Basin Analysis: Principles and Application to Petroleum Play Assessment

Third Edition

Philip A. Allen[†] and **John R. Allen**
[†]*Department of Earth Science & Engineering, Imperial College London*

WILEY-BLACKWELL

A John Wiley & Sons, Ltd., Publication

Library of Congress Cataloging-in-Publication Data
Allen, P. A.
  Basin analysis : principles and application to petroleum play assessment / Philip A. Allen, Department of Earth Science & Engineering, Imperial College London & John R. Allen. – Third edition.
     pages cm
  Includes bibliographical references and index.
  ISBN 978-0-470-67376-8 (pbk.) – ISBN 978-0-470-67377-5 (hardback) – ISBN 978-1-118-45030-7 (epub)   1. Sedimentary basins.
2. Petroleum–Geology.   3. Petroleum–Prospecting.   I. Allen, John R. (John Richard), 1953–   II. Title.
  QE571.A45 2013
  552'.5–dc23

                                                2013001792

A catalogue record for this book is available from the British Library.

Wiley also publishes its books in a variety of electronic formats. Some content that appears in print may not be available in electronic books.

Cover image: The image on the front cover is an overlay of the faults and salt (black) structures in the Gulf of Mexico and surrounding coastal plain, with an image of the dispersal of sediment from the Mississippi plume. The tectonic map is from Rowan, M.G., Jackson, M.P.A. & Trudgill, B.D., Salt-related fault families and fault welds in the northern Gulf of Mexico, Bulletin of the American Association of Petroleum Geologists, 83, 1454–1484, Figure 1, page 1456. AAPG © 1999. Reprinted by permission of the AAPG whose permission is required for further use. The colour image shows mixing of the turbid Mississippi plume with the clear blue water of the Gulf of Mexico, and is from the NASA Earth Observatory, http://earthobservatory.nasa.gov.
Cover design by Nicki Averill Design & Illustration

Set in 9/11 pt Minion by Toppan Best-set Premedia Limited
Printed and bound in Malaysia by Vivar Printing Sdn Bhd

2   2014

Contents

Preface to the third edition

The first edition of *Basin Analysis* appeared in 1990, and the second edition in 2005. In a geometrical series, we should publish the third edition in 2020, but the widespread adoption of Basin Analysis courses in undergraduate and postgraduate curricula, and the speed of advancement and breadth of engagement of the subject make an earlier updating necessary. We continue to believe that knowledge of the basic principles of the thermomechanical behaviour of the lithosphere, the dynamics of the mantle, and the functioning of sediment routing systems provide a sound background for studying sedimentary basins, and are a prerequisite for the exploitation of resources contained in their sedimentary rocks. There are therefore no significant departures from the underlying philosophy followed in the first two editions.

Basin analysis increasingly represents a vehicle for the adoption of integrated approaches to geoscientific problems. Integration is very easy to talk about and very difficult to do. To focus on sedimentary basins dynamically is therefore a way in which diverse specialists can contribute to a wider mission. The big mistake in the study of sedimentary basins is to believe that it is the preserve of a particular discipline. Sedimentologists are tempted to believe that they hold the key to the understanding of sedimentary basins through their expertise in the depositional environments and stratigraphic architectures of sedimentary rocks. Likewise, geophysicists may assert that they hold the key through their understanding of the continuum mechanics of the way the lithosphere deforms and its thermal consequences. And structural geologists will no doubt defend their position that the key ingredient of basin analysis is the understanding of the nature of the brittle deformation of the crust and the evolution of its contractional and extensional fault arrays. To tackle basin analysis effectively in the future, there needs to be a 'mind-set' that has the courage to stray into adjacent disciplines, and an openness to share the significance of one's own specialism. We aim to help develop this mind-set through reading this text.

The third edition is renamed *Basin Analysis: Principles and Application to Petroleum Play Assessment* to make clear that we choose to show the outworking of a basin analysis approach to the exploration for hydrocarbons. It is lengthened and published in larger format, allowing us to expand the content of some chapters in recognition of the growth of certain areas. In order to allow as fluent a text as possible, we have also removed some of the quantitative though often simplified work to Appendices. These Appendices contain details of quantitative derivations of concepts widely applied in basin analysis, and practical exercises that give hands-on experience of a wide variety of physical problems. The Appendices allow the student to gain additional skill in quantitative problem solving, but the selection of topics is necessarily incomplete and somewhat arbitrary. The key generic learning points from carrying out the calculations are emphasised. A rigorous treatment of principles used in basin analysis is the

hallmark of this book, but despite the inclusion of 60 Appendices it is not a quantitative manual. We aim to illustrate important foundational principles in basin analysis rather than attempt an encyclopaedically descriptive coverage.

We set the scene for sedimentary basin analysis in Part 1 by outlining the compositional and rheological zonation of the Earth before providing a brief background to the geodynamics relevant to the formation and evolution of basins. Sedimentary basins are then classified in terms of this geodynamical background. The following chapter on the physical state of the lithosphere aims to give the essentials of the mechanics that underpin the way the lithosphere extends, shortens, bends and undergoes heating and cooling.

Part 2 of the book makes use of this information in considering the two main mechanisms for sedimentary basin formation, namely stretching of the continental lithosphere and flexure associated with mountain building, which are dealt with in Chapters 3 and 4. In each of these chapters the basic geological and geophysical observations of the rift–drift suite of basins and of foreland basin systems are presented before physical models are outlined and assessed. We have expanded the treatment of the effects of mantle dynamics to reflect the current emphasis on deep-to-surface connections and spend some time assessing whether mantle flow can be recognised in surface dynamic topography, and its implications for basin development. The last chapter of Part 2 is an updated treatment of basins associated with strike-slip tectonics.

Part 3 concerns the basin-fill in two chapters; firstly through an integrated view of the sediment routing system that links sediment erosion with transport and deposition from source-to-sink, and secondly through a process-orientated view of stratigraphy. These two interlinked chapters conform to the general philosophy of the textbook in highlighting the role of Earth surface dynamics in controlling the sedimentary systems feeding and occupying basins. The emphasis on the basin-fill continues with a rigorous treatment of the burial history and thermal history of sedimentary basins. Chapter 9 focuses on the loss of porosity during burial and the techniques necessary to correctly extract the time-depth trajectory of horizons or units within the basin-fill. The following chapter evaluates the various processes affecting thermal history from a modelling viewpoint, followed by an evaluation of a wide range of temperature proxies measured in basin sedimentary rocks.

We finish in Part 4, as before, with the application of basin analysis principles to the assessment of petroleum systems and plays. Firstly, we describe the building blocks of the petroleum play, which in essence covers the current state of understanding of the natural life cycle of petroleum from its biological precursors, generation and migration, through to the stages of entrapment, alteration and dissipation. Relative to the second edition of *Basin Analysis*, this is now presented in a condensed and updated form. We believe that many of the fundamentals of the scientific understanding of these key processes have not significantly changed over the past decade, but, nevertheless, in updating this section, we wanted to bring attention to the advances made in a number of important areas. Two areas that stand out for particular mention are the plays associated with gravity-driven foldbelts, which have become a major exploration focus over the past 10 to 15 years, and the plays associated with sand injectites, which have been conceived and have grown in importance over the past decade.

An addition in this third edition is a new chapter that begins by showing, through a small number of examples, how all of the elements come together to produce some classic conventional plays in truly world-class petroleum provinces that are of enormous economic importance. The examples chosen are the Niger Delta, the Campos Basin of Brazil, the pre-salt of the Brazilian Santos Basin, and the Northwest Shelf of Australia. In addition, in recognition of a major shift in focus that has occurred over the past decade, a number of unconventional petroleum plays are described. There is no doubt that the production contribution and economic importance of unconventional petroleum is growing as the rate of discovery of conventional petroleum resources has moved into long-term decline. Many unconventional gas resources indicate the operation of quite unconventional petroleum systems. While oil sands and heavy oils represent conventional oils that have been severely altered, unconventional gas plays such as basin-centred tight gas, shale gas and gas hydrates indicate processes at work that vary significantly from conventional models of petroleum formation and occurrence. Finally, the principles of basin analysis, and many of the geological understandings developed in the search for petroleum, can be applied in the new emerging area of carbon dioxide geosequestration. Although the sector is in its infancy, geological assessment has an important role to play in identifying and evaluating potential subsurface sites of CO_2 storage, and it is fitting that this edition of *Basin Analysis* concludes with a brief description of this potentially important new area of application.

There is an increasingly impressive number of books that address important aspects of basin analysis. In the field of geodynamics and heat flow, we are pleased to acknowledge in particular the mighty *Geodynamics* (second edition 2002) by Donald Turcotte and Gerald Schubert, *Dynamic Earth* (1999) by Geoff Davies, with the all-important subtitle 'Plates, Plumes and Mantle Convection', *The Solid Earth* (second edition 2005) by Mary Fowler, *Isostasy and Flexure of the Lithosphere* (2001) by Tony Watts, the recently published (2011) *Heat Generation and Transport in the Earth* by Claude Jaupart and Jean-Claude Mareschal, and *Crustal Heat Flow* (2001) by G.R. Beardsmore and J.P. Cull. Valuable information is found in *Quantitative Thermochronology* (2006) by Jean Braun, Peter van der Beek and Geoffrey Batt. In the field of Earth surface dynamics, *Geomorphology: The Mechanics and Chemistry of Landscapes* (2010) by Bob Anderson and Suzanne Anderson is a *tour de force*, and further important contributions are made by Jon Pelletier's *Quantitative Modelling of Earth Surface Processes* (2008), and *Tectonic Geomorphology* (2001) by Doug Burbank and Bob Anderson. Andrew Miall's *The Geology of Stratigraphic Sequences* (second edition 2010) continues to provide a wealth of information on stratigraphic ground-truth, and *Sedimentology and Sedimentary Basins* by Mike Leeder (second edition 2011) places basin analysis in a firm context of sedimentary processes. *Diagenesis: A Quantitative Perspective* (1997) by Melvin Giles remains an important source of information on compaction and diagenesis. *Tectonics of Sedimentary Basins* (second edition 2012) edited by Cathy

Busby and Antonio Azor contains a compilation of chapters of direct relevance to basin analysis. *Petroleum Geoscience* by Jon Gluyas and Richard Swarbrick (2004) is a valuable source of information and useful handbook for the practising petroleum geoscientist that focuses on the geoscience fundamentals important in the various stages of the Exploration & Production 'value chain', from frontier exploration through to field appraisal, development and production. In the petroleum area, it is timely to also recognise the *American Association of Petroleum Geologists* for the remarkable role they have played in the continuing education of the petroleum industry workforce, and in particular the excellent publications that have contributed so much to the understanding of petroleum genesis and occurrence that we have attempted to synthesise in this book. Finally, and importantly, we have benefitted greatly from Magnus Wangen's quantitative treatment of an eclectic mix of topics of relevance to basin analysis in *Physical Principles of Sedimentary Basin Analysis* (2010).

We are grateful to a large number of past and present colleagues, friends and students who have provided input for the third edition, have commented on drafts of individual chapters, or have simply inspired us with their conversations and published work, including Rhodri Davies, Saskai Goes, Alex Whittaker, Sanjeev Gupta, Gary Hampson, Howard Johnson, Lidia Lonergan, Chris Jackson, Al Fraser, Amy Whitchurch and Nikolas Michael of Imperial College London, Hugh Sinclair (Edinburgh), Andy Carter (UCL), Alex Densmore (Durham University), Nicky White (Cambridge), John Armitage (Paris), Robert Duller (Liverpool), Niels Hovius (Potsdam), Sébastien Castelltort and Guy Simpson (Geneva University), Kerry Gallagher (Rennes), Peter Burgess (Royal Holloway), Chris Paola (Minnesota), Paul Heller (Wyoming), Andrew Miall (Toronto), Andrew Hurst (Aberdeen), Ian Lunt and Hans-Morten Bjornseth (Statoil) and Andy Morrison (Pura Vida Energy). PAA gratefully acknowledges the interaction with Australian geoscientists during the tenure of a Royal Society International Travel award, especially Mike Sandiford and Andrew Gleadow at Melbourne and Louis Moresi at Monash. PAA is especially pleased to acknowledge Lidia Lonergan who has co-taught a Basin Analysis course in Imperial College London for several years.

There will not be a fourth edition.

Biographies

Philip Allen graduated with a Bachelor's degree in Geology from the University of Wales, Aberystwyth, and a PhD from Cambridge University. He held lectureships at Cardiff and Oxford, and professorships at Trinity College Dublin, ETH-Zürich and Imperial College London. He is a process-oriented Earth scientist with particular interests in the interactions and feedbacks between the solid Earth and its 'exosphere' through the critical interface of the Earth's surface.

John Allen has over 30 years of experience in the international oil and gas industry as a petroleum geologist, exploration manager, senior exploration advisor, and business strategist with British Petroleum (BP) and BHP Billiton, as well as several years of experience as a non-executive director. He is currently based in Melbourne, Australia.

Philip A. Allen
John R. Allen

PART 1

The foundations of sedimentary basins

CHAPTER ONE
Basins in their geodynamic environment

Summary

Sedimentary basins are regions of prolonged subsidence of the Earth's surface. The driving mechanisms of subsidence are related to processes originating within the relatively rigid, cooled thermal boundary layer of the Earth known as the lithosphere and from the flow of the mantle beneath. The lithosphere is composed of a number of plates that are in motion with respect to each other. Sedimentary basins therefore exist in a background environment of plate motion and mantle flow.

The Earth's interior is composed of a number of compositional and rheological zones. The main compositional zones are between crust, mantle and core, the crust containing relatively low-density rocks overlain by a discontinuous sedimentary cover. The mechanical and rheological divisions do not necessarily match the compositional zones. A fundamental rheological boundary is between the lithosphere and the underlying asthenosphere. The lithosphere is sufficiently rigid to comprise a number of relatively coherent plates. Its base is marked by a characteristic isotherm (c.1600 K) and is commonly termed the thermal lithosphere, which encloses a mechanical lithosphere. The upper portion of the thermal lithosphere is able to store elastic stresses over long time scales and is referred to as the elastic lithosphere. The continental lithosphere has a strength profile with depth that reflects its composition, temperature and water content. A weak, ductile zone exists in the lower crust below a brittle–ductile transition, but the strength of the underlying lithospheric mantle is uncertain. The oceanic lithosphere lacks this low-strength layer, its strength increasing with depth to the brittle–ductile transition in the upper mantle.

The relative motion of plates produces deformation, magmatism and seismicity concentrated along oceanic plate boundaries. Continental lithosphere is more complex, exhibiting seismicity and deformation far from plate boundaries, and with a heat flow and geotherm that is strongly influenced by radiogenic self-heating. Plate boundary forces and elevation contrasts strongly influence the state of stress of lithospheric plates.

Sedimentary basins have been classified principally in terms of the type of lithospheric substratum (i.e. continental, oceanic, transitional), their position with respect to the plate boundary (intracontinental, plate margin) and type of plate motion nearest to the basin (divergent, convergent, transform). The formative mechanisms of sedimentary basins fall into a small number of categories, although all mechanisms may operate during the evolution of a basin:

- Isostatic consequences of changes in crustal/lithospheric thickness, such as caused mechanically by lithospheric stretching, or purely thermally, as in the cooling of previously upwelled asthenosphere in regions of lithospheric stretching.
- Loading (and unloading) of the lithosphere causes a deflection or flexural deformation and therefore subsidence (and uplift), as in foreland basins.
- Viscous flow of the mantle causes non-permanent subsidence/uplift known as dynamic topography, which can most easily be recognised in the domal uplifts of the ocean floor at volcanic hotspots.

From the point of view of lithospheric processes there are two major groups of basins: (i) basins due to lithospheric stretching and subsequent cooling, belonging to the rift–drift suite; and (ii) basins formed primarily by flexure of continental and oceanic lithosphere.

1.1 Introduction and rationale

Maps of the global or plate-scale distribution of sediment thickness reveal strong variations (Fig. 1.1). It can be seen at both the global scale (Fig. 1.1) and the plate or continental scale (Fig. 1.2) that much of the area of the continental interiors is devoid of any sedimentary cover, with Precambrian crystalline rocks exposed at the surface. Elsewhere, the greatest sedimentary thicknesses are found in particular geological settings such as at extensional continental margins and fringing the world's great collisional mountain belts. These regions of large sedimentary thickness have undergone extensive and prolonged subsidence (Bally & Snelson 1980). The complexities of geological history have resulted in a patchwork of currently subsiding active basins and their ancient counterparts. Sedimentary basins, ancient and modern, are the primary archive of information on the evolution of the Earth over billions of years.

Basin Analysis: Principles and Application to Petroleum Play Assessment, Third Edition. Philip A. Allen and John R. Allen.
© 2013 John Wiley & Sons, Ltd. Published 2013 by John Wiley & Sons, Ltd.

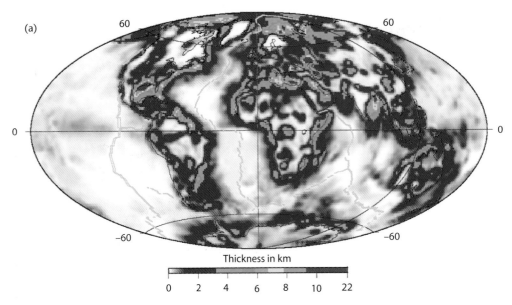

(a)

Thickness in km

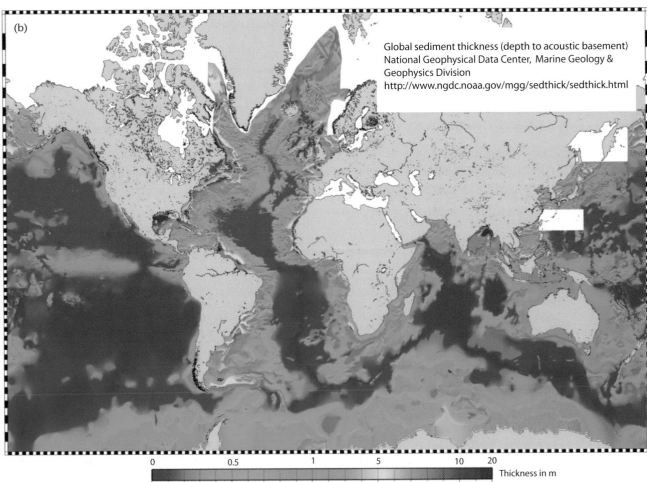

(b)

Global sediment thickness (depth to acoustic basement)
National Geophysical Data Center, Marine Geology &
Geophysics Division
http://www.ngdc.noaa.gov/mgg/sedthick/sedthick.html

Thickness in m

Fig. 1.1   (a) Global sediment thickness, from Laske & Masters (1997) based on the digital database of Gabi Laske at the University of San Diego, California (http://mahi.ucsd.edu/Gabi). (b) A higher-resolution map of the total sediment thickness in the ocean, from Divins (2003).

(a)

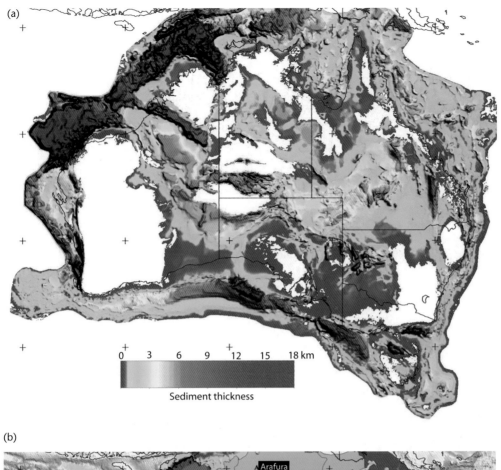

0 3 6 9 12 15 18 km

Sediment thickness

(b)

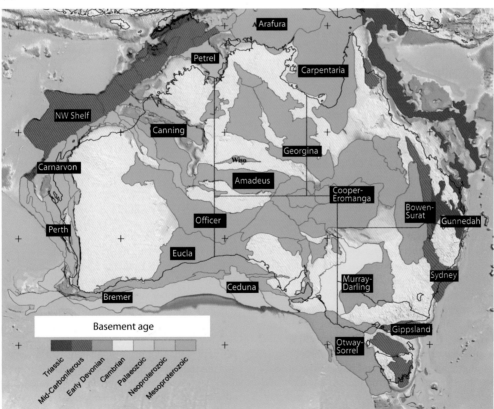

Arafura

Petrel

Carpentaria

NW Shelf

Canning

Carnarvon

Georgina

Wiso

Amadeus

Cooper-
Eromanga

Officer

Bowen-
Surat

Gunnedah

Perth

Eucla

Ceduna

Murray-
Darling

Sydney

Bremer

Basement age

Gippsland

Otway-
Sorrel

Triassic Mid-Carboniferous Early Devonian Cambrian Palaeozoic Neoproterozoic Mesoproterozoic

Fig. 1.2   (a) Sediment thicknesses, Australia. Note that the greatest sediment thicknesses are in continental margin basins superimposed on Paleozoic basement rocks. In addition, there are high sediment thicknesses in the continental interior related to structural inversion of intracontinental mountain belts (e.g. Petermann and Alice Springs orogenies) bordering the Officer and Georgina basins. This tectonic activity compartmentalised large cratonic basins. Note also that large areas of old cratons have no sedimentary cover. (b) Outlines of ancient and modern sedimentary basins shown according to basement age. Note that the bulk of the basement ages are Precambrian (green). Younger basement rocks underlie basins of the continental margins (e.g. NW Shelf, Ceduna Basin in south), and the terranes east of the Tasman Line. Both (a) and (b) fromFrOG Tech Pty Ltd. (2005).

The location of sedimentary basins and their driving mechanisms are intimately associated with the motion of discrete, relatively rigid slabs, which together represent the cooled thermal boundary layer of the Earth. The outer shell of the Earth comprises a relatively small number of these thin, relatively rigid plates, which are in a state of motion with respect to each other. Such motions set up plate boundary forces that may be transferred considerable distances into the interior of the plates, so that sedimentary basins exist in a background environment of stress set up by plate motion.

The lithospheric plates are the surface manifestation of a slow thermal convection in the mantle, and are subject to differential thermal stresses along their bases. The mantle and lithosphere therefore do not operate as independent systems. We see spectacular evidence for the interaction of mantle processes and the lithosphere in the volcanic and topographic expression above mantle flow structures, some of which may have risen from the core–mantle boundary. We also discern, though less spectacularly, the effects on mantle flow caused by the subduction of cold slabs of oceanic crust at ocean–continent boundaries.

Deep Earth processes involving the thermomechanical behaviour of the lithosphere and the flow of the underlying mantle are coupled to Earth surface processes of erosion, sediment and solute transport and deposition in sedimentary basins. This coupling between 'deep' and 'surface' is the fundamental basis for the practice of broad, integrative thinking in basin analysis, and underpins the understanding of sedimentary basins as geodynamical entities. It is also the framework for the study of petroleum systems in sedimentary basins (Fig. 1.3). The connectedness between deep and surface geodynamics is emphasised throughout this text, and the fruits of an improved understanding derived by studying such connections are illustrated in the application to the exploration of hydrocarbons in Chapters 11 and 12.

Two key, dovetailed concepts therefore underlie this necessity of integration in basin analysis:

1. The dynamics of the solid Earth results in tectonic processes at various scales that control the generation of space in which sediment may accumulate for long periods of time. Tectonic processes determine the bulk strain and strain rate at which the basin and its structures form, and also control thermal history, magmatism and burial history. At a smaller scale, tectonic processes generate fault arrays that provide the template for spatial patterns of uplift and subsidence, and determine the pathways of sediment transport from source to sink.
2. Sediment is derived by weathering and erosion and released into dispersal systems *en route* to long-term depositional sinks (Allen 2008). The 'erosional engine' of the sediment routing system is powered by tectonic processes causing uplift of rocks, and the deposition of sediment to form stratigraphy is strongly modulated by the spatial patterns and rates of tectonically generated subsidence. The functioning of sediment routing systems determines the gross depositional environments in a basin and its stratigraphic geometries and has important feedbacks into tectonic deformation, structural reactivation and exhumation.

1.2   Compositional zonation of the Earth

There are three main compositional units: the crust, mantle and core (Fig. 1.4).

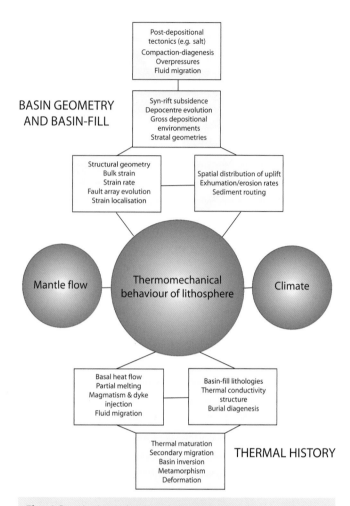

Fig. 1.3   The basis for integration in basin analysis is the coupling between deep and surface geodynamics, the effects of which cascade through the various aspects of basin analysis to the application to petroleum systems.

1.2.1   Oceanic crust

The crust is an outer shell of relatively low-density rocks. The oceanic crust is thin, ranging from approximately 4–20 km in thickness, 6–10 km being 'normal', and with an average density of about 2900 kg m^{-3}. It comprises a number of layers that reflect its mode of creation: an upper veneer (layer 1) of unconsolidated or poorly consolidated sediments, generally up to 0.5 km thick; an intermediate layer 2 of basaltic composition, consisting of pillow lavas and associated products of submarine eruptions; and a layer 3 of gabbros and peridotites that comprise the parent rocks, which upon differentiation give rise to the basalts of layer 2. The oceanic crust was formerly thought to be distinctly layered in terms of velocity of seismic waves, but more recent views are that it possesses a more gradual and continuous increase in velocity with depth.

Ocean crust is being created as new oceanic seafloor at 2.8 km^2 yr^{-1}, and the current area of ocean crust is 3.2×10^8 km^2, roughly 60% of the surface of the Earth. At this rate of production, the area of ocean crust would double in a further 100 Myr. Yet the lifetime of ocean crust is short – the oldest oceanic crust in today's oceans is as young

(a)

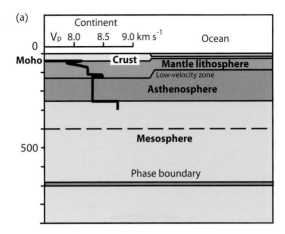

(b)

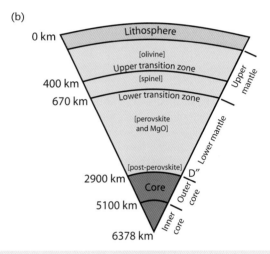

Fig. 1.4 The main compositional (a) and rheological (b) boundaries of the Earth. The most important compositional boundaries are between crust, mantle and core. There are also compositional variations within the continental crust and compositional variations caused by phase changes in the mantle. V_p is the velocity of P-waves. The main rheological boundary is between the lithosphere and the asthenosphere. P-wave velocities increase markedly beneath the Moho, but decrease in a low-velocity zone in the asthenosphere. The lithosphere is rigid enough to act as a coherent plate. The perovskite to post-perovskite transition at >2600 km is associated with the D″ seismic discontinuities.

Fig. 1.5 Age of the ocean crust in the Atlantic Ocean. Colour code for the age of the oceanic crust is red for zero at the mid-ocean ridge and blue for Jurassic. Reproduced with permission from Mueller *et al.* (2008). Note that the age of the ocean floor closely mimics the depth of the ocean floor below sea level (see Chapter 2).

Table 1.1 Physical properties of the Earth and its compositional zones

Properties of the Earth	
Equatorial radius	6.378137×10^6 m
Volume	1.0832×10^{21} m³
Mass	$5.9736 \ 10^{24}$ kg
Mean density	5.515×10^3 kg m⁻³
Surface area	5.10×10^{14} m²
Land area	1.48×10^{14} m²
Mean land elevation	875 m
Continental area (incl. continental shelves)	2.0×10^{14} m²
Water area	3.62×10^{14} m²
Mean ocean depth	3794 m
Mean surface heat flow	87 mW m⁻²
Age of Earth	4.55 Ga
Crust	
Mass	2.36×10^{22} kg
Mean thickness of continental crust	40 km
Mean thickness of oceanic crust	6 km
Mean continental heat flow	65 mW m⁻²
Mean oceanic heat flow	101 mW m⁻²
Zero-pressure density (granodiorite)	2700 kg m⁻³
Zero-pressure density (basalt and gabbro)	2950 kg m⁻³
Mantle	
Volume	9.06×10^{20} m⁻³
Mass	4.043×10^{24} kg
Zero-pressure density (peridotite)	3250 kg m⁻³
Core	
Volume	1.77×10^{20} m³
Mass	1.883×10^{24} kg

as Jurassic in age (c.150 Ma) (Fig. 1.5) and the average age of oceanic crust is about 100 Ma. This is because the oceanic crust caps the gravitationally unstable oceanic lithosphere; as a result, it is consumed by subduction.

1.2.2 Continental crust

The continental crust is thicker, ranging from 30 to 70 km, but with an average thickness of 35–40 km (Table 1.1). It was originally thought to be divided into two layers, each with a distinct composition and density: (i) an upper layer with physical properties similar to those of granites, granodiorites or diorites, overlain by a thin, discontinuous veneer of sedimentary rocks. This so-called 'granitic layer' has a thickness of between 20–25 km and a density of 2500–2700 kg m^{-3}. The term 'granitic' is, however, misleading, since average densities are greater than that of granite. (ii) A lower layer of primarily basaltic composition, but the pressure and temperature at depths in excess of 25 km imply that the rocks are granulites, or their high-pressure, high-temperature equivalents, eclogites or amphibolites. The density of this lower layer is 2800–3100 kg m^{-3}. These layers may not in reality be well defined, and instead a more continuous variation of composition with depth may exist.

Information on the density of crustal rocks has been obtained largely by observations on seismograms of the speed of seismic waves passing through the various layers, combined with laboratory experiments on rock materials. The existence of a low-velocity crust was discovered by the geophysicist Mohorovicic. At the crust–mantle boundary, seismic **P** (longitudinal) wave velocities increase markedly; this abrupt increase in velocity may reflect a corresponding increase in rock density (Fig. 1.4). This horizon is known as the Mohorovicic discontinuity or Moho. The Moho varies in depth considerably. The continental crust thickens under orogenic belts, thins under zones of rifting and attenuates completely at continental margins (Fig. 1.6).

In some regions, particularly the attenuated margins of continents, the crust is intermediate in character and thickness between typical oceanic and continental varieties. This may be due to the injection of dense intrusions, to metamorphism, or to other processes accompanying stretching. In particular, the depth to the Moho may be abnormally great due to the igneous underplating of the crust associated with high temperatures in the asthenosphere. The presence of several kilometres of underplate emplaced in the Early Tertiary over the head of the Iceland plume has been interpreted from seismic experiments carried out on the northwest European continental margin.

Since the upper part of the continental crust is thick and gravitationally stable, it is not subducted, and contributes to the much greater average age of the continental crust (10^9 years). However, the denser lower continental crust and the underlying lithospheric mantle may undergo recycling into the Earth's interior (see Chapter 2).

1.2.3   Mantle

The *mantle* is divided into two layers: the upper and lower mantle. The upper mantle extends to about 680 km ± 20 km and is punctuated by phase transitions. The lower mantle extends to the outer limit of the core at 2900 km, with an increasing density with depth.

Although there are of course no *in situ* measurements of the composition of the mantle, it can be estimated from the chemistry of volcanic and intrusive rocks derived by melting of the mantle, from tectonically emplaced slivers of mantle rock preserved in orogenic belts known as ophiolites, from nodules preserved in volcanic rocks, from minerals brought to the surface explosively in kimberlites, and, importantly, from the remote but sophisticated measurement of the mantle using seismic waves. The main constituent of the mantle is thought to be olivine, mostly the Mg-rich variety forsterite.

Olivine is known to undergo phase changes to denser structures at pressures equivalent to depths of 390–450 km and c.700 km in the Earth. At 390–450 km olivine is thought to change to spinel via an exothermic reaction involving a 10% increase in density. At c.700 km spinel changes to perovskite (magnesium silicate) and magnesium

oxide in an endothermic reaction. These phase changes can be recognised by variations in the velocity of **S**-waves (McKenzie 1983) and may determine the scale of convection in the mantle (Silver *et al.* 1988). The high-pressure form *post-perovskite* is stable at temperatures above 2500 K and pressures above 120 GPa, corresponding to a depth in excess of 2600 km in the innermost mantle. The perovskite to post-perovskite transition may be responsible for the seismic discontinuity known as D″ (Peltier 2007) (Fig. 1.4).

1.3   Rheological zonation of the Earth

The mechanical or rheological divisions of the interior of the Earth do not necessarily match the compositional zones. One of the rheological boundaries of primary interest to students of basin analysis is the differentiation between the lithosphere and the asthenosphere. This is because the vertical motions (subsidence, uplift) in sedimentary basins are principally a response to the deformation of this uppermost rheological zone of the Earth, or to the guiding through the lithosphere of stresses transmitted from the mantle.

1.3.1   Lithosphere

The *lithosphere* is the rigid outer shell of the Earth, comprising the crust and the upper part of the mantle. It is of particular importance to note the difference between the *thermal* and *elastic* thicknesses of the lithosphere. It is generally believed (e.g. Parsons & Sclater 1977; Pollack & Chapman 1977) that the base of the lithosphere is represented by a characteristic isotherm (1100–1330 °C, or 1600 K) at which mantle rocks approach their solidus temperature. The solidus temperature of peridotite in K is 1500 + 0.12p, where the pressure p is in MPa. This defines the *thermal lithosphere* (Figs 1.7, 1.8). Heat flow in the thermal lithosphere is by conduction, whereas heat flow in the underlying asthenosphere is by convection. The temperature gradient in the convecting mantle is the adiabatic gradient, which is approximately 0.5 K km^{-1} in the shallow upper mantle.

Typical thicknesses of lithosphere under the oceans vary from c.5 km at mid-ocean ridges to c.100 km in the coolest parts of the oceans. The lower boundary of the lithosphere is poorly defined under continents, depths of 100 km to 250 km being typical. However, the precise seismological mapping of the base of the thermal lithosphere is difficult since it does not correspond to a compositional boundary. Instead, it corresponds to a rheological change in peridotite, which at about 1600 K becomes sufficiently weak to convect. A thermal boundary layer separates the convecting asthenosphere from the mechanically strong lithosphere above. Stepwise increases in velocities of **S** and **P** waves with depth through the lithosphere suggest that it also contains minor compositional boundaries within it.

It is puzzling that continental subcrustal lithosphere can extend to great depths of >200 km, since it should become gravitationally unstable relative to the underlying hot asthenosphere (see Chapter 2). This thick subcrustal lithosphere may avoid delamination by being depleted by the extraction of melts, resulting in harzburgites. Harzburgites, the Mg-rich refractory residuum of peridotite after the extraction of basaltic melts, are not only less dense, but also have a higher (~500 K) melting temperature than unmodified peridotite, allowing these cratons to retain both their long-term buoyancy and their thickness. The main period of growth of cratonic lithosphere was in the Archaean, but since then there have been cycles of enhanced

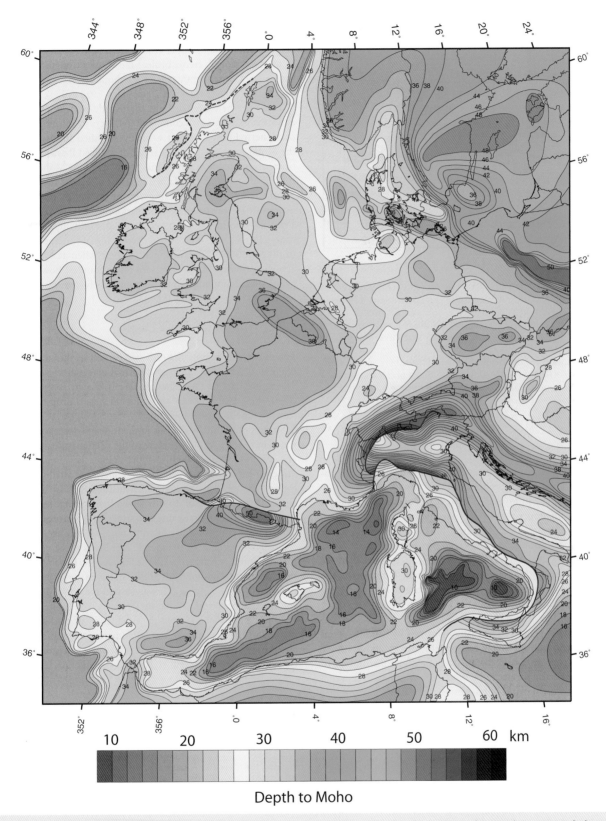

Fig. 1.6 Depth to the Moho, western Europe, from Dèzes & Ziegler (2002). Note that the Moho is deep under orogenic belts where the crust has been thickened tectonically, such as the Pyrenees (Spain), Alps (France, Switzerland, Austria), Dinarides (Croatia) and Apennines (Italy) and under the cratonic Scandinavian shield. The crust is thin in the basins floored with oceanic crust in the eastern Mediterranean and the Atlantic Ocean. The crust is also thin in regions of stretching of the continental lithosphere, as can be seen particularly in the North Sea region, and in the Rhine-Bresse-Rhone region of the West European Rift system. Free download from EUCOR-URGENT, http//comp1.geol.unibas.ch/downloads/Moho_net/euromoho1_3.pdf.

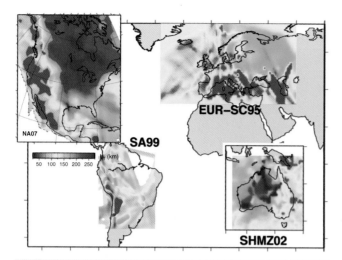

Fig. 1.7 Thickness of the thermal lithosphere (1200 °C isotherm) from a conversion of the **S**-wave velocity model of Bedle & van der Lee (2009) to temperature. The temperature conversion follows Goes and van der Lee (2002) and assumes a constant peridotitic composition for the mantle lithosphere. Note (especially for North America and Europe) that there is a thick, cold core of the plate, shown by blue colours, up to 250 km thick, surrounded by relatively thin, hot lithosphere shown by red colours (≪100 km thick), such as the Basin and Range province of west-central USA, and the Mediterranean region of southern Europe.

magmatism, judged from the frequency of age of detrital zircons (Condie 1998; Hawkesworth *et al.* 2009).

The rigidity of the lithosphere allows it to behave as a coherent plate, but only the upper part of the lithosphere is sufficiently rigid to retain elastic stresses over geological time scales (say 10^7 years). Below this upper *elastic lithosphere* creep processes efficiently relax elastic stresses, so there is a physical and conceptual difference between the elastic lithosphere and the thermal lithosphere. The lithosphere below the upper elastic portion must therefore be sufficiently weak to relax elastic stresses but sufficiently rigid to remain a coherent part of the surface plate.

The lithospheric plates can be easily deformed by bending about a horizontal axis, but are highly resistant to torsion about steeply inclined axes. This latter property of strength allows the motion of plates over the Earth's surface to be modelled, assuming no internal deformation, except at plate boundaries. But how do the oceanic and continental lithosphere compare in terms of their long-term (>1 Myr) flexural strength? Different views exist on this problem (Jackson 2002; Burov & Watts 2006).

On the one hand, oceanic plates are stronger because they consist of more mafic mineral assemblages and contain fewer intrinsic weaknesses such as old fault systems, but on the other hand are thinner and hotter and therefore bend more easily than continental plates. The strongest part of the oceanic lithosphere occurs in the mantle between 20 and 60 km depth, below which it becomes increasingly ductile (Fig. 1.9). In contrast, the continents contain quartz, which shows ductile flow at lower temperatures than olivine, and contain heterogeneities, but are thicker and cooler than oceanic plates. The strength profile is potentially complex. In one model (jelly sandwich)

both the upper crust and mantle are strong, separated by a mid-lower crustal aseismic ductile zone that has been invoked as a level of detachment of major upper crustal faults (e.g. Kusznir & Park 1987). In another model (crème brûlée) the elastic strength is confined to the crust. Consequently, major earthquakes are confined to this brittle crustal layer (Jackson 2005) (Figs 1.9, 1.10).

There are heterogeneities in the mantle part of the lithosphere, although they are small compared to the crust. Seismological studies of western Europe suggest a highly stratified lithosphere beneath the Moho. In particular, a 'channel' of reduced **P**-wave velocities has been interpreted between 10 and 20 km below the Moho. This 10 km-thick layer cannot be explained in terms of partial melting since the solidus temperature is far in excess of the actual temperature; the hydration (serpentinisation) of peridotites has been postulated as a possible mechanism. Whatever the cause, this upper low-velocity channel may serve as a zone of decoupling of the upper lithosphere from the lower portion of the lithosphere when acted upon by tangential tectonic forces. There are few examples, however, where a process of decoupling can be unambiguously demonstrated at these levels.

1.3.2   Sub-lithospheric mantle

The underlying region, the *asthenosphere*, is weaker than the lithosphere and is able to undergo deformation relatively easily by flow. The upper part of the asthenosphere is known as the low-velocity zone, where **P** and **S**-wave transmission speeds drop markedly, presumably due to partial melting.

Studies of minute variations in the transmission speeds of seismic waves have allowed the structure of the deep mantle to be visualised and mapped (Nolet 2008), a topic known as *seismic tomography*. Zones of faster than average seismic velocity are attributed to propagation through denser rock, which in turn is most likely due to a cooler temperature. Zones with slower than average seismic velocity are likewise thought to be due to warmer temperatures. Variations in temperature are probably caused by large-scale convection.

Far from being inert, the mantle represents a vast volume of rock that dynamically interacts with the lithosphere. Instabilities rise from the core–mantle boundary as *plumes* of hot material that impinge on the base of the overlying lithosphere and may have a major role in continental break-up (Burke & Dewey 1973; White & McKenzie 1988). The thermal effects of the subduction of cold oceanic lithosphere and the insulating effects of supercontinental assemblies are also thought to be recognisable in the thermal structure of the mantle (Gurnis *et al.* 1996; O'Neill *et al.* 2009) (Chapter 5).

1.4   Geodynamic background

1.4.1   Plate tectonics, seismicity and deformation

Plate tectonics is a kinematic theory that describes the motion of the lithosphere as comprising a relatively small number of rigid plates that deform at their boundaries. These plate boundary regions are very narrow in comparison with the size of individual plates. In the oceans, earthquakes are strongly concentrated in narrow bands corresponding to mid-ocean ridges, subduction zones and transform fault zones (Barazangi & Dorman 1969) (Fig. 1.11). Elsewhere, the oceanic crust is essentially aseismic. Consequently, the oceans behave as described by plate tectonic theory.

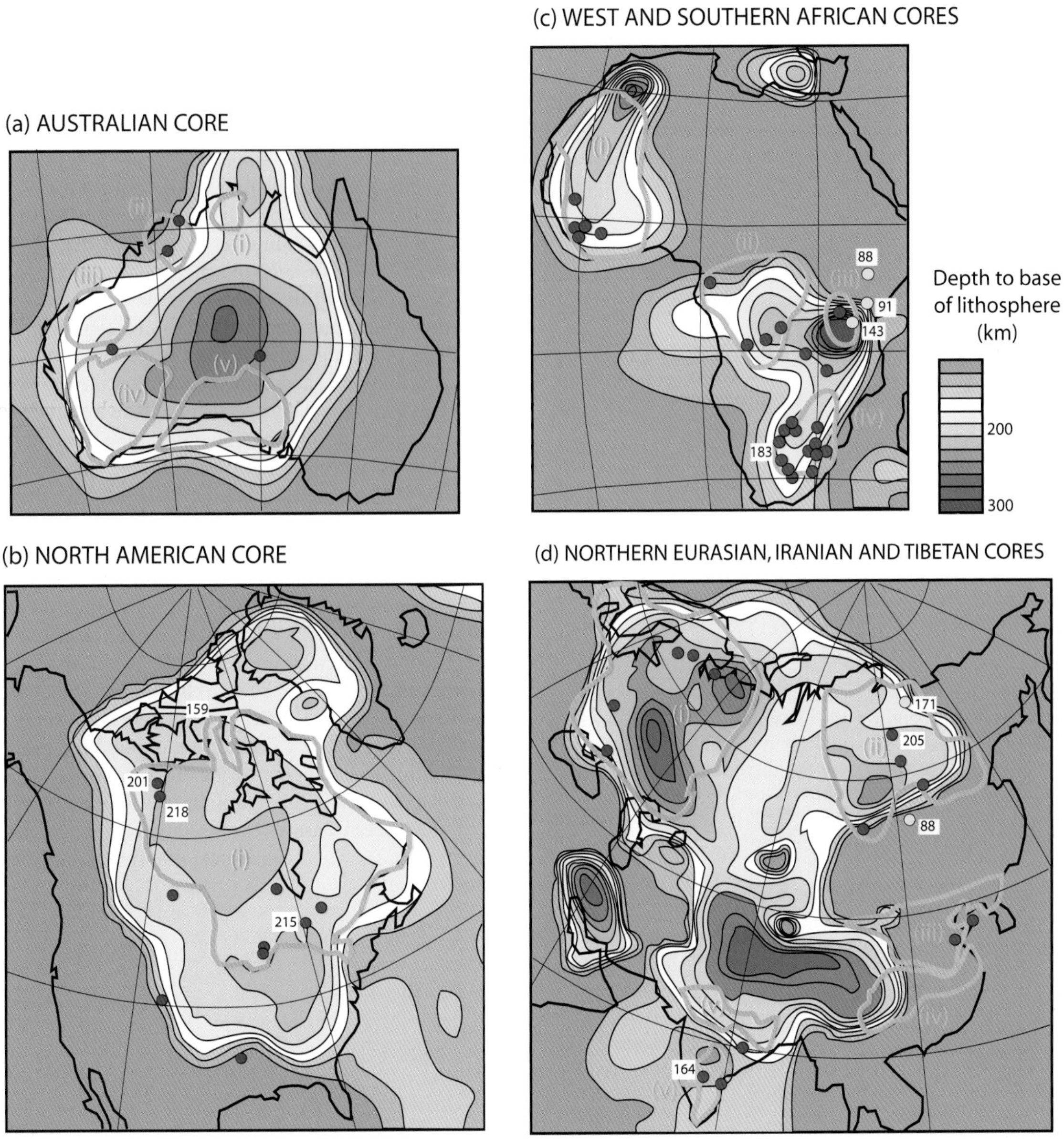

(c) WEST AND SOUTHERN AFRICAN CORES

(a) AUSTRALIAN CORE

(b) NORTH AMERICAN CORE

(d) NORTHERN EURASIAN, IRANIAN AND TIBETAN CORES

Depth to base
of lithosphere
(km)

200

300

Fig. 1.8 Lithospheric thickness derived from **S**-wave velocities (Preistley & McKenzie 2006). The yellow circles show the locations of kimberlites and alkali basalts lacking diamonds, with white boxes showing lithospheric thickness derived from nodule mineralogies. The red circles show the locations of kimberlites bearing diamonds, with the lithospheric thickness where indicated. The yellow lines show the boundaries of the cratons. (a) Australia: i, Darwin craton; ii, Kimberley craton; iii, Pilbara block; iv, Yilgarn block; v, Gawler craton. (b) North America: i; North American craton. (c) Africa: i, West African craton; ii, Angolan craton; iii, Tanzanian craton; iv, Kalahari craton. (d) Eurasia: i, East European craton; ii, Siberian craton; iii, North China craton; iv, Yangtze Block of South China craton; v, Himalayan cratons; vi, Dharwar craton. Diamond occurrences show that the lithospheric thickness beneath cratons exceeds 250 km. Reprinted from Preistley & McKenzie (2006), with permission from Elsevier.

In the continents, however, earthquakes are very widely distributed (Fig. 1.11). Geodetic measurements, principally from global positioning system (GPS) networks, confirm that active deformation is taking place over very extensive areas rather than being confined to plate boundary zones (England 1987). Continental plates clearly undergo deformation a long way from plate boundaries, in the form of intracontinental orogenies causing structural reactivation and basin development in continental interiors. In such cases, the driving force may be the excess potential energy of elevated continental crust

or the far-field transmission of in-plane stresses from a plate boundary. Plate tectonics theory is therefore a far less adequate description of the deformation of the lithosphere in continental regions. Instead, the continents may be better approximated by thin viscous sheets that deform in a ductile fashion; at a large scale, the deformation is continuous rather than discrete. This view of a diffuse, continuous deformation of continental lithosphere is supported by geodetic measurements of regions such as the Aegean Sea, eastern Mediterranean. Whatever the best descriptions for deformation in the oceanic and continental lithosphere, the nature and rates of relative plate motion govern many aspects of the geodynamic environment of basins.

The fact that earthquake epicentres occur at depths as great at 650 to 700 km along some plate boundaries suggests that a process exists that is capable of transferring brittle material to depths normally associated with deformation by flow. This process of plate subduc-

tion is responsible for both the relative youth of the oceanic crust and the distribution of earthquake epicentres.

1.4.2   The geoid

The Earth's outer fluid envelope, the oceans, flows to a level that is determined by Earth's gravitational potential field, which is controlled by the distribution of density with depth in the solid Earth. The water of the oceans accumulates to a height that is not a perfect reflection of the oblate spheroid shape of the Earth. Instead, mean sea level, representing the observed geoid height, shows very large undulation-like departures from the idealised shape (Fig. 1.12). These departures, or geoid anomalies, may result from a complicated set of contributing factors, but the fact that they vary significantly from one plate to another suggests that there must be an important long-wavelength sub-lithospheric contribution. This sub-lithospheric source of geoid anomalies presumably derives from the density heterogeneity of the mantle, which may itself be due to the history of convective flow and slab subduction.

The equivalence in terms of gravitational potential of different sections of oceanic and continental lithosphere is an important question. For example, a 125 km-thick continental lithosphere with an average density of $2750 \, kg \, m^{-3}$ should produce a (+6 m) geoid anomaly, consistent with an isostatically compensated continental lithospheric column with a surface elevation of 1 km. Such a column is also likely to be in potential energy balance with a mid-ocean ridge with a water depth of 2.5 km. Bearing in mind the mean elevations of the continents and ocean basins (Table 1.1 and §1.4.3), this suggests that continents achieve a mean elevation that is balanced by the adjacent ocean basins. The isostatic balance between adjacent columns of oceanic and continental lithosphere, and the horizontal

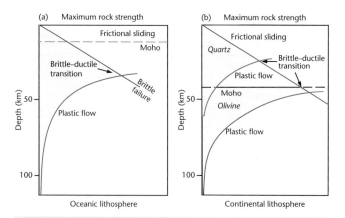

Fig. 1.9 Strength profiles of the continental and oceanic lithosphere. Strength also depends on whether the lithosphere is in tension or compression, and on the presence of volatiles such as water (Jackson 2005). The model for the continental lithosphere is called 'jelly sandwich', but an alternative view for the continental lithosphere is that the upper mantle is not strong and is aseismic, which is termed 'crème brûlée' (compare Burov & Watts 2006 and Jackson 2002).

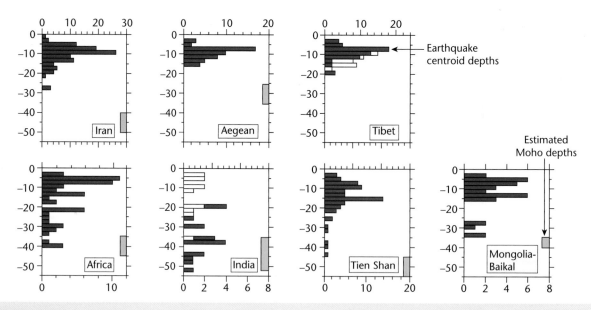

Fig. 1.10 Depth of occurrence of earthquakes in several regions of the world. In the top row, the distribution of earthquakes shows that they must originate only from the upper part of the crust, at depths of less than 20 km. In the bottom row, earthquakes occur over a larger depth range of up to 40–50 km. However, comparison with the estimates of Moho depth suggests that even these deep earthquakes may originate from the crust. In other words, there is little indication that the mantle is seismogenic. Consequently, the mantle may contribute little to the elastic strength of the continental plates (Jackson *et al.* 2008).

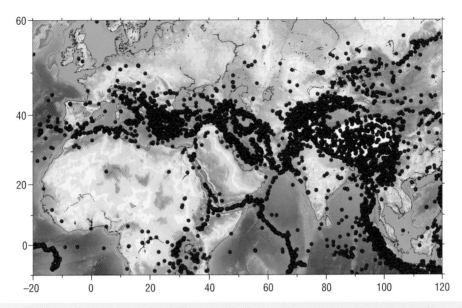

Fig. 1.11 Earthquake epicentres, shown in red circles, are distributed over a very wide mountainous region, whereas they are concentrated in narrow zones in the oceans. From Jackson (2005).

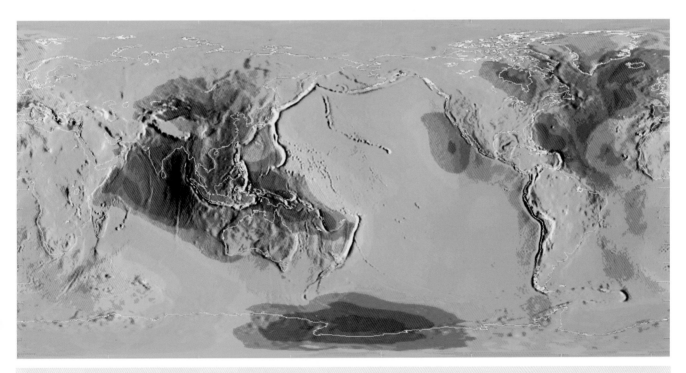

Fig. 1.12 The geoid height anomaly: the geoid relative to a reference ellipsoid, showing two major geoid highs and lows per circumference of the Earth. One geoid high is situated across much of Africa, western Eurasia and the North Atlantic. Another occupies the western Pacific area. These long-wavelength geoid height anomalies may be almost permanent features of the mantle (Torsvik *et al.* 2008).

forces between them, is given in more detail in Chapter 2 and in Appendices 2, 3 and 4.

1.4.3 Topography and isostasy

The mean elevation of the land is just 875 m, whereas the mean depth of the ocean basins is 3794 m (Table 1.1). These mean elevations reflect the buoyancy of the underlying continental and oceanic lithosphere. Where the oceanic lithosphere is extremely thin at mid-ocean ridges, the water depth is commonly between 2 and 3 km. This is the water depth at which we expect the top of a column of normal asthenosphere to lie. In some regions the mid-ocean ridge crest shallows and may become emergent, as in Iceland, suggesting that the underlying asthenosphere is hotter and therefore more buoyant than normal. Either side of the elevated mid-ocean ridges, the ocean basins deepen with a square root of crustal age relationship (see §2.2.7).

Continental topography extends to the edge of the continental shelves, at approximately 200 m present-day water depth, and up to the highest mountain peak at c.8 km. There is no relationship between the age of the continental crust and its elevation. Topography is primarily controlled by the history of deformation and erosion, with an upper limit on the height of mountain ranges provided by rock strength (Willett & Brandon 2002). That is, once a critical elevation is reached, rocks are prone to collapse under gravitational forces.

Simple calculations of isostasy underpin many aspects of basin analysis. Isostasy is the way in which hydrostatic equilibrium (Archimedes principle) influences the support of the oceanic and continental plates by the mantle. For equilibrium to be maintained, we can equate the surface forces due to the differing rock columns under continental and oceanic lithosphere (see Appendix 2). A familiar parameterisation for isostasy is of a mountain belt with a 'root', the excess mass in the elevated continental crust being compensated by the mass deficit at depth in the continental root.

The presence of continental blocks of varying thickness and topography adjacent to oceanic lithosphere of varying thickness causes pressure differences between adjacent columns of rock, which result in horizontal surface forces that act in addition to the lithostatic term (§2.1.1) (Appendices 3 and 4). These deviatoric or in-plane stresses may be compressive or tensile. For a continental crustal block adjacent to mantle-density oceanic lithosphere, the tensional horizontal deviatoric stress in the continent is of the order of −10 to −100 MPa, but if the continental block is 70 km thick, as in some mountain ranges, the tensile deviatoric stress may reach −150 MPa. Thickened continental lithosphere, as in continental plateaux such as Tibet and the Altiplano, consequently pushes against the adjacent platforms and ocean basins, leading to extension in the elevated regions.

Elevated continental plates surrounded by passive margins should therefore be in deviatoric tension. Continental plates surrounded by convergent boundaries, on the other hand, experience compressional deviatoric stresses roughly orthogonal to the orientation of the convergent boundary, as seen in the Indo-Australian plate (Hillis & Reynolds 2000) (Fig. 1.13).

1.4.4 Heat flow

A global heat flow map shows that the oceans have high surface heat flows, especially at the mid-ocean ridges (Fig. 1.14). The average surface heat flow value for the ocean crust is 101 mW m^{-2}, nearly double the continental average (Table 1.1). Fig. 1.14b demonstrates that the surface heat flow of the ocean floor shows the same pattern as the ocean bathymetry, with the same root-age dependence. This

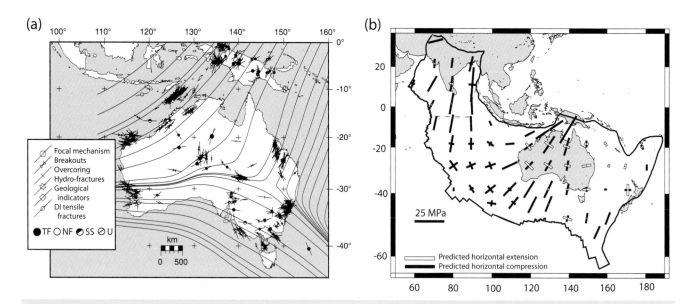

Fig. 1.13   (a) Stress trajectories based on the Australian stress map (Hillis & Reynolds 2000): NF, normal fault; SS, strike-slip; TF, thrust fault; U, unknown. (b) Stresses predicted by an elastic numerical model of the forces acting on the Indo-Australian plate, after Coblentz *et al.* (1998). Ridge push forces are balanced by fixing the collisional segments of the northeastern margin of the plate (Himalayas, New Guinea and New Zealand). Bars indicate orientation and magnitude of extensional (open) and compressive (solid) deviatoric stresses. Note that compressive deviatoric stresses align themselves normal to the convergent boundaries. Maximum deviatoric compressive forces are <50 MPa.

(a)

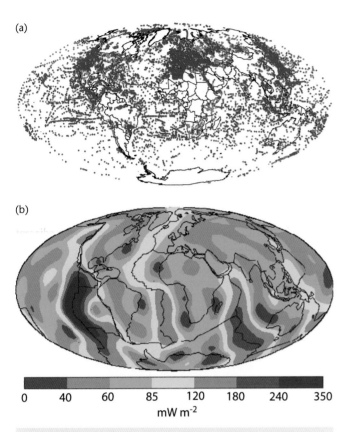

(b)

Fig. 1.14   (a) Global surface heat flow database, based on the compilation of Pollack *et al.* (1993) and additional data from Davies and Davies (2010), making over 38,000 measurements. (b) Contoured surface heat flow (degree 12 spherical harmonic), from Pollack *et al.* (1993).

shows that as the ocean lithosphere moves away from the spreading centre, it cools, thickens and sinks.

On the continents, there is considerable variability in surface heat flow values, but the variability is primarily explained by variations in the radiogenic heat production of continental rocks (§2.2.3). It is common for over 50% of the surface heat flow to be caused by radiogenic heating. Consequently, although the global average continental surface heat flow is $65\,\mathrm{mW\,m^{-2}}$, the basal heat flows under the continents must be much smaller, typically between 20 and $30\,\mathrm{mW\,m^{-2}}$. High concentrations of radiogenic elements in upper crustal rocks causes self-heating and the elevation of the geotherm above the nonradiogenic case. Consequently, although the oceanic geotherm may be linear due to conduction, the geotherm in continental lithosphere is typically curved (§2.2.2). Elevated geotherms may play an important role in controlling continental seismicity and deformation.

1.4.5   Cycles of plate reorganisation

The relative motion of plates with constructive, conservative and destructive plate margins creates a continually changing picture of continental splitting, ocean basin creation, ocean closure and continental collision. This cycle of plate motion involving the birth and closure of oceans is termed the Wilson cycle since it is based on early ideas of the opening and closing of the Atlantic Ocean by John Tuzo

Wilson (Wilson 1966). Many sedimentary basins can be fitted into a particular phase of the Wilson cycle.

A related concept is that of the aggregation of supercontinents and their dispersal in response to convective flow in the mantle (Anderson 1982; Gurnis 1988). Rodinia, Gondwana and Pangaea are all examples of supercontinental assemblies, with a repeat time scale of the order of 300 Myr. Supercontinents are thought to act as insulators of the mantle, hindering it from losing heat as it does so effectively through the oceanic lithosphere. Over a period of time (about $10^{8}\,\mathrm{yr}$), sub-plate temperatures would rise as the buoyant supercontinent is trapped on a geoid high between two adjacent cold downwellings. Eventually, the lateral spread of heat would allow the plate to move off the geoid high and to settle over a downwelling at a geoid low, causing the widespread flooding of the continental surface. The time scale of the supercontinental cycle therefore depends primarily on the time taken to incubate the sub-plate mantle sufficiently to inflate the supercontinent (Grigné *et al.* 2007), cause break-up, and lead to migration of continental fragments to adjacent geoid lows. The Wilson cycle is therefore driven fundamentally by mantle circulation (Gurnis, 1988).

Although very generalised, a wide range of basin types can be placed in the geodynamical context of supercontinental assembly and dispersal. Extension of the supercontinent triggers the formation of the rift-drift suite of sedimentary basins, comprising continental rifts, continental rim basins, cratonic basins, failed rifts, protooceanic troughs, and passive margins. The subduction phase associated with convergence triggers the formation of ocean trenches, accretionary wedge basins and fore-arc basins trenchwards of the magmatic arc, a range of basins in the retro-arc region, and strike-slip basins linked to oblique convergence. Continental collision is typified by foreland basin systems, including wedge-top deposition (De Celles & Giles 1996; Ford 2004). Consequently, certain basin types are common during particular phases of the Wilson cycle. In addition, the frequency of occurrence of basin types will depend on their longterm preservation by being shielded from tectonic recycling. This is an important consideration for Proterozoic basins, whose preservation potential is enhanced by incorporation into relatively rigid, stable, cratonic cores.

The dispersal and reassembly of plates may take place in two modes, termed introversion and extroversion (Nance & Murphy 2003; Murphy & Nance 2008). In the case of introversion, plates disperse from the supercontinental core, producing new interior oceans bordered by extensional trailing edges of the continental plates, and convergent boundaries along their external perimeter. Subsequent subduction of the interior ocean causes it to close and to result in reassembly of the plates in essentially the same orientation as in the original super-assembly. Extroversion, on the other hand, involves the continued expansion of the continental plates away from the supercontinental core, and the eventual closure of the external ocean by subduction along the advancing edges of the dispersing continental plates. Introversion and extroversion therefore control the large-scale setting for basin development at the time scale of the supercontinental cycle.

1.5   Classification schemes of sedimentary basins

Ideally, classifications are theories about the basis of natural order rather than dull catalogues compiled only to avoid chaos (Gould

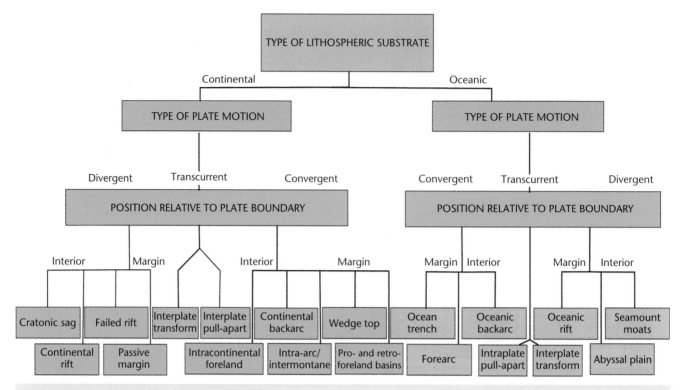

Fig. 1.15   Classification scheme for sedimentary basins based on their plate tectonic setting. Modified from scheme used by Kingston *et al.* (1983a). AAPG © 1983. Reprinted by permission of the AAPG whose permission is required for further use.

1989, p. 98, quoted in Ingersoll & Busby 1995, p. 2). In this sense, classification schemes for sedimentary basins should both reveal something of the underlying mechanisms for basin development and reflect the natural variability of the real world.

Classification schemes of sedimentary basins based on plate tectonics have much in common. Their lineage derives from Dickinson's influential work in 1974, which emphasised the position of the basin in relation to the type of lithospheric substrate, the proximity of the basin to a plate margin, and the type of plate boundary nearest to the basin (divergent, convergent, transform) (Fig. 1.15). The evolution of a basin could then be explained by changing plate settings and interactions.

Bally (1975) and Bally and Snelson (1980) differentiated three different families of sedimentary basins based on their location in relation to megasutures, which in this context can be defined to include all the products of orogenic and igneous activity associated with predominantly contractional deformation. The boundaries of megasutures are often associated with subduction, whether it be of slabs of oceanic lithosphere (Benioff or B-type subduction) or of relatively buoyant continental lithosphere (Amferer or A-type subduction) and may also be the sites of important wrench tectonism along transform faults. Ingersoll and Busby (1995) developed the classifications of Dickinson (1974) and Ingersoll (1988) to recognise 26 different types grouped into classes of divergent settings, intraplate settings, convergent settings, transform settings and hybrid settings. Within these settings, Ingersoll (2011) suggested that there were numerous variants depending on sediment supply and geological inheritance.

1.5.1   Basin-forming mechanisms

The goal of categorising a sedimentary basin and thereby gaining some predictive insights into the hydrocarbon potential of frontier basins is common to industry classifications (e.g. Huff 1978; Klemme 1980; Kingston *et al.* 1993a, b). Although such classifications undoubtedly have their uses, particularly in predicting the presence of key elements of the petroleum play (Chapter 11), they have the effect of scrambling some of the essential differences and similarities between basins from the point of view of geodynamical mechanisms.

Ingersoll and Busby (1995) and Ingersoll (2011) recognised seven subsidence mechanisms, operating to different degree in their various basin types (Fig. 1.16), which can be summarised as:

1. crustal thinning, such as caused primarily by stretching or surface erosion;
2. lithospheric thickening, such as caused by cooling following stretching or accretion of melts derived from the asthenosphere;
3. sedimentary and volcanic loading causing isostatic compensation;
4. tectonic (supracrustal) loading causing isostatic compensation;
5. subcrustal loading caused by subcrustal dense loads such as magmatic underplates or obducted mantle flakes;
6. mantle flow primarily due to the subduction of cold lithospheric slabs;
7. crustal densification due to changing P-T conditions or intrusion of high-density melts.

Subsidence mechanisms

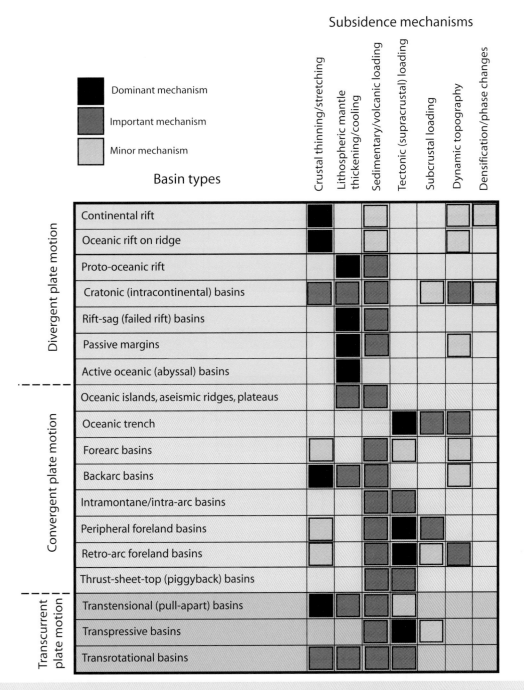

Fig. 1.16 Matrix of main basin types and the principal mechanisms of subsidence (adapted from Ingersoll & Busby 1995 and Ingersoll 2011). Note that subsidence mechanisms operate in more than one basin type, and that several mechanisms may operate in a single basin type. Some basins are 'polyhistory' and may pass through phases characterised by different sets of mechanisms.

From the point of view of fundamental lithospheric processes, the major mechanisms for regional subsidence and uplift (*isostatic*, *flexural* and *dynamic*) can be summarised as follows (Fig. 1.17):

- *Isostatic* consequences of changes in crustal and lithospheric thickness; the thickness changes may be brought about purely thermally by *cooling* of lithosphere, for example following mechanical stretching. Thickness changes causing *thinning* may be caused by mechanical stretching, subaerial erosion or at depth by removal

(or delamination) of a deep lithospheric root. Mechanical *thickening* of crust and lithosphere, as in zones of continental convergence, generally causes isostatic uplift. Thickening of the lithosphere by cooling, however, causes subsidence.
- *Loading* and unloading at the surface and in the subsurface, including the far-field effects of in-plane stresses; *loading* of the lithosphere may take place on a small scale in the form of volcanoes or seamount chains, and on a large scale in the form of mountain belts, causing *flexure* and therefore subsidence. The

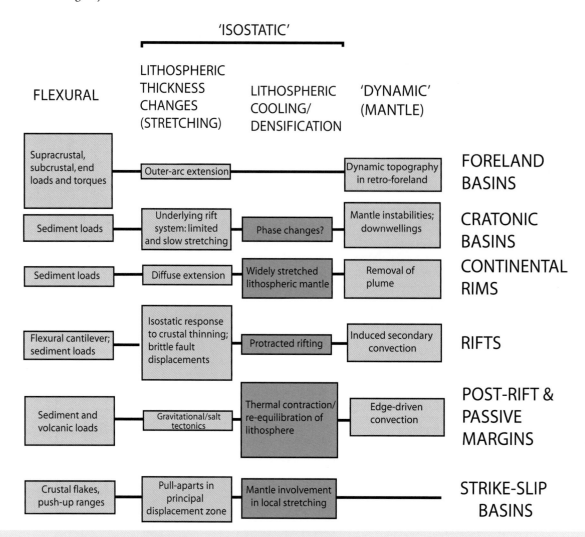

Fig. 1.17 Genetic basin classification based on flexural, isostatic and dynamic mechanisms, and their importance in a number of common basin types. Size of box reflects roughly the importance of the mechanism in the formation and evolution of the basin type. Cratonic basins, continental rims, rifts, failed rifts and passive margins belong to a suite of basins associated with continental extension. Foreland basins are associated with flexure, whereas strike-slip basins are controlled by the structural geology of zones of transcurrent deformation.

sediment infilling a basin also acts as a sedimentary load, amplifying the primary driving mechanism.

• *Dynamic* effects of asthenospheric flow, mantle convection and plumes; subsidence or uplift are caused by the buoyancy effects of changes in temperature in the mantle. Since these temperature changes are transmitted by viscous flow, the surface elevation changes may be termed *dynamic*.

For a given basin type, some or all of the mechanisms given above may have a major or minor role (Fig. 1.17). In addition, a given mechanism operates in more than one basin type. The classification scheme is therefore a matrix. Rather than attempting an encyclopaedic coverage of all basin types, we focus on the main lithospheric processes in Part 2 of this book. After an introductory chapter on the fundamentals of lithospheric mechanics (Chapter 2), basin-forming processes are considered by investigating basins primarily caused by

lithospheric stretching (Chapter 3), and basins primarily caused by flexure (Chapter 4). Chapter 5 discusses the role of mantle-lithosphere interactions in basin development. Basins related to strike-slip deformation are considered in Chapter 6.

Basins formed by stretching or thinning of the continental lithosphere fall within an evolutionary sequence (Kinsman 1975; Veevers 1981) (§3.1.1). The early stages of the sequence correspond to the development of intracontinental sags (cratonic basins) and continental rim basins, which lack clear evidence of brittle stretching, and continental rifts, which comprise clear extensional fault systems that may or may not be associated with topographic doming. Such rifts may evolve into oceanic spreading centres or may be aborted to form failed rifts or aulacogens. With seafloor creation and drifting of the continental edge away from the spreading centre, passive margin basins develop. The sequence has been termed the *rift–drift suite* of sedimentary basins. The mechanisms of interest within this evolutionary sequence are therefore primarily the thermal and mechanical behaviour of the lithosphere under tension, and the thermal contraction of the lithosphere following stretching.

Basins formed by flexure are mainly linked to plate convergence (Chapter 4). Flexure of oceanic lithosphere as it approaches subduction zones is responsible for the formation of deep oceanic trenches. It was the investigation of the deflection of the oceanic lithosphere at arc–trench boundaries that provided much of the framework for the general theory of lithospheric flexure. Flexure of the continental lithosphere in continental collision zones gives rise to *foreland basin systems*. Flexure of the lower or subducting plate generates pro-foreland basins, whereas flexure of the upper or overriding plate generates retro-foreland basins (Naylor & Sinclair 2008). Where subduction zone roll-back takes place, as in the Adriatic-Apennines region of Italy, the lower plate is flexed to produce a pro-foreland basin, but the retro-foreland region is extensional. Foreland basin pairs may also be associated with intracontinental mountain building where there is no subduction, as in the late Precambrian and Paleozoic of central Australia.

Processes within the mantle have an important role in basin development (Chapter 5). Mantle flow structures, probably originating at the core–mantle boundary, impinge on the base of the lithosphere and spread out laterally over a length scale of 10^3 km. They are instrumental in continental splitting and the formation of new ocean basins, and have a major role in igneous underplating and isostatic regional uplift. Present-day *hotspots* over ascending limbs of mantle convection systems are characterised by topographic doming and commonly rifting. Basins located over downward limbs of convection systems, *cold-spot basins*, appear to be broad, gentle sags. The onset of subduction of cold oceanic slabs at ocean–continent boundaries causes far-field tilting of the continental plate towards its margin and therefore has a potentially major effect in forearc, intra-arc and retro-arc settings. In addition, it is believed that supercontinental assemblies in the geological past have experienced very long-wavelength topographic doming caused by elevated sub-lithospheric temperatures generated by an overlying insulating lid. Mantle flow instabilities associated with steps in the base of the lithosphere, or associated with the drip-like removal of gravitationally unstable mantle lithosphere, may also have important impacts on surface topography.

Different basin types have a typical time span of existence (Ingersoll & Busby 1995), after which they may be uplifted and eroded, locally inverted, or modified into another basin type to become polyhistory in nature. Time spans may range over 4 orders of magnitude (Woodcock 2004). Oceanic trench basins have the shortest life span (0.1–0.8 Myr) due to rapid subduction velocities and early incorporation into an accretionary wedge. Trench-slope, intra-arc, extensional back-arc and strike-slip basins also have relatively short life spans in the range 2–25 Myr, reflecting their tectonically active settings. Rift, forearc, back-arc and foreland basins have longer time spans in the range 4–125 Myr. The longest life spans (60–440 Myr) are found in cratonic, passive margin and ocean basins, since they are situated in regions of low strain rate, experiencing prolonged thermal subsidence.

CHAPTER TWO
The physical state of the lithosphere

Summary

Knowledge of the behaviour of the lithosphere is essential if we are to understand the initiation and development of sedimentary basins. Lithospheric processes are responsible for the highly dynamic nature of the tectonics taking place near the surface of the Earth. The Earth's outer layer or lithosphere can be regarded as a thermal boundary layer between the cool atmosphere or oceans and the hot interior. The lithosphere is therefore a thermal entity, but it also has a physical significance. The upper portion of the lithosphere is sufficiently rigid that it is able to store and transmit stresses. Since the deformations caused by applied forces are generally recoverable, this outer zone is known as the elastic lithosphere.

Applied forces of whatever origin cause stresses that result in deformation or strain. The simplest view of applied forces is of those due simply to the weight of an overlying rock column, known as *lithostatic stress*. The difference between the actual stress and the lithostatic stress is a tectonic contribution known as *deviatoric stress*. Deviatoric stresses can be either tensile or compressive and are commonly transmitted from plate boundaries or originate from the gravitational potential energy due to elevation.

In an elastic solid there is a clear relationship between the stresses and resultant strains. The exact relationship depends on material properties known as Young's modulus and Poisson's ratio. Where only one of the principal axes is non-zero, a *uniaxial* state of stress is said to occur, and the relation between stress and strain is called *Hooke's law*. If there are two non-zero components of principal stress, we have the condition termed *plane stress*. In an analogous fashion, *uniaxial strain* and *plane strain* refer to the coordinate system of principal strains. In a state of stress where all the principal stresses are equal, an *isotropic* state of stress, the fractional volume change caused by isotropic compression is given by the *bulk modulus* or its reciprocal, the *compressibility*. Since rocks have a finite compressibility, they increase in density with depth in the Earth, but they also expand due to heating. The net effect on rock density is critical to understanding the long-term gravitational stability of the continental lithosphere.

The lithosphere is able to bend or flex. The shape of the flexure of the lithosphere depends on its rigidity and on the nature of the applied force or load causing the bending. Bending is accompanied by longitudinal stresses and bending moments in the plate. The bending moment is related to the local radius of curvature by a coefficient called *flexural rigidity*. The bending stresses around the outer arc of flexed lithosphere may be sufficient to cause faulting or fault reactivation.

In order to understand the mechanical behaviour of the Earth it is necessary to understand its thermal structure, since this determines its rheology (Gk. *rheos*, stream, flow). In the lithosphere, heat transfer is predominantly by conduction, whereas in the mantle, convection is extremely important. Convection is the natural result of differential heating above a critical Rayleigh number, and estimates for Earth's mantle strongly suggest that convection must be taking place. A thermal boundary layer develops along the upper surface of a convecting material because heat is lost by conduction to the surface. This cool boundary layer, the lithosphere, detaches and sinks due to gravitational instability along subduction zones. For conduction, the heat flux is related to the temperature gradient by a coefficient, the *thermal conductivity*. Heat fluxes in the continents are determined primarily by conduction from radioisotopic heat sources, whereas in the oceans they reflect cooling of newly created oceanic lithosphere. The variation of temperature with depth is known as the *geotherm*. Measurements of continental heat flows indicate a linear relationship between radiogenic heat generation and surface heat flow, with the intercept indicative of the 'reduced' or basal heat flow from the mantle.

Since radiogenic heat-producing elements (HPEs) are concentrated in the upper crust, they can easily be mobilised by surface processes of erosion and sedimentation, which potentially has an important effect on the continental geotherm. The crust beneath a self-heating basin is hotter because of the effect of the sedimentary blanket. This heating may be sufficient to reactivate old faults or cause new tectonic deformation. Erosion and sediment deposition also have a potentially important transient effect on the underlying geotherm. The time scale for the recovery of the geotherm following an instantaneous erosion or deposition event is of the order of 10^6–10^7 years. Rapid exhumation, where the upward advection of hot crustal rocks outweighs the effect of conductive cooling, causes a marked curvature in the geotherm, with important implications for the interpretation of thermochronometers such as zircon and apatite fission track.

The reference equipotential surface for the Earth is termed the *geoid*. Some geoid anomalies probably represent density differences in the Earth set up by plate tectonic processes. The analysis of surface gravity data (*Bouguer anomalies*) provides important information on the way in which topography is compensated isostatically. Loads are supported by the flexural strength of the lithosphere, the degree of compensation

Basin Analysis: Principles and Application to Petroleum Play Assessment, Third Edition. Philip A. Allen and John R. Allen.
© 2013 John Wiley & Sons, Ltd. Published 2013 by John Wiley & Sons, Ltd.

depending on the flexural rigidity of the plate and the wavelength of the load. Techniques such as admittance and coherence allow the flexural rigidity to be estimated from a correlation of the Bouguer gravity anomaly with the spectral wavelength of the topography.

Mantle convection is thought to take place by means of a thermally activated creep. Crustal rocks may also behave in a ductile manner by pressure-solution creep, or they may deform in a brittle way by fracturing (*Byerlee's law*). The occurrence of earthquake foci and the results of laboratory experiments on rock mechanics suggest that at depths of about 15–25 km and temperatures of c.300 °C the continental crust starts to become ductile or plastic, but the rheology of the crust is complex owing to its compositional heterogeneity. The onset of ductility in the middle crust may serve mechanically to decouple the upper crust from the lower crust and mantle lithosphere. The lower lithosphere may deform as an elastic solid on short time scales but viscously on longer time scales. The time scale of the viscous relaxation of stresses is not fully understood but may be estimated from studies of postglacial rebound. In the case of a flexed plate, plastic deformation may take place if a critical elastic curvature is exceeded.

The key to understanding the deformation of the lithosphere is the profile of lithospheric strength with depth. Oceanic and continental lithospheres have different strength profiles. Strength depends on composition, temperature, volatile content such as water, and presence of heterogeneities acting as weaknesses. The strong part of the lithosphere not only supports loads elastically, but also determines the depth of occurrence of earthquakes. There are two competing models for the strength profile of the continental lithosphere. The 'jelly sandwich' model envisages a strong, brittle, seismogenic upper crust and mantle lithosphere separated by a weak, ductile lower crust, whereas the 'crème brûlée' model proposes that there is one strong, seismogenic layer restricted to the crust.

Sedimentary basins represent a physical deformation of the lithosphere. In order to understand how basins are initiated and evolve through time, it is necessary to have some appreciation of the physical state of the lithosphere. This physical state can be thought of in terms of its thickness, mineralogy, thermal structure, and consequently its strength when acted upon by tangential forces or its rigidity when flexed by orthogonal forces. This chapter introduces some fundamental ideas on the physical state of the lithosphere and how this determines its deformation or flow. In the succeeding chapters in Part 2, the particular cases of the stretching (extension) of the lithosphere (Chapter 3) and the bending (flexure) of the lithosphere (Chapter 4) are considered before investigating the large-scale interaction of the lithosphere and mantle in Chapter 5 and processes in zones of strike-slip deformation in Chapter 6.

2.1   Stress and strain

2.1.1   Stresses in the lithosphere

Body forces on an element of a solid act throughout the volume of the solid and are directly proportional to its volume or mass. For example, the force of gravity per unit volume is the product of ρ, the density, and g, the acceleration of gravity. The body forces on rocks within the Earth's interior depend on their densities, but density is itself a function of pressure. If we normalise rock densities to a zero-pressure value (Appendix 1), typical mantle rocks would have densities of about 3250 kg m^{-3}, ocean crust (basalt and gabbro) would have densities of about 2950 kg m^{-3}, whereas continental granites and diorites would be in the range of 2650 to 2800 kg m^{-3}. Sedimentary rocks are highly variable, ranging from 2100 kg m^{-3} for some shales to 2800 kg m^{-3} for very compact marbles (Tables 2.1, 2.2).

Surface forces act only on the surface area bounding a volume and arise from the interatomic stresses exerted from one side of the surface to the other. The magnitude of the force depends on the surface area over which the force acts and the orientation of the surface. The normal force per unit area on horizontal planes increases linearly with depth. That due to the weight of the rock overburden is known as *lithostatic stress* or *lithostatic pressure*. This concept of surface forces in the Earth's interior is made use of in considering the way in which hydrostatic equilibrium (Archimedes principle) influences the support of the oceanic and continental plates by the mantle, an important application known as *isostasy*. Appendix 2 deals with the isostatic balance for a range of geological settings. We will examine a simple application of isostasy in studying the bending of lithosphere in §2.1.5.

Normal surface forces can also be exerted on vertical planes (Fig. 2.1). If the normal surface forces, σ_{xx}, σ_{yy} and σ_{zz}, are all equal and

Table 2.1 Physical (thermal) properties of common minerals (after Goto & Matsubayashi 2008)

Mineral	Density ρ (kg m^{-3})	Thermal conductivity K (W m^{-1} K^{-1})	Specific heat c (J kg^{-1} K^{-1})	Thermal diffusivity κ (10^{-6} m^2 s^{-1})
Quartz	2648	7.69	741	3.92
Albite	2620	2.2	776	1.08
Anorthite	2760	1.68	745	0.82
Orthoclase	2570	2.32	707	1.28
Muscovite	2831	2.32	796	1.03
Illite	2660	1.85	808	0.86
Smectite	2608	1.88	795	0.91
Chlorite	2800	5.15	818	2.25
Calcite	2710	3.59	820	1.62
Seawater	1024	0.59	3993	0.15
Mud (grain)	2731	3.4	758	1.64

Table 2.2 Physical and thermal properties of common rocks (in part from Turcotte & Schubert 2002; Blackwell & Steele 1989; Jaupart & Mareschal 2011, p. 416, p. 422, thermal conductivities and diffusivities at room temperature)

	Zero-pressure density kg m^{-3}	Young's modulus E 10^{11} Pa	Poisson's ratio v	Thermal conductivity K W m^{-1} K^{-1}	Thermal diffusivity κ m^2 s^{-1} (x10^{-6})	Coefficient of thermal expansion α 10^{-5} K^{-1}
Sedimentary						
Shale	2100–2700	0.1–0.7	0.1–0.2	1.2–3.0	0.8	
Clay/siltstone				0.80–1.25		
Sand				1.7–2.5		
Sandstone	1900–2500	0.1–0.6	0.1–0.3	1.5–4.2	1.3	3.0
Limestone	1600–2700	0.5–0.8	0.15–0.30	2.0–3.4	1.2	2.4
Dolomite	2700–2800	0.5–0.9	0.1–0.4	3.2–5.0	2.6	
Halite			0.15	5.4–7.2		13
Metamorphic						
Gneiss	2600–2850	0.4–0.6	0.15–0.25	2.1–4.2		
Amphibolite	2800–3150		0.4	2.1–3.8		
Marble	2670–2750	0.3–0.8	0.2–0.3	2.5–3.0	1.0	
Quartzite					2.6	
Igneous						
Basalt	2950	0.6–0.8	0.2–0.25	1.3–2.9	0.9	
Granite	2650	0.4–0.7	0.2–0.25	2.4–3.8	1.0–1.6	2.4
Diabase	2900	0.8–1.1	0.25	2.0–4.0		
Gabbro	2950	0.6–1.0	0.15–0.20	1.9–4.0		1.6
Diorite	2800	0.6–0.8	0.25–0.30	2.8–3.6	1.2	
Granodiorite	2700	0.7	0.25	2.0–3.5		
Mantle						
Peridotite	3250			3.0–4.5	1.7	2.4
Dunite	3000–3700	1.4–1.6		3.7–4.6		
Miscellaneous						
Water				0.6	0.15	
Ice	917		0.31–0.36	2.2	1.2	5.0
Soil				0.17–1.10	0.2–0.6	

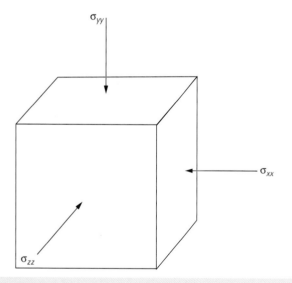

Fig. 2.1 Normal surface forces acting on vertical and horizontal planes. After Turcotte & Schubert (2002).

they are also equal to the weight of overburden, the rock is said to be in a *lithostatic state of stress*. The normal surface forces, σ_{xx}, σ_{yy} and σ_{zz}, are rarely equal when a rock mass is being subjected to tectonic forces. In such a case, the total horizontal surface force (normal stress) acting on a continent, for example, would be made up of two components, a lithostatic term and a tectonic contribution known as a deviatoric stress ($\Delta\sigma_{xx}$),

$$\sigma_{xx} = \rho_c gy + \Delta\sigma_{xx} \qquad [2.1]$$

where ρ_c is the density of the continent. Normal stresses can be either *tensile* when they tend to pull on planes or *compressive* when they push on planes. Horizontal deviatoric stresses may result from uplift producing excess potential energy (Appendices 3, 4) or may be transmitted from plate boundaries, when they are commonly referred to as *in-plane* or *intraplate stresses*.

Surface forces acting parallel to a surface are known as *shear stresses*. Examples are provided by a thrust sheet with a lower fault plane that experiences a frictional resistance, or the gravitational sliding of a rock mass down an inclined plane.

Stress components can be generalised at any point in a material by using the x, y, z coordinate system. At any point we can envisage three mutually perpendicular planes on which there are no shear stresses. Perpendiculars to these planes are known as *principal axes of stress* and can be labelled

σ_1 = maximum principal axis
σ_2 = intermediate principal axis
σ_3 = minimum principal axis,

where convention is that σ is positive for compressional stress and negative for extensional stress. There are certain states of stress that can be described by use of the principal axes notation:

uniaxial stress has a finite σ_1, and $\sigma_2 = \sigma_3 = 0$

biaxial or *plane stress* has $\sigma_1 > \sigma_2$ and $\sigma_3 = 0$

triaxial stress is the general state $\sigma_1 > \sigma_2 > \sigma_3$.

If the principal stresses, σ_1, σ_2 and σ_3, are identical, the state of stress is isotropic and any principal stress is equal to the pressure. In such a case, any set of orthogonal axes qualifies as a principal axis coordinate system. This is known as a *hydrostatic state of stress*. Where the state of stress is not isotropic, the pressure is equal to the mean of the normal stresses

$$p = \frac{1}{3}(\sigma_1 + \sigma_2 + \sigma_3) \qquad [2.2]$$

Subtraction of the mean stress (that is, pressure) from the normal stress component reveals the *deviatoric normal stresses*. It is common for two of the principal stresses to be non-zero, giving the state of *plane stress*. Such a system is suggestive of the horizontal stresses in the lithosphere caused by tectonic processes.

The state of stress of the lithosphere is determined principally by lateral changes of gravitational potential caused by variations in density and elevation of the surface, and by transmission from plate boundaries, basal traction and underlying mantle flow structures. The stress field caused by lateral variations in gravitational potential is called an *ambient lithospheric stress state* (Dahlen 1981), and a plate-scale mean gravitational potential energy can be defined (Coblentz & Sandiford 1994) (Fig. 2.2). Departures of gravitational potential in a lithospheric column from the mean plate-scale value reveals the state of stress at that column, which has important implications for continental tectonics and the development of sedimentary basins. Slow-moving plates surrounded by mid-ocean ridges, such as Africa and Antarctica, are most likely to display an ambient lithospheric stress state since the magnitudes of plate boundary and basal tractional forces are likely to be small.

The gravitational potential energy per unit area of a column of rock above a depth y is given by the integral of the vertical stress σ_{yy} from y to the surface h (Molnar & Lyon-Caen 1988). For Africa, the plate-scale mean gravitational potential energy per unit area is $2.4 \times 10^{14}\,\mathrm{N\,m^{-1}}$. This is similar to the potential energy of cooling oceanic lithosphere at a water depth of nearly 4.5 km and continental lithosphere with an elevation of just 70 m (Fig. 2.2). Most of the continental area of Africa (mean elevation 875 m) therefore has a gravitational potential energy greater than the plate-scale mean, implying that it is in deviatoric tension. Highly elevated regions such as the Ethiopian swell, East African Rift and southern Africa have large horizontal extensional stresses of 15 MPa, 9 MPa and 8 MPa respectively – enough to cause continental stretching (see Chapter 3). In other plates, where there is a convergent margin, such as the northern boundary of the Indo-Australian plate, the state of stress is strongly influenced by plate boundary effects (Hillis & Reynolds 2000) (Fig. 1.13).

The global *in situ* stress field was compiled in the World Stress Map (Zoback 1992), and is continually updated and re-released. Stress orientations in the crust are based on focal mechanism solutions of earthquakes from a range of seismogenic depths, borehole break-outs mostly from sedimentary basins <4 km deep, fault slip data from surface outcrops, and fracture orientations from engineering activities generally restricted to the top 1 km. These various data sources are of different reliability and originate from different depth ranges.

Some continental plates (North and South America, western Europe) show horizontal stress orientations parallel to the direction of plate motion, indicating that the forces driving and/or resisting plate motion are also responsible for the regional stress orientations. However, in the Indo-Australian plate (Hillis & Reynolds 2000) stress orientations do not parallel the N to NNE absolute motion direction and appear to be controlled by its complex convergent boundary stretching from the Himalayas to New Guinea in the north and its largely transcurrent boundary in New Zealand in the east, combined with the effects of ridge-push along its long southern boundary (Fig. 1.13).

The magnitude and orientation of intraplate stress of lithospheric plates is therefore most satisfactorily explained by a combination of topographic (gravitational potential energy) and plate boundary forces (Coblentz *et al.* 1998). This intraplate stress field potentially controls the reactivation of old tectonic structures (Hand & Sandiford 1999), the compartmentalisation of sedimentary basins into smaller depocentres, and may be the source of deviatoric stresses driving lithospheric extension (Chapter 3).

2.1.2 Strain in the lithosphere

Strain is the deformation of a solid caused by the application of stress. We can define the components of strain by considering a rock volume with sides δx, δy and δz, which changes in dimensions but not in shape, so that the new lengths of the sides after deformation are $\delta x - \varepsilon_{xx}\delta x$, $\delta y - \varepsilon_{yy}\delta y$ and $\delta z - \varepsilon_{zz}\delta z$, where ε_{xx}, ε_{yy} and ε_{zz} are the strains in the x, y and z directions (Fig. 2.3). As long as the deformation of the volume element is relatively small, the volume change, or *dilatation*, is simply the sum of the strain components $(\varepsilon_{xx} + \varepsilon_{yy} + \varepsilon_{zz})$. Volume elements may also change their position without changing their shape, in which case the strain components are due to *displacement*.

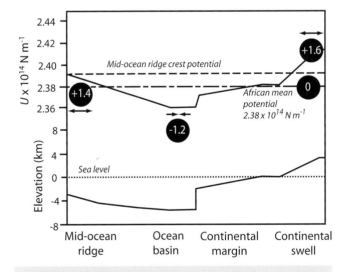

Fig. 2.2 Calculated potential energy U along a transect from an elevated continental swell, an Atlantic-type continental margin, deep ocean basin and mid-ocean ridge, with values of potential difference between calculated values and plate-scale mean potential for Africa in black circles. Profile at base is bathymetry and topography along transect. Modified from Coblentz & Sandiford (1994, p. 832, fig. 1). Calculations of U use a continental crustal density of $2750\,\mathrm{kg\,m^{-3}}$.

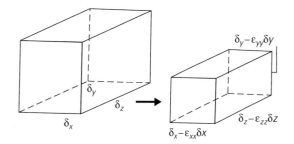

Fig. 2.3 A rectangular block that changes its dimensions but not its shape: this is a deformation involving no shear. After Turcotte & Schubert (2002).

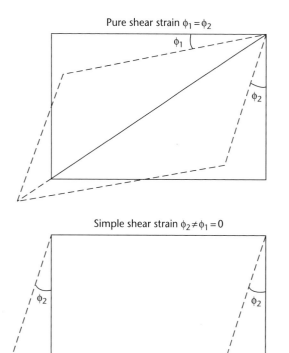

Fig. 2.4 Difference between pure shear strain (no solid body rotation) and simple shear strain (solid body rotation is $\phi_2/2$). After Turcotte & Schubert (2002).

Shear strains, however, may distort the shape of an element of a solid. In the case of a two-dimensional rectangular element that is distorted to a parallelogram, the shear strain is dependent on the amount of rotation of the sides of the rectangular element. Thus, shear strain in the two dimensions, x and y, is determined by the angles through which the sides of the rectangle are rotated (Fig. 2.4).

If the angles through which the sides of the rectangle are rotated (ϕ_1 and ϕ_2 in Fig. 2.4) are not equal, *solid body rotation* is said to have occurred (Fig. 2.5). Solid body rotations do not involve changes in the distances between neighbouring elements of a solid and therefore do not reflect strain.

The deformation of any element can now be described according to the shear strain and the solid body rotation. If no solid body rota-

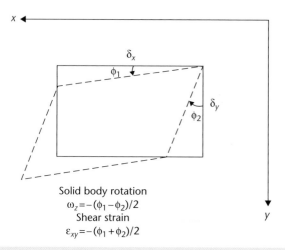

Fig. 2.5 Deformation of a rectangle into a parallelogram by a strain field involving shear. After Turcotte & Schubert (2002).

tion occurs, $\phi_1 = \phi_2$, and the deformation is a result only of shear strains, it is known as *pure shear*. If there is solid body rotation, but $\phi_1 = 0$, the element has undergone *simple shear*. Fig. 2.5 illustrates the two circumstances. We shall see how models of pure shear (e.g. uniform extension with depth) and simple shear (e.g. asymmetrical extension associated with trans-lithospheric shear zones) have been applied to the formation of extensional basins (Chapter 3).

As in the case of shear stresses, shear strains can be described with reference to a coordinate system that is orientated such that shear strain components are zero. Such a system contains the *principal axes of strain*. The fractional changes in length along the directions of the principal strain axes are the *principal strains*. In the three-dimensional case the condition of isotropic strain is satisfied by

$$\varepsilon = (\varepsilon_1 + \varepsilon_2 + \varepsilon_3)/3 = \Delta/3 \qquad [2.3]$$

where Δ is the dilatation and ε is the mean normal strain.

It is very rare, however, for strain to be homogeneous. *Deviatoric strain* components are strains that are the difference between the actual strain and the mean normal strain. Deviatoric strains invariably result from the operation of tectonic processes. Their analysis therefore greatly aids the interpretation of lithospheric deformation.

In an analogous fashion to the treatment of states of stress, we can define different states of strain:

- *Uniaxial strain* is where there is only one non-zero component of principal strain, that is, assuming the non-zero axis to be ε_1, $\varepsilon_2 = \varepsilon_3 = 0$.
- *Plane strain* is where only one of the principal strain components is zero, for example $\varepsilon_3 = 0$ and ε_1 and ε_2 are non-zero. It is a common starting point for studying lithospheric deformation, since it can be assumed that the strain in the direction of an infinite plate will be zero.

Strain in the lithosphere commonly results from tectonic activity. This tectonic activity, such as the slip on faults, is manifested at the surface as a co-seismic surface strain. Although these surface strains may be very small, creep-like displacements, the strains resulting

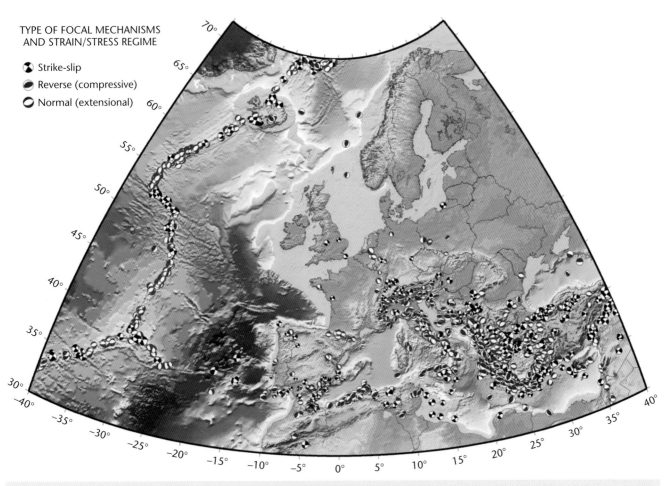

TYPE OF FOCAL MECHANISMS
AND STRAIN/STRESS REGIME

- Strike-slip
- Reverse (compressive)
- Normal (extensional)

Fig. 2.6 Focal mechanism solutions for earthquakes in the Eurasia-Mediterranean area, overlain by topography, comprising 1608 earthquakes with Mw values between 2.6 and 7.5. Reprinted from Olaiz *et al.* (2009), with permission from Elsevier.

from single large earthquakes may be measured in several metres of displacement at the surface, causing railway tracks, fences and roads to be visibly offset. Repeated co-seismic displacements allow the strain rate to be calculated, with units of $[T]^{-1}$. The first ground motion associated with earthquakes (known as first motion studies or focal mechanism studies) allows the sense of motion to be identified as extensional, contractional or strike-slip (Fig. 2.6).

Classical geodesy involves the construction of benchmarks in a geodetic network and the use of triangulation to measure surface strains with a theodolite. However, tectonic displacements can now be measured from satellites in space, especially using the global positioning system (GPS) (Figs 2.6, 2.7). Synthetic aperture radar interferometry is able to make radar backscatter images of the Earth's surface before and after a tectonic movement, thereby allowing the displacement to be accurately quantified (Fig. 2.8). Radar interferometry allows the distribution of surface strain to be imaged at scales from long plate boundaries to kilometre-scale segments of single faults.

2.1.3 Linear elasticity

It is important to know the relationship between the stress and the strain in a piece of lithosphere. These relationships reflect the basic

flow laws of Earth materials (§2.3). *Elastic* materials deform when they are subjected to a force and regain their original shape and volume when the force is removed. For relatively low temperatures, pressures and applied forces, almost all solid materials behave elastically. The relation between stress and elastic strain is linear.

However, at high temperatures and pressures or high levels of stress, rocks do not behave elastically. In near-surface regions where temperature and pressures are low, rocks deform by brittle fracture at high levels of stress. Deeper in the Earth, high temperatures and pressures cause the rock to deform plastically under an applied force, with no fracturing. Brittle materials that have exceeded their yield strength, and plastic materials, do not regain their original shape when the force is removed.

Since much of the lithosphere behaves as a strong material over geological (i.e. >10^6 yr) periods of time, it is able to bend under surface loads, to store the elastic stresses responsible for earthquakes, and to transmit stresses over large horizontal distances. This fundamental property of the lithosphere is crucial to an understanding of the formation of sedimentary basins.

The theory of linear elasticity underpins a great deal of thought on lithospheric mechanics and often constitutes the basic assumption in models of lithospheric behaviour. In a linear, isotropic, elastic solid the stresses are linearly proportional to strains, and mechanical

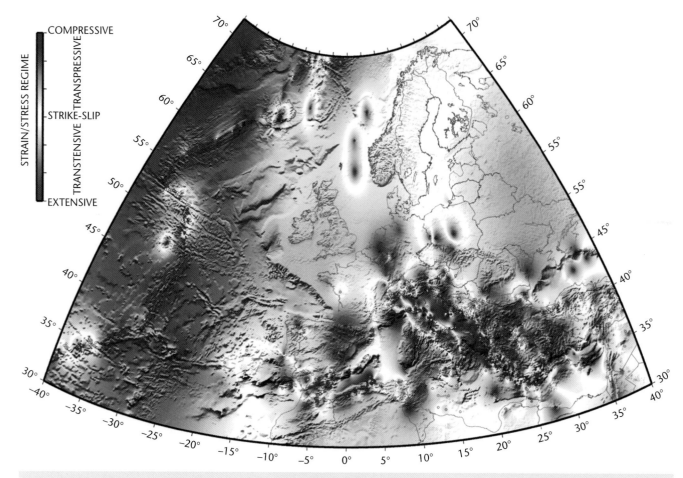

Fig. 2.7   Map of continuous stress-strain for the western Eurasian plate. Colour coding is based on calculations of the ratio of the maximum horizontal strain divided by the maximum vertical strain. Note the high amount of extensional strain in the Aegean Sea. Note also the change from the compressional Adriatic domain to the extensional Tyrrhenian domain in Italy. Reprinted from Olaiz *et al.* (2009), with permission from Elsevier.

properties have no preferred orientation. The principal axes of stress and the principal axes of strain coincide. The relation between the principal strain and the components of principal stress can be stated as follows.

$$\varepsilon_1 = \frac{\sigma_1}{E} - \frac{v\sigma_2}{E} - \frac{v\sigma_3}{E} \qquad [2.4]$$

$$\varepsilon_2 = -\frac{v\sigma_1}{E} + \frac{\sigma_2}{E} - \frac{v\sigma_3}{E} \qquad [2.5]$$

$$\varepsilon_3 = -\frac{v\sigma_1}{E} - \frac{v\sigma_2}{E} + \frac{\sigma_3}{E} \qquad [2.6]$$

The exact partitioning of stresses to give a resultant strain is clearly strongly influenced by E and v, which are material properties known as *Young's modulus* and *Poisson's ratio* respectively. In general terms, a principal stress produces a strain component σ/E along the same axis and strain components $-v\sigma/E$ along the two other orthogonal axes.

Where only one of the principal stresses is non-zero (uniaxial stress), a shortening in the direction of the applied compressive stress will be accompanied by an extension in the two orthogonal directions (Fig. 2.9), and vice versa. Under these conditions where, let us say $\sigma_2 = \sigma_3 = 0$ and $\sigma_1 \neq 0$, there is a simple relation along the axis of uniaxial stress

$$\sigma_1 = E\varepsilon_1 \qquad [2.7]$$

This well-known relationship is called *Hooke's law* (Fig. 2.9).

There is a fractional volume change or dilatation due to uniaxial stress, but the contraction in the direction of uniaxial stress is compensated by expansion by half as much in the two other orthogonal directions.

In an isotropic state of stress, all the principal stresses and principal strains are equal. The pressure under these conditions is related to the dilatation by K, the *bulk modulus*, or its reciprocal β, the *compressibility*. These parameters therefore give the fractional volume change during isotropic compression under a given pressure. If there is a volume change, in order that matter is conserved, there must be

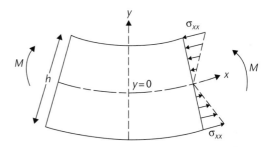

showing that as *v* approaches 1/2 the bulk modulus tends to infinity, that is, the material becomes essentially incompressible.

2.1.4 Flexure in two dimensions

Since the lithosphere behaves elastically, it is able to bend when force systems or loads are applied to it. We will return to this topic in considerable detail when considering the initiation and maintenance of foreland basins (Chapter 4). The aim here is to provide a brief background of the way the lithosphere responds by flexure to these applied force systems. A full analysis is provided by Turcotte and Schubert (2002), and the derivation of the general flexure equation is given in Appendix 5. The concepts involved in the flexure of an elastic solid may be briefly summarised as follows:

- Flexure results from vertical forces, horizontal forces and torques (bending moments) in any combination. Horizontal loads are commonly neglected, perhaps unwisely, in many geodynamical problems.
- The bending moment is the integration of the fibre (normal) stresses on notional cross sections of the plate acting over the distance to the mid-plane of the plate (Fig. 2.10). The bending moment is related to the local radius of curvature of the plate by a coefficient called the *flexural rigidity*. Flexural rigidity is proportional to the cube of the equivalent *elastic thickness*. When applied to the lithosphere, the equivalent elastic thickness does not represent a real physical discontinuity.
- A general flexural equation (see eqn. [2.10]) can be derived that expresses the deflection of the plate in terms of the vertical and horizontal loads, bending moment and flexural rigidity. This equation can readily be adapted for use in the study of geological problems.

There will be different bending stresses under situations where the plate is pinned at one or both ends, and according to whether it is point loaded (for example, at its free end) or uniformly loaded along its length, or loaded in some other fashion. However, the general flexural equation provides the basic starting point for more specific analyses.

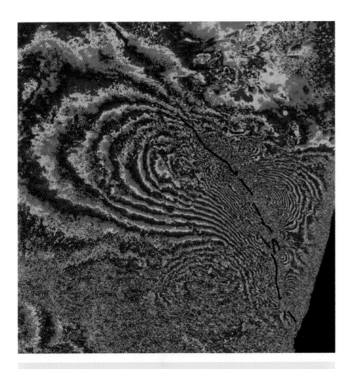

Fig. 2.8 Radar interferometric image of the surface displacement resulting from the Landers earthquake, California. Each colour cycle represents 2.8 cm of displacement. The surface rupture of the fault is shown in black. From Massonnet *et al.* 1993, reproduced in colour at http://southport.jpl.nasa.gov/scienceapps/dixon/report3.html. Reproduced with permission of Nature Publishing Group.

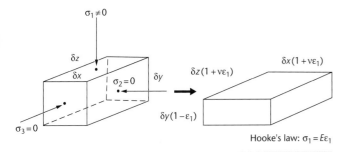

Hooke's law: $\sigma_1 = E\varepsilon_1$

Fig. 2.9 Deformation under a uniaxial stress. Contraction in the direction of the compressive stress σ_1 is compensated by extension in the two orthogonal directions. Hooke's Law is the relation between stress and stress along the principal axis of compressive stress. After Turcotte & Schubert (2002).

an increase in density (see §2.1.6 and Appendix 1). Such a density increase $\delta\rho$ is given simply by

$$\delta\rho = \rho\beta p \qquad [2.8]$$

where *p* is the pressure, ρ is the density of the solid element, and β the compressibility. In terms of the previously defined Young's modulus *E* and Poisson's ratio *v*,

$$K = \frac{1}{\beta} = \frac{E}{3(1-2v)} \qquad [2.9]$$

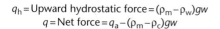

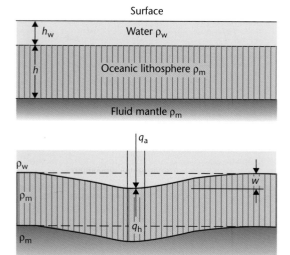

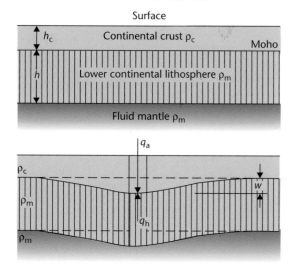

Fig. 2.11 Model for calculating the upward-acting hydrostatic restoring force q_h on an oceanic plate overlain by water, deflected by an applied load q_a. After Turcotte & Schubert (2002).

Fig. 2.12 Model for calculating the upward-acting hydrostatic restoring force q_h on the base of some continental crust where the deflection caused by the applied load q_a is assumed to be filled with material of the same density as the continental crust. This approximates the case of a fully compacted sediment-filled basin on continental lithosphere. After Turcotte & Schubert (2002).

When an applied load flexes a plate, the deflected region is filled either with water, as in the case of oceanic lithosphere or a starved continental basin, or with sediment, as in the case of most basins adjacent to hinterlands undergoing erosion. This infilling material has a lower density than the mantle that is being replaced (Figs 2.11, 2.12). The density difference can be denoted by $\Delta\rho$. The magnitude of the restoring force on the base of the deflected plate can be estimated by considering a balance of pressure (ρgh) under the region of maximum deflection and under the unaffected region. This upward restoring force is $\Delta\rho gw$, and the net vertical force acting on the plate is the applied load less this restoring hydrostatic force. The general flexural equation (Appendix 5) therefore becomes

$$D\frac{d^4w}{dx^4} + P\frac{d^2w}{dx^2} + \Delta\rho gw = q_a(x) \qquad [2.10]$$

where $\Delta\rho$ is ($\rho_m-\rho_w$) for a purely water-filled basin (Fig. 2.11) and $\Delta\rho$ is ($\rho_m-\rho_s$) for a fully sediment-filled basin (Fig. 2.12). The analytical solution of the general flexure equation in the context of a foreland basin is given in §4.2.

The lithosphere has a different flexural response according to the spatial distribution of the load $q_a(x)$. If the wavelength of a load of a certain mass, for example excess topography, is sufficiently short, the vertical deflection of the lithosphere is small, and the lithosphere can be regarded as infinitely rigid for loads of this scale. However, if the wavelength of a load of the same mass is sufficiently long, there is an effective isostatic response approaching hydrostatic equilibrium, and the lithosphere appears to have no rigidity. These two situations must be regarded as end members. The degree of compensation of the topographic load is the ratio of the deflection of the lithosphere to its maximum or hydrostatic deflection. The response of the lithosphere to an applied load, such as the excess mass in a mountain belt,

is identical to its response to sediment loads in a basin. The degree of compensation for sediment loads of varying wavelength is dealt with in §2.1.5 and revisited in §9.3.2.

For reasonable values of plate thickness and flexural rigidity it is found that horizontal forces applied at the end of a plate are generally inadequate to cause buckling. Horizontal forces as buckling agents may, however, be much more important where the lithosphere has been strongly thinned in regions of high heat flow or where it is rheologically layered and decoupled mechanically (§4.4).

The general features of basins such as oceanic trenches and foreland basins adjacent to mountain belts can be explained by flexural models. Applications of the foregoing discussion of elasticity and flexure to sedimentary basin analysis are given in Chapter 4.

2.1.5 Flexural isostasy

The operation of flexure can be recognised in a number of settings in nature. The clearest examples are isostatic rebound during deglaciation, and the downflexing of moats adjacent to oceanic seamounts and island chains. Other important examples are the formation of foreland basins associated with mountain building, and the flexing of ocean plates ahead of subduction zones, whereas less obvious cases are the subsidence beneath sediment loads such as the Amazon fan. In all these cases the lithosphere supporting tectonic, volcanic and sedimentary loads is strong enough to store elastic stresses. Instead of the Airy isostatic model, in *flexural isostasy*, loads are compensated regionally by a lithosphere approximating an elastic sheet overlying a fluid substratum.

The problem of flexural isostasy can be tackled by considering the way the lithosphere supports periodic topography, as given in

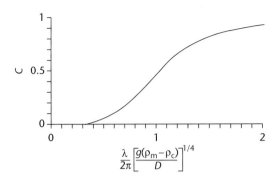

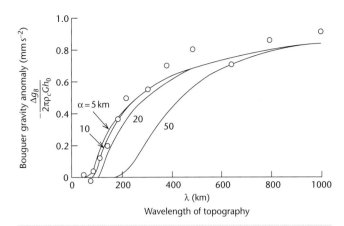

Fig. 2.13 Dependence of the degree of compensation C on the non-dimensional wavelength of periodic (sinusoidal) topography. D is the flexural rigidity, λ is the wavelength of the load, ρ_m and ρ_c are the mantle and crustal or infill densities, and g is the acceleration due to gravity. From Turcotte & Schubert (2002, p. 123, fig. 3.26), © Cambridge University Press, 2002.

Fig. 2.14 Spectral technique for the estimation of flexural rigidity from the wavelength of the Bouguer gravity anomaly. Correlation of Bouguer gravity with topography (admittance) for the United States (Dorman & Lewis 1972) compared with the gravity formula for loading by periodic topography, for different values of the flexural parameter α. The best agreement is with $\alpha = 20\,\text{km}$, or flexural rigidity $D = 10^{21}\,\text{Nm}$, which with $E = 70\,\text{GPa}$ and $v = 0.25$, gives an equivalent elastic thickness Te of 6 km. After Turcotte & Schubert (2002).

Appendix 6. This simplified analysis can then be extended to the case of a sedimentary basin of a certain wavelength and amplitude.

The way in which the lithosphere supports loads flexurally, or the degree of compensation C, is related to the rigidity of the plate undergoing flexure and also the wavelength of the load. With a sinusoidal, periodic load, the degree of compensation is the magnitude of the maximum deflection compared to the maximum deflection expected from a plate with zero flexural rigidity (Airy isostasy), or expressed differently

$$C = \frac{(\rho_m - \rho_s)}{\rho_m - \rho_s + \frac{D}{g}\left(\frac{2\pi}{\lambda}\right)^4} \qquad [2.11]$$

where $(2\pi/\lambda)$ is termed the *wave number* (e.g. Watts 1988) (Fig. 2.13).

The same idea of compensation as a function of wavelength can be used to investigate the flexural support for the sediment loads in sedimentary basins (Chapter 9). The sediment load is assumed to have a characteristic wavelength. Taking the example of a sedimentary basin 200 km wide ($\lambda/2 = 200\,\text{km}$) with a sinusoidal sediment load and an underlying lithosphere of flexural rigidity $10^{24}\,\text{Nm}$, $(\rho_m - \rho_s) = 800\,\text{kg m}^{-3}$, the degree of compensation C is about 0.12. This suggests that the lithosphere behaves very rigidly to this wavelength of load. Changing the wavelength of the load such that $\lambda/2$, the width of the basin, is now 400 km, $C = 0.68$, indicating that the sediment load is only weakly supported. In this case of large compensation, Airy-type isostasy is approached. It must be stressed, however, that rarely can C be estimated accurately in geological situations.

The very common negative Bouguer anomalies extending well beyond the topography of mountain belts suggest that they are flexurally compensated. In present-day settings, the extent of this compensation can be estimated from the correlation of Bouguer gravity anomalies with topography as a function of its wavelength, known as *admittance* (Fig. 2.14) (Dorman & Lewis 1972). Topography with a wavelength of less than 100 km is not compensated, whereas topography with a wavelength of greater than 1000 km is fully compensated. From Fig. 2.14, it can be seen that different curves are predicted for different flexural rigidities. The best fit for data from western

United States is for a flexural rigidity of just $10^{21}\,\text{Nm}$ (equivalent elastic thickness of c.6 km).

A related spectral technique is the conversion of both the topography and Bouguer gravity field by a fast Fourier transform (Forsyth 1985). The *coherence* function is the square of the correlation coefficient between the topographic and gravity signals. A coherence of 1 indicates complete compensation on a plate with no flexural strength, and a coherence of 0 indicates that the load is completely supported by the strength of the plate. On a plot of coherence versus wavenumber ($k = 2\pi/\lambda$), the rollover from a coherence of 1 to 0 gives a measure of the flexural rigidity or equivalent elastic thickness. Lowry and Smith (1994) mapped flexural rigidity variations across the Basin and Range/Colorado Plateau/Rocky Mountain area using coherence analysis. They found that areas of low flexural rigidity (average of $9 \times 10^{21}\,\text{Nm}$, Te = 10 km) correlated with areas of high surface heat flows in the Basin and Range. The highest flexural rigidities were in the Rocky Mountains (average of $3 \times 10^{23}\,\text{Nm}$, Te = 33 km), with areas of Archaean cratons reaching elastic thicknesses of 77 km.

The good correspondence of flexural rigidity with heat flow indicates that the flexural rigidity of the continental lithosphere is related to the thickness of the elastic layer overlying a temperature-dependent ductile layer. This elastic layer thickness also contains almost all earthquakes on the continents (Maggi *et al.* 2000; Jackson *et al.* 2008) (Fig. 1.10), so there is a close correspondence between the seismogenic crust, the thickness of the strong elastic layer and the effective elastic thickness (§4.3). A global map of elastic thickness based on coherence is found in Audet and Bürgmann (2011) (Chapter 4).

2.1.6 Effects of temperature and pressure on rock density

Since rocks are compressible (they can change in volume), they change in density due to changes in both temperature and pressure.

Owing to their dependence on pressure, rock densities are commonly expressed as a zero-pressure value. For example, the rock type comprising the mantle is peridotite, which has a zero-pressure density of 3250 kg m^{-3}, whereas the gabbros and basalts that comprise the oceanic crust have zero-pressure values of 2950 kg m^{-3}, and the granites making up most of the continental crust have zero-pressure values of about 2650 kg m^{-3} (Table 2.2).

To work out the increase in density with increasing depth within the Earth (Appendix 1), we assume an isotropic state of stress, in which case the pressure is related to the dilatation Δ and the reciprocal of the compressibility β (known as the bulk modulus K), as given in eqn. [2.9]. Since the density change $\delta\rho$ of a solid element with volume V is related to the fractional change of volume δV, which is equal to the dilatation Δ, and since there is no change in mass, the isothermal change in density, in terms of the pressure, the initial density and the compressibility, is given by eqn. [2.8].

The density of rock increases with depth below the surface of the Earth due to isothermal compression. For example, at 1 GPa, equivalent to a depth of 38.5 km in granite, the density of granite increases from its zero-pressure value of 2650 to 2730 kg m^{-3}. Assuming the mantle lithosphere to be peridotitic in composition, peridotite increases in density from 3282 kg m^{-3} at 1 GPa (under 38.5 km of granite) to 3477 kg m^{-3} at 7 GPa, equivalent to a depth of 205 km, close to the base of the thermal lithosphere in continental interiors.

We can incorporate the effects of thermal expansion on rock density. When there is no change in pressure (isobaric), the density of a rock depends on its density at a reference temperature such as 0 °C, denoted ρ^*, the volumetric coefficient of thermal expansion α_v – the fractional change in volume with temperature *at a constant pressure* – and the temperature relative to the reference temperature ΔT

$$\rho = \rho^* (1 - \alpha_v \Delta T) \qquad [2.12]$$

Taking the volumetric coefficient of thermal expansion as 2.4×10^{-5} K^{-1} for granite, a geotherm (§2.2) using a basal heat flow of 30 mW m^{-2}, a radiogenic heat production of 2.5×10^{-6} W m^{-3}, an exponential depth constant of 10 km, a thermal conductivity of 3.0 W m^{-1}°C^{-1}, a surface temperature of 0 °C, and a density at a reference temperature of 0 °C of 2650 kg m^{-3}, the density at a new temperature of 430 °C, which is approximately the Moho temperature (depth of 35 km), is 2620 kg m^{-3}. Comparing the density decrease caused by isobaric thermal expansion of −30 kg m^{-3} and the density increase caused by the isothermal compression of +80 kg m^{-3}, it can be seen that the compression effect dominates in the granitic crust.

Turning to peridotite, and using the same volumetric thermal expansion coefficient of 2.4×10^{-5} K^{-1} and the same geotherm, and a density at 0 °C of 3250 kg m^{-3}, the density at a depth of 35 km is 3216 kg m^{-3}, and the density at a depth of about 200 km is 3088 kg m^{-3}. The density decrease caused by the isobaric thermal expansion between the Moho and the base of the thermal lithosphere is therefore −128 kg m^{-3}. This can be compared with the density increase caused by the isothermal compression of +195 kg m^{-3}. The density increase caused by the increasing pressure with depth therefore also dominates in the peridotitic mantle lithosphere (Fig. 2.15).

The relative effects on rock density of changes in temperature and pressure can also be assessed from their contribution to the state of stress. For the situation of zero horizontal strain, and where the rock behaves in a linear elastic fashion, the horizontal thermal stress

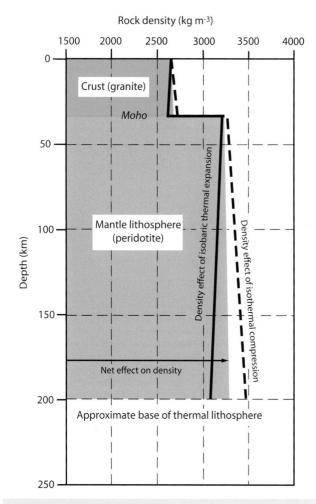

Fig. 2.15 Effects of temperature and pressure on the density of rocks with depth in the Earth. Densities are calculated for a geothermal gradient determined by a basal heat flow of 30 mW m^{-2}, a radiogenic heat production in the crust of 2.5×10^{-6} W m^{-3}, a depth constant for radiogenic heat production a_r of 10 km, thermal conductivity of 3 W m^{-1} K^{-1}, compressibility of 3×10^{-11} and 1×10^{-11} Pa^{-1} for granite and peridotite respectively, volumetric coefficient of thermal expansion of 2.4×10^{-5} K^{-1} for both granite and peridotite, Poisson's ratio of 0.225 and 0.25 for crustal granite and mantle peridotite respectively, and zero-pressure densities of 2650 and 3250 kg m^{-3} for granite and peridotite respectively. The trade-off between the thermal expansion due to increasing temperature with depth versus the increased density due to increased pressure with depth can be gauged by a dimensionless number N_T. For the most plausible range of parameter values, the effect of compression exceeds thermal effects.

relative to the horizontal lithostatic stress is given by a dimensionless number (Wangen 2010, p. 61) N_T, given by

$$N_T = \frac{E \alpha G}{3 v \rho_b g} \qquad [2.13]$$

where G is the linear geothermal gradient ($\Delta T = Gy$), E is Young's modulus, v is Poisson's ratio, and ρ_b is the bulk density of the rock column above depth *y*. For the case of a 30 km-thick granitic crust overlying a peridotitic mantle, a linear geothermal gradient of 20 °C km^{-1}, $E = 60$ GPa, and $v = 0.25$, N_T is approximately 0.6, showing that density changes due to thermal expansion and contraction are a significant fraction of the effects of pressure. When $G > 30$ °C km^{-1}, the thermal and pressure effects are approximately equal.

Taking into account the dual effects on density of increasing temperature with depth and increasing pressure with depth, it is clear that where the continental lithosphere is thick, the deep subcrustal part is prone to gravitational collapse into the hot asthenosphere since it is denser. Melt segregation from the thick continental subcrustal lithosphere under continental interiors (cratons) is probably required to reduce the density enough to prevent delamination of continental roots (O'Hara 1975; Jordan 1979). Melting selectively removes garnet, which is heavy, and reduces the iron content of the residual mantle. Changes in density of the residual mantle rock due to melting depends on the percentage of melt removed (Preistley & McKenzie 2006), 20% melt removal, for example, reducing mantle density from 3390 to 3340 kg m^{-3}. This has a significant effect on buoyancy and stability of the mantle lithosphere of continental cores.

2.2 Heat flow

2.2.1 Fundamentals

To understand the mechanical behaviour of the Earth it is necessary to know something of its thermal structure, since rock rheologies commonly depend on temperature, itself a function of depth. The temperature distribution of the Earth must reflect the inputs and outputs of heat to the Earth system. In other words, there is a heat transfer or flow, achieved by processes of conduction, convection and radiation. The essential differences between these processes are as follows. Conduction is a diffusive process whereby kinetic energy is transferred by intermolecular collisions. Convection, on the other hand, requires motion of the medium to transmit heat. Electromagnetic radiation, such as that of the Sun, can also transmit heat, but it is of relatively minor importance in the Earth's *internal* heat budget. It is, however, of paramount importance in determining Earth's *surface* heat budget.

The processes of conduction and convection are of differing importance in different zones. In the lithosphere, heat is transported primarily through conduction, whereas in the mantle, convection of heat from the Earth's deep interior is dominant. Convection is a much faster process of heat transfer than conduction.

The fundamental relation for conductive heat transport is given by *Fourier's law*. It states that the heat flux *Q* is directly proportional to the temperature gradient, and takes the mathematical form

$$Q = -K\frac{dT}{dy} \qquad [2.14]$$

where *K* is the coefficient of thermal conductivity, *T* is the temperature at a given point in the medium, and *y* is the coordinate in the direction of the temperature variation, generally vertical for heat flow through the lithosphere to the Earth's surface.

The heat flux at the Earth's surface gives a good indication of processes within the interior. Temperature measurements can be made on land in caves, mines, and, better still, in deep boreholes, and

allow the heat flux to be calculated as long as the thermal conductivity *K* is known. *K* can be measured in the laboratory on rock samples by subjecting them to a known heat flux and measuring the temperature drop across the sample (Beardsmore and Cull 2001) (Table 2.1). Temperature measurements of the ocean floor can also be made by penetrating seafloor sediments with a temperature probe. The same probe contains a heater that enables the *in situ* thermal conductivity to be calculated. The wide range of values of thermal conductivity for sedimentary rocks is due, to a large extent, to large variations in porosity (see also §2.2.6). Heat flow is in units of mW m^{-2} or cal cm^{-2} s^{-1}. Surface heat fluxes are sometimes expressed in heat flow units where 1 HFU is equivalent to 10^{-6} cal cm^{-2} s^{-1} or 41.84 mW m^{-2}. Thermal conductivity is measured in units of watts per metre per degree centigrade (W m^{-1} °C^{-1}), and occasionally in cal cm^{-1} °C^{-1}.

Earth's average surface heat flow (87 mW m^{-2}) corresponds roughly to one household light bulb (100 W) over an area of a tennis court. However, heat flow measurements of the oceans and continents reveal important variations (Table 2.2, Sclater *et al.* 1980a, 1981, §1.4.4). Regions of high heat flow on the continents generally correspond to active volcanic areas, such as the Andes, to those areas underlain by highly radiogenic crust, or to regions of extensional tectonics, such as the Basin and Range province of western USA. Continental collision zones typically have low to normal surface heat flows. In areas devoid of active tectonics and volcanicity, the heat flow appears to be inversely correlated to the age of the rocks (Jessop & Lewis 1978) (Table 2.3). This can be explained by the decreasing abundance with age of the radioactive heat-producing isotopes of uranium, thorium and potassium (§2.2.3). Surface heat flows are strongly influenced by the underlying rock type. Granites produce large amounts of radiogenic heat, whereas basalts and peridotites produce almost no radiogenic heat (§2.2.3) (Table 2.4). In the oceans the surface heat flows are related not to the concentration of radioisotopes but to the age of the seafloor. Newly created oceanic crust cools by conduction as it travels away from the mid-ocean ridge, thereby explaining this relationship (§2.2.7). Mean oceanic surface heat flows (101 mW m^{-2}) are higher than their continental counterparts (65 mW m^{-2}) (Table 2.3). Approximately 60% of the heat loss of the Earth takes place through the ocean floor (Parsons 1982).

Since radioactive isotopes decay to stable daughter products, there must be a steadily decreasing heat production with time from radioactive decay, known as *secular cooling*. The rate at which heat is being transferred to the Earth's surface is therefore also decreasing with time, in turn slowing down the mantle convection system. Analysis of the abundance of the heat-producing radioisotopes and their stable daughter products suggests that heat production was twice the present value 3000 million years ago (Table 2.4, Fig. 2.16), with ^{238}U and ^{232}Th taking over from ^{235}U and ^{40}K as the main heat producers because of the latters' relatively short half-lives.

Heat travels down temperature gradients (eqn. [2.14]). The lithosphere is subjected to changes in temperature with time, particularly associated with tectonic processes. Consequently, it is important to be able to develop a rule for the heat loss or gain of an element of lithosphere (Appendix 7). Once such a rule is established, it can be modified to account for three-dimensional variations in heat flow.

2.2.2 The geotherm

The variation of temperature with depth is called the *geotherm*. In Appendices 7 and 8, we solve the one-dimensional conduction

Table 2.3 Global mean surface heat flow. Data from Pollack *et al.* (1993) and Stein & Stein (1992) in brackets. Continental and oceanic surface heat flows depend on surface geology and age as well as asthenospheric temperatures

	Mean surface heat flow	
	mW m^{-2} and standard error	Heat flow units (HFU)
Continents:	65	1.55
Cenozoic (sedimentary/metamorphic) (0–65 Ma)	63.9 ± 0.9	1.53
Cenozoic (igneous) (0–65 Ma)	97.0 ± 5.6	2.32
Mesozoic (sedimentary/metamorphic) (65–251 Ma)	63.7 ± 1.3	1.52
Mesozoic (igneous) (65–251 Ma)	64.2 ± 3.0	1.53
Paleozoic (sedimentary/metamorphic) (251–544 Ma)	61.0 ± 1.2	1.46
Paleozoic (igneous) (251–544 Ma)	57.7 ± 2.6	1.38
Proterozoic (544–2500 Ma)	58.3 ± 1.4	1.39
Archaean (2000–3800 Ma)	51.5 ± 2.4	1.23
Oceans:	101	2.41
Cenozoic (0–65 Ma)	89.3 (125.2) ± 2.8	2.13
Mesozoic (65–251 Ma)	44.6 (51.0) ± 2.8	1.07
Worldwide	87	2.08

Table 2.4 Typical concentrations of radioactive elements from typical rock types comprising continental and oceanic crust and undepleted mantle (from Fowler 1990). ^{238}U and ^{235}U have half-lives of 4.47×10^9 yr and 7.04×10^8 yr, ^{232}Th has a half-life of 1.40×10^{10} yr, and ^{40}K has a half-life of 1.25×10^9 yr. Because of their shorter half-lives, ^{40}K and ^{235}U were the most important heat producers in the distant past, whereas today heat is produced mostly by ^{238}U and ^{232}Th

	Granite	Tholeiitic basalt	Alkali basalt	Depleted peridotite (harzburgite)	Average continental upper crust	Average oceanic crust	Undepleted 'fertile' mantle
Concentration by weight							
U (ppm)	4	0.1	0.8	0.006	1.6	0.9	0.02
Th (ppm)	15	0.4	2.5	0.04	5.8	2.7	0.10
K (%) (10^{-19} W kg^{-1})	3.5	0.2	1.2	0.01	2.0	0.4	0.02
Heat generation (10^{-18} W kg^{-1})							
U	3.9	0.1	0.8	0.006	1.6	0.9	0.02
Th	4.1	0.1	0.7	0.010	1.6	0.7	0.03
K	1.3	0.1	0.4	0.004	0.7	0.1	0.007
Total	9.3	0.3	1.9	0.020	3.9	1.7	0.057
Density (kg m^{-3})	2.7	2.8	2.7	3.2	2.7	2.9	3.2
Heat generation (10^{-6} W m^{-3})	2.5	0.08	0.5	0.006	1.0	0.5	0.02

equation to derive the geotherm. It is commonly said that the mean geothermal gradient in the continental lithosphere is 30 °C km^{-1}, based on the gradient calculated in deep boreholes and coal mines. But extrapolation of this rate downwards from the surface would result in temperatures of 900 °C at a Moho depth of 30 km, and of 3750 °C at the base of a 125 km-thick lithosphere. Comparison with the solidus for peridotite demonstrates that this cannot be the case, since downward extrapolation implies that the mantle lithosphere would undergo wholesale melting, yet it is able to transmit S-waves.

If we continue to assume that the heat flow is by conduction in one direction from a hot mantle to a cold Earth surface, with a value of 30 mW m^{-2}, through lithosphere with a thermal conductivity of 3 W m^{-1} K^{-1}, giving a mean geothermal gradient of 10 K km^{-1}, the geotherm would closely approach the solidus for peridotite at the base of the lithosphere (Fig. 2.17). However, this geothermal gradient is considerably lower than observations suggest in the shallow crust and in sedimentary basins.

We conclude that the conduction geotherm given by eqn. [2.14] does not apply throughout the continental lithosphere. The solution to this problem is that the continental lithosphere generates its own heat by radiogenic decay. This self-heating causes the geotherm to be curved, with a higher geothermal gradient at shallow depths and a lower geothermal gradient in the rest of the continental lithosphere. The framework for the understanding of radiogenic heat production is given in §2.2.3.

The radiogenic heat production cannot be assumed to be uniform with depth since the abundances of the heat-producing radiogenic isotopes vary strongly according to rock type (Table 2.4). The heat generation of the crust, being 'granitic' in composition, far exceeds that of the mantle. Despite the fact that the very large volume of the mantle means that it contributes 80% of the Earth's radiogenic heat, it is the crustal contribution that is of importance in basin analysis and that determines the continental geotherm in the lithosphere. The question is how the radiogenic heat generation in the continental crust is modelled (Appendix 9) (§2.2.3).

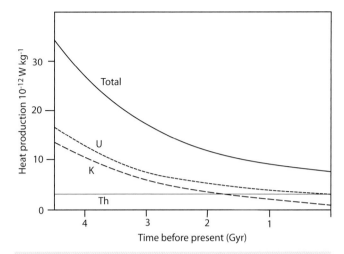

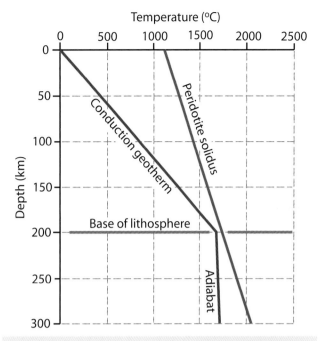

Fig. 2.16 Rate of heat production in W kg^{-1} due to the decay of radiogenic isotopes of U, K and Th, and the total heat production from all isotopes. Uranium concentrations in the mantle are 31×10^{-9} kg kg^{-1}, but heat production is 9.81×10^{-5} W kg^{-1}: Thorium has a higher concentration of 124×10^{-9} kg kg^{-1}, but a lower heat production of 2.64×10^{-5} W kg^{-1}: K is highly concentrated at 31×10^{-5} kg kg^{-1}, but has a low heat production rate of 3.48×10^{-9} W kg^{-1}. The secular reduction of heat production over time is due to the various half-lives of the isotopic systems, which vary from c.10^9 to 10^{10} yrs. From Turcotte & Schubert (2002). © Cambridge University Press, 2002.

Fig. 2.17 The geotherm for the lithosphere and the solidus for peridotite. The geotherm approaches the solidus at the base of the lithosphere, where partial melting may take place. The geothermal gradient is calculated using 1D conduction with a thermal conductivity of 3.0 W m^{-1} K^{-1}, a basal heat flow of 30 mW m^{-2}, and a surface temperature of 0 °C. Pressure is calculated using a granitic 40 km-thick crust with density 2750 kg m^{-3} and a mantle lithosphere extending to a depth of 200 km with density 3300 kg m^{-3}. The geotherm is adiabatic in the asthenosphere and taken as 0.5 °C km^{-1}. The solidus for peridotite is given by $T = 1500$ (K) $+ 0.12p$ where pressure (p) is in MPa. Note that the geotherm approaches the solidus near the lithosphere–asthenosphere boundary, leading to partial melting.

In some cases, it is possible that heat is conducted in more than one direction, as may be common in regions of large lateral variations in surface temperatures. Such lateral variations might also arise where the lithosphere is stretched over a relatively narrow zone, as is common in strike-slip basins (Chapter 6), so that there is both upward and lateral loss of heat by conduction.

In the following sections, we make some general observations about radiogenic self-heating of the crust and its effect on the continental geotherm (§2.2.3) before discussing the effect of erosion and sedimentation on the geotherm (§2.2.4 & §2.2.5), and the effect of variable thermal conductivity on the geotherm of the sedimentary basin-fill (§2.2.6). The thermal history of the basin-fill is dealt with in Chapter 10.

2.2.3 Radiogenic heat production

Radiogenic heating is caused by the spontaneous decay of the elements uranium (^{235}U and ^{238}U), thorium (^{232}Th) and potassium (^{40}K). Partial melting of mantle rock, for instance at a mid-ocean ridge, depletes the residuum in incompatible elements such as U, Th and K, which are concentrated in the basaltic melt (Table 2.4). Consequently, ocean crust is concentrated in radiogenic elements relative to 'fertile' (undepleted) mantle, and even more concentrated relative to depleted residual mantle rock. Processes causing growth of continental crust further concentrates the incompatible elements, so concentrations of radiogenic elements are an order of magnitude higher in continental crust than in oceanic tholeiitic basalts. It therefore follows that the radiogenic self-heating of the lithosphere may be highly variable.

In areas devoid of active tectonics and volcanicity, continental heat flow values appear to be strongly correlated with the type of underlying crust. In NE North America, for example, the higher heat flows over the Appalachians (average 58 mW m^{-2}) than over the North American Shield (average 29 mW m^{-2}) may be explained by the different thicknesses of underlying tonalitic crust (Pinet *et al.* 1991) containing the radiogenic HPEs uranium, thorium and potassium (§2.2). In general, granitic terranes have high surface heat flows, whereas basic and ultrabasic igneous rocks and many sedimentary rocks are associated with low surface heat flows.

The contribution of radiogenic heating to the continental geotherm is generally approached in two ways (Fig. 2.18): (i) the assumption of a slab or small number of slabs of uniform radiogenic heat production (in W kg^{-1} or W m^{-3}) (solutions in Appendix 9); and (ii) the assumption that there is an exponential reduction of the radiogenic heat production A from a surface value A_0 at a rate given by the depth constant a_r:

$$A = A_0 \exp(-y/a_r) \qquad [2.15]$$

At a depth of a_r km, A has reduced to $1/e$ of its surface value A_0.

(a)

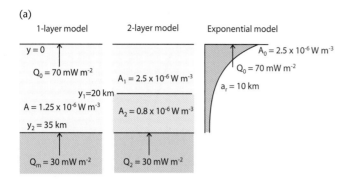

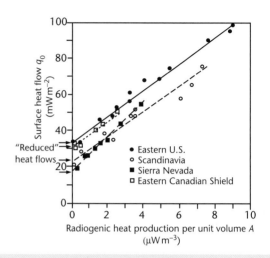

Fig. 2.19 Surface heat flow versus radiogenic heat production for a number of different geological provinces, showing a linear regression. The intercept on the y-axis gives a reduced, basal or mantle heat flow. See text for a discussion of the interpretation of these plots. Data from Roy *et al.* (1968).

(b)

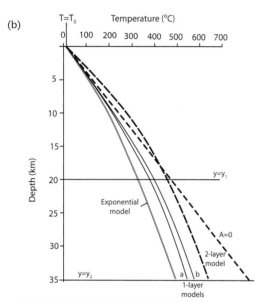

Fig. 2.18 Geotherms for the continental crust. (a) Model set-up: one-layer model with a constant internal heat generation for a 35 km-thick crust with (i) a surface heat flow of 70 mW m^{-2}, and (ii) a basal heat flow of 30 mW m^{-2}; a two-layer model with a highly radiogenic 20 km-thick upper crust and a weakly radiogenic 15 km-thick lower crust; and a model with an exponentially decreasing radiogenic heat production with depth. (b) Resultant geotherms, with a linear geotherm for zero radiogenic heat production (A = 0) and surface heat flow of 70 mW m^{-2} shown for comparison.

The heat flow q at any depth y is given by:

$$q = -q_m - A_0 a_r \exp(-y/a_r) \qquad [2.16]$$

where q_m is the basal heat flow from the mantle in W m^{-2}, and A_0 is the radiogenic heat production in W m^{-3}. This can be expressed in terms of the surface heat flow q_0

$$q_0 = q_m + a_r A_0 \qquad [2.17]$$

From eqn. [2.17] it is clear that when the surface heat flow q_0 is plotted against the radiogenic heat production A_0, the linear regression has a slope of a_r and an intercept of q_m. It is therefore possible to obtain an estimate of the basal heat flow and the differentiation

(that is, distribution with depth) of heat production from such plots (Fig. 2.19).

However, the interpretation of q_0–A_0 plots is not straightforward (Fig. 2.20). First, the distribution of radiogenic heat production may not be exponential with depth. Second, secular cooling causes a reduction in A_0, which causes a flattening of the regression and makes estimation of q_m problematical. Third, lateral conductive heating of a 'cold' terrane by an adjacent 'hot' terrane reduces the variation in the suite of A_0 values and reduces the slope of the linear regression. For these reasons, some care should be exercised in estimating a_r and q_m. A large number of q_0–A_0 plots, however, indicate that the basal heat flow is commonly less than half of the surface heat flow. Since the global average for the continents is 65 mW m^{-2}, the average basal heat flow from the mantle must be approximately 30 mW m^{-2}.

An alternative method for estimating the vertical distribution of HPEs in the upper part of the lithosphere, which is particularly valuable where the depth distribution of HPEs is variable, is to calculate a depth distribution parameter a that integrates over depth the product of radiogenic heat production $A(y)$ and the depth y (Fig. 2.20) (Sandiford & McLaren 2002). The variable a therefore differentiates between shallow (small a) and deep (large a) sources of radiogenic heat production. The effects of crustal growth, tectonic deformation and surface processes of erosion and sedimentation can then be viewed in a-q_c space, where q_c is the crustal radiogenic heat production (Fig. 2.21). q_c is simply the integration of the radiogenic heat production over the crustal thickness y_c.

The upward emplacement of granitic melts causes no significant change in the total crustal heat production, but a reduction in the depth distribution, since radiogenic material is moved to shallower depths (Fig. 2.21a). Tectonic deformation may cause thinning or thickening of the radiogenic layer. In the former, extension causes a reduction in the total crustal heat production and in most cases an upward movement of the HPEs, causing a reduction in the depth distribution parameter. On the other hand, shortening causes an

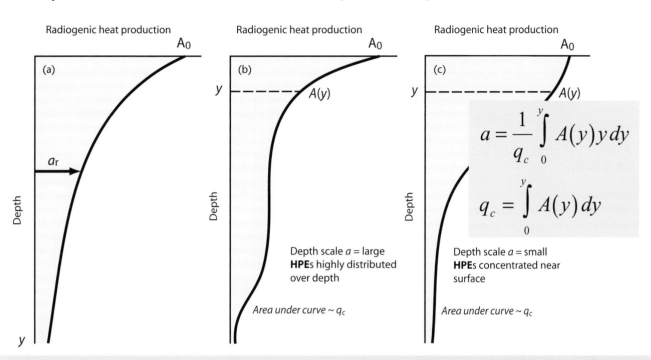

Exponential distribution

Constant q_c: variable depth distribution a

$$a = \frac{1}{q_c} \int_0^y A(y)\,y\,dy$$

$$q_c = \int_0^y A(y)\,dy$$

Fig. 2.20   The depth distribution of heat-producing elements (HPEs) can be approximated by an exponential function with a depth constant a_r and a surface heat production A_0, or can be denoted by a variable depth distribution parameter a and crustal heat production q_c. The depth distribution parameter a incorporates the effects of deep versus shallow sources of HPEs, where the value of q_c is identical to the area under the curves in the three profiles shown.

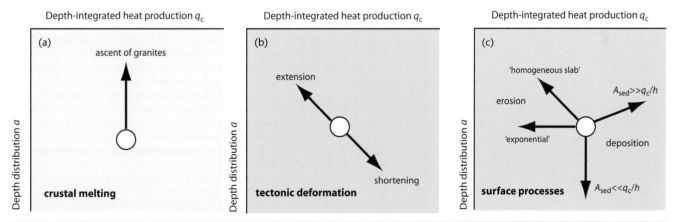

Fig. 2.21   The effects of crustal growth, surface processes and tectonic deformation can be recognised in the q_c-a space (after Sandiford & McLaren 2002). The depth distribution of HPEs a and the depth-integrated heat production q_c are shown in Fig. 2.20. Upward migration of radiogenic melts raises the depth distribution of HPEs while keeping the total heat production the same (a). Tectonic extension and shortening affects both the depth distribution and the total heat production, as shown in (b). Surface processes of erosion and deposition have impacts depending on the magnitude and distribution of HPEs in the crust (c), but in general erosion reduces a and q_c, whereas deposition increases a and q_c. Reprinted with permission from Elsevier.

increase in the total crustal heat production and a deepening of the distribution of HPEs, leading to an increase in the depth distribution parameter (Fig. 2.21b). Surface processes of erosion and deposition may have varied effects (Fig. 2.21c). Erosion reduces the thickness of the radiogenic layer, causing a drop in the value of q_c, accompanied by a reduction in the depth distribution a. Deposition has the opposite effect, but the precise response in q_c-a space depends on whether the heat production of the sediment is large or small in relation to the ratio q_c/a.

2.2.4   Effect of erosion and sediment blanketing on the geotherm

Since radiogenic elements are concentrated in the upper part of the lithosphere, the crustal radiogenic heat production is likely to be strongly affected by the processes of erosion and sedimentation. Deep erosion may potentially strip highly radiogenic crust from a piece of continental lithosphere, thereby reducing its radiogenic heat production, whereas sedimentation may deposit highly radiogenic sediment on top of a normal section of radiogenic crust.

We can initially consider this as a steady-state problem by comparing the geotherm under eroded crust versus under a sediment blanket (see also Appendix 13). Transient effects are considered in §2.2.5 and Appendix 12. The original thickness of the radiogenic crust is y_c. We treat the eroded crust as comprising a slab of radiogenic material that has been thinned erosively by h_e. In this case, the temperature at a depth $0 < y < y_c$ is given by:

$$T = T_0 + \frac{\{q_m + A(y_c - h_e)\}}{K} y - \frac{A}{2K} y^2 \qquad [2.18]$$

where T_0 is the surface temperature, q_m is the basal heat flow, K is the thermal conductivity, and A is the uniform radiogenic heat production in the slab.

The eroded sediment is deposited as a uniform layer of thickness h_b above normal crust comprising an initial radiogenic slab of thickness y_c, giving a total thickness of $(h_b + y_c)$ for the slab. We neglect, for simplicity, the effect of sediment porosity, and assume that the thermal conductivity of the basin-fill is the same as the thermal conductivity of the adjacent and underlying crust (see §2.2.6 for the impact of thermal conductivity variation). The geotherm through and under the basin is

$$T = T_0 + \frac{\{q_m + A(y_c + h_b)\}}{K} y - \frac{A}{2K} y^2 \qquad [2.19]$$

The difference in temperature at any depth between the eroded crust and the basin is linearly dependent on the depth y and the amount of erosion and deposition. It is given by

$$\Delta T = \frac{A}{K} y (h_b + h_e) \qquad [2.20]$$

where h_b is the thickness of the basin-fill and h_e is the depth of erosion. If K is $3\,W\,m^{-1}\,^{\circ}C^{-1}$, A is $2.5 \times 10^{-6}\,W\,m^{-3}$, and $(h_b + h_e)$ is $10\,km$, the temperature difference at the base of the sedimentary basin is about $40\,^{\circ}C$ and at a depth of $y = 10\,km$ it is about $80\,^{\circ}C$ (Fig. 2.22).

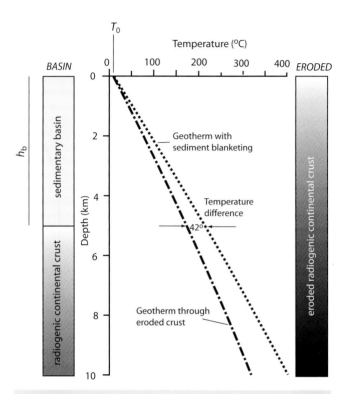

Fig. 2.22   Steady-state crustal geotherms resulting from erosion and sedimentation, using a basal heat flow of $30\,mW\,m^{-2}$ and a radiogenic heat production with $A_0 = 2.5 \times 10^{-6}\,W\,m^{-3}$ and a thermal conductivity of $3\,W\,m^{-1}\,K^{-1}$. The surface temperature T_0 is $10\,^{\circ}C$. The steady-state geotherm following erosion by $5\,km$ (long-short dash) shows the effect of stripping off $5\,km$ of highly radiogenic upper crust. The steady-state geotherm resulting from sediment blanketing (short dashes) has elevated temperatures caused by the presence of a homogeneous $5\,km$-thick basin filled with radiogenic sediment. The temperature difference between the geotherms for the eroded crust and the blanketed crust is $42\,^{\circ}C$ at the base of the sedimentary basin. Upper crustal rocks under the sedimentary basin are therefore likely to undergo deformation or reactivation of old tectonic structures as a result of the thermal effects of surface processes (see also Sandiford & McLaren 2002).

The effect of erosion and sedimentation can be made more realistic by invoking a two-layer slab model (Appendices 9 and 12) where each slab has its own radiogenic heat production and thermal conductivity, representing the porous basin-fill and the underlying crust. For reasonable parameter values, it can be seen that, at steady state, temperatures are strongly elevated in and under the sedimentary basin, with its radiogenic infill and normal thickness upper crustal radiogenic slab compared to the adjacent eroded crust. The radiogenic structure of the crust and basin-fill are therefore critical to the prediction of the continental geotherm. The deposition of a radiogenic, self-heating sedimentary infill has an important impact on basin-fill and crustal temperatures, and thereby influences the thermal maturation of lithologies within the basin and potentially the deformation of upper crustal rocks (Hand & Sandiford 1999).

2.2.5 Transient effects of erosion and deposition on the continental geotherm

The preceding discussion treats the geotherm as a steady-state problem. However, the geotherm may be affected by time-dependent processes, such as an abrupt change in erosion and sedimentation rate at the surface of the Earth. We assume that the initial geothermal gradient is linear, and that there are no additional radiogenic heat sources. We firstly consider the transient effects on subsurface temperatures following an instantaneous change in erosion rate or sedimentation rate (following Wangen 2010, p. 181–186) (Appendix 12).

The instantaneous removal of rock causes warm crust to be subjected to a cold surface temperature T_0. Over time, the warm crust cools by conduction to the surface of the Earth. The time scale of this cooling depends on the diffusivity of crustal rocks. The depth to which the temperature perturbation is felt is the diffusion distance of the form $\sqrt{\kappa t}$ (see §2.2.7). A similar situation arises with instantaneous deposition, but there is an instantaneous cooling followed by a transient heating with the same time scale.

The temperature following erosion by an amount l (Fig. 2.23) is the sum of the steady-state solution (first two terms on right-hand side) and a transient solution (third term on right-hand side)

$$T(y,t) = T_0 + G(y+l) + Gl\,erfc\left(\frac{y}{2\sqrt{\kappa t}}\right) \qquad [2.21]$$

and, similarly, the temperature following sediment blanketing by a thickness h (Fig. 2.24) is

$$T(y,t) = T_0 + G(y-h) + Gh\,erfc\left(\frac{y}{2\sqrt{\kappa t}}\right) \qquad [2.22]$$

where T_0 is the surface temperature and G is the geothermal gradient, y is zero at the new surface, and $erfc$ is the complementary error function. Figs 2.23 and 2.24 show that after ~1 Myr, the geotherm is close to its original linear gradient before the erosion event and before the blanketing event.

Erosion and sedimentation, however, may be rapid, but not instantaneous. Even at fast erosion rates of $1\,mm\,yr^{-1}$, it should take 1 million years to erode 1000 m of upper crust. Consequently, it is valuable to consider what effect the speed of erosion and deposition have on the underlying geotherm. To do so requires the Earth's surface to be treated as a *moving boundary*.

When rocks are exhumed rapidly, they move towards the surface of the Earth by advection while at the same time cooling by conduction. During high rates of exhumation due to erosion, hot rocks are brought closer to the surface of the Earth but are unable to cool quickly enough to reach the linear steady-state geotherm due to conduction. That is, the temperature field is determined by two effects: the conductive cooling at a rate given by the thermal conductivity and temperature gradient, and advection of rock towards the surface at the exhumation rate. When the competing rate effects of conduction (or diffusion) and advection are involved in a problem, it is customary to use a *Péclet number*. We assume that the rate of exhumation E is steady over a sufficiently long period of time so that a stable thermal state exists, governed by the balance of advection and conduction (Willett & Brandon 2002).

We firstly non-dimensionalise the vertical depth coordinate and the temperature coordinate, so that $y^* = y/L$ and $T^* = T/L$, where L is the depth to a lower boundary, which might, for instance, be the

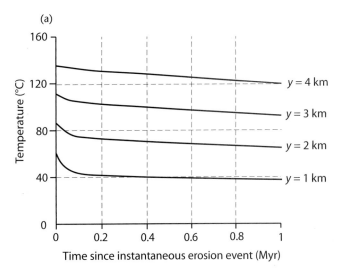

(a)

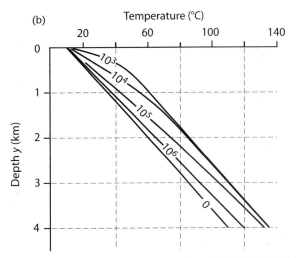

(b)

Fig. 2.23 Effect of instantaneous erosion. A thickness of $l = 1\,km$ is removed instantaneously by erosion of crust with a surface temperature T_0 of $10\,°C$, geothermal gradient $25\,°C$ per km, and thermal diffusivity κ of $10^{-6}\,m^2\,s^{-1}$. (a) Temperature trajectories for different depths below the surface after erosion, showing a decreasing rate of cooling by conduction following the erosion event. (b) Geotherms as a function of time following erosion, and the linear geotherm prior to erosion ($t = 0$). After c.$10^6\,yr$, the geotherm has almost returned to its steady-state gradient.

depth to the base of the crust. Secondly, we make use of the Péclet number, which measures the efficiency of advective versus conductive heat flow:

$$Pe = \frac{EL}{\kappa} \qquad [2.23]$$

and the diffusivity $\kappa = K/\rho c$, where K is the thermal conductivity ($W\,m^{-1}\,K^{-1}$), ρ is the rock density ($kg\,m^{-3}$), and c is the specific heat ($W\,kg^{-1}\,K^{-1}$). The dimensionless form of the temperature is given by Braun *et al.* (2006, p. 79)

$$T^*(y^*) = \frac{1 - \exp(-Pey^*)}{1 - \exp(-Pe)} \qquad [2.24]$$

(a)

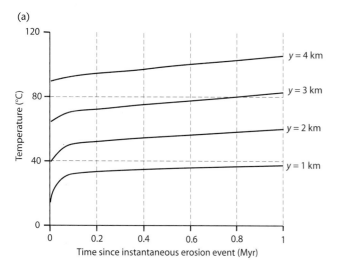

(b)

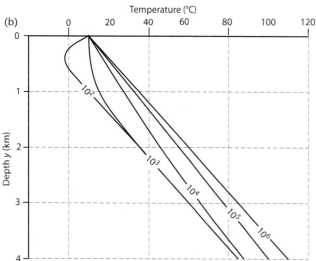

Fig. 2.24 The effect of instantaneous deposition on the crustal geotherm, with a steady-state geotherm G of 25 °C per km, surface temperature T_0 of 10 °C, thermal diffusivity of 10^{-6} m² s⁻¹, and depositional thickness h of 1 km. (a) Temperature trajectories for different depths, showing an instantaneous cooling followed by recovery. (b) The geotherm following instantaneous cooling due to sedimentation at times of 10^2 to 10^6 years. It takes in the order of 10^6 years to regain the linear geotherm with gradient G.

and is shown in Fig. 2.25. It can be seen that at Péclet numbers of greater than 1 the geotherm is significantly curved. It is therefore important to know what Péclet numbers are likely to occur in nature. Taking $\rho = 2800$ kg m⁻³, $c \sim 1000$ W kg⁻¹ K⁻¹, and $K = 2.5$ W m⁻¹ K⁻¹, the diffusivity of crustal rocks is approximately $\kappa = 10^{-6}$ m² s⁻¹. Using eqn. [2.23], the Péclet number for the crust is >1 for exhumation rates of greater than 1 mm yr⁻¹. Rates of this order are common in tectonically active settings. The effect on the geotherm of advection results in a surface heat flow 'anomaly' compared to regions undergoing conduction only. At $E = 2$ mm yr⁻¹, the surface heat flow is approximately 2.5 times the surface heat flow in an adjacent region where there is no advection ($Pe = 0$). This has profound consequences for the interpretation of surface heat flow data, and for the

correct interpretation of thermochronometric data (see §10.4). If these differences in surface heat flow could be confidently attributed to differential rates of exhumation, the surface heat flow 'anomaly' would allow the curvature of the geotherm under the exhuming region to be estimated.

The transient form of the geotherm following a rapid change in the exhumation rate, as may be caused by the initiation of a major tectonic event, or following a rapid change in the sedimentation velocity, as may occur through internally-forced delta switching or externally forced increases in sediment discharge caused by catchment processes, is solved in Appendix 12.

2.2.6   Effect of variable thermal conductivity

Fourier's law (eqn. [2.14]) states that the heat flow is related to the temperature gradient by a coefficient known as the thermal conductivity. Whereas the thermal conductivity is not expected to vary greatly in the crystalline crust, it is highly variable in the sedimentary basin-fill. This variability derives from the wide range of sedimentary lithologies in the basin, their pore-filling fluid, and their diagenetic phases (Giles 1997; Allen & Allen 2005) (see Chapters 9 and 10).

Ignoring, for the moment, lithological variations, thermal conductivities of sediments vary as a function of depth because of their porosity loss with burial (§9.2). Neglecting also radiogenic heat production within the basin-fill, the geotherm in a basin with different thermal conductivities of sedimentary layers is given by

$$T_y = T_0 + (-q)\left\{\frac{l_1}{K_1} + \frac{l_2}{K_2} + \frac{l_3}{K_3} + \dots\right\} \qquad [2.25]$$

where l_1 to l_n are the thicknesses of the layers with thermal conductivities K_1 to K_n, and $l_1 + l_2 + l_3 \dots$ must of course be equal to y. The heat flow q is shown negative since heat travels vertically towards the surface in the opposite direction to the depth coordinate y.

Typical thermal conductivities of the minerals comprising framework grains and diagenetic products, and the thermal conductivity of pore-filling water, are given in Table 2.2. Methods of calculating the bulk thermal conductivity of different stratigraphic units are outlined in §10.3.1. The bulk thermal conductivity of most sedimentary rocks ranges between 1.5 W m⁻¹ K⁻¹ (shales) and 4.5 W m⁻¹ K⁻¹ (sandstones). The individual conductivities of framework, matrix and pore fluid are also dependent on temperature, so thermal conductivity values (Table 2.2) are commonly quoted as surface temperature values (Brigaud & Vasseur 1989). The general trend is that non-argillaceous rocks have higher conductivities than argillaceous rocks, and that conductivity increases with increasing porosity.

A fundamental result of maintaining a constant heat flux through a heterogeneous basin-fill with variable thermal conductivity is that the geothermal gradient must vary with depth. A worked example is given in Appendix 13. Taking a basin composed of the stratigraphy given in Table 2.5, the geotherm is highly variable (Fig. 2.26). Comparison with the linear conduction profile using Fourier's law with $q_m = 30$ mW m⁻² and $K = 3.0$ W m⁻¹ K⁻¹ emphasises the importance of this approach to the prediction of thermal history. Note the strong insulating effect of the marine shales and the highly conducting effect of quartzites and evaporites (anhydrite).

As we have seen in §2.2.3, the sediments of the basin-fill are also radiogenic. Variations are likely to exist in the internal heat production of the different sedimentary lithologies. We can assume that these different sedimentary layers are the same units that we used

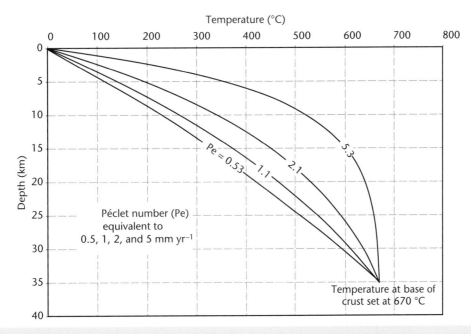

Fig. 2.25 Curves show geotherms for different values of the exhumation rate, shown as a Péclet number. At high exhumation rates, the upward advection of hot rock towards the surface outweighs the conductive cooling, causing highly curved geotherms. Note that in the case of high exhumation rate, the geothermal gradient changes from 40–60 °C km^{-1} in the upper 5 km of the crust to <10 °C km^{-1} in the lower crust. For apatite fission track thermochronometers, the high geothermal gradient at high Péclet numbers expected in rapidly exhuming orogens has the effect of reducing the true depth of erosion compared to estimates using 'normal' or 'standard' geotherms. For U-Th/He thermochronometers, the geothermal gradient used to convert cooling ages to depths of erosion is highly susceptible to the effects of erosion and variable surface topography (Braun *et al.* 2006).

Table 2.5 Bulk thermal conductivities of stratigraphic layers in a model basin-fill (surface temperature values from Brigaud & Vasseur 1989) and radiogenic heat production values

Lithology	Depth range (m)	Bulk thermal conductivity (W m^{-1} K^{-1})	Density (kg m^{-3})	Internal heat production A ($\times$ 10^{-6} W m^{-3})
Evaporite	0–500	6.0	2960	2.0
Dolomite	500–1500	3.2–3.5	2870	1.0
Shales	1500–2300	1.2–3.0	2100–2700	4.0
Quartzite	2300–3500	7.0	2650	0.5
Granite	>3500	2.4–3.8	2650	5.0

for calculations of bulk thermal conductivity. For a uniform radiogenic heat production extending to the base of the crust and a uniform thermal conductivity

$$T = T_0 + \left(\frac{q_m + A y_c}{K}\right) y - \frac{A}{2K} y^2 \qquad [2.26]$$

must be modified to account for n number of layers of thermal conductivity K_1 to K_n, radiogenic heat production A_1 to A_n, and thickness l_1 to l_n

$$T_y = T_0 + \left\{ \left(\frac{q_m + A_1 y_c}{K_1}\right) l_1 + \left(\frac{q_m + A_2 y_c}{K_2}\right) l_2 + \cdots \right\} \\ - \left\{ \frac{A_1}{2K_1} l_1^2 + \frac{A_2}{2K_2} l_2^2 + \cdots \right\} \qquad [2.27]$$

where y_c becomes the sum of the sedimentary layers of thickness $l_1 + l_2 + \ldots l_n$ if the geotherm is calculated for the basin-fill only. The

solution for a heterogeneous basin-fill using eqn. [2.27] and the data in Table 2.5 gives the geotherm shown in Fig. 2.26, and demonstrates that the linear conduction geotherm may be a very poor approximation of the temperature field in parts of sedimentary basins.

2.2.7 Time-dependent heat conduction: the case of cooling oceanic lithosphere

Many problems involve heat flows that vary in time. An obvious example is the heat flow associated with the intrusion of an igneous body, but the example used here is the cooling of the oceanic lithosphere and its consequent subsidence. This process finds great application in sedimentary basins experiencing a period of cooling following stretching (Chapter 3). Heat sources *within* the medium are unimportant in the time-dependent problem of the cooling of oceanic lithosphere (that is, $A = 0$).

At the crest of an ocean ridge, hot mantle rock injected in dykes and extruded as lava flows is suddenly subjected to a cold surface

(a)

Thermal stratigraphy   *n*

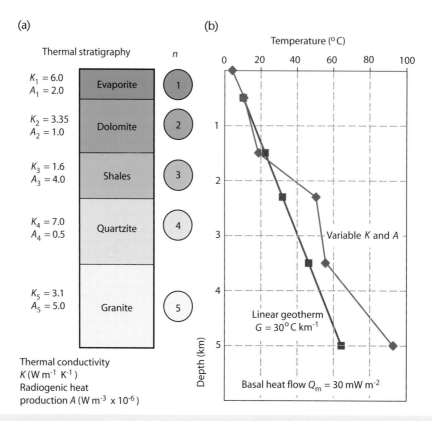

$K_1 = 6.0$
$A_1 = 2.0$ — Evaporite — 1

$K_2 = 3.35$
$A_2 = 1.0$ — Dolomite — 2

$K_3 = 1.6$
$A_3 = 4.0$ — Shales — 3

$K_4 = 7.0$
$A_4 = 0.5$ — Quartzite — 4

$K_5 = 3.1$
$A_5 = 5.0$ — Granite — 5

Thermal conductivity
K (W m^{-1} K^{-1})
Radiogenic heat
production A (W m^{-3} × 10^{-6})

(b)

Fig. 2.26   Stratigraphy with variable thermal conductivity (a) used to calculate the geotherm in (b), illustrating the insulating and conducting properties of common rock types.

temperature and then continues to lose heat to the cold seawater as the seafloor spreads away from the ridge. The initial cooling can be treated as instantaneous. We know from the 1D conduction equation (Appendix 7) where $A = 0$ that

$$\frac{\partial T}{\partial t} = \kappa \frac{\partial T^2}{\partial y^2} \qquad [2.28]$$

where the diffusivity $\kappa = K/\rho c$, and K is the thermal conductivity, ρ is the density, and c is the specific heat or thermal capacity. Eqn. [2.28] is a simple diffusion equation. We can define a characteristic time scale τ as the time necessary for a temperature change to propagate a distance l in a material with a thermal diffusivity κ (units of [L]^2 [T]$^{-1}$):

$$\tau = \frac{l^2}{\kappa} \qquad [2.29]$$

The length scale $l = \sqrt{\kappa\tau}$ gives the distance that a temperature change propagates in time τ. It is known as the *thermal diffusion distance*.

Let the surface plates move away from the ridge along the coordinate x with a velocity u (Fig. 2.27). The cooling rocks form the oceanic lithosphere and the boundary between this relatively rigid upper layer and the easily deformed mantle is an isotherm with a value of about 1600 K (1300 °C). The thickness of the oceanic litho-

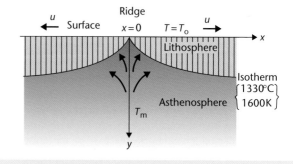

Fig. 2.27   Schematic diagram of the cooling oceanic lithosphere at a mid-ocean ridge. The oceanic plate moves away from the ridge at a velocity u. Its age is therefore determined by x/u, where x is the horizontal distance from the ridge crest. After Turcotte & Schubert (2002), © Cambridge University Press, 2002.

sphere is clearly a function of its age, where age can be expressed in terms of x/u.

Using a time-dependent instantaneous cooling model, the isotherms below the seafloor as a function of age are parabolic (Fig. 2.28). The surface heat flow as a function of age is based on the 1D conduction equation. It is compared with actual heat flow measurements of the ocean floor in Fig. 2.29 (Sclater *et al.* 1980a). The presence of regions of poor correspondence between the predicted and

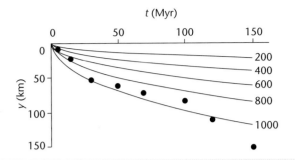

Fig. 2.28 Calculated isotherms for an oceanic lithosphere that is instantaneously cooled. The values of the isotherms are $T-T_s$°K. The dots are the estimated thicknesses of the oceanic lithosphere in the Pacific. From Leeds *et al.* (1974).

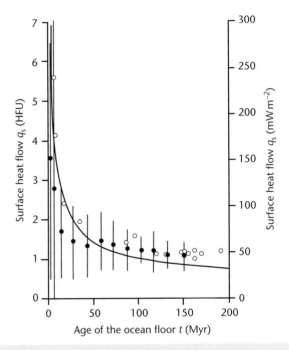

Fig. 2.29 Comparison of measured ocean heat flows (mean and standard deviation) and those predicted using the instantaneous cooling model, as a function of age. Black circles, data from Sclater *et al.* (1980a). Open circles, data from sediment-covered regions of the Atlantic and Pacific oceans, from Lister *et al.* (1990).

observed surface heat flow measurements is probably due to temperature changes associated with hydrothermal circulation of seawater through the oceanic crust. These effects become less important with age as impermeable sediments blanket the ocean floor. For large ages (>80 Ma), however, an additional heat source appears to be recognisable, which may be mantle convection beneath the lithospheric plates.

The solution for the temperature as a function of age has the form

$$\theta = erfc\left(\frac{x}{2\sqrt{\kappa t}}\right) \qquad [2.30]$$

where it can be seen that the denominator in the bracket is twice the thermal diffusion distance, *erfc* is the complementary error function, x is the horizontal distance from the mid-ocean ridge crest, t is the age of the oceanic crust, and θ is a dimensionless temperature ratio given by

$$\theta = \frac{T - T_a}{T_s - T_a} \qquad [2.31]$$

where T_a is the initial temperature (the asthenospheric temperature), T_s is the constant temperature of the space into which the ocean lithosphere is emplaced (the seawater temperature), and T is the temperature at time t.

The cooled oceanic material forms a thermal boundary layer equivalent to the thickness of the new oceanic lithosphere. The choice of the definition of the boundary layer is largely arbitrary, but if we define it as the thickness to where $\theta = 0.1$, it is found that the thermal boundary layer is 2.32 times the thermal diffusion distance $\sqrt{(\kappa t)}$ or $\sqrt{(\kappa x/u)}$. The thickness of the oceanic lithosphere at an age of 50 Myr is therefore approximately 92 km, taking $\kappa = 10^{-6}\,\text{m}^2\,\text{s}^{-1}$. The oceanic lithosphere should increase parabolically in thickness with age, and therefore also with distance from the ridge crest.

As it moves away from the ridge crest, the oceanic lithosphere cools and contracts. This increases the density of the oceanic lithosphere and causes a higher lithostatic stress on the underlying mantle. An isostatic balance shows that the seafloor must subside. Any column through the oceanic lithosphere can be balanced isostatically to give the depth of the ocean floor as a function of distance from the ridge crest, or age of the oceanic crust (Fig. 2.27).

The observed heat flows in the oceans, such as the Atlantic and Pacific oceans (Parsons & Slater 1977), and the observed bathymetry, are in general agreement with a model of instantaneous cooling of new oceanic material and its loss of heat through time by conduction, resulting in subsidence. The general form of the relationship between the bathymetry of the ocean floor and its thermal age for oceanic lithosphere less than 70 Myr old is

$$h = h_{ridgecrest} + Ct^{1/2} \qquad [2.32]$$

where h is the depth of the ocean floor, $h_{ridgecrest}$ is commonly in the region of 2.5 km, t is the age of the oceanic lithosphere, and C is a coefficient approximately equal to 0.35. This is commonly termed a *root-age relationship*. If $h_{ridgecrest} = 2.5$ km, and $C = 0.35$, the ocean depth h at $t = 20$ Myr is 4 km, and at $t = 50$ Myr is 5 km. The heat flow Q ($10^{-3}\,\text{W}\,\text{m}^{-2}$) for oceanic lithosphere younger than 120 Ma follows a similar but *inverse root-age* relationship:

$$Q = 473t^{-1/2} \qquad [2.33]$$

Since the age of the oceanic lithosphere is directly related to its distance from the ridge crest by the spreading rate, the oceanic bathymetry increases gradually and the heat flow decreases gradually away from the site of spreading. The isostatic balance and the root-age relationship are affected by the presence of marine sediments overlying the ocean crust. However, in the deep sea the cover of pelagic sediments is generally only a relatively thin veneer.

2.2.8 Convection, the adiabat and mantle viscosity

Whereas the thermal structure of the lithosphere is dominated by conduction, that of the mantle is determined primarily by convection. The lithosphere simply serves as a thermal boundary layer exhibiting high temperature gradients. Extensional basins, other than thin-skinned varieties, involve upwelling of convecting asthenosphere. We therefore present here a very brief account of the thermal structure of the upper mantle. Further information on the impact of mantle dynamics on the Earth's surface and basin development is given in Chapter 5.

In the interior of a vigorously convecting fluid, the mean temperature increases with depth along an *adiabat*. The *adiabatic temperature gradient* is the rate of temperature increase with depth caused by compression due to the overlying rock column. It is roughly linear in the mantle (Appendix 14). The compressional pressure forces cause a decrease in volume and therefore an increase in density. The relationship between the density and pressure changes is given by the *adiabatic compressibility*, β_a. For a solid, it is somewhat smaller than the isothermal compressibility β given in eqn. [2.9].

The purely adiabatic expressions for the variations of density and pressure with depth in the mantle do not perfectly match observed values based on seismic velocities. In particular, at about 400 km there is a density discontinuity thought to be due to the phase change of olivine. This change to the denser spinel structure is thought to occur at a pressure of 14 GPa or 140 kbar and a temperature of 1900 K. The phase change is *exothermic*, causing heating of the rock by c.160 K. The olivine–spinel phase change probably enhances mantle convection rather than blocks it. There is a second discontinuity in density at about 700 km, but its origin is less clear – it is probably also due to a change in mantle composition to perovskite. There is a sharp change in **P**-wave velocity at about 700 km. Close to the outer limit of the core is a seismic discontinuity known as D″ that is related to the change to a denser mineral structure known as post-perovskite (see §1.2.3). This change probably influences the formation of instabilities in the mantle at the core–mantle boundary, some of which may rise through the mantle as vertical conduits or plumes.

If a substance is *incompressible*, its volume is incapable of contracting, so adiabatic heating cannot take place. Rocks, however, are sufficiently compressible that adiabatic temperature changes are extremely important under the large pressure changes in the mantle. The adiabatic geotherms in the upper mantle underlying oceanic and continental lithospheres are different since continental lithosphere has its own near-surface radioactive heat source. Below this near-surface layer the heat flow is assumed to be constant at about 30 mW m^{-2}. It is clear from the shape of the geotherm that the continental lithosphere as a thermal entity extends to about 200 km below the surface (§1.2).

The *onset of convection* is a threshold above which the influence of the temperature difference in the fluid layer exceeds that of the viscous resistance to flow. The temperature difference is conventionally evaluated as the fluid temperature at a point as a departure from the temperature expected from a purely conduction profile. The onset of convection is marked by a critical dimensionless number known as the *Rayleigh number, Ra*. A Rayleigh number analysis of Earth's mantle suggests that it must be fully convecting (Appendix 40) (more details in Chapter 5).

There are also large *lateral* temperature heterogeneities in the mantle, such as where cold lithospheric plates are subducted at ocean trenches. These lateral temperature variations are extremely important in providing a driving force for mantle convection. In the case of a descending cold lithospheric slab, the low temperatures of the plate cause it to be denser than surrounding mantle, providing a gravitational body force tending to make the plate sink. A second

factor is the distortion of the olivine–spinel phase change since the pressure at which this phase change occurs depends on temperature. The upward displacement of the phase change in the cold descending plate provides an additional downward-acting body force, helping to drive further plate motion. These two processes are often referred to as *trench-pull*. If trench-pull forces are transmitted to the oceanic plate as a tensional stress in an elastic lithosphere with a thickness of 50 km, the resultant tensional stress would be as high as 1 GPa. The forces arising through the elevation of the ridge crest relative to the ocean floor constitute a *ridge-push* force, also helping to drive plate motion. Ridge-push is an order of magnitude smaller than trench-pull, but since the latter is countered by enormous frictional resistances, the two agents of net trench-pull and ridge-push may be comparable.

One of the fundamental differences between fluids and solids is their response to an applied force. Fluids deform continuously under the action of an applied stress, whereas solids acquire a finite strain. Stress can be directly related to strain in a solid, but in fluids applied stresses are related to *rates* of strain, or, alternatively, velocity gradients. In Newtonian fluids there is a direct proportionality between applied stress and velocity gradient, the coefficient of proportionality being known as *viscosity*. (In non-Newtonian fluids there may be complex relationships between applied stresses and resultant deformations, see §2.3.) The problem of direct relevance to sedimentary basin analysis is the viscous flow in the mantle. One way that the viscosity of the mantle can be estimated is by studying its response to loading and unloading (see also Chapter 4 and Appendices 30, 31).

Mountain building is an example where the crust–mantle boundary is depressed through loading, but orogeny is so slow a process that the mantle manages to constantly maintain hydrostatic equilibrium with the changing near-surface events. In contrast, the growth and melting of ice sheets is very rapid, so that the mantle adjusts itself dynamically to the changing surface load; the way in which it does so provides important information on mantle viscosity. The displacement of the surface leads to horizontal pressure gradients in the mantle that in turn cause flow. In the case of positive loading, fluid is driven away from the higher pressures under the load, the reverse being true on unloading. The surface displacement decreases exponentially with time as fluid flows from regions of elevated topography to regions of depressed topography. If w is the displacement at any time and w_m is the initial displacement of the surface, the form of the exponential decrease in surface topography with time is

$$w = w_m \exp(-t/\tau_r) \qquad [2.34]$$

where τ_r is the characteristic time for the exponential relaxation of the initial displacement. It is given by

$$\tau_r = (4\pi\mu)(\rho g \lambda) \qquad [2.35]$$

where μ is the viscosity and λ is the wavelength of the initial displacement. Mantle viscosity can therefore be estimated from postglacial rebound if this relaxation time can be found (Appendix 30).

Elevated beach terraces that have been dated (usually by ^{14}C) provide the basis for quantitative estimates of the rate of postglacial rebound. Large lateral variations in rebound rates are found. The former Lake Agassiz in North America has a number of lake shoreline terraces that have been uplifted by flexural rebound accompanying deglaciation (Watts 2001) (Appendix 31). The Baltic region of Scandinavia is also renowned for its raised beaches. After correcting the

(a)

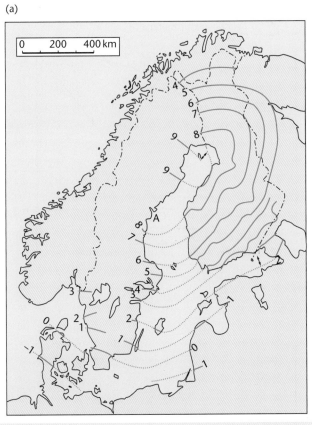

(b)

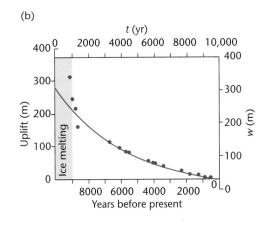

Fig. 2.30 Mantle viscosity can be estimated through studies of postglacial isostatic rebound. (a) Present-day rates of uplift occurring in Scandinavia (after Flint 1971; material reproduced with permission of John Wiley & Sons, Ltd.). Contours are in mm yr^{-1}. (b) Postglacial uplift of the mouth of the Angerman River, Sweden, over the last 10,000 years compared with the exponential relation in eqn. [2.38] with a constant viscosity of 10^{21} Pa s (after Turcotte & Schubert 2002; © Cambridge University Press, 2002). Angerman River mouth in the Gulf of Bothnia is marked by an 'A'.

uplift of Swedish beaches for absolute (eustatic) changes in sea level, a relaxation time of $\tau_r = 4400$ years is found. Assuming a reasonable wavelength of the displacement for the Scandinavian glaciation to be $\lambda = 3000$ km, the viscosity of the mantle is estimated to be $\mu = 1.1 \times 10^{21}$ Pa s (10^{22} poises) (Fig. 2.30). However, this approximate analytical solution does not take into account: (i) the flexural rigidity of the elastic lithosphere; and (ii) the depth-dependency of mantle viscosity.

2.3 Rock rheology and lithospheric strength profiles

2.3.1 Fundamentals on constitutive laws

Rocks in the convecting mantle, the mantle lithosphere and the crust are subjected to different temperatures, pressures, volatile contents and strain rates, which causes them to deform in different ways. For example, in the crust, joints and faults are evidence that rocks can behave as *brittle* materials, that is, they behave elastically up to a limit, beyond which they fail by fracturing. Alternatively, the widespread occurrence of folds suggests that rocks can also behave in a *ductile* manner. Ductile deformation in crustal rocks takes place by pressure-

solution creep (Rutter 1976, 1983), whereby dissolution of minerals in zones of high pressure and their precipitation in areas of low pressure causes creep even at low temperatures and pressures. In the mantle, however, convection is thought to take place by a thermally activated creep that depends exponentially on both temperature and on pressure. When the effective viscosity is stress-dependent, it is analogous to the flow of a power-law fluid in a channel. Creep also takes place in the mantle lithosphere where it can relax elastic stresses. In this case, the rheology is a combination of an elastic and viscous behaviour – a *viscoelastic* rheology. Some of these rheologies (flow laws) are discussed in further detail with reference to the deformation of the mantle (§2.3.2) and the continental crust (§2.3.3).

A key concept in rock rheology is that there is a change in deformation behaviour under changing pressures. When confining pressures approach a rock's brittle strength a transition takes place from brittle (elastic) behaviour to plastic behaviour (Fig. 2.31a). This transition is at the rock's yield stress σ_0. The elastic strain can be regarded as recoverable, but the deformation in the plastic field is not recoverable when the stress or load is removed (Fig. 2.31b). If the deformation continues indefinitely without any addition of stress above σ_0, the material is said to exhibit an *elastic–perfectly plastic* behaviour (Fig. 2.31c). Laboratory studies of the mantle rock dunite show that it conforms closely to the elastic–perfectly plastic rheology (Griggs

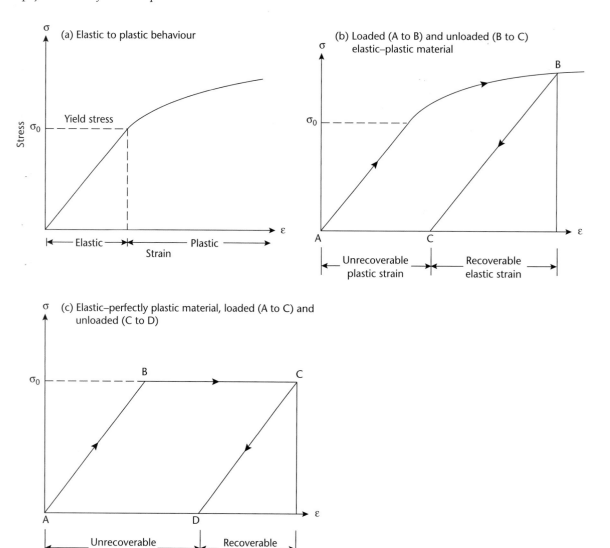

Fig. 2.31   Deformation of solids with different constitutive laws. Stress-strain trajectories for: (a) a solid showing a transformation from elastic to plastic behaviour; (b) loading and unloading of an elastic–plastic material. Unloading of the material once it has passed into the plastic field results in an unrecoverable deformation or plastic strain; and (c) a material with an elastic–perfectly plastic rheology. Plastic strain continues indefinitely without any addition of stress above the yield stress.

et al. 1960). A typical depth at which the brittle–plastic transition would occur for dunite is about 17 km; below this the rock should yield plastically under large deviatoric stresses. The importance of the brittle–ductile transition for the strength profile of the lithosphere is discussed in §2.3.4.

2.3.2   Rheology of the mantle

It was initially believed that at the high temperatures and low strain rates of the upper mantle, Newtonian (linear) flow would take place. Alternatively, the mantle may act as a power-law fluid, in which case the mantle's viscosity would be stress-dependent. For a power-law fluid the strain rate or velocity gradient is proportional to the power n of the stress (Fig. 2.32). Velocity gradients and therefore strain rates

are greater near the walls of the channel where shear stress is a maximum, whereas a core region experiences small strain rates. This plug-flow appearance of the velocity profiles for large n is a direct consequence of the stress dependency of the effective viscosity μ_{eff}, which changes from low at the walls to high at the centre of the flow. This must be true because

$$\mu_{eff} = \tau \bigg/ \frac{du}{dy} \qquad (2.36)$$

that is, the viscosity μ is the coefficient linking the shearing stresses to the resultant velocity gradient or strain rate.

The idea can be applied to shear flow in the asthenosphere, which can be treated as having a heated lower boundary and an upper

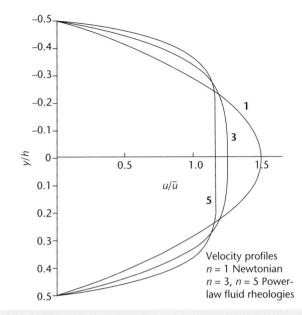

Fig. 2.32 Velocity profiles in a channel of thickness h for power-law fluid rheologies with $n = 1$ (Newtonian), $n = 3$ and $n = 5$. Distance from the channel wall is expressed by y/h. The velocity u is scaled by the average velocity $\bar{u}$. After Turcotte & Schubert (2002).

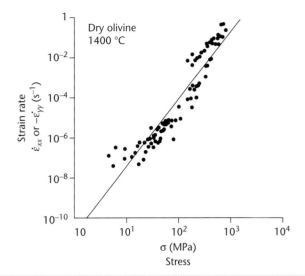

Fig. 2.33 Observed dependence of strain rate on stress for olivine at a temperature of 1400 °C (after Ashby & Verall 1977). Dry olivine obeys an approximate cubic power rheology in these laboratory experiments. After Turcotte & Schubert (2002), © Cambridge University Press, 2002.

cooler boundary at the base of the rigid lithosphere. Shear in the asthenospheric velocity profile is concentrated in zones close to the lower boundary of the asthenosphere where the fluid is hottest and viscosity the smallest. The upper part of the asthenosphere, on the other hand, tends to behave like a rigid extension of the overlying lithosphere as a result of this temperature and stress dependency of viscosity. Frictional heating can also have important consequences for shear flow of a fluid with a strongly temperature-dependent viscosity.

The fact that seismic shear waves can be propagated through the mantle suggests that it is an elastic solid, yet it also appears to flow like a viscous fluid, enabling, for example, postglacial rebound to take place. A material that behaves as an elastic solid on short time scales, but viscously on long time scales, is known as a viscoelastic or *Maxwell* material.

In a Maxwell material, the rate of strain $\dot{\varepsilon}$ is the sum of the linear elastic strain rate $\dot{\varepsilon}_e$ and a linear viscous strain rate $\dot{\varepsilon}_f$. The elastic strain of a material under a uniaxial stress is

$$\varepsilon_e = \sigma/E \qquad [2.37]$$

as will be recalled from Hooke's law (eqn. [2.12]), where E is Young's modulus. The rate of strain is therefore the time derivative of ε_e, $d\varepsilon_e/dt$. The rate of strain in a viscous Newtonian fluid subjected to a deviatoric normal stress σ is a velocity gradient $\delta u/\delta x$. The effective viscosity is taken as the ratio between stress and twice the strain rate. Therefore

$$\frac{d\varepsilon_f}{dt} = -\frac{\partial u}{\partial x} = \frac{\sigma}{2\mu} \qquad [2.38]$$

(where the minus sign simply indicates a tensional strain by convention). Since the total strain is the sum of the elastic and viscous fluid strains, the total strain rate is

$$\frac{d\varepsilon}{dt} = \frac{1}{2\mu}\sigma + \frac{1}{E}\frac{d\sigma}{dt} \qquad [2.39]$$

This is the fundamental rheological equation relating strain rate, stress and rate of change of stress for a *Maxwell viscoelastic* material. When a strain is initially rapidly applied to the viscoelastic medium, the time-derivative terms in eqn. [2.38] dominate and the material behaves elastically. Subsequently, if there is no change in the strain, the stress relaxes to 1/e of its original value in a time $2\mu/E$, which is known as the *viscoelastic relaxation time*. For the asthenosphere, this relaxation time is of the order of 30 to 40 years, that is, longer than the period of seismic waves but shorter than the duration of postglacial rebound. The stress relaxation time is a strong function of temperature, rheological parameters and initial stress, and its range of values for the lithosphere is controversial. Some workers believe that the relaxation time is very short ($<10^6$ yr) while others believe it to be sufficiently long (20–30 Myr) to have a major impact on sedimentary basin geometry and subsidence. A viscoelastic rheology with a long relaxation time was used in early models of foreland basin evolution in particular (Beaumont 1978, 1981; Quinlan & Beaumont 1984).

Most views of mantle rheology come from laboratory studies. Since the mantle is composed primarily of olivine, laboratory studies of the creep of olivine at high temperatures are particularly relevant to mantle rheology. The strain rate in dry olivine $\dot{\varepsilon}_{xx}$ (or $-\dot{\varepsilon}_{yy}$) at 1400 °C as a function of stress appears to follow a cubic power-law rheology reasonably well (Ashby & Verall 1977) (Fig. 2.33). Other rocks deform in a non-linear way at high temperatures in the laboratory, with slightly different coefficients in the power law (ice $n = 3$, halite 5.5, limestone 2.1, dry quartzite 6.5, wet quartzite 2.6, basalt 3). It is reasonable therefore to assume that the mantle has a power-law rheology. This power-law effect is likely to be small, however, compared with the temperature dependence of mantle rheology.

2.3.3   Rheology of the continental crust

As every field geologist knows, crustal rocks exposed at the Earth's surface are strongly fractured and faulted at a range of scales. The occurrence of brittle failure depends on the stress overcoming the frictional resistance of the rock (Atkinson 1987 and Scholz 1990 for more details). This resistance to brittle failure increases with pressure and therefore depth.

It has been known for a considerable time that there is a roughly constant ratio between the frictional force on a potential failure plane F and the normal stress N. This ratio is equal to the coefficient of friction f, or the tangent of the angle of sliding friction ϕ

$$\frac{F}{N} = f = \tan\phi \qquad [2.40]$$

Byerlee (1978) found a large range of friction coefficients for different rock types at low pressures, but at moderate pressures (5–100 MPa, 50–1000 bar) the correlation was poor, and at high pressures (200–2000 MPa, 2–20 kbar) there was no rock type dependence at all. *Byerlee's law* states

$$F = aN + b \qquad [2.41]$$

where the coefficients a and b are 0.6 and 0.5 kbar at normal pressures of >2 kbar (200 MPa). Byerlee's law has the same form as the Navier-Coulomb failure criterion

$$\tau_c = C + \sigma\tan\phi \qquad [2.42]$$

where the shear stress acts in the direction of and balances the frictional force on a fracture plane, τ_c is the critical shear stress for failure, σ is the normal stress, ϕ is the angle of internal friction (in the range 0.7 to 1.7 for compact, coherent rocks), and C is the strength that exists even at zero normal stress (pressure) called *cohesion*. The shear stress needed to produce failure increases as the confining pressure increases. For initiation of a new fracture σ should be higher than for slip on an existing fracture.

Byerlee (1978) suggested that at normal pressures greater than 200 MPa (2 kbars), C is 50 MPa (0.5 kbars) and μ is 0.6 for a wide variety of rock types (Fig. 2.34). Where in the crust would we expect to find normal pressures of 200 MPa? The weight of a column of rock of density 2800 kg m^{-3}, height 1 km, and g = 10 m s^{-2}, is 28 MPa. We would expect to find 200 MPa of pressure at depths of 7 to 8 km in the crust. Earthquake foci are found at depths up to 15 km in the continental crust, equivalent to about 420 MPa (4.2 kbar). Between the surface and 15 km, therefore, we should expect brittle deformation on unlubricated faults. Below 15 to 20 km, however, earthquakes are much less common, and the rheology is ductile or 'plastic'. Depths of about 10–15 km correspond with a temperature of 300 °C in continental crust.

Although faulting of the brittle upper crust is very familiar, there is also observational evidence that near-surface rocks deform in both a plastic and fluid-like manner. The texture of many folded rocks suggests that the deformation responsible for the folding was the result of diffusive mass transfer. But the lower temperatures involved preclude the thermally activated diffusion processes found in the

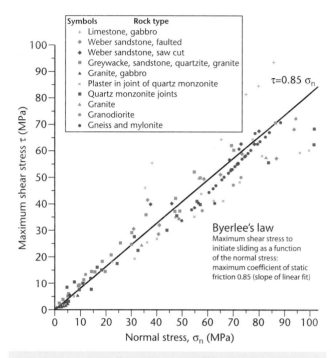

Fig. 2.34   Byerlee's law for the maximum shear stress τ to initiate sliding as a function of the normal stress σ_n for a variety of rock types. The linear fit defines a coefficient of static friction of 0.85 (Byerlee 1977). Reproduced with kind permission from Springer Science+Business Media.

mantle. Instead, the diffusive processes take place through *pressure solution* whereby material is transported in solution from areas of high intergranular pressure and stress and precipitated in regions of low pressure and stress, leading to creep. Point and line contacts between grains in a sandstone would represent such high-pressure/stress zones. The presence of water around the grains acts as the solvent and facilitates the transport of dissolved material. The pressure-solution process also leads to compaction (§9.2). Strain rate is linearly proportional to applied stress in pressure-solution creep, so the deformation is equivalent to that of a Newtonian fluid. It explains viscous folding of rocks at quite low temperatures.

Although pressure solution probably dominates as the main ductile mechanism at low temperatures in the upper crust, the rheology of the crust is likely to be complex as a result of the many compositional changes taking place within it. The layering of the continental crust affects its strength as a function of depth (§2.3.4). In particular, a ductile region in the middle crust is thought to be the location of detachment of major extensional faults (Kusznir *et al.* 1987). This topic of decoupling in mid-crustal regions is dealt with in greater depth in considering extensional basins in Chapter 3.

In summary, the consideration of rock deformation textures, laboratory studies of quartz-bearing rock and continental seismicity suggests that a seismogenic upper crust characterised by discontinuous-frictional (brittle) faulting passes below 10–15 km (c.300 °C) depth into an aseismic continuous quasi-plastic region where creep takes place in mylonitic fault zones (Sibson 1983; Scholz 1990). In the deep crust at temperatures greater than 450 °C, fully ductile continuous deformation is thought to dominate in gneissose shear zones (Grocott & Watterson 1980). The vertical zonation of

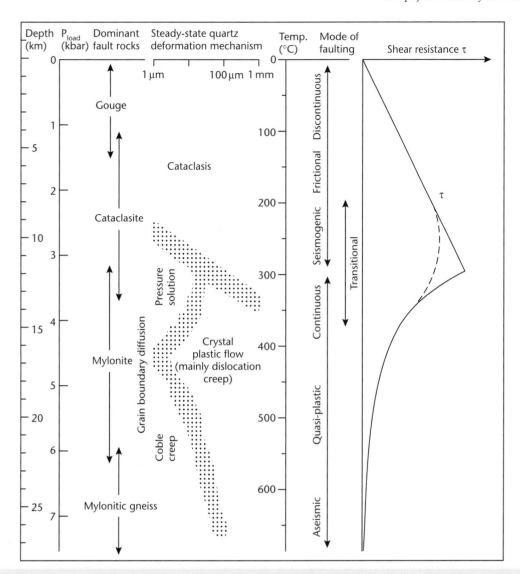

Fig. 2.35 Sibson's (1983) conceptual model for a major fault zone in the continental crust, showing the dominant deformation mechanisms.

deformation mechanism envisaged by Sibson (1983, p. 744) is shown in Fig. 2.35.

2.3.4 Strength profiles of the lithosphere

Lithospheric strength is controlled by lithospheric rheology, which itself is dependent on the heat flow or geothermal gradient, the composition of lithospheric rocks and the presence of volatiles. The thickness of the crust and lithosphere and the geothermal gradient therefore interact to produce distinctive strength profiles. We should expect these profiles to be dramatically different between oceanic and continental lithosphere (Lynch & Morgan 1987).

The lithosphere can deform by brittle (faulting) and by ductile mechanisms, each with its own yield strength. The yield stress (strength) of the lithosphere at any depth is conventionally calculated as the lesser of the two yield strengths for brittle failure and ductile creep (Brace & Kohlstedt 1980). The integral of the yield stress profile

over depth is the total yield stress of the lithosphere, equivalent to the horizontal deviatoric force required to cause non-elastic strain. At one or several points in the profile, the brittle and ductile yield strengths may be equal. These points or depths are known as the *brittle–ductile transition*.

The brittle yield strength is assumed to be related to the pressure and therefore to the depth in the crust. It is independent of strain rate, temperature and rock type. The ductile yield stress, however, is strongly dependent on strain rate, temperature and rock composition. Let us firstly take the relatively straightforward case of the oceanic lithosphere (Fig. 2.36).

The Anderson theory of faulting expresses the dip of a fault in terms of the coefficient of static friction. The tectonic stress required to cause faulting under tension is then

$$\Delta\sigma_{xx} = \frac{-2f_s(\rho gy - p_w)}{(1+f_s^2)+f_s}$$ [2.43]

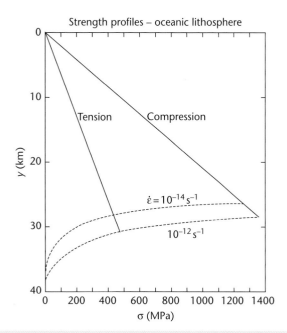

Fig. 2.36 Strength profile for the oceanic lithosphere calculated using eqn. [2.43] for tension and eqn. [2.44] for compression. The dashed lines are the stresses associated with creep in the lithosphere at the strain rates indicated. After Turcotte & Schubert (2002), © Cambridge University Press, 2002.

and under compression is

$$\Delta\sigma_{xx} = \frac{2f_s(\rho gy - p_w)}{(1+f_s^2) - f_s}$$ [2.44]

where $\Delta\sigma_{xx}$ is the deviatoric stress, f_s is the static coefficient of friction, ρgy is the lithostatic stress, and p_w is the pore pressure. To find the failure stress of the oceanic crust, we assume the pore pressure is hydrostatic, so $p_w = \rho_w gy$, $\rho = 3300\,kg\,m^{-3}$ and $\rho_w = 1000\,kg\,m^{-3}$, and $f_s = 0.6$. The results for tension and compression are sketched in Fig. 2.36.

But how far downwards in the oceanic lithosphere does the Anderson theory extend? Power-law creep as a function of temperature is given by an equation of the form

$$\dot{\varepsilon} = A\sigma^n \exp\left(\frac{-E_a}{RT}\right)$$ [2.45]

For olivine, $A = 4.2 \times 10^5\,MPa^{-3}\,s^{-1}$, E_a is the activation energy = $523\,kJ\,mol^{-1}$, and $n = 3$. Taking strain rates of $10^{-12}\,s^{-1}$ and $10^{-14}\,s^{-1}$, the stress as a function of depth for a geothermal gradient of $25\,K\,km^{-1}$ can be found (Fig. 2.36). We then assume that the lower of the frictional stress and the creep stress determines the strength of the oceanic lithosphere. The change from a frictional stress to a power-law stress takes place at a depth of c.28 km for the conditions given above. This is the brittle–ductile transition. Note that its depth depends on the geothermal gradient and strain rate, as well as on the 'sign' of the deviatoric stress (tension or compression).

The same concepts apply to the continental lithosphere, but geothermal gradients are likely to be more complex because of the internal heat generation of radiogenic minerals.

The effects of variations in geothermal gradient are illustrated in Fig. 2.37 using basal heat flows of $25\,mW\,m^{-2}$ representing a cold

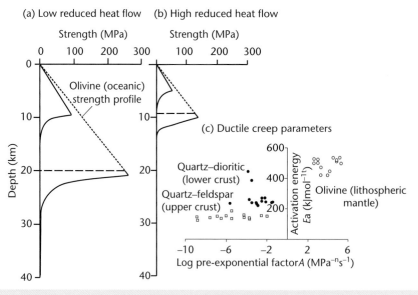

Fig. 2.37 Strength profiles for the continental lithosphere for different geothermal gradients, using (a) a low reduced heat flow of $25\,mW\,m^{-2}$ for a cold continental shield area, and (b) a higher reduced heat flow of $59\,mW\,m^{-2}$ for an area undergoing extension. $K = 2.5\,W\,m^{-1}\,K^{-1}$, $A_0 = 2.1 \times 10^6\,W\,m^{-3}$, $T_0 = 15\,°C$, universal gas constant $= 8.31451\,J\,mol^{-1}\,K^{-1}$, strain rate is $10^{-12}\,s^{-1}$, $n = 3$. For the olivine rheology, $A = 104\,MPa^{-3}\,s^{-1}$, $E_a = 500\,kJ\,mol^{-1}$; for brittle failure under tension, $f_s = 0.6$, $\rho = 3300\,kg\,m^{-3}$. For the lower crustal rheology, $A = 100\,Pa^{-3}\,s^{-1}$, $E_a = 300\,kJ\,mol^{-1}$; and for brittle failure under tension, $f_s = 0.6$ and $\rho = 2750\,kg\,m^{-3}$. (c) Plot of activation energy E_a versus the pre-exponential factor A for a range of rock types typical of the lithospheric mantle, lower crust and upper crust. Derived from table 1 in Fernandez & Ranalli (1997).

continental shield, and 59 mW m^{-2} representing the high geothermal gradients typical of the Basin and Range province, USA (Lynch & Morgan 1987) ($K = 2.5\,$W m^{-1} K^{-1}, $A_0 = 2.1\,\mu$W m^{-3}, $a_r = 10\,$km). There are a number of points to note. First, there are two high-strength brittle regions in the lithosphere, one in the upper crust and one in the upper mantle, each underlain by brittle–ductile transitions. Second, the brittle–ductile transition occurs at a much shallower level in the crust for the hot geotherm than for the cold geotherm. Third, the area to the left of the strength profile, representing the total yield strength of the lithosphere, is much greater for the cold lithosphere. We should expect cold lithosphere to be 'stronger' than hot lithosphere when flexed or subjected to horizontal extensional or compressional deviatoric stresses. The total yield strength of the lithosphere is considerably smaller (about half) under tension than under compression.

Oceanic lithosphere has a yield strength that is a strong function of its age. The entire oceanic crust is brittle, and the yield strengths of oceanic lithosphere older than about 10 Ma under both extension and compression are considerably higher than any likely imposed tectonic forces. Continental lithospheric strength, on the other hand, is strongly dependent on the basal heat flow or geothermal gradient. For a low geotherm (basal heat flow of 25 mW m^{-2}) the strength under extension is an order of magnitude greater than any likely imposed forces; consequently, cold continental lithosphere is unlikely to undergo significant deformation. At higher geotherms the strength under extension is much reduced, making the continental lithosphere susceptible to extensional deformation. For example, the actively extending crust in the Basin and Range province, USA, is 30 km thick with a basal heat flow of 70 mW m^{-2} (Lachenbruch & Sass 1978), whereas in the eastern USA, which is tectonically inactive, crustal

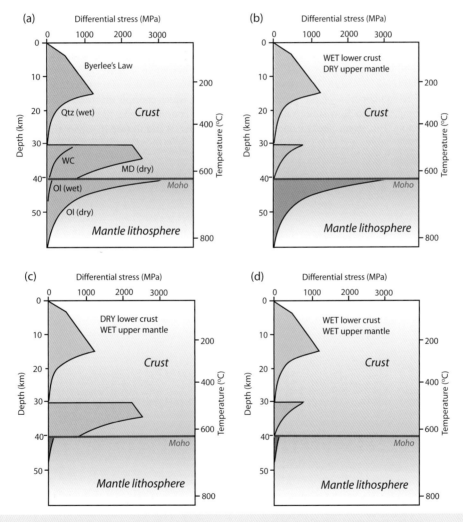

Fig. 2.38 Strength profiles: differential stress (rock strength) of the continental lithosphere (after Mackwell *et al.* 1998; Jackson 2002) based on the presence or absence of water. The Moho is at 40 km, and the geotherm is for a surface heat flow of 70 mW m^{-2}. The strain rate is 10^{-15} s^{-1} in (a) to (d). (a) Rheologies are wet quartz (Qtz) in the upper crust, Maryland diabase (MD) or undried granulite (WC) in the lower crust, and olivine (Ol) in the lithospheric mantle. The frictional strength of the upper crust is given by Byerlee's law for quartz. In (b), a wet (weak) lower crust and dry (strong) upper mantle constitutes the jelly sandwich model of lithospheric strength. In (c), a dry (strong) lower crust and wet (no strength) upper mantle may represent conditions in continental cratons. In (d), a wet (weak) lower crust and a wet (weak) upper mantle causes almost all strength to reside in the seismogenic upper crust, known as the crème brûlée model.

thickness is 40 km and basal heat flow is 33 mW m⁻². A global comparison of tectonic activity and basal heat flows shows that regions with high heat flows (>60 mW m⁻²), such as the Rhine Graben (western Europe) and Shansi Graben (China), undergo rifting, whereas regions with lower heat flows (<50 mW m⁻²) do not show signs of active extensional deformation (Kusznir & Park 1987, p. 42, fig. 8).

Crustal composition is an important factor, particularly at high geothermal gradients. The quartz rheology of the upper crust is weaker than the plagioclase rheology of the lower crust, which is in turn much weaker than the olivine rheology of the mantle. The thicknesses of upper and lower crust are therefore important in determining strength profiles. Crustal thickening has the effect of weakening the lithosphere, since the weaker quartz-plagioclase rheologies may replace the stronger olivine rheology at the depths normally associated with the lithospheric mantle. At crustal levels, a strong plagioclase or ultramafic lower crustal rheology may be replaced by a weaker quartz rheology during thickening. Thickened crust in zones of convergence may therefore be prone to extensional collapse.

During extension of the lithosphere at a finite strain rate, an increase in geothermal gradient weakens the lithosphere, whereas the thinning of the crust simultaneously strengthens it. The balance between net strengthening and net weakening is strongly affected by the strain rate. At high strain rates ($\dot{\varepsilon}$ of 10^{-13} to 10^{-14} s⁻¹) there is less opportunity for heat loss and re-equilibration of the geothermal gradient, leading to net weakening, whereas slow strain rates ($\dot{\varepsilon}$ of 10^{-16} s⁻¹) promote net strengthening. Fast strain rates therefore lead to intense localised extension and high surface heat flows, whereas slow strain rates generate broad regions of extension with low surface heat flows.

In conclusion, the distribution of rock strength in the lithosphere has a profound influence on its mechanical behaviour, including during basin development. Conventional views are that there are two high-strength layers corresponding to the brittle upper crust (quartz) and brittle upper mantle (olivine), separated by a weak lower crust (diabase or granulite), causing it be termed the *jelly sandwich model* (Chen & Molnar 1983) (Fig. 2.38b). An alternative is that there is essentially a strong crust with a dry lower crust overlying a wet, weak mantle, so lithospheric strength resides almost entirely within a single strong crustal layer within which earthquakes would be confined (Fig. 2.38c). In this *crème brûlée model*, the continental lithosphere is more like the oceanic lithosphere in its strength profile. The presence of volatiles such as water, even in very low quantities, is known to dramatically lower the creep strength of rocks (Maggi et al. 2000). Consequently, small concentrations of water, perhaps related to migration of metasomatic fluids or derived from subduction, might lower the strength of the continental lithospheric mantle. In contrast, the lower crust under continental cratons (Fig. 2.38d) is most likely of an anhydrous granulitic composition, with a permeability barrier to underlying wet mantle and where volatiles have been driven off. It is clear, therefore, that the lithospheric strength profile, which controls the storage of elastic stresses, the deformation of the continents and the mechanical behaviour during extension (Chapter 3) and flexure (Chapter 4), is potentially complex.

PART 2

The mechanics of sedimentary basin formation

CHAPTER THREE
Basins due to lithospheric stretching

Summary

Cratonic basins (intracontinental sags), continental rim basins, rifts, failed rifts, proto-oceanic troughs and passive continental margins fall within an evolutionary suite of basins unified by the process of lithospheric stretching. Rifts are a clear manifestation of lithospheric stretching, demonstrated by the shallow depth to the Moho, high surface heat flows, volcanic activity, seismic activity with predominantly extensional focal mechanism solutions, negative Bouguer gravity anomalies and commonly elevated rift margin topography.

Cratonic basins show little evidence of crustal stretching but subside over very long periods of time, probably due to cooling of relatively thick continental lithosphere. Continental rim basins are similar in overlying essentially unstretched crust, but are coeval with the drift phase of adjacent continental margins. Discrete, localised continental rifts appear to form on normal thickness crust and extend slowly over long periods of time. Rift zones evolve as linked fault arrays that control sediment dispersal during the syn-rift phase. At higher strain rates, localised rifts may evolve into passive margins. Wide extended domains with supra-detachment basins occur on previously thickened crust that extends quickly over a short period of time. Local anomalies in the ductile lower crust may be amplified to produce core complexes within these wide extended terranes.

At high levels of stretching the continental lithosphere attenuates sufficiently to allow the creation of new marine basins (proto-oceanic troughs) floored with oceanic crust. Passive continental margins evolve principally by slow cooling following break-up. They are in general seismically inactive, and their tectonics is dominated by gravity-driven collapse, salt movement and growth faulting. Heat flows are near normal in mature examples. Passive continental margins can be categorised according to their magmatic activity, their nourishment with or starvation of sediment, and the importance of salt and gravitational tectonics. There may be considerable asymmetry between conjugate margins, especially in the ductile lower crust.

Early investigations suggested that rifting fell into two end-member idealisations. *Active rifting* involves the stretching of the continental lithosphere in response to an active thermal process in the asthenosphere, such as the impingement of a hot mantle plume on the base of the lithosphere. *Passive rifting*, on the other hand, involves the mechanical stretching of the continental lithosphere from unspecified distant extensional forces, with passive upwelling of asthenosphere. Subsidence in rifts is an isostatic response to the stretching of the continental lithosphere. The post-rift stage of failed rifts and the drift stage of passive continental margins are due to thermal contraction during cooling of the stretched lithosphere and upwelled asthenosphere. The sediment loads are supported flexurally during this long phase of cooling. Dynamical extensional models of the lithosphere incorporate a rheologically layered lithosphere. These numerical models make predictions of strain rate evolution as a function of the changing temperature field and viscosity within this layered lithosphere.

The simplest formulation of continental extension is the 'reference' uniform stretching model and its derivatives. It involves uniform stretching with depth, instantaneous extension, one-dimensional heat transport by conduction, no magmatic activity, no internal radiogenic heat sources and the operation of Airy isostasy throughout. The reference uniform stretching model makes important first-order predictions. Initial fault-controlled subsidence is dependent on the initial crustal to lithospheric thickness ratio and on the stretch factor β. The subsequent thermal subsidence has the form of a negative exponential, and depends on the stretch factor β alone. The uniform stretching model serves as a useful approximation for subsidence and paleotemperature in rifted basins such as the North Sea, and in sediment-starved passive margins such as the Bay of Biscay and Galicia margin of the eastern central Atlantic. However, sediment-nourished passive margins require the post-rift subsidence to be modelled with flexural rather than Airy isostatic support. Volcanic margins require the effects of melt segregation, igneous underplating and transient dynamic uplift to be accounted for.

Subsequent modifications to the reference uniform stretching model have investigated: (i) the effects of a protracted period of stretching at low strain rate; (ii) depth-dependent stretching; (iii) extension by simple shear along trans-crustal detachments; (iv) elevated asthenospheric temperatures; (v) magmatic activity; (vi) secondary small-scale convection; (vii) radiogenic heat sources; (viii) greater depths of lithospheric necking; (ix) flexural support; and (x) phase changes. These modifications affect the predictions of syn-rift and post-rift subsidence, and in particular predict elevated syn-rift topography in the form of topographic swells or rift margin uplifts.

Dynamical approaches to lithospheric extension involve plane stress or plane strain models with various boundary conditions and initial conditions. Dynamical models are especially instructive in explaining why rifts remain narrow, with small bulk extensional strains, or develop into oceanic spreading centres. One set of dynamical models indicates that at moderate to low initial strain rates, extension is limited by an

Basin Analysis: Principles and Application to Petroleum Play Assessment, Third Edition. Philip A. Allen and John R. Allen.
© 2013 John Wiley & Sons, Ltd. Published 2013 by John Wiley & Sons, Ltd.

increase in viscosity in the mantle lithosphere below the Moho. In contrast, at high initial strain rates, complete rifting results from the concentration of the extensional force on a progressively thinner high-strength layer within the lithosphere. The strength profile, extent of decoupling or coupling between upper and lower strong layers of the lithosphere, and the effects of strain softening in accelerating thinning are all important facets of numerical models of continental extension. Analogue models are able to replicate the narrow, localised rifts and extensive tilted fault block terranes observed in passive margins as a function of strain rate. Analogue models also show that gravity spreading of weak, previously thickened crust leads to wide extended terrains and exhumed ductile lower crust in core complexes.

The amount of extension (stretch factor β) and the strain rate history can be estimated by a number of techniques. The thermal subsidence history of stratigraphy penetrated in boreholes allows the amount of lithospheric stretching and strain rate history to be estimated. Crustal stretching can be calculated from imagery of the depth to the Moho using gravity and seismic data. Stretch factors can also be estimated from forward sequential tectonostratigraphic modelling using well-constrained crustal and basin profiles derived from deep seismic reflection and refraction data.

3.1   Introduction

3.1.1   Basins of the rift–drift suite

Continental sags (cratonic basins), rifts, failed rifts, continental rim basins, proto-oceanic troughs and passive margins form part of an evolutionary sequence of basins unified by the processes of lithospheric extension (Dietz 1963; Dewey & Bird 1970; Falvey 1974; Kinsman 1975; Veevers 1981). Two linked mechanisms explain the majority of observations in these basins: (i) brittle extension of the crust, causing extensional fault arrays and fault-controlled subsidence; and (ii) cooling following ductile extension of the lithosphere or negative dynamic topography sustained by downwelling mantle flow, leading to regional sag-type subsidence.

Basins of the rift–drift suite occur in an evolutionary sequence leading eventually to the formation of new ocean basins. They can be visualised as falling in an existence field with axes of the bulk strain due to stretching (stretch factor) and the extensional strain rate (Fig. 3.1):

- *Cratonic basins* (*syn.* intracratonic sags) lack evidence for widespread extensional faulting but experience long-lived sag-type subsidence; both bulk strain and strain rate are low.
- *Continental rim basins* are located on essentially unstretched continental lithosphere and experience slow sag-type subsidence coeval with the late syn-rift and drift phases of the adjacent passive margin.
- *Rifts* are characterised by well-developed extensional faulting. Tensile deviatoric stresses are sufficient to overcome rock strength. Bulk strain and strain rate vary widely from narrow-slow, localised rifts, to wide-fast, diffuse extensional provinces and *supradetachment basins*.
- *Failed rifts* occur where the brittle stretching stops before reaching a critical value necessary for the formation of an ocean basin, and subsequent subsidence takes place due to cooling.
- *Proto-oceanic troughs* occur where the stretching has rapidly attenuated the lithosphere to allow a new ocean basin to form; bulk strain and strain rate are both very high.
- *Passive margins* are dominated by broad regional subsidence due to cooling following complete attenuation of the continental lithosphere. The passage from rift to passive continental margin takes place at the *rift–drift transition*.

Some rifts are at high angles to associated plate boundaries (Burke 1976, 1977). Some of these rifts appear to be linked to arms of triple junctions associated with the early stages of ocean opening – these failed rifts are termed *aulacogens*. Others are aligned at high angles to associated collision zones and are termed *impactogens* or *collision grabens*. Whereas aulacogens are formed contemporaneously with the ocean opening phase, impactogens clearly post-date this period, being related temporally with collision. The Western European Rift System, containing the Rhine Graben, has been cited as an example of an impactogen (Sengör *et al.* 1978), and collision in the Grenville orogeny has also been invoked as a cause of the Keweenawan Rift in North America (Gordon & Hempton 1986).

In this chapter, the mechanisms of lithospheric extension are focused upon as a means of explaining the rift–drift suite of sedimentary basins. This suite is characterised by a distinctive set of geological, geophysical and geomorphological observations that collectively include elevated heat flows, volcanicity, extensional faulting, thinned crust and regional sag-type subsidence. Embedded within the overarching theme of lithospheric extension are complexities caused by the total amount of stretching, the strain rate and the thermal and mechanical properties of the continental lithosphere undergoing extension.

3.1.2   Models of continental extension

Early investigations of lithospheric extension suggested that rifting fell into two end-member classes (Sengör & Burke 1978; Turcotte 1983; Morgan & Baker 1983; Keen 1985; Bott 1992): active and passive (Fig. 3.2).

In *active rifting* deformation is associated with the impingement on the base of the lithosphere of a thermal plume or sheet. Conductive heating from the mantle plume, heat transfer from magma generation or convective heating may cause the lithosphere to thin. If heat fluxes out of the asthenosphere are large enough, relatively rapid thinning of the continental lithosphere causes isostatic uplift. Tensional stresses generated by the uplift may then promote rifting.

In *passive rifting* unspecified tensional stresses in the continental lithosphere cause it to fail, allowing hot mantle rocks to penetrate the lithosphere. Crustal doming and volcanic activity are only secondary processes. McKenzie's (1978a) widely cited model for the origin of sedimentary basins belongs to this class of passive rifting. If passive rifting is occurring, rifting takes place first and doming may follow but not precede it. Rifting is therefore a passive response to a regional stress field. It is not easy to determine whether a given rift is either active or passive, since for small mantle heat flows the amount of uplift may be minimal. In addition, it should be understood that active and passive models are idealised abstractions that represent 'end members'. Real-world cases may exhibit aspects of each (Khain

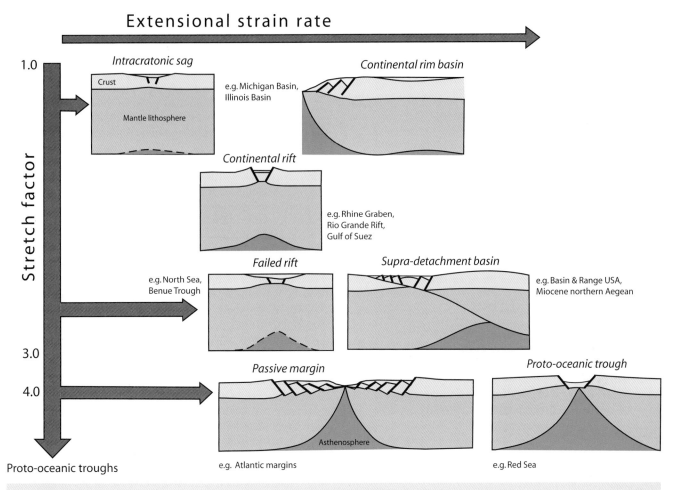

Fig. 3.1   Basins of the rift–drift suite in terms of increasing stretch factor and extensional strain rate.

1992). The East African and Ethiopian rifts (Fig. 3.3) are good candidates for active rifting whereas the wide Basin and Range province of southwestern USA (Fig. 3.4) is a candidate for passive rifting.

The syn-rift subsidence during stretching of the lithosphere is caused by brittle extension of the crust. Several mechanisms have been postulated as influences on the subsidence characterising the post-rift (or syn-drift) phase of passive margin development (§3.3.2). It is universally agreed that the main mechanism for post-rift subsidence is cooling following lithospheric thinning: the upwelling of asthenosphere (McKenzie 1978a) is followed by thermal contraction (for detailed treatment see Appendices 17 and 19). Lithospheric stretching, however, may be accompanied by important magmatism, producing dyke swarms, plutons and extensive basaltic sheets (Royden *et al.* 1980; White & McKenzie 1989). Emplacement of large volumes of basaltic melt into the crust (or along its base) should produce transient uplift, followed by subsidence as the extruded, intruded and underplated material cools. The result, long after emplacement, is determined by the density and thickness of the igneous additions to the lithosphere (Appendix 27). Commonly, these igneous additions have a higher density than the crust but lower density than the mantle. Assuming that the bulk of the igneous accretion replaces lithospheric mantle, passive margins with large amounts

of magmatic activity should remain relatively elevated compared to non-magmatic margins.

It has also been postulated that subsidence may result from phase changes in lower crustal or mantle lithosphere rocks (Podladchikov *et al.* 1994; Kaus *et al.* 2005). Mineral phase changes may cause a rapid increase or decrease in the density of the lithosphere. The change from gabbro to eclogite may cause basin subsidence due to a density increase (Ahern & Dikeou 1989; Stel *et al.* 1993), whereas a density decrease, causing uplift, may result from the change from garnet to plagioclase-lherzolite. Phase changes are most likely to be important where the stretching is rapid (Armitage & Allen 2010).

The growing sediment prism acts as a load on the lithospheric substrate and is supported isostatically. This may be by lithospheric flexure (Watts *et al.* 1982; Beaumont *et al.* 1982), as long as the wavelength of the load is sufficiently short. Historically, there was much debate as to whether flexural support involved an elastic plate or a viscoelastic (Maxwell) plate (§2.3) overlying a weak substratum (Watts *et al.* 1982). Since active faulting and high heat flows accompany the early stages of rifting it has been assumed that an Airy isostatic model is most applicable during this period, although a small but finite flexural strength (Chapter 4) has been inferred for the active rifting phase in East Africa (Ebinger *et al.* 1989, 1991).

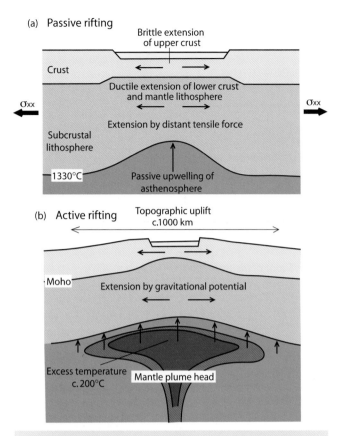

(a) Passive rifting

Brittle extension
of upper crust

Crust

Ductile extension of lower crust
and mantle lithosphere

σ_{xx}

Extension by distant tensile force

Subcrustal
lithosphere

1330°C

Passive upwelling of
asthenosphere

(b) Active rifting

Topographic uplift
c.1000 km

Moho

Extension by gravitational potential

Excess temperature
c. 200°C

Mantle plume head

Fig. 3.2 Active and passive rifting end-member idealisations. (a) Passive rifting driven by a distant tensile deviatoric force σ_{xx} causing thinning of the lithosphere and passive upwelling of hot asthenosphere. (b) Impingement on the base of the lithosphere of a mantle plume causes long-wavelength topographic doming and gravitationally driven extension of the lithosphere.

(Goetze & Evans 1979; Ranalli 1995), which is the profile of lithospheric strength as a function of depth, underpins the rheological modelling of lithospheric extension (see §2.3.4).

The *rheology of the lithosphere* and its response to extensional deformation is most strongly controlled by its temperature and its mineralogy, and is strongly affected by the presence of volatiles such as water. A material can only maintain stresses over geological time if the ratio of actual temperature to its melting temperature (known as the *homologous temperature*) is less than about 0.4. This implies that only the upper parts of the crust and upper mantle are capable of supporting elastic stresses over long periods of time. Most dynamical models of lithospheric extension use laboratory data for the dominant mineral comprising crust and mantle – that is, quartz + feldspar for the crust, and olivine for the mantle. On a graph of stress at which failure takes place ($\sigma_1 - \sigma_3$) versus depth in the continental lithosphere, two brittle–ductile transitions and two stress-supporting regions are commonly predicted, one in the upper-mid crust, and the other in the upper mantle lithosphere (Fig. 2.38). Since the quartz + feldspar rheology of the continental lithosphere is weak at high temperatures, zones of orogenic thickening are prone to collapse by extension – an idea originally proposed by Tapponnier and Molnar (1976).

We examine the results of numerical models of lithospheric extension in §3.5. One of the major results of numerical modelling is that the changing temperature field of lithospheric particles during extension results in a changing viscosity of the mantle lithosphere over time. Consequently, dependent on the initial strain rate, the lithosphere may extend rapidly into a run-away state, or may stop extending (Newman & White 1999). This may explain dynamically the strain rate and bulk strain history of narrow rifts, wide rifts and passive margins (Fig. 3.5). The compilation in Fig. 3.5 shows a clustering of examples as relatively narrow, localised rifts (<100 km), but with extensional strain rates varying over two orders of magnitude (from $<10^{-16}\,s^{-1}$ to $10^{-14}\,s^{-1}$).

3.2 Geological and geophysical observations in regions of continental extension

3.2.1 Cratonic basins

'Intracratonic basins', 'cratonic basins', 'interior cratonic basins' and 'intracontinental sags' are circular- to oval-shaped crustal sags, located on stable continental lithosphere (Sloss & Speed 1974; Sloss 1988, 1990). We restrict the use of the term 'cratonic basin' to those basins located some distance from stretched or convergent continental margins, distinct from rifts where a history of continental extension is unequivocal, but located on a variety of crustal substrates, irrespective of whether they are crystalline shields *sensu stricto*, accreted terranes, or ancient foldbelts and rift complexes (Holt *et al.* 2010; Allen & Armitage 2012). Many cratonic basins overlie thick continental lithosphere (Hartley & Allen 1994; Downey & Gurnis 2009; Crosby *et al.* 2010).

Cratonic basins are characterised by predominantly shallow-water and terrestrial sedimentation. Their subsidence history is prolonged, occasionally marked by an initial stage of relatively fast subsidence, followed by a period of decreasing subsidence rate (Nunn & Sleep 1984; Stel *et al.* 1993; Xie & Heller 2009; Armitage & Allen 2010), somewhat similar to that of ocean basins (Sleep 1971). Cratonic basins generally lack well-developed initial rift phases, though this

Post-rift sediments, however, are gently dipping and of wide extent, suggesting that flexure takes over at some stage after the end of rifting. Watts *et al.* (1982) believed that the characteristic pattern of stratigraphic onlap on the eastern Atlantic and other continental margins suggested an increasing rigidity of the lithosphere with time – the expected result of an elastic lithosphere heated during the rifting stage and subsequently cooling.

In the last decade or two, emphasis has shifted away from active and passive rifting models and towards an understanding of the dynamics of stretching a piece of continental lithosphere with a particular strength and viscosity structure (§3.5) (Houseman & England 1986; Sonder & England 1989; Newman & White 1999; Huismans & Beaumont 2002, 2008). Increasingly, numerical model predictions are examined alongside high-resolution 3D seismic reflection data sets. The fundamental infrastructure of these numerical models is the description of the rheology of the continental lithosphere (Fernandez & Ranalli 1997).

The *strength of the lithosphere* can be judged from laboratory experiments (e.g. Kohlstedt *et al.* 1995), though there is the problem of 'scaling-up' laboratory experiments to the strain rates experienced by rocks in the lithosphere. The concept of the *strength envelope*

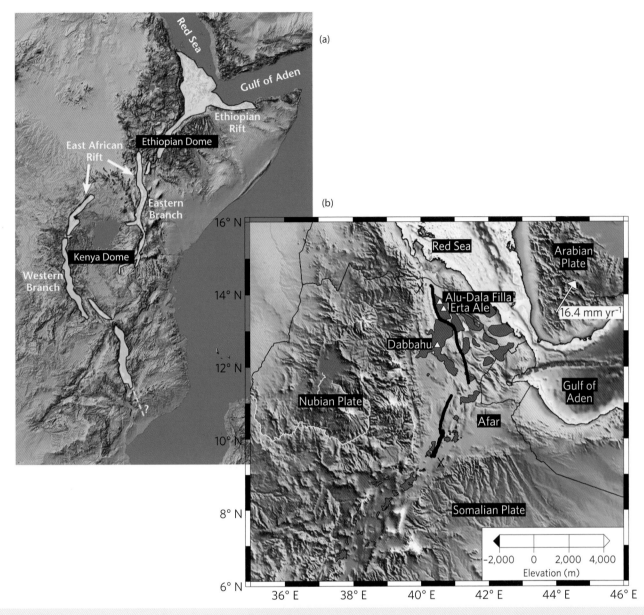

Fig. 3.3 (a) Topographic swells of east and northeast Africa, with location of main rift zones. (b) Topography of the Ethiopia-Afar region with magmatism in red. Reproduced from Bastow & Keir (2011), with permission of Nature Publishing Group.

may be due in part to the poor seismic imaging of the base of cratonic basins preserved on land. In some cases, cratonic basins are connected by a rift or failed rift zone to the ocean, as in the Neoproterozoic Centralian Superbasin of Australia (Walter *et al.* 1995; Lindsay 2002), the Lower Paleozoic Illinois and Oklahoma basins of USA (Braile *et al.* 1986; Kolata & Nelson 1990; Leighton *et al.* 1990) and the Mesozoic phase of the Chad Basin of north-central Africa (Burke 1976). This geometry suggests that many cratonic basins lie at the tips of failed rifts extending into the continental plate at a high angle from the extensional plate margin, which may be the site of former triple junctions (Burke & Dewey 1973). The association of cratonic basins with continental stretching is supported by the local occurrence of rifts below the sag-type basin-fill. Stretch factors are, however, very low beneath the cratonic basin, for example 1.05 to 1.2

in the Hudson Bay (Hanne *et al.* 2004) (Fig. 3.6) and <1.6 in the West Siberian Basin (Saunders *et al.* 2005).

Cratonic basins are large, with surface areas ranging from the relatively small Anglo-Paris Basin (10^5 km^2), through to the large Hudson Bay (1.2×10^6 km^2), Congo (1.4×10^6 km^2) and Paraná basins (1.4×10^6 km^2) to the giant Centralian Superbasin (2×10^6 km^2) and West Siberian Basin (3.5×10^6 km^2) (Sanford 1987; Leighton & Kolata 1990, p. 730; Walter *et al.* 1995, p. 173; Vyssotski *et al.* 2006; Crosby *et al.* 2010). In cross section, they are commonly simple saucers, with sediment thicknesses typically less than 5 km, and rarely <6–7 km (as in the West Siberian (Fig. 3.7), Illinois and Paraná basins). However, in some cases, the circular planform shape is a result of later compartmentalisation of a previously more extensive platform or ramp, as in the cratonic basins of North Africa, such as

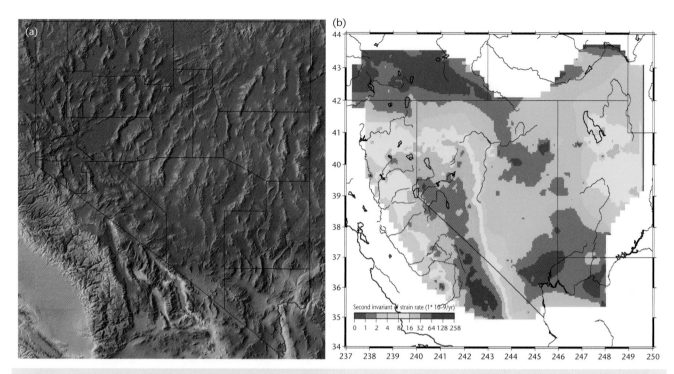

Fig. 3.4 (a) Topography of the Basin and Range province, southwestern USA, showing a large number of extensional fault blocks and basins over an extensive area of approximately 1000 km × 1000 km. Digital elevation model from National Geophysical Data Center, National Oceanic and Atmospheric Administration (NOAA), http://www.ngdc.noaa.gov/. (b) Map of crustal strain rate along the plate boundary between the Pacific and North American plates, derived from GPS measurements of horizontal station velocities (arrows). Strain rate is contoured in units of $10^{-9}\,yr^{-1}$. The range of values shown is from 0 to $3600 \times 10^{-9}\,yr^{-1}$, equivalent to 0 to $11 \times 10^{-14}\,s^{-1}$. A value of $72 \times 10^{-9}\,yr^{-1}$ (yellow tone) corresponds to $23 \times 10^{-16}\,s^{-1}$. Note the very high rates of deformation along the plate boundary zone and the lower strain rates in the Basin and Range province responsible for the N–S extensional fault blocks seen in part (a). From Map 178, Nevada Bureau of Mines and Geology, A geodetic strain rate model for the Pacific-North American plate boundary, western United States, by Kreemer, C., Hammond, W.C., Blewitt, G., Holland, A.A. & Bennett, R.A.

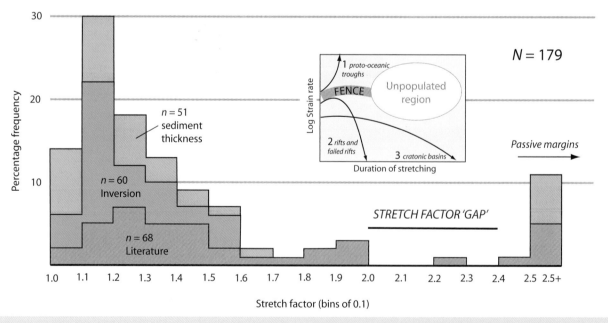

Fig. 3.5 Rift basins plotted in an existence field of stretch factor and strain rate. Literature-based data (grey) were obtained by MSc Petroleum Geoscience students at Imperial College London in 2010 and 2011. Other estimates were obtained from sediment thickness information (particularly ICONS database) and from inversion of thermal subsidence data (Wooler *et al.* 1992; White 1994). There is a scarcity of rift basins with stretch factors in the range 2.0 to 2.5.

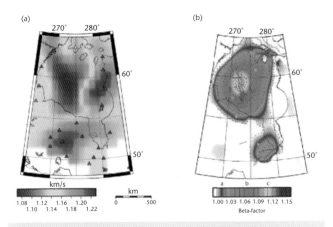

Fig. 3.6 Hudson Bay Basin, Canada. (a) Seismic velocity anomalies based on tomographic reconstructions (Pawlak *et al.* 2011), representative of the mid-crust, showing a slowing in the region beneath the main sedimentary depocentre; and (b) calculated stretch factors based on modelling of sediment thickness (from Hanne *et al.* 2004).

Al Khufra, Murzuk and Ghadames (Selley 1972, 1997; Boote *et al.* 1998) and the latest Proterozoic–Early Ordovician 'Sauk' sequence of east-central North America (Sloss 1963, 1988). They experience a long duration of subsidence, measured in hundreds of millions of years, implying slow sediment accumulation rates. The cratonic basins of North America, for instance, accumulated sediment at rates of 20 to 30 m Myr^{-1} (Sloss 1988), which is extremely slow compared to rifts, failed rifts, young passive margins, foreland basins and strike-slip basins (Allen & Allen 2005), but relatively fast compared to the adjacent platforms. Laterally equivalent platformal areas, such as the Transcontinental Arch of USA, accumulated c.1 km of sediment between Cambrian and Permian, at a rate of 3–4 m Myr^{-1} (Sloss 1988).

Although cratonic basins are very long-lived, it is important to recognise that the basin-fill is commonly composed of a number of different megasequences (or *sequences* of Sloss 1963), some of which may be associated with entirely different mechanisms of formation, such as strike-slip deformation, flexure and unequivocal stretching. Consequently, it is important, wherever possible, to extract the cratonic basin megasequence from the polyhistory basin-fill (Kingston *et al.* 1983a) for analysis. In other cases, basins have remained as

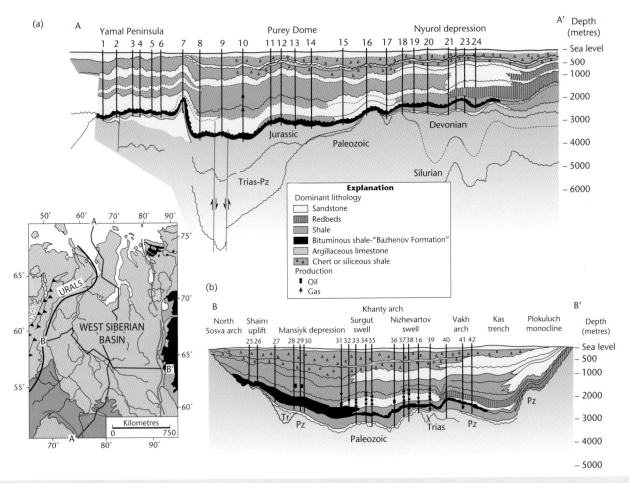

Fig. 3.7 Cross-sections of West Siberian Basin, A-A' and B-B', along transects shown in the inset geological map. From Vyssotski *et al.* (2006) and Saunders *et al.* (2005).

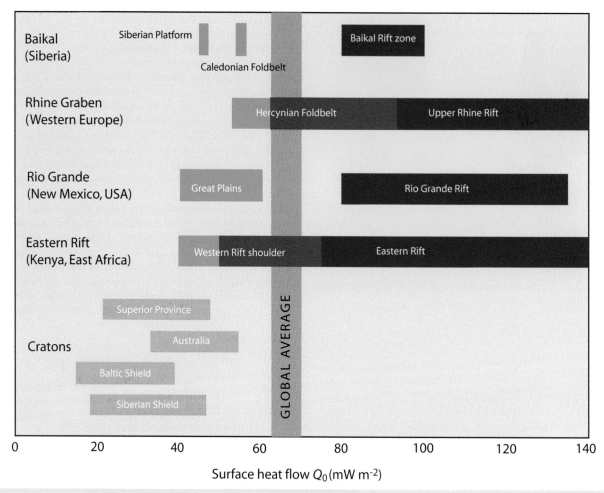

Fig. 3.8 Surface heat flows in some continental rifts and surrounding regions, compared with some cratonic basins and the global continental surface heat flow average.

cratonic basins throughout their history but have existed long enough to have been strongly affected by several tectonic mechanisms of subsidence and uplift. Consequently, there may be a primary mechanism for basin formation, and different secondary mechanisms for later modification (Allen & Armitage 2012).

3.2.2   Rifts

Rifts (classified as 'interior fracture basins' by Kingston *et al.* 1983a, b) are areas of crustal extension, and seismic studies show them to overlie thinned crust. Regions of rifting at the present day are characterised by negative Bouguer gravity anomalies, high surface heat flow and volcanic activity, all of which suggest that in addition to crustal extension, a thermal anomaly exists at depth. The essential observations in zones of continental extension are summarised here in terms of heat flow, seismicity, crustal thickness, gravity data, faults, surface topography, bulk strain and strain rates.

The presence of active volcanoes and elevated heat flows in rift zones demonstrates active thermal processes. However, the measured values of heat flow are often difficult to interpret, owing to complications caused by shallow magmatic intrusions, groundwater convection and spatial/temporal variability of conductive sediments and rocks. In general, rift zones have surface heat flows of 90–110 mW m^{-2}. This is up to a factor of 2 higher than in surrounding unstretched terranes (Fig. 3.8). Values are higher in volcanic rifts such as the Eastern Rift, Kenya, and lower in non-volcanic rifts such as those of Malawi, Tanganyika and the Jordan–Dead Sea rift zone of the Middle East. Volcanic activity has a major influence on rift sedimentation, as exemplified by the Eastern Rift of Kenya (Ebinger & Scholz 2012) and the Rio Grande Rift of southwestern USA (Mack & Seager 1990).

Rifts are characterised by high levels of earthquake activity. In the continental lithosphere, earthquake epicentres commonly delineate active rift zones or reactivated orogenic belts. Focal mechanism solutions in general indicate normal dip-slip faulting with orientations roughly parallel to the long-axis orientation of the rift. In some continental rifts, such as the Rhine Graben, strike-slip focal mechanisms dominate dip-slip solutions by 3:1 (Illies 1977). Although earthquakes are common in regions of continental rifting, they typically have moment magnitudes of up to 5.0 (Rhine Graben) or 6.0 (East African Rift), with shallow focal depths of <30 km, indicating that the earthquakes are located in the brittle upper crust.

Seismic studies show that the Moho is elevated beneath rift zones. The southern Rhine Graben is an example (Fig. 3.9). The Moho reaches a depth of 24 km near the Kaiserstuhl volcano due west of Freiburg, Germany, directly beneath the centre of the graben. The Moho is dome-shaped, deepening to the north, NW and NE to about 30 km. Approximately 3 km of syn-rift sedimentary rocks are found in the graben, so the continental crust has been thinned from 30 km to 21 km, that is, by a factor of 1.4. In the North Sea failed rift, the crust (pre-Triassic) is greater than 31 km thick beneath the Shetland Platform and Scandinavian Shield, but is less than 16 km thick beneath the Viking Graben (Klemperer 1988). The continental crust has therefore been thinned by a factor of approximately 2 immediately beneath the deepest part of the Viking Graben (Fig. 3.10). An important observation, however, is that some regions of extensive, diffuse extension, such as the Basin and Range province of SW USA, are located on previously thickened crust. The Moho was therefore anomalously deep at the onset of extension, and extension/thinning has brought the Moho back to 'normal' depth. This is also the case for the Tibetan Plateau, which is undergoing active extension and overlies crust as much as 70 km thick.

Rift zones have characteristic gravity signatures – typically a long-wavelength Bouguer gravity low with sometimes a secondary high located in the centre of the rift zone. The conventional explanation is that rift zones have anomalously hot material in the mantle beneath the rift, producing a mass deficit and therefore a negative gravity anomaly. The mass deficit is commonly associated with slow shear wave velocities, as below the Rio Grande Rift of southern USA (Fig. 3.11). The subsidiary gravity high is thought to be due to the intrusion of dense magma bodies within the continental crust. Regions of widespread, diffuse extension, such as the Basin and Range province of SW USA, show a series of gravity highs corresponding to basement blocks, and c.20 km-wide gravity lows corresponding to sedimentary basins. The gravity lows most likely reflect the mass deficit of light basin sediments.

Rift zones are typified by normal dip-slip faults with a variable number of strike-slip faults depending on the orientation of the rift axis in relation to the bulk extension direction. Consequently, the central Death Valley Basin is close to orthogonal to the extension direction and is typified by dip-slip normal faults, whereas the northern Death Valley Basin is more oblique and has faults with important strike-slip motion (Burchfiel & Stewart 1966).

Faults in rifts are not infinite in extent: instead there is a displacement-length relationship, with most of the slip being taken up on a small number of interacting major fault segments. Fault displacement dies out towards the tips of fault segments. A single fault segment, whose length is commonly between 10 and 100 km, has a maximum cumulative displacement (d)-length (L) relationship (Schlische 1991; Dawers & Anders 1995)

$$d \approx kL \qquad\qquad [3.1]$$

where k is a constant of proportionality that varies between 0.01 and 0.05 for many fault systems (Fig. 3.12).

The growth of populations of interacting faults is affected by a stress feedback that promotes the growth of a small number of major through-going border faults at the expense of smaller displacement faults that become dormant (Schlische 1991; Dawers & Anders 1995; Cowie *et al.* 2000). The evolution of the fault array has a profound effect on the topography and river drainage of rifted regions (Gupta *et al.* 1999; Cowie *et al.* 2000, 2006) (Fig. 3.13), and therefore on the gross depositional environments and sediment thicknesses of the basin-fill (Gawthorpe & Leeder 2000). Relay ramps between fault segments along the basin edge are commonly the entry points of river systems into the hangingwall depocentres. Footwall evolution, however, controls the establishment of the main drainage divide and therefore the size and sediment discharge of river catchments draining into the rift (Cowie *et al.* 2006). The Jurassic Brent–Strathspey–Statfjord fault array system in the North Sea Basin (McLeod *et al.* 2000), the Neogene fault array of the Gulf of Suez in eastern Sinai (Sharp *et al.* 2000) and the modern fault array of the Lake Tanganyika Basin (Rosendahl *et al.* 1986) are all excellent examples. Studies of such evolving fault arrays during the Jurassic syn-rift phase of the North Sea Basin suggest that linkage takes place rapidly, perhaps over 3–4 Myr (McLeod *et al.* 2000). Most major border faults dip steeply inwards towards the basin centre and are planar as far as they can be imaged. However, some rift-bounding faults are low-angle and listric, taking up very large amounts of horizontal extension, as in the supra-detachment basins of SW USA. Metamorphic rocks may be unroofed from <25 km depth in footwall 'core complexes' (Wernicke 1985).

The stratigraphic patterns of rift basins reflect this underlying pattern of fault array evolution (Gawthorpe & Leeder 2000). The common evolution from small, disconnected faults to full linkage produces a stratigraphic theme of isolated continental, hydrologically closed basins with lakes evolving into open rifts connected to the ocean, containing shallow-marine or deep-marine sediments fringed by fan deltas. The Miocene evolution of the Gulf of Suez is a classically documented example of such a rift evolution (Garfunkel & Bartov 1977; Evans 1988; Sharp *et al.* 2000). The concept of fault linkage also underpins the interpretation of extensional fault basins recognised on seismic reflection profiles (Prosser 1993).

Currently or recently active rift zones typically have elevated rift flank topography bordering a depositional basin. There may be two length scales of surface uplift. The best examples of the large length scale (several hundred kilometres) are the >3 km-high topographic swells of Ethiopia and East Africa (Baker *et al.* 1972; King & Williams 1976) (Fig. 3.14). Other domal uplifts are found in northern Africa, such as those in the Tibesti and Hoggar regions. These swells are commonly associated with widespread volcanic activity. Whereas the large domes of eastern and northeastern Africa are currently undergoing rifting, the smaller domes of north-central Africa are not. At a smaller length scale (<100 km) are the linear rift flank uplifts associated with border fault arrays. The <1 km-high highlands bordering the Gulf of Suez are a good example. Border fault footwalls involve upward tectonic fluxes, leading to enhanced denudation. In the southern Rhine Graben, tectonically driven exhumation of the rift flank has resulted in 2 to 3 km of erosion, exposing Hercynian crystalline basement in the Vosges of Alsace (France) and the Black Forest

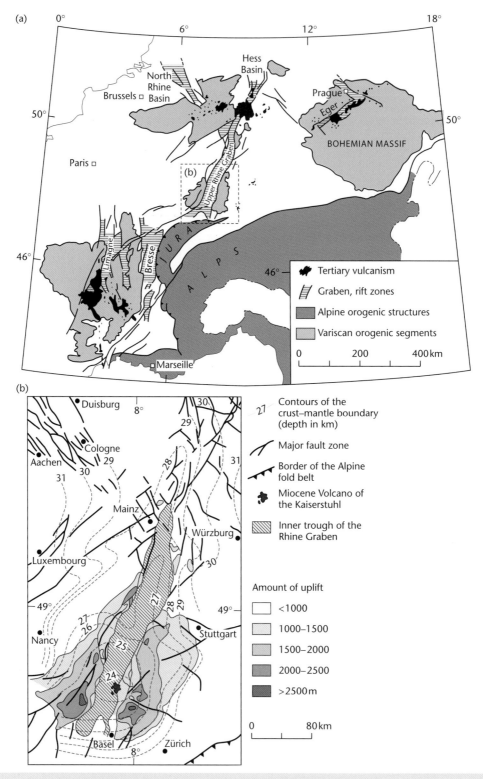

Fig. 3.9 (a) Location and main elements of the late Eocene–Recent Western European Rift System, with sites of Tertiary volcanicity. (b) Depth to the Moho below sea level (in km), showing a mantle bulge in the southern Rhine Graben centred on the Kaiserstühl volcano (Illies 1977). The largest amounts of denudation are found on the rift flanks above the shallow mantle.

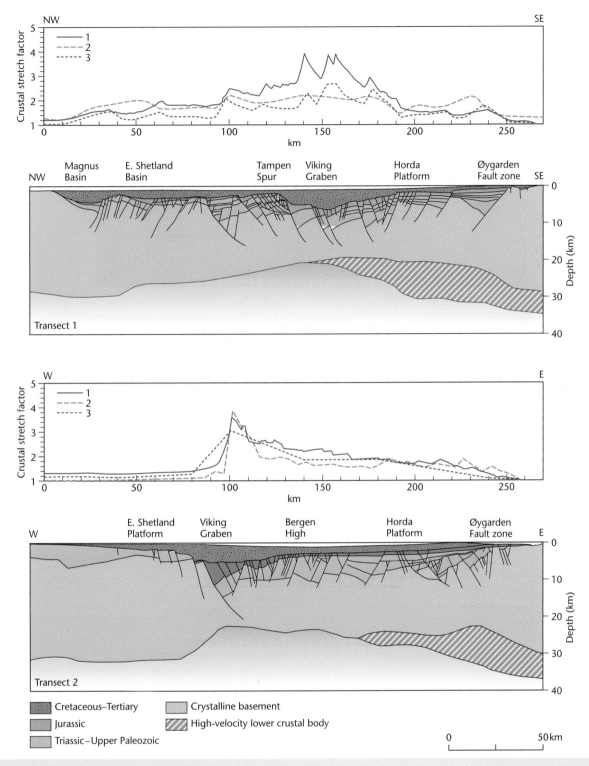

Fig. 3.10 Two crustal transects across the Viking Graben, northern North Sea (from Christiansson *et al.* 2000; Skogseid *et al.* 2000), based on the integration of all available geological and geophysical information, with estimates of crustal stretch factor, based on: (1) crustal thinning, assuming an initial crustal thickness of 36 km; (2) reverse modelling; and (3) forward tectonostratigraphic modelling.

(a)

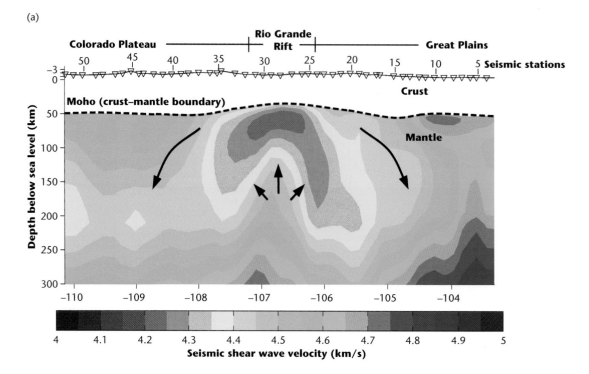

(b)

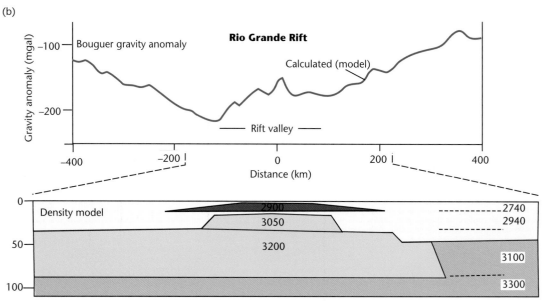

Fig. 3.11   (a) Seismic shear wave velocities (km s⁻¹) beneath the Rio Grande region, USA, indicating the existence of hot mantle, perhaps associated with a mantle upwelling. The mass deficit is responsible for a large-wavelength gravity low. Image from RISTRA programme, Wilson *et al.* (2005). (b) Gravity profile (c.33°N) and density model for the Rio Grande Rift of New Mexico, after Ramberg (1978). The secondary gravity high is thought to be due to the presence of dense igneous bodies beneath the rift. Densities shown in kg m⁻³.

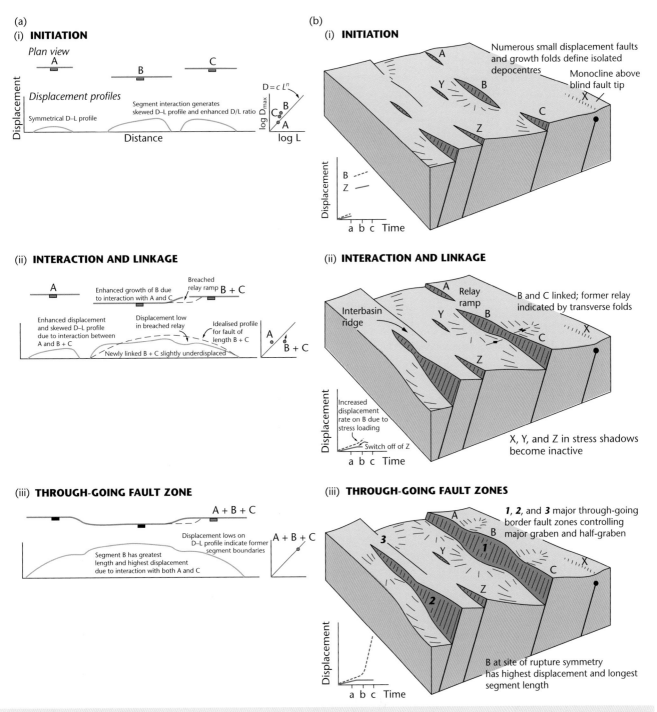

Fig. 3.12 (a) Displacement-length relationship of normal faults. Fault segments grow from an initiation stage (i) with individualised displacement-length relationships, to interaction and linkage (ii), and then to a through-going fault zone (iii), where there is one displacement-length profile that may contain remnant information of the original fault segments. (b) 3-D expression of fault interaction and growth from initiation, interaction and linkage, to a through-going fault zone. Reproduced from Gawthorpe & Leeder (2000), with permission of John Wiley & Sons, Ltd.

(a)

(i)   Fault propagation and interaction prior to linkage

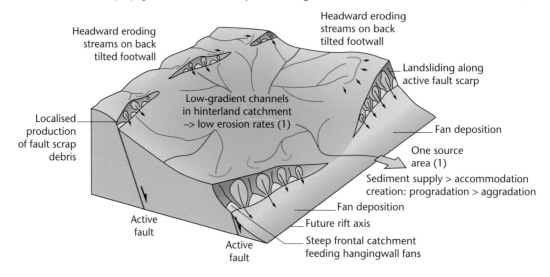

(ii)   Fault linkage and slip rate increase

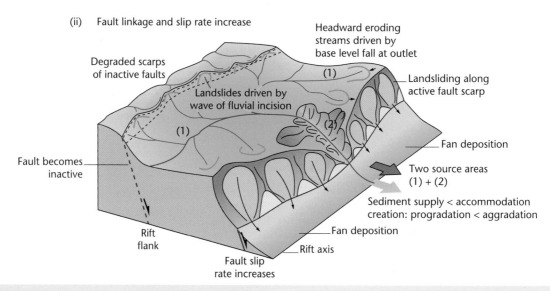

Fig. 3.13   (a) Schematic diagrams showing effects of fault linkage on surface topography and drainage in the fault propagation and interaction stage (i) and after fault linkage, which transfers extension to the main border fault (ii). (b) Numerical model results from fault linkage/surface processes model, showing evolution of erosional footwall. Reproduced from Cowie *et al.* (2006), with permission of John Wiley & Sons, Ltd.

of Germany. Regions of extensive, diffuse extension are associated with plateau-type topography, such as the Basin and Range, USA, and especially Tibet. In the first case, shallow subduction of relatively buoyant oceanic lithosphere beneath the North American plate, and in the second case, thickening of continental lithosphere during India-Asia collision, are the driving forces for extensive topographic uplift and extension.

Extensional basins vary greatly in their duration of subsidence, total extensional strain, and therefore in their strain rate (Fig. 3.5). Friedmann and Burbank (1995) recognised two distinct families of basins, which could be differentiated according to their strain rate,

total extensional strain (or stretch factor β) and the dip of the master faults (Fig. 3.15):

- Discrete continental rifts located on normal thickness crust (such as the Rhine Graben, Baikal Rift, Rio Grande Rift) extend slowly ($<1\,mm\,yr^{-1}$) over long periods of time (10–>30 Myr), with low total extensional strain (generally <10 km). Master fault angles are steep (45–70 degrees). Seismicity suggests that crustal extension takes place down to mid-crustal levels. At higher strain rates, narrow rifts may evolve through increased stretching into passive margins.

(b)

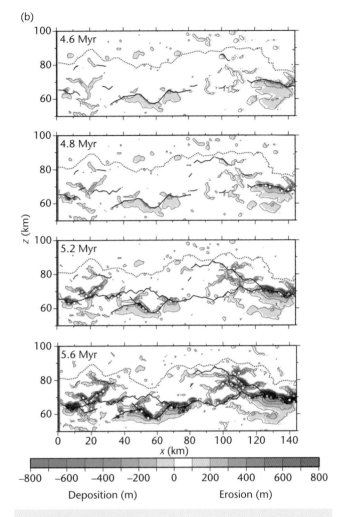

Fig. 3.13 *Continued*

3.2.3 Failed rifts

Failed rifts are those basins in which rifting has been aborted before the onset of seafloor spreading and passive margin development. Their rift phase is identical to that outlined in the previous paragraphs (§3.2.2). During cooling, failed rifts widen and post-rift sedimentary rocks onlap the previous rift shoulders, producing a *steer's head* geometry. A sedimentary evolution from non-marine to shallow-marine in the syn-rift phase and deeper-marine in the post-rift phase is typical.

The Benue Trough of central-western Africa and the North Sea are two excellent examples of aborted rifting. The Benue Trough is 1000 km long, 100 km wide, and is filled with <5 km of fluvial, deltaic and marine Cretaceous sedimentary rocks. At the southwestern end of the failed rift, the Tertiary Niger Delta has built a wedge of fluvial, deltaic and submarine fan deposits 12 km thick. A similar focusing of river drainage and delta growth along failed rift arms is found in the Lena Delta of the giant West Siberian Basin (Vyssotski *et al.* 2006) and the Brent Delta of the Viking Graben of the North Sea Basin (Morton *et al.* 1992).

In the northern North Sea a major period of rifting took place in the Middle Jurassic. At this time sediment was dispersed longitudinally along the graben. In the N–S oriented Viking Graben fluviatile deposits pass northwards into deltaic and shallow-marine deposits of the Brent Group (Morton *et al.* 1992). The Mid-Cretaceous saw the end of the rift phase and sediment onlapped the graben shoulders onto the East Shetland Platform in the west and the Norwegian Platform in the east. Thick deposits of Cretaceous chalks, Paleogene submarine fan sandstones and basinal shales, and Neogene mudstones typify this post-rift phase. In the southern part of the North Sea, the basin was filled by major sediment supplies from rivers draining the European mainland, producing mega-clinoforms due to westward delta progradation (Overeem *et al.* 2001; Huuse & Clausen 2001) (Fig. 3.16). The southern-central North Sea therefore shows the tendency for shallowing due to basin filling during the later stages of the thermal subsidence phase of failed rifts.

3.2.4 Continental rim basins

Continental rim basins or 'sag basins' (Huismans & Beaumont 2002) are thought to evolve over unstretched to slightly stretched continental lithosphere at the same time as passive margin development. They occur during the formation of new ocean basins where each continental margin subsides to produce a wide, shallow basin inboard of the ocean-facing margin (Veevers 1981, 2000; Favre & Stampfli 1992). The broad sag-type subsidence is suggested to be due to cooling following plume activity driving continental rifting (Fig. 3.17), but similar effects might be caused by thermal relaxation of upwelled asthenosphere relayed laterally from brittle upper crustal stretching associated with passive margin formation, or by cooling following convective thinning of the mantle lithosphere during continental attenuation (Huismans & Beaumont 2002). The hallmark of continental rim basins is that their centres are located in the order of 10^3 km inboard from the continental edge, their basin-fills are contemporaneous with the late syn-rift and drift phase of adjacent passive margins, their subsidence rates are low, their continental basement is typified by minimal amounts of brittle faulting, and magmatism is absent.

• Supra-detachment basins occur within wide extended domains with previously thickened crust. They typically extend quickly ($<20 \, \text{mm} \, \text{yr}^{-1}$) over short periods of time (5–12 Myr) with a high amount of total extensional strain (10–80 km). Master faults (detachments) are shallow in dip (10 to 30 degrees), but may have originated at higher angles. Local anomalies in the ductile lower crust are amplified to produce core complexes (Wernicke 1985).

The location of continental rifts is commonly determined by the existence of old fundamental weaknesses in the lithosphere. A number of examples of this phenomenon are the opening of the modern Atlantic Ocean along the Paleozoic Iapetus suture (Wilson 1966), the Cretaceous separation of southeastern Africa from Antarctica along a Paleozoic failed rift (Natal Embayment) (Tankard *et al.* 1982) and Cenozoic rifting in the East African Rift system, which follows Precambrian structural trends (McConnell 1977, 1980).

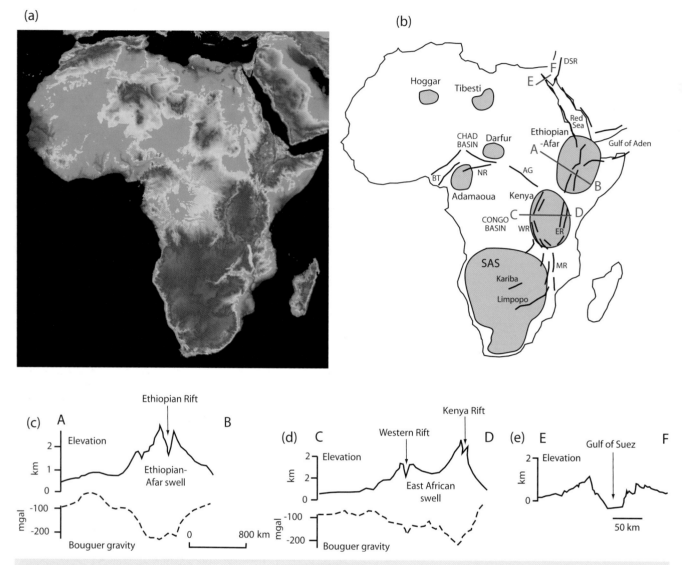

Fig. 3.14 (a) Topographic image of Africa (reproduced with the permission of NASA). (b) The major domal uplifts of Africa (Afar and East African domes) are due to uplift over hotspots in the mantle. SAS, Southern Africa swell. Also shown are the smaller topographic uplifts of central Africa and the main rift systems: AG, Abu Gabra Rift; BT, Benue Trough; DSR, Dead Sea Rift; MR, Malawi Rift; ER and WR, Eastern and Western Rifts; NR, Ngaoundere Rift. Topographic and Bouguer gravity profiles across the Afar (A–B) and East African (C–D) swells are shown in (c) and (d). (e) Topographic profile along E–F showing rift flank uplift across the Gulf of Suez (after Ebinger *et al.* 1989).

Sag-type subsidence commonly produces extensive, shallow-marine basins, which may accumulate widespread carbonate and organic-rich deposits, with a semi-restricted connection to the open ocean. Continental rim basins therefore have considerable economic importance (Hillgärtner *et al.* 2003).

3.2.5 Proto-oceanic troughs

At high amounts of stretching, continental rifts or backarc basins may evolve into new oceanic basins through a stage known as a proto-oceanic trough. Proto-oceanic troughs are characterised by young oceanic crust and very high surface heat flows. The transition from stretched continent to new ocean basin can be seen in north-eastern Africa and Arabia. The southern Red Sea contains young (<5 Ma) oceanic crust along its 50 km-wide axial zone, with flanking shelves underlain by stretched continental lithosphere. To the south the Red Sea undergoes a transition to the continental Afar Rift, and to the north into the continental Gulf of Suez Rift. The sedimentary evolution of the Red Sea area involves Oligo-Miocene syn-rift deposition of continental and shallow-marine sediments. As stretching continued through the Miocene, thick evaporites formed in the periodically isolated, proto-oceanic trough. During the Pliocene to Holocene, the Red Sea accumulated pelagic foraminiferal-pteropod oozes in deep water.

At the transition from rift basin to youthful ocean basin, subsidence commonly outpaces sediment supply, leading to the deposition

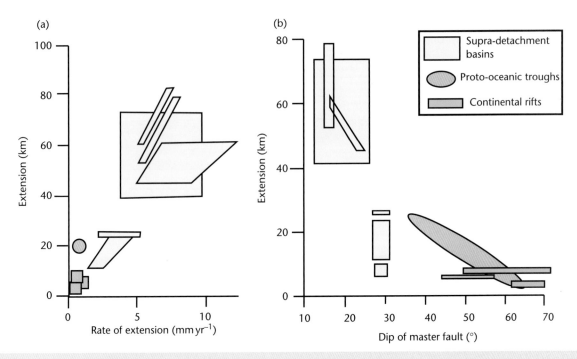

Fig. 3.15 Rifts, supra-detachment basins and proto-oceanic troughs in terms of their strain rate, total extensional strain, and dip of master faults. Based on Friedmann & Burbank (1995), reproduced with permission of John Wiley & Sons, Ltd.

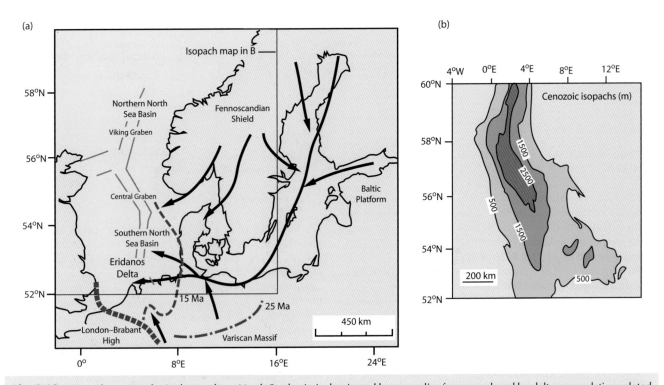

Fig. 3.16 Post-rift stratigraphy in the southern North Sea basin is dominated by megaclinoforms produced by delta progradation related to major river systems draining Europe, such as the Eridanos Delta. (a) Drainage system for the Eridanos Delta, with the shoreline shown at 25 Ma and 15 Ma (after Overeem *et al.* 2001; reproduced with permission of John Wiley & Sons, Ltd.); (b) Cenozoic isopachs (excluding Danian) in the North Sea (after Huuse & Clausen 2001; reproduced with permission of John Wiley & Sons, Ltd.).

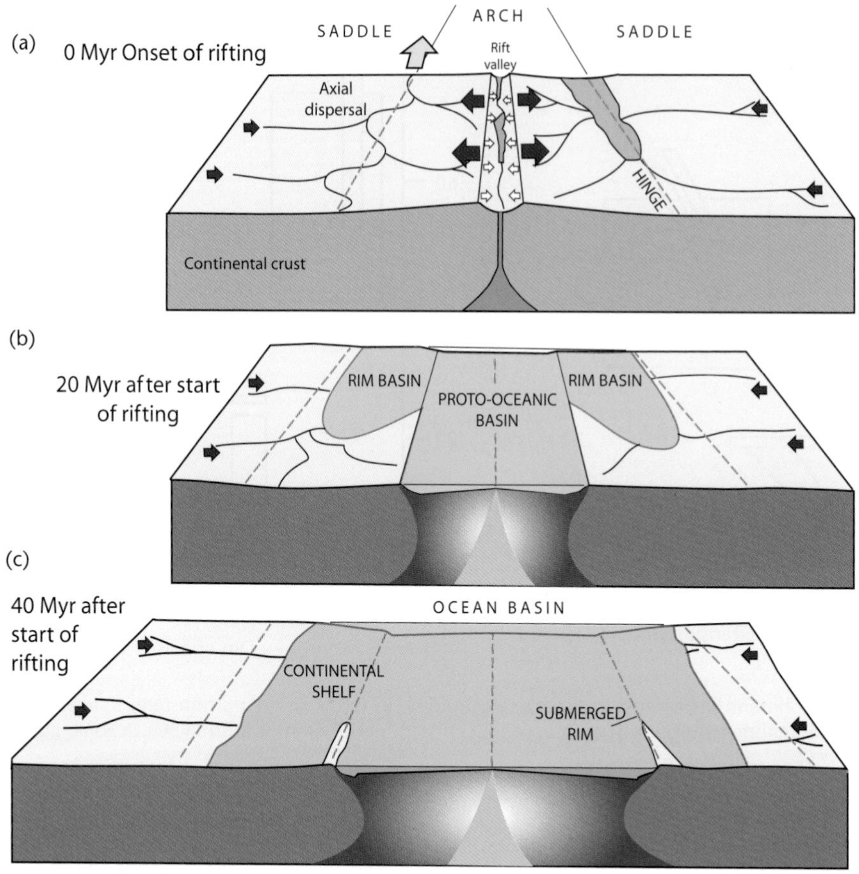

Fig. 3.17 Formation of continental rim basins (Veevers's model), as a component in the evolution of a continental margin and spreading centre. (a) Onset of rifting with rift valley and rift flank uplift, with small transverse river systems in rift axis, larger axial systems running down hinge line. (b) 20 Myr after the start of rifting, flooding of rim basins with restricted ocean circulation, adjacent to new ocean basin, with transverse continental drainage. (c) 40 Myr after the start of rifting, rim basins fully connected to ocean over flooded rift margin. From Veevers (1984), with permission of Oxford University Press.

of a triad of distinctive facies associations indicative of sediment starvation:

- *Evaporites.* The intermittent connection of developing rifts with the sea during the incipient stage provides ideal conditions for the formation of thick evaporites. Such evaporites occur along the margins of the Atlantic Ocean (Emery 1977; Rona 1982) and under the Red Sea (Lowell & Genik 1972).
- *Black organic-rich shales.* High organic productivity and restricted marine circulation may allow the preservation of organic-rich shales. Such conditions are likely to prevail where youthful ocean basins contain submarine sills restricting the throughput of water.
- *Pelagic carbonates.* In new ocean basins with little particulate sediment supply, deep-water pelagic carbonates may directly overlie the foundered pre-rift 'basement' or newly created seafloor. The faulted basement topography controls the type of deposit, with uplifted fault block edges and seamounts accumulating shallow-water carbonates, and intervening troughs being the sites of fine-grained oozes. This pattern of sedimentation related to fault block shoulders and troughs has been interpreted

from the Triassic–Jurassic of the Tethyan realm of southern Europe (Bernoulli & Jenkyns 1974).

3.2.6 Passive continental margins

The formation of juvenile oceanic spreading centres, as in the 20 Myr-old Red Sea–Gulf of Aden, and then mature (>40 Myr) ocean basins is accompanied by the development of passive margins on the adjacent stretched continental lithosphere (review in Bond *et al.* 1995). Fully developed passive margins, such as those bordering the Atlantic Ocean, are characterised by extensional faulting, large-scale gravitational tectonics (slumps, slides, glide-sheets) and salt tectonics.

Passive continental margins involve strongly attenuated continental crust stretched over a region of 50–150 km, and exceptionally as much as 400–500 km (Keen *et al.* 1987), overlain by thin or thick sediment prisms. They are in general seismically inactive, and in mature examples heat flows are normal. Passive continental margins (also known as *Atlantic-type margins*) are characterised by seaward-thickening prisms of marine sediments overlying a faulted basement

Table 3.1 Conjugate margins of the Atlantic

Western margin	Eastern margin	Start of main rifting and duration
Southern Grand Banks	Iberia/Galicia	Valanginian (137 Ma) 15–25 Myr
Flemish Cap	Goban Spur	Barremian (127 Ma) 15–20 Myr
Labrador	SW Greenland	Barremian (127 Ma) 40–65 Myr

with syn-rift sedimentary sequences, often of continental origin. The post-rift seaward-thickening sediment prisms consist predominantly of shallow-marine deposits. Seismic reflection sections show some passive margins to be underlain by linked listric extensional fault systems that merge into low-angle sole faults. The post-rift or drifting phase, in contrast, is typically dominated by gravity-controlled deformation (salt tectonics, mud diapirism, slumps, slides, listric growth faults in soft sediments) (Rowan *et al.* 2004).

Passive margins overlie earlier rift systems that are generally sub-parallel to the ocean margins, or less commonly at high angles to the ocean margin (as in the case of failed arms of triple junctions such as the Benue Trough, Nigeria), or along transform fault zones (e.g. Grand Banks and Gulf of Guinea). The early syn-rift phase of sedimentation is commonly separated from a later drifting phase by an unconformity (the 'breakup' unconformity of Falvey 1974). Some passive margins exhibited considerable subaerial relief at the end of rifting (leading to major unconformities), as in the case of the Rockall Bank, northeastern Atlantic, whereas in others, the end of rifting may have occurred when the sediment surface was in deep water, as in the Bay of Biscay and Galicia margin of Iberia (Montadert *et al.* 1979; Pickup *et al.* 1996).

Since passive margins represent the rifted edges of a piece of continental lithosphere, now separated by an ocean basin, it is possible to identify the original matching margins on either side of the ocean. These are known as *conjugate margins*. They are particularly well developed on either side of the northern Atlantic (Table 3.1).

Comparison of conjugate margins is informative regarding the geometry of extension prior to ocean basin development. For example, deep seismic reflection profiles show some conjugate pairs of passive margin to be symmetric, with seaward-dipping rotated fault blocks, whereas other deep profiles suggest the presence of a flat-lying or landward-dipping detachment or shear zone, producing a markedly asymmetrical pattern (Fig. 3.18). The profiles across the Labrador and SW Greenland margins show that although the brittle upper crust has been extended symmetrically, the lower crustal extension is particularly asymmetrical. Some margins show thin sediment covering wide regions of highly faulted upper crust, commonly separated from underlying serpentinised upper mantle by a horizontal detachment (e.g. Iberia, Galicia and SW Greenland margins). Other margins with thick sediment prisms consist of one or two major tilted crustal blocks and lack a horizontal detachment (e.g. Labrador margin).

Passive margins can be categorised according to:

- the abundance of volcanic products;
- the thickness of sediments, from sediment nourished to sediment starved;
- the presence or absence of gravitationally driven and salt tectonics in the post-rift phase (§11.7.2).

Volcanically active margins (Fig. 3.19) are characterised by extrusive basalts, lower crustal igneous accretions, and significant uplift at the time of break-up. Continental extension and ocean spreading are thought to be intimately related to mantle plume activity (Fig. 3.20). Volcanic activity, generally tholeiitic, occurring at the time of break-up (White & McKenzie 1989), is commonly associated with subaerial emergence, as in the northern North Atlantic in the Early Tertiary (e.g. Skogseid *et al.* 2000). *Non-volcanic margins* lack evidence of high thermal activity at the time of break-up.

Sediment-starved margins have relatively thin 2–4 km-thick sediment veneers draping large arrays of rotated syn-rift fault blocks above a sub-horizontal detachment, as in the Bay of Biscay (Fig. 3.20b). *Sediment-nourished margins* comprise very thick (<15 km) post-rift sedimentary prisms overlying a small number of tilted upper crustal fault blocks and a wide region of mid-lower crustal extension, as in the Baltimore Canyon region of the Eastern Seaboard of North America (Fig. 3.20c), and the Labrador margin.

Extensional faulting dies out in the post-rift phase with only minor reactivation of older normal faults. Growth faults are common in areas of high sedimentation rate (e.g. off the Niger Delta, African coast). Gravitational mass movements, from small slumps to gigantic slides, are very important during the post-rift drifting phase (§11.7.2.2). The continental slope and rise off southwestern and southern Africa were subject to major slope instability during the Cretaceous and again in the Tertiary (Dingle 1980). The slide units are over 250 m thick and can be traced for 700 km along strike and for nearly 50 km down paleoslope. The recent 3500 km^3 Storegga slide (8150 yr BP) off the Norwegian margin (Bondevik *et al.* 1997) is a reminder of the ongoing gravitational instability of continental margins.

The sediment supply to the passive margin prism comes principally by rivers eroding the continental land surface, but significant quantities of sediment may be accreted to the lower continental slope by thermohaline current-driven drifts. Major river systems build out large embankments and submarine fans that may extend directly onto oceanic crust, as in the case of the Amazon fan.

Evaporites typify the first marine incursions during the incipient ocean phase of the proto-oceanic trough or young passive margin. When buried under an overburden of passive margin sediments these evaporites become mobile, producing sheets and diapirs (Fig. 3.21). The Brazilian continental margin and the western Grand Banks, Newfoundland, show examples of major diapiric activity. The movement of subsurface salt and its penetration of the seabed causes submarine bathymetry that is important in the guiding of gravity flows to perched basins and the deep sea. Basins perched on the gravitationally driven post-rift thrust wedge may be 'filled and spilled' in a cascading process of sediment filling. Structures such as salt diapirs, whose growth affects seabed topography, interact with the incisive power of submarine channelised currents (Mayall *et al.* 2010). Incisive currents flowing over slow-growing structures tend to cut through, whereas less powerful currents are deflected by fast-growing seabed structures. Clark and Cartwright (2009) referred to such channel-topography interaction as 'confinement' and 'diversion'.

The Atlantic margin shows great variety in the nature of the passive margin prograding wedge. The Senegal margin of western Africa contains a thick carbonate bank extending over the stretched continental crust. Further to the SE the Niger has built a thick deltaic clastic wedge, provoking growth faulting and mud diapirism.

(a)   Symmetric (pure shear)

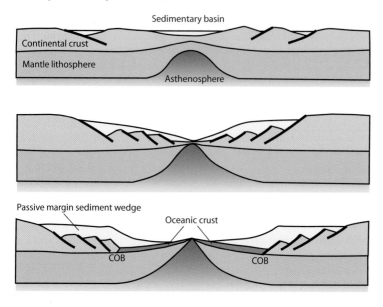

(b)   Asymmetric (simple shear)

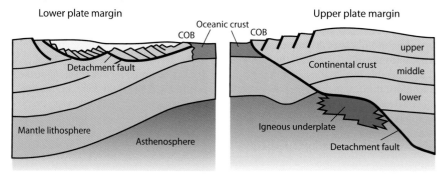

Fig. 3.18   Conjugate margins based on deep seismic information (after Lister *et al.* 1986; Louden & Chian 1999). (a) Symmetric margin (pure shear), and (b) asymmetric margin (simple shear) with a lithospheric detachment fault. COB is continent–ocean boundary.

Further south off Gabon and remote from the Niger Delta, oceanic muds overlie thick diapirs of evaporite. Off the coast of southwestern Africa a 'normal' siliciclastic margin exists with seaward-prograding clinoforms reaching far out into the basin and overlying the oceanic crust. This latter type is also common to the highly sediment-nourished North American Atlantic margin, where a well-developed coastal plain and continental shelf extend to 200 m water depth, with a continental slope descending to the abyssal plain of the Atlantic Ocean at water depths in excess of 4 km.

3.3   Uniform stretching of the continental lithosphere

3.3.1   The 'reference' uniform stretching model

Falvey (1974) proposed that the subsidence histories of various continental rift basins and continental margins could be explained quali-

tatively by extension in both crust and subcrustal lithosphere. The crust was assumed to fail by brittle fracture and the subcrustal lithosphere to flow plastically. The isostatic disequilibrium caused by the crustal extension leads to a compensating rise of the asthenosphere, and consequent regional uplift. Partial melting of upwelled asthenosphere leads to volcanism and further upward heat transfer. Eventually, after the continental lithosphere has been extended and thinned to crustal levels, new oceanic crust is generated as the rift evolves into a continental margin.

McKenzie (1978a) considered the quantitative implications of a passive rifting or mechanical stretching model, assuming the amount of crustal and lithospheric extension to be the same (*uniform stretching*). The stretching is symmetrical, no solid body rotation occurs, so this is the condition of *pure shear*. He considered the instantaneous and uniform extension of the lithosphere and crust with passive upwelling of hot asthenosphere to maintain isostatic equilibrium (Fig. 3.22). The initial surface of the continental lithosphere is taken to be at sea level, and since the lithosphere is isostatically compen-

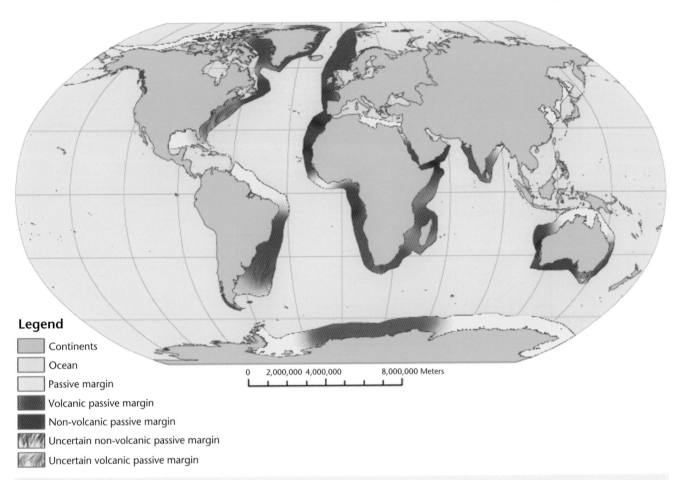

Legend

- Continents
- Ocean
- Passive margin
- Volcanic passive margin
- Non-volcanic passive margin
- Uncertain non-volcanic passive margin
- Uncertain volcanic passive margin

0   2,000,000 4,000,000   8,000,000 Meters

Fig. 3.19 Global distribution of passive margins, categorised as volcanic or non-volcanic, from http://en.wikipedia.org/wiki/File:Globald.png.

sated throughout (Airy model), the subsidence or uplift consequent upon mechanical stretching can be obtained (Appendices 19, 21).

A key component of the reference uniform stretching model is the assumption that the base of the plate remains at the same temperature during the stretching and subsequent cooling of the continental lithosphere. Consequently, the heat flow increases instantaneously (by a factor β) as a result of the mechanical stretching, since by Fourier's law it is proportional to the temperature gradient, and subsequently declines (Appendix 23). The steady-state geotherm before stretching is linear. Following stretching, the new, elevated geotherm relaxes as the stretched lithosphere cools and thickens, so that the post-rift subsidence is driven by a time-dependent transient cooling (Fig. 3.23). Much of the mathematics of the reference uniform stretching model is geared to calculating this transient component (Appendix 22).

The results of McKenzie's (1978a) quantitative model of uniform stretching can be summarised as follows:

- The total subsidence in an extensional basin is made up of two components: an initial *fault-controlled subsidence*, which is dependent on the initial thickness of the crust y_c compared to the initial thickness of the lithosphere y_L, and the amount of stretch-

ing β; and a subsequent *thermal subsidence* caused by relaxation of lithospheric isotherms to their pre-stretching position, which is dependent on the amount of stretching alone.

- Whereas the fault-controlled subsidence is modelled as instantaneous, the rate of thermal subsidence decreases exponentially with time. This is the result of a decrease in heat flow with time (Fig. 3.24a). The heat flow reaches 1/e of its original value after about 50 Myr for a 'standard' lithosphere, so at this point after the cessation of rifting, the dependency of the heat flow on β is insignificant.

The time scale of the thermal subsidence following stretching is determined by a diffusive thermal time constant of the lithosphere. For an initial lithosphere thickness of 125 km and a thermal diffusivity of lithospheric rocks of $10^{-6}\,m^2\,s^{-1}$, the time constant used by McKenzie (1980a) is $\tau = y_L^2 / \pi^2 \kappa$, which is equal to 50 Myr. After two lithospheric time constants (c.100 Myr) thermal subsidence curves for all values of the stretch factor are similar (Fig. 3.24b).

The elevation change of the surface of the Earth at the time of the onset of stretching is a trade-off between the effect of crustal stretching (causing subsidence by faulting) and the effect of the stretching of the subcrustal lithosphere (causing uplift by thermal expansion).

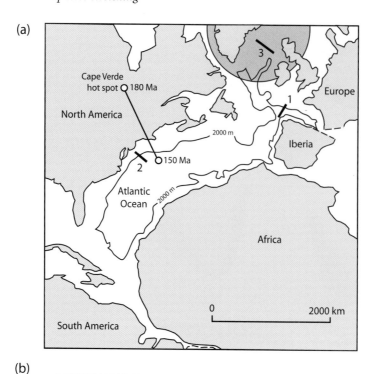

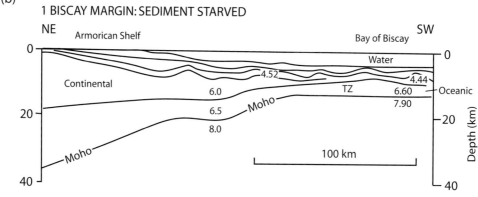

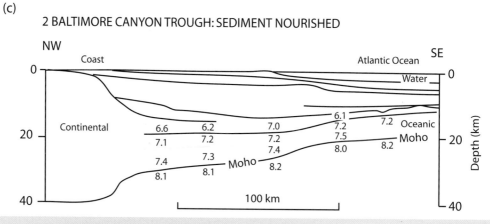

Fig. 3.20   (a) Location of margins in the central–north Atlantic region on a Middle Jurassic reconstruction (170 Ma), shortly after the onset of seafloor spreading. (b) Biscay margin, which is sediment starved. (c) Baltimore Canyon Trough margin, which is thickly sedimented. (d) Hatton bank margin, which is characterised by important magmatic activity. Shaded area shows extent of extrusive basalts. Moho is overdeepened due to the presence of an igneous underplate. TZ, ocean–continent transition zone; OC, ocean crust; numbers are seismic wave velocities in km s[-1]. After White & McKenzie (1989), reproduced with permission of the American Geophysical Union.

(d)

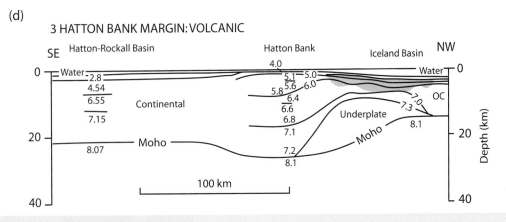

3 HATTON BANK MARGIN: VOLCANIC

Fig. 3.20 *Continued*

(a)

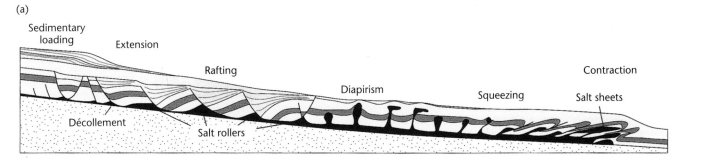

(b)

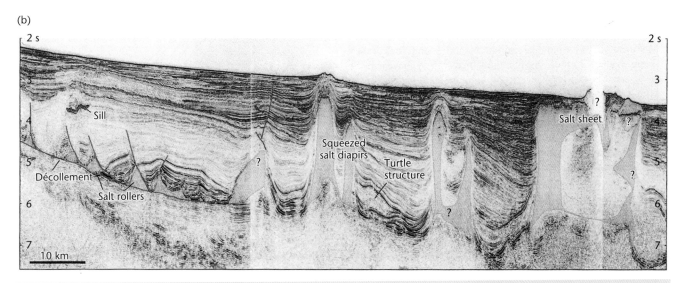

Fig. 3.21 (a) Structures commonly found on passive margins related to a salt detachment, dominated by sediment loading and extension at the rear, diapirism in the centre and fold-thrust deformation at the front (after Fossen 2010, p. 391). (b) Salt diapirs originating from a detachment level, Espirito Santo Basin, offshore Brazil (CGG Veritas, in Fossen 2010, p. 391). Haakon Fossen, 2010 © Cambridge University Press.

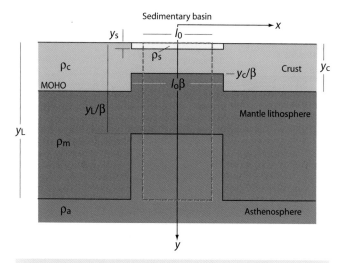

Fig. 3.22 Set-up for the reference uniform stretching model (McKenzie 1978a). The crust and subcrustal lithosphere stretch horizontally and thin vertically, uniformly with depth. Sediment, crustal, mantle lithosphere and asthenospheric densities are ρ_s, ρ_c, ρ_m and ρ_a. The crust and lithosphere have an initial thickness of y_c and y_L, and the zone of stretching has an initial width l_0. The crust and subcrustal (mantle) lithosphere stretch uniformly by a factor β.

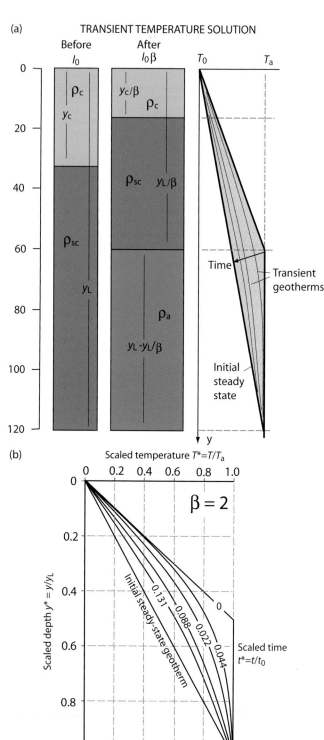

Fig. 3.23 (a) The initial steady-state geotherm is linear, due to conduction through a lithosphere of initial thickness y_L. The geotherm at time $t = 0$ immediately after stretching is linear from the surface to a new depth y_L/β. Shown for $\beta = 2$ are the transient geotherms at different times during the subsequent cooling. (b) Temperature versus depth for different times following stretching, for a stretch factor of $\beta = 2$. Temperature is scaled by the asthenospheric temperature T_a; depth is scaled by the initial lithospheric thickness y_L; time is scaled by a diffusive time constant of the lithosphere $t_0 = y_L^2/\kappa$, as in eqn. 2.29 (§2.2.7).

At a y_c/y_L ratio of 0.12, corresponding to a crustal thickness of 17 km and a lithosphere thickness of 125 km, there is no surface elevation change (Fig. 3.25). Consequently, regions with thick crusts should experience larger amounts of fault-controlled subsidence than those with thin crusts. Similarly, regions with thick subcrustal lithosphere should experience greater and more prolonged post-rift thermal subsidence than those with thin subcrustal lithospheres.

The assumptions, boundary conditions and development of the model are elaborated in Appendices 17 to 23.

Modifications of the boundary conditions and assumptions of the uniform stretching model are discussed in §3.4. For convenience, these assumptions and boundary conditions are listed below (some are implicit rather than stated):

- stretching is uniform with depth;
- stretching is instantaneous;
- stretching is by pure shear;
- the necking depth is zero;
- Airy isostasy is assumed to operate throughout;
- there is no radiogenic heat production;
- heat flow is in one dimension (vertically) by conduction;
- there is no magmatic activity;
- the asthenosphere has a uniform temperature at the base of the lithosphere.

3.3.2 Uniform stretching at passive continental margins

The uniform stretching model has been applied to the formation of passive margins (e.g. Le Pichon & Sibuet 1981). The assumptions are identical to those given above (§3.3.1): stretching of the whole lithosphere occurs instantaneously at time $t = 0$ (or within a period of 20 Myr, following Jarvis & McKenzie 1980; see also §3.4.1), heat

(a)

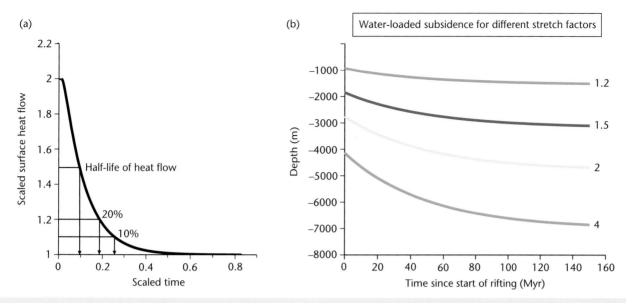

(b)

Fig. 3.24 (a) Surface heat flow, scaled by the surface heat flow prior to stretching, versus time scaled by the diffusive time constant of the lithosphere, for a stretch factor of 2. Note the exponential reduction in heat flow over time. The half-life for the surface heat flow is at 0.1 of the scaled time, that is, 0.1 t_0. (b) Water-loaded subsidence due to cooling, plotted against time since the start of rifting, as a function of the stretch factor, shown for β values of 1.2, 1.5, 2, and 4.

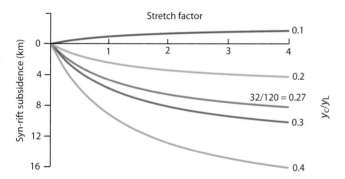

Fig. 3.25 Syn-rift subsidence as a function of the crustal/lithosphere thickness ratio y_c/y_L for stretch factors β up to 4, using the uniform stretching model. Crustal, mantle and sediment densities are 2700 kg m^{-3}, 3300 kg m^{-3} and 2000 kg m^{-3} respectively. At a crust/lithosphere thickness ratio of 0.12 (corresponding to a crust of 15 km in a lithosphere 125 km thick), there is neither uplift nor subsidence during rifting. For thinner crusts, uplift occurs, and for thicker crusts, subsidence occurs. Since crustal thicknesses are typically 30–35 km, the syn-rift phase should normally be characterised by subsidence.

production from radioactivity is ignored, local (Airy) isostatic compensation is maintained throughout, and the continental surface is initially at sea level. The initial fault-controlled subsidence (Appendix 21) is determined by the stretch factor and y_c/y_L ratio. However, in the case of passive margins, it is important to investigate the subsidence $S(\infty)$ at time $t = \infty$. Just as the initial subsidence S_i is a linear function of $(1 - 1/\beta)$, so $S(\infty)$ can also be expressed as a linear function of $(1 - 1/\beta)$. By introducing new parameters ρ'_L and ρ'_c as the average densities of the mantle part of the lithosphere and crust at time $t = \infty$, the final subsidence $S(\infty)$ can be expressed in terms of

the average densities of the crust and lithosphere, the initial thicknesses of the crust and lithosphere and the stretch factor (Appendix 28).

The difference between the initial subsidence S_i and the final subsidence $S(\infty)$ is the subsidence caused by the progressive return to thermal equilibrium, i.e. that due to thermal contraction on cooling. The latter is termed thermal subsidence S_t and of course $S_t = S(\infty) - S_i$.

As an example of this kind of analysis, it is possible to examine seismic reflection profiles across the northern Bay of Biscay and Galicia continental margins (Montadert *et al.* 1979) (Fig. 3.20b) and to test whether the uniform stretching model accurately predicts the observed subsidence. Active extensional tectonics started in the Late Jurassic–Early Cretaceous (c.140 Ma). Extension extended outwards from a central region, creating new fault blocks that were progressively tilted. By the time oceanic crust was emplaced (120 Ma) the subsiding trough had reached a depth of about 2.4 km, and simultaneously active tectonics ceased, giving way to thermal subsidence. The Bay of Biscay is relatively starved of sediment, minimising the effects of sediment loading compared to, for example, the US Atlantic continental margin (Fig. 3.20c).

The following constants were chosen (Parsons & Sclater 1977) to fit the Biscay and Galicia data (Le Pichon & Sibuet 1981):

initial lithospheric and crustal thicknesses; $y_L = 125$ km and $y_c = 30$ km (from refraction data)
mantle density at 0 °C; $\rho^*_m = 3350$ kg m^{-3}
crustal density at 0 °C; $\rho^*_c = 2780$ kg m^{-3}
water density; $\rho_w = 1030$ kg m^{-3}
volumetric coefficient of thermal expansion; $\alpha_v = 3.28 \times 10^{-5}$ °C^{-1}
temperature at base of lithosphere; $T_m = 1333$ °C

Using these constants, the initial fault-controlled subsidence simplifies to $S_i = 3.61(1 - 1/\beta)$ km, the final subsidence becomes

$S(\infty) \sim 7.83(1 - 1/\beta)$ km, so the thermal subsidence is the difference between $S(\infty)$ and S_i, giving $S_t \sim 4.21(1 - 1/\beta)$ km.

However, since the Bay of Biscay margin is 120 Myr old rather than being infinitely old, S_{120} is somewhat smaller than $S(\infty)$. As a result, the total subsidence at 120 Myr is $S_{120} \sim 7.23(1 - 1/\beta)$ and the thermal subsidence at 120 Myr is $S_{t120} \sim 3.64 (1 - 1/\beta)$.

Mid-ocean ridge crests are generally at about 2.5 km water depth, suggesting that zero-age oceanic lithosphere under 2.5 km of water is in equilibrium with a 'standard' continental lithospheric column. Therefore, during rifting, the asthenosphere should theoretically not be able to break through the thinned continental lithosphere as long as S_i is less than 2.5 km. Using the equation for the initial subsidence, the stretch factor required to produce 2.5 km of subsidence is 3.24; the crust by this time will be reduced in thickness to 9.25 km and will most likely be highly fractured – it is probable therefore that the asthenosphere would break through when this depth was reached. This represents the continent–ocean transition.

In the Bay of Biscay the estimated total subsidence at 120 Myr since rifting (S_{120}) for $\beta = 3.24$ is 5.2 km, and the final subsidence $S(\infty)$ for an infinitely large β is 7.8 km. One should therefore expect to find continental crust in the Bay of Biscay at water depths of 5.2 km or shallower in the absence of sedimentation. Oceanic crust should not be found at shallower water depths.

The extension estimated from fault block geometries in the upper crust is relatively high (approximately $\beta = 2.6$) based on migrated seismic reflection profiling, indicating that the crust is substantially thinned and close to the value at which the asthenosphere could break through. Along the seismic profile the depth to the surface of the continental block is 5.2 km, which shows that the model very satisfactorily explains the main features of the Biscay margin.

A regional synthesis suggests that water depth S_w varies linearly with the thinning of the continental crust, following a relation close to $S_w = 7.5(1 - 1/\beta)$, which is almost identical to the model prediction of $7.23(1 - 1/\beta)$.

It is perhaps surprising that the Biscay and Galicia data fit the simple uniform stretching model so well. This is probably because during phases of rapid stretching, and in the absence of large sediment loads, the lithosphere is compensated on a local scale rather than responding by flexure.

The eastern US and Canadian passive margin, however, is very thickly sedimented, with over 10 km along most of the margin (Fig. 3.26), and considerably more in areas such as the Baltimore Canyon, where a deep offshore well (COST B-2) was drilled in March 1976 (Poag 1980). The subsidence history of thickly sedimented margins such as this is profoundly affected by the *sediment load*. The sediment load is supported by the rigidity of the plate, and the borehole records (and seismic reflection profile data) need to be *backstripped* to obtain the tectonic subsidence (§9.4).

3.4 Modifications to the uniform stretching model

It is clear that there are a number of observations in regions of continental extension that suggest that the assumptions in the uniform stretching model should be re-examined. We firstly consider the effect of finite (protracted) periods of rifting, since rifts typically have syn-rift phases lasting 20–30 Myr or longer. We then briefly consider other possible modifications to the reference uniform stretching model. Modifications are dealt with under the following headings:

- *Protracted periods of stretching* cause slowly extending lithosphere to cool during the phase of stretching.
- *Non-uniform (depth-dependent) stretching*: the mantle lithosphere may stretch by a different amount to the crust.
- *Pure versus simple shear*: the lithosphere may extend along trans-crustal or trans-lithospheric detachments by simple shear.
- *Elevated asthenospheric temperatures*: the base of the lithosphere may be strongly variable in its temperature structure due to the presence of convection systems such as hot plumes.
- *Magmatic activity*: the intrusion of melts at high values of stretching modifies the heat flow history and thermal subsidence at volcanic rifts and some passive margins.
- *Induced mantle convection*: the stretching of the lithosphere may induce secondary small-scale convection in the asthenosphere.
- *Radiogenic heat production*: the granitic crust provides an additional important source of heat.
- *Depth of necking*: necking may be centred on strong layers deeper in the mid-crust or upper mantle lithosphere.
- *Flexural compensation*: the continental lithosphere has a finite elastic strength and flexural rigidity, particularly in the post-rift thermal subsidence phase.

3.4.1 Protracted periods of rifting

The reference model assumes instantaneous rifting of the lithosphere followed by thermal subsidence as the lithosphere re-equilibrates to its pre-extension thickness. This is an attractive assumption since it gives a simple, well-defined initial condition for the thermal calculations. Jarvis and McKenzie (1980) revised this one-dimensional model by allowing for protracted periods of stretching. If the duration of stretching is large compared with the diffusive time scale of the lithosphere ($\tau = y_L^2/\pi^2\kappa$), which is 50–60 Myr for a standard lithosphere, some of the heat diffuses away before the stretching is completed. In the general case, if the time taken to extend the continental lithosphere by a factor β is less than τ/β^2 Myr, the results are similar to those of the uniform stretching model with instantaneous stretching. However, the sensitivity of the results depends on the value of the stretch factor and whether total subsidence or heat flow is being considered:

- Considering total subsidence, if $\beta < 2$, the duration of extension must be less than τ/β^2 Myr for the instantaneous stretching model to be a reasonable representation. If $\beta > 2$, the duration of extension must be less than $\tau(1 - 1/\beta)^2$.
- Considering heat flow, if $\beta < 2$, the duration of extension must be less than τ Myr for the instantaneous stretching model to be a reasonable representation. If $\beta > 2$, the duration of extension must be less than $\tau(2/\beta)^2$.

Although stretching is known to be short-lived in some extensional provinces, such as the Pannonian Basin (<10 Myr to reach a stretch factor of 3) (Sclater *et al.* 1980b), many sedimentary basins appear to have undergone more protracted periods of rifting, considerably in excess of $60/\beta^2$ Myr. The Paris Basin, for instance, rifted in the Mid-Permian and sedimentation was restricted to elongate rift troughs until close to the end of the Triassic. The rift phase was therefore close to 60 Myr in duration. The Triassic (Carnian–Norian, 212–200 Ma) continental red beds and evaporites of the Atlantic margin of northeastern USA and Canada were deposited in fault-bounded rifts, but seafloor spreading did not commence until the

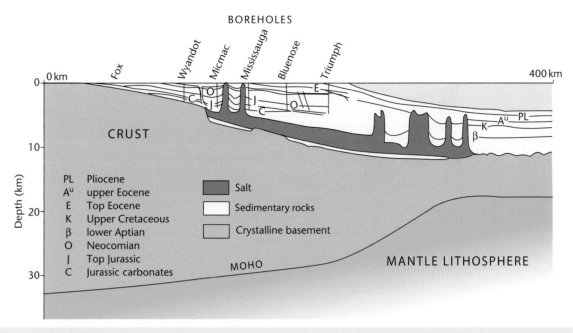

Fig. 3.26   Composite cross-section of the Scotian margin and basin. The sediment–basement interface is based on wells that penetrate it, in addition to seismic reflection and refraction observations (Jansa & Wade 1975). After Beaumont *et al.* (1982), reproduced with permission of John Wiley & Sons, Ltd.

Bajocian at about 170 Ma. The stretching in the Red Sea (Cochran 1983) appears to have occurred diffusely through a combination of extensional block faulting and dyke injection over an area of the order of 100 km wide. This phase of diffuse extension has lasted for 20 to 25 Myr in the northern Red Sea and is still occurring. A phase of rifting took place in the North Sea in the Permo-Triassic. There was another phase in the Jurassic that lasted until the Early Cretaceous. The duration of this major phase of extension was of the order of 50 Myr.

Taking the North Sea as an example, the stretch factor for the Jurassic extensional phase rarely exceeds 1.5, except along the axis of the Viking Graben. Using $\tau = 50$ Myr and $\beta = 1.5$, the critical duration of rifting from Jarvis and McKenzie (1980) is about 27 Myr. Consequently, on this basis, we should expect the protracted duration of extension during the Jurassic phase in the North Sea to have had significant effects on the heat flow and subsidence history. Despite this, the North Sea appears to be satisfactorily explained by the uniform stretching model (Sclater & Christie 1980; Wood & Barton 1983).

Although the concept of instantaneous extension is useful for modelling of sedimentary basins that form by extension at high strain rates ($>10^{-15}$ s^{-1}), at lower strain rates the thinning lithosphere and asthenosphere cool during its slow upward advection (Jarvis & McKenzie 1980; Armitage & Allen 2010; Allen & Armitage 2012) (Fig. 3.27). During slow stretching, the lithosphere is subject to heat conduction, upward advection due to the lateral extension, and internal heating due to radioactive decay. The geotherm is then solved through time as the lithosphere extends and cools. The composition of the upwelled asthenosphere is assumed to be identical to that of the lithospheric mantle it replaces, so that there is no additional downward-acting load due to a change in chemical composition and therefore density (cf. Kaminski & Jaupart 2000).

At very low rates of extension (strain rates of 10^{-16} s^{-1}) the heat loss by conduction and the upward advection of warm lithosphere are similar in value. In this case the lithosphere is less thermally buoyant to counter faulting in the crust, resulting in syn-rift subsidence that is more prolonged and greater than in the reference model. Subsequent thermal subsidence is less than in the faster strain rate reference model. The overall subsidence profile therefore has a more constant slope, with a slight elbow when the stretching ceases (Fig. 3.27).

In problems where there are competing effects of diffusion (heat conduction) and advection, it is conventional to use a Péclet number:

$$Pe = \frac{v y_L}{\kappa} \qquad [3.2]$$

where v is the mean upward velocity, y_L is the thickness of the lithosphere and κ is the thermal diffusion coefficient. Since the strain rate $\dot{\varepsilon}$ (in 1D) is $\delta v / \delta y$, eqn. [3.2] becomes

$$Pe = \frac{\dot{\varepsilon} y_L^2}{2\kappa} \qquad [3.3]$$

We assume that for $Pe > 10$, upward advection dominates, whereas if $Pe < 1$ then diffusion dominates. For a strain rate of 10^{-15} s^{-1} upward advection dominates as $Pe = 20$. If the strain rate is 10^{-17} s^{-1}, $Pe = 0.2$ and thermal diffusion dominates, such that as the lithosphere is extended, it cools as material is slowly advected upwards (Figs 3.28, 3.29). For strain rates of the order of 10^{-16} s^{-1}, $Pe = 2$, which means that the thermal diffusion and upward advection of material are comparable, and the assumption of instantaneous extension is no longer an accurate representation.

Returning to the Viking Graben example, we can calculate the Péclet number in order to assess the effects of protracted rifting.

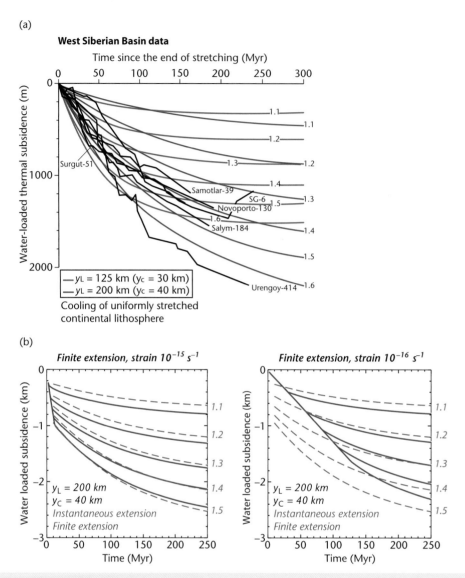

Fig. 3.27   (a) Water-loaded subsidence for a continental lithosphere with an initial thickness of 200 km, compared with the reference model for $y_L = 125$ km, with subsidence curves from the West Siberian Basin superimposed (red). Subsidence is calculated assuming that the lithosphere is isostatically compensated locally and that the initial configuration is at sea level. We assume that density remains constant within the crust at 2900 kg m^{-3}, and varies linearly due to thermal expansion within the lithosphere. The density at the base of the lithosphere is assumed to be 3400 kg m^{-3}, and the temperature is assumed to be 1330 °C. Simple instantaneous extension of a piece of 200 km-thick lithosphere produces subsidence due to thermal relaxation that continues for more than 250 Myr. (b) Effect on subsidence of finite times of rifting (Armitage & Allen 2012). Water-loaded subsidence for instantaneous extension is compared with subsidence for finite extension at strain rates of 10^{-15} and 10^{-16} s^{-1}. The base of the lithosphere is initially at 200 km, at a temperature of 1330 °C, and the crust is initially at 40 km thick. Slow upward advection of rock causes heat loss by conduction during the protracted rift phase.

If $\beta = 1.5$, the width of the stretched crust is 150 km (initial width 100 km), the lithospheric time constant $\tau = 50.2$ Myr ($y_L = 125$ km), $\kappa = 10^{-6}$ m^2 s^{-1}, and the duration of stretching is 50 Myr, the strain rate becomes 3.17×10^{-16} s^{-1}, and the Péclet number is 2.5. This indicates that the effects of protracted rifting should be significant.

3.4.2   Non-uniform (depth-dependent) stretching

Two families of models have been proposed to deal with the possibility of non-uniform stretching:

- *discontinuous* models, in which there is a discontinuity or decoupling between two layers with different values of the stretch factor β;
- *continuous* models, where there is a smooth transition in the stretching through the lithosphere.

The extent of coupling or decoupling of rheologically contrasting layers, such as the mid-upper crust and lower crust, is a fundamental factor in controlling the evolution of stretched continental lithosphere (Huismans & Beaumont 2002, 2008).

(a)

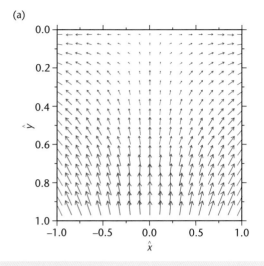

(b)

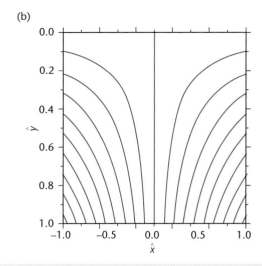

Fig. 3.28 Advection of rock in a piece of stretched continental lithosphere during finite periods of extension, showing (a) the vectors of the velocity field in dimensionless depth versus dimensionless width, and (b) the flowlines. Vertical velocity is set to zero at the surface where $y = 0$. From Wangen (2010). © Magnus Wangen 2010, reproduced with the permission of Cambridge University Press.

(a)

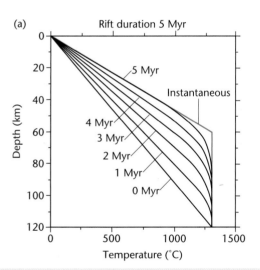

(b)

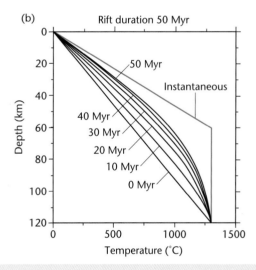

Fig. 3.29 Effect of finite rifting on the geotherm, from Wangen (2010), for rift durations of 5 Myr (a) and 50 Myr (b). Note the gradual (rather than instantaneous) increase in temperature due to the transient thermal effects, especially at long rifting times.

Critically, both sets of models make a first-order prediction – that zones of continental stretching should be characterised by elevated rift margin topography.

3.4.2.1 Discontinuous stretching with depth

Extension may not be uniform with depth because of the changing rheological properties of the lithosphere. If the lithosphere extends inhomogeneously and discontinuously, there must be a depth d at which the upper and lower parts of the lithosphere are decoupled. This zone of detachment or shear is where listric faults in the overly-

ing brittle zone sole out (Montadert *et al.* 1979; Kusznir *et al.* 1987). Structural evidence (e.g. seismic studies of the Basin and Range province, Bay of Biscay and northwestern European continental shelf) demonstrates that some steep faults near the surface become listric into near-horizontal detachments where a transition to ductile behaviour takes place. The focal depths of earthquakes in old cratons further suggest that while the upper part of the lithosphere has relatively high strength and is seismically active, the lower part is aseismic, probably due to the operation of ductile deformation mechanisms (Sibson 1983; Scholz 1988). In some instances at least, there may therefore be decoupling of the upper and lower zones at about mid-crustal levels, allowing the two rheological layers to extend by different amounts, giving a non-uniform discontinuous stretching.

The initial subsidence and thermal subsidence for the case of depth-dependent extension, where the zone above d extends by β_c and the zone below d extends by ductile deformation by a different amount β_{sc}, is given by Royden and Keen (1980). If the lower zone stretches by ductile deformation more than the brittle upper zone, uplift should occur if the depth to decoupling approximates the crustal thickness ($d \sim y_c$). This uplift occurs at the same time as extension, and is an attractive feature of the model in view of the updoming characteristic of many present-day rift systems such as in East Africa.

As an example of discontinuous depth-dependent stretching, in the centre of the Pattani Trough, Gulf of Thailand, the stretch factors for crust β_c and lithosphere β_L derived from subsidence plots were 2.35 and 1.90, indicating that crustal thinning was 20% greater than lithospheric thinning in the graben region (Hellinger & Sclater 1983). However, there are few examples where depth-dependent stretching can be demonstrated convincingly (Shillington *et al.* 2008).

The temporal relationship between faulting and rift flank uplift and the wavelength and amplitude of the rift flank uplift provide important constraints on the lithospheric stretching model. For example, if pre-rift sedimentary rocks are preserved in the graben but eroded from the flanks, it is a good indication that crustal uplift did not precede rifting. Footwall uplift may be due to co-seismic strain along border faults, in which case the wavelength of the uplift (c.10 km) will be smaller than the fault spacing. However, if the rift flank uplift has a much larger wavelength, it might imply that the subcrustal lithosphere was extended over a larger region than the more confined crustal extension. At first the crustal thinning, causing subsidence, outstrips the uplift from subcrustal thinning in the graben area, but the reverse is true beyond the graben edge. Later, after extensional tectonics has ceased, both flanks and graben should subside due to cooling and thermal contraction of the upwelled asthenosphere.

Regions initially at sea level that are uplifted, such as graben flanks, are subject to erosion and theoretically should subside to a position below sea level after complete cooling. But a second effect is the added crustal uplift caused by isostatic adjustment to the removed load. These two processes govern how much erosion will take place before the rift flanks subside below sea level and erosion effectively stops. In the southern Rhine Graben, the present-day surface uplift of the rift flanks is about 1 km. The erosional exhumation of footwall rocks is, however, of the order of 2.5 km (Illies & Greiner 1978). A further implication of rift flank erosion is an increase in the area of subsidence, so non-uniform extension models incorporating erosional effects predict larger subsiding basins than uniform extensional models. Regions that are not initially at sea level prior to rifting, such as the Tibet Plateau and the Basin and Range province, have a different, lower base level for erosion.

A comparison of the stratigraphies generated by the uniform (one-layer) stretching model and the two-layer (depth-dependent) stretching model from the passive margin off New Jersey, USA, is shown in Fig. 3.30. Both models show a well-developed coastal plain, hinge zone and inner shelf region underlain by a thick sequence of seaward-dipping strata. The main difference is in the stratigraphy of the coastal plain, the one-layer model over-predicting syn-rift sediment thickness. The two-layer model, however, explains the lack of syn-rift (Jurassic) stratigraphy by the lateral loss of heat to the flanks of the rift, causing uplift and subaerial emergence.

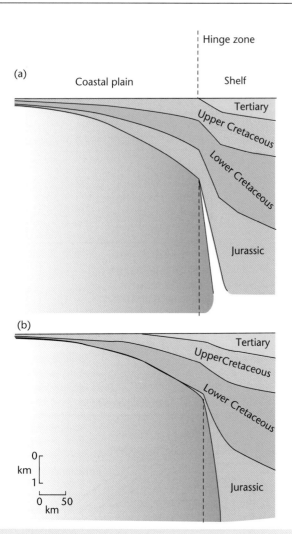

Fig. 3.30 Synthetic stratigraphy along profiles crossing the coastal plain and shelf off New Jersey, constructed using the flexural loading model of Watts & Thorne (1984). (a) One-layer uniform stretching model. (b) Two-layer model in which the lithosphere and crust are thinned by equal amounts seaward of the hinge zone, but only the mantle lithosphere is thinned landward of the hinge zone. The lithospheric thinning promotes early uplift of the zone landward of the hinge line, and helps to explain the absence of Jurassic strata from this region (after Steckler *et al.* 1988).

3.4.2.2 Continuous stretching with depth

The implication of different amounts of stretching in the crust and mantle lithosphere is that there must be a surface or zone of discontinuity separating the regions with different values of β. Although such models are successful in explaining the common rift flank uplift that accompanies extension (e.g. Nova Scotian and Labrador continental margin, East African Rift, Red Sea, Rhine Graben), they have a number of requirements:

- the existence of an intra-lithospheric discontinuity, which although evident in some settings (e.g. Biscay margin), is by no means universally 'proven';

- a mechanism by which the mantle is detached and stretched by a different amount to the overlying crust, and a means of solving the attendant space problem in the mantle.

If the stretching is non-uniform but *continuous* with depth, these objections are removed (Rowley & Sahagian 1986). It is possible that the mantle may respond to extension as a function of depth, the strain rate decreasing as the extension is diffused over a wider region. This can be modelled by considering a geometry of an upward-tapering region of stretching (Fig. 3.31). If ϕ is the angle between the vertical and the boundary of the stretched region in the mantle lithosphere, the amount of stretching depends on the depth beneath the crust and angle ϕ. The variation of β_{sc} with depth can then be integrated from the base of the crust to the base of the lithosphere to obtain estimates of initial and total subsidence in a similar fashion to the uniform stretching case. For large values of ϕ, the initial subsidence is increased but the amount of post-extension thermal sub-sidence is decreased. The wider zone of mantle stretching results in uplift of the rift shoulders, and the horizontal length scale of the uplift is an indication of the value of ϕ. In the Gulf of Suez region, the horizontal width over which the topographic uplift occurs is ~250 km. Since the thickness of the subcrustal lithosphere is likely to be approximately 90 km, the taper angle ϕ is given by $\phi = \tan^{-1}(250/90) = 70°$. In the Rhine Graben, ϕ is approximately $\tan^{-1}(80/90)$ ~40°.

A point in the rift shoulder region should therefore initially experience uplift, followed by a comparable amount of subsidence; if erosion has occurred, the final elevation of the point will be below its initial height. The same general pattern is observed where the crustal stretching varies from a minimum at the rift margin to a maximum at the rift centre (Fig. 3.31) (White & McKenzie 1988). The implications of stretching the mantle over a wider region than the crust (but with equal total amounts of extension) is that stratigraphic onlap should occur over previous rift shoulders during the post-rift phase, a feature commonly found in 'rift-sag' or 'steer's head' type basins.

3.4.3 Pure versus simple shear

The lithosphere may extend asymmetrically where the zone of ductile subcrustal stretching is relayed laterally from the zone of brittle crustal stretching (Buck *et al.* 1988; Kusznir & Egan 1990) (Figs 3.32, 3.33). This is the situation of *simple shear*. Wernicke (1981) proposed such a model, based on studies of Basin and Range tectonics, envisaging that lithospheric extension is accomplished by displacement on a large-scale, gently dipping shear zone that traverses the entire lithosphere. Such a shear zone transfers or 'relays' extension from the upper crust in one region to the lower crust and mantle lithosphere in another region. It necessarily results in a physical separation of the zone of fault-controlled extension from the zone of upwelled asthenosphere.

Wernicke (1981, 1985) suggested that there are three main zones associated with crustal shear zones (Fig. 3.33): (i) a zone where upper crust has thinned and there are abundant faults above the detachment zone; (ii) a 'discrepant' zone where the lower crust has thinned but there is negligible thinning in the upper crust; and (iii) a zone where the shear zone extends through the subcrustal (mantle) lithosphere.

Since crustal thinning by fault-controlled extension causes subsidence, but subcrustal thinning produces uplift, we should expect subsidence in the region of thin-skinned extensional tectonics but tectonic *uplift* in the region overlying the lower crust and mantle thinning (the discrepant zone). Subsequent asthenospheric cooling may result in one of two things. Thermal subsidence of the region above the discrepant zone may simply restore the crust to its initial level. However, if subaerial erosion has taken place in the meantime, thermal subsidence will lead to the formation of a shallow basin above the discrepant zone. The basement of the basin should be unfaulted. However, beneath the zone of thin-skinned extensional tectonics there should be minimal thermal subsidence.

Stretching of the lithosphere combined with unloading along major detachment faults can result in the unroofing of mantle rocks in their footwalls. These are called *core complexes* and *gneissic domes* (Fig. 3.34). The so-called 'turtlebacks' of the Death Valley region and the Whipple Mountains of Nevada are examples (Lister and Davis 1989).

Tectonic unloading may also result in flexural uplift of adjacent footwall areas along major detachment faults (Fig. 3.35). Kusznir

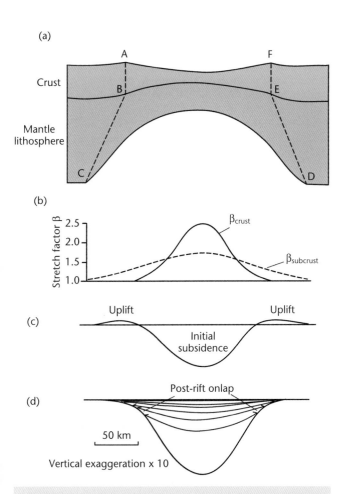

Fig. 3.31 Basin filling pattern resulting from continuous depth-dependent stretching (Rowley & Sahagian 1986; White & McKenzie 1988). (a) Geometry of a tapering region of extension in the subcrustal lithosphere. (b) Stretch factors in the crust and subcrustal lithosphere as a function of horizontal distance. (c) Initial subsidence and uplift immediately after stretching, showing prominent rift flank uplift. (d) Total subsidence 150 Myr after rifting, showing progressive onlap of the basin margin during the thermal subsidence phase, giving a steer's head geometry.

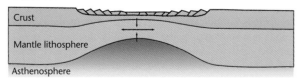

(a)   PURE SHEAR

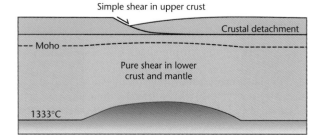

(b)   SIMPLE SHEAR

(c)   SIMPLE SHEAR–PURE SHEAR

Fig. 3.32 Models of strain geometry in rifts (Coward 1986; Buck *et al.* 1988). (a) Pure shear geometry with an upper brittle layer overlying a lower ductile layer, producing a symmetrical lithospheric cross section with the initial fault-controlled subsidence spatially superimposed on the thermal subsidence. The ductile stretching may be accompanied by dilation due to the intrusion of melts (Royden *et al.* 1980). (b) Simple shear geometry with a through-going low-angle detachment dividing the lithosphere into an upper plate or hangingwall, and a lower plate or footwall. Thinning of the lower lithosphere is relayed along the detachment plane, producing a highly asymmetrical lithospheric cross-section (after Wernicke 1981, 1985). Initial fault-controlled (syn-rift) subsidence is spatially separated from the thermal subsidence. (c) Hybrid model of simple shear in the upper crust on listric (shown) or planar faults, and pure shear in the ductile lower crust and mantle lithosphere (Kusznir *et al.* 1991).

et al. (1991) refer to this as the *flexural cantilever effect*. They use a model of simple shear in the crust and pure shear in the lithospheric mantle. The scale of the flexural cantilever effect depends on the depth at which the detachment soles out. The model has been successfully used to explain footwall uplift and erosion in the Jeanne d'Arc Basin, Grand Banks of Newfoundland, and Viking Graben, North Sea (Marsden *et al.* 1990; Kusznir *et al.* 1991), and the Tanganyika Rift of East Africa (Kusznir & Morley 1990).

Models involving large-scale simple shear do not explain basins that have a thermal subsidence spatially superimposed on a fault-controlled subsidence, as in the North Sea (Klemperer 1988). Such examples are more suggestive of pure shear. However, it is possible that the upper crust may deform by simple shear and the lower crust and subcrustal mantle lithosphere by pure shear, with a mid-crustal detachment (Coward 1986) (Fig. 3.32). Symmetrical (pure shear) and asymmetrical (simple shear) geometries were both outcomes of

the numerical experiments of Huismans and Beaumont (2002, 2008), dependent on the rheological structure of lithospheric layers and the strain rate of stretching (see §3.5).

3.4.4   Elevated asthenospheric temperatures

The reference uniform stretching model envisages an asthenosphere with a laterally uniform temperature (1330 °C), which rises passively to fill the region of lithospheric stretching. Some workers have suggested that continental stretching is in some cases associated with mantle plume activity (Spohn & Schubert 1983; Houseman & England 1986). Such activity may raise the local asthenospheric temperatures by as much as 200 °C (Fig. 3.36). Plume heads typically have diameters in the region of 1000 km (Griffiths & Campbell 1990) (Chapter 5). Laboratory experiments indicate that starting plumes may generate as much as 600 m of surface uplift (Hill 1991) – insufficient to solely explain the 3 km-high swells of eastern Africa. However, mantle plume activity may drive continental extension by elevating the lithosphere, including its surface, thereby giving it excess potential energy compared to its surroundings (Houseman & England 1986).

The driving stress caused by the uplift may exceed a threshold value, causing *run-away or accelerating extension*, eventually leading to ocean crust production. Alternatively, the driving stress may be considerably lower than (about half of) the threshold value, causing *negligible extension*, as in many low-strain continental rifts. If the driving stress is intermediate in value, the extension is thought to be *self-limited* by the cooling of the ductile portion of the lithosphere. This produces aborted rift provinces such as the North Sea. These ideas are further developed in §3.5.

One of the complexities of this type of model is the *time scale of the plume*. Removal of the plume head at early, intermediate or late stages has major effects on basin development. Another aspect is the *volcanicity* generated by the anomalously high asthenospheric temperatures (§3.4.5). Mantle plumes are therefore commonly associated with very high volume basaltic igneous provinces such as the Karoo, Deccan and Parana examples.

Plume activity has been invoked as a particularly effective mechanism of generating new ocean basins, such as the Atlantic. The opening of the northern Atlantic in the Paleocene has been related to the impingement of the Icelandic plume on the base of the lithosphere (White & McKenzie 1989). Mantle plume effects may have been common during supercontinental break-up of Rodinia (c.850 Ma), Gondwana (c.550 Ma) and Pangea (c.250 Ma).

3.4.5   Magmatic activity

Although melting is not predicted at low to moderate values of stretching in the reference uniform stretching model, it is well known that rift provinces are associated with minor (e.g. Western Rift, Africa; North Sea) to major (Eastern Rift, Africa; Rio Grande) volcanism. Most importantly, continental break-up at high values of stretching is commonly associated with vast outpourings of flood basalts, indicating major melting of the asthenosphere by adiabatic decompression. Melts liberated by decompression are assumed to separate from their residue and to travel upwards to either be erupted at the surface, or to be emplaced as igneous bodies in the crust. Underplating of the crust by igneous bodies can cause uplift of the surface (Brodie & White 1994) (Appendix 27). McKenzie and Bickle (1988) showed how the amount of melt generated

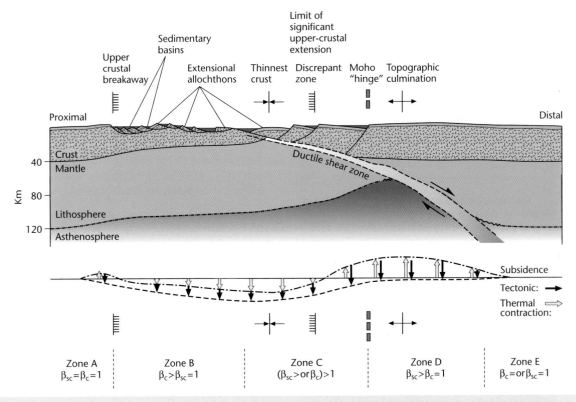

Fig. 3.33 Simple shear model of the entire lithosphere, developed from the Basin and Range province of SW USA (Wernicke 1985). This geometry takes of the order of 10–15 Myr to develop. Mid-crustal rocks in the hangingwall may initially pass through greenschist or amphibolite metamorphic conditions in the ductile shear zone, followed by uplift, cooling and deformation in the brittle field. β_c and β_{sc} refer to the stretch factors of the crust and subcrustal lithosphere respectively.

depends on the potential temperature of the asthenosphere (defined as the temperature the asthenosphere would have if brought to the surface adiabatically without melting) and the amount of stretching (Chapter 5).

Melting is greatly facilitated by the presence of volatiles and by elevated asthenospheric temperatures, as would be expected over a plume head (Chapter 5). The high plume temperatures provide long-wavelength (c.10^3 km) dynamic support for the Earth's surface. Pulsing in this dynamic support (§5.2.5) may cause periodic uplift, the formation of erosional landscapes, and the transport of sediment to the ocean, followed by subsidence, as has been interpreted for the Iceland plume at 55 Ma (White & Lovell 1997; Hartley *et al.* 2011). Pulses of hot material are thought to have travelled outward from the main conduit of the Iceland plume in a channel below the lithosphere (Rudge *et al.* 2008), producing an annulus of uplift and subsidence like a ripple travelling outwards from a stone thrown into a pond.

In summary, melting, igneous underplating and dynamic support from a hot asthenosphere affect the subsidence experienced at the Earth's surface during the period of lithospheric stretching (Fig. 3.36). Dynamic effects from the mantle are discussed further in Chapter 5.

3.4.6 Induced mantle convection

Models involving an active asthenospheric heat source should predict *uplift before rifting*. At present there is no broad consensus on the temporal relationships between uplift and extension associated with lithospheric stretching. However, Steckler (1985) suggested that in the Gulf of Suez, the rift flank uplifts were *not* formed as a precursor doming event prior to rifting, but rather formed *during* the main phase of extension. The rift appears to have initiated (by Miocene times) at near sea level, since the tilted fault blocks associated with early rifting experienced both subaerial erosion and marine deposition. The early Miocene topography of the Gulf of Suez region was subdued, and stratigraphic thicknesses are uniform over the area (Garfunkel & Bartov 1977). However, at the end of the early Miocene, 8–10 Myr after the onset of rifting, a dramatic change took place – there is a widespread unconformity, and conglomerates appear at the rift margins suggesting major uplift and unroofing at this time.

In the Gulf of Suez, the lithosphere must have extended by 2.5 times as much as the crust to explain the uplift and rift subsidence. This brings the uniform extension model seriously into doubt in terms of its ability to predict the lithospheric heating. How then does one explain the additional amount of heating that the lithosphere under the Gulf of Suez has undergone? This extra heat may have resulted from convective flow induced by the large temperature gradients set up by rifting. Numerical experiments confirm that secondary small-scale convection should take place beneath rifts (Buck 1984, 1986). Convective transport should heat the lithosphere bordering the rift, causing uplift of rift shoulders concurrent with extension within the rift itself (Fig. 3.37). If this mechanism is correct, it removes the need for additional active, sub-lithospheric heat sources.

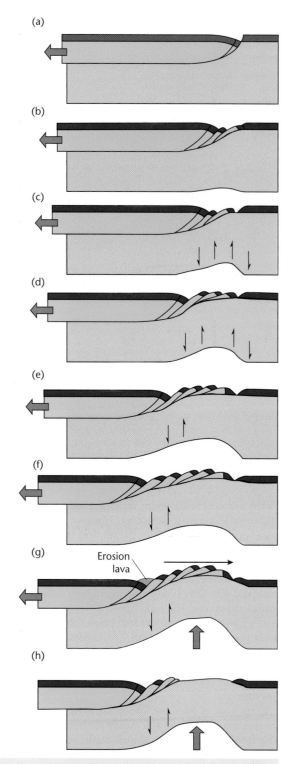

(a)

(b)

(c)

(d)

(e)

(f)

(g) Erosion lava

(h)

Fig. 3.34 Sequential evolution of a piece of continental lithosphere undergoing simple shear and isostatic compensation, showing the development of flat detachment faults, tilted listric fault blocks and metamorphic core complexes. Based on Wernicke & Axen (1988), from Fossen (2011, p. 339). Reproduced with the permission of Cambridge University Press.

It also removes the requirement for two-layer extension to explain high heat flows at rift margins.

Secondary small-scale convection may also affect plates adjacent to ocean–continent margins. Convective flow in the mantle may be triggered by the step in plate thickness at the ocean–continent transition (§5.2.4) as the plate migrates relative to the underlying mantle. The dynamic topography caused by these edge effects may extend c.10^3 km into the continental plate, and may also be responsible for widespread melting of the asthenosphere, producing trailing-edge magmatic provinces, such as Victoria, Australia. Further discussion of dynamic topography is found in Chapter 5.

3.4.7 Radiogenic heat production

Although the very large volume of mantle contributes 80% of the Earth's radiogenic heat, it is the crustal contribution that determines the continental geotherm and which is of importance in basin analysis. The crustal radiogenic heat production can be modelled either as a series of slabs of different internal heat production, or as an exponential that decays with depth (§2.2.3). As a rule of thumb, the radiogenic heat production may be of roughly equal importance to the basal heat flow in determining the continental geotherm.

In practice, radiogenic heat production makes little difference to the shape of the subsidence curves predicted by the reference uniform stretching model, though it potentially strongly affects paleotemperature estimations. It is therefore very important to include radiogenic contributions to the heat flow in modelling of thermal indicators in sedimentary basins, such as vitrinite reflectance (Chapter 10).

3.4.8 Flexural compensation

It is likely that at low to moderate values of stretching, the lithosphere maintains a finite strength during basin development (Ebinger *et al.* 1989), so that it responds to vertical loads by flexure (Chapter 4). The degree of compensation (between the end members of Airy isostasy for a plate with no flexural rigidity and zero compensation for an infinitely rigid plate) depends on the flexural rigidity of the plate and the wavelength of the load (§2.1.5). There are two situations where flexure is likely to be important: (i) tectonic unloading by extension along major detachments, leading to a regional cantilever-type upward flexure (Kusznir & Ziegler 1992), causing the footwall region to be mechanically uplifted; and (ii) downward flexure under the accumulating sediment loads in the syn-rift and especially post-rift phase. Flexural compensation of sediment loads should be particularly important in passive margin evolution because of the secular increase in flexural rigidity, and where sediment loads are relatively narrow, as in narrow rifts and some pull-apart basins.

3.4.9 The depth of necking

Necking is the very large-scale thinning of the lithosphere caused by its mechanical extension. Necking should take place around one of its strong layers (Braun & Beaumont 1987; Weissel & Karner 1989; Kooi *et al.* 1992). The *depth of necking* is defined as the depth in the lithosphere that remains horizontal during thinning if the effects of sediment and water loading are removed. For the reference uniform stretching model, the necking depth is implicitly 0 km. That is, all depths below the surface experience an upward advection during thinning (when sediment/water loading is removed). However, the

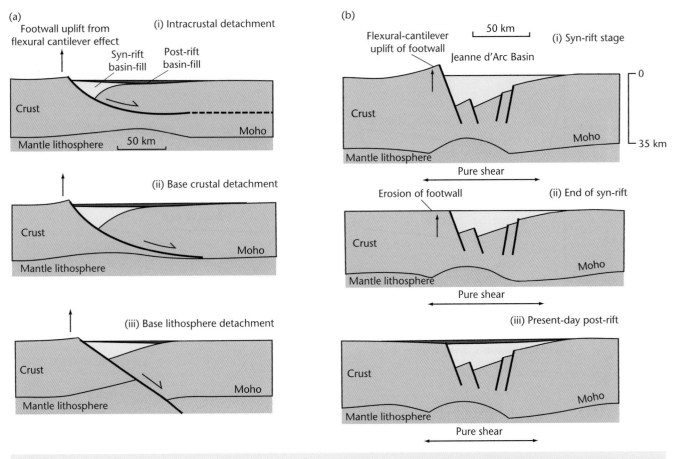

Fig. 3.35 Sedimentary basin geometry and crustal structure predicted by a simple-shear/pure-shear model including the flexural cantilever effect (Kusznir & Egan 1990; Kusznir *et al.* 1991). (a) Crustal structures after 100 Myr and 30 km extension, for an intra-crustal detachment (i), a base-crustal detachment (ii), and a base-lithosphere detachment (iii). Equivalent elastic thickness Te = 5 km. (b) Sequential development of the Hibernia-Ben Nevis profile of the Jeanne d'Arc Basin, showing flexural uplift and erosion of the unloaded footwall of the main detachment fault. The total amount of extension is 18 km, initial fault dip = 60°, initial crustal thickness is 35 km, and Te = 10 km.

necking depth may be deeper in the shallow mantle lithosphere. In this case, there is advection of material upwards from below the necking depth, but advection of material downward from above the necking depth.

If the necking depth is in the strong mantle lithosphere, there is a regional flexural uplift, causing a pronounced rift shoulder. We would expect a deep necking depth where the lithosphere is cold and strong, with a strong subcrustal mantle, such as in the Transantarctic Mountains and Red Sea region (Cloetingh *et al.* 1995). On the other hand, if the necking depth is within the upper-mid crust, there is a downward regional flexure, promoting subsidence of the rift margins. This should occur where the lithosphere is weak, or where the crust is thickened, as in the Pannonian Basin of eastern Europe. The level of necking therefore controls the amount of rift shoulder denudation, and, consequently, the sediment delivery to the basin during the syn-rift phase (van Balen *et al.* 1995; ter Voorde & Cloetingh 1996). There is theoretically an equilibrium depth of necking where there is no net flexural response. For a sediment-filled basin and initial crustal thickness of 33 km within a 100 km-thick lithosphere, this

equilibrium depth is about 10 km. Odinsen *et al.* (2000) used a necking depth of 18 km for their analysis of the northern North Sea, which therefore can be viewed as a relatively deep necking depth, promoting regional flexural uplift.

3.4.10 Phase changes

Although in general the lithosphere does not melt during slow extension without the activity of an underlying plume, mineral composition may still change due to decompression (Podladchikov *et al.* 1994; Kaus *et al.* 2005). Decompression may cause mantle rocks to cross the transition from a garnet- to a plagioclase-lherzolite, which takes place at a maximum depth of approximately 50 km. It would cause a reduction in density by c.100 kg m^{-3} and therefore surface uplift. For this to occur, extension must be rapid, extending the crust to breakup (β~4) within 10 Myr.

The gabbro–eclogite phase change may cause subsidence due to the associated increase in density of around 500 kg m^{-3} (Joyner 1967; Haxby *et al.* 1976), and has been used to explain the formation of the

(a)

(b)

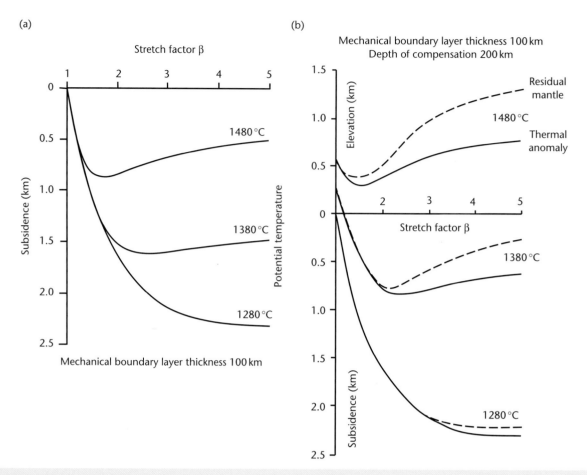

Fig. 3.36 Uplift and subsidence associated with plume activity at a spreading margin (after White & McKenzie 1989). (a) Subsidence at the time of rifting as a function of the stretch factor for potential temperatures of 1280 °C (normal), 1380 °C and 1480 °C. Each curve incorporates the effects of lithospheric thinning and crustal additions of melts caused by decompression of the mantle. (b) The effects of the reduced density of the abnormally hot asthenosphere (thermal anomaly) and the reduced density of the depleted mantle from which melt has been extracted (residual mantle) for three potential temperatures, as in (a). Plume activity causes the lithosphere to be elevated well above the level expected for an asthenosphere of normal temperature. The depth of compensation of 200 km is typical of the depth over which anomalously hot mantle is likely to extend. Reproduced with permission of the American Geophysical Union.

Williston and Michigan basins in North America (Haxby *et al.* 1976; Ahern & Dikeou 1989; Baird *et al.* 1995). Although a large proportion of the lower crust may be composed of gabbro, to generate this phase change, a large upwelling of warmer material to the base of the crust (Baird *et al.* 1995) and the presence of water (Ahrens & Schubert 1975) are required. This combination is probably rare where extension of the continental lithosphere takes place away from ocean subduction zones.

3.5 A dynamical approach to lithospheric extension

3.5.1 Generalities

Dynamic approaches to the modelling of continental extension use the constitutive laws of lithospheric materials to describe the three-dimensional deformation of the continental lithosphere under extension (§2.3). A number of *plane strain* models making a range of

assumptions of lithospheric rheology (Keen 1985, 1987; Buck 1986; Braun & Beaumont 1987; Dunbar & Sawyer 1989; Lynch & Morgan 1990) show, in general, that the style of deformation is controlled by strain rate, initial geotherm and rheological structure. Consequently, any initial heterogeneities causing variability in the mechanical and thermal properties of the lithosphere are likely to be highly influential in determining the resulting deformation. These initial perturbations that cause lateral strength variations might be thickness variations, pre-existing deep faults, thermal anomalies or rheological inhomogeneities (Fernandez & Ranalli 1997).

Plane stress models approximate the lithosphere to a thin viscous sheet, in which the rheological properties of the sheet are vertically averaged (examples are England & McKenzie 1982, 1983; Houseman & England 1986; Sonder & England 1989; Newman & White 1999). A single power-law rheology is used in these models to describe lithospheric deformation.

Dynamic models require *boundary conditions* on the margins of the extending lithosphere (Appendix 18). The choice of boundary conditions has important implications for the evolution through

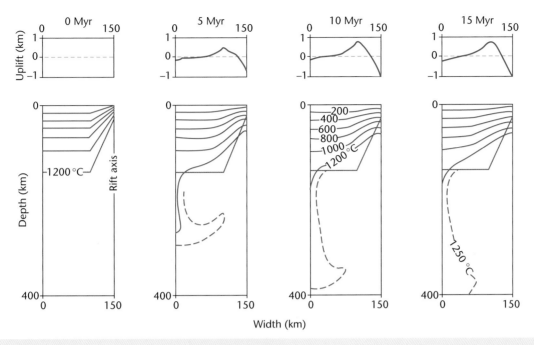

Fig. 3.37   Development of small-scale convection beneath rifts from the numerical experiments of Buck (1984). The rift is assumed to be symmetrical, with an initial half-width of 50 km. The average stretch factor is 1.6 and the internal temperature of the asthenosphere is 1300 °C. Comparison of the 1200 °C isotherm for the four time steps (0 Myr, 5 Myr, 10 Myr and 15 Myr) shows how the rift flank is progressively heated through time, causing rift shoulder uplift. Top boxes show the topography caused by combined convection and conduction. The rift flank uplift builds substantially between 5 and 10 Myr, with a maximum elevation of approximately 800 m and a width of 500–600 m. Reprinted from Buck (1984), with permission from Elsevier.

time of parameters in the model. For example, the investigator may choose a constant velocity, a constant strain rate, a constant stress, a constant heat flux or a constant basal temperature boundary condition. The latter is used in the reference uniform stretching model. Use of this boundary condition causes the heat flow from the asthenosphere to increase as stretching proceeds, since by Fourier's law, the flux is proportional to the temperature gradient, which must increase during lithospheric thinning if the basal temperature is held constant.

When regions of stretched lithosphere are plotted in terms of their bulk strain (or stretch factor) or width, and their extensional strain rate (Fig. 3.5) it is clear that there is great variability, with strain rate varying over two orders of magnitude and width varying over one order of magnitude.

Brun (1999) proposed that there are two different modes of continental extension (Fig. 3.38):

1. *Narrow rifts* with normal thickness crust (e.g. Rhine Graben; Gulf of Suez; Baikal; East Africa) are due to relatively small amounts of extension ($\beta < 2$). Narrow rifts may evolve into passive margins (e.g. Eastern Seaboard, USA; Biscay-Armorican margin) alongside oceanic spreading centres. In crust with normal thickness (30–40 km) and geothermal gradient (30 °C km^{-1}), the process of lithospheric stretching is by a very large-scale, localised thinning of the lithosphere or *necking* (Kooi et al. 1992). At early stages of lithospheric necking, rifts are 30–40 km wide, with the Moho elevated immediately beneath the rift. Some early rifts, such as those of East Africa, are wider (60–70 km). At late stages of necking, passive margins develop,

with widths of 100–400 km and with the Moho shallowing from 30–40 km under the undisturbed lithosphere to 8–10 km at the continent–ocean boundary.

2. *Wide extended domains* where extension follows earlier crustal thickening by c.20 Myr or more, providing time for the thickened brittle crust to spread gravitationally under its own weight over a weak layer in the lower crust (Jones et al. 1996). The width of the extending region can be as wide as the region of previously thickened crust – as much as 1000 km in the Basin and Range, USA. As a result of spreading, crustal thicknesses return to 'normal' (c.30 km), at which point extension stops. Local zones of exhumed ductile lower crust are termed *core complexes*. Core complexes are common in the Aegean (Greece) (Lister et al. 1984) and the Basin and Range province (Lister & Davis 1989), and have been interpreted in orogenic belts as diverse as the Alps of Austria–Switzerland (Frisch et al. 2000), the Variscan of west-central Europe (Burg et al. 1994a) and the Pan-African of western Saudi (Blasband et al. 2000). Flat-lying detachments associated with core complexes have been interpreted as initially low-angle faults that traversed the entire lithosphere. More recent studies suggest that they are initially steep and progressively rotated to a very shallow dip during extension, with a flat Moho undisturbed by any trans-lithospheric faults/shear zones.

An important set of questions immediately springs to mind: what controls the duration and total stretching of a piece of continental lithosphere? Do rifts stay narrow because the driving force for extension is removed? Or is it because the mantle lithosphere gains in strength and self-limits extension (England 1983)? The first

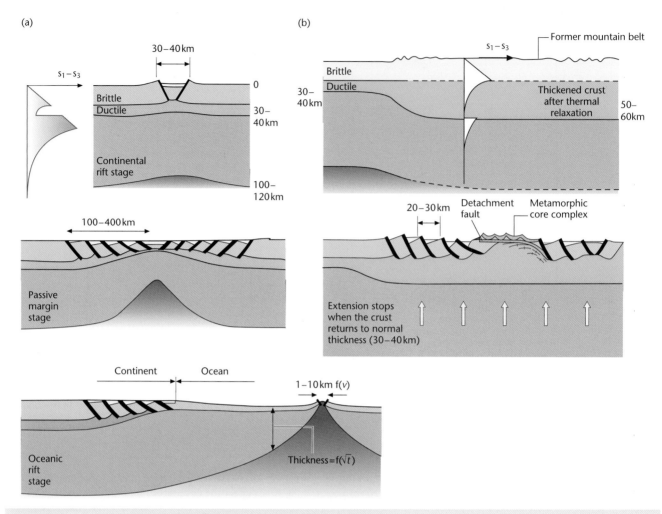

Fig. 3.38   Geological and structural differences between (a) narrow and (b) wide rifts, after Brun (1999). Wide rifts typically develop where the lithosphere has been previously thickened, with a Moho at >50 km. Extension leads to the unroofing of metamorphic core complexes. Reprinted by permission of the Royal Society.

possibility can be termed '*force-controlled extension*' and the second possibility '*viscosity-controlled extension*'.

Huismans and Beaumont (2002, 2008) also generated a complex and variable set of lithospheric geometries during extension by varying the rheological make-up of a layered lithosphere in numerical simulations. Two frictional plastic layers, representing upper and lower lithosphere, are separated by a weak ductile layer. Decoupling is possible between the upper and lower lithospheric layers across this ductile layer, with a wet quartz rheology above and a strong dry olivine rheology below. The brittle crust and upper mantle are allowed to strain-soften during deformation (that is, during deformation, the rock becomes weaker). Numerical outcomes were strongly dependent on the strength profiles, extension velocity and coupling between upper and lower lithospheric layers (Figs 3.39, 3.40):

- *AS* – asymmetric upper lithosphere rifting and symmetric lower lithospheric rifting were produced at a *low* rifting velocity where the upper and lower lithospheric layers were *decoupled*.

- *AA* – fully asymmetric rifting of both layers was produced at a *low* rifting velocity with *coupled* upper and lower lithospheric layers.

- *SS* – fully symmetric rifting of both upper and lower lithospheric layers took place at a *high* rifting velocity, for both *decoupled* and *coupled* cases.

Consequently, pure (symmetric) and simple shear (asymmetric) deformation, uniform and depth-dependent stretching, and variations in extension velocity and strain rate, are seen to be different outcomes dependent on the rheologies used, the extent of strain softening, and the coupling between different lithospheric layers.

3.5.2   Forces on the continental lithosphere

The forces responsible for stretching of the continental lithosphere in the reference uniform stretching model are not specified. Horizontal tensile stresses, however, must result from one or a combination of sources: (i) in-plane stresses transmitted from a plate boundary

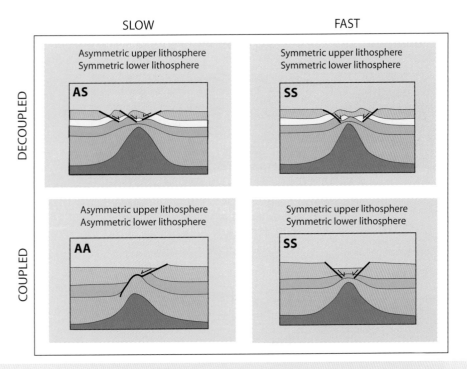

Fig. 3.39 Outline of results of numerical thermomechanical model of the extension of the continental lithosphere, redrafted and colour added, from original in Huismans & Beaumont (2002, 2008). The occurrence of symmetrical (pure shear) and asymmetrical (simple shear) of the upper and lower parts of the lithosphere is plotted in terms of the rate of extension and the coupling or decoupling of the lithospheric layers. The decoupled models include a weak, ductile lower crust with a wet quartz rheology. Note that this weak layer undergoes strong shearing. Fast extensional velocities promote pure shear.

or from basal traction; and (ii) stresses resulting from potential energy differences arising from buoyancy contrasts in the continental lithosphere. The most important cause of potential energy variations is elevation of the continental surface.

Imagine a slab of continental lithosphere acted upon by *distant extensional forces*. Changes in the distant extensional force should cause the region to experience variations in the strain rate at the same time, regardless of location. If the crust varies in thickness across our slab of lithosphere, *buoyancy forces* caused by these thickness variations should interfere with the distant extensional forces. Where the crust is thick, buoyancy forces assist the distant force to promote extension. But where the crust is thin, buoyancy forces oppose the distant extensional forces and resist extension. Could extension therefore be stopped by thinning the crust to a critical point?

It is also possible that buoyancy forces may operate in the absence of a strong distant extensional force. For example, if the crust is thick, or if the lithosphere is elevated above a mantle plume, the buoyancy forces alone may drive extension. A further source of extensional stresses is the application of shear stresses to the base of the lithosphere, for example from the convectional motion in the mantle, with extension occurring above the divergent flow of a convection cell. However, these stresses are likely to be relatively minor because the asthenosphere has a very low viscosity.

Lateral transmission of mechanical energy through the lithosphere as a result of plate collision has been proposed as a cause of continental rifting by Molnar and Tapponnier (1975). For example, the Baikal Rift of central Asia may be influenced by plate collision along the Himalayan front in the south, and a further convergent plate boundary exists in the Pacific on the east, both boundaries being roughly 3000 km distant from the rift itself. The Rhine Graben has the collision boundary of the Alps in close proximity in the south, and the opening Atlantic (1000 km distant) and subsiding North Sea (500 km distant) in the west and north. However, the frequency of collision boundaries compared to the large number of modern and ancient rifts is negligible. It therefore seems inconceivable that collision events have a *primary* role to play in providing the deviatoric in-plane forces necessary for the rifting of the continents. A numerical model of a plate of constant thickness loaded horizontally by an indenter (Neugebauer 1983), using limits of lithospheric stress under compression and extension (Brace & Kohlstedt 1980), indicates that the predicted stress regime falls considerably short of that required of a self-sustaining mechanism for rifting in the Baikal and Rhine rifts.

In addition to the forces applied at the edge of a continental plate by relative plate motion, there is an additional set of buoyancy forces set up by crustal thickness contrasts (England & McKenzie 1983; England 1983). The elevation of the continental lithosphere causes pressure differences to exist between it and neutrally elevated lithosphere. The relative importance of this driving force compared to that caused by applied forces at the plate boundaries can be gauged by an *Argand number* ***Ar*** (see also §4.5.6). If the Argand number is small, that is, the effective viscosity is large for the ambient rate of strain, the deformation of the continental lithosphere will be entirely due to boundary forces. If ***Ar*** is large, however, the effective viscosity

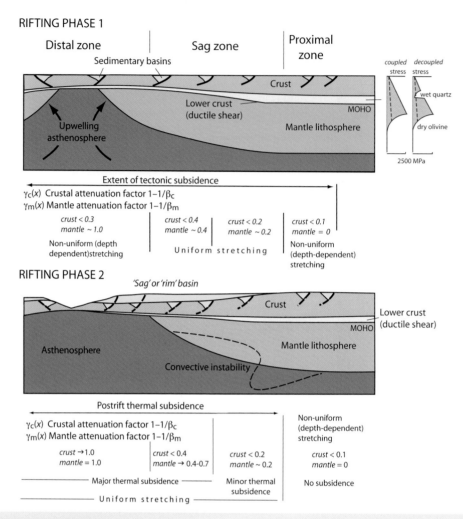

Fig. 3.40 Summary of depth-dependent (non-uniform) stretching resulting from extension of continental lithosphere in the numerical thermomechanical model of Huismans & Beaumont (2002, 2008). Three zones can be differentiated: a distal continental margin, a sag zone, and a proximal zone. During phase 1, mantle thinning is concentrated in the distal zone and decreases progressively towards the proximal zone. However, crustal thinning is concentrated in the sag zone due to shearing of the lower crust and brittle deformation of the upper crust. With further extension (rifting phase 2), a post-rift sag or continental rim basin forms due to thermal subsidence, whereas brittle deformation of the crust continues in more distal positions. Convective instabilities may form at the base of the mantle lithosphere.

will appear to be small and the medium will not be strong enough to support the elevation contrasts, so the forces due to crustal thickness changes will dominate deformation. Numerical modelling using this approach explained the tectonic styles of major crustal shortening in the Himalayas and extension in Tibet (England & Houseman 1988, 1989).

3.5.3   Rheology of the continental lithosphere

The link between the forces on the lithosphere and its deformation is the rheology of lithospheric materials (§2.3). The rheology of the lithosphere controls its deformation under any set of initial and boundary conditions. Consequently, rheology underpins all dynamical models of basin development. Key to the correct treatment of rheology is the concept of the strength envelope (Goetze & Evans 1979; Ranalli 1995; Fernandez & Ranalli 1997) (§2.3.4). The conti-

nental lithosphere is widely regarded as rheologically layered, so that at its simplest, it is a four-layer model. The high-strength regions of stabilised or 'standard' continental lithosphere (with a Moho temperature of about 600 °C) occur in the mid-crust and mantle lithosphere immediately beneath the Moho. Two low-strength regions also occur – in the lower crust, and at greater depths in the mantle lithosphere.

Information on rock rheology necessary for basin modelling essentially comes from laboratory experiments of rock deformation. Ignoring, for the moment, brittle failure, the constitutive laws derived from these experiments express strain rate as a function of temperature, pressure and deviatoric stress. Although laboratory experiments may not be fully representative of conditions in the lithosphere, the strain rate of rocks in the lithosphere is thought to obey a *power-law creep* relationship, as follows:

$$\dot{\varepsilon} = A\tau^n \exp(-E_a/RT) \qquad\qquad [3.4]$$

where τ is the deviatoric stress, T is the absolute temperature, R is the universal gas constant ($8.3144\,\mathrm{J\,mol^{-1}\,K^{-1}}$), and A, Q and n are constants dependent on the type of material undergoing deformation (Fernandez & Ranalli 1997). The activation energy E_a ranges from about $500\,\mathrm{kJ\,mol^{-1}}$ for dry olivine to about $160\,\mathrm{kJ\,mol^{-1}}$ for quartz. The power exponent n is in the range 3 to 5. Eqn. [3.4] is in fact the common form of many temperature-activated processes, for example the maturation of organic matter to form petroleum.

It is a simple matter to calculate the difference in strain rate for dry olivine in the uppermost mantle lithosphere for a 20 °C temperature change from 600 °C to 580 °C (873 K to 853 K). We assume that the driving deviatoric stress τ, exponent n and material coefficient A are constant. A temperature change of just 20 °C (from 600 °C to 580 °C) causes over an order of magnitude change in the strain rate. This is because *viscosity* is highly dependent on temperature.

Although it is possible to measure strain rates of the surface of the Earth using techniques such as global positioning system geodesy (e.g. Clarke *et al.* 1998; Bennett *et al.* 1997; Kreemer *et al.* 2003), it is not known how representative this is of longer time scales and of the lithosphere as a whole. Whereas deformation of the upper crust is by brittle faulting, it is not known whether this fault-related upper crustal extension follows the same strain rate pattern as the ductile lower crust and lithospheric mantle. In some parts of the world (such as the Basin and Range province of SW USA, and the Aegean region of Greece and Turkey) upper crustal extension is diffuse and widespread, being distributed on very many individual faults. This suggests that extension of the ductile layers of the lithosphere may control the brittle extension of the upper crust. The strain rate of ductile deformation is determined by viscosity. It is open to question whether the amount of extension varies as a function of depth, but

in the general case it can be assumed that strain and strain rate are independent of depth, representing the case of *pure shear*. The entire lithosphere is then assumed to extend at a rate determined by the depth-integrated viscosity.

3.5.4　Numerical and analogue experiments on strain rate during continental extension

Using a relatively simple rheological model of a low-viscosity crust (not temperature-dependent) and a power-law creep lithospheric mantle, Newman and White (1999) showed the strain rate history of a piece of extending continental lithosphere assuming pure shear. A number of scenarios are possible, depending on the initial strain rate/magnitude of the distant driving force (Fig. 3.41).

- At *high initial strain rate*, complete rifting of the lithosphere takes place before any significant heat loss. The strain rate increases through time because the same distant driving force acts on progressively thinner lithosphere.
- At *lower initial strain rate*, strain rate rises at first because of the same lithospheric thinning effect, but mantle cooling becomes important and reduces the strain rate. The time taken for the strain rate to become negligible (less than about $10^{-17}\,\mathrm{s^{-1}}$) depends on the initial strain rate – it is short for high initial strain rates and long for smaller initial strain rates. In other words, extension continues longer where the initial strain rate is slower.

How likely is it that the mantle lithosphere will cool sufficiently to significantly increase viscosity and stop extension? Let us take eqn. [3.4] and derive an expression for the temperature change at the

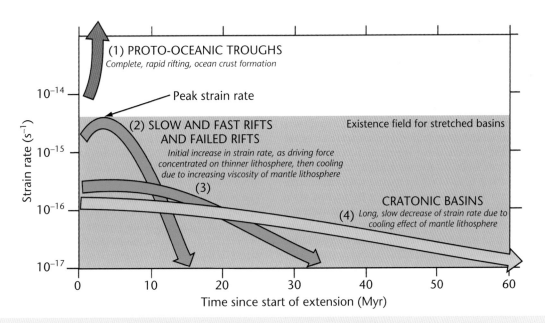

Fig. 3.41 Variation of the strain rate with time based on a set of 1D numerical experiments by Newman & White (1999), adapted by Allen & Armitage (2012) and reproduced with permission of John Wiley & Sons, Ltd. Each experiment uses the same rheology, but the driving force varies, resulting in different initial strain rates. At very high initial strain rates (1), complete, rapid rifting of the lithosphere occurs. In (2) and (3) the strain rate rises at first because the same driving force acts on a progressively thinner lithosphere. Strain rate then falls because of the effect of cooling on viscosity in the mantle lithosphere. In (4) initial strain rates are low, so the mantle cooling effect dominates. Apart from passive margins, peak strain rates in extensional sedimentary basins rarely exceed $2 \times 10^{-15}\,\mathrm{s^{-1}}$.

Moho of a piece of continental lithosphere extending with a stretch factor β required to cause a fall in the strain rate from an initial value $\dot{\varepsilon}_0$ to a new value $\dot{\varepsilon}_1$

$$\log_e \dot{\varepsilon}_0 - \log_e \dot{\varepsilon}_1 = \frac{E_a \Delta T}{R T_{MOHO}^2} - n \log_e \beta \qquad [3.5]$$

where T_{MOHO} is the temperature at the Moho, and the other parameters are defined above. The temperature change, ΔT, required to lower the strain rate by an order of magnitude is then

$$\Delta T = \{2.3 + n \log_e \beta\} \frac{R T_{MOHO}^2}{E_a} \qquad [3.6]$$

If the lithospheric stretch factor β is 1.2, $E_a = 500 \, kJ \, mol^{-1}$, $T_{MOHO} = 580 \, °C \, (853 \, K)$, $n = 3$, $R = 8.314 \, J \, mol^{-1} \, K^{-1}$, and the temperature change required to lower the strain rate by an order of magnitude is 34 °C. If the stretch factor β is 2.5, the answer is 61 °C. This is because, with the higher stretch factor, the driving force is concentrated on a thinner slab of lithosphere, which causes the deviatoric stress to increase by a factor β as the lithosphere thins. This effect offsets the increase in viscosity caused by cooling.

How long should it take to accomplish the temperature reduction causing an order-of-magnitude fall in the strain rate? To answer this question requires us to know something about the geothermal gradient in the lithospheric mantle G_m, the thermal conductivity of crust K_c and mantle lithosphere K_m, the radiogenic heat production of the crust A_c, the thermal time constant of the lithosphere τ, the initial thickness of the crust y_{c0}, and initial Moho temperature T_o. The expression for the time period required, t, simplifies to

$$t = K (\dot{\varepsilon}_0)^{-1/2} \qquad [3.7]$$

where the coefficient K is equal to 4×10^{-7} for the following parameter values: initial geotherm $G_m = 25 \, °C \, km^{-1}$, lithospheric time constant $\tau = 60 \, Myr$, initial Moho temperature $T_o = 580 \, °C$, thermal conductivity of crust $K_c = 2.5 \, W \, m^{-2} \, K^{-1}$, thermal conductivity of mantle $K_m = 3.0 \, W \, m^{-2} \, K^{-1}$, crustal radiogenic heat production $A_c = 1.1 \, \mu m \, W \, m^{-3}$, initial crustal thickness $y_{c0} = 33 \, km$, and initial lithospheric thickness $y_{L0} = 120 \, km$.

The striking result is that the time taken to reduce the strain rate by an order of magnitude depends on the inverse square root of the initial strain rate. That is, if the initial strain rate is high, the time taken to reduce it by an order of magnitude is short. If the initial strain rate is low, the time taken to reduce the strain rate by an order of magnitude is long. If the initial strain rate is $3 \times 10^{-15} \, s^{-1}$, the strain rate reduces to an order of magnitude smaller than this initial value in 7 Myr. When the initial strain rate is $3 \times 10^{-16} \, s^{-1}$, this time scale becomes 23 Myr. This compares favourably with the observed time scale of rifting known from the geological record.

If the initial strain rate is less than c.$10^{-16} \, s^{-1}$, it may take as much as 60 Myr for the extension to stop. It is likely that during this long period of time the driving force may be removed, for example, by changes in relative plate motion, or by changes in the convectional motion in the mantle (see also Chapter 5).

The subsidence history of sedimentary basins allows the stretch factor and peak strain rate of the underlying continental lithosphere to be estimated. White (1994) and Newman and White (1999) developed this technique and applied it to literally thousands of borehole subsidence records. They discovered that there was a strong relation-

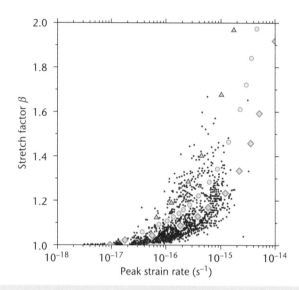

Fig. 3.42 Relation between peak strain rate and stretch factor for 2195 separate rifting episodes (small dots). Model results are for activation energies E_a of 300 kJ mol^{-1} (triangles), 500 kJ mol^{-1} (circles) and 1200 kJ mol^{-1} (diamonds). After White (1994) and Newman & White (1999).

ship between the peak strain rate and the final stretch factor in all of the sedimentary basins studied (Fig. 3.42).

If continental extension is controlled by a distant driving force, we would expect large variations in this relationship between peak strain rate and stretch factor, since continental lithosphere is heterogeneous. However, if viscosity controls extension, we should expect just the relationship observed. Differences caused by variations in the magnitude of the distant driving forces are much smaller than the effects of viscosity (or values of the activation energy E_a, which may vary significantly).

If the viscosity-controlled extension model is correct, previously stretched lithosphere should be *stronger* than 'normal' lithosphere. That is, for a given peak strain rate, we should expect a smaller stretch factor in the case of the previously stretched lithosphere. This appears to be supported by the data of Newman and White (1999). However, there is still an unanswered question: do previously rifted pieces of lithosphere tend to rift again? The geological record suggests 'yes'. The numerical model suggests perhaps 'no'.

Considerable insights have also been gained from analogue experiments. Analogue models are experiments conducted with materials of carefully chosen viscosities, strengths and densities, scaled in such a way to represent the deformation of a layered lithosphere at the appropriate strain rates. These models are informative about lithospheric stretching under a variety of conditions. Different results are obtained dependent on the strength profile of the lithospheric model and the scaled strain rate of the analogue experiment. The analogue experiments replicate many of the different styles of continental extension, particularly narrow, localised rifts versus extensive tilted fault block terranes (Brun 1999) (Fig. 3.38).

- For a simple two-layer model (brittle upper layer over a ductile lower layer), at low strain rates (and therefore low ductile strengths), deformation remains localised in an asymmetric graben; at high strain rate, the faulted zone is much wider, with highly tilted fault blocks.

- For a simple two-layer model with constant strain rate, small brittle layer thicknesses (therefore low brittle strength) produced extensively faulted models, whereas high brittle strength resulted in localised rifts.
- Where the two-layer analogue model is able to extend under its own weight, faulting invades the whole model, producing a wide rift with highly tilted fault blocks rather than horsts and graben.

These results therefore suggest that ductile and brittle strengths and spreading under gravity play major roles in style of deformation. Horsts and graben appear to typify regions of slow strain rate, whereas multiple tilted fault blocks are typical of high strain rate regions.

In four-layer models (two high-strength zones, corresponding to the upper crust and upper mantle lithosphere), low strain rates produce a narrow rift whose delimiting faults sole out into the brittle–ductile interface. Whereas a single zone of necking develops at low strain rates, at higher strain rates multiple necking (that is, *boudinage*) causes a widening of the extending zone, as in the passive margin stage. A local, low-viscosity heterogeneity beneath the brittle–ductile interface (such as caused by the thermal effects of a large pluton, or by partial melting) causes an increase in stretching and allows the ductile layer to exhume to form core complexes. Initially steep normal faults are rotated into flat-lying detachments. In wide rifts, therefore, local anomalies are prone to amplification to produce strong variations in extension and exhumation.

3.6 Estimation of the stretch factor and strain rate history

An estimate of the amount of extension that has taken place can be obtained from a number of methods (Fig. 3.43). Knowledge of the strain rate history or total stretch factor is important as a basis for predicting the geothermal gradient and heat flow history of basin sediments, as well as in showing the role of rift structures in accommodating basin sediments.

3.6.1 Estimation of the stretch factor from thermal subsidence history

It has previously been shown (§3.3) that the 'reference' uniform stretching model predicts initial uplift or subsidence depending on the ratio of crustal to lithospheric thickness y_c/y_L and the stretch factor β. If y_c/y_L is known for a particular basin, the fault-controlled initial subsidence could in theory be used to estimate the amount of stretching (see equations in Appendix 21). However, the extreme variations in syn-rift thicknesses make this a very unreliable technique.

The rate of thermal subsidence in a basin generated by uniform stretching is dependent on the stretch factor. The post-rift thermal subsidence phase of sedimentary basins also lacks the extreme thickness variations characteristic of syn-rift deposits. The standard technique is the construction of a set of subsidence curves for different values of the stretch factor β (Appendix 19). Curves can be constructed for a water-loaded basin, where the density of the infill is $\rho_w = 1000\,\mathrm{kg\,m^{-3}}$. In this case the post-rift sediment thicknesses

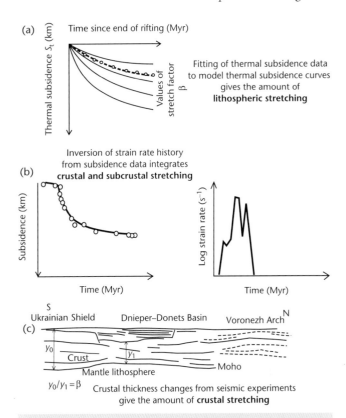

Fig. 3.43 Methods for calculation of the amount of stretching. (a) Calculation of the lithospheric stretch factor from thermal subsidence curves. (b) Inversion of strain rate history of the lithosphere from subsidence curves. (c) Calculation of the crustal stretching from mapping of the Moho.

need to be decompacted and backstripped to reveal the water-loaded tectonic subsidence history (Appendices 56, 57 and §9.4). The observed backstripped post-rift subsidence can then be compared with the model curves to reveal the best-fitting stretch factor (Appendix 19). Alternatively, model curves can be constructed that show the thermal subsidence for different values of the stretch factor β for a sediment-loaded basin. Construction of such curves requires estimation of the bulk density of the sediment column as a function of time. The model sediment-loaded curves can then be compared with the decompacted total subsidence curve derived from borehole data or outcrop sections.

Thermal subsidence data can also be inverted to obtain the stretch factor, as shown in Appendix 20. A workflow showing the steps taken from decompacting a stratigraphic succession, through inversion of the stretch factor, to estimating paleotemperatures of selected horizons, is given in Appendix 58.

3.6.2 Estimation of the stretch factor from crustal thickness changes

In some circumstances the attenuation of the crust can be estimated from deep seismic (refraction and wide-angle reflection) results. For example, the Moho rises by 5 km in the southern part of the Rhine Graben, and 3 km of sediment has accumulated. The crust has therefore thinned by 8 km (Emter 1971; Mueller *et al.* 1973). Similarly, the nearby Limagne Graben (France) contains 2 km of sediments, and the Moho rises to within 24 km of the surface (Hirn & Perrier 1974). Since the crust is about 30 km thick in the Massif Central, which separates the Rhine–Bresse rift system from the Limagne Graben, the amount of crustal attenuation in these two cases can be estimated to be between 1.2 and 1.3.

The North Sea is a failed rift that underwent rifting in the Jurassic–Early Cretaceous. Regional deep seismic reflection profiles and gravity profiles in the Viking Graben–Shetland Platform area have been used to estimate the depth to the Moho (Barton 1986; Klemperer 1988). More recently, the deep structure of the Viking Graben and adjacent areas of the northern North Sea (60–62°N) has been investigated based on an integrated study of deep seismic reflection and refraction data, gravity and magnetic data (Christiansson *et al.* 2000) (Fig. 3.10). Where unaffected by Caledonian crustal roots and by Mesozoic stretching, the Moho lies at a depth of 30–32 km. It shallows to about 20–22 km beneath the Viking Graben. Since the graben is filled with approximately 10 km of Triassic to Cenozoic sedimentary rocks, the thickness of crystalline and Paleozoic crustal rocks in the Viking Graben is only about 10 km. This represents a stretch factor of over 3. In contrast, the East Shetland and Horda platforms have stretch factors of less than 1.5. These values reflect the cumulative effects of several post-Caledonian stretching events.

The estimates of crustal stretching derived from mapping of the Moho have been compared with estimates from subsidence analysis (Giltner 1987; Badley *et al.* 1988; Christiansson *et al.* 2000) and forward tectonostratigraphic modelling (Odinsen *et al.* 2000). The former are somewhat higher than the stretch factors derived from the other two methods, which emphasises the care that needs to be taken in the interpretation of β estimates. A similar exercise in the Rockall, mid-Norway and SW Barents Sea areas also showed large discrepancies between the crustal stretching estimated from deep seismic imagery and subsidence analysis (Skogseid *et al.* 2000).

3.6.3 Estimation of the stretch factor from forward tectonostratigraphic modelling

Stretch factors have been estimated along transects of the northern North Sea where the crustal profile, stratigraphic thicknesses and faults are very well constrained (Odinsen *et al.* 2000). Forward tectonostratigraphic models simulate crustal structure and basin development through time using a rheological model of the lithosphere, with a finite flexural strength based on Braun and Beaumont (1987) and Kooi *et al.* (1992). The forward model has the advantage of discriminating the stretching from each rifting event. Odinsen *et al.* (2000) compared the stretching in the Permo-Triassic (maximum of $\beta = 1.41$ in the Viking Graben) versus Jurassic (maximum of $\beta = 1.53$ in the Viking Graben) rifting events in the northern North Sea. The calculated distribution of stretch factors along two well-constrained crustal transects (Fig. 3.10) matches well earlier estimates based on 1D subsidence analysis (Giltner 1987; Badley *et al.* 1988).

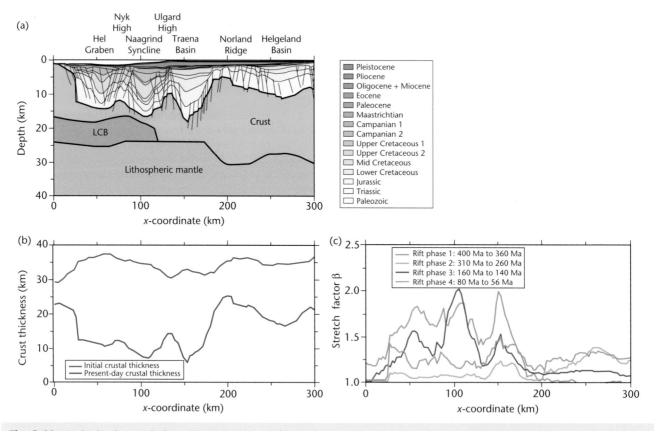

Fig. 3.44 Multiple phases of rifting, Voring margin, mid-Norway. (a) Sedimentary basin, crust and lower crustal body (LCB) across a 300 km-wide transect. (b) Thickness of present-day crust and estimate of its initial thickness. (c) Calculated stretch factors along the profile for four rifting episodes. From Wangen (2010, p. 247). © Magnus Wangen 2010, reproduced with the permission of Cambridge University Press.

3.6.4 Inversion of strain rate history from subsidence data

White (1993, 1994) has taken the novel approach of calculating the strain rate history from the subsidence history of sedimentary basins using inverse methods. The stretch factor β can be viewed as representative of the total strain integrated over the time scale of extension. The stretch factor is therefore related to the vertical strain rate $\dot{\varepsilon}_v(t)$ by

$$\beta = \exp\left(\int_0^{\Delta t} \dot{\varepsilon}_v(t)\, dt\right) \qquad [3.8]$$

where Δt is the duration of stretching. If the strain rate is constant, $\beta = \exp(\dot{\varepsilon}_v \Delta t)$. In the inverse method, the subsidence as a function of time is solved iteratively to give the strain rate as a function of time $\dot{\varepsilon}_v(t)$. Multiple stretching episodes can be resolved by this technique.

3.6.5 Multiple phases of rifting

Many extended provinces have experienced a number of distinct rifting episodes. The Voring margin, offshore mid-Norway, for example, extended in the Devonian (400–360 Ma), Permian (310–260 Ma), Late Jurassic (160–140 Ma) and finally in the Late Cretaceous and Paleocene (80–56 Ma), which led to break-up and the formation of the North Atlantic Ocean (Lundin & Doré 1997; Wangen *et al.* 2007). The strain rate history therefore shows multiple peaks. The crustal stretch factors for each of these stretching phases can be plotted across a basin transect (Fig. 3.44). If the crustal thickness before the first rifting phase can be constrained, it allows the stretch factors for each phase and the total crustal stretch to be calculated. The initial thickness of the crust can be determined if Airy isostasy is assumed and the water depth, sediment thickness, sediment bulk density and crustal thickness are known from the present-day profile (see Chapter 9).

The total (cumulative) crustal stretch is

$$\beta_{max} = \beta_1 \beta_2 \beta_3 \beta_4 \qquad [3.9]$$

where 1 to 4 are the rift phases. The cumulative stretch factor at any point in the geohistory t can also be found by a comparison of the crustal thickness initially and at the time of interest:

$$\beta_{max}(t) = \frac{y_{c,0}}{y_c(t)} = \frac{y_{c,0}}{y_{c,0} - f_w W_d(t) - f_s(t) y_s(t)} \qquad [3.10]$$

where $y_{c,0}$ is the initial crustal thickness, $y_c(t)$ is the crustal thickness at time t, f_w is a density factor for water equal to $(\rho_m - \rho_w)/(\rho_m - \rho_c)$, f_s is a density factor for bulk sediment $(\rho_m - \rho_w)/(\rho_m - \rho_b(t))$, and $\rho_b(t)$ is the average bulk density of the sediment columns as a function of time.

Table 3.2 Results of estimates of crustal stretch factor for multiple stretching phases of the Voring margin, mid-Norway (from Wangen 2010, p. 251) © Magnus Wangen 2010, reproduced with the permission of Cambridge University Press

Phase i	y_i	$\rho_{b,i}$	Wd_i	$y_{c,i}$	$f_{s,i}$	$\beta_{max,i}$	β_i
1	1	2200	0	35028	2.2	1.00	1.18
2	1800	2200	300	29690	2.2	1.18	1.07
3	2300	2250	550	27670	2.1	1.27	1.77
4	10000	2550	950	15660	1.5	2.24	1.96
5	15000	2560	1050	8000	1.5	4.38	–

Input data and calculated stretch factors for the four rift phases recognised in the Voring margin are given below. Note that the cumulative stretch factor increases over time and reaches $\beta_{max} > 4$ by the time of ocean crust formation in the Paleocene.

CHAPTER FOUR
Basins due to flexure

Summary

Flexure is the long-wavelength deflection of a lithosphere of finite strength caused by the application of an external force system. A general flexural equation can be established for the case of a thin elastic plate overlying a weak fluid subjected to vertical applied forces, horizontal forces and torques or bending moments. The general flexural equation can then be used in a range of geodynamical situations by applying different boundary conditions.

Flexure of the lithosphere is most clearly demonstrated at the margins of former ice caps, at oceanic islands, seamount chains, ocean trenches and foreland basins. Lithospheric flexure also supports sediment loads in most sedimentary basins.

The flexural basins associated with ocean–continent and continent–continent plate margins are particularly well represented in the geological record. In an Andean-type setting the flexural basin on the subducting oceanic plate is the ocean trench, and the flexural basin on the upper plate is a retro-arc foreland basin. In Alpine-type settings the flexural basin on the subducting plate is called a peripheral or pro-foreland basin, whereas the flexural basin on the overriding or indenting plate is termed a retro-foreland basin. Some foreland basins are related to subduction zone roll-back (subduction zone retreat) and are associated with prominent backarc extension. The formation and evolution of foreland basins is intimately related to the processes of tectonic shortening and exhumation of the adjacent orogenic wedge.

The deflection of the oceanic lithosphere along seamount chains such as the Hawaiian Islands can be explained by either the flexure of a continuous plate loaded by a vertical applied force (representing the excess mass of the seamount chain), or by the flexure of a plate broken beneath the vertical applied force. Some useful and simple expressions for the geometry of the deflection can be derived that involve the amplitude and width of the deflection, and the location and height of the forebulge. In particular, the wavelength of the deflection is dependent on the *flexural rigidity* (units of Nm) of the plate, or *flexural parameter* (units of km). The maximum deflection is dependent on the flexural rigidity and the magnitude of the applied load. Flexure of continental lithosphere at sites of plate convergence requires the tectonic and sedimentary load to be treated as spatially widely distributed.

The oceanic lithosphere appears to become stronger as it ages and cools, but it does not weaken in its ability to support loads with time since loading. This suggests that any viscous relaxation that takes place does so very quickly (less than 10^5–10^6 yr). The factors determining the flexural rigidity of the continental lithosphere are less clear, and there is no relationship between flexural rigidity and its age. A number of factors may be responsible, such as differences in geothermal gradient caused by strong variations in the crustal radiogenic heat production, decoupling between a strong upper crust and strong underlying mantle lithosphere during bending, and plastic yielding at high curvatures. The lithosphere approaching collision zones may also be segmented, producing a stepped, compartmentalised deflection. Strong spatial variations in flexural rigidity have been measured in the forelands of many mountain belts. At a global scale, continental cores are very rigid, whereas continental megasutures marking the accretion of old mountain belts and arcs, plate margins and regions of rifting, are weak.

Although linear elasticity suggests that whole-lithosphere buckling should not take place under horizontal (in-plane) loading, analogue experiments and numerical models using a rheologically layered lithosphere consistently reproduce long-wavelength (several hundred kilometre) antiforms and synforms. Sedimentary basins located at buckle synforms may be difficult to discriminate from back-bulge depositional zones. In-plane stresses may also invert older tectonic structures, causing large sedimentary basins to be subdivided into several separate depocentres.

Orogenic wedges act as primarily vertical loads on the deflected plate in collision zones. The movement of a load over a foreland plate forces a wave-like deflection ahead of it. The orogenic wedge is viewed as a dynamic unit with a critical surface slope in which the gravitational and deviatoric forces strive to achieve a steady state. Disturbances of equilibrium by externally imposed forces (such as changes in convergence rate) may result in major changes in the rates of shortening or extension of the wedge. These processes in turn influence the erosion of the mountain belt to provide detritus for the foreland basin, and the configuration of the load driving the deflection of the overriden plate. Orogenic wedges are strongly influenced by the strength of the basal detachment. Salt-based orogens typically have very low surface slopes.

Analogue models using scaled physical experiments are successful at visualising major crustal structures produced during convergence, including major crustal buckles, conjugate shears and bivergent wedges. They also show the strong impact of localised and distributed patterns of erosion on the exhumation paths of particles in the wedge, and the teleconnections between opposite flanks of the wedge driven by erosion.

Basin Analysis: Principles and Application to Petroleum Play Assessment, Third Edition. Philip A. Allen and John R. Allen.
© 2013 John Wiley & Sons, Ltd. Published 2013 by John Wiley & Sons, Ltd.

Early numerical models focused on diagnostic basin stratigraphy, and treated erosion, sediment transport and deposition as a diffusional problem. Later three-dimensional plane strain numerical models illustrate the complex links between plate convergence, tectonic fluxes within bivergent wedges and exhumation. The mechanics of these wedges can be investigated using the Argand number and Ampferer number. Two dynamical end members are the indentation model and the subduction model. Plane strain numerical models show that erosion of the surface of the orogenic wedge is of major importance to wedge dynamics as well as to the filling of adjacent foreland basins. Such models now provide a mechanical explanation for the topography, distribution of high-grade metamorphic rocks and thermochronology of rocks in orogenic belts.

A distinction can be drawn between mountain building associated with subduction, in which case there is a high relative velocity between the orogenic wedge and the lower plate, and mountain building far from plate boundaries, associated with smaller amounts of shortening. These intracontinental orogenic belts (e.g. Tien Shan) and their flexural basins (Tarim, Junggar) typify the interior of Asia at the present day.

Thickness changes of continental lithosphere during mountain building cause lateral (in-plane) forces that are important in influencing gravitational collapse. Erosion is an important component in the processes affecting the gravitational potential energy of lithospheric columns and thereby may play a role in causing distinct cycles of contractional deformation and extensional collapse.

Tectonic deformation and surface processes of erosion and deposition may be coupled at a range of scales, from individual anticlines, to arrays of folds, to whole orogens. The vigour of surface processes therefore has a direct link to structural geometries and loci of fault activity, exhumation rates, basin styles and gross depositional environments.

In Chapter 4 we focus on the process of lithospheric flexure, with the goal of understanding better the long-wavelength behaviour of the lithosphere when acted upon by applied force systems or loads. This is of relevance in a number of situations in basin analysis. Firstly, flexure of the lithosphere may be the primary cause of the subsidence or uplift in a sedimentary basin. The most important example is the downflexing of a continental plate by force systems set up during mountain building. This produces *foreland basins*. Secondly, the lithosphere may undergo a long-wavelength flexure in supporting sediment loads. This takes place in basins of all types (Chapter 1), including those due to stretching of the lithosphere (Chapter 3).

The most obvious geological scenario for long-wavelength bending of the lithosphere is downflexing by the growth of major ice masses, such as the Laurentide and Scandinavian ice caps during the Pleistocene, and the flexural rebound accompanying melting (§4.1.1). A similar effect is produced by the magmatic and topographic growth of oceanic seamount chains such as the Hawaiian-Emperor chain of the Pacific, and volcanic islands such as the Canaries, which flexes down the adjacent ocean floor producing a bathymetric moat (§4.1.2). Oceanic lithosphere is also flexed down by the accumulating sediment load off major river deltas, such as that of the Amazon fan (§4.1.3). The bending of oceanic plates as they approach subduction zones (§4.1.4), and the bending of continental plates in zones of continental collision (§4.1.5), are dynamic examples of lithospheric flexure of direct relevance to basin analysis.

The theory of the bending of an elastic beam provides the foundation for the study of the flexure of the lithosphere. The reader is referred to §2.1.3 and §2.1.4 for some of the fundamentals of flexure of linear elastic materials. §2.3 contains background to more complex rheologies. The theory of the ideal behaviour of linear elastic plates when loaded by vertically acting loads (§4.2) and horizontal loads (§4.4) is applied to a range of geological situations, but principally to the case of the foreland basins associated with mountain building, where sedimentary basins are intimately linked to the dynamics of orogenic wedges and their fold-thrust belts (§4.5). The interplay of tectonics, climate and lithospheric flexure controls the stratigraphic infill of foreland and wedge-top basins (§4.6).

A short note on terminology is required. The deformed shape of the lithosphere caused by flexure in response to applied loads is called the deflection w. The amplitude and wavelength of the deflection are determined by the material properties and thickness of the elastic part of the plate and the mass and spatial distribution of the load. The material properties and thickness of the elastic plate are encapsulated in the flexural rigidity D (Nm), which can also be expressed as an elastic thickness Te (m), or a flexural parameter α (m).

4.1   Basic observations in regions of lithospheric flexure

Both oceanic and continental lithosphere is capable of a long-wavelength bending termed flexure. Flexure of the oceanic lithosphere takes place at ocean trenches, seamount chains and individual oceanic volcanic islands. Flexure of the continental lithosphere takes place at sites of rifting, strike-slip faulting, at passive margins and most emphatically at sites of plate convergence. The sedimentary basins caused by flexure of continental lithosphere located adjacent to zones of tectonic shortening commonly occur as pairs, separated by an orogenic belt or magmatic arc (Fig. 4.1). Flexural basins are elongated along the tectonic strike, with an asymmetrical cross section deepening towards the orogenic belt or magmatic arc. On the continental lithosphere, Bouguer gravity anomalies are negative, indicating the presence of a mass deficit at depth, caused by the downward penetration of relatively light 'granitic' material replacing mantle. Flexural basins on continental lithosphere are close to strongly eroding source areas, and are typically filled with large thicknesses of syn-orogenic sediment. Ocean trenches on the other hand may be well-nourished or starved of sediment depending on the proximity of major sediment routing systems. Heat flows in regions of lithospheric flexure are close to normal, or slightly reduced due to crustal and lithospheric thickening during plate convergence.

4.1.1   Ice cap growth and melting

Flexure and flexural rebound associated with the growth and melting of ice caps is a classic situation where long-wavelength bending of the continental lithosphere, and the time scale of the response, can be investigated.

The growth of ice caps causes a downflexing of the underlying lithosphere, with a proglacial lake occupying the flexural depression (Fig. 4.2a). Such proglacial lakes are typically 100–200 km across.

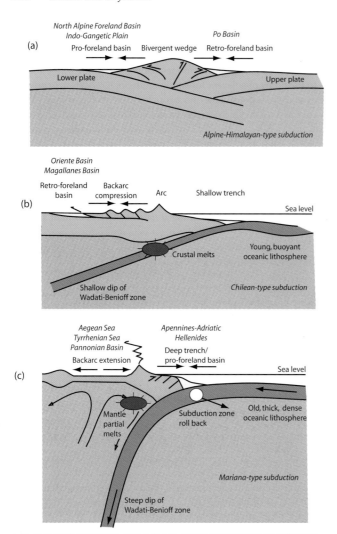

Fig. 4.1   Flexure at sites of plate convergence. Schematic illustration of (a) pairs of pro-foreland and retro-foreland basins at sites of continent–continent collision, as in the Alps; (b) retro-foreland basins and ocean trenches at sites of ocean–continent convergence, as in the Andes (especially the Chilean transect) and Sevier belt of the USA, where subduction is or was shallow; and (c) pro-foreland basins and backarc extension related to subduction zone roll-back, as in the Aegean and Tyrrhenian seas, and steep subduction of the oceanic plate, as in the Marianas. Based partly on Uyeda & Kanamori (1979) and Stern (2002).

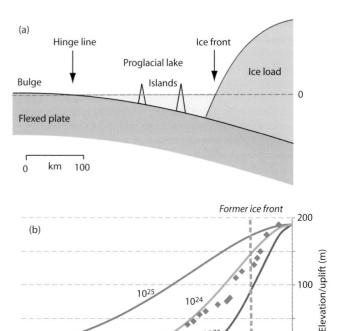

Fig. 4.2   Proglacial lakes and postglacial rebound. (a) Schematic diagram showing a proglacial lake with islands and shorelines occupying a flexural depression generated by the load of an ice cap. (b) The elevation of paleolake shorelines of Lake Algonquin as a function of horizontal distance from the maximum extent of the North American ice cap, with best-fitting curves for flexural rigidity. Part (a) is adapted from Watts (2001).

Beach ridges around the margin of the lake and rimming islands uniquely define the position of past lake shorelines. Today, the beach ridges are found at different elevations, with a systematic trend with distance from the former ice front (Fig. 4.2b).

The Lake Algonquin (Canada) 12 ka shoreline is now elevated at 200 m at locations close to the former ice cap, and at intermediate locations there is a progressive reduction of the amount of postglacial uplift (Colman *et al.* 1994). The trend in elevation of the beach ridges can be used to constrain the flexural rigidity of the Canadian plate, providing a best fit of 70–100 km for the elastic thickness. The flexural rebound has taken place over the last 12 kyr, which demonstrates that flexural adjustment is rapid (Turcotte & Schubert 2002). A

similar time scale for flexural adjustment to changing loads is found in the case of the raised beaches of the Baltic resulting from deglaciation of the Scandinavian ice cap (Flint 1971; Mitrovica 1996).

4.1.2   Oceanic seamount chains

The Hawaiian-Emperor seamount chain is relatively narrow compared to the length of the Pacific plate, so acts like a line load (Fig. 4.3). The ocean crust is deflected downward (c.3 km) by the 8 km-thick volcanic load of the seamount chain, producing a moat with a maximum water depth of 4 km and an outer rise. Using the velocity-density model shown in Fig. 4.3, the free-air gravity anomaly is calculated and compared with a gravity model at different values of the elastic thickness. An excellent fit is obtained for an elastic thickness of 40 km (Watts & ten Brink 1989). The 'basin' or 'moat' is filled with marine sediment and debris flow/rock avalanche material derived by collapse of the volcanic edifice.

The loads on the Pacific plate are of two types: (i) a *driving load* made of volcanic material that is replacing ocean water; and (ii) a *sediment load* infilling the basin or 'moat' either side of the seamount chain. The sediment replaces ocean water, so is a positive downward-acting load. These surface loads are opposed by an upward-acting hydrostatic restoring force.

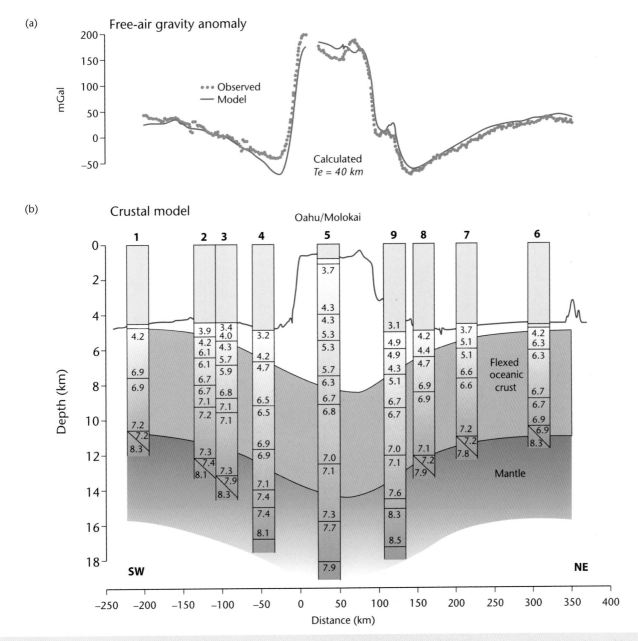

Fig. 4.3 Hawaiian-Emperor seamount chain, Pacific Ocean, showing velocity-density model and free-air gravity anomaly, and best-fitting deflection for an elastic thickness of 40 km. Adapted from Watts & ten Brink (1989). Red line is bathymetry/topography across Oahu. Reprinted by permission of Macmillan Publishers Ltd.

Flexure under oceanic seamount loads can also be imaged using seismic reflection experiments. Images from a profile across Tenerife, Canary Islands, show the structure of the infilling stratigraphy of the two 'moats' (Fig. 4.4). The early (oldest) seismic sequences show very little thickening towards the volcanic edifice, but the youngest sequence shows pronounced thickening towards the island and thinning towards the outer rises (bulges). This indicates that the main phase of flexural subsidence is very recent (Watts 2001).

4.1.3 Flexure beneath sediment loads

When sediment is deposited in the ocean, it displaces water and acts as a distributed load on the underlying lithosphere. The degree of compensation of a periodic load is given in §2.1.5. The flexure under the sediment load depends on the magnitude of the sediment load, the wavelength of the load, and the flexural rigidity of the lithosphere.

In the case of the Amazon fan, which has accumulated on the Brazilian continental margin since the mid-Miocene, the ocean crust is buried by a thick (<12 km) wedge of sediment, which extends c.200 km across the margin. Using a density model for the sediment and underlying crust, the free-air gravity anomaly can be used to constrain the most likely value of flexural rigidity (Te of 31 km – see §2.1.4). The sediment load of the Amazon fan has flexed down the Brazilian margin by approximately 2 km compared to adjacent sea-floor (Watts *et al.* 2009).

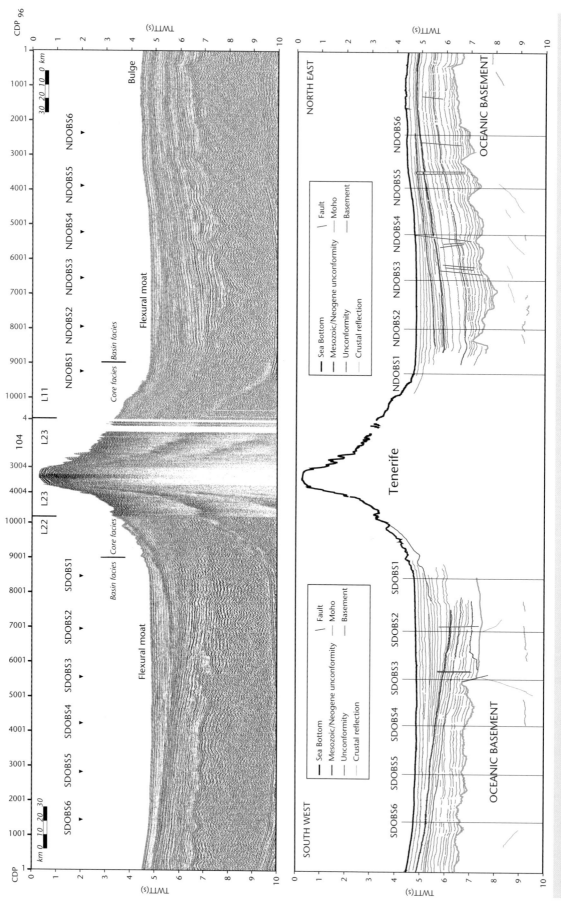

Fig. 4.4 Cross-section of Tenerife, Canary Islands, based on seismic reflection data, showing deflection of the ocean crust and the structure of the sedimentary infill of the moat on each side of the island line load. Youngest stratigraphy shows thickening towards load, indicating active volcanic island growth and flexure of the oceanic plate. Reprinted from Watts *et al.* (1997), with permission of Elsevier.

4.1.4   Ocean trenches

At ocean–ocean and ocean–continent convergent boundaries, there is strong evidence that the oceanic lithosphere is bent to high curvatures. Seaward of the ocean trench is a positive free-air gravity anomaly (Watts & Talwani 1974) that corresponds to relatively elevated seafloor. The trench and accretionary prism are characterised by a large negative free-air gravity anomaly, and the island arc by a large positive anomaly. These anomalies can be explained by the bending of the oceanic plate by a force system associated with subduction. The shape of the plate at depth can be imaged from the foci of the earthquakes that define Benioff-Wadati zones (Isacks *et al.* 1969; Isacks & Molnar 1971).

The ocean trenches of the world are remarkably uniform in their bathymetry and wavelength (Fig. 4.5). They are generally 50 to 100 km in width and 2 to 4 km below the general level of the adjacent ocean floor. They are asymmetrical in cross section, with a steeper limb flanking the overriding plate, and a shallow slope on the side of the down-going plate. Most trenches are relatively starved of sedi-

ment (e.g. Tonga), whereas others are filled with sediment (e.g. Lesser Antilles, and the northern part of the Burmese Sunda trench) and therefore lack a pronounced bathymetric expression. The key control is the availability of a major source of sediment; for example, the Ganges-Brahmaputra and Irrawaddy systems are responsible for the filling of the Burma trench, while the same trench in Sumatra is relatively deficient in sediment fill.

Cross-sections of the Andes indicate that where the dip of the down-going slab is shallow, as in Ecuador, there is very little volcanism and weak but widespread seismicity (Pilger 1984). Deformation extends from the active plate margin huge distances (up to 700 km) into the centre of the plate, giving rise to crustal shortening and earthquake activity in fold-thrust belts facing the craton. Where the plate is being subducted at a high angle, there is active recent volcanism, seismicity is restricted to a narrow Benioff zone, and a well-developed longitudinal median sedimentary trough is developed. The cause of the locally shallow angle of the down-going slab may be the subduction of relatively buoyant oceanic ridges or of very young oceanic lithosphere.

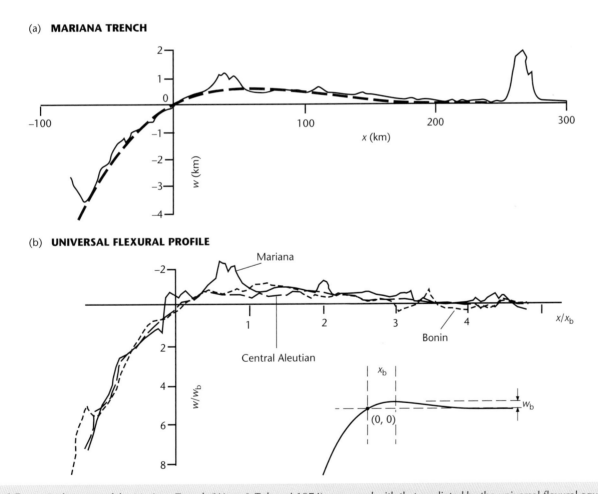

Fig. 4.5 (a) Bathymetry of the Mariana Trench (Watts & Talwani 1974) compared with that predicted by the universal flexural equation with x_b taken as 55 km and w_b as 0.5 km (after Turcotte & Schubert 1982, p. 130; © Cambridge University Press, 2002). (b) Non-dimensional comparison of the bathymetric profile of the Mariana, Bonin and Central Aleutian oceanic trenches with a universal flexural profile (after Caldwell *et al.* 1976; reprinted with permission of Elsevier). Deflection w is scaled by the maximum elevation of the bulge w_b, and the horizontal coordinate x is scaled by the horizontal distance to the crest of the bulge x_b. The x-coordinate is zeroed at the first node of the deflection x_0 where $w = 0$ (inset).

4.1.5   Mountain ranges, fold-thrust belts and foreland basins

Elongate to arcuate basins caused primarily by flexure are found at ocean–continent and continent–continent collision zones (Fig. 4.1) where they are located on continental lithosphere (Dickinson 1974). In an Andean-type setting (ocean–continent), the basin on the lower plate is the ocean trench, and the basin on the upper plate is a foreland basin, commonly termed of *retro-arc type*. In a Himalayan-type or Alpine-type setting (continent–continent), the foreland basin on the lower plate is of *peripheral* or *pro-foreland* type, and the basin on the upper plate is of *retro-foreland* type (Willett *et al.* 1993). Some orogenic belts contain large *intramontane basins*, as is classically displayed in the Andes of South America (Horton *et al.* 2001). However, these intramontane basins are caused by localised extensional and strike-slip tectonics rather than predominantly by lithospheric flexure. Foreland basins may be partitioned by the uplift of crustal blocks cored by basement rocks (Schwartz & DeCelles 1988), as in the Laramide uplifts of the Rocky Mountains of the western interior of the USA. Such situations are termed *broken forelands*.

Foreland basin systems are dynamically linked to adjacent convergent orogenic belts. These convergent orogens are zones of considerable (hundreds of kilometres) crustal shortening, achieved by translations along thrust faults and by ductile thickening. The thrust belts characteristic of orogenic belts are typically parallel to the strike of the orogenic belt, but in detail comprise salients and re-entrants (known as syntaxes). The Appalachians of eastern North America and the Himalayas of India-Pakistan are Paleozoic and Cenozoic examples respectively. Many orogenic belts are also characterised by ductile extensional detachment faults and shear zones that facilitate rapid rates of exhumation of high-grade rocks. The South Tibet detachment system of the Himalayas (Burg *et al.* 1984) and the Simplon fault of the Swiss Alps (Mancktelow 1985) are examples.

Some thrust belts and foreland basins are related to *subduction zone roll-back*, or *subduction zone retreat* (Royden 1993), where the rate of subduction exceeds the rate of plate convergence (see §4.5.2). Subduction roll-back produces short, highly arcuate thrust systems such as the Apennines and Carpathians of Europe. Foreland basins developed in such systems may be associated with regions of retro-arc extension (e.g. Pannonian Basin) or the creation of new oceanic basins (Apennines-Tyrrhenian Sea, Malinverno & Ryan 1986; Doglioni *et al.* 1998).

There are a number of well-documented orogenic belts and foreland basin systems related to Cenozoic orogenesis, such as the Pyrenees of Spain, Alps of France and Switzerland, Apennines of Italy, Himalayas of India and Pakistan, and Andes of South America. These examples provide benchmarks of the close linkage between lithospheric flexure, convergent tectonics and basin development. As an example, the Himalayas are the world's highest and largest orogenic belt. A classic foreland basin (Burbank *et al.* 1996) with >6 km of sedimentary fill is found to the south of the Main Boundary Thrust of the Himalayan system (Fig. 4.6). The Himalayas are 250 to 350 km wide and extend for thousands of kilometres around the site of indentation of India with Asia. Convergence between the Indian indenter and Asia is currently c.60 mm yr^{-1} (Bilham *et al.* 1997). Shortening in the Himalayan fold-thrust belt is in excess of 700 km (DeCelles *et al.* 1998). Focal mechanism solutions of earthquakes (Molnar & Chen 1982) show that the Indian shield in the south is undergoing weak extensional faulting, the Lower Himalaya is experiencing thrust-fault deformation, and the High Himalaya has a

mixture of extensional and strike-slip deformation. Seismic imaging shows that the Moho descends from about 35 km under the Indian shield, via a number of steps, to 70 km under Tibet. Although it is unlikely that the Indian plate slides smoothly under Asia, the gravity data in the Ganga Basin and Himalayas (Lyon-Caen & Molnar 1985) can be explained by the flexure of the Indian plate beneath the topographic load of the mountain belt (Fig. 4.6).

Foreland basins are therefore emphatically syn-orogenic. The bulk of the sediments in foreland basin systems (DeCelles & Giles 1996) are found in a *foredeep*, extending from the thrust front across the foreland plate. The foredeep is classically asymmetrical, with subsidence rates at a maximum close to the thrust front. Sedimentation is dominated by sediment delivery from the orogenic belt, but significant sediment discharges may come from the opposite side of the basin, particularly early in basin development (Yong *et al.* 2003). Proximal foredeep deposits are progressively incorporated into the orogenic wedge as the thrust belt migrates over the pro-foreland. Sediments also accumulate in *wedge-top* or *piggyback basins* located on top of the deforming thrust wedge. These basins are commonly involved in thrust deformation, and may be eroded as the orogenic wedge is exhumed (see §4.6.4).

4.2   Flexure of the lithosphere: geometry of the deflection

Since the oceanic lithosphere has a relatively simple thermal structure and is relatively simple rheologically, it is therefore the appropriate starting place for the consideration of flexure. Early measurements of free-air gravity anomalies in the ocean were highly influential in explaining the bathymetry of the seafloor, particularly adjacent to oceanic islands and seamount chains. Where there is negligible sediment loading, the bathymetry of the ocean floor and its free-air gravity anomalies can be explained by a model involving a line load acting on either a continuous plate (§4.2.1) or on the end of a broken plate (§4.2.2). The effect of a load system that has a significant spatial distribution is discussed in §4.2.3.

4.2.1   Deflection of a continuous plate under a point load (2D) or line load (3D)

Some general concepts of flexure can be appreciated by considering the bending of lithosphere under loads acting a long way from the edge of the plate. A geological situation of this type is the flexure of the lithosphere under the load of mid-plate oceanic islands such as the Hawaiian archipelago (Vening-Meinesz 1941, 1948; Watts & Cochran 1974). The Hawaiian ridge is a long (thousands of kilometres) line of volcanic islands of about 150 km width. The ridge therefore approximates a line load. The ridge is flanked by a depression, the Hawaiian Deep, and then an outer rise (Fig. 4.3). These three morphological elements (load, basin and outer rise or forebulge) are a constant theme in flexural problems.

In a 2D profile, a point load (such as a single oceanic volcano) is equivalent to a line load in 3D (such as an oceanic seamount chain). It is useful to start a consideration of flexure with a point load, since a more realistic, distributed load can be approximated by the superimposition of the point load solutions (§4.2.3).

We assume that a vertically acting load V_0 is emplaced at $x = 0$, and that the deflection goes to zero ($w \rightarrow 0$) a long way from the load ($x \rightarrow \infty$) (Appendix 29). It is also assumed that there are no horizon-

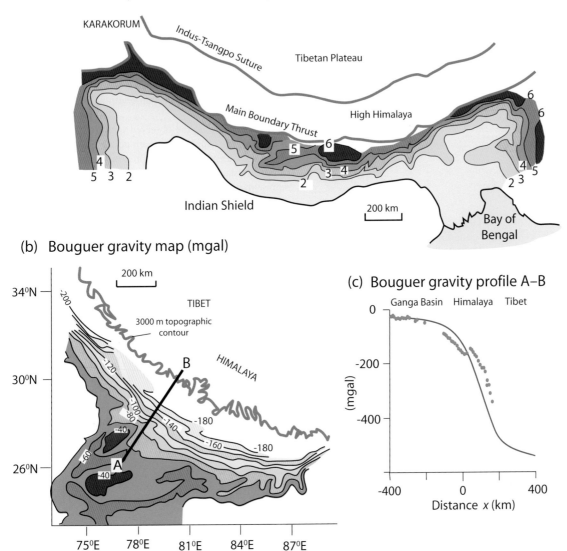

(a) Himalayan foreland basin isopachs (km)

KARAKORUM

Indus-Tsangpo Suture

Tibetan Plateau

Main Boundary Thrust

High Himalaya

Indian Shield

200 km

Bay of Bengal

(b) Bouguer gravity map (mgal)

200 km

34°N

TIBET

3000 m topographic contour

B

HIMALAYA

30°N

-200

-120

-100

-80

-40

-180

-140

-160

-180

26°N

-60

A

-40

75°E   78°E   81°E   84°E   87°E

(c) Bouguer gravity profile A–B

Ganga Basin   Himalaya   Tibet

0

-200

(mgal)

-400

-400   0   400

Distance *x* (km)

Fig. 4.6 The Himalaya and Ganga Basin. (a) Isopach map of sediment thicknesses in the Himalayan foreland basin, based on Raiverman *et al.* (1983). (b) Contours of Bouguer gravity anomaly over the Ganga Basin. (c) Observed Bouguer gravity anomalies along profile A–B, compared with the anomalies computed assuming the topography is locally compensated (Airy model) by thickening of the crust beneath central India. Note that there is an apparent mass excess in the Himalaya and a mass deficit over the Ganga Basin (after Lyon-Caen & Molnar 1985). (b) and (c) reproduced with permission of John Wiley & Sons, Inc.

tal applied loads ($P = 0$), in which case, following Appendix 5, the general flexural equation is

$$D\frac{d^4w}{dx^4} + \Delta\rho gw = 0 \tag{4.1}$$

where w is the deflection, x is the horizontal scale, D is the flexural rigidity, and $\Delta\rho$ is the difference in density between mantle ρ_m and infilling material ρ_i.

The general solution for a fourth-order differential equation such as this requires four boundary conditions. For one side of a symmetrical deflection (Fig. 4.7) we already have

$V = V_0/2$ at $x = 0$, and $V = 0$ at $x > 0$

$w \rightarrow 0$ as $x \rightarrow \infty$

We can add a third boundary condition for a continuous (unbroken) plate:

$dw/dx = 0$ at $x = 0$

and as a fourth we use a vertical force balance between the vertical force ($V_0/2$ for half of the symmetrically flexed plate) and the upward restoring force of the mantle, which gives

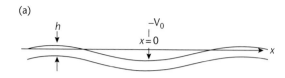

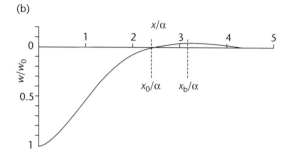

Fig. 4.7 (a) Deflection of a continuous elastic plate under a line load. (b) Theoretical deflection of the elastic lithosphere under a line load applied at the centre of an infinitely extensive plate. Parameters are defined in the text. The deflection w is scaled against the maximum deflection w_0. The horizontal distances are scaled against the flexural parameter α. Bending moments and horizontal in-plane forces are zero. From Turcotte & Schubert (2002), © Cambridge University Press, 2002.

$$\frac{V_o}{2} = (\rho_m - \rho_i) g \int_0^\infty w(x)\,dx \qquad [4.2]$$

The solution using these boundary conditions is

$$w(x/\alpha) = w_{max} \exp(-x/\alpha)(\cos x/\alpha + \sin x/\alpha) \qquad [4.3]$$

where the characteristic flexural length scale (or *flexural parameter* of Walcott 1970), is given by

$$\alpha = \left\{ \frac{4D}{\Delta \rho g} \right\}^{1/4} \qquad [4.4]$$

and the maximum deflection in terms of the vertical load is

$$w_{max} = \frac{V_0}{2} \frac{1}{\Delta \rho g \alpha} \qquad [4.5]$$

From this solution, several useful and simple expressions emerge for the geometry of the deflection (Fig. 4.7).

The half-width of the depression (x_0) can be found since it is defined by the horizontal distance from the maximum deflection ($x = 0$) to the point where the deflection is zero ($w = 0$). Setting $w = 0$, eqn. [4.3] gives

$$x_0 = \frac{3\pi\alpha}{4} \qquad [4.6]$$

The distance from the line load ($x = 0$) to the highest part of the forebulge (x_b) can be found since at the forebulge crest, the slope of the deflection is zero. That is, $dw/dx = 0$ at $x = x_b$. Eqn. [4.3] then gives

$$x_b = \pi\alpha \qquad [4.7]$$

The height of the forebulge w_b above the datum of zero deflection can also be found using the condition that $x = \pi\alpha$ at the point where $w = w_b$. As a result, eqn. [4.3] reduces to

$$w_b = -w_0 \exp(-\pi) = -0.0432\, w_0 \qquad [4.8]$$

If the deflection under a line load approximates a sedimentary basin, the half-width of the basin is represented by x_0 if base level is taken as $y = 0$. However, the half-width enclosed by the deflection is represented by x_b.

By comparison of the observed with the theoretical bathymetric or free-air gravity profiles, something can be said about the thickness of the equivalent elastic lithosphere (defined in §2.1.4) under the Hawaiian Islands, always assuming the theory to be an adequate representation of the geodynamics (Watts & Cochran 1974). For example, the bathymetric profile suggests that the crest of the outer rise is about 250 km from the line load of the Hawaiian Islands, i.e. $x_b = 250$ km. Assuming the moat to be water-filled, $\Delta\rho = 2300\, \text{kg m}^{-3}$. With a gravitational acceleration of $10\, \text{m s}^{-2}$, the flexural parameter from eqn. [4.4] is 80 km, which gives a flexural rigidity of 2.4×10^{23} Nm. If Young's modulus E is 70 GPa and Poisson's ratio $v = 0.25$ this gives an equivalent elastic thickness Te of 34 km.

4.2.2 Deflection of a broken plate under a line load

If the Hawaiian plate were broken under the Hawaiian Islands, the boundary conditions would need to be modified (Walcott 1970). In this case, we would consider the deflection of a semi-infinite elastic plate subjected to a line load applied at its end (Fig. 4.8) (Appendix 29). Assuming that no external torque is applied at $x = 0$, simple expressions describing the geometry of the deflection can be obtained as follows.

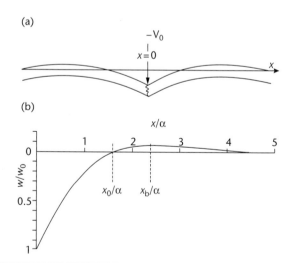

Fig. 4.8 (a) Deflection of a broken elastic plate under a line load applied at its end. Horizontal in-plane forces and applied bending moments are again zero. (b) Theoretical deflection of the broken plate under a line load applied at its end. Note that the half-width of the basin is narrower than for the unbroken plate, and that the elevation of the forebulge is greater than for the unbroken plate. Vertical and horizontal axes are scaled by w_0 and flexural parameter α respectively. From Turcotte & Schubert (2002), © Cambridge University Press, 2002.

The maximum deflection for a broken plate of the same flexural rigidity and under the same vertical load is twice that of an unbroken plate (Turcotte & Schubert 2002, p. 126). If the maximum deflection is known, the deflection as a function of x for a broken plate loaded at its end is given by (Turcotte & Schubert 2002, p. 127–129)

$$w(x) = w_0 \exp(-x/\alpha)\cos(x/\alpha) \qquad [4.9]$$

where the maximum deflection is given by

$$w_{max} = \frac{V_0}{\Delta \rho g \alpha} \qquad [4.10]$$

The half-width of the basin (at $w = 0$) is

$$x_0 = \frac{\pi \alpha}{2} \qquad [4.11]$$

showing that the basin is narrower for the case of a broken plate. The distance to the crest of the forebulge (where $dw/dx = 0$) is

$$x_b = \frac{3\pi \alpha}{4} \qquad [4.12]$$

showing that narrower forebulges characterise broken plates. Finally, the height of the forebulge (where $x_b = 3\pi\alpha/4$) is given by

$$w_b = w_0 \exp(-3\pi/4)\cos(3\pi/4) = -0.067 w_0 \qquad [4.13]$$

indicating a considerably larger forebulge amplitude for a broken plate.

Returning to the case of the Hawaiian Islands with $x_b = 250$ km, eqn. [4.12] gives 106 km for the flexural parameter, and flexural rigidity is therefore 7.26×10^{23} Nm. The equivalent elastic thickness with $E = 70$ GPa and $v = 0.25$ is 49 km.

Seismic refraction studies (Shor & Pollard 1964) suggest that the Moho is deflected downwards by approximately 10 km under the centre of the Hawaiian Islands. If 10 km is the maximum deflection w_0, the height of the outer rise above the undeflected seafloor can be found from eqn. [4.8] for the unbroken plate and eqn. [4.13] for the broken plate. The results are 432 m and 670 m respectively. Measurements of the bathymetry of the seafloor surrounding the Hawaiian Islands indicate that the outer rise is elevated above regional by about 500 m (Chase *et al.* 1970), but this does not permit us to tell whether the continuous or broken plate best fits the observational data.

A good fit was found between the observed free-air gravity anomalies across the Hawaiian-Emperor seamount chain and those calculated for the continuous elastic plate if $D = 5 \times 10^{22}$ Nm and for the broken elastic plate if $D = 2 \times 10^{23}$ Nm. These values are close to the flexural rigidities estimated from the wavelength of the deflection given above.

Another geological situation is of the bending of oceanic lithosphere at arc-trenches. This configuration is similar to that of an end load on a broken plate, but in this case we cannot ignore the bending moments acting on the subducting plate. Subducting oceanic slabs are subject to a large number of forces. There is a large force acting downwards due to the negative buoyancy of the cold oceanic lithosphere. This may be enhanced at depths of 200 to 300 km where the olivine to spinel phase change takes place. The oceanic slab also experiences resistance to its downward motion at its tip, which may have a horizontal and rotational (torque) component. It also experi-

ences resistance along its upper contact with the overriding plate, which also has a horizontal component, as has been suggested for the Taranaki-Wanganui region of New Zealand, where the Pacific plate becomes 'locked' against the Australian plate (Stern *et al.* 1992).

It is possible to obtain some estimates of the geometrical aspects of the flexural depression by expressing the trench profile in terms of the height of the forebulge (w_b) and the half-width of the forebulge ($x_b - x_0$) (Fig. 4.9). The solution for the latter is

$$x_b - x_0 = \pi\alpha/4 \qquad [4.14]$$

In other words, the half-width of the forebulge is a direct measure of the flexural parameter. A forebulge is well developed seaward of deep-sea trenches in the northwestern Pacific (Watts & Talwani 1974). The observed bathymetric profile across the Mariana Trench (Fig. 4.5) suggests that the half-width of the forebulge (measured on the flank facing the trench) is about 55 km. Eqn. [4.14] then gives 70 km for the flexural parameter. Using $E = 70$ GPa, $v = 0.25$, and $\Delta\rho = 2300$ kg m^{-3}, from eqn. [4.4] the flexural rigidity of the Pacific plate at the Marianas trench using the broken plate model is 1.35×10^{23} Nm, equivalent to an elastic thickness of 28 km.

Both the deflection w and the height of the forebulge w_b are functions of the flexural parameter and bending moment. Dividing the expression for w by the expression for w_b (i.e. w/w_b) eliminates the unknown parameters and gives a universal flexure profile of the type shown in Fig. 4.9. This universal profile applies to any two-dimensional elastic flexure under end loading, so can be applied to a variety of geological contexts.

As an example of its application, let us return to the case of oceanic trenches, such as the Mariana, Bonin and Central Aleutian trenches (Fig. 4.5). For the Mariana Trench, taking horizontal distances from the point $x = 0$ at the oceanward intersection of the plate surface with sea level, the distance to the crest of the forebulge is approximately 55 km and its height is taken as 0.5 km (Fig. 4.5a). The three trenches are shown in terms of dimensionless depth versus dimensional horizontal distance in Fig. 4.5b. Although the fit between the theoretical deflection and the observed bathymetry is not perfect, particularly where volcanoes have built cones in the forebulge region, the correspondence is generally good, suggesting that the lithosphere at ocean trenches can indeed be modelled in this way.

The same approach can initially be taken to the flexure of the continental lithosphere. As in the case of the oceanic lithosphere, the universal flexural profile of Fig. 4.9 should be applicable, but since regions of continental collision are characterised by mountains acting as source areas for large volumes of detritus, basins occupying the flexural depressions tend to be rapidly filled with sediment. As a result, $\Delta\rho$ in the flexural parameter (eqn. [4.4]) is the difference between mantle and sediment densities, $\rho_m - \rho_s$.

4.2.3 Deflection of a continuous plate under a distributed load

Flexural basins on the continental lithosphere, such as foreland basins, are generally loaded by a force system that has a spatial distribution far removed from a line-load approximation. Consequently, it is necessary to consider the solution of the general flexural equation for the case of an applied load $q_a(x)$. A distributed load can be treated as a series of discrete point loads, and the sum of the solutions for the individual point loads superposed to obtain a solution for the

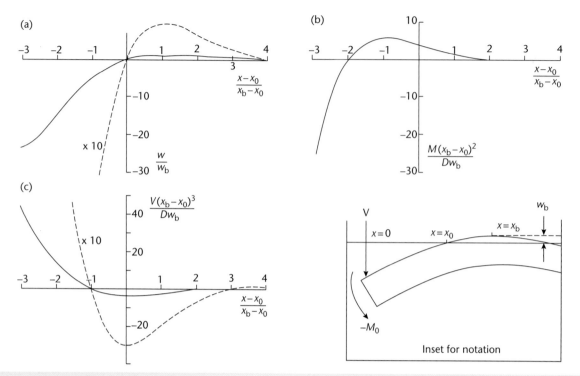

Fig. 4.9   Universal solution for the deflection of an elastic lithosphere under a vertical end load and a bending moment, approximating to the situation at ocean trenches. (a) Dependence of the non-dimensional displacement w/w_b on the non-dimensional position $(x - x_0)/(x_b - x_0)$. Dashed line shows vertical exaggeration of 10:1 to show geometry of forebulge. (b) The non-dimensional bending moment $(M(x_b - x_0)^2/Dw_b)$ versus non-dimensional position. Note that the maximum bending moment is found roughly one third of the distance from $x = x_0$ to the maximum deflection. (c) The non-dimensional vertical shear force $(V(x_b - x_0)^3/Dw_b)$ as a function of non-dimensional position, showing that it reaches a maximum at the point of maximum deflection. Dashed line is at vertical exaggeration of 10:1. From Turcotte & Schubert (2002), © Cambridge University Press, 2002. Inset shows notation.

distributed load (Wangen 2010, p. 293). A semi-numerical solution for the total deflection from a series of discrete point loads is provided in Appendix 33. The total deflection $w(x)$ is then the sum of the deflections from $i = N$ point loads located with their centres at $x_{c,i}$

$$w(x) = \sum_{i=1}^{N} \frac{V_i}{2\Delta\rho g\alpha} f(|x - x_{c,i}|/\alpha) \qquad [4.15]$$

where the function f is given by

$$f(|x - x_{c,i}|/\alpha) = \exp(-|x - x_{c,i}|/\alpha)$$
$$(\cos\{|x - x_{c,i}|/\alpha\} + \sin\{|x - x_{c,i}|/\alpha\}). \qquad [4.16]$$

$V_i/2\Delta\rho g\alpha$ is the ith vertically acting load counteracted by the upward restoring force from the mantle, and α is the flexural parameter (Fig. 4.10).

Jordan (1981) made an early attempt to forward model the deflection of the continental lithosphere in the western interior of the USA during the Late Jurassic–Cretaceous Sevier orogeny using a distributed load system on a linear elastic plate. The crustal loads were approximated by rectangular blocks of density ρ_c, with their centres s situated at a distance x km, half-width a, and height above sea level h (Fig. 4.10) (Appendix 32). These crustal blocks float on a fluid mantle with density ρ_m. The deflection of each load block w_i is calculated using a solution of the universal flexure equation. The sum

of the deflections from all of the rectangular crustal blocks gives the total deflection. Jordan (1981) moved the crustal blocks to represent the tectonic load of a fold-thrust belt at different stages in the evolution of the Sevier orogen. This type of two-dimensional modelling has been applied to the Tertiary Bermejo foreland basin of Argentina (Cardozo & Jordan 2001) and the Triassic Sichuan foreland basin of China (Yong *et al.* 2003).

Since the total deflection is the superposition of the deflections from discrete point loads, the extent of mutually constructive and destructive interference of the discrete deflections depends on the wavelength of the distributed load. The effect of the wavelength of a periodic load on flexural compensation is discussed in §2.1.5. The concept is identical to the case of a distributed system of discrete point loads. As the spatial distribution becomes more spread out, the total deflection approaches that expected from an Airy isostatic balance.

4.2.4   Bending stresses

The flexure of an elastic plate causes bending stresses acting normally to an imaginary cross section of the plate σ_{xx}. We know from a consideration of linear elasticity (§2.1.4 and Appendix 5) that the magnitude of the longitudinal bending stress depends on the curvature of the plate (or radius of curvature R), which in turn depends on the applied force and its material properties

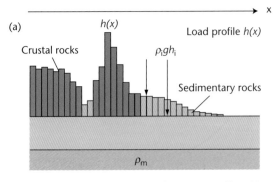

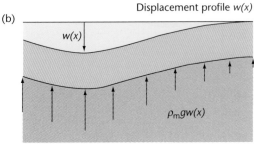

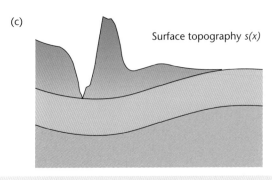

Fig. 4.10 Flexure under a distributed load. (a) The individual deflections caused by load blocks of given height *h*, density ρ and width δx are solved analytically, and summed to give the displacement profile *w(x)* in (b). The upward restoring force is shown by arrows. The distributed load then 'sinks' into this displacement profile to give the resultant topographic profile *s(x)* (c). Based on Jordan (1981), Cardozo & Jordan (2001) and Yong et al. (2003).

$$\sigma_{xx} = E\frac{y}{R} \qquad [4.17]$$

where *E* is Young's modulus and *y* is the distance from the mid-plane of the plate. As a reminder, the integration of the bending stresses over the cross-section gives the bending moment *M*, and the bending moment is related to the curvature by a coefficient, the flexural rigidity *D*. Longitudinal bending stresses, or fibre stresses, may be large enough to cause faulting of the outer arc of a flexed plate (below).

The solutions for the bending stress of a plate undergoing flexure are given in Appendix 38. The maximum fibre stress is found around the outer arc of the plate, and for a broken plate loaded at its end is given by

$$\sigma_{xx} = -6M/h^2 \qquad [4.18]$$

where *h* is equivalent to the effective elastic thickness of the plate T_e. A negative fibre stress denotes tension. The bending moment in the region of the forebulge M_b, at low confining pressures, can be related to parameters that can be easily measured, such as those describing the overall shape of the deflection (e.g. w_b, x_o and x_b), as follows:

$$M_b = -\left(\frac{\pi^2}{8}\right)\left\{D\frac{w_b}{(x_b - x_0)^2}\right\} \qquad [4.19]$$

where *D* is the flexural rigidity ($= Eh^3/12(1 - v^2)$), w_b is the maximum upward deflection of the forebulge, x_0 is the position of the first node, x_b is the position of the forebulge crest, and $(x_b - x_0)$ is the width of the inward facing flank of the forebulge. Using typical parameter values, the tensile stress around the outer arc of a flexed plate is -100 MPa – well beyond the brittle strength of carbonate rocks near the surface (c. -30 MPa, Paterson 1958).

The bending stresses on the upper surface of a flexed continuous plate show a maximum compressive stress at $x = \pi\alpha/4$ and a maximum tensile stress at $x = \pi\alpha/2$ (Fig. 4.11). Since the first node is at $x = 3\pi\alpha/4$, this implies that the outer part of the basin and the forebulge are in tension. For a broken plate, the entire basin and forebulge are in tension, with a maximum tensile stress at $x = \pi\alpha/4$. This may explain the occurrence of extensional faults of small displacement on the basinward-facing slope of foreland basins imaged on seismic reflection profiles (Bradley & Kidd 1991), and the conjugate sets of normal faults and joints cutting carbonates of the underlying passive margin megasequence in the Apulian foreland of the Apennines of northern and central Italy (Billi & Salvini 2003). Flexural bending stresses may also be important in the reactivation of pre-existing faults formed in the passive margin phase.

4.3 Flexural rigidity of oceanic and continental lithosphere

4.3.1 Controls on the flexural rigidity of oceanic lithosphere

The lithosphere can be regarded as a thermal boundary layer losing heat to the atmosphere and oceans by conduction. The oceanic lithosphere thickens as a function of age, being ~6 km at ridge axes and thickening to about 100 km under the oldest (Jurassic) ocean floor (§2.2.7). The ability of the oceanic lithosphere to support loads should also therefore be a function of its age. However, analysis of the flexure of the oceanic lithosphere at the Hawaiian-Emperor seamount chain (Watts 1978) suggested that equivalent elastic thicknesses were large at the old (75–80 Ma) end of the chain near the northernmost Emperor seamount, but low at the young (0–5 Ma) end near Hawaii, despite the fact that the age of the underlying oceanic crust (80–100 Ma) differs very little along the length of the chain. In other words, volcanoes deflected oceanic lithosphere of different ages *at the time of loading*.

A plot of 139 estimates of the elastic thickness of the oceanic lithosphere against age of the lithosphere at the time of loading (Fig. 4.12) (Watts 2001, p. 242) shows that oceanic flexural rigidity expressed as an equivalent elastic thickness (Te) is approximated by the depth to the 300–600 °C isotherm, with a central isotherm of 450 °C (Karner et al. 1983). This relationship is exponential with time, similar to that of the oceanic bathymetry using a cooling plate

(a)

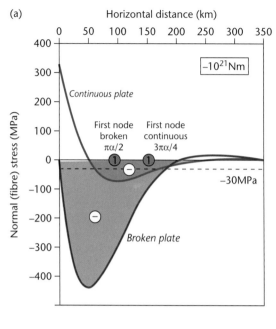

(b)

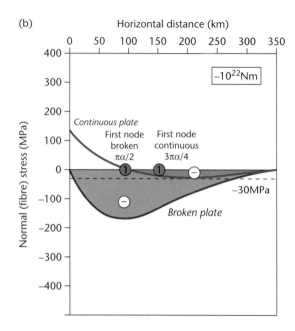

(c)

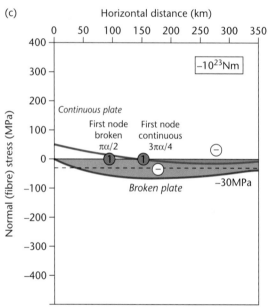

Fig. 4.11   The distribution of fibre stresses (normal stresses due to bending) for a continuous plate (red) and a broken plate (blue) for flexural rigidities of (a) 10^{21} Nm, (b) 10^{22} Nm and (c) 10^{23} Nm. A vertical load is applied at $x = 0$. For the continuous plate, the maximum tensile stress (negative by convention) is situated at $\pi\alpha/2$, compared with the position of the first node at $3\pi\alpha/4$, so the maximum tensile stress is in the outer part of the basin, where the plate dips down towards the maximum deflection. This is where we would expect to see down-to-the-basin normal faults, or normal reactivation of older structures. For the broken plate, the tensile stress is at a maximum at $\pi\alpha/4$ whereas the first node is at $\pi\alpha/2$, so the maximum tensile stress is located at half way between the maximum deflection and the first node. The dashed line shows the yield strength of 30 MPa, applicable to carbonates in the bulge region (that is, at the surface), showing that yield strengths are likely to be overcome by bending stresses, especially for weak, highly curved plates.

model such as that shown in Fig. 2.27 (Parsons & Sclater 1977). When depth of earthquake occurrence is plotted against age of oceanic lithosphere it is evident that seismic activity is also confined to lithosphere cooler than 600 °C (Maggi *et al.* 2000) (Fig. 4.13), strongly supporting the view that the elastic portion of the ocean lithosphere is defined by a characteristic isotherm.

Although investigation of the oceanic lithosphere allows us to lay down some fundamental concepts in flexure, it is of little importance in basin analysis, since most sedimentary basins of interest are formed on continental lithosphere. It is therefore important to understand what controls the flexural rigidity of continental lithosphere.

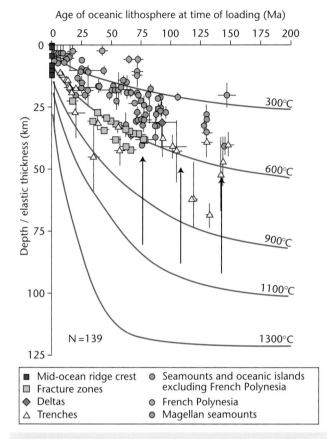

Age of oceanic lithosphere at time of loading (Ma)

Legend:
- ■ Mid-ocean ridge crest
- □ Fracture zones
- ◆ Deltas
- △ Trenches
- ● Seamounts and oceanic islands excluding French Polynesia
- ● French Polynesia
- ● Magellan seamounts

Fig. 4.12 Plot of the flexural rigidity of the oceanic lithosphere (expressed as equivalent elastic thickness Te), as a function of its age at the time of loading (after Watts *et al.* 1982, updated in Watts 2001; © Cambridge University Press). The equivalent elastic thickness estimates are shown for mid-ocean ridges and fracture zones, seamounts and oceanic islands, trenches and deltas. Isotherms are for a cooling plate model (Parsons & Sclater 1977), showing that most data fall within the range 300°C to 600°C. The general relationship is that equivalent elastic thickness increases with the age of the oceanic lithosphere at the time of loading.

4.3.2 Flexure of the continental lithosphere

The continental elastic lithosphere responds differently to the oceanic elastic lithosphere at a range of scales of deformation. The continents appear to accumulate strains over long periods of geological time, with wide plate boundary zones of diffuse deformation such as found in the Himalayas and Tibet, whereas the oceanic lithosphere remains relatively intact over its short lifetime of up to about 180 Myr, with narrow, well-defined plate boundary zones (review in Molnar 1988). The continents are consequently exceedingly complex, particularly in their inherited fabric and strength profiles (§2.3) (Vauchez *et al.* 1998). The flexure of continental lithosphere is therefore controversial. There are no simple measurable parameters that can be confidently correlated with the observed flexural rigidity of the continental lithosphere (McNutt *et al.* 1988). The change through time ('secular evolution') of the flexural rigidity of continental lithosphere has in particular been widely discussed.

The elastic plate thickness (or flexural rigidity) of continental lithosphere has been calculated in a wide variety of tectonic settings. Flexural rigidities have been estimated from late glacial rebound (Crittenden 1963; McConnell 1968; Walcott 1970; Nakiboglu & Lambeck 1983), from regions of lithospheric extension (Banks & Swain 1978; Bechtel *et al.* 1987; Weissel & Karner 1989; Ebinger *et al.* 1991; Kusznir *et al.* 1991; Upcott *et al.* 1996), at passive continental margins (Watts 1988; Keen & Dehler 1997), and especially from the flexure of the continental lithosphere at foreland basins (Stewart & Watts 1997) (Table 4.1).

A plot of elastic thickness of the continental lithosphere versus thermal age, or versus the age at the time of loading (Watts 2001), shows no global relationship (Fig. 4.14). Values appear to range between <5 km and 110 km, with no clear modes at particular depth ranges, although rift provinces tend to have values of Te of <30 km. There must be other factors contributing strongly to the value of the flexural strength of the lithosphere. One factor is the present-day geothermal gradient and surface heat flow (Artemieva 2006), which are strongly influenced by the concentration of radiogenic HPEs in the crust (Willett *et al.* 1985; Pinet *et al.* 1991) (Fig. 4.15). If continental Te is determined by the depth to a characteristic isotherm, variations in the continental geotherm would explain variations in the value of continental Te. However, observations on the depth to the base of the elastic lithosphere (Willett *et al.* 1985) imply a very wide spread of temperature from 300 to 900°C, and Audet and Bürgmann (2011) found no relationship between surface heat flow and elastic thickness calculated from the spectral coherence of gravity with topography (Fig. 4.16).

Other mechanisms affecting the flexure of the continental lithosphere are the possible *decoupling* of a strong upper crust from a strong underlying mantle during plate bending (Fig. 4.17) (Burov & Diament 1995; Lavier & Steckler 1997). Where the lithosphere is young and hot, it is prone to decoupling, thereby explaining the low Te in regions such as the Apennines and Alps. Where the lithosphere is old and cold, the lithosphere does not decouple, which explains the high Te in regions such as the Himalayas. Lavier and Steckler (1997) stress the role of sediment blanketing in raising temperatures high enough to cause weakening, yielding and therefore decoupling.

The concept of a yield strength envelope (§2.3.4) suggests that if stresses caused by loading exceed the strength of the lithosphere, it will *yield* rather than flex elastically. The yielding of a flexed plate should be related to its curvature (Goetze & Evans 1979). Initially, as curvature increases (beyond c.10^{-7} m^{-1}) the upper and lower parts of the plate should yield, thereby reducing its elastic thickness. Measurements at ocean trenches indicate curvatures from 1×10^{-6} to 8×10^{-7} m^{-1}, suggesting that some yielding should be taking place (McNutt 1984).

A correlation has been proposed between the elastic plate thickness and the surface curvature in map view of the thrust belts of mountain ranges (McNutt & Kogan 1987). Highly arcuate mountain belts are associated with low elastic thicknesses (Alps, Carpathians) whereas long linear belts tend to be supported by a strong plate (Urals, Appalachians). The highly arcuate belts are commonly associated with backarc extension (e.g. Pannonian Basin), itself related to the dip of the subducting slab (Uyeda & Kanamori 1979). The plan view shape is related to the cross-sectional curvature of the flexed plate; steeply dipping highly curved plates are flexurally weaker, and shallowly dipping plates are stronger (McNutt & Kogan 1987). This implies that the equivalent elastic thickness is not so

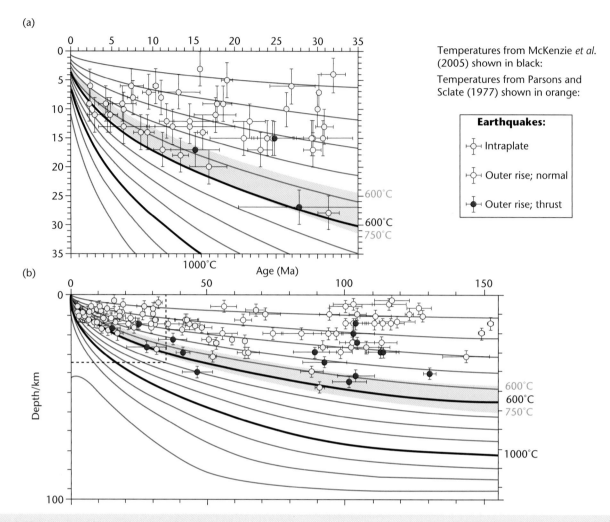

(a)

Temperatures from McKenzie *et al.* (2005) shown in black:

Temperatures from Parsons and Sclate (1977) shown in orange:

Earthquakes:

⊢◌⊣ Intraplate

⊢◌⊣ Outer rise; normal

⊢●⊣ Outer rise; thrust

(b)

Fig. 4.13 Depth of occurrence of earthquakes in the ocean lithosphere, from Maggi *et al.* (2000). The depths of earthquakes are shown in relation to the temperature of their occurrence based on the cooling models of McKenzie *et al.* (2005) and Parsons & Sclater (1977). Almost all earthquakes take place at temperatures of less than 600 °C. Shaded area shows temperature range of 600–750 °C.

much related to a characteristic isotherm, but is more due to the extent to which the elastic lithosphere has flexed under high bending stresses.

It could be argued that the curvature of thrust belts is more strongly influenced by variations in the mechanical properties of the stratigraphy and shallow basement being deformed, rather than scaling on the rigidity of the underlying plate. Sinclair (1996) therefore plotted equivalent elastic thickness estimates versus the curvature of the foreland basin's outer margin using carefully screened examples (see discussion in Zweigel & Zweigel 1998). A linear relationship was found, supporting the idea that weak plates correlate with arcuate mountain belts and arcuate foreland basins (McNutt *et al.* 1988).

There is no convincing evidence that the continental lithosphere underlying mountain belts and their foreland basins experiences long time scale viscoelastic relaxation (McNutt *et al.* 1988; cf. Beaumont 1981; Willett *et al.* 1985). Field observations supporting interpretations of viscoelastic behaviour with long relaxation times (>10 Myr)

are based on migration of forebulge positions and proximal foreland basin deepening (Quinlan & Beaumont 1984; Tankard 1986). These features can be explained by other mechanisms on an elastic plate (§4.6.5).

The age of the continental lithosphere at the time of loading appears to set an upper bound to the maximum possible equivalent elastic thickness that can be observed. However, many Te values fall well below this level, suggesting that some weakening has taken place, but regardless of plate age. As introduced above, this mechanism of weakening may be related to *decoupling* or to *yielding*, reducing the effective elastic thickness from the purely elastic case by a factor of up to 2.

Values of flexural rigidity can be estimated from gravity data and its spectral properties (Lowry & Smith 1994; Simons *et al.* 2000; Kirby & Swain 2009; Perez-Gussinye *et al.* 2004, 2009; Audet & Bürgmann 2011). In some cases, gravity data indicate significant spatial variations in flexural rigidity. For example, the foreland of the Alpine chain of Europe shows strong variations in the estimated flexural

Table 4.1 Estimates of equivalent elastic thickness (Te) of the continental lithosphere at sites of mountain belts and foreland basins, compiled from Garcia-Castellanos and Cloetingh (2012, Table 8.1, pp. 158–160) and Watts (2001, Table 6.2, pp. 251–253)

Region	Load/orogen	Basin	Te (km)	Error (±km)
South Iberia	Betics W	Guadalquivir	10.0	5.0
South Iberia	Betics E	Guadalquivir	2.5	2.5
Central Iberia	Iberian Central System	Duero/Tajo/Madrid	7.0	5.0
North Iberia	Pyrenees, Iberian Chain, Coastal Catalan Range	Ebro Basin	22.5	10.0
Western European Platform	Pyrenees	Aquitaine Basin	25.5	10.0
Western European Platform	Western Alps	North Alpine Basin	25.0	15.0
Western European Platform	Eastern Alps	North Alpine Basin	35.0	12.0
North Adriatic	Alps/Apennines	Po Basin west	20.0	5.0
North Adriatic	Alps/Apennines	Po Basin east	10.0	8.0
North Adriatic	Apennines	Adriatic	28.0	3.0
Central Adriatic	Apennines	Adriatic	6.5	3.5
Adriatic	Apennines/Dinarides	Apennines-Dinarides	14.0	10.0
South Adriatic	Apennines		28.0	3.0
South Adriatic	Apennines/Hellenides		11.0	8.0
Moesian Platform	South Carpathian	Getic Depression	11.0	2.0
East European Platform	SE Carpathian	East Carpathian Basin	12.0	5.0
East European Platform	Carpathian		35.0	10.0
East European Platform	NE Carpathian	East Carpathian Basin	18.0	7.0
East European Platform	NW Carpathian	West Carpathian Basin	12.0	4.0
East European Platform	East Caucasus	North Caucasian Basin	55.0	15.0
East European Platform	West Caucasus	North Caucasian Basin	70.0	15.0
East European Platform	Urals		75.0	25.0
East European Platform	Urals		120.0	20.0
Scandinavia	Caledonian	Caledonian Basin	87.0	20.0
Scandinavia	Caledonian	Caledonian Basin	39.0	31.0
Central Mongolia			7.0	3.0
South Mongolia	Altai-Gobi		15.0	5.0
West Mongolia			40.0	10.0
Siberian Platform	Verkhoyansk	Verkhoyansk Basin	50.0	10.0
Arabian Shield	Zagros		50.0	25.0
Arabian Shield	Oman Mountains	North Oman Basin	13.0	3.0
Turkmenistan-Iran	Kopet Dag	Trans-Caspian Basin	25.0	5.0
Asia	Karakorum		121.0	10.0
South Tarim	Kunlun	Tarim Basin	44.0	25.0
North Tarim	Tien Shan	Tarim Basin	53.0	20.0
Dzungarian		Junggar Basin	12.5	12.5
Tadjik Depression	Pamir		15.0	5.0
China	Indosinian/Longmenshan	Sichuan Basin	48.0	5.0
Indian Shield	Karakorum		99.0	10.0
Indian Shield	East Himalaya	Ganges	90.0	15.0
Indian Shield	West Himalaya	Ganges	34.0	6.0
Indian Shield	Himalaya	Ganges	25.0	5.0
West Taiwan	Taiwan	West Taiwan	15.0	6.5
East Papua New Guinea			20.0	10.0
West Papua New Guinea			75.0	10.0
New Zealand		Wanganui Basin	17.0	8.0
East Australia	South Island New Zealand		22.5	12.5
North Australia	Banda Arc-Timor		79.1	10.0
Northwest Australia	Banda Arc-Timor		55.0	25.0
East Antarctic			62.0	37.0
South American Platform	Burdwood Bank	South Falkland Basin	13.0	7.0
South American Platform	East Andes (Argentina)		14.0	
South American Platform	Andes (Argentina)	Puna Basin	14.0	
South American Platform	Andes (Argentina)	Puna Basin	20.0	
South American Platform	Andes (Bolivia)	Chaco Basin	30.5	0.5
South American Platform	Andes (Peru)		40.0	15.0
South American Platform	Andes (Ecuador)		25.0	20.0

Continued

Table 4.1 *Continued*

Region	Load/orogen	Basin	Te (km)	Error (±km)
South American Platform	Andes (Colombia)	Llanos Basin	70.0	
South California	North Transverse Ranges		50.0	5.0
South California	South Transverse Ranges		10.0	5.0
North American Platform	Appalachians		105.0	25.0
North American Platform	Appalachians		70.0	15.0
North American Platform	Ouachita orogen	Anadarko Basin (Oklahoma)	50.0	10.0
North American Platform	Ancestral Rocky Mountains	Paradox Basin	25.0	
North American Platform		Idaho-Wyoming Basin	22.0	
North American Platform	Arctic/Alaska	Colville Trough	65.0	5.0
Northwest Africa	East Atlas	Ouarzazate Basin	20.0	1.0
Northwest Africa	West Atlas	Ouarzazate Basin	9.0	1.0

Reproduced with permission of John Wiley & Sons, Ltd.

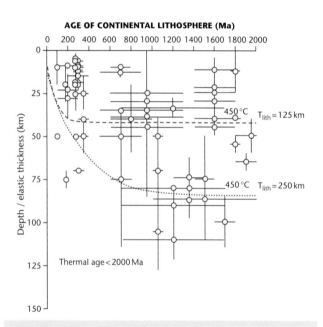

Fig. 4.14 Plots of flexural rigidity expressed as equivalent elastic thickness versus age of the continental lithosphere, for 66 foreland basins and six late glacial rebound sites (after Watts 2001; © Cambridge University Press). Note that there is no correlation for this continental dataset.

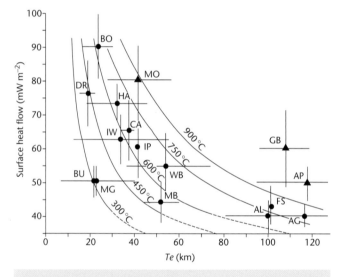

Fig. 4.15 Plot of the equivalent elastic thickness Te versus surface heat flow for a number of sites of flexure. The solid curves indicate depths to isotherms at a given surface heat flow for a steady-state thermal model. The plot demonstrates the very wide range in temperature (300 °C to >900 °C) for the base of the elastic lithosphere. Flexure sites are as follows: AG, Lake Agassiz; AL, Lake Algonquin; FS, Fennoscandia; CA, Caribou Mountains; IP, Interior Plains; GB, Ganges Basin; IW, Idaho–Wyoming thrust belt; MB, Michigan Basin; BU, Boothnia Uplift; MG, Midcontinent gravity high; HA, Lake Hamilton; BO, Lake Bonneville; DR, North Great Dividing Range; AP, Appalachian foreland basin; MO, Molasse Basin; WB, Williston Basin. Full details in Willett *et al.* (1985). Reproduced with permission of Nature Publishing Group.

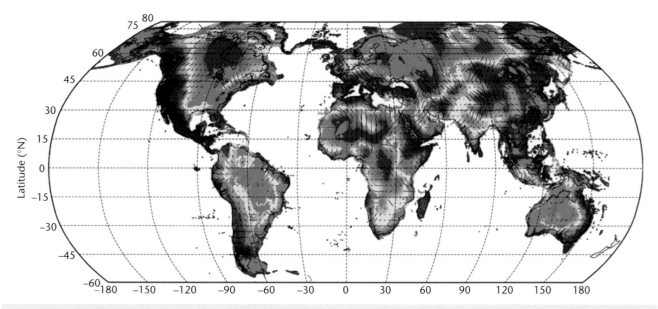

Fig. 4.16 Global map of the elastic thickness of the continental lithosphere calculated from the coherence between Bouguer gravity and topography (from Audet & Bürgmann 2011, fig. 1). Colour scale is from 0 to 150 km for elastic thickness, with blue tones for high Te and red tones for low Te. Small black arrows are direction of anisotropy. Grey areas are where Te estimates are unreliable. Continents have rigid cores with Te > 100 km. Reprinted by permission of Macmillan Publishers Ltd.

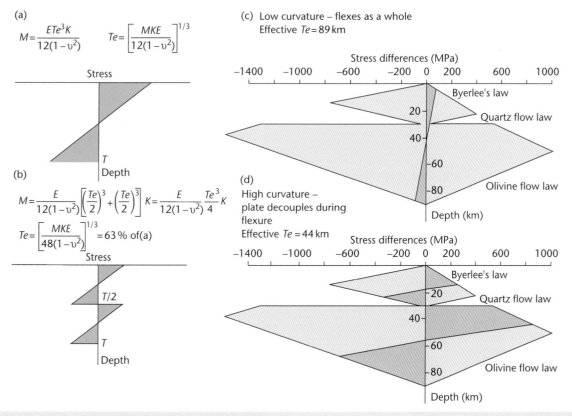

Fig. 4.17 The effects of decoupling at high curvatures illustrated by cross-sections of stress in the flexed lithosphere (after McNutt *et al.* 1988). (a) Distribution of fibre stresses in an elastic plate of thickness T showing extension in the upper half of the plate and compression in the lower half. (b) A purely elastic plate that is decoupled at $T/2$ so that the upper and lower portions flex independently with the same radius of curvature as in (a). The equivalent elastic thickness is 63% of that in (a). M is the bending moment, K the radius of curvature, E Young's modulus, and v Poisson's ratio. (c) and (d) Rheologically layered continental lithosphere showing the failure envelope under extension (positive stress) and compression (negative stress). The strength is limited by frictional sliding (Byerlee's law) in the upper crust and uppermost mantle and by ductile flow in the lower crust and lower lithosphere. The purple-shaded areas represent the stress resulting from flexure superimposed on the failure envelope. In (c) the plate is flexed at a very low curvature. The fibre stresses rarely exceed the failure criterion of the plate (there is a small amount of yielding at the top and bottom) and the equivalent elastic thickness (89 km) is almost exactly the same as the thickness of the elastic plate. In (d) the plate is flexed to a high curvature. This leads to the failure criterion being exceeded, and a decoupling zone forms in the lower crust. The equivalent elastic thickness calculated for the total bending moment sustained in both the crust and the mantle is just 44 km. Decoupling therefore has strong potential mechanical implications for flexure. Reproduced with permission of the American Geophysical Union.

rigidity, with low values in western Switzerland, in close proximity to the Rhine Graben (Stewart & Watts 1997). A reduction of lithospheric strength was attributed to the thermal weakening of the plate in the region of the Rhine Rift.

Spatial variations in the flexural rigidity of the foreland plate may affect its response to moving tectonic loads (Schedl & Wiltschko 1984) during their migration history. The most likely scenario involves the closure of an ocean basin, initially causing the loading of the attenuated continental margin prior to full collision, when normal thickness continental lithosphere is flexed (Stockmal *et al.* 1986). Stewart and Watts (1997) identified such a strike-normal increase in flexural rigidity with distance from the Andean orogenic belt in Ecuador using the Bouguer gravity field. In such a situation, with continued convergence the orogenic wedge should initially deflect a weak lower plate, and then a progressively stronger lower plate. Although this is conceptually reasonable, there is no evidence of an increase in flexural rigidity of the European foreland plate during Alpine orogenesis from the marine (late Eocene–early Oligocene) 'flysch' stage to the predominantly non-marine (late Oligocene–Miocene) 'Molasse' phase (Allen *et al.* 2001).

Plate flexure may be non-cylindrical due to spatial variations in the load and spatial variations in the properties of the flexed plate, including the presence of prominent fault zones segmenting the subducting plate. In the foreland of the Apennines of Italy, there are offsets in the positions of the depocentres and forebulges delimiting distinct slab segments (Royden & Karner 1984; Royden *et al.* 1987) (Fig. 4.18). The stepped deflection is thought to be due to segmentation of the lithosphere within the subduction zone. The Apennine thrust belt mimics the deeper segmentation. In the segment with a forebulge far to the northeast, the thrust belt has also advanced relatively far to the northeast. In contrast, the lesser-travelled thrust units are correlated with forebulges relatively to the southwest.

4.4   Lithospheric buckling and in-plane stress

Long-wavelength folding or *buckling* of the lithosphere has been suggested for the Indian Ocean (Fleitout & Froideveaux 1982), but the wide applicability to the continents is in some doubt. Long-wavelength folding of the continental lithosphere has now been suggested from several continents (central Australia, central Asia, Canada, western Europe) (Lambeck 1983; Cloetingh *et al.* 1999; Fernandez-Lozano *et al.* 2011). The typical wavelengths observed are 50–600 km. If lithospheric buckling of significant amplitude takes place, it is important to recognise it and to discriminate it from the effects of lithospheric flexure due to the predominantly vertically applied loads in mountain belts.

4.4.1   Theory: linear elasticity

We have previously seen that the lithosphere flexes under an applied load system that may include vertically applied distributed loads, torques and horizontal loads (§2.1.4, §4.2). The question has not yet been addressed of whether the lithosphere is able to buckle under large horizontal, in-plane forces. We firstly approach this problem using an elastic beam under horizontal compression, and then extend the analysis to a lithosphere with an upward restoring force.

It is possible to find out if the lithosphere should buckle under an in-plane stress (Turcotte & Schubert 2002, p. 118). We take the following parameter values: Young's modulus $E = 70\,\mathrm{GPa}$, Poisson's

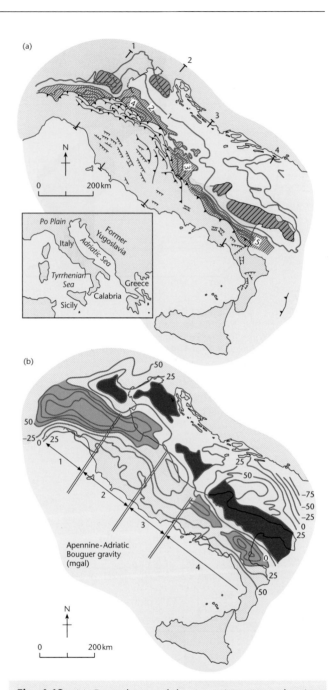

Fig. 4.18   (a) General map of the Apennine system showing depth to the base of the Pliocene in the Adriatic and Po basins at a 1 km contour interval. There are four distinct outer rise/forebulge segments recognised in the basal Pliocene surface (shown in wide diagonal hatch). The narrow diagonal hatch shows the Apennine foreland basin and parts of the basal Pliocene surface that is below 2 km in depth. (b) Simplified map of the Bouguer anomaly gravity field (milligals). Contour interval is 25 mgal. The four morphological outer rises correspond to Bouguer gravity highs, suggesting that they are maintained by regional flexure. The dark shaded areas represent anomalies of greater than 0 mgal in sectors 1 to 3 and 50 mgal in sector 4: light shading shows gravity anomalies less than −50 mgal in sectors 1 to 3 and 0 mgal in sector 4. These sectors are believed to be bounded by major tears in the subducted plate, segmenting it at depth. After Royden *et al.* (1987), compiled from Ogniben *et al.* (1975), Morelli *et al.* (1975) and Bertotti *et al.* (2001).

ratio $v = 0.25$, mantle density $\rho_m = 3300\,kg\,m^{-3}$ and ocean water density $\rho_w = 1000\,kg\,m^{-3}$, and elastic plate thickness $h = 50\,km$. This gives a critical stress for buckling of 5.3 GPa, which is far too high for the brittle part of the lithosphere to withstand without faulting. If we reduce the elastic thickness and cover the elastic lithosphere by a sedimentary layer, thereby reducing the density difference $\Delta\rho$, the critical stress for buckling reduces (by about 75%) to slightly more realistic values. However, the wavelength at such reduced critical stresses is $\ll 50\,km$. This suggests that although thin elastic layers embedded in surrounding rocks may deform by buckling, large-scale, whole-elastic lithosphere folding should not occur in nature. In the following paragraphs we consider the conditions where buckling may take place.

4.4.2 Lithospheric buckling in nature and in numerical experiments

Unrealistically high stresses are required to initiate buckling of the entire elastic lithosphere when it is treated using linear elasticity. Yet observations suggest that long-wavelength buckles are common in the continental lithosphere, and develop early during phases of short-ening in distant mountain belts. Sedimentary basins of central Australia, central Asia, the Himalayan perimeter, and Iberia have all been interpreted in terms of long-wavelength lithospheric buckling (Fernandez-Lozano *et al.* 2011). Basins with a previous extensional history are also believed to have been modified by in-plane stresses causing buckling (Ziegler *et al.* 1995).

Lithospheric folding is largely controlled by rheological and thermal structure. The initial crustal thickness and thermal state of the lithosphere are therefore very important in determining the effectiveness of far-field compressive stresses (Cloetingh & Burov 1996). Consequently, different responses to compressive forces are expected in the relatively warm Alpine foreland of Europe compared to the relatively cold lithosphere of the Russian Platform, because of their different strength profiles. In addition, buckling can be initiated at much lower compressive in-plane stresses with an elastic-ductile rheology than with a linear elastic rheology.

Cloetingh *et al.* (1999) recognised different types of long-wavelength folding (Fig. 4.19):

- *Regular or periodic folding* is inferred from undulations of the 'basement' imaged by seismic reflection and refraction, variations of Bouguer gravity anomalies and periodic vertical tectonic movements. Most cases or regular folding occur in young, weak lithosphere strongly affected by horizontal loads. Long-wavelength warping is occasionally detected by geomorphic observations such as the tilting of river terraces. Numerical experiments suggest a number of different behaviours based on the rheological structure prior to folding. Where a strong upper crust overlies a weak lower crust and mantle, as in very young lithosphere, folds develop along the top and bottom of the upper crust monoharmonically (Fig. 4.19). If the upper mantle is strong, the folds in the upper crust may develop independently from the folds at the top of the upper mantle, producing a decoupled, biharmonic pattern. The short wavelength (30–60 km) corresponds to crustal folds, whereas the longer wavelength (200–350 km) corresponds to mantle folds. Where the combined crust and upper mantle are strong and mechanically welded, as in very old lithosphere, monoharmonic folds once again develop, but with a greater wavelength (>500 km) than in the 'young lithosphere' case.

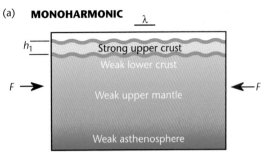

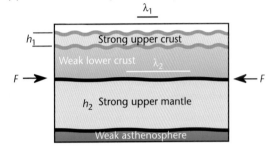

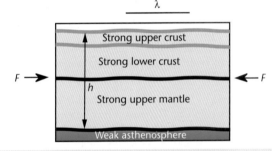

Fig. 4.19 Different styles of periodic buckling by a compressional force F; h_1 and h_2 are the thicknesses of the competent crust and mantle respectively, which fold with wavelengths λ_1 and λ_2. Where the lower crust is very weak, the upper crust and mantle lithosphere fold with different wavelengths ($\lambda_1 < \lambda_2$), producing decoupled or biharmonic folding (b). Where there is a single coherent layer of coupled crust and mantle, as in both young (<150 Ma) and very old (>1000 Ma) lithospheres, monoharmonic folding develops (a and c). After Burov *et al.* (1993) and Cloetingh *et al.* (1999).

- *Irregular or aperiodic folding* generally occurs in very young, weak lithosphere, where the deformation is strongly affected by such factors as erosion, lateral spreading of the sedimentary basin-fill, inhomogeneities in the crust (such as pre-existing sedimentary basins) and non-linear rheological behaviour. Both wavelengths and amplitudes may vary unsystematically along the plate. They may also vary with progressive shortening.

Based on numerical plane strain models using a non-linear elastic–plastic (crust) and temperature-dependent power-law creep, Burg and Podladchikov (2000) predicted that lithospheric buckles, flanked by sedimentary basins, should spontaneously evolve into regions of asymmetric folding and thrusting. For example, the Kashmir and Peshawar basins either side of the Hazara-Kashmir syntaxis in northern Pakistan are thought to have been initiated by buckling. The Miocene to Recent Central and South Adriatic basins, located between the Apennines and the Dinarides of southern Europe, are also thought to be due to lithospheric buckling of the Adriatic crust (Bertotti *et al.* 2001) (Fig. 4.18).

A persistent problem is how buckles are maintained over geological time scales, since in the absence of continuing compression, they should collapse because of the density contrasts between crust and mantle at the Moho. Only very strong crustal rheologies, expected in cold, strong cratons such as Australia or the Russian Platform, would be expected to withstand these pressure differences at the Moho for geological time scales of >20 Myr. In weak lithospheres, the buckles may collapse within 10 Myr.

Since basins due to lithospheric buckling are likely to be found in the foreland of continental collision zones, they may be difficult to discriminate from flexural basins driven by essentially vertically applied load systems in mountains belts. Despite some similarities in wavelength and amplitude, the main discriminating factors between the two types of basin are:

- Flexural foreland basins are very deep (<10 km) close to the thrust front and taper strongly towards the foreland. Beyond the flexural forebulge in the back-bulge region, amplitudes of the deflection are very small (≪200 m). Migration of the orogenic wedge causes major stratigraphic onlap of the pro-foreland plate through time.
- Basins due to lithospheric buckling appear to form within a train of uplifts and depressions extending for ≫1000 km from the plate boundary. Depressions may be 10^2–10^3 m deep, with sediment derived from both margins, producing a symmetrical stratigraphic stacking.

Flexural foredeeps and basins due to lithospheric buckling should not therefore be confused. However, it may be very difficult to discriminate between some back-bulge depositional zones and some basins representing synforms due to lithospheric buckling.

4.4.3 Origin of intraplate stresses

The state of stress of a plate is fundamental to its tectonic deformation, reactivation (inversion) of older structures and basin development. This state of stress results from forces acting along its boundaries and from gravitational body forces due to lateral density and elevation differences. The calculation of a continent's state of stress therefore depends on factors such as the presence of mid-ocean ridges, subduction zones and collisional boundaries, as well as its topography. Estimates of intraplate stress, even for relatively simple plates such as Africa and Indo-Australia, are therefore difficult (Richardson 1992; Coblentz & Sandiford 1994; Coblentz *et al.* 1998). For present-day stress fields, the modelled intraplate stresses can be compared with observations compiled on the World Stress Map (Heidbach *et al.* 2010; www.world-stress-map.org).

In some plates, such as North America and South America, the stress field (orientation of maximum compressive or tensile stress) and focal mechanism solutions of earthquakes suggest that intraplate stresses are dominated by the effect of ridge-push forces. A plate surrounded by mid-ocean ridges should be in compression if ridge-push dominates (Zoback 1992), and the orientation of the maximum compressive stress should be roughly parallel to the direction of relative plate motion. In both Africa and the Indo-Australian plate, the stress of a large part of the plate is affected by the complexities of a convergent margin opposite a mid-ocean ridge, with different seg-

ments involving subduction, strike-slip deformation and collision. Consequently, there is little correlation between the orientation of ridge-push, the relative plate motion and the observed stress field (Richardson 1992; Hillis & Reynolds 2000). In Africa, intraplate stress orientations are also strongly affected by topography, leading to horizontal NE–SW tension in regions such as East Africa and Ethiopia.

The presence of intraplate compression sufficient to cause structural inversions, basin compartmentalisation and long-wavelength buckling therefore depends on the interaction of ridge-push forces driving the plate towards opposite convergent boundaries, mediated by the important effects of lateral density and topographic variations. The effectiveness of intraplate stresses in causing deformation may also depend on the rheological pre-conditioning of the intraplate crust, perhaps influenced by changes in its thermal structure associated with sediment blanketing (Sandiford & Hand 1998; Hand & Sandiford 1999; Sandiford & McLaran 2002) (§2.2.4).

4.5 Orogenic wedges

4.5.1 Introduction to basins at convergent boundaries

The main focus of this chapter is on basins generated primarily by flexure of the continental lithosphere. These foreland basin systems are common in regions of plate convergence, but the 'megasutures' (Bally & Snelson 1980) of convergent boundaries comprise a wide range of tectonic and magmatic products as well as sedimentary basins.

Convergent boundaries can be broadly classified as ocean–ocean, ocean–continental and continent–continent types. Convergent margins involving the subduction of oceanic crust are characterised by magmatic/volcanic arcs. The main components of convergent arc-related systems are, from overridden oceanic plate to overriding plate (Dickinson & Seely 1979) (Fig. 4.20):

- An outer rise on the oceanic plate, recognised as an arch in the abyssal plain. This is the flexural forebulge of the descending oceanic plate.
- A trench or deep trough, commonly >10 km deep, situated oceanward of the arc (review in Underwood & Moore 1995). The bathymetric expression of the trench much depends on the sediment supply into it and, associated with this, the rate of encroachment from the arc of the accretionary wedge. The trench is the bathymetric expression of the flexed oceanic plate.
- A subduction complex composed of tectonic stacks of fragments of oceanic crust, its pelagic cover and arc-derived turbiditic sediments, together with perched or accretionary basins ponded on top of the accretionary wedge. The subduction complex makes up the inner slope of the trench. Where accretion rates are high, the subduction complex may rise to shelf depths or even become emergent.
- A forearc basin between the ridge or terrace formed by the subduction complex and the volcanic arc (review in Dickinson 1995).
- The magmatic arc caused by partial melting of the overriding plate and possibly subducted plate when the latter reaches between 100 and 150 km depth. The volcanism is predominantly andesitic. Small intra-arc basins may form by extensional or strike-slip tectonics or in collapsed calderas (Smith & Landis 1995).

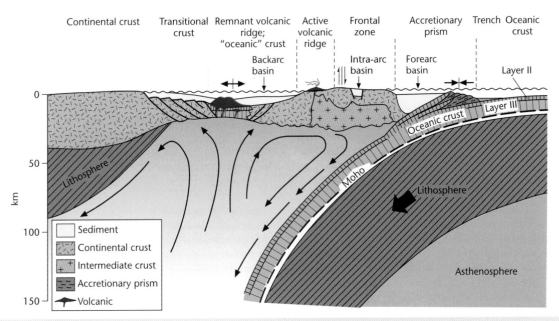

Fig. 4.20 A convergent ocean–arc boundary showing the location of the trench, accretionary wedge, forearc basin, intra-arc and backarc basins, modified from Dickinson & Seeley (1979). The intra-arc of intra-massif basin and the forearc basin both contain deep-marine to non-marine sediments. The accretionary basin and subduction complex contains tectonic slices of abyssal plain, slope and trench deposits, together with ophiolites and metamorphics.

- The backarc region floored by oceanic or continental lithosphere (review in Marsaglia 1995). Where the lithosphere is oceanic, the backarc region typically undergoes extension. Backarc basins are some of the most rapidly extending regions of the Earth's crust today, a prime example being the Aegean Sea of the eastern Mediterranean. Where the lithosphere is continental, as in Andean-type margins, the backarc (or retro-arc) region is typically a zone of flexural subsidence related to major fold-thrust tectonics along the arc boundary.

Some convergent margins consist of oceanic arcs and extensional backarcs and some consist of magmatic arcs on continental crust with active backarc compression (Cordilleran-type) (Molnar & Atwater 1978). These two possibilities are exemplified by the western and eastern margins of the Pacific Ocean respectively (Uyeda & Kanamori 1979; Tamaki & Honza 1991; Stern 2002) (Fig. 4.1).

In the western Pacific the convergent margins are generally subducting oceanic lithosphere of Mesozoic age (i.e. ~100 Ma) and backarc extension is widespread. In the eastern Pacific the subducting lithosphere is much younger (<50 Ma). Here the arcs are located on the overriding continental plates, with fold-thrust belts and retro-arc foreland basins in the backarc region. Since oceanic lithosphere cools and thickens with age, the older Mesozoic lithosphere of the western Pacific should be more gravitationally unstable than the <50 Ma lithosphere of the eastern Pacific. The older lithosphere should therefore subduct more rapidly and may exceed the convergence rates of the plates (see §4.5.2). This should cause an oceanward migration of the subducting hinge together with the forearc elements. This process has been called roll-back (e.g. Dewey 1980). It leads to extension in the region behind the rolled-back forearc, that is, backarc spreading. Others believe that backarc extension is related to secondary mantle

convection above the subducted oceanic plate (Fig. 4.1) (Toksöz & Bird 1977; McKenzie 1978b).

There are three families of arc systems (Dewey 1980): (i) extensional arcs where the velocity of roll-back exceeds the oceanward velocity of the overriding plate, producing backarc extensional basins. As the arc migrates oceanward the forearc region becomes isolated from continental sediment sources and consequently is starved of major sediment supply. Examples are the Mariana and Tonga arcs, eastern Indonesia. (ii) Neutral arcs where there is a balance between the rates of roll-back and oceanward movement of the overriding plate, producing well-developed subduction complexes but no backarc extension. Examples are the Alaska–Aleutian and Sumatran (western Indonesia) arcs. (iii) Compressional arcs where the subducting crust is young and the velocity of roll-back is low. This places the forearc region in compression, causing thrusting in both overriden oceanic and overriding continental crust. Examples are the Canadian–western USA Cordillera and the Peruvian Andes.

Arc behaviour may change through time as the age of the subducted oceanic lithosphere changes. If the age of the subducted oceanic lithosphere gets progressively older, for example, backarc basins may form, then be closed, and margins change from western Pacific type to Cordilleran type.

Suturing of two continental plates produces a complex amalgam of intense structural deformation, regional metamorphism, plutonism and basin formation (see summaries in Dewey 1977; Hsu 1983). The ocean closing process may involve a number of variations in the manner of continent–continent docking: (i) the continental plate may initially collide with an arc-backarc system before continent–continent suturing, causing inversion of extensional backarc basins, and choking of subduction complexes with continental fragments; (ii) the continental plates may compress a number of microplates in the collision zone; and (iii) the collision may be highly irregular or oblique, triggering diachronous orogenic activity and major strike-slip displacements. The lithospheric shortening, thickening, metamorphism and plutonism involved in continental collision are accompanied by the formation of sedimentary basins in a number of settings. Foreland basins form both in front of (pro- or peripheral foreland basins) and behind (retro-foreland basins) the overriding plate. Intermontane basins are found within the megasuture. Extensional and strike-slip basins are located in shear zones produced by 'escape tectonics' from the collision, or as impactogens (Chapter 3).

4.5.2 The velocity field at sites of plate convergence

Subduction of plates is the fundamental driver for plate motion and deformation of the lithosphere. In the presence of water, subduction also causes melting and the building of magmatic arcs. The dynamics of subduction, and the nature of the arc systems at the plate boundary, are controlled by a force balance on the descending plate. This force balance is made up of: (i) *driving forces*, such as the gravitational body forces of the cold slab, and compositional effects from phase changes; and (ii) *resistive forces*, due to the viscous resistance of the mantle. The balance between driving and resistive forces results in a net force (Fig. 4.21).

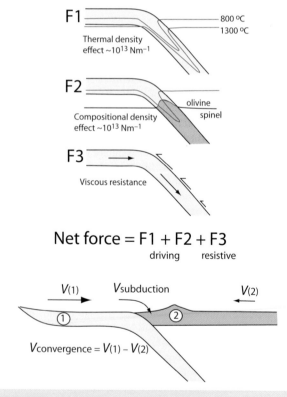

Fig. 4.21 The forces on subducting lithosphere, due to the density effect of negatively buoyant, cold lithosphere (F1), the olivine–spinel phase change (F2), and the viscous resistance exerted along the top surface of the descending slab (F3). The two plates have velocities $V_{(1)}$ and $V_{(2)}$, and Plate 1 subducts at $V_{subduction}$. From Sandiford (http://jaeger.earthsci.unimelb.edu.au/msandifo/Publications/Geodynamics/geodynamics.html).

The velocity of convergence between two plates V_{con} is taken up by a combination of a subduction velocity V_{sub} and internal deformation of the overlying plate V_{def}, which may be extensional (as in the Aegean and the Tyrrhenian Sea) or contractional (as in the Andes). If V_1 is the velocity of the subducting plate, and V_2 is the velocity of the overlying plate

$$V_{con} = V_1 - V_2 = V_{sub} + V_{def} \qquad [4.20]$$

The velocity of subduction results from the action of gravity on the density contrast between the cold slab and the hotter mantle that surrounds it. The thermal density effect and the compositional density effect are usually about the same in magnitude. Subduction velocities are generally 0–$10\,cm\,yr^{-1}$. At velocities higher than this, viscous resistance forces probably act in a negative feedback on descent velocity. The convergence velocity, on the other hand, may be smaller or greater than the subduction velocity, in which case the difference is taken up in internal deformation of the upper plate.

A number of scenarios are possible within this framework (Fig. 4.22):

- Tension results if the convergence velocity is less than the velocity of subduction ($V_{sub} > V_{con}$), causing roll-back, or trench 'suction', as in the Tyrrhenian Sea, Italy.
- Compression results when the convergence velocity is more than the subduction velocity ($V_{sub} < V_{con}$), leading to contractional orogens linked to ocean plate subduction, such as the Andes.
- When the convergence velocity far exceeds the subduction velocity ($V_{sub} \ll V_{con}$), continental collision takes place, as in the Himalayas.

Basin development in regions of convergent tectonics is controlled at the large scale by these relative velocity scenarios.

4.5.3 Critical taper theory

Three possible mechanisms for the driving forces responsible for crustal shortening have been proposed: (i) *gravity sliding*; (ii) *gravity spreading*; and (iii) *horizontal deviatoric push*. Gravity sliding requires a potential energy gradient for sliding under gravity alone. The gravitational force acting on the gradient must overcome the resistance to movement for gravity sliding to occur. Physical models suggest that the angle need only be small (a few degrees) where the effective normal stress on the inclined surface is reduced by the presence of high pore fluid pressures (Hubbert & Ruby 1959). However, field studies, particularly in the Rocky Mountains (e.g. Dahlstrom 1970), suggest that many thrusts dip in the opposite sense to the mass transport direction. Gravity sliding is therefore not likely to be the sole or dominant process in mountain belt tectonics. In the discussion that follows, we therefore concentrate on models involving horizontal deviatoric push and gravitational spreading. The most commonly used model involves a force balance in a tapered orogenic wedge (Fig. 4.23).

Price (1973) believed that orogenic wedges behave as plastic entities as a result of the interfingering of a large number of thrust sheets into a mechanically interdependent system. Gravity acting on an orogenic 'high' would provide the driving force for a spreading along thrust faults. The horizontal shear stress produced by gravity is given by

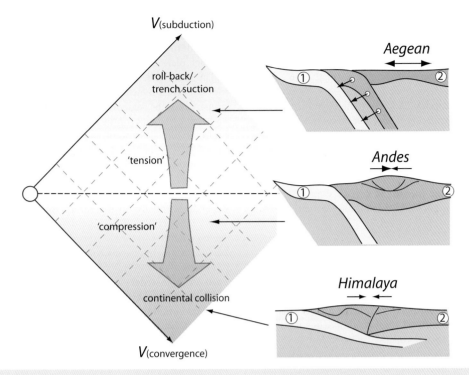

Fig. 4.22 The convergence velocity (V1 – V2) and subduction velocity determine the state of tension and compression in the zone of convergence. High subduction velocities promote subduction zone roll-back, as in the Aegean and Tyrrhenian seas. High convergence velocities promote compressive orogens such as the Himalaya. From Sandiford (http://jaeger.earthsci.unimelb.edu.au/msandifo/Publications/Geodynamics/geodynamics.html).

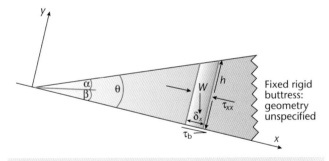

Fig. 4.23 Simplified model of an accretionary wedge (after Platt 1986). α is the surface slope, β is the basal slope and θ is the taper angle ($\alpha + \beta$). A small segment of the wedge of basal length δx and height h is subjected to a body force W, to a basal traction $\tau \delta x$, and to 'push' forces from the rear, produced by the longitudinal deviatoric stresses τ_{xx}. Frontal accretion, underplating, erosion and changes in basal shear stress (for example by changes in the rate of convergence/subduction) place the wedge out of equilibrium. The wedge responds by shortening and thickening or by collapsing by extension.

$$\tau = \rho g h \alpha \qquad [4.21]$$

where ρ = density, g = gravitational acceleration, h = depth below surface, and α = surface slope of wedge (Elliott 1976). However, although this strain system is capable of producing local shortening which must be compensated elsewhere by extension in order to maintain a constant angle of the wedge, it cannot explain the evidence of large-scale net shortening in orogenic belts. Nor is any explanation given for the prior existence of a gravitational 'high' at the onset of spreading.

Horizontal compressional forces (deviatoric 'pushes') exerted along convergent plate boundaries may be primarily responsible for the dynamics of orogenic wedges. The convergent orogen can be regarded as a wedge-shaped prism resting on a rigid slab with a rigid buttress at the rear (Chapple 1978). The wedge behaves as a single mechanically continuous, dynamic unit, with the longitudinal force applied at the rear of the wedge counterbalanced by resistance to sliding on its base. The geometry attained is one of a wedge tapering along its length, with a dynamic balance between the gravitational forces arising from the slope of the wedge, the push from the rear and the basal shear force or 'traction'. This balance can be expressed as follows (Platt 1986):

$$\tau_b = \rho g h \alpha + 2K\theta \qquad [4.22]$$

where the additional term on the right-hand side is due to the horizontal 'push', K is the yield strength of the wedge, and θ is the angle at the front of the wedge (Fig. 4.23). Davis *et al.* (1983) also accounted for the internal strength and pore fluid pressure of the wedge. They introduced the term 'critical taper' to describe the surface slope that would produce a wedge in a state of yield throughout. Davis *et al.* (1983) therefore expressed the basal shear stress in terms of the gravitational force and the material properties of the wedge as follows:

$$\tau_b = \rho g h \alpha + (1-\lambda)K\rho g h \theta \qquad [4.23]$$

where λ is the ratio of pore fluid to lithostatic pressure, and the other terms are essentially the same as in eqn. [4.22].

Different patterns of deformation in the orogenic wedge result from externally imposed changes in its geometry, such as frontal accretion versus basal accretion or underplating (Platt 1986):

- *Frontal accretion* is the accumulation of material at the tip of the wedge, thereby lengthening the wedge. The response, if longitudinal deviatoric stresses are large enough, is internal shortening of the wedge. This may take the form of out-of-sequence thrusting or backthrusting.
- *Underplating* of material to the underside of the wedge, causing the wedge to thicken and increase in surface slope. The wedge may respond by extension, lengthening the wedge and lowering surface slopes.

Other factors influencing the shape of the orogenic wedge are: (i) *erosion* or removal of material from the top of the wedge, which encourages renewed shortening; and (ii) *changes in basal shear stress*: an increase in τ_b, caused, for example, by an increase in the rate of subduction, leads to shortening and thickening; a decrease in τ_b may cause extension (Dahlen 1984). If subduction ceases, τ_b vanishes and the orogenic wedge should collapse by extension at the rear. Platt (1986) explains the uplift of high-pressure rocks to surface positions in the rear of orogenic belts as due to this process of extension (Fig. 4.24).

The implications for basin development of this dynamic model are clear. Variations in rate of subduction, magnitude of deviatoric compression or material properties of the wedge may cause large temporal variations in the load configuration and therefore in the deflection of the plate. In particular, lengthening and contraction of the wedge will have an impact on the position of the forebulge on the downgoing slab relative to the orogenic front and on stratigraphic geometries in the foreland basin-fill.

The critical taper model is limited to cases of purely brittle-frictional (Coulomb) deformation. Many mountain belts, such as the Andes (Isaaks 1988) and Himalayas (Le Pichon *et al.* 1992), have a low-gradient toe area, and a steeper slope leading to a broad plateau region. These topographic features can be explained by using a model including the effects of temperature-dependent ductile creep (Williams *et al.* 1994). The basal surface of the wedge, the décollement zone, may have different rheological properties to the material within the wedge, because of the effects of dynamic crystallisation (Rutter & Brodie 1988). This keeps deformation focused on the décollement. At the toe of the wedge, deformation within the wedge and in the basal décollement is dominated by frictional sliding and brittle failure, giving the classic tapered profile. However, at greater depths and higher temperatures the lower part of the wedge becomes ductile. This increases the ratio of applied basal shear stress to the strength of material in the lower part of the wedge, forcing an increase in the taper and a steeper surface slope. At even greater depths, the décollement zone itself may also become ductile, decreasing the ratio of applied basal shear stress to the strength of the lower part of the wedge and causing the surface profile to flatten into a plateau.

Abrupt changes in the wedge taper and the surface slope are brought about by abrupt changes in either the applied basal shear stress in the décollement zone or by the strength of the material in the lower part of the wedge. Such abrupt changes are most likely caused by a change in the deformational mechanism from brittle to

ductile. A model using a brittle–ductile critical taper explains the east–west topographic profile across the Andes (near 20°S) from the sub-Andean foothills to the steep slopes of the Eastern Cordillera and the high plateau of the Altiplano.

The likelihood of orogenic wedges having brittle–ductile behaviour depends on factors such as the thermal environment, material properties of the wedge and décollement materials, convergence rate and elastic thickness of the lithosphere. Some wedges may be almost perfect frictional Coulomb wedges, such as in Taiwan, whereas others may exhibit significant ductile behaviour.

The orogenic wedge may be entirely erosional, providing sediment to adjacent foredeeps, or partly depositional, with prominent wedge-top basins. Whether the wedge-top is erosional or depositional depends essentially on the surface and basal slopes of the wedge (Ford 2004) (Fig. 4.25). These slopes are themselves influenced by the regional flexure of the underlying plate and the strength of the basal detachment of the wedge.

Flexure of the foreland plate accommodates part or all of the thickened orogenic wedge, so we can think of tectonic shortening-related uplift and flexural-related subsidence as in a trade-off. Depending on this trade-off, the wedge-top may be depositional or erosional. We need to define a surface slope to the wedge (α), and a basal slope (β), and the two added together give the taper angle. In Fig. 4.25a, the wedge-top is half erosional and half depositional, and the foredeep is depositional. The depositional zone that lies on the buried thrust front, but which is not ponded by thrust-related structural highs can be called an *inner foredeep*. This is to distinguish it from a foredeep proper located outboard of the frontal thrust and from structurally confined wedge-top basins inboard of the frontal thrust.

Three different behaviours are envisaged (Fig. 4.25b–d):

- Where the basal slope is high (6°) and the surface slope is also high (2.5°), an extensive erosional wedge-top and a relatively simple, narrow depositional wedge-top zone (inner foredeep) result. There is a wide zone of active sedimentation filling the flexural deflection in the foredeep proper. The North Alpine Foreland Basin of Switzerland is an example.
- Where the basal slope is very high (9°), but the surface slope is very low (zero to 0.5°), the thrust wedge almost entirely fills the deflection, with active sedimentation in a series of wedge-top basins. Shortening takes place across the wedge and there is little or no frontal advance. The northern sector (Ferrara–Romagna) of the Apennines of Italy is an example.
- Where the basal slope is relatively low (4°) and the surface slope is low (0.5°), there is a broad region of wedge-top deposition in several depocentres separated by eroding thrust-related culminations, and a moderately wide foredeep proper. The wedge continues to advance over the foreland. The south-central Pyrenees of northern Spain is an example.

When data on the surface and basal slopes of orogenic wedges are compiled (Ford 2004) it is evident that low surface slopes are commonly associated with weak detachments caused by the presence of salt (Fig. 4.26). Salt-based orogens have surface slopes of always less than 1°, but basal slopes vary over a wide range. Salt is so weak that it cannot sustain high surface slopes. Classic examples of salt-based orogenic wedges and fold-thrust belts are the northern Apennines of Italy, the external South Pyrenean units, the Salt Range of Pakistan, and the Zagros of Iran.

The theoretical relationship between surface and basal slopes is given by:

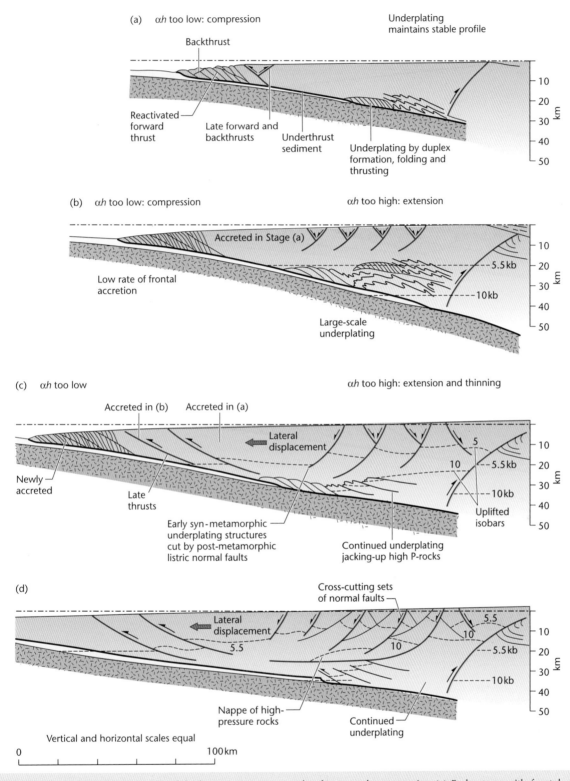

Fig. 4.24 Platt's (1986) evolutionary model of an accretionary wedge from youth to maturity. (a) Early stage with frontal accretion dominant. The gravitational effect of the surface slope (first term on right-hand side of eqn. [4.23]) is too low in the frontal region, which therefore shortens and thickens internally. (b) Large-scale underplating is the dominant mode of accretion, so that the rear of the wedge extends by extensional faulting and possibly by ductile flow near the base of the wedge. The deeper parts of the wedge may also undergo high-pressure metamorphism. (c) Continued underplating and resultant extension has lifted the high-pressure rocks towards the surface. Extension towards the rear of the wedge promotes some shortening (late thrusting) at the front. (d) In the mature stage, underplating and extension have brought the high-pressure rocks to levels accessible to erosion. The prism is now 300 km long, comparable to the Makran wedge of Pakistan. After Platt (1986).

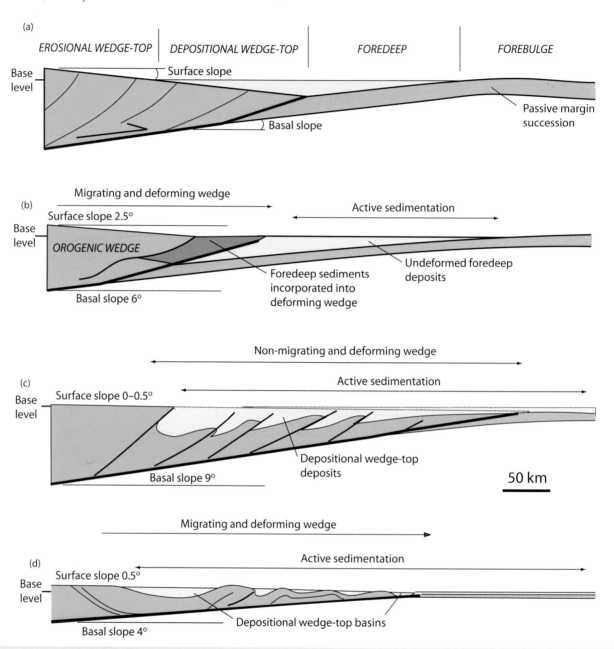

Fig. 4.25 Depositional versus erosional wedge tops, after Ford (2004). (a) The extent of the depositional and wedge-top zones depends on the geometry of the critically tapered wedge combined with the flexural response of the underlying lithosphere, resulting in a basal slope and a surface slope. (b) Basal slope of 6° and surface slope of 2.5°, showing an erosional wedge-top and depositional foredeep, with some foredeep sediment incorporated tectonically into the frontal part of the wedge. This situation may reflect a moderate basal strength of the wedge and a moderate flexural rigidity of the underlying plate. (c) Basal slope of 9° and surface slope of zero to 0.5°, with the wedge filling the deflection, and active sedimentation in wedge-top basins. This situation may reflect a weak underlying lithosphere. (d) Basal slope of 4° and surface slope of 0.5°, causing a narrow taper angle, with well-developed piggyback (wedge-top) basins within a deforming fold-thrust belt. The active sedimentation includes the deforming wedge and the foredeep. This situation characteristically results from the presence of a weak basal detachment comprising salt. Reproduced with permission of John Wiley & Sons, Ltd.

$$\alpha + \beta = \frac{\beta + (\tau_0/\rho g H)}{1 + (1 - \lambda)(2/(\csc\phi - 1))} \qquad [4.24]$$

where ρ is the average density of the wedge, H the depth to the salt detachment, τ_0 is the basal shear stress on the salt, ϕ is the angle of internal friction of the wedge, and λ is the pore fluid pressure (here assumed to be zero), and *csc* means the cosecant trigonometric function. High values of strength of the basal detachment result in high taper angles with high surface slopes. Low strength of the basal detachment, such as caused by the presence of salt (yield strength of c.1 MPa), results in a narrower taper and particularly in a very low surface slope.

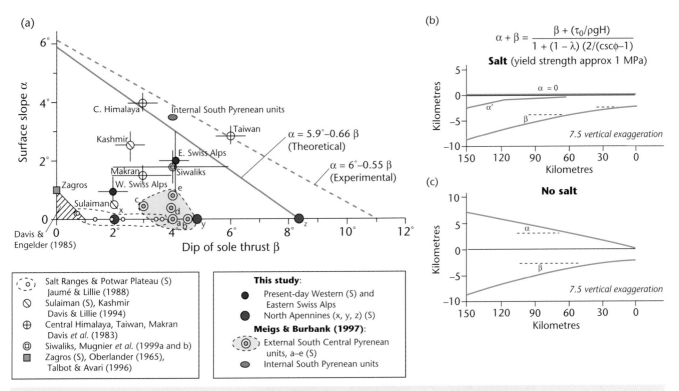

Fig. 4.26 Plot of orogenic wedges in terms of surface and basal slopes (a) with examples of theoretical and experimental curves for salt based (b) and non-salt based (c) orogenic wedges, from Ford (2004). Low surface slopes typify salt-based orogens. (S) in key to (a) indicates a salt basal detachment. Reproduced with permission of John Wiley & Sons, Ltd.

The surface and basal slopes can also be plotted in terms of the pore fluid pressure, expressed as the ratio of the fluid to lithostatic pressures on the basal detachment, λ_b, and on the angle of internal friction, ϕ, denoting the rock strength of the orogen (Dahlen 1984; Willett & Schlunegger 2010). The wedge is said to be *subcritical* when basal sliding is not possible without further internal thickening, and *stable* when the wedge slides on its basal detachment without internal reorganisation. Stable wedges are steeper and higher tapered than subcritical wedges and occur at higher rock strengths of the wedge and lower pore fluid pressures. Actively deforming wedges are therefore favoured by higher pore fluid pressures on the basal detachment.

The effect of basal detachment strength on surface slope of the orogenic wedge has obvious implications for the preservation of wedge-top basins, the erosion rate of the orogen and sediment delivery to the foreland basin system, and also for the approximation of the distributed load driving flexure.

The thrust wedge of the orogen and the foreland basin system are mechanically coupled (Sinclair 2012) and linked by the surface processes of erosion, sediment transport and deposition. During the evolution of the orogen and foreland basin system, there is an accretionary flux into the wedge, as foreland basin sediments are deformed and uplifted, F_a, and an erosional flux, F_e, that transmits mass from eroding mountain hinterlands to the basin in sediment routing systems (Fig. 4.27).

4.5.4 Double vergence

We have so far considered the mechanics of critically tapered orogenic wedges as a Coulomb model (Davis *et al.* 1983; Dahlen 1984;

Platt 1986). One of the conspicuous features of mountain belts and oceanic accretionary prisms is their doubly vergent nature. That is, thrust displacements are directed outwards from the core of the orogen on both flanks. Bivergent wedges can also be produced in numerical experiments involving the subduction of mantle lithosphere beneath a crustal layer with a Coulomb-plastic rheology (Willett 1992; Beaumont *et al.* 1996a). The basic model set-up (Fig. 4.27c) is a laterally uniform layer undergoing a plane strain deformation caused by coupling across a detachment to two underlying, converging rigid plates. One underlying plate slides (subducts) beneath the other at point S. At the base of the upper layer, the velocity field to the left of S is uniform and positive, acting from left to right, whereas to the right of S the velocity is zero. The upper layer has a Coulomb yield strength with a coefficient of friction ϕ that is larger than the coefficient of friction in the weak detachment layer ϕ_d. Model runs show that deformation spreads upwards from S, initially as two conjugate shear zones, bounding a nearly triangular region of minimal internal deformation that is uplifted as a block. Subsequently, shear zones develop on the pro-wedge side of S, producing a long, tapered pro-wedge with a shallow surface slope, and a more steeply tapered retro-wedge. This is supported by the obviously different topographic slopes on the different flanks of mountains belts such as the Southern Alps of New Zealand and Alps of central-western Europe.

The thickened upper layer must be balanced isostatically, giving a crustal root. If the orogen is supported flexurally, flanking foreland basins will develop. As the root develops, however, heating may cause the rheology of crustal materials to change from a Coulomb-plastic flow to a thermally activated power-law viscous flow. This weakens support for the orogenic wedge, promoting the development of low surface slopes bounding a plateau. If heating continues, the orogen

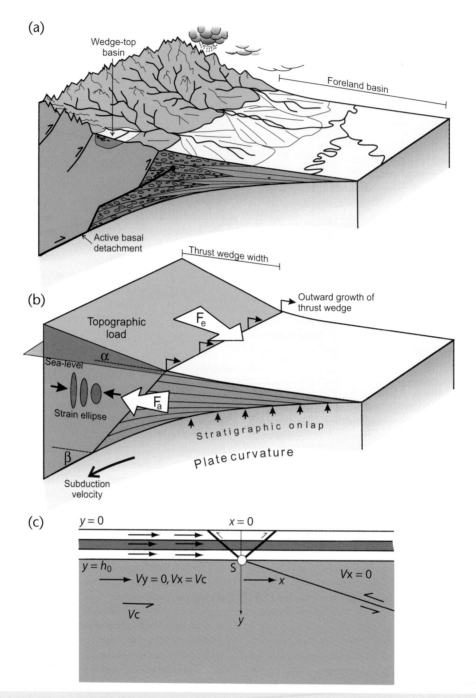

Fig. 4.27 Coupled thrust wedge/foreland basin systems, from Sinclair (2012) and reproduced with permission of John Wiley & Sons, Ltd. (a) Geological and topographic characteristics showing thrust wedge with ramp-flat staircase geometry of basal detachment. (b) Quantifiable parameters include the erosional (F_e) and tectonic (F_a) accretionary fluxes, the mean surface slope of the wedge α, and the mean slope of the basal detachment β. (c): geodynamic model, showing crustal deformation focused at a point S, where mantle of the plate on the left, moving with a constant tangential velocity of convergence V_c, detaches and is underthrusted. V_x and V_y are the horizontal and vertical velocities respectively. Overlying crust has initial thickness h_0 to the left of S. The tangential velocity to the right of S is zero. There is no stress on the upper surface. There is no requirement for a 'bulldozer' backstop in this geodynamic model. Inset modified from Willett *et al.* (1993).

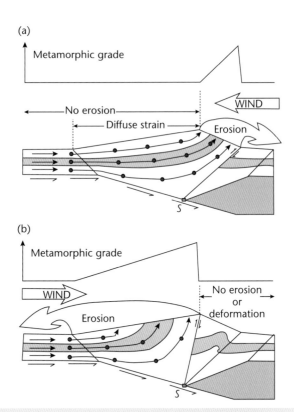

(a)

Metamorphic grade

No erosion

Diffuse strain

WIND

Erosion

S

(b)

Metamorphic grade

WIND

Erosion

No erosion
or
deformation

S

Fig. 4.28 Effects of erosion on a bivergent wedge (after Willett *et al.* 1993). Precipitation concentrated on the retro-wedge (a), and pro-wedge (b) increases erosion, causing an advection of high-grade metamorphic rocks from the middle and lower crust (and even mantle lithosphere) to the surface. Passive shaded layer shows middle crust. Lines are material trajectories, and dots show progressive equal-time positions of points originally aligned vertically.

may collapse by extension in order to achieve a low taper angle. Orogenic collapse (Dewey 1988) drives material to the flanks, which continue to behave as Coulomb-plastic wedges. Consequently, extension in the rear of an orogen should be accompanied by contraction (thrusting) towards the flanks.

The requirement of the wedge to alter its taper angle also depends on the convergence velocity. If the convergence velocity falls, the viscous-based orogen is required to achieve a new geometry with a lower taper, which can best be achieved by extension. If the convergence velocity stops, the wedge must extend until there is no surface slope.

The removal of material from the surface of the wedge by erosion has a strong feedback to the generation of mass in the orogenic wedge caused by convergence (Fig. 4.28). We should expect high rates of erosion to require high tectonic fluxes into the wedge to replace the material lost at the surface. Consequently, erosion appears to 'suck up' rocks from depth. This can be recognised in the exhumation patterns of mountain belts, which commonly have lower crustal (and even mantle) rocks exposed in their cores. If erosion is asymmetrical because of climatic differences on each side of the mountain belt, the wet, windward side of the orogen should be characterised by high exhumation rates of high-grade metamorphic rocks, whereas the dry leeward side should continue to be draped by upper crustal rocks (Fig. 4.28). This pattern is seen well in the Southern Alps of New

Zealand (Koons 1989; Norris *et al.* 1990), the central Himalaya (Fielding 2000), the European Alps (Beaumont *et al.* 1996a; Schlunegger & Willett 1999) and Taiwan (Lin 2000; Dadson *et al.* 2003).

We reconsider these scenarios posed by numerical models in relation to exhumation and erosion, sediment flux and basin development in §4.6.3.

4.5.5 Analogue models

Analogue experiments give insights into the mechanics of crustal deformation in orogenic wedges (Davy & Cobbold 1991). Analogue experiments are scaled to be representative of the lithosphere. Materials are therefore chosen to represent the rheological layering and strength profile of the lithosphere. A simple system comprises a brittle upper sand layer simulating the crust, and a lower layer of high-viscosity power-law silicone representing the lower crust, above a basal low-viscosity fluid representing the mantle. This basal fluid layer allows the analogue experiment to be isostatically balanced. In a four-layer system, a brittle upper mantle lithosphere layer and ductile lower mantle lithosphere layer are constructed beneath a brittle upper crust and ductile lower crust.

Shortening in analogue experiments is normally produced by a piston and commonly results in buckling followed by bivergent thrusting (Cobbold *et al.* 1993; Burg *et al.* 1994b). The buckles have a wavelength controlled by the relative thicknesses of the brittle and ductile layers. Scaled up to the real world, such large buckles should be obvious as major crustal-scale antiforms with a wavelength of c.200 km, such as the Penninic culmination of the Central Alps of Europe (Burg *et al.* 2002). The thrusts originate from the inflection points of the major buckles, and so are controlled in location and spacing by the presence of the early first-order buckles. Further shortening results in amplification of the thrusted buckles, giving rise to imbricated brittle wedges.

Indentation sandbox models involve a rigid indenter with the same height as the undisturbed sand (representing undeformed crust), which allows sand to spread laterally outwards over the indenting plate (Koons 1989, 1990). With progressive deformation, a doubly vergent wedge forms, with thrusts propagating away from the indenter to form an outboard wedge or retro-wedge, and towards and over the indenter as an inboard wedge or pro-wedge. The outboard wedge is decoupled along a low-friction Mylar sheet at the base, and thrusts rise from this surface. However, within the inboard wedge, thrusts propagate up through the wedge material itself at an angle corresponding to the angle of internal friction of the sand ϕ. Thrusts in the inboard wedge are therefore shallower than in the outboard wedge. Sandbox experiments are also capable of reproducing the conditions of mantle subduction, where a sand cover is dragged by an underlying Mylar sheet (Malavieille 1984). Once again, a doubly vergent wedge is produced centred over the point of detachment, the majority of thrusts facing the subducting plate.

The effects of erosion and sedimentation can also be studied using analogue experiments (Bonnet *et al.* 2008; Konstantinovskaya & Malavieille 2005). Some experiments, for instance, indicate a teleconnection from erosion of the retro-wedge and frontal accretion of the pro-wedge (Hoth *et al.* 2007, 2008), indicating that the pro- and retro-wedges and their foreland basins are mechanically coupled (Fig. 4.29). The distributed or localised nature of erosion also has an effect of the deformation of the bivergent wedge in analogue experiments: localised, focused erosion of the pro-wedge, for example, prohibited out-of-sequence displacement, whereas distributed

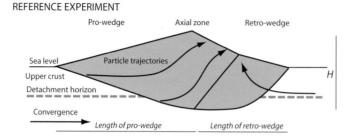

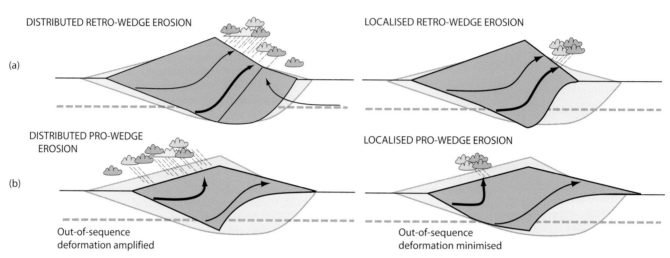

Fig. 4.29 Summary of results of analogue models investigating the effects of focused (localised) and distributed rainfall on the pro-wedge (b) and retro-wedge (a), after Hoth *et al.* (2006). Particle trajectories are shown in arrows. Thicker arrows indicate significant influence of erosion. Erosion drives larger fluxes towards surface and shortens length of retro- or pro-wedge compared to reference experiment.

erosion amplified out-of-sequence thrusting. Fold geometries in foreland fold-thrust belts and the large-scale geometry of the wedge are also affected by the presence of, rate of accumulation and competence of syntectonic sediment loads (Storti & McClay 1995; Duerto & McClay 2009). At the scale of the critically tapered wedge, high sedimentation rate reduces the number of thrust sheets, decreases the internal shortening and decreases the critical taper angle. At a smaller scale, low competency results in tight, asymmetric detachment folds, whereas higher competence results in an array of clearly foreland verging folds. Consequently, different tectonic styles result from the different impacts of syntectonic pro-wedge sedimentation:

- The absence of syntectonic pro-wedge sediment causes an uplifted, outcropping fold-thrust belt.
- An underfilled pro-foreland basin results in long-displacement blind thrust sheets.
- An overfilled pro-foreland basin results in short-displacement blind thrust sheets.

These results are particularly interesting in light of the results of numerical models of the interaction of erosion and deposition with tectonic deformation in orogenic wedges and foreland fold-thrust belts (§4.5.6).

Analogue experiments are therefore good at allowing a visualisation of the development of major crustal features such as thrusted antiforms and bivergent wedges. The tectonic wedges created in ana-

logue experiments have geometries as predicted by critically tapered wedge mechanics (§4.5.3) and are very similar to the wedges produced in numerical simulations using similar boundary conditions (see §4.6.4).

4.5.6 Numerical approaches to orogenic wedge development

There are three force systems that control deformation in the orogenic wedge:

F1: the compressive force integrated over the thickness of the crust
F2: the gravitational force (or increased potential energy) due to crustal thickening
F3: the basal traction force integrated across the base of the deforming zone

The tendency for the thickened crust to spread out laterally is given by the ratio **F2/F1**, which is known as the *Argand number* (see also §3.5.2) (England & McKenzie 1982). For a linear viscous material

$$Ar = \frac{\rho g h^2}{\mu V_p} \qquad [4.25]$$

where ρ is the density, g is the acceleration of gravity, μ is the viscosity, h is the length scale (thickness) of the mechanical model, and V_p is the horizontal velocity at the base of the crust. For a typical crust with $\rho = 2800\,\text{kg m}^{-3}$, $\mu = 10^{23}\,\text{Pa s}$, $h = 35\,\text{km}$, and $V_p = 0.05\,\text{m yr}^{-1}$, the 'viscous' Argand number is 5. For a Coulomb-plastic material, the Argand number becomes (Willett 1999)

$$Ar = \frac{2}{\tan \phi} \qquad [4.26]$$

where ϕ is the angle of internal friction, typically 15° for crustal rocks. Using this value, the 'Coulomb' Argand number is 7.

If $Ar \gg 1$, even small increases in thickness are counteracted by gravity, redistributing the thickening away from the plate boundary and broadening the zone of deformation. If Ar is small, however, the crustal thickening is localised.

The ratio **F3/F1** is known as the *Ampferer number* (Ellis *et al.* 1995; Ellis 1996). It reflects the strength of the crust-mantle coupling. The length scale of deformation in subduction models scales on the Ampferer number. If Am is nearly 1, the model crust is strongly coupled to the underlying mantle, which means that the basal tractive forces are transmitted strongly into the crust, causing thickening. If Am is close to zero, there is weak coupling, and the basal tractive forces are inefficient at causing crustal thickening.

Numerical models of orogenic wedge development have been developed from two starting points (Figs 4.30, 4.31):

1. *Indentation models* assume that forces are transmitted laterally from a rigid indenter into a less rigid continental plate approxi-

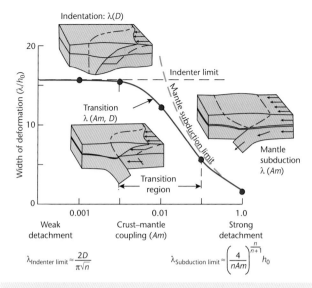

Fig. 4.31 Deformation width λ normalised by the initial crustal thickness h_0, versus the crust-mantle coupling Am. The normalised indenter lateral length scale is $D/h_0 = 25$. For a given indenter length scale, the combined model width of deformation λ is controlled by indenter mechanics for weakly coupled (detached) systems ($Am = 0$), and by mantle subduction mechanics for strongly coupled ($Am \sim 1$) systems. After Ellis (1996).

mated by a thin viscous sheet (England *et al.* 1985; England & Houseman 1986). As mimicked by plasticine analogue models (e.g. Peltzer & Tapponnier 1988), deformation of the viscous sheet spreads out over a strike-normal distance related to the width of the indenter. If λ is the length scale of the deformation and D is the strike-parallel width of the indenting body

$$\lambda = \frac{2D}{\pi\sqrt{n}} \qquad [4.27]$$

where n is the power-law exponent of the viscous sheet. In this model, if D is 3000 km, which is the approximate lateral scale of the India–Asia collision, and the rheology of the viscous sheet is linear ($n = 1$), the length scale of the deformation should be in the region of 2000 km. For $n = 2$ and $n = 3$, λ reduces to 1350 km and 1100 km respectively. In the indentation model, crust (and mantle lithosphere) is thickened in the relatively weak continental lithosphere against the boundary with the rigid indenter. Gravity acts on the thickened layer, which causes λ to increase as convergence continues. Eventually, very large regions of crustal uplift such as the Tibetan Plateau are produced. The forces on the thickened crust are quantified by the Argand number in indentation models.

2. *Mantle subduction* models invoke the transmission of forces from an underlying subducting mantle lithosphere to the base of the crust. These forces act across a crust–mantle boundary that must therefore undergo considerable sub-horizontal shear. This zone of sub-horizontal decoupling is termed a *detachment*. The width of the region of crustal thickening depends critically on the strength of the coupling along the detachment. This coupling is quantified by the Ampferer number. If the original

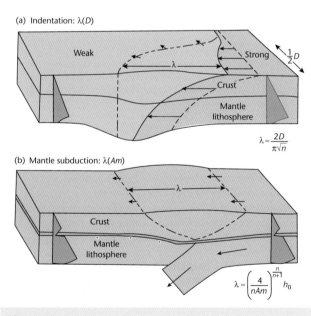

Fig. 4.30 Schematic illustration of two end-member models of continental deformation resulting from convergent continent–continent tectonics (after Ellis 1996). In (a), indentation over a lateral length scale D causes thickening over a width λ. Half of the indenter is shown. Thickening occurs in the crust and mantle lithosphere in the relatively weak indented plate. (b) Continental deformation is driven mostly by mantle subduction. Crust in both plates is weak, with a strength minimum at the Moho, which decouples crustal deformation from underlying mantle subduction. The width of deformation λ depends on the strength of the mantle-crust coupling, given by the Ampferer number Am.

thickness of the crust is h_0, the length scale of the deformation in mantle subduction models is given by

$$\lambda = \left(\frac{4}{nAm}\right)^{\frac{n}{n+1}} h_0 \qquad [4.28]$$

If the crustal layer is initially 35 km thick, we adopt a linear viscous rheology ($n = 1$) and a very strong coupling between the crust and mantle ($Am = 1$), the width of deformation is just 70 km. If there is a very weak coupling between crust and mantle ($Am = 0.1$), λ becomes c.220 km. Both of these estimates are considerably smaller than in the indentation model, suggesting that the length scale of deformation in an orogenic belt is an indicator of the underlying mechanics.

Comparing the indentation and mantle subduction models (Ellis 1996), the increase in the width of deformation increases with continued convergence, but at different rates. The value of λ is initially much lower in the mantle subduction case, but the growth of λ with convergence increases more rapidly than in the indentation model (Fig. 4.31).

An orogenic model involving both indentation and mantle subduction processes suggests that there may be different evolutionary paths for orogenic belts:

- For situations where the crust is strongly coupled to the mantle lithosphere (Am high), crustal thickening should initially be controlled by mantle subduction, especially where D is large. In the case of orthogonal relative plate motion, there will be very limited strike-slip or escape tectonics.
- For situations where the crust is weakly coupled to the mantle lithosphere (Am low), mechanics will be dominated by indentation, especially for small indenter length scales (D). Even during orthogonal convergence, strike-slip tectonics and lateral escape should be widespread.
- Since λ grows faster for mantle subduction than for indentation, an orogen initially dominated by mantle subduction mechanics may, with continued convergence, evolve into an orogen dominated by indenter-style mechanics.

4.5.7 Low Péclet number intracontinental orogens

There are a number of orogenic belts occurring far from any plate margin, where conventional ocean–continent or continent–continent subduction is not taking place. The intracontinental mountain belts of inland China, such as the Tien Shan, are examples (Fig. 4.32). These intracontinental orogenies, and their equivalent from the geological past, such as the Petermann and Alice Springs orogenic belts of central Australia, do not involve large amounts of shortening, but nevertheless are typified by high topography and well-developed flexural basins.

Questions immediately arise as to the source of the compressional forces that cause mountain building far from plate margins. One possibility is that horizontal compression is transmitted from a plate boundary far into the continental plate (Cloetingh *et al.* 1999). Other possibilities are that intraplate deformation is driven by lateral contrasts in gravitational potential energy, with or without the thermal effects of sediment blanketing (Stüwe & Barr 2000) (§4.5.8).

Intraplate orogenies are typified by relatively low rates of tectonic advection of rock (<1 mm yr^{-1}) compared to 'plate boundary orogenies' dominated by subduction (>10 mm yr^{-1}), and have consequently

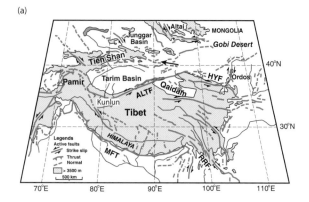

(a)

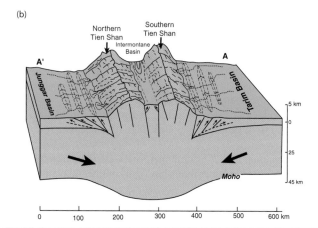

(b)

Fig. 4.32 (a) Location map of the intracontinental orogenic belts of central Asia, principally the Tien Shan, Altyn Tagh and Kunlun ranges, with intervening basins such as the Tarim, Junggar and Qaidam. ALTF, Altyn Tagh Fault; RRF, Red River Fault; HYF, Haiyuan Fault; MFT, Himalayan Main Frontal Thrust; (b) Schematic cross-section of the Pleistocene to Holocene Tien Shan. The Tien Shan is an example of a low Péclet number orogeny. Reprinted from Fu *et al.* (2003), with permission of Elsevier.

been termed 'low Péclet number' orogenies (Sandiford 2002). At low convergence rates, the thermal effects of tectonic advection in the crust are outweighed by the cooling effect of conduction (thermal Pe <1), allowing for progressive strengthening of the orogen over time, and with little potential for gravitational collapse.

4.5.8 Horizontal in-plane forces during convergent orogenesis

Lateral forces are caused by the difference between the gravitational potential energy of mountains relative to adjacent lowlands. The potential energy is the integration over depth in the lithosphere of vertically acting stresses (Dalmayrac & Molnar 1981; Jones *et al.* 1998). The magnitude of these horizontal, in-plane forces can be visualised by plotting σ_{yy} against depth y (Appendix 39). In contrast, isostatically supported topography results from a balance between σ_{yy} at a depth of compensation between the lithospheric column under consideration and a reference lithospheric column. Consequently, the relationship between surface elevation and lateral intraplate forces may change during the evolution of a mountain belt in response to

(a) Topography

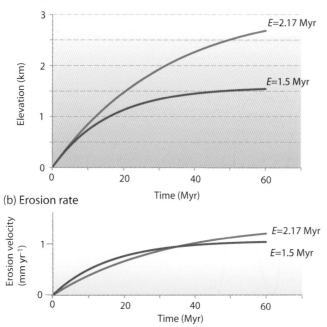

(b) Erosion rate

Fig. 4.33 Topography (a) and erosion velocity (b) for a mountain belt subject to a uniform thickening strain for different values of an Erosion constant *E* (see Appendix 39 for details). Elevation and erosion rate both become asymptotic after c.40 Myr, suggesting that orogenies have a characteristic time scale. Calculations based on algorithms in Stüwe & Barr (2000).

tectonic thickening, convective thinning and detachment of subcrustal lithosphere, and erosional thinning of the crust (Sandiford & Powell 1990; Stüwe & Barr 2000).

The elevation H and horizontal in-plane force F_h are plotted for a number of scenarios involving variation in crustal and subcrustal lithospheric thickening strains, ε_c and ε_L, respectively, in Fig. A39.1. If the crust thins, for example by erosion outpacing tectonic thickening, $\varepsilon_c < 1$, and if the lithosphere thickens, for example by tectonic shortening, $\varepsilon_L > 1$.

We now assume that an orogen grows at a constant vertical strain rate and that the surface is eroded at a rate proportional to elevation (Stüwe & Barr 1998, 2000). A possible pathway over time in the $\varepsilon_c - \varepsilon_L$ plane is:

- Early in orogenic evolution, convergent tectonics is accompanied by an increasing in-plane force pushing outward from the elevated region as the crust thins by erosion, causing the crustal thickening strain to be less than the subcrustal lithosphere thickening strain, $\varepsilon_c < \varepsilon_L$.
- As topography increases, erosion rate also increases, and the crust continues to thin; the negatively buoyant subcrustal lithosphere dominates the lateral in-plane force, causing it to decrease in magnitude and then reverse in sign. An in-plane force now pushes from the foreland towards the elevated region, despite the existence of high topography.
- In a final stage, a steady-state elevation is achieved as the orogen collapses inwards gravitationally, as may occur in the Andes.

This trend of increasing and positive lateral force F_h followed by diminishing F_h and reversal of sign is strongly affected by the erosion rate. Slow erosion promotes very high in-plane forces of long duration, whereas fast erosion produces much reduced in-plane forces acting towards the lowland, of short duration.

It is emphasised that the lateral force is calculated as the integration over depth of σ_{yy}, but there is always a positive push outwards from the orogen at shallow depths because of the elevation of the mountain above sea level (Appendices 4, 39). This may lead to the formation of shear zones to accommodate the varying lateral forces as a function of depth in the orogen.

Furthermore, in a convergent orogen, the lateral in-plane force acts in combination with a horizontal deviatoric, tectonic force. The effective horizontal force is therefore the difference between the two (Sonder & England 1989; Zhou & Sandiford 1992). During a cycle of convergent tectonics, the effective horizontal force may increase, then decrease, then increase again, giving distinct contractional deformation episodes separated in time (Stüwe & Barr 2000). At reasonable strain rates, the orogen builds to the required thickness and topography after 30–40 Myr (Fig. 4.33). This is similar to the duration of many foreland basin systems (§4.6).

4.6 Foreland basin systems

4.6.1 Introduction

Foreland basins are examples of basin evolution controlled primarily by a moving tectonic load. Deposition in foreland basin systems can be understood by investigating the effects of a moving load system on the deflection of an elastic foreland plate (Jordan 1981 and Schedl & Wiltschko 1984 for early treatments). The static deflection is described for continuous and broken plate models in §4.2. For a moving load system, the deflection simply moves as a wave through the foreland plate ahead of the load system. If the load system moves relative to the foreland plate, so does the deflection.

The shape of the deflection is an asymmetric 'low' (foredeep) close to the load and a broad, low-amplitude, uplifted forebulge far from the load. Since the supracrustal load in a convergent mountain belt moves closer to a point fixed to the lower, pro-foreland plate through time, and since we know from the solution of the flexural equation (eqns. [4.3] and [4.9] for continuous and broken plates) that the deflection increases towards the load, it follows that subsidence rates in the pro-foreland basin should accelerate through time, until the same fixed point is overridden by the thrust wedge, and sedimentation ceases. The original location of the point fixed to the foreland plate at the onset of flexure determines the precise subsidence history followed (Allen *et al.* 1991; Crampton & Allen 1995). Consider the following three locations (Fig. 4.34):

- *Point A* is originally located beyond the flexural forebulge in the back-bulge region (i.e. for a line load on the end of a broken plate, at $x/\alpha > 5$). It will initially experience negligible uplift/subsidence, but as convergence continues, will be 'dragged' through the flexural forebulge, causing slow but prolonged rock uplift and erosion. The erosion will continue until $x = \pi\alpha/2$, producing an unconformity with a large chronostratigraphic gap.
- *Point B* is originally situated within the forebulge region ($\pi\alpha/2 < x < 3\pi\alpha/2$). It will immediately experience rock uplift and erosion. As convergence continues it will be 'dragged' through

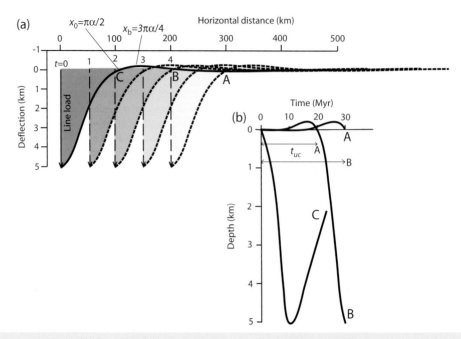

Fig. 4.34 Migration of a flexural wave under a moving load system. (a) Movement of the deflection across a foreland plate as a result of load migration. (b) Points A, B and C experience different subsidence histories depending on their initial location with respect to the load. Point A spends 30 Myr in a backbulge or forebulge position. Point B is initially uplifted in the forebulge, then subsides in the foreland basin. Point C experiences basin subsidence, followed by uplift as it is incorporated into the thrust wedge. t_{uc} denotes the time spent in the erosional region and therefore the duration of the unconformity.

the first node of the deflection where $w = 0$, and will then experience accelerating subsidence as the point becomes buried by foreland basin sediments. The chronostratigraphic gap should be smaller than for Point A.

- *Point C* is originally located close to the orogenic wedge inboard from the first node and the flexural bulge. It will, from the outset, experience subsidence. The rate of subsidence will increase through time as the point becomes closer to the maximum deflection. Eventually, with continued convergence, the point is likely to become incorporated into the orogenic wedge itself so that the foreland basin stratigraphy can only be viewed in allochthonous thrust units. There should be a conformity at the basal surface of the pro-foreland basin megasequence.

The key prediction for a foreland basin generated by a moving load is therefore that the basin-fill is underlain by an unconformity with a variable and predictable stratigraphic gap (Crampton & Allen 1995; Allen *et al.* 2001). This is a megasequence boundary caused by flexural forebulge uplift. Points located close to the load start subsiding earlier than distal points. At any one time, proximal locations subside faster than distal locations. The overall form of the subsidence rate at any one location is convex-up, indicating accelerating subsidence through time (e.g. Homewood *et al.* 1986; Vergés *et al.* 1998; Xie & Heller 2009) (§9.5).

4.6.2 Depositional zones

In the simplest terms, foreland basins develop at the front of active thrust belts where the bulk transport direction is towards the evolving basin. Because the thrust load is inherently mobile, the foreland basin itself becomes involved in the deformation. To what

extent the basin becomes dissected or becomes completely detached depends on a number of variables, including the propagation rate of the thrust tips, availability of subsurface easy-slip horizons underlying the basin, and the angle of convergence. The foreland basin system (DeCelles & Giles 1996) contains four depositional zones (Fig. 4.35):

- on moving thrust sheets as a *thrust-sheet-top* (Ori & Friend 1984), *wedge-top* or *piggyback basin*, which receive sediment from the eroding orogenic wedge;
- ahead of the active thrust system in a *foredeep*, which is supplied with sediment from both the continental foreland and the orogenic wedge;
- on the *flexural forebulge* if accommodation is available, for example because the foreland lithosphere is submerged below sea level as a result of negative dynamic topography (§5.2);
- as a shallow, broad *back-bulge* basin filled with shallow-marine and continental sediments; ongoing convergence should cause the back-bulge depozone to be uplifted and eroded in the flexural forebulge.

Individual foreland basins may contain examples of these four depozones, but the importance of particular depozones varies according to the geodynamics of each case. In the North Alpine Foreland Basin of Switzerland and eastern France, the foreland basin system was dominated by a well-developed foredeep and minor wedge-top basins (Allen *et al.* 1991; Sinclair 1997). The first clastic wedges (lower Oligocene), composed essentially of turbidites, were shed partly into ponded basins located on top of thrust sheets and partly overspilled into foredeeps. As the foreland basin evolved through the Oligocene, thrust-sheet-top basins became far

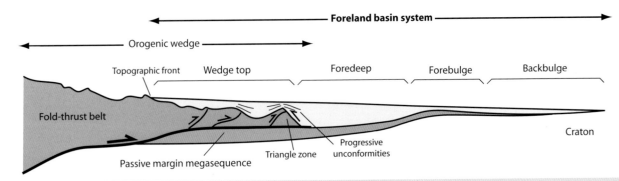

Fig. 4.35 The four depositional zones of a foreland basin system, as envisaged by DeCelles & Giles (1996). Reproduced with permission of John Wiley & Sons, Ltd.

less conspicuous features of the inner margin of the basin. Post-depositional tectonics (late Miocene–Pliocene) detached the entire basin in western Switzerland, using Triassic salt as an easy-slip horizon, as deformation progressed into the Jura province (Homewood *et al.* 1986; Burkhard & Sommaruga 1998).

In other settings, wedge-top deposition was particularly important in basin evolution. A lucid picture of a linked system of inner thrust-sheet-top basins and outer foredeeps is provided by the Apenninic chain of Italy (Ricci Lucchi 1986 for synthesis; Artoni 2007) (Fig. 4.36). In the northern Apennines, the entire system of basins migrated eastwards onto the foreland. Sedimentation was initially dominated by turbidites and some hemipelagics. As deformation continued, the inner thrust-sheet-top basins became uplifted at the end of the Miocene and cannibalised to provide erosional detritus for the foredeep. Sediments range from continental coarse clastics to shelfal mixed carbonate-siliciclastics and turbiditic deep-water sandstones. The Apenninic foreland basin depocentre has now extended into the Adriatic Sea where penecontemporaneous thrust deformations have produced submarine structural culminations. These seafloor highs have been subsequently denuded by slope failure and submarine erosion (Ori *et al.* 1986). Three depositional seismic units have been recognised in the Plio-Pleistocene of the Adriatic section of the foreland basin comprising initially turbiditic and hemipelagic deposits and subsequently deltaic deposits shed from the Apennines and Southern Alps via the Po sediment routing system.

Tectonics has a primary influence on sediment dispersal patterns. Uplifting thrust fronts may not act as major sediment suppliers but may instead form barriers or guides to basinward sediment transport (Ramos *et al.* 2002; Clevis *et al.* 2004; Vergés 2007; Barrier *et al.* 2010; Allen *et al.* 2013). Structural topography is commonly provided by lateral or oblique ramps (Fig. 4.37), as in the southern Pyrenees, or sediment is funnelled along through-going strike-slip faults unrelated to the boundaries of thrust sheets, such as the Sillaro Line and Forli Line in the Italian Apennines. The apices of major fluviatile systems in the southern Pyrenees are located at structural lows or re-entrants in the thrust front, whereas small, locally developed fans with highly restricted drainage basins typify the structural salients in the thrust front (Hirst & Nichols 1986; Barrier *et al.* 2010). This same bimodal picture of small, locally sourced fans coexisting with large megafans sourced from deep within the orogen is also found in the Andes (Horton & DeCelles 1997) and Himalayas (Gupta 1997).

Back-bulge and forebulge deposition are more common in retro-foreland basin systems than in pro-foreland types. Retro-foreland basins such as the Cretaceous Rocky Mountains (Sevier) foreland

basin in western USA and the series of basins east of the Andes differ in important ways from pro-foreland basins (Naylor & Sinclair 2008; Sinclair *et al.* 2005). Their main distinguishing characteristic is that they commonly evolve from regions of backarc extension, and the composition of the sedimentary fill reflects the large amounts of plutonic and volcanic rocks in the orogenic belt. In addition, since they are located on the upper plate, the velocity between retro-wedge and foreland plate is small, restricted to the period of orogenic growth (Naylor & Sinclair 2008). Excellent examples are the Magallanes Basin of Argentina (Biddle *et al.* 1986), the basins of the northern Argentinian Precordillera, such as the Bermejo Basin (Cardozo & Jordan 2001) and the Central Andean foreland basin system of eastern Bolivia, northernmost Argentina, Paraguay and southwestern Brazil (Horton & DeCelles 1997).

Foreland basins contain a gross stratigraphic evolution related to the geodynamical controls on subsidence and sediment supply. The oldest deposits of foreland basins are commonly predominantly fine-grained, turbiditic sediments that accumulated in sub-shelf water depths, commonly termed 'flysch', which pass distally into shallow-water carbonates deposited close to the flexural forebulge (Dorobek 1995; Sinclair 1997; Allen *et al.* 2001). The younger deposits of foreland basins are, in contrast, predominantly shallow-water or continental and typify the term 'Molasse'. This kind of vertical megasequence is found in the North Alpine Foreland Basin of Switzerland (Matter *et al.* 1980; Sinclair *et al.* 1991; Allen *et al.* 1991; Schlunegger *et al.* 1997a, b) (Fig. 4.38) in the form of a basal deepening-up trend, followed by two shallowing-upwards megasequences. The shoreline unit at the top of the Lower Marine Molasse (Fig. 4.38) can be thought of as the pivot point between an early *underfilled* stage (Covey 1986; Sinclair 1997) and a later steady-state stage. During the underfilled stage the topography of the orogenic wedge was most likely relatively subdued, sediment delivery rates were low and the orogenic advance rate relatively high, which collectively caused deep-water conditions in the foreland basin (Sinclair & Allen 1992). After the mountain belt had grown to a steady state, rapid erosion counterbalanced tectonic uplift, the advance rate was slow, and the basin was filled to the spill-point with detritus. During this phase any excess sediment was removed from the foreland basin by fluvial and/or shallow-marine processes. Present-day analogues are the overfilling and sediment export to the ocean of the Indo-Gangetic foreland basin and the Po Basin of northern Italy.

Key processes in the foreland that affect sedimentation are the reactivation of crystalline basement uplifts, such as the Laramide

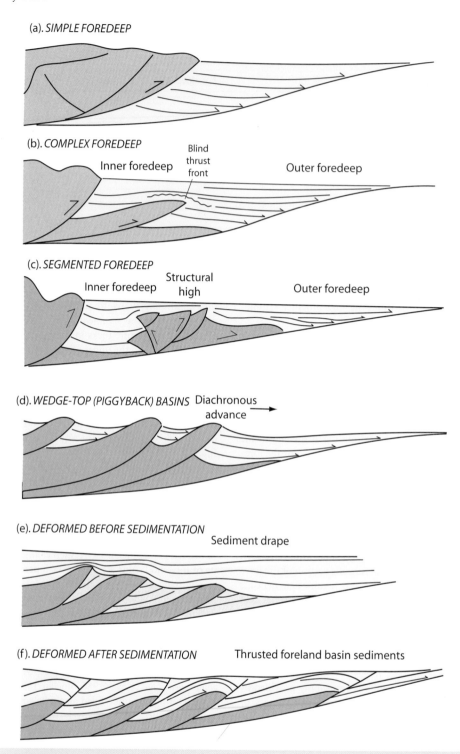

(a). *SIMPLE FOREDEEP*

(b). *COMPLEX FOREDEEP*

Inner foredeep   Blind thrust front   Outer foredeep

(c). *SEGMENTED FOREDEEP*

Inner foredeep   Structural high   Outer foredeep

(d). *WEDGE-TOP (PIGGYBACK) BASINS* Diachronous advance

(e). *DEFORMED BEFORE SEDIMENTATION*

Sediment drape

(f). *DEFORMED AFTER SEDIMENTATION*   Thrusted foreland basin sediments

Fig. 4.36 Foreland basin/thrust belt interactions, based mainly on seismic records from the Apennines and foreland of Italy (modified from Ricci-Lucchi 1986). Basins may be simple, asymmetrical wedges with stratigraphic onlap onto the foreland plate, as in the North Alpine Foreland Basin of Switzerland (a). Basins may be complex as a result of tectonic advance into the basin, causing an inner foredeep separated from an outer foredeep by buried (blind) thrusts (b) or by more complex zones of faulting (c). The wedge-top may be disconnected from the foredeep, consisting of an array of piggyback basins separated by thrust culminations, as in the Eocene of the southern Pyrenees (d). Some basins are draped over a previously deformed substrate (e), whereas others are dissected and shortened after the period of foreland basin deposition (f). Reproduced with permission of John Wiley & Sons, Ltd.

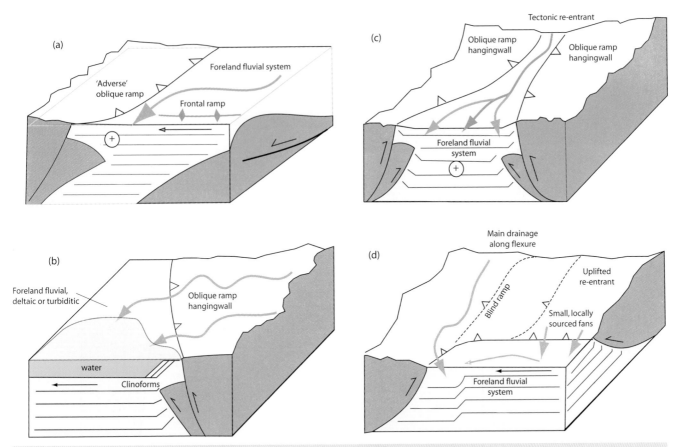

Fig. 4.37 Sediment dispersal patterns by rivers at mountain fronts, adapted from Vergés (2007), based on Pyrenean examples. (a) River guided by frontal ramp flows axially, and breaks through into outer foreland basin when faced with adverse topographic slope of oblique ramp hangingwall. (b) River originating on oblique ramp hangingwall flows down topography into fluvial system or marine embayment. (c) River system constrained by tectonic re-entrant between two oppositely verging oblique ramps, flows transversely into foreland fluvial system. (d) River flows down flexure associated with major oblique ramp, whereas smaller systems drain small catchments on thrust-related hangingwalls and flow longitudinally.

structures in the Rocky Mountain area of North America (Beck *et al.* 1988) and the Sierras Pampeanas of Argentina (Jordan & Allmendinger 1986). Basement grabens in the foreland plate may be inverted during compression, such as the Birmingham Graben during the Lower Paleozoic Appalachian orogeny (Bayona & Thomas 2003), or the foreland plate may be faulted by the outer-arc extensional stresses caused by flexure (§4.2.4) (Bradley & Kidd 1991). These are examples of important effects superimposed on the larger wavelength first-order signature of flexure.

4.6.3 Diffusive models of mountain belt erosion and basin deposition

The convergence of an orogenic thrust wedge over a foreland plate can be considered mechanically as a distributed vertical load system translating laterally over a flexed plate. Erosion of the thrust wedge and deposition in the foreland basin has been treated as a diffusional process (Flemings & Jordan 1989; Sinclair *et al.* 1991). In diffusional problems, the transport rate or flux is proportional to the topographic gradient, and the rate of erosion or sedimentation is proportional to the curvature.

Sinclair *et al.* (1991) proposed that the simplest model of mountain belt erosion and foreland basin sedimentation must contain four parameters (Fig. 4.39): (i) the equivalent elastic thickness of the foreland plate, T_e; (ii) the rate of thrust-front advance, A; (iii) the surface slope angle of the orogenic wedge, α; and (iv) the effective transport coefficient, K, representing the constant of proportionality between topographic slope and sediment transport rate. The orogenic wedge is modelled as a critically taped system (Davis *et al.* 1983; Platt 1986; Boyer 1995) migrating over an elastic plate that flexurally supports tectonic and sedimentary loads.

From the theory of diffusion, the rate of erosion in the mountain belt or rate of deposition in the basin is given by

$$\frac{\partial h}{\partial t} = K \frac{\partial^2 h}{\partial x^2} \qquad [4.29]$$

where K is the effective transport coefficient or diffusivity. Deposition takes place (positive change in h) when the topographic curvature is positive. Erosion takes place (negative change in h) where topographic curvature is negative. In each increment of time, the thrust wedge moves over the foreland plate, so the total change in distributed load $\Delta q(x)$ is made of two components

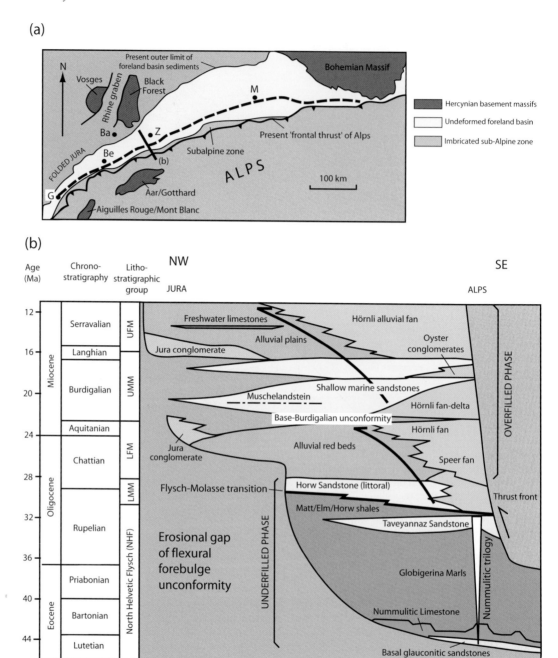

Fig. 4.38 The stratigraphy of the North Alpine Foreland Basin in the vicinity of Zürich, Switzerland (modified from Sinclair *et al.* 1991). (a) Map of North Alpine Foreland Basin: M, Munich; Z, Zürich; Be, Bern; B, Basel; G, Geneva. (b) The stratigraphy is made of a basal deepening-up sequence from shallow-water Nummulitic limestones to hemipelagic marls and turbidites derived from the Alpine orogen (the Nummulitic Trilogy), followed by two grand shallowing-up and coarsening-up cycles. Coarse clastic wedges fringe the Alpine thrust front, whereas the feather-edge of the basin along the Jura margin was relatively passive, with the exception of some local pockets of conglomerate. Reproduced with permission of John Wiley & Sons, Inc.

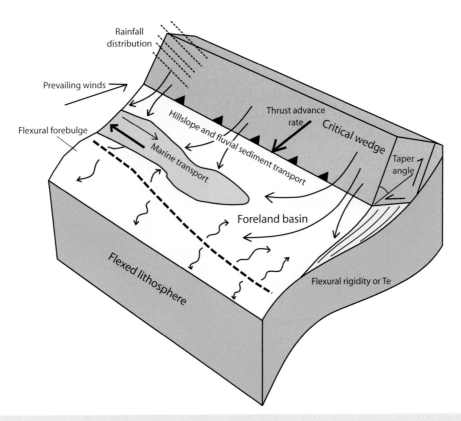

Fig. 4.39 Parameters necessary to describe a diffusive mountain belt system are the advance rate and surface taper of the orogenic wedge, flexural rigidity of the foreland plate, and sediment transport coefficients (Sinclair *et al.* 1991). The distinction between relatively local transport on hillslopes and far-field transport in river systems can either be tackled by attributing two different effective diffusivities to these geomorphic subsystems, or treating the fluvial transport as advective (Johnson & Beaumont 1995).

$$\Delta q(x) = \left(\frac{\partial h}{\partial t} + \frac{\partial T}{\partial t} \right) g\rho_s \qquad [4.30]$$

where the first partial derivative is the topographic change due to erosion or deposition, and the second partial derivative is the change in the tectonic load $T(x)$ in that time step. The new deflection is calculated using flexural isostasy.

A sensitivity analysis of the four parameters critical to the diffusion model (Sinclair *et al.* 1991; Allen & Allen 2005) shows that the flexural rigidity of the foreland plate is a first-order control on the geometry of the flexed plate and foreland basin, as expected from point and line-load approximations (§4.2). However, a greater effect is caused by large variations in the transport coefficient K (Fig. 4.40). For the same flexural rigidity, small transport coefficients ($100\,\mathrm{m^2\,y^{-1}}$) result in narrow, marine basins, whereas high transport coefficients ($800\,\mathrm{m^2\,y^{-1}}$) result in broad, terrestrial basins. The effect of large sediment loads is to force the flexural forebulge far onto the foreland, perhaps even burying it in sediment derived from the orogenic wedge.

Diffusional models have been particularly informative in showing the stratigraphic consequences of changes in the load configuration. For example, a slowing of the advance rate of the orogenic wedge, combined with an increase in its surface taper angle, was interpreted by Sinclair *et al.* (1991) as the cause of distal unconformities in the North Alpine Foreland Basin of Switzerland, without the need to invoke viscoelastic relaxation of the lithosphere (cf. §2.3.2).

The effects of episodic thrusting have specifically been investigated (Flemings & Jordan 1989, 1990) using a model of a mountain belt as a fault-bend fold over a crustal-scale ramp (Suppe 1983). Different transport coefficients were used for transport processes in rivers erosively incising the mountain belt and for alluvial rivers in the foreland basin. The change from quiescence to renewed thrusting was accompanied by a trapping of coarse sediment near the thrust front and the movement basinwards of the forebulge. Continued thrusting and sediment delivery to the basin forced the forebulge away from the orogen, causing stratigraphic onlap over an erosional unconformity. Other studies (Heller *et al.* 1988) suggest that during periods of tectonic activity, coarse clastic sediment wedges are trapped close to the thrust front ('syntectonic' conglomerates), whereas during quiescence sediment is spread further from the thrust front, onlapping the basin margin ('antitectonic' conglomerates) (further discussion in §8.3.2) (Fig. 4.41). Other approaches involving the dynamics of catchment erosion and sediment dispersal (Densmore *et al.* 2007; Armitage *et al.* 2011) show a similar behaviour of proximal trapping

(a)

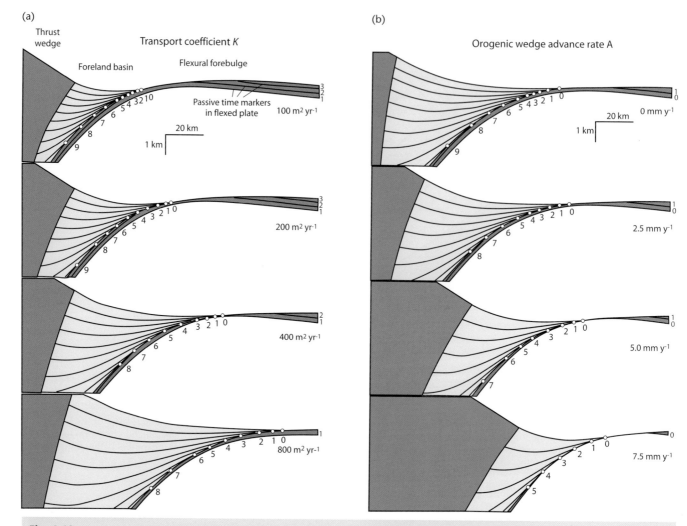

(b)

Fig. 4.40   Sensitivity tests of two of the parameters in Sinclair *et al.*'s (1991) diffusion model for mountain belt denudation and foreland basin evolution. The reference model involves a transport coefficient $K = 400\,m^2\,yr^{-1}$, slope angle = 1.5°, advance rate = 2.5 mm yr^{-1}, and equivalent elastic thickness of 20 km. In the sensitivity tests, one parameter is varied while the others are held constant at their reference values. Lines in basin are at equal-time steps (chrons). Numbered positions on top of foreland plate are pinch-out (stratigraphic onlap) positions. Lines within foreland plate are passive markers to show extent of erosion of foreland plate in flexural forebulge. (a) Effect of variations in transport coefficient K, showing major impact on basin depth, basin width and pinch-out migration rate. (b) Effect of variations in the orogenic advance rate, showing major impact on state of filling of the basin. Marine underfilled basins are favoured by low transport coefficients, low tapers and fast advance rates, whereas continental overfilled basins are favoured by high transport coefficients, high tapers and slow advance rates, irrespective of flexural rigidity. Reproduced with permission of John Wiley & Sons, Inc.

of conglomerate wedges during a transient period in response to an increase in slip-rate on border faults, followed by progradation of a conglomerate sheet as steady state is attained (Clevis *et al.* 2004; Allen & Heller 2012). The landscape response time to a change in tectonic forcing is c.10^6 years.

An episodically moving thrust sheet above a basal detachment has a strong impact on the stratigraphic patterns in the overlying basin (Clevis *et al.* 2004). Movement at moderate rates (~5 mm yr^{-1}) over low-angle (2–6°) faults causes sufficient uplift of rocks to counterbalance flexural subsidence, causing fluvial systems to prograde and spill sediment into neighbouring basins. This effect increases with rate of movement and angle of fault, potentially causing fluvial incision in the piggyback basin-fill.

These different model results emphasise the importance of the 'fine detail' of the different numerical models. Nevertheless, diffusional models have been helpful in enabling a link to be made between orogenic events and large-scale basin evolution.

4.6.4   Coupled tectonic-erosion dynamical models of orogenic wedges

Although a coupling between mountain belt denudation, exhumation, flexure and deposition was made in the diffusional models discussed in §4.6.3, the coupling was through the arbitrary choice of an effective transport coefficient or bulk diffusivity. This transport

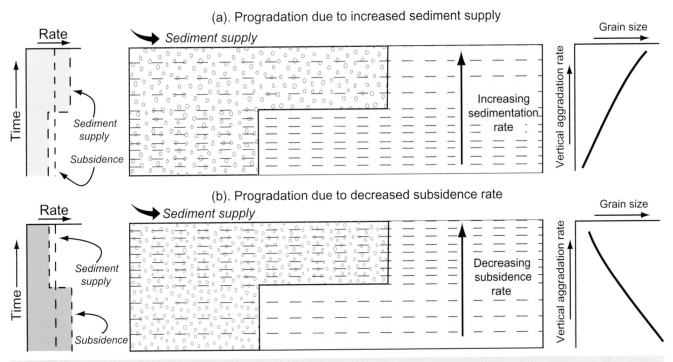

Fig. 4.41 Two contrasting situations leading to the progradation of gravels into a sedimentary basin. In (a), an increase in sediment supply causes a thickening of depositional increments and an upward coarsening of grain size. In (b), a reduction of accommodation caused by tectonic subsidence causes gravel progradation but thinner depositional increments. These two idealised cases are referred to as 'syntectonic' and 'antitectonic' gravels by Heller & Paola (1992). Reproduced from Allen & Heller (2012), with permission of John Wiley & Sons, Ltd.

coefficient encapsulates all of the different factors influencing erosion, deposition and sediment transport. In addition, tectonic fluxes have been crudely approximated (e.g. fault-bend fold over a crustal-scale ramp) or left unspecified by use of a critical taper model. Rarely, tectonic folds are allowed to grow by co-seismic slip increments and to be influenced by surface erosion (Ellis & Densmore 2006).

An early attempt to consider the mountain belt in planform, and to incorporate the important geomorphic elements (hillslopes, channels) in the sediment routing system, was the study by Koons (1989) of the Southern Alps of New Zealand. However, the channels were fixed and did not interact with the diffusive hillslopes in this early model. Subsequent models have incorporated the dynamics of stream incision, hillslope erosion and sediment transport (see Chapter 7) on a deforming crustal template. A commonly used example at the orogenic scale is a 2D plane strain thermomechanical model with a crust of thickness h, with a strength profile determined by a brittle Coulomb-type upper layer and a temperature-dependent power-law lower layer (Willett 1992; Willett *et al.* 1993; Fullsack 1995). Below the crust are two mantle plates – one representing the subducting foreland lithosphere, and the other representing the overriding upper plate. Convergence between the two plates induces a shear force or traction on the base of the crustal layer, driving deformation. We have previously considered this type of model in terms of Argand number and Ampferer number (§4.5.6).

The tectonic development of the orogenic wedge, its isostatic compensation and its erosional unroofing, are clearly crucial to the analysis of foreland basins. On the basis of plane strain numerical models, we can consider the tectonic evolution of a convergent boundary such as the Alps of Europe as following three stages (Beaumont *et al.* 1996a; Pfiffner 2000; Pfiffner *et al.* 2002; Kempf & Pfiffner 2004):

1. Initially, tectonics are dominated by subduction of the pro-lithosphere due to the negative buoyancy of the down-going slab (slab-pull effect). Subduction complexes form along a relatively thin shear zone above the subducting oceanic crust. Horizontal propagation of the accretionary prism is likely to be rapid, but the pro-foreland basin is likely to be underfilled with turbiditic 'flysch'.

2. In an intermediate stage, the subduction–collision transition, the subduction load decreases by the subduction of more buoyant continental lithosphere (or because of slab break-off), causing the pro-lithosphere to flexurally rebound. Parts of the pro-lithosphere detach and form part of a bivergent wedge instead of being subducted. Activity of a retro-shear should begin when the pro-lithosphere is no longer fully subducted during this transition phase.

3. In a final collision stage, when there is a negligible subduction load, well-defined pro- and retro-wedges form. The entire crustal pro-layer is incorporated into a critical-tapered orogenic wedge. Denudation of the bivergent wedge should be particularly strong during this phase. Flexural basins (pro- and retro-foreland basins) are supplied with large amounts of basins, causing filling and overfilling in the 'Molasse' phase.

Since the tendency for thickened crust to spread laterally is strongly affected by the gravitational force, denudation of the orogen will hinder lateral tectonic progradation. If the rate of crustal thickening equals the rate of denudation, the orogen should not grow in size. We can refer to this as a *steady-state orogen*. The same idea applies to critically tapered Coulomb wedges (§4.5.3). If the wedge taper is reduced, for example by erosion, the Argand number falls, preventing lateral tectonic progradation. Only after internal deformation of the wedge has restored gravitational forces to a threshold level can the wedge continue to prograde laterally.

Erosion at the surface can be modelled as either a function of elevation, relief or slope (Schlunegger & Willett 1999) or related to a set of geomorphic and climatic rules (Kooi & Beaumont 1994, 1996; Beaumont *et al.* 1996b; Willett 1999). As an example of the latter, erosion is commonly modelled through calculation of the rate of river incision, which is taken as proportional to stream power (§7.4.2, Appendix 44). The efficiency of erosion versus tectonic uplift can be gauged from a dimensionless number, termed *Erosion number* by Willett (1999)

$$Ne = \frac{kL}{u} \qquad [4.31]$$

where u is the rate of tectonic uplift, L is the half-width of the uplifting block of crust, and k is a coefficient proportional to the bedrock incision efficiency and precipitation rate with units time^{-1}. In the coupled tectonic-erosion model, Ne can be expressed in terms of the ratio between the accretion mass fluxes due to convergence (numerator) and the erosional mass flux through the upper surface (denominator), giving

$$Ne^* = \frac{4kL^2}{V_p h} \qquad [4.32]$$

The Erosion number strongly controls the topographic profile and the exhumation pattern (Fig. 4.42a, b). If Ne tends to infinity, erosion planes off the topography ruthlessly despite finite rates of tectonic uplift. The Erosion number Ne strongly controls the time taken for a mountain belt to achieve steady state and its maximum elevation. Coupled models therefore now allow specific precipitation patterns, such as wet orographic flanks (high Ne) and dry rain-shadow flanks (low Ne), to be investigated as mechanisms for particular patterns of exhumation and topography (Fig. 4.42c).

Do surface processes of erosion and deposition affect tectonic deformation in orogens and fold-thrust belts? Scaled analogue experiments using sandboxes indicate that erosion plays an important role in controlling structural style, localisation of deformation and exhumation trajectories (Konstantinovskaya & Malavieille 2005; Bonnet *et al.* 2008) (Fig. 4.43).

It is generally accepted that mass redistribution related to surface processes modifies the state of stress of the lithosphere, which may induce internal deformation and an overall isostatic adjustment (Beaumont *et al.* 2000). Prolonged erosion in rapidly deforming regions can lead to advection of hot, mechanically weak rocks towards the surface, causing enhanced localisation and exhumation through a feedback process, as proposed in regions such as the Nanga Parbat of the Himalayas and the Southern Alps of Nevw Zealand (Zeitler *et al.* 2001a, b; Finlayson *et al.* 2002; Koons *et al.* 2003). The spatial scale

at which interaction can take place between surface processes of erosion and deposition and tectonic deformation is critical to the role of river incision on plateau flank and mountain peak uplift (Molnar & England 1990; Montgomery 1994).

Although erosion and deposition have been demonstrated to be important in the evolution of natural orogens, that is, at very large spatial scales (Beaumont *et al.* 1992; Avouac & Burov 1996; Willett 1999; Pfiffner 2000), there are reasons that the coupling between deformation and surface processes is more problematic at smaller length scales (<100 km). For example, rock strength and flexural rigidity causes flexural compensation of the small loads associated with river valleys, which would therefore not be experienced by the deforming crust. The plastic deformation of the crust is therefore important for the operation of coupling at intermediate spatial scales.

The surface processes of erosion and deposition may interact with three-dimensional deformation in a number of settings. Here, we are primarily interested in the possible coupling between surface processes and tectonics in fold-thrust belts.

Previous studies of the interaction of tectonic deformation and surface processes were restricted to 2D. In the simplest 3D studies, the 3D deformation was imposed kinematically and uncoupled from surface processes. The substrate was regarded as rigid and unresponsive to the patterns of erosion and deposition at the surface. Some studies of long-lived escarpments and orogenic wedges fall into this category (Tucker & Slingerland 1994; Willett *et al.* 2001). Other models constructed tectonic deformation kinematically (either as an initial condition, or allowed to evolve subsequently) and introduced a flexural isostatic response to erosion and deposition (e.g. Kooi & Beaumont 1994; Johnson & Beaumont 1995; Garcia-Castellanos 2002) mediated by the flexural rigidity. To understand how surface processes influence internal deformation requires full (e.g. Braun 1993; Braun & Beaumont 1995) or simplified thin-sheet 3D mechanical models (Medvedev & Podlachikov 1999a, b).

It has been recognised for some time that foreland basin megasequences show a temporal change from deep-water deposition in an underfilled basin to shallow-marine/continental deposition in a filled basin (Covey 1986; Allen *et al.* 1991; Sinclair 1997). The underfilled 'flysch' stage was attributed to the combination of a fast advancing orogenic wedge with low rates of sediment yield from a largely submarine, low-angle wedge. The filled 'Molasse' stage was attributed to a slowing of frontal advance and high sediment yield from a largely subaerial, steep wedge (Sinclair & Allen 1992; Allen & Allen 2005). These relationships are supported by a geodynamic model of an orogenic wedge that evolves during progressive shortening that includes the coupling between erosion of the wedge and its deformation (Simpson 2006b). The emergence of the wedge above sea level is associated with a number of effects in the numerical model: increased exhumation rates of the orogen, reduced thrust-front advance rates, a narrowing of the orogen width, and a shift from slow-deposition in piggyback basins to rapid deposition in an extensive foredeep.

Consequently, the main effects of a wedge becoming subaerial are: (i) erosion rates are expected to increase; and (ii) the gravitational load of water will decrease as water depths reduce during surface uplift of the wedge, reducing to zero on emergence. The geodynamical model used to generate the results in Fig. 4.44 is an elasto-plastic rheology (sediment) with a frictional basal décollement above an elastic plate (T_e 10 km) situated above an inviscid substrate. The

(a)

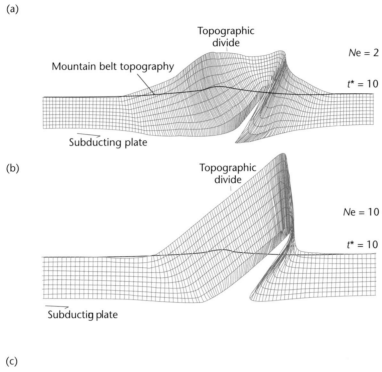

(c)

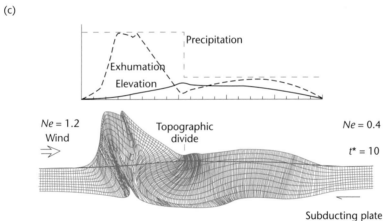

Fig. 4.42 Variations in coupled tectonic-erosion model for *Ne* = 2 (a) and *Ne* = 10 (b) for a dimensionless time of $t^* = 10$ ($t^* = tV_p/h$) (after Willett 1999; reproduced with permission of the American Geophysical Union). In both cases, elevation and exhumation have reached steady state. Note the greater amount of erosional removal of rock (exhumation) in the high-erosion case. (c) Numerical model with orographic focusing of precipitation such that the pro-wedge has an erosion number *Ne* of 0.4 and the retro-wedge an erosion number of 1.2, which closely reflects the precipitation patterns in the Southern Alps of New Zealand.

mechanical model is flexurally compensated, and erosion and deposition are by simple diffusion. During steady convergence, the wedge starts as a classic fold-thrust belt with regularly spaced thrust-related anticlines and intervening wedge-top basins. As the wedge becomes partially subaerial, and deformation localises at the rear, causing rapid exhumation, accompanied by a high discharge of sediment to fill a broad foredeep. Piggyback basins are uplifted and eroded and the thrust-front advance rate slows.

Simpson (2006c) introduced a dimensionless parameter reflecting the efficiency of surface processes given by

$$\tilde{\kappa} = \frac{\kappa}{L^2 \dot{\varepsilon}} \qquad [4.33]$$

where κ is the surface process diffusivity, L is the length scale of the initial thickness of the elasto-visco-plastic layer, and $\dot{\varepsilon}$ is the initial

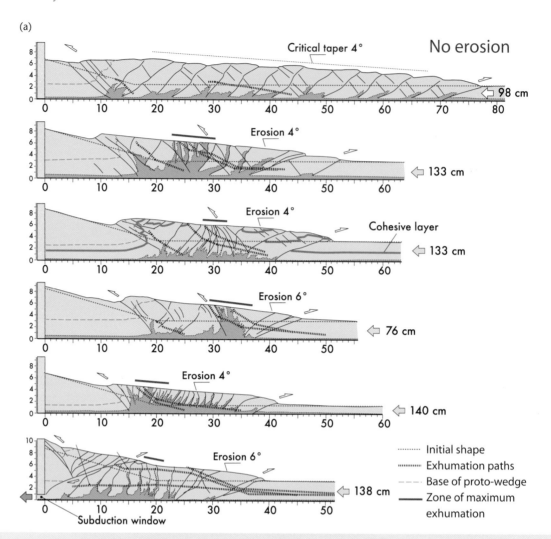

(a)

Fig. 4.43 Results of scaled sandbox analogue experiments to investigate the effect of erosion rate on the development of fold-thrust belts, from Konstantinovskaya & Malavieille (2005). Sequences begin with the no-erosion case, for low (a) and high (b) basal friction experiments. The no-erosion cases show an imbricated stack of uniformly spaced foreland-verging thrusts. The cases with erosion show the total amount of shortening on the right of each panel. In the low friction experiments, erosion causes exhumation to be concentrated at the mid-rear of the wedge, associated with clusters of closely spaced subvertical faults. In the high basal friction experiments, exhumation takes place along inclined thrusts in the middle part of the wedge. Erosion accounts for up to 50% of the total area of the initial model. Reproduced with permission of John Wiley & Sons, Inc.

imposed strain rate. The non-dimensional diffusivity is the ratio of the deformation time scale $(1/\dot{\varepsilon})$ to the time scale for surface mass transport (L^2/κ). Low values of $\tilde{\kappa}$, where tectonic deformation is essentially unaffected by surface processes, promote the formation of an array of thrust-related folds defining regularly spaced wedge-top basins, with exhumation peaking at each fold culmination. At very high values of $\tilde{\kappa}$, an extensive foredeep is formed, with no wedge-top deposition. Exhumation is concentrated in an intense region of shortening at the rear of the wedge (Fig. 4.45). Transitional states exist between these two extremes.

In Simpson's models, erosion and deposition influence the deforming plate by modifying the distribution of vertical surface loads and

by changing the plate thickness and therefore the strength distribution. The surface evolution processes are modelled as transport-limited, consisting of two parts: a concentrative fluvial transport (as in channels) and a dispersive transport (as on hillslopes) (Smith & Bretherton 1972; Simpson & Schlunegger 2003):

$$\frac{\partial h}{\partial t} = \frac{\partial}{\partial x}\left[\left(\kappa + cq^n\right)\frac{\partial h}{\partial x} \right] + \frac{\partial w}{\partial t} \qquad [4.34]$$

where h is the surface topography, w is the deflection of the plate, κ is a hillslope diffusivity, q is the surface fluid flux related to rainfall,

(b)

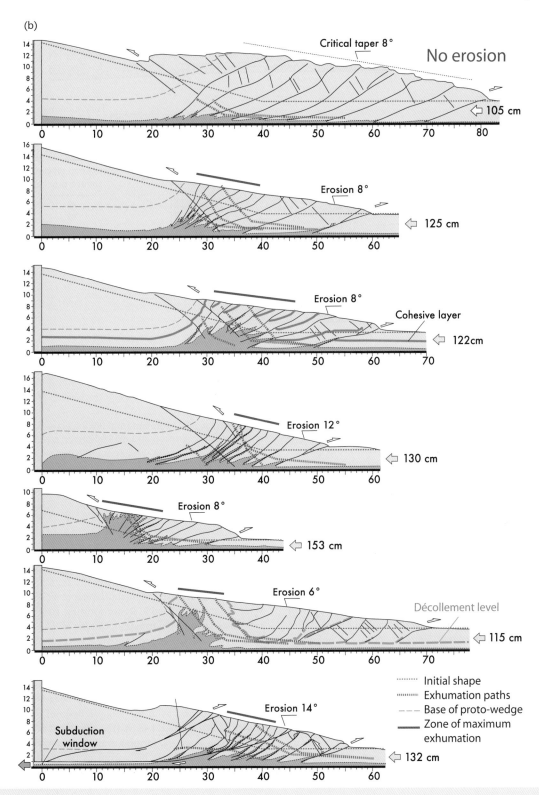

Fig. 4.43 *Continued*

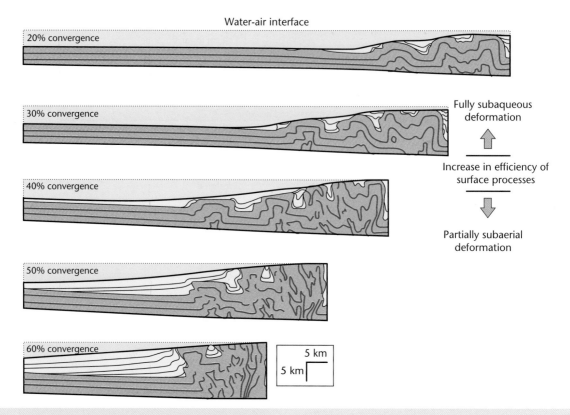

Fig. 4.44 Evolution of an orogenic wedge, from Simpson (2006b), using a thin layer of sediment on a flexing elastic lithosphere above a weak (inviscid) substrate, which undergoes progressive convergence and shortening. Erosion and deposition are governed by diffusion. The emergence of the wedge above sea level tends to reactivate tectonic deformation in the orogen, a reduction of orogen width and a stalling of frontal propagation. This leads to a cessation of wedge-top basin development and the promotion of thick foredeep deposition. Reproduced with permission of John Wiley & Sons, Ltd.

c is a fluvial erosion coefficient, n is a fluvial erosion power exponent, and $\partial w/\partial t$ is the 'tectonic' uplift/subsidence rate. Inspection of eqn. [4.34] shows that the first bracketed term represents erosion/deposition, whereas the second bracketed term accounts for uplift or subsidence caused by deformation of the underlying plate. When eqn. [4.34] is nondimensionalised (Simpson 2004a), an important parameter emerges – the ratio of a time scale for deformation to a time scale for fluvial erosion defined as

$$R = \frac{c\alpha^n L^{n-1}}{\dot{v}} = \frac{\tau_d}{\tau_e} \qquad [4.35]$$

where L is the characteristic length scale, $\dot{v}$ is the imposed boundary displacement rate, the deformation time scale is $\tau_d = L/\dot{v}$, and the fluvial erosion time scale $\tau_e = L^{2-n}/(c\alpha^n)$, where α is the rainfall rate in excess of infiltration. Consequently, when R is large, fluvial processes are fast relative to the rate of deformation, and the reverse is true where R is small.

Simpson (2004b) also varied the initial regional topographic slope (denoted by β) between 0% and 2%. His numerical model indicates that when β and/or R are low (that is, the initial regional topography is flat and/or relative rates of deformation compared to fluvial processes are low), folding is two-dimensional with small-scale drainage

networks transporting sediment locally from anticlines to adjacent basins. This may be characteristic of the Zagros, for example. When β and/or R are relatively high, the development of a large-scale transverse river network cuts across growing structures and strongly influences tectonic deformation, causing amplification and localisation of folding. The largest transverse rivers are therefore associated with the greatest structural and topographic relief of doubly plunging folds.

The interaction of tectonic deformation and surface processes is ideally studied where deformation propagates into the margins of foreland basins, as in the Junggar Basin of north-central China (Fu *et al.* 2003) (Figs 4.46, 4.47). The thrust-related anticlines are cut through by rivers, forming transverse canyons, as is commonly observed in fold-thrust belts from as far afield as the Apennines, Zagros, Pyrenees, Swiss and French Alps, Himalayas and central Andes (Oberlander 1985; Alvarez 1999; Simpson 2004a, b).

4.6.5   Modelling aspects of foreland basin stratigraphy

Flexure plays a major role in the generation of accommodation for wedge-top and foredeep deposition of stratigraphy. The typical large-scale geometry of foreland basin stratigraphy is of wedge-shaped

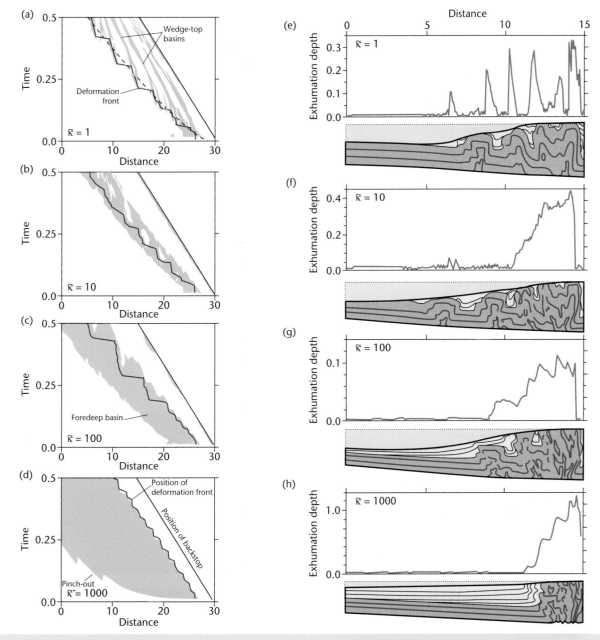

Fig. 4.45 Effect of surface processes on orogenic wedge and foreland basin evolution, after Simpson (2006c). Left: time-distance charts showing chronostratigraphy at different values of the dimensionless transport coefficient $\tilde{\kappa}$, from 1 to 1000. Wedge-top deposition at low values of $\tilde{\kappa}$ evolves into extensive foredeep deposition at high $\tilde{\kappa}$. The time scale is dimensionless, and the horizontal distance is scaled by the original thickness of the deforming layer. Right: shallow exhumation of arrays of folds at low values of $\tilde{\kappa}$ evolves to deep exhumation of wedge at high $\tilde{\kappa}$. The exhumation depth and the horizontal distance are both scaled by the original thickness of the deforming layer. Reproduced with permission of John Wiley & Sons, Ltd.

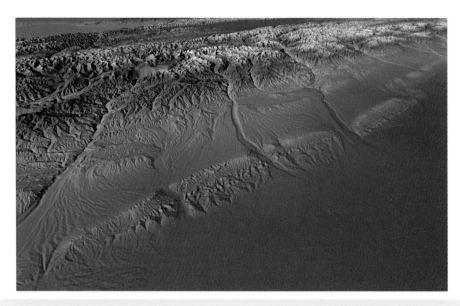

Fig. 4.46 Oblique digital elevation model of the front of the northern Tien Shan at the margin of the Junggar Basin, central-northern China, showing the interaction between tectonic deformation and transverse rivers and fans.

(a)

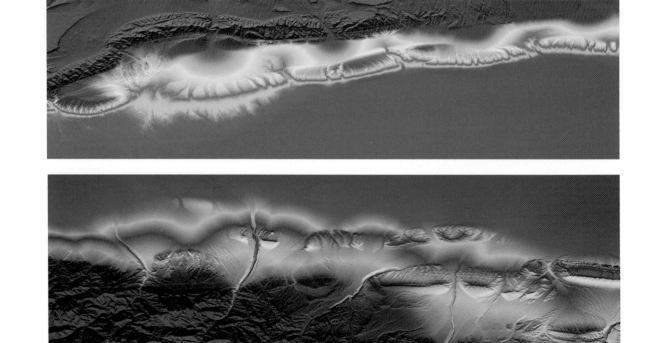

Fig. 4.47 (a) Digital false-colour topographic images of northern Tien Shan in the region of Kuitan, showing frontal thrust-related folds and surface drainage. (b) Geological map showing deformation of Quaternary. (c) Structural cross-section. C, Carboniferous; P-T, Permo-Triassic; J1, J2, J3, Jurassic; K, Cretaceous; E, Eocene; N1, N2, Neogene; Q, Quaternary. Reprinted from Fu *et al.* (2003), with permission from Elsevier.

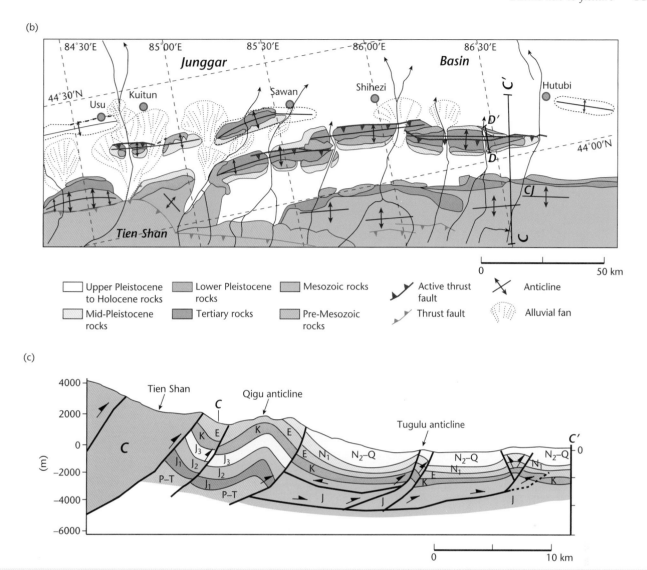

(b)

Upper Pleistocene to Holocene rocks
Mid-Pleistocene rocks
Lower Pleistocene rocks
Tertiary rocks
Mesozoic rocks
Pre-Mesozoic rocks
Active thrust fault
Thrust fault
Anticline
Alluvial fan

(c)

Fig. 4.47 *Continued*

units, thick close to the orogenic load and thinning onto the foreland into a 'feather-edge'. This is a reflection of the lateral gradient in subsidence rate from the centre of the load to the peripheral bulge. Superimposed on this is commonly a spatial translation (with respect to the underlying plate) of stratigraphic units ahead of a moving load, a consequence of continued convergence being accommodated in the orogenic belt. This effect causes a general onlap of successively younger stratigraphy onto the foreland. Numerous case studies exist that demonstrate this simple geometrical style, such as the Magallanes Basin of southern South America (Fig. 4.48).

The deflection of a foreland plate must depend on the flexural rigidity of the flexed lithosphere, the nature of the distributed loads (topographic/thrust loads, subsurface loads, horizontal end forces, bending moments, sediment and water loads) and the presence of pre-existing heterogeneities. As a consequence, it is vital in considering foreland basin stratigraphy to consider the previous geological history of the lithosphere. The most common scenario is the *Wilson cycle* (§1.4.5).

The Wilson cycle implies that during continental collision, the overriding plate flexes an inherited passive margin structure in the underlying plate. Two points immediately stand out: (i) since the lithosphere has been previously heated and stretched (§3.1) it will possess a flexural rigidity or equivalent elastic thickness that reflects this history, the 'memory' diminishing with time since rifting; and (ii) the first loads will be emplaced on a pre-existing continental margin bathymetry rather than on an imaginary flat surface representing the 'regional' elevation (Fig. 4.49).

Stockmal *et al.* (1986), using a depth-dependent plate rheology, proposed four end-member models to account for the transition from passive margin to foreland basin. Whereas at early stages the thermal age of the passive margin is an important, even dominant control on basin development, this effect becomes less important as

(a)

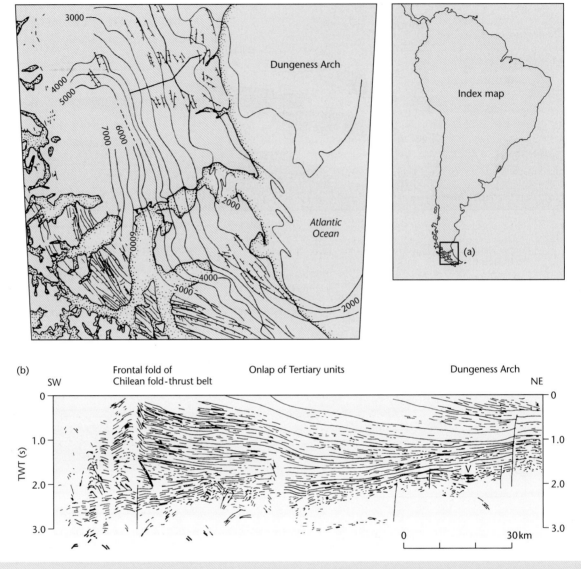

Fig. 4.48   The Magallanes Basin, southern South America (after Biddle *et al.* 1986). (a) Isopachs (in m) of sedimentary fill of Magallanes Basin based on a structural contour map drawn on top of the Tobifera volcanics (representing the last stages of rifting prior to foreland basin flexure (Bruhn *et al.* 1978; Gust *et al.* 1985). Isopachs show a clear thinning onto the Dungeness Arch. (b) Line drawing interpretation of seismic reflection record from the frontal folds of the Chilean fold-thrust belt to the flank of the Dungeness Arch. Stratigraphic units show strong onlap above the Tobifera volcanics (V). The clinoforms are thought to be fan deltas derived from the Andean mountain belt that prograded into deep-marine shales.

the overthrust wedge is progressively emplaced on stronger unstretched lithosphere. At this stage, the thickness of the overthrust load is of greatest importance. The maximum foreland basin depths prior to post-deformational erosion vary from ~3 km for a low topography (e.g. Zagros) to ~16 km for a high topography (e.g. Himalayas). Post-deformation erosion and tectonic thinning (extension) can cause massive unroofing of up to 40 km, bringing high-grade metamorphic rocks to the surface. Very thick overthrust wedges can develop with little topographic expression if they are emplaced

on a deep oceanic bathymetry. Further information on the development of orogenic wedges and their numerical modelling is found in §4.5.

4.6.5.1   The flexural forebulge unconformity

Passage of the flexural forebulge can cause a complex arrangement of unconformities (Price & Hatcher 1983). Initial uplift in the forebulge region is followed by subsidence as the bulge moves onto the

Passive margin stage

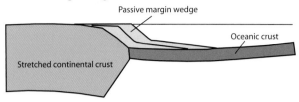

Early convergent stage

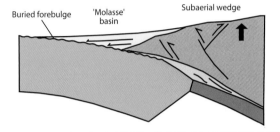

Late convergent stage

Fig. 4.49 Model involving orogenic loading of a previously stretched continental margin during the early stages of convergence (Stockmal *et al.* 1986 – AAPG © 1986 and reprinted by permission of the AAPG whose permission is required for further use; Watts 1992), modified by Allen *et al.* (1991). The first orogenic loads are emplaced on a weaker lithosphere at considerable water depths.

craton. The unconformities should migrate in a time-transgressive manner onto the craton (Crampton & Allen 1995).

Consider a flexural forebulge unconformity generated by an orogenic load moving across a linear elastic foreland plate of flexural rigidity *D*. If the advance of the orogen is steady, the stratigraphic gap is dependent on: (i) the time-average erosion rate on the forebulge; and (ii) the duration of uplift in the forebulge region, given by

$$\tau = \frac{x_2 - x_1}{V} \tag{4.36}$$

where x_1 and x_2 are the first node and second node of the deflection respectively, which is equivalent to the width of the forebulge, and *V* is the advance rate of the orogenic system. The width of the forebulge on a continuous elastic plate scales on the flexural parameter α (eqn. [4.4]), with $x_2 = 7\pi\alpha/4$ and $x_1 = 3\pi\alpha/4$. Hence

$$\tau = \frac{x_2 - x_1}{V} = \frac{\pi\alpha}{V} \tag{4.37}$$

where the flexural parameter is given by

$$\alpha = \left\{ \frac{4D}{\Delta\rho g} \right\}^{1/4} \tag{4.38}$$

and $\Delta\rho$ is the difference in density between the mantle and the material filling the deflection. The model stratigraphic gap should increase from zero at the point of conformity corresponding to the first node of the deflection, to a maximum at a distance corresponding to the initial position of the second node of the deflection (Fig. 4.50). In other words, this zone of increasing stratigraphic gap is equivalent to the initial width of the forebulge. In the case of inherited bathymetry from the passive margin stage, the zone of increasing stratigraphic gap is telescoped because the initial uplift of the forebulge takes place below sea level. The model prediction of Crampton and Allen (1995) compares favourably with the observed spatial distribution of stratigraphic gap at the base of the Tertiary foreland basin megasequence in the Alps of Switzerland (Fig. 4.51).

In some cases, such as the Devonian (Acadian) peripheral bulge to the Appalachian orogen, the forebulge unconformity system shows an orogenward migration that has been attributed to viscoelastic relaxation of the lithosphere (Tankard 1986). A series of splaying unconformities that overstep systematically toward the orogen indicates contemporaneous uplift and orogenward movement of the forebulge at this time.

The location and geometry of a flexural forebulge may also be strongly influenced by pre-existing lithospheric heterogeneities (Waschbusch & Royden 1992). The flexural forebulge may position itself over weak zones in the lithosphere, and then episodically migrate to another weak zone, instead of migrating steadily. Reactivation of old faults may also strongly disrupt the uplift, subsidence and migration history of the forebulge region. Extensional reactivation of normal basement faults by outer-arc bending stresses (Bradley & Kidd 1991) or inversion of normal basement faults by the compressional stresses generated during orogeny (Meyers *et al.* 1992; Ussami *et al.* 1999; Gupta & Allen 2000) have both been described.

4.6.5.2 Foreland basin isopachs and pinch-outs

The migration of depocentres and of feather-edge pinch-outs of stratigraphy gives an impression of the mobility of the distributed loads and/or variations in the lithospheric response. An early modelling exercise was carried out on the Cretaceous retro-foreland basin in the western United States (Jordan 1981). The progressive eastward shift of depocentres through time in the Idaho–Wyoming region was predicted using the palinspastically restored thrust load configuration during the Sevier orogeny. Careful matching of predicted with observed stratigraphy allowed a flexural rigidity of about 10^{23} Nm to be estimated for the Cretaceous lithosphere. There was no necessity to invoke a changing flexural rigidity with time, suggesting that the lithosphere behaved elastically over the modelled 70 Myr time span and that lateral variations in rigidity were also insignificant. Similar modelling exercises have been carried out in the Neogene Argentinian foreland of South America (Cardozo & Jordan 2001) and in the Triassic Longmen Shan foreland basin of Sichuan, China (Yong *et al.* 2003).

Some of the Alpine foreland basins of Europe also show a clear relationship between load mobility and depocentre migration

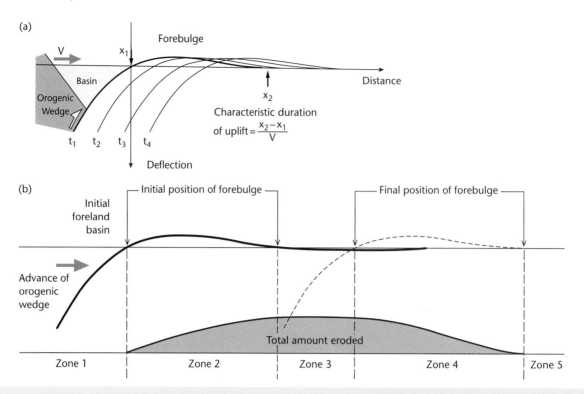

Fig. 4.50 (a) Flexural profiles at times t_1 to t_4, to show the duration of uplift in the flexural forebulge region, using a broken elastic plate. (b) Geometry of the flexural forebulge unconformity. The existence of inherited bathymetry causes zones 1 and 2 to be translated further onto the foreland plate, and zone 2 becomes narrower. After Crampton & Allen (1995).

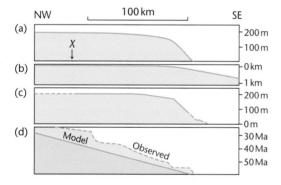

Fig. 4.51 Fit between model predictions for the forebulge unconformity and observations in eastern Switzerland, using the linear elastic model of Crampton & Allen (1995). (a) The modelled total erosion at the forebulge unconformity versus distance from the initial position of the orogenic wedge, with Te = 10 km, orogenic advance rate 8 mm yr⁻¹ and erosion proportional to elevation with coefficient 0.02. Point *X* is initially 200 km from the orogenic front. Note the rapid increase in erosional gap from 0 to 200 m across the zone of inherited shelf bathymetry shown in (b). (c) The observed erosion versus approximate restored distance from the orogen (based on Herb 1988), showing close similarity to the modelled erosion. (d) As a test, the modelled onlap of sedimentary rocks overlying the unconformity are compared with the observed onlap along a restored section (based on Herb 1988, Sinclair *et al.* 1991 and Crampton & Allen 1995).

(Puigdefàbregas *et al.* 1986; Homewood *et al.* 1986; Puigdefàbregas *et al.* 1992; Vergés *et al.* 1998; Burkhard & Sommaruga 1998). In the North Alpine Foreland Basin of western Switzerland, a classical pro-foreland basin, pinch-outs and depocentres of stratigraphic units migrated onto the European craton during the Oligo-Miocene, corresponding to the main collisional phase of Alpine orogenesis. A palinspastic restoration of thrust units over the same time period coupled with a time-bracketing technique for dating thrust movements allowed fault tip propagation and shortening rates to be estimated (Homewood *et al.* 1986; Sinclair & Allen 1992). There was a close correspondence between the rates of depocentre and pinch-out migration and of thrust-related shortening in western Switzerland. This feature initially suggests, as in the Cretaceous Sevier belt, that the lithospheric response is constant through time.

If the pinch-out migration rate far exceeds the rate at which the mountain belt advances, the foreland basin should progressively widen with time (Fig. 4.52). This might result from a number of processes. For example, it would be expected if the mountain belt loaded a progressively stronger elastic lithosphere as it overrode first the passive margin, then the unstretched craton. It might also result from the forelandward shift of the centre of gravity of the load by erosion and deposition of sediment in the basin. On the other hand, if the pinch-out migration rate slows relative to the rate of advance of the mountain belt, or if the forebulge is 'dragged' inwards toward a stationary mountain front, the foreland basin should narrow (Fig. 4.52). Although this kind of offlapping relationship has been interpreted to result from stress relaxation in a viscoelastic lithosphere (Quinlan & Beaumont 1984; Tankard 1986), a similar effect can be

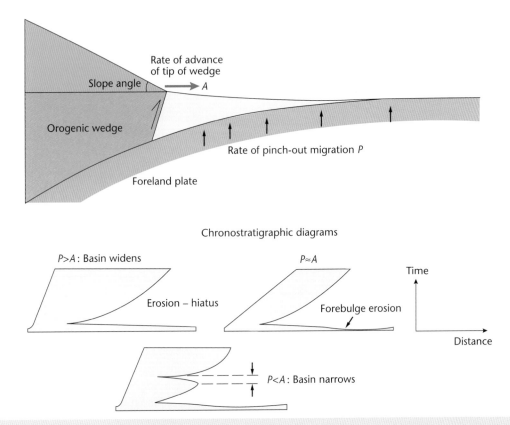

Fig. 4.52 Influence of the rates of advance of the tip of the orogenic wedge and of pinch-out migration on the foreland basin width. Note that onlap onto the foreland plate takes place even in the case of a time-constant basin width. A narrowing of the basin has been interpreted as due to viscoelastic relaxation, but can also result from reorganisation of the orogenic load on an elastic plate.

produced by thickening of the adjacent orogenic wedge loading an elastic plate (Flemings & Jordan 1989; Sinclair *et al.* 1991).

4.6.5.3 Underfilling and overfilling

The topography caused by thrusting in mountain belts is eroded to provide detritus to fill the foreland basin. Moretti and Turcotte (1985) first applied a diffusion equation to this process, and Flemings and Jordan (1989) and Sinclair *et al.* (1991) have specifically applied 2-D diffusion modelling to the stratigraphy of foreland basins. The coupling between orogenic wedge development and basin filling has been extended to 3-D by Johnson & Beaumont (1995) and Clevis *et al.* (2004).

The 2-D diffusion model provides a starting point for considering the controls on foreland basin stratigraphy. Sinclair *et al.* (1991) modelled the stratigraphy of the North Alpine Foreland Basin in the Central Alps using a diffusion model for erosion, a thrust-wedge model based on the concept of a critical taper (§4.5.3 for details),

and a rheology of a linear elastic plate for the flexed European lithosphere. The amount of sediment transported to the foreland basin is proportional to the slope, the transport coefficient (K) varying over several orders of magnitude according to climatic and hydrologic setting. Flemings and Jordan (1989), for example, suggest that K for fluvial transport is c.10^4 m^2 yr^{-1}, whereas it is only 10^{-2} m^2 yr^{-1} for the decay of isolated hillslopes. The values obtained from a study of the modern sub-Andean foreland were of the order 10^4 m^2 yr^{-1} (Flemings & Jordan 1989). Marr *et al.* (2000) estimated the transport coefficient to be 10^4 m^2 yr^{-1} and 10^5 m^2 yr^{-1} for gravel and sand respectively. The magnitude of the rate of transport of sediment into the basin clearly determines whether, for a given flexural response, it is *underfilled* or *overfilled* (Covey 1986).

Sinclair *et al.* (1991) simulated movement of the forebulge, both vertically and horizontally, and the attendant unconformities produced in the basin-filling stratigraphy purely by varying the rate of thrust belt propagation and wedge thickening coupled with diffusive erosion. It was not necessary to vary the flexural rigidity of the European plate through time, nor to invoke a non-elastic rheology

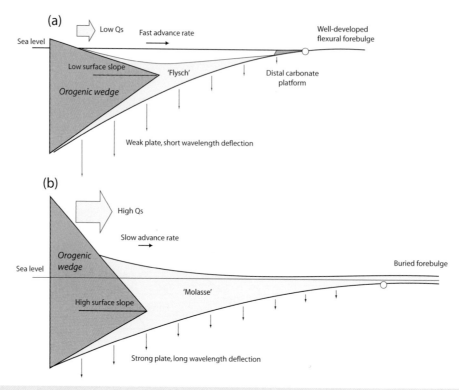

Fig. 4.53 Highly schematic illustration of the factors favouring underfilled 'flysch' basins (a) compared to overfilled 'Molasse' basins (b). Pro-foreland basin systems commonly evolve from flysch to Molasse stages.

for the flexed plate. Subtle unconformities ('sequence boundaries') associated with abrupt basinward shifts or progressive offlap could be produced by thickening the load and slowing its propagation rate. Erosion of the load then caused onlap as the centre of gravity of the load migrated forelandwards through redistribution of sediment from wedge to basin. Rapid advance of the thrust tip encour-

aged underfilling of the basin, whereas lower values of thrust belt migration rate promoted overfilling, the sedimentation keeping pace with the creation of new space in the basin (Fig. 4.53). These combinations of factors are supported by the results of numerical models of coupled surface processes and wedge tectonics (Simpson 2006a).

CHAPTER FIVE
Effects of mantle dynamics

Summary

Plate tectonics operates in the upper thermal boundary layer of an underlying mantle convection system. Flow within the mantle takes place at different spatial scales, including flows induced by the subduction of cold lithospheric slabs, incubation under continental lids, small-scale convection driven by edge effects, lower-mantle megaplumes, and upper-mantle hotspots. This flow and the consequent mass anomalies at depth result in a dynamic topography at the Earth's surface that is important in basin analysis.

The starting point for an analysis of mantle flow is a consideration of buoyancy forces set up by density heterogeneities in the mantle. Buoyancy forces are counteracted by viscous resistance. The force balance provides fundamental information about the scaling between parameters and allows the derivation of dimensionless groups that describe various aspects of the flow and thermal characteristics of a fluid layer, such as the Rayleigh number. Consideration of the Rayleigh number for the mantle indicates that convection must be occurring. This convective flow efficiently transports heat and drives high plate velocities, but is laminar.

Numerical and physical experiments illustrate the convection patterns of upwellings and downwellings produced by heating a fluid layer from below, from the surface and with internal radiogenic heating. Models using highly temperature-dependent viscosity structures in the mantle suggest that large aspect ratio convection cells are stable, such as must underlie the Pacific plate. Detailed three-dimensional velocity models of the mantle (seismic tomography) also show mass anomalies that must be sustained by flow. These mass anomalies show that the effects of plate tectonics, such as mid-ocean ridges and subducting slabs, can be recognised deep into the Earth's interior.

Measurements of the Earth's gravity field reveal geoid height anomalies. An underlying convecting system should produce variations in the height of the geoid over upwelling and downwelling limbs. At long wavelength, the observed geoid of the Earth shows major zones of positive and negative geoid anomaly. Viscous flow models of the mantle suggest that mass excesses in the upper mantle caused by the remnants of cold subducted slabs generate geoid lows, whereas hotspots and mid-ocean ridges are correlated with the presence of hot megaplumes in the lower mantle and are associated with geoid highs.

The surface topography on land, and the bathymetry beneath the sea, produced by mantle flow, is known as dynamic topography. It represents a deflection of the surface of the Earth caused by the presence of 'blobs' of buoyancy in the mantle. It can be recognised, for example, in regions of plate subduction due to the cooling effect of the oceanic slab, beneath supercontinents due to heating caused by the presence of an insulating lid, near the edges of continents due to small-scale convection triggered by the ocean–continent boundary, and, of course, in the form of topographic 'hotspot' swells in the ocean associated with magmatism. Some of the sub-plate flow structures responsible for topographic doming and magmatism are thought to be plumes originating from the core–mantle boundary. There is evidence of unsteadiness in plume activity, with pulses of buoyancy causing relatively rapid uplift of the Earth's surface followed by subsidence. Such pulses may cause cycles of erosional landscape development and delivery of sands to the deep sea, followed by draping with marine sediments.

Mantle flow is commonly associated with melt generation, igneous underplating and surface magmatism. The presence of plumes may cause sufficient elevation of asthenospheric temperatures to result in adiabatic decompression and the formation of flood basalt provinces (large igneous provinces – LIPs). The North Atlantic igneous province and the Ethiopian Afar regions are examples. LIPs are commonly connected to currently active volcanic hotspots by tracks of extinct volcanoes.

Mantle dynamics affects basin development by causing topographic uplift and the export of particulate sediment from erosional landscapes to sedimentary basins. Mantle dynamics also causes negative dynamic topography in the form of sag-type basins of the continental interiors. In geological history, cratonic basins have initiated preferentially at times of continental dispersal away from supercontinental assemblies. Their long, slow subsidence has been attributed to negative dynamic topography over cold downwellings, to thermal contraction of previously stretched and heated thick continental lithosphere, and to sediment loading following an earlier stage of stretching.

Dynamic uplift of the continental surface produces domal features that promote centrifugal river drainage patterns. The periods of dynamic uplift can potentially be recognised by the location of erosive knickzones in the river long profile. Time-varying dynamic topography may also affect the position of continental drainage divides and the direction of river discharge to the ocean, as in the reversal of Amazon drainage from the Pacific to the Atlantic.

Dynamic topography is also a primary factor in the history of long-term sea-level change and the extent of continental flooding. In particular, times of extensive uplift of the seafloor in the form of superswells, as in the Pacific during the Cretaceous, may be primarily responsible for elevated sea levels, rather than reflecting increases in spreading rate in the mid-ocean ridge system.

Basin Analysis: Principles and Application to Petroleum Play Assessment, Third Edition. Philip A. Allen and John R. Allen.
© 2013 John Wiley & Sons, Ltd. Published 2013 by John Wiley & Sons, Ltd.

5.1 Fundamentals and observations

5.1.1 Introduction: mantle dynamics and plate tectonics

Although plate tectonics has been highly successful as an explanation of the relative motion of plates and the deformation at their boundaries, there is little integration of mantle dynamics into what is essentially a kinematic theory. For example, volcanic hotspots such as those of Hawaii and Iceland, which are related to the ascent of deeply rooted mantle plumes, do not fit into plate tectonic theory. Within the solar system, the operation of plate tectonics is unique to Earth, but plume volcanism occurs on Mars and Venus as well as on a number of moons. The surface of Mars appears to be a single static plate acting as a 'frozen lid'; Venus is a convecting planet beneath a thick viscous lid, but no mobile plates, and the tectonics is spatially continuous and unstable. On Earth, therefore, we have a particular problem in understanding the interaction of internal thermal convection with an upper thermal boundary layer made of a discrete number of plates in relative motion.

The three-dimensional mapping of seismic velocity variations in the mantle (seismic tomography) now provides unprecedented detail on small but systematic density variations that reflect tempera-ture (and/or chemical) heterogeneities. These heterogeneities reveal a picture of deep penetration of cold subducted lithospheric slabs into the mantle, and of large hot upwellings that seem to rise from a highly dynamic core–mantle boundary (Fig. 5.1). Subducted slabs of lithosphere appear to both penetrate and to queue at the 670 km discontinuity. The residues of the subducted slabs that penetrate the discontinuity appear to descend to the core–mantle boundary, where they are reheated before being resurrected as a mantle plume. Mantle plumes appear to travel upwards and impinge on the base of the lithosphere, spreading out like a mushroom, and uplifting the overlying plate. Extension of the lithosphere over the plume head may lead to continental splitting (Chapter 3), the formation of new spreading centres and further subduction of lithosphere.

Convection systems in the mantle are the engines for the surface tangential motions of plates. Subducting slabs are cold downwellings, and spreading oceanic ridges are upwellings (Davies & Richards 1992), so the lithospheric plates must be regarded as an integral part of the convection system rather than being dragged passively by basal traction over the convection system. In other words, plate tectonics is the surface expression of convection.

Plate tectonic theory explains a wide range of geological and geophysical observations in the oceans and at plate boundaries. It has been far less successful in explaining the topography, seismicity, neo-

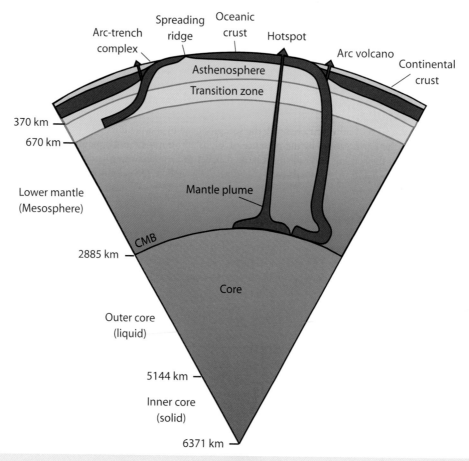

Fig. 5.1 Schematic section through a segment of the Earth showing the subduction of lithospheric slabs that are laid out along the 670 km discontinuity or penetrate to a graveyard above the core–mantle boundary (CMB). A mantle plume is shown rising from the CMB to produce a hotspot in the overlying oceanic plate. Modified from Stern (2002) and reproduced with permission of John Wiley & Sons, Inc.

tectonics and subsidence history of the continental interiors. This is partly due to the complex rheology of the continental lithosphere, and partly due to the lack of integration of mantle dynamics into a satisfactory explanation of continental deformation and basin development. Stratigraphers have long recognised the supreme importance of vertical movements ('epeirogeny') of the continents in generating the stratigraphic sequences of the world's cratons (Sloss 1963). In particular, the major transgressions of the continents that have taken place through geological time (e.g. Cambrian–Early Ordovician and Late Cretaceous, Bond 1979) cannot be explained purely by a eustatic sea-level rise (Bond 1976, 1978) and must involve a major component of widespread relative sea-level rise most likely related to mantle processes (§5.4.3).

It is increasingly recognised, therefore, that the near-surface motion of lithospheric plates needs to be understood by reference to the mantle convection system (Anderson 1982). In Chapter 5, we look briefly at the evidence for flow in the mantle, with the specific goal of understanding the significance of this flow for subsidence within continental sedimentary basins. The reader is referred to §2.2.8 for introductory material on mantle viscosity, convection in the mantle and the adiabatic temperature gradient. The rheology of mantle rocks is dealt with in §2.3.2.

5.1.2 Buoyancy and scaling relationships: introductory theory

The buoyancy contrasts of the interior of the Earth, which are the forces responsible for its flow, derive essentially from lateral (horizontal) density contrasts. These density contrasts may be thermal or compositional in origin.

Buoyancy results from gravity acting on a density contrast $\Delta\rho$ in a volume V. It is a force B given by

$$B = -gV\Delta\rho = -g\Delta m \qquad [5.1]$$

where Δm is the mass anomaly resulting from the density contrast and the minus sign is because gravity and weight are positive downwards whereas buoyancy is conventionally positive upwards (Davies 1999, p. 212). Consequently, the buoyancy force depends strongly on the volume of the density contrast. The density contrast resulting from temperature differences is given by the volumetric coefficient of thermal expansion α_v (eqn. [2.12]). Using parameter values for mantle rock, a temperature difference of 1000 °C causes just a 3% change in density. Nevertheless, large buoyancy forces in the mantle may result from the subduction of cold slabs of lithosphere (negative buoyancy) or from the presence of plume heads (positive buoyancy).

A simple consideration of the buoyancy forces associated with the motion of plates explains convection in the mantle (Turcotte & Oxburgh 1967) (see §5.1.3). A plate represents a cooled thermal boundary layer, so is cold and tends to sink due to negative buoyancy. The sinking is opposed by viscous stresses caused by the resulting flow of the mantle, with the stresses increasing with the velocity of the sinking. Consequently, a balance is reached between the viscous resistance and the negative buoyancy at a certain velocity u, given by Davies (1999, p. 216)

$$u = D\left(\frac{g\rho\alpha_v T\sqrt{\kappa}}{4\mu}\right)^{2/3} \qquad [5.2]$$

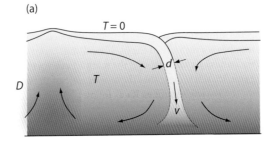

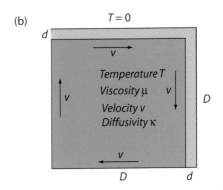

Fig. 5.2 Sketch (a) and model set-up (b) of a convective flow driven by subduction at velocity v of a plate thickness d in a fluid layer of thickness D, viscosity μ and temperature T. After Davies (1999, p. 214), © Cambridge University Press, 1999.

where D is the depth of the vertically descending slab, and ρ, μ, κ and T are the density, viscosity, thermal diffusivity and temperature of the interior of the fluid mantle (Fig. 5.2). Using parameter values of $D = 3000$ km (thickness of the whole mantle), $\rho = 4000$ kg m^{-3}, $T = 1400$ °C, $\mu = 10^{22}$ Pa s, $\alpha = 2 \times 10^{-5}$ °C^{-1} and $\kappa = 10^{-6}$ m^2 s^{-1}, the downward velocity is 90 mm yr^{-1}. This is close to the known velocities of plate motion (Lithgow-Bertelloni & Richards 1998).

The same force balance allows the thickness of the lithosphere at the point of subduction d to be estimated, treating it as a diffusive length scale (see §2.2.7) equal to $\sqrt{\kappa t}$ (or $\sqrt{\kappa D/u}$), and the surface heat flux to be estimated, using Fourier's law (see §2.2.2), which states $q = KT/d$ where K is the thermal conductivity. Using the same parameter values as above, and $K = 3$ W m^{-1} K^{-1}, we obtain 33 km for the thickness of the oceanic lithosphere and 130 mW m^{-2} for the surface oceanic heat flow. These values are consistent with observations of oceanic plates, suggesting that a basic theory of mantle flow driven by buoyancy explains at first order the velocities of plates.

Combining eqn. [5.2] and the expression for diffusion length ($d = \sqrt{\kappa t}$) gives an expression with two dimensionless groups that embeds how the different parameters scale:

$$\left(\frac{D}{d}\right)^3 = \frac{g\rho\alpha TD^3}{4\kappa\mu} \qquad [5.3]$$

For example, eqn. [5.3] can be used to ask 'what would be the lithospheric thickness, heat flow and plate velocity if the mantle viscosity were 10 times lower at some early stage in the evolution of the Earth'? Keeping other parameter values constant, the plate thickness would reduce to 15 km, the heat flow would increase to 275 mW m^{-2}, and the plate velocity would increase to c.400 mm yr^{-1}.

The right-hand side of eqn. [5.3] is a dimensionless group that contains a great deal of information about convection of a fluid layer. Removing the numerical factor 4, it is the *Rayleigh number*. The Rayleigh number is a measure of the likelihood and vigour of convection. For the parameter values above, the mantle has a **Ra** of c.3×10^6. Rewriting eqn. [5.3] it is clear that the length scale of the convection depends on the Rayleigh number **Ra** as follows

$$\frac{d}{D} = kRa^{-1/3} \qquad [5.4]$$

where the coefficient of proportionality k is about 1.6. In other words, if **Ra** increases, the cooled boundary layer of lithosphere d reduces in thickness relative to the depth of the convecting layer D. At **Ra** $= 3 \times 10^6$, d is about 1% of the thickness of the convecting layer.

In a similar fashion, other parameter groups can be scaled against **Ra**. Instead of investigating length scales, we could choose velocities, in which case

$$\frac{u}{U} \sim Ra^{2/3} \qquad [5.5]$$

where U is the characteristic velocity of the problem equal to κ/D. This ratio u/U is a *Péclet number*, which for parameter values typical of the mantle is approximately 9000.

The heat flow q can also be expressed in terms of its scaling with the Rayleigh number. Modifying eqn. [5.4] with Fourier's law gives

$$q = \left(\frac{KT}{D}\right)Ra^{1/3} \qquad [5.6]$$

where the scaling variable (KT/D) expresses the heat that would be conducted across the entire fluid layer in the absence of convection. The total heat flow (in the presence of convection) q versus the conductive heat flow KT/D is known as the *Nusselt number*, **Nu**. Consequently, eqn. [5.6] becomes **Nu** $\sim$ **Ra**$^{1/3}$. For the mantle, the Nusselt number is about 100, indicating that heat flow by convection is two orders of magnitude more efficient than that by conduction.

Accepting that the mantle is convecting in some way, it is possible to estimate whether the flow is laminar or turbulent. This is given by the Reynolds number **Re**

$$Re = \frac{ud}{v} \qquad [5.7]$$

where u is the velocity of the flow, d is the length scale and v is the kinematic viscosity equal to μ/ρ. The kinematic viscosity of the mantle is difficult to estimate, since it is dependent on the temperature, but let us take a value of $10^{17} \, \mathrm{m^2 \, s^{-1}}$ as appropriate for the upper mantle. The flow velocity can be estimated from flow laws for incompressible Newtonian fluids; let us take $u\rho = 10^{-7} \, \mathrm{m \, s^{-1}}$ as an order of magnitude estimate ($1 \, \mathrm{m \, yr^{-1}}$ is $3.17 \times 10^{-8} \, \mathrm{m \, s^{-1}}$, so u is about $30 \, \mathrm{mm \, yr^{-1}}$ in this calculation). Using eqn. [5.7], **Re** is c.7×10^{-19}. This is much smaller than the critical **Re** for turbulence, indicating that the convecting flow in the mantle is laminar.

In summary, a simple quantitative model of convection involving a fluid layer with a cooled thermal boundary layer undergoing subduction allows the scaling relationships between parameter groups to be investigated. The Rayleigh number predicts the onset and vigour of convection, the Péclet number is a measure of the velocity of the plate as a cooled thermal boundary layer, the Nusselt number indicates the efficiency of heat flow by convection, and the Reynolds

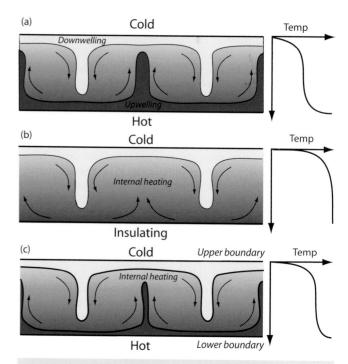

Fig. 5.3 Various modes of heating of a fluid layer, with temperature profiles. (a) Layer of fluid heated from below and cooled at the top. (b) Insulating (no flux) boundary below and internally self-heated by radiogenic decay. (c) Hot lower boundary and internally heated fluid layer. See text for explanation. After Davies (1999, p. 226) © Cambridge University Press, 1999.

number is a measure of turbulence. For the mantle, **Ra** $\approx 3 \times 10^6$, **Pe** ≈ 9000, **Nu** ≈ 100, and **Re** $\approx 10^{-18}$, indicating a strongly convecting layer driving high plate velocities and vigorous heat loss, but laminar flow.

5.1.3　Flow patterns in the mantle

The standard starting point for studies of convection is a layer of fluid heated from below and cooled at the top (Fig. 5.3a). Fluid close to the cold upper surface is cooled, forming a cold thermal boundary layer that is the lithosphere or plate. Likewise, fluid close to the hot lower surface is heated, forming a hot thermal boundary layer. Above a critical Rayleigh number, hot material rises from the lower boundary layer and reinforces the downward motion of fluid from the upper boundary layer, forming a series of rotating cells. The convection cells have a width that scales on the thickness of the fluid layer (Bercovici *et al.* 2000).

Differences in the flow patterns are expected depending on the precise conditions at the lower and upper boundaries and within the fluid layer itself. For example, if the fluid layer is self-heated by radiogenic heat production, and the lower boundary is insulating (involving no heat flow) (Fig. 5.3b), then cold downwellings would originate from the cooled upper boundary layer, but any upwellings would be a passive response rather than involving positively buoyant material. In other words, the fact that upwelling is occurring, as at mid-ocean ridges, does not mean that the ascending mantle material is hotter than average. The upwelling material is simply being displaced by the descending cold mantle material.

Perhaps the most characteristic feature of simple convection where a constant-viscosity layer is heated from below and cooled at the top is that it is symmetrical, consisting of sinking cold currents with an equal and opposite velocity and temperature to rising warm currents. In between the upper and lower thermal boundary layers, where temperature gradients are strong, is a well-mixed fluid at the average temperature of the bottom and top boundaries. The Earth's lithospheric plates comprise the upper thermal boundary layer. If the layer convects more vigorously (**Ra** increases), more of the layer will have a homogeneous temperature, the thermal boundary layers will be thinner and the temperature gradients through them will be higher, driving larger heat fluxes. Fluid in the gravitationally unstable thermal boundary layers must eventually either rise or sink, producing upwellings or downwellings, which therefore must set up a flow of material in the thermal boundary layer from divergent zones above upwellings to convergent zones at downwellings. The thermal boundary layers must thicken in the direction of motion. For example, hot fluid arriving above an upwelling zone travels towards a downwelling and cools to the surface as it does so, before becoming gravitationally unstable enough to sink. The simple theory of convection therefore explains many of the essential features of plate motion (§1.4).

The simplest type of convection is two-dimensional, with counter-rotating cylinders or rolls in the third dimension. The horizontal dimensions of the individual convecting cells in a fluid layer of vertical thickness D, heated from below, is approximately equal to D. This horizontal dimension is also the length of the horizontal currents in the thermal boundary layers. In the simplest terms, the length of the horizontal currents is determined by the amount of cooling necessary to induce gravitational instability and downwelling. At a higher Rayleigh number (**Ra** ~ 10^5) in fluids of constant viscosity, a three-dimensional pattern develops (Busse & Whitehead 1971), which may be bimodal or spoke-like, with linear upwellings joining at a vertex. Fluids with temperature-dependent viscosity produce three-dimensional polyhedral patterns of squares, triangles and hexagons.

The volume of mantle undergoing convection is a *spherical shell* whose outer surface area is over three times the area of the inner surface. This promotes a highly three-dimensional circulation and enhances the effects of layering in mantle viscosity. Spherical shell models (Bunge *et al.* 1996; Davies & Davies 2009) generate near-surface circulations that are very long in the horizontal dimension, with rising spoke-like columns and veins, and descending sheets (Fig. 5.4).

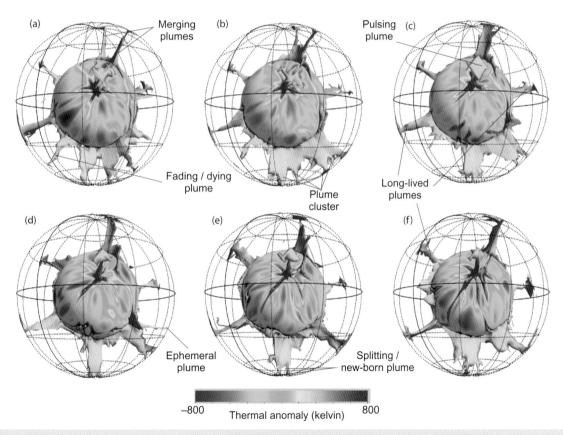

Fig. 5.4 Temporal evolution of an isochemical, incompressible mantle at a Rayleigh number of 1.4×10^9, forming a spherical shell, from Davies & Davies (2009). Each simulation, (a) to (f), is spaced 35 Myr apart. The temperature scale is the temperature in excess of or below the lateral average. Each snapshot shows a radial surface just above the core–mantle boundary and a hot iso-surface 500 K hotter than the average for their depth. Hot upwelling structures (plumes) are prominent. The majority are long-lived and migrate slowly, whereas others are more mobile and short-lived. Over time, new plumes form and old plumes fade and die. Smaller plumes may coalesce, and larger plumes may pulse in their intensity. Reprinted with permission from Elsevier.

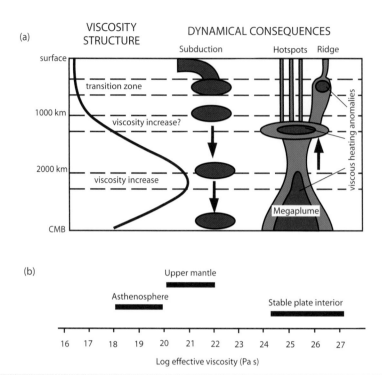

Fig. 5.5   (a) Schematic diagram showing the subduction of cold slabs and the development of megaplumes resulting in mid-ocean ridges and hotspots at the surface, based on seismic tomography of the mantle, adapted from Cadek *et al.* (1995) and reproduced with permission from Elsevier. CMB, core–mantle boundary. (b) Estimates of effective viscosities for stable plate interiors, upper mantle, and asthenosphere.

5.1.3.1   Effects of internal heating

We expect internal heating of the fluid layer to affect the spatial pattern of convection as well as the vigour. The simplest situation of internal heating would be a uniform distribution of heat production with depth, an isothermal upper boundary, and a zero heat flux through the insulating lower boundary (Fig. 5.3b). Because of the zero heat flux at the bottom boundary, no active upwellings occur, and instead a passive upward flow takes place to compensate for the active downwelling of cold material from the upper thermal boundary layer. If the bottom boundary is allowed to conduct a heat flux, and there is an internal heat generation (Fig. 5.3c), the upper boundary must conduct outwards both the basal heat as well as the internal heat. Consequently, the upper thermal boundary layer develops a greater temperature drop than the basal boundary. This causes the upper boundary to develop more numerous or more vigorous downwellings than upwellings rising from the lower boundary. Internal heating therefore breaks down the symmetry between upwellings and downwellings. Since the Earth is known to have very significant internal sources of heat relative to the basal heat flow from the core, we should expect downwellings of an upper thermal boundary layer to dominate the pattern of convection, and upwellings should be relatively weaker (Bercovici *et al.* 2000). This is precisely the situation seen at the Earth's surface.

5.1.3.2   Effects of a temperature-dependent viscosity

The viscosity of mantle materials obeys a temperature-dependent Arrhenius-type law (see also §10.2). Since the absolute temperature occurs in the denominator of the Arrhenius exponent, there are very large viscosity variations at lower temperatures. This is why the viscosity variation in the upper mantle may be extreme (Fig. 5.5a), from perhaps 10^{21} Pa s in its lower part to 10^{18} Pa s in the asthenosphere and 10^{25} Pa s in the lithosphere, representing a variation of viscosity of seven orders of magnitude in a depth range of just 200 km (Fig. 5.5b). This makes the lithosphere much stronger and potentially less mobile than the rest of the mantle. It in turn forces the underlying mantle to heat up, increasing the temperature contrast between the hot interior and the cold upper boundary. Since material in the upper thermal boundary layer must cool a great deal to become gravitationally unstable and sink relative to its cold, immobile surroundings, the horizontal extent of the convection cell may become elongated, as demonstrated in both laboratory and numerical experiments (Ratcliff *et al.* 1997). This would explain the very large lateral distance of the spreading ridge to the subducting margin of the Pacific plate. Large aspect ratio (long-wavelength) convection cells have also been imaged by seismic tomography (Su & Dziewonski 1992). With a stronger viscosity-temperature dependence, convection experiments show that the cold upper thermal boundary layer may form a stagnant, rigid lid over a convecting interior, as is believed occurs in the planet Mars.

Convection theory therefore goes some way towards explaining the occurrences of divergent and convergent boundaries in the lithosphere together with the long aspect ratio of plates required. Convection theory also predicts that there should be topographic variations at the surface due to upwellings (high topography) and downwellings (low topography). This topographic signal needs to be separated from the topographic effects of density and thickness variations in the lithosphere. The topography remaining after the removal of isostatic effects originating within the lithosphere from the observed topography is commonly referred to as *dynamic topography* and is discussed in further detail in §5.2.

5.1.4   Seismic tomography

Very detailed velocity models of the mantle can be constructed using seismic tomography. Seismic tomography is a technique using many measurements of seismic wave arrivals at a network of recording stations to construct a best-fitting three-dimensional model of the **S**-wave velocity structure of the upper mantle or the **P**-wave velocity structure of the lower mantle (Dziewonski & Woodhouse 1987; Su *et al.* 1994; Kárason & van der Hilst 2000; Nolet 2008). By computing slices at different depths, it is possible to see how the velocity structure varies as a function of depth within the mantle. The velocity structure most likely reflects density differences, which in turn reflect temperature differences, but may also reflect compositional heterogeneity. Near the surface ($y = 50$ km), the velocity structure is dominated by the presence of the continental and oceanic plates, the continental shields being fast and the mid-ocean ridges being slow. The effects of the overlying plates are lost beneath a certain depth in the mantle, but this depth is surprisingly large. A distinctive 'slow' seismic structure can be related to overlying oceanic ridge systems down to at least 1000 km (Su *et al.* 1994; Cadek *et al.* 1995). The remnants of old lithospheric slabs have also been detected from their effect on seismic velocities down to more than 1000 km (Wen & Anderson 1995). The correlation with plate tectonics is lost at 1500–1700 km, but at about 2000 km, the surface tectonic pattern can once again be weakly recognised from seismic tomography. In this region of the lower mantle, major mantle plumes and remnants of old lithosphere can be detected from seismic heterogeneities. This suggests that plumes originate close to the core–mantle boundary. It also suggests that this region is a graveyard for deeply subducted lithospheric remnants. Two 'megaplume' structures can be detected in the lower mantle below 2000 km. These major upwellings in the lower mantle are correlated with an abundance of hotspots on the Earth's surface, and are thought to be long-lived, stable flow structures. The presence of lower-mantle megaplumes, hotspots, mid-ocean ridges, subduction zones and remnant lithospheric slabs have been explained as the dynamic consequences of a mantle with a strong viscosity stratification (Fig. 5.5a, b).

5.1.5   Plate mode versus plume mode

Davies (1999) distinguished between a type of convection driven by the upper cooled thermal boundary layer ('plate mode') and a type driven by processes at the hot lower thermal boundary layer ('plume mode'). The recognition of different modes of convection helps to solve the puzzle of the relationship between the movement and boundaries of plates and the circulation in the convecting mantle.

The 'plate mode' is the mode of convection driven by the negative buoyancy of subducting oceanic plates. Plates are part of the underlying convection system, and their negative buoyancy drives convection, so they are active rather than passive components in an integrated plate-mantle system. Plates are sufficiently strong, owing to the temperature-dependence of rheology (§2.3), that they normally resist 'dripping' downwards by their negative buoyancy. Since we know that subduction is commonplace, there must be a factor causing downward flow of plates, such as the existence of zones of weakness (for instance, major faults). The position of plate boundaries therefore controls the locations of passive upwellings at spreading centres and downwellings at subduction zones. In this sense, surface plate motion 'organises' deeper mantle flow, a phrase coined by Brad Hager.

The life cycle of an oceanic plate begins at a spreading ridge, where mantle cools by conduction and thickens as it moves away horizontally (§2.2.7). The plate returns to the mantle, where it subducts and heats up by absorbing heat from its surroundings, thereby cooling the Earth's interior. The mantle loses most of its internal heat by this plate cycle, the remainder (about 10%) being lost by conduction through the continental lithosphere. The dominance of mantle flow organised by plate movement is seen in the topography associated with this plate-scale flow – the mid-ocean ridges elevated at kilometres above the adjacent ocean floor.

The 'plume mode', in contrast, is driven by the upwelling of positively buoyant material from the lower thermal boundary layer of the mantle. Many mantle plumes appear to be fixed in position relative to each other and are unrelated to the motion of the plates and to present-day plate boundaries. Although there are between 40 and over 100 volcanic hotspots (Burke & Wilson 1976; Crough & Jurdy 1980; Morgan 1981) (Table 5.1), not all of these are associated with plumes originating from the core–mantle boundary. Topographic swells in the ocean, however, such as the 2000 km-wide Hawaiian swell, can only be satisfactorily explained by the presence of a column of buoyant material beneath the lithosphere, and the association with volcanic activity supports the idea that this buoyant material is hot. Many postulated plumes have a present-day distribution correlated with the two major geoid highs (§5.1.6) in the Pacific and under Africa (Burke & Torsvik 2004; Burke *et al.* 2008) (Fig. 5.6). Other hotspots are within 1000 km of the edge of continents and may be related to a smaller scale of convective circulation driven by the step in the base of the lithosphere (King 2007) (Figs 5.6, 5.7) (§5.2.4).

Plumes transport heat from the interior of the Earth to the lithosphere. If the plume is envisaged as a vertical cylinder with radius r, with material flowing at velocity u, then the buoyancy flux B is

$$B = g\Delta\rho\pi r^2 u \qquad [5.8]$$

where $\Delta\rho$ is the density deficit between the plume material and the ambient mantle. Buoyancy flux therefore scales on the discharge of plume material and its density deficit. The buoyancy fluxes of the world's major volcanic hotspots are shown in Table 5.1. The buoyancy flux is closely related to the topographic expression of the swell – the largest swells are supported by the largest buoyancy fluxes, Hawaii being the largest. Estimates of the total heat transport by all known plumes (Davies 1988; Sleep 1990) suggest that plumes account for about 6% of global heat flow, which is similar to estimates of the heat transport out of the core (Stacey 1992). This supports the notion that plumes originate from the thermal boundary layer at the base of the mantle. Hotspots confidently related to mantle plumes have on average twice the buoyancy flux of hotspots that are candidates for edge-driven convection.

A physical manifestation of the plume mode is the flood basalt provinces, from which volcanic hotspot tracks emerge, such as the Chagos–Laccadive Ridge extending southwards from the Deccan Traps of western India to the present-day volcanic centre of Réunion Island in the Indian Ocean (Fig. 5.8). Flood basalts (§5.3.2, §5.3.3), or Large Igneous Provinces (LIPs), may extend up to 2000 km across, and may be several kilometres thick, so total volumes of extrusive eruptions range up to 10 million km³. The LIPs and tracks are

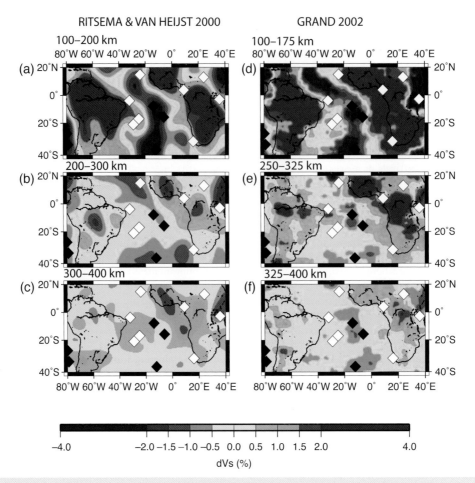

RITSEMA & VAN HEIJST 2000   GRAND 2002

Fig. 5.6 S-wave tomography of the South America-Atlantic-Africa region from Ritsema & van Heijst (2000) (left) and Grand (2002) (right) for three depth slices, with hotspots shown as black and white diamonds from Sleep (1990), from King (2007). White diamonds are located where edge-driven convection is favourable geometrically, whereas black diamonds are in unfavourable positions geometrically. Colour scheme is based on the percentage departure from a mean velocity dVs.

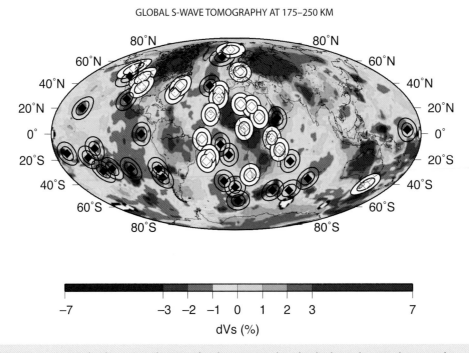

GLOBAL S-WAVE TOMOGRAPHY AT 175–250 KM

Fig. 5.7 Global S-wave tomography from Grand (2002) for the 175–250 km depth slice, showing hotspots from Sleep (1990), with circles of radius 660 km and 1000 km drawn around the centre of the hotspot. Candidates for edge-driven convection (EDC) are those where the circle intersects a blue (fast) anomaly, shown in white. After King (2007).

Table 5.1 World's major hotspots, with buoyancy flux, after Sleep (1990), Turcotte and Schubert (2002), and Davies (1988), classified according to whether the upwelling is related to edge-driven convection or deeper plumes, based on: (i) Courtillot *et al.* (2003); (ii) Montelli *et al.* (2003); and (iii) King (2007). Those marked (4) are tentatively assigned to EDC based on their spatial proximity to a continental edge, but some of this category may be plume-related. Based on this classification, the average buoyancy flux for plume-generated hotspots is 1.7 Mg s^{-1} and for EDC is 0.86 Mg s^{-1}

Hotspot	Flux (Mg s^{-1}) Sleep 1990/Turcotte & Schubert 2002 p. 261/Davies 1988	Location-type 1 – Courtillot *et al.* 2003; 2 – Montelli *et al.* 2003; 3 – King 2007
Afar, African plate	1.2/1.2/-	Plume-generated[1,2,3]
Ascension, South American plate	-/0.9/-	Plume-generated[2,3]
Australia, East, Indo-Australian plate	0.9/0.9/-	Edge-driven[4]
Azores, Eurasian plate	1.1/1.1/-	Plume-generated[2,3]
Balleny, Antarctic plate		Edge-driven[4]
Baja, eastern Pacific	0.3/0.3/-	Edge-driven[4]
Bermuda, North American plate	1.1/1.3/-	Edge-driven[3]
Bouvet, African plate	0.4/0.4/-	Plume-generated[1]
Bowie seamount, Pacific plate	0.3/0.6/-	Edge-driven[3]
Canary, African plate	1.0/1.0/-	Edge-driven?[3] Plume-generated[2]?
Cape Verde, African plate	1.6/1.0/-	Edge-driven[3]
Caroline Islands, Pacific plate	1.6/1.6/-	Plume-generated[1]
Comoros, African plate		Edge-driven[4]
Crozet, Antarctic plate	0.5/0.5/-	Plume-generated[2,3]
Darfur, African plate	-/0.4/-	Plume-generated[1]
Discovery, African plate	0.5/0.6/-	Plume-generated[1]
East African, African plate	-/0.6/-	?Edge-driven[4]
Ethiopian, African plate	-/1.0/-	?Edge-driven[4]
Easter, Nazca plate	3.3/3.3/-	Plume-generated[1,2,3]
Eifel, Eurasian plate		Plume-generated[2]
Fernando, South American plate	0.5/0.7/0.9	Edge-driven[3]
Galapagos, Nazca plate	1.0/1.0/-	Plume-generated[1]
Great Meteor seamount (New England), African plate	0.5/0.4/0.4	Plume-generated[1]
Hawaii, Pacific plate	8.7/7.4/6.2	Plume-generated[1,2,3]
Hoggar, African plate	0.9/0.6/0.4	Edge-driven[3]
Iceland, Eurasian plate	1.4/1.4/-	Plume-generated[1,2,3]
Jan Mayen, Eurasian plate		Edge-driven[3]
Juan de Fuca/Cobb seamount, Pacific plate	0.3/0.3/-	Edge-driven[4]
Juan Fernandez, Nazca plate	1.6/1.6/1.7	?Edge-driven[4]
Kerguelen, Antarctic plate	0.5/0.4/0.2	Plume-generated[1,2,3]
Lord Howe, Indo-Australian plate		Edge-driven?[4]
Louisville, Pacific plate	0.9/2.0/3.0	Plume-generated[1,2,3]
Macdonald seamount, Pacific plate	3.3/3.6/3.9	Plume-generated[1]
Madeira, African plate		Edge-driven[3]
Marquesas Islands, Pacific plate	3.3/4.0/4.6	Plume-generated[1]
Marion, Antarctic plate		Plume-generated[1]
Martin	0.5/0.6/0.8	
Meteor, African plate	0.5/0.4/0.4	Plume-generated[1]
Mount Cameroon, African plate		Edge-driven[3]
Pitcairn, Pacific plate	3.3/2.5/1.7	Plume-generated[1]
Raton, North American plate		Edge-driven[3]
Réunion, African plate	1.9/1.4/0.9	Plume-generated[1,2,3]
St Helena, African plate	0.5/0.4/0.3	Edge-driven[3]
Samoa, Pacific plate	1.6/1.6/-	Plume-generated[1,2,3]
San Felix, Nazca plate	1.6/2.0/2.3	Edge-driven[4]
Socorro, Pacific plate		Edge-driven[3]
Tahiti, Pacific plate	3.3/4.6/5.8	Plume-generated[2,3]
Tasman, Central, Indo-Australian plate	0.9/0.9/-	Edge-driven[3]
Tasman, East, Indo-Australian plate	0.9/0.9/-	Edge-driven[3]
Tibesti, African plate	-/0.3/-	Edge-driven[3]
Trindade, South American plate		Edge-driven[3]
Tristan, African plate	1.7/1.1/0.5	Plume-generated[1,2,3]
Vema seamount, African plate	-/0.4/-	Edge-driven[3]
Yellowstone, North American plate	1.5/1.5/-	Edge-driven[3]

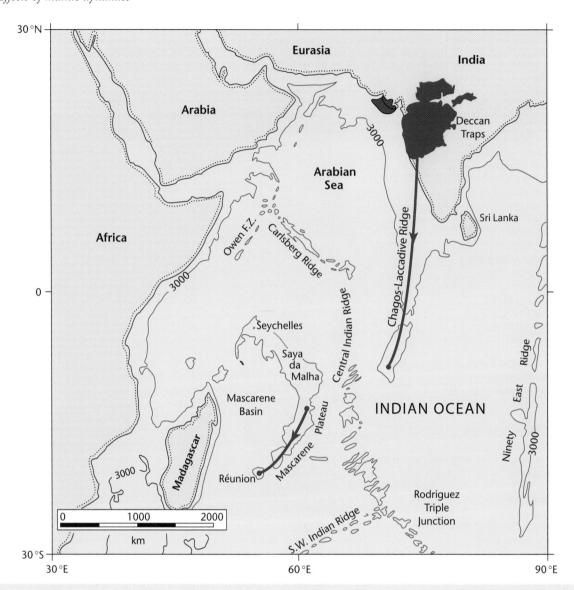

Fig. 5.8 The path of the Réunion hotspot track in the Indian Ocean and the Deccan flood basalt province. After White & McKenzie (1989), reproduced with permission of the American Geophysical Union.

explained (Morgan 1981; Richards *et al.* 1989; Griffiths & Campbell 1990) as due to the arrival of a plume head, followed by a plume tail as the overlying plate moves away from the conduit. The track shows the path of the overlying plate over the plume tail, ending at the present-day volcanic centre (Fig. 5.9).

It is unlikely that material of normal mantle composition can generate >1 million km³ of extrusive basalts from melting of a plume head. An additional important factor may be the presence of a higher basaltic composition in the plume head, perhaps due to entrainment of previously subducted oceanic crust (Hoffman & White 1982), which would lower the solidus temperature. Other possibilities are higher plume temperatures, or lithospheric thinning (Chapter 3) caused by the impingement of the plume on the base of the plate.

In summary, the presence of topographic swells, and LIPs connected to active volcanic centres by hotspot tracks, can best be explained by the rise of long-lived columns of buoyant mantle mate-

rial 200–300 °C hotter than the surrounding mantle. The rise of these mantle plumes most likely takes place from the lower hot boundary layer of the mantle and contributes to the 'plume mode' of mantle convection. Since the distribution of hotspots shows no relation to present-day plate boundaries, the plate mode and the plume mode of convection do not appear to be strongly coupled (Stefanik & Jurdy 1994).

5.1.6   The geoid

Measurements of the Earth's gravity field also provide much information about the dynamics of the mantle. Numerical simulations of convection in the upper mantle (McKenzie *et al.* 1980) show that upwellings and downwellings should be accompanied by variations in the sea surface (amplitude of up to 10 m) and in the gravity anomaly (amplitude of up to 20 mgal) (Fig. 5.10). The geoid is the

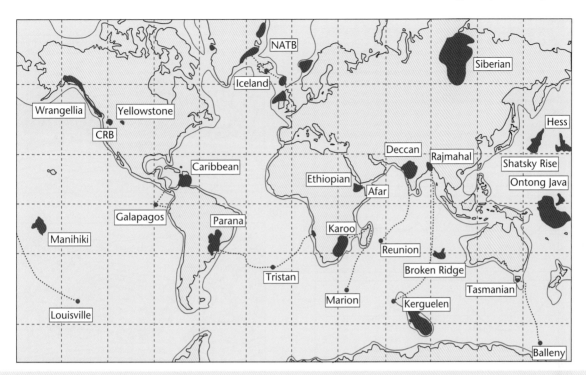

Fig. 5.9 Map of oceanic and continental flood basalt provinces and hotspot locations. Dashed lines connect flood basalt provinces with present-day volcanic centres (volcanic hotspots), representing hotspot tracks. The Chagos–Laccadive Ridge joining the Deccan with Réunion is discussed in the text. Hotspots are concentrated on and around geoid highs (Fig. 5.7). CRB, Columbia River Basalt province; NATB, North Atlantic Tertiary Basalt province. From Duncan & Richards (1991) (fig. 11.12 of Davies 1999), reproduced with permission of John Wiley & Sons, Inc.

gravitational equipotential surface around the Earth and has the form of an oblate spheroid (§1.4.2). Departures from the reference geoid due to the non-uniform distribution of mass within the Earth are called geoid height anomalies. Upwellings have positive gravity anomalies and an elevated sea surface (positive geoid height anomaly), whereas the reverse is true of the downwellings. This is at first contradictory. We should expect a density excess at depth to cause a geoid high and a positive free-air anomaly, and a density deficit should produce a geoid low and a negative free-air anomaly. However, the free surface is deflected upwards over the rising limbs of the convection cells, and downwards over the descending limbs. This outweighs the effects of the density differences. The net result is that rising limbs are associated with geoid highs and positive free-air anomalies.

A global map of the geoid anomaly (Lemoine *et al.* 1998) (Fig. 1.12) shows maximum geoid anomalies of approximately 100 m, with highs in the southwestern Pacific (+80 m) and northern Atlantic (+60 m), and lows in the Indian Ocean (−100 m), Antarctica and Southern Ocean (−60 m) and the western Atlantic Ocean and adjoining American continent (−40 m). Although some of the geoid pattern is related to plate tectonic processes such as subduction (see §5.2.2), many of the geoid height anomaly features cannot be explained in this way and reflect deeper processes in the mantle.

It was recognised at an early stage (e.g. Runcorn 1967) that zones of plate convergence are generally associated with highs in the observed long-wavelength geoid. For example, the major subduction zones of the Earth all have geoid highs (Chase 1979). The long-wavelength (degree <10[1]) geoid anomalies are broadly comparable to those expected from the excess density of seismically active, cold subducted slabs. However, the geoid anomaly depends not only on the 'driving' density contrasts at depth, but also on the viscosity structure of the mantle. This is because the density contrasts, such as due to chemical layering, set up flows causing a deformation of the boundaries of any density contrast, the amplitude of which depends on the viscosity structure. Incorporating dynamic effects, the presence of high-density material in the upper mantle (such as due to subducted lithosphere) should lead to long-wavelength geoid highs. But the presence of dense material in the lowermost mantle should lead to long-wavelength geoid lows. Consequently, long-wavelength geoid highs can be associated with high-density material (old slabs) in the upper mantle and low-density material (megaplumes) in the lower mantle. Hager (1984) found that the geoid anomalies (at degree <10) associated with subduction could best be explained by a dynamic model in which the density effects of subducting slabs penetrated deep into the lower mantle (through the 670 km discontinuity) and in which the viscosity contrast between upper and lower mantle was of the order of 50 to 100.

Subtracting the estimated effects of subduction (the so-called 'slab geoid') from the observed geoid reveals the *residual geoid* (Crough

---

[1] Wavelength is normally expressed in terms of spherical harmonic degree *l*. For degree *l*, there are *l* wavelengths in the circumference of the Earth. At degree 2, there are just two hemispherical highs and lows.

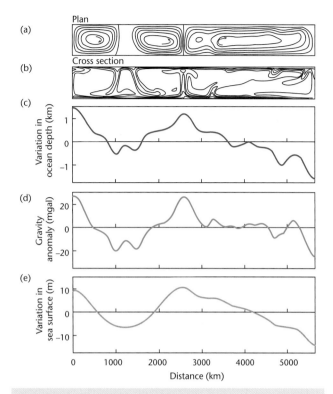

(a) Plan

(b) Cross section

(c)

(d)

(e)

Fig. 5.10 Computer modelling of convection in an upper mantle with constant viscosity heated from below. (a) and (b) Temperature field contoured at 100 °C intervals, showing three upwellings and two downwellings in plan and cross-section; (c) variation in depth of the ocean floor; (d) variation in the gravity anomaly; and (e) variation in height of the sea surface (geoid height anomaly) resulting from the convection pattern in (a) and (b). Reproduced from McKenzie *et al.* (1980) by permission from Macmillan Publishers Ltd.

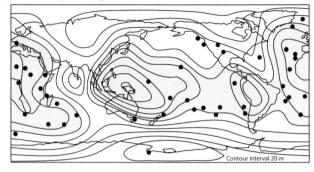

(a) Observed geoid: Degree 2–10

Contour interval 20 m

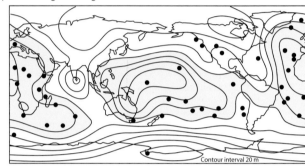

(b) Residual geoid: Degree 2–10

Contour interval 20 m

Fig. 5.11 Observed (a) and residual (b) long-wavelength geoid at degree 2–10 (Lerch *et al.* 1983; Hager 1984), contoured at a 20 m interval. Geoid highs are shaded. The residual geoid is obtained by subtracting a dynamically consistent slab geoid from the observed geoid. The distribution of surface hotspots, sites of intraplate volcanism, and anomalous plate volcanism (black dots) shows a correlation with geoid highs.

& Jurdy 1980; Hager 1984) (Fig. 5.11). It is noticeable that there is a residual geoid high along a nearly continuous E–W band, with a maximum in the Pacific Ocean and a secondary peak over Africa. This pattern may reflect megaplume activity in the lower mantle. The correlation of residual geoid highs with the location of hotspots supports this view.

In summary, the pattern of geoid anomalies is highly diagnostic of the processes of slab subduction and mantle convection. The challenge for us is how to make the connection between deep sublithospheric processes and basin development on the continents.

5.2 Surface topography and bathymetry produced by mantle flow

5.2.1 Introduction: dynamic topography and buoyancy

The term '*dynamic topography*' is defined as 'the vertical displacement of the Earth's surface generated in response to flow within the mantle' (Richards & Hager 1984). It is therefore distinct from isostatic topography generated by near-surface density contrasts such as due to crustal thickness changes.

Nevertheless, as pointed out by Davies (1999), there is some confusion over what dynamic topography really is and how it differs from the topography resulting from a simple isostatic balance. If we accept that the oceanic tectonic plates constitute an upper, cooled, thermal boundary layer, then because these plates are actively participating in a global convection system, the surface elevation of the oceanic plates represents dynamic topography. That is, the increasing bathymetry of the ocean floor from the spreading centre to the abyssal plain (§2.2.7) is dynamic topography dominated by the effect of cooling of the upper thermal boundary layer. The bathymetry of the ocean floor is, however, calculated using an isostatic balance, neglecting the viscous stresses set up by three-dimensional variations in buoyancy. Ambiguously, therefore, dynamic topography results from a purely static balance.

The broad oceanic swells and their continental counterparts are thought to be associated with underlying upwellings of mantle flow structures and clearly constitute 'dynamic topography'. The less easily demonstrable topographic changes associated with smaller scales of convection are also clearly 'dynamic'.

Other situations exist where cooling takes place, but where it is not related to the Earth's convection system, such as following mechanical stretching of the lithosphere, or surrounding a large intrusion in the crust. The topographic effects of these processes can also be calculated by an isostatic balance, but such topography is not 'dynamic'.

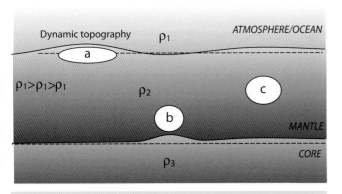

Fig. 5.12 Schematic diagram showing effect of buoyant blobs on the upper and lower surfaces of a fluid layer, which are assumed to be free to deflect vertically. A less dense fluid (such as the atmosphere or oceans) overlies the upper surface, and a denser fluid (core) underlies the lower surface. Adapted from Davies (1999, p. 234, fig. 8.5), © Cambridge University Press, 1999.

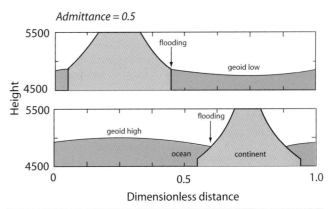

Fig. 5.13 Topography of continents and elevation of sea surface, showing degree of continental flooding when the continent is positioned over a geoid high (top), and geoid low (bottom). Admittance is 0.5, maximum geoid is 100 m. The topography of the continent is the result of isostasy, dynamic topography and seawater loading. The continent experiences a greater amount of flooding when it is positioned over a geoid low. After Gurnis (1990a), reproduced with permission of American Association for the Advancement of Science.

Turning to other mechanisms, topography may result from the subduction of cold oceanic slabs into the hot mantle, setting up a large-scale flow, in which case the topography is 'dynamic'. But the regional bending of the oceanic plate beneath the ocean trench, and the regional bending of the retro-arc region of the continental plate in an ocean–continent convergence zone, are due to flexure of the elastic lithosphere (Chapter 4), are not linked to the underlying convection system, and are not 'dynamic'.

The fact that the bathymetry of oceanic plates can be explained by a cooling plate model (§2.2.7) shows that there is no recognisable influence of deeper mantle convection in the elevation of the ocean floor. The submarine topography is explainable using the plate mode of convection (§5.1.5). Numerical convection models can produce topography that matches observed seafloor bathymetry well; this topography is due to cooling in the upper thermal boundary layer of the convection system. The dynamic topography associated with subduction, however, is much more difficult to model, since numerical models include an artificial, local weakening of the viscosity structure to simulate a subduction zone fault, and the predicted topography is highly sensitive to details of the deeper viscosity structure and of the plate thickness approaching the subduction zone, both of which are poorly constrained.

The underpinning concept for the role of mantle dynamics in forming surface topography is that buoyancy drives convective flow and causes a deflection of the surfaces of the fluid layer undergoing motion. The distribution within the fluid layer of the sources of buoyancy, which we can call 'blobs', is critical to the topographic deflection (Fig. 5.12) (Davies 1999). For example, if the blob is close to the upper surface, its buoyancy must be opposed by a gravitational force acting on the upward deflected surface. The downward-acting weight of the upward deflected surface counterbalances the upward-acting buoyant force of the blob. This is a static force balance with no momentum. If the blob is located close to the lower boundary, the lower boundary is deflected upwards but the upper surface is barely deflected since it is too far away to be affected by the viscous stresses transmitted by the blob. If the blob is situated in the centre of the fluid layer, it may deflect both the upper and lower layers. The

dynamic topography of the Earth's surface therefore carries important information about the distribution of buoyancy and therefore about convection in the interior of the Earth.

As introduced above, when a viscous fluid is disturbed by the presence of a parcel of material with positive buoyancy, viscous stresses are set up that cause the free surface to move upward (Fig. 5.12). The response time of the mantle to a disturbance is dependent on its viscosity (estimated from glacial rebound studies, see §2.2.8, §4.1.1) and the wavelength of the parcel of positive buoyancy (Zhong *et al.* 1996). For mantle flow related to subduction, with a wavelength of $1–3 \times 10^3$ km (Gurnis 1992, 1993; Burgess & Gurnis 1995) the time scale is 10^4 yr, and for larger wavelengths of anomaly ($>5 \times 10^3$ km), such as expected under insulating supercontinental assemblies (Anderson 1982; Gurnis 1988), the response time is up to 10^5 yr (Gurnis *et al.* 1996). The mantle can therefore be considered as responding effectively instantaneously to the disturbance by a positive buoyancy parcel.

The dynamic topography associated with the long-wavelength heterogeneities identified by seismic tomography should theoretically be large (>1000 m, Hager & Clayton 1989) in order to explain the undulations in the geoid. However, dynamic topography of this scale has not been detected (Le Stunff & Ricard 1995). A number of studies suggest that the global scale variations in dynamic topography, corrected for effects within the lithosphere, are in the range 300–500 m (e.g. Cazenave *et al.* 1989; Nyblade & Robinson 1994). However, variation in the degree-2 geoid is about 50 m (Hager *et al.* 1985).

The ratio of geoid to dynamic topography is termed *admittance*. Although a continent may be located over a geoid high, which therefore has elevated water levels in the ocean, the greater dynamic topography in this region may cause subaerial emergence rather than flooding (Fig. 5.13). The admittance is therefore important in studying continental flooding histories (Gurnis 1990a). For low admittance, flooding takes place preferentially over geoid lows. At high values of admittance, the continent may be flooded preferentially over geoid highs. The flooding history of the North American continent during the Phanerozoic (maximum of 30% of the continental surface) (Bond 1979) indicates that since the geoid should lie within the range 0–100 m, the amplitude of dynamic topography must be up to 150 ± 50 m.

The recognition of the amplitude and wavelength of dynamic topography at the Earth's surface at the present day is made difficult by the isostatic compensation of strong crustal and lithospheric thickness and density variations. If these isostatic effects could be removed from the Earth's topographic field, we should obtain the dynamic component attributable to mantle flow. When this 'residual' topography is mapped globally, it shows a pattern of long wavelengths of up to several thousand kilometres and heights of up to 1 km (Fig. 5.14) (Steinberger 2007). Secondly, we could estimate

dynamic topography using a density model of the mantle derived from tomographic studies (§5.1.4) (Moucha *et al.* 2008). Such estimates are broadly in line with the estimates derived from 'residual' topography.

In the following paragraphs, we examine the dynamic topography associated with different geodynamic situations. First, we consider the subduction of cold oceanic slabs, particularly those that subduct at shallow angles (Mitrovica *et al.* 1989; Gurnis 1993; Spasojevic *et al.* 2008). Second, we look at the very large-scale dynamic topography associated with supercontinental assemblies (Anderson 1982). Third, we reduce scale and speculatively consider small-scale convection associated with edges and steps beneath the continental lithosphere (King & Anderson 1998). Fourth, we evaluate the dynamic topography associated with mantle plumes and superplumes (Lithgow-Bertelloni & Silver 1998).

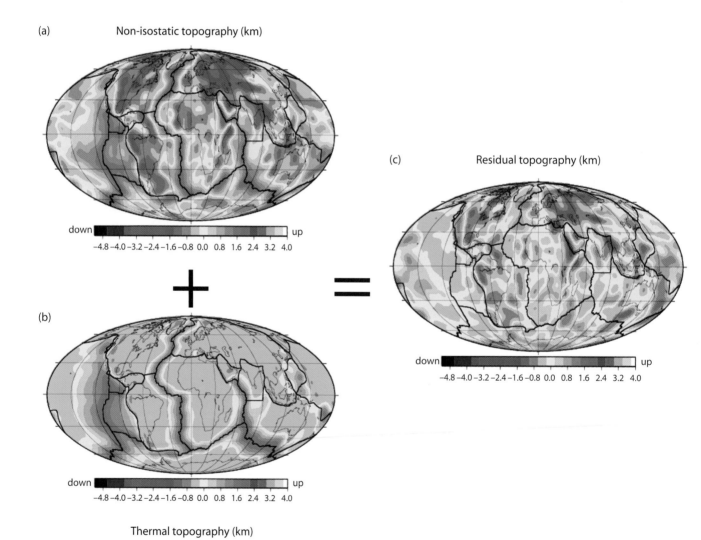

Fig. 5.14 Global dynamic topography based on residual topography after removal of isostatic effects of crustal and lithospheric thickness changes and thermal topography due to the cooling of the ocean lithosphere (Steinberger *et al.* 2001). Part (a) is the non-isostatic topography after removing isostatic effects from the actual topography. Part (b) is the thermal topography calculated from the age_2.0 ocean floor age grid of Müller *et al.* (2008a) for ages less than 100 Ma. Part (c) is the residual topography calculated for an assumed global seawater cover, equivalent to dynamic topography thought to result from mantle circulation.

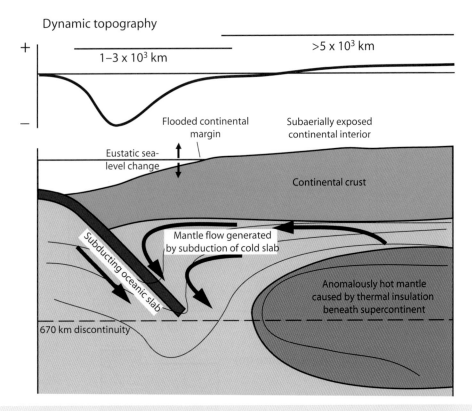

Fig. 5.15 Schematic diagram to show generation of dynamic topography from subduction of a cold oceanic slab (wavelength $1-3 \times 10^3$ km) and from mantle insulation beneath a supercontinent (wavelength $>5 \times 10^3$ km) (after Burgess *et al.* 1997).

5.2.2 Dynamic topography associated with subducting slabs

If dynamic topography results from the thermal effects of flow in the mantle, we should be able to recognise dynamic topography behind ocean trenches where cold lithospheric slabs disturb the mantle temperature field (Parsons & Daly 1983; Mitrovica & Jarvis 1985; Gurnis 1992) (Fig. 5.15). The ramp-like tilting of the continents towards their oceanic active margins indicated by the preservation of extensive (>1000 km) wedge-shaped stratigraphic packages (see §5.4) has been taken to reflect dynamic topography above shallowly subducting ocean slabs on a number of continents, such as western North America (Mitrovica *et al.* 1989; Burgess *et al.* 1997), the Russian Platform (Mitrovica *et al.* 1996) and eastern Australia (DiCaprio *et al.* 2009).

Although the amplitude of dynamic topography behind subduction zones is controversial, the wavelength is thought to extend 1000–2000 km from the ocean trench into the continental plate. Consequently, dynamic topography may be recognised as a realm of subsidence extending far beyond the location of retro-arc foreland basins (Mitrovica *et al.* 1989; Catuneanu *et al.* 1997; Burgess & Moresi 1999).

Dynamic topography depressions of approximately 500 m are predicted by geoid models over subducting slabs (Hager & Clayton 1989). However, the measurement of dynamic topography behind trenches is problematic, since the dynamic topography must be separated from other forms of subsidence. One approach is to compare such regions with areas unaffected by slab-related dynamic topography. This can be done by a comparison of the distribution of topog-

raphy (hypsometry) behind ocean trenches compared with a world average (Gurnis 1993). This suggests that regions within 1000 km of ocean trenches are depressed by 400–500 m in the western Pacific (Gurnis 1993). A second approach is to remove the isostatic effects of lithospheric thickness changes and thermal subsidence. A study of backarc regions of the southwestern Pacific (Wheeler & White 2000) indicated a maximum of 300 m of negative long-wavelength (c.10^3 km) dynamic topography (Fig. 5.16).

Since there must be major uncertainties in the estimation of dynamic topography behind ancient subduction zones, a first approach to the modelling of dynamic topography in basin analysis is to use an expression for the geometric form of the dynamic topography by assuming that it is made of two components (Gurnis 1992; Coakley & Gurnis 1995; Burgess & Gurnis 1995): (i) an exponential component with an exponent λ and maximum deflection f_m; and (ii) a linear tilt with a maximum gradient α in m km^{-1} and a maximum distance from the trench at which tilting occurs η. Combining these components gives

$$f(x) = f_m\left(e^{-x/\lambda}\right) + \alpha(\eta - x) \qquad [5.9]$$

where x is the horizontal orthogonal distance from the trench.

With eqn. [5.9] we are able to approximate the dynamic topography as a function of distance from the trench axis and to consider its impact on the surface elevation of the retro-arc region of a continent as a thought experiment. We initially take the following parameter values: $\lambda = 200$ km, $f_m = 2000$ m, $\alpha = 0.5$ m km^{-1}, and $\eta = 2000$ km. If we superimpose on this distribution of dynamic topography $f(x)$ a realistic deflection due to flexure in the retro-arc

(a) Predicted dynamic topography (m)

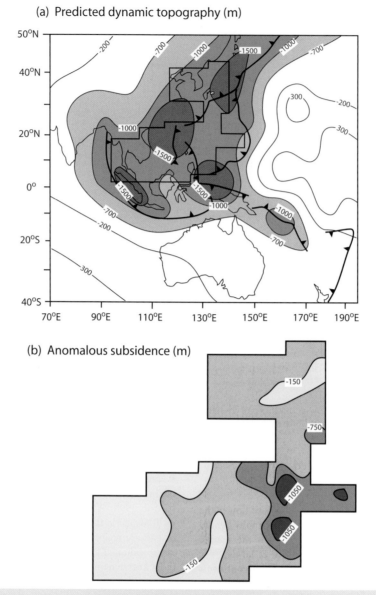

(b) Anomalous subsidence (m)

Fig. 5.16 (a) Predicted dynamic topography in SE Asia (Lithgow-Bertelloni & Gurnis 1997) using a history of subduction over the past 180 Myr. (b) The 'anomalous' subsidence after removal of isostatic effects related to basin formation by Wheeler & White (2000), showing that the observed dynamic topography is significantly less than predicted by slab models.

region $w(x)$, it is apparent that the dynamic topography is sufficient to exceed the effects of flexural forebulge uplift (Allen & Allen 2005, p. 179). Consequently, the back-bulge and flexural forebulge regions are zones of (transient) subsidence (Fig. 5.17a), as envisaged by DeCelles and Giles (1996). The parameter values used are similar to those used to model the Phanerozoic evolution of the North American craton (Burgess & Gurnis 1995). If, however, we take parameter values that more closely reflect recent estimates of the amplitude of dynamic topography behind trenches ($\lambda = 200$ km, $f_m = 500$ m, $\alpha = 0.2$ m km^{-1}, and $\eta = 2000$ km), the dynamic topography is insufficient to exceed the magnitude of forebulge uplift. Consequently, the flexural forebulge remains an erosional zone in the retro-arc region (Fig. 5.17b). If we reduce the maximum distance of tilt to 1000 km, which might result from a steepening of slab dip (see below), the

dynamic topography reduces to such low values that the back-bulge region is most likely non-depositional while the forebulge is erosional (Fig. 5.17c). Clearly, the amplitude and wavelength of dynamic topography is extremely important to the preservation of forebulge and back-bulge stratigraphy (see also Burgess & Moresi 1999).

The horizontal distance into the upper plate affected by dynamic topography should be related to the dip of the subducting slab (Mitrovica *et al.* 1989). Shallow subduction angles of <45° are required to produce deflections of the continental plate >500 km from the ocean trench for a range of initial slab temperatures relative to the mantle of −200 K to −800 K. These initial temperature contrasts most likely reflect the age of the oceanic plate at the point of subduction. The vertical amount of dynamic topography, however, is determined by the temperature contrast between the

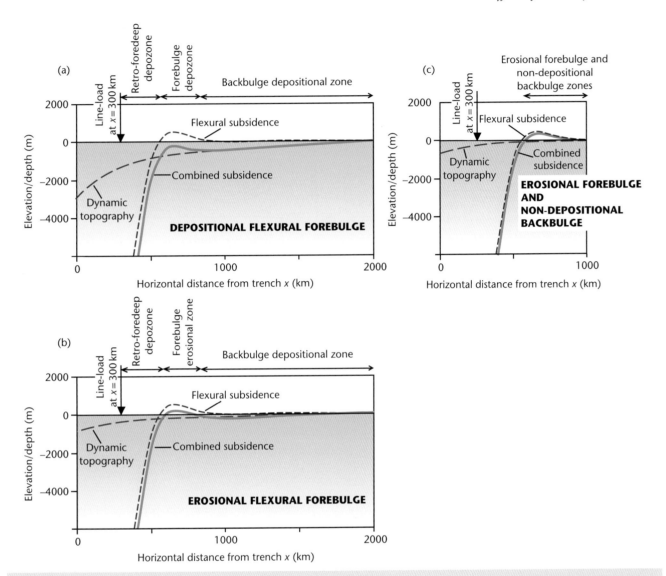

Fig. 5.17 Combination of flexural subsidence and dynamic topography behind subduction zones. In all cases, the flexural subsidence is that due to a line load at $x = 300$ km on a continuous plate with a flexural rigidity of 5×10^{23} Nm with a maximum deflection w_0 of 10 km. $\Delta\rho$ in the flexural parameter is 1300 kg m^{-3}. (a) Dynamic topography calculated using $f_m = 2000$ km, $\lambda = 200$ km, $\alpha = 0.5$, and $\eta = 2000$ km. The dynamic subsidence is large enough to overwhelm flexural uplift in the flexural forebulge region. (b) Dynamic topography calculated using $f_m = 500$ km, $\lambda = 200$ km, $\alpha = 0.2$, and $\eta = 2000$ km. Dynamic subsidence is sufficient to cause extensive backbulge deposition, but the flexural forebulge is erosional. (c) Dynamic topography calculated using $f_m = 500$ km, $\lambda = 200$ km, $\alpha = 0.2$, and $\eta = 1000$ km. Dynamic subsidence is insufficient to offset the effects of flexural forebulge uplift, and the backbulge region experiences such limited subsidence that it is likely to be non-depositional.

slab and the surrounding mantle, the flexural rigidity of the lithosphere, as well as the slab dip (Mitrovica *et al.* 1989). On cessation of subduction, any transient dynamic topography should be removed, the time scale of the uplift depending only on the initial temperature contrast between the slab and the mantle. 25 Myr is a typical recovery time for 95% of the uplift. There are therefore essentially two distinct modes of subduction, shallow and steep, that determine the dynamic subsidence of the continental interior behind ocean–continent boundaries, with secondary effects from plate age, rate of subduction, flexural rigidity, and viscosity structure of the upper mantle.

The evolution of dynamic topography through time may be complex. This is because the age of the oceanic slab, the velocity of plate subduction, and the angle of subduction change through time. The dips of slabs soon after the initiation of subsidence are believed to be close to vertical (Gurnis & Hager 1988). Over a period of perhaps 50 Myr, the slab shallows as a horizontal pressure gradient in the upper mantle develops (Stevenson & Turner 1977). The two most steeply dipping subduction zones in the Pacific, the Mariana and Kermadec, corresponding to young oceanic lithosphere, are both nearly vertical, whereas the oldest slab, Japan, has a dip of just 45°. Shallow slab dips may also be due to the attempted subduction

(a) EVOLUTION OF OCEAN–CONTINENT CONVERGENCE

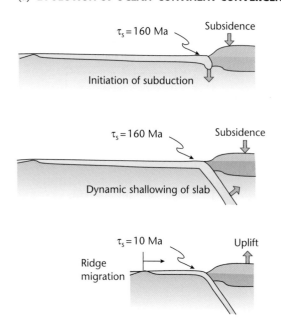

(b) DYNAMICALLY CONTROLLED STRATIGRAPHY

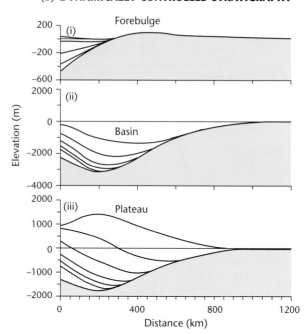

Fig. 5.18   Three stages in the evolution of dynamic topography over a subduction zone (a) and the associated chronostratigraphic surfaces in the region of the continental plate affected by dynamic topography (b). (i) An old (160 Ma) slab initially dips vertically, then penetrates the mantle by 100 km. The chronostratigraphic surfaces are separated by 2 Myr if the slab descent rate is 50 mm yr^{-1}. The slab then shallows in dip (ii). Chronostratigraphic surfaces are shown for each 10 degree decrease in slab dip. In (iii), the age of slab at the ocean trench progressively decreases to 10 Ma, causing uplift (no erosion). After Gurnis (1992), reproduced with permission of American Association for the Advancement of Science.

of relatively buoyant mid-ocean ridges and plateaus (e.g. Peru trench).

One scenario is the subduction of progressively younger and more buoyant lithosphere as an ocean basin closes (Fig. 5.18). In this case, the negative dynamic topography (subsidence) at the initiation of steep subduction of an old plate would be restricted to a narrow zone close to the trench bordered by a bulge on the continental plate. Subsequent rapid shallowing of dip of the slab would drive the bulge towards the continental interior. As the ocean basin is eventually closed, the subduction of young oceanic lithosphere generates long-wavelength uplift, producing an extensive plateau in the retro-arc region.

Some of these features can be identified in the subduction history of the Farallon plate beneath North America (Conrad *et al.* 2004; Forte *et al.* 2007; Spasojevic *et al.* 2008) (Fig. 5.19). In the Tertiary, a very large tract of western USA was uplifted (e.g. Colorado Plateau) as the Farallon plate shallowed in dip and decreased in age (Cross 1986), leading to high gravitational potential energy and extensional collapse (Jones *et al.* 1996). A similar model has recently been applied to the tilting of Laurentia in the Ordovician to explain subsidence in the Michigan Basin (Coakley & Gurnis 1995) and to the entire North American craton during the Phanerozoic (Burgess & Gurnis 1995; Burgess *et al.* 1997). Different timings of subduction, different slab angles and ages, and different locations of subduction, are capable of producing the alternate periods of uplift and subsidence demonstrated in the six transgressive/regressive sequences (supersequences) of Sloss (1963) (Fig. 8.6).

The Farallon slab has also been used to model mantle convection backwards in time from the present-day tomography (Conrad & Gurnis 2003; Liu *et al.* 2008). These results clearly show the extent of subsidence in the Cretaceous Western Interior Seaway of west-central North America (Fig. 5.19).

A point that is rarely addressed but which is of major importance is that dynamic subsidence is transient rather than permanent. When the flow responsible for dynamic topography is removed, the continental interior should rebound to its non-dynamic position, at a rate determined by mantle viscosity. In the case of mantle plumes, igneous underplating of the crust causes a permanent isostatic subsidence, whereas dynamic uplift over the plume head is transient. Burgess and Gurnis (1995) correctly conclude that another mechanism must be present to allow long-term preservation of thick stratigraphic sequences in the continental interiors.

5.2.3   Dynamic topography associated with supercontinental assembly and dispersal

The history of continental migration is clearly marked by phases of continental aggregation into supercontinental assemblies, followed by break-up and dispersal before aggregation takes place once again (Anderson 1982; Kerr 1985) (§1.4.5). This recurring signal in the geological record is marked by collisional orogenesis during continental assembly, and extrusive volcanism and rifting during break-up (Condie 2004). Rodinia, Gondwana and Pangaea are all examples of

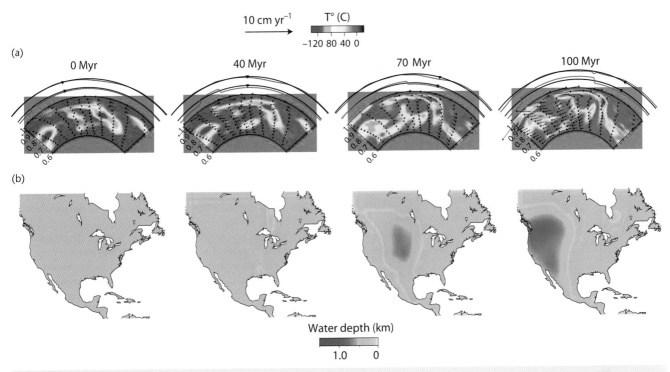

Fig. 5.19 (a) Cross-sections of the mantle (at 41°N) with velocity vectors (black) superimposed on the temperature field, due to subduction of the Farallon ocean slab from the left of the model and its passage beneath North America. The shallow subduction of the slab can be seen in the blue colours indicating cool temperatures. (b) Predicted continental flooding using an initially flat continent at 100 Ma and a eustatic sea-level correction from Haq & Al-Qahtani (2005). This model has a viscosity structure of 10^{20} Pa s for the upper mantle, 10^{21} Pa s for the transition zone and 1.5×10^{22} Pa s for the upper mantle. From Spasojevic *et al.* (2009), reproduced with permission of John Wiley & Sons, Inc.

supercontinental assemblies in the geological past. For example, the Laurentian (North American) plate broke away from a late Proterozoic Gondwanan supercontinent at about 575 Ma, then underwent accretions in the Late Devonian–Carboniferous (c.360 Ma) marking the initial assembly of the Pangaean supercontinent. By the Permian (c.250 Ma) Pangaea was a huge continental mass. In the Triassic (c.220 Ma) it started to split, with fragments dispersing rapidly in the Jurassic and Early Cretaceous. The time scale of the cycle above is of the order of 350 Myr. It should not escape the reader's attention that these aggregation-dispersal cycles roughly correspond with the first-order sea-level cycles of Vail *et al.* (1977a) and Haq *et al.* (1987) (see §5.4.3, §8.2.1).

Continents, and especially supercontinental assemblies, provide a different and more complex boundary condition for the heat loss of the Earth, since they introduce important lateral temperature variations across the continent–ocean boundary, and insulate the underlying mantle (Grigné *et al.* 2007). It is instructive to pause to speculate why the time scale of 300–400 Myr is characteristic of the 'supercontinental cycle' (Nance *et al.* 1988), since at present there is no theoretical basis for this regularity.

A number of time scales may be relevant. The diffusion time scale τ_κ can be written

$$\tau_\kappa = D^2/\kappa \qquad [5.10]$$

and represents the time it would take for a fluid layer of vertical dimension D to cool by conduction if convection ceased. With $\kappa = 10^{-6}$ m^2 s^{-1} and $D = 3000$ km, this conductive time scale is very large (many times the age of the Earth) and does not provide much information about the convective process. On the other hand, the convective time scale τ_c is given by

$$\tau_c = D/u = \tau_\kappa Ra^{-2/3} \qquad [5.11]$$

and represents the time it takes for fluid to cross the fluid layer of depth D at the convective velocity u. If $\mathbf{Ra} = 3 \times 10^6$, the convective time scale becomes 14 Myr, equivalent to a steady convective velocity of c.0.2 m yr^{-1}, whereas at $\mathbf{Ra} = 10^6$, $\tau_c = 130$ Myr and $v = 0.02$ m yr^{-1}, showing the strong dependency on Rayleigh number.

Instead, consider an insulating lid dominated by heat conduction to represent a slab of continental lithosphere above a fluid mantle (Fig. 5.20). The heat flux through the lid can be written

$$q = K_L \frac{T - T_0}{d} \qquad [5.12]$$

where K_L is the uniform thermal conductivity of the continental lid, and T and T_0 are the temperatures at the base and top of the lid of thickness d. If K is the thermal conductivity of the fluid layer of thickness h, and T_0 is set to zero, we obtain a thermal impedance term known as the Biot number

$$B = \frac{K_L}{K} \frac{h}{d} \qquad [5.13]$$

(a)

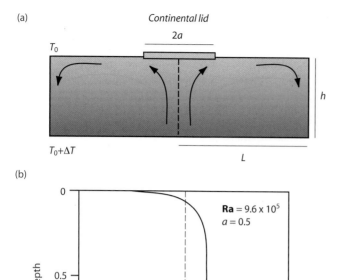

(b)

Fig. 5.20 (a) Geometrical parameters for convection beneath a supercontinent of half-width a acting as a lid conducting heat to the surface. There is no internal heating. Incubation of heat beneath the lid generates an upwelling, which feeds an elongate cell of width L. (b) Temperature profile beneath the continental lid from experimental work (Guillou & Jaupart 1995). Two dimensionless numbers characterise the flow: a Ra of c.10^6 and a dimensionless lid width of 0.5 ($2a/h$, where a is the half-width of the lid and h is the depth of the fluid layer). After Jaupart & Mareschal (2011, p. 268), reproduced with the permission of Cambridge University Press, 2002.

where **B** is small for a highly insulating lid. Note that the dimensionless number is the product of a dimensionless thermal conductivity and a dimensionless length scale. For the Earth K_L/K is slightly less than unity, and since d/h is approximately 10 under thick continental cores, **B** must be somewhat less than 10, which is relatively small. That is, the heat flux through the base of the thick continental lithosphere is small (c.10 mW m^{-2}) compared with the average heat flux through the oceanic plates driven by the convective cells in the mantle (100 mW m^{-2}). A wide insulating lid therefore has the effect of warming the mantle beneath it, but also of warming the entire interior of the Earth.

The presence of a wide, highly insulating lid also affects the circulation patterns in the mantle. For low continental extents (37% of the Earth's surface is covered with continents at the present day) of <40%, the flow pattern simply scales on the size of the continent or supercontinent (Lenardic & Moresi 1999; O'Neill *et al.* 2009). The fact that flood basalts are associated with supercontinental rifting and

break-up suggests that the effects of the continental insulator are coupled in some way to processes in the deep mantle. It is possible that heating beneath the continental lid causes localisation of plumes, which transmit buoyancy-related viscous stresses, which in turn drives rifting, basin development and voluminous extrusive activity (Hill 1991; Courtillot *et al.* 2003). Alternatively, plumes may initiate from the lower thermal boundary layer at the core–mantle boundary in regions devoid of cold subducted slabs, which descend close to continental margins rather than under supercontinents (O'Neill *et al.* 2009). Heating beneath supercontinents could therefore be partly the result of the absence of the cooling effect of subduction, and partly the result of the relatively rapid advection of heat in plumes and/or relatively slow passive upward advection. The presence of the lid prevents this heat from escaping to the surface, as it does under the oceans.

The size of the supercontinent influences the flow structures in the heated mantle beneath. The sub-plate mantle may become organised into elongated convective cells, with aspect ratios determined by the width of the lid and by the Rayleigh number. When the Rayleigh number is greater than about 10^6, the simple cells become unstable, producing small-scale flow structures superimposed on the larger cell structure (Jaupart & Mareschal 2011, p. 268). This may be important for generating periodic uplift and subsidence of the continental surface (see §5.2.4 and §5.2.5).

Since the elongate large-scale cells beneath an insulating lid have a size dependent on the width of the insulating lid (Guillou & Jaupart 1995; Grigné *et al.* 2007; Phillips & Coltice 2010), the largest supercontinental assemblies are likely to have the greatest impact on the elevation of mean temperatures of the convecting fluid layer (Fig. 5.21). Small continents have difficulty in preventing the effects of cooling *via* subduction at their margins from affecting sub-plate temperatures. On the other hand, large supercontinents successfully prevent the ingress of subduction-related cooling. For very large continents, however, inefficient lateral advection of heat may cause drip-type instabilities to form in association with small-scale convection cells with a wavelength of <1000 km (O'Neill *et al.* 2009). Such instabilities may be important in individualising cratonic basin depocentres (see §5.4.2).

Gurnis (1988) conducted numerical experiments on the feedbacks between supercontinental assembly and dispersal, the geoid and mantle convection, the main results of which have been supported by experiments in a spherical geometry (Phillips & Bunge 2007). In experiments involving a single plate overlying a convecting mantle, a geodynamical cycle involves: (i) a phase of sub-plate heating over the upwelling, when the horizontal velocity of the plate is zero and the average horizontal stress is also zero. While the plate is stationary, the average temperature along its base rises (1 °C per 2.5 Myr for approximately 450 Myr gives a temperature increase of 180 °C) and its topography rises. (ii) Plate migration: as the sub-plate mantle is heated, the plate experiences significant tension. Its maximum topography corresponds to the time of maximum horizontal extensional stress. The plate moves rapidly off the hot upwelling, relaxing the extensional stresses and reducing its surface topography as it does so. (iii) Plate settling over a downwelling, when the average stress along its base becomes compressive. The topographic range experienced by the surface of the plate during this cycle is approximately 3 km. A plate located over an upwelling could accumulate sufficient extensional stresses to break, the two continental fragments dispersing rapidly towards adjacent downwellings, and eventually (150 Myr after

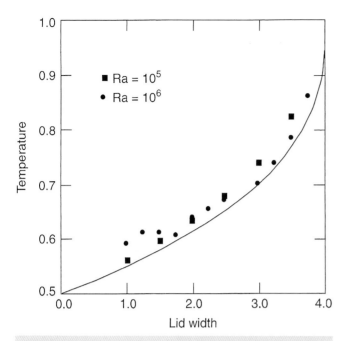

Fig. 5.21 Effect of lid width on the mean temperature of a convecting fluid layer, for two Rayleigh numbers of 10^5 and 10^6. Based on Grigné *et al.* 2007 (Jaupart & Mareschal 2011, p. 270), reproduced with the permission of Cambridge University Press, 2011.

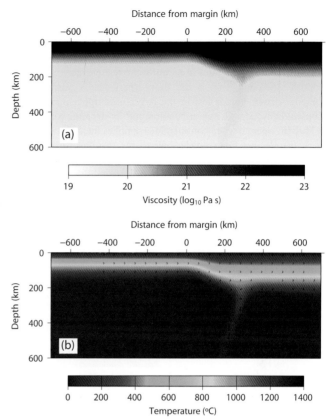

Fig. 5.22 Edge-driven convection or 'corner flow' associated with mantle circulation beneath a continental margin. The flow is generated after 40 Myr from a 100 km step in the thermally and compositionally defined lithosphere. (a) Viscosity, (b) temperature and flow vectors. Density is a function of temperature and melt depletion, and there is no pressure dependence. The model has an initial uniform density distribution. Density increases during evolution of the model by cooling of the ocean lithosphere. From John Armitage (pers. comm.).

break-up) colliding with each other again to form a new supercontinent. After about 450 Myr, the supercontinent has insulated the mantle for long enough to develop a new upwelling under the plate. Assembly of continental fragments and incubation of the mantle beneath the supercontinent is therefore very slow, whereas break-up and dispersal is relatively fast.

The time scale for a supercontinental cycle therefore depends primarily on the long time taken to incubate the sub-plate mantle sufficiently to inflate the supercontinent, cause break-up, and lead to the relatively rapid migration of continental fragments to adjacent geoid lows.

5.2.4 Dynamic topography associated with small-scale convection

It has been proposed (McKenzie 1967; Richter 1973; McKenzie *et al.* 1974) that there is a small scale of convection confined to the upper 650–700 km of mantle, but there is a lack of clear topographic expression and gravity field of the upwelling and downwelling limbs of such a convection system, or of lower lithosphere 'dripping' downwards to drive small-scale convection. It is possible, however, that small-scale convection may take place accompanying lithospheric stretching (Buck 1986; Keen & Boutilier 1995) or as an edge-driven effect at ocean–continental boundaries (Vogt 1991; King & Anderson 1998; Conrad *et al.* 2004). Edge-driven convection may be responsible for some LIPs situated close to continental margins, such as that of Victoria, Australia (King & Anderson 1995).

Steps in the base of the continental lithosphere, such as at the margins of cratons, and at the continental–ocean boundary (Fig.

5.22), may represent significant heterogeneities that interact with mantle flow, driving a pulsating secondary circulation with a downwelling fixed close to the ocean–continent boundary and an upward flow of hot mantle displaced about 600 km from the continental edge (Elder 1976; Vogt 1991). The Bermuda Rise is thought to be the seafloor bathymetric expression of such an upwelling, whereas the anomalously deep Nova Scotian margin of the northwestern Atlantic Ocean has been suggested to represent a cold downwelling of an edge-driven convection system (Conrad *et al.* 2004).

Pioneering models of edge-driven convection (King & Anderson 1995) required a shallow mantle beneath a continental lid that was hotter than under thinner lithosphere, as would occur under incubating supercontinents (Gurnis 1988). The small-scale edge-driven flow at a cratonic or continental margin interacts with the large-scale mantle flow. When this large-scale flow is small, the edge-driven effects stand out clearly, and model runs showed an increasing dominance of the edge-driven flow over time (King & Anderson 1998).

A second scenario is that the thick lithosphere is moving relative to the underlying upper mantle so that there is a shear between the two. If the mantle moves from the craton to the ocean relative to the lithosphere, that is, the flow experiences a negative step, the edge-driven flow maintains itself, but if the relative motion is in the opposite direction, that is the flow experiences a positive step, the edge-driven flow is lost in the larger-scale upper-mantle flow. Consequently, edge-driven circulation and its dynamic topography is likely to operate when continental assemblies incubate an underlying mantle that flows slowly towards the margins. Qualitatively, elevated bathymetry is expected in the oceanic lithosphere c.10^3 km from the continental edge, while the continental lithosphere adjacent to the continent–ocean boundary is expected to experience slow subsidence due to cold downwelling. Such edge-driven down-

welling is expected to initiate during the period of continental inflation and extensional break-up. The negative dynamic topography associated with the downwelling under the continental or cratonic edge would take the form of low-angle ramp-like tilting towards the ocean, or sub-circular sags, depending on the three-dimensional nature of the edge-driven convection (for further discussion see §5.4.2).

Small-scale convection may also develop under continental lithosphere without the effect of a step at the continental edge (Petersen *et al.* 2010). The horizontal scale of the temperature variations modelled in the shallow mantle beneath the continental lithosphere is hundreds of kilometres, representing a considerably smaller length scale than expected for larger flow structures originating at the core–mantle boundary (Fig. 5.23). Fluctuations in the dynamic topogra-

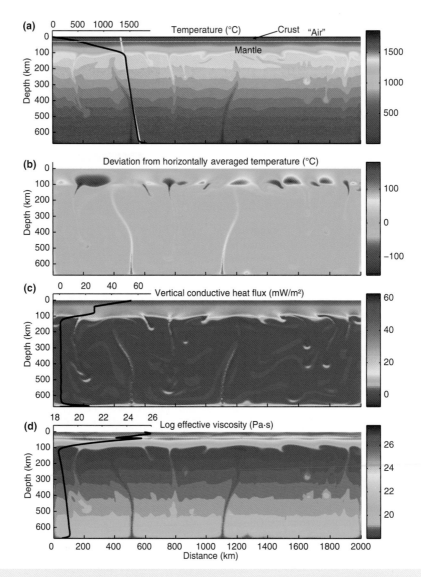

Fig. 5.23 Small-scale convection from a numerical convection model (Petersen *et al.* 2010). (a) Temperature. Black line shows averaged model geotherm, and grey line is the isentropic (adiabatic) profile for a potential temperature of 1315 °C. (b) Deviation from the horizontally average temperature. (c) Vertical conductive heat flux; black line is horizontal average. (d) Effective viscosity; black line is horizontal average. Note the development of small-scale instabilities, causing surface variations in heat flow and topography that may be important in the generation of stratigraphic sequences. Reproduced with permission of American Association for the Advancement of Science.

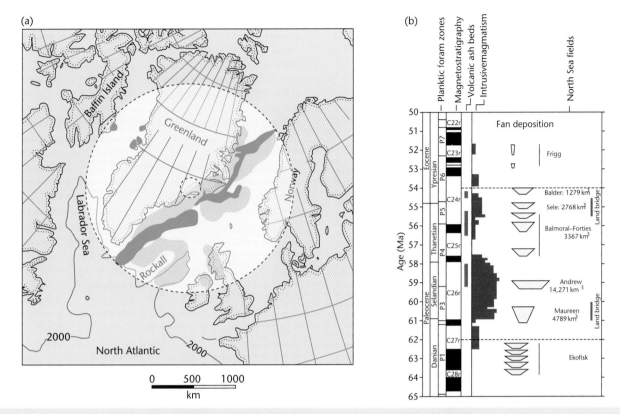

Fig. 5.24 (a) Reconstruction of the northern North Atlantic region just after the onset of ocean spreading (magnetic anomaly 23), showing the extent of the Iceland plume (White and McKenzie 1989; reproduced with permission of the American Geophysical Union). Solid shading shows position of extrusive volcanic rocks, and light shading shows extent of Early Cretaceous igneous activity. Uplift over the plume head, shown by circular area, can be recognised in thermochronological data such as apatite fission tracks. (b) Pulsing of the Iceland plume is thought to have caused episodic deposition of sands in the deep sea (White and Lovell 1997; reprinted by permission from Macmillan Publishers Ltd.). Periods of deep-sea fan deposition correspond to times of high levels of intrusive igneous activity.

phy of the continental surface of this length scale may explain the stratigraphic sequences of sedimentary basins.

5.2.5 Pulsing plumes

The periodic emplacement of sand beneath the deep waters of the North Sea during the Early Tertiary (White & Lovell 1997) (Fig. 5.24b) was interpreted as a result of pulsating activity of the nearby Iceland plume. Apart from causing variations over time in the volumes of erupted volcanic rocks (O'Connor *et al.* 2002), pulsating plumes might be expected to cause periodic phases of uplift of the seafloor (or continental surface), leading to the generation of erosional unconformities, tilting of sedimentary strata, and perhaps even to climate changes caused by the shallowing or occlusion of oceanographic gateways (Jones *et al.* 2001; Shaw-Champion *et al.* 2008; Hartley *et al.* 2011; Poore *et al.* 2011). Pulsing plumes may also cause phases of igneous underplating beneath the crust, resulting in relative uplift during intrusion followed by subsidence following crystallisation (McLennan & Lovell 2002).

The transmission of thermal pulses of the early Iceland plume is thought to have produced diachronous and transient relative uplift as they spread out from the main plume conduit (Rudge *et al.* 2008). The radially spreading mantle material is thought to flow in a channel of low viscosity beneath the plate, with instabilities giving rise to

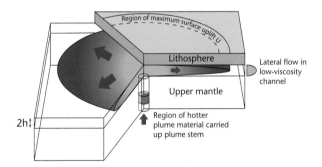

Fig. 5.25 Model for radial, channel-like flow associated with the impingement on the base of the lithosphere of a plume. After Rudge *et al.* (2008), reprinted with permission from Elsevier.

relatively rapid pulses of uplift and subsidence (Fig. 5.25). The dating of the transient uplift suggests that the uplift-subsidence motion (150–250 m – greater than can be accounted for by eustatic mechanisms) took place on the time scale of about 10^6 yr – smaller than the periods conventionally associated with convection, but similar to the periodicity of volcanic activity found in some hotspot chains (O'Connor *et al.* 2002).

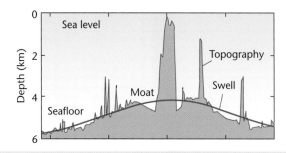

Fig. 5.26 Bathymetric profile of the Hawaiian Ridge at Oahu (after Watts 1978) showing that the hotspot swell is elevated about 2 km above the adjacent seafloor. Reproduced with permission of the American Geophysical Union.

The mechanisms that cause and control temporal variations in mantle plumes are not understood. Of the hypotheses that have been proposed, most have difficulty reproducing the correct frequency of multiple pulses, or need to invoke unrealistic excess temperatures and heat flows in the convecting upper mantle. It is currently uncertain whether variations in plume activity are related to changing fluxes of subducted cold material to the core–mantle boundary, to secondary instabilities within the plume conduit, or to interactions with phase transitions and rheological variations.

5.2.6   Hotspots, coldspots and wetspots

There is a smaller wavelength of thermal activity in the mantle compared to the very long wavelength (degree 2–10) of the geoid. This shorter wavelength pattern is shown by the spatial distribution of hotspots (Fig. 5.9) (see §5.1.5). Hotspots are at their most obvious in the ocean, where they are associated both with volcanism and strongly elevated topography. The Hawaiian Islands, Iceland and Canaries are all excellent examples. The recognition of hotspots on the continents is far more problematic. A key feature of hotspots is that they are assumed (perhaps incorrectly) to be fixed relative to a stationary reference frame in the underlying mantle, so absolute plate motion causes a hotspot track to be formed on the moving plate.

Initially, it is instructive to look at hotspot swells in the ocean (Table 5.1). Topographic uplifts over hotspots in the ocean are large, with swells elevated >1 km above the seafloor (Fig. 5.26). The topographic doming over a hotspot must be maintained by excess buoyancy. The buoyancy flux of eqn. [5.8] can be expressed in the alternative form (Davies 1988; Sleep 1990)

$$B = \Delta\rho WE u_p \qquad\qquad [5.14]$$

where $\Delta\rho$ is the density contrast of the uplift, WE is the cross-sectional area of the swell, and u_p is the plate velocity. The buoyancy fluxes of the Earth's present-day hotspots (Table 5.1) range over a factor of 20, with Hawaii at 8.7 Mg s^{-1} the largest. The average buoyancy flux for the 37 currently active hotspots is in the region of 1.35 Mg s^{-1}. It can immediately be appreciated that there is an inverse relationship between the cross-sectional area of the swell and the plate velocity for the same buoyancy flux from the mantle.

The Cape Verde swell typifies plume-related temperature anomalies and bathymetric relief (Courtney & White 1986) (Fig. 5.27). The best-fitting convection model to explain the temperature distribution involves a plume neck of 150 km diameter feeding a mushroom-like

head of hot mantle material 1500 km across. The temperature anomaly in the plume head at the base of the plate is generally in the region of 100 °C, with the neck of the plume in excess of 300 °C above ambient asthenospheric temperatures. The centre of the Cape Verde swell is elevated by 1900 m relative to the normal oceanic depth; most of this relief can be attributed to dynamic uplift over the convecting plume head.

Geoid anomalies suggest that plume upwellings are spaced at about 2500–4000 km in the upper mantle beneath the oceans (McKenzie *et al.* 1980), but their distribution beneath the continental lithosphere is more speculative. The highspots associated with alkaline volcanism in north-central Africa, such as the Hoggar and Tibesti domes, may be related to activity in the underlying mantle (Thiessen *et al.* 1979; Sahagian 1980). Doming over hot plumes has also been interpreted as responsible for particular centrifugal paleo-drainage patterns (Cox 1989) (§5.4.1), as well as for the eruption of vast piles of basalts (§5.3.1, §5.3.2).

Hotspots should also produce swells on the continental lithosphere. Taking the globally 'typical' hotspot ($B = 1.35$ Mg s^{-1}) and a continental plate ($\Delta\rho = 2700$ kg m^{-3}) with a track velocity of 20 mm yr^{-1}, the cross-sectional area of an average topographic dome is about 800 km^2. For a dome elevated at 1 km above the regional elevation, we would expect a width of c.1000 km. For a fast-moving plate (say 100 mm yr^{-1}), the cross-sectional area reduces to just 160 km^2 for the same buoyancy flux. Since slow-moving plates are likely to have better developed hotspot swells than fast-moving plates, hotspot swells should be more significant and more easily recognised at times of continental assembly before rapid plate dispersal.

Fluid dynamical work (Griffiths & Campbell 1990; Hill 1991) suggests that low-viscosity plumes may initiate from within the Earth and ascend as a spherical pocket of fluid (plume head) fed by a pipe-like conduit (plume tail) continuously supplying buoyant, hot material to the head region (Fig. 5.28). It is likely that enlargement of the plume occurs by the entrainment of material heated by the ascent of the plume from the core–mantle boundary. Flood basalt provinces have been interpreted as originating through the melting of the heads of newly started plumes, whereas oceanic island chains represent the tracks of the relatively long-lived plume tails as the plate migrates over the mantle.

Fluid dynamical experiments also suggest that the diameter of a new plume head varies according to the volume flux and temperature excess of the source material provided to the plume head, and the thermal and viscosity properties of the lower mantle into which the plume starts to ascend (Griffiths & Campbell 1990). The plume head then grows by entrainment as it ascends through the mantle, so that the plume head diameter grows as a function of the distance travelled. By the time the plume head has penetrated into the upper mantle it should have cooled to only 100–200 °C above the ambient temperature.

Upon nearing the surface of the Earth, the plume head spreads out into a disc of hot material with positive buoyancy (Fig. 5.28a). This produces a dynamic surface uplift. The scaling of laboratory experiments suggests that the timing and magnitude of the surface uplift depend strongly on the viscosity of the upper mantle. The results from laboratory experiments in which a plume head is sourced from the core–mantle boundary with a buoyancy flux of 3×10^4 N s^{-1} and a source temperature excess of 300 °C, lower mantle dynamic viscosity of 10^{22} Pa s (kinematic viscosity of 2.5×10^{18} m^2 s^{-1}) ascending into an upper mantle with a viscosity of 3×10^{20} Pa s, is shown in Fig. 5.28b. The surface is initially weakly uplifted while the plume head

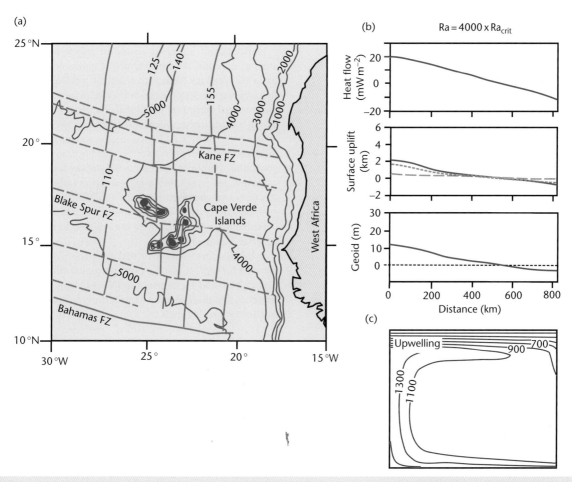

Fig. 5.27 Cape Verde hotspot. (a) Bathymetric map (in m) of the Cape Verde Rise, with isochrons in Ma. Fracture zones are shown in dashed lines. (b) Heat flow, surface uplift and the geoid for a convection cell at 4000 times the critical Rayleigh number for convection. The total uplift is shown as a conductive component (dashed) and a convective component (dotted). (c) Temperature distribution responsible for the results shown in (b). After Courtney & White (1986), reproduced with permission of Oxford University Press.

is entirely within the lower mantle (−25 Myr). When the plume head enters the low-viscosity zone of the upper mantle (−3 Myr) the surface uplift takes place rapidly, reaching a maximum elevation of 600 m after further ascent to just beneath the lithosphere. At this stage the plume head has a diameter of 1300 km. A number of factors may increase the maximum elevation of the topographic dome: (i) penetration of the hot plume into the cold lithosphere; and (ii) a volume increase caused by melting. The release of large amounts of basalts by melting of the plume can only take place once the plume head has reached shallow depths. It should be noticed that this is only possible a number of millions of years after the maximum surface uplift. The development of smaller-scale gravitational instabilities over the cap of the plume as cold, dense material is squeezed between the ascending plume head and the Earth's surface may facilitate the last-stage ascent of the plume and produce the high surface uplifts seen today in locations such as East Africa, and the large outpourings of basalts in the geological past, such as the Siberian and Deccan (India) Traps (Fig. 5.8).

Hotspot swells in the oceans are well documented. We might also expect topographic uplift of the continents over sub-lithospheric upwellings. Such uplifts are likely to be of the order of 1 km in height and perhaps 1000 km in width. The domes of Africa, such as the Hoggar of the Saharan region, and the Bié and Namibian domes of southwest Africa (Al-Hajri *et al.* 2010), are good candidates for uplift over upwellings. Over geological time, all the continents must have passed over mantle upwellings, leaving hotspot tracks. Some parts of continents must have a veritable criss-cross pattern of hotspot tracks (Morgan 1983). Within a band of approximately 10^3 km width, the continental surface should be transiently elevated, leading to erosion and the development of regional unconformities. Such unconformities would be recognisable long after the transient dynamic uplift has ceased. If we ignore flexural uplift due to mountain building (Chapter 4), and dynamic topography related to plate subduction (§5.2.2) and supercontinental assembly (§5.2.3), involvement in hotspot uplift may be a major cause for erosional unconformities in cratonic sedimentary sequences.

If there are upwellings beneath continents, we should also expect there to be downwellings, which would be cold and characterised by negative dynamic topography. The Congo Basin may be situated over one such downwelling (Hartley & Allen 1994; Downey & Gurnis 2009; Crosby *et al.* 2010, but compare Buiter *et al.* 2011). It is situated adjacent to the East African upwelling. The fact that many

(a)

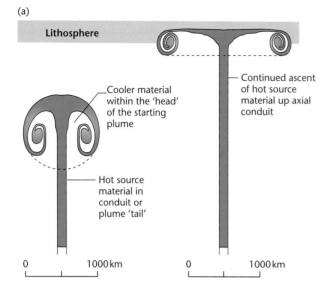

(b)

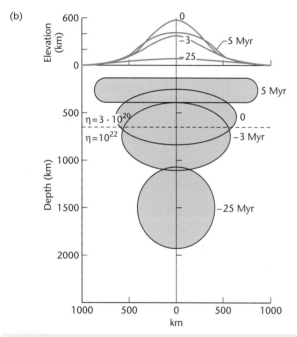

Fig. 5.28 (a) Schematic diagram of the ascent of a plume through the mantle and its mushrooming into a disc beneath the lithosphere, with darker shading indicating higher temperatures, after Hill (1991) and reprinted with permission from Elsevier. (b) The dimensions of a starting plume together with the predicted uplift, based on laboratory experiments. During the lateral spreading of the head, the input of material from the source is discontinued, simulating the carrying away of the head from the source region by plate motion. After Griffiths & Campbell (1990), reproduced with permission of the American Geophysical Union.

sedimentary basins have sedimentary sequences deposited over long periods of time raises the possibility that small-scale convection systems are pinned to the lithospheric plates (see §5.2.4). If so, these convection systems would have upwellings that do not move significantly relative to the plate, in contrast to the well-developed hotspot

tracks caused by migration of a plate over a quasi-stationary system of mantle upwellings.

Hotspots and highspots should of course be important source areas of erosional detritus. Peripheral uplifts around a sedimentary basin may feed the basin with sediment and load the lithosphere. In this sense, some cratonic basins may be more passive receptacles for the erosional detritus of peripheral or annular uplifts than areas with strong mechanical subsidence (see Sahagian 1993 for a similar view).

The presence of volatiles reduces the solidus for mantle rocks and therefore promotes melt generation, and hydration decreases density, generating buoyancy. Consequently, topographically elevated volcanic hotspots may reflect mantle wetspots rather than plume heads. A wetspot might result from the subduction of oceanic crust, for example. However, volcanic seamount chains such as Hawaii have persisted for 100 Myr, so there is a problem in the required volume of water-rich mantle rock.

It has also been suggested that the buoyancy of topographically elevated hotspots might be a compositional effect caused by depletion in iron after melt extraction (Morgan *et al.* 1995). It is unlikely, however, that this effect can explain the entire buoyancy represented by the topographic swell.

5.3 Mantle dynamics and magmatic activity

Igneous activity is a common feature of some continental rifts and passive margins at the point of break-up. Igneous activity is caused by adiabatic decompression due to one or a combination of lithospheric stretching, elevated asthenospheric temperatures and presence of volatiles. Hot plumes derived from the core–mantle boundary have a profound effect on asthenospheric temperatures, melt generation by adiabatic decompression and surface uplift. Plume heads occupy areas of the asthenosphere generally 1000–2000 km across with temperature anomalies of 100–200°C, and in the oceans are responsible for bathymetric swells 1–2 km above the surrounding seafloor. The elevation of sub-lithospheric temperatures over a mantle plume may generate large amounts of melting and the upward migration of melts to form igneous underplates and extrusive basalts. Transient surface uplift is caused by the dynamic effects of hot asthenospheric flow, whereas permanent uplift results from igneous underplating of the crust. Consequently, volcanically active passive margins and continental rifts stand elevated topographically compared to non-volcanic equivalents. The amount of melt generated and its composition is related to the plate thickness, excess temperature, stretch factor and percentage of volatiles.

It is beyond doubt that some regions of continental stretching are unrelated to any 'active' thermal processes. However, in other instances, continental stretching and break-up are associated with voluminous basaltic volcanism, as in the Deccan of India, the Karoo of southern Africa and the Parana of South America. Magmatically active rifts and continental margins are commonly associated with anomalously high temperatures in the asthenosphere caused by the presence of mantle plumes (White & McKenzie 1989).

The long chains of volcanic islands (e.g. Hawaiian-Emperor seamount chain) and bathymetric swells in the oceans are associated with the eruption of extensive basalts whose geochemistry indicates that they originate by melting of mantle elevated above the normal temperature of the asthenosphere. Such high temperatures are thought to be due to the ascent of hot material from the lower thermal boundary layer at the base of the mantle. Two sorts of uplift pattern and igneous activity should result from plume activity, depending on whether the lithosphere overlies a plume head or plume tail. Plume head regions should be areally extensive and equant (1500 km to 2500 km across), giving rise to LIPs, whereas plume tail provinces should be narrow (<300 km wide) and linear, that is, they should be hotspot tracks.

5.3.1 Melt generation during continental extension

Igneous activity in sedimentary basins is diagnostic of basin-forming mechanisms. In the reference uniform stretching model of McKenzie (1978a), no melts are generated because the geotherm does not intersect the solidus for crustal and mantle materials. However, it is well known that rift provinces are associated with minor (e.g. Western Rift, Africa, North Sea) to major (Eastern Rift, Africa, Rio Grande) volcanism. Most importantly, continental break-up at high values of stretching is commonly associated with vast outpourings of flood basalts, indicating major melting of the asthenosphere by adiabatic decompression. Melts liberated by decompression are assumed to separate from their residue and to travel upwards to either be erupted at the surface, or to be emplaced as igneous bodies in the crust. McKenzie and Bickle (1988) showed how the amount of melt generated depends on the potential temperature of the asthenosphere – defined as the temperature the asthenosphere would have if brought to the surface adiabatically without melting – and the amount of stretching (Fig. 5.29). For example, if a 100 km-thick lithosphere is stretched by a factor of 2, we would expect a melt thickness of about 2 km for a potential temperature of 1400 °C. At higher values of stretching ($\beta = 5$), we would expect a melt thickness of 10 km for the same potential temperature. The normal potential temperature of the asthenosphere is 1280 ± 30 °C, so the examples above refer to an excess temperature of the order of 100 °C attributable to plume activity.

The conditions at the site of partial melting are reflected in the composition of the erupted basalts. Stretching at infinitely high values at a normal asthenospheric potential temperature of 1280 °C produces melt with the composition of a mid-ocean ridge basalt (MORB). However, increasing the potential temperature to 1480 °C causes an increase in MgO and a decrease in Na_2O. Consequently, we would anticipate the generation of first alkali basalts and then tholeiitic basalts as stretching increases over an asthenosphere with a potential temperature of 1480 °C.

The presence of a hot asthenosphere affects the subsidence experienced at the Earth's surface caused by lithospheric stretching. This is mainly caused by two effects: the addition of igneous bodies beneath the crust (see below), and the dynamic uplift from the mantle plume.

The density of igneous rock generated by adiabatic decompression of the mantle depends on the potential temperature at the site of partial melting, and ranges between 2990 and 3070 kg m^{-3} (at a depth of 10 km) for potential temperatures of 1280 °C and 1480 °C respectively. This range of density is mid-way between the density of the continental crust and the density of the mantle, so there is a strong likelihood that melts will be trapped and underplated beneath the crust. Since these igneous additions replace lithospheric mantle (density 3300 kg m^{-3}), the net effect is uplift relative to the depth expected from uniform stretching of the lithosphere without magmatic activity (Appendix 27).

Assuming Airy isostasy, the amount of underplating X required to generate an amount of surface uplift U can be found by balancing

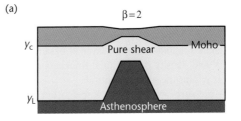

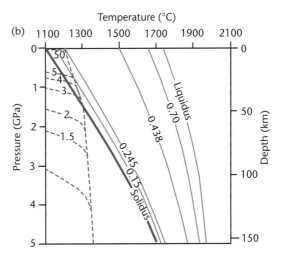

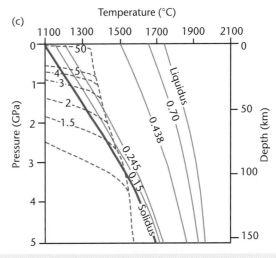

Fig. 5.29 Melt generation by adiabatic decompression during pure shear extension, based on McKenzie & Bickle (1988), White & McKenzie (1989) and Latin & White (1990). (a) Uniform stretching of the lithosphere by a stretch factor of 2. (b) Adiabatic upwelling due to stretching at different values of β from 1.5 to 50. Potential temperature of the asthenosphere 1280 °C, mechanical boundary layer thickness of 100 km, kinematic viscosity 4×10^{15} m^2 s^{-1}. Curves between the liquidus and the solidus show the melt fraction by weight. Only small amounts of melt can be produced at high values of stretching. (c) As for (b), but with a potential temperature of 1480 °C, showing that large amounts of melt can be liberated at relatively low values of the stretch factor.

two lithospheric columns. In the undisturbed column, crust of thickness y_c and density ρ_c occurs within a lithosphere of thickness y_L and subcrustal density ρ_m. In the underplated lithospheric column, crust of the same thickness is underlain by a layer of basaltic underplate with thickness X and density ρ_x. The total thickness of this lithospheric column is y_L plus the surface uplift U.

Balancing the pressures at a depth of compensation at the base of the lithosphere yields

$$U = X\left(1 - \frac{\rho_x}{\rho_m}\right)$$ [5.15]

The density of the mantle lithosphere can be taken as 3300 kg m^{-3} and the density of the basaltic underplate as 3000 kg m^{-3}. The rock uplift for an underplate thickness of 5 km is therefore 455 m, or approximately one tenth of the underplate thickness. This is also the surface uplift without any erosion. We can incorporate erosion in the simplest way as follows (Brodie & White 1994). Consider a piece of lithosphere underplated with basalt of thickness X that is eroded so that the surface elevation change is zero (Fig. A27.1). The crust therefore has a thickness of y_c minus the denudation D. We can once again perform an isostatic balance, giving

$$D = X\left(\frac{\rho_m - \rho_x}{\rho_m - \rho_c}\right)$$ [5.16]

where D is the total denudation of crust caused by the emplacement of a basaltic underplate of thickness X. The total denudation for an underplate thickness of 5 km if there is no change in the surface elevation is 2.5 km, if crustal rocks of density 2700 kg m^{-3} are eroded. We therefore have the maximum erosion case and the no-erosion case. Reality is probably in between. Clearly, magmatic underplating can have significant effects on surface uplift and on the delivery of erosional products to neighbouring sedimentary basins (White & Lovell 1997). The uplift caused by magmatic underplating is permanent.

The circulation in the mantle plume maintains anomalously high temperatures that cause a dynamic uplift, as described above. The combined effect of underplating and dynamic support is to elevate the continental plate to sea level or above during the rifting process for excess asthenospheric temperatures of 100–150 °C. This is entirely supported by the observation that the basalts erupted at rifted continental margins are commonly subaerial rather than submarine, and flow for large distances downslope to produce extensive flood basalt provinces.

5.3.2   Large igneous provinces

It is known that hotspot tracks emerge from LIPs, or flood basalt provinces (Morgan 1981). The tracks are marked by volcanoes with an age progression along the chain, highly suggestive of the relative movement of the oceanic plate across a plume tail. The end of the track is a currently active volcano. Correspondingly, large equant igneous provinces are interpreted as the result of impingement of a plume head on the base of the lithosphere. The largest submarine flood basalt province is the Ontong Java Plateau east of Papua New Guinea, whereas the Siberian traps of northwestern Russia, the Karoo of southern Africa, and the Parana of South America are celebrated continental flood basalt provinces from the geological record (Duncan & Richards 1991).

5.3.3   The northern North Atlantic and the Iceland plume

The northern Atlantic contains extensive basalts covering c.500 × 10^3 km^2 on both sides of the ocean in Greenland and in northwest Britain, Ireland and the Faroes (Fig. 5.24). The basalts range from 52–63 Ma in age, with a main eruption phase at 59 Ma (Mussett *et al.* 1988). The thermal anomaly responsible for these igneous provinces is now located beneath Iceland. The Iceland region continues to be dynamically supported by the underlying plume over an area of anomalously shallow seafloor with a radius of 1000 km.

In the subsurface seismic record, extrusive basalt sheets can be recognised as seaward-dipping reflectors along both the east Greenland and Rockall–Faroes–Norwegian margin. Volumetrically smaller basaltic volcanism can be seen in onshore exposures such as the Tertiary Igneous province of northwest Scotland and northeast Ireland. Thick (<8 km) prisms of accreted igneous rocks with seismic velocities >7.2 km s^{-1} have been imaged across the Hatton Bank and Voring Plateau beneath thinned continental crust, most likely representing underplated igneous rocks. White and McKenzie (1989) estimate the combined underplate and extrusive basalt volume as <10^7 km^3.

The Icelandic plume has been identified as the 'smoking gun' for the widespread surface uplift of the North Atlantic area in the Early Tertiary (Paleocene). The regional surface uplift caused by underplating and dynamic support is thought to have promoted high rates of denudation and transport of particulate load into the deep-sea environments of the North Sea and NW Atlantic margin basins. High rates of denudation in large areas of northwestern Europe during the Early Tertiary can be inferred from apatite fission track analysis (papers in Doré *et al.* 2002). In the neighbouring sedimentary basins, such as the Faroes–Shetland Basin and the Rockall–Porcupine troughs, sandstones were episodically delivered from this exhuming continental landmass. White and Lovell (1997) suggested that the main pulses of submarine fan deposition correspond to the main pulses of magmatic activity during the time period 62 to 54 Ma.

5.3.4   The Afar region, Ethiopia

The Afar region of northeast Africa is currently undergoing extension, accompanied by voluminous intrusive and extrusive magmatism. It illustrates the magmatism associated with continental break-up and the transition to a new ocean basin, during continental rifting (Ebinger & Casey 2001). The uppermost mantle beneath the Horn of Africa (to a depth of 400 km) is seismically (P- and S-waves) slow (7.4–7.5 km s^{-1}) (Fig. 5.30), suggesting that the mantle is hot and most likely subject to partial melting. It has been suggested that the mantle beneath Ethiopia is the hottest on Earth (Bastow *et al.* 2008). One view is that the Ethiopian region is the location of a plume stem originating from a northward continuation of the African super-plume (Furman *et al.* 2006) – a deep, long-lived, thermal structure with a footprint on the core–mantle boundary of 2000 × 4000 km and a height of 2000 km. An alternative view is that it is part of a single large plume centred on Turkana (Ebinger & Sleep 1998), and a further possibility is of two plumes rising beneath southern Ethiopia and Afar (Rogers *et al.* 2000).

A characteristic feature of continental rift provinces overlying hot mantle is that a large (~80%) part of the extension is achieved by the

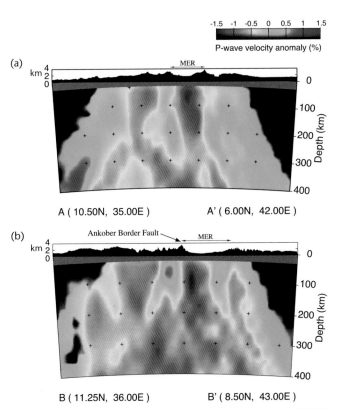

-1.5 -1 -0.5 0 0.5 1 1.5

P-wave velocity anomaly (%)

(a)

km

A (10.50N, 35.00E)   A' (6.00N, 42.00E)

(b)

Ankober Border Fault   MER

km

B (11.25N, 36.00E)   B' (8.50N, 43.00E)

Fig. 5.30 Vertical cross-sections of the P-wave velocity model, with contours in P-wave velocity anomaly (%), for (a) the continental rifting part of the Ethiopian region, and (b) the region of transition to ocean crust towards Afar. The low-velocity zone extends from 75–400 km depth and is thought to be the northward continuation of the African superplume. The very slow velocities are thought to be due to partial melting in the upper mantle. The Main Ethiopian Rift (MER) is situated towards the eastern flank of the low-velocity structure, not above its centre, suggesting that pre-existing tectonic structures are important in localising melt migration at the base of the lithosphere. From Bastow *et al.* (2008), reproduced with permission of John Wiley & Sons, Inc.

intrusion of dykes, as occurs below the Baikal rift (Thybo & Nielsen 2009). This means that stretch factors are relatively low, which affects the timing and volume of decompression melts generated by thinning during advanced stages of plate break-up. Magma intrusion also reduces plate strength, thereby causing later localisation of stretching. The Main Ethiopian Rift is consequently displaced towards its eastern edge from the locus of the underlying low-velocity zone.

The final stretching leading to a new ocean basin caused voluminous (2 km thick) basaltic volcanism due to decompression melting of the asthenosphere at 29–31 Ma. This is just prior to, or at the same time as the onset of rifting in the Red Sea and Gulf of Aden. These flood basalts of the NW Plateau directly overlie the low-velocity zone in the upper mantle. Active volcanism continues to the present day in the Afar Depression and parts of the Main Ethiopian Rift. Low-temperature thermochronometry suggests that Ethiopia experienced surface uplift at 30–20 Ma. Extension in the Main Ethiopian Rift started at ~20 Ma in the south of the region, but migrated to the

north at ~10 Ma. The coincidence of the onset of topographic uplift and the onset of extrusive volcanism at 30 Ma suggest that they both relate to the arrival of a hot mantle flow structure beneath the area, though whether it was a plume head (Griffiths & Campbell 1990; Hill 1991) followed by a narrow conduit or tail, is unknown. Early dyking controlled the orientation of later border faults, which accommodated most slip some millions of years later.

5.4   Mantle dynamics and basin development

Mantle dynamics influence basin development in a number of possible ways. Principally, mantle processes control dynamic topography (§5.2), which has an important impact on the location and elevation of source regions for sediment and on the location and rate of subsidence of sedimentary basins. However, extracting the component of dynamic topography from the total topographic signal of the surface of the Earth and from its sedimentary archive is challenging. The most promising situations are areas where other basin-forming mechanisms are absent or subdued, such as in continental interiors. Cratonic basins (§5.4.2) therefore offer the possibility of inverting mantle processes from the stratigraphic record.

Topography supported by convective flow in the mantle occurs at different wavelengths and amplitudes. The dynamic topography due to the African superplume (Nyblade & Robinson 1994) is approximately 500 m over a large part of the African continent, whereas the dynamic topography related to the convective cells responsible for hotspots is likely to be 0.5–1 km over a much shorter wavelength of 500–1000 km. In addition, a number of domal uplifts occur close to the ocean–continent transition in western Africa, such as the Namibian dome (Al-Hajri *et al.* 2010).

5.4.1   Topography, denudation and river drainage

It is a challenging problem to estimate the dynamic topography of the continental surface in the geological past. One approach is to map the stratigraphic gaps of unconformities attributable to dynamic support, and to convert denudation estimates to amounts of topographic uplift. Phases of uplift also correspond to pulses of increased sediment delivery to offshore sedimentary depocentres (Walford *et al.* 2005). This procedure has been carried out, for example, in the late Cenozoic domal uplifts in southwestern Africa (Al-Hajri *et al.* 2010) (Fig. 5.31). Whereas a phase of Oligocene uplift is attributed to the initiation of the African superplume, a Pliocene unconformity suggests the rapid emergence of a number of smaller domes around the western perimeter of the superplume.

Surface uplift of the continental surface affects the course of river systems draining the elevated region. Centrifugal drainage patterns may develop over domal uplifts (Cox 1989). River long profiles potentially contain information about the uplift history experienced by the drainage area (Roberts & White 2009; Pritchard *et al.* 2009), since erosional landscapes respond to relative uplift by the upstream passage of a wave of erosion recognised by a knickzone (Fig. 5.32) (Appendix 47). When applied to African examples (Fig. 5.31), where topography is expected to reflect convective circulation (Burke & Gunnell 2008), a coherent picture of surface uplift, denudation and sediment supply to the ocean is obtained. In the magmatic domes of Hoggar and Tibesti of Saharan Africa, and the amagmatic domes of Bié, Namibia and South Africa, uplift started 30–40 millions of years ago. A further application to the rivers of the Colorado Plateau

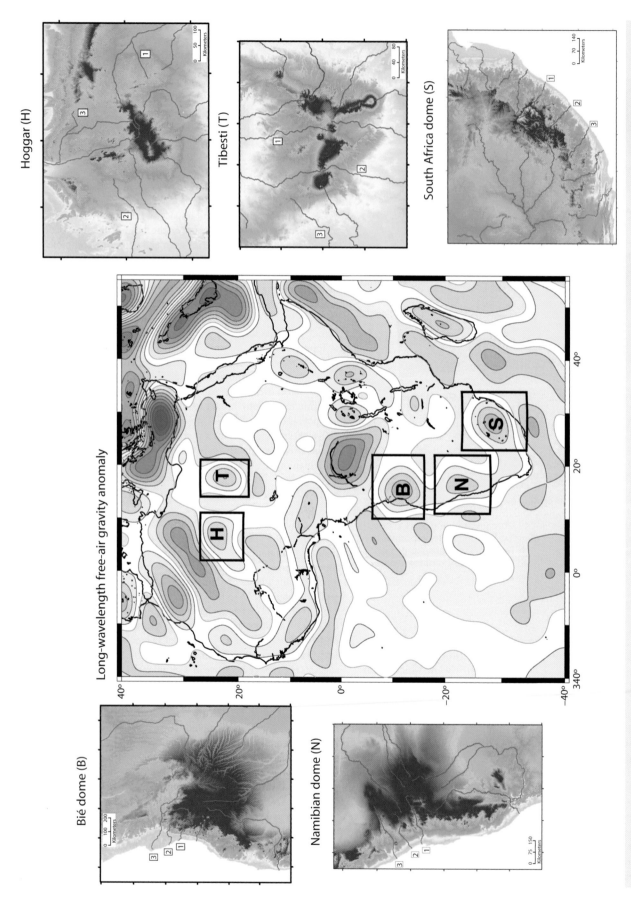

Fig. 5.31 Long-wavelength free-air gravity map of Africa, from Roberts & White (2010), showing positive (red) and negative (blue) anomalies, contoured at 5 mGal. The long-wavelength gravity anomalies may represent convective circulation beneath the lithosphere. Boxed positive anomalies are topographic domes where rivers have been analysed in order to estimate the timing of uplift events from knickpoints. H, Hoggar; T, Tibesti; B, Bié dome; N, Namibian dome; S, South Africa dome. Rivers whose long profiles were used are shown in the inset maps. Reproduced with permission of the American Geophysical Union.

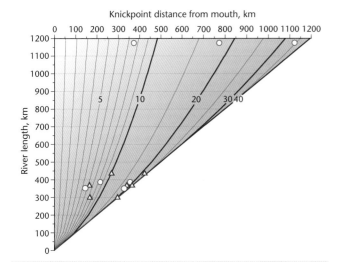

Fig. 5.32 Look-up chart for estimating the timing of uplift events from the position of knickpoints along the river profile. Erosion parameters are advective coefficient $v = 50\,\mathrm{m^{1-m}\,Myr^{-1}}$, exponent for distance $m = 0.5$, and erosional diffusivity $\kappa = 10^5\,\mathrm{m^2\,Myr^{-1}}$. Numbered curves are isochrons in Myr, which give the timing of the uplift event for a given knickpoint distance from the river mouth. Shown on look-up chart are three rivers from the Bié dome, southwest Africa (open circles), and three rivers from the South Africa dome (open triangles), each with 2 or 3 knickpoints. From Roberts & White (2010, fig. 15), reproduced with permission of the American Geophysical Union. See also Appendix 47.

region, which experienced uplift over the Yellowstone hotspot, also yielded valuable information on the timing and amount of dynamic topography (Roberts *et al.* 2012).

It is argued that rapid changes in Miocene shoreline positions and a change in flow direction of the paleo-Amazon River took place as a result of surface elevation changes driven by the subduction of oceanic slabs beneath the South American continent (Shephard *et al.* 2010) (Fig. 5.33). The impact of mantle dynamics on drainage network is also seen in the case of the Yellowstone hotspot (Beranek *et al.* 2006; Wegmann *et al.* 2007).

The rate of motion of an overlying plate relative to a mantle flow structure should determine the amplitude, wavelength and longevity of topographic uplift. Let us take two contrasting situations. In the first case of a rapidly moving plate relative to the mantle, such as the northward movement of the Australian plate in the Pliocene to Recent (Heine *et al.* 2010), the uplift is short-lived and relative sea-level changes, recognised by shoreline migration, regionally variable. In the second case of a slowly moving plate relative to the mantle, shoreline migration may be uniform over very large distances, giving the impression of a eustatic control.

5.4.2 Cratonic basins

Among the many hypotheses for the formation of cratonic basins (Allen & Armitage 2012 for summary), a common theme is the thermal contraction of the lithosphere following heating (Sleep & Snell 1976; Xie & Heller 2009). Cratonic basins are described and discussed in relation to lithospheric stretching in §3.2.1. Models of stretching of a continental plate assume that the basal boundary is

isothermal (McKenzie 1978a) or experiences a constant heat flow (Kaminski & Jaupart 2000). Instead, the base of the lithosphere can be treated as the boundary between a 'stagnant' or insulating continental lid and a convecting underlying mantle (Sleep 2009). In this case the uplift and subsidence of the Earth's surface constitutes dynamic topography.

Dynamic topography may therefore be manifested on the continents at the present day, apart from the distribution of hotspots and highspots (§5.2.6), in the regions of sag-like subsidence of cratonic basins (Fig. 5.34). The Eyre Basin of Australia (Alley 1998; Veevers *et al.* 1982), the Congo Basin of Africa (Downey & Gurnis 2009) and the Hudson Bay of Canada (Hanne *et al.* 2004) are potentially examples. Some workers have suggested that basins of this type, such as the Congo Basin, are located over downwellings of convective cells in the mantle (Hartley & Allen 1994; Heine *et al.* 2008; Downey & Gurnis 2009). This would explain the lack or paucity of evidence for stretching of the continental crust under some cratonic basins (Allen & Armitage 2012). If cratonic basins are located over downwellings, these convective cells must be attached to the overlying plate, since the locus of subsidence stays fixed relative to the continental crust for protracted periods of time, despite relative motion of plates with respect to the stationary reference frame of the sub-plate mantle. Other workers attribute sag-type subsidence to the ponding of plume material in a closed region beneath the continental plate caused by thinning (Sleep 2009).

The subsidence histories revealed by stratigraphic thicknesses penetrated in boreholes in cratonic basins can be matched by predictions from a range of models, including slow strain rate uniform stretching of an initially thick lithosphere (Armitage & Allen 2010) as well as those invoking dynamic effects transmitted from an underlying convecting mantle.

5.4.3 The history of sea-level change and the flooding of continental interiors

Since the oceanic lithosphere is the cooled thermal boundary layer of the underlying mantle convection system, its root-age trend for ocean bathymetry should play an important role in controlling long-term global sea level through its effect on ocean basin volume. This effect may operate in a number of ways: (i) changes over time in the temperature of the upper mantle beneath the oceanic plates could conceivably result in long-term global sea-level changes. As an illustration of this potential, it is known that the depth of the mid-ocean ridge crests, where unaffected by plumes, varies broadly over a depth range of 1 km. A modest temperature change in the upper mantle of just 40–50 °C would generate this order of variation in seafloor topography. If the thermal structure of the mantle changed significantly over geological time, such changes might be manifested in variations in the extent to which continents are flooded. (ii) Growth in the size of a newly formed ocean basin would cause a temporal increase in the average age of the ocean floor, and therefore an increase in the volume of the ocean basin, with no change in the production rate of new seafloor (Xu *et al.* 2006). This process has been used to explain the Cenozoic sea-level fall as a result of the progressive aging of the world's ocean basins.

If the root-age depth of the ocean floor of the form $d = a + bt^{1/2}$ (where d is seafloor depth, t is age, and a and b are constants), or *thermal topography*, is subtracted from the observed seafloor bathymetry (Davies & Pribac 1993), the mid-ocean ridges are removed, and a broad swell is observed in the central and western Pacific, with a

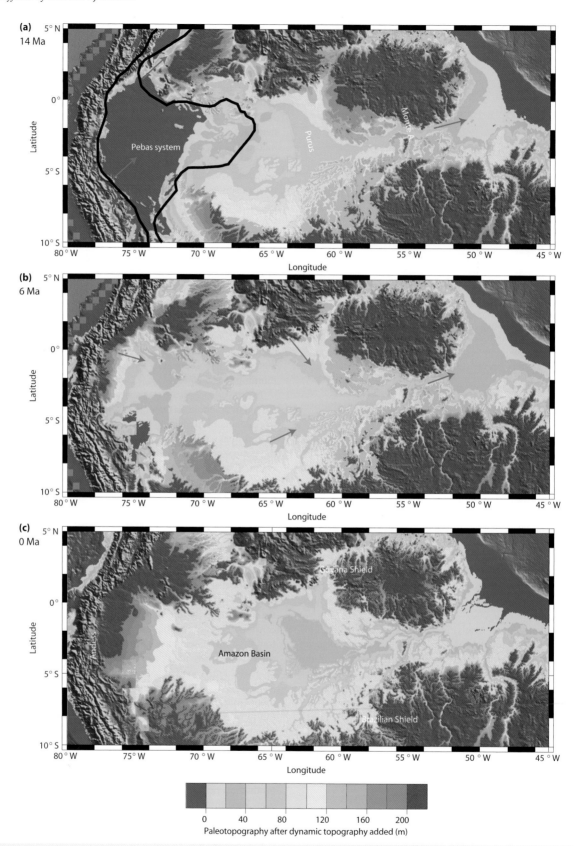

Fig. 5.33 Paleogeography of the Amazon Basin region at 14 Ma (a), 6 Ma (b) and present day (c), showing the reversal of paleodrainage of the Amazon River in the late Miocene. Paleotopographies include calculation of dynamic topography caused by subduction of the Farallon, Phoenix and Nazca plates. Red arrows show sediment transport directions. From Shephard *et al.* (2010). Reprinted by permission from Macmillan Publishers Ltd.

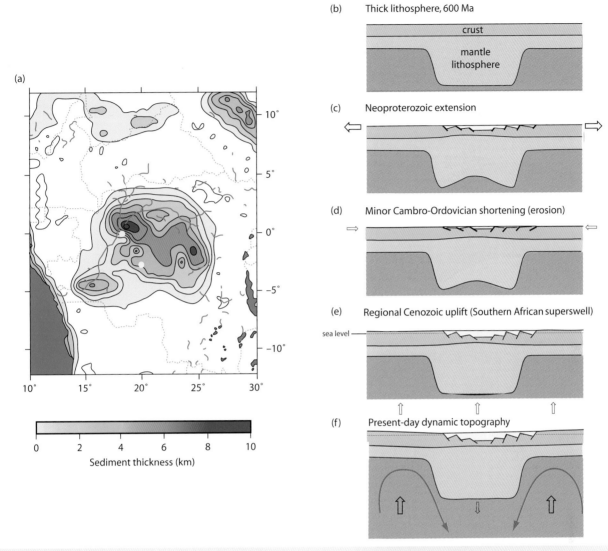

Fig. 5.34 Mantle circulation may be involved in the evolution of some cratonic basins, such as the Congo Basin (after Downey & Gurnis 2009). Convective circulation in the mantle may be responsible for the late Cenozoic to present-day subsidence of the basin and the uplift of the surrounding annulus (Hartley & Allen 1994; Crosby *et al.* 2010; Downey & Gurnis 2009). (a) Sediment isopachs showing roughly circular basin outline with subsidiary depocentres. Yellow circles are borehole locations. (b) to (f) show sequential evolution from Neoproterozoic to present-day. Reproduced with permission of John Wiley & Sons, Inc.

broad low observed between Australia and Antarctica (residual topography, Fig. 5.14). The south Pacific feature, several thousand kilometres in extent and hundreds of metres in elevation, is known as a *superswell* (McNutt & Fischer 1987; McNutt 1998; Adam & Bonneville 2005). Superswells correspond to regions of slow seismic velocities from tomographic studies, and with an anomalously high geoid. These observations suggest that superswells overlie relatively warm mantle. They are regions that have escaped the cooling effects of subduction for long (100–200 Myr) periods of time.

Superswells undoubtedly formed in the past. In the south Pacific region, seafloor updoming in the Cretaceous was accompanied by the eruption of flood basalts (Ontong Java Plateau) due to the spawning of new plumes in the heated region below the superswell (Davies & Pribac 1993) (Fig. 5.35). The importance of this observation is that

the uplift of the superswell most likely reduced the volume of the ocean basins and displaced water over the continental shelves, causing the sea-level highs characteristic of this period of Earth history (Bond 1978). Assuming a 5000 km-wide superswell of roughly circular planform shape, with an average bathymetric elevation above the regional bathymetry of 500 m, gives a volume of displaced water of $10 \times 10^6 \, km^3$. Since the area of ocean surface is $360 \times 10^6 \, km^2$, and neglecting isostatic compensation, the global water depth increase is approximately 25–30 m, from the effect of one superswell alone. If the effects of thermal expansion of the Cretaceous ocean is also included, the continents are expected to have been inundated extensively during the Cretaceous, as is confirmed from the transgressive nature of marine stratigraphy of this age (Bond 1976, 1978) (Fig. 8.23).

(a) Extension

Subduction   Warming - upflow   Cooling

Bottom thermal boundary layer

(b) Extension

Plumes

(c) Extensive intraplate volcanism   Extension

Plumes

Fig. 5.35 Possible evolution of the Pacific mantle and superswell during the Cretaceous, after Davies & Pribac (1993). In (a) a region of mantle between two subduction zones warms up and causes a thermal doming of the Earth's surface, causing extension. The warming promotes the formation of new plumes that cause igneous accretions to the overlying plate (b). Generations of plumes and their igneous products are formed and transported away through plate motion (c). Note that Davies & Pribac (1993) invoke a large region of mantle heating that spawns plume activity rather than a 'superplume'. Reproduced with permission of the American Geophysical Union.

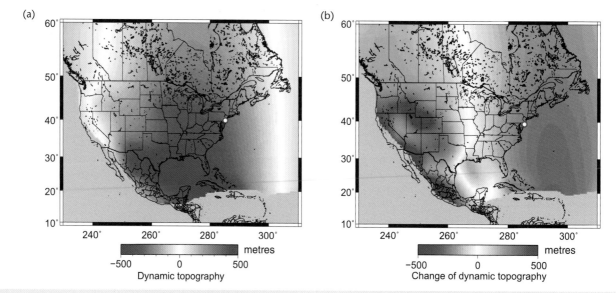

Fig. 5.36 Effect of dynamic topography on estimates of eustatic sea level, after Spasojevic *et al.* (2008). (a) Predicted dynamic topography in North America and the western Atlantic Ocean based on a mantle convection model, with the red circle showing the location of the New Jersey coastal plain. (b) Difference between predicted dynamic topography at 50 Ma (early Eocene) and the present day. The results in (a) and (b) are consistent with the position of paleoshorelines on the coastal plain of eastern USA, and explain some of the discrepancies in Fig. 8.23. Reproduced with permission of the American Geophysical Union.

The long-term trends in global (eustatic) sea level (10–100 Myr duration) and the associated inundations of continental margins are the subject to debate in terms of the methodology used to construct a 'global' cycle chart of sea-level change (Carter 1998; Miall 1992) and in terms of the mechanisms for such long-term trends. With the assumption that these long-term trends are eustatic and therefore globally synchronous (Vail *et al.* 1977a; Haq *et al.* 1987), a mechanism of variation in the spreading rate of mid-ocean ridges is conventionally invoked (Hays & Pitman 1973; Pitman 1978). Doubts have subsequently arisen over whether the history of spreading ridge activity, judged from the age distribution of oceanic crust, matches the known flooding history of continents over the last c.200 Myr (Rowley 2002). In addition, this model neglects the impact of subduction on dynamic topography of adjacent continents (Mitrovica *et al.* 1989; Gurnis 1990b) (see §5.3), which would loosen the connection with spreading rate variations. The continental platforms that have undoubtedly been affected by variations in dynamic topography over geological time are also the source regions for data supporting a global eustatic model.

Convection experiments have been used to predict dynamic topography back through time (Conrad & Gurnis 2003; Moucha *et al.* 2008) (Fig. 5.36). They allow for the cooling effect on the mantle of the descent of cold slabs, such as the Farallon plate (Forte *et al.* 2007), which has resulted in a trend in dynamic topography in the New Jersey area, a region providing critical data for determinations of global sea level in the Cretaceous to present interval (Miller *et al.* 2005). Over the same time interval, there was little relative sea-level change on the conjugate African margin, which was unaffected by the subduction of the Farallon plate, demonstrating that variations in dynamic topography form an important component of the total relative sea-level history of areas providing critical data for the global cycle chart.

In assessing the flooding history of the continents, dynamic topography is likely to modify and even dominate the effects of eustatic change. Variations of 100 m in dynamic topography over periods of tens of millions of years are predicted by spherical shell mantle flow models. By including the effects of dynamic topography, eustatic change cannot be solely due to changes in spreading ridge volumes. Spreading ridge variations, and the attendant long-term geochemical fluxes of the Earth, therefore, do not track eustatic sea level in any simple way.

In summary, from the viewpoint of the basin analyst, one of the most important effects of mantle flow is its impact on sea level and continental inundation.

CHAPTER SIX
Basins associated with strike-slip deformation

Summary

Sedimentary basins generally form by localised extension along a strike-slip fault system that may be related to either divergent, convergent or oblique relative plate motion. Less commonly, loading resulting from local crustal thickening may cause flexural subsidence. Although strike-slip basins form in a wide variety of geodynamical settings, such as oceanic and continental transforms and arc and suture collisional boundaries, they are best known from intracontinental and continental margin environments.

In simple systems, the orientation of the strike-slip fault in relation to the plate motion vector is important in determining whether divergent (transtensile) or convergent (transpressive) strike-slip takes place. This guide breaks down, however, in complex regions of continental convergence such as Turkey.

Zones of strike-slip tectonics are characterised by active seismicity on strike-parallel and strongly oblique faults, with zones of infrequent large earthquakes along locked segments, and frequent small earthquakes along unlocked segments. Some of the world's best known and most hazardous faults are strike-slip. Heat flows are generally low, suggesting that major strike-slip fault zones are weak and therefore generate little frictional heat. Geodetic surveys and paleomagnetic results show that small and large crustal blocks commonly rotate about a sub-vertical axis during strike-slip deformation. Characteristic geomorphic features result from the lateral displacement of adjacent terranes.

Sedimentary basins in zones of strike-slip deformation are diverse and complex. Some are clearly thin-skinned and related to extensional or contractional detachments in the weak lower crust. Others involve thinning of the mantle lithosphere and experience elevated surface heat flows. A number of different types of basin can be discriminated on the basis of kinematic setting, chief of which are fault-bend basins, overstep basins, transrotational and transpressional basins.

The bulk of the shear strain is accommodated in a central *principal displacement zone* (PDZ), which may be linear to curvilinear in plan view and steeply inclined in cross section. The PDZ commonly branches upwards into a splaying system of faults producing a *flower structure*. Some fault zones, such as the Garlock Fault in California, penetrate to great depths, terminating in the middle crust, whereas others link at depth with relatively shallow low-angle detachments belonging to orogenic wedges. Strike-slip zones are characterised by *en echelon* arrangements of faults and folds that are orientated in a consistent pattern with respect to the strain ellipse. The most important fractures are termed *Riedel shears*, but extension fractures are also formed. *En echelon* folds may form, with axes roughly at right angles to the extension fractures. The exact pattern of faults and folds produced in any particular fault zone depends on the local geological fabric and the youth or maturity of the fault system.

The precise structural pattern is controlled by a number of factors, including: (i) the kinematics (convergent, divergent, parallel) of the fault system; (ii) the magnitude of the displacement; (iii) the material properties of the rocks and sedimentary infills in the deforming zone; and (iv) the configuration of pre-existing structures.

The PDZ is characteristically segmented. The individual segments may be linked, both in plan view and cross-sectional view, by *oversteps*. If the sense of an along-strike overstep is the same as the sense of fault slip, a *pull-apart basin* is formed; if the sense of the overstep is opposite to that of fault slip, a *push-up range* is formed. Pull-apart basins appear to develop in a continuous evolutionary sequence at releasing bends with increasing offset, from narrow 'spindle-shaped' basins to 'lazy S' and 'lazy Z' basins to 'rhomboidal' basins and eventually into ocean-floored basins. However, the weakness of major strike-slip zones may cause transform-normal extension, producing markedly asymmetrical strike-slip basins. A global compilation suggests that pull-apart basins have a linear length-width relationship. This has implications for the kinematic models of their formation and evolution, and favours the idea of coalescence of individual scale-dependent depocentres during progressive deformation.

Strike-slip deformation is associated with arc–continent collision and continent–continent collision. In the case of the Caribbean–South American plate boundary zone, strike-slip basins were activated diachronously due to the dextral interplate motion as the Caribbean arc elongates eastwards. Pull-apart basin stages are phases in a complex geological history involving flexural foreland basin formation, inversion and E–W extension. In continent–continent collision zones, such as Himalaya-Tibet and central Asia, strike-slip tectonics is a response to lateral extrusion towards the Pacific and oblique collision of the Indian indenter.

Numerical modelling of basins at stepovers using a basal forcing are successful at generating the broad features of pull-apart basins, such as the Dead Sea Basin. Analogue models replicate many of the key features of natural pull-apart basins and emphasise the importance of small amounts of transtension compared to pure strike-slip.

Basin Analysis: Principles and Application to Petroleum Play Assessment, Third Edition. Philip A. Allen and John R. Allen.
© 2013 John Wiley & Sons, Ltd. Published 2013 by John Wiley & Sons, Ltd.

Thermal and subsidence modelling is poorly developed in strike-slip basins, largely on account of their complex structural history. In basins involving lithospheric thinning, the uniform extension model has been applied with modifications for the lateral loss of heat through the basin walls during the extension. Other basins appear to form over zones of thin-skinned extension, with no mantle involvement. These basins, such as the Vienna Basin in the compressive Alpine-Carpathian system, are cool and lack a well-developed phase of post-extension thermal subsidence.

6.1 Overview

6.1.1 Geological, geomorphological and geophysical observations

Basins associated with strike-slip deformation are generally small and complex compared to cratonic sags (Chapter 3), passive margins (Chapter 3) and foreland basins (Chapter 4). They are intimately linked to the detailed structural evolution of an area, and mechanical models have been relatively slow to appear because of the extreme complexity of this history of deformation. Nevertheless, basins in zones of strike-slip deformation have been generated in analogue 'sandbox' experiments, and numerical three-dimensional models are available for basin development in particular structural settings.

Strike-slip deformation occurs where principally lateral movement takes place between adjacent crustal or lithospheric blocks. In pure strike-slip, the displacement is purely horizontal, so there is no strain in the vertical dimension y. This is the situation of *plane strain*. The vertical stress in strike-slip faulting is the vertical lithostatic stress $\sigma_{yy} = \rho g y$. The horizontal stresses are the deviatoric stresses, one compressional and the other extensional. The vertical stress is always the intermediate stress. Two conjugate strike-slip faults are anticipated from this state of stress, one right-lateral (dextral) and the other left-lateral (sinistral), inclined at an angle φ to the principal stress σ_{xx} (Fig. 6.1).

In reality, movement in strike-slip zones is rarely purely lateral, and displacements are commonly *oblique*, that is, involving a certain amount of normal or reverse dip-slip movement. Oblique slip may therefore characterise any strike-slip zone, but particular zones may experience a net contraction while others may suffer net extension. The stress regimes responsible for these two variants on pure strike-slip deformation are known as *transpressive* and *transtensile*.

Major strike-slip deformation and associated basin formation takes place in a wide range of geodynamical situations. Strike-slip zones may be associated with entire plate boundaries such as the San Andreas Fault system of California and the Alpine Fault system of New Zealand, microplate boundaries, intraplate deformations or small fractures of limited displacement. Sylvester (1988), drawing considerably on Woodcock's (1986) genetic scheme (Fig. 6.2), proposed a classification of strike-slip faults into *interplate* and *intraplate* varieties (Table 6.1). He recommended use of the term 'transform' fault for deep-seated interplate types and 'transcurrent' fault for intraplate strike-slip faults confined to the crust. Of greatest importance to students of basin analysis are the strike-slip faults cutting continental lithosphere where uplifting and eroding source areas for sediment are available.

The development of regions of extension and shortening along strike-slip systems has been related to the relative orientation of the plate slip vectors and the major faults (Mann *et al.* 1983). In the case of the San Andreas and Dead Sea strike-slip systems, basins develop where the PDZ is divergent with respect to the plate vector (Fig. 6.3). In contrast, uplifts or push-up blocks such as the Transverse Ranges, California, and the Lebanon Ranges of the Levant, occur where the PDZ is convergent with respect to the plate vector. This simple relationship is unlikely to apply where deformation takes place on many faults enclosing rotating crustal blocks. It is a very poor guide to patterns of uplift and subsidence in complex regions of continental convergence such as Turkey (Sengör *et al.* 1985).

Strike-slip zones are characterised by extreme structural complexity. Individual strike-slip faults are generally linear or curvilinear in plan view, steep (sub-vertical) in section, and penetrate to considerable depths, perhaps decoupling crustal blocks at the base of the seismogenic crust (that is, at 10–15 km). In contrast to regions of pure extension or contraction, strike-slip zones possess prominent *en echelon* faults and folds, and faults with normal and reverse slip

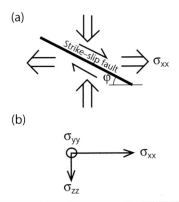

Fig. 6.1 Plane strain approximation for strike-slip faults. (a) Conjugate strike-slip faults at an angle φ to the principal stress σ_{xx}. (b) Principal stresses are related by $\sigma_{zz} > \sigma_{yy} > \sigma_{xx}$.

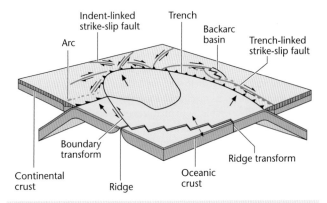

Fig. 6.2 Genetic classification of major classes of strike-slip fault according to plate tectonic setting (after Woodcock 1986). See also Table 6.1. Reproduced with permission from the Royal Society.

Table 6.1 Sylvester's (1988) classification of strike-slip faults, slightly modified

1 Interplate 'transforms' (deep-seated, delimiting plate)
   1.1 *Ridge transform faults*
      Displace segments of oceanic crust with similar spreading vectors
      e.g. Romanche fracture zone (Atlantic Ocean)
   1.2 *Boundary transform faults*
      Separate different plates parallel to the plate boundary
      e.g. San Andreas (California), Alpine Fault (New Zealand), Central Range Fault Zone (offshore Trinidad)
   1.3 *Trench-linked strike-slip faults*
      Accommodate horizontal component of oblique subduction
      e.g. Atacama Fault (Chile), Median Tectonic Line (Japan)
2 Intraplate 'transcurrent' faults (confined to crust)
   2.1 *Indent-linked strike-slip faults*
      Bound continental blocks in collision zones
      e.g. North Anatolian Fault (Turkey), East Anatolian Fault (Turkey), Altyn Tagh Fault (Mongolia-China), Kunlun Fault (Tibet), Red River Fault (SE Asia)
   2.2 *Intracontinental strike-slip faults*
      Separate allochthons of different tectonic styles
      e.g. Garlock Fault (California)
   2.3 *Tear faults*
      Accommodate different displacement within a given allochthon or between the allochthon and adjacent structural units
      e.g. Asiak fold-thrust belt (Canada)
   2.4 *Transfer faults*
      Linking overstepping or *en echelon* strike-slip faults
      e.g. Southern and Northern Diagonal faults (eastern Sinai, Israel)

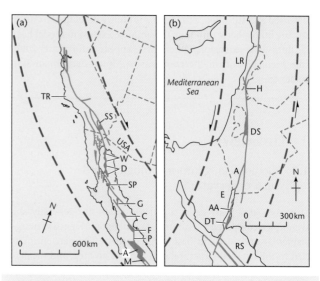

Fig. 6.3 Regions of compression and extension along strike-slip boundaries between rigid continental plates related to the orientation of the fault zone with respect to the plate slip vector. After Mann *et al.* (1983). (a) Pacific–North American plate boundary. The dashed lines are theoretical interplate slip lines from Minster *et al.* (1974). The major zone of compression is the push-up block of the Transverse Ranges where the dextral San Andreas Fault zone has a 'convergent' orientation with respect to the interplate slip line. The prominent pull-apart basins, however, are situated relatively to the south where the fault zone has a 'divergent' orientation with respect to the interplate slip line. TR, Transverse Ranges; SS, Salton Sea pull-apart at a right step between the San Andreas and Imperial faults. Pull-aparts in the Gulf of California include: W, Wagner Basin; D, Delfin Basin; SP, San Pedro Martir Basin; G, Guaymas Basin; C, Carmen Basin; F, Farallon Basin; P, Pescadero Basin complex; A, Alarcon Basin; M, Mazatlan Basin. (b) Arabia–Sinai (Levant) plate boundary zone (Garfunkel 1981; Ben-Avraham *et al.* 1979). Theoretical interplate slip lines are from Le Pichon & Francheteau (1978). The prominent area of compression is the Lebanon Ranges push-up block where the sinistral Dead Sea Fault zone is 'convergent' with respect to the interplate slip lines. Most of the pull-apart basins are in the Dead Sea and Gulf of Aqaba regions where the fault zone is locally 'divergent' with respect to the interplate slip lines. Pull-aparts include: H, Hula Basin; DS, Dead Sea Basin; A, Arava Fault trough; E, Elat Basin in the northern Gulf of Aqaba; AA, Arnona-Aragonese Basin; DT, Dakar-Tiran Basin. Dead Sea Fault Zone extends from Lebanon Ranges (LR) in the north to Red Sea (RS) in the south. © *Journal of Geology* 1983. Reproduced with the permission of University of Chicago Press Journals.

commonly coexist. The vergence direction of folds and the mass transport indicators from thrusts associated with transpression are distinctively poorly clustered or apparently random.

Zones of major strike-slip tectonics are marked by important seismicity (Fig. 6.4). In California, the relative velocity between the Pacific and North American plates is 47 mm yr⁻¹. Much of this displacement is taken up on the right-lateral (dextral) San Andreas Fault. However, as is the case with many other major translithospheric strike-slip faults, the San Andreas Fault occurs within a broader zone of faulting that may stretch laterally for 500 km. Much of the present-day seismicity in the San Andreas system originates from faults oblique to the main fault zone (e.g. the NE-trending Garlock Fault and Big Pine Fault) (Nicholson *et al.* 1986a). These oblique faults define crustal blocks that are rotating about a vertical axis. Seismicity dies out below about 15 km, indicating the brittle–ductile transition. Focal mechanism solutions in strike-slip zones are commonly highly variable, with mixtures of strike-slip, extension and compression (e.g. North Anatolian Fault, Turkey, Taymaz *et al.* 1991), reflecting the complexities of deformation within a zone of overall strike-slip deformation.

Paleomagnetic studies support the idea that crustal blocks commonly undergo rotations about vertical axes (Fig. 6.5). The amount and scale of rotation varies greatly. The western Transverse Ranges of southern California have experienced net clockwise rotations of 30°–90°, for example, whereas in the Cajon Pass region (Fig. 6.6) there has been no significant rotation since 9.5 Ma. Small blocks can rotate rapidly, such as the Imperial Valley area, which has rotated by

35° in the last 0.9 Ma. The existence of such rotating blocks supports the view that the crustal blocks deform like a set of dominoes (Freund 1970). There is an obvious physical implication of the existence of rotating blocks, which is that the blocks must detach on their boundary faults at some level in the crust or upper mantle (Terres & Sylvester 1981; Dewey & Pindell 1985).

Pure strike-slip plate boundary faults should fall along a small circle drawn from the pole of rotation that defines the relative motion between the two plates. This is true for the San Andreas system along the boundary of the Pacific and North American plates (Fig. 6.4). However, seismicity is distributed unevenly along the fault. In some

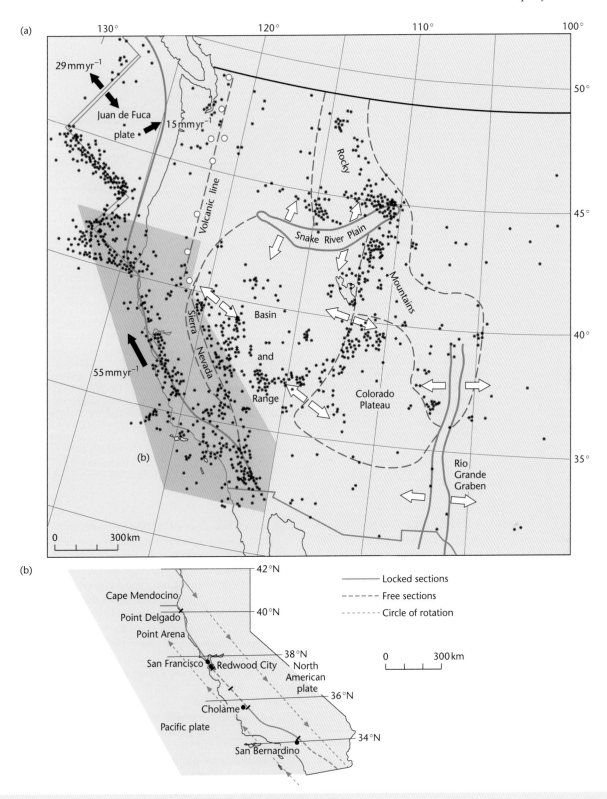

Fig. 6.4 The San Andreas system, California. (a) Distribution of seismicity (stars). Solid arrows are relative plate velocities; open arrows are stress directions from focal mechanism studies. (b) Surface trace of the locked and unlocked sections of the San Andreas Fault. Thin dashed lines are small circles to the pole of rotation for the motion of the Pacific plate relative to the North American plate. Compiled from Jennings *et al.* (1975), Nicholson *et al.* (1986a) and Turcotte & Schubert (2002). © Cambridge University Press, 2002.

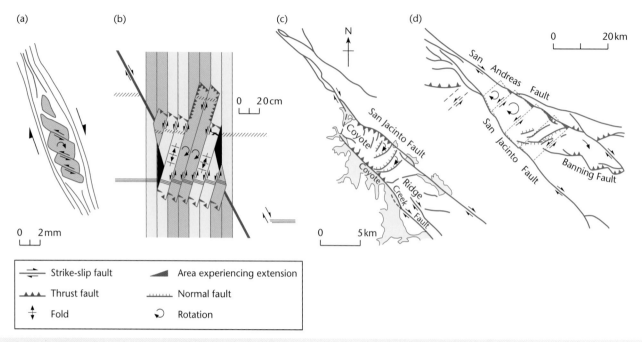

Fig. 6.5 Examples of block rotation by strike-slip faulting at scales from mm to km (after Nicholson *et al.* 1986a). (a) Fracture and rotation of a feldspar crystal along cleavage planes in a ductile matrix. (b) Rotation of a hard surface soil layer as a result of the Imperial Valley earthquake of 1979. The ruled lines are the evenly spaced furrows of a ploughed field. (c) Rotating blocks defined by secondary cross-faults between an overlapping right step from the Coyote Creek Fault to the San Jacinto Fault. (d) Block model for rotation near the intersection of the San Jacinto and San Andreas faults inferred from geology and seismicity. Reproduced with permission of John Wiley & Sons, Inc.

regions there are very few historical earthquakes. The fault in these 'locked' regions appears to accumulate strain and then release it in major earthquakes. The 1906 San Francisco moment magnitude 7.6 earthquake is an example. It produced a 4 m surface rupture all the way along a 200 km-long locked section from Redwood City to Cape Mendocino (Fig. 6.4). Elsewhere, small earthquakes are very abundant. In these 'free' sections, frequent small earthquakes and aseismic creep release the accumulating strain.

The heat flow measured above major strike-slip faults, such as those of the San Andreas system, is not significantly elevated compared to background values (Fig. 6.6) (Lachenbruch & Sass 1980, 1988). The heat flow associated with the Dead Sea Transform is also comparable to average continental values (Ben-Avraham *et al.* 1978). When slip on a fault occurs under a large stress, as given by Byerlee's law, significant frictional heating takes place. Repeated movements on a frictional strike-slip fault should generate large amounts of heat and contribute to elevated heat flows (Molnar 1992). The fact that observed heat flows are not significantly elevated suggests that major strike-slip faults are relatively weak structures set in a background of strong upper crust. Measurements of the principal stresses close to strike-slip faults in California also suggest that there is little shear stress resisting strike-slip fault motion (Zoback *et al.* 1987). Low heat flows also indicate that the localised extension along strike-slip zones does not in general involve significant mantle upwelling.

Whereas most of California has low to moderate surface heat flows, the Salton Trough-Imperial Valley in the extreme south has high surface heat flows (>2.5 HFU, Lachenbruch & Sass 1980) and known geothermal resource potential, as a northward continuation of the Gulf of California tectonic-thermal system (Figs 6.6, 6.7). The region is active seismically, with right-lateral slip on the main Imperial Valley Fault. A seismic refraction and gravity study of the region (Oliver *et al.* 1981; Fuis *et al.* 1984) shows that the Imperial Valley is a gravity high, despite the presence of a 3.7–3.8 km-thick sedimentary basin (Fig. 6.7). This is most likely due to the presence of intruded mafic basement domed up under the Imperial Valley where the extension is greatest.

Some of the world's best known and most hazardous faults, such as the San Andreas and North Anatolian faults, are of the strike-slip variety. They commonly have startling geomorphic expression and stand out strongly in satellite images. The accumulation of individual slip events over long periods of time commonly results in a linear trough along the fault zone. Streams draining adjacent uplands and their interfluves are commonly offset laterally, demonstrating long-term strike-slip displacement. Since alluvial fans accumulate downstream of the range front, their apices may also be displaced laterally, and stream channels on the fan surface may thus become beheaded. Geomorphic evidence for strike-slip tectonics is classically displayed along the San Andreas system of California (Wesson *et al.* 1975; Keller *et al.* 1982) and along the Hexi Corridor of north-central China (Li & Yang 1998) (Fig. 6.8).

The evolution of some orogenic belts, notably the North American Cordillera, involves very large magnitude lateral displacements of

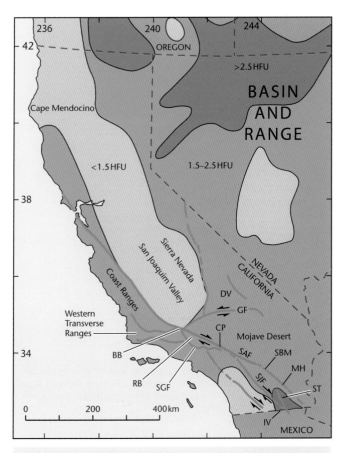

Fig. 6.6 Fault systems and surface heat flow of California and adjacent regions. High heat flows are found in the extensional Basin and Range province and in the Salton Trough–Imperial Valley area close to the Mexican border. However, coastal California and most of southern California have low heat flows despite the presence of major strike-slip faults. BB, Big Bend; CP, Cajon Pass; DV, Death Valley; GF, Garlock Fault; IV, Imperial Valley; MH, Mecca Hills; RB, Ridge Basin; SAF, San Andreas Fault; SBM, San Bernardino Mountains; SGF, San Gabriel Fault; SJF, San Jacinto Fault; ST, Salton Trough. After Lachenbruch & Sass (1980). Reproduced with permission of the American Geophysical Union.

terranes. Up to 1500 km of relative lateral motion is thought to have occurred in both the Mesozoic North American Cordillera and in the Tertiary India–Asia collision (Molnar & Tapponnier 1975). Clearly, strike-slip processes offer the possibility of immense lateral translations of crustal terranes (Styron *et al.* 2011). Strike-slip movement may result from oblique convergence or from regional tectonic extrusion towards an 'open' plate boundary (§6.3.4) (Kreemer *et al.* 2003).

6.1.2 Diversity of basins in strike-slip zones

Since the history of deformation is invariably complex in strike-slip zones, the associated sedimentary basins are correspondingly

complex. They commonly form in areas of localised extension caused by the geometry and kinematic history of the fault configuration, or in areas of net shortening, where flexural loading may drive subsidence. Characteristically, a basin experiences both extension and shortening during its life span (part of what Ingersoll (1988) calls the 'Reading cycle', after Reading (1980)), or one part of the basin may experience shortening while another part is undergoing extension. Rift basins and foreland basins may undergo a phase of strike-slip deformation due to changes in the stress field set up by relative plate motion. Strike-slip basins are therefore emphatically syntectonic and there is little evidence of substantial thermally driven subsidence. This latter feature may be due at least in part to the small size of strike-slip basins. Their narrowness (usually less than 50 km wide) causes extreme heat loss to the sides as well as vertical conduction to the overlying sea or atmosphere.

A number of different types of basins form in strike-slip zones. Bucket terms for these varied basins are 'pull-apart basins' (Burchfiel & Stewart 1966; Crowell 1974a) and 'strike-slip basins' (Mann *et al.* 1983). The term 'pull-apart basin' is also used for a sub-class of basins formed in local transtensional settings (Ingersoll & Busby 1995). Other terms in usage are 'rhombochasms' (Carey 1976), 'wrench grabens' (Belt 1968) and 'rhomb grabens' (Freund 1971).

We can identify two broad types of strike-slip basin in terms of thermal and subsidence history: (i) strike-slip basins with mantle involvement; these can be thought of as 'hot' or hyperthermal basins; and (ii) strike-slip basins that are relatively thin-skinned; they can be regarded as 'cold' or hypothermal basins (Fig. 6.9).

A more detailed breakdown of basins in strike-slip zones on the basis of the kinematic setting and geometry of bounding faults has also been proposed. The advantage of such a scheme is that individual phases of basin evolution can potentially be discriminated, but the disadvantage is that many basins show characteristics of more than one 'type'. A slightly modified version of the scheme proposed by Nilsen and Sylvester (1995) is as follows:

- *Fault-bend basins* commonly develop at bends in the main strike-slip fault where localised extension takes place (Aydin & Nur 1982). A type example is the Ridge Basin, California, in the San Andreas system. The Vienna Basin, Austria, is another basin developed at a releasing bend, but within a broader compressional orogenic setting.
- *Overstep basins* form between the ends of two sub-parallel strike-slip fault segments, which may merge at depth into one single master fault. A type example is the Dead Sea Basin along the Gulf of Aqaba-Dead Sea Transform, Middle East.
- *Transrotational basins* form as triangular gaps between crustal blocks undergoing rotation about a sub-vertical axis. A type example is the Los Angeles Basin of southern California.
- *Transpressional basins* are elongate depressions parallel to the regional strike of folds and faults in zones of oblique convergence. These basins subside primarily by flexure caused by the supracrustal loads of tectonically shortened regions along the strike-slip zone. A type example is the Ventura Basin of California.

The sedimentary fill of strike-slip basins reflects their varied structural history, with highly asymmetrical longitudinal and transverse distribution of sedimentary facies, and abundant and complex unconformities. Subsidence rates are commonly extremely high, but subsidence may be relatively short-lived and inversion events are common. In many cases, it may be difficult to separate the signature

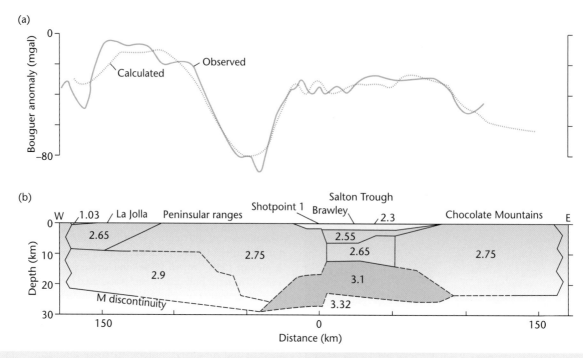

Fig. 6.7 Salton Trough–Imperial Valley region, southern California (after Fuis *et al.* 1982). (a) Gravity profile and (b) gravity density model for a ENE cross section of Imperial Valley from La Jolla to Chocolate Mountain. Solid lines on density model are derived from seismic refraction, whereas those that satisfy gravity data are shown dashed. A dense volume of sub-basement compensates gravitationally for the sedimentary rocks of the Salton Trough. Data compiled by the U.S. Geological Survey.

of strike-slip related subsidence from other mechanisms of subsidence in complex fault zones, as, for example, along the Caribbean–South American plate boundary (Escalona & Mann 2011).

6.2   The structural pattern of strike-slip fault systems

6.2.1   Structural features of the principal displacement zone (PDZ)

Strike-slip faults are linear to curvilinear in plan view and generally possess a principal displacement zone (PDZ) along which the bulk of the shear strain is accommodated. However, changes in the orientation of the fault and/or the influences of the local geological fabric may cause deformation to extend beyond the PDZ into juxtaposed crustal blocks (Fig. 6.10). In cross section, the PDZs of large strike-slip faults are steeply inclined and commonly grade upwards from narrow well-defined zones cutting igneous and metamorphic basement rocks at depth, to a braided, more diffuse deformation in the overlying sedimentary cover. This upward branching effect has led to the fault splays being christened '*flower structures*' or '*palm tree*

structures' (Fig. 6.11). Some strike-slip faults link at depth with low-angle detachments, as occurs in foreland fold and thrust belts such as the Vienna Basin (Royden 1985) and in regions of regional extension such as the Basin and Range, USA (Cheadle *et al.* 1985). In the latter case, the Garlock Fault, a major strike-slip fault at a high angle to the San Andreas trend in California (Figs 6.4, 6.6), appears to terminate downwards on a low-angle surface situated in the middle crust (9–21 km depth) according to deep seismic reflection profiling (COCORP).

En echelon arrangements of faults and folds are commonly associated with strike-slip displacements (Fig. 6.10). These structures are consistently arranged both in orientation and sense of strain with respect to the PDZ. They are distinct from the *oversteps* between different segments of the PDZ (§6.2.2). *En echelon* arrangements have been produced in model studies involving the deformation of clay, loose sand or artificial materials (e.g. Riedel 1929; Cloos 1955; Harris & Cobbold 1984; Dooley *et al.* 2004; Wu *et al.* 2009). Similar patterns have been observed in 'natural' environments where alluvium has been affected by seismic disturbance, as in the 1975 earthquake in Imperial Valley, California (Sharp 1976). These experimental results and limited natural occurrences suggest that five sets of fractures are associated with a shear displacement (Fig. 6.12):

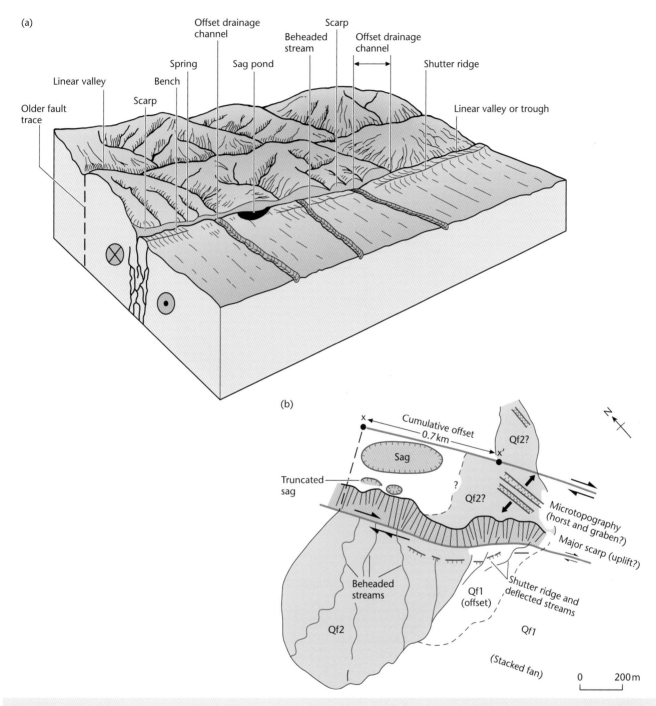

Fig. 6.8 Geomorphic features associated with strike-slip faults in the San Andreas system of California. (a) A linear trough along the fault, sag ponds, shutter ridges, offset ridges and drainages, springs, scarps and beheaded streams are typical geomorphic features indicative of strike-slip faulting. The older, abandoned fault trace displays analogous but erosionally degraded features. Modified from Wesson *et al.* (1975). (b) A large fan is offset from its catchment system, producing beheaded streams, shutter ridges and extensional fault scarps. Modified from Keller *et al.* (1982), reproduced with permission of John Wiley & Sons, Ltd.

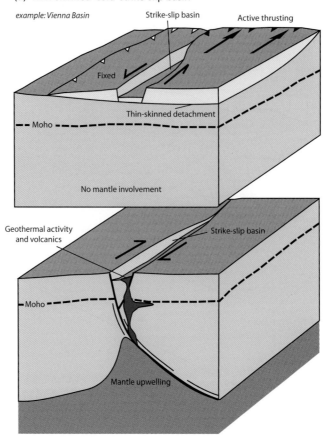

(a) Thin-skinned 'cold' strike-slip basin

example: Vienna Basin

Strike-slip basin

Active thrusting

Fixed

Thin-skinned detachment

Moho

No mantle involvement

Geothermal activity and volcanics

Strike-slip basin

Moho

Mantle upwelling

(b) Hot' strike-slip basin with mantle involvement

example: Salton Trough

Fig. 6.9 Schematic diagram of broad classes of strike-slip basins. (a) Thin-skinned strike-slip basins, which are hypothermal. (b) Strike-slip basins with mantle involvement, which are hyperthermal.

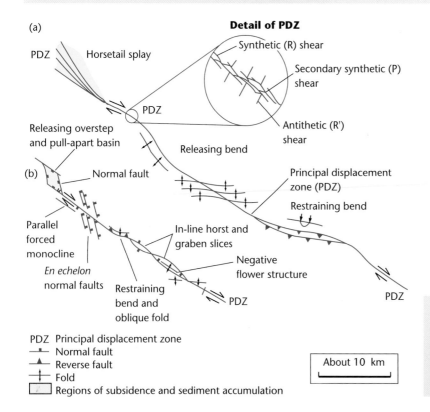

(a)

PDZ   Horsetail splay

Detail of PDZ

Synthetic (R) shear

Secondary synthetic (P) shear

PDZ

PDZ

Releasing overstep and pull-apart basin

Releasing bend

Antithetic (R') shear

(b)   Normal fault

Principal displacement zone (PDZ)

Restraining bend

Parallel forced monocline

In-line horst and graben slices

En echelon normal faults

Negative flower structure

Restraining bend and oblique fold

PDZ

PDZ

PDZ   Principal displacement zone
—•— Normal fault
—▲— Reverse fault
—+— Fold
▨ Regions of subsidence and sediment accumulation

About 10 km

Fig. 6.10 (a) The plan view arrangement of structures associated with an idealised right-lateral (dextral) strike-slip fault. (b) Adaptation to a slightly divergent setting with the predominance of pull-aparts, *en echelon* normal faults and graben slices within the PDZ. Modified from Christie-Blick & Biddle (1985).

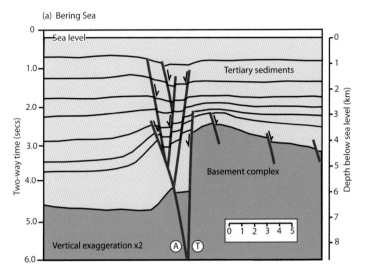

(a) Bering Sea

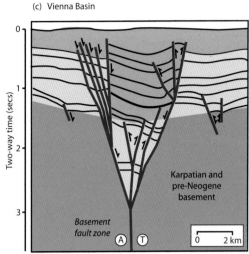

(c) Vienna Basin

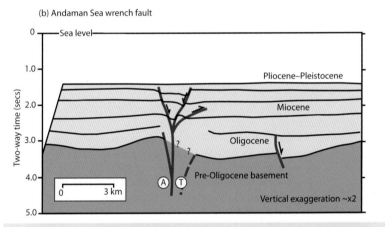

(b) Andaman Sea wrench fault

Fig. 6.11 The major characteristics in cross-sectional view of an idealised strike-slip fault. (a) Major fault zone in Bering Sea, and (b) Andaman Sea wrench fault (modified from Christie-Blick & Biddle 1985). (c) Vienna Basin, Austria, showing an asymmetric negative flower structure from seismic data, in Hinsch *et al.* (2005), reinterpreted by Wu *et al.* (2009, p. 1617, fig. 9; reprinted with permission from Elsevier). A, displacement away from reader; T, displacement towards reader.

1. Synthetic strike-slip faults orientated at small angles to the regional shear couple. These are frequently termed *Riedel (R) shears*. The sense of offset is the same as that of the PDZ.
2. Antithetic strike-slip faults orientated at high angles to the regional shear couple. These are termed *conjugate Riedel shears* or *R′*. The sense of offset is opposite to that of the PDZ.
3. Secondary synthetic faults or *P shears* with a sense of offset similar to that of the PDZ.
4. *Tension fractures* related to extension in the strain ellipse.
5. Faults parallel to the PDZ and shear couple, or the *Y shears* of Bartlett *et al.* (1981).

En echelon folds develop with their axial traces parallel to the long axis of the strain ellipse, indicating shortening perpendicular to the extension demonstrated by the tension fractures (Fig. 6.12). Geological examples are invariably more complicated than the situ-ation shown in Fig. 6.12, because the geological fabric influences fault orientations and also because faults and folds become rotated during progressive deformation so that the present fault configuration represents a cumulative picture over time. However, the idealised pattern is useful as a predictor of fault and fold occurrences if the regional shear direction is known, or, alternatively, as a predictor of the latter where observations are possible on the resultant fracture and fold pattern.

The offset pattern on strike-slip faults in cross-section can be exceedingly complex. The sense of displacement may vary along one fault from horizon to horizon, and within a flower structure faults of opposite displacement occur together (Fig. 6.11). Additionally, a given fault in one cross-section may commonly switch in dip in another cross-section. These characteristics make strike-slip fault zones distinctive compared to regional extensional and contractional fault systems.

(a)

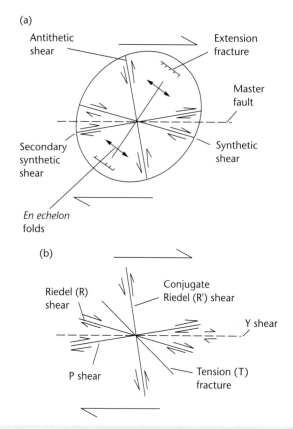

(b)

Fig. 6.12 The angular relations between structures that form in an idealised right-lateral simple shear, compiled from clay models and from geological examples (after Christie-Blick & Biddle 1985). (a) Fractures and folds superimposed on a strain ellipse for the overall deformation. Terminology of structures from Wilcox *et al.* (1973). (b) Riedel shear terminology, modified from Tchalenko & Ambraseys (1970) and Bartlett *et al.* (1981).

The precise structural pattern is controlled by a number of factors including (Christie-Blick & Biddle 1985): (i) convergent, divergent or simple strike-slip (parallel) kinematics; (ii) magnitude of the displacement; (iii) material properties of the rocks and sedimentary infills in the deforming zone; and (iv) configuration of pre-existing structures giving a geological fabric.

1 Convergent, divergent and parallel kinematics

Convergent strike-slip causes the development of many reverse faults and *en echelon* folds. Depending on the obliquity of the strike-slip, this pattern may grade into a fold-thrust belt, as in the western Transverse and Coast Ranges of California (Figs 6.4, 6.6). Upward-splaying flower structures have an overall antiformal structure caused by the net shortening. They are commonly known as *positive flower structures*. Folds are less well developed in divergent strike-slip settings, taking the form of flexures associated with extensional faulting. Flower structures take on an overall synformal structure in regions of net divergence, and are consequently known as *negative flower structures*. The Tekirdag depression in the Marmara Sea, Turkey, situated along a restraining bend of the North Anatolian Fault, is situated in a broad negative flower structure (Okay *et al.* 1999) (Fig. 6.13).

The scale of the distribution of convergence or divergence varies from regional to local. For example, regional divergent strike-slip may take place where major fault strands are oblique to interplate slip vectors, as in the southern part of the Dead Sea Transform (Fig. 6.3). On a much more local scale, divergent strike-slip may take place at releasing fault oversteps and fault junctions. Divergent strike-slip faults also develop on a local scale where crustal blocks rotate between bounding wrench faults.

A detailed study of the rotations and oversteps associated with a strike-slip displacement following an earthquake was carried out by Terres and Sylvester (1981). The deformation of a recently ploughed carrot field in Imperial Valley following the earthquake of 15 October 1979 is shown in Fig. 6.5b. This very small-scale example shows many of the features predicted from model experiments, with both extension and contraction occurring simultaneously. The topsoil broke up into elongate blocks along the furrows, and these blocks moved relative to each other along conjugate Riedel shears (R') or antithetic shears. Extension occurred along the long axis of the strain ellipse and folding along the short axis. The elongate blocks delimited by shears, which Dewey (1982, p. 382, fig. 10) has termed '*Riedel flakes*', show clockwise rotations in this example of right-lateral strike-slip. A much larger rotated block with dimensions of 20 km by 70 km, the Almacik flake along the North Anatolian Fault, Turkey, is also thought to be due to Riedel flaking (Sengör *et al.* 1985) (Fig. 6.14). Here, clockwise rotation of the tectonostratigraphic domains of 110° has taken place since the initiation of faulting. The right-lateral strike-slip along the North Anatolian PDZ has produced families of thrusts, Riedel and P shears.

Rotations may also take place along straight segments of adjacent strike-slip faults, causing geometrical space problems. These space problems are resolved by gaps and overlaps occurring at fault block corners known as transrotational basins. Fig. 6.15 (see also Fig. 6.5d) shows how regions of extension and contraction may alternate along the strike-slip fault.

2 Magnitude of the displacement

Laboratory experiments indicate that there is a sequential development of structural features in strike-slip zones (Tchalenko 1970 and Wilcox *et al.* 1973 for clay models, Dooley *et al.* 2004 and Wu *et al.* 2009 for sandbox models, Bartlett *et al.* 1981 for rock samples under confining pressure). In the rock sample experiments the zone of deformation first of all expands rapidly due to the development of folds and fractures. Weakening of the deforming zone soon, however, stabilises the spread of the zone of deformation, and later structures are concentrated in a central core zone (Odonne & Vialon 1983). All early-formed structures are rotated by later deformations, so their orientation is dependent on the magnitude of the displacement.

Some of the features observed in experiments can be found in natural occurrences, whereas other features are difficult to match. An increasing complexity from discontinuous faults and folds along low-displacement boundaries to through-going PDZs along high-displacement boundaries is commonly observed. Active fault systems that are accompanied by sedimentation may, however, be buried faster than they deform. In this way, the uppermost, younger sediments may record less deformation than the older stratigraphy.

3 Material properties in the deforming zone

The lithologies of rocks in the strike-slip zone and the rates and pressure-temperature conditions of deformation all control the individual structural character of the fault zone. These factors vary from

(a)

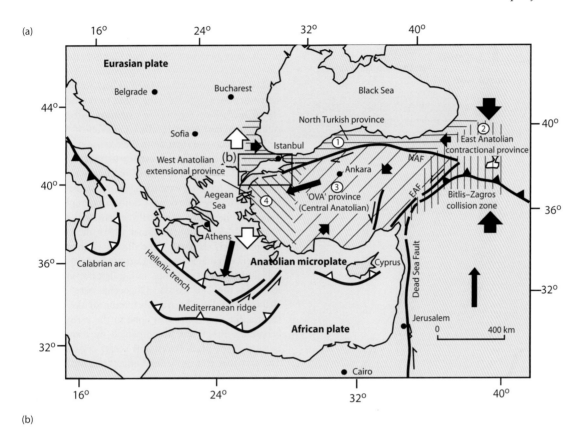

(b)

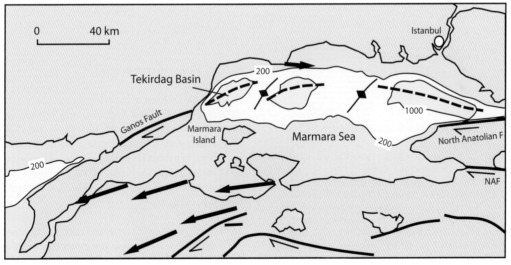

Fig. 6.13 (a) Tectonics of the Eastern Mediterranean (Okay *et al.* 1999). The four main tectonic regimes of Turkey (shown by numbers in circles) are caused by the post late Serravalian (c.12 Ma) westward escape of the Anatolian block from the east Anatolian convergence zone onto the oceanic lithosphere of the eastern Mediterranean Sea (after Sengör *et al.* 1985). (1) Weakly active North Turkish province characterised by limited E–W shortening; (2) East Anatolian contractional province of N–S shortening, situated mostly to the east of the meeting point of the East Anatolian and North Anatolian strike-slip faults; (3) the Central Anatolian 'Ova' province with NE–SW shortening and NW–SE extension, containing large, roughly equant-shaped complex basins termed 'ovas'; (4) West Anatolian extensional province characterised by N–S extension. The arrows are roughly proportional to the magnitude of the total strain. EAF, East Anatolian Fault; NAF, North Anatolian Fault. (b) Detail of the Marmara Sea region along the North Anatolian Fault zone (Okay *et al.* 1999), showing selected bathymetric contours in m. Fault plane solutions are from Taymaz *et al.* (1991). Arrows are displacement vectors derived from GPS using a fixed station at Istanbul (Straub & Kahle 1995). The Tekirdag Basin is located in a broad negative flower structure. Reproduced with permission of John Wiley & Sons, Inc.

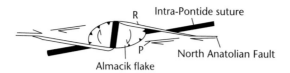

Fig. 6.14 Interpretation of the Almacik flake along the North Anatolian Fault in Turkey (Sengör *et al.* 1985), based on the idea of 'Riedel flaking'. R and P shears are shown.

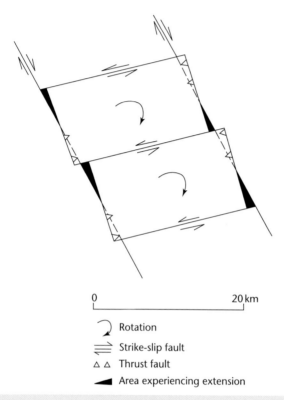

Fig. 6.15 Model of rotation of blocks near the intersection of the San Andreas and San Jacinto faults, California, based on earthquake hypocentral locations and first-motion studies (Dibblee 1977; Nicholson *et al.* 1986b). During a large earthquake, one of the bounding faults moves by right-lateral slip. Continued aseismic motion between the bounding faults is accommodated by clockwise rotations of small blocks and by minor left-lateral movements on faults defining these rotating blocks. Rotation causes overlaps (small reverse faults) and gaps (small normal faults) to form along the major bounding right-lateral faults. Viewed on a crustal scale, these rotating flakes would be detached from the underlying lower crust/mantle by thrust faults that would merge with the associated left-lateral strike-slip faults. Reproduced with permission from the Society of Economic Paleontologists and Mineralogists, Tulsa, Oklahoma.

one strike-slip zone to another, but they also may vary in time within an individual strike-slip fault system. Deformation of different stratigraphic units, strain-weakening or strain-hardening effects, uplift and erosion or sedimentation and burial, or changes in heat flow related to crustal thinning and/or volcanism can all have significant effects on the evolution of the fault zone.

4   Pre-existing geological fabric

Continental strike-slip fault systems commonly intersect or make use of older heterogeneities. For example, the Great Basin, western USA, underwent Laramide shortening (NE–SW trend), then two phases of extension, one related to backarc processes (axis of extension orientated NE–SW), the other to the development of a right-lateral megashear (change in extension direction to NW–SE). The right-lateral faults and, locally, conjugate left-lateral faults associated with this clockwise rotation in the extension direction may be related to the existence of older crustal structures.

The Great Basin of the Basin and Range province, SW USA, is an example of regional extension, so the strike-slip motion is taking place in a transtensile regime. Some topographic lows such as Death Valley change from principally extensional where it is orientated roughly N–S, to transtensional where the orientation changes to approximately NW–SE. The Rhine Graben of western Europe, however, originated in a background environment of continental collision north of the Alps in late Eocene times. Early extensional faults in the late Eocene–Oligocene were subsequently used as strike-slip faults. The transformation of the Upper Rhine Graben from a rift to a transform zone in the early Miocene (Rotstein & Schaming 2011) was accompanied by transpressional uplift and erosion of the Black Forest and Vosges crystalline massifs (Fig. 3.9). The fact that transpressional effects diminish northwards suggests that the source of the compression was the Alpine orogen to the south.

Pre-existing structures that significantly influence the location and orientation of folds and faults during strike-slip are termed '*essential*' structures. On the other hand, pre-existing structures that exert no control on later deformation are termed '*incidental*' (Christie-Blick & Biddle 1985, p. 13).

6.2.2   Role of oversteps

In §6.1.2 the location of overstep basins between the tips of two subparallel strike-slip fault segments was introduced. An *overstep* or *stepover* is a structural discontinuity between two approximately parallel overlapping or underlapping faults. Oversteps are extremely important in determining the location of regions of subsidence and uplift along a strike-slip system. We consider a numerical model of basin development at an overstep in §6.4.1. There appear to be two basic types (Fig. 6.16):

1. Oversteps along the strike of faults, that is, observed in plan view; faults are continuous in the down-dip direction.
2. Oversteps along the dip of faults, that is, observed in cross-section; faults are otherwise continuous in plan view.

Both types of overstep may occur along the same fault or fault zone. If the sense of an along-strike overstep is the same as the sense of fault slip, a *pull-apart basin* is formed (Aydin & Nur 1985). If, on the other hand, the sense of the overstep is opposite to that of fault slip, a *push-up range* is formed. Down-dip oversteps are probably as important as along-strike oversteps, but are more difficult to identify and map. Seismicity studies on active strike-slip faults such as the Calaveras Fault, California, show a distinct offset of the trends of hypocentres, breaking the fault into an upper segment (2–7 km depth) and a lower segment 2 km away (4–10 km depth). This suggests that a down-dip overstep is present (Reasenberg & Ellsworth 1982).

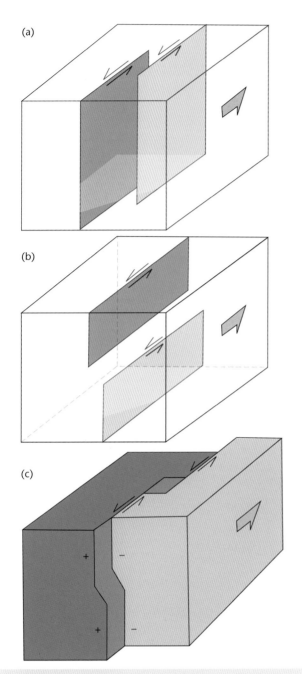

Fig. 6.16 Oversteps on strike-slip faults. (a) Along-strike oversteps in which faults are continuous in the down-dip direction. (b) Down-dip oversteps in which the faults are continuous in plan view. (c) Combination of along-strike and down-dip oversteps, with a pull-apart basin located at the along-strike overstep. After Aydin & Nur (1985).

6.3 Basins in strike-slip zones

In general, subsidence in strike-slip zones tends to occur where the deformation is accompanied by a component of divergence (Fig. 6.3), such as at bends and oversteps in the fault or in gaps between rotating flakes. Uplift, on the other hand, tends to occur where there is a component of convergence (Fig. 6.3), although flexure of an underlying block by an overriding block may cause subsidence. Uplift takes

place at convergences of faults into a fault junction or at bends in fault traces. These zones of uplift provide sourcelands of detritus available to fill nearby basins. Bends, oversteps and fault junctions associated predominantly with extension and subsidence are termed '*releasing*', while those associated with shortening and uplift are termed '*restraining*' (Crowell 1974b). Parallel-sided basins in which oceanic crust occurs are termed *rhombochasms* (Carey 1976).

The causes of fault bends, oversteps and junctions are varied. Some bends may result from pre-existing crustal heterogeneities, as in the case of strike-slip faults propagating along older extensional fractures. This appears to be the case in Jamaica (Mann *et al.* 1985), which lies entirely within a 200 km-wide seismic zone of strike-slip deformation between the North American and Caribbean plates (Escalona & Mann 2011). Extensional faults at the east of the Paleocene–Eocene Wagwater rift graben have been reactivated as reverse faults along a restraining bend in the Neogene strike-slip system. Other bends may be due to deformation of initially straight faults as a result of incompatible slip at a fault junction (e.g. the 'big-bend' of the San Andreas Fault, Figs 6.4, 6.5), rotation of adjacent blocks (southern San Andreas Fault), or intersection of the fault with a zone of greater extensional strain (e.g. western end of North Anatolian Fault) or contractional strain (e.g. near junction of North Anatolian and East Anatolian faults in eastern Turkey, Fig. 6.13).

Fault branching and oversteps may develop by a number of mechanisms, including segmentation of curved fault traces, intersection of weak zones orientated obliquely to the direction of strike-slip, and changes in stress fields caused by fault interaction. The precise reasons for the formation of oversteps and branches are obscure (Aydin & Nur 1985).

Strike-slip basins may be *detached* and therefore thin-skinned. Examples are known from both areas of pronounced regional shortening, such as the Vienna Basin and the St George Basin in the Bering Sea, Alaska, and in areas of regional extension, such as the West Anatolian extensional province of Turkey and the Basin and Range of the USA.

6.3.1 Geometric properties of pull-apart basins

An assessment of the width, length and depth (sediment thickness) of pull-apart basins is a useful starting point for the consideration of their kinematics and geological evolution. When viewed in two dimensions, there is a well-established relationship between length and width of pull-apart basins (Aydin & Nur 1982). In three dimensions, basin depth scales on along-strike length (Hempton & Dunne 1984).

Pull-apart basins have an aspect ratio (length to width) of approximately 3:1 (Fig. 6.17) (Table 6.2) (Aydin & Nur 1982; Basile & Brun

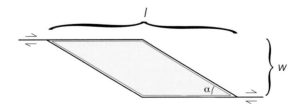

Fig. 6.17 Notation for a pull-apart basin at an overstep between two parallel strike-slip fault segments. After Gürbüz (2010).

Table 6.2   Geometrical properties of pull-apart basins; data from Aydin & Nur (1982), Hempton & Dunne (1984), Crowell (1974a, 1982) and Gürbüz (2010)

Name of pull-apart basin	Strike-slip fault	Displacement *l* (km)	Width (overstep) *w* (km)	Sediment thickness *d* (km)
Dead Sea Transform, Arabian plate				
Dead Sea Basin	Dead Sea Rift	100	25	7.5
Hula Basin	Dead Sea Rift	20	7	1.5
Lake Kinneret	Dead Sea Rift	17	5	
Elat Basin	Gulf of Elat	45	10	
Aragonese Basin	Gulf of Elat	40	9	
Tiran-Dakor	Gulf of Elat	65	8	
Karkom	Paran Fault	18	6	
Alpine Fault Zone, New Zealand				
Hanmer Plains	Hope Fault	4.5	2.7	0.85
Glynnwye Lake	Hope Fault	0.3	0.09	0.76
San Andreas Fault system, California–Nevada				
Cholame Valley	San Andreas Fault	17	3	
Koehn Lake	Garlock Fault	40	11	
Searleys Valley	Garlock Fault	1.6	0.38	
Elsinore Lake	Elsinore Fault	12	3	
Brawley Basin	Imperial Fault	10	7	
Ridge Basin	San Gabriel Fault	40		4.5
Hog Lake basin	San Jacinto Fault	0.68	0.17	
Hemet Basin	San Jacinto Fault	22	5	
Caribbean area				
Cariaco Basin	El Pilar Transform	125	35	
Magdalen Basin		200	100	6.7
La Gonzales Basin	Bocono Fault, Venezuela	23	6.2	
Merida-Mucuchies Basin	Bocono Fault, Venezuela	6.2	1.7	
Lake Valencia	Valencia Fault	30	11.5	
Casanay Basin	El Pilar Fault	3	1.2	
Lake Izabal	Polochic Fault (Guatemala)	80	30	
Motagua Valley	Motagua Fault	50	20	
Anatolia, Turkey				
Tekirdag (Marmara Sea)	North Anatolian Fault	26	10	2.02
Zahan Basin	East Anatolian Fault			
Hazar Basin	East Anatolian Fault	5	3	
Erzincan	North Anatolian Fault	48	16	3.90
Niksar Basin	North Anatolian Fault	15	12	
Susehri Basin	North Anatolian Fault	23	6	
Carrington		12		0.85
Others				
Hornelen (Norway)		70		7
Bovey (England)	Sticklepath Fault	16		1.25
Petrockstow		4	1	0.67
Chuckanut		50		6
Lefors		12		1.2
Iran	Dasht-e Bayaz Fault (Iran)	12	5	
Vienna (Austria)		80		5.2
Death Valley (eastern California)	Black Mountain Fault	28		2.75

1999 from experimental results). As slip along the fault increases, so does the basin width; narrow grabens evolve to wider basins flanked by terraced sidewall fault systems, as seen in analogue experiments (Dooley & McClay 1997; McClay & Dooley 1995). The acute angle between the main strike-slip fault and the basin-bounding faults of the pull-apart basin is commonly 30° (Fig. 6.17). This figure is probably dependent on the rheology of the deforming crustal materials and the overlap and/or separation of the strike-slip faults, but numerical experiments suggest that the key parameter is the thickness of the brittle layer (Petrunin & Sobolev 2006, 2008).

The length l along the main strike-slip fault, and the width of the pull-apart basin w, are related by

$$\log l = c_1 \log w + \log c_2 \qquad [6.1]$$

where c_1 is approximately 1, and c_2 is between 2.4 and 4.3 (Aydin & Nur 1982, $n = 62$), from which it can be seen that $l = w^{c_1} c_2$. Since $c_1 \sim 1$, and $c_2 \sim 3$, l/w is approximately 3. Gürbüz (2010) added data from the North Anatolian Fault Zone that fell on a similar trend (Fig. 6.18). The true thickness of basin deposits d scales on basin length l in a relatively small data set of nine ancient and seven modern pull-apart basins (Hempton & Dunne 1984)

$$d = 0.8l + 0.26 \qquad [6.2]$$

This relationship suggests that basin depth depends on the amount of stretching associated with strike-slip displacement. Combining all width, length and depth data, Gürbüz (2010) derived the following regression

$$d = 0.1104l - \left(8.7550 \cdot 10^{-2}\right)w \qquad [6.3]$$

for data in the range $l < 10^5$ m, $w < 10^4$ m and $d < 10^4$ m. This relationship must break down at high values of displacement (along-strike length), since basin depths seldom exceed 10 km (Zak & Freund 1981). An upper limit to basin depth despite further increases in displacement is also a feature of numerical models of pull-apart basin evolution (Petrunin & Sobolev 2008) (see §6.4.1).

The implications of these geometrical properties are discussed in §6.3.2.

6.3.2 Kinematic models for pull-apart basins

A number of kinematic models exist for the development of pull-apart basins (Fig. 6.19). All models involve a process leading to local extension and subsidence bounded by strike-slip fault segments, accommodated in the brittle domain of the lithosphere (see §6.4.1). The main kinematic models are:

1 Overlap of side-stepping faults
The simplest model is based on pioneer field studies of active strike-slip basins such as the Dead Sea and the Hope Fault Zone in New Zealand. Discontinuous parallel fault segments that have a horizontal separation develop oversteps. As the master faults lengthen, producing more overlap, the basin lengthens while its width remains fixed by the original separation of the parallel master faults (Fig. 6.19a). A fixed basin width with a lengthening displacement implies highly variable l/w ratios, which is not found in global compilations (Fig. 6.18a) (§6.3.1), so this mechanism may be restricted to short increments of displacement. The model has been widely applied to the San Andreas Fault system in the Gulf of

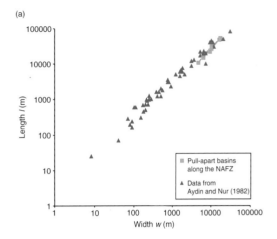

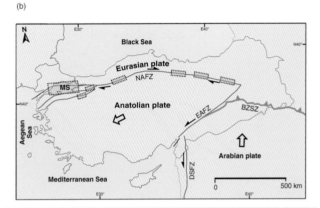

Fig. 6.18 (a) Plot of pull-apart basin width-length (displacement) relationship using the dataset of Aydin & Nur (1982) supplemented by the Turkish data of Gürbüz (2010). Turkish pull-apart basins are shown in the map in (b). MS, Marmara Sea; NAFZ, North Anatolian Fault zone; EAFZ, East Anatolian Fault zone; DSFZ, Dead Sea Fault zone; BZSZ, Bitlis–Zagros suture zone. From west to east along the NAFZ, the pull-apart basins are Tekirdag, Yenisehir, Pamukova, Bolu and Yenicaga, Tasova-Erbaa and Niksar, Susehri and Erzincan.

California-Salton Trough area (Crowell 1974b), and the Dead Sea Fault system (Garfunkel *et al.* 1981). This configuration of lengthening, parallel master faults has been used to simulate pull-apart development using elastic dislocation theory (Rodgers 1980) (see §6.3.3). This theory predicts that when the overlap is about equal to the separation, the pull-apart develops two depocentres in regions of local extension separated by a zone of secondary strike-slip faulting in the basement. As the overlap increases, the depocentres become more widely spaced and the intervening zone of strike-slip faulting broader. Although there are instances where field observations fit the theoretical elastic dislocation model rather well (e.g. Cariaco Basin, Schubert 1982, and Vienna Basin, Hinsch *et al.* 2005), basins that are filled with sediment more rapidly than the upward propagation of faults from the basement may develop fault systems at strong variance to those predicted by elastic dislocation theory. The presence of two distinct depocentres was also noted in

(a) Overlap of side-stepping faults

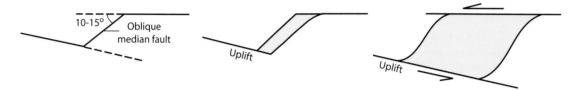

(b) Slip on divergent fault segments

(c) Nucleation on *en echelon* fractures

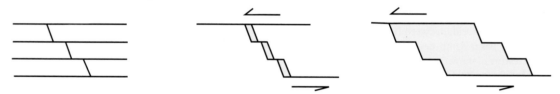

(d) Coalescence of scale-dependent basins

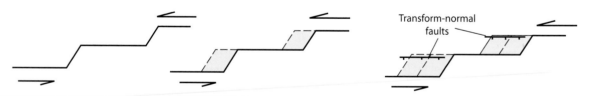

(e) Transform-normal extension

Fig. 6.19 Models of pull-apart basin development (after Mann *et al.* 1983 (reproduced with permission of University of Chicago Press) and Ben-Avraham & Zoback 1992). (a) Pull-apart opening between left-stepping and sinistral master strike-slip faults. Fault separation and basin width remain constant through time whereas fault overlap and basin length increase with the amount of strike-slip displacement. The elastic dislocation model of Rodgers (1980) applies to this kind of pull-apart mechanism. (b) Pull-apart opening across an oblique median fault and non-parallel master faults. Note that compression and uplift occur on one side of the pull-apart while an extensional 'gap' develops on the other. (c) Pull-apart formation by nucleation from extensional fractures, based on shear box experiments (Koide & Bhattacharji 1977). (d) Pull-apart formation by coalescence of small-scale similar sub-basins, as suggested by Aydin & Nur (1982). This allows the widening of pull-aparts with increased offset, in contrast to (a). (e) Pull-apart basin formation caused by simultaneous strike-slip motion on a transform fault and transform-normal extension (after Ben-Avraham & Zoback 1992). The pull-aparts are markedly asymmetric towards the transform fault.

sandbox experiments involving small amounts of transtension (Wu *et al.* 2009), but it does not appear to be a common feature of pull-apart basins.

2 Slip on divergent fault segments

Detailed field mapping of some active pull-apart basins suggests that bounding strike-slip faults may be non-parallel (Freund 1971) (Fig. 6.19b). If non-parallel, non-overlapping faults are slightly divergent and connected by a short oblique segment, continued strike-slip may open up a basin along one side of the oblique fault and cause compression along the other side. Large-scale examples are the V-shaped wedges between strike-slip faults in central Asia (An Yin 2010).

3 Nucleation on *en echelon* fractures or Riedel shears

Shear box experiments suggest that pull-apart basins may be structurally analogous to *en echelon* extensional fractures produced in clay materials (Fig. 6.19c). As deformation continues, the shear fractures join to form a pull-apart basin. A similar process of formation from rotated large-scale tension gashes or Riedel shears has also been proposed. Nucleation on extensional sigmoidal fractures and subsequent growth should produce highly variable *l/w* ratios that are not sup-

ported by observational evidence (Fig. 6.18a), so this explanation cannot have broad applicability.

4 Coalescence of adjacent pull-apart basins into larger system

The linear relationship between basin length *l* (fault overlap) and basin width (fault separation *w* based on a global compilation) (§6.3.1) (Aydin & Nur 1982; Gürbüz 2010), irrespective of absolute size, may be due to the coalescence of adjacent pull-apart depocentres into a single larger basin with increasing offset, or to the formation of new fault strands parallel to existing ones (Fig. 6.19d). Coalescence of scale-dependent basins should result in the merging of a series of different depocentres whose individuality may still be recognisable in sediment isopachs.

5 Transform-normal extension

Markedly asymmetric, large strike-slip basins bounded by a transform segment on one side and sub-parallel normal faults on the other may be due to transform-normal extension (Ben-Avraham & Zoback 1992) (Fig. 6.19e). Examples of highly asymmetric basins are the Gulf of Elat (Dead Sea Transform) (Ben-Avraham 1985) and the Cariaco Basin (offshore Venezuela) (Fig. 6.20). In the Gulf of Elat, the sense

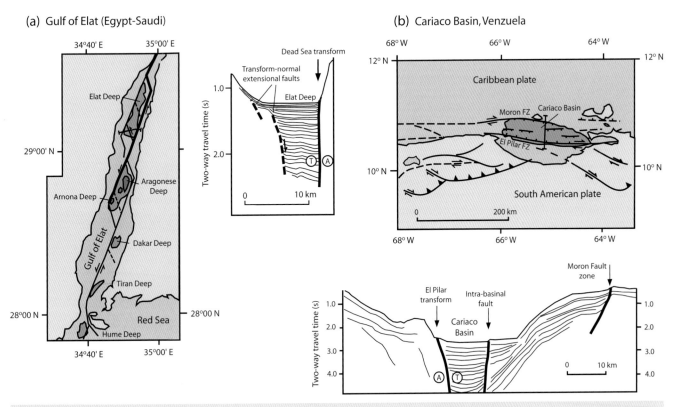

Fig. 6.20 (a) Bathymetry and structure of the Gulf of Elat showing segments of the Dead Sea Transform and location of main topographic depressions. Inset shows cross-sectional geometry across the Elat Deep derived from seismic reflection surveys. Asymmetry of the basin-fill is towards the transform fault. (b) Tectonic map and seismic profile across Cariaco Basin, northern Venezuela (Schubert 1982), showing basin asymmetry towards El Pilar Transform. A, displacement away from reader; T, displacement toward reader. Modified from Ben-Avraham and Zoback (1992).

of asymmetry switches along the PDZ, the deeper side of the sub-basin always being located against the strike-slip Dead Sea Transform, with the shallower flank adjacent to normal faults. This suggests that subsidence occurs due to extension approximately normal to the transform at the same time as strike-slip displacement is taking place, in contrast with kinematic models of deformation in strike-slip zones where extension and compression directions are at 30 to 45° to the strike-slip fault plane (Fig. 6.12). This style of deformation may be due to the weakness of strike-slip faults (see also §6.3.3). A model of stress orientation close to weak transform faults embedded in a strong crust (Zoback *et al.* 1987) suggests that far-field transtensional stresses cause fault-normal extension, and transpressional stresses cause fault-normal compression.

6.3.3 Continuum development from a releasing bend: evolutionary sequence of a pull-apart basin

Analysis of neotectonic pull-apart basins in their embryonic stage suggests that they are associated with: (i) releasing bend geometries, in which master faults do not overlap but are connected by oblique median strike-slip faults; and (ii) non-parallel master faults (see also (1) and (2) above). Gentle restraining bends appear to produce narrow *spindle-shaped* pull-apart basins like the Clonard Basin, Haiti at the left-step in the Enriquillo–Plantain Garden Fault Zone. In contrast, sharp restraining bends produce multiple staggered pull-apart basins, as in the Salton Trough of California. Since embryonic pull-apart basins appear to occur on releasing bends of through-going faults, they are unlikely to be analogous to the tension gashes or Riedel shears produced in shear box experiments (see (3) above).

Continued offset produces basin shapes known as 'lazy S' (between sinistral faults) and 'lazy Z' (between dextral faults) (Mann *et al.* 1983) (Fig. 6.21), representing a transitional stage between spindle-shaped basins between master faults with no overlap and rhomboidal basins between overlapping master faults. S- and Z-shaped pull-apart basins are particularly common where the master faults are widely separated (>10 km) and their strikes are non-parallel. The Death Valley Basin, California, has a pronounced Z-shape.

Lengthening of the S- or Z-shaped basins produces rhomb-shaped pull-apart basins. Length to width ratios (overlap to separation) l/w increase by this process because the separation remains fixed by the width of the releasing bend. Rhomboidal pull-apart basins are characterised by overlapping master faults and deep depocentres at the ends of the pull-apart basin, separated by a shallow sill(s). The Cariaco Basin of the Venezuelan borderlands is a good example, with two sub-circular deeps in excess of 1400 m and a shallow sill at 900 m. Multiple deeps may be arranged diagonally across the rhomboidal pull-apart, as is the case in the Gulf of Aqaba in the northern Red Sea, the deepest segment of the sinistral Dead Sea Fault system (Ben-Avraham *et al.* 1979). The rapid subsidence between the overlapping master faults generally greatly exceeds sedimentation, leading to deep-marine or lacustrine environments in the basin. The presence of prominent depocentres at the ends of rhomboidal pull-apart basins does not support the mechanism of simple extension between overlapped master faults (see (1) above), but is predicted by the elastic dislocation model (Rodgers 1980) and is replicated under transtension in sandbox experiments (Wu *et al.* 2009). The depocentres may be smaller compartments in the basin floor of the larger

pull-apart basin, that is, the rhomboidal basin is in a coalesced stage of development (see (4) above).

With continued offset, rhomboidal pull-apart basins may develop into narrow oceanic basins where the length (overlap) greatly exceeds the basin width (separation). The Cayman Trough along the bound-

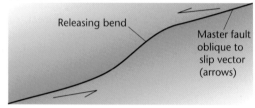

(a) Initial fault geometry

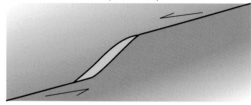

(b) Basin nucleation–spindle-shaped basins

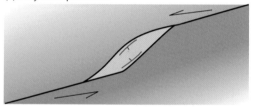

(c) Lazy S-shaped basin

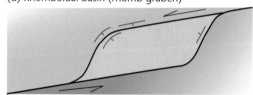

(d) Rhomboidal basin (rhomb graben)

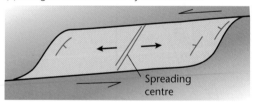

(e) Elongate basins floored by oceanic crust

Fig. 6.21 Continuum model of pull-apart development, after Mann *et al.* (1983). (a) Pull-apart begins life on a releasing fault bend. The size of the bend controls the pull-apart basin width. (b) Spindle-shaped basin. Continued offset produces lazy S-shaped basin (c), then a rhomb-shaped basin (d), then, finally, a long, elongate trough floored with oceanic crust (e). Most pull-aparts do not reach the stage represented by (e); instead they tend to be terminated after the basin length has reached about three times the original width of the releasing bend (or fault separation). © *Journal of Geology* 1983. Reproduced with permission of University of Chicago Press Journals.

ary between the Caribbean and North American plates is a type example. The large *l/w* ratio suggests that coalescence is unimportant at this stage.

6.3.4 Strike-slip deformation and pull-apart basins in obliquely convergent orogens

Oblique convergence causes strike-slip deformation in a number of fold-thrust belts and orogenic belts. The Caribbean–South America plate boundary contains a number of pull-apart basins, which developed diachronously from west to east from the Paleogene to the present day in the overall context of arc–continent collision. Basin development continues today in the Columbus Basin of eastern offshore Trinidad (Escalona & Mann 2011). Other basins along this strike-slip zone are, from west to east, the Falcon Basin, Bonaire Basin, Cariaco Basin and Carupano Basin (Figs 6.22, 6.23). The eastward displacement of the Caribbean plate relative to South America produces a major E–W dextral strike-slip zone with a component of southward overthrusting, which has successively displaced the path

of the paleo-Orinoco River to the south (Fig. 6.22). The dextral strike-slip motion due to the eastward elongation of the Caribbean arc and the southward thrust propagation results in a three-stage evolution of the plate boundary zone: (i) initial arc–continent collision, in which the Caribbean arc and forearc terranes are overthrusted onto the passive margin of northern South America, causing a flexural foreland basin to form; (ii) cessation of deformation in the fold-thrust belt as tectonic activity translates eastwards and isostatic rebound, structural inversion and strike-slip deformation is associated with slab break-off of the northward-subducted oceanic lithosphere attached to the South American plate. A slowing of subsidence in the foreland basins is accompanied by overfilling with continental sediment; and (iii) a final stage of arc–continent collision is marked by the complete break-off of the oceanic lithosphere of the South American plate, and E–W extension of the Caribbean arc as it elongates eastwards, producing deep half-grabens bounded by oblique normal faults. The periods of dominance of strike-slip tectonics are therefore phases in a longer, more complex tectonic history (Figs 6.24, 6.25).

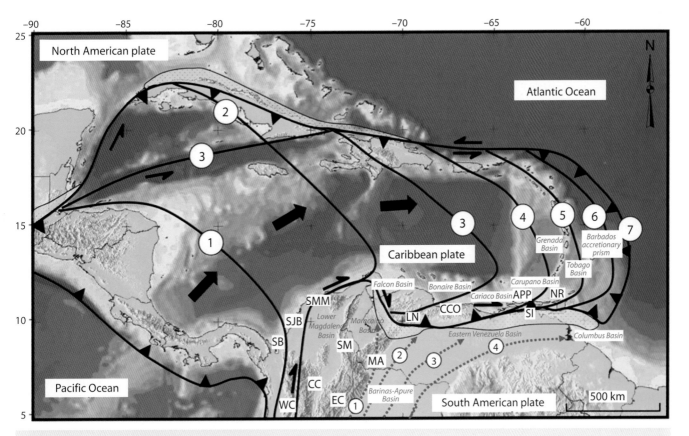

Fig. 6.22 Diachronous eastwards movement of the Caribbean plate, with numbers showing the locations of the leading edge, from Late Cretaceous (1) at 80 Ma to Recent (7) at 0 Ma, from Escalona & Mann (2011). Stippled yellow areas are foreland basins caused by diachronous oblique convergence. Dashed blue lines are the paleo-courses of the Orinoco River between Paleocene (1) at 60 Ma to Recent (4) at 0 Ma. Abbreviations: WC, Western Cordillera; CC, Central Cordillera; EC, Eastern Cordillera; SB, Sinu Belt; SJB, San Jacinto Belt; SM, Santander massif; MA, Merida Andes; LN, Lara nappes; CCO, Cordillera de la Costa; APP, Araya-Paria Peninsula; SI, Serrania del Interior; NR, Northern Range of Trinidad. Main sedimentary basins are also labelled. Reprinted with permission from Elsevier.

(a) Mesozoic–Paleogene thicknesses in TWT (msecs)

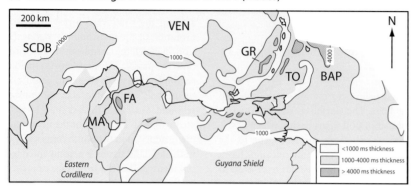

(b) Neogene thicknesses in TWT (msecs)

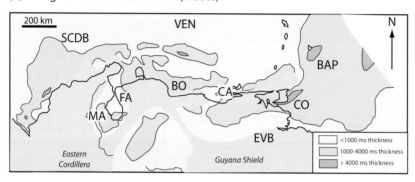

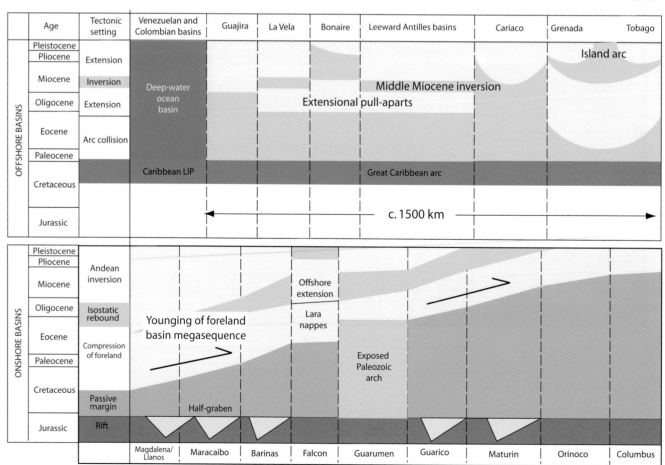

Fig. 6.23 Isopachs based on two-way travel time (in milliseconds) for (a) the Mesozoic–Paleogene, and (b) the Neogene of the Caribbean-South American plate boundary zone. Modified from maps in Escalona and Mann (2011). SCDB, South Caribbean Deformed Belt; MA, Merida Andes; F, Falcon Basin; BO, Bonaire Basin; CA, Cariaco Basin; EVB, East Venezuela Basin; VEN, Venezuelan Basin; CO, Columbus Basin; GR, Grenada Basin; TO, Tobago Basin; BAP, Barbados Accretionary Prism. Reprinted with permission from Elsevier.

Fig. 6.24 Basin development in the offshore and onshore areas of the margin of the Caribbean and South American plates, modified from Audemard & Serrano (2001) in Escalona & Mann (2011). Grey tone indicates stratigraphic gaps. The onshore basins are dominated by the eastward migration of a wave of flexural subsidence caused by tectonic shortening along the transpressional plate margin. The offshore basins lack this flexural signature, and instead subside by local extension, commonly with a mid-Miocene phase of inversion. Reprinted with permission from Elsevier.

Strike-slip faults are a prominent feature of regions of oblique continent–continent convergence, such as Himalaya-Tibet and the Chinese interior. They are an important component of arc-parallel deformation related to the active deformation of the Himalayan system (Fig. 6.26) (Styron *et al.* 2011). India–Asia collision was characterised by an early stage (55–30 Ma) of shortening, lateral extrusion of SE Asia (32–17 Ma), followed by extensive strike-slip deformation producing conjugate faults and N–S trending extension. The zones of NE-trending and NW-trending dextral strike-slip faults cover large regions of central Asia (An Yin 2010). Conjugate sets of strike-slip faults producing V-shaped wedge blocks, opening to the east, are found in Tibet, Mongolia, Afghanistan and SE Asia (Taylor *et al.* 2003). They lie at 60–75° to the maximum compressive stress, open in the direction of the topographic gradient, and individual faults are hundreds of kilometres long. They appear to be typical of continental extrusion (Peltzer & Tapponnier 1988) rather than plate boundary transform movement.

6.4 Modelling of pull-apart basins

6.4.1 Numerical models

The complexity of the structural geology of the PDZ makes numerical modelling approaches to the understanding of basins in strike-slip zones difficult. Although there is no 'reference' model, as in the case of the uniform stretching of the continental lithosphere (Chapter 3), the subsidence of pull-apart basins is essentially a result of the stretching and thinning of crustal materials between the main strike-slip faults. The quantitative modelling of basin development in strike-slip zones has followed a different route to the case of simple extension. These approaches are based more on the details of deformation around interacting faults, crustal rheology and strain localisation than about the large-scale processes of lithospheric stretching and subsequent cooling described in Chapter 3. Furthermore, different quantitative models for the formation of basins in strike-slip zones may not be easily discriminated from the surface displacement fields recognised in geomorphic and geodetic studies (Gomberg & Ellis 1994).

The *en echelon* faults typical of strike-slip zones can be modelled as surfaces of discontinuity that produce a neighbouring displacement field using *dislocation theory*, but with no interaction between the faults (Rodgers 1980) (Fig. 6.27b). The faults can also be viewed as large-scale fractures that interact to produce a local stress field (Segall & Pollard 1980). Two-dimensional plane stress models for pull-apart basins such as the latter calculate the strain around two *en echelon* fault tips in an elastic medium. The 2-D plane stress

model produces unrealistic, deep depocentres adjacent to the fault tips characterised by very high subsidence rates, together with unrealistic vertical uplifts above the fault tip. These effects result from the plane stress approximation, which does not allow stresses to develop in the vertical direction within the plate; consequently, stresses cannot resist deformation near the fault tips. The advantage of a three-dimensional model is that vertical stresses near the fault tips resist deformation, which dampens subsidence and uplift patterns in this region (Fig. 6.27a).

In the three-dimensional linear elastic model proposed by Katzman *et al.* (1995), the deformation is driven by a shear from below to simulate relative plate motion. The upper crust (c.15 km) is penetrated fully by faults that detach above the weak lower crust. Pull-apart basins develop close to two *en echelon* vertical faults 150 km long, spaced 10 km apart, with 0–20 km overlap (Fig. 6.28). The shear zone at the base of the plate is allowed to vary between depths of 10 and 40 km. Two geometrical factors have important effects on pull-apart basin development:

1. *Shear zone width*: the basal shear causes slip to taper towards the fault tips, the rate of tapering decreasing with wider shear zones. As the shear zone width increases, the basin widens, lengthens and becomes shallower. Although the surface deformation varies strongly as a function of the shear zone width, the necking profiles in cross-section do not significantly vary.
2. *Fault overlap*: full grabens develop between the overlapping faults over the entire overlap zone, becoming half-grabens just outside the overlapping zone. Uplift decreases with the amount of overlap, reducing to zero at an overlap of c.20 km. Less overlap causes greater rotation about a vertical axis.

The Dead Sea Basin is an ideal test for this three-dimensional numerical model. It is located along the left-lateral Dead Sea Transform at the plate boundary between the African and Arabian plates (Garfunkel *et al.* 1981; Aydin & Nur 1982). The basin is narrow (c.7–18 km), deep (c.10 km), roughly symmetric, and extends along the strike of the bounding faults for 132 km (Fig. 6.29). Gravity and seismic reflection data (ten Brink & Ben-Avraham 1989) show that the Moho is not elevated beneath the basin. It is therefore thought to be uncompensated isostatically (as expected from its c.10 km wavelength – see §2.1.5). The flanks of the basin are not topographically elevated compared to the regional elevation. The Dead Sea Basin is subsiding over a wide area, over several tens of kilometres in length. The extent of the basin subsidence is in harmony with the model predictions using 20–40 km-wide shear zones. However, the symmetric full-graben nature of the basin over a 50 km N–S

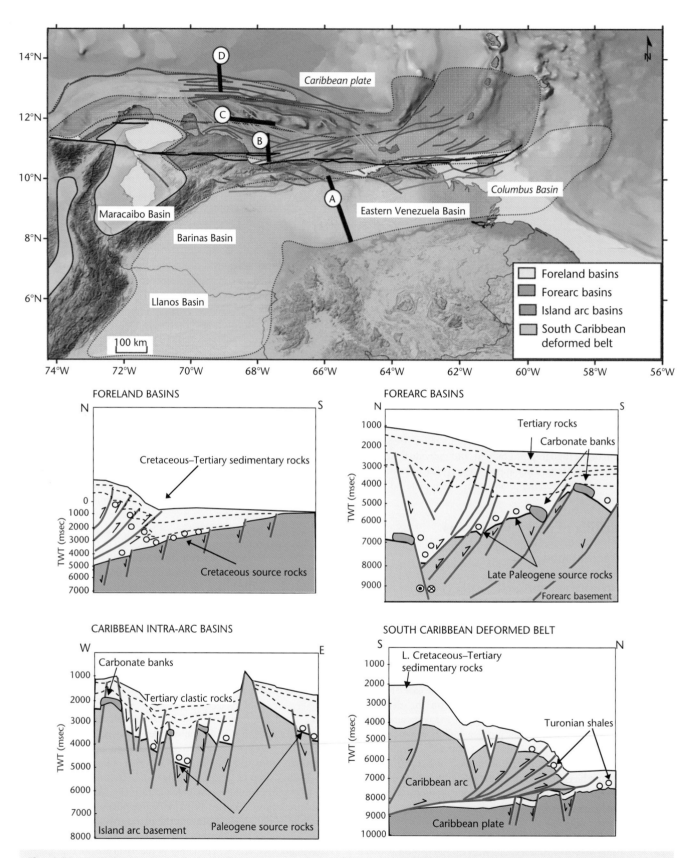

Fig. 6.25 Main basin types and petroleum provinces in the Caribbean–South America strike-slip zone. Profile A typical of foreland basins. The Caribbean arc and forearc are thrusted over the South American passive margin generating a flexural foreland basin. The fold-thrust belt runs along the modern Caribbean coastline; Profile B typical of forearc basins in the Caribbean arc; Profile C typical of deep Caribbean island arc-related basins; Profile D typical of the South Caribbean Deformed Belt, a zone of north-vergent thrusting linked to the southward subduction of the Caribbean plate beneath northern South America. Yellow circles show source rocks. After Escalona & Mann (2011), reprinted with permission from Elsevier.

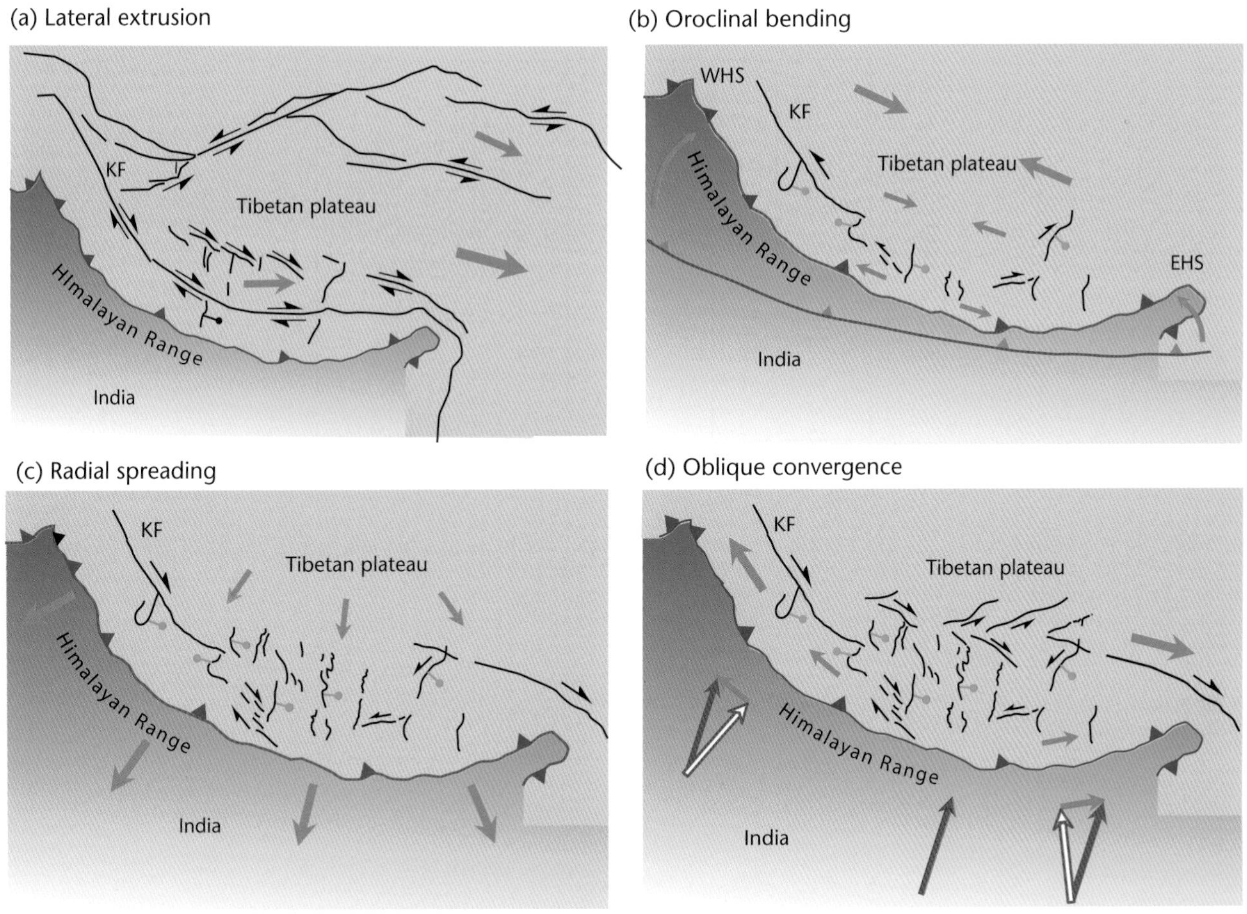

(a) Lateral extrusion

(b) Oroclinal bending

(c) Radial spreading

(d) Oblique convergence

Fig. 6.26 Models for large-scale response to indentation of India in the Himalayan-Tibet region, from Styron *et al.* (2011). (a) Lateral extrusion to the east; (b) regional oroclinal bending; (c) radial spreading from Tibet; (d) oblique convergence causing northwestwards and eastwards escape. KF, Karakorum Fault; EHS, Eastern Himalayan Syntaxis.

distance, and the lack of topographic uplifts suggest that fault overlap must be greater than 20 km (twice the fault spacing). Katzman *et al.* (1995) suggested that the overlap distance increased during basin development.

Numerical modelling of the deformation associated with strike-slip faulting ideally requires the *rheology* of crustal and subcrustal materials to be specified. An example is a continental crust with a quartz- or plagioclase-dominated rheology, overlying an olivine-dominated mantle (Fig. 6.30). A constant left-lateral transform motion (6 mm yr^{-1}, to simulate the Dead Sea Transform, Garfunkel & Ben-Avraham 1996) is applied to the continental block with seeds causing the nucleation of two fault strands separated by 12 km (Petrunin & Sobolev 2006, 2008). Surface heat flow varies from 40 to 70 mW m^{-2}, spanning the global average. In the brittle crust, the strain is able to localise by a plastic softening process, which controls the rate at which extension and subsidence is distributed between the active fault segments. The viscosity of the lower crust is also allowed to vary.

There are a number of first-order conclusions from these rheologically tuned numerical models:

- Deep basins can form with thick brittle layers and viscous lower crust. Subsidence rate depends on the strength of the lower crust.

A weak lower crust reduces subsidence rate and may even cause uplift after a certain amount (70–90 km) of strike-slip displacement.

- Maximum basin thickness depends strongly on the geotherm, high surface heat flows (60 mW m^{-2}) being associated with slow subsidence rates. Initial subsidence rates are fast, reducing over time, and resulting in uplift after a certain amount of strike-slip displacement.
- Rapid strain softening of the crust promotes rapid subsidence.

Petrunin and Sobolev (2006, 2008) introduced the term 'brittle brick stretching' to describe the way in which brittle deformation is localised on the bounding faults to create pull-apart basins. Assuming that the free surface of crust subsides as the brittle 'brick' extends, the amplitude of subsidence h_s depends on the initial thickness of the brittle layer h_0 by a conservation of mass

$$h_s = h_0 \left(1 - \frac{l_0}{l} \right)$$

[6.4]

where l_0 and l are the initial and current lengths of the layer. Isostatic compensation of the sediment load is not accounted for, as is likely where the wavelength of load is small (§2.1.5) and where the

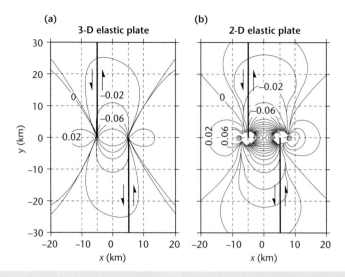

Fig. 6.27   Comparison between 2-D and 3-D plane stress models for pull-apart basins (after Katzman *et al.* 1995). A sinistral displacement discontinuity of 1 km is defined across two *en echelon* strike-slip faults. Vertical displacement is shown for (a) a 3-D elastic plate, and (b) a 2-D thin elastic plate with elastic thicknesses of 15 km. Contour interval 0.02 km, Poisson's ratio 0.25 and Young's modulus 75 GPa. The 2-D plane stress model produces unrealistic deformation around the fault tips. Reproduced with permission of the American Geophysical Union.

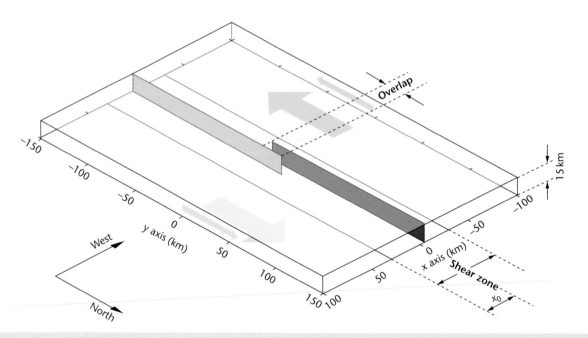

Fig. 6.28   Three-dimensional block diagram showing the set-up for the numerical pull-apart basin model (Katzman *et al.* 1995). Shaded planes define zones of weakness having zero shear strength (faults). Shaded arrows indicate the displacements defined as boundary conditions. The base of the model in the vicinity of the faults (within the shear zone) is stress free and thus freely deforming. The total width of the shear zone, $2x_0$, and the overlap distance between the fault segments, are varied in the model. Reproduced with permission of the American Geophysical Union.

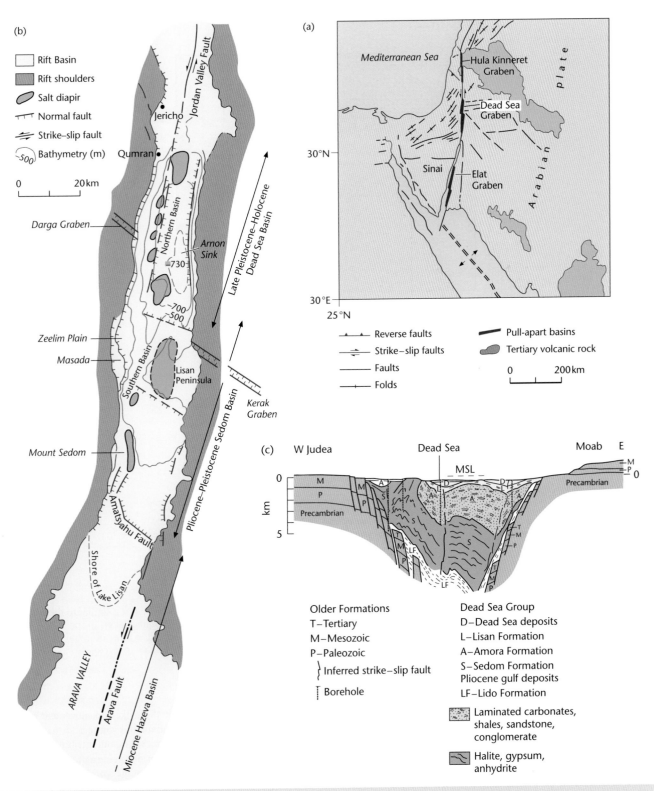

Fig. 6.29 (a) Location of Dead Sea Basin. (b) Structural interpretation, showing main faults, basins and salt diapirs (from Garfunkel *et al.* 1981). (c) Cross-section of Dead Sea Basin between Mount Sedom and Moab (from Zak & Freund 1981). MSL, Mean sea level. (b), (c) Reprinted with permission from Elsevier.

(a) Model set-up

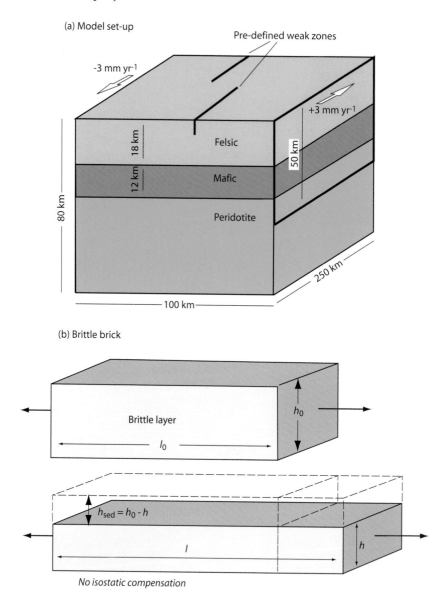

(b) Brittle brick

No isostatic compensation

Fig. 6.30 (a) Model set-up for 3-D numerical model of Petrunin & Sobolev (2008), showing rheological layering similar to the situation in the Dead Sea transform, with a strike-slip velocity of 6 mm yr^{-1}. Fault seeds (predefined weak zones) are placed in the upper 10 km of felsic crust with a spacing of 12 km. (b) Brittle brick model of Petrunin & Sobolev (2008, fig. 12). A brittle layer of initial thickness h_b and length l_0 is stretched to thickness h and length l, allowing sediment in the pull-apart basin to accumulate to thickness h_{sed}. Reprinted with permission from Elsevier.

continental lithosphere is rigid. Stronger crust has a thicker brittle layer, so sedimentary basins should be deeper on relatively cold lithosphere for the same amount of strike-slip displacement. Once again, taking the Dead Sea Basin as an example, seismicity occurs at depths of up to 18–22 km, which therefore approximates the thickness of the brittle layer h_0. The basin extends over a N–S distance of 132 km between bounding faults, and 60–65 km of strike-slip displacement is thought to have occurred during the early Miocene (25–14 Ma). Further strike-slip displacement took place in the Pliocene–Pleistocene-Holocene (last 4.5 Myr). Removing the early Miocene deformation alone therefore gives $l = 132$ km and $l_0 < 72$ km. Eqn. [6.4] gives a sediment thickness of 9 km, which is much greater than the thickness of preserved Hazeva Formation in the Arava Valley (Fig.

6.29). However, we can consider the strike-slip movement over the last 4.5 Myr, during which the Sedom and younger formations were deposited. We take l_0 once again as 132 km, but attribute only 50 km of strike-slip displacement to this period, giving a predicted sediment thickness of 7.5 km. This is close to the combined thickness of the Pliocene to Holocene sediments. The model subsidence rate is 1–2 mm yr^{-1}, close to the estimates of present rates (0.8–0.9 mm yr^{-1}). The main point to be appreciated from these modelling results is that sediment thicknesses in pull-apart basins are controlled by the thickness of the brittle layer undergoing stretching and the rate of strike-slip displacement.

If isostatic compensation is included, eqn. [6.4] can be modified to

$$h_s = h_{uc}\left(1 - \frac{l_0}{l}\right)\frac{(\rho_{lc} - \rho_{uc})}{(\rho_{lc} - \rho_s)} \qquad [6.5]$$

where the subscripts uc, lc and s refer to upper crust, lower crust and sediments (Fig. 6.31). This is likely to be more important when the crust is weak, allowing viscous flow in the lower crust to achieve isostatic compensation. Taking $h_{uc} = 15$ km, and ρ_{uc}, ρ_{lc} and ρ_s as 2750 kg m^{-3}, 2850 kg m^{-3} and 2500 kg m^{-3}, and using the displacement data for the late Miocene (see above), h_s is 2 km, similar to the preserved sediment thicknesses. Using the displacement data for the last 4.5 Myr, h_s becomes 1.6 km, much smaller than the brittle brick prediction of 7.5 km.

A comparison of the brittle brick model, the fully isostatically compensated model, and observations on basin depth are shown in Fig. 6.32.

6.4.2 Sandbox experiments: pure strike-slip versus transtension

Most models of pull-apart basins consider the case of pure strike-slip motion, that is, the slip vector is on average parallel to the PDZ marked by major basement-cutting faults. Pull-apart basins also occur where the motion is obliquely compressional (e.g. Karakorum, Himalaya, Tibet) (§6.3.4), but especially where the motion is obliquely transtensional (e.g. southern Dead Sea system, Vienna Basin). Numerical models show that even small amounts of transtension (5°) can cause the area of subsidence to broaden out significantly, by a factor of 2 to 3 (ten Brink *et al.* 1996). Scaled (ratio of 10^{-5}) sandbox models also show that the fault geometries and surface areas of pull-apart basins developed in pure strike-slip are different to those associated with transtension (Dooley *et al.* 2004; Wu *et al.* 2009) (Fig. 6.33). Whereas pull-apart basins in both settings are elongate, sigmoidal or rhomb-shaped grabens, the transtensional pull-apart basins are wider, with two distinct depocentres instead of a centrally located singular depocentre (Fig. 6.34). With pure strike-slip, a deep, narrow elongate rhomb-shaped, flat-bottomed basin is formed, flanked by shallow terraces. The fault structure is essentially a narrow negative flower structure. In contrast, two depocentres with opposite polarity, separated by an intra-basinal high, forming a much wider region of subsidence, typify transtension. Basement faults form a broad, complex, upward-diverging negative flower structure.

The introduction of transtension therefore strongly affects the overall dimensions of the pull-apart basin, being both wider and longer than in pure strike-slip, without appreciably changing the length-width aspect ratio. The Vienna Basin, for example, is situated within the sinistral transform system that detaches at depth at the base of the Alpine-Carpathian orogenic wedge, has two depocentres bordered by *en echelon* faults (Hinsch *et al.* 2005) (Fig. 6.34), and consists of a broad, asymmetric flower structure, showing many similarities with the transtensional pull-apart basins produced in sandbox experiments (Wu *et al.* 2009). Although dual depocentres are also found in pure strike-slip models, their development is thought to be enhanced in transtension. The Cariaco Basin, Venezuela (Schubert 1982), is a further example. Along the Dead Sea Transform, the southernmost basins, such as the Gulf of Aqaba, are currently forming under 2–5° transtension, compared to pure strike-slip in the north, thereby allowing a direct comparison of the size and fault geometries in the two settings. As in sandbox experiments, the transtensional

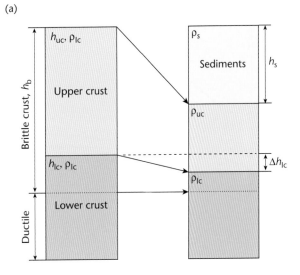

(a)

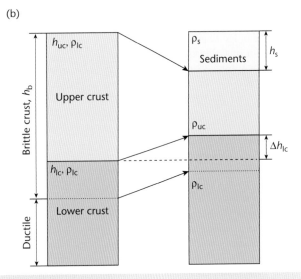

(b)

Fig. 6.31 Contrasting isostatic results for (a) brittle brick stretching (BBS) model where the lower boundary of the brittle layer is located within the lower crust and is fixed because of the high viscosity of the lower crust, and (b) the lower boundary of the brittle layers is located within the lower crust and complete Airy isostatic compensation is achieved. Note the difference in thickness of sediments in the basin. Reprinted from Petrunin & Sobolev (2008, p. 396, figs 12 and 13), with permission from Elsevier. Subscripts uc and lc refer to upper crust and lower crust. See text for discussion.

Gulf of Aqaba pull-apart basin is larger in surface area (both length and width) compared to the pure strike-slip main Dead Sea Basin, and depocentres are at both ends rather than centrally located inside a terraced basin margin (Fig. 6.35).

6.4.3 Application of model of uniform extension to pull-apart basins

The local extension in pull-apart basins causes very rapid subsidence and sediment accumulation, over 10 km of sediments accumulating

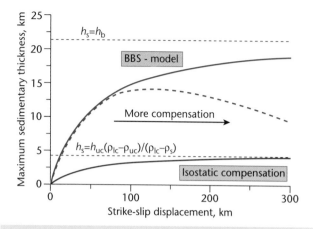

Fig. 6.32 Comparison of maximum sediment thickness versus displacement length for the brittle brick model, the full compensated model and (dashed) observations from the Dead Sea area. Reprinted from Petrunin & Sobolev (2008, fig. 15) with permission from Elsevier.

in a period of less than 5 Myr in some examples such as the Miocene Ridge Basin of California (Link & Osborne 1978). It is debatable whether the subsidence of pull-apart basins can be approximated by a model of lithospheric extension (McKenzie 1978a, §3.3.1), in which fault-controlled subsidence is followed by a thermal subsidence phase. Late Paleozoic pull-apart basins of eastern Canada (such as the 100×200 km Magdalen Basin, Gulf of St. Lawrence area) are thought to have undergone a rift, then thermal subsidence history (Bradley 1983). However, the generally small size of pull-apart basins implies that lateral temperature gradients must be large. Lateral heat conduction to the basin walls becomes an important source of heat loss; the narrower the basin the greater the lateral cooling. The critical basin width below which lateral heat loss to the sides becomes important appears to be about 100 km (Steckler 1981) to 250 km (Cochran 1983). In comparison, most strike-slip basins are 10–20 km in width. Since narrow basins cool rapidly during extension, the subsidence is greater than predicted for the rift stage by the uniform extension model (§3.3.1). Rapid early subsidence related to lateral heat loss may help to explain the sediment starvation and deep bathymetries characteristic of the early phases of pull-apart basin development. The subsequent post-rift thermal subsidence should be correspondingly less. The Tekirdag depression along the North Anatolian Fault has been active in the Pliocene–Quaternary and has a basin floor at maximum depths of 1150 m, suggesting that basin subsidence has far outstripped sediment supply for the last <4 Myr (Okay *et al.* 1999). The Ridge Basin of California was initiated in the late Miocene and was initially occupied by a deep-marine embayment before continental conditions prevailed in the Pliocene (Link & Osborne 1978). These two examples therefore support the idea of early underfilled stages in pull-apart basins, caused by high subsidence rates that outpace sediment supply.

6.4.4 Pull-apart basin formation and thin-skinned tectonics: the Vienna Basin

It has previously been observed that the common block rotations in zones of strike-slip deformation require a detachment beneath the

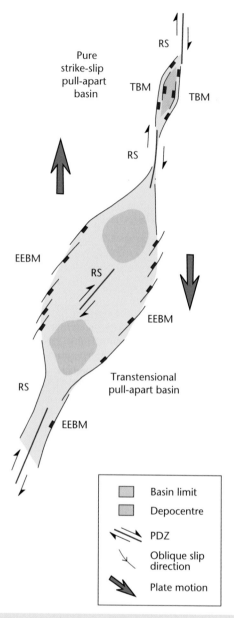

Fig. 6.33 Pull-apart basins in a dextral strike-slip system based on sandbox experiments, showing at top a small basin bordered by a narrow terraced margin (TBM) in the case of pure strike-slip, and at bottom a larger basin with two depocentres and a wide margin with en echelon normal faults (EEBM) for the case of trantension. RS, Riedel shears. After Wu *et al.* (2009, Fig. 11), reprinted with permission from Elsevier.

brittle upper crust. Upper crustal faults are believed to sole out into these detachment zones, which may be rheologically weak zones in the mid-lower crust, or may be related to lithospheric extension or to lithospheric convergence. The stretching of the upper crust without associated mantle upwelling is suggested by the generally low surface heat flows of many strike-slip zones. The narrowness of strike-slip basins, and of the topographic features associated with strike-slip deformation, indicate that they are essentially uncompensated isostatically. Such cases can be referred to as thin-skinned

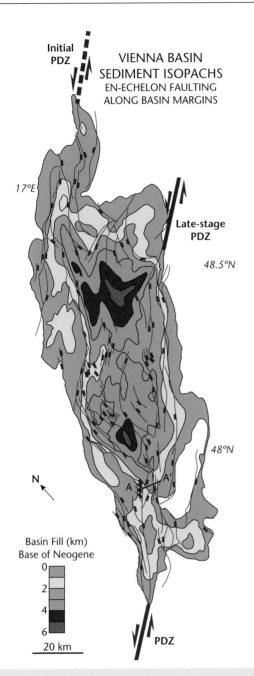

Fig. 6.34 Isopachs of Neogene fill of Vienna Basin, and faults, showing dual depocentres and *en echelon* faulting along basin margin, typical of transtensional strike-slip motion. Cross-section along transect A-A' is shown in Fig. 6.11c. From Hinsch *et al.* (2005), in Wu *et al.* (2009, Fig. 9), reprinted with permission from Elsevier.

stension (Figs 6.34, 6.36). The NE–SW trending major fault systems active during sedimentation in the Vienna Basin do not significantly disrupt the underlying autochthonous cover rocks of the European plate and appear to pass into flat detachments (at c.10 km depth) using older thrust planes (Royden 1985). Since the extension is thin-skinned, the lower crust and mantle are unaffected by the extension near the surface. This is supported by a number of lines of evidence: (i) the uniformly low heat flows of 45–60 mW m^{-2} through the Vienna Basin and adjacent regions (Cermak 1979; Royden 1985); (ii) subsidence curves derived from boreholes show no recognisable thermal subsidence phase; and (iii) low thermal gradients and low levels of organic maturity in basin sediments.

Other pull-apart basins are situated in regions where the lithosphere has been thinned. These basins have high geothermal gradients, high surface heat flows, and correspondingly high levels of organic maturity in the basin-fill. The Salton Trough area of the northern Gulf of California is an example. Pull-apart basins with mantle involvement are likely to behave more like extensional basins than their thin-skinned counterparts.

Within the Carpathian orogenic system, the Vienna Basin is situated near the thrust front, and the sinistral strike-slip deformation is related to the relative movement between active and inactive nappes in the Alpine-Carpathian belt. As the distance from the thrust front increases, extension may reach progressively greater depths, ultimately affecting the entire lithosphere. This may explain why the Pannonian Basin, situated far to the south of the arcuate Carpathian suture, involves mantle lithosphere extension and is situated above elevated asthenosphere (Fig. 6.36). The Pannonian Basin therefore experiences elevated surface heat flows, in contrast with the thin-skinned Vienna Basin located close to the orogenic thrust front.

6.5 Characteristic depositional systems

The sedimentary fills of strike-slip basins have a number of features in common (Miall 2000). Basin geometries are deep but relatively narrow, with high syndepositional relief causing conglomerates and breccias to be banked up against faulted basin margins. Sedimentation rates are rapid. Lateral facies changes are also rapid, so that marginal breccias may pass laterally directly into lacustrine mudstones. Fault movements cause syndepositional unconformities to form in individual basins and different stratigraphies to develop in closely adjacent basins, making correlation difficult. Basin sediments are commonly offset from their source, as may be proved by a mismatching between size of depositional system and drainage area, or between the petrography of basin sediments and that of hinterland geology. In modern basins there may be offsets of geomorphological features such as rivers, alluvial fans or submarine canyons.

The best-known intracontinental transform is the San Andreas system, and one of the best-documented pull-apart basins in this system is *Ridge Basin*, California. It shows many of the elements indicated above. The basin was initiated in the Miocene and continued to accumulate sediment during the Pliocene, after which it was uplifted. It contains over 13.5 km of sediment (maximum stratigraphic thickness, but not the true thickness measured in any one location) deposited at an estimated rate of 3 mm yr^{-1}, and is located between the San Andreas and San Gabriel faults (Fig. 6.37). During the late Miocene the San Gabriel Fault was a major active strand of the San Andreas system. The Ridge Basin formed to the east of a

strike-slip basins. Thin-skinned strike-slip basins should exhibit a distinctive subsidence and thermal history compared to basins involving mantle upwelling.

The 200 km by 60 km Vienna Basin contains up to 6 km of Miocene sedimentary rocks and formed adjacent to the coeval Carpathian thrust belt. The Vienna Basin is an excellent example of a rhombohedral pull-apart that developed on top of the allochthonous thrust terranes of the Alpine-Carpathian system in a background environment of lithospheric shortening, but in a local environment of tran-

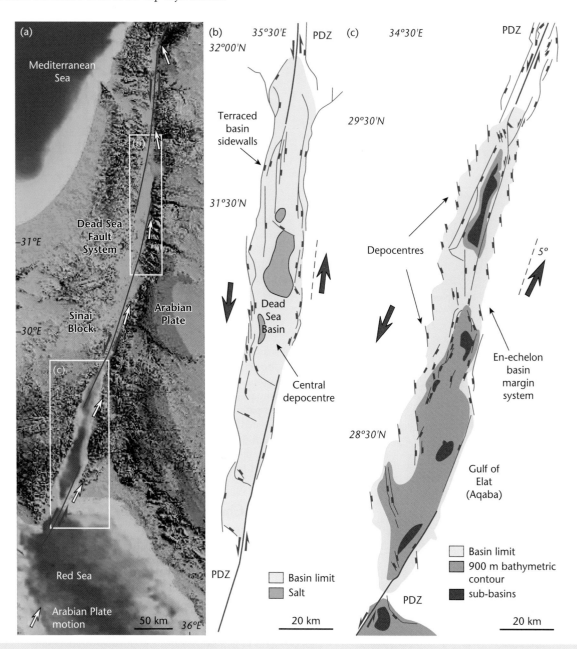

Fig. 6.35 Pull-apart basins in the Dead Sea Fault system. (a) Digital elevation model showing topography and main faults. The Dead Sea pull-apart basin in the north (b) is pure strike-slip, whereas the Gulf of Elat (Aqaba) in the south (c) is transtensional with $\alpha = 5°$ (Wu *et al.* 2009, p. 1618, fig. 10). Reprinted with permission from Elsevier.

releasing bend in the fault. During the late Miocene–Pliocene over 60 km dextral strike-slip took place along the San Gabriel Fault, but in the Pleistocene slip was transferred to the San Andreas Fault along the northeastern flank of the basin.

The sedimentary fill of the basin (Crowell & Link 1982; Link 1982) is made up of a basal non-marine unit (Mint Canyon) overlain by the 2.2 km-thick upper Miocene Castaic Formation, consisting of marine mudstones and turbidites. The Castaic Formation is overlain by the 9–11 km-thick, mostly non-marine Ridge Basin Group, with marine deposits in the lowermost 600 m. The Ridge Basin Group comprises marginal breccias along the active western fault scarp

(Violin Breccia), central lacustrine deposits, chiefly mudstones (8 km), of the Peace Valley Formation, fluviatile clastic wedges of the Ridge Route Formation (9 km) along the eastern margin of the basin, and a basin-wide, final basin filling of alluvial sands and gravels (1.1 km) of the Hungry Valley Formation. The marginal alluvial cones and talus of the Violin Breccia were derived from the SW and pass very rapidly (within 1.5 km) into lacustrine shales and siltstones of the Peace Valley Formation, or sandstones of the Ridge Route Formation. The thick clastic wedges of the Ridge Route Formation were shed from source areas to the NE of the basin, but the younger Hungry Valley Formation was derived from the N, NW and W. This

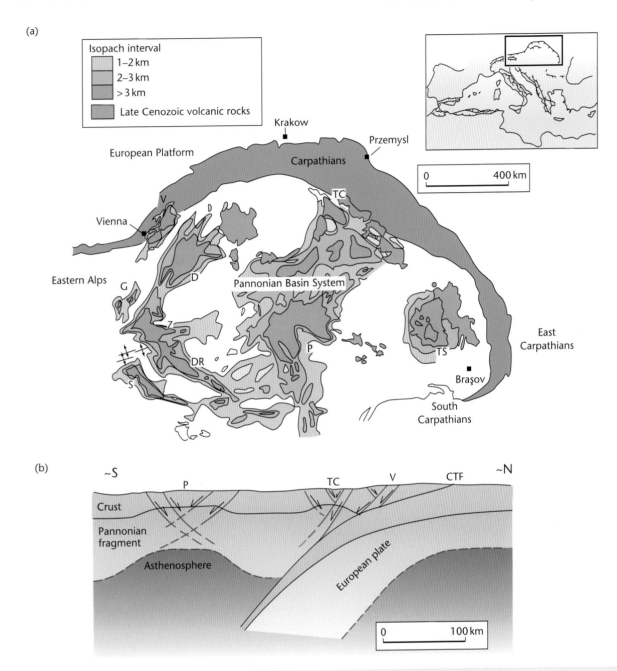

Fig. 6.36 Map (a) and schematic cross section (b) to show the relationship between the Vienna Basin to other basins in the Carpathian-Pannonian system. The Vienna Basin is situated at a left step (releasing bend) in a sinistral strike-slip system accommodating relative movement between active and inactive nappes in the Alpine-Carpathian orogenic system. The Vienna and Transcarpathian basins are located on the leading (thin) edge of the Pannonian lithosphere and above the deflected European plate. The Pannonian Basin, however, is located entirely on the Pannonian lithosphere, where it overlies asthenosphere. Extension in the Pannonian Basin therefore involves mantle, and the basin is consequently 'hot' compared to the 'cool' Vienna Basin (Royden 1985; © SEPM Society for Sedimentary Geology 1985). CTF, Carpathian Thrust Front; V, Vienna Basin; P, Pannonian Basin; TC, Transcarpathian Basin; TS, Transylvanian Basin; S, Sava Basin; DR, Drava Basin; D, Danube Basin; G, Graz Basin; Z, Zala Basin. © SEPM Society for Sedimentary Geology (1985).

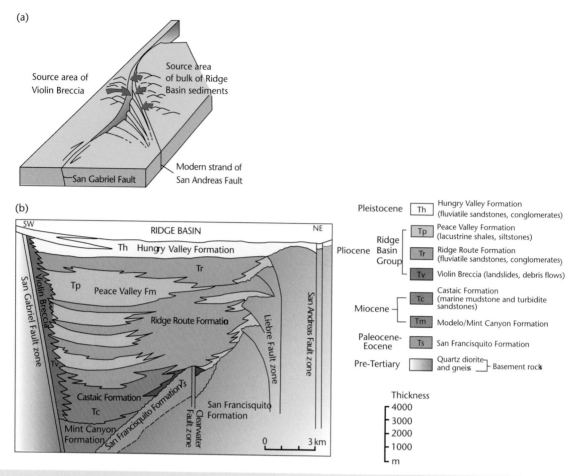

Fig. 6.37 Ridge Basin, California. (a) General tectonic and depositional setting for the Ridge Basin as a pull-apart on the releasing bend of the San Gabriel Fault (Crowell 1974b). (b) Generalised cross-section showing the sedimentary facies and principal faults. The fine-grained lacustrine depocentres (Peace Valley) and landslide/debris flow breccias (Violin Breccia) occur close to the San Gabriel Fault zone (after Link & Osborne 1978 and Crowell & Link 1982).

demonstrates the complexity of sourceland switching in strike-slip basins. Dextral strike-slip along the San Gabriel Fault displaced the source region for the Violin Breccia northwestward with time. As a result, the successive alluvial fans that form the Violin Breccia become younger northwestward, and form an overlapping or shingled pattern. Within the axial part of the basin, sediments were transported southeastward down the axis of the basin concurrently with northwestward migration of the depocentre.

The Ridge Basin has been compared, in terms of its structural and sedimentological development, with the larger Hornelen Basin, Norway, and the smaller Little Sulphur Creek group of basins, southern California (Nilsen & McLaughlin 1985). Each basin is characterised by marginal fans located tight up against the active strike-slip fault, axial lacustrine facies and streamflow-dominated fans along the opposite margin. These streamflow-dominated fans contribute most of the sediment to the basin, sometimes filling the basin completely and spreading alluvial deposits across to the active fault scarp to interfinger with the talus fans. Rates of deformation, catchment and fan development are commonly highly asymmetrical in basins of this type (Allen & Hovius 1998).

The present-day submarine equivalents of the Ridge Basin are found in the California Borderland basins. This area, to the west of the San Andreas Fault, is underlain by an arc complex formed during subduction of the Pacific plate in the Mesozoic to early Cenozoic (Howell & Vedder 1981). A large number of small basins filled with submarine fans formed during the Paleogene, and Oligocene dextral strike-slip faulting fragmented the region into *en echelon* ridges and rhomboidal basins (Fig. 6.38). Sedimentation is

Fig. 6.38 Strike-slip basins of the California Borderland (Moore 1969; Junger 1976). (a) Map view. (b) Interpretations of seismic reflection profiles (Howell *et al.* 1980), vertical exaggeration ×10. Q, Pleistocene and Holocene sediments; Tp, middle and late Pliocene sediments; Tpm, Miocene and early Pliocene sediments; Tm, early to middle Miocene sediments; Tmo, cherty, calcareous and siliceous shale of late Oligocene to middle Miocene; Tmv, Miocene volcanics; Kl-To, Late Cretaceous to Oligocene sediments. Reproduced with permission of John Wiley & Sons, Ltd.

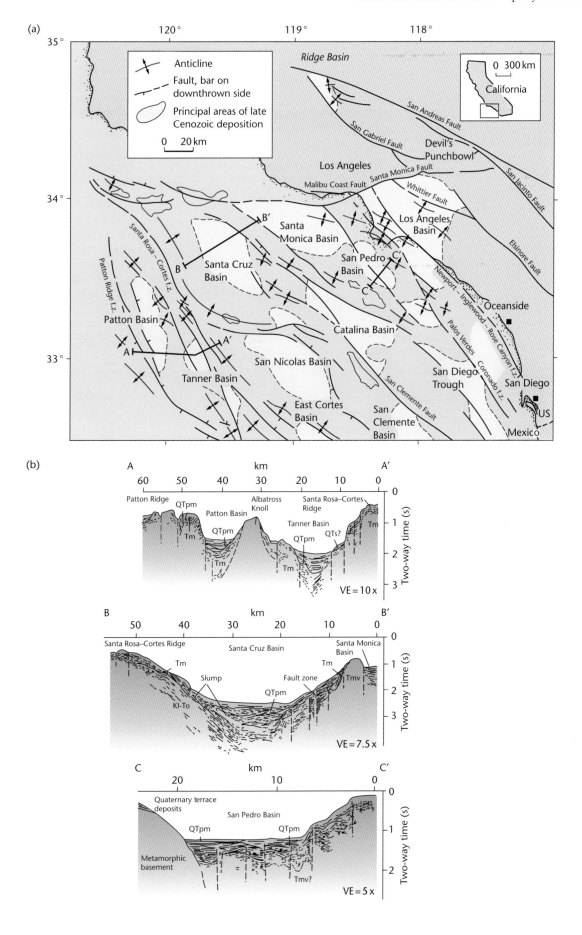

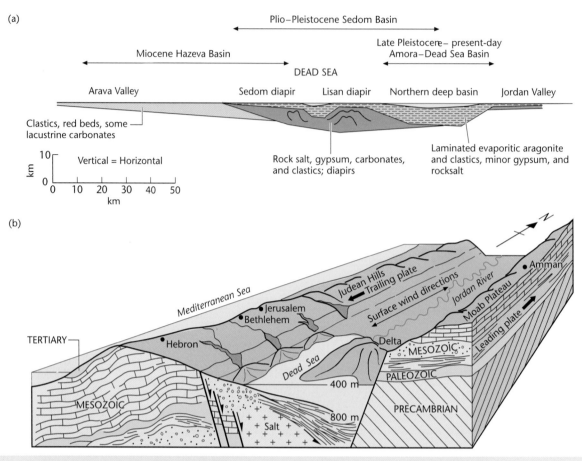

Fig. 6.39 Cross section (a) and 3-D view (b) of the Dead Sea-Arava depression. (a) The northward migration of the basin depocentres. Early Miocene (25–14 Ma) strike-slip of about 60–65 km opened up the Arava Basin, which was filled with about 2 km of red beds during a pause in the strike-slip displacement. Later movement in the last 4.5 Myr has allowed the deposition of >4 km of marine to lacustrine rock salt of the Sedom Formation, followed by 3.5 km of lacustrine evaporitic carbonates and clastics (after Zak and Freund 1981). Reproduced with permission from Elsevier. (b) Today, sediment enters the northern basin via high discharge ephemeral streams feeding fan deltas along the semi-arid western edge of the basin (Judean Hills), and the perennial, axial-flowing Jordan River. The Jordan has a drainage area in a humid region, and deposits a fine-grained delta at its mouth. Autochthonous supply to the lake is of aragonite and gypsum precipitates (Manspeizer 1985).

dominated by turbidites fed from the nearby American coast into a background environment of fall-out of fine-grained terrigenous and pelagic material. The region has several well-studied deep-sea fans (e.g. La Jolla, Navy) characterised by important submarine slides (e.g. Sur submarine slide on Monterey fan, described by Normark & Gutmacher 1988). Sedimentation rates increase towards the coast, so that the Los Angeles and Ventura basins located close to the continental source contain as much as 8 km of Neogene sediment, whereas the more offshore basins such as the Tanner and Patton basins (Fig. 6.38) have much reduced sedimentary thicknesses (Howell *et al.* 1980).

The sedimentation in a classic pull-apart in an arid climate is well illustrated by the Dead Sea. Movement along the Dead Sea Fault

commenced in the Miocene in response to the opening of the Red Sea. It has continued to move up to the present day. The basin contains over 10 km of fluvial clastics, lacustrine limestones and evaporites. Like the Ridge Basin, depocentres have moved considerably, producing a highly diachronous fill. The Miocene Hazeva Formation consists of continental clastics and some lacustrine carbonates and is found in the Arava Valley in the south. The Pliocene–early Pleistocene Sedom Formation consists mainly of lacustrine salts, gypsum, carbonates and some clastics, and occurs in the central section. The Pleistocene to Recent Amora and younger formations consist of laminated evaporitic (gypsum) and aragonite sediments which are accumulating today in the modern Dead Sea in the northern sector of the basin (Fig. 6.39) (Zak and Freund 1981).

PART 3

The sedimentary basin-fill

CHAPTER SEVEN
The sediment routing system

Summary

The sediment routing system is the integrated, dynamical system connecting regions of erosion, sediment transfer, temporary storage and long-term deposition, from source to sink. Sediment routing systems, ancient and modern, are characterised by complex internal dynamics, but are also sensitive to changes in external mechanisms such as climate and tectonics. They typically comprise a number of *segments*, or compartments, each governed by a distinctive set of erosional, transportational and depositional processes. The zones between segments involve complex sediment transfers and may act as staging areas for sediment storage prior to down-system redistribution. The storage of sediment in 'transient states', such as alluvial valleys and shelf-edge prisms, increases the transit time of sediment through the sediment routing system segment, and potentially buffers signals propagated from source to sink.

Particulate sediment is fed from source to sink from upland regions undergoing weathering, regolith production and erosion, accompanied by the release of solutes. The regolith is a zone of physical mobility and biologically mediated chemical changes, and ranges in thickness from zero to >100 m. The bulk of particulate sediment and dissolved loads is transported in the hydrological cycle, principally by run-off. The flux of sediment, or sediment yield, varies strongly from place to place on the Earth's surface, and also varies strongly depending on the size of the area monitored and the time scale of the observations. Very few rivers are in a pristine state, that is, without interference by the activities of humans. The modelling or prediction of sediment yields in the erosional engines of sediment routing systems is consequently difficult. Dissolved loads depend on weathering reactions with the geological substrate, but also on inputs from precipitation and changes caused by evaporation. The chemistry of river water entering the oceans profoundly affects global biogeochemical cycles, especially the long-term carbon cycle.

Erosion rates at a point can be assessed using a range of techniques based primarily on the measurement of the thermal history of minerals and detrital grains. Thermochronometric methods allow the cooling history of a mineral or grain to be determined, and converted to long-term erosion rate using a geothermal gradient. Catchment-averaged erosion rates can be estimated from the abundance of cosmogenic nuclides in bulk stream samples. In basin analysis, there are strong limitations on the accurate estimation of catchment-averaged erosion rates in paleo-catchments that have since disappeared, and therefore considerable uncertainty in how to approximate sediment effluxes of upland catchments in geological sediment routing systems.

The erosional engine of sediment routing systems is dominated, in non-glaciated regions, by the coupled process systems of hillslopes and channels. In regions of vigorous erosion, hillslopes are dominated by landsliding. Hillslope evolution can be modelled as a diffusive process, where the diffusivity depends on climate, vegetation and biological factors, and rock and regolith strength. Valley spacing is set by the rate of tectonic uplift of rock and rate of precipitation, so valleys are deeply entrenched but narrowly spaced on the wet, tectonically highly active western flank of the Southern Alps of New Zealand, but broad and less entrenched on the dry, less tectonically active eastern flank. Rates of bedrock incision by rivers are commonly treated as controlled by a discharge-slope product, termed the *stream power rule*.

Long-range sediment transport through alluvial rivers approximates uniform flow down an inclined plane, which causes a shear stress to be exerted on the bed, resulting in sediment transport. Long-range transport and deposition in rivers can be treated diffusively, with the diffusivity determined by river power and the intermittency of sediment-transporting events. There is commonly a sharp transition along an alluvial river between a gravel-rich zone and a sand-rich zone, termed the gravel front. The dispersal and preservation of gravel in sedimentary basins is a sensitive indicator of tectonic activity. A key concept in understanding the delivery of sediment from the erosional engine to the depositional sink of the basin is the time scale of the frequency of the external forcing relative to the diffusive time scale of the basin, or equilibrium time. Large alluvial systems may buffer high-frequency sediment discharge signals, with important implications for stratigraphic architecture.

Sediment is transported by rivers to the ocean, so the coastal zone is a moving boundary that migrates in response to sediment discharge fluctuations, subsidence and eustatic change. The coast and shelf is typically a staging area that feeds sediment into down-system sinks by processes such as longshore drift, offshore advection by plumes, tides, waves and storms, and sediment gravity flows. Sediment accumulates over long time periods in a range of continental and marine depositional systems dependent on the availability or creation of accommodation.

Sediment is selectively extracted from the mobile surface flux at the Earth's surface to build stratigraphy. This selective extraction causes a characteristic down-system reduction in grain size. The down-system trend is diagnostic of the probability density function of grain size in the sediment supply, the magnitude of the sediment discharge and the distribution of accommodation in the basin.

Basin Analysis: Principles and Application to Petroleum Play Assessment, Third Edition. Philip A. Allen and John R. Allen.
© 2013 John Wiley & Sons, Ltd. Published 2013 by John Wiley & Sons, Ltd.

The sediment routing system concept requires that deposited sediment is traced back to erosional source regions, which can be done using a range of thermochronological, mineralogical and geochemical tracers. The time period between the cooling through a characteristic isotherm and the stratigraphic age of the detrital grains, known as the lag time, is useful in understanding the functioning of sediment routing systems. The footprint or 'fairway' of the sediment routing system, if mapped, gives critical information on the volumetric budget over specified geological time intervals.

Erosional and depositional landscapes respond transiently when perturbed by external mechanisms such as tectonics and climate. This transient behaviour has been studied particularly in coupled catchment-fan systems and in regions of evolving extensional fault arrays. Response times for the achievement of steady-state conditions following a perturbation are commonly c.10^6 yr. When allied with a downstream-fining model, increases in slip-rate on basin border faults are recognised by a rapid fining and retrogradation, followed by a slow recovery of progradation and coarsening, producing a wedge-like stratigraphic unit. In contrast, step change increases in precipitation result in thin sheet-like gravel bodies extending far into the basin. Many stratigraphic trends may reflect similar transient behaviour in response to different types of external forcing.

The magnitude of the sediment discharge into a basin (Qs) relative to the space made available for sediment deposition (accommodation) controls the extent of basin under- or over-filling and gross depositional environments, but also influences sediment dispersal pathways. At low Qs, sediment underfills predominantly marine or lacustrine troughs and commonly flows axially from spaced input points controlled by tectonic structures. At intermediate Qs, tectonic structures continue to deflect and guide sediment dispersal pathways, but basin compartments are commonly filled. At high Qs, sediment discharges bury active or defunct tectonic structures, and sediment dispersal is transversely down the regional paleoslope.

7.1 The sediment routing system in basin analysis

The sedimentary character of basins is determined by the balance between tectonic subsidence causing the generation of long-term accommodation and sediment influx (see also Chapter 8). Sedimentary basins should not therefore be regarded as passive receptacles, but instead as dynamic entities occupied by complex physical and biological process systems. This chapter concerns the geomorphological and biogeochemical process systems operating in source regions of sediment production and export, as well as in the depositional sinks of sedimentary basins. These dynamic environments connected by the transfer of mass fluxes of particulate sediment and dissolved loads from source to sink are termed *sediment routing systems*.

Sediment routing systems, ancient and modern, are integrated dynamical systems connecting erosion in mountain catchments to downstream deposition (Allen 1997; Somme *et al.* 2009; Allen & Heller 2012) (Fig. 7.1). They represent a vigorous way in which the Earth recycles mass, and their dynamics are fundamental to the global response to mountain building and climate change (Whipple 2009), as well as providing thoroughfares for the transmission of chemical signals from mountains to the ocean (Meybeck 1987; Hay 1998; Galy *et al.* 2007). Sediment routing systems therefore participate strongly in many geochemical cycles, such as that of carbon, including the drawdown of atmospheric CO_2 mediated by rates of silicate weathering (Raymo & Ruddiman 1992), the delivery of particulate organic carbon to the ocean and its removal from the short-term carbon cycle by rapid burial in deltas and sediment fans (Galy *et al.* 2007; France-Lanord & Derry 1997) and the catalysing of ocean anoxia in sheltered and semi-enclosed seas by changes in freshwater discharge by rivers (Beckmann *et al.* 2005).

The upland catchments acting as source regions for sediment release mass fluxes into the transportational and depositional regimes of sedimentary basins (Figs 7.1, 7.2). But this release is neither uniform nor steady. In particular, the sediment and solute effluxes of upland catchments are time-dependent or *transient* in relation to the forcing mechanisms for environmental change, and signals are transformed by propagation through sediment routing systems with their own internal dynamics. Understanding these transformations is a major goal of Earth surface dynamics research and a key to the interpretation of stratigraphy (Chapter 8).

Sedimentary basins accommodate particulate sediment to build stratigraphy by a selective extraction from surface fluxes. This selective extraction from surface fluxes leaves the downstream flux depleted in coarser grain sizes, thereby generating a distinctive downstream-fining trend (Fedele & Paola 2007; Duller *et al.* 2010; Whittaker *et al.* 2011) and depositional styles dependent on the resulting grain-size mix (Strong *et al.* 2005).

The goal of the basin analyst is to identify in surface outcrops or in borehole stratigraphy the different sediment routing systems comprising a sedimentary succession. These paleo-sediment routing systems are building blocks of stratigraphy in the dynamical sense that *teleconnections* exist coherently within them (Hoth *et al.* 2008). That is, a change in climate in a sediment source region may be transmitted as a signal of discharge through the sediment routing system and is manifested in a change in depositional style, such as bed thickness, geometry, or grain size. Secondly, depositional products may be linked to a distinct area of provenance in terms of tracers, such as those provided by geochemistry, mineralogy, geochronology, thermochronology or petrography. And, thirdly, sediment routing systems are amenable to a volumetric or mass balance, since the system is closed in terms of the sediment budget. A 'footprint' or 'fairway' therefore links component parts of the sediment routing system.

The quantification of ancient sediment budgets is critical in explaining and predicting gross depositional environments and large-scale stratigraphic trends in depositional basins (Marr *et al.* 2000; Strong *et al.* 2005), in constraining geodynamic models of topographic growth and decay (Willett & Brandon 2002) and in evaluating the impact of climate change on Earth surface processes (Whipple 2009). Depositional sinks may be sourced from one catchment or from several, causing a mixing of provenance indicators. Mixing of sediment fluxes may also take place where a supply from an onshore catchment is joined by a longshore supply from a littoral cell (Covault *et al.* 2011) (Fig. 7.3).

The stratigraphic archive of sediment routing systems may consist of stratigraphic cycles forced externally or driven by internal dynamics. In addition, most sediment routing systems are divided into

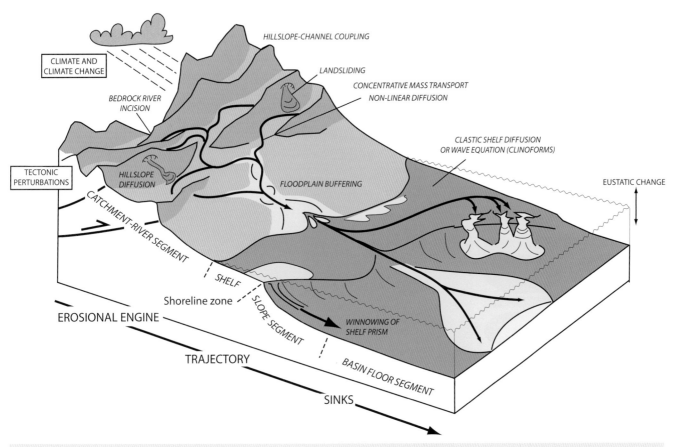

Fig. 7.1 Concept of the sediment routing system from source to sink. The sediment routing system is driven and perturbed by tectonics and climate. It comprises a number of morphodynamic, mutually coupled subsystems or segments. The erosional engine feeds sediment to a predominantly transportational system (the trajectory) and to long-term sinks. From Allen & Heller (2012, p. 113, fig 6.2), modified from Somme et al. (2009).

segments or *compartments* characterised by a distinct set of physical processes (Somme *et al.* 2009; Carvajal & Steel 2012), separated from other segments by moving boundaries or energy 'barriers' or 'fences' (Fig. 7.2). An example is the boundary between the terrestrial river-dominated segment and the continental shelf segment, with a moving boundary represented by the coast and its deltas (Swenson *et al.* 2000; Swenson *et al.* 2005; Kim *et al.* 2006). These energy fences are potentially strong filters or barriers for sediment flux signals generated in upland catchments. Each segment may possess zones of temporary storage of sediment, as in the continental shelf prisms of the Golo system of eastern Corsica (Somme *et al.* 2011)(Fig. 7.4) or even permanent storage on a geological time scale, depending on the pattern of tectonic subsidence and accommodation generation. We term these *transient states* and *absorbing states* respectively (Malmon *et al.* 2003; Allen & Heller 2012) (Fig. 7.2).

The structure of this chapter is to initially consider the processes acting in the erosional engine. These processes cause regolith production by weathering and removal by erosion. Distinctive erosional landscapes dominated by hillslopes and downcutting channels reflect the climate, vegetation, bedrock lithologies and tectonic forcings affecting upland catchments, and are collectively responsible for the volume of sediment released, its probability density function of grain size, as well as the spacing of outlets into depositional basins.

The development of landscape as a complex response to ongoing tectonics is investigated in the context of gravel-rich systems, which are known to be particularly sensitive recorders of tectonic change (Jordan *et al.* 1988; Allen & Heller 2012). Catchment-fan systems are particularly well developed in the extensional tilted fault block landscapes of the Basin and Range province, USA (Allen & Densmore 2000; Densmore *et al.* 2007), but also along the active margins of foreland basins in regions of contractional mountain building (Arenas *et al.* 2001; Lopez-Blanco 2002; Jones *et al.* 2004; Charreau *et al.* 2009; Barrier *et al.* 2010).

Having investigated the functioning and sensitivity of sediment routing systems, the reader is in a better position to appreciate the full set of parameters controlling stratigraphic architectures and variability of depositional style in the basin-fill discussed in Chapter 8.

7.2 The erosional engine

7.2.1 Weathering and the regolith

Weathering is the decay and disintegration of rock *in situ* at the Earth's surface. A weathering mantle forms between pristine bedrock

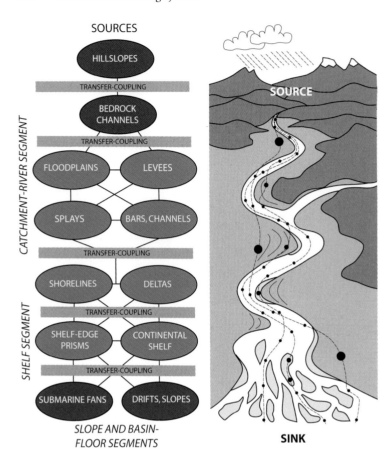

SOURCES

HILLSLOPES

TRANSFER-COUPLING

BEDROCK CHANNELS

TRANSFER-COUPLING

FLOODPLAINS   LEVEES

SPLAYS   BARS, CHANNELS

TRANSFER-COUPLING

SHORELINES   DELTAS

TRANSFER-COUPLING

SHELF-EDGE PRISMS   CONTINENTAL SHELF

TRANSFER-COUPLING

SUBMARINE FANS   DRIFTS, SLOPES

CATCHMENT-RIVER SEGMENT

SHELF SEGMENT

SLOPE AND BASIN-FLOOR SEGMENTS

SOURCE

SINK

Fig. 7.2 Right: sediment is transferred from a source region to a sink along trajectories (dashed lines). Some trajectories involve short times with brief periods of storage in the sediment routing system (small circles) whereas others involve long transit times with prolonged periods of storage in transient states (large circles). Long periods of storage of sediment in transient states implies buffering of sediment supply signals. Left: sources (orange), sites of temporary storage (transient states) in green and sites of permanent storage (absorbing states) in blue. The transfer of sediment between macrogeomorphic segments may be complex and involve dynamic feedbacks. From Allen & Heller (2012, p. 112, fig 6.1).

and the land surface, known as *regolith*. Water moves up and down in the regolith by capillary action and gravity respectively (Fig. 7.5a), generating a typical hillslope weathered profile (Fig. 7.5b). The presence of an enveloping and reactive moisture film is critical to the weathering of bedrock and regolithic materials.

The first indication of weathering of bedrock is the presence of *fractures*, from which chemical weathering spreads out. Rock becomes increasingly chemically weathered and fractured and becomes mobilised by downslope gravitational processes, commonly mediated by the activities of animals and plants. Strongly weathered rock where the original rock structure is recognisable is called *saprolite*, whereas when the weathered material has been more strongly moved, it is referred to as *mobile regolith*. The mobility or strain of regolith increases upwards from zero in unaltered bedrock to a dilated, expanded volume in the mobile regolith. The density of fractures also increases upwards, suggesting that they are part of the expansion process.

Fracturing of rock during weathering reduces particle size, enhancing transportability, and reduces bulk rock strength. Fractures also serve as routes for air and water to pass through rock, and increase the surface area exposed to chemical weathering. Fractures and cracks commonly result from expansion and contraction associated with near-surface temperature changes. In tectonically active regions, fractures also result from deformation at depth, so that rock may arrive at the surface pre-fractured, where the distribution of fractures may have an important impact on weathering and erosion (Molnar *et al.* 2007).

7.2.1.1   Chemical weathering

Weathering is conventionally divided into mechanical and chemical categories. Mechanical weathering involves: (i) processes causing a volumetric change in the rock mass, including exfoliation, insolation and hydration weathering; and (ii) processes causing a volumetric change through the introduction of material, commonly water, but also salts, into pores, void spaces and fissures in the rock mass, such as freeze-thaw processes and salt weathering. Although these processes of mechanical weathering may be locally dominant, chemical weathering is of greater global importance. The fundamental importance of weathering, whether physical or chemical, for basin analysis is that it provides the starting particulate and dissolved material for release into sediment routing systems.

Chemical weathering involves the chemical breakdown of bedrock and the formation of new mineral products. A detailed treatment of the processes in chemical weathering is found in Anderson and Anderson (2010, p. 183–202), and a very brief overview is provided here. The main chemical weathering processes are:

- *Solution* involves the action of water as a solvent. For example, gypsum ($CaSO_4$) dissolves easily in water. The tendency of a mineral to dissolve in water is expressed by its *equilibrium solubility*. This is affected by the temperature and pH of the local environment. As an example, quartz (SiO_2) has a low but finite solubility below a pH of 10, and is highly soluble in very alkaline waters above this value. Alumina (Al_2O_3) is only soluble in

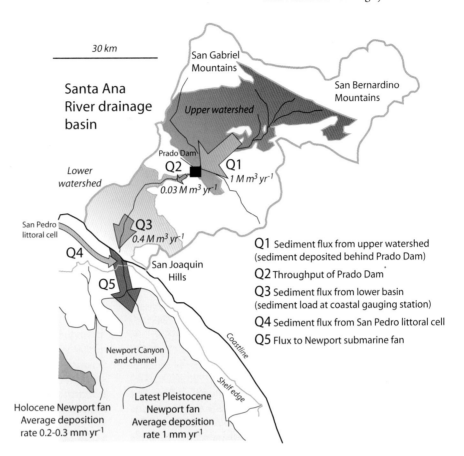

30 km

Santa Ana River drainage basin

San Gabriel Mountains

San Bernardino Mountains

Upper watershed

Lower watershed

Prado Dam

Q2

Q1
1 M m³ yr⁻¹

0.03 M m³ yr⁻¹

San Pedro littoral cell

Q3
0.4 M m³ yr⁻¹

Q4

Q5

San Joaquin Hills

Q1 Sediment flux from upper watershed (sediment deposited behind Prado Dam)

Q2 Throughput of Prado Dam

Q3 Sediment flux from lower basin (sediment load at coastal gauging station)

Q4 Sediment flux from San Pedro littoral cell

Q5 Flux to Newport submarine fan

Coastline

Shelf edge

Newport Canyon and channel

Holocene Newport fan Average deposition rate 0.2-0.3 mm yr⁻¹

Latest Pleistocene Newport fan Average deposition rate 1 mm yr⁻¹

Fig. 7.3 Sediment routing system of the Santa Ana River and Newport submarine fan, California, with sediment fluxes shown in terrestrial compartment and sediment accumulation rates in submarine sinks. Derived from Covault *et al.* (2010, p. 249, fig.1) and Warrick & Rubin (2007).

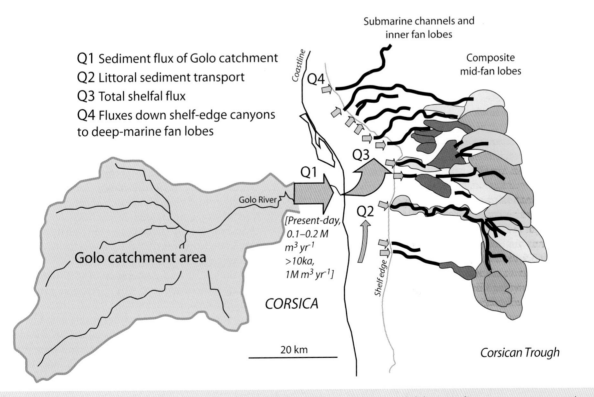

Submarine channels and inner fan lobes

Composite mid-fan lobes

Q1 Sediment flux of Golo catchment

Q2 Littoral sediment transport

Q3 Total shelfal flux

Q4 Fluxes down shelf-edge canyons to deep-marine fan lobes

Coastline

Q4

Q3

Q1

Golo River

Q2

Golo catchment area

[Present-day, 0.1–0.2 M m³ yr⁻¹ >10ka, 1M m³ yr⁻¹]

CORSICA

Shelf edge

20 km

Corsican Trough

Fig. 7.4 The Golo sediment routing system is fed from a catchment in eastern Corsica, and delivers sediment to a narrow continental shelf before export to deep-marine lobes, with age of lobes colour-coded (Somme *et al.* 2011). Reproduced with permission of John Wiley & Sons, Ltd.

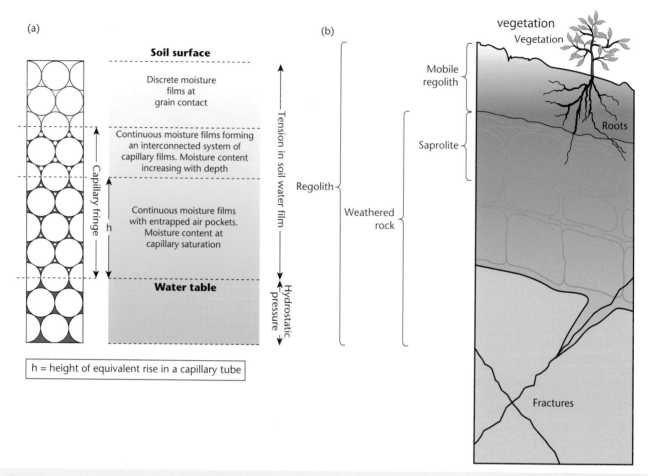

Fig. 7.5 (a) The zones of moisture in the regolith, showing the extent of the capillary fringe (after Carson 1969; reproduced with permission from Methuen/Routledge). (b) Schematic diagram of the weathered profile on a hillslope, showing a low-strength, mobile zone near the surface, often termed soil, which consists of materials that have been detached from the weathered rock below; the underlying weathered rock consists of saprolite (strongly weathered but immobile rock retaining the bedrock structure) above fractured, strong, less-weathered rock (after Anderson & Anderson 2010, p. 163; © Robert S. Anderson, Suzanne P. Anderson 2010, published by Cambridge University Press, translated with permission). Regolith structure therefore depends on strength and mobility.

conditions seldom found in nature, below a pH of 4 and above a pH of 9. As a result, alumina accumulates as a residue during weathering, whereas silica may be slowly leached. Calcium carbonate ($CaCO_3$), in contrast, has a steadily decreasing solubility in alkaline waters. However, the low solubility of $CaCO_3$ in pure water is rarely applicable in the natural environment because dissolved CO_2 in water causes $CaCO_3$ to be replaced by calcium bicarbonate $Ca(HCO_3)_2$, which is highly soluble (see 'Acid hydrolysis'). The solubility of a particular mineral is commonly affected by the presence of other ions from other sources. This makes calculation of the equilibrium solubility of mixtures of minerals and solutes difficult, as in the chemical weathering of common aluminosilicates that comprise much of the continental crust.

- *Oxidation and reduction* involves the gain or loss of charge by the addition (reduction) or loss (oxidation) of negatively charged electrons. The oxygen dissolved in water is the most common oxidising agent. Oxidation results in the formation of oxides and hydroxides, as in the oxidation of sulphides such as iron pyrite (FeS_2) under anaerobic conditions to produce sulphuric acid and iron hydroxide. The oxidation of organic matter in soils by bacteria produces CO_2 and therefore generates acidity. The acidity is then used in the hydrolysis of minerals (see 'Acid hydrolysis'). The tendency for oxidation and reduction to take place is indicated by the *redox potential* (Eh), measured in millivolts.

- *Hydration* involves the absorption of water into the crystal lattice, making it more porous and therefore more susceptible to weathering. A common example is the transformation of the iron oxide hematite to the hydrated iron hydroxide limonite.

- *Acid hydrolysis* is the reaction of a mineral with acidic weathering agents, where the acidity is mainly derived from the dissociation of atmospheric CO_2 in rainwater and soil zones by respiration of plant roots and bacterial decomposition of plants, producing in both cases carbonic acid (H_2CO_3). Hydrolysis involves the replacement of metal cations in the crystal lattice such as K^+, Na^+, Ca^{2+} and Mg^{2+} by the hydrogen or hydroxyl ions of water. The released cations combine with further hydroxyl ions, commonly to form *clay minerals*. Examples are the hydrolysis of albite (plagioclase feldspar $NaAlSi_3O_8$) to *kaolinite* ($Al_4Si_4)O_{10}(OH)_8$, and the hydrolysis of orthoclase feldspar ($KAlSi_3O_8$) to *illite* ($K_2Al_4(Si_6Al_2O_{20})(OH_4)$). Such reactions in general produce a clay mineral residue plus the release of silica, metal cations and bicarbonate ions in solution. Acid hydrolysis involving CO_2 is commonly termed *carbonation*. Carbonation dominates the weathering of limestones.

Chemical weathering of rock with the average composition of the continental crust (Taylor & McLennan 1995) can be compared with the solute loads of the world's rivers (Stumm & Morgan 1996) (see §7.2.2). The most abundant solute is bicarbonate, HCO_3^-, generated by the dissolution of limestone ($CaCO_3$) and the acid hydrolysis of aluminosilicates such as feldspars.

The mineralogical composition of the regolith is determined not just by the type and intensity of chemical weathering processes, but also by the parent bedrock. There are considerable differences in the way basalts and granites weather under the same climatic regime. Clay minerals are particularly diagnostic of both weathering processes and parent material. A primary factor is the extent of leaching, which strips minerals of their metal cations and eventually of their silicon and iron, leading to a stable aluminium-rich residue (gibbsite).

A large throughput of water is necessary for advanced stages of leaching, leading to regoliths dominated by gibbsite, kaolinite and aluminium oxides and hydroxides. This is only accomplished in regions of high precipitation rates such as the equatorial and humid tropical regions (Fig. 7.6). The process is commonly termed *laterisation*. Where leaching is less intense, the cations released by the breakdown of bedrock promote the formation of cation-bearing clay minerals such as those of the illite and smectite groups. These zones of moderate leaching typify the temperate zones of Asia, Europe and North America (Fig. 7.6). In arid and semi-arid climates, there are very low rates of chemical weathering and little accumulation of weathering products except for carbonates and salts as hard crusts and concretions.

The clay mineral assemblage is not only a function of climate, but also of position within the regolith profile. This is because the flux of water generally decreases with depth in the regolith, caused by a downward reduction in permeability. Residual minerals such as kaolinite and gibbsite are consequently found at the top of weathering profiles, whereas smectite and illite may be found at deeper levels in the same regolith where leaching is less intense.

The thickness of the regolith depends on the trade-off of two rate effects:

- rate of bedrock weathering, which is strongly controlled by climatic factors such as temperature and rainfall;
- rate of removal by denudation, which is controlled by climatic, tectonic and topographic factors (§7.3).

Regoliths can only attain great thicknesses (over 100 m) in localities with humid, warm climates and subdued topography. Even in these situations, regolith growth is self-limiting since the development of a thick, weathered mantle decreases permeability and reduces the throughput of water to the weathering front in contact with pristine bedrock. Although studies suggest that weathering rates are higher under some finite thickness of regolith than on bare bedrock, the rate of regolith production is commonly assumed to decay exponentially with regolith thickness (Anderson & Humphrey 1990):

$$E_w = k_w \exp(-m_w R) \qquad [7.1]$$

where E_w is the rate of descent of the bedrock–regolith weathering front, k_w is a coefficient representing the pristine bedrock weathering rate (units of $L\ T^{-1}$), m_w is a weathering rate decay constant (equal to $1/R_0$, where R_0 is the depth at which the weathering rate reduces to $1/e$ of its surface value k_w), and R is the regolith thickness. Under such circumstances of low bedrock weathering under thick regoliths, weathering activity may be restricted to intense leaching of the soil zone. The length scale R_0 is likely to be of the order of 1 m.

Regolith production rates are difficult to measure directly. Techniques in *cosmogenic radionuclide dating*, however, allow reliable estimates to be made. Cosmogenic radionuclides (such as ^{10}Be and ^{26}Al) are produced *in situ* by cosmic radiation, at a rate P_0 dependent on latitude and altitude (Lal 1991; Bierman 1994). The principle is that the production rate in a bare piece of bedrock, or in a sedimentary deposit such as a river terrace undergoing no erosion or deposition, P decays exponentially with depth below the surface y (Fig. 7.7), so that

$$P = P_0 \exp(-y/y^*) \qquad [7.2]$$

where y^* is a rate constant equal to 0.5–0.7 m for most common lithologies (that is, the production rate decays downwards to P_0/e in a distance y^*). The rate of change of concentration with time is then the production rate $P(t)$ minus the rate of decay λN

$$\frac{\partial N}{\partial t} = P(t) - \lambda N \qquad [7.3]$$

where N is the concentration of cosmogenic radionuclides per unit volume of rock, and λ is a decay constant analogous to a half-life.

Regolith shields the underlying bedrock from bombardment by cosmic rays, reducing radionuclide production rates at the regolith–bedrock interface. If the thickness of regolith H at a site is constant over time (losses at the surface balancing production of regolith from bedrock), the rock to regolith conversion rate is

$$E = \frac{\{P_0 \exp(-H/y^*)\} y^*}{N} \qquad [7.4]$$

where E is the bedrock lowering rate. Application of this technique, for example in the Wind River Range of USA (Small *et al.* 1999), showed that bedrock to regolith production rates were of the order of just $10 \times 10^{-6}\ m\ yr^{-1}$. This is much lower than typical erosion rates in tectonically active areas (§7.3).

In steep mountains with humid climates, although the rate of regolith production is high, it is removed by hillslope erosion. The combination of high regolith production caused by climatic factors, and high topographic slopes caused by tectonic factors, produces optimum conditions for the erosional sediment efflux of the landscape.

Since chemical weathering depends on temperature (as well as the other factors noted above), it is commonly assumed that chemical weathering rates for silicates are approximated by an Arrhenius-type relationship (White & Blum 1995):

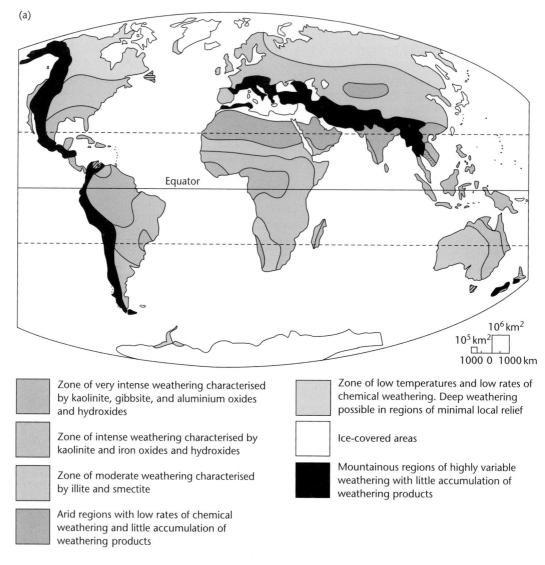

(a)

Equator

10^6 km^2
10^5 km^2
1000 0 1000 km

Zone of very intense weathering characterised by kaolinite, gibbsite, and aluminium oxides and hydroxides

Zone of intense weathering characterised by kaolinite and iron oxides and hydroxides

Zone of moderate weathering characterised by illite and smectite

Arid regions with low rates of chemical weathering and little accumulation of weathering products

Zone of low temperatures and low rates of chemical weathering. Deep weathering possible in regions of minimal local relief

Ice-covered areas

Mountainous regions of highly variable weathering with little accumulation of weathering products

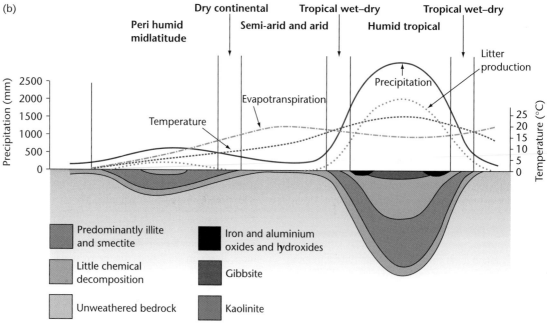

(b)

Peri humid midlatitude

Dry continental

Semi-arid and arid

Tropical wet–dry

Humid tropical

Tropical wet–dry

Litter production

Precipitation

Evapotranspiration

Temperature

Precipitation (mm)

Temperature (°C)

Predominantly illite and smectite

Little chemical decomposition

Unweathered bedrock

Iron and aluminium oxides and hydroxides

Gibbsite

Kaolinite

Fig. 7.6 The global pattern of weathering, modified from Strakhov (1967). (a) Map of major weathering zones. (b) Latitudinal zonation showing regolith thickness and type.

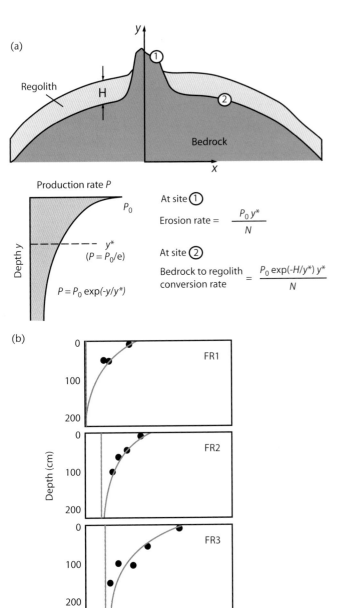

Fig. 7.7 (a) Calculation of bedrock erosion and regolith production rates using cosmogenic radionuclides (after Burbank & Anderson 2001; reproduced with permission of John Wiley & Sons, Ltd.). At site 1, bedrock erosion rate depends on the production rate of the radionuclide at the sample site P_0, the depth rate constant y^*, and the concentration of radionuclides per unit volume of rock N. At site 2, bedrock is converted to regolith at a rate that is also affected by the thickness of the regolith H. (b) Profiles of the concentration of ^{10}Be in three river terraces (FR1, FR2 and FR3) along the modern Fremont River, Utah (Repke *et al.* 1997; reproduced with permission from Elsevier), showing a roughly exponential reduction with depth.

$$r_T = A \exp(-E_a/RT) \qquad [7.5]$$

where r_T is the rate of chemical weathering (as a function of temperature), A is a pre-factor that varies with the chemical species being weathered, E_a is the activation energy (in kJ mol^{-1}), R is the universal gas constant, and T is absolute temperature (K). In gross terms this

temperature dependence is shown by the latitudinal global pattern of weathering shown in Fig. 7.6.

In summary, there are a number of factors controlling the rate of chemical weathering:

- *Organic activity* in soils generates soil acidity (CO_2) through decomposition, aids the retention of water through the build-up of organic matter, and biological activity promotes permeability.
- *Climate* controls weathering reactions through the effect on chemical kinetics of temperature (Arrhenius-type equation), imparting a latitudinal gross pattern of chemical weathering rates. Rainfall controls vegetation type, biological activity in the soil and the activity of water as a solvent.
- *Fractures* facilitate mechanical break down and ingress of fluids catalysing chemical weathering; they originate in the regolith, but also through tectonic strains.
- *Kinetics of mineral reactions*; chemical weathering requires pore waters to be undersaturated with respect to the mineral being weathered. Once saturation is achieved, an increase in the flow rate will not cause further reaction. Whereas the chemical weathering of evaporites is determined by the rate of removal of the saturated solution by flushing, in relatively insoluble minerals such as silicates the rate of chemical weathering is determined by the kinetics by which ions are detached from crystal surfaces.
- *Bedrock composition* controls the stability of the mineral components to weathering through their degree of polymerisation, as illustrated by the *Goldich series*. Monomer silicates like olivine are most easily weathered, and framework silicates such as quartz are most resistant.
- *Topography* influences the rate of removal of regolith by erosion, which invigorates chemical weathering by subjecting new, fresh bedrock to weathering. Topography also controls drainage and therefore rates of flushing. Flow rates on mountainous slopes are high, but flat, poorly drained slopes have very low flow rates.
- *Time* is required for chemical changes to take place. Weathering profiles are seldom in equilibrium with surface conditions.

7.2.2 Terrestrial sediment and solute yields

The bulk of particulate and dissolved solids are transported from the continents by running water. This flux connecting land and ocean reservoirs of the hydrological cycle is known as *run-off*. It can be viewed most simply in the context of a simple box model for the global water cycle. There is a net gain in the fluxes between the land and atmosphere (precipitation exceeds evaporation over the land areas), which is compensated by a run-off flux from the land to the ocean reservoirs. Although the quantity of water in the continental part of the hydrological cycle is small compared to the ocean reservoir, the flux of water from land to ocean is considerable.

Rain is not simply channelled through rivers to the ocean. The surface water balance contains run-off as one component

$$P = E + T + \Delta S + \Delta G + R \qquad [7.6]$$

where P is precipitation, E is evaporation, T is transpiration, ΔS is the change in storage of water in the soil, ΔG is the change in storage of water as groundwater and R is run-off – the overland flow across the land surface as rills and gullies, streams and rivers.

The relative importance of these parameters depends on climatic, topographic and geological setting, with important regional

Table 7.1　Run-off of world's major rivers

River	Catchment area (km²)	Annual precipitation (mm y⁻¹)	Discharge at river mouth (m³ s⁻¹)	Run-off coefficient C_r
Amazon	6,150,000	1490	200,000	0.69
Nile	2,715,000	832	317	0.004
Lena (Russia)	2,430,000	355	16,200	0.59
Orinoco	945,000	1300	34,900	0.9
Murray (Australia)	910,000	582	698	0.04
Seine (France)	78,600	711	685	0.39

variations. The ratio of streamflow measured from a river and the precipitation falling over the drainage basin is called the *run-off coefficient C_r*. The run-off coefficient varies strongly, principally according to climatic setting and the extent of human interference through irrigation and damming. The drainage basins listed in Table 7.1 provide some representative examples. The humid tropical (Amazon, Orinoco) and cold climate (Lena) drainage basins have high run-off coefficients compared to arid-region catchments (Nile, Murray).

Major variations in run-off can be predicted by comparing global maps of precipitation versus evaporation:

- Mean annual precipitation is highest in the tropics where humid air rises because of convectional instability, in mountainous regions because of orographic cooling, and between 35° and 60°N and S associated with atmospheric instability causing storms.
- Mean annual evaporation is highest where there is a heat source, a plentiful source of water, and low moisture contents in the air; these conditions are best satisfied in the subtropical oceans. Evaporation rates over dry continental areas are commonly much lower because of the scarcity of liquid water.

Precipitation exceeds evaporation in equatorial and high temperate to polar latitudes, whereas there is a net deficit in the subtropics. More than half of the global run-off occurs in South America, where it is concentrated in the equatorial regions. The Amazon River alone contributes 15% of the total annual global run-off.

7.2.2.1　Sediment yield

The erosion of the terrestrial surface results in a flux of sediment from source regions to depositional sinks. The fluxes of sediment through the process subsystems of erosion, transport and deposition are dependent not only on the processes operating within the subsystems, but also on the linking processes between the subsystems. This is particularly true of the transfer of sediment from the erosional zone to the transport zone. The functioning of channel-hillslope processes is introduced in §7.4, and the long-range transport of particulates in sediment routing systems is considered in §7.5. Here, we initially concentrate on a more descriptive approach to the global patterns of sediment production from erosional areas and its delivery to sedimentary basins.

The denudation of a drainage basin with area A_d results in a discharge of sediment and solute to the ocean or to lakes. If D is the total discharge of sediment and solutes at the exit of the drainage basin, and there is no change in the storage of sediment within the drainage basin over time, the long-term average denudation of the drainage basin per unit area is simply

$$\frac{\partial h}{\partial t} = \frac{(1-\phi)}{\rho} \frac{D}{A_d} \qquad [7.7]$$

where h is the elevation, ϕ is the porosity of rocks of density ρ in the catchment undergoing weathering and erosion, and the discharge D is measured as mass per unit time (kg yr⁻¹). Measurement of both the particulate and solute loads in rivers is required to calculate the total average denudation of the catchment. If D_s denotes the particulate sediment discharge, eqn. [7.7] can be rewritten as

$$Y = \frac{D_s}{A_d} \qquad [7.8]$$

where Y is termed the *sediment yield*, with units of kg m⁻² yr⁻¹. Sediment yield therefore has the units of a *mass flux*. For example, the Amazon River at its mouth has a solid load of 1150 Mt yr⁻¹ and a solute load of 223 Mt yr⁻¹ (1 mega ton is 10⁹ kg). The drainage area is 6,150,000 km². The sediment yield is therefore 187 t km⁻² yr⁻¹, whereas the total average denudation rate is 223 t km⁻² yr⁻¹. For a rock density of 2700 kg m⁻³ and an average porosity of 5%, the catchment-wide average denudation rate (eqn. 7.7) is 79 mm kyr⁻¹. Although this figure is very low, it should be remembered that certain parts of the Amazon drainage basin in the Andes are experiencing very rapid denudation while large parts of the floodplain are undergoing no denudation at all.

The ratio between particulate and dissolved denudation rates is variable (§7.2.2). Globally, rivers transport about four times as much suspended particulate load as dissolved load (Meybeck 1987; Milliman & Meade 1983; Summerfield 1991).

There are two strategies for evaluating catchment-averaged sediment yields and mechanical denudation rates based on the sediment exported from the catchment: use of the rate of filling of artificial reservoirs and natural lakes, and calculation of the volume of sediment accumulated in sedimentary basins (Fig. 7.8).

Einsele and Hinderer (1997) studied the filling of artificial reservoirs and lakes in order to estimate sediment yields in the contributing drainage basins. In open lakes and reservoirs the solutes are assumed to be flushed through, but the particulate load is mostly trapped in the water body, allowing a sediment budget to be calculated. For example, the Tarbela reservoir on the Indus River near Islamabad is an artificial lake extending 80 km upstream from the dam, covering an area of 625 km². It has a mean depth of 22 m, giving it an initial storage capacity of 13.9 km³. The drainage area in the predominantly semi-arid northwestern Himalaya is 171,000 km². The rate of filling of the reservoir (about 2% per year) indicates that the average sediment yield is 1170 t km⁻² yr⁻¹. The present-day mechanical denudation rate for the Indus catchment is therefore just over 400 mm kyr⁻¹.

The High Aswan reservoir in southern Egypt is 500 km long and covers an area of 5000 km², with a storage capacity of about 130 km³. The drainage area feeding the reservoir is 1.839×10^6 km². After 25

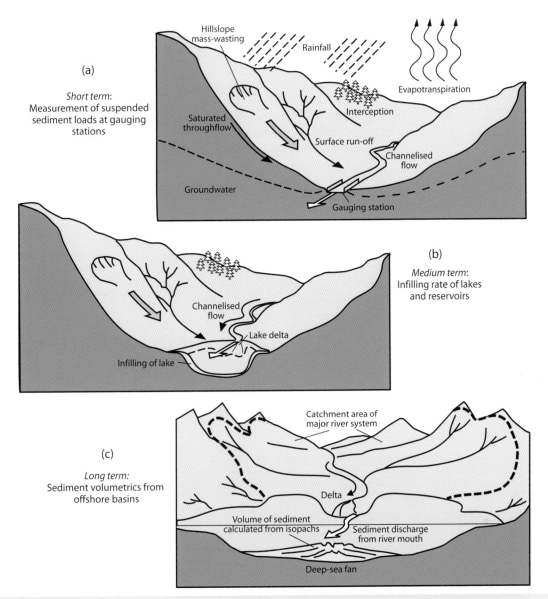

Fig. 7.8 Methods of estimating sediment yield. In (a) the short-term sediment yield is the discharge of sediment in kg yr⁻¹ measured at a gauging station divided by the contributing catchment area. In (b) the medium-term sediment yield is the infill rate of a lake basin or artificial reservoir, divided by the contributing catchment area. In (c) the long-term sediment yield is the solid sediment volume derived from sediment isopachs of dated intervals of the stratigraphy divided by the paleocatchment area.

years of operation the reservoir had collected 2800 to 3300 × 10⁶ t of sediment. Three other reservoirs in SE Sudan upstream of the Aswan High Dam have accumulated about the same amount of sediment as the main Aswan reservoir, giving a total of about 6000 × 10⁶ t. The annual accumulation rate is therefore 240 × 10⁶ t yr⁻¹. The sediment yield for the upper and middle reaches of the Nile River system is c.130 t km⁻² yr⁻¹, and the mechanical denudation rate (using $\rho = 2750 \, \text{kg m}^{-3}$ and $\lambda = 0.05$) is 45 mm kyr⁻¹, an order of magnitude less than in the Himalayan Indus catchment.

An allied strategy is to estimate the volume of sediment in the depositional zone of the sediment routing system using isopachs and cross-sections derived from borehole and seismic reflection data. The average solid phase accumulation rate for Asian sedimentary basins since the beginning of the Tertiary has been calculated after correction of stratigraphic thicknesses for the effects of compaction (Métivier *et al.* 1999) (Fig. 7.9). The Bengal Basin, for example, contains a vast amount of Upper Tertiary sediments (locally <21 km in thickness) deposited primarily from southwestward prograding deltas feeding a large deep-sea cone. The Bengal Basin has accumulated 12 × 10⁶ km³ solid volume during the Tertiary, with a rate of 0.45 × 10⁶ km³ My⁻¹ over the last 2 Myr. The Bengal Basin has received sediment from the Ganges, Brahmaputra and the rivers of the eastern platform of India (Mahanadi, Krishna and Godavari), which have a total drainage area of c.2.266 × 10⁶ km². The average mechanical denudation rate for these Indian subcontinent catchments during the Quaternary is about 200 mm ky⁻¹, with an average sediment yield of

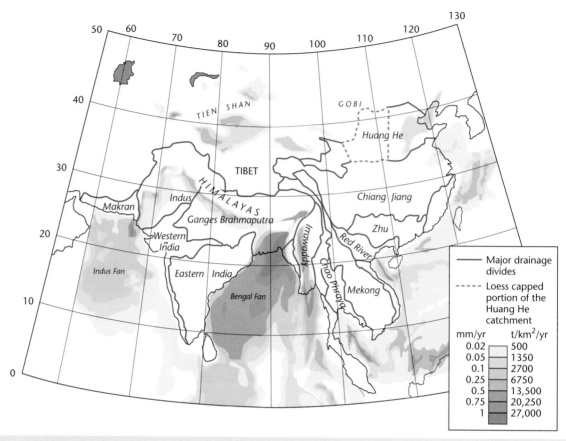

Fig. 7.9 Sediment mass (t km² yr⁻¹) accumulated in basins in south and southeastern Asia during the last 2 Myr, digitised from a large number of sources by Métivier *et al.* (1999), and equivalent sedimentation rate (mm yr⁻¹). The sediment was delivered to the ocean by the river systems of Asia. The limits of the present-day Ganges-Brahmaputra and Indus sediment routing systems are superimposed.

546 t km⁻² yr⁻¹ (using $\rho = 2750$ kg m⁻³). The Ganges, Brahmaputra, Mahanadi, Krishna and Godavari rivers have a combined river mouth discharge of 0.49×10^6 km³ Myr⁻¹. This is in excellent agreement with the estimate derived from the volume of preserved stratigraphy, and indicates that present-day discharges in large, buffered systems may be closely representative of longer-term (geological) rates (Einsele *et al.* 1996; Métivier *et al.* 1999; Métivier & Gaudemer 1999).

Longer-term trends in sediment discharge to the ocean rely on the mapping of sediment thicknesses derived from seismic data (Métivier *et al.* 1999; Clift 2006). Correlations are made between trends in sediment discharge, climatic and tectonic changes in the sediment source regions, and major river drainage reorganisation. The sediment budget of the Indus Fan, in particular, has been used in this way (Clift *et al.* 2001, 2002).

7.2.2.2   Global patterns of sediment yield

The total amount of sediment discharged to the ocean is estimated to be c.20×10^9 t yr⁻¹ (Milliman & Syvitski 1992; Walling & Webb 1996) or c.16×10^9 t yr⁻¹ (Syvitski & Milliman 2008). Since the land surface of the Earth is 1.48×10^8 km², the globally averaged sediment yield is 108–135 t km⁻² yr⁻¹, but there is enormous variability around this globally averaged figure (Fig. 7.10). Some of the highest sediment yields come from rugged oceanic islands such as

Java, New Guinea, Taiwan and New Zealand (Table 7.2). Even higher values (>50,000 t km⁻² yr⁻¹) come from Chinese rivers draining the easily erodible loess region. At the other extreme, there are some very large river systems with sediment yields of less than 1 t km⁻² yr⁻¹, such as the Yenesei and Dneiper basins of the former Soviet Union.

The main point evident from the numerous published sediment yield maps (e.g. Walling & Webb 1983; Milliman & Meade 1983; Lvovich *et al.* 1991) are that sediment yields are highest in an arc around the Pacific and Indian Ocean margins from Pakistan to Japan. This area is characterised by Cenozoic mountain building, steep topography, and generally high temperatures and annual rainfalls. Low sediment yields typify desert regions and the cold, formerly glaciated low-relief regions of Eurasia and Canada. The key question is what controls this wide variability in sediment yield. If the answer to this question were known, it would assist the understanding of the likely mass fluxes to sedimentary basins in the geological past.

Sediment yield models commonly use the sediment loads of rivers as an index of sediment yield (Milliman & Meade 1983; Summerfield & Hulton 1994; Berner & Berner 1997; Hay 1998; Hovius 1998). These sediment load data are strongly biased towards measurements made at the mouths of the world's major rivers, although Milliman and Syvitski (1992) have emphasised the influence of small mountainous streams on the global sediment budget. However, the

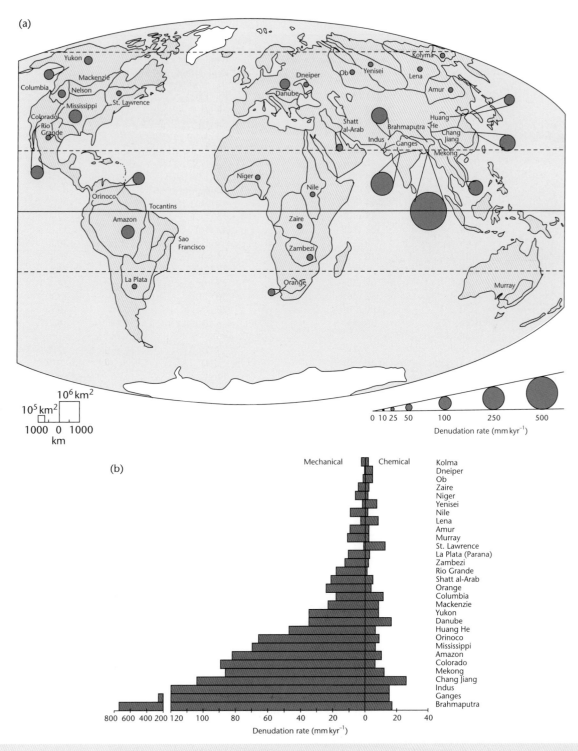

Fig. 7.10 Global patterns of sediment yield. (a) Estimated total denudation rates for major externally drained basins. Size of circles is proportional to denudation rate. (b) Histogram comparing mechanical and chemical denudation in the world's major externally drained basins. After Summerfield & Hulton (1994), reproduced with permission of the American Geophysical Union. The solute contributions of the major rivers in terms of silicate, carbonate and evaporite weathering is shown in Fig. 7.17 (§7.2.2) and Table 7.2.

Table 7.2 Sediment and solute yields of the world's major rivers (after Summerfield & Hulton 1994, and sources cited therein)

River basin	Area 10⁶ km²	Sediment yield t km⁻² yr⁻¹ (equivalent mechanical denudation rate, mm kyr⁻¹)	Solute yield t km⁻² yr⁻¹ (equivalent chemical denudation rate, mm kyr⁻¹)	Chemical denudation as % of total
Amazon	5.98	221 (82)	29 (11)	11.6
Amur	2.04	28 (10)	6 (2)	17.6
Brahmaputra	0.64	1808 (670)	49 (18)	2.6
Chang Jiang (Yangtze)	1.73	281 (104)	72 (27)	20.4
Colorado	0.70	239 (89)	19 (7)	7.4
Columbia	0.67	48 (18)	32 (12)	40.0
Danube	0.79	94 (35)	45 (17)	32.4
Dneiper	0.54	2 (1)	12 (4)	85.7
Ganges	0.98	694 (257)	42 (16)	5.7
Huang He	0.79	127 (47)	18 (7)	12.4
Indus	0.93	323 (120)	42 (16)	11.5
Kolyma	0.65	9 (3)	4 (1)	30.8
La Plata (Parana)	2.86	30 (11)	9 (3)	23.1
Lena	2.45	7 (3)	22 (8)	75.9
Mackenzie	1.77	62 (23)	23 (9)	27.1
Mekong	0.76	232 (86)	36 (13)	13.4
Mississippi	3.20	189 (70)	20 (7)	9.6
Murray	1.14	30 (11)	6 (2)	9.7
Nelson	1.24	–	16 (6)	–
Niger	2.16	19 (7)	4 (1)	17.4
Nile	3.63	28 (10)	3 (1)	9.7
Ob	2.98	6 (2)	11 (4)	64.7
Orange	0.89	65 (24)	11 (4)	14.5
Orinoco	0.92	179 (66)	23 (9)	11.4
Rio Grande	0.63	48 (18)	4 (1)	7.7
Sao Francisco	0.62	11 (4)	–	–
Shatt al-Arab	0.89	56 (21)	14 (5)	20.0
St Lawrence	1.05	2 (1)	34 (13)	94.4
Tocantins	0.76	–	–	–
Yenisei	2.55	5 (2)	18 (7)	78.3
Yukon	0.84	94 (35)	23 (9)	19.7
Zaire	3.63	14 (5)	6 (2)	30.0
Zambezi	1.41	34 (13)	6 (2)	15.0

Reproduced with permission of the American Geophysical Union.

sediment loads at river mouths are affected by the extent to which sediment is stored in river floodplains. The extent of storage is likely to increase in larger drainage basins with substantial depositional and transportational components of the sediment routing system. This storage effect explains the well-established inverse relationship between drainage basin area and sediment yield (Fig. 7.11) (Milliman & Meade 1983; Milliman & Syvitski 1992). This inverse relationship has, however, been questioned as a fundamental property of natural drainage systems, and has been attributed at least in part to human impact (Walling & Webb 1996).

There are a large number of problems with the use of data on sediment loads measured at river mouths (Milliman & Meade 1983). The most obvious is that the effects of human activities, such as land use, dam construction and irrigation schemes, must be accounted for. There are now very few 'pristine' rivers with discharges of water and sediment unaffected by the activities of humans (Syvitski et al. 2005; Syvitski & Milliman 2007). In addition, sediment loads are invariably suspended load only, with the bedload contribution ignored. This is unlikely to be a problem in the downstream reaches of large rivers, but may introduce a significant error in short, steep mountainous streams. Finally, the suspended load sediment data

have been collected using different methods over protracted periods of time and are difficult to compare.

Glaciated terrains appear to have a different and more complex relation between sediment yield and catchment size than unglaciated river basins (Hallet et al. 1996). Sediment delivery from the upland part of glaciated catchments may be particularly high when pluvial conditions promote the rapid evacuation of loose morainic material from a previously glaciated region. Some currently glaciated basins, such as the steep and wet catchments of Alaska and the Southern Alps of New Zealand, have very high yields, in excess of most non-glaciated settings.

7.2.2.3 Controls on sediment yield

Numerous attempts have been made to correlate sediment yield with topographic and climatic factors. The resulting relationships invariably work reasonably well at restricted spatial scales, or in certain prescribed geomorphic settings, but involve a large amount of scatter at the global scale. A multivariate regression including anthropogenic factors is discussed in §7.2.3.

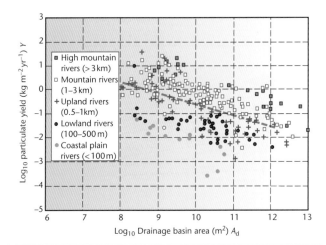

Fig. 7.11 Plot of drainage basin area versus particulate sediment yield for the elevation classes of rivers of Milliman & Syvitski (1992). The regression of all the river data has the form $Y = \exp(9.18 - 0.46\ln A_d)$, where Y is yield, and A_d is the area of the drainage basin. After Hay (1998), reprinted with permission from Elsevier.

Rates of precipitation: a number of authors have investigated the link between sediment yield and mean annual precipitation, rainfall variability and specific run-off (water discharge per square kilometre of drainage basin, in mm yr^{-1}). Any relationship between precipitation and run-off variables and sediment yield must be strongly mediated by the effects of vegetation (Leeder *et al.* 1998). There is some consensus that sediment yield reaches a maximum in semi-arid areas where vegetation cover is sparse, with a second maximum where mean annual precipitation exceeds 1000 mm. Since most hillslope processes are sensitive to the intensity of rainfall, however, the rainfall peakedness (ratio of average monthly precipitation and maximum monthly precipitation) may be more important in governing sediment yield (Fournier 1960). The effects of vegetation cover can be seen where catchments have been disturbed by human activities. Sediment yields from cleared agricultural land is higher (by factors of up to 100) than from forested catchments.

Topographic effects: erosion rate can be linked to a variety of topographic parameters such as mean or maximum elevation, large-scale or local relief, slope and drainage basin area. Various correlations have been claimed for a range of different data sets. For example, these comprise the sediment discharges of 280 rivers, including small mountainous streams (Milliman & Syvitski 1992), a data set of 285 drainage basins of varying size and climatic zone (Allen 1997; Hovius 1998), a subset of the world's largest 29 rivers (Summerfield & Hulton 1994), a compilation of sediment yields in central Europe (humid temperate climate) up to 1981 based on reservoir and lake studies (Schröder & Theune 1984), and a selection of mid-latitude drainage basins (Ahnert 1970; Pazzaglia & Brandon 1996).

The bewildering range of correlations with topographic and climatic parameters claimed by different authors suggests that erosion rate data cannot be compressed onto one simple plot. A multivariate analysis of sediment yield data reveals that a combination of environmental and topographic factors only explains about half of the variance in global sediment yield data (Hovius 1998). However, when plotted against a proxy for tectonic uplift rate (Fig. 7.12), tec-

tonically inactive and tectonically active settings are discriminated. For example, tectonically inactive cratonic settings are characterised by very low sediment yields of <100 t km^{-2} yr^{-1}, whereas currently or recently tectonically active contractional mountain belts have sediment yields of 100–10000 t km^{-2} yr^{-1}. Pinet and Souriau (1988) similarly noted different relationships for young, tectonically active orogens and old, tectonically inactive landscapes. This suggests that physical insights based on the different sets of processes dominating in low-relief and high-relief areas must be included in the analysis (see BQART equations, §7.2.3).

Montgomery and Brandon (2002) suggested a non-linear regional-scale relation between erosion and mean slope derived from a digital elevation model (DEM):

$$E = E_0 + \frac{KS}{\left[1 - (S/S_c)^2\right]} \qquad [7.9]$$

where E is the erosion rate, E_0 is a background erosion rate due to chemical weathering (c.0.01 mm yr^{-1} based on the mean chemical denudation rate for the world's 35 largest drainage basins, Summerfield 1991), S is the mean slope, S_c is a limiting hillslope gradient for landsliding, and K is a rate constant. Here, mean slope is derived from the range of values within a 10 km-diameter analysis window around points spaced every 2 km, using a 10 m-resolution DEM. In the Olympic Mountains of Washington, USA, $E_0 = 0.016 - 0.059$ mm yr^{-1}, $K = 0.6$ mm yr^{-1}, and $S_c = 40°$. Eqn. [7.9] can be expressed in terms of a mean local relief, derived from the same 10 m-resolution DEM with a 10 km-diameter analysis window, giving

$$E = E_0 + \frac{KR_z}{\left[1 - (R_z/R_c)^2\right]} \qquad [7.10]$$

where R_z is the mean local relief, and R_c is the limiting local relief. Data from low-relief, tectonically inactive areas show a linear relation between erosion and mean local relief, as previously suggested by Ahnert (1970). However, this linear relation does not hold for high-relief, tectonically active areas. In such cases, eqn. [7.10] gives a good fit with $E_0 = 0.01$ mm yr^{-1}, $R_c = 1500$ m, and $K = 2.5 \times 10^{-4}$ mm yr^{-1} (Fig. 7.13). The mean local relief R_z appears to reach a maximum of c.1500 m ($R_c = 1500$ m), and is very rarely >2000 m, suggesting that rock strength may control the maximum relief attainable in a mountain belt.

There are therefore essentially two different sorts of landscape in terms of erosion rate. In *low-relief landscapes*, such as low-lying shield areas and alluvial plains, thick regoliths may develop. The rate of removal of regolith is determined mainly by the erosivity of the transport processes, rather than by the availability of loose, easily transportable material. This may also be the case in regions of extreme aridity where streamflow processes are negligible. In these *transport-limited* circumstances, parameters such as mean annual rainfall and specific run-off may offer the best correlation with sediment yield. Where the removal of loose material by streams is efficient, the rate of removal of sediment may, however, be limited by the rate at which loose material is supplied by hillslope weathering. These circumstances are therefore *weathering-limited*, and hillslope erosion processes determine the erosion rate, which depends linearly on mean slope or local relief. These conditions may be typical of low-relief, mid-latitude, temperate zone landscapes. In *high-relief landscapes*, however, physical transport processes of hillslope erosion by landsliding and channelised flow are almost always capable of

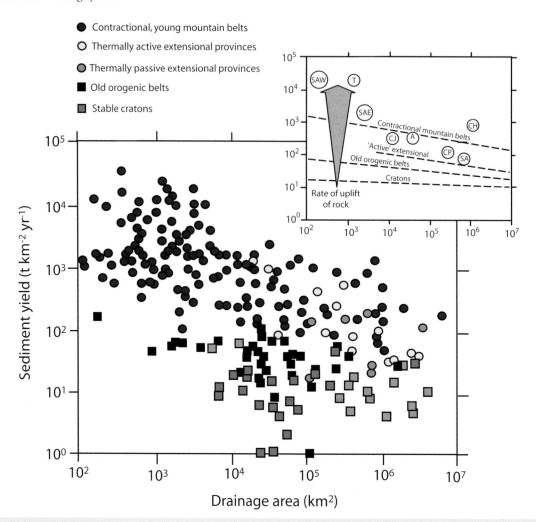

Fig. 7.12 Plot of sediment yield versus drainage basin area for five categories of tectonic setting, derived by Hovius (1995) using the sediment yield data of Milliman & Syvitski (1992). Inset shows approximate trends for various tectonic settings in the same existence field of drainage area (in km^2) versus sediment yield (in t km^{-2} yr^{-1}). SAW, Southern Alps New Zealand western flank; SAE, Southern Alps New Zealand eastern flank; T, Taiwan; CJ, Central Japan; A, Alps; CP, Colorado Plateau; SA, Southern Africa; CH, Central Himalaya.

removing regolith above a critical rate of rock uplift. Erosion rates then stabilise at a certain mean local relief determined by rock and soil strength. It is estimated that rock uplift rates of $1\,mm\,yr^{-1}$ are required to sustain combined potential chemical and physical erosion in mountainous areas (Koons 1995). In high-relief landscapes, therefore, the key process is the landsliding of critically steep hillslopes and efficient removal of debris by streams.

7.2.2.4   Transport of dissolved loads

Chemical weathering results in the release of ions in solution. Precipitation must infiltrate to deep levels in the regolith in order to promote chemical breakdown and the flushing of solutes into the groundwater system. This process of infiltration and flushing depends on climate and topography. In semi-arid regions, water is drawn to the surface during dry periods, leading to reprecipitation of solutes, and chemical weathering is retarded. On steep slopes, water runs off hillslopes quickly and does not spend long periods in contact with bedrock and regolith. In low-lying, gently sloping areas, the regolith

is poorly drained, so flushing rates are low. Chemical fluxes are highest on gentle upland slopes, where both infiltration and flushing are moderately high. Although particulate sediment yields are far higher than solute yields in mountainous areas, solute yields are relatively higher in lowland regions.

The groundwater system delivers solutes to rivers. The solute discharge of rivers may be modified by non-denudational inputs such as precipitation, wind-blown dust, salt aerosols (particularly close to the ocean), mineralisation of organic matter, plant metabolism and human pollution (Gaillardet *et al.* 1999). Outputs include those of evaporation and losses to groundwater and soils. Consequently, the dissolved loads of rivers cannot be simply or directly linked to rates of chemical denudation.

The chemistry of water in the hydrological cycle strongly reflects the reactions taking place with water and soil, with plants and with decomposing organic matter, through dilution or concentration caused by additions of precipitation and losses by evaporation, and particularly through the weathering of rock. Waters from calcareous catchments contain high amounts of *total dissolved solids* (TDS),

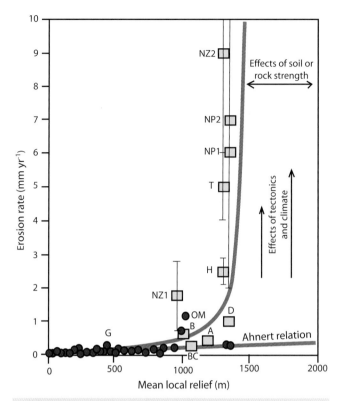

Fig. 7.13 Plot of erosion rate versus mean local relief using data compiled from Ahnert (1970), Summerfield and Hulton (1994) and Pazzaglia and Brandon (1996). Solid line is model fit using eqn. [7.10] with $E_0 = 0.01\,mm\,yr^{-1}$, $R_c = 1500\,m$, and $K = 2.5 \times 10^{-4}\,mm\,yr^{-1}$. Solid squares are from tectonically active regions; NZ1 and NZ2, Southern Alps of New Zealand; H, central Himalaya; NP1 and NP2, the Himalayan portion of the Indus River Basin; T, Taiwan; BC, British Columbia Coast Range; OM, Olympic Mountains; D, Denali portion of the Alaska Range; A, European Alps. Ganges (G) and Brahmaputra (B) also shown for reference. After Montgomery and Brandon (2002), reprinted with permission from Elsevier.

particularly Ca^{2+} and HCO_3^-, but small particulate loads. Waters draining different igneous, metamorphic and sedimentary rock types have a varied chemical composition and higher TDS.

It is customary to plot river and lake water on a graph with axes of TDS (mg l^{-1}) and the cationic ratio (Fig. 7.14)

$$\frac{Na^+}{Na^+ + Ca^{2+}}$$

or the anionic ratio

$$\frac{Cl^-}{Cl^- + HCO_3^-}$$

Waters are consequently classified in terms of the source of the solute load (Gibbs 1970):

- Dominated by *atmospheric precipitation*, with low salt concentrations because of dilution (20–30 mg l^{-1}), and a chemistry dominated by Na^{2+}, so the cationic ratio is nearly 1. This class of water

characterises rivers draining areas of low relief with well-weathered bedrock and plentiful rainfall, such as the tropical rivers of Africa and South America.
- Dominated by *weathering reactions* in rock and soil, with higher TDS and a wide spread of cationic ratio depending on the reactions involved. There is a low ratio in limestone terrains where waters are dominated by calcium bicarbonate. In the Amazon Basin, 85% of the solute load is derived from a relatively small area of intense chemical weathering in the Andes, whereas in the lowland part of the catchment the river the control is by precipitation.
- Dominated by *evaporation* and subsequent precipitation, with high salt concentrations (TDS 1000–2000 mg l^{-1}), reflecting the precipitation of $CaCO_3$ and the relative concentration of Na$^+$ and Cl$^-$ through evaporative losses. This is common in the rivers of hot and arid climates such as the Jordan River of the Near East. These typical river compositions evolve to the composition of seawater.

The chemistry of river input to the ocean therefore varies from continent to continent and from region to region in response to variations in the control of precipitation, weathering and evaporation superimposed on the signature provided by bedrock type (§7.2.2.6). Rivers in arid Kazakhstan have solute concentrations of 6000–7000 mg l^{-1}, compared to 10 mg l^{-1} for rivers in the humid Amazon Basin, where solutes are diluted by large amounts of run-off.

The measurement of the solute concentration or 'salinity' of a water sample is conventionally in units of mg l^{-1}. That is, the solutes are measured as a concentration. Consequently, the transport rate of solutes is the product of the water discharge (m^3 y^{-1}) and the solute concentration expressed as TDS (mg l^{-1}). The solute discharge from the world's rivers, and the solute flux from river catchments, reflect the competing effects of concentration and run-off.

7.2.2.5 Effects of bedrock weathering on solutes

Study of unpolluted catchments composed of one bedrock lithology illuminates, after correction for atmospheric salts, the typical solute loads of rivers draining known rock types (Meybeck 1987) (Fig. 7.15). The solute loads of rivers draining large multi-lithologic catchments (such as the Amazon, Stallard & Edmond 1983), or even the entire land surface of the Earth (Holland 1981), can therefore be estimated with knowledge of the percentage of each rock type comprising the region being investigated. The calculated global value for solute discharge to the ocean can then be compared with the measured average composition of rivers (Livingstone 1963; Meybeck 1979). More than 80% of the total dissolved load of the world's rivers is made of four ions: HCO_3^-, SO_4^{2-}, Ca^{2+} and SiO_2 (Stumm & Morgan 1996)(Fig. 7.16).

The geographic origin of river-borne material has an important impact on the TDS and suspended loads, expressed as percentages of the total input to the oceans (Table 7.3). The dissolved and suspended loads from cold, temperate, tropical and arid regions shows that the tropical zone dominates the world's run-off. It is therefore the major source of silica and organic carbon. The dissolved silica reflects the high rates of chemical weathering in tropical regions. Despite occupying 17% of the Earth's surface, the arid regions contribute a minute amount of run-off. However, the total ions, though small, are four to five times the volume of run-off, indicating higher ionic concentrations in arid-region rivers. Cold regions, despite occupying 23% of

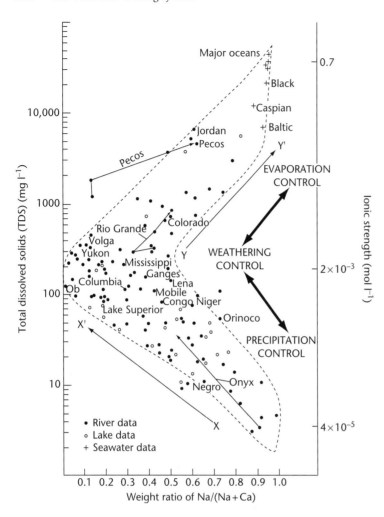

Fig. 7.14 Gibbs's (1970) scheme for global water chemistry, wherein the total dissolved solids (TDS) and ionic strength of surface waters are plotted against the cationic weight ratio of Na/(Na + Ca). The arrows connecting data points show the geochemical evolution of river waters from source downstream. Rivers plotting along the trend from X to X', such as the McKenzie (Arctic Canada) and Ganges (India), occur in regions with highly active weathering processes. Rivers falling along the trend from Y to Y', such as the Jordan (Middle East), Rio Grande and Colorado (arid southwestern North America), occur in areas experiencing high amounts of evaporation, and evolve towards the composition of seawater. Reproduced with permission of American Association for the Advancement of Science.

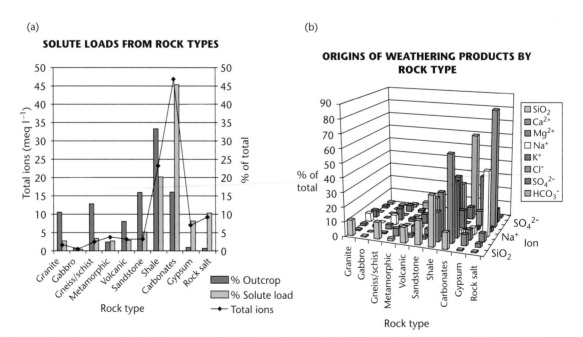

Fig. 7.15 (a) Solute loads (meq/l) from a range of common rock types compared to their outcrop area. (b) Distribution of eight major ions according to rock type, expressed as a percentage of the total solute load. Data from Meybeck (1987).

Table 7.3 Dissolved and suspended loads of different climatic regions

Geographic region	Area (%)	Run-off (%)	Dissolved SiO₂ (%)	Ions (%)	TOC (%)	Suspended matter (%)
Cold regions	23.4	14.7	5.4	15.5	17.5	2.7
Temperate	22.4	27.5	19.9	39.9	28.5	56.5
Tropical	37.0	57.2	73.6	41.8	52.0	34.2
Arid	17.2	0.65	1.0	2.8	1.3	6.6

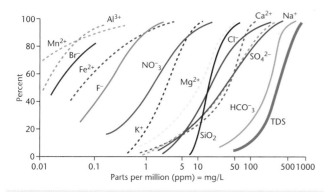

Fig. 7.16 Cumulative probability curves for concentration of chemical species in rivers. Dashed, cations. Solid, anions. TDS, total dissolved solids. From Anderson & Anderson (2010, p. 186, fig. 7.22; reproduced with permission of John Wiley & Sons, Ltd.) and Stumm & Morgan (1996, fig. 15.1).

Table 7.4 Rock types at the Earth's surface by percentage

Class	Rock type	Percentage (by area)
Plutonic	Granite	10.4
	Gabbro and ultrabasics	0.6
Metamorphic	Marble	0.4
	Amphibolite	1.9
	Mica-schist	1.5
	Gneiss	10.4
	Quartzite	0.8
Volcanic	Basalt	4.15
	Andesite	3.0
	Rhyolite	0.75
Sedimentary	Quartz-arenite	12.6
	Arkose (feldspathic arenite)	0.8
	Greywacke (lithic, argillaceous arenite)	2.4
	Shale	33.1
	Limestone and dolomite	15.9
	Evaporite (gypsum and halite)	1.3

the Earth's surface, contribute less than 3% of the suspended matter to the ocean.

7.2.2.6 The composition of waters draining different rock types

The surface of the Earth is covered with different rock types (Table 7.4). The weathering of these rock types produces distinct chemistries of solutes in the run-off. (Atmospheric contribution of salts must be accounted for and care must be taken to exclude all anthropogenically affected waters.) If we combine the typical water analyses of major rock types with their relative abundance on the surface of the continents, we get a global picture of the solute delivery to the ocean from rivers (corrected for atmospheric salts). The solutes derived from different rock types is shown in Fig. 7.15a, b. The impact of different rock types on the solute loads of the world's major rivers is shown in Fig. 7.17 (Gaillardet *et al.* 1999).

Taking *granite* as a standard, the chemical weathering rates of the world's major rock types (Table 7.5) shows that evaporites and carbonates weather most rapidly. Volcanic and non-crystalline metamorphic rocks and shales weather at moderate rates, while granites, gneisses, gabbros and sandstones have low rates of weathering. Chemical sediments, such as carbonates and evaporites (which have high yields of solutes), and shales (which are very extensive) therefore dominate the solute fluxes of rivers draining to the ocean. Crystalline rocks have a minor contribution to global weathering in terms of solutes.

In terms of individual elements (Fig. 7.15b), *silica* and *potassium* originate mostly by weathering of silicates. *Sodium* is derived mostly

from the weathering of the salt halite (NaCl), the remainder coming from the weathering of silicates. *Magnesium* originates from both dolomites, silicates and to a lesser extent Mg-rich evaporites. *Calcium* is mainly derived from the weathering of carbonates. *Chloride* is complicated by the contamination from the atmosphere, but is strongly controlled by the dissolution of halite. *Sulphate* comes from the weathering of pyrite, evaporitic sulphate and organic sulphur compounds. *Bicarbonate* is derived from atmospheric and soil CO_2 as well as from the dissolution of carbonate rocks.

7.2.2.7 Relation between solute and suspended load

The global average of solute load is roughly one fifth of the global average of suspended load, but the ratio varies from continent to continent and from drainage basin to drainage basin. A key determinant is the topographic setting of the river. In mountainous regions, the particulate sediment yield is much higher than the solute yield, whereas in lowland regions the two are roughly in balance (Fig. 7.18).

The relationship between suspended and solute load for the world's major rivers is found in Table 7.6 and Fig. 7.10.

7.2.3 BQART equations

Sediment discharge to the ocean is controlled by a range of geomorphic and geologic factors, but is also increasingly conditioned by the activities of man (Syvitski *et al.* 2005; Syvitski & Milliman 2007).

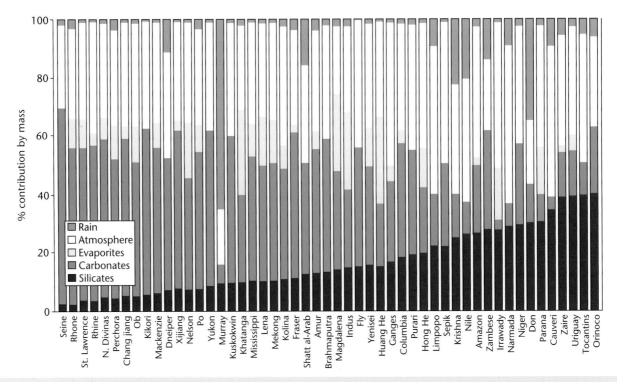

Fig. 7.17 Contributions of various bedrock and atmospheric sources of solutes in the world's major rivers, ranked left to right according to the percentage contribution of silicate weathering. The bars show the fraction of the total solute concentration in mg l^{-1} from each source. The contribution from rain is mostly Na and Cl from sea salt. After Gaillardet *et al.* (1999, fig. 5) and Anderson & Anderson (2010, fig. 7.23).

Table 7.5 Weathering rates of major rock types, relative to granite

Rock type	Weathering rate
Granite	1
Gneiss/schist	1
Gabbro	1.3
Sandstones	1.3
Volcanics	1.5
Shales	2.5
Other metamorphic	5
Carbonates	12
Gypsum	40
Rock salt	80

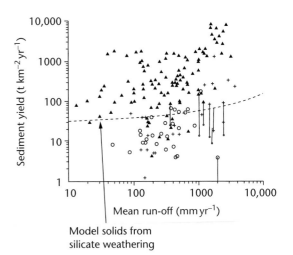

Fig. 7.18 Relation between mean run-off and sediment yield for rivers throughout the world. Open circles, coastal plain (0–100 m at headwaters) and lowland (100–500 m); crosses, upland rivers (500–1000 m); solid triangles, mountain (1–3 km) and high mountain (>3 km). Model curve is prediction for solids derived by silicate weathering (after Stallard 1995), showing that mountainous catchments provide high solid yields for a given run-off, whereas lowland rivers provide low solid yields for the same run-off.

Consequently, there has been a concerted effort to incorporate anthropogenic factors into an equation that would explain sediment discharge at the present day. This global predictor is called BQART.

The evolution of the BQART predictor begins with the recognition that the sediment discharge of rivers Q_s relates to their maximum relief R and basin area A (Milliman & Syvitski 1992; Mulder & Syvitski 1996), whereby

$$Q_s = \alpha A^{0.41} R^{1.3} \qquad [7.11]$$

and α is a constant of proportionality. Although this regression does not explicitly take into account precipitation, climate, vegetation and

Table 7.6 Water discharge, suspended and solute loads, and mean elevation of world's major rivers

River	Water discharge ($m^3 s^{-1}$)	Suspended load (Mt y^{-1})	Solute load (Mt y^{-1})	Mean elevation (m)
Amazon	200,000	1,150.0	223	426
Brahmaputra	19,300	520	61	2,734
Columbia	7,930	15	35	1,329
Colville	492	520	6	469
Danube	6,660	70	60	501
Dneiper	1,650	2.1	11	152
Fraser	3,550	20	11	1,140
Ganges	11,600	524	75	890
Indus	7,610	250	41	1,855
Irrawaddy	13,600	260	92	758
Jana	920	3	1	703
Lena	16,200	12	88	602
Mackenzie	9,830	125	64	634
Magdalena	6,980	220	28	1,203
Mekong	14,900	160	60	1,062
Mississippi	18,400	400	125	656
Murray	698	30	9	266
Niger	6,020	32	10	429
Nile	317	125	18	662
Ob	12,200	16	50	301
Orange	2,890	91	12	1,241
Orinoco	34,900	150	39	456
Parana	18,000	112	56	564
Po	1,490	18	10	793
Rio Grande	95	30	2	1,279
Shatt al-Arab	1,460	103	18	669
St. Lawrence	14,300	4	59	265
Xi Jiang	9,510	80	132	670
Yangtze (Chang Jiang)	28,500	480	226	1,688
Yellow (Huang He)	1,550	120	22	1,885
Yenisei	17,800	13	65	749
Yukon	6,180	60	34	741
Zaire	40,900	32.8	36	740
Zambezi	6,980	48	15	1,033

bedrock geology, it was assumed by Syvitski and Morehead (1999) that climatic factors, including temperature, were embedded into the coefficient and exponents of eqn. [7.11]. Taking basin-averaged temperature as a parameter, they modified eqn. [7.11] to derive an estimate of sediment discharge as a function of area (A), relief (R) and temperature (T) – the ART equation:

$$Q_s = c_1 A^{c_2} R^{c_3} e^{kT} \qquad [7.12]$$

where c_1, c_2 and c_3 are empirically derived coefficients based on climate characteristics, T is temperature, and k is another coefficient related to the strength of physical versus chemical weathering. The ART equation applies to pristine conditions, unaffected by the various activities of man. But eqn. [7.12] accounts for relatively little of the variance of a global compilation of data from nearly 500 rivers, suggesting that anthropogenic factors need to be considered. Consequently, the sediment discharge was expressed in a modified equation that explicitly accounted for climatic factors, the BQART equation of Syvitski and Milliman (2007):

$$Q_s = 0.02 BQ^{0.31} A^{0.5} RT \qquad [7.13]$$

for basin-averaged temperatures of >2 °C, and

$$Q_s = 0.04 BQ^{0.31} A^{0.5} R \qquad [7.14]$$

for basin-averaged temperatures of <2 °C, and climatic and human factors are embedded in the new variable B. Units are in kg s^{-1} for Q_s, km^2 for A, km for R, and °C for T. B is given by

$$B = IL(1 - T_E)E_h \qquad [7.15]$$

where I is a glacier erosion factor, L is a basin-wide lithology factor, T_E is the trapping efficiency of lakes and man-made reservoirs, and E_h is a human-influenced soil erosion or 'anthropogenic' factor. The derivation of the BQART equation and the attribution of values to the variables, as well as a dimensionless analysis of the river discharge data, are given in Fig. 7.19 and Appendix 41. Use of BQART accounts for substantially more of the variance in the global sediment load data.

The sediment discharge of the BQART equation can be divided by basin area to derive the average yield for suspended particulate matter, since measured riverine sediment discharges do not account for dissolved loads or bedload. Consequently

$$Y_s = 0.02 BQ^{0.31} A^{-0.5} RT \qquad [7.16]$$

for basin-averaged temperatures of greater than 2° C.

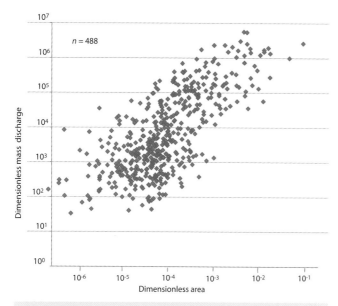

Fig. 7.19 Plot of 488 rivers in terms of dimensionless suspended sediment discharge and dimensionless catchment area, using the BQART equations and database of Syvitski & Milliman (2007) (§7.2.3), non-dimensionalised in Appendix 41.

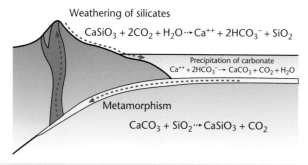

Fig. 7.20 Long-term carbon cycle involving chemical weathering of silicate rocks in mountainous regions, precipitation of carbonate in the ocean, burial, metamorphism and outgassing of CO_2 from volcanoes, with the main chemical reactions involved. From Anderson & Anderson (2010, p. 200, fig. 7.29), © Robert S. Anderson, Suzanne P. Anderson 2010, published by Cambridge University Press, translated with permission.

7.2.4   Chemical weathering and global biogeochemical cycles

Chemical weathering of rocks plays an important role in global biogeochemical cycles. Of particular interest is the role of chemical weathering in regulating climate through its effect on the drawdown of atmospheric carbon dioxide (Raymo *et al.* 1988; Raymo & Ruddiman 1992). Chemical weathering affects the carbon cycle by its participation in a long-term cycle involving the use of atmospheric CO_2 in silicate weathering, the run-off of bicarbonate in solution, the precipitation of carbonate in the oceans, and the volcanic outgassing of CO_2 from metamorphosed rocks (Fig. 7.20).

The general reaction for the chemical weathering of silicate rocks, taking a calcium silicate (the pyroxene wollastonite) as the starting material is

$$CaSiO_3 + 2H_2CO_3 = Ca^{2+} + 2HCO_3^- + SiO_2 + H_2O \qquad [7.17]$$

where the carbonic acid H_2CO_3 is derived by the dissolution of CO_2 in water, as occurs in the regolith where soil CO_2 is dissolved in rainwater. The bicarbonate ion is released and is transported as a solute in the world's rivers (Figs 7.15, 7.16). On entering the ocean, it combines with Ca^{2+} to precipitate the mineral calcium carbonate $CaCO_3$, commonly mediated by biological activity

$$Ca^{2+} + 2HCO_3^- = CaCO_3 + H_2CO_3 \qquad [7.18]$$

The result of eqn. [7.17] and eqn. [7.18] is to transfer CO_2 gas from the atmosphere to calcium carbonate solid on the seafloor. If this process were to continue for long periods of time, atmospheric concentrations of CO_2 would reduce well below those known to have existed in the past. However, CO_2 is returned to the atmosphere by metamorphism of rocks and by the outgassing of volcanoes. Taking a mixture of SiO_2 and $CaCO_3$, representing the mineralogy of typical pelagic sediments of the ocean floor, carbon dioxide is released, leaving the calcium silicate mineral that we started with:

$$CaCO_3 + SiO_2 = CO_2 + CaSiO_3 \qquad [7.19]$$

Variations between the weathering-driven drawdown flux and the tectonically driven emission flux of carbon over geological time may be linked to past changes in climate. One view is that tectonic uplift of mountain belts increases chemical weathering, draws down CO_2, leading to cooling through the greenhouse effect (Raymo *et al.* 1988; Raymo & Ruddiman 1992). Alternatively, increases in rates of seafloor spreading and volcanic outgassing linked to metamorphism at subduction zones increases atmospheric CO_2 concentrations, leading to increased chemical weathering, which stabilises climate change (Berner 2004). These contrasting views continue to be debated.

The carbon cycle is also affected by the removal or oxidation of organic carbon. In densely vegetated regions, organic carbon is transported in particulate form, and is buried in thick sedimentary accumulations offshore from river mouths (Galy *et al.* 2007; France-Lanord & Derry 1997; Hilton *et al.* 2008, 2011). This organic material, which has drawn down atmospheric CO_2 as a result of photosynthesis, may either be buried for long time scales to produce hydrocarbons (Chapter 10), or may be oxidised, returning CO_2 to the atmosphere. The fluxes of particulate organic carbon estimated for major river systems in southern Asia suggest that this factor may be important in the global carbon cycle.

7.3   Measurements of erosion rates

7.3.1   Rock uplift, exhumation and surface uplift

There is some confusion in the literature regarding the terminology for various rates involved in tectonics and denudation (for discussion see England & Molnar 1990). A given point on the Earth's surface experiences vertical changes in its position relative to some reference datum such as the centre of the Earth caused by: (i) the uplift of underlying crustal rocks caused by tectonic or isostatic processes; (ii) denudation or deposition at the surface of the Earth; and (iii) compaction of the underlying sediment pile. All of these factors may contribute to *surface uplift* (or subsidence). Neglecting compaction and deposition in upland areas, surface uplift is normally the net result of rock uplift and denudation.

Now consider a crustal section through a mountain range with an average elevation h and a crustal thickness h_c, shown in Fig. 7.21a. If we instantaneously erode this crustal section and do not thicken the

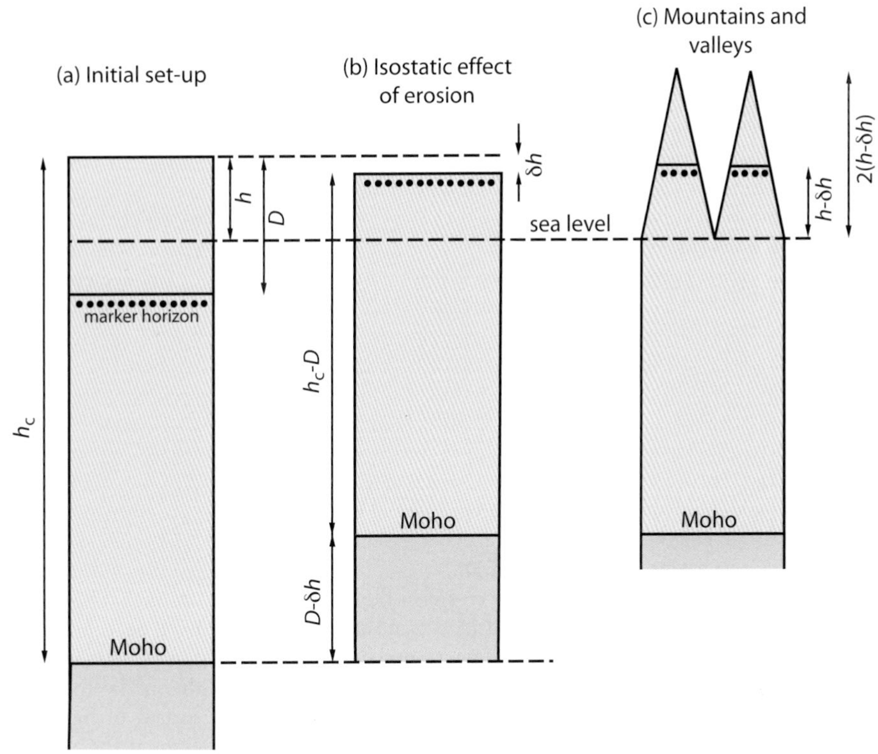

(c) Mountains and valleys

(a) Initial set-up

(b) Isostatic effect of erosion

Fig. 7.21 Highly schematic crustal cross-sections to illustrate the possible isostatic effects of erosion on the elevation of mountain peaks and valleys, modified from Molnar & England (1990). The initial set-up in (a) shows a surface elevation of h and a marker horizon at a depth D. Erosion of amount D in (b) brings the marker horizon to the surface, resulting in a change in surface elevation Δh. In (c), rivers excavate deep valleys to sea level, which results in an uplift of the mountain peaks to an elevation $2(h-\Delta h)$, while the marker horizon remains at an elevation of $h-\Delta h$. Airy isostasy is assumed throughout. Reproduced with permission of Nature Publishing Group.

crust tectonically to compensate for the loss of crust by erosion, it will respond isostatically to the unloading by erosion, as shown in Fig. 7.21b. The line with black circles below it is a reference horizon in the crust. Note that it has been uplifted between (a) and (b). This is the *uplift of rock*. Note also that all of the crust above the reference horizon has been eroded between (a) and (b). This is the *exhumation* or *denudation* D. As a result of the erosion and the isostatic adjustment, there is a change in the mean elevation of the surface Δh. In this case it is a surface lowering, so the surface uplift is negative. It is essential to recognise the differences between the uplift of rock, the denudation and the surface uplift.

Now imagine that the Earth's surface is carved into mountains and valleys, as shown in Fig. 7.21c. In this case the mean elevation is the same as in Fig. 7.21b, but the highest mountain peaks are considerably higher than the mean elevation initially. In this sense, erosion builds mountains. The isostatic balance for the situation described above is as follows. The pressure at depth h_c in (a) is simply $\rho_c h_c g$. The pressure at the same depth after erosion (b) is $\rho_c(h_c - D)$ $g + \rho_m(D - \Delta h)g$. Equating the two pressures and simplifying slightly, we obtain

$$\Delta h = D\frac{(\rho_m - \rho_c)}{\rho_m} \qquad [7.20]$$

If $\rho_c = 2800\,\mathrm{kg\,m^{-3}}$ and $\rho_m = 3300\,\mathrm{kg\,m^{-3}}$, the elevation change is only 15% of the total denudation. The elevation of the surface above sea level is $h - \Delta h$.

In Fig. 7.21c valleys are carved down to sea level, and the highest peak is therefore at an elevation of $2(h - \Delta h)$. If the denudation is 4 km, the elevation change from eqn. [7.20] is 0.454 km. If the mountain range is initially at an elevation of 3 km, the highest peak will

now be at 5 km, with deep valleys incised to sea level. Although this concept has been successfully applied to the Laramide uplifts of the Rocky Mountain region and Sierra Nevada, USA (Small & Anderson 1995, 1998), and the Himalayas (Molnar & England 1990; Burbank 1992; Montgomery 1994), it probably exaggerates the uplift of mountain peaks as an isostatic result of surface erosion.

It is also important to discriminate between local rates and regional rates of surface uplift and denudation. Whereas a local rate is measured at a point, and may be influenced by very specific tectonic and geomorphic processes such as slip on a fault or the occurrence of a landslide, isostatic responses depend on regional rates of denudation.

7.3.2 Point-wise erosion rates from thermochronometers

There is a wide range of techniques used to estimate erosion rates at a point. Any individual technique is useful for a particular time range, using particular materials. A large number of techniques are applicable only to Quaternary materials (see Noller *et al.* 2000 for an excellent compilation). For example, dendrochronology (tree rings) can only be used up to about 10 ka, and radiocarbon dating (^{14}C) has a useful time range of only up to 35 ka. U–Th radioisotopic dating, amino acid racemisation, thermoluminescence and optically stimulated luminescence (OSL) all have upper age limits of c.300 ka. All of these techniques, therefore, are restricted to Quaternary materials and are routinely used in geomorphological studies. The time range is significantly extended by the use of *in situ* cosmogenic radionuclides (e.g. ^{10}Be and ^{26}Al), and even further by the use of thermochronometric methods such as He diffusion during U–Th decay and apatite/zircon fission track analysis. Dating techniques making use of radiometric clocks with higher closure temperatures and longer half-lives, such as ^{40}Ar–^{39}Ar (half-life 1.25×10^9 yr), ^{87}Rb–^{87}Sr (half-life 1.4×10^9 yr), ^{40}K–^{40}Ar (half-life 4.89×10^9 yr), ^{235}U–^{207}Pb (half-life 7.04×10^8 yr) and ^{147}Sm–^{143}Nd (half-life of 1.06×10^{11} yr), provide estimates of very long-term (10^6–10^7 Myr) erosion rates. It needs to be stressed that the numerical values for erosion rate derived from different techniques with different temporal sensitivities cannot be directly compared (but see §7.3.5).

Since this book concerns basin analysis, we are mostly focused on the techniques providing 'geological' rather than 'geomorphological' rates of erosion. Use of a number of radiometric clocks with different closure temperatures (Table 7.7) allows cooling histories to be estimated. Combined with assumed or estimated geothermal gradients, these cooling histories can be used to calculate denudation rates over geological time periods. As an example, a cooling curve based on several different thermochronometers is shown for the Central Alps of Switzerland (Hurford 1986; Hurford *et al.* 1989) (Fig. 7.22). Multiple thermochronometers show the different thermal evolution of footwall and hangingwall blocks across a major extensional shear zone in the Central Alps (Mancktelow 1985).

A fundamental tool in the assessment of erosion is the relationship between the cooling age of a sample and its present-day elevation (Fig. 7.23). Samples might be recovered from a vertical drill hole, but samples from field campaigns are generally obtained from sloping valley sides. The isotherms in these regions are likely to be curved by the effect of irregular topography and geotherms may also be affected by high rates of rock advection at high Péclet number (§2.2.4 and §2.2.5). Ignoring these complications for the moment, the different ages of differently elevated samples are assumed to result from the time since passing through the same characteristic isotherm. Consequently, the expectation is that the highest elevated samples have the oldest cooling ages (Fig. 7.23), and that the slope of the age-elevation relationship is indicative of the rate of denudation.

The age-elevation relationship may be approximately linear, but may also show a break in slope reflecting the non-linear increase in annealing rate of fission tracks, or diffusion rate of helium, with temperature (Fig. 7.24). Thermochronological ages therefore rapidly decrease in the partial annealing zone (or partial retention zone for U–Th/He). Rock uplift of the crust may preserve a fossil partial annealing zone (PAZ) at a shallower depth or higher elevation, thereby revealing the amount of denudation that has taken place.

Conversion of a cooling history to a denudation history is not necessarily straightforward. The geothermal gradient is rarely well constrained, though a near-surface value of 25–30 °C km^{-1} is often assumed. The geothermal gradient may be affected by changes in the basal heat flow, the internal radiogenic heat production, the advection of heat from magmas and hot fluids, and the 'bending' of isotherms by the upward exhumation of rock towards an erosional surface of high relief. This latter effect is especially important to account for with low-temperature chronometers (Fig. 7.25). Rapid erosion rates (>5 mm yr^{-1}), associated with high thermal Péclet numbers (§2.2.5), may cause very high geothermal gradients (60–100 °C km^{-1}) to exist in the exhuming rock body (Stüwe *et al.* 1994; Mancktelow & Grasemann 1997), and geothermal gradients may be very different between mountain peaks and valley floors (Fig. 7.25).

Further details on the use of low-temperature thermochronometers to estimate catchment erosion rates, together with information on the techniques used, are deferred to §7.3.4. The use of thermochronometry on detrital minerals in the basin-fill is discussed in §7.6.1 and §10.4.3.

7.3.3 Catchment-scale erosion rates from cosmogenic radionuclides

We have previously learnt that the accumulation of cosmogenic radionuclides in rocks at the Earth's surface depends on latitude, altitude and depth below the surface (§7.2.1.1). If a rock is brought

Table 7.7 Closure temperatures of some common isotopic systems (in part from Braun *et al.* 2006, p. 5, table 1.1)

Mineral and dating technique	Closure temperature (°C)
Hornblende (K–Ar)	500 ± 50
Monazite (U–Pb)	525 ± 25
Muscovite (Rb–Sr)	500 ± 25
Muscovite (K–Ar)	350 ± 50
Biotite (K–Ar)	300 ± 50
Titanite (fission track)	265–310
Zircon (fission track)	240 ± 20
Biotite (Rb–Sr)	275 ± 25
Zircon (U–Th/He)	200–230
K-Feldspar (K–Ar)	150–350
Titanite (U–Th/He)	150–200
Apatite (fission track)	110 ± 10
Apatite (U–Th/He)	75 ± 5

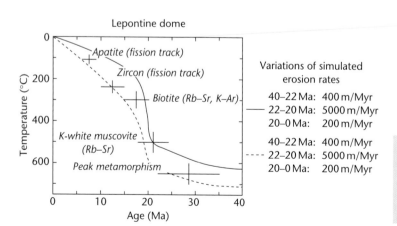

Fig. 7.22 Example of the use of a range of radiometric and thermochronometric techniques from the Lepontine dome, Central Alps, to constrain thermal (cooling) history (Schlunegger & Willett 1999). Solid and dashed lines are alternative model curves based on erosion data shown to right.

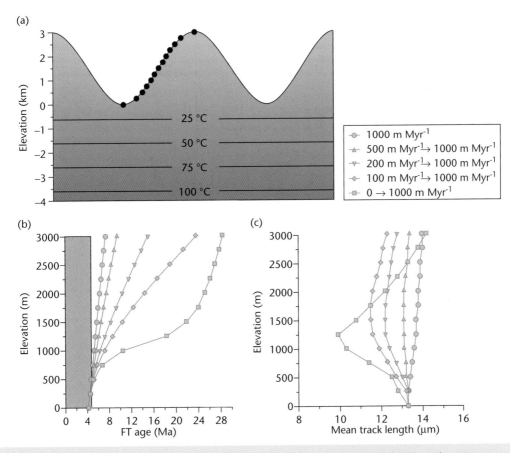

Fig. 7.23 Predicted apatite fission track age–elevation relationships for different denudation histories (from Braun *et al.* 2006, p. 10, fig. 1.5). (a) Simple model of thermal structure beneath a varying surface topography, with geothermal gradient 25 °C km⁻¹ and a surface temperature at sea level of 10 °C. Black dots are surface samples for which age and mean track length are predicted. Isotherms close to the topographic surface are increasingly unrealistic. (b) and (c) Predicted fission track ages and mean track lengths as a function of elevation for six different denudation histories with up to 1 km Myr⁻¹ (values in box) prior to 4.8 Ma followed by rapid exhumation for all samples since 4.8 Ma (shaded box in (b)). Reproduced with permission from Cambridge University Press.

closer to the surface by erosion, it experiences a greater amount of cosmic-ray bombardment. The cosmogenic radionuclide abundance is therefore partly a function of the rate of erosion.

The concentration of cosmogenic nuclides has been used to estimate the age of bedrock surfaces (Burbank *et al.* 1996) and depositional surfaces such as fluvial infill terraces (Repke *et al.* 1997). They can also be used to estimate the erosion rate at a point, ε, assuming the rate to be steady. A point in an exhuming mass of rock will experience increasing dosages of cosmogenic radiation P (production rate of new nuclides such as ^{10}Be per unit mass of quartz) as it gets closer to the land surface, where it experiences a surface production rate P_0. At the surface the concentration C is the integration over time of the production rate history

$$C = \int_0^\infty P_0 e^{-\varepsilon t/y^*} = P_0 y^*/\varepsilon \qquad [7.21]$$

where $t = y^*/\varepsilon$ is the time it takes to pass through a depth zone y^* thick at a velocity ε. If nuclides decay during this process of exhumation, eqn. [7.21] becomes

$$C = \frac{P_0}{(\varepsilon/y^*) + \lambda} \qquad [7.22]$$

where λ is the inverse half-life of the nuclide. For very long half-life, or rapid erosion rate, eqn. [7.22] reduces to eqn. [7.21]. If the decay constant of production with depth y^* is typically 0.7 m, and the surface production rate P_0 is six atoms per gram of quartz per year), a concentration of ^{10}Be of 0.6×10^6 atoms per gram gives a steady erosion rate of just 0.00525 mm yr⁻¹ (about 5 microns per year). Allowing for nuclide decay during the exhumation process, taking $\lambda = 5 \times 10^{-7}$ (yr⁻¹) for ^{10}Be, using eqn. [7.22], gives an erosion rate of 0.0049 mm yr⁻¹ for the same other parameter values. Bedrock erosion rates using this method have been obtained from a wide range of tectonic and climatic settings (Bierman 1994; Biermann & Caffee 2002) (Fig. 7.26).

If samples are taken of many detrital quartz grains derived from the catchment, it is possible to estimate a catchment-wide mean erosion rate (Brown *et al.* 1995; Granger *et al.* 1996). However, since altitude and topographic shielding affect the cosmic flux, the best results come from catchments with little (less than a few hundred metres) relief. The distribution of quartz-bearing lithologies should be relatively uniform throughout the area of the catchment. Different erosional processes (bedrock landsliding through to soil creep) also affect cosmogenic dosages. This effect can be accounted for by considering the abundance of cosmogenic radionuclides in different size classes of

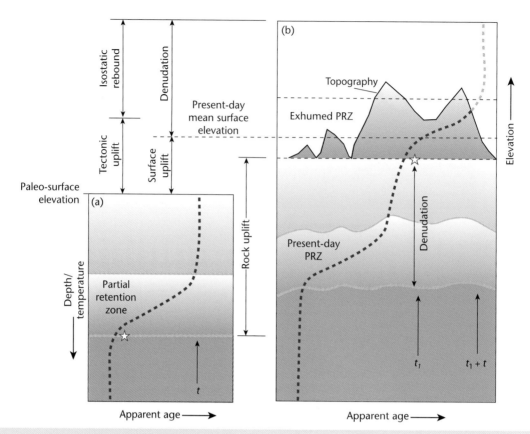

Fig. 7.24   Different types of uplift, denudation and thermochronological age–elevation relationship, showing the position of the partial retention zone (PRZ) for U–Th/He or partial annealing zone (PAZ) for the apatite fission track. From Braun *et al.* (2006, p. 11, fig. 1.6), modified from Fitzgerald *et al.* (1995). (a) Initial thermochronological age-depth profile developed under conditions of tectonic stability of duration *t*. Thermochronological ages decrease through the PRZ. (b) At time t_1, the crust starts to be uplifted and partially eroded, causing the old PRZ to be exhumed. A sample (star) originally near the base of the pre-existing PRZ will record the age of the onset of erosion t_1, and its elevation compared to the original depth will give the amount of rock uplift. © Jean Braun, Peter van der Beek, Geoffrey Batt 2006, published by Cambridge University Press, translated with permission.

the total detrital sediment load. Using this technique, a 3 km² catchment in Puerto Rico was found to have a catchment-wide average denudation rate of 40 mm ky⁻¹, using the concentrations of ¹⁰Be (Brown *et al.* 1995). On a larger scale, Schaller *et al.* (2001) used ¹⁰Be concentrations in quartz grains in river terraces and bedload from a number of European river basins to estimate catchment-wide denudation rates of 20–100 mm ky⁻¹ over the last c.10⁴ years.

Clearly, catchment-averaged erosion rates are of great interest to the basin analyst, since the rate can be used to estimate sediment discharges into adjoining basins. We are effectively 'letting nature do the averaging', both spatial and temporal. The spatial averaging results from sediment being derived from throughout the catchment, mixed and deposited at the sampling point. The temporal averaging is the time taken for grains to pass through the depth range y^* at the average velocity ε. For the data in the preceding paragraph, this averaging time is $0.7/5 \times 10^{-6}$ yr, or 0.14 Myr. At faster erosion rates, the time average reduces to tens of thousands of years.

This technique has been used in a number of settings. Granger *et al.* (1996) compared the average catchment erosion rate for fan lobes deposited at the margins of ancient Lake Lahontan in Nevada using: (i) the volume of dated fan material; and (ii) ¹⁰Be and ²⁶Al concentrations in fan sediments. The average catchment erosion rates (0.58×10^{-3} mm yr⁻¹ and 0.3×10^{-3} mm yr⁻¹) for the two contributing catchments and fans were very similar using the two methods. The cosmogenic nuclide dating method has also been used in more mountainous and wet settings, including Sri Lanka (Blanckenburg *et al.* 2004).

The measurement of cosmogenic nuclide concentrations in river terraces allows catchment-averaged erosion rates to be estimated for the time period of terrace construction. This can be done if the concentration of nuclides is recognised as due to inheritance during terrace formation, plus a component accumulated since abandonment of the active terrace. If flights of terraces can be analysed, a detailed picture of the denudational lowering of the contributing

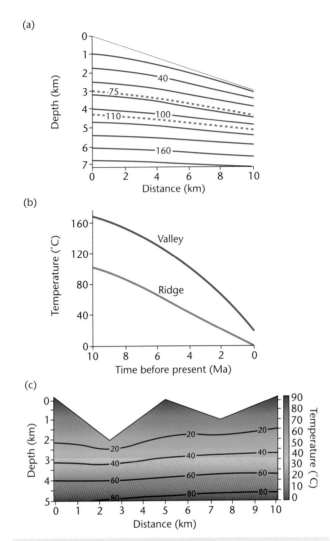

Fig. 7.25 Thermal structure beneath a steadily eroding ridge and valley, from Anderson & Anderson (2010, p. 156, fig. 6.45). (a) Thermal structure beneath ridge and valley topography, and the temperature versus time experienced by samples emerging at the ridge crest and the valley bottom (b). Isotherms for the base of the PRZ (75 °C) and PAZ (110 °C) are dashed. A sample collected from the valley floor would pass through the closure temperature for apatite at about 4.5 Ma and through the closure temperature for U–Th/He at just over 2 Ma, whereas a sample collected from the ridge would be close to the closure temperature for apatite at >10 Ma, and pass through the U–Th/He closure temperature at 7 Ma (after Safran 2003, fig.2). (c) Steady-state temperature field beneath a ridge and valley topography, for a uniform basal heat flow and no radiogenic contribution, contoured in intervals of 20 °C. Note the expansion of isotherms beneath the ridge and compression under the valleys, and that the effect of the irregular topography dies out within a depth equivalent to the topographic relief.

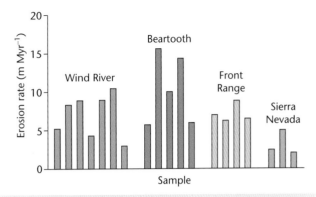

Fig. 7.26 Erosion rates estimated from cosmogenic radionuclide analysis on samples obtained from bedrock in mountains in the western United States. Radionuclides used are ^{10}Be. Erosion rates range from c.5–15 m Myr^{-1} (after Small *et al.* 1997).

catchments can be obtained, as has been achieved in the River Meuse of Germany (Schaller *et al.* 2004) (Fig. 7.27).

7.3.4 Catchment erosion rates using low-temperature thermochronometers

An explanation of the use of apatite and zircon fission track analysis in constraining burial history is given in §10.4.3 and §10.5.2. Here, we focus on the use of fission tracks in apatite and zircon, and on helium produced during the decay of uranium to thorium (U–Th/He) in constraining catchment erosion rates.

Fission track analysis is a relatively modern (since the 1960s, Naeser 1967) thermochronological technique based on the atomic damage caused to minerals by their spontaneous decay, almost exclusively by the fission of ^{238}U. The fission of ^{238}U produces a trail of damage to the lattice known as a *fission track*. Fission tracks become visible when the crystal is chemically etched and viewed at high magnification.

It is safe to assume that, statistically, fission takes place at a constant rate. The concentration of the parent isotope ^{238}U in minerals such as apatite ($Ca_5(PO_4)_3(F,Cl,OH)$), zircon ($ZrSiO_4$) and sphene ($CaTiO(SiO_4)$) is high enough to produce plenty of tracks, but low enough so that the crystal is not completely criss-crossed by tracks, making measurement difficult.

A *fission track age* (Hurford & Carter 1991) can be determined in a similar way to other radiometric methods, that is, through knowledge of the relative abundance of the parent and daughter products. In the case of fission track dating, the abundance of the parent is the original number of ^{238}U atoms – this can be estimated by irradiating the sample in a nuclear reactor. The abundance of the daughter product is proportional to the number of spontaneous fission tracks per unit volume of the apatite crystal.

There are a number of ways in which fission tracks can be measured and analysed. The most common is to measure the lengths of between 50–150 individual horizontal tracks that have been etched just below the surface of the polished section of the crystal. These are known as *confined tracks*. The data are normally shown as a histogram of track length with a mean and standard deviation. Importantly, fission tracks are metastable features that fade or *anneal*, which causes track lengths to shorten. Annealing is mostly controlled by

(a)

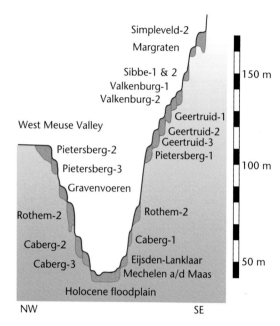

(b)

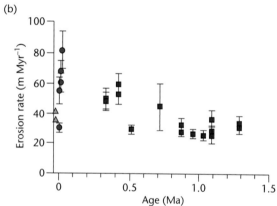

Fig. 7.27 Use of cosmogenic radionuclides to estimate basin-averaged erosion rates, after Schaller *et al.* (2004, figs 2 and 6). (a) Terraces along the River Meuse of western Europe date back to more than 1.3 Ma. The basin-averaged erosion rate of the contributing catchment area can be estimated from the concentrations of ^{10}Be in the alluvial cover of these terraces. (b) A two-fold acceleration of erosion rate from the early to middle Pleistocene, from 20–40 m Myr^{-1} to 30–90 m Myr^{-1}, is suggested by the ^{10}Be data. Reprinted with permission from Elsevier.

temperature (and also time), so track length distributions reveal information about thermal history. The higher the temperature experienced by the apatite crystal, the greater the annealing.

Individual minerals have a closure temperature below which fission tracks are preserved. For *apatite*, the closure temperature is 110 °C ±10 °C, but at geological time scales there is a range of temperature of at least 60 °C below the closure temperature where significant annealing takes place: this is termed the *partial annealing zone* (PAZ). Some annealing may even take place at room temperature, but it is generally regarded as negligible at these temperatures.

For *zircon*, the PAZ is between 200 and 350 °C, but the annealing behaviour of zircon is not as well known as for apatite. Because of the different closure temperatures of apatite and zircon, apatite fission track analysis is particularly useful in thermal studies of the upper few kilometres of the crust, whereas zircon fission track analysis is more informative for deeper levels of burial.

In summary, fission tracks are continuously formed at all temperatures, but are only preserved for geological time scales below the closure temperature. Their abundance therefore gives a fission track age. The track length distribution gives a record of the annealing history, which can be related to the time-temperature trajectory of the crystal.

Apatite fission track distributions are sensitive to different tectonic histories (Gallagher *et al.* 1998) involving a different trajectory and speed of heating and cooling. Commonly, the fission track age bears no relation to any specific thermal or tectonic event. Only where rapid cooling takes place, such as caused by rapid exhumation of a rift flank or orogenic belt, are many long fission tracks recorded, and the fission track age is a good indication of the timing of the cooling event. If a phase of heating is followed by cooling, as in basin development followed by inversion, the histogram of track lengths is typically bimodal (Fig. 10.18).

Apatite fission track analysis is routinely used in the quantification of denudation in a wide range of tectonic situations, including convergent mountain belts, rift flanks and passive margin escarpments (useful summary in Gallagher *et al.* 1998). The rate of cooling derived from fission track thermochronometry is very valuable in estimations of long-term sediment effluxes of erosional hinterlands, and in the timing of distinct tectonic exhumation events. For example, the onset of rapid exhumation in the Central Alaska Range was dated as 6 Ma on the basis of the recognition of two distinct populations of fission tracks (Fitzgerald *et al.* 1995). Some highly annealed tracks formed during a period of slow cooling, whereas other relatively unannealed tracks formed during a rapid exhumation event.

The accumulation of ^{4}He from the decay of uranium and thorium to stable lead can also be used as a thermochronological tool (Wolf *et al.* 1996). The basis for U–Th/He dating is that ^{4}He nuclei (α particles) are produced by the decay of ^{238}U, ^{235}U and ^{232}Th. The helium produced diffuses from the crystal over time, escaping through its surface faces. The rate of diffusion scales on temperature with an Arrhenius-type relationship

$$\frac{\kappa}{a^2} = \frac{\kappa_\infty}{a^2}\exp(-E_a/RT) \qquad [7.23]$$

where κ is the diffusivity, κ_∞ is the diffusivity at an infinite temperature, and a is the diffusion domain radius, which may be the physical grain dimension. Diffusivities are measured in the laboratory for certain minerals with different chemical compositions, grain size and shape. He diffusivity is best understood in apatite. Laboratory data suggest that the closure temperature in apatite for the retention of helium is c.80 °C (Zeitler *et al.* 1987), and more recent estimates are 68 ± 5 °C (Reiners 2002), so U–Th/He methods extend the temperature range of sensitivity to lower temperatures compared to apatite fission track analysis. This gives additional data on the thermal evolution of the upper 2–3 km of the crust or sedimentary basin-fill.

One method of investigating the relationship between temperature history and He diffusion is to study the He age distribution in boreholes where the temperature distribution with depth is known. He ages are predicted to decrease rapidly downhole, because of the effect

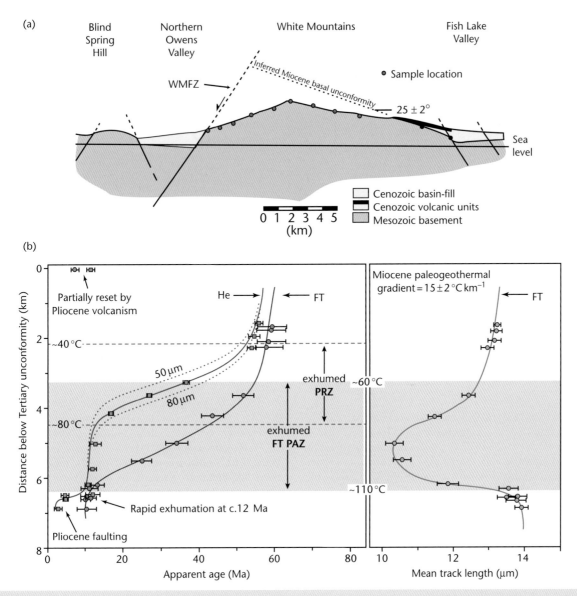

Fig. 7.28 (a) Cross-section of the northern White Mountains, east-central California and west-central Nevada, showing the tilted fault block geometry. The main border fault (White Mountains Fault Zone, WMFZ) has a total of 8 to 9 km normal displacement, resulting in a 25° eastward tilt recorded by Tertiary volcanic rocks. (b) Integrated apatite fission track and U–Th/He data showing an exhumed partial annealing zone for apatite fission tracks and an exhumed partial retention zone for U–Th/He. Samples from below the exhumed PRZ are invariant with age and directly date the time of inception of footwall cooling caused by extensional faulting. After Stöckli *et al.* (2000).

of temperature on diffusion rates (Wolf *et al.* 1998). Consequently, there is a *helium partial retention zone* (HePRZ), which extends from approximately 40 °C to 70 °C. This is conceptually similar to the PAZ in fission track analysis. The existence of a HePRZ was confirmed by Stöckli *et al.* (2000) who found a HePRZ overlying an apatite PAZ in samples taken from the White Mountains of California.

Cooling ages from (U–Th)/He diffusion are most simply related to the time of cooling of a sample through its closure temperature. Rapid cooling of samples during exhumation may reveal a fossil or paleo-HePRZ, as we have seen with apatite fission track analysis. Stöckli *et al.* (2000) recognised a distinct break in slope at 12 Ma

when apparent He ages were plotted against depth, signifying the onset of rapid exhumation caused by extensional faulting in the White Mountains of eastern California (Fig. 7.28). This event is barely recognisable from the distribution of apatite fission track apparent ages.

Although the benefits of a low-temperature thermochronometer such as U–Th/He are clear, there are drawbacks: (i) the isotherms beneath evolving mountainous topography may be strongly warped, making estimates of exhumation problematical (Braun 2005; Reiners *et al.* 2005); (ii) helium concentrations may be strongly affected by thermal events near or at the surface of the Earth, such as forest fires

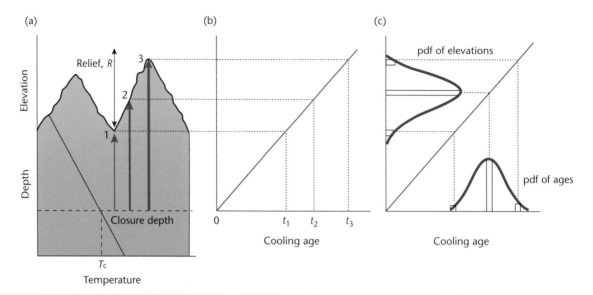

Fig. 7.29 Use of elevation profiles to estimate exhumation rate in a steady-state landscape where the erosion rate of the ridge and valley floor are the same, after Brewer *et al.* (2003, fig. 2; reproduced with permission of John Wiley & Sons, Ltd.) and Anderson & Anderson (2010, p. 150, fig. 6.36). (a) Three samples, labelled 1 to 3, from a valley side, span a range of topographic elevation *R*, giving three cooling ages t_1 to t_3 (b). The long-term erosion rate is $R/\Delta t$. (c) Samples of detrital grains from a stream bed draining this upland area have a pdf of cooling ages, which can be mapped onto a pdf of elevations in the catchment, assuming that the geothermal gradient and the catchment hypsometry are known.

(Mitchell & Reiners 2003; Heffern *et al.* 2008) or even high daytime temperatures; and (iii) helium ages are prone to being reset during even small amounts of burial of detrital grains in sedimentary basins.

Thermochronometric tools can be used to constrain catchment erosion by use of the *age–elevation relationship* of fission tracks (Figs 7.29, 7.30) (see also §7.3.2) (Brewer *et al.* 2003; Stock *et al.* 2006). The exhumation rate can be estimated from the slope of the cross-plot between elevation and age (Fig. 7.29). In the simplest scenario, the highest sample from the mountain peak is assumed to be eroding at the same rate as the lowest sample in the valley bottom. The vertical spacing of the highest and lowest samples and the difference in the cooling ages therefore defines the erosion rate. The long-term erosion rate ε is then given by

$$\varepsilon = R/\Delta t \qquad [7.24]$$

where Δt is the difference between the youngest and the oldest cooling ages, and R is the relief between mountain and valley. If the elevation difference is 2.5 km, and the difference in cooling ages is 10 Myr, the long-term average erosion rate is 0.25 mm yr^{-1}.

The age–elevation relationship of thermochronometers can also be used to estimate where sediment comes from in an upland catchment (Stock *et al.* 2006) (Fig. 7.30). The U–Th/He ages of detrital sediment grains collected from the mouth of a catchment can be presented as a pdf (probability density function), and assumed to contain spatially averaged information on erosion rate. The simplest model for spatial averaging is that the sediment is derived uniformly across the catchment. The pdf of measured U–Th/He in detrital apatites can be compared with the pdf predicted by a convolution of the observed catchment hypsometry with the local age–elevation relationship. In this way, each elevation in the catchment is assigned an AHe age. If the measured pdf matches the predicted pdf perfectly, then erosion is viewed as uniform (Brewer *et al.* 2003; Ruhl & Hodges

2005), but mismatches of the pdfs indicate localised erosion, glacial modification of the catchment landscape (Brocklehurst & Whipple 2004) or sediment storage (Fig. 7.30). Results from two catchments in the Sierra Nevada of eastern California show that one (Inyo Creek) is eroding uniformly, whereas the other (Lone Pine Creek) has a strong mismatch of the measured and predicted pdf. The results suggest that low-temperature thermochronometers are valuable for assessing whether erosion is uniform, with sediment efficiently flushed from the catchment, or highly non-uniform as a result of localised erosion, and/or sediment storage. However, identification of the precise factors causing the mismatch of pdfs is challenging.

Erosion rates may also be calculated by measuring the isotopic signatures of river water or sediment loads, in comparison to the isotopic composition of different source lithologies in the catchment. A spatially uniform erosion of the catchment would yield isotopic signatures in water or sediment that reflected the isotopic ratio of the parent lithology and the area as a percentage of the catchment occupied by this lithology. Departures from this uniform model should reflect different erosion rates in the catchment. The total denudation rate for the whole catchment can then be obtained by summing the effects of the source areas with different lithologies. The isotopic ratios of Sr and Nd in detrital clay minerals from Nepalese catchments in the Himalayas have been used in this way (Galy *et al.* 1996).

7.3.5 Erosion rates at different temporal and spatial scales

It is evident from the discussion in §7.3.2, §7.3.3 and §7.3.4 that erosion rates can be estimated at different temporal resolutions. Consequently, an erosion rate estimate derived for a very short time resolution provides very different information to the erosion rate derived for a long time resolution. For example, the rate of bedrock lowering caused by a single debris flow in an upland channel gives a local

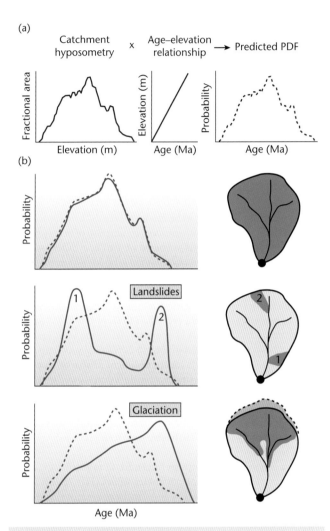

Fig. 7.30 Cooling age pdfs can be used to determine where sediment comes from in an upland catchment (after Stock *et al.* 2006, p. 726,fig. 1). (a) A predicted pdf is made from a catchment hypsometry and age–elevation relationship. (b) The predicted pdf will be closely mimicked by the measured pdf derived from low-temperature thermochronological methods such as U–Th/He if sediment is derived uniformly from the catchment. Local derivation of grains from discrete, localised events such as landslides will cause strong departures from the predicted pdf, whereas a systematic non-uniform derivation, for example from a glaciated upper catchment, will cause a skew in the measured pdf towards higher elevations and therefore towards older ages (Brocklehurst & Whipple 2004).

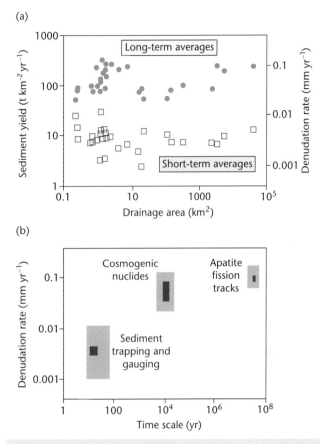

Fig. 7.31 Erosion rate estimates are dependent on the time scale of the methodology used (after Kirchner *et al.* 2001, figs 1 and 2; Anderson & Anderson 2010, p. 143, fig. 6.26). (a) [10]Be-based cosmogenic methods produce long-term averages that are much higher than short-term estimates based on sediment gauging station and reservoir trapping data, irrespective of drainage area. (b) These measurements are in turn different from the geological time scale represented by apatite fission track-based estimates.

erosion rate over a time period of minutes to hours. At the other extreme, cooling histories obtained from apatite fission track studies provide estimates of erosion rates at the time scale of millions of years (Fig. 7.31).

There is considerable value in compiling the variation of erosion rates derived from methods with different temporal resolution (Dadson *et al.* 2003). The Taiwan orogen has rock uplift rates of 5–7 mm yr⁻¹, mean annual precipitation of 2.5 m yr⁻¹ and strong earthquake activity, which together promote rapid rates of landsliding, debris flows and fluvial bedrock incision (Hovius *et al.* 2000).

Erosion rates at the decadal time scale have been calculated from the suspended sediment discharges of Taiwanese rivers at 130 gauging stations, from the rate of filling of 12 major reservoirs, and from the annual discharge of sediment into the ocean by rivers. The 30-year annual average erosion rate of Taiwan is 3.9 mm yr⁻¹ using suspended load data, with significant regional differences (up to 60 mm yr⁻¹ in the tectonically active SW Taiwan). Discharges of sediment into the ocean (average of 384 Mt yr⁻¹ for the period 1970–2000) support the gauging station data. Erosion rates for the Holocene (<10 ka) have been derived from dated river terraces cut into bedrock using ¹⁴C in wood or plant fragments in the alluvial veneer. Incision rates were calculated by dividing the radiocarbon age by the elevation of the bedrock terrace above the present-day river. Holocene erosion rates range from 1.5 mm yr⁻¹ to 15 mm yr⁻¹, with broadly the same pattern as the rates obtained from sediment discharges. Long-term erosion rates derived from low-temperature thermochronometry (apatite fission track) are 3.0–6.0 mm yr⁻¹ and 1.5–2.5 mm yr⁻¹ in the metamorphic core of the mountain belt and in the fold-thrust belt of SW

Taiwan respectively. A comparison of these erosion rates suggests that despite the difference in temporal resolution, the metamorphic core of the Central Range of Taiwan has consistent values of 3–6 mm yr^{-1} across the temporal scales considered. However, large variations in Holocene rates are thought to reflect the more localised impact of growing tectonic structures (e.g. in the fold and thrust belt of SW Taiwan). At decadal scales, individual major earthquakes are responsible for major spatial and temporal variations in erosion rate.

We saw in §7.3.2 that erosion rates can be estimated from the sediment volumes in neighbouring sedimentary basins. This enables sediment volumes of known age to be used to calculate erosion rates at the temporal resolution of the stratigraphy. The spatially and temporally variable sediment effluxes of channel-hillslope systems in individual catchments are smoothed as the sediment is transported through major fluvial systems with large storage capacities in floodplains, channel bars and lakes. Consequently, erosion rates (sediment yields) derived from the sediment discharge at the mouths of large river systems such as the Ganges–Brahmaputra are a buffered record of the contributing inputs (Métivier *et al.* 1999). In addition, the sediment preserved as marine stratigraphy is a time-integrated record of the sediment discharge from the land area over the time periods resolvable from stratigraphic dating tools. In general, therefore, longer time scale and larger spatial scale 'geological' estimates of erosion rate are strongly smoothed relative to the characteristics of the individually erosive geomorphic events (see also §8.4.1).

7.4   Channel-hillslope processes

The engine for the sediment routing system, termed the *clastic factory* by Leeder (1999), is the coupled process systems of hillslopes and rivers. The topography generated by tectonics is degraded through the action of this coupled system. Rivers typically become entrenched in position, cutting down like cheese wires through the regionally uplifting bedrock (Burbank *et al.* 1996). The intervening hillslope system provides erosional material by soil creep, overland flow, gullying, debris flows, landslides and rockfalls. If the rate of hillslope erosion is slower than the rate of valley lowering, hillslopes steepen and eventually become susceptible to landsliding involving bedrock. The amplitude of hillslopes is therefore set by the effect of rock strength on landsliding. The fluvial system transports away material derived from hillslope erosion through its channel networks. This evacuation of valley bottoms is particularly rapid where the stream is incising into bedrock. Evacuation may be significantly delayed where high-magnitude trigger events such as earthquakes cause an instantaneous discharge from hillslopes, blocking valleys with landslide and debris-flow dams. The efflux of the hillslope-bedrock channel system sets a boundary condition for far-field sediment dispersal.

In mountain ranges landslide activity, triggered by rainstorms and co-seismic shaking (Densmore & Hovius 2000), is critical to the clearing of hillslopes (Densmore *et al.* 1997). A number of rapidly uplifting mountain belts (Southern Alps of New Zealand, Karakorum of the western Himalayas) are believed to exhibit a steady-state topography where the tectonic rate of uplift of rock is balanced by the landslide-dominated erosional sediment efflux. In the Southern Alps of New Zealand, the flux of material due to landsliding estimated over a 60-year period (several mm yr^{-1}) is roughly equivalent to the sediment output of the main rivers draining the region (Hovius *et al.* 1997).

In the sections that follow, we present a physically based but necessarily incomplete account of the processes operating in hillslope-channel systems. These processes fashion landscapes and deliver sediment to basins. Rather than attempt a comprehensive review of such a rapidly growing field of study, we highlight some basic modelling approaches to the evolution of hillslopes (§7.4.1), the incision of bedrock rivers (§7.4.2) and the far-field transport of sediment in alluvial rivers (§7.5). We introduce some additional concepts in numerical landscape models as a way of integrating many of the ideas found in the preceding sections in §7.6.3.

7.4.1   Modelling hillslopes

The evolution of hillslopes can be modelled as a diffusive process, in a manner similar to the one-dimensional conduction of heat (Chapter 2). Sediment diffusion models are based on two first-order assumptions: (ii) the *sediment continuity equation* (also known as the *Exner equation*), which states that the spatial variation in the sediment transport rate is proportional to the vertical erosion or aggradation rate of the substrate, assuming there to be no changes in the concentration of suspended sediment (see Allen 1997, p. 180–181) (Appendix 43); and (ii) that the flux is proportional to the local gradient, which is a basic notion underpinning all diffusional problems.

The derivation of solutions for the sediment discharge, slope and maximum relief of hillslopes undergoing diffusion is given in Appendix 42. The shape of diffusional hillslopes is parabolic, with a maximum slope at the river channel and a minimum slope at the hillslope crest. The characteristic time scale for diffusional hillslopes is of the familiar form $\tau = L^2/\kappa$ that we have seen previously in problems of heat flow (§2.2).

We can use the relationships in Appendix 42 to predict the shape of hillslopes under a range of parameter values (Fig. 7.32b). If two channels located 100 m apart ($L = 50$ m) incise at a rate of 0.5 mm yr^{-1}, typical of tectonic uplift rates, and a diffusion coefficient applicable to the slow processes of rainsplash and soil creep is 50×10^{-3} m^2 yr^{-1}, the maximum gradient of the hillslope, as given by eqn. [A42.5] is 0.5 (tan^{-1} 0.5 is 27°). The time constant from eqn. [A42.6] is 50 kyr.

Diffusional theory can also be used to study the degradation of fault scarps (Hanks *et al.* 1984), river terraces (Avouac 1993) or independently dated wave-cut lake margin scarps, such as those of Pleistocene Lake Bonneville, Utah. The scarp profile evolves over time by an erosional smoothing of the upper portion and a depositional smoothing of the lower portion. The slope of the mid-point of the scarp decays as the square root of time, as in the thermal evolution of a piece of newly created oceanic crust (§2.2.7). Diffusivities in the present-day arid landscapes of the American west were estimated as 5–100 m$^2 \times 10^{-3}$ yr^{-1} (Burbank & Anderson 2001). However, a large number of studies suggest that we cannot always assume that the diffusivity κ is constant and independent of x on hillslopes and scarps of various origins. Furthermore, the upper parts of hillslope profiles may be strongly affected by the regolith production rate (§7.2), so that they are weathering-limited and therefore not strictly determined by linear diffusion (Rosenbloom & Anderson 1994).

If the channel incision rate is large compared to the diffusive mass wasting of the hillslope, a critical slope will be reached, after which landsliding takes place. This is why mountains with high channel incision rates, such as the Southern Alps of New Zealand and the Finisterre Range of Papua New Guinea, have hillslopes dominated by landsliding. Landslide-dominated hillslopes are straight rather than parabolic (Anderson 1994; Densmore *et al.* 1997; Densmore & Hovius 2000). The critical slope for the onset of landsliding depends on bedrock strength, vegetation cover and fluid pressures in the regolith and bedrock. It is likely to vary between c.30° and 60°. For example, Ellis *et al.* (1999) found a critical slope for landsliding of

just 34°. For the hillslope system introduced above, this critical slope would be achieved at a channel incision rate of 1.2 mm yr⁻¹. For effective diffusivities in the range 10 to 100×10^{-3} m² yr⁻¹, the channel incision rates required to initiate landsliding at critical slopes between 40° and 60° are between 0.2–3.5 mm yr⁻¹. This range covers the tectonic uplift rates of recent mountain belts. The rates of rock uplift along the Alpine Fault in the Southern Alps of New Zealand is over 5 mm yr⁻¹, with rates of 1 mm yr⁻¹ along the main drainage divide of the Southern Alps (Koons 1989). The Central Range of Taiwan and the Finisterre Range, Papua New Guinea, have similar tectonic rates of rock uplift.

The flanks of mountain belts are commonly sculpted into transverse valleys and interfluve ridges with a characteristic spacing (Hovius 1996). The transverse river systems draining linear mountain belts appear to maintain a near constant aspect ratio (Fig. 7.33; Table 7.8)

$$\frac{W}{S} \approx 2.2 \qquad [7.25]$$

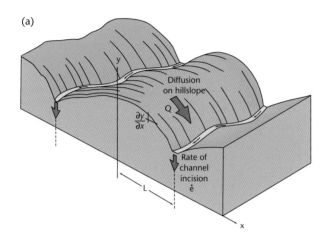

(a)

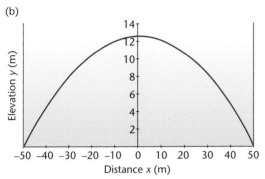

(b)

Fig. 7.32 (a) Schematic showing hillslopes and bedrock channels, with notation. (b) Solution (parabolic) for the hillslope profile using the parameter values given in the text.

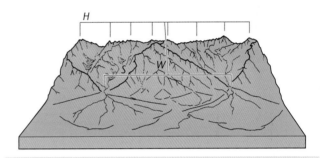

Fig. 7.33 Schematic illustration of two river catchments in a mountainous area with an average elevation of the axial ridge *H* and half-width *W*. The catchments have a transverse spacing *S* at the mountain front, as seen from the locations of the apices of the alluvial fans. Table 7.8 gives data from Hovius (2000).

Table 7.8 Mountain belt geometry and drainage spacing (after Hovius 2000)

Mountain belt	H (m)	W (km)	a (m km⁻¹)	L (km)	n	S (km)	R
Southern Alps	2653	21.1	123	198	19	10.99	1.92
Finisterre Range	3640	25.5	137	187	15	13.36	1.91
Maoke Range	4038	36.4	108	298	19	16.56	2.20
Barisan Range	1914	28.0	66	439	33	13.72	2.04
Central Range	3292	25.0	124	243	20	12.29	1.95
Kirgizskiy Khrebet	4101	34.7	92	202	14	15.52	2.23
Northern Tien Shan	4950	38.1	84	269	15	19.18	1.99
Apennines	1949	39.2	48	547	33	17.65	2.22
Sierra Nevada	3192	84.0	36	422	12	38.36	2.19
Peruvian Andes	5797	81.9	70	1000	25	40.00	2.05
Central Himalayas	7822	158	49	1610	12	139.02	1.17

Above: H, mean height of the culminations of the drainage divide; *W*, average half-width of the mountain belt, measured from the main drainage divide to the mountain front; *a*, mean gradient of the enveloping slope of the range flank; *L*, length of section over which spacing of drainage outlets was measured; *n*, number of streams draining section measured; *R*, aspect ratio of transverse drainages for the section (*R* = *W/S*); *S*, lateral spacing of main catchments at the mountain front. Note that the value of *R* in the central Himalayas is markedly different to the tight clustering of *R* around 2 for the other mountain belts. Anticlines in the Himalayan foothills cause drainage deflection and capture, which amalgamates transverse streams into larger catchments with a wider spacing.
Reproduced with permission of John Wiley & Sons, Ltd.

where W is the length of the main transverse trunk stream, and S is the width of the catchment or outlet spacing. In the Southern Alps of New Zealand, the transverse landslide-dominated catchments on the wet western flank have transverse spacings of 5 to 10 km and catchment lengths of 20 to 25 km. Maximum hillslope relief is about 1500 m. This geomorphological sculpture is controlled by the tectonic vertical uplift rate of rocks ($V(x)$) (Appendix 45).

On the western flank of the Southern Alps of New Zealand (Koons 1989) (Fig. 7.34), where the lateral distance from interfluve to interfluve $L = 3000$ m, the maximum topographic relief $y_{max} = 1500$ m, and the vertical tectonic uplift rate of rock V is approximately 5 mm yr^{-1}, the effective erosional hillslope diffusivity must be 15 m^2 y^{-1}. It is considerably higher than the value of 0.01 m^2 yr^{-1} estimated for the semi-arid to arid hillslopes in the western United States

(Rosenbloom & Anderson 1994). The effective erosional hillslope diffusivity encompasses a whole range of geomorphological processes acting on the hillslope, and therefore may lump together the effects of mean annual precipitation (over 12 m yr^{-1}) and rainfall intensity, rock strength, vegetation and temperature-dependent weathering. The value calculated here reflects the high activity of rapid mass wasting events. The mean gradient of the hillslopes from the diffusion model is y_{max}/L, which for the Southern Alps examples is 0.5, or 27°. This corresponds closely with measured mean values.

Inspection of eqn. [A42.12] shows that the transverse spacing of hillslopes must always be small in order to balance high rates of tectonic uplift of rock in a steady-state landscape. On the drier eastern flank of the Southern Alps (<1 m yr^{-1} mean annual precipitation) the tectonic uplift rates reduce from 1 mm yr^{-1} at the main

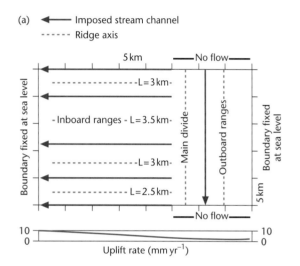

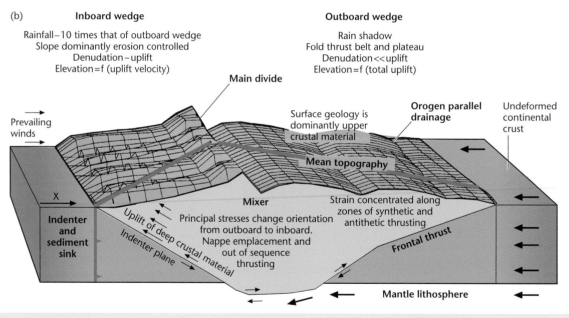

Fig. 7.34 The doubly vergent orogenic wedge of the Southern Alps of New Zealand has deeply etched, narrowly spaced, transverse drainages on the wet western flank where tectonic uplift rates are high. On the drier eastern flank, rivers are much less deeply entrenched and are broadly spaced. (a) Boundary and initial conditions for the numerical model in Koons (1989). (b) Plate velocities outside the wedge, tectonic transport directions within the wedge, and surface topography for a doubly vergent wedge with precipitation from the left. The marked asymmetry in denudation promotes a strong asymmetry in rock uplift within the wedge. Modified from Koons (1989).

divide to zero at the east coast of South Island. On this flank, braided river valleys are broadly spaced and less deeply entrenched into bedrock.

From eqn. [A42.12], the aspect ratio of y_{max}/L can also be written as $VL/2\kappa$. This is a dimensionless ratio of a flow rate of rock across a section of the transverse drainage versus an effective erosional diffusivity. It has the form of a *Péclet number*. If the *erosional Péclet number* is greater than about 0.6, it is unlikely that erosional processes can keep pace with tectonic uplift rates to produce a steady-state landscape. The erosional Péclet number for the wet western flank of the Southern Alps, using the parameter values above, is 0.5.

7.4.2 Bedrock river incision

The products of hillslope erosion are transported away by channelised flows (rivers). Channelised flow takes place wherever a stream power threshold is locally exceeded (Montgomery & Dietrich 1988). There is a fundamental distinction, however, between rivers incised into bedrock (this section) and alluvial rivers with beds and banks of sediment (§7.5). Bedrock rivers are those that lack a coherent bed of active alluvium (Howard 1987). Consequently, bedrock rivers have a transport capacity that is larger than that required to transport all of the available sediment. They incise by abrasion by the sediment load, plucking of blocks from the bed, and cavitation (Hancock *et al.* 1998). Bedrock rivers are the dominant channel type in mountainous topography.

There are a number of interrelated ways of approximating the rate of bedrock incision. A common approach makes use of the idea that bedrock incision is controlled by an area-slope or discharge-slope product. This is often termed the *stream power rule* (Appendix 44). A related view (Howard 1994) is that the rate of bedrock incision is proportional to the shear stress on the bed. Alternatively, the rate of incision could be related to the excess stream power available to erode bedrock in the channel bed above that required for sediment transport (Seidl & Dietrich 1992). The modelling of bedrock incision

using the area-slope or discharge-slope product is illustrated in Appendix 46.

Hillslopes and bedrock channels are dynamically linked, but they have different response times. The hillslope response time is L^2/κ (eqn. [A42.6]). The response time of the channel system must be influenced by the length of the channel and the average velocity of the knickpoints generated when the channel adjusts to its new base level. Such knickpoints migrate headwards (Appendix 47), but slow down in their upstream velocity as the stream power decreases due to a reduction in the contributing drainage area. Anderson (1994) calculated channel system response times assuming the stream power rule as typically 10^3–>10^4 yr. This is significantly longer than the response time for landslides, but is comparable to the hillslope response time. This suggests that hillslopes characterised by the high effective diffusivities produced by landslides are always in equilibrium with the incising channels. Less steep hillslopes dominated by slow diffusion, however, may not be in equilibrium with channel incision and may display transient rather than steady-state morphologies.

We saw in §7.3.1 that deep incision by rivers causes a removal of mass from the area undergoing erosion, which drives an isostatic compensation (Molnar & England 1990). Consequently, the high mountain peaks fringing deeply incised valleys may experience a surface uplift. We can now restate this possibility in terms of a bedrock river incision rule. Assuming the isostatic uplift to be due to a local (Airy) compensation rather than to a regional flexure (Appendix 2), the rate of uplift of the mountain peaks is

$$\frac{\partial y_p}{\partial t} = \frac{1}{2}\left(\frac{\rho_c}{\rho_m}\right)c_1 x^2 S \qquad [7.26]$$

where ρ_c and ρ_m are the crustal and mantle densities respectively, y_p is the elevation of the mountain peak, x is the horizontal coordinate, and S is the slope. Using $c_1 = 0.5\,\text{m}^{-1}\,\text{Myr}^{-1}$, profiles along the Arun and Karnali Rivers in the Nepalese Himalaya show the elevation of mountain peaks some distance downstream from the drainage divides in the Tibetan Plateau (Montgomery 1994) (Figs 7.35, 7.36).

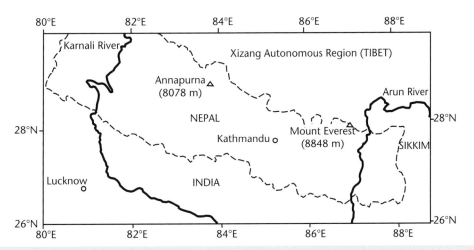

Fig. 7.35 The Himalayas and Tibetan Plateau, with the location of the Arun and Karnali rivers.

(a)

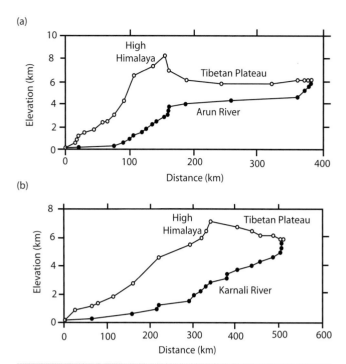

(b)

Fig. 7.36 Long profiles of the mountain peaks (open circles) and valley bottoms (filled circles) along the Arun (a) and Karnali (b) rivers, showing the high elevation of the summits c.200 km from the headwater regions, suggesting that bedrock river incision rates produce an isostatic uplift of the peaks. After Montgomery (1994), reproduced with permission of the American Geophysical Union.

The greatest relief is on the edge of the Himalayan plateau, where the Karnali River has incised 4500 m below the surrounding peaks. As much as 20–30% of the present elevation of the Himalayan peaks can be explained by isostatic compensation.

7.5   Long-range sediment transport and deposition

7.5.1   Principles of long-range sediment transport

A fundamental characteristic of alluvial channels is that they have sufficient sediment to equal or exceed the transport capacity. The transport of sediment in river systems has been the subject of modelling by a generation of civil engineers, geomorphologists, and, latterly, geologists. Our purpose here is not to attempt a full description of long-range sediment transport by rivers. For more comprehensive treatments the reader is referred to Parker (1978a, b), Paola *et al.* (1992) and Dade and Friend (1998). Bridge (2003) and Mackey and Bridge (1995) provide much useful information on rivers and fluvial deposits. The fact that river profiles can be matched by solutions deriving from the field of thermal conduction (Fig. 7.37) is an encouragement that diffusion may operate at the large scale in fluvial systems.

Various forms of diffusion equation have been applied to long-range fluvial transport. Here, we concentrate on an assessment of long-range fluvial transport of a mixture of gravel and sand. The quantitative development of this group of models can be found in

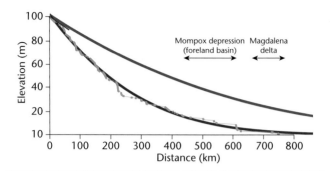

Fig. 7.37 Magdalena River long profile, and model curve based on heat conduction solution, as explained in Appendix 48. The two curves are for different slopes close to $x = 0$. The red line is an almost perfect fit to the longitudinal profile. Blue line, slope at $x = 0$ of 0.167×10^{-3}; red line, slope at $x = 0$ of 0.333×10^{-3}.

Paola *et al.* (1992) and Marr *et al.* (2000). It is well known in modern mixed gravel–sand fluvial systems that there is a relatively steep gravelly proximal zone with an abrupt change to a lower gradient sandy zone (Sambrook-Smith & Ferguson 1995). To be able to predict the movement of the 'gravel front', or 'gravel–sand transition' would be of considerable benefit in basin analysis.

Most models of sediment transport in rivers make the same set of assumptions (Paola 2000):

(1) It is assumed that the long-term flow of water in a river approximates that of a steady uniform flow down an inclined plane. For a flow of depth h and density ρ on a slope $\delta y / \delta x$, the downslope component of the fluid weight on a unit area of the river bed is $\rho g h \delta y / \delta x$. This downslope acting force must be opposed by an equal and opposite drag force exerted on the fluid by the unit area of bed – this is the shear stress τ_0 (Fig. 7.38a). Consequently, the force balance gives

$$\tau_0 = -\rho g h \frac{\partial y}{\partial x} \qquad [7.27]$$

Rivers are anything but steady and uniform. However, eqn. [7.27] can be used as an approximation for long-term river behaviour. It works best for shallow, high-gradient streams and worst for deep, low-gradient streams. If the flow depth is large compared to the channel width, h should be replaced by the *hydraulic radius R* in eqn. [7.27] (Fig. 7.38b).

(2) Second, we make use of an equation that expresses the resistance to flow in the channel. A fluid moving over its bed and banks experiences frictional losses of energy known as *flow resistance*. Where the bed of the river is rough, as is always the case in natural rivers, the energy losses should in some way be related to the length scale of the roughness of the bed. This roughness can be expressed in a number of ways. A common method is to use the *Darcy-Weisbach friction factor f*

$$f = \frac{8\tau_0}{\rho_f u^2} \qquad [7.28]$$

where u is the flow velocity, and the other parameters are explained above. The friction factor (or similar forms such as the Chézy coefficient C and Manning's n) varies strongly according to the grain size of the sediment on the river bed and is especially affected by the

(a)

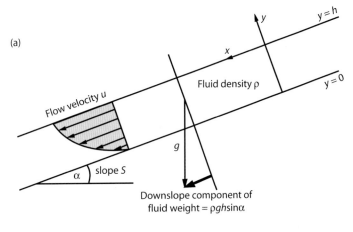

(b)

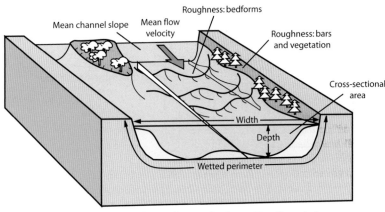

Discharge = Cross-sectional area x Mean flow velocity
Hydraulic radius = Cross-sectional area/Wetted perimeter

Fig. 7.38 (a) Notation for steady uniform flow down an inclined plane of slope sin α. (b) The wetted perimeter and hydraulic radius of a river. The presence of bar forms, pools, dunes and ripples, vegetation and pebbles all contribute to the friction factor of the river.

presence of bedforms such as ripples and dunes and of macroforms such as bars, chutes and pools.

(3) Third, we conserve the discharge of water through the system. Consider a slice of width B of an alluvial basin of length L, which has active channels on its surface with cumulative width b, of flow depth h and containing flows of velocity u (Fig. 7.39). Let β be the fraction of the section width B occupied by channels, so $\beta = b/B$. The discharge of water Q_w in the channels occupying the width of floodplain B is bhu, or averaged across the floodplain is βhu. Consequently,

$$\beta = \frac{Q_w}{hu} \qquad [7.29]$$

(4) Finally, we make use of the sediment continuity equation, modified by the use of a Shields dimensionless shear stress τ^*,

$$\tau^* = \frac{\tau_0}{(\rho_s - \rho_f)gD} \qquad [7.30]$$

where D is the median grain size, and ρ_s and ρ_f are the sediment and fluid densities respectively. There appear to be strong limits on the value of the dimensionless shear stress, so that it can be treated as a constant, at about 1.4 times the critical shear stress at the threshold of particle motion in coarse-grained braided rivers, and between 1

and 2 in alluvial sand-bed rivers (Paola & Seal 1995; Dade & Friend 1998; Parker *et al.* 1998). It is probably a constant, because if it is too high, the stream erodes its banks and widens its bed, thereby reducing the shear stress per unit area of stream bed (Parker 1978a, b). There is a sharp jump in τ^* at the gravel–sand transition or 'gravel front' in many rivers. The modified sediment continuity equation where rivers are flowing over a template experiencing tectonic subsidence $\sigma(x)$ therefore has the form

$$\frac{\partial y}{\partial t} = \frac{\partial}{\partial x}\left(\kappa \frac{\partial y}{\partial x}\right) - \sigma(x) \qquad [7.31]$$

where the effective diffusivity or transport coefficient κ of the alluvial system is dependent principally on the discharge of water, a friction (flow resistance) factor, a dimensionless sediment transport rate and the dimensionless shear stress (modified from Paola 2000):

$$\kappa = I_f\left(\frac{q_s^*}{\tau^{*3/2}}\right)\frac{c_d^{1/2}q_w}{C_0}\left(\frac{1}{s-1}\right) \qquad [7.32]$$

where s is the submerged specific weight of the sediment $(\rho_s - \rho_f)g$, c_d is a drag coefficient, q_w is the average water discharge per unit width, q_s^* is a dimensionless sediment discharge per unit width

(a)

(b)

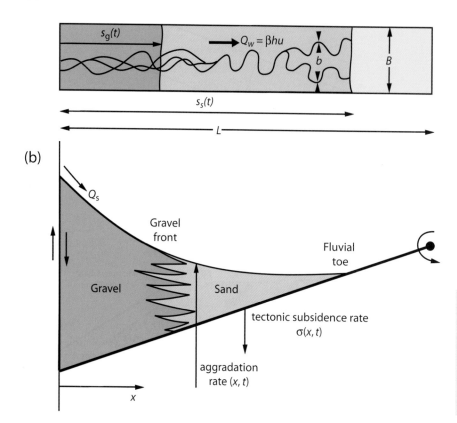

Fig. 7.39 Map view (a) and cross-section (b) of a slice of alluvial basin of length L, width B, containing channels with cumulative width b undergoing tectonic subsidence $\sigma(x)$ approximated by a linear tilt from a distant hinge. Note the break in slope at the gravel front and the position of the fluvial toe. After Marr *et al.* (2000), reproduced with permission of John Wiley & Sons, Ltd.

determined by τ^* and a sediment transport formula, τ^* is the dimensionless shear stress, C_0 is the volume of sediment concentration in the deposit, proportional to porosity, and I_f is a term to represent transport rate fluctuations, termed intermittency. Eqn. [7.32] is a useful formulation of the effective fluvial diffusivity for present-day rivers, but has limited applicability in geological studies. An alternative formulation (Smith & Bretherton 1972; Simpson & Schlunegger 2003), where sediment-hydrodynamic constraints are lacking is:

$$\kappa = \kappa_l + c_f (\alpha x)^m \qquad [7.33]$$

where κ_l is the linear hillslope diffusivity, α is the mean annual precipitation, x is the horizontal coordinate along the direction of concentrative flow, which causes the water discharge q_w to increase downstream, m is an exponent, which is normally assumed to equal 2, and c_f is a fluvial transport coefficient, assumed to be equal to 10^{-6}.

Eqn. [7.32] and eqn. [7.33] are functionally similar, since q_w approximates αx, the drag coefficient, the submerged specific weight of the sediment and the dimensionless shear stress can be assumed to be constant, and κ_l is negligible compared to the second term on the right-hand side of eqn. [7.33]. Typical values for κ used by Marr *et al.* (2000) are $0.01\,\mathrm{km^2\,yr^{-1}}$ in the gravel regime and $0.1\,\mathrm{km^2\,yr^{-1}}$ in the sand regime.

The characteristic time scale for a mixed diffusive-advective ('concentrative') system, as in eqn. [7.33], has the form

$$\tau = \frac{L^2}{\kappa_l + c_f(\alpha x)^m} \qquad [7.34]$$

where L is the system length. Since κ_l is negligible compared to the concentrative term, and x equates to L as the length term, the characteristic time scale becomes

$$\tau = \frac{1}{c_f \alpha^2} \qquad [7.35]$$

showing that the response time is negatively proportional to precipitation, but is independent of system size. Response time is negatively proportional to precipitation since higher rates of precipitation cause a more rapid catchment response to perturbation (see also §7.6.3).

As stated in the paragraphs above, in the context of a diffusive-concentrative approach, the simplest approximation of the time required for equilibrium to be achieved with a set of forcing conditions, or *basin response time* T_{eq}, is L^2/κ, where L is the basin length, as in the diffusional problems previously encountered. In a two-diffusion model, the gravel and sand regimes may have their own response times. For a basin length of 100 km and κ for gravel of $0.01\,\mathrm{km^2\,yr^{-1}}$, the response time is $10^6\,\mathrm{yr}$, whereas with κ for sand, we have a response time of $10^5\,\mathrm{yr}$. An important parameter should be the period of the forcing T compared to the basin response time T_{eq}. Marr *et al.* (2000) considered cross-sections of modelled basin stratigraphy under different conditions of forcing, using the total sediment flux Q_s, the water discharge Q_w, the gravel fraction f_g, and the rate of tectonic subsidence σ. When the forcing has a rapid sinusoidal change (for example, in sediment supply), $T \ll T_{eq}$, the time for equilibrium to be re-established is almost constant for a fixed gravel fraction, suggesting that the basin response time is controlled by the fluvial system rather than the frequency of the forcing. The value of this *intrinsic* response time is determined by the gravel fraction (fast – $<10^5$ yr – with low gravel fractions and slow – 10^5–10^6 yr – for high gravel fractions). In the simulations found in Figs 7.40, slow forcing is $T = 10^7$ yr, whereas fast forcing is $T = 10^5$ yr. The important aspects for basin analysis are as follows.

With *slow forcing* (Fig. 7.40a):

- The movements of the sand toe and gravel front are in phase with variations in sediment flux, with proximal and distal accumulation of sediment both occurring at the time of high sediment flux. Retreat of the fluvial toe at times of low sediment flux produces distal unconformities.
- During periods of increased subsidence, sediment is trapped in proximal regions, and distal areas are starved, causing a retreat of both the gravel front and the fluvial toe and the creation of distal unconformities. During periods of reduced subsidence rate, the gravel front and fluvial toe prograde since accommodation is reduced. Both proximal and distal accumulation are therefore out of phase with subsidence rate.
- The position of the gravel front moves outwards when the gravel percentage (for a constant sediment flux) increases, producing a vertical coarsening-up in a vertical section.

For *rapid forcing* (Fig. 7.40b):

- Rapid changes in sediment flux produce gravel and sand regions with markedly different slopes, and the position of the gravel front is out of phase with the forcing. Reduced sediment supply causes a reduction in proximal slopes, the cutting of proximal unconformities and the progradation of the gravel front. In contrast, reduced sediment supply causes the fluvial toe to retreat. During periods of increased sediment supply, proximal slopes increase, which causes the gravel front to retreat, while the fluvial toe progrades slightly.
- Rapid variations in subsidence rate have little effect on the gravel front.
- The position of the gravel front and proximal accumulation are in phase with the gravel fraction.

It is clear, therefore, that the stratigraphy of an alluvial system exhibits a complex response to the various forcing mechanisms. The response of the system depends on the time scale of the forcing compared to the basin response time T/T_{eq}. Proximal and distal unconformities, variations in proximal and distal accumulation, and movements of the gravel front and fluvial toe may be in phase or out of phase with external forcing.

Heller *et al.* (1988) and Paola *et al.* (1992) recognised that gravel prograded away from or aggraded close to the mountain front depending on the balance between the sediment supply driven by tectonic uplift, and the accommodation caused by tectonic subsidence. Progradational sheets define 'syntectonic' gravels whereas stacked proximal wedges define 'antitectonic' gravels. These stratigraphic geometries were replicated in numerical experiments of catchment-fan systems subjected to changes in tectonic forcing (Allen & Densmore 2000; Densmore *et al.* 2007). At step increases in fault slip-rate, fan systems backstepped before transient progradation, suggesting that the 'antitectonic' mode is a transient response before resumption of steady-state conditions. Incorporation of grain size in these numerical time-dependent models revealed that the transient backstepping is associated with grain-size fining in the hangingwall basin close to the fault plane (Armitage *et al.* 2011) (§7.6.3).

The response times of alluvial systems have also been approximated from measurements of the sediment discharge leaving the catchment. River basins may have extensive floodplain areas in their downstream portions, which act as buffers to changes in any forcing variables, such as base-level change at a river mouth, or changes in sediment flux near their headwaters, and therefore may have long response times. If the system is assumed to be diffusive in character, and the discharge of water varies systematically with floodplain width W, the mass effective diffusivity κ of the channel-floodplain system is

$$\kappa = \frac{Q_s}{W\left\langle \dfrac{\partial y}{\partial x} \right\rangle} \qquad [7.36]$$

where Q_s is the sediment discharge, and $\langle\rangle$ denotes the spatial average of the slope. If L is the downstream length of the river-floodplain system and H the maximum relief between its upstream and downstream ends, the response time becomes

$$\tau = \frac{L^2}{\kappa} = \frac{L^2 W \langle \partial y / \partial x \rangle}{Q_s} = \frac{LWH}{Q_s} \qquad [7.37]$$

The large Asian river systems have typical values of $L\sim10^6$ m, $W\sim10^5$ m, $H\sim1$–2×10^2 m (slopes of 10^{-3} to 10^{-4}) and sediment discharges Q_s of 10^{7-8} m^3 y^{-1} (Métivier & Gaudemer 1999). The characteristic response time is therefore in the region of 10^5 to 10^6 years. Castelltort and Van Den Driessche (2003) carried out a similar analysis on 93 of the world's major rivers, and found that response times varied between 10^4 yr to more than 10^6 yr. (Fig. 7.41). Contrary to the result in eqn. [7.35], the response time in this case depends on the scale of the channel-floodplain system. As we saw in the forward modelling of alluvial stratigraphy above, large alluvial systems therefore should strongly buffer any variations in sediment supply with frequencies of less than 10^{5-6} years. This has strong implications for the detection of high-frequency driving mechanisms in the stratigraphy of sedimentary basins (Chapter 8).

7.5.2 Sediment transport in marine segments of the sediment routing system

Sediment transport by rivers is generally treated as a diffusion problem in which the patterns of deposition and erosion are determined by the curvature of the topography. A question is whether the same approach can be used for the shoreline, the continental shelf and the deep sea. Some authors have extended a linear diffusion approach to sediment dynamics in the ocean (Jordan & Flemings 1991; Johnson & Beaumont 1995; Rivenaes 1992, 1997), whereas others have used a diffusivity dependent on water depth (Kaufman *et al.* 1991). Models such as these are successful in producing a bathymetric rollover analogous to a shelf-edge and continental slope (see also §8.3.3).

A number of lines of evidence suggest that at long time scales, which statistically amalgamate the effects of individual meteorologically or seismically triggered sediment transport events, the sediment transport rate scales on water depth. For example, the orbital velocity of water particles under waves decreases with depth, so wave-generated shear stresses scale on water depth. Tidal currents also, at

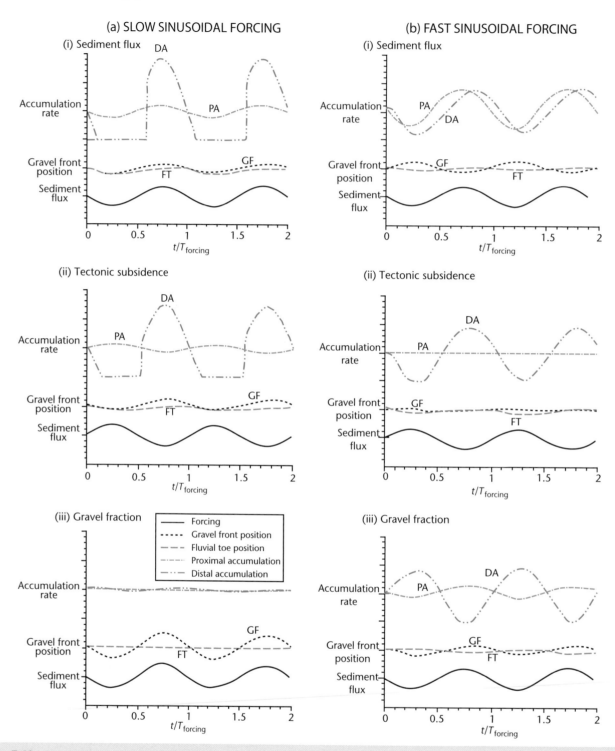

Fig. 7.40 Patterns in alluvial stratigraphy under slow sinusoidal forcing ($T > T_{eq}$) (a) and fast sinusoidal forcing ($T < T_{eq}$) (b) of sediment flux (i), tectonic subsidence (ii) and gravel fraction (iii). Graphs show time variation of the position of the gravel front (GF) and fluvial (FT) toe, proximal (PA) and distal accumulation (DA) and the forcing variable. Further details can be found in Marr *et al.* (2000). From Marr *et al.* (2000), reproduced with permission of John Wiley & Sons, Ltd.

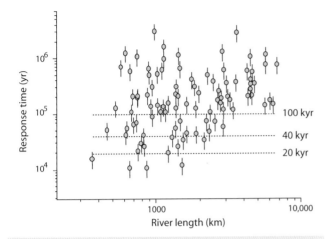

Fig. 7.41 Response times of some modern drainage basins, from a dataset compiled by Castelltort and Van Den Driessche (2003) using the relations given in eqns [7.43] and [7.44]. Hydrological and geomorphological data from Hovius (1998). Response time is plotted against river length. Vertical bars show uncertainties based on the unknown bedload sediment transport rates of the rivers shown. Note that the majority of response times are $>10^5$ yr, and some are $>10^6$ yr.

a broad scale, decrease in basal shear stress in deeper water (Ericksen & Slingerland 1990), though the effects of local flow constrictions between islands and shallows may have overriding importance for sediment transport rates (Mitchell *et al.* 2011).

The ubiquitous presence of clinoforms in seismic reflection profiles of coastal stratigraphy (Pirmez *et al.* 1998), representing progradation of submarine slopes in pro-delta and related regions along the coastline, suggests that the morphodynamics might be treated as advective and wave-like rather than diffusive (Paola 2000, p. 148). Although the sediment dynamics at river entry points to the ocean involve processes such as inertial jets, buoyancy-driven plumes, and gravity-driven particulate flows, the overall migration of the pro-delta slope may be wave-like (Figs. 8.30, 8.31). The elevation of the seabed below sea level η as a purely geometrical function of horizontal distance x and time t is given by the wave equation

$$\eta(x,t) = A + A\sin(kx - \sigma t) \qquad [7.38]$$

where σ is the radian frequency ($2\pi/T$) and k is the wave number ($2\pi/L$), T is the wave period, L is the wavelength and A is the amplitude of the clinoform. If t is held constant, eqn. [7.38] describes the shape of the clinoforms, whereas if x is held constant, eqn. [7.38] gives the elevation of the seabed over time. Inserting reasonable parameter values for a relatively small subaqueous delta, consider clinoforms 100 m high (amplitude of 50 m) and 25 km long (wavelength of 50 km), advancing with a time-averaged celerity $c = L/T$ of 5 m yr^{-1}. The period of the clinoform wave is 10,000 years, and the elevation change of a point along the x direction, in the absence of eustatic change, corresponds with a cycle of water depth change of height 100 m and duration 10 kyr.

The celerity of the wave-like clinoforms is sustained by a sediment feed from large rivers. The block of sediment deposited per unit width during the progradation of clinoforms by one wavelength is

$2AL$, which takes place in one period T. The solid discharge rate per unit width is therefore $(1/C_0)(2AL/T)$, where C_0 is the volume fraction of solid sediment in the clinoforms. If the volume fraction of solid sediment is 0.8 (average porosity of 0.2), and taking the same clinoforms as above with a celerity of 5 m yr^{-1}, the solid discharge per unit width is 625 m^2 yr^{-1}. To build clinoforms of this size and advance rate (celerity) along a delta front of 100 km width would require a sediment supply of 62.5×10^6 m^3 yr^{-1}, or 165 Mt yr^{-1} for a sediment density of 2650 kg m^{-3}, equivalent to the suspended sediment load of a major river (Table 7.6).

The celerity (phase velocity) can also be expressed dynamically, since it scales directly on the sediment flux variation with depth, $c = (1/C_0)\,dq_s/dh$. The change in seabed elevation over a time step t then becomes

$$\frac{\partial \eta}{\partial t} = \frac{1}{C_0}\left(\frac{dq_s}{dh}\right)\frac{\partial \eta}{\partial x} = c\frac{\partial \eta}{\partial x} \qquad [7.39]$$

The equilibrium time for the wave-like progradation of the delta clinoforms across a shelf of width L is L/c. The sedimentation rate, neglecting compaction or tectonic subsidence, can be obtained from eqn. [7.39]. The seabed slope is a maximum at half the clinoform height, which for the parameter values above is 0.006284. The seabed elevation change due to sedimentation is therefore 5×0.006 m yr^{-1} or 30 mm yr^{-1}.

The Gargano and Po subaqueous deltas of the Adriatic coast of eastern Italy are examples of clinoform progradation in the Holocene, fed by large sediment discharges from river catchments (Cattaneo *et al.* 2003). The Gargano delta has prograded during the late Holocene highstand over the last 5 kyr at rates of sediment accumulation of <15 mm yr^{-1}, as muddy clinoforms c.35 m thick, and containing a volume of 180 km^3 of sediment, delivered primarily by river sediment sources to the west (Correggiani *et al.* 2001). It has prograded 15–20 km in the last 5000 years, giving a celerity of 3–4 m yr^{-1}. These values are concordant with those expected of a wave-like movement of the delta as outlined above.

7.5.3 Depositional sinks: sediment storage

In the sediment routing system concept, sediment is transported from source to sink, but there are a number of possible sinks available for short-term storage and long-term preservation. These sinks are occupied by depositional systems, which are sets of depositional environments linked by the process of sediment routing. Depositional systems and their sedimentary products reflect the integration of autogenic (internal) and allogenic (external) controls. Sedimentary basins with different driving mechanisms (Chapters 3 to 6) have distinctive assemblages of depositional systems and facies. More comprehensive treatments of sedimentary facies and depositional systems are found in Reading's *Sedimentary Environments* (1996), Einsele's *Sedimentary Basins* (2000), Leeder's *Sedimentology and Sedimentary Basins, Second Edition* (2011) and Miall's *Geology of Stratigraphic Sequences* (2010). Below, we provide a brief overview of the common terrestrial and marine depositional systems filling sites of storage in the sediment routing system.

Sediment routing systems may be efficient conveyors of sediment from the erosional engine to the deep sea, with little intervening storage, or they may contain 'transient states' where sediment is stored for long periods of time. The Holocene Santa Ana Basin in California (Romans *et al.* 2009) has a steep, incised catchment occupied by the Santa Clara River, which is linked along the sediment routing system via a submarine canyon with the deep-sea Heuneme Fan (Fig. 7.3). From 7 ka to 2 ka the Santa Clara River fed directly to the Heuneme Fan by sediment being routed across the narrow shelf by a canyon with its head at the river mouth. However, at 2 ka the Santa Clara River shifted position away from the canyon head, causing a sediment supply from a littoral cell sweeping southeastwards along the California coast to be tapped and delivered to the deep-sea fan. This temporal change illustrates a 'connected' mode where fluvial sediment supply is routed directly to the deep sea, and a 'disconnected' mode where the fluvial sediment source is cut off. In the connected mode, the Santa Ana Basin system is a good example of 'reactive' dynamics (Allen 2008), in which response times are very short and signals are propagated largely unchanged from source to sink.

In the late Quaternary Golo source-to-sink system of eastern Corsica and the adjacent offshore region of the Mediterranean Sea, the Golo River supplies sediment to a relatively narrow shelf (10 km wide), and to a deep-marine trough filled with turbiditic deposits comprising submarine fan lobes (Somme *et al.* 2011) (Fig. 7.4). Sediment stored in the onshore fluvial system comprises ~13% of the total sediment budget. At the time scale of 10^3–10^4 years, there is a strong mismatch between sediment accumulation rates in the deep sea and variations in the fluvial supply, indicating that sediment must be stored transiently along the fairway of the sediment routing system. The main staging area is the shallow continental shelf where clinoformal prisms of sediment accumulated in several phases in the Quaternary. The Golo system shows that there are no simple or direct teleconnections from source to sink due to the complexities of fluvial incision and alluviation, storage in coastal and shelfal prisms, and the timing of sediment release to the deep sea. The Golo system has more 'buffered' properties (Allen 2008) than the Santa Ana system, and sediment supply signals from the erosional engine are transformed or shredded during propagation from source to sink.

7.5.3.1 Continental depositional systems

Terrestrial depositional systems include the deposits of alluvial fans and fan deltas, rivers, deserts, lakes, slope wastage, and ice sheets and glaciers. Basins with *through drainage* are dominated by well-established river systems and perennial lakes, whereas basins with *internal drainage* are characterised by ephemeral river systems, generally shallow and short-lived lakes, continental sabkhas and deserts. The fraction of the continental surface occupied by internally drained (endorheic) basins varies from continent to continent (Fig. 7.42a), dependent on climatic zone and topography. Externally drained (exorheic) river basins range from small, steep catchments, generally along tectonically active plate margins associated with subduction, to very large, low-gradient catchments flowing from intercontinental drainage divides to distant trailing-edge plate margins (Fig.7.42b). Rivers are the conveyors of sediment to the ocean, representing the most important flux in the sediment routing system. Large rivers tend to enter the ocean with a roughly regular spacing. Their sediment discharge to the ocean may directly reflect the erosional effluxes of contributing upland catchments, or may be modulated by sea-level change at the river mouth, in which case valley incision during base-level lowering strongly impacts sediment delivery offshore. These two situations might be termed 'conveyor belt' versus 'vacuum cleaner' modes (Fig. 7.43) (Blum & Hattier-Womack 2009). The high sedi-

ment discharges of rivers build thick submarine wedges and fans along the continental margin.

Alluvial basin types may be dominated by transverse (e.g. Atlantic coastal plain of North America) or longitudinal (e.g. Po system in northern Italy, Ganges of northern India) drainage systems. River systems may change from longitudinal to transverse during the evolution of the foreland basin systems of mountain belts, as in the southern Pyrenees (Whitchurch *et al.* 2011). Alluvial basins can also be classified according to the nature of proximal, medial and distal elements. Tectonically active basin margins commonly have alluvial fans as proximal elements. Medial elements include braidplains and high-sinuosity alluvial systems of transverse or longitudinal type. Distal elements may be lake margins, terminal fans and sabkhas, deltas and estuaries.

Alluvial fans form where rivers emerge from valleys onto an unconfined plain or major trunk valley, building semi-conical depositional landforms (Bull 1977; Blair & McPherson 1994). They are sensitive indicators of the erosional unroofing of hinterland terrains and of the accommodation generation in neighbouring basins (Whipple & Trayler 1996; Allen & Hovius 1998; Allen & Densmore 2000; Allen & Heller 2012). Fans vary greatly in area. The classical arid-region fans of the American southwest, such as those of Death Valley, California, have radial distances of 1 to 7 km, slopes of 2 to 7 degrees, and are dominated by debris-flow processes. Other depositional systems are low-gradient, fluvial megafans with radial distances of over 100 km, such as that of the Kosi River in the Himalayan foreland (Gupta 1997) and the megafans of the Andean retroforeland (Horton & DeCelles 1997; Horton *et al.* 2001). Megafans depend on an amalgamation of catchments in the mountainous hinterland to supply large quantities of sediment and water. Where fans enter a deep sea, fjord or lake, their surface processes may be dominated by subaqueous mass flows and turbidity currents (Nemec & Steel 1988; Collella & Prior 1990).

Deserts occur in both Arctic areas, where sediment is derived from wasting glaciers, and internally drained tropical zones where large sand seas (ergs) are concentrated. Areas of low rainfall occur as two discontinuous belts around latitudes of 20–30° and are associated with persistently high atmospheric pressures. Deserts also occur in the centres of large continental masses such as the Taklimakan and Gobi deserts of central Asia. The occurrence of aeolian deposits in the stratigraphic record is therefore, to a large extent, a reflection of an ancient climatic zone.

Deserts include a variety of environments of deposition, including giant sand seas, stony wastelands, interdune sabkhas with temporary lakes, dried-up river courses, and expanses of wind-blown silt and clay known as *loess*. The largest present-day ergs occur in topographic basins in intracontinental positions, such as the Saharan examples (Rub al Khali erg in Saudi Arabia is 560,000 km^2 in area) and those of central Australia. Ancient examples such as the Permian Rotliegend Sandstone of northern Europe similarly occupied large continental sags. Sediment is exported from deserts by wind to form loess deposits thousands of kilometres away from the source, and is blown into the ocean to form an aeolian contamination to pelagic sediments.

Lacustrine depositional systems are highly sensitive to climate and generally occupy tectonically controlled depressions. There is a great diversity of lake basins and lake waters. They can be divided into lakes that have an outlet and are hydrologically open and those that lack an outlet and are hydrologically closed (Fig. 7.44) (Allen & Collinson 1986; Talbot & Allen 1996). Their hydrological status determines lake

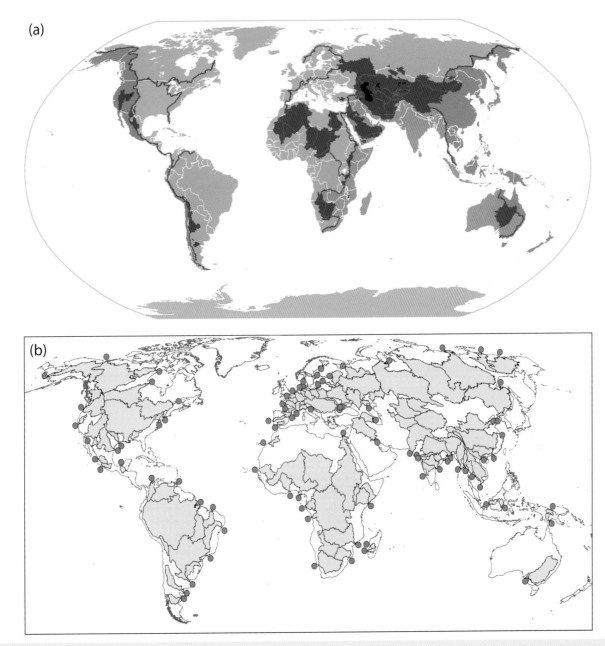

Fig. 7.42 (a) Global map of river drainage, with colour coding based on the ocean to which rivers flow. Black colour is internal (endorheic) drainage. Black lines are continental drainage divides. (b) The world's c.100 largest river drainage basins with circles at their entry points into the ocean or large lakes. The world's largest sediment discharges into the oceans are along trailing-edge passive margins.

water chemistry and therefore the terrigenous clastic versus chemical and biochemical sedimentation in the lake. Hydrologically closed lakes commonly form in the centres of endorheic (internal drainage) regions such as the Chad Basin, north-central Africa. They are characterised by a dominance of chemical and biochemical sediments, and evaporites are common. Through-flowing lakes typify many continental rift systems such as in East Africa and the Baikal Rift of central Asia. They are dominated by terrigenous clastic input and, where river input is negligible, by alkaline earth carbonates.

Lacustrine depositional systems may change in character rapidly as a result of climatic fluctuations. A large number of present-day lakes are shrunken, saline remnants of much larger and more dilute ancestors that existed during the pluvial period coinciding with the last glacial maximum (Flint 1971). Examples are the Dead Sea and precursor Lake Lisan; Great Salt Lake, USA, and precursor Lake Bonneville; and Lake Eyre, Australia, and precursor Lake Dieri. The expanded lakes, such as Lake Lisan, typically accumulated alkaline earth carbonates with an abundant flora of diatoms, whereas the

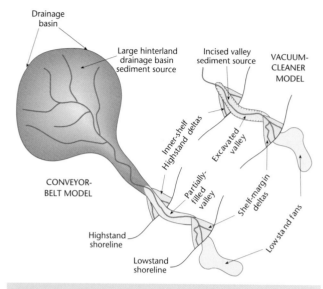

Drainage
basin

Large hinterland
drainage basin
sediment source

Incised valley
sediment source

VACUUM-
CLEANER
MODEL

Inner-shelf
Highstand deltas

Excavated
valley

Partially-
filled
valley

Shelf-margin
deltas

Lowstand fans

CONVEYOR-
BELT MODEL

Highstand
shoreline

Lowstand
shoreline

Fig. 7.43 Conveyor belt versus vacuum cleaner models for sediment supply, from Blum & Hattier-Womack (2009, p. 25, fig. 11). The conveyor belt model involves a large inland drainage basin feeding large amounts of sediment directly to the shelf margin, whereas the vacuum cleaner model produces smaller quantities of sediment by excavating an incised valley.

present-day remnant has a lake floor covered with gypsum and halite. An analysis of Holocene/Quaternary lake levels and salinities suggests that minima and maxima recur at a frequency that is most likely a result of orbital eccentricities of the Earth around the Sun. The lake-level variations therefore appear to be a complex response to Milankovitch forcing.

About 10% of the present Earth's surface is covered with ice and a further 20% is affected by permafrost. During some periods of Earth history, such as the Neoproterozoic, Ordovician, Permo-Carboniferous and late Cenozoic, ice and permafrost cover was much more extensive (Hambrey 1994). Glacial environments are complex and laterally extremely variable. Moving ice is found in a number of settings, principally as polar ice sheets and as coastal ice shelves, and as valley glaciers and their marine outlets (tidewater glaciers). Glacial sedimentary facies reflect the various processes involved in ice movement and melting (Eyles *et al.* 1985; Ehlers 1996). On land, sediment is deposited by direct deposition or lodgement under active ice, some from reworking by later sediment gravity flows at the ice terminus, and some is deposited by subglacial streams or as deltas at their exit points. Sediment is also dumped from ice at its contact with lakes and the ocean as glacilacustrine and glacimarine sediment. Rain-out diamictites are caused by the melting of sediment-loaded floating ice, which distributes a wide grain-size range of sediment, including large *dropstones*, over the lake or sea bed.

Upland catchments acting as engines for sediment routing systems may be strongly modified by glaciation. Cycles of glaciation may control the long-term trends in sediment efflux to neighbouring basins (Brocklehurst & Whipple 2007; Dühnforth *et al.* 2008). A common feature recognised in the sediment output of mountain ranges affected by glaciation and deglaciation is a peak in sediment

discharge early in the deglaciation phase as glacially milled sediment is rapidly evacuated from upland catchments by hillslope and fluvial processes (Hinderer 2001; Kuhlemann *et al.* 2002).

7.5.3.2   Coastal and nearshore depositional systems

Coastal and nearshore areas are commonly staging areas for the temporary deposition of particulate sediment before further transport to the absorbing state of the deep sea. In addition, coastal and nearshore sites may accumulate very great thicknesses of sediment over geological time periods in the form of deltaic deposits. The storage of sediment as prisms or wedges at the edge of the continental shelf during highstands is an important process affecting the dynamics of source-to-sink systems.

The geomorphology and oceanography of the Earth's siliciclastic coastlines reveals an exceedingly complex interplay between fluvial input on the one hand, and basinal parameters such as wave energy, tidal range and storm regime on the other. Deltaic coasts are strongly influenced by the dynamics of the outflowing river water, whereas non-deltaic coastal systems are dominated by waves, tides and their combination (Hayes 1979; Galloway 1975; Postma 1990) (Fig. 7.45). The dispersion of the outflow into the receiving water body falls into three categories: inertia-dominated jets, friction-dominated jets, and buoyancy-dominated outflows (Wright & Coleman 1974; Wright 1977). The dynamics of the outflow control the transport distance of particulates of different size in the estuary, lake or ocean.

Deltas develop where river systems debouch into the ocean, inland seas and lakes. Where tidal and wave energies are low, distributary channels are able to build out into the sea unhindered by coastal erosion, producing a typical 'birdsfoot' pattern, as shown by the Mississippi Delta, USA. Where wave energies are strong compared to the river inflows and tides, the sediment delivered to the sea is moulded into curved ridges at the delta's front and some is redistributed along the shore as beaches and spits. Deltas of this type, such as the Senegal (West Africa) or the Grijalva (Gulf of Mexico) are roughly arc-shaped and prograde slowly because of the destructive nature of approaching waves. Deltas strongly affected by tides have tidal channels cutting deep into the coastline, with associated tidal sand ridges or shoals elongated in the same direction as the tidal current pathways, such as the Ganges–Brahmaputra Delta in the Bay of Bengal, and the Mahakam Delta, Indonesia.

Wave-dominated shorelines occur on delta fronts and on non-deltaic coasts. They are dominated by beaches (directly attached to the land) and barrier islands separated from the land by a shallow lagoon. Tidal inlets are few and short-lived on wave-dominated barrier islands. Because of the limited connection to the open sea, the lagoons behind such barriers are prone to abnormal salinities (e.g. Padre Lagoon, Gulf of Mexico). Storms are capable of breaching the barrier, however, producing washover channels and fans. *Wave- and tide-influenced shorelines* comprise barrier islands highly dissected by tidal inlets and associated tidal deltas. Wave-influenced tidal inlets are highly mobile, migrating in the direction of net longshore sediment transport. They therefore progressively replace the wave-built barrier with coalesced tidal inlet sequences. When tidal range is large (mesotidal to macrotidal, 2 to >4 m) estuaries dominate the coastal geomorphology (Dalrymple *et al.* 1992). Fine-grained sediments fringe the estuary in the form of intertidal and supratidal flats, whereas sands dominate the central zone. Tidal flats dissected

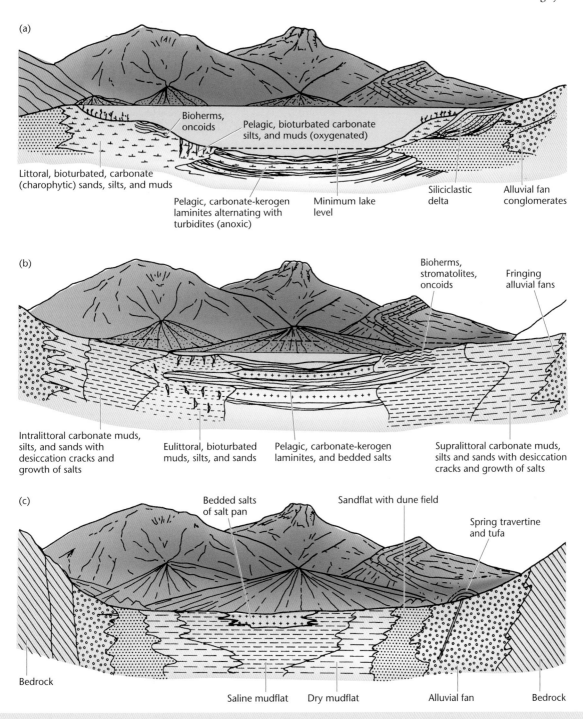

(a)

Bioherms, oncoids

Pelagic, bioturbated carbonate silts, and muds (oxygenated)

Littoral, bioturbated, carbonate (charophytic) sands, silts, and muds

Pelagic, carbonate-kerogen laminites alternating with turbidites (anoxic)

Minimum lake level

Siliciclastic delta

Alluvial fan conglomerates

(b)

Bioherms, stromatolites, oncoids

Fringing alluvial fans

Intralittoral carbonate muds, silts, and sands with desiccation cracks and growth of salts

Eulittoral, bioturbated muds, silts, and sands

Pelagic, carbonate-kerogen laminites, and bedded salts

Supralittoral carbonate muds, silts and sands with desiccation cracks and growth of salts

(c)

Bedded salts of salt pan

Sandflat with dune field

Spring travertine and tufa

Bedrock

Saline mudflat

Dry mudflat

Alluvial fan

Bedrock

Fig. 7.44 Idealised depositional environments and facies in lacustrine systems. (a) Open freshwater lake with through drainage. (b) Hydrologically closed perennial salt lake. (c) Hydrologically closed ephemeral salt pan. After Eugster and Kelts (1983), modified from Allen and Collinson (1986), reproduced with permission of John Wiley & Sons, Ltd.

by highly sinuous tidal creeks can be extremely extensive on some low-wave-energy mesotidal to macrotidal coasts.

Carbonate sediments are produced in great abundance in shallow, warm waters where the biological and physico-chemical conditions are optimal for carbonate precipitation and fixation. Arid shorelines with low terrigenous input are characterised by deposition of car-

bonates and evaporites. The Trucial Coast, Persian Gulf, is an example of a modern carbonate-rich marginal marine *sabkha*. Arid shorelines may also be dominated by siliciclastic sedimentation, as in Baja California, the Gulf of Elat and some parts of the Arabian Gulf. Here the sabkhas are composed of siliceous sands and muds with a possible admixture of carbonate grains.

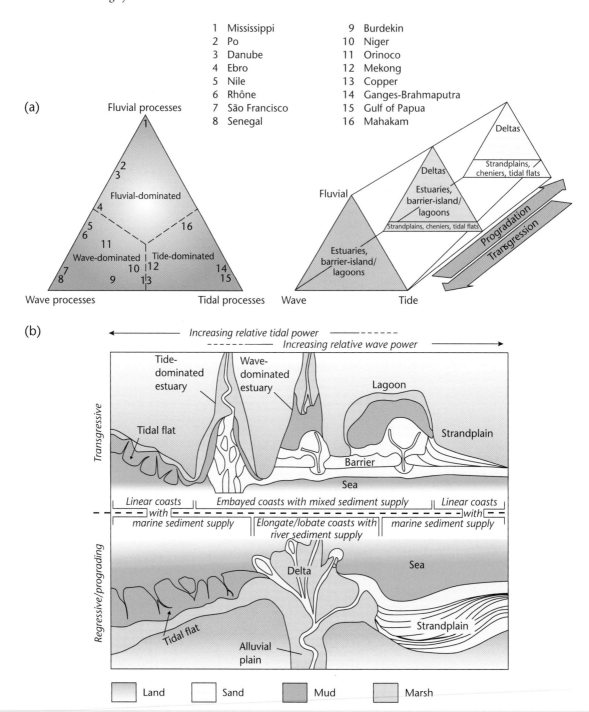

1 Mississippi   9 Burdekin
2 Po   10 Niger
3 Danube   11 Orinoco
4 Ebro   12 Mekong
5 Nile   13 Copper
6 Rhône   14 Ganges-Brahmaputra
7 São Francisco   15 Gulf of Papua
8 Senegal   16 Mahakam

Fig. 7.45   Delta classification according to fluvial discharge, wave power and tidal range, following Galloway (1975), Postma (1990), Reading and Collinson (1996). (a) Ternary diagram and its modification into a Swiss Toblerone, to account for the different morphological features found during progradation and transgression of the coast. (b) Plan views of transgressive/retrogradational and regressive/progradational coasts under varying conditions of wave power, tidal range and marine and fluvial sediment supply (based on Heward 1981 and Boyd *et al.* 1992).

7.5.3.3 Continental shelf depositional systems

Continental shelf systems are extremely complex and are highly sensitive to sea-level fluctuations. Modern continental shelves can be classified according to the nature of the sediment (relict, palimpsest, modern) and the hydraulic regime (wave-, tide-, storm- or oceanic-current-dominated) (Swift 1974; Johnson & Baldwin 1986). Classifying shelves according to the dominant shelf currents (Swift *et al.* 1986), tide-dominated shelves occupy 17% of the present-day shelf area, storm-dominated 80%, and shelves dominated by intruding ocean currents a very small percentage.

Much of the continental shelf between the latitudes of 30°S and 30°N is an area of high organic productivity and is covered not by river-derived or relict siliciclastic sediments, but by organic carbonate material (Fig. 7.46). There are two major categories of *subtropical carbonate shelf* (Ginsburg & James 1974). (i) *Rimmed shelves* sheltering protected shelf lagoons. Their margins often fall precipitously into the abyssal depths. Some rimmed shelves are attached to continental areas as in the Great Barrier Reef of Australia. Others are now *isolated platforms*, as in the Bahamas. (ii) *Open shelves*, such as Yucatan, western Florida and northern Australia, slope gently towards the continental edge and are termed *ramps*. Because of the lack of a protective rim, they are strongly affected by storm waves and tidal currents.

Reefs are *biogenic* constructions on the seafloor and can generally be divided into a: (i) reef core comprising skeletons of reef-building organisms and a lime-mud matrix; (ii) reef flank of bedded reef debris; and (iii) inter-reef of subtidal shallow-marine carbonates (or siliciclastics). Where reefs form a natural breakwater on the windward sides of shelves or islands, however, they protect a back-reef environment from wave attack.

7.5.3.4 Deep-sea depositional systems

The seabed of northwest Africa provides an informative example of an integrated system of slope and deep-marine processes and products (Weaver *et al.* 1992; Masson 1996; Frenz *et al.* 2009; Talling *et al.* 2007) (Fig. 7.47). Catastrophic collapse of volcanic islands in the Canaries has repeatedly produced debris avalanches that load the adjacent seabed and trigger large debris flows. Some of these debris flows have yielded turbidity currents that travelled downslope to the Madeira abyssal plain at water depths of over 5000 m. Other flows originate from the Agadir Canyon and are routed for thousands of kilometres to the deep ocean. Turbidites *en route* to the deep sea commonly interact with seabed topography caused by sedimentation, gravitational tectonics and salt movement.

There are four fundamentally different environments of deposition of clastic sediments in the deep sea: (i) *slope aprons*, which accumulate between the shelf and basin floor, and vary in width from <1 km to >200 km. (ii) *Submarine fans* that build oceanward at the base of the shelf slope (Fig. 7.48a,b) or directly from river input on steep coasts as fan-deltas (Fig.7.8c). They receive sediment from river mouths, from alternative feeder systems such as submarine canyons or from reworking of shelf prisms and from littoral cells. (iii) *Deep-water sediment drifts* formed by the deep thermohaline circulation of the oceans, composed of fine-grained *contourites* (Fig.7.8d). (iv) *Basin plains*, which are flat, relatively deep areas that act as the ultimate sediment traps for clastic sediments eroded from the continents and from submarine highs.

Sedimentation in the open sea beyond the influence of the continental land masses is controlled by two major factors: the fertility of the surface water and the presence of the '*calcite compensation depth*' (CCD), below which carbonate from the skeletons of marine organisms is dissolved. The biological productivity of the oceans is highly concentrated in near-surface waters where light allows photosynthesis of phytoplankton. A small percentage (1%) of the carbon fixed by photosynthesis accumulates in dead organic tissues on the ocean bed. In low-oxygen conditions, this organic matter is converted into black, laminated, organic-rich shales known as *sapropels*. Highest productivity is where there is an ample supply of nutrients, as in upwelling zones (see also §11.3). Nutrient supply may also be strongly influenced by climate change, with increased productivity during glacial periods when nutrients such as iron are blown into the ocean from expanded desert regions. Beckmann *et al.* (2005) found that shallow shelves and semi-enclosed seas were prone to anoxia at times of high river discharge in the Cretaceous.

Above the CCD, calcareous oozes derived from micro-organisms such as foraminifera predominate. Below this, the silica skeletons of radiolaria and diatoms produce siliceous oozes. There are also regions of seafloor dominated by red and brown clays derived from volcanoes, meteorites and dust blown from continents. The CCD has varied in its depth in the ocean during geological time, commonly as a result of variations in temperature and productivity, as at the Eocene–Oligocene boundary (Kennett & Shackleton 1976).

7.5.4 Downstream fining

Over many repeated sediment transport events, a portion of the mobile sediment flux is extracted to build stratigraphy. Stratigraphy is therefore a strongly filtered record of this succession of sediment transport events. The selective extraction of mass from the mobile sediment flux during downstream transport produces a characteristic trend in preserved grain size (Fig. 7.49). This grain-size trend in turn affects fluvial style and depositional environments (Strong *et al.* 2005). Critical to the prediction of downstream fining (Fedele & Paola 2007; Duller *et al.* 2010; Whittaker *et al.* 2011; Allen & Heller 2012) is the influence of the spatial distribution of tectonic subsidence $\sigma(x)$ on the downstream trend in the long-term sediment transport rate $q_s(x)$. This downstream trend in $q_s(x)$ reflects the interplay between the volumetric discharge of the sediment supply and the volume extracted to fill accommodation.

The impact of deposition is to reduce the sediment discharge q_s in the downstream direction from its initial value $q_s(0)$ at $x = 0$ (Appendix 50). Eventually, the sediment supply is exhausted at a depositional length L_d. At each point in the downstream direction $x < L_d$ the solid sediment volume extracted by deposition as a fraction of the surface sediment flux is $(1 - \varphi)\sigma(x)/q_s(x)$, where φ is the porosity of the deposited sediment. The mean grain size of the deposited stratigraphy therefore depends on the grain-size characteristics in the supply, and the downstream extraction controlled by the spatial distribution of deposition. Granulometric trends are therefore potentially valuable in constraining parameters describing the volumetric budget of sediment routing systems (Whittaker *et al.* 2011; Parsons *et al.* 2012).

Since both $q_s(x)$ and $\sigma(x)$ are determined by the tectonics and climate of the basin and its hinterlands, different basin types are expected to contain sediment routing systems with characteristic rates of downstream fining. For example, it is possible to discriminate

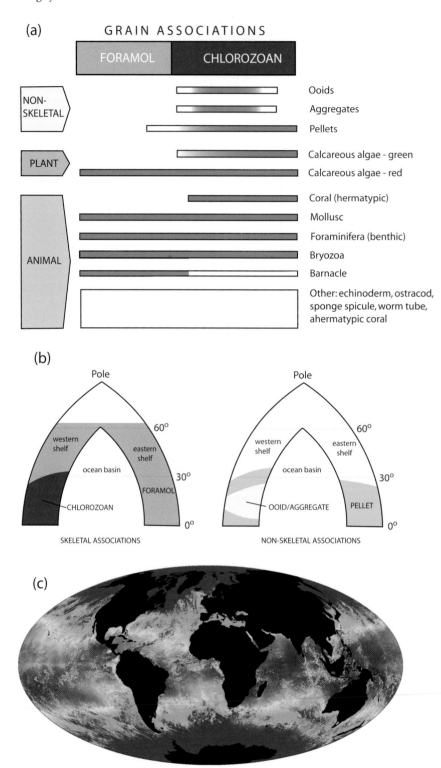

Fig. 7.46 Distribution of warm-water and temperate-water carbonate sediments on the world's continental shelves. (a) Principal carbonate grain types and their associations, from Lees & Buller (1972). Shading in bars indicates whether component is 'dominant' (colour) or 'important'. (b) Distribution of main carbonate grain associations on the continental shelves of ideal oceans, one hemisphere shown only, from Lees & Buller (1972). The ooid/aggregate association is shown as restricted to the western side of the ocean. Carbonate productivity is linked to sea surface temperatures, shown in (c). Not shown are cold water reefs and mounds, which also occur, often in the bathyal water depths along continental margins.

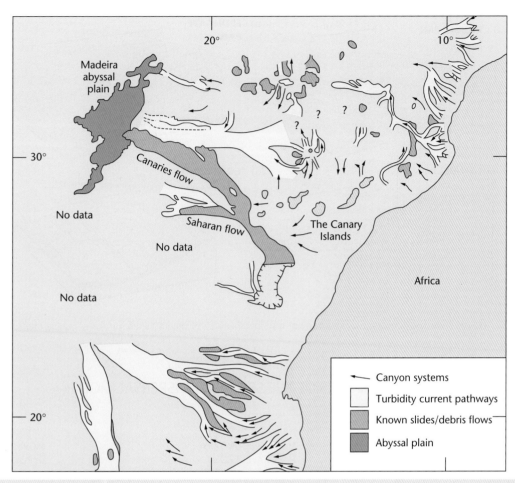

Fig. 7.47 Sediment routing systems offshore NW Africa, showing the transport paths of turbidity currents and giant debris flows. The Canaries debris flow originated by the loading of the seabed by a rock avalanche caused by the catastrophic collapse of a nearby volcanic seamount. The Saharan debris flow, however, was due to failure of the continental margin. After Weaver *et al.* (1992), reprinted with permission from Elsevier.

between catchment-fan systems filling small piggyback basins in thrust belts, which have rapid rates of downstream fining, from large, coalesced fluvial systems spreading sediment from the cores of mountains belts to distant foredeeps, which have low rates of downstream fining (Duller *et al.* 2010; Whittaker *et al.* 2011; Allen *et al.* 2013).

7.6 Joined-up thinking: teleconnections in source-to-sink systems

In order to understand sediment routing systems dynamically, a number of workflows need to be completed. First, it is essential to locate the source area for the sediment occupying depositional sinks. The chief method of doing so is to link the petrography, mineralogy, or chemistry of detrital grains to a source region using specific tracers. Of growing importance is the use of thermochronometric and radiometric information contained in the detrital grains, which helps in the location of the source area, but provides additional information on its rate of exhumation and age. Second, the 'footprint' or 'fairway' of the sediment routing system needs to be delimited or mapped so that a volumetric budget can be assessed. The volumetric budget is an important tool in evaluating the size of the total sediment reservoir and its partitioning by grain size. Third, documentation of downstream trends in grain size, facies, depositional environments and depositional styles enable the functioning of the sediment dispersal system and its selective extraction to build stratigraphy to be evaluated. These downstream trends are symptomatic of the external forcing of the sediment routing system by tectonics and climate, as well as the internal (unforced) dynamics. They are discussed in §7.5.4.

7.6.1 Provenance and tracers; detrital thermochronology

For many years a range of techniques have been applied to the study of the provenance of sand-grade sediment, including analysis

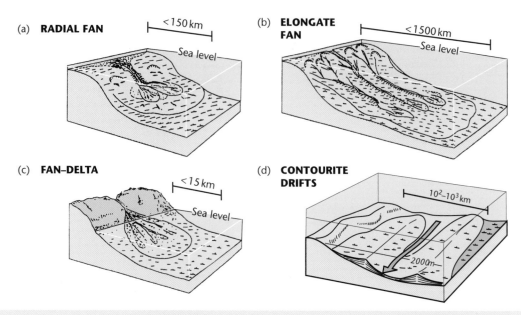

Fig. 7.48   Main types of deep-sea depositional environment for siliciclastics. All diagrams are vertically exaggerated.

ESSENTIALS OF THE SEDIMENT DISPERSAL MODEL

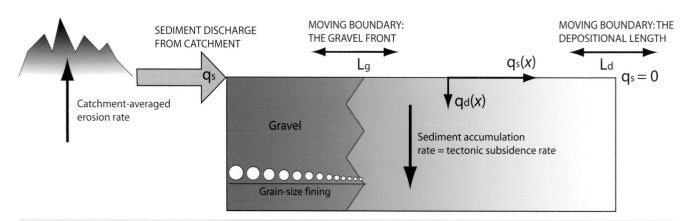

Fig. 7.49   Fundamentals of the downstream fining model. Catchment erosion generates a sediment supply to a neighbouring basin $q_s(x = 0)$. Part of this surface flux is extracted to build stratigraphy, forming a depositional flux $q_d(x)$, and the remainder passes downstream as a surface flux $q_s(x)$. Grains are selectively sorted by size to join the depositional flux, causing a downstream fining. Note that the fining is not a result of abrasion.

of light-fraction petrography (Zuffa 1985) and heavy minerals (Mange & Maurer 1991). Bulk geochemistry, isotopes and thermochronometers are increasingly used to constrain both the location and rock type of source areas and their exhumation history. Such integrated studies allow unprecedented insights into orogenic evolution, basin development and even global climate change. The most intensively studied mountain range is the Himalayas, perhaps because of its global importance to Cenozoic climate change and ocean water chemistry. Thermochronological studies linking the cooling history of crustal materials in the Himalayan chain to their burial and preservation in the foreland basin sediments of the Siw-

laliks are found in Harrison *et al.* (1993), Najman *et al.* (1997), DeCelles *et al.* (1998, 2001, 2004), Najman and Garzanti (2000), Huyghe *et al.* (2001), White *et al.* (2002), Bernet *et al.* (2006) and Szulc *et al.* (2006).

Samples taken *in situ* from bedrock in mountain belts provide invaluable information on the cooling/exhumation history of that location, but in rapidly exhuming orogens such as the Himalayas, Taiwan and the Southern Alps, the information is restricted to the last few millions of years, dependent on the thermochronometer used. However, detrital material found in the stratigraphy of foreland basin systems provides a longer record of the evolution of the mountain belt. This record is based on the mineralogical, chemical and isotopic characteristics of detrital sediment sourced from the eroding mountain belt, including their thermochronometers, such as Ar/Ar data on detrital micas (Copeland & Harrison 1990; Harrison *et al.* 1993), fission tracks in zircon (Cerveny *et al.* 1988; Brandon & Vance 1992; Bernet *et al.* 2004) and in apatite (Lonergan & Johnson 1988; Glotzbach *et al.* 2011).

A valuable aid in the interpretation of the evolution of mountain belts and their sedimentary basins is the concept of lag time (Fig. 7.50). Lag time is the difference between the stratigraphic age of a sediment and that sample's thermochronological or cooling age. Although the lag time must involve a period of transport and perhaps intermittent storage, this is generally assumed to be small (Brandon & Vance 1992), so the lag time provides information on the velocity of exhumation. Consequently, systematic variations in lag time can be interpreted in terms of variations in denudation rate, which might be driven by climate or tectonics.

Systematic variations in lag time are expected during the large-scale, long-term evolution of a mountain belt and its foreland basin systems (Beaumont *et al.* 1999; Willett *et al.* 2001; Willett & Brandon 2002; Whitchurch *et al.* 2011). During the early stage of orogenesis, the tectonic mass influx is thought to outpace the sediment efflux at the surface, leading to crustal thickening, surface uplift and the construction of topography – the *constructional* phase. As hillslopes and bedrock rivers increase the erosion rate, the mountain belt enters a steady state, or dynamic equilibrium, in which the tectonic influx roughly balances the erosional efflux. During the steady-state phase, the orogen ceases to grow. Orogens are then thought to enter a *destructional* phase, when tectonic influx into the orogen decreases, topography reduces, and the erosional efflux exceeds the tectonic influx. This evolution should be accompanied by a reduction and then increase in lag time (Fig. 7.50).

Samples from modern rivers integrate the fission track characteristics of the entire catchment area, though in proportions related to the details of the location and erosion rate of sediment sources and the details of downstream sorting. The most reliable detrital thermochronology samples will therefore be from areas where additional information is available on provenance, such as heavy-mineral petrography (Ruiz *et al.* 2004), geochemical/isotopic data (Spiegel *et al.* 2004; Szulc *et al.* 2006) or U–Pb geochronology (Carter & Bristow 2000; Rahl *et al.* 2003; Bernet *et al.* 2006).

A relatively large number of detrital grains are needed in order to detect different age groups in the population of detrital grain ages. To avoid potential problems of resetting of thermochronological information during basin subsidence and burial, argon dating of white mica and zircon fission track techniques are most widely used. It is important to deconvolve the detrital grain-age distribution into its component distributions, as shown for sediments in the Karnali River of western Nepal (Fig. 7.51). A number of techniques are available for the deconvolution, including that of Stewart and Brandon (2004), which can be downloaded free from www.geology.yale.edu/ ~brandon/Software/FT_PROGRAMS/BinomFit/index.html.

7.6.2 Mapping of the sediment routing system fairway

The sediment routing system fairway is the total spatial extent of the individual thread-like sediment transport paths from source to sink. Although detrital sediment is commonly linked to source areas using provenance studies, it is rare for the sediment routing system fairway

to be specifically mapped. It is clearly problematic to do so where the source and the sink are separated by long distances of non-deposition. But where the fairway can be recognised and mapped, invaluable information can be gained on the sediment budget and the downstream fractionation of sediment in terms of grain size. We illustrate the concept of the sediment routing system fairway using an example where the feeder paleovalleys and fluvial segment are well preserved, and passes downstream into coastal and deep-marine environments.

The middle to late Eocene Escanilla Formation of the south-central Pyrenees is an example where a sediment routing system fairway has been mapped (Michael 2013). Two main sources in the Sis and Pobla-Gurb areas supplied sediment to a fluvial system that flowed longitudinally to the west, guided by active tectonic structures in the southern foreland of the Pyrenean mountain belt (Bentham *et al.* 1993; Puigdefábregas *et al.* 1992; Vincent 2001; Beamud *et al.* 2003). The Escanilla fluvial system passed into marine environments in the Ainsa and Jaca basins (Fig. 7.52).

Mapping of the fairway of the sediment routing system allows the surface areas of the different segments to be visualised, and sediment isopach information allows this to be converted to volume. Over the 7.9 Myr duration of the mid–late Eocene system, $3–4 \times 10^3 \, km^3$ of sediment was released from catchments eroding the Pyrenees, suggesting a physical erosion rate of $0.2–0.3 \, mm \, yr^{-1}$ in the mountain belt, which is consistent with estimates derived from thermochronology. Since downstream transects constrain the grain-size fractions (conglomerate, sandstone, siltstones and mudstones), the volumetric partitioning of the Escanilla system can be approximated. Taking the sediment routing system from source to deep-marine sink, gravel comprises c.12%, sandstone c.23% and mud and silt 65% of the total sediment released. Although the Escanilla is a relatively small paleo-sediment routing system, it provides valuable generic information on the volumetric discharge and grain-size mix fluxed from upland catchments during orogenesis.

7.6.3 Landscape evolution models and response times

The significance of numerical landscape evolution for basin analysis is that such models allow the sediment supply to basins to be simulated in terms of tectonic, climatic and geomorphic sets of rules. The incorporation into a physically realistic numerical model of feedbacks between the various processes that shape a landscape also allows system behaviour, such as the response time to a change in a driving mechanism or forcing variable, to be studied. Numerical landscape models range in their spatial and temporal scales, from whole contractional orogens and passive margin mega-escarpments (length scales of c.100 km and time scales of c.10^7 yr) to individual extensional fault blocks or thrust-related anticlines (length scales of c.10 km, time scales of c.10^6 yr), and even to the fault scarps generated by single seismic events (length scale of c.100 m, time scales of 10^3–10^5 yr). The resolution (of the elevation of topography) required by the landscape evolution model varies accordingly, from centimetres in the case of a degrading fault scarp, to c.100 m for an orogen. The spatial and temporal scale and topographic resolution commonly affect the way in which tectonic and geomorphic processes are dealt with in the numerical model. Burbank and Anderson (2001) provide a useful summary. Many of the algorithms used in numerical landscape models are identical to or similar to those given in §7.4 and §7.5. We concentrate here on the intermediate scale of extensional fault blocks and anticlinal folds in fold-thrust belts.

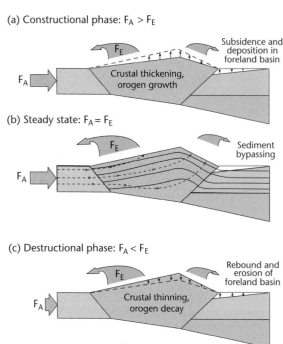

(a) Constructional phase: $F_A > F_E$

(b) Steady state: $F_A = F_E$

(c) Destructional phase: $F_A < F_E$

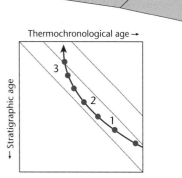

Fig. 7.50 Evolutionary stages of a collisional mountain belt in terms of the lag time of detrital zircons sampled from the fore-land sediments (after Jamieson & Beaumont 1998; Willett & Brandon 2002). (a) Constructional phase where the tectonic mass influx F_A is greater than the erosional mass efflux F_E, leading to crustal thickening, orogenic growth and a temporal increase in denudation rate. (b) Steady-state phase, where $F_A = F_E$, the orogen ceases to grow and denudation rates become constant. (c) The destructional phase, where $F_A < F_E$, causing crustal thinning, topographic decay and a temporal decrease in denudation rates. These phases are reflected in a trend of decreasing then increasing lag time (inset). 1, 2 and 3 represent the constructional, steady-state and destructional phases.

7.6.3.1   Tilted extensional fault blocks

Early attempts to model the evolution of simple fault-bounded ranges used a single planar fault with uniform slip, diffusional modi-fication of the tectonically generated topography using a single, con-stant value of κ, and flexural compensation (King *et al.* 1988; Stein *et al.* 1988). Later models incorporated channels and hillslopes that strongly interacted with transport rules for the disposal of sediment derived from hillslope erosion, and hillslopes were allowed to fail by landsliding (Densmore *et al.* 1998; Ellis *et al.* 1999). Such models not only simulated very realistic landscapes in the Basin and Range prov-ince of SW USA, but also allowed system behaviour to be better

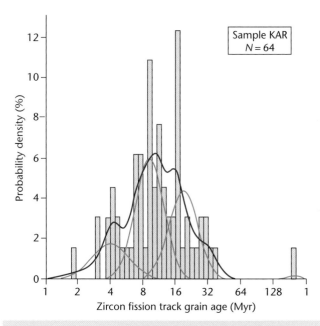

Fig. 7.51   Detrital zircon fission track ages of sediment from the present-day Karnali River, western Nepal. There are 64 single grain ages (histogram), from which a probability-density plot can be made (red line). The grain ages were deconvolved into five component sub-populations using a binomial best-fitting tech-nique (green lines). From Stewart & Brandon (2004).

evaluated. These tectonic-geomorphic systems involve steep trans-verse drainage systems over uplifting footwall blocks, with fans in the neighbouring hangingwall basins (Whipple & Trayler 1996; Allen & Hovius 1998) (Fig. 7.53). The area of the catchment acting as a source terrain for sediment A_1, relative to that of the hangingwall fan A_2, is an indicator of how this relatively simple tectonic-geomorphic system functions (Fig. 7.53a).

Tectonics plays a major role in the creation of uplifting footwalls as source regions for sediment, but also in determining the spatial pattern and rate of hangingwall subsidence, and thereby in control-ling fan thickness and fan progradation distance. Not surprisingly, plots of fan area versus catchment area from examples in the arid SW USA (Bull 1962; Whipple & Trayler 1996; Allen & Hovius 1998) show that areas with high fault displacement rates and areas with low fault displacement rates are discriminated by their value of A_2/A_1 (Fig. 7.53b). This implies that coarse-grained fan bodies are stacked against active range-bounding faults, whereas fans coalesce and pro-grade basinwards where fault displacement rates are smaller.

The growth of fault arrays, from fault initiation, to fault linkage, to the development of a through-going border fault system (§3.2) has profound effects on depositional systems (Gawthorpe and Leeder 2000) (Fig. 7.54). Abrupt changes in the rate of slip on range-bounding extensional faults, as may be produced by the process of segment linkage, can be recognised by changes in mean catchment erosion rate and mean fan deposition rate, but there is a delay in the achievement of a new steady state under the new fault slip-rate condi-tions (Allen & Densmore 2000; Densmore *et al.* 2007; Allen 2008; Allen & Heller 2012). The stratigraphic response during this transient stage, $<10^6$ yr in duration, is abrupt retrogradation accompanied by grain-size fining, followed by gradual progradation (Densmore *et al.* 2007; Armitage *et al.* 2011) (Fig. 7.55). Because of this transient

(a)

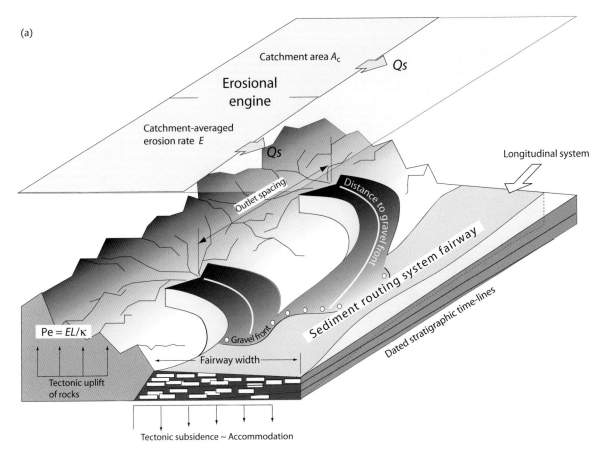

(b)

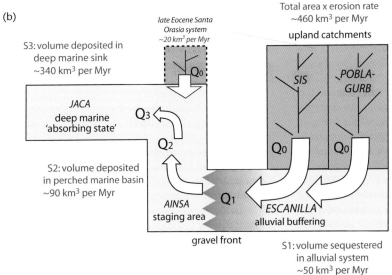

Fig. 7.52 (a) Schematic illustration of the mapping of the terrestrial compartment of a paleo-sediment routing system fairway, based on the mid–late Eocene Escanilla system of the south-central Pyrenees, Spain. (b) Plan view of a schematic representation of the Escanilla paleo-sediment routing system, showing sediment fluxes from upland catchments and sequestered volumes per Myr in the terrestrial segment (S1), delta, shelf and slope segment of the Ainsa Basin (S2) and deep marine segment of the Jaca Basin (S3). Part of S3 is made of continental deposits following basin-filling. The Santa Orasia source was operative only in the latest Eocene.

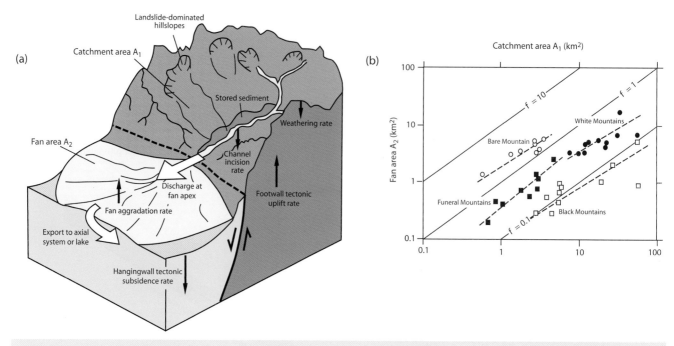

Fig. 7.53 (a) Schematic block diagram for a simple sediment routing system comprising an uplifting erosional footwall and a depositional hangingwall basin occupied by a fan (after Allen & Hovius 1998 and Allen & Densmore 2000). (b) The relationship between fan area (A_2) and catchment area (A_1), for fans in the arid southwestern USA, showing that individual regions with particular slip rates on the border fault are discriminated in terms of their value of A_2/A_1.

behaviour, rapid ($<10^6$ yr) changes in the tectonic 'boundary conditions', such as changes in slip-rate on bounding faults, are unlikely to be easily recognised in basin stratigraphy. Step changes in climatic parameters, such as mean annual precipitation, have a different stratigraphic response. A doubling of precipitation produces a coarse-grained spike that extends down-system far from the depositional apex (Fig. 7.55).

The deforming crustal template in regions of extension strongly interacts with erosion and drainage development. As footwalls emerge, river systems are etched into their flanks either side of a catchment divide. Individual *en echelon* fault segments interact at their tips in a variety of geometrical arrangements, most important of which is the *relay zone* (Larsen 1988). These relay zones have an important role in focusing sediment delivery to hangingwall basins from larger-than-normal catchments and are therefore associated with fan and fan-delta development (Leeder & Gawthorpe 1987; Gupta *et al.* 1999; Cowie *et al.* 2006). Such sedimentary bodies adjacent to relay zones are potentially important as hydrocarbon reservoirs in rift provinces. Two types of catchment appear in model runs (Cowie *et al.* 2006): small *frontal catchments* draining proximal footwalls contribute little sediment by volume to rift sedimentary basins, whereas large *hinterland catchments* that evolve by drainage reorganisation back to the main drainage divide, control the location of major hangingwall depocentres.

7.6.3.2   Thrust-related anticlines

At a similar wavelength to the extensional fault blocks of the Basin and Range are the thrust-related folds of contractional tectonic provinces such as the Zagros of Iran and western Pakistan (Mann & Vita-Finzi 1988). Folds are approximately 10 km in wavelength and

2–3 km in amplitude. There are a number of questions that are especially relevant to this tectonic situation: as hangingwall rocks are transported up the thrust ramp, how is the topographic profile and river drainage development related to the tectonic paths of hangingwall rocks? What is the effect of progressively unroofing rocks with markedly different erodibilities? And what effects do the contractional tectonics have on antecedent rivers?

In areas with high rates of tectonic uplift of rocks, the topographic relief is essentially controlled by the ratio U/c_v, where U is the tectonic uplift rate of rock and c_v is the efficiency of bedrock incision expressed as a velocity (Appendices 44, 45). Streams are known to change their gradients markedly over substrates of different erodibility (Hack 1973). Tucker and Slingerland (1996) estimated bedrock erodibilities c_v to vary over an order of magnitude in the arid landscapes of the Zagros. Using a numerical landscape evolution model, they successfully simulated the progressive unroofing of stratigraphy with strongly varying erodibility related to the growth of a series of tectonic folds. Sediment flux from the fold-thrust belt reflects both tectonic growth of the fold structures but also the extent of exposure of resistant versus weak lithologies. For a landscape dominated by bedrock channels, the time required to reach equilibrium (90% of equilibrium sediment flux) with a given rate of tectonic uplift is proportional to the rock erodibility c_v, the uplift rate U and the spatial scale (width) of the uplifting region L,

$$\tau = k \frac{U^{\left(\frac{1}{n}-1\right)} L}{c_v^{1/n}} \qquad [7.40]$$

where n is the exponent for slope in eqn. [A44.6], commonly assumed to be 2/3. Calibrated against model output from a landscape

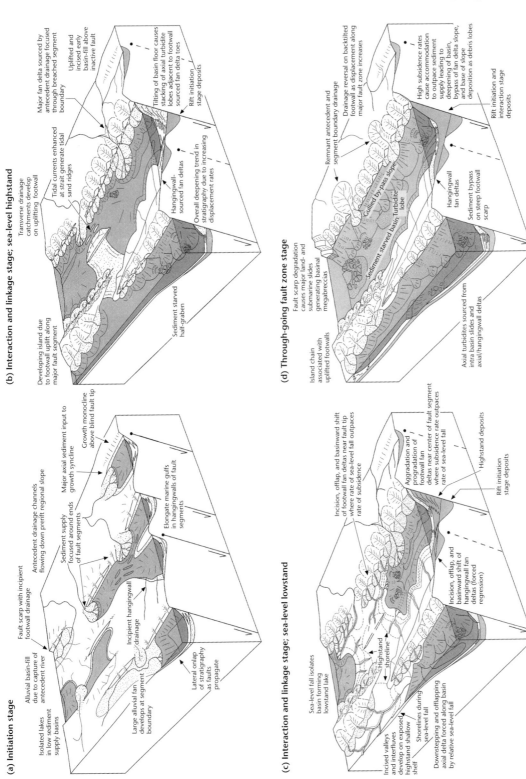

(a) Initiation stage

Fault scarp with incipient footwall drainage

Antecedent drainage channels flowing down prefit regional slope

Sediment supply focused around ends of fault segments

Major axial sediment input to growth syncline

Growth monocline above blind fault tip

Alluvial basin-fill due to capture of antecedent river

Isolated lakes in low sediment supply basins

Incipient hangingwall drainage

Large alluvial fan develops at segment boundary

Elongate marine gulfs in hangingwalls of fault segments

Lateral onlap of stratigraphy as faults propagate

(b) Interaction and linkage stage; sea-level highstand

Transverse drainage catchments develop on uplifting footwall

Tidal currents enhanced at strait generate tidal sand ridges

Major fan delta sourced by antecedent drainage focused through breached segment boundary

Uplifted and incised early basin-fill above inactive fault

Tilting of basin floor causes stacking of axial turbidite lobes adjacent to footwall sourced fan delta toes

Rift initiation stage deposits

Developing island due to footwall uplift along major fault segment

Hangingwall-sourced fan deltas

Overall deepening trend in stratigraphy due to increasing displacement rates

Sediment starved half-graben

(c) Interaction and linkage stage; sea-level lowstand

Sea-level fall isolates basin forming lowstand lake

Incised valleys and interfluves develop on exposed highstand shelf

Shorelines during sea-level fall

Downstepping and offlapping axial delta forced along basin by relative sea-level fall

Highstand shoreline

Incision, offlap, and basinward shift of footwall fan deltas near fault tip where rate of sea-level fall outpaces rate of subsidence

Aggregation and progradation of footwall fan deltas near center of fault segment where subsidence rate outpaces rate of sea-level fall

Incision, offlap, and basinward shift of hangingwall fan deltas (forced regression)

Highstand deposits

Rift initiation stage deposits

(d) Through-going fault zone stage

Remnant antecedent and segment boundary drainage

Drainage reversal on backtilted footwall as displacement along major fault zone increases

High subsidence rates cause accommodation to outpace sediment supply leading to deepening of basin, bypass of fan delta slope, and base of slope deposition as debris lobes

Rift initiation and interaction stage deposits

Gullied by-pass slope

Sediment starved basin turbidite lobe

Hangingwall fan deltas

Sediment bypass on steep footwall scarp

Fault scarp degradation causes major land- and submarine slides generating basinal megabreccias

Island chain associated with uplifted footwalls

Axial turbidites sourced from intra basin slides and axial/hangingwall deltas

Fig. 7.54 (a) Schematic block diagram showing the depositional systems related to stages in fault growth and linkage in coastal/marine environments, after Gawthorpe & Leeder (2000). (a) Fault initiation stage. (b) Interaction and linkage stage of fault growth at relative sea-level highstand. (c) Interaction and linkage stage of fault growth at relative sea-level lowstand. (d) The through-going fault zone stage of fault growth. Reproduced with permission of John Wiley & Sons, Ltd.

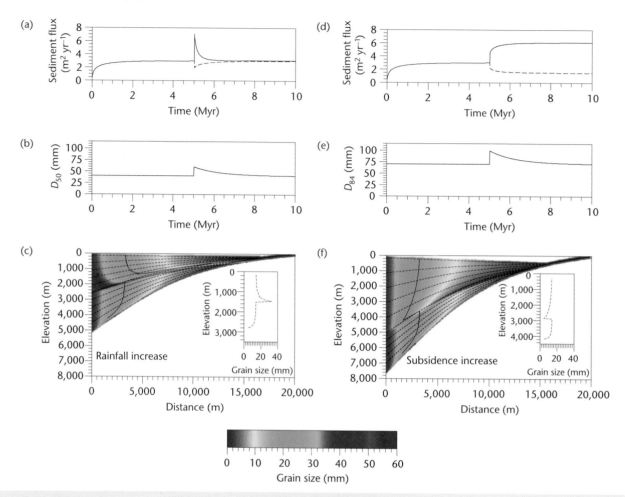

Fig. 7.55 Response of a catchment-fan system to a step change in mean annual precipitation accompanied by a coarsening of mean grain size D_{50} (a to c) and in the slip rate of a vertical fault accompanied by an increase in the coarse tail D_{84} (d to f) (after Armitage *et al.* 2011). Step change takes place after 5 Myr of model spin-up. (a) Sediment flux out of catchment, showing effect of two-fold increase (solid line) and two-fold decrease (dashed line) of precipitation at 5 Myr. (b) Change in D_{50} at 5 Myr due to two-fold increase in precipitation; (c) Stratigraphic pattern in half-graben, colour coded for grain size, showing coarse-grained spike extending far down-system due to increased sediment flux. Inset shows grain-size profile at a distance of 5 km from the depositional apex. (d) Sediment flux out of catchment for a two-fold increase in slip rate on fault (solid line) and two-fold decrease in slip rate (dashed line) at 5 Myr. (e) Change in coarse tail of grain size (D_{84}) at 5 Myr due to increase in tectonic slip rate; (f) Stratigraphic pattern in half-graben, showing retrogradation and fining at 5 Myr. Inset shows grain-size profile at 5 km from depositional apex. Initial conditions are mean annual precipitation of $1\,myr^{-1}$, slip rate $1\,mm\,yr^{-1}$, $D_{50} = 40\,mm$, $D_{84} = 70\,mm$. Solid line in stratigraphy shows 20 mm grain size; dashed lines (chrons) are at 1 Myr intervals. The sediment efflux shows a transient behaviour as landscapes adjust to the new climatic and tectonic boundary conditions. The time-scale of this transient response is of the order of 1 Myr.

evolution model used to simulate the topography of the Zagros (using $c_v = 0.02$ to $0.2\,myr^{-1}$, $L = 40\,km$, and $U = 1\,mm\,yr^{-1}$), the coefficient of proportionality k is 7–10. The geomorphic response time τ for sediment flux is between c.0.1 Myr and 4 Myr for the weak and resistant lithologies respectively. For the resistant lithologies, the geomorphic response time is very long (>1 Myr). In this case, it is most likely that periods of different thrust displacement rate in the fold-thrust belt would not be discernible in the stratigraphic record of neighbouring foreland and thrust-sheet-top basins. On the other hand, small thrust-related anticlines composed of weak lithologies should respond quickly to changes in tectonic boundary conditions, and produce a recognisable pulse of sediment in the basin. Fig. 7.56 shows the evolution of sediment flux through time as rocks with different erodibility are unroofed.

It is apparent from a casual inspection of topographic maps that some streams cut transversely through folds, whereas others are deflected around their tips (Fig. 7.57). For example, in the Marche region of the Italian anticlines, a series of rivers drain to the Adriatic coast by cutting straight through the points of maximum tectonic uplift in NW–SE oriented folds (Alvarez 1999). The impact of growing tectonic structures on drainage patterns has been discussed by a number of authors (Jolley *et al.* 1990; Burbank & Vergés 1994; Talling *et al.* 1995; Burbank *et al.* 1996; Gupta 1997; Vergés 2007; Barrier et al. 2010). One of the major impacts for basin analysis is that sediment entry points into basins may be shifted by growing tectonic structures.

The simplest approach to understanding this problem is to consider a river that is in equilibrium between channel incision and

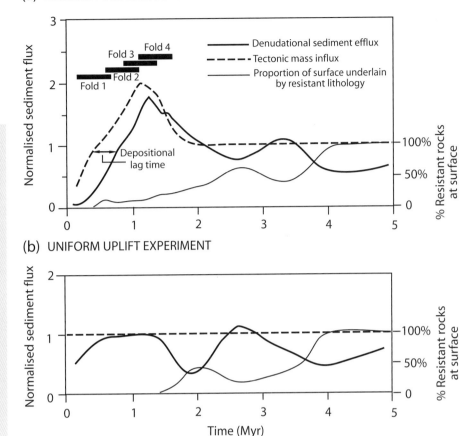

(a) ZAGROS FOLD BELT SIMULATION

(b) UNIFORM UPLIFT EXPERIMENT

Fig. 7.56 Landscape evolution model results of the unroofing of folds in the Zagros fold-thrust belt. (a) Tectonic influx from uplift of rock (dashed line) and sediment efflux out of the model grid (solid line) versus time. In this simulation, four anticlinal folds grow progressively in time. Note the lag time between tectonic influx and depositional response of <0.5 Myr during the growth phase of the folds. The thin solid line shows the percentage of resistant rocks exposed at the land surface as a function of time. Note that the decrease then increase of sediment efflux is due to this variation in erodibility of outcropping rocks. (b) Sediment efflux in which the variations are solely due to lithological variations in the stratigraphy being unroofed. Tectonic influx is spatially and temporally uniform. After Tucker and Slingerland (1996), reproduced with permission of John Wiley & Sons, Ltd.

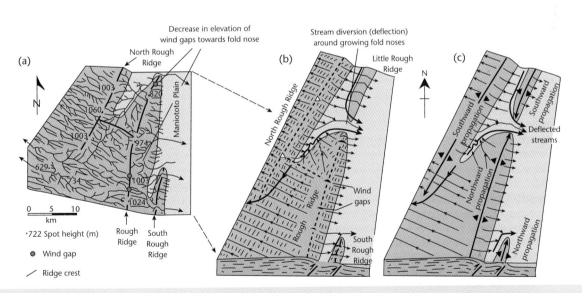

Fig. 7.57 Evolution of landscape in the face of growing folds, exemplified by the Southern Alps of central Otago, New Zealand (after Jackson *et al.* 1996). (a) Drainage network, fold crests, plunging fold noses, water gaps, wind gaps and their elevations in the vicinity of Rough Ridge. Note the decrease in elevation of the wind gaps towards the fold nose. (b) Cartoon of drainage pattern, showing clear diversion of drainages around the nose of each growing fold. (c) Interpretation of fold propagation of same region of Rough Ridge and the resultant drainage development. Reprinted with permission from Elsevier.

tectonic uplift rates. For bedrock streams, we can modify eqn. (A44.6) for the case of equilibrium to give

$$\frac{U}{c_v} = \left(\frac{Q_w}{Q_*}\right)^m S^n \qquad [7.41]$$

where Q_* is a characteristic channel discharge (equal to the total area of the catchment times the average precipitation rate), Q_w is the water discharge, S is the slope, c_v is the rock erodibility, and U is the tectonic uplift velocity. If the uplift rate of rocks in the fold crest region is high ($U = 0.01\,\mathrm{m\,yr^{-1}}$) and the stratigraphy is strongly resistant ($c_v = 0.025\,\mathrm{m\,yr^{-1}}$), the dimensionless ratio U/c_v is high (0.4), indicating that a stream is likely to be deflected. If, however, the uplift rate of rocks in the fold crest region is low ($U = 0.001\,\mathrm{m\,yr^{-1}}$) and the rocks are weakly resistant to erosion ($c_v = 0.25\,\mathrm{m\,yr^{-1}}$), U/c_v is low (0.004), indicating that the discharge-slope product of the stream may be sufficient to cut through the growing anticline.

As a growing fold emerges, the width of the fold should increase. Field studies (e.g. Burbank *et al.* 1996; Jackson *et al.* 1996) suggest that if the elevation of the entrance and exit of the fold are fixed, the widening of the fold may cause a decrease in stream gradient and stream power, leading to defeat of the stream by the growing structure (see eqn. 7.41). Alternatively, aggradation upstream of the entrance to the fold may cause avulsion of the stream to a lower position of the floodplain, effectively diverting the stream away from the growing fold. Deflected (defeated) streams are commonly captured by adjacent streams, which increases their discharge, allowing them to incise through the growing fold. The interaction of river drainage, erosion, topography and tectonic displacements is currently a rich area of research.

Coupled tectonic-erosion models at the scale of whole orogens, discussed in Chapter 4, involve the description of the tectonic deformation of the lithosphere in zones of convergence, coupled with a surface landscape of hillslopes and active channels and depositional basins (Beaumont *et al.* 1992; Willett *et al.* 1993; Beaumont *et al.* 1996b; Kooi & Beaumont 1994, 1996). As we have seen (§4.5), orogenic wedges with an asymmetry of climate (precipitation) on windward and leeward flanks are associated with an asymmetry of exhumation of deep crustal rocks. A structurally much simpler situation at a similar spatial and temporal scale is provided by the classic mega-escarpments found along segments of passive margins (Gilchrist & Summerfield 1990). The high escarpments of southern India (Western Ghats) (Gunnell 1998), Namibia–South Africa (Gallagher & Brown 1997), Brazil (Gallagher *et al.* 1994) and SE Australia (Seidl *et al.* 1996) are good examples. Landscape evolution models of passive margin mega-escarpments are found in Kooi and Beaumont (1994) and Braun and Sambridge (1997).

7.6.4   Interaction of axial and longitudinal drainage

The sediment supply to a basin, and the regional paleoslopes set up by tectonic uplift and subsidence, together control the flow paths of rivers and the boundaries between transverse systems and axial or longitudinal systems. In an extensional half-graben, sediment may be supplied transversely from rapidly uplifting footwalls and slowly rotating hangingwalls. In foreland basin systems, the transverse supply is predominantly from the orogenic wedge, and to a less extent from the flexural forebulge. In the wedge-top region, sediment may be shed backwards from growing contractional folds and thrust culminations. The moving boundaries between transverse and longitudinal transport are represented by major facies, grain size and architectural changes. Their movement is a record of environmental changes driven by tectonics and climate. The dynamics of moving boundaries is a familiar problem encountered in the investigation of shoreline migration (Kim *et al.* 2011).

Coalesced alluvial fans derived from the footwall block typify tectonically active basins (Whipple & Trayler 1996; Allen & Hovius 1998; Gawthorpe & Leeder 2000). Progradation of the bajada is controlled by the mass balance determined by sediment supply and accommodation, but is also influenced by erosion at the toes of the fans by axial rivers, known as 'toe-cutting' (Leeder & Mack 2001; Pépin *et al.* 2010; Perez-Arlucea *et al.* 2000). Toe-cutting is an effective way of boosting sediment discharges of the axial system.

We let the sediment discharges from hangingwall and footwall fans be q_h and q_f, and the axial river discharge to be q_a. The transfers from footwall and hangingwall fans to the axial system are q_{fa} and q_{ha}. The half-graben subsides by a simple rotation with a basin width w_b at a maximum rate s at $x = W_b$, reducing to zero at $x = 0$. The lengths of the fans can be calculated using a sediment mass balance, assuming that the fluvial system remains constant in surface area and there is no rotation or extension across the half-graben. Taking the hangingwall fan, the sediment below the fan surface is made up of that above base level and that below base level, accommodated by tectonic subsidence. However, if the fan slope and maximum elevation do not change over time, the sediment deposited is equal to the increment in part below base level, so that

$$w_h = \sqrt{2\frac{W_b}{S_{max}}(q_h - q_{ha})} \qquad [7.42]$$

Similarly, for the footwall-derived fans, the mass balance gives

$$w_f = 2\frac{(q_f - q_{fa})}{S_{max}} \qquad [7.43]$$

From eqn. [7.42] and eqn. [7.43] it is clear that the positions of the fan toes are determined by the mass balance. Inserting reasonable values for a Basin and Range half-graben, let $W_b = 20\,\mathrm{km}$, and $S_{max} = 1\,\mathrm{mm\,yr^{-1}}$, the widths of the fans scale on the sediment discharge terms. The total sediment budget is $q_T = W_b S_{max}/2$. The fraction delivered from each margin is therefore given by eqn. [7.42] and eqn. [7.43] scaled by q_T.

If the fraction of the total budget supplied from each margin remains constant, the fan toes remain stationary, despite changes in the maximum subsidence rate. Inspection of eqn. [7.42] shows that the length of hangingwall-derived fans remains constant when the net discharge $q_h - q_{ha}$ scales with $S_{max}/2$. In other words, when S_{max} doubles through an increase in slip-rate on a bounding fault, the length of the fan stays fixed if the net discharge increases by 50%. If sediment discharge increases through a climate change, with no change in accommodation driven by fault-related subsidence, the fan length must change in response. This suggests that climatically driven movement of the fan toe should be more variable than damped (coupled) tectonically driven changes.

The width of the axial tract should vary according to the tectonic or climatic forcing. For the coupled tectonic model, the tract remains almost stationary, but an increase in the net discharge forced by climate results in fan toe progradation from each side of the basin, and the eventual meeting of the oppositely derived fans.

The concept above can be extended to the case of an array of folds bordering wedge-top basins. The basement surface is assumed to subside in a sinusoidal pattern, and we neglect horizontal advection of rock during tectonic shortening. Accommodation is created by an

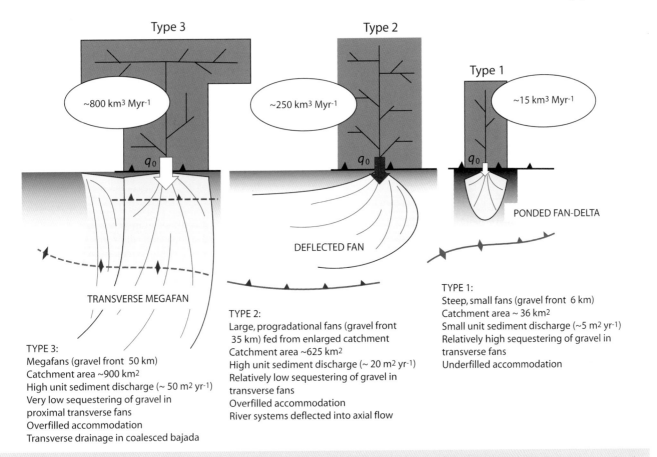

Fig. 7.58 Three scenarios of sediment dispersal in continental environments away from a mountain front, based on the ratio of sediment supply q_s to accommodation $\sigma(x)$ based on Allen *et al.* (2013). Right: a small, steep fan or fan-delta supplied by a small catchment in the external fold-thrust belt builds into an underfilled wedge-top basin at low $q_s/\sigma(x)$. Centre: a larger catchment reaching the axial zone of the mountain belt fills the wedge-top basin. Sediment is guided axially by fold-thrust anticlines to feed downsystem depocentres, at intermediate values of $q_s/\sigma(x)$. Left: large catchments oversupply the wedge-top region, burying defunct thrust-related structures, dispersing sediment to the foredeep down the regional paleoslope as large transverse fans, at high values of $q_s/\sigma(x)$. Reproduced with permission of John Wiley & Sons, Ltd.

increase in the amplitude of the sine curve, keeping wavelength constant. The new accommodation up to a transverse fan length from the proximal margin is

$$h_0 S_{max} \int_0^{wf} \sin\left(\pi + \frac{2\pi x}{\lambda}\right) dx \qquad [7.44]$$

which when evaluated gives

$$w_f = \frac{\lambda}{2\pi}\left\{\cos^{-1}\left(-\frac{(q_f - q_{fa})}{g_0 \lambda S_{max}} + 1\right) - \pi\right\} \qquad [7.45]$$

For example, with a wavelength λ of 40 km (basin width of 20 km), a maximum subsidence rate S_{max} of 0.1 mm yr^{-1}, a net sediment discharge to the fan of $q_f - q_{fa}$ of 400 m^2 yr^{-1}, and an amplitude $h_o = 1$ km, the proximal fan length is c.12 km. If using Excel, eqn. [7.45] needs to be solved by taking arccos(x) = cos^{-1}(x) and using an on-line calculator, or by making a look-up table of cos(x) in Excel, but this is slow.

We can make use of the idea of an axial sediment discharge. For a closed budget, the axial discharge is that remaining after sediment

extraction to build fringing fans, that is, $q_T = q_h + q_f$, and $q_{fa} + q_{ha} = q_a$. The axial budget is broken down into that deposited and that exported downslope q_e. The fraction of the total discharge exported through the axial system is a key parameter in understanding the dynamics of sediment routing systems in tectonically active regions. For $q_e/q_T \sim 0$, we anticipate perfectly filled basins with little far-field transport. For $q_e/q_T \gg 0$, we anticipate major longitudinal downslope export of sediment guided by the tectonic grain. When there is marked asymmetry in the sediment supply combined with a high total discharge, we anticipate extensive systems that bury underlying tectonic structures and follow the regional transverse paleoslope.

Sediment dispersal patterns are strongly affected by the balance between the sediment supply q_s and the distribution of accommodation $\sigma(x)$ (Allen *et al.* 2013). Where this ratio is low, steep depositional systems such as fan deltas build into underfilled lacustrine or marine basins. At intermediate values of $q_s/\sigma(x)$, larger depositional systems fill structurally confined basins, but sediment dispersal is guided by tectonic structures into an axial flow. At higher values of the ratio, large transverse megafans bury defunct tectonic structures and follow the regional paleoslope from the mountain belt to the flexural foredeep (Fig. 7.58).

CHAPTER EIGHT
Basin stratigraphy

Summary

The stratigraphy of a sedimentary basin reflects the interplay between sediment supply and accommodation generation. Accommodation generation is controlled by tectonic subsidence and uplift, global (eustatic) sea-level change and compaction, all of which affect the position of base level. Sediment supply depends on the erosion rate and size of catchments serving as feeder systems. Variations in these parameters in space and time result in a characteristic packaging of stratigraphic units in the basin-fill.

A simple 1-D experiment of superimposing a sinusoidal eustatic variation on a steady tectonic subsidence shows that the timing of the peak in relative sea level, analogous to the occurrence of the deepest-water facies in a sedimentary cycle, and the timing of the lowest relative sea level when the smallest water depths occur, depends on the balance between the sinusoidal variation and the background tectonic subsidence. Consequently, we should expect genetic stratigraphic units to have boundaries that are inherently diachronous. Variation in the sediment supply results in stratigraphic cycles that are limited by the availability of accommodation, or are limited by the magnitude of the sediment supply. Over- or under-supply of sediment to fill accommodation results in stratigraphic architectures involving basinward progradation, aggradation or hinterlandward retrogradation. The position over time of the shoreline, the shelf-edge or the clinoform rollover may be descending or ascending, retrogradational or aggradational.

Stratigraphic architectures can be broken down into a hierarchy of genetic units or cycles, from long-term megasequences ($\sim 10^7$–10^8 yr) and tectonostratigraphic units ($\sim 10^6$–10^7 yr), depositional sequences ($\sim 10^6$ yr) down to metre-scale cycles of c.10^3–10^4 yr duration attributed to orbital forcing or unforced internal dynamics. Depositional sequences are recognised on the basis of key stratigraphic surfaces such as unconformities and their lateral conformities, maximum flooding or maximum regression. Within depositional sequences, packages of stratigraphy exhibit characteristic stratal terminations such as downlap, onlap and erosional truncation that are diagnostic of the stage during a cycle of relative sea level in which they were deposited.

Forcing mechanisms for the generation of stratigraphic cycles may be changes in tectonic boundary conditions, such as flexure, in-plane stress, slip on faults, growth of folds, and continental dynamic topography, or changes in base level caused by eustatic mechanisms. The most important long-term eustatic mechanisms are changes in the volume of the ocean basins caused by the trade-off between the changing hypsometry of the mid-ocean ridge system, and the occurrence of dynamic uplift of the ocean floor by deep mantle processes associated with subduction, oceanic hotspots and superswells. Shorter-term mechanisms for changing the volume of ocean water are thermal expansion of the global ocean, losses through subduction, and fluctuations caused by the growth and decay of land-based ice caps. Stratigraphic cycles, however, may be generated by feedbacks within sediment routing systems, causing an unforced autocyclity. Unforced cyclicity may be convolved with external orbitally forced cyclicity in sedimentary successions, such as the well-documented Triassic Latemar of the Italian Dolomites.

Dynamical approaches to stratigraphy attempt to model the complexities of the sediment routing system from the viewpoint of the stratigraphic product. These models allow a visualisation of stratigraphic architectures under known forcing conditions. Other models investigate the functioning of the sedimentary systems, or components of the sedimentary systems, by examination of the physics. The investigation of the formation of prograding clinoforms where rivers enter the ocean, using the shoreline as a moving boundary, is an example of the latter, and is a prerequisite for the understanding of the shoreline and shelf-edge trajectories in stratigraphic successions.

Stratigraphy is a selective, partial and 'filtered' record of past events. The transformation of sediment routing systems into stratigraphy, therefore, is complicated by the incompleteness of the stratigraphic record, and the difficulties posed by upscaling in time. Measured rates of sediment accumulation reduce with the length of time interval considered, suggesting that there is a power-law magnitude-frequency distribution of stratigraphic record and the intervening gaps. The incompleteness of the stratigraphic record suggests that the transfer of sediment from surface flux to stratigraphic record is analogous to a signal passing through a gate. Gates might be related to the exceeding of a threshold, a random sampling of sediment fluxes, or a regularly repeating selection.

Basin Analysis: Principles and Application to Petroleum Play Assessment, Third Edition. Philip A. Allen and John R. Allen.
© 2013 John Wiley & Sons, Ltd. Published 2013 by John Wiley & Sons, Ltd.

Previous chapters have provided some insights into the possible physical causes of the main classes of sedimentary basin. The long-term subsidence required for the maintenance of sedimentary basins is generated by these large-scale physical mechanisms. The nature of the stratigraphy in sedimentary basins is a result of variations in the interplay between sediment supply and tectonic subsidence in time and space, but also on the internal dynamics of sediment routing systems.

In this chapter we examine the main controls on the stratigraphy of sedimentary successions, starting by using simple physical principles. We consider the difficulties involved in understanding stratigraphy from the viewpoint of modern depositional systems, and discuss the incomplete recording of erosional and depositional landscapes in the 'filtered' stratigraphic record. A number of topics are chosen to illustrate the role played by external (allogenic) and internal (autogenic) processes in generating distinctive stratigraphic geometries.

8.1 A primer on process stratigraphy

8.1.1 Introduction

We use the term *process stratigraphy* informally as the science of the recognition and interpretation of the genetic structure of stratigraphy. *Genetic stratigraphy* (Galloway 1989; Homewood *et al.* 1992) and *dynamic stratigraphy* (Matthews 1974; Cross 1990) are equally appropriate terms, and *time stratigraphy* (Wheeler 1964) is an allied concept. The term *sequence stratigraphy* has connotations with the paradigm of global sea-level control on stratigraphic architectures, though modern definitions of the term reduce the emphasis on this aspect (Catuneanu *et al.* 2009). Miall (2010) provides a comprehensive treatment of the historical background of stratigraphic approaches, the descriptive stratigraphic framework, and the mechanisms for the formation of stratigraphic cycles and architectures.

Process stratigraphy makes use of systematic branches of stratigraphy (biostratigraphy, magnetostratigraphy, chemostratigraphy, chronostratigraphy, lithostratigraphy) but is not itself a taxonomic exercise. The fundamental aim of process stratigraphy is to understand the driving mechanisms for the range of stratigraphic architectures in time and space found in sedimentary basins.

The history of 'sequence stratigraphy' dates back to the mid-1900s (Miall 2010) when sedimentary successions were recognised to be divisible into packages of rock separated by unconformities of inter-regional extent (Sloss 1950, 1963). An analysis of the geometry of stratigraphic units deduced from their seismic reflection character (Payton 1977) led industry Earth scientists to believe that eustatic changes of sea level were the primary control on the development of stratigraphic packages. In 1987 a 'global' sea-level chart was published (Haq *et al.* 1987) on the basis of the apparent recognition of distinctive stratigraphic boundaries (unconformities and correlative conformities) of the same age in widely separated locations. A second wave of sequence stratigraphic thinking (Wilgus *et al.* 1988) introduced ideas on accommodation in sedimentary basins and on the relationship between stratigraphic packaging and cycles of relative sea level. Since the late 1980s there has been substantial criticism of some of the new stratigraphic terminology, of its over-reliance on eustasy as a driving mechanism and on the presumed global synchroneity of key stratigraphic surfaces such as sequence boundaries (see Miall 1991, 1992 for a trenchant view). More recently, numerical

modelling and physical experiments (Harbaugh *et al.* 1999; Burgess *et al.* 2002; Swenson *et al.* 2005; Paola *et al.* 2009; Martin *et al.* 2009) have been directed at examining in detail some of the controls and feedbacks involved in the generation of stratigraphic packages, including those due to unforced internal dynamics.

The sequence stratigraphy practised in the decades prior to the turn of the century was therefore underpinned by two linked but separable concepts (Carter 1998). The first derives largely from the recognition of geometrical relationships between seismic reflectors and key seismic surfaces, commonly unconformities, as elaborated in Payton (1977). The second concept is the integration of the timing and amplitude of these relationships recognised in seismic reflection data into a global sea-level curve (Vail *et al.* 1977b; Haq *et al.* 1987; Posamentier & Vail 1988). Whereas the methodological or descriptive aspects of the breakdown of stratigraphy derived from seismic reflection data remain intact, there continues to be vociferous debate of inter-regional correlation and the global sea-level chart.

Process stratigraphy specifically focuses on the dynamics of the controls on stratigraphic patterns: (i) the volume and granulometry of the sediment supply; and (ii) the rate of generation and spatial distribution of accommodation. The interplay between these factors at different time and spatial scales is responsible for the stratigraphic architectures found in sedimentary basins. The key idea of process stratigraphy is that space is made available for sediment to accumulate, termed *accommodation*. Sediment fluxes cause an underfilling, filling or overfilling of this available space. Process stratigraphy is therefore essentially a mass or volume balance exercise. Accommodation generation and sediment supply vary strongly in space and in time, causing a complex four-dimensional packaging of stratigraphic units. This process stratigraphic approach is similar to the 'standardised' approach that emphasises the basic building blocks or genetic units of stratigraphy as a response to changing accommodation and sediment supply rather than simply base-level change (Catuneanu 2006; Catuneanu *et al.* 2009).

A number of texts and reviews provide overviews of sequence or process stratigraphy, such as Emery and Myers (1996), Homewood *et al.* (1992, 2000), Miall (2000, 2010), Coe (2003), Helland-Hansen and Hampson (2009) and Catuneanu *et al.* (2009). High-resolution sequence stratigraphy, which focuses on the structure of stratigraphy below the level of the depositional sequence, is colourfully examined by Van Wagoner *et al.* (1990). The application of sequence stratigraphy to carbonate reefs and platforms is treated by Schlager (1992), Sarg (1988) and Schlager (2005). The results of physical experiments designed to understand stratigraphic geometries and their controls is found in Martin *et al.* (2009).

8.1.2 Accommodation, sediment supply and sea level

Eustasy is global sea level measured from the sea surface to a fixed datum, such as the centre of the Earth. The controls on eustatic sea level are discussed in §8.2.2.5.

Relative sea level is sea level measured relative to a moving datum, often a distinctive horizon such as a bed, unit or boundary within a sediment pile, or the lower contact with basement. Relative sea level is therefore affected by processes such as tectonic uplift and subsidence, compaction and eustasy.

Water depth is the vertical distance between the sea surface and the seabed. Although water depth commonly changes during a change in relative sea level, it may also be influenced by the sediment input into a sedimentary basin with no relative sea-level change. Eustasy, relative sea level and water depth are all distinctive concepts (Fig. 8.1).

Accommodation – the space made available for sediment to accumulate – is controlled by *base level*, since sediment can only accumulate long-term up to base level. Base level may be a graded stream profile on land, or a graded shelf profile on the continental shelf.

We can write a change in accommodation ΔA as

$$\Delta A = \Delta E + \Delta S + \Delta C \qquad [8.1]$$

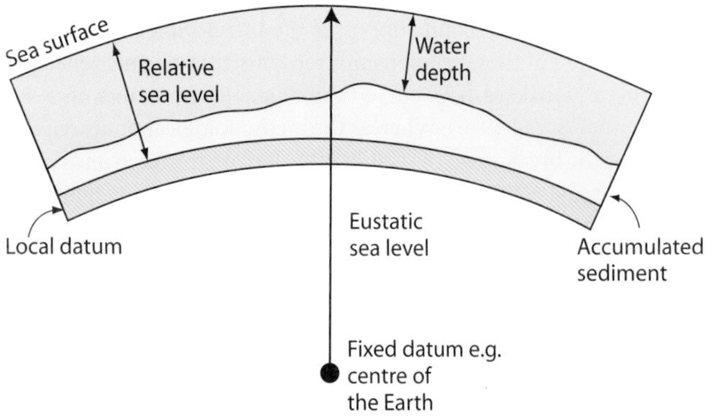

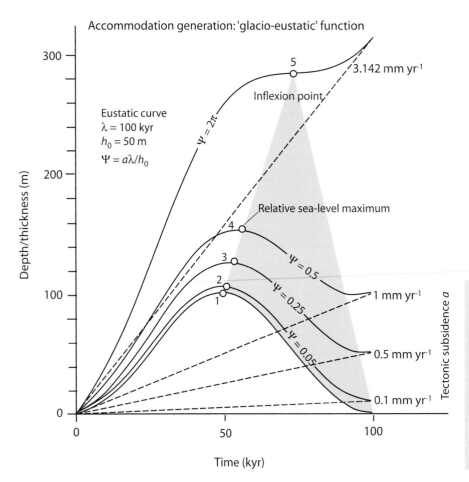

Fig. 8.1 Definitions of terms used in process stratigraphy (after Jervey 1988; Emery & Myers 1996, reproduced with permission of John Wiley & Sons, Ltd.): eustatic sea level, relative sea level and water depth.

where A is accommodation, E is eustasy, S is subsidence, and C is compaction. A change in water depth ΔW can also be written

$$\Delta W = \Delta A - \Delta D = (\Delta E + \Delta S + \Delta C) - \Delta D \qquad [8.2]$$

where ΔD is the amount of sediment deposited.

8.1.3 Simple 1D forward models from first principles

Sloss (1962) argued that transgressions and regressions of the shoreline were controlled (at least in part) by the relative magnitudes of the rate of sea-level change and rate of subsidence. We can verify this from a simple 1D model using first principles.

Any forward model must make use of the relationship among the key parameters in eqn. [8.2]. For example, eustatic sea level can be treated as two end-member types: (i) a rapid rise followed by a slow fall; and (ii) slow rise in sea level followed by a rapid fall. In initial forward modelling studies and conceptual exercises, such as the early offerings of the Exxon Group (Vail *et al.* 1977a, b), sediment supply was considered to be constant through the cycle of relative sea-level change.

In the discussion that follows, we build an analytical method for considering the stratigraphic cycles generated under a sinusoidal eustatic variation in a basin with a background tectonic subsidence rate and with a sediment supply coupled to the relative sea-level variation. We ignore compaction and the isostatic response to water and sediment loads. Although the assumptions and algorithms are undoubt-

Fig. 8.2 Variation in relative sea level through a cycle of eustatic change with wavelength λ and amplitude h_0 in a basin with a linear tectonic subsidence rate a. The dimensionless parameter Ψ varies from 0.05 to 2π, corresponding to tectonic subsidence rates of 0.1 to $\pi\,\mathrm{mm\,yr^{-1}}$. Increasing values of Ψ cause the relative sea-level maximum (open circle) to be delayed in the cycle. For the 'glacial' eustatic parameters used, tectonic subsidence must be $> \pi\,\mathrm{mm\,yr^{-1}}$ in order for the relative sea-level fall (shaded area) to disappear.

edly a crude simplification of the complexities of nature, a simple 1-D forward model is the prerequisite for all process stratigraphic thinking.

The variation of relative sea level through a cycle of eustatic change with wavelength λ and amplitude h_0 in a basin with a linear tectonic subsidence rate a is given in Appendix 51 (Fig. 8.2), and is a quantitative version of the pioneering early work of Barrell (1917). A number of different relative sea-level scenarios result from the variations in the magnitude of the eustatic change compared to the subsidence rate. This can be expressed in terms of the dimensionless parameter $\psi = a\lambda/h_0$. If the amplitude of the sinusoidal eustatic change far exceeds the change in tectonic subsidence (ψ is small), the relative sea-level variation is also near-sinusoidal, and the maximum relative sea level is very close to the eustatic peak. As the tectonic subsidence rate increases relative to the amplitude of eustatic change (ψ increases), the relative sea-level curve becomes more asymmetrical, and the peak of relative sea level becomes delayed in the eustatic cycle. At a critical tectonic subsidence rate ($\psi = 2\pi$), there is no relative sea-level fall, but instead an inflexion point in relative sea level significantly following the eustatic peak. In Fig. 8.2, relative sea level is equivalent to accommodation since the curves begin at zero water depth rather than at some point on a graded profile.

For a given eustatic period λ, the time period to the relative sea-level inflexion point decreases with increasing values of the dimensionless parameter $\psi = a\lambda/h_0$. Relative sea level increases monotonically when $\lambda a/2\pi h_0 > 1$, or $\psi/2\pi > 1$.

Clearly, the timing of the peak in relative sea level, analogous to the occurrence of maximum water depths in a stratigraphic succession, and the timing of the lowest relative sea level, analogous to the smallest water depths in a stratigraphic succession, depends on the interplay of tectonic subsidence and eustatic sea level, even with no sediment supply variations. If two adjacent areas with different histories of tectonic subsidence are correlated, we should expect different timings of the maximum and minimum water depths.

Eustatic sea-level change takes place at a wide range of periods and amplitudes. The dominant signal for the glacial-interglacial variations of the Pleistocene is a period of 100 kyr and amplitude of 50 m (variation in height of 100 m) (Carter 1998, 2005). We call this a typical 'glacio-eustatic cycle'. On the other hand, cycles of eustatic change inferred from stratigraphy, such as the mid–late Eocene cycles in the Nummulitic Limestone of Switzerland (Allen *et al.* 2001) and the Oligocene stratigraphy of the New Jersey coastal plain (Kominz and Pekar 2001), imply cycles of longer duration (c.0.5 Myr to 2 Myr) and lower amplitude (10 to 20 m). The fact that these cycles exist in stratigraphy deposited during climatic conditions when there was limited ice suggests that mechanisms other than glacio-eustasy were responsible. We call these variations 'non-glacial cycles'. In unforced cycles (§8.2.3) no eustatic change is required, and the facies and water depth changes are produced by internal dynamics driving variations in sediment transport rate.

Substitution of parameter values in eqn. [A51.5] shows that the critical tectonic subsidence rate for a monotonic rise in relative sea level (for 'glacial cycles' with $h_0 = 50$ m and $\lambda = 100$ kyr) is 3.14 mm yr^{-1}. Such rates of tectonic subsidence are rare, which implies that eustatic cycles of this type are likely to always produce relative sea-level falls. If relative sea-level falls are accompanied by erosion, we should expect to find erosional unconformities in stratigraphy formed under the influence of 'glacio-eustatic' cycles. If, however, the eustatic variation is 20 m amplitude with a wavelength of 400 kyr, the critical tectonic subsidence rate becomes 0.3 mm yr^{-1}, and if the amplitude and wavelength are 10 m and 1 Myr, the critical tectonic subsidence rate becomes 0.06 mm yr^{-1}. These last values are below the rate of subsidence experienced in most sedimentary basins. In situations of slow 'non-glacial' eustatic change, relative sea level may monotonically rise under a background tectonic subsidence.

It can also be seen from Fig. 8.2 that the triangular field of relative sea-level fall narrows to a point at approximately $t = 50,000$ yr on the curve $a_{\text{crit}} = 3.14$ mm yr^{-1}. Clearly, when correlating different locations in a basin with spatially variable tectonic subsidence rate, we should expect significant diachroneity of the onset of erosion due to relative sea-level fall and of flooding during relative sea-level rise. The delay in the onset of erosion is a quarter of a eustatic wavelength, or $\lambda/4$. The difference in the timing of flooding is also $\lambda/4$. Since diachroneity of key stratigraphic surfaces is to be expected from a globally synchronous eustatic change, it is curious that the inference of stratigraphic synchroneity is seen as a key test for a eustatic control. A critique of global correlation is found in Miall (1991, 1994) and reviewed in Miall (2010).

Sediment supply controls how much of the accommodation is filled (Fig. 8.3). In the simplest scenario, the sediment supply rate s could be considered a constant velocity, but it is more reasonable to assume that sediment supply rate is coupled in some way to the rate of change of relative sea level. It would also be simple to formulate sediment supply as coupled to the rate of relative sea-level variation but with a certain time lag to account for the response time of sediment delivery systems to any base-level change (Chapter 7). The sediment accumulated at any point within a relative sea-level cycle, and the remaining water depth representing unfilled accommodation, can be calculated using eqns [A51.9] and [A51.10].

At high rates of sediment supply relative to the rate of relative sea-level change, accommodation is rapidly filled, leading to bypassing of the excess sediment and the cutting of disconformities or unconformities. At low sediment supply rates, accommodation is never completely filled during a cycle of relative sea-level change, causing increasing water depths over time. The variation in water depth through a stratigraphic cycle, the thickness of sediment preserved, the nature of flooding surfaces and disconformities and their time span, are all related to the interplay of eustatic variation, tectonic subsidence rate and sediment supply rate. This is a fundamental reason why stratigraphic cycles are so variable.

Some conclusions from this analysis from first principles of cycle development are as follows (in all of the analysis, water depth and relative sea level are assumed to be zero at the start of a cycle of relative sea-level change):

Cycle thickness: if previously deposited sediment is easily eroded during a relative sea-level fall, stratigraphic cycle thicknesses depend on two factors:

1. Where sediment supply rates are high, cycle thickness depends on the tectonic subsidence over the period of time between the onset of the cycle and the time of relative sea-level lowstand. This is between π and $3\pi/2$ through a eustatic cycle of duration 2π. If $a = 0.1$ mm yr^{-1} and $\lambda = 10^5$ yr, cycle thicknesses, ignoring compaction, should be between 5 m and 7.5 m.
2. Where sediment supply rates are low compared to the tectonic subsidence rate, cycle thickness depends only on the sediment accumulated over the duration of the eustatic cycle. For $s_0 = 0.1$ mm yr^{-1} and $\lambda = 10^5$ yr, the cycle thickness is c.6 m.

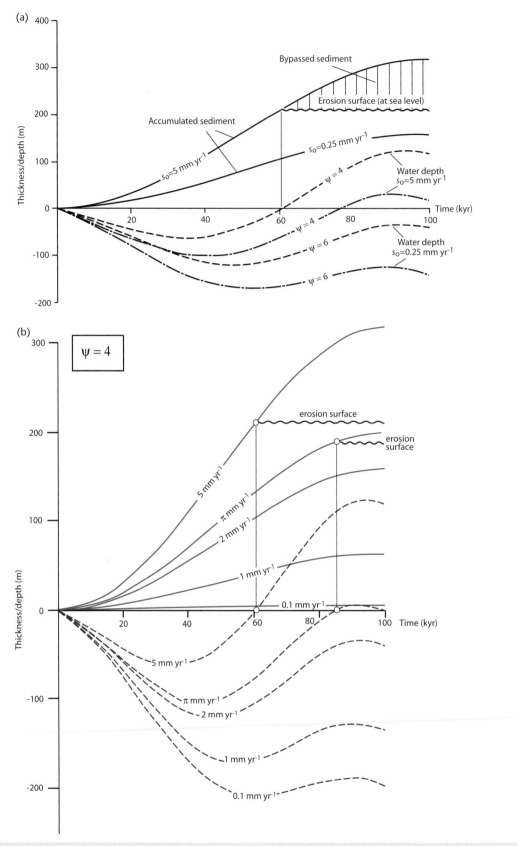

Fig. 8.3 Water depth and sediment accumulation during relative sea-level cycles with variable sediment supply rate coupled to the rate of change of relative sea level. (a) Accumulated sediment (solid lines) and water depth (dashed lines) (for Ψ of 4 and 6) for maximum sedimentation velocities s_0 of 0.25 and 5 mm yr^{-1}. When $s_0 = 5$ mm yr^{-1} and Ψ = 4, the basin is filled with sediment by 60 kyr, after which sediment is bypassed over an erosion surface. At Ψ = 6, the basin remains water-filled until nearly 80 kyr. At the lower maximum sedimentation velocity s_0 of 0.25 mm yr^{-1}, the basin remains water-filled throughout the cycle of relative sea-level change. (b) Accumulated sediment (solid blue lines) and water depth (red dashed lines) for a range of vertical sedimentation velocities s_0 with Ψ = 4. The evolution of water depth through a cycle of relative sea level, sedimentary facies and the occurrence of erosional bypass surfaces are all critically dependent on the sediment supply.

We can therefore get similar stratigraphic cycle thicknesses under two contrasting scenarios: by erosion during a relative sea-level fall, and by underfilling of the available accommodation. The first is obviously a case of *accommodation-limited* stratigraphy. The latter is a case of *sediment-supply-limited* stratigraphy.

Sedimentary facies and water depth variation: a common feature of model output is that stratigraphic cycles deepen-up. How then do we generate the common *shallowing-up* cycles of the stratigraphic record? If we believe that eustatic change is important, the most straightforward answer is that the preservation of the shallow-water upper portion of a stratigraphic cycle must be due to the predominance of relative sea-level cycles possessing inflexion points, but insignificant relative sea-level falls. In many continental and shallow-marine environments, sediment supply is easily capable of filling accommodation. The generation of shallowing-upward cycles therefore depends critically on the value of ψ, rather than on the sediment supply rate. Alternatively, if there is no eustatic change, and the cycles are unforced, the common pattern of shallowing-up cyclicity may be due to variations in flooding and progradation driven by internal dynamics of the sedimentary system under a background tectonic subsidence (§8.2.3).

Progradation, aggradation and retrogradation: sediment cannot accumulate for long periods of time above the graded profile. Consequently, within a relative sea-level cycle, sediment may be bypassed downstream or into deeper water. High sediment supply in a basin with adequate accommodation will cause sedimentary facies belts to *aggrade*. If sediment supply is less than the accommodation, facies belts will *retrograde*. If sediment supply exceeds accommodation, facies belts will *prograde*. Net retrogradation, net progradation and net aggradation depend on the balance between the accumulated sediment over time and the accommodation generation. Sediment supply variations, caused by climate change or internal dynamics, are potentially extremely important in influencing stratigraphic packaging, without recourse to eustatic change.

Sediment accumulation may be limited by either a low sediment supply or low rate of accommodation generation, that is, *sediment-supply-limited*, or *accommodation-limited*. This produces two distinct stratigraphic trends and facies assemblages. In the first case, new water depth is generated with every cycle of relative sea level. After several cycles of relative sea level, marine shorelines will be strongly retrogradational or transgressive, and basinal locations should be dominated by deep-water shale deposition. In the second case, if sediment supply is high but accommodation is limited, the shoreline regresses immediately, causing the progradation of deltaic and coastal plain sediments.

The depositional architecture is therefore very sensitive to the interplay between sediment supply and accommodation generation. We can envisage a continuum of geometrical possibilities where the sediment supply rate varies in relation to the accommodation. For example, for a morphology involving low-gradient topsets and steeper-gradient clinoforms (Fig. 8.4), the increase in topset accommodation ΔV_{ta} (as a volume) is the product of the relative sea-level change and the topset area. If the sediment supply is greater than ΔV_{ta}, the clinoforms will prograde, the *offlap break* or coastline migrating strongly offshore. However, if the sediment supply is progressively reduced, the offlap break first climbs, then becomes near-vertical, and then migrates landward when sediment supply is less than ΔV_{ta}. We further develop the problem of clinoform migration and the trajectory of the shoreline in §8.3.3.

8.2 Stratigraphic cycles: definition and recognition

The cyclical nature of stratigraphy at a range of scales has been recognised for many decades, from the recognition of small-scale millimetre- or centimetre-scale rhythms of laminated sediments in lake and ocean sediments, to metre-scale cycles in the facies, grain size or bed thickness of sediments and sedimentary rocks, through to meso-scale, c.10^1–10^2 m-thick sedimentary packages bounded by unconformities and surfaces recording rapid deepening, to large-scale 10^2–10^3 m-thick units that can be correlated across an entire continent. There is therefore a hierarchy of scale, and a corresponding hierarchy of driving mechanisms for cyclicity (Fig. 8.5).

8.2.1   The hierarchy from beds to megasequences

8.2.1.1   Supersequences and megasequences

The importance of rock-stratigraphic units traceable over wide areas of the North American continent and bounded by unconformities of 'inter-regional scope' has long been recognised (Sloss 1950, 1963, 1988). Six such *supersequences* (termed 'sequences' by Sloss) occupying the Phanerozoic were defined, with durations of 50 to 120 Myr (Fig. 8.6). Each supersequence is thinner and represents a shorter period of deposition at the centre of the craton compared to its margin, so that the bounding unconformities increase in duration from the craton edge (10 Myr gap) to the craton centre (150 Myr gap). The supersequences have been interpreted as due to the effects of cycles of subduction around the edges of the North American plate (§5.2) (Burgess & Gurnis 1995; Burgess *et al.* 1997).

Hubbard *et al.* (1985), using primarily a subsurface seismic reflection database, introduced a similar concept of *megasequences*, representing the stratigraphic packages deposited during distinct phases of plate motion. Consequently, a megasequence may correspond to a period of continental extension with syn-rift and post-rift components, or to a period of convergent plate motion characterised by flexure of the continental lithosphere. Megasequences are also bounded by extensive, but not global, major unconformities. Since sedimentary basins are deformations of the lithosphere under a set of plate-scale driving forces, the concept of the megasequence is of primary importance in basin analysis.

The package of stratigraphy associated with a recognisable phase of tectonic deformation is commonly referred to as a *tectonostratigraphic unit* (e.g. Yong *et al.* 2003). Although some authors view the tectonostratigraphic unit as synonymous with the megasequence (Nikishin & Kopaevich 2009), it is best used for a level of tectonically controlled stratigraphic packaging below the status of the megasequence (Fig. 8.5).

8.2.1.2   Depositional sequences

The primary meso-scale units of stratigraphy are conventionally termed *depositional sequences*. They are coherent packets of strata that are genetically related and which can be traced for considerable distances across a basin. Depositional sequences can be recognised in the subsurface using seismic stratigraphic methods, the tops and bases being marked by bounding unconformities or laterally correlatable conformities (Mitchum *et al.* 1977) (Fig. 8.7). Within a depositional sequence, individual strata can exhibit a variety of geometrical

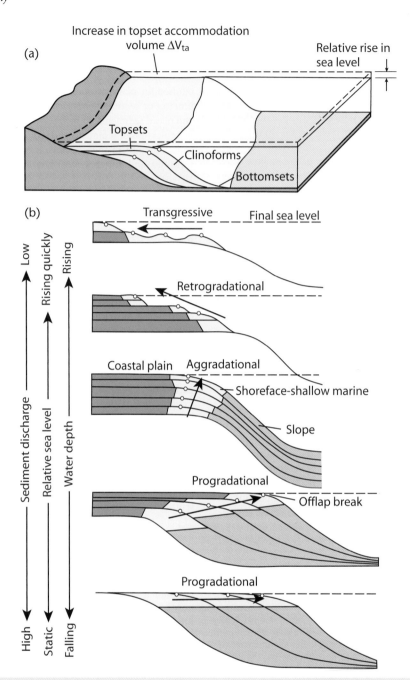

Fig. 8.4 Large-scale architecture of depositional units in relation to accommodation and sediment supply (after Galloway 1989). (a) A rise in relative sea level causes an increase in topset accommodation volume ΔV_{ta}, equal to the product of the relative sea-level rise and the topset area. (b) Stratigraphic patterns change from transgressive to retrogradational, aggradational and progradational as the sediment supply increases relative to the topset accommodation. White circles approximate position of beach (or offlap break). AAPG © 1989. Reprinted by permission of the AAPG whose permission is required for further use.

relationships to the depositional boundary (Dunbar & Rogers 1957). Widespread recognition of these geometrical relationships was only possible with the availability of high-quality seismic reflection data. As well as helping to define a depositional sequence boundary, these types of discordant relationship also furnish clues as to the origin of the unconformity. Onlap, downlap and toplap indicate non-depositional hiatuses, whereas truncation indicates an erosional

hiatus, or it may be the result of structural disruption (Vail *et al.* 1977a).

Meso-scale stratigraphic units may also be defined in terms of the trend of the shoreline, called *transgressive–regressive cycles* (Johnson & Murphy 1984; Embry & Johannessen 1992) or in terms of the recognition of the maximum inundation of the land (*maximum flooding surfaces*) rather than by their erosional unconformities,

HIERARCHY OF GENETIC UNITS

DRIVING MECHANISMS

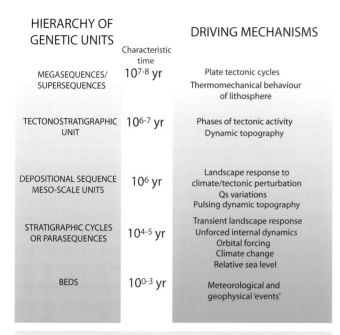

Fig. 8.5 Hierarchy of genetic stratigraphic units from beds to megasequences, with the most likely formative mechanisms at the appropriate scale.

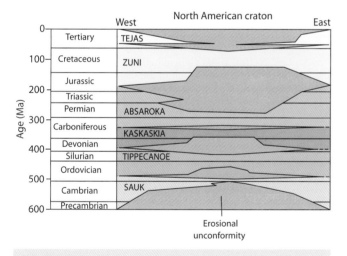

Fig. 8.6 The supersequences of the North American craton of Sloss (1963).

termed *genetic sequences* (Frazier 1974; Galloway 1989) (Fig. 8.7). These different approaches place varying emphasis on the recognition of distinct phases in a cycle of relative base-level change and on the characteristic surfaces representing the onset of base-level fall, the end of base-level fall, the end of regression and the end of transgression.

In essence, the different approaches to sequence stratigraphy can be summarised as follows (Fig. 8.8). Sequence stratigraphic models

as derivatives of early seismic-stratigraphy-based methods require the identification of stratigraphic packages related to particular stages of a cycle of relative sea level, termed highstand, lowstand and transgressive. Emphasis is given to the recognition of sequence boundaries located at the onset or end of relative sea-level fall, and their correlative conformities (Hunt & Tucker 1992). A stratigraphic model based on transgressive–regressive cycles requires the recognition of shallowing (regressive) and deepening (transgressive) trends, and the sequence boundary is a *maximum regressive surface*. In contrast, the genetic sequence model (Galloway 1989) is based on the recognition of highstand, lowstand and transgressive deposits, with a sequence boundary defined at the maximum flooding surface.

Maximum flooding surfaces represent the period of greatest inundation of the basin margin (Fig. 8.9). Although the depositional sequence boundary (Mitchum *et al.* 1977) can be easily recognised on seismic reflection sections by a downward shift in the position of coastal onlap, it is commonly difficult to recognise in log, core and outcrops. The maximum flooding surface, which is also easily recognised on seismic reflection sections as the reflector with the most landward penetration, is also easily recognised in electrical logs, cores and outcrop, since it is commonly associated with the deepest-water or condensed sedimentary facies (Loutit *et al.* 1988). The position of the maximum flooding surface within depositional sequences is strongly determined by the minima in sediment supply. The use of maximum flooding surfaces as a tool in subsurface mapping is well illustrated by the case study of Partington *et al.* (1993) in the North Sea Basin.

Stratigraphic geometries and correlations can be indexed well by the tracking of certain moving boundaries in the sediment routing system, particularly the shoreline and the shelf-edge break (Fig. 8.10). These moving boundaries are characterised by trajectories, which over time may be ascending, descending or horizontally migrating (Henriksen *et al.* 2009; Helland-Hansen & Hampson 2009; Henriksen *et al.* 2011). Migration of the shoreline break and shelf–slope break leads to the formation of small and large clinoforms respectively, which are particularly well imaged on seismic reflection profiles (Fig. 8.11), but may also be recognised from outcrop sections in cases of exceptional exposure (Fig. 8.12). In the case of outcrop sections, the trajectory of the shoreline break is traced through the proxy of sedimentary facies. Ascending regressive trajectories are typical of high sediment supplies and adequate accommodation causing a rising of base level, whereas descending regressive trajectories commonly result from falling base level, commonly known as *forced regression*.

8.2.1.3 The 'global' sea-level chart

Using seismic reflection results, a team of geologists and biostratigraphers from Exxon constructed a chart of relative sea level through time (Vail *et al.* (1977b), updated and improved by Haq *et al.* (1987, 1988). These charts, colloquially known as the Vail and Haq curves, are based on the concept of depositional sequences and the kinds of baselap and toplap at sequence boundaries. The Vail-Haq curves are composed of *cycles of relative change of sea level*. In essence, the Haq *et al.* (1987) curve is presented as having two components: (i) a *long-term* eustatic curve defining 'first-order cycles' or 'megacycles' of period >100 Myr; it is probably equivalent to the sea-level change resulting from net changes in mid-ocean ridge volumes, subduction losses of water, and dynamic topography of the ocean floor and

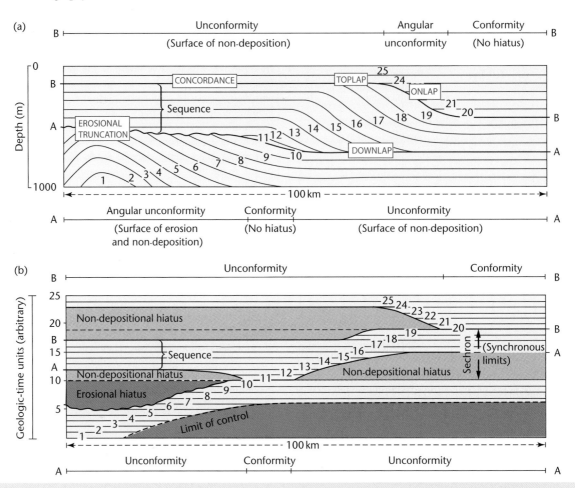

Fig. 8.7 The basic concept of the depositional sequence as outlined by Vail *et al.* (1977a). (a) Generalised stratigraphic section of a depositional sequence. A sequence boundary A changes from an angular unconformity in the left half of the diagram to a conformity in the centre and to a non-depositional unconformity on the right. The sequence boundary B passes from a non-depositional unconformity on the left to an angular unconformity in the centre and a lateral conformity on the right. Unconformities are dated at the points where they have become conformable. Units 1 to 25 represent strata deposited during successive time intervals. (b) Generalised chronostratigraphic section of the same stratigraphic sequence as in (a). The depositional sequence between surface A and B ranges in age from the beginning of chron 11 until the end of chron 19. Chronostratigraphic charts of this type are sometimes termed 'Wheeler diagrams', following Wheeler (1958). Boxes: the geometrical relationships of strata to a depositional sequence boundary or to any other surface within a depositional sequence. Relations to upper surface include (1) erosional truncation, (2) toplap (commonly non-depositional rather than erosional) and (3) concordance. Relations to lower surface include (1) onlap where the overlying strata are near-horizontal and the surface is inclined, (2) downlap where the overlying strata are inclined and (3) concordance. AAPG © 1977. Reprinted by permission of the AAPG whose permission is required for further use.

continental shelves (see §8.2.2.5, §5.4.3); and (ii) a *short-term* eustatic curve, composed of higher frequencies defining 'second-order cycles' or 'supercycles' of duration 8–10 Myr and 'third-order cycles' or 'cycles' of duration 1–5 Myr.

The chronostratigraphical basis for the global sea-level chart of Haq *et al.* (1987) is a combination of radiometric dates, magneto-stratigraphy (geomagnetic polarity reversals) and biostratigraphy, but there is a limit to the time resolution of stratigraphic boundaries, which makes the question of their global synchroneity problematic (Miall 1991, 1994). The Vail-Haq global sea-level chart remains controversial. It is evident from a consideration of process stratigraphy that the timing of depositional sequence boundaries is determined

by the interacting rate effects of sediment accumulation, tectonic subsidence and eustatic change. For example (§7.5), large river catchments are strongly buffered systems that have long response times to a base-level change such as eustatic rise or fall. Since different catchments have different response times (10^4–10^6 years), a *synchronous* base-level change should result in a *diachronous* stratigraphic response. A eustatic change at a point in time should only produce a *partial* or *apparent* synchroneity in basins with different tectonic and sediment supply histories (see also Parkinson and Summerhayes 1985 for a thought experiment along these lines). The search for global synchroneity is therefore fundamentally futile. There is also some doubt as to whether 2nd, 3rd, 4th ... *n*th-order cycles can

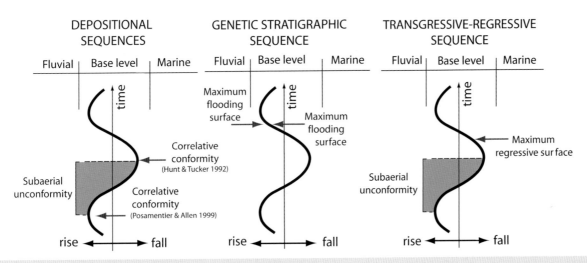

CYCLE OF RELATIVE SEA-LEVEL CHANGE	MESO-SCALE MODEL	DEPOSITIONAL SEQUENCE	DEPOSITIONAL SEQUENCE	DEPOSITIONAL SEQUENCE	GENETIC SEQUENCE	TRANSGRESSIVE–REGRESSIVE
	END OF TRANSGRESSION	HST Highstand	Early Highstand	HST Highstand	HST Highstand MFS	HST Highstand
	END OF REGRESSION	TST Transgressive	TST Transgressive	TST Transgressive	TST Transgressive	TST Transgressive
	END OF BASE-LEVEL FALL	Late LST Lowstand	LST Lowstand	LST Lowstand	Late LST Lowstand	MRS
	ONSET OF BASE-LEVEL FALL	Early Lowstand	Late Highstand	Falling Stage CC	Early Lowstand	RST Regressive
		HST Highstand CC	Early Highstand	HST Highstand	HST Highstand	

Fig. 8.8 Terminology and relationship of stages of deposition to a cycle of relative sea level, following Catuneanu *et al.* (2009). The main distinction is between depositional sequences, genetic stratigraphic sequences and transgressive–regressive sequences, which place different emphasis on key stratigraphic surfaces. LST, TST, HST and RST stand for lowstand systems tract, transgressive systems tract, highstand systems tract and regressive systems tract. CC, correlative conformity; MFS, maximum flooding surface; MRS, maximum regressive surface. Reprinted with permission from Elsevier.

actually be discriminated. It is possible that stratigraphic cyclicity exists in a broad continuum of frequency (Wilkinson *et al.* 1998), and that some apparent cyclicity is an illusion caused by the statistical techniques used (Vaughan *et al.* 2011).

In summary, we follow Carter (1998) in believing that the Haq *et al.* (1987) curve is a 'noisy' amalgam of a wide range of local sea-level signals, and should not be used as a global benchmark. Consequently, its use as a chronostratigraphic tool by assuming *a priori* that a certain stratigraphic boundary has a globally synchronous and precise age, which it is therefore safe to extrapolate into a basin with poor age control, is hazardous.

8.2.1.4 Systems tracts

Depositional sequences can be subdivided into smaller units of stratigraphy that have distinct stacking patterns of chronostratigraphic increments. These smaller units are termed *systems tracts* (Van Wagoner *et al.* 1988), and they are themselves composed of *parasequences*, or *paracycles*. The tracts of depositional systems (Brown & Fisher 1977) have been related to specific intervals of cycles of relative sea level (Van Wagoner *et al.* 1988; Posamentier *et al.* 1988; Posamentier & Vail 1988) (Fig. 8.13). In idealised form these are:

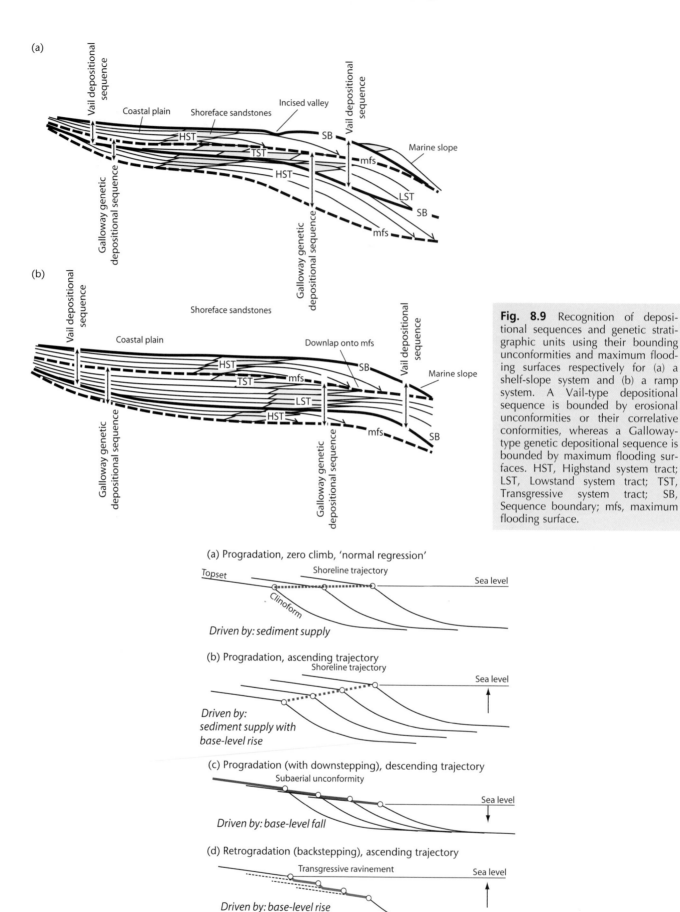

Fig. 8.9 Recognition of depositional sequences and genetic stratigraphic units using their bounding unconformities and maximum flooding surfaces respectively for (a) a shelf-slope system and (b) a ramp system. A Vail-type depositional sequence is bounded by erosional unconformities or their correlative conformities, whereas a Galloway-type genetic depositional sequence is bounded by maximum flooding surfaces. HST, Highstand system tract; LST, Lowstand system tract; TST, Transgressive system tract; SB, Sequence boundary; mfs, maximum flooding surface.

Fig. 8.10 Trajectory of the shoreline associated with progradation and retrogradation, showing ascending and descending trends. Adapted from Catuneanu *et al.* (2009), reproduced with permission from Elsevier.

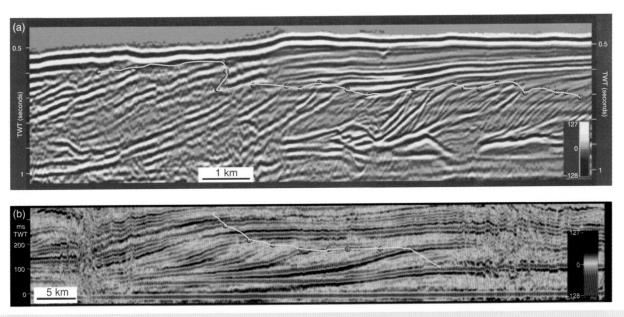

Fig. 8.11 Examples of seismic dip sections showing clinoforms and shelf-edge trajectories, from (a) Oligo-Miocene of the mid-Norwegian shelf, and (b) Triassic (Anisian) of the Barents Shelf. Rollover positions are small red circles, joined by yellow lines showing the trajectory. Reproduced from Helland-Hansen & Hampson (2009) with permission of John Wiley & Sons, Ltd.

- *Lowstand systems tract.* When relative sea-level fall is rapid, no space is available for further sedimentation, the former shelf is incised by streams (producing incised valley systems), and sedimentation is transferred to the basin floor and slope. Base-of-slope fans (*lowstand fans*) are nourished by sediment bypassed through the shelf and slope by valleys and canyons. *Slope fans* result from deposition on the middle or base of the continental slope and may be coeval with the basin-floor fan. *Lowstand wedges* are characterised by onlap onto the slope and at the same time by progradation and downlap onto the previous basin floor or slope fans. They are deposited at times of low but very slowly changing sea level, particularly during a slow relative rise. Lowstand fans, slope fans and lowstand wedges are all associated with underlying erosional sequence boundaries. When the relative sea-level fall is gradual, sedimentation may be progressively shifted basinwards to the shelf-edge where both onlap in a landward direction and downlap in a basinward direction take place. The base of the *shelf margin* systems tract is therefore a non-erosional sequence boundary. On a ramp margin rather than a shelf-break margin, slopes are too low to allow for significant submarine canyon formation and turbidite deposition (Van Wagoner *et al.* 1988). Since there is little bypass of sediment to the basin floor, falling relative sea level may cause the deposition of a set of downstepping prograding wedges, known as *forced regressive wedges* (Posamentier *et al.* 1992) in a *forced regressive systems tract* (Hunt & Tucker 1992; Posamentier & James 1993).
- *Transgressive systems tract.* During a rapid relative sea-level rise, the underlying lowstand or shelf margin system tracts are transgressed (the *transgressive surface*). Where the transgressive surface is erosional, it is called a *ravinement surface*. Sets of parasequences making up the transgressive system tract are commonly retrogradational, that is, they back-step onto the basin margin,

with strong onlap in a landward direction and downlap onto the transgressive surface in a basinward direction. As the rate of relative sea-level change slows down, the sets of parasequences change from being retrogradational to being aggradational, the surface at which this occurs being that of the *maximum flooding*. *Condensed sections* occur in the basin during times of transgression.
- *Highstand systems tract.* After the maximum flooding, the relative sea-level rise slows, and sets of aggradational parasequences are succeeded by progradational parasequences with clinoform geometries. These highstand systems tract parasequences onlap onto the underlying sequence boundary in a landward direction and downlap onto the top of the transgressive systems tract or lowstand systems tract in a basinward direction, so that there is a prominent *downlap surface* below a highstand system tract. Marine condensed sequences occur during the early stages of a highstand systems tract before major progradation takes place (Loutit *et al.* 1988). Highstands offer the possibility of thick subaerial deposition of fluviatile sediments.
- *Regressive systems tracts* are defined by the basinwards migration of the shoreline, commonly associated with clinoform progradation over a downlap surface (Embry & Johannessen 1992). Regression may be associated with a subaerial unconformity in fluvial environments, cut during base-level fall, and a maximum regressive surface in marine environments when the rate of accommodation generation (during base-level rise) exceeds the sediment supply to the coastline.

The position within a cycle of relative sea level has a strong effect on the timing of sand delivery to the deep sea, particularly adjacent to the entry points of rivers into the ocean (Henricksen *et al.* 2011) (Fig. 8.14). Descending shelf-edge trajectories are linked to sediment

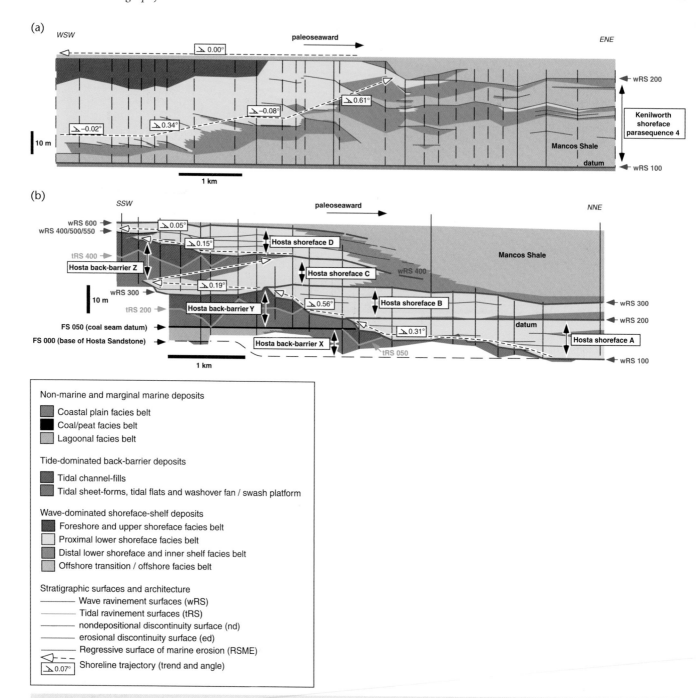

Fig. 8.12 Shoreline trajectories revealed from outcrop datasets: (a) dip-oriented outcrop facies panel for the regressive wave-dominated delta of the Kenilworth tongue (Hampson 2000; reproduced with permission of John Wiley & Sons, Ltd.); (b) dip-oriented facies panel for the transgressive mixed wave-tide dominated barrier island system of the Hosta Sandstone (Sixsmith *et al.* 2008).

bypass of the incised river valley and continental shelf and delivery of sediment to basin-floor accumulations. Ascending trajectories are associated with progradation of fluvio-deltaic systems and condensation of the basin floor.

The methodology of systems tract recognition has been extended to carbonate-dominated systems (e.g. Sarg 1988; Schlager 1992, 2005). Although there are significant differences between siliciclastic and carbonate systems, the underlying principles are similar. The

effects of relative sea-level change on the diagenetic history of carbonate reservoir rocks are discussed in §11.5.6.

8.2.1.5 Smaller genetic stratigraphic units

The resolution of most seismic reflection sections generally allows only depositional sequences and megasequences to be recognised. However, log, core and outcrop data show that stratigraphy is marked

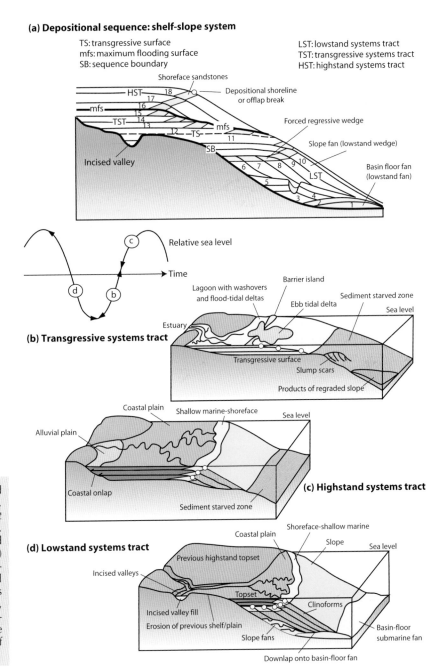

(a) Depositional sequence: shelf-slope system

TS: transgressive surface
mfs: maximum flooding surface
SB: sequence boundary

LST: lowstand systems tract
TST: transgressive systems tract
HST: highstand systems tract

(b) Transgressive systems tract

(c) Highstand systems tract

(d) Lowstand systems tract

Fig. 8.13 Characteristic systems tracts and their relation to the relative sea-level curve, according to the Exxon Group (Posamentier *et al.* 1988), modified by Emery and Myers (1996). These block diagrams are extremely idealised and have a strong vertical exaggeration. (a) Arrangement of systems tracts within a depositional sequence in a coastal plain–continental shelf-slope system. Numbers are relative ages of chrons. (b), (c) and (d) show transgressive, highstand and lowstand systems tracts respectively. Open circles are depositional shoreline or offlap break. Reproduced with permission of John Wiley & Sons, Ltd.

by variability at higher resolution. The larger stratal geometries of depositional sequences can be linked at high resolution with their typical facies assemblages (Van Wagoner *et al.* 1990; Posamentier & Chamberlain 1993).

The typical small-scale cycles seen in core and at outcrop consist of an upward-shallowing trend followed by an abrupt deepening. These cycles are known as *parasequences* or *paracycles* (Van Wagoner *et al.* 1990), defined as relatively conformable units of genetically related beds or bed-sets bounded by marine flooding surfaces or their correlative surfaces (Figs 8.15, 8.16). The boundaries of parasequences, as in depositional sequence boundaries, are thought to be

synchronous surfaces. Parasequences thus defined are placed at a lower level than depositional sequences, but there is no further implication of hierarchical level (3rd, 4th, 5th-order, etc.). They have long been recognised in a variety of paleoenvironmental settings ranging from fluvial to deep-marine, and in both siliciclastic- and carbonate-dominated systems. Parasequences are small-scale but fundamental building blocks of stratigraphy, generally a few metres to <15 m in thickness. Their lateral extent varies according to the uniformity of basin tectonic subsidence and the nature of the external or internal forcing. In the Proterozoic Rocknest platform of northwest Canada (Grotzinger 1986) 140 to 160 progradational shallowing-up cycles

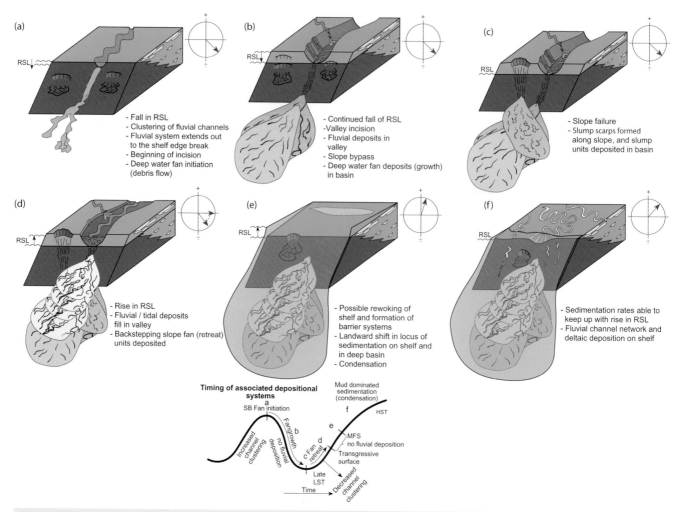

Fig. 8.14   The evolution of a prograding system linked to a river sediment supply, based on the Neogene of offshore mid-Norway (a to f) and eastern Venezuela (g and h), after Henriksen *et al.* (2011). In (a) to (f) the style of the shelf-edge trajectory is shown by the red circle, with ascending (upper half of circle) and descending (lower half of circle) trajectories shown by arrows. (a) Descending shelf trajectory with incision of shelf and start of basin floor deposition; (b) and (c) descending shelf-edge trajectory with major river incision of shelf and sediment bypass to the basin floor; (d) turnaround from descending to ascending shelf-edge trajectory and landward shift in locus of basin floor sedimentation towards shelf slope; (e) high-angle ascending shelf-edge trajectory and drowning of shelf; and (f) low-angle ascending shelf-edge trajectory where sedimentation rates keep up with relative sea-level rise, allowing fluvial-deltaic progradation, but condensation on the shelf slope and basin floor. Block diagrams in (a) to (f) occur on the points indicated on the curve of relative sea level in the inset diagram. (g) and (h): Simplified depositional models showing evolution of high-relief (g) to low-relief (h) margins, based on the Neogene of offshore eastern Venezuela. The early Pliocene high-relief margin shows sediment bypass of the slope through base-Pliocene canyon systems, and long run-out of turbidites. The late Pliocene–Pleistocene low-relief margin shows prograding sediments filling pre-existing bathymetry, with gravitationally driven tectonics, and short run-out distances of turbidites. Reproduced with permission of John Wiley & Sons, Ltd.

can be traced for enormous distances (>100 km) both perpendicular and parallel to strike, whereas other field studies suggest that parasequences are commonly laterally impersistent.

The glacially forced sea-level changes of the late Quaternary have a similar frequency to those required to explain some parasequences. The time scale of the fluctuation is also reminiscent of the frequency of climatic change attributed to eccentricities in the orbital motion of the Earth (Hays *et al.* 1976; Imbrie 1982). These variations are commonly termed *Milankovitch* cycles (see also §8.2.2.5). Time-series analysis of the periodicity of stratigraphic cycles, such as in the

Alpine Triassic, have led to different claims about the identification of Milankovitch-type forcing (e.g. Goldhammer *et al.* 1990; Zühlke *et al.* 2003), and some cycles may be entirely unforced (Burgess *et al.* 2001).

Parasequences deposited during greenhouse periods, when glacio-eustatic changes are expected to have been minimal, may be unforced, or forced at a longer frequency than typical 'icehouse' Milankovitch cycles. For example, cycles of relative sea-level change interpreted in Eocene 'greenhouse' stratigraphy in the Swiss Alps appear to have a frequency of c.0.4–1.0 Myr (Gupta & Allen 1999; Allen *et al.* 2001).

(g) Early Pliocene high-relief margin, underfilled basin

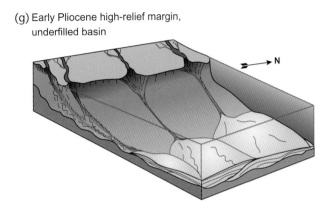

(h) Late Plio-Pleistocene low-relief margin, overfilled basin

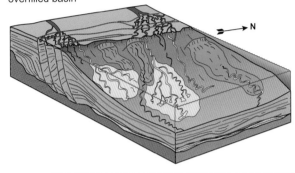

Fig. 8.14 *Continued*

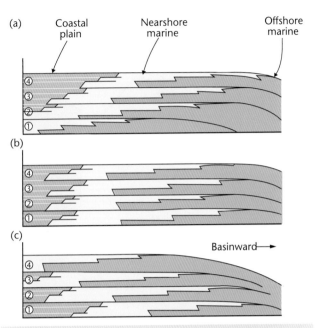

Fig. 8.15 Stacking patterns of parasequence sets (after Van Wagoner *et al.* 1988) are indicative of the trend in time of sediment supply and relative sea-level change or accommodation. (a) Progradational parasequence set with a basinward migration of the shoreline, characteristic of the highstand systems tract and lowstand prograding wedge. (b) Aggradational parasequence set showing no movement of the shoreline, characteristic of the shelf-margin systems tract. (c) Retrogradational parasequence set, with a landwards migration of the shoreline, characteristic of the transgressive systems tract.

The Oligocene stratigraphy of the New Jersey coastal plain exhibits cycles of c.0.5–2.0 Myr duration (Kominz & Pekar 2001). The lower amplitude and longer frequency of greenhouse cycles may also influence the transport of sediment across the continental shelf and its delivery to the deep sea (Burgess & Hovius 1998; Somme *et al.* 2009).

Although the high-frequency (Milankovitch) pacemaker of environmental change can be recognised in the stratigraphic successions of lakes, the stacked paleosols of floodplains and the pelagic sediments of the deep sea, its recognition is more in doubt in systems involving long-term transient responses of erosional and depositional landscapes. As discussed in §7.6.3, high-frequency signals may be 'shredded' during propagation through sediment routing systems involving sediment transport and storage, making the recognition of Milankovitch forcing problematical. It is probable, therefore, that Milankovitch-band forcing of stratigraphic cyclicity is well preserved in environments of relative stasis, but less easy to decipher in highly active sediment transport threads.

8.2.2 Forcing mechanisms

In Chapters 3 to 6 we introduced the main mechanisms for the formation and development of sedimentary basins. In this section we consider in more detail some of the mechanisms responsible for long-term stratigraphic patterns in the basin-fill. They can be divided into:

- tectonic mechanisms that control the spatial and temporal pattern of subsidence and the evolution of sediment routing systems;
- eustatic mechanisms that essentially control accommodation and set base level;
- climatic mechanisms that affect sediment supply to basins;
- internal dynamics unforced by external controls.

8.2.2.1 Tectonic mechanisms: flexure

The subsidence in stretched basins comprises: (i) a fault-controlled initial subsidence caused by mechanical stretching of the upper brittle layer of the lithosphere; and (ii) a thermal subsidence caused by the cooling and contraction of the upwelled asthenosphere (Chapter 3). These mechanisms are amplified by sediment and water loading (§9.4). In the period of active stretching or rifting, the lithosphere is generally viewed as being in a state of purely local (Airy) isostasy. That is, the lithosphere behaves as a very weak support for any superimposed loads in the active rift. However, the presence of gently dipping post-rift sediments onlapping the basin margin (Figs 8.17, 8.18a) suggests that a broadly distributed subsidence characterises the post-rift stage. The way in which the lithosphere distributes loads in the post-rift stage is by flexure. Whereas the stretching event determines the first-order depth and size of the basin, the flexural response of the lithosphere has a strong influence on depositional sequences in the post-rift stage.

If the lithosphere behaves *elastically* over geological time periods, flexural rigidity may depend on its thermal age (§4.3.2). As a result

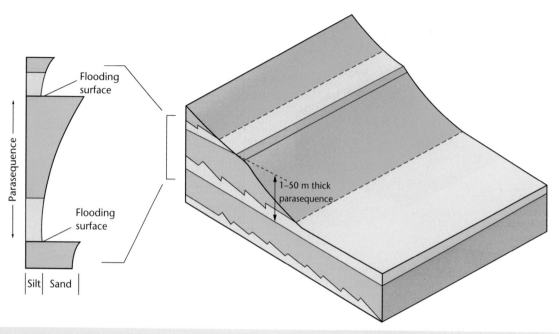

Fig. 8.16   A shallow-marine parasequence involving upward coarsening, representing a shallowing of water depth over time. The parasequence is bounded by flooding surfaces. After Emery and Myers (1996), reproduced with permission of John Wiley & Sons, Ltd.

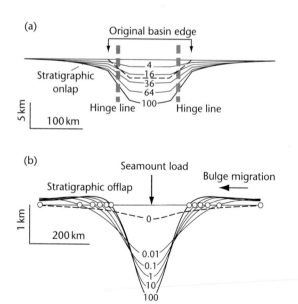

Fig. 8.17   Contrast of elastic and viscoelastic behaviours. (a) Stratigraphy of a model stretched basin with $\beta = 2.0$, overlying an elastic lithosphere where the equivalent elastic thickness (Te) is the depth to the 450 °C isotherm. Te increases with thermal age, causing stratigraphic onlap (steer's head geometry). (b) Response of a viscoelastic lithosphere, which weakens rapidly following loading, causing stratigraphic onlap and a gradually narrowing basin. Vertical exaggeration of x 10. Numbers are time in Myr since rifting. After Watts *et al.* (1982), reproduced by permission of the Royal Society.

of lithospheric cooling, basin stratigraphy should reflect an increase in flexural rigidity with time by a progressive overstepping of younger strata at the margin of the basin. If the lithosphere behaves *viscoelastically* (and the time constant of the viscous relaxation is long), stresses are relaxed by a viscous flow, causing the flexural rigidity to decrease with time. For the same tectonic (cooling plate) driving mechanism, this different basement response produces a pattern of stratigraphic offlap, with the youngest sediments restricted to the basin centre. The widths of the basins produced on elastic and viscoelastic lithosphere are therefore markedly different (Fig. 8.17). For the elastic lithosphere, the post-rift sediments strongly overstep the syn-rift sediments giving a *'steer's head'* geometry. In contrast, on a viscoelastic plate, overstep of post-rift sediments is minimal because of stress relaxation with time.

The pattern of stratigraphic onlap has been modelled in detail for the passive margin of eastern North America (Watts 1982). Sedimentation is assumed to keep pace with subsidence, maintaining a constant bathymetric profile through time. Fig. 8.18a shows the initial strong onlap of sediments onto the basement at the transition from fault-controlled Airy-type subsidence to flexural-controlled subsidence. However, lateral heat flow causes thermal uplift on the coastal plain, abruptly terminating onlap. By about 16 Myr after rifting, flexural subsidence outstrips thermal uplift and the sediments again progressively onlap basement.

The large-scale wedge-shaped geometry of foreland basin stratigraphy is a reflection of the lateral gradient in subsidence rate from the centre of the load to the peripheral bulge. The spatial translation (with respect to the underlying plate) of stratigraphic units ahead of a moving load, a consequence of continued convergence being accommodated in the orogenic belt, causes a general onlap of successively younger stratigraphy onto the foreland (Fig. 8.18b). The

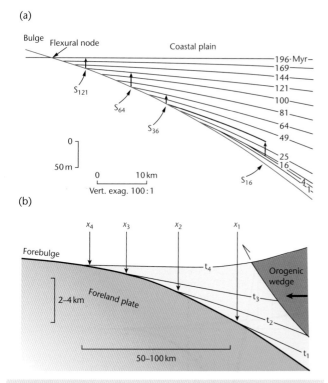

Fig. 8.18 Impact of flexure on stratigraphic onlap on passive margins and in foreland basins. (a) Coastal plain and shelf stratigraphy using a model of a passive margin in which the tectonic subsidence is due to thermal contraction following stretching. Sediments rapidly infill the continental shelf, keeping a near-constant bathymetry with time. The sedimentary load flexes a cooling plate that increases in rigidity with time since heating. The initial lithospheric thickness y_L is 125 km, initial crustal thickness y_c 31.2 km, coefficient of thermal expansion α_v 3.4 × $10^{-5}\,°C^{-1}$, mantle temperature 1333 °C, initial densities of 2800 and 3300 kg m^{-3} for crust and mantle lithosphere respectively, and the uniform density of the infilling material is 2500 kg m^{-3}. The stretch factor is 3.0 and the equivalent elastic thickness is given by the depth to the 450 °C isotherm. Solid boundaries are stratigraphic units with ages in Myr since the end of rifting. Effects of compaction have been ignored. The arrows show the amounts of coastal aggradation at given time periods. After Watts (1982). (b) Diagrammatic illustration of a foreland basin showing stratigraphic onlap. Locations x_1 to x_4 show the successive positions of pinch-outs corresponding to the chronostratigraphic lines t_1 to t_4. Reproduced by permission of the Royal Society.

distal margins of foreland basins are therefore typified by retrogradational stacking patterns causing a time-average onlap of the foreland plate. Further details on the stratigraphy of foreland basins, including the basal unconformity, can be found in §4.6.5.

The stratigraphic onlap associated with flexure on passive margins, such as the continental margins of the Atlantic, and on the distal flank of foreland basins, may be very widespread, but is not global.

8.2.2.2 Tectonic mechanisms: fault blocks and fault array evolution

The three-dimensional evolution of accommodation during the development of a stretched piece of continental lithosphere plays a

major role in determining the stratigraphy of rifts (see §7.6.3.1). A single fault block and basin bounded by a vertical fault in an elastic domain is associated with a tectonic displacement field that decays away from the fault plane (Fig. 8.19). Uplift of the footwall block causes erosion to be concentrated close to the fault plane, and, similarly, subsidence in the hangingwall decays with distance from the fault plane. Modelling of simple catchment-fan systems suggests that changes in climate and tectonic slip-rate on the border fault result in transient responses in the sediment routing system with a time scale of c.10^6 years. A typical external driving mechanism for an increase in slip on the border fault is the interaction of an array of faults leading to the localisation of slip on a small number of major structures (Schlische 1991; Cowie *et al.* 2006) (Fig. 3.12a).

The evolution of a fault array requires individual faults to interact and join, which results in relay zones or transfer zones between fault segments (Fig. 3.12b), which are thought to strongly influence the entry points of rivers into the hangingwall basin (Gawthorpe & Hurst 1993). The evolution of a fault array from rift initiation to rift climax (Prosser 1993) commonly results in accommodation generation outpacing sedimentation in hangingwall depocentres, leading to basin deepening and relative sea-level rise. However, at the same time, adjacent footwall regions experience tectonic uplift, causing relative sea-level fall and erosion. If an observer were to walk up a relay ramp from hangingwall basin to footwall crest (Fig. 3.12b), they would therefore record first relative sea-level rise, then relative sea-level fall along the same time line. Correlation of the relative sea-level rise from the hangingwall across the sedimentary basin would not reflect eustatic change, only the timing and magnitude of tectonically driven subsidence.

Depositional environments and sedimentary facies, and the position of grain-size fronts such as the fan toe, reflect the interplay of the generation of accommodation by fault array evolution and sediment supply by erosion of uplifting footwalls and rotating hangingwalls. Catchments located in the rapidly uplifting footwall and the slowly rotating hangingwall of an extensional half-graben system compete in their effect on displacing axial river systems (Kim *et al.* 2011) (§7.6.4). The fronts of oppositely sourced fan systems act as moving boundaries, so that fan progradation or retrogradation may result primarily from the variation of sediment supplies available to the footwall and hangingwall fans rather than simply reflecting base-level change driven by tectonics (Fig. 8.20).

8.2.2.3 Tectonic mechanisms: changes of in-plane stress

Variations in regional stress fields acting within inhomogeneous lithospheric plates may cause vertical movements large enough to have a major impact on stratigraphy. In-plane stresses acting on a deflected plate may enhance or reduce the curvature of the deflection. In-plane forces may also produce long-wavelength lithospheric buckles where the plate is rheologically layered, with a weak lower crust (§4.4).

An examination of the possible effects of changing horizontal (i.e. in-plane) stresses on a passive continental margin with an overlying sedimentary load and an age-dependent elastic thickness (Cloetingh *et al.* 1985; Karner 1986) shows that the application of an in-plane force may cause a vertical motion of the continental lithosphere. A change from a tensile stress to a compressive stress produces a net uplift of the basin margin, forcing a stratigraphic offlap, whereas the reverse produces a net subsidence and enhanced stratigraphic onlap.

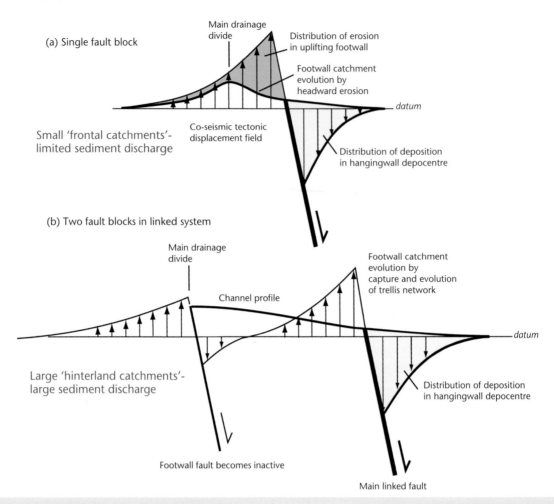

Fig. 8.19 Tectonic displacement field associated with a single normal fault (a) showing distribution of denudation in footwall and sedimentation in hangingwall basin. Footwall catchments are small and steep and deliver relatively small amounts of sediment to the hangingwall, which tends to be underfilled. Two faults in a linked system (b) may cause the development of large hinterland catchments discharging high sediment supplies into the adjacent, overfilled rift basin.

These patterns of offlap and onlap might erroneously be attributed to eustatic sea-level change. The effects of in-plane stresses are most important when the lithosphere is flexurally weak, as in the early stage of stretching. Changes in horizontal stresses also have effects towards the basin centre, but because the total subsidence is large in these areas, the contribution of in-plane stresses to the deflection is masked.

Accepting that large-wavelength lithospheric folds exist (§4.4), the critical question is whether in-plane stresses of significant magnitude change in sign over a sufficiently short time period to significantly influence stratigraphic packaging, and whether any such changes can be 'inverted' from the sedimentary record of basins. Vertical motions due to changes of in-plane stress combined with the effects of dynamic topography may be highly influential in controlling the stratigraphic packaging of many of the world's cratons.

8.2.2.4   Dynamic topography

Uplift and subsidence caused by dynamic effects in the mantle are discussed in §5.2. Dynamic uplift of the continental surface may

produce domal uplifts incised by rivers, as has been recognised in Africa (Cox 1989; Pritchard *et al.* 2009; Roberts & White 2010). Dynamic uplifts may also be linked to increased sediment delivery to the ocean, and to progradation of clinoforms off major river entry points. An example is the enhanced rates of sediment delivery by the Zambezi River to its coastal delta caused by pulses of dynamic uplift of southern Africa in the late Cenozoic (Walford *et al.* 2005). Dynamic uplift in the ocean, however, affects the volumetric capacity of the ocean basins and may therefore have a eustatic effect, as discussed in §5.4.3 and §8.2.2.5.

8.2.2.5   Mechanisms affecting global sea levels

Changes in basement height relative to a datum (sea level or 'regional') caused by large-scale mechanisms such as lithospheric thickness changes, flexure and mantle dynamics may be extremely widespread, but they are not global. Global sea-level changes (absolute changes relative to the centre of the Earth) result either from changes in the amount of water in the oceans or from changes in the volume of the ocean basins.

(a) Extensional

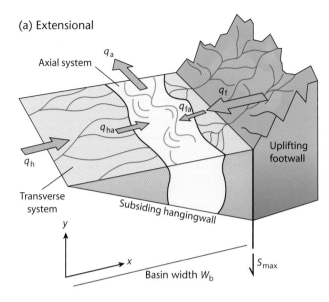

(b) Contractional

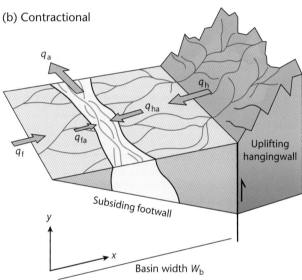

Fig. 8.20 Transverse and axial sediment dispersal in a half-graben (a) or wedge-top basin (b), modified from Kim *et al.* (2011). The position of fan toes, the width of the axial system, and the axial sediment flux depend on the two transverse supplies relative to accommodation in the basin. q_h and q_f are sediment discharges from the hangingwall and footwall blocks respectively; q_{ha} and q_{fa} are the sediment discharges from the hangingwall-derived fans and the footwall-derived fans to the axial system respectively; q_a is the axial system discharge. The basin, of width W_b, subsides with a maximum velocity S_{max} adjacent to the border fault.

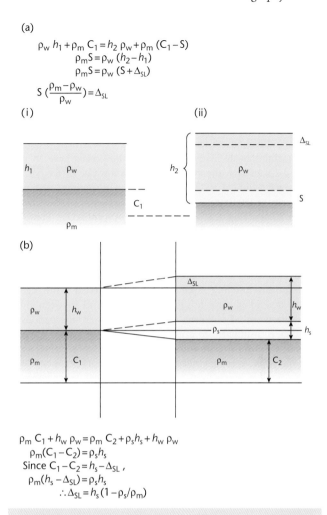

$$\rho_w h_1 + \rho_m C_1 = h_2 \rho_w + \rho_m (C_1 - S)$$
$$\rho_m S = \rho_w (h_2 - h_1)$$
$$\rho_m S = \rho_w (S + \Delta_{SL})$$
$$S\left(\frac{\rho_m - \rho_w}{\rho_w}\right) = \Delta_{SL}$$

$$\rho_m C_1 + h_w \rho_w = \rho_m C_2 + \rho_s h_s + h_w \rho_w$$
$$\rho_m(C_1 - C_2) = \rho_s h_s$$
Since $C_1 - C_2 = h_s - \Delta_{SL}$,
$$\rho_m(h_s - \Delta_{SL}) = \rho_s h_s$$
$$\therefore \Delta_{SL} = h_s(1 - \rho_s/\rho_m)$$

Fig. 8.21 (a) Isostatic effect of a change in water depth in the ocean. Since the wavelength of the water load is very large, the compensation can be regarded as local (Airy). (i) Initial ocean of depth h_1. (ii) An increase in water depth to h_2 results in an increase in sea level of Δ_{SL}. (b) Derivation of the sea-level change resulting from the deposition of sediment in the ocean. If the density of mantle is 3300 kg m⁻³ and the density of ocean sediment is 2500 kg m⁻³, the sea-level change is 0.24 of the sediment thickness h_s. Consequently, if 25 mm of sediment is deposited every 1000 yr, the sea-level change is 6 mm per 1000 yr.

In considering absolute sea-level change, it is important to incorporate the effects of isostasy (§2.1.1). An increase in water depth in the ocean basins causes subsidence of the ocean floor due to the excess weight of the water at the new sea level (Fig. 8.21a). A decrease in water depth causes uplift of the ocean floor due to the removal of part of the water column. The displacement of the oceanic basement relative to the centre of the Earth by changes in the water load would be found throughout the oceans, although the effect would only be observed at places such as oceanic islands. These regions would experience the same displacement as the oceanic basement, and would therefore act as a 'dipstick', indicating the sea-level change. The isostatic effects of changes in absolute sea level are given in Appendix 52.

We can examine global sea-level change in terms of five possible causes:

1. *Continuing differentiation of lithospheric material as a result of plate tectonic processes.* The volume of water in the oceans may be added to by contributions from mid-ocean ridge and island arc volcanism. Counteracting this, water may be removed by hydrothermal alteration of new crust and at subduction zones.

These processes are probably roughly balanced. There is no evidence for appreciable, long-term changes in the chemical composition of the oceans during the past 1500 Myr, suggesting that the volume of seawater has not greatly increased as a result of Phanerozoic volcanism.

2. *Changes in the volumetric capacity of the ocean basins caused by sediment influx or removal.* The total sediment discharged to the oceans is $16–20 \times 10^9$ t yr^{-1} (Milliman & Syvitski 1992; Walling & Webb 1996). If we ignore the volume of sediment produced autochthonously in the oceans, and the oceans have an area of 3.2×10^8 km^2, the average accumulation rate of terrestrially derived sediment is 25 mm per 1000 yr (assuming a sediment density of 2500 kg m^{-3}). This would cause a sea-level rise of 6 mm per 1000 yr (Fig. 8.21b) (derivation in Appendix 53). However, sediment in the ocean is removed by tectonic accretion and subduction at active margins, and continued spreading creates new ocean floor. It is uncertain how much sediment is removed at subduction zones and eventually melted to contribute to arc volcanism, and how much is tectonically accreted to the edge of the overriding plate. It is likely that the balance between influx and removal of sediment, when averaged over long periods of time, is insufficient to cause rates of sea-level change of more than c.1 mm per 1000 years (Pitman 1979).

3. *Changes in the volumetric capacity of the ocean basins.* Long-term (first-order) variations of sea-level, which are commonly assumed to be global, have been thought to reflect variations in the mean rate of spreading at mid-ocean ridges (Hays & Pitman 1973), causing global sea levels to rise and fall in response to faster and slower rates of ocean crust formation (Pitman 1978). This view is based on the observation that the volume of the present world ridge system (c.1.6×10^8 km^3) is a significant proportion, over 10%, of the volume of ocean water (c.1.35×10^9 km^3). Because the bathymetry of the ocean is dependent on age of the ocean crust, variations in spreading rate would cause major changes in ridge volumes, the faster spreading ridges being hotter and therefore more voluminous. Bearing in mind that a change in ocean depth of magnitude Δh produces a change in sea level relative to the centre of the Earth of magnitude $\Delta_{SL} = 0.7\Delta h$, the sea-level fluctuations resulting from mid-ocean ridge volume changes can be estimated if the shape of the ocean basins is known from the *hypsometric curve* (Pitman 1979).

This argument hinges on the existence of a distribution of ocean crust ages supporting a correlation of increased spreading rate with elevated global sea levels, which is not settled (Rowley 2002; Conrad & Lithgow-Bertelloni 2007). But, more seriously, it requires the effects of subduction to be taken into account. Subduction of cold oceanic slabs sets up a mantle circulation resulting in dynamic topography at the Earth's surface (Hager 1980; Gurnis 1990) that may compensate for, or perhaps even balance, the increase in ridge volume during increases in spreading rate.

Dynamic topography, however, affects all of the ocean basins, including those adjacent to non-subducting margins. If this is the case, then the long-term relative sea-level changes recorded on continental margins might not reflect global (eustatic) change but instead the variable effects of dynamic topography (Moucha *et al.* 2008). Since continental margins are not stable over long geological time scales, including those used to construct and calibrate the so-called 'global' sea-level chart of Vail *et al.* (1977b) and Haq *et al.* (1987), including the New Jersey margin of NE

USA and the northwest African margin, any correlation between long-term trends on the 'global' cycle chart and ocean spreading rate are in danger of being spurious.

Vertical movements on continental margins caused by mantle circulation are thought to cause dynamic topography variations of c.100 m over time scales of 30 Myr (Moucha *et al.* 2008) (Fig. 8.22). These values are similar to those attributed to global sea-level variations caused by ocean spreading rate changes, making the connections between spreading rate, volume of ocean basins and continental flooding unnecessary. The variability of dynamic topography from place to place would also help to explain the large differences in estimates of global sea levels over the last 100 Myr (Fig. 8.23).

4. *Thermal expansion and contraction of the oceanic water reservoir.* Changes in the average temperature of the oceanic reservoir, caused by climate change, should cause changes in its volume by thermal expansion and contraction and, therefore, a *thermosteric* sea-level change. However, the reservoir of ocean water is very large, so there is a characteristic time scale for heating and cooling to manifest in sea-level change. In broad terms, the importance of this effect can be gauged from the thermal expansion coefficient of ocean water, but the world oceans are heterogeneous in their temperature structure, making prediction of thermosteric sea-level change complex. Current estimates (Antonov *et al.* 2005; Marcelja 2010) are that the sea-level rise resulting from climate-driven warming of the world's oceans is about 0.4 mm each year over the last half century. Thermosteric sea-level rise is therefore approximately 30–40 mm over a roughly 100-year period up to the present day, compared to an observed total rise of 100–200 mm over the same 100-year time period. That is, the thermosteric rise is ~20–30% of the total observed rise. Sea-level rises due to thermal expansion are therefore rapid geologically, and comparable in rate of sea-level rise to the effect of melting of major ice caps (Warrick & Oerlemans 1990), but are much smaller in magnitude than those due to the complete melting of land-based ice caps (see 5).

5. *Changes of available water by abstraction in and melting of polar ice caps and glaciers.* Large sea-level changes can be caused by abstraction of water from the oceans into land-based ice sheets, and, of course, the subsequent release on melting. Ice shelves are unimportant since the floating ice displaces its own mass of water. These changes are called *glacio-eustatic* (Fairbridge 1961; Donovan & Jones 1979). Total melting of the Antarctic land ice, some 2 to 3×10^7 km^3, would result in an increase in water depth of between 60 and 75 m. Melting of the much smaller Greenland ice cap would probably cause an additional 5 m increase in water depth. Allowing for the water-loaded depression of the ocean floor discussed above (Fig. 8.21), the actual sea-level rise associated with the complete melting of today's land-based ice sheets is expected to be in the region of 50 m. Calculations of former ice sheet volumes are speculative. Assuming that former ice sheets reached an equilibrium maximum thickness, and delineating their maximum limits from Quaternary geology, the sea-level fall resulting from locking up of water in Pleistocene ice sheets is thought to have been about 100 m. The total change in sea level corresponding to the removal of Pleistocene-scale ice sheets is therefore about 150 m and liable to have major impacts on basin stratigraphy. In geological terms, the formation and melting of ice caps is a rapid process (c.10 mm yr^{-1}) (Hays *et al.* 1976; Imbrie 1982). The

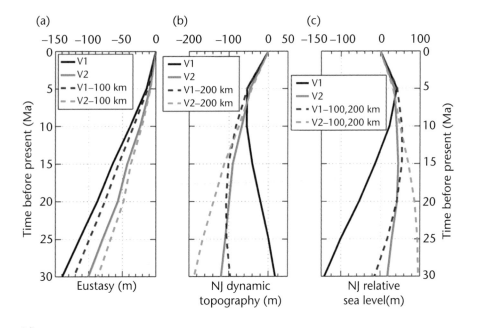

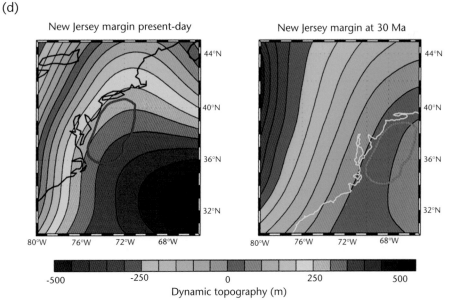

Fig. 8.22 Dynamic topography at the New Jersey margin, at the present-day and projected backwards to 30 Ma, after Moucha *et al.* (2008). (a), (b) and (c) show eustasy (relative to the present-day), dynamic topography and relative sea level of the New Jersey margin over the last 30 Myr for different viscosity models of the mantle, V1 and V2. The suffixes '100 km' and '200 km' refer to the depths above which all lithospheric heterogeneity is set to zero. (d) Maps of dynamic topography at the New Jersey margin at the present-day and at 30 Ma, for model V2_200 km. The blue and orange lines enclose the area used for integration of dynamic topography in order to calculate the curves in (a, b and c). Reprinted with permission from Elsevier.

amplitude of sea-level change resulting from major glacial-interglacial cycles (100 m) is of the same order as the variation in the long-term 'global' sea-level curve.

The lithosphere also responds to changes in the water and ice loads; these *glacio-isostatic* adjustments (Clark *et al.* 1978; Peltier 1980) operate at a rate dependent on mantle viscosity. At any high-latitude locality close to the former ice front, there is therefore a competition in the history of deglaciation between rates of isostatic rebound and rates of eustatic sea-level rise (Belknap *et al.* 1987).

Late Quaternary and Holocene sea-level fluctuations following the Pleistocene glaciation are relatively well understood compared to more ancient sea-level changes. Sea-level curves derived from oxygen

isotopes measured in benthonic foraminifera recovered in deep-sea cores (Shackleton 1977; Shackleton & Opdyke 1973) and from tropical coral terraces (Fairbanks & Matthews 1978; Aharon 1983) show a high-frequency fluctuation, with eight sea-level maxima (and eight minima) in the last 120,000 years. These sea-level changes vary from 20 m to 180 m in total height. They have a long primary period of about 10^5 years, and shorter secondary periods of about 40,000 and 20,000 years.

The Serbian physicist Milankovitch first suggested that periodic variations in solar forcing might be caused primarily by orbital fluctuations of the Earth. For example, changes in the shape of the elliptical path of the Earth's orbit around the Sun (or *eccentricity*) have a periodicity of approximately 100 kyr, and may be responsible for the

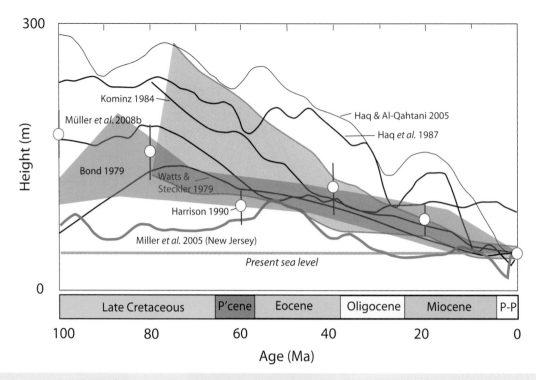

Fig. 8.23 A comparison of estimates of sea-level changes in the last 100 Myr, based on Spasojevic *et al.* (2008, fig. 1). The curve labelled Watts and Steckler (1979) is based on the subsidence history of boreholes from the continental margins of eastern North America. The lines labelled Vail *et al.* (1977b), Haq *et al.* (1987) and Haq & Al-Qahtani (2005) are based on patterns of coastal onlap recognised on seismic reflection lines. The ranges labelled Harrison (1990) and the red-shaded area labelled Bond (1979) are from estimates of the amount of flooding of the continental area of North America. The curve labelled Miller *et al.* (2005) is based on data obtained from backstripping of wells drilled in the New Jersey coastal plain. Kominz (1984) is based on calculations of mid-ocean ridge volumes, with the maximum and minimum values shown by the area shaded blue.

glacial cycles characteristic of the Pleistocene. The Earth also wobbles in its spin about its axis of rotation (*precession*), with the axis of rotation varying in its inclination with a periodicity of c.21 kyr. The axis of rotation also leans with respect to the plane of the Earth's orbit around the Sun (*obliquity* of the ecliptic), the angle of lean varying on a period of c.41 kyr. These three different frequencies interact to produce a complex signal that may act as a pacemaker for climate change, causing eustatic variations responsible for stratigraphic parasequences. Unfortunately, the precision of dating ancient sequences is generally insufficiently precise to allow reliable comparison of cycle durations with Milankovitch frequencies (see also §8.2.1.3 and §8.2.1.5).

8.2.3 Unforced cyclicity

Parasequences generated by allocyclic mechanisms, such as eustatic sea-level change, are commonly assumed to be driven by Milankovitch orbital forcing (e.g. Goldhammer *et al.* 1990; Osleger 1990). A convincing demonstration of orbital forcing should include a spectral analysis of the cyclicity. However, such statistical analyses sometimes show that apparently ordered patterns of cycles are indistinguishable from patterns produced by random depositional processes (Drummond & Wilkinson 1993; Wilkinson *et al.* 1996, 1997; Vaughan *et al.* 2011). It is possible, therefore, that many high-frequency cycles may be produced by an unforced autocyclicity.

An early theory for the generation of shallowing-up cycles in peritidal carbonates envisaged a subtidal *carbonate factory*, with land-

ward transport to produce progradational supratidal and intertidal facies belts (Ginsburg 1971). Progradation was thought to reduce the area of subtidal carbonate production. Continued tectonic subsidence would therefore eventually outpace sediment production, leading to flooding. Two-dimensional (Demicco & Spencer 1989; Hardie *et al.* 1991; Goldhammer *et al.* 1993) and three-dimensional (Burgess *et al.* 2001) numerical models successfully simulate stacked autocyclic parasequences in peritidal carbonates. The internal dynamics of the numerical model developed by Burgess *et al.* (2001) involves an advective landward transport of sediment from subtidal areas driven by prevailing winds. This advection causes accretion of carbonate on the seaward flanks of the islands, causing them to prograde (Fig. 8.24). A less important seaward sediment transport takes place by a gradient-dependent diffusive transport. Laterally impersistent, peritidal cycles (<2.5 m thick) are produced by the landward nucleation, detachment and seaward migration of intertidal–supratidal islands, leaving new subtidal zones in their rear. Such cycles resemble the thin, laterally impersistent, shallowing-up peritidal cycles commonly reported from field studies (Hardie & Shinn 1986; Shinn 1986; Pratt & James 1986; Adams & Grotzinger 1996).

The role of non-orbital forcing in the formation of the classic Middle Triassic carbonate cycles of the Latemar carbonate platform of the Alps (Fig. 8.25) is controversial. The carbonate succession has long been interpreted as an excellent example of Milankovitch-band orbital forcing (Hardie *et al.* 1986; Hinnov & Goldhammer 1991). Improved biostratigraphy and radiometric ages (Brack *et al.* 1996; Mundil *et al.* 1996) allowed the time duration of the Latemar cyclic

(a)

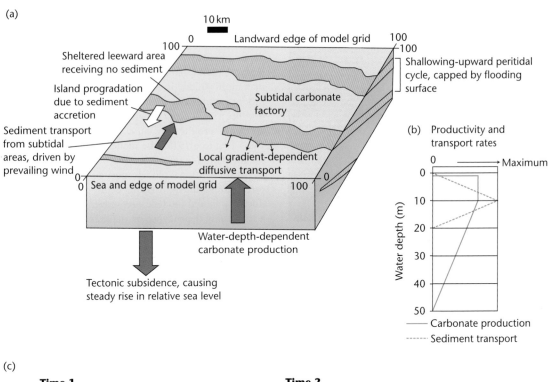

(b) Productivity and transport rates

(c)

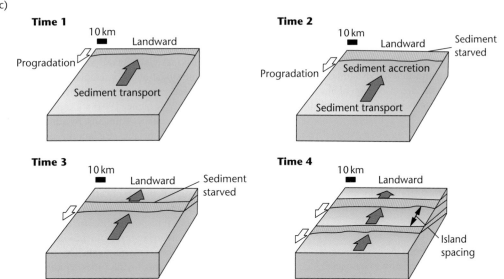

Fig. 8.24 Model for the generation of unforced high-resolution cyclicity in peritidal carbonates, after Burgess *et al.* (2001). (a) Illustration of different processes involved in generating prograding inter- and supra-tidal islands and autocyclic shallowing-upward cycles. (b) Depth-dependent carbonate productivity relationship and sediment transport rate. (c) Different stages in the evolution of prograding islands. Time 1: landward transport of sediment causes accretion of inter- and supra-tidal flat. Time 2: continued accretion drives inter- and supra-tidal flat progradation, which causes a sediment-starved leeward side to develop. Time 3: sediment-starved lee subsides and prograding island forms, allowing a new subtidal carbonate factory to develop. Time 4: a second island system develops as the entire process is repeated. Reproduced with permission of John Wiley & Sons, Ltd.

succession to be revised, making an orbital forcing unlikely. On the other hand, a spectral analysis (Preto *et al.* 2001) of the cyclicity continues to support a pure orbital model. It has subsequently been suggested that the shorter shallowing-up cycles (c.4 kyr) are autocyclic, whereas longer-term bundles of cycles may correspond to orbital forcing, including precession, obliquity and short eccentricity (Zühlke *et al.* 2003). The Latemar example shows that very high-precision chronostratigraphy is required to test for allocyclic orbital forcing versus unforced autocyclity.

Unforced autocyclicity commonly emerges from numerical approaches to sediment routing systems including complex internal feedbacks. For example, in catchments coupled to fans across a basin-

Fig. 8.25 Latemar cycles of the Alpine Triassic. From Bechstädt *et al.* (2003)

bounding fault, an unforced cyclicity or oscillation known as *tintinnabulation* (ringing like bells) was produced in numerical models (Humphrey & Heller 1995). Feedbacks in a coupled system between an erosional catchment and a depositional fan also generated variations in the amount of fanhead incision without external forcing (Pépin *et al.* 2010). Coupling of fluvial and shallow-marine sediment dynamics also explains many of the first-order features of subaerial and subaqueous delta progradation (§8.3.3).

It is likely that external forcing and internal dynamics both take place to a varying degree, and that the deconvolution of any external signal from a system with its own internal dynamics may be difficult to achieve.

8.3   Dynamical approaches to stratigraphy

The groundwork for many numerical stratigraphic models was provided by Sloss (1962) who identified the four main variables affecting stratigraphic geometries:

- the quantity of sediment supply from the basin margins;
- the rate of subsidence in the basin relative to base level;
- the dispersal of sediment by various transport processes;
- the composition and texture of the sediment.

It is regrettable that much early sequence stratigraphic literature ignored this groundwork, and instead focused on eustatic change as the exclusive control on stratigraphic geometries. A number of approaches to quantitative modelling of the basin-fill have simultaneously developed (review in Paola 2000). This section can give a mere snapshot of the wide diversity and complexity of the modelling approaches to stratigraphy. We do so by reviewing some of the fundamental and commonly used principles upon which more sophisticated approaches are based.

8.3.1   Carbonate stratigraphy

The building of carbonate stratigraphy is fundamentally based on a notion of carbonate productivity as a function of water depth. This water-depth dependence is caused by the attenuation of light with depth, but the effects of turbidity and temperature may also be important (Lerche *et al.* 1987). Pomar (2001) suggested that the depth–productivity curve is also strongly affected by the type of biota. Euphotic biota produce a strong depth dependence, with a maximum productivity in very shallow waters; oligotrophic biota produce a midwater depth maximum, and photo-independent biota produce a weak depth dependence (Fig. 8.26). Armed with a carbonate depth–productivity curve (e.g. Bice 1991; Bosscher & Schlager 1992; Demicco 1998), the modeller can readily simulate carbonate sedimentation in one dimension.

Two-dimensional models must account for physical transport of carbonate sediment. A strong depth dependence of carbonate productivity should produce steep-edged carbonate platforms, whereas a less-strong dependence on depth, or vigorous offshore transport, should flatten the carbonate platform slope into that of a ramp (Fig. 8.27). Euphotic organisms such as corals and green algae should produce steep-edged platforms, whereas oligophotic organisms such as the larger foraminifera and red algae should produce ramps. Carbonate depositional systems commonly possess clinoforms whose shape depends on the effects of lateral sediment transport, carbonate productivity, eustatic variation and accommodation generation by tectonic subsidence (Lawrence *et al.* 1990). Continued subsidence of clinoform geometries produced by a strongly depth-dependent productivity curve may cause drowning of the platform/ramp (Schlager 1981). This may occur spontaneously, without external forcing, or during a eustatic rise superimposed on the background tectonic subsidence (Dorobek 1995; Galewsky 1998; Allen *et al.* 2001).

Two-dimensional geometrical models rely on sediment filling available accommodation on a template subject to tectonic subsidence. Geometrical models (e.g. Aigner *et al.* 1989, 1990; Lawrence *et al.* 1990; Shuster & Aigner 1994) are capable of dealing with both carbonate and siliciclastic sediments. By adjusting parameter values, realistic simulations of the stratigraphy of given basins can be achieved. Mixed carbonate-siliciclastic sedimentation can also be simulated in the geometrical model SEDPAK (Kendall *et al.* 1991a, b) (http://sedpak.geol.sc.edu/).

More dynamic treatments of carbonate systems, such as the three-dimensional model of Burgess *et al.* (2001), combine physical processes of sediment transport (diffusion, advection) with a carbonate depth–productivity relationship to produce a dynamic mosaic of carbonate depositional environments. This model is capable of simulating the common small-scale cycles typical of peritidal sequences. Dynamic models involving the transport of carbonate sediments inevitably share much in common with models designed for siliciclastic systems.

8.3.2   Siliciclastic stratigraphy

Since the accumulation of siliciclastic sediment is not a straightforward function of depth, simulations of siliciclastic stratigraphy are commonly two-dimensional geometrical models in which the sediment fills available space up to an equilibrium profile. This is the graded stream profile on land, the shoreface profile, and the graded continental shelf (e.g. Jervey 1988; Burgess & Allen 1996; Ross *et al.* 1995). Commonly, these models are a combination of a dynamic treatment of thermal subsidence and flexural response to sediment loads and a geometrical treatment of the physical processes of sediment transport (Fig. 8.28) (Burgess & Allen 1996; Steckler 1999).

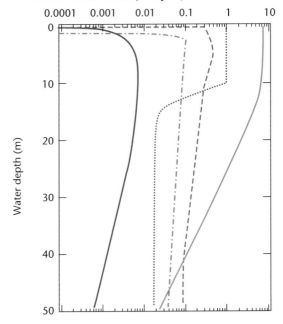

(a) **PRODUCTION RATE (mm yr⁻¹)**

(b) **NORMALISED PRODUCTION RATE**

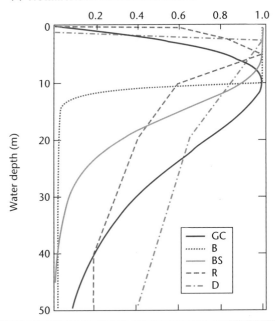

Fig. 8.26 Carbonate productivity versus depth from various authors (a), normalised by the maximum production rate in (b), after Paola (2000). GC, Gildner and Cisne (1990); B, Bice (1991); BS, Bosscher and Schlager (1992); R, Read *et al.* (1991); D, Demicco (1998). Reproduced with permission of John Wiley & Sons, Ltd.

Within this spectrum, some models (Ross *et al.* 1995) consider the grain-size distribution of the sediment by partitioning sand and mud deposition within different depositional environments. Others (Kendall *et al.* 1991a, b) include the effects of sediment compaction, or incorporate a viscous relaxation time in calculations of isostatic support (Steckler 1999). However, the physical laws of erosion, sedi-

ment transport and deposition are not explicitly treated in these geometrical models.

Process-driven simulations require models based on equations governing sediment-water fluid dynamics constrained by conservation of energy and momentum. The Sedsim modelling package (Stanford University, Lee *et al.* 1992; CSIRO, http://www.csiro.au/products/Sedsim.html) is an example of this approach. However, dynamic sediment-hydrodynamic simulations do not capture the long-term effects that control stratigraphic preservation.

Other dynamic approaches simplify sediment erosion, transport and deposition approximated by physical laws that allow a mass or volume balance (Paola & Martin 2012). For example, fluvial systems can be treated as diffusive (Paola 2000). When combined with tectonic subsidence, stratigraphy can be simulated as a function of the diffusive sediment transport coefficient as well as tectonic parameters controlling accommodation and source-area erosion (Flemings and Jordan 1989; Sinclair *et al.* 1991; Paola *et al.* 1992).

A key problem, however, is the formulation of the diffusive transport coefficient. While retaining the essential diffusive character of sediment transport, a number of approaches have been developed to deal with the problem of the transport coefficient (see also §7.5.1 and §7.5.2):

- a two-diffusion model uses separate transport coefficients for gravel and sand that are different by a factor of 10 (Marr *et al.* 2000);
- transport coefficient values may be made to vary across a continuum according to the percentages of sand and mud (Rivenaes (1992, 1997); or
- transport coefficients can be assigned to multiple grain-size classes (Granjeon & Joseph 1999).

These latter two approaches may be more valid for the continental shelf than for fluvial systems. Simpson and Schlunegger (2003) used a bulk diffusive approach to sediment flux from eroding catchments, but invoked a mixed diffusive-advective algorithm for the transport coefficient, which recognised the important effect of flow concentration in the downstream direction within the drainage net.

A depth-dependent diffusivity has been invoked for sediment transport on the continental shelf (Kaufman *et al.* 1991), which accounts for the decrease of wave energy with depth. A more dynamic approach considers the details of individual sediment transport events summed over time (Niedoroda *et al.* 1995). Although individual transport events may be advective, the aggregate effect may be broadly diffusive. As in fluvial systems, evaluating the diffusive sediment transport coefficients as a function of meteorological and oceanographical conditions is problematic. Numerical code needs to be adapted in moving from a storm-dominated continental shelf (Niedoroda *et al.* 1995) to a tidally dominated continental shelf (Ericksen & Slingerland 1990). Shallow-marine stratigraphy can also be modelled as a wave-like progradation producing clinoforms (Kim *et al.* 2011) (§8.3.3).

Dynamical forward models are successful in simulating many of the high-resolution features seen in stratigraphy at outcrop and in borehole records, including the sand:shale ratio in the deposited layers. The evolution of the sand:shale ratio and stratigraphic geometry through a cycle of progradation and retrogradation for a fluvial-dominated delta is shown as an example in Fig. 8.29 (Cross *et al.* 1993; Granjeon & Joseph 1999). During progradation, sand is concentrated in a high-energy, narrow upper shoreface, whereas during

(a)

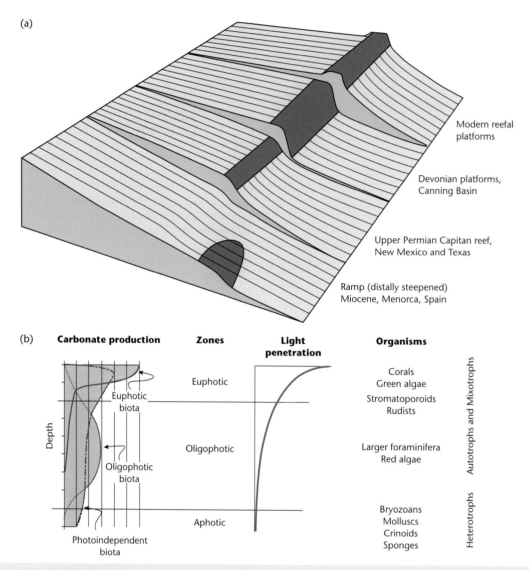

Modern reefal
platforms

Devonian platforms,
Canning Basin

Upper Permian Capitan reef,
New Mexico and Texas

Ramp (distally steepened)
Miocene, Menorca, Spain

(b)   **Carbonate production**   **Zones**   **Light penetration**   **Organisms**

Euphotic

Euphotic
biota

Oligophotic

Oligophotic
biota

Aphotic

Photoindependent
biota

Depth

Corals
Green algae
Stromatoporoids
Rudists

Larger foraminifera
Red algae

Bryozoans
Molluscs
Crinoids
Sponges

Autotrophs and Mixotrophs

Heterotrophs

Fig. 8.27 Carbonate depositional geometries, modified after Pomar (2001). (a) Range of morphologies from ramps to rimmed platforms. (b) Carbonate productivity in relation to the type of biota. Variations in the dominant biota control the water depth range of maximum productivity. Ramps may be distally steepened where oligophotic organisms dominate. Reproduced with permission of John Wiley & Sons, Ltd.

retrogradation driven by a relative sea-level rise, sand is distributed more widely due to erosional ravinement of the underlying deltaic deposits. The same model can be used to produce a three-dimensional simulation at the basin scale.

Fully coupled models do not assume that sediment supply is determined by some empirical formula related to elevation, slope or relief, nor that it is always available in sufficient quantity to fill accommodation, but instead incorporate surface process laws into a sediment routing model on a tectonically active template (e.g. Johnson & Beaumont 1995; Clevis *et al.* 2003, 2004). These models provide powerful insights into the linkage between catchment development, sediment routing and depositional geometries in a fully coupled system containing geodynamical representations of tectonic uplift and basin subsidence.

8.3.3 Shelf-edge and shoreline trajectories; clinoform progradation

Evaluation of the trajectory of the shoreline and shelf-edge is a means of understanding the dynamic controls on seismic facies and sequence stratigraphic geometries (Helland-Hansen & Gjelberg 1994; Helland-Hansen & Martinsen 1996; Steel & Olsen 2002). Trajectories can be recognised along dip-oriented seismic reflection profiles from the movement of the break in slope of the shoreface, the rollover of subaerial and subaqueous deltas, and the shelf–slope break (Mellere *et al.* 2002; Pirmez *et al.* 1998; Steel & Olsen 2002). The clinoforms associated with the shoreface have a scale of five to tens of metres, whereas subaerial delta clinoforms are 2–40 m in amplitude, subaqueous delta clinoforms range from 20 to 100 m, and

(a)

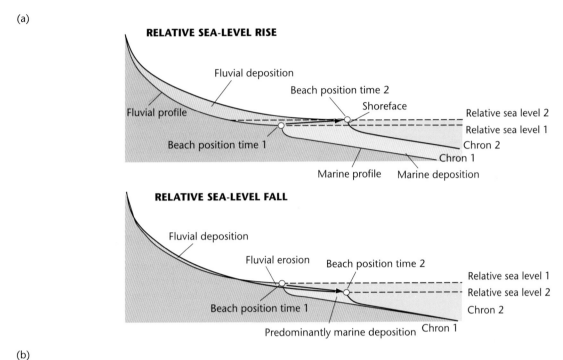

(b)

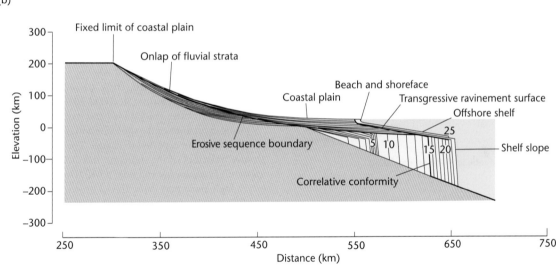

Fig. 8.28 Example of a geometrical 2D model for stratigraphy at a passive margin (after Burgess and Allen 1996), showing how depositional sequences and bounding unconformities can be simulated. (a) Movement of the fluvial and marine profile during relative sea-level rise and fall. (b) Computer-generated stratigraphy with an erosive sequence boundary and its distal marine conformity, and a transgressive ravinement surface generated during relative sea-level rise. Details in Burgess & Allen (1996). Small numbers in (c) are chrons.

the shelf slope clinoform may be measured in hundreds of metres (Fig. 8.30).

As introduced in §8.2.1.2, the shoreline and shelf-edge trajectories may show both a descending or ascending regressive trajectory (Fig. 8.10). The shoreline trajectory is determined by the interplay between sediment supply, short-term (<0.5 Myr) eustatic changes, subsidence/uplift and the effects of basin physiography on coastal sediment hydrodynamics. Based on outcrop studies, flat or descending shelf-edge trajectories are commonly associated with high fluvial sediment supplies, and aggradational or ascending shelf-edge trajectories are associated with wave-dominated shallow-marine conditions (Steel & Olsen 2002). The shelf-edge trajectory is thought to be driven by a longer-term relative sea-level change (0.5–3 Myr) than that affecting the shoreline (Carvajal & Steel 2009). The distance between the shelf-edge and the shoreface changes during the evolution of compound clinoforms (Fig. 8.31).

(a) PROGRADATION PHASE

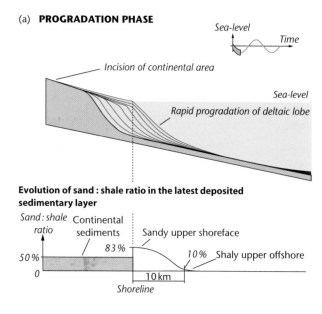

(b) RETROGRADATION PHASE

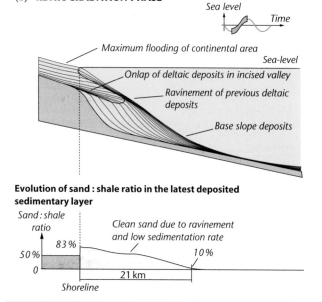

Fig. 8.29 Computer-generated two-dimensional transects through a fluvial dominated delta, showing the sand:shale ratio in the preserved sediment during the progradational phase (a) and retrogradational stage (b). After Granjeon and Joseph (1999).

across the continental shelf also depends on the amplitude and frequency of eustatic sea level, with differences between icehouse and greenhouse conditions (Somme *et al.* 2009).

Supply of coarse sediment to clinoforms at the shelf-edge is commonly attributed to the repeated transit across the shelf of shoreface deltas, driven by relative sea-level change or by sediment supply. For example, shoreface deltas may migrate rapidly across the shelf if fed by high sediment fluxes of large rivers. There is therefore a characteristic shelf transit time (Burgess & Hovius 1998) that depends on the depth-averaged sediment discharge, clinoform height and shelf width. This transit time can be estimated from seismic reflection dip-oriented profiles:

- For flat or descending shelf-edge trajectories developed during stable or falling relative sea level, shelf transit times are rapid (seldom exceeding 10^5 yr), even for low-supply systems and wide shelves.
- During glacial periods, eustatic sea-level excursions of c.100 m should be effective at transporting sand across the shelf, reducing transit times relative to greenhouse periods.

Sediment is delivered to the deep sea by transport through the coastal-shelf-slope segment of the sediment routing system to the deep-sea segment. This transfer can be essentially directly from river input, via a canyon incised into the shelf, to the deep sea. The small sediment routing systems of the California Borderlands are examples of this rapid throughput from mountain to deep sea (Covault *et al.* 2010, 2011). The alternative way of delivering sediment to the deep sea is by the arrival at the shelf-edge of shoreface deltas that have crossed the continental shelf. These are called shelf-edge deltas, and may take on different forms (Steel *et al.* 2008), depending on the local conditions of relative sea-level fall and sediment supply.

During excursions of the shoreline and shelf-edge, sediment transport processes are expected to change. For example, during a transgressive trajectory of the shoreline, the shelf is transgressed, thereby impacting wave energy and tidal power. During a descending trajectory, a high angle may cause fluvial processes to dominate, whereas a small angle may favour tide and wave activity, due to resonance, funnelling and shallowing effects in coastal seas.

We return to the idea of clinoforms as an example of the use of a dynamical approach to stratigraphy. As introduced above, sediment delivered to the ocean by rivers is stored in deltaic systems (§7.5.3.2), commonly with a compound clinoform geometry consisting of a couplet of subaerial and subaqueous prograding deltas (Fig. 8.31) (Nittrouer *et al.* 1996). The subaerial delta consists of low-gradient fluvial and delta plain topsets and a shoreface foreset, whereas the subaqueous data comprises a shallow-marine topset, foreset and deeper-marine bottomset. Each delta has a clinoform *rollover*, marking the point of gradient change from topset to foreset, but its position is highly variable. Progradation rates of the subaerial and subaqueous delta are also highly variable. In the Ganges–Brahmaputra system, the subaerial delta is prograding very slowly, but the subaqueous delta is prograding much more rapidly at 10 m yr^{-1} (Kuehl *et al.* 1997). There must therefore be a sediment partitioning between the subaerial and subaqueous clinoforms. In general, rivers entering marine basins with high energy (tides, waves) have well-developed subaqueous deltas, as in the Bay of Bengal, Papua New Guinea and Amazon shelf, whereas those entering low-energy marine basins, such as the Gulf of Mexico, have poorly

Progradation of *subaqueous deltas* and an ascending shelf-edge trajectory requires a strongly positive sediment budget, which is controlled by the migration of shoreface deltas across the shelf. This may occur during times of falling relative sea level (Muto & Steel 2002), but also during rising relative sea level and highstand (Burgess & Hovius 1998) where high sediment influx from major rivers fills the available accommodation generated by eustatic rise, as in offshore eastern Venezuela and the Niger Delta regions. The transit of deltas

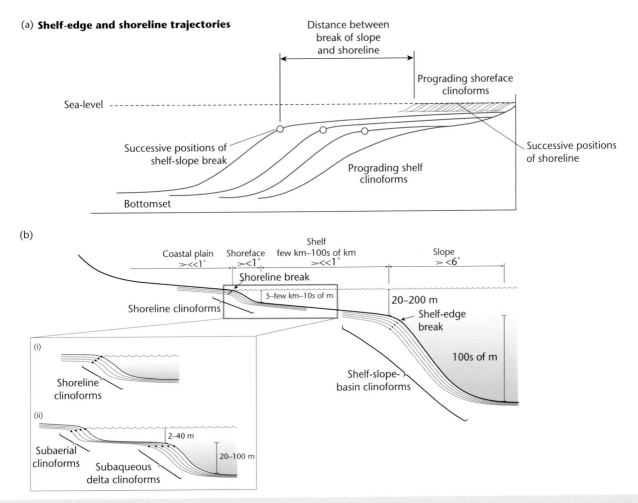

Fig. 8.30 (a) Schematic diagram illustrating shoreline and shelf-edge trajectories for the case of prograding shoreface and shelf-slope clinoforms (after Henriksen *et al.* 2011; reproduced with permission of John Wiley & Sons, Ltd.), and (b) dimensions of clinoforms associated with the shoreface, subaerial deltas, subaqueous deltas and the shelf slope (after Helland-Hansen & Hampson (2009), reproduced with permission of John Wiley & Sons, Ltd.).

developed subaqueous deltas. An increase in the strength of shallow-marine sediment transport relative to the river discharge favours growth of the subaqueous delta and causes a greater separation, both vertically and laterally, between the shoreface and the rollover (Fig. 8.31) (Swenson *et al.* 2005).

Compound clinoform development can be modelled in a basin with a steady water depth and river water and sediment fluxes, treating the shoreface as a moving boundary (Swenson *et al.* 2005). Neglecting the depositional porosity and compaction, the change in bed elevation η takes the familiar form of a divergence of the depth-averaged sediment flux q_s

$$\frac{\partial \eta}{\partial t} = -\frac{\partial}{\partial x}(\beta q_s) \qquad [8.3]$$

where β is the fraction of the transport surface occupied by channels ($\beta = 1$ seaward of the shoreline). Since the sediment flux depends on the integration over time of countless river floods,

coastal storms and tidal cycles that are unknown, these hydrological/meteorological events are captured by a single 'characteristic' event with a specified magnitude and recurrence interval that is responsible for the long-term evolution of the subaerial and subaqueous deltas. A time series of river discharge can be replaced by an impulse function with floods of specified magnitude (sediment and water discharges q_{sf} and q_{wf}), frequency (T_f) and duration ($T_f I_f$), where I_f is the time fraction or *intermittency* of floods (Fig. 8.32). In between the characteristic events is a 'dead time' during which no sediment transport is assumed to take place. Similarly, in the ocean, a characteristic event might be a coastal storm with frequency T_s and duration $I_s T_s$ that generates waves (breaker depth h_b) and offshore-directed currents (velocity v_o) controlling the long-term evolution of the subaqueous delta. Terrestrial floods and coastal storms may occur at the same time, may be independent or may be phase shifted.

Three different sets of rules are required for sediment hydrodynamics:

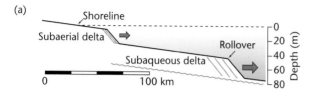

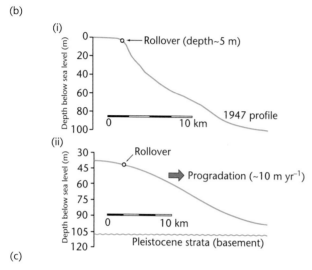

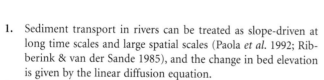

Fig. 8.31 (a) Schematic of compound (subaerial and subaqueous) clinoforms based on Nittrouer *et al.* (1996), (b) the bathymetric profiles of the Mississippi River delta (in 1947) (i) and the Ganges-Brahmaputra subaqueous delta (ii), showing profiles shapes and rollover positions (from Swenson *et al.* 2005, fig. 2). (c) Model definition sketch for delta progradation and development of compound clinoforms (from Swenson *et al.* 2005, fig.3). q_{so} and q_{wo} are the depth-averaged sediment and water discharges entering the model from rivers at $x = 0$. D is the depth of the water body, h_b is the breaker depth of waves, and H_b is wave height. Reproduced with permission of the American Geophysical Union.

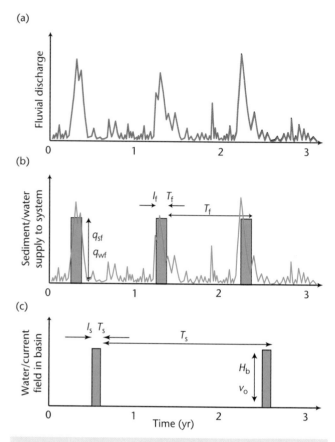

Fig. 8.32 The characteristic event approach of Swenson *et al.* (2005, fig.4). (a) Time record of fluvial discharge. (b) Approximation of the time series in (a) by a periodic impulse function of specified amplitude and frequency. The flood has a periodicity T_f, the time fraction during which flooding is active, or intermittency, I_f, so the duration of individual floods is $I_f T_f$. q_{sf} and q_{wf} are the sediment and water discharges associated with the river floods. (c) Equivalent representation of the wave current field of the sea that acts on the sediment supply to produce clinoform geometries. H_b is the wave height of breakers, and v_0 is the strength of downwelling currents produced by coastal set-up (that is, 'storm surge'). Shaded rectangles are the 'characteristic events' used in modelling of a storm-influenced coastal shelf offshore a river entry point. Reproduced with permission of the American Geophysical Union.

1. Sediment transport in rivers can be treated as slope-driven at long time scales and large spatial scales (Paola *et al.* 1992; Ribberink & van der Sande 1985), and the change in bed elevation is given by the linear diffusion equation.
2. Sediment and water dynamics in the surf zone are complex, but a long-term equilibrium is treated as a vertical step with the height of the breaker depth h_b.
3. In the shallow sea, wave-current interaction drives sediment transport beyond the surf zone. The suspended sediment flux is treated as a combination of current- and slope-driven and is estimated using a non-linear advection-diffusion equation in

which the diffusivity is water-depth dependent and the advective term is a depth- and slope-dependent wave-like celerity (details in Swenson *et al.* 2005).

The depth-dependent wave- and current-driven sediment dynamics below the surf zone generates a clinoform rollover and a low-gradient subaqueous delta foreset (Fig. 8.33). Using these sets of rules and a moving boundary solution at the shoreface, model runs show that the magnitude and frequency of floods and storms control the partitioning of sediment between and growth rates of subaerial and subaqueous deltas. Subaqueous and subaerial delta progradation may even be phase shifted, posing major problems for conventional sequence stratigraphic models that invoke external controls for stratigraphic architectures.

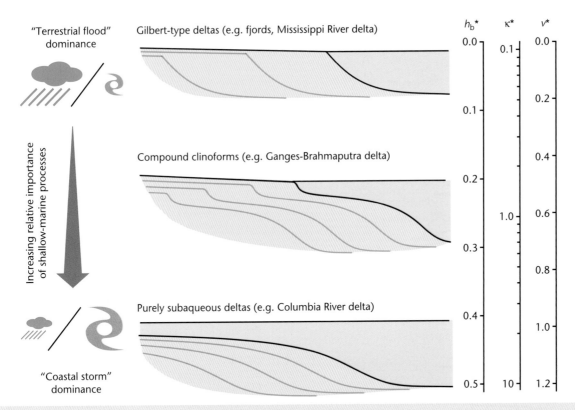

Fig. 8.33 Clinoform geometries based on a spectrum from dominance of terrestrial (river) floods to dominance by coastal storms. River dominance favours Gilbert-type deltas, compound clinoforms are found in intermediate positions, and purely subaqueous sigmoidal deltas are found in regions of coastal storm dominance. h_b^*, κ^* and v^* are dimensionless parameters. h_b^* is the fraction of the basin depth affected by high wave energies based on the water depth of breaking waves (high for 'shallow' basins and low for 'deep'), and κ^* and v^* are the slope and current-driven diffusivities controlling the shallow-marine sediment flux. These dimensionless parameters collectively quantify how flood and storm parameters and basin geometry combine to control clinoform behaviour. From Swenson *et al.* (2005), reproduced with permission of the American Geophysical Union.

8.4   Landscapes into rock

8.4.1   Stratigraphic completeness

"Continuity and steadiness of sedimentation are troublesome notions"

Peter M Sadler, 1981, *Journal of Geology*, 89, p. 569

It is self-evident that if we extrapolate rates of deposition measured in modern environments to the long time scales of the geological past, stratigraphic sections are less thick than predicted, so must contains gaps or hiatuses (Reineck 1960; Newell 1962). So how complete is the stratigraphic record? It is a selective and partial record of events that took place in the past, rather than a faithful recorder of everything that happened. The extraction from surface fluxes to build the stratigraphic record involves a transfer function that may be negative, implying erosion or non-deposition, or positive, involving deposition and preservation.

"Time transforms sediment routing systems into geology, and like history, selectively samples from the events that actually happened to create a narrative of what is recorded"

Philip A Allen, 2008, in 'From Landscapes into Geological History', *Nature*, 451, 274–276

Owing to the occurrence of erosion and non-deposition, which produces gaps in the stratigraphic record, the completeness of the stratigraphic record is a function of the time interval over which accumulation rates are measured (Sadler 1981; Sadler & Strauss 1990). Stratigraphic incompleteness clearly influences the reliability of the sedimentary archive to say meaningful things about environmental changes in the past, but also brings into question how representative modern depositional environments are of the fragmentary mosaics constituting the stratigraphic record.

Consider as a thought experiment a randomly generated succession of 100 beds of variable thickness, deposited at equal time intervals (Fig. 8.34). Consequently, the true sediment accumulation rate varies over time. If this randomly generated time series of sediment accumulation rate is sampled at a range of chosen time intervals t^*, it is clear that the sediment accumulation rate generated varies greatly according to the value of t^*. When $t^* = 5$ time units, implying a high resolution in the sampling, the rate appears to slowly increase unsteadily then slowly decrease unsteadily over the total of 100 time units. When $t^* = 10$ time units, a double peak of sediment accumulation rate appears; with $t^* = 20$ time units, a pronounced single cycle of increase then decrease is observed; and with $t^* = 50$ units, implying low resolution in the sampling, the rate varies little, but slightly decreases over time. Trends and cycles therefore appear and disappear depending on the time interval of the sampling relative to the

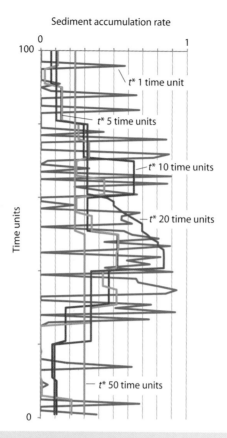

Sediment accumulation rate

*t** 1 time unit

*t** 5 time units

*t** 10 time units

*t** 20 time units

*t** 50 time units

Fig. 8.34 Randomly generated sediment accumulation rate over time for 100 time units representing 100 depositional events (blue). Other curves are generated by sampling the 100 events with different time resolution *t**, at 5, 10, 20 and 50 time units. The trends in sediment accumulation rate depend on the sampling resolution.

This can be appreciated from another simple thought experiment. Consider a stratigraphic succession deposited over a 10 Myr period, as shown in Fig. 8.35, with periods of steady deposition at a rate of 0.1 mm yr^{-1} interspersed randomly with periods of non-deposition. The total thickness accumulated is 400 m, which gives an overall sediment accumulation rate of 0.04 mm yr^{-1}, whereas the individual beds were deposited at 0.1 mm yr^{-1}. The background subsidence rate or drift is therefore 0.04 mm yr^{-1}. 40% of the time period of 10 Myr is represented by deposition and 60% by a non-depositional gap.

The long-term sediment accumulation rate is 0.04 mm yr^{-1} and the total sediment deposited in the 10 Myr interval is 400 m. If the sampling interval is 10 Myr, $t'/T = 1$, indicating that the stratigraphic succession is 'complete' at this scale of resolution. If, however, the sample interval is reduced, it can be seen from inspection of Fig. 8.35 that stratigraphic gaps become increasingly apparent. Stratigraphic successions are therefore less 'complete' at higher resolution. Sediment accumulation rates within each time interval are also increasingly higher than the background 'drift' at higher sampling resolution.

Completeness at a given time scale t^* can be modelled from:

- the duration of the stratigraphic section, which is generally known from biostratigraphic, paleomagnetic and geochronological data;
- the long-term net accumulation rate or 'drift', which is derived from sediment thickness;
- the unsteadiness of the sedimentation rate, which is best modelled statistically in terms of wavelength and amplitude (Schwarzacher 1987; Sadler & Strauss 1990).

Completeness is then a positive function of time scale, drift and wavelength of fluctuations, and an inverse function of standard deviation and amplitude of sedimentation rate fluctuations. In other words, thick stratigraphic sections of long duration with slow fluctuations in sedimentation rates tend towards being 'complete', whereas thin stratigraphic sections of short duration with rapid, high-amplitude fluctuations tend towards being 'incomplete'.

The significance for basin analysis is that different basin types, with different histories of subsidence and sediment supply, should be characterised by different levels of stratigraphic completeness. If the rate of sediment accumulation is plotted in logarithmic space against the time span for which they were determined, there is a clear inverse relationship with considerable scatter, but with a number of clusters or 'modes' (Sadler 1981) (Fig. 8.36a) that reflect the common ways of estimating accumulation rates. The variability is caused by differential compaction, non-uniformity at a given location, and unsteadiness in time, which reflect in some complex way the sediment hydrodynamics of the depositional environment and its uplift/subsidence history. Consequently, accumulation rate versus time-span relationships should vary according to depositional environment and basin type (for example, fluvial, lacustrine, carbonate platforms and reefs, siliciclastic shelves, deep-marine minibasins and abyssal plains). Environments above wave base characterised by limited accommodation have similar trends in accumulation rate versus time span, especially at time spans >1 kyr, suggesting that they have a common influence – tectonic processes of stretching, cooling and flexure. In contrast, sediments deposited below wave base, where accommodation is not limited, have a different relationship between accumulation rate and time span (Fig. 8.36b).

Sadler (1981, p. 580) proposed that completeness should be calculated as the long-term rate of sediment accumulation of the whole

total time T represented by the randomly generated sedimentary succession.

The question of stratigraphic completeness can therefore only be addressed if we specify at what time interval t^* the absence or presence of stratigraphic record is to be judged, within a longer time span for the duration of the entire stratigraphic succession T. A 'complete' stratigraphic section is therefore one where there is a record in each time interval t^*, so that any hiatuses present have a duration of less than t^*.

Peter Sadler formalised this problem of stratigraphic completeness (Sadler 1981). Take a stratigraphic section where $t^* = T$. The completeness must be equal to 1, since the section contains a preserved sedimentary record. But, as the time intervals become finer, the likelihood arises that some intervals t^* will fall within gaps, so completeness reduces, and becomes a minimum when there is only one time interval left in the whole stratigraphic section that contains a sedimentary record. The fraction of time intervals that have left a record is then t^*/T. The stratigraphic completeness therefore depends on the time interval t^*, so sections might be regarded as 'highly complete', or 'poorly complete' dependent on the length of the time interval t^* alone. Completeness has the character of a probability.

The history of sediment accumulation derived from dated sections of rock of known thickness becomes an increasingly distorted record of the true rate of accumulation with lower values of completeness.

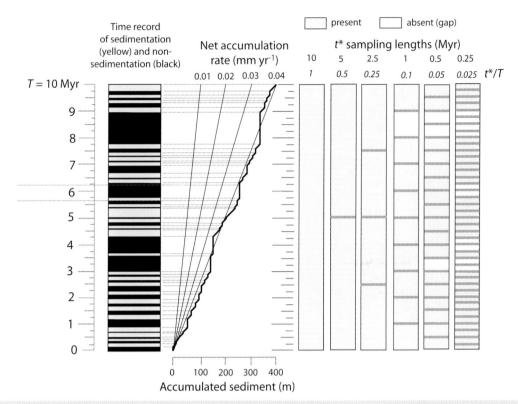

Fig. 8.35 A random distribution of record and gap in a 10 Myr-long stratigraphic succession is sampled at different resolution t^*, resulting in different values of completeness. The stratigraphic succession is built by the accumulation of depositional events with a sedimentation rate of 0.1 mm yr⁻¹, separated by periods of non-deposition, resulting in a net sediment accumulation rate or 'drift' of 0.04 mm yr⁻¹. By decreasing the sampling interval t^* from 10 Myr to 0.25 Myr, erosional gaps become apparent, illustrating the dependency of stratigraphic completeness on t^*.

section S compared to the average rate of accumulation at the given time scale t^*, denoted S^*. The measured deposition rate decreases as a power-law function of the interval of time over which it is measured. The expected completeness is then

$$(S/S^*) = \left(\frac{t^*}{T}\right)^{-m} \qquad [8.4]$$

where the exponent m is the gradient of the trends in Fig. 8.36a. The value of m differs in different depositional environments and basin types (Fig. 8.36b) and is in the range $-1 < m < 0$. The peak in relative frequency of m values according to depositional setting is as follows:

Fluvial	−0.70
Siliciclastic shelf	−0.40
Carbonate platform	−0.45
Lacustrine	−0.35
Marine minibasins	−0.20
Abyssal plain carbonate ooze	−0.20

Consider a carbonate stratigraphic succession 1000 m thick, deposited in a time period of 12 million years, with time intervals of 1 Myr provided by ammonite stratigraphy. The expected completeness is therefore $(1/12)^{0.45} = 0.33$, meaning that only one third of the chosen time intervals will be represented by a sedimentary record. In comparison, let us take a marine basin 1000 m thick of duration 12 million years, with time intervals of 2 Myr; the expected completeness

is $(2/12)^{0.2} = 0.7$, meaning that the stratigraphic succession is 70% complete at the time resolution of 2 Myr.

Rates of sediment accumulation also appear to be scale-dependent in individual basins (Schumer & Jerolmack 2009). The power-law dependence of measured accumulation rate on time interval may be a result of a power-law distribution of erosional gaps, or 'waiting times' between recorded events (Schumer & Jerolmack 2009; Schumer *et al.* 2011). The longer the time interval, the greater the likelihood that longer hiatuses will be sampled, so the length distribution of hiatuses is 'heavy tailed'. If the tail of the distribution of hiatal period length is described by a parameter γ, the observed sediment accumulation rate decreases as t^* raised to the power $(\gamma - 1)$ over several orders of magnitude (Schumer & Jerolmack 2009). The scaling exponent is in the range $0.1 < \gamma < 0.5$ for time scales up to 10^4 yr for fluvial and marine environments (Jerolmack & Sadler 2007). At time scales greater than 10^4 years, γ is close to unity, which represents a constant deposition rate, implying that the sediment accumulation is dependent on the steady drift of tectonic subsidence alone at long time scales.

The concept of stratigraphic completeness is critical for the interpretation of trends in sediment accumulation rate. For example, an increase in sedimentation globally over the last 5 Myr, combined with an increase in grain size, has been attributed to global climate change affecting erosion rate (Hay *et al.* 1988; Zhang *et al.* 2001; Molnar 2004). But it is possible that rates estimated for the last 5 Myr are compared with rates derived from thicknesses of sediment more than 5 Myr old, where time resolution is of lower resolution. The idea of

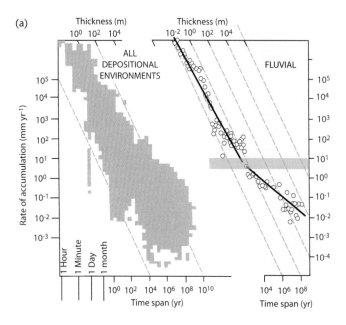

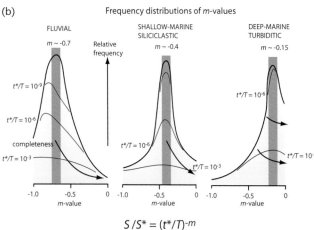

$$S/S^* = (t^*/T)^{-m}$$

S sediment accumulation rate of whole succession ('drift')
S* Sediment accumulation rate of sample interval
T Time duration of whole succession
t* Time duration of sample interval
m Completeness exponent

Fig. 8.36 (a) 'Sadler' plot for all depositional environments, showing (left) relationships between bed thickness, time span of observation, and rate of accumulation, and (right) a subset for fluvial environments, showing a decrease in net rate of accumulation for longer time spans. Note the different gradients of the best-fit lines through the data. (b) Value of the completeness exponent *m* for fluvial, shallow-marine siliciclastic and deep-marine turbiditic depositional environments and processes. Arrows indicate trend with increasing completeness. Redrawn from Sadler (1981), republished with permission of University of Chicago Press.

a global, climatically driven increase in erosion rates has been challenged on the basis that:

• grain-size coarsening can be demonstrated to be strongly diachronous over the last ~10 Myr rather than showing an abrupt increase at 5 Ma (Charreau *et al.* 2009);

• oceanic ^{10}Be/^{9}Be evidence indicates stable weathering rates over the last 10 Myr (Willenbring & von Blanckenburg 2010);

• measured deposition rate is known to decrease as a power-law function of the time over which it is measured.

The increase in sediment accumulation rate is therefore most likely due to the recording of longer hiatuses in the time period before 5 Ma rather than being driven by a real change in erosion rate.

With this view of stratigraphic completeness, the stratigraphic record is seen to be strongly filtered by the time scale and statistical properties of sediment dynamics over the Earth's surface relative to the slower rates and lower statistical variability of accommodation generation. Only a small fraction of the sediment transport events taking place on the surface of the Earth survive this stratigraphic filter (Schumer & Jerolmack 2009; Schumer *et al.* 2011).

The selective preservation of stratigraphy by the 'stratigraphic filter' described above is essentially a random or stochastic process. But the question can be raised of whether stratigraphic completeness can also be approached by approximating the transfer function between present-day fluxes and stratigraphic archive by simple analytical models. This gives the capability of evaluating the way in which sediment discharge signals are modified through propagation in sediment routing systems, and the stratigraphic product of this modification. We can think of the signal as being blocked or allowed to pass through different types of gate (as in electronic engineering and neurobiology). Gating is one form of signal transformation where a signal with certain probability-density characteristics is selectively sampled to generate a transformed record with different probability-density characteristics (Fig. 8.37); other transformations are a relaxation over time from a step change in a driving mechanism, a buffering of a spike-like driver into a lower-amplitude and longer-wavelength signal, or the damping of a periodic or alternating driver.

8.4.2 Gating models

The transformation of surface sediment dynamics into a stratigraphic record, represented by the metaphor 'from landscapes into rock' can be viewed as a complex process of selection that we term gating. Even in very simple systems, the actual deposition rate at a point on the bed is very different from the net sediment accumulation over longer time scales. Two examples below illustrate this idea.

The simplest forms of gating can be visualised from simplified deterministic approaches at a range of spatial and temporal scales. At the small scale, the migration of an array of *ripple marks* under a steady current causes alternate periods of erosion and deposition at a fixed point on the bed, since deposition takes place on the lee side of the ripple in the separation zone, whereas erosion and sediment transport take place on the stoss side (Fig. 8.38). If the length of the lee side is x_l, the period of time spent in the depositional mode is x_l/V_x, where V_x is the horizontal velocity (celerity) of the ripple train. Periods of deposition, lasting x_l/V_x, recur periodically at intervals of x_r/V_x, where x_r is the spacing of the ripple mark, resulting in a sawtooth pattern of erosion and deposition from a steady, continuous sediment transport rate.

This sawtooth pattern of alternating deposition and erosion does not produce a stratigraphic record unless there is a background 'drift'. Such a drift is produced by bedform climb under conditions of a finite vertical (fall-out) sediment flux. When the ripple marks climb subcritically, a portion of the bedform height is top-truncated during

Transformation in distance or time

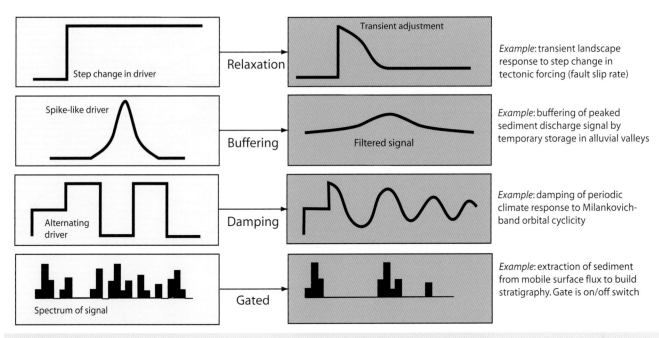

Fig. 8.37 Schematic diagram illustrating various forms of signal transformation applicable to sediment routing systems.

the passage of the array of ripple marks. Only when the ripple marks begin to climb at angles above critical are stoss sides preserved. Consequently, the lee-side gate continues to operate but is superimposed by a process of erosional top truncation at subcritical angles of climb, and ceases to operate in the case of supercritically climbing current ripples. The fraction of the bedform height preserved as depositional cross-sets depends on the angle of climb, which in turn reflects the balance between the horizontal and vertical fluxes. A second gate is therefore operative, dependent on angle of climb. The stratigraphic preservation from a simple array of migrating bedforms is therefore a question of completeness, and the time period represented by a set or coset of cross-stratified deposits is a complex amalgam of periods of erosion and deposition due to the operation of 'lee-side' and 'bedform-climbing' gates.

Another simplified example, at a larger time and spatial scale, is the variability in sediment accumulation as a sinusoidal variation about a longer-term background subsidence rate or 'drift'. If sediment is deposited during a time of net positive accommodation it is momentarily preserved, but long-term preservation is only achieved when the deposits of a time interval survive erosion during the younger time interval (Fig. 8.39). This is another example of top truncation. The chance of passing through this sinusoidal gate depends essentially on the background tectonic subsidence rate a. If the period of the sinusoidal variation in sediment accumulation is 100 kyr and the amplitude 50 m, the fraction of time represented by preservation is less than 10% for a tectonic subsidence rate of 0.2 mm yr^{-1}. Increasing the tectonic subsidence rate to 1 mm yr^{-1} causes the fraction of time represented by preservation to increase to c.40%. Inspection of Fig. 8.39 shows that the fragments preserved in stratigraphy are located very early in the rising limb of the sediment accumulation (accommodation) curve.

There are alternative gating models. For example, consider a limited and fixed volume gating of each sediment-transporting event that takes place. In this case, irrespective of the sediment supply, if $q_s(x)$ is above some threshold value, a fixed increment of deposition will take place. The time pattern of sediment accumulation will depend on the temporal distribution of sediment-transporting events above a certain threshold magnitude. More realistically, a variable volume may be gated up to a volume determined by the tectonically generated accommodation. These scenarios are best considered visually (Fig. 8.40). The impact on the sediment accumulation curve and on stratigraphic completeness of the different gating models is considerable.

The selective extraction from surface fluxes to build underlying stratigraphy using a transfer function from the active surface layer to the inactive substrate or bed (made use of in §7.5.4) is an example of a gating model where the volume of sediment preserved depends on the spatial distribution of long-term accommodation generation. The extraction process depletes the surface flux in particles of a certain grain size, leading to characteristic down-system grain-size trends in stratigraphy (Fedele & Paola 2007; Duller *et al.* 2010; Whittaker *et al.* 2011; Allen & Heller 2012; Parsons *et al.* 2012). This is an example of an accommodation-controlled gating process that has predictable outcomes for grain size.

Gating is therefore the selective opening (activation) and closing of gateways allowing transmission of signals, at its simplest like an on-off switch (Fig. 8.40). Gating 'samples' the probability-density characteristics of meteorological or geophysical events to selectively generate stratigraphy (Fig. 8.41). Even with a steady driver, such as a continuous sediment-transporting flow over a granular bed, there is a complex gating process that eventually results in a cross-stratified coset of sand deposited with a net accumulation rate far different to

(a) ZERO ANGLE OF CLIMB

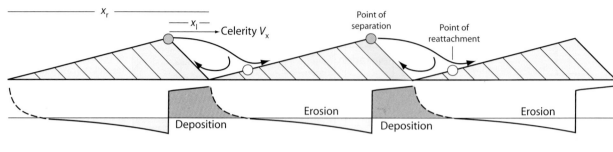

(b) SUBCRITICALLY CLIMBING

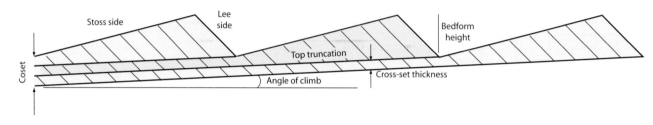

(c) CRITICALLY CLIMBING

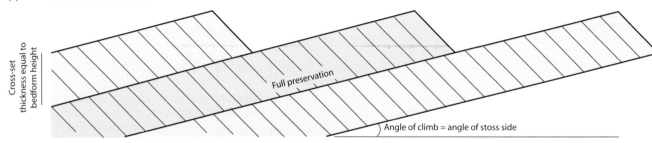

Fig. 8.38 The partial recording of time in the deposits of a migrating train of current ripples with length of lee side x_l, ripple spacing x_r and celerity V_x. (a) Alternate periods of erosion and deposition take place due to the passage of a field of periodic bedforms; (b) top truncation takes place during the migration of subcritically climbing current ripples; (c) full preservation takes place at a critical angle of climb equal to the slope of the stoss side. The net accumulation rate of a coset of cross-strata is therefore much lower than the rate of deposition by avalanching of the lee slope.

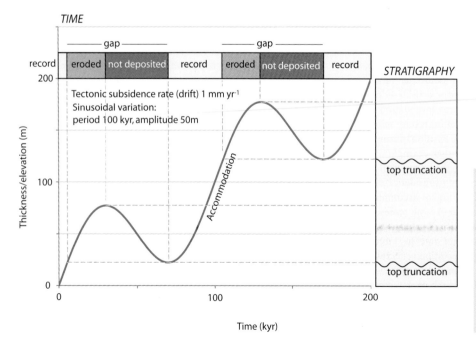

Fig. 8.39 Relative sea level, equivalent to accommodation, for a sinusoidal sea-level change superimposed on a background tectonic subsidence or drift, based on the pioneering work of Barrell (1917). It is assumed that sedimentation continually keeps pace with accommodation generation. Erosional and non-depositional gaps are present due to loss of accommodation, represented in the stratigraphic column by unconformities.

GATE: ON-OFF SWITCH

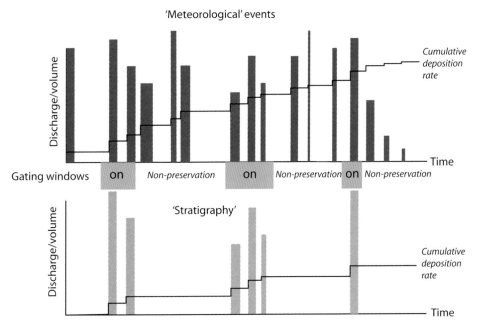

'Meteorological' events

Fig. 8.40 The operation of a gate as an on-off switch, showing the transformation of the meteorological signals or impulses into stratigraphy. The gated signals result in a lower net sediment accumulation rate. An on-off gate might be sporadic subsidence events associated with seismic slip on faults, or the cutting of channels by floods, followed by their filling with sediment.

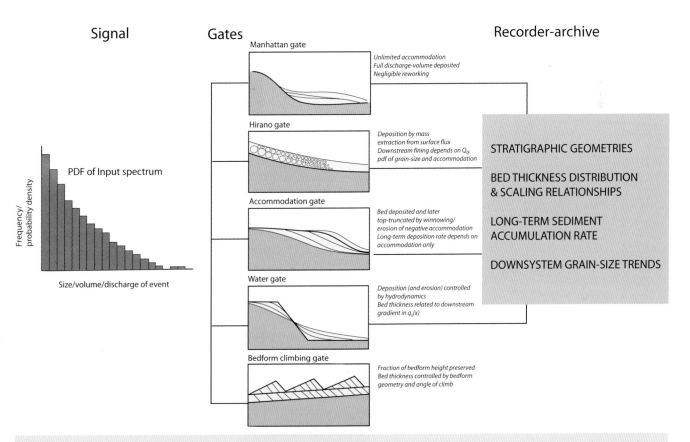

Fig. 8.41 Various forms of gating causing a transformation of the input spectrum (impulse function) into the stratigraphic record.

the local rate of avalanching down the lee face of a ripple. Signals are more likely to be unsteady, such as storms on the continental shelf, or river floods. In this case, it is valuable to think of the size and repeat time (intermittency) of *characteristic events* (Swenson *et al.* 2005; Fig. 8.32), as discussed in §8.3.3.

8.4.3 Hierarchies and upscaling

Nature exhibits complicated behaviour, but displays striking simplicity in its forms, whether in the shape of a wind-blown dune, a river meander or a snowflake. This combination of 'complicated behaviour arising from a simple form is a classic symptom of a complex system' (Werner 1999, p. 102). Such complexities arise from the non-linearities and feedbacks in natural systems that are open to exchanges of energy across their boundaries, so that they are sensitive to external forcings. Complex systems become more organised as they evolve, and by the interactions within them, new behaviours emerge that were not previously evident. These self-organised new behaviours are described by new dynamical variables. Variables with short (fast) time scales may become *slaved* to longer time scale (slow) variables, losing their status as independent variables. The best example of this is the slaving of small-scale and fast variables determining the physics of motion of individual sand grains over the floor of a desert to the larger-scale and slower variables determining the motion of large barchan dunes, or the even larger-scale and slower-still motion of large draas with 200 m-high slip-faces.

Complex natural systems can be investigated by identifying the different dynamical levels from small to large, thereby forming a hierarchical system of interacting levels (Fig. 8.42). Returning to the movement of bedforms (Werner 1999), a hierarchical model can be formulated. At the lowest level, the important dynamical variables are the physics of fluid flow around individual sand grains. Dynamics from the lower level are abstracted to a higher level in the hierarchy to explain the shape and velocity of sand ripples and dunes. The dynamics of single bedforms are in turn abstracted to a higher level to explain the position of bedform fields with repeating patterns of regularly spaced bedform crests. The dynamical variables therefore correspond to a certain temporal scale, and the different levels of the complex natural system interact with different aspects of the external environment.

In basin analysis, and particularly in the study of basin stratigraphy, we are also faced with a problem of moving from the short-term

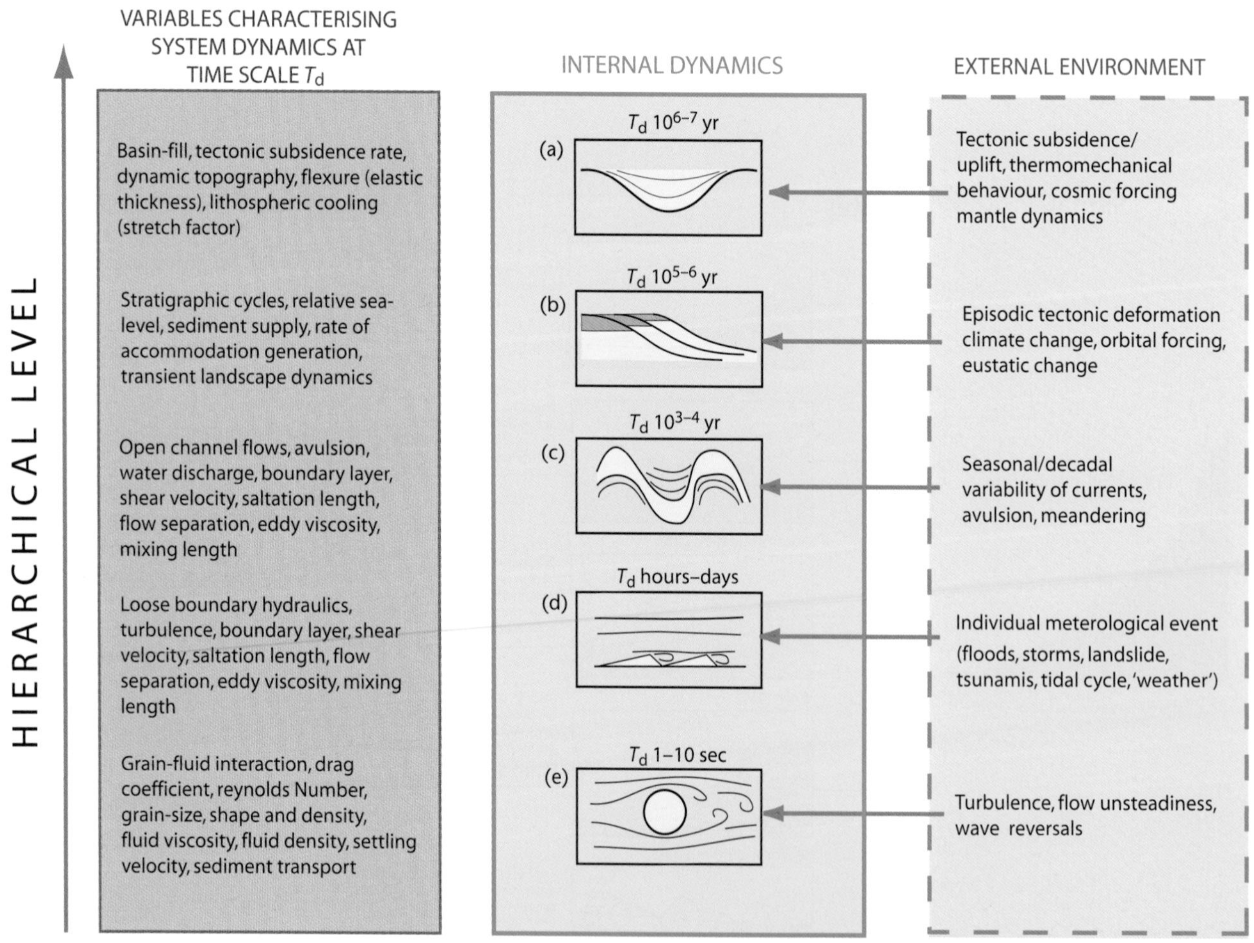

Fig. 8.42 Hierarchical levels: the external environmental forcings and internal dynamical variables at different time scales T_d. See discussion in text.

scale of individual sediment-transporting 'meteorological' or 'geophysical' events to the long-term time scale of the stratigraphy of the entire basin-fill (Fig. 8.42). Hierarchical modelling suggests that this upscaling from short-term to long-term cannot be achieved by simply using the same dynamical parameters for longer time periods, but instead that we should be searching for new sets of dynamical variables. These new variables will operate at a hierarchical level with different external forcings than at the short-term level.

The physics of fluid flow around individual grains, or the interaction between grains in a highly concentrated flow, constitutes the important dynamics at the shortest time scale of seconds. The important dynamical variables are fluid and grain densities, fluid viscosity, and instantaneous velocity; a classic dimensionless group is the Reynolds number, and the underlying concept is of flow resistance (drag), settling and viscous shear stresses. The external environment is individual 'geophysical' events, such as a landslide or a river flood.

At a longer temporal scale, the physics just described is slaved to dynamical variables that describe the movement of trains of bedforms, such as sand dunes in an open channel flow or estuary, that grow to equilibrium over hours, days and years, and are described by dynamical variables referring to the outer flow rather than to individual grains, such as flow depth and time-averaged speeds. The underlying concept is net sediment transport of bedforms and bedform fields, and the external environmental forcings are tides, waves, currents and gravity flows acting repeatedly over time scales longer than that of individual events.

At the next hierarchical level are morphodynamic forms, such as river meanders and delta lobes, which grow and evolve over periods of thousands of years. The dynamical variables are the discharge and grain-size distribution of the sediment supply, the average slope, and the floodplain or delta accretion rate. Feedbacks in this hierarchical level may induce important autocyclicity,

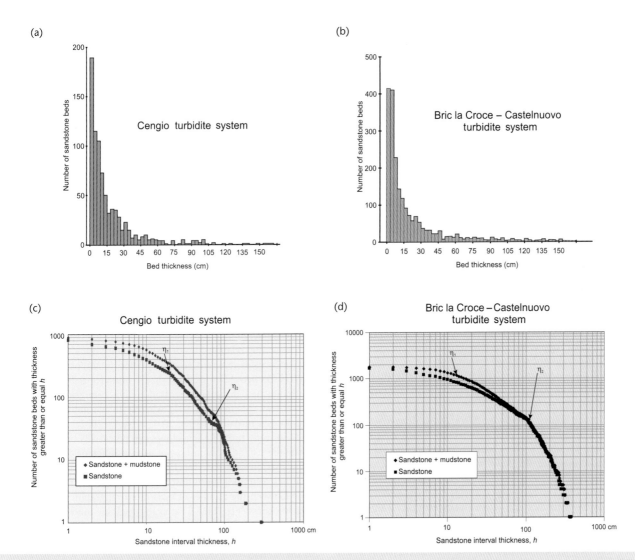

Fig. 8.43 Scaling relationships of turbidites. (a) and (b) show the magnitude-frequency distributions of turbidite bed thickness from two turbiditic successions in the Piedmont area of northern Italy (Felletti & Bersezio 2010). (c) and (d) are plots of the logarithm of the number of sandstone beds thicker than *h* versus sandstone bed thickness *h*. Segmentation of the distribution is marked by breaks in slope, giving thickness thresholds that split the distribution into statistically different sub-populations. The slope of the segments gives the scaling exponent β in the power law for turbidite bed thickness. Reprinted with permission from Elsevier.

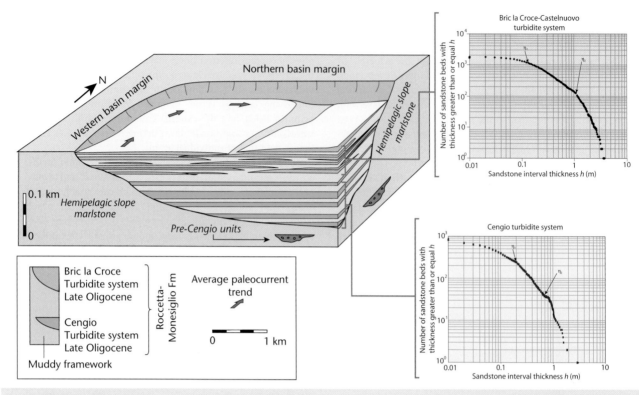

Fig. 8.44 The effect of basin confinement on the scaling relationships of turbidites, Piedmont Basin northern Italy. After Felletti & Bersezio (2010, fig. 10), reprinted with permission from Elsevier.

resulting in, for example, avulsion or delta lobe switching, but external forcings might include orbital forcing of climate or changes in the local subsidence rate.

At a higher level, these large morphodynamic elements are slaved to longer-term variables controlling the preservation or erosion of depositional units with distinctive three-dimensional geometries, producing stratigraphic cycles and stacked cycles with time scales of thousands to hundreds of thousands of years. The dynamical variables are sediment supply, water depth, tectonic processes generating accommodation, eustatic sea level forced from the external environment by Milankovitch-band orbital variations, episodic (in time and space) tectonic deformation, and some forms of dynamic topography such as pulsing plume behaviour.

Finally, at the highest level in the hierarchy, at time scales of tens to hundreds of millions of years, sedimentary basins are controlled by dynamical variables such as long-term patterns of subsidence/uplift, and sediment supply. Important variables and concepts relate to the amount of stretching, the flexural rigidity, and the amount of crustal shortening. This hierarchical level is driven by external factors such as mantle dynamics and relative plate motion, resulting in a thermomechanical evolution of the lithosphere.

Upscaling from short to long time scales requires the identification of *new* sets of dynamical variables and *new* forcing mechanisms from the external environment, rather than the extrapolation of the same variables from a low hierarchical level to a higher level. For the basin analyst, this means that the focus must be on high hierarchical levels if the objective is to understand broad stratigraphic patterns and the stratigraphic completeness of the basin-fill.

8.4.4 Magnitude-frequency relationships

We conclude this chapter by referring to a fundamental property of stratigraphy – the scaling relationships of its component beds; that is, the statistical distribution of thickness, horizontal width and three-dimensional volume (Malinverno 1997). Beds of sandstone, for example, are likely to have different scaling relationships according to their different depositional settings.

The scaling relationships of turbidite sandstones have been a particular focus of attention, since they comprise important deep-water targets for petroleum exploration (Drinkwater & Pickering 2001). Turbidite bed thickness distributions are commonly regarded as power laws (Malinverno 1997), whereas an alternative model is of a log-normal mixture (Talling 2001; Sylvester 2007). The techniques used to study turbidite sandstones can potentially be used to investigate the scaling relationships of beds in different depositional settings (Malinverno 1997; Sinclair & Cowie 2003).

It has commonly been observed that there is an inverse relationship between the number of beds in a stratigraphic succession and bed thickness (Fig. 8.43). That is, thin beds are much more frequent than thick beds. If the statistical distribution is a power law, a straight line will be obtained on a log-log plot of $N(h)$ versus h, where $N(h)$ is the number of beds whose measured thickness is greater than h, which is proportional to bed thickness h raised to an exponent $-\beta$, and a is a constant representing the total number of beds measured in the data set. That is

$$N(h) = ah^{-\beta} \qquad [8.5]$$

If data plot as a straight line, the power-law exponent β can be determined by visual inspection or by calculation of the least squares coefficient R^2 (Carlson & Grotzinger 2001; Sinclair & Cowie 2003). Power-law models are poor, however, at accounting for thin beds in the least squares fitting (Sylvester 2007). Significant departures from a straight line on log-log plots is commonly interpreted in terms of *segmented* power laws (Rothman *et al.* 1994; Sinclair & Cowie 2003), made of two different populations of beds, or of a truncated power law. Breaks in slope between segments with different β on the log-log graph occur at threshold bed thicknesses η_1, η_2, etc. The values of these threshold thicknesses may reflect the different factors affecting the emplacement of turbidites, such as flow rheology, different source areas or different basin-floor geometry.

Log-normal distributions for turbidite bed thickness, or log-normal mixtures of, for example, thin- and thick-bedded populations, appears to satisfy some bed thickness field data, such as from the extremely well-documented Miocene Marnoso Arenacea of Italy (Talling 2001). The bimodality of turbidite bed thicknesses may reflect two modes of transport and depositional mechanics, such as dilute and highly concentrated variants, or slow and rapid deposition rates.

The bed thickness distribution of turbidites may also be used to assess the degree of basin confinement. Confined turbidite flows are forced to aggrade vertically, thereby changing the bed thickness distribution relative to unconfined flows that are free to expand. This transition from ponded to unconfined can be recognised in the bed thickness statistics of turbidites from the Tertiary Piedmont Basin of northern Italy (Felletti & Bersezio 2010) (Fig. 8.44). The effects of basin confinement were also explored by Sinclair and Cowie (2003) from Tertiary flysch basins around the Alpine arc of France and Switzerland.

The intense treatment of turbidite sandstone thickness distributions and scaling properties with bed length and volume, suggests that a similar approach should be adopted in other depositional settings, such as fluvial, aeolian, coastal and shallow-marine, where the mechanics of sediment transport and deposition by air and water flows should be reflected in the power-law or log-normal mixture distributions.

CHAPTER NINE
Subsidence history

Summary

The present-day stratigraphic units of the basin-fill have been reduced in thickness by the progressive effects of compaction over time. The stratigraphic column is therefore not a direct recorder of sediment accumulation rates in the past. The burial of sedimentary layers in a basin causes porosity loss, which may be due to mechanical compaction, physico-chemical changes such as pressure solution, and cementation. The processes involved in mechanical compaction are well understood in relation to clay-rich soils. There is a well-known relationship between the effective vertical stress and the void ratio for normal consolidation. Overpressuring can be predicted from a dimensionless parameter representing the ratio of hydraulic conductivity to rate of sediment accumulation. Burial of low-permeability sediments accompanied by active sedimentation leads to overpressures in the pore fluid at depths of >2 km, where hydrofracturing may take place.

In studying the subsidence in sedimentary basins it is common to assume that there is no change in the solid volume during burial, so that mechanical compaction of framework grains due to increasing compression from the overlying sediment-water column solely drives porosity reduction.

A quantitative analysis of subsidence rates through time relies primarily on the decompaction of stratigraphic units to their correct thickness at the time of interest. Two other corrections must also be made in order to plot subsidence relative to a fixed datum such as present sea level. These are: (i) corrections for the variations in depositional water depth through time; and (ii) corrections for absolute fluctuations of sea level ('eustasy') relative to the present sea-level datum.

The decompaction of stratigraphic units requires the variation of porosity with depth to be known. Some sedimentary formations exhibit a linear relation between porosity and depth, but, self-evidently, a linear relationship cannot hold at large depths since porosities would become negative. Estimates of porosity from borehole logs (such as the sonic log) for a wide range of different lithologies suggest that normally pressured sediments exhibit an exponential reduction of porosity with depth. There are a number of different formulations of this exponential relationship, with different pre-exponential factors and exponents. Strong deviations from the expected porosity-depth curve are found in overpressured units.

Information on changing paleobathymetry through time may come from sedimentary facies and distinctive geochemical signatures, but principally from micropaleontological studies. Benthic microfossils are especially useful. Eustatic corrections are hazardous to apply, and a simple transferral from the Vail-Haq curve is not recommended. Correction for the widely accepted long-term eustatic variation of sea level is probably justified.

The sediment deposited in a marine basin replaces water, and so drives further subsidence of the basement. The exercise of partitioning the subsidence due to tectonics, and that due to sediment loading, is termed *backstripping*. If the lithosphere is in local Airy isostasy, the decompacted subsidence, corrected for paleobathymetric and eustatic variations, can be simply used to calculate the tectonic component. This requires the average bulk density of the sediment column as a function of time to be calculated. However, if the lithosphere supports the sediment load by a regional flexure, the separation of the tectonic and sediment contributions is more complex. The flexural loading of the sedimentary basin can be accounted for if both the flexural rigidity and spatial distribution of the sediment load are known.

Sedimentary basins dominated by particular mechanisms of formation have characteristic backstripped subsidence histories. The backstripped tectonic subsidence, therefore, can potentially be inverted to provide vital clues to the thermomechanical behaviour of the lithosphere and of mantle processes. Flexural basins and stretched basins have significantly different subsidence histories.

Basin Analysis: Principles and Application to Petroleum Play Assessment, Third Edition. Philip A. Allen and John R. Allen.
© 2013 John Wiley & Sons, Ltd. Published 2013 by John Wiley & Sons, Ltd.

9.1 Introduction to subsidence analysis

Improvements in the dating of stratigraphic units and in estimates of past depositional water depths, largely brought about by advances in micropaleontology, have allowed the development of quantitative techniques in geological analysis of sedimentary basins. Van Hinte (1978) termed this quantitative approach '*geohistory analysis*'. The first qualitative attempts at plotting subsidence/uplift and paleowater depth as a function of time date back at least to Lemoine's *Géologie du Bassin de Paris* published in 1911. Quantitative geohistory analysis was developed in the 1970s, principally in response to a vastly improved commercial paleontological database. It is now more commonly referred to as burial or subsidence analysis.

Subsidence analysis aims at producing a curve for the tectonic subsidence and sediment accumulation rates through time. In order to do this, three corrections to the present stratigraphic thicknesses need to be carried out:

1. *Decompaction*: present-day stratigraphic thicknesses must be corrected to account for the progressive loss of porosity with depth of burial.
2. *Paleobathymetry*: the water depth at the time of deposition determines its position relative to a datum (such as present-day sea level).
3. *Absolute sea-level fluctuations*: changes in the paleo-sea level relative to today's also needs to be considered (see §8.2.2).

Having made these corrections, comparisons between boreholes or other sections are readily made possible. In addition, the subsidence curves give an immediate visual impression of the nature of the driving force responsible for basin formation and development (§9.5).

The time-depth history of any sediment layer can be evaluated, therefore, if the three corrections above can be applied. Such a time-depth history can also be tested from independent methods. These fall into the main classes of organic thermal indicators, thermochronometric indicators and mineralogical thermal indicators, all of which are discussed in §10.4.

The addition of a sediment load to a sedimentary basin causes additional subsidence of the basement. This is a simple consequence of the replacement of water ($\rho_w = 1000\,\mathrm{kg\,m^{-3}}$), or less commonly air, by sediment ($\rho_s \sim 2500\,\mathrm{kg\,m^{-3}}$). The total subsidence is therefore partitioned into that caused by the tectonic driving force and that due to the sediment load. The way in which this partitioning operates depends on the isostatic response of the lithosphere (Appendices 2 and 6; §2.1.5). The simplest assumption is that any vertical column of load is compensated locally (Airy isostasy). This implies that the lithosphere has no strength to support the load. Alternatively, the lithosphere may transmit stresses and deformations laterally by a regional flexure. The same load will therefore cause a smaller subsidence in the case of a lithosphere with strength sufficient to cause flexure. The technique whereby the effects of the sediment load are removed from the total subsidence to obtain the tectonic contribution is called *backstripping*. Backstripped subsidence curves are useful in investigating the basin-forming mechanisms (§9.5).

Burial history (Chapter 9) and thermal history (Chapter 10) can be used to determine the oil and gas potential of a basin and to estimate reservoir porosities and permeabilities. Burial history curves from a number of locations can also be used to construct paleostructure maps at specific time slices. Combined with information on thermal maturity, this can be a powerful tool in evaluating the timing of oil migration and likely migration pathways in relation to the development of suitable traps (see Chapter 11 for fuller discussion).

9.2 Compressibility and compaction of porous sediments: fundamentals

The compressibility of elastic rocks was introduced in §2.1, and further information is given in Appendix 54. Here, we focus on the compressibility of the *porous* rocks and sediments that comprise the basin-fill. Sediments turn into sedimentary rocks by a process of consolidation, which involves compaction of the solid framework and the occlusion of pore space. A layer of fine-grained sediment will compact considerably, even under its own weight, or with relatively small overburdens. The compaction of fine-grained deltaic sediments at major river mouths is an example. We therefore firstly consider some general principles of compressibility and compaction that derive principally from the field of soil mechanics. We then shift emphasis to the long-term burial history of sedimentary basins involving much greater depths and pressures.

Progressive burial of sediment by overlying layers during basin evolution causes a number of physical and chemical changes to the basin-fill. In this chapter we are primarily concerned with the loss of porosity, increase in bulk density and decrease in stratigraphic layer thicknesses that accompany burial.

It is initially important to define some interrelated and overlapping terms:

- *Compressibility* refers to the elastic response of a solid material allowing a reduction of volume caused by an increase in pressure or stress. Under an isotropic compression the pressure is related to the fractional volume change by the *bulk modulus K* and its reciprocal the *compressibility β*. A decrease in volume implies an increase in bulk density in the sample or stratigraphic unit. Compressibility is an important concept for the change in thickness of layers of chemical sediments such as salt.
- *Consolidation* is a term usually applied to soils and young sediments and refers to the decrease in volume by a loss of water under static loading, which results in an increase in strength. Soils follow a consolidation line (Appendix 54), which is a relationship between porosity or void ratio (§9.3) and the logarithm of effective stress (§9.2.1).
- *Compaction* is the change in dimensions of a volume of sediment by a reduction of the pore space between a solid framework as a result of loading. The loading commonly takes the form of the gravitational weight of an overlying column of water-saturated sediment. Compaction is assumed to cause a vertical shortening of the sediment layer with no change in the horizontal directions, so is a *uniaxial strain*.
- *Porosity loss* refers to the loss of pore volume that commonly accompanies burial, and may or may not be related to volumetric strains. For example, cementation of a sandstone may result in a loss of porosity but may not affect the volume occupied by the sedimentary rock, and therefore involves no strain.

Compaction and porosity loss are affected by three sets of interrelated processes (Giles 1997 for summary):

1. Mechanical compaction, which is the mechanical rearrangement and compression of grains in response to loading. Mechanical compaction dominates in the cool upper portions of sedimentary basins.
2. Physico-chemical compaction due to processes such as *pressure solution*, which is particularly important in carbonates.
3. Cementation, which involves the filling of pore space by chemical precipitation, which is related to temperature rather than to loading. Chemical compaction becomes increasingly important in the warm lower portions of sedimentary basins, the cement arresting further mechanical compaction.

For a given lithology, observations show that there is a general exponential reduction in porosity and increase in bulk density with depth. The following sections therefore aim to explain this trend and to highlight important consequences for basin analysis.

The total volume of a sedimentary rock is made of a solid volume and pore volume. During burial the total volumetric strain is therefore made of a change in pore volume and a change in volume of the solid phase. Compaction causes a major reduction in the pore fluid volume accompanied by a small reduction of the solid volume due to compression, whereas cementation increases the solid volume at the expense of the pore fluid volume. Changes in the solid volume during mineral transformation and cementation are commonplace. For example, the transformation of aragonite to calcite causes an increase in 8% by volume, but the dehydration of gypsum to anhydrite causes a reduction of 37.5% in volume, and dehydration reactions in shales such as the illitisation of K-feldspar and kaolinite causes a volume decrease of 9.6%. This means that the common starting point to the study of the loss of porosity during burial, the assumption that the solid volume remains constant (since the compressibilities of many rock-forming minerals are small) (Schneider *et al.* 1993) is prone to error.

9.2.1 Effective stress

General relationships for the burial of sedimentary layers can be obtained using basic principles of soil mechanics. For a water-saturated clay, for example, the overlying weight on a layer of sediment is supported jointly by the fluid pressure in the pores and the grain-to-grain mechanical strength of the clay aggregates. The *effective stress* is the combination of the vertical stress and the fluid pressure:

Effective stress σ' = vertical compressive stress σ − fluid pressure p

$$[9.1]$$

This well-known relationship is known as *Terzaghi's law* (Terzaghi 1936; Terzaghi & Peck 1948). As the amount of gravitational compaction increases, the effective stress also increases. Since the vertical compressive stress σ is determined by the weight of the overlying water-saturated sediment column,

$$\sigma = \bar{\rho}_b g y \qquad [9.2]$$

(where $\bar{\rho}_b$ is average water-saturated bulk density, g is acceleration due to gravity, and y is depth), this increasing vertical load must be divided between p and σ.

If the ratio of fluid pressure to overburden pressure $p/\sigma = \lambda$, it must vary between zero where the fluid pressure is non-existent, to

1 where the sediment layer is effectively 'floating' on the highly pressured pore-filling fluid. We can therefore write

$$p = \lambda \sigma = \lambda \bar{\rho}_b g y \qquad [9.3]$$

and

$$\sigma' = (1 - \lambda)\bar{\rho}_b g y \qquad [9.4]$$

As the sediment load is increased, the extra vertical stress is taken up initially by an increase in pore fluid pressure so that both p and λ increase. With time, however, water is expelled from the pores, reducing fluid pressures but increasing effective stresses. The increase in σ' results in compaction of the grains supporting the stresses, reducing porosity. The lowest pressure that the pore fluid can attain is that due to the hydrostatic column, that is

$$p = \rho_w g y \qquad [9.5]$$

in which case

$$\lambda = \frac{\rho_w}{\bar{\rho}_b} \qquad [9.6]$$

This is called the normal pressure since the pore fluids are at a pressure equivalent to the head of a static body of water and the grain-to-grain contacts are supporting the stratigraphic layer. If $\lambda > \rho_w/\bar{\rho}_b$, the pore fluids are at a higher pressure than hydrostatic.

Assuming that water is free to be expelled from the sediment pore space and is not trapped, increasing burial should lead to the equilibrium state where p is hydrostatic. If this is the case, from eqns [9.4] and [9.6]

$$\sigma' = (\bar{\rho}_b - \rho_w) g y \qquad [9.7]$$

which states that the magnitude of the stress causing compaction is a function of depth and the difference between the water-saturated sediment and water densities.

Drawing on soil mechanics and reservoir engineering literature (Burland 1990; Zimmerman *et al.* 1986; Audet & McConnell 1992), the effective vertical stress is related to porosity or *void ratio e* (Fig. 9.1) (see also Appendix 54, eqn. [A54.3]).

Sedimentary layers are capable of compacting under the action of gravity on their mass. The gravitational compaction of fine-grained sediments is particularly important in delta regions where suspended sediment accumulates rapidly. It can be estimated from a consolidation line (Fig. 9.1 and Appendix 54), assuming hydrostatic conditions. The vertical effective stress is the integration of the specific weight of the porous sediment over depth. Using $C = 0.2$, $e_0 = 1.5$, $\sigma'_0 = 1\,\text{kPa}$, and $(\rho_s - \rho_f)g = 10^4\,\text{Pa m}^{-1}$ in eqn. [A54.3], the reduction of porosity with depth and as a function of effective vertical stress is shown in Fig. 9.2. We further develop ideas on porosity-depth relationships in the following sections.

9.2.2 Overpressure

The Terzaghi law is a statement of the importance of fluid pressures, but it is not an explanation of the physical processes leading to normally pressured and overpressured sediments. Audet and McConnell

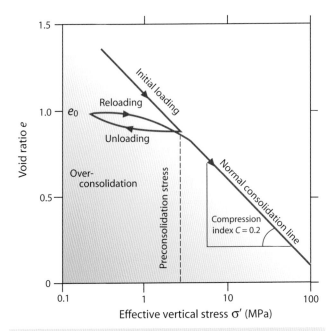

Fig. 9.1 The logarithm of effective vertical stress versus the void ratio, as obtained from an oedometer test (a 1-D compression test), showing the normal consolidation line with a surface void ratio $e_0 = 1$ (porosity is 0.5) and a compression index C of 0.2. The compression index is the slope of the straight normal consolidation line. The stress that the sample has previously been subjected to is the pre-consolidation stress. The path followed during initial loading, unloading and reloading of a compacted clay-rich soil is shown by the over-consolidation red line. From Wangen (2010, fig 4.2). © Magnus Wangen 2010, reproduced with the permission of Cambridge University Press.

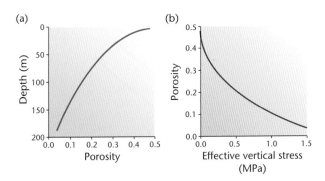

Fig. 9.2 (a) Porosity as a function of depth for water-saturated sediment that follows the normal consolidation line (Fig. 9.1). (b) Porosity of the same sediment as a function of the effective vertical stress. From Wangen (2010, p. 92, fig 4.3). © Magnus Wangen 2010, reproduced with the permission of Cambridge University Press.

(1992, 1994) proposed a dimensionless *sedimentation parameter* α (Fig. 9.3) given by

$$\alpha = \frac{k_1}{V_0}\left(\frac{\rho_s}{\rho_f}-1\right) \qquad [9.8]$$

where k_1 is the hydraulic conductivity of the sediments (typical values of 10^{-7} to 10^{-9} mm s^{-1}, Olsen 1960), which is dependent on permeabil-

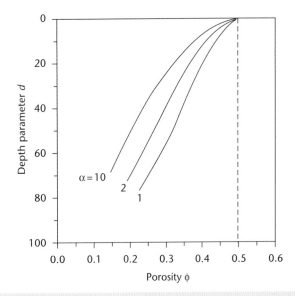

Fig. 9.3 Effect of the sedimentation parameter α on the porosity-depth curve. Porosity is shown versus a dimensionless depth for a time corresponding to 20 Myr after the start of gravitational compaction. For the parameter values used to calculate the porosity-depth curves, $d = 10$ corresponds to c.1 km burial depth, and maximum burial depths are 7–8 km. Note that at high values of the sedimentation parameter α, the curve is 'normal' or hydrostatic, with increasing effects of overpressuring at lower values of α. The initial porosity ϕ_0 is 0.5. After Audet and McConnell (1994), reprinted with permission from Elsevier.

ity, fluid viscosity and density, as well as clay mineralogy, V_0 is the time-averaged sedimentation rate (typical values of 0.1 to 1 mm yr^{-1}), and ρ_s and ρ_f are the sediment and pore fluid densities respectively (typically 2700 and 1050 kg m^{-3}). The sedimentation parameter is the ratio of the rate at which pore fluid moves through the sediments versus the rate at which new sediment accumulates as an overburden at the depositional surface. Consequently, if $\alpha \gg 1$, fluid is expelled from the compacting sediment relatively fast, leading to normally pressured pore fluids. In contrast, if $\alpha \ll 1$, pore fluid is retained in the compacting sediment, probably caused by fast sediment accumulation rates, leading to *overpressured* pore fluid.

Porosity reduction normally involves the expulsion of pore fluids, but a pressure gradient is required to drive the expulsion. Overpressure is therefore a basic requirement of the efficient expulsion of pore fluids from a volume of sedimentary rock. From eqn. [9.8], overpressures should be particularly developed in stratigraphic intervals that were rapidly deposited.

To develop this discussion further, we make use of two expressions, one for the Darcy velocity (see §9.3) and the other for the pressure gradient in a transformed coordinate system:

$$v_D = -\frac{k}{\mu}\frac{\partial p}{\partial y} \qquad [9.9]$$

and

$$\frac{\partial p}{\partial y} = (1-\phi)\frac{\partial p}{\partial \zeta} \qquad [9.10]$$

where p is the overpressure, e is the void ratio, v_D is the Darcy velocity, ζ is the transformed vertical coordinate, such that $dy = d\zeta/(1 - \phi)$ (see §9.3.3 and eqn. 9.25), k is the Darcy permeability, and μ is the fluid viscosity. The transformed depth ζ is a net (solid) depth or thickness with porosity removed. At the base of the sedimentary basin we can assume that there is no flow of pore fluid at all, as pore water is sealed into the very low-permeability rocks. We can also assume that the maximum velocity is at the surface through the least compacted sediments. But the basin also experiences a sedimentation rate, denoted V_0. Solution for the Darcy flux, assuming the porosity to vary with the porosity-free depth ζ (eqns [9.9] and [9.10]) shows that the maximum velocity at the surface is identical to the sedimentation rate $V_0/(1 - \phi)$, that is, pore fluid is buried with the sediments everywhere in the basin (Wangen 2010, p. 389) and cannot be expelled quicker than it is incorporated as new pore fluid.

The basic equation for overpressure build-up can be solved using a function for void space decreasing with depth, measured as porosity-free rock, similar to the exponential reduction of porosity with depth explained in §9.3.2. The void space therefore decreases as a function of the thickness of porosity-free rock as follows:

$$e = e_0 \exp\left(-\frac{\zeta^* - \zeta}{\zeta_0}\right) \qquad [9.11]$$

where e is the void ratio, e_0 is the void ratio at the surface, ζ^* is the total basin thickness in the porosity-free coordinate, ζ_0 is the depth constant for the reduction of porosity in the ζ-depth coordinate, and $\zeta^* - \zeta$ is the porosity-free depth from the surface. The 'real' depth (the depth including porosity) is given by

$$y(\zeta) = (\zeta^* - \zeta) + \zeta_0(e_0 - e) \qquad [9.12]$$

where e is the void ratio at the given position in the ζ-depth coordinate. If the surface porosity is 0.5 ($e_0 = 1$), and ζ_0 is 2500 m in a basin of porosity-free depth (ζ^*) of 10,000 m, the real depth of the base of the sedimentary basin is 12,454 m, using eqns [9.11] and [9.12].

The overpressure for a basin experiencing a constant rate of sediment deposition, with a void ratio given by eqn. [9.11] (Wangen 2010, p. 392) depends on an exponent of the permeability (typically $n = 3$) and inversely on the gravity number N_g

$$N_g = \frac{k_0 \Delta\rho g}{\mu V_0} \qquad [9.13]$$

where $\Delta\rho$ is the density difference between sediment and pore fluid, k_0 is the surface permeability, μ is the fluid viscosity, and V_0 is the surface sedimentation velocity (Fig. 9.4). Most importantly, overpressure is proportional to deposition rate and inversely proportional to surface permeability (Fig. 9.5). The overpressure increases with depth and time, which potentially leads to an exceedance of the lithostatic pressure at depths of greater than about 2 km. *Hydrofracturing* acts as a valve to reduce the overpressure by increasing the permeability of sedimentary rocks.

9.3   Porosity and permeability of sediments and sedimentary rocks

Sediments and sedimentary rocks are composed of solid framework particles and pore space occupied by pore fluid and pore-filling minerals that have grown since the time of deposition. The pore space

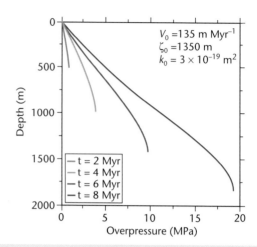

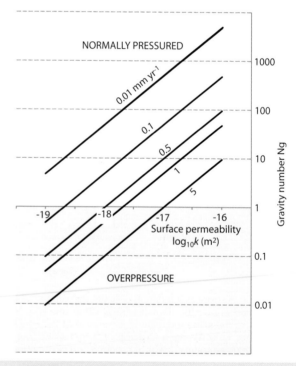

Fig. 9.4   The overpressure versus depth in a basin as a function of time, for a surface net sedimentation rate V_0 of 0.135 mm yr^{-1}, depth decay constant for the downward reduction in void ratio, ζ_0, of 1350 m, and a surface permeability k_0 of 3×10^{-19} m^2, giving a gravity number N_g of 1. The overpressure approaches and exceeds lithostatic after 6–8 Myr (where the characteristic time is 10 Myr), potentially leading to hydrofracturing. From Wangen (2010, fig 12.5b). © Magnus Wangen 2010, reproduced with the permission of Cambridge University Press.

Fig. 9.5   Variation of the gravity number N_g for different values of the surface permeability k_0 and sedimentation rate V_0. At $N_g \ll 1$, overpressures are expected to build up.

controls the total amount of fluid stored in a sediment or sedimentary rock, and the size and arrangement of the pores controls the ability of the rock to transmit fluids by a slow permeable flow.

The volume fraction of void space in a porous medium is the porosity ϕ. The volume of void space relative to the solid volume of a sample is the void ratio e. The two are related by

Table 9.1 Porosity and porosity-depth data for common lithologies (Sclater & Christie 1980; Hansen 1996 in brackets, from sonic log porosities)

Lithology	ϕ_0	y_0 (m)	c (km^{-1})
Shale	0.63 (0.71)	1960 (1961)	0.51 (0.51)
Sandstone	0.49	3703	0.27
Chalk	0.70	1408	0.71
Shaly sandstone	0.56	2464	0.40

$$\phi = \frac{e}{e+1}$$
$$e = \frac{\phi}{1-\phi}$$

[9.14a,b]

If porosity is 0.5, the void space is the same as the solid volume, so $e = 1$.

Pores may be connected with each other, allowing flow, or they may be unconnected. The term *effective porosity* is used to describe the porosity of connected pores. Porosity commonly reduces with depth of burial and is modelled as a negative exponential (Athy 1930)

$$\phi(y) = \phi_0 \exp(-y/y_0)$$

[9.15]

where ϕ_0 is the surface porosity, and y_0 is a characteristic depth at which the porosity has diminished to $1/e$ of its surface value. The typical values of the depth constant y_0 for common lithologies in the North Sea were given by Sclater and Christie (1980), who expressed y_0 as its reciprocal in km^{-1}, denoted c (Table 9.1). Porosity-depth measurements are discussed in §9.3.2.

The flux of fluid (volume per cross-sectional area per time) v_D flowing through a porous medium of length l in response to a pressure difference Δp is the well-known Darcy's law

$$v_D = -\frac{k}{\mu}\frac{\Delta p}{l}$$

[9.16]

where k is the permeability and μ is the fluid viscosity. The typical permeabilities of common rock types are given in Fig. 9.6. The Darcy flux can also be expressed as an average fluid velocity

$$v_D = \phi v_f$$

[9.17]

since fluid only flows through a fraction of the cross-sectional area of the sample given by the porosity. Taking the pressure difference across a 20 cm-long sediment sample of 5 MPa, a permeability of 1 mD (1 Darcy x 10^{-3} or 0.986923 × 10^{-15} m^2), and fluid viscosity of 10^{-3} Pa s, the Darcy flux is 0.025 mm s^{-1} and the average fluid velocity is 0.01 mm s^{-1}. This is equivalent to 36 mm hour^{-1} or ~300 m yr^{-1}.

Since permeability is of such profound importance for the flow rate of a fluid through a porous medium, and permeability is affected by the size and arrangement of pores, it would be very valuable to have a simple model for permeability as a function of porosity. Such a model for well-sorted sands is described in Appendix 55. There is commonly, however, very wide variation between porosity and permeability in subsurface rocks. Log$_{10}$ permeability versus linear porosity expressions are commonly used for a range of different lithologies (Fig. 9.7).

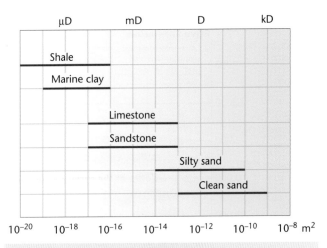

Fig. 9.6 Typical permeabilities of sediments and rocks. From Wangen (2010, p. 13, fig 2.9). © Magnus Wangen 2010, reproduced with the permission of Cambridge University Press.

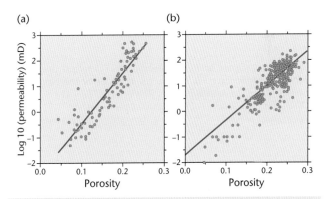

Fig. 9.7 Permeability in log$_{10}$ values versus porosity (after Wangen, 2010, Fig. 2.12), with least squares linear best fits, data from Bloch *et al.* (2002) (a) and Dutton *et al.* (2003) (b). © Magnus Wangen 2010, reproduced with the permission of Cambridge University Press.

9.3.1 Measurements of porosity in the subsurface

As a prerequisite for a further discussion of compaction and porosity loss during burial, it is important to have some understanding of how porosity can be estimated in the subsurface. Porosity can be directly measured on core and sidewall cores recovered from a borehole, but such direct measurements tend to be concentrated on zones of known economic interest, such as reservoir intervals (see Fig. 9.8). The distribution of porosity with depth in a borehole must therefore be obtained by remote methods, principally from the interpretation of downhole electrical logs. For example, sonic, neutron and density logs are sensitive to lithology and porosity.

The *sonic log* is a recording of the time taken (interval transit time, Δ_t) for a compressional sound wave emitted from a sonic sonde to travel across 1 ft (c.25 cm) of rock surrounding the borehole to a receiver. The interval transit time is the inverse of velocity, which is a function of lithology and porosity. When lithology is known from other data such as a drill cuttings log, Δ_t may be used to calculate porosity from the *Wyllie time-average equation*:

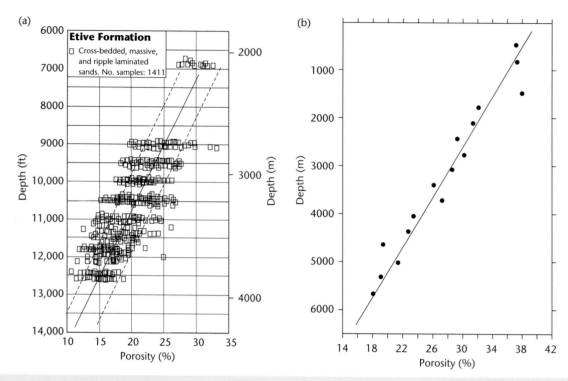

Fig. 9.8   Examples of a linear relationship between porosity and depth. (a) Cross-bedded, massive and ripple-laminated sandstones from the Jurassic Etive Formation, Brent Group of the North Sea, from Giles (1997). Dashed lines give the 90% confidence interval for porosity about the regression line. Plot is based on 1411 samples. Reproduced with kind permission of Springer Science+Business Media B.V. (b) Tertiary sands, southern Louisiana, based on over 17,000 cores, averaged every 1000 ft (c.300 m), after Blatt (1982). The reduction in sand porosity with depth is thought to be due to the compaction of ductile rock fragments.

$$\Delta_t = \Delta_{t\,ma}(1-\phi) + \phi(\Delta_{t\,f}) \qquad [9.18]$$

where $\Delta_{t\,ma}$ is the transit time through the solid rock matrix (51.3–55.5 ms/ft for sandstones, 43.5–47.6 ms/ft for limestones), and $\Delta_{t\,f}$ is the transit time through the pore space fluid (dependent on fluid salinity, 189 ms/ft for freshwater). Rearranging eqn. [9.18], a solution is obtained for porosity:

$$\phi = \frac{\Delta_t - \Delta_{t\,ma}}{\Delta_t - \Delta_{t\,f}} \qquad [9.19]$$

It has been found that in uncompacted, geologically young sands, the Wyllie time-average equation needs to be corrected by a compaction factor.

The principle of the *density log* is that gamma rays are emitted by a radioactive source in the logging tool, and are scattered and lose energy as a result of collisions with electrons in the rock surrounding the borehole. The number of scattered gamma rays recorded at a detector, also on the logging tool, depends on the density of electrons in the formation. This electron density is virtually the same as the formation bulk density for most common minerals, although for some evaporite minerals such as rock salt and sylvite (KCl), and coal, there is a significant difference.

The *formation density* (FDC) sonde is normally calibrated in freshwater-filled limestone formations, so that the log reads the actual bulk density for limestone and for freshwater. The bulk density is a function of the average density of the substances making up the

rock surrounding the borehole, that is, both the rock matrix and fluid-filled pores, and the relative volumes occupied, as shown by the equation

$$\rho_b = \phi\rho_f + (1-\phi)\rho_{ma} \qquad [9.20]$$

where ρ_f is the average density of the fluid occupying the pore space, which is a function of temperature, pressure and salinity, and ρ_{ma} is the average density of the rock matrix. By rearranging eqn. [9.20] we very easily arrive at a means to determine porosity:

$$\phi = \frac{\rho_{ma} - \rho_b}{\rho_{ma} - \rho_f} \qquad [9.21]$$

ρ_{ma} may itself be decomposed into its constituent parts. When clay minerals are present, a correction that accounts for the density and amount of clay may be needed in order to avoid inaccurate porosity determination,

$$\rho_b = \phi\rho_f + V_{clay}\rho_{clay} + (1-\phi-V_{clay})\rho_{ma} \qquad [9.22]$$

where ρ_{clay} is the average density of the clay, and V_{clay} is the fraction of the total rock occupied by clay. The clay correction is substantial when ρ_{ma} and ρ_{clay} are very different. This is usually at shallow depths where the clay is particularly uncompacted. Hydrocarbons may also lower the density log recording. Generally, the presence of oil in the invaded zone pore space has a negligible effect on the density log, but residual gas has a considerable effect that must be corrected for.

The principle of the *neutron log* is that neutrons emitted from a radioactive source collide with nuclei in the rock surrounding the borehole and are captured. A detector, or detectors, counts the returning neutrons. Because neutrons lose most energy when they collide with a hydrogen nucleus in rocks surrounding the borehole, the neutron log gives a measure of its hydrogen content, that is, its liquid-filled content. It can be used, therefore, to determine porosity. The most commonly used neutron tool is the *compensated neutron log* (CNL). The CNL is calibrated to read true porosity in clean limestones. Its unit of measurement, as presented on the log, is therefore the 'limestone porosity unit'. When measuring in lithologies other than limestone, for example in a quartzose sandstone, a correction needs to be made.

The neutron log responds to *all* the hydrogen present in the formation, including the water bound up in clay minerals. As a result, it is very sensitive to clay content. In shales, the neutron log normally shows a very high reading, and needs correction. Although oil has a hydrogen content close to water and has only a small effect on the CNL, the hydrogen content of gas is considerably lower. As a result, the neutron response is low in rocks that contain gas within the depth of investigation of the tool (generally <30 cm).

Of all the porosity tools, the sonic log is most widely used (e.g. Magara 1976; Ware & Turner 2002). This is largely because the neutron and density logs are normally only run in the deeper zones of hydrocarbon interest in a borehole, and the sonic log may therefore be the only porosity log available in the shallow sections of the borehole. Porosity from the sonic log can be determined if the lithology is known. The most common method of obtaining porosity from the density and neutron logs is to cross-plot them.

Precise porosity determinations can be made on samples of the formation penetrated by a borehole. During a logging run, small sidewall cores can be obtained by firing a sleeved 'bullet' into the formation. Intervals of reservoirs that are cored provide more promising material for porosity and permeability determination. Samples are drilled out of the slabbed core and analysed in the laboratory. Porosity trends with depth can then be obtained from a particular geological province if the same reservoir is cored at a wide range of different depths.

9.3.2 Porosity-depth relationships

Observations on the porosity-depth relationship of sedimentary rocks come from rock deformation experiments and from measurements from natural subsurface settings.

Rock deformation experiments show that compaction rate is a strong function of mineralogy. Clays compact easily and quartz-rich lithologies compact relatively slowly. There are important deviations from this behaviour. For example, sandstones with ductile lithic clasts compact more quickly than the linear relationship between log porosity and effective stress.

Direct measurements on core material and remote measurements from downhole logging devices such as the sonic log provide an enormous database on the relation between present depth and porosity (Fig. 9.9). A number of factors affect the porosity-depth relationship, chief of which are: (i) gross lithology, shales compacting quickly compared to sandstones; (ii) depositional facies, which controls grain size, sorting and clay content and therefore initial (surface) porosity; (iii) composition of framework grains: for example, pure quartz arenites differ from lithic arenites containing ductile fragments; (iv) temperature strongly affects chemical diagenesis, such as quartz

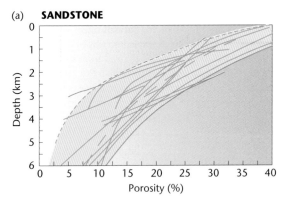

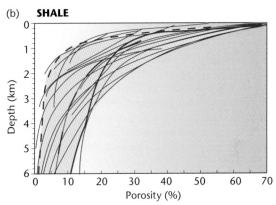

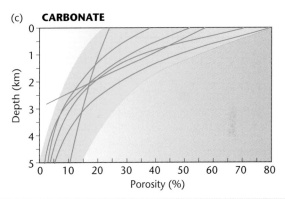

Fig. 9.9 Compilation of porosity-depth curves for sandstones (a), shales (b) and carbonates (c). Sources of datasets in Giles (1997). Note that shales compact early compared to sandstones. The porosity-depth relation for carbonates varies according to grain types and amount of cementation. Reproduced with kind permission from Springer Science+Business Media B.V.

cementation, clay mineral growth and pressure solution; and (v) time: porosity loss may require sufficiently long periods of time.

The simplest trend recognised between porosity and depth is a linear trend of the form

$$\phi = \phi_0 - ay \qquad [9.23]$$

where ϕ and ϕ_0 are the porosity at depth y and the initial porosity respectively, and a is an empirically derived coefficient. This linear relationship appears to fit data carefully chosen from specific sedimentary facies within a certain stratigraphic unit of given geological age from one basin, for example, cross-stratified, massive and ripple cross-laminated sandstones of the Jurassic Etive Formation of the North Sea (Fig. 9.8a). In this example, the slope of the linear regression gives $a = 0.09\,km^{-1}$. The values derived from sonic log data from shales in the Norwegian sector of the North Sea (Hansen 1996) was $\phi_0 = 0.62$, and $a = 0.00018\,km^{-1}$. However, a linear relationship self-evidently cannot apply at large depths, since porosities would have to become negative. A more widely used porosity-depth relation therefore has the form of a negative exponential (Athy 1930; Hedberg 1936), which produces an asymptotic low porosity with increasing depth.

For normally pressured sediments, the variation of porosity ϕ with depth y can be expressed in a slightly modified form of the Athy relationship in eqn. [9.15]

$$\phi(y) = \phi_0 \exp(-cy) \qquad [9.24]$$

where c is a coefficient determining the slope of the ϕ-depth curve, y is the depth, and ϕ_0 is the porosity at the surface (Fig. 9.10). In other words, the surface porosity declines to $1/e$ of its original surface value at a depth of $1/c\,km$. On a depth versus log porosity graph, the value of c is the inverse of the rate of change of porosity with depth. The coefficient c can therefore be estimated if a number of porosity measurements can be made, for example from a sonic log from a representative borehole in the basin. This relationship has been applied to a range of different lithologies, each with its own value of c (Sclater & Christie 1980; Halley and Schmoker 1983; Hansen 1996) (Table 9.1).

Porosity-depth curves, such as those in Fig. 9.9, and in large databases (Baldwin & Butler 1985), are based on a statistical analysis of large numbers of samples derived from normally pressured as well as overpressured lithologies that have been subjected to differing amounts of cementation and from different geothermal gradients (Dzevanshir *et al.* 1986; Scherer 1987). We have already seen that porosity reduction resulting from compaction is best viewed as related to the logarithm of effective stress. This may explain a lot of the scatter in porosity-depth data (Gluyas & Cade 1997). Although potentially useful, these large regressions lack any underlying physical principles.

The evolution of porosity with depth can also be investigated in a physically more meaningful way by using Audet and McConnell's (1992, 1994) sedimentation parameter α. Numerical solution of the equation governing the evolution of porosity as a function of depth below the depositional surface y and time t, as a function of the sedimentation parameter, shows that clay strength (compressibility) and permeability (hydraulic conductivity) strongly affect overpressure and the evolution of porosity in a sedimentary basin. The curve of porosity versus depth is strongly affected by the sedimentation parameter (Fig. 9.3). As α varies from 1 to 10, the sediment becomes more compacted at a given depth, shown by a dimensionless depth parameter d. This is because at $\alpha = 10$ pore fluid is easily expelled, leading to near hydrostatic pore fluid pressures and significant compaction, whereas when $\alpha = 1$ pore fluid motion is inhibited, causing reduced porosity loss in the sediment.

In summary, there is a proliferation of porosity-depth relations, each based on a similar principle of porosity destruction under the increasing effective stresses experienced during burial, but differing in detail. It seems likely that each 'standard' porosity-depth curve may be valuable if carefully calibrated and then used only within the region of calibration. For example, coefficients in the expressions should be evaluated for carefully selected porosity data in which grain size/sorting, composition, geothermal gradient and geological age are constrained.

9.3.3   Porosity and layer thicknesses during burial

To calculate the thickness of a sediment layer at any time in the past, it is necessary to computationally remove overlying sediment layers and allow the layer of interest to decompact, following a chosen porosity-depth relationship. In so doing, we keep solid mass constant and monitor the changes in total volumes (including porosity) and therefore stratigraphic thicknesses (Appendix 56).

One procedure is to transform the vertical coordinate into a Lagrangian coordinate so that a chosen horizon does not move in the new coordinate system over time (see Wangen 2010, p. 95), as introduced in §9.2.2. We call this Lagrangian vertical coordinate ζ. The relationship between the ζ coordinate and the 'real' depth coordinate y (Fig. 9.11) is

$$d\zeta = (1 - \phi)dy \qquad [9.25]$$

so $d\zeta$ is the solid (porosity-free) volume in the vertical interval dy. The ζ coordinate is measured from the base of the sedimentary basin or the base of a chosen sedimentary layer, whereas the y coordinate

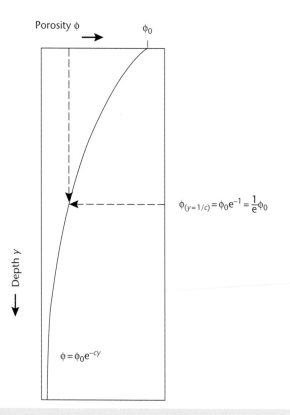

Fig. 9.10   Schematic diagram illustrating the use of the porosity-depth coefficient c. If a porosity-depth curve is known and it is exponential, c can be found by determining the depth at which the porosity ϕ has decreased to $1/e$ of its surface value ϕ_0. This should be repeated for all lithologies.

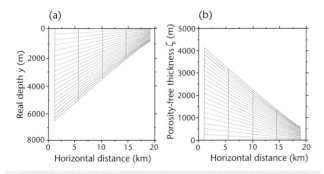

Fig. 9.11 The Lagrangian coordinate system for considering burial in a sedimentary basin. A 2-D basin in the normal y-coordinate system (a) compared with the basin in terms of porosity-free depths measured from the basement (b). See also Fig. A56.3. During burial, the porosity-free coordinate system travels with the subsiding basement ($\zeta = 0$), whereas the porosity-free distance to a stratigraphic horizon ζ_i, or the distance to the surface of the basin ζ^*, increases. Modified from Wangen (2010, fig. 3.17). © Magnus Wangen 2010, reproduced with the permission of Cambridge University Press.

is measured downwards from the surface of the basin, such as from sea level.

When expressed in the ζ coordinate system, the exponential porosity-depth relationship of Athy (1930) is modified to

$$\phi = (\phi_0 - \phi_{min})\exp\left(-\frac{(\zeta^* - \zeta)}{\zeta_0}\right) + \phi_{min} \qquad [9.26]$$

where ϕ_0 is the surface porosity, ϕ_{min} is the porosity at great depth, ζ is the current total porosity-free thickness of the basin, and ζ_0 is the characteristic depth constant in the ζ coordinate system, analogous to y_0 in eqn. [9.15]. The 'real' depth of the chosen horizon below the top surface of the basin is then given by

$$y = \frac{1}{1 - \phi_{min}}\left\{(\zeta^* - \zeta) + \zeta_0 \ln\left(\frac{1 - \phi}{1 - \phi_0}\right)\right\} \qquad [9.27]$$

from which it can be seen that the real depth y is a function of the ζ-depth below the surface, equal to $\zeta^* - \zeta$.

The layer thickness is

$$dy = \frac{1}{1 - \phi_{min}}\left\{d\zeta + \zeta_0 \ln\left(\frac{1 - \phi_1}{1 - \phi_2}\right)\right\} \qquad [9.28]$$

where dy is $y_2 - y_1$, $d\zeta = \zeta_2 - \zeta_1$, and the subscripts 1 and 2 refer to the top and bottom boundaries of the chosen stratigraphic unit.

Let us take parameter values as follows: for the entire basin-fill we assume the same lithology with $\phi_0 = 0.5$, $\phi_{min} = 0.03$ and $\zeta_0 = 1350\,m$; the total porosity-free basin thickness is $\zeta^* = 3700\,m$. The layer thickness for the entire basin-fill using $d\zeta = 3700\,m$ in eqn. [9.28] is 4692 m. In other words, a solid volume of height 3700 m becomes a total volume, including pore space, of 4692 m.

Now let the basin-fill total thickness increase over time as a result of a steady accumulation of sediment at the rate of $V_0\,m\,Myr^{-1}$. Consequently, $\zeta^* = V_0 t$, where t is the time since the start of subsidence. Despite the fact that the sediment accumulation rate is constant, the

burial curve (real depth versus time) is concave-up, due to the effects of progressive compaction. This forms the basis for the calculation of burial history for a basin-fill of variable lithology under an unsteady sediment accumulation rate.

The decompacted depth of a stratigraphic layer at any time t can be written in terms of the void ratio

$$s(t) = \sum_i \{1 + e_i(t)\}d\zeta_i \qquad [9.29]$$

where i is the number of stratigraphic layers in the basin at time t, $e_i(t)$ is the void ratio of the ith layer at time t, and $d\zeta_i$ is the net (porosity-free) thickness of the ith layer. Eqn. [9.29] can be used where the net (porosity-free) thicknesses are known (as in a forward model), or can be calculated from present-day stratigraphic thicknesses (Appendix 57).

The step-by-step procedures in carrying out a decompaction are given in Appendix 56, using the solution given in Sclater and Christie (1980) and summarised in Allen and Allen (2005), followed by a method using the solution in Wangen (2010). Further experience in decompaction can be gained by carrying out the practical exercise that takes you from stratigraphic thicknesses through to paleotemperatures (Appendix 58).

The sources of the data points for the burial history curve are the stratigraphical boundaries of presumed known age defining stratigraphical units of known present-day thickness. All depths are, however, in relation to a present-day datum, normally taken as mean sea level. Consequently, it is necessary to correct the decompacted subsidence curve for firstly the difference in height between the depositional surface and the regional datum (*paleobathymetric correction*), and, secondly, for past variations in the ambient sea level compared to today's (*eustatic correction*). Finally, the sediment weight drives basement subsidence. In order to calculate the true tectonic subsidence, it is necessary to remove the effects of the excess weight of the sediment compared to water. This is known as *backstripping* (§9.4.1, Appendix 57).

9.4 Subsidence history and backstripping

In the previous section we saw how it was possible to account for the progressive compaction of a sedimentary column during burial in a sedimentary basin. The decompacted sediment thicknesses allow a graph of sediment accumulation rate versus time to be constructed. However, this sediment accumulation rate graph is not the same as the subsidence rate of a chosen geological horizon relative to a constant datum. Nor does it directly give us the magnitude of the tectonic driving force as a function of time. To gain insights into subsidence rate and tectonic driving force we need to make a number of modifications or 'corrections' to the sediment accumulation rate curve derived by decompaction of the stratigraphic column. In this section we perform these modifications on a 1D stratigraphic column. We then briefly consider the use of 2D models.

9.4.1 Backstripping techniques

As a reminder, the total subsidence of the basement underlying a sedimentary basin is made of three parts at any time t: the sediment deposited $y(t)$, the water depth $W_d(t)$, and the change in global sea level $\Delta_{SL}(t)$. We can perform an isostatic balance between a

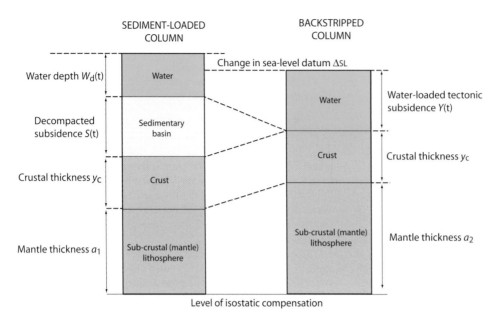

SEDIMENT-LOADED
COLUMN

BACKSTRIPPED
COLUMN

Fig. 9.12 Isostatic balance to illustrate backstripping of sediment load. From Wangen (2010, p. 243, fig 7.30). © Magnus Wangen 2010, reproduced with the permission of Cambridge University Press.

lithospheric column through the sedimentary basin, and a column in which the sediment load has been removed and has been replaced with water (Fig. 9.12) to obtain the tectonic or backstripped subsidence $Y(t)$. The isostatic balance down to a depth of compensation in the ductile asthenosphere is

$$\rho_w g W_d(t) + \bar{\rho}_b g y(t) + \rho_c g y_c + \rho_m g a_1 = \rho_w g Y(t) + \rho_c g y_c + \rho_m g a_2$$

[9.30]

where y_c is the crustal thickness, and a_1 and a_2 are the mantle lithosphere thicknesses before and after removal of the sediment load, and ρ_w, $\bar{\rho}_b$, ρ_c and ρ_m are the densities of water, bulk sediment, crust and mantle respectively.

Global sea-level change $\Delta_{SL}(t)$ changes the datum at the top of the lithospheric column. Equating the thicknesses in the two columns, and removing y_c since crustal thickness does not change, and g since it occurs in each term, we have

$$W_d(t) - \Delta_{SL}(t) + y(t) + a_1 = Y(t) + a_2$$

[9.31]

Combining eqns [9.30] and [9.31] gives the general solution for tectonic subsidence

$$Y(t) = W_d(t) + y(t)\left(\frac{\rho_m - \bar{\rho}_b(t)}{\rho_m - \rho_w}\right) - \frac{\rho_m}{\rho_m - \rho_w}\Delta_{SL}(t)$$

[9.32]

When performing this calculation for many different layers in the basin-fill, it is necessary to know the sea-level change, water depth change, sediment bulk density and decompacted sediment thickness of each component layer. It is best performed by computer, and then repeated for each subsequent time in basin evolution. The chart constructed is the basis for the calculation of paleotemperature (Chapter 10) and hydrocarbon expulsion history (Chapter 11).

A commentary on the parameters required to carry out a full backstripping is provided as follows:

Sediment bulk density: the average bulk density of the basin-fill depends on the bulk densities and thicknesses of the component

stratigraphic units. When the basin is made of i units with different bulk density $\rho_{b,i}$ and thickness y_i, the average bulk density of the basin-fill is simply the summation given by

$$\bar{\rho}_b(t) = \frac{\sum_i \rho_{b,i}(t)\Delta y_i(t)}{\sum_i \Delta y_i(t)}$$

[9.33]

The bulk density of each layer $\rho_{b,i}$ is made of the combination of the density of water filling porosity ρ_w and the density of the solid phase (grains and cements) $\rho_{s,i}$

$$\rho_{b,i}(t) = \phi_i(t)\rho_w + (1 - \phi_i(t))\rho_{s,i}$$

[9.34]

Following previous usage, the average basin-fill bulk density can also be expressed in the porosity-free coordinate system in terms of the void ratio e, since porosity can be modelled as an Athy-type exponential

$$\bar{\rho}_b(t) = \frac{\sum_i (e_i(t)\rho_w + \rho_{s,i})d\zeta_i}{\sum_i (1 + e_i(t))d\zeta_i}$$

[9.35]

The advantage of using the ζ-depth coordinate system is that the only variable is porosity, since solid layer thickness $d\zeta_i$ and the sediment framework density $\rho_{s,i}$ for each layer are constants in the Lagrangian coordinate system.

Assuming the basin to be composed of a uniform lithology, the average basin porosity is the integration of the porosity over the depth range of the entire basin-fill of thickness y, which when evaluated gives

$$\bar{\phi} = \frac{\phi_0 y_0}{y}(1 - \exp(-y/y_0))$$

[9.36]

where y_0 is the depth scale for the exponential reduction of porosity, and ϕ_0 is its surface value. The average bulk density, likewise, is a

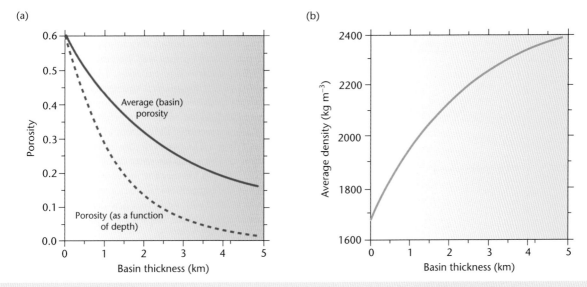

Fig. 9.13 Average basin porosity and porosity as a function of depth based on an Athy-type porosity-depth relationship (a) and average bulk density (b) of the sedimentary basin. From Wangen (2010, p. 245, fig. 7.31). © Magnus Wangen 2010, reproduced with the permission of Cambridge University Press.

modification of eqn. [9.34], using the average porosity from eqn. [9.36] and assuming that the density of the sediment framework (matrix) does not vary, is given by

$$\bar{\rho}_b = \bar{\phi}\rho_w + (1-\bar{\phi})\rho_s \qquad [9.37]$$

The average porosity of the basin-fill and the average bulk sediment density of the basin-fill are shown in Fig. 9.13, using the following parameter values: $\rho_w = 1000\,\mathrm{kg\,m^{-3}}$, $\rho_s = 2650\,\mathrm{kg\,m^{-3}}$, $\phi_0 = 0.6$, and $y_0 = 1350\,\mathrm{m}$. As the basin thickness (or depth) increases, the average porosity of the basin-fill decreases and the average bulk density increases. The same trajectory affects individual layers in the basin-fill, which become less porous, denser and thinner over time during burial.

Paleobathymetric corrections: the estimation of water depth for a given stratigraphic horizon is generally far from easy, yet it is essential in order to accurately study subsidence history. As an example of the potential problems, consider two sedimentary basins A and B (Fig. 9.14). In basin A the water depth is 5 km. If the basement subsides tectonically by 1 km over a certain time period, and during this time sediment is supplied to the basin such that it becomes filled to the brim, local isostasy shows that about 15 km of sediment will have accumulated at the end of the time period. In basin A, therefore, a stratigraphic thickness of 15 km reflects a very small (1 km) driving subsidence. In basin B, however, a tectonic subsidence of 1 km takes place in a basin that has its depositional surface already at sea level. The resultant sediment thickness is barely 3 km at the end of the same time period. Clearly, for the same rate of tectonic subsidence, enormously different stratigraphic thicknesses can result, depending on the initial and ensuing paleobathymetry.

Information on paleobathymetry comes from a number of sources, chief of which are benthic microfossils, other faunal and floral assemblages, sedimentary facies and distinctive geochemical signatures. Without paleontology, it is very difficult to constrain paleobathymetry. A compilation of papers illustrating the use of

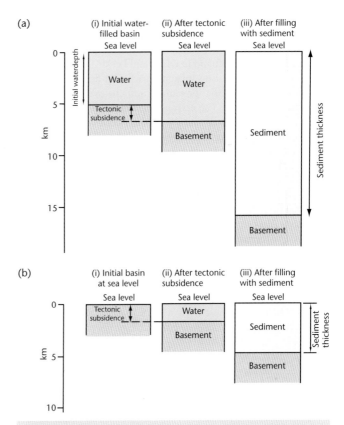

Fig. 9.14 Schematic illustration of the effects of initial water depth on sediment thickness. (a) Sedimentary basin with an initial water depth of 5 km (i) undergoes 1 km of tectonic subsidence (ii) and is filled with sediment to a depth of 15 km (iii). (b) Sedimentary basin initially at sea level (i) undergoes 1 km of tectonic subsidence (ii) and accumulates c.3 km of sediment (iii).

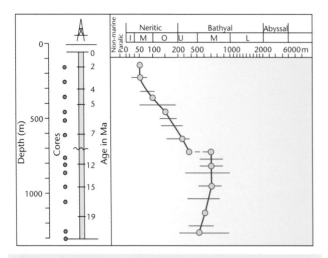

Fig. 9.15   An early example of the use of micropaleontological data to estimate paleobathymetric changes through time (after van Hinte 1978). Dots indicate core data, from which paleobathymetric estimates are made. As much paleontological information as possible is used to guide the choice of the position of the paleobathymetric curve through the depth ranges indicated by the horizontal lines. The mid-points of the ranges need not be used. AAPG © 1978. Reprinted by permission of the AAPG whose permission is required for further use.

micropaleontological data in assessing sea-level change and stratigraphic architectures is found in Olson and Leckie (2003).

Although some organisms inhabit a particular depth range as an adaptation to hydrostatic pressure, most paleodepth estimates are indirectly obtained. For example, qualitative estimates can be obtained by a comparison with the modern occurrences of certain species or assemblages and by recognition of the ecological trends of benthonic and planktonic organisms through time. Estimates can also be obtained by quantitative methods, using ratios of, for example, plankton/benthos, arenaceous/calcareous foraminifera, percentage of radiolarians or ostracods, or, alternatively, using species dominance and diversity, morphological characters and so on. These many techniques allow the paleontologist to make meaningful interpretations of environmental factors such as the chemical environment (salinity, pH, oxygen and CO_2 contents, nutrient availability), the physical environment (temperature, light, energy level, type of substrate, turbidity) and the biological environment. As much information as possible needs to be synthesised to produce a reliable depth estimate. An example given in Fig. 9.15 shows the type of depth information that can be obtained paleontologically, together with the ranges over which the estimate may span.

Sedimentary and geochemical data are, by comparison, far less useful. Sedimentary facies reflect supply and depositional process and are not therefore particularly diagnostic of depth. Although some structures, such as wave ripple marks, are restricted to particular depth ranges (<200 m), and quantitative techniques are available to constrain the depth range at which wave ripples formed (Allen 1984), the sedimentary facies more likely will provide a back-up to the paleontological observations in marine sediments. In continental environments the potential for using sedimentary facies is greater. If shoreline facies can be identified, for example, the calculation of floodplain slopes from fluviatile sediments can give valuable information on heights above sea level. Structures such as desiccation

cracks indicate subaerial emergence, and wave mini-ripples indicate exceptionally shallow water depths. The most obvious geochemical data relate to the carbonate dissolution depth (CCD) below which calcareous material is dissolved. Because most of this material is in the form of calcareous microfossils, it falls within the realm of the paleontologist. Some mineral species such as glaucony and phosphates may provide useful information on paleowater depth, but estimates are likely to be far from precise.

Eustatic corrections: the concept of sea-level change relative to a reference datum is known as *eustasy*, and the absolute, global sea-level variation relative to this reference datum is a eustatic change. We saw in §8.2.2.5 (Fig. 8.21) that any increase (or decrease) of water in the ocean must be compensated isostatically. If an ocean with initial water depth h_1 is filled with water to a new depth h_2, the sea-level rise Δ_{SL} compensated for isostatic depression of the ocean floor is

$$\Delta_{SL} = \left(\frac{\rho_m - \rho_w}{\rho_m}\right)(h_2 - h_1) \qquad [9.38]$$

The new elevation of sea level relative to a sea level datum, caused by an increase in the water column h, is termed *freeboard*, and is given by

$$f = \frac{(\rho_m - \rho_w)h}{\rho_m} \qquad [9.39]$$

Inserting reasonable values for the density terms for an oceanic basin ($\rho_m = 3300 \, \text{kg} \, \text{m}^{-3}$ and $\rho_w = 1030 \, \text{kg} \, \text{m}^{-3}$) gives $f = 0.69h$.

At present, there is no consensus on a global eustatic curve that could be used to make corrections of decompacted subsidence data in order to extract the tectonic subsidence. The 'global' sea-level curve (Haq *et al.* 1987) is thought to be an amalgam of local records (Carter 1998). In some cases, where the tectonic subsidence and paleowater depth history are well constrained, the eustatic signal may be retrievable. Detailed analysis of a number of borehole records from the Atlantic coastal plain in northeastern USA and the adjacent continental slope has used a tectonic model of slow cooling following rifting, to isolate the eustatic component (Miller *et al.* 1998). The long-term eustatic curves for the Cenozoic, which show a long-term lowering of 150–200 m since 65 Ma, are similar to the estimates based on ocean ridge volume changes (Pitman 1978; Kominz 1984), but substantially less than the Haq *et al.* (1987) estimate of a c.300 m lowering (Fig. 8.25). Significant differences between various estimates of eustatic sea level may result from dynamic topography (§8.2.2.4, §8.2.2.5).

Bearing in mind these uncertainties, it is advisable firstly to decompact ignoring any possible global sea-level fluctuations. Following this, the sea-level changes associated with first-order (long-term) cycles can be included, following Kominz (1984), Kominz *et al.* (1998) and Miller *et al.* (1998). It is not advisable to *a priori* invoke the short-term eustatic excursions proposed by Haq *et al.* (1987).

Isostatic models of compensation: we have seen above that the sediment and water load above a horizon of interest in a sedimentary basin causes an isostatic effect so that the total subsidence observed is made of a tectonic driving force component and a sediment/water load component. Watts and Ryan (1976) were the first to propose the isolation of the tectonic driving force by removal of the isostatic effects of the sediment load, and called the technique *backstripping*. Further details are given in Appendix 57.

Since the lithosphere underlying sedimentary basins has flexural strength, it is likely that sediment loads of appropriate wavelength are compensated flexurally rather than in a local Airy fashion. In order to understand the principle of flexural backstripping we return to the problem of the deflection of the lithosphere under a periodic load found in §2.1.5 and Appendix 34. The degree of compensation for a periodic load (eqn. [2.11]) is determined principally by the flexural rigidity D and the wave number $2\pi/\lambda$. The relationship between degree of compensation and wave number is shown in Fig. 2.13. Taking the fill of a sedimentary basin 200 km wide ($\lambda/2 = 200$ km) with a sinusoidal sediment load and an underlying lithosphere of flexural rigidity 10^{24} Nm, $(\rho_m - \rho_s) = 800$ kg m^{-3}, the degree of compensation C is about 0.12. This suggests that the lithosphere behaves very rigidly to this kind of wavelength of load. Changing the wavelength of the load such that $\lambda/2$, the width of the basin is now 400 km, $C = 0.68$, indicating that the sediment load is only weakly supported. In this case of large compensation, Airy-type isostasy is approached.

Flexural backstripping can therefore only be carried out if there is knowledge of the flexural rigidity of the underlying lithosphere and of the spatial distribution of the sediment load. The procedure is normally carried out sequentially on a number of sediment layers, where the isopachs of each layer are known from seismic reflection data. The flexure is then given by

$$Y(k) = \frac{(\rho_s - \rho_w)S(k)}{(\rho_m - \rho_s)}\Phi(k) \qquad [9.40]$$

where $Y(k)$ is the frequency domain equivalent of the flexure, $S(k)$ is the frequency domain equivalent of the sediment thickness, and $\Phi(k)$ is a wave number function equivalent to the compensation C outlined above.

Flexural backstripping can be extended to three dimensions by replacing the wave number by a 3D wave number

$$k = \sqrt{(k_x^2 + k_z^2)} \qquad [9.41]$$

where k_x is the wave number in the x direction, and k_z is the wave number in the z direction.

An example of the use of flexural backstripping from the Valencia Trough of the western Mediterranean is given in Fig. 9.16.

Regional seismic reflection lines allow the structure of the crust and basin-fill to be defined in detail. This allows the subsidence history to be deconvolved into a number of rifting episodes (§3.6.5), and also for the stretch factor to be calculated across the crustal profile for each episode. The 2D profile is treated as a number of columns where we have information on sediment thickness and its bulk density over time, together with paleowater depth. For each column, crustal thickness and stretch factor can be calculated using the Airy isostatic solutions given above. Analysis of 2D profiles such as those through the Voring passive margin offshore mid-Norway, shows large variations in crustal thickness, stretch factor and backstripped tectonic subsidence across the 300 km-wide profile (Wangen 2010, p. 246–259).

9.5 Tectonic subsidence signatures

One of the benefits of a complete decompaction and backstripping procedure is that the subsidence history of basins can be compared without the complications of different paleobathymetric, eustatic, compactional and isostatic effects (Xie & Heller 2009). This enables the tectonic driving force for subsidence, and for basins due to lithos-

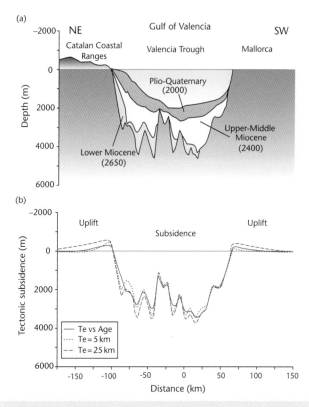

Fig. 9.16 Application of flexural backstripping technique to the Valencia Trough, a young rift basin in the western Mediterranean between the Spanish mainland and the Balearic Islands, which extended in the early Miocene. (a) Stratigraphic cross section with estimated densities (kg m^{-3}) of the Plio-Quaternary, upper/middle Miocene and lower Miocene. (b) Backstripped tectonic subsidence using three models for the equivalent elastic thickness. After Watts and Torné (1992), reproduced with permission of the American Geophysical Union.

pheric stretching, the stretch factor β, to be evaluated at different positions within a single basin and between different basins. It also enables the tectonic subsidence history of sedimentary basins to be discriminated and compared with theoretical curves.

Stretching of the continental lithosphere and flexure of the lithosphere are by far the most important mechanisms for prolonged and widespread subsidence. Stretching and flexure produce entirely different subsidence profiles:

- Stretching of the continental lithosphere produces a rapid syn-rift subsidence followed by an exponentially decreasing post-rift subsidence due to thermal relaxation. Although a range of subsidence history possibilities arise from the duration of the stretching, three-dimensional heat flow, induced convection in the zone of upwelled asthenosphere, maintenance or removal of a hot source in the mantle, lateral relay during simple shear, and depth-dependent stretching (Chapter 3), nevertheless, the diagnostic signature of continental stretching is an early, rapid, fault-controlled phase of subsidence followed by a 'concave-up' phase as the lithosphere cools. The amount of syn-rift and post-rift subsidence depends essentially on the amount of stretching β.

• Flexure of the lithosphere under a moving load, such as an orogenic wedge, typically produces an accelerating subsidence through time in pro-foreland settings. This is the simple result of the propagation of a flexural wave with a maximum deflection under the load across the pro-lithosphere (Chapter 4). Consequently, locations on the foreland plate distant from the load are initially uplifted in a forebulge region, producing an unconformity, and then subside at an increasing pace as they become involved in foredeep subsidence. The subsidence history of retro-foreland basins, however, is approximately linear, since the the mountain belt load does not move rapidly across the upper plate (Naylor & Sinclair 2008). The characteristic tectonic subsidence history is controlled by tectonic progradation taking place during an early phase of orogenic growth. The magnitude and geometry of flexural subsidence depend on the flexural rigidity of the underlying plate and the magnitude and spatial distribution of the applied load.

The typical subsidence signatures of stretched basins and foreland basins are compared in Fig. 9.17a, b. A first-order conclusion from such a comparison is that: (i) stretched basins contain rift to post-rift megasequences with a duration of 10^1–>10^2 Myr, with syn-rift tectonic subsidence rates of typically <0.2 mm yr^{-1} and exponentially decreasing post-rift tectonic subsidence rates of <0.05 mm yr^{-1}; and (ii) pro-foreland basin megasequences are typically 20–40 Myr in duration and involve convex-up subsidence signatures with maximum tectonic subsidence rates of 0.2–0.5 mm yr^{-1}. Thicknesses of foreland basin stratigraphy range up to 10 km. Retro-foreland basin subsidence is in general slower (<0.05 mm yr^{-1}) and more protracted (40–80 Myr) than in pro-foreland basins, and subsidence curves are linear rather than convex-up (Naylor & Sinclair 2008; Sinclair 2012). Strike-slip basins may be particularly difficult to discriminate on the basis of tectonic subsidence signature, since basins in strike-slip zones are strongly related to the kinematics of border fault movement. Basins may have components of thermal relaxation, lateral heat loss and flexure from nearby push-up ranges contributing to the tectonic subsidence signature. Tectonic subsidence rates are commonly very high compared to all other basin types (>0.5 mm y^{-1}), but strike-slip megasequences may be short in duration (c.10 Myr) due to the complexities of deformation within the PDZ.

Cratonic basins are typified by very long periods (>10^2 Myr) of slow subsidence interrupted by regional unconformities, with tectonic subsidence rates of 0.01–0.04 mm yr^{-1} (Fig. 9.18). If periods of accelerated subsidence and negative subsidence (unconformities) are identified and removed from the backstripped subsidence history, the background or 'reduced' subsidence signature may be diagnostic of

(a)

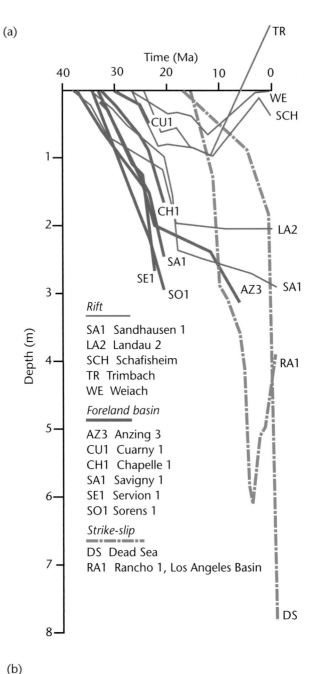

(b)

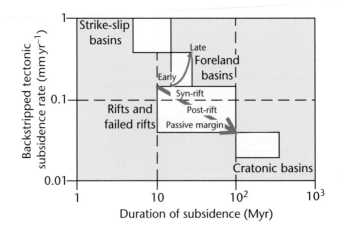

Fig. 9.17 (a) Comparison of the typical subsidence histories of foreland basins, rift and strike-slip basins, using decompacted subsidence curves. Thick solid curves are boreholes in the flexural North Alpine Foreland Basin of Switzerland and southern Germany. Thin solid curves are the Tertiary rift phase of boreholes in the Rhine rift and its southerly continuation in northern Switzerland. The Swiss boreholes (TR, WE and SCH) have experienced Neogene uplift. Dash-dot lines are two strike-slip basins. (b) Plot of duration of subsidence versus typical tectonic subsidence rate, allowing foreland basins, rift, failed rift, passive margin and strike-slip basins to be discriminated.

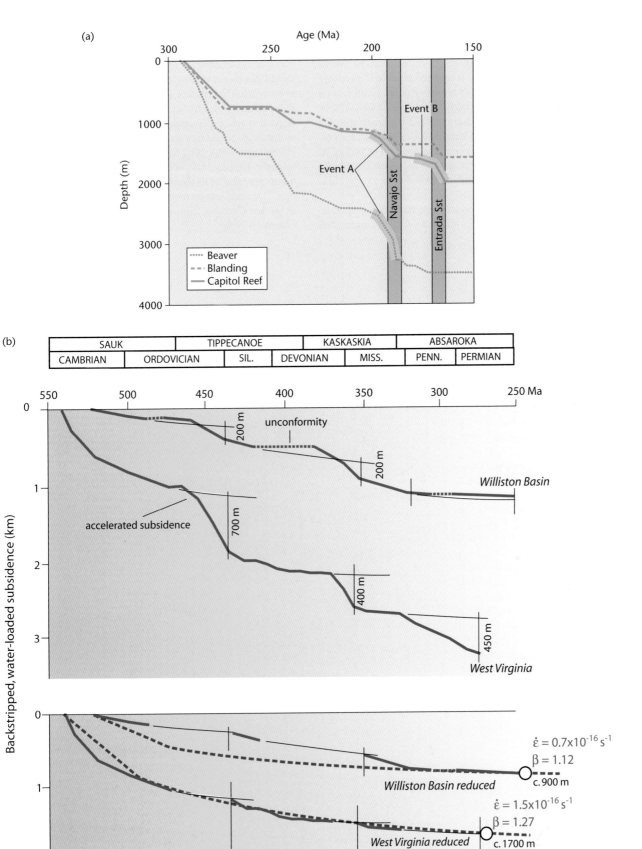

Fig. 9.18 Subsidence histories showing long-term drift superimposed by phases of accelerated subsidence and uplift. (a) Periods of convex-up accelerated subsidence ascribed to flexural subsidence in the retro-foreland of the Colorado Plateau region, western USA (Allen *et al.* 2000). (b) Residuals superimposed on long-term drift from the North American craton (Armitage & Allen 2010). Top: Total backstripped tectonic subsidence for two locations showing periods of accelerated subsidence and unconformities. Base: Same localities with residuals removed to give a long-term drift with model fits (red dashed lines) for extensional strain rates and stretch factors of $0.7 \times 10^{-16}\,s^{-1}$ and $\beta = 1.12$ (Williston) and $1.5 \times 10^{-16}\,s^{-1}$ and $\beta = 1.27$ (West Virginia) respectively.

the tectonic driving mechanism. In the cratonic basins of North America, Armitage and Allen (2010) recognised a 'reduced' signature that could be matched by a model of low extensional strain rate ($\dot{\varepsilon}$ of 0.7 to $1.5 \times 10^{-16}\,s^{-1}$) and low stretch factor (β of 1.12–1.27). In other cases, prolonged background subsidence in cratonic areas may be caused by dynamic topography (Burgess & Gurnis 1995; Burgess *et al.* 1997; Burgess & Moresi 1999) (see Chapter 5). The timing of the onset of cratonic basin subsidence appears to be linked to break-up of major supercontinental assemblies (Allen & Armitage 2012, p. 605, fig. 30.2).

In many basins more than one subsidence mechanism may operate. This may be particularly the case in the retro-foreland regions of major ocean–continent convergence zones. In these regions flexural subsidence due to loading in retro-fold and thrust belts may be added to by dynamic subsidence associated with subduction of cold oceanic slabs. Allen *et al.* (2000), for example, recognised two convex-up flexural events superimposed on a very long-term, linear to slightly concave-up subsidence curve in the Mesozoic of the Colorado Plateau area, USA (Fig. 9.18a). The addition of an in-plane stress related to continental collision to the post-rift history of stretched basins may also be recognised by 'anomalous' increases in subsidence rate. This combination of mechanisms has been invoked, for example, in the late Cenozoic history of the North Sea failed rift (van Wees & Cloetingh 1996).

CHAPTER TEN
Thermal history

Summary

Subsidence in sedimentary basins causes thermal maturation in the progressively buried sedimentary layers. Indicators of the thermal history include organic, geochemical, mineralogical and thermochronometric measurements. The most important factors in the maturation of organic matter are temperature and time, pressure being relatively unimportant. This temperature and time dependency is described by the *Arrhenius equation*, which states that the reaction rate increases exponentially with temperature; the rate of the increase, however, slows with increasing temperature. The cumulative effect of increasing temperature over time can be evaluated by integrating the reaction rate over time. This is called the *maturation integral*. It can be related directly to measurable indices of burial.

Paleotemperatures are controlled by the basal heat flow history of the basin (which in turn reflects the lithospheric mechanics), but also by 'internal' factors such as variations in thermal conductivities, heat generation from radioactive sources in the continental crust and within the sedimentary basin-fill, effects of sediment deposition rate producing thermal blanketing, regional water flow through aquifers and surface temperature variation. Heat flow heterogeneities may be caused by the presence of salt diapirs and fractures permitting advective heat transport by fluids, and by the heating effects of igneous sills and dykes.

Thermal conductivity models of the basin-fill can be developed from knowledge of framework mineralogy and porosity. The effect of a heterogeneous basin-fill, assuming a constant heat flow, is an irregular rather than linear geotherm, particularly where there is a wide range of insulating (for example marine shale) and conducting (for example salt) lithologies. Radiogenic heat production is greatest where the underlying basement is granitic, and where the basin-fill contains 'hot' shales. Radiogenic heat production is particularly important in deep, long-lived basins. Advection of fluids along regional aquifers has profound consequences for heat flow in basins and can locally override basal heat flow contributions. Advective heat transport depends on the temperature of the pore fluids, but also on rock porosity. Although compactionally driven fluid movement is slow and thermally relatively ineffective, gravitationally driven flow through aquifers is very important. Recharge areas of water in topographically elevated areas around the basin margin, such as in foreland basins, rifts and intracratonic sags, displaces basinal brines and strongly affects the temperature history of basin sediments. Major climatic changes of long frequency cause temperature changes to be propagated through the upper part of the basin-fill that may affect thermal indicators.

Subsurface temperatures can be estimated from borehole measurements, with a correction applied to account for the cooling that takes place during the circulation of drilling fluids. These corrected formation temperatures allow near-surface geothermal gradients to be calculated.

Vitrinite reflectance is the most widely used organic indicator of thermal maturity. Other organic and mineralogical indicators are also used. Apatite fission track analysis is now a well-established thermochronological tool, and the diffusion of helium during U–Th decay is an increasingly used technique. Both thermochronometers allow the timing of thermal events as well as the maximum paleotemperature to be assessed.

Vitrinite reflectance measurements plotted against depth – termed R_o profiles – provide useful information on the thermal history of the basin. The 'normal' pattern is a sublinear relationship between log R_o and depth, indicating a continuous, time-invariant geothermal gradient. R_o profiles with distinct kinks between two linear segments (doglegs) indicate two periods of different geothermal gradient separated by a thermal event. R_o profiles with a sharp break or jump (offsets) indicate the existence of an unconformity with a large stratigraphic gap. The R_o profile in basins that have undergone continuous subsidence intersect the surface at values of 0.2 to 0.4% R_o. Inverted basins that have lost the upper part of the basin-fill by crustal uplift and erosion have profiles intersecting the surface at higher values of R_o. The offset of the R_o profile from the 'normal' profile can be used to estimate the amount of denudation, but care needs to be taken with the possible effects of thermal conductivity variations in the basin and the correct choice of surface temperature at the time of maximum paleotemperature of the basin-fill.

Burial of thermochronometers in sedimentary basins, such as apatite fission track and U–Th/He, may cause partial or full annealing. Quartz cementation is also temperature-dependent, taking place at 90–120°C, and typically at depths of 2500–4500 m. Quartz cementation arrests further mechanical compaction.

Studies of present-day heat flows and ancient geothermal gradients suggest that thermal regime closely reflects tectonic history. In particular, *hypothermal* (cooler than average) basins include ocean trenches and outer forearcs and foreland basins. *Hyperthermal* (hotter than average) basins include oceanic and continental rifts, some strike-slip basins with mantle involvement, and magmatic arcs in collisional settings. Mature passive margins that are old compared to the thermal time constant of the lithosphere tend to have near-average heat flows and geothermal gradients.

Basin Analysis: Principles and Application to Petroleum Play Assessment, Third Edition. Philip A. Allen and John R. Allen.
© 2013 John Wiley & Sons, Ltd. Published 2013 by John Wiley & Sons, Ltd.

10.1   Introduction

Subsidence in sedimentary basins causes material initially deposited at low temperatures and pressures to be subjected to higher temperatures and pressures. Sediments may pass through diagenetic, then metamorphic regimes and may contain indices of their new pressure-temperature conditions. Thermal indices are generally obtained from either dispersed organic matter, temperature-dependent chronometers such as fission tracks in apatite and zircon or from mineralogical trends. A great deal of effort has been spent in attempting to find an analytical technique capable of unambiguously describing thermal maturity, and an equal amount of effort attempting to correlate the resulting proliferation of indicators.

Numerical values of the organic geochemical parameters are dependent on time, thermal energy and type of organic matter (e.g. Weber & Maximov 1976 for an early contribution). The evolution of clays and other minerals is controlled by temperature and by chemical and petrological properties. The scale of maturation to which a given organic or mineralogical phase can be calibrated is that of *coal rank*. Any analytical technique must be able to make use of very small amounts of dispersed organic matter in order to be valuable in basin analysis. Vitrinite reflectance and elemental analyses enable coal rank to be related to hydrocarbon generation stages. Thermochronological tools such as apatite fission track analysis and the diffusion of He during U–Th decay offer the important advantage of providing information on thermal evolution instead of solely on maximum paleotemperature reached.

The objective in this chapter is to describe the use of a number of thermal indicators in constraining and calibrating the thermal evolution of the basin-fill. The implications for the generation of hydrocarbons and for the diagenesis of reservoir rocks are developed in Chapter 11. Fundamentals on heat flow are given in §2.2.

10.2   Theory: the Arrhenius equation and maturation indices

It is now believed that the effects of depth *per se* on the maturation of organic matter are of minor importance, the most important factors being *temperature* and *time*. Pressure is relatively unimportant. Philippi (1965) assessed the effect of pressure by studying hydrocarbons in two Californian basins. In the Los Angeles Basin, hydrocarbons were generated at about 8,000 ft (~2.4 km) whereas in the Ventura basin, generation did not take place until about 12,500 ft (3.8 km) of burial. Since pressure or stress is directly related to depth of burial ($\sigma = \rho g h$) this suggests that pressure does not play a major role in hydrocarbon generation. However, the generation of hydrocarbons in the two basins took place at the same temperature, strongly suggesting that subsurface temperature was the overriding control.

The relationship between temperature and the rate of chemical reactions is given by the *Arrhenius equation*:

$$K = A \exp(-E_a/RT) \qquad [10.1]$$

where K is the reaction rate, A is a constant sometimes termed the *frequency factor* (it is the maximum value that can be reached by K when given an infinite temperature), E_a is the activation energy, R is the universal gas constant, and T is the absolute temperature in kelvin. The constants in the Arrhenius equation can be estimated

from compilations of organic metamorphism (e.g. Hood *et al.* 1975; Shibaoka & Bennett 1977). The activation energies of each individual reaction involved in organic maturation are not known, but for each organic matter type a distribution of activation energies may be established from laboratory and field studies. For example, a distribution of activation energies for the maturation of vitrinite from 159 to 310 kJ mol^{-1}, centred on 226 kJ mol^{-1}, was suggested by Burnham and Sweeney (1989).

The Arrhenius equation suggests that reaction rates should increase exponentially with temperature, so that a 10 °C rise in temperature causes the reaction rate to double. This result is widely known, but it is less widely realised that the rate of increase in reaction rate slows down with increasing temperature, so at 200 °C the reaction rate increases by a factor of 1.4 for a 10 °C rise in temperature (Robert 1988). Clearly, both time and temperature influence organic maturation, a view supported by the occurrence of shallower oil generation thresholds as the sediments containing the organic matter become older (Dow 1977). Connan (1974) believed that the threshold of the principal zone of oil generation was related to the logarithm of the age of the formation, further supporting a time-temperature dependence obeying the laws of chemical kinetics.

The cumulative effect of increasing temperature can be evaluated from the *maturation integral*, the reaction rate integrated over time,

$$C = A \int_0^t \exp(-E_a/RT) + C_0 \qquad [10.2]$$

where C_0 is the original level of maturation of the organic material at the time of deposition ($t = 0$). The maturation integral for any nominated horizon can be calculated if the decompacted burial history (Chapter 9), heat flow through time, and thermal conductivities of the sediments and basement are known or can be assumed. For the less mathematically inclined, eqn. [10.2] shows that when paleotemperatures are plotted on an exponential scale, the area under the curve from deposition to a given time is proportional to the maturation integral at that time at the nominated horizon (plus the value of C_0). Some authors believe that the maturation integral is related to measurable values of vitrinite reflectance (see §10.4.2) (Royden *et al.* 1980; Falvey & Middleton 1981).

Hood *et al.* (1975) devised an artificial maturation parameter, the *level of organic metamorphism* (LOM), based on a rank progression of coals, from lignites to meta-anthracites. Hood's diagram shows the relationship between the 'effective heating time' and the maximum temperature attained. The effective heating time is defined as the length of time the temperature remains within a 15 °C range of the maximum temperature. This method, although not stated explicitly by Hood *et al.* (1975), is based on the first-order chemical kinetics outlined above.

Another application of the Arrhenius relationship is the *time-temperature index* (TTI) (Lopatin 1971; Waples 1980). This index is based on the view that the reaction rate doubles for every 10 °C rise in temperature over the entire range from 50 °C to 250 °C. Since the method assumes that the reaction rate continues to double in 10 °C intervals over the entire temperature range to 250 °C, it tends to overestimate maturity. The reaction cannot continue indefinitely because the materials undergoing thermal maturation are used up.

Other techniques, such as those of Tissot (1969), Tissot and Espitalié (1975) and Mackenzie and Quigley (1988), have been developed that enable masses of petroleum generated during thermal maturation of organic matter to be calculated. The Mackenzie and Quigley model is described in relation to petroleum source rocks in §11.4.

10.3 Factors influencing temperatures and paleotemperatures in sedimentary basins

Chapter 2 contains some basic concepts about heat flow, and the specific problem of one-dimensional (vertical) heat flow in basins due to stretching is addressed in Chapter 3. Here, we are concerned with the various 'internal' factors that influence the temperatures within sedimentary basins: (i) variations in thermal conductivity, most commonly due to lithological heterogeneity; (ii) internal heat generation; (iii) convective/advective heat transfer within fractured and unfractured sediments; and (iv) surface temperature changes.

10.3.1 Effects of thermal conductivity

The distribution of temperature with depth (geotherm) in the continents is primarily determined by conductive heat transport. We know the relation between heat flux and temperature gradient, as given by Fourier's law (eqn. 2.29). This law states that conductive heat flux is related to the temperature gradient by a coefficient, K, known as the coefficient of _thermal conductivity_. If two measurements of temperature are known, one T_y at depth y and another T_0 at the surface ($y = 0$), Fourier's law can be restated as

$$q = -K(T_y - T_0)y \qquad [10.3]$$

which by rearrangement becomes

$$T_y = T_0 + \left(\frac{-qy}{K}\right) \qquad [10.4]$$

where q is the heat flux (negative for y increasing downwards). We are here initially ignoring internal heat production within the sedimentary pile.

Ignoring also for the moment lithological variations, thermal conductivities of sediments vary as a function of depth because of their porosity loss with burial (§9.3). Eqn. [10.4] can be modified to account for the different thermal conductivities of the sedimentary layers,

$$T_y = T_0 + (-q)\left\{\frac{l_1}{K_1} + \frac{l_2}{K_2} + \frac{l_3}{K_3} + \cdots\right\} \qquad [10.5]$$

where l_1 to l_n are the thicknesses of the layers with thermal conductivities K_1 to K_n, and $l_1 + l_2 + l_3 \ldots$ must of course be equal to y. Falvey and Middleton (1981) recommended the use of a function that assumed an exponential relation between porosity and depth

$$K = K_{min} - \{(K_{min} - K_0)\exp(-\gamma y)\} \qquad [10.6]$$

where K_{min} is the thermal conductivity deep in the sedimentary section, K_0 that at the sediment surface, and γ is a constant for a given section. Since K varies with depth, temperature gradients must also vary with depth in order to maintain a constant heat flow. If present-day heat flow can be calculated from a borehole by measurement of conductivities and surface and bottom hole temperatures, eqns [10.4] and [10.6] can be used to find the temperature at any

depth. If paleoheat flow is then assumed to be constant with depth, the temperature history of any chosen stratigraphic level can be estimated. The assumption of a constant heat flow with depth is a condition of any one-dimensional steady-state heat conduction model. Measurements in some sedimentary basins such as the North Sea failed rift (Andrews-Speed _et al._ 1984), however, suggest that this is not a good assumption, deep circulation of water most likely being responsible for the departure from the steady-state assumption (see §10.3.4).

A fundamental requirement in the estimation of geotherms, temperatures and paleotemperatures in sedimentary basins is therefore the bulk thermal conductivity of the different sedimentary layers making up the basin-fill. Thermal conductivities can be measured in the laboratory (Carslaw & Jaeger 1959; Sass _et al._ 1971) and _in situ_ (Beck _et al._ 1971). The bulk thermal conductivity of most sedimentary rocks ranges between $1.5\,\mathrm{Wm^{-1}K^{-1}}$ (shales) and $4.5\,\mathrm{Wm^{-1}K^{-1}}$ (sandstones) (Table 10.1). These estimates depend mostly on the mineralogy of the framework grains, the type and amount of material in the matrix (commonly clay minerals), and the porosity and fluid content (commonly water) (Brigaud & Vasseur 1989). The individual conductivities of framework, matrix and pore fluid are also dependent on temperature. The general trend is that non-argillaceous rocks have higher conductivities than argillaceous rocks, and that conductivity increases with increasing porosity. If measurements of the thermal conductivities of the different components in a rock can

Table 10.1 Density and thermal properties of some common minerals and rock types. Thermal conductivities are given at surface temperatures (in part from Brigaud and Vasseur 1989).

Rock type	Density (kg m⁻³)	Thermal conductivity K (Wm⁻¹K⁻¹)	Volumetric coefficient of thermal expansion α_v (10⁻⁵ K⁻¹)
Minerals common in sediments and sedimentary rocks			
Water (pore fluid)	1000	0.6	
Quartz	2650	7.7	
Calcite	2710	3.3	
Dolomite	2870	5.3	
Anhydrite	2960	6.3	
Kaolinite	2630	2.6	
Chlorite	2780	3.3	
Illite/smectite	2660	1.9	
Sedimentary rocks			
Shale	2100–2700	1.2–3.0	
Sandstone	1900–2500	1.5–4.2	3
Limestone	1600–2700	2.0–3.4	2.4
Dolomite	2700–2850	3.2–3.5	
Metamorphic rocks			
Gneiss	2600–2850	2.1–4.2	
Amphibolite	2800–3150	2.1–3.8	
Igneous rocks			
Basalt	2950	1.3–2.9	
Granite	2650	2.4–3.8	2.4
Gabbro	2950	1.9–4.0	1.6
Peridotite	3250	3.0–4.5	2.4

Reproduced with permission of John Wiley & Sons, Ltd.

be made, the bulk thermal conductivity of the sedimentary layer can be estimated. The overall thermal conductivity structure of the basin-fill can then be estimated from a knowledge of the mineralogy, porosity and fluid content of the stratigraphy filling the basin. The availability of large 'continuous' subsurface data sets from hydrocarbon exploration boreholes has revolutionised the ability to make such estimates.

The *effective thermal conductivity* of a clean quartzose sandstone with pore-filling water should decrease with increasing porosity, since the pore fluid is insulating. However, the effective thermal conductivity of a clean quartzose sandstone may be almost invariant with depth (Fig. 10.1b). This is due to the decrease in conductivity

of the framework quartz grains with increasing temperature, which offsets the increase in conductivity due to compaction (porosity loss) (Palciauskas 1986).

The effects of clays in a sandstone, for example as pore-filling authigenic cements, is to decrease the bulk thermal conductivity of the argillaceous sandstone, since clays have an insulating effect (Fig. 10.1a). Feldspar and most clays do not show a marked effect of temperature on thermal conductivity, so the effect of compaction commonly dominates. A clay-water mixture (shales) increases in conductivity rapidly with depth because of compaction, whereas a feldspar-water mixture, because it compacts similarly to a sand, increases in conductivity much more slowly with depth (Fig. 10.1c).

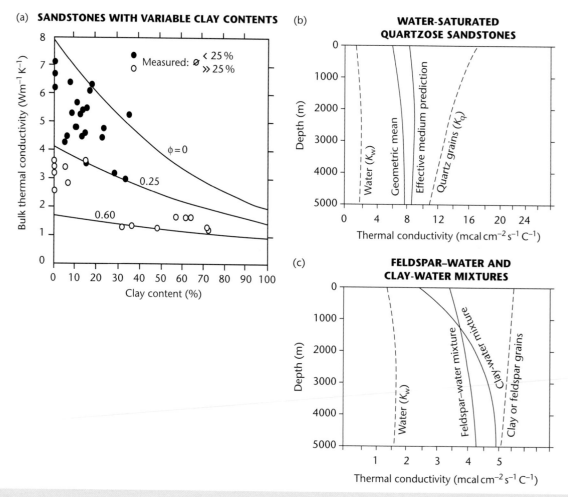

Fig. 10.1 The thermal conductivities of sedimentary rocks. (a) Influence of mineralogy on thermal conductivity of water-saturated sandstones with a variable clay content, using thermal conductivities of $7.7\,W\,m^{-1}\,K^{-1}$, $2.0\,W\,m^{-1}\,K^{-1}$ and $0.6\,W\,m^{-1}\,K^{-1}$ for quartz, clay and water respectively (after Brigaud & Vasseur 1989, reproduced with permission of John Wiley & Sons, Ltd.). (b) Thermal conductivity of water-saturated quartzose sandstone as a function of depth. The effective medium prediction and the empirical relation closely agree, demonstrating a negligible increase in thermal conductivity with depth, despite the fact that the quartz grains decrease in thermal conductivity considerably with depth. (c) Feldspar-water and clay-water mixtures, showing that the thermal conductivities increase markedly with depth, especially for clay-rich sediments. This is principally due to the effects of compaction. In (b) and (c) the temperature gradient is $30\,°C\,km^{-1}$ and the surface temperature $20\,°C$. After Palciauskas (1986), reproduced with permission from Editions Technip.

The bulk conductivity of a sediment layer can therefore be thought of as being made up of the contributions of the pore fluid and the grain conductivities. Assuming a geometric mean model for the two-phase media of solid and fluid (Woodside & Messmer 1961), the bulk conductivity is

$$K_{bulk} = K_s^{(1-\phi)} K_w^{\phi} \qquad [10.7]$$

where K_s and K_w are the thermal conductivities of sediment grains and water respectively, and ϕ is the porosity, assumed to be filled with water. An alternative method, termed the *effective medium* theory calculates an effective bulk thermal conductivity for a randomly inhomogeneous medium made of constituents with volume fractions V_i and thermal conductivities K_i. The basic result of the theory is

$$K^{-1} = \sum_{i=1}^{n} 3V_i (2K + K_i)^{-1} \qquad [10.8]$$

This expression is particularly useful where mixed components are present in the sediment layer. For example, for the water-quartz mixture mentioned above, if the quartz framework ($K_q = 5.4\,\mathrm{W\,m^{-1}\,^{\circ}C^{-1}}$ at $T = 100\,^{\circ}C$) occupies 0.7 of the rock volume, and water ($K_w = 0.7\,\mathrm{W\,m^{-1}\,^{\circ}C^{-1}}$ at $T = 100\,^{\circ}C$) occupies 0.3 of the rock volume, the bulk conductivity from effective medium theory (eqn. [10.8]) is approximately $3.3\,\mathrm{W\,m^{-1}\,^{\circ}C^{-1}}$. From the general result in eqn. [10.7], the bulk conductivity is approximately $2.9\,\mathrm{W\,m^{-1}\,^{\circ}C^{-1}}$.

A fundamental result of maintaining a constant heat flux through a heterogeneous basin-fill is that the geothermal gradient must vary with depth. As an example, consider the thermal conductivity structure derived from the borehole shown in Fig. 10.2a. Keeping the heat flux at $63\,\mathrm{mW\,m^{-2}}$, the geotherm varies as in Fig. 10.2b. Clearly, the presence of a heterogeneous basin-fill negates the assumption of a linear conduction geotherm. The implications for the interpretation of maturity profiles are discussed in §10.5.

10.3.2 Effects of internal heat generation in sediments

Heat generation by radioactive decay in sediments may significantly affect the heat flow in sedimentary basins (Rybach 1986). Although all naturally occurring radioactive isotopes generate heat, the only significant contributions come from the decay series of uranium and thorium and from ^{40}K (Table 10.2). As a result, heat production varies with lithology, generally being lowest in evaporites and carbonates, low to medium in sandstones, higher in shales and siltstones and very high in black shales (Rybach 1986; Haack 1982; Rybach & Cermak 1982).

In the continents, crustal radioactivity may account for a large proportion (20–60%) of the surface heat flow (§2.2.3). For a purely conductive, one-dimensional (vertical) heat flow, the temperature at any depth y is determined by the surface temperature T_0, the basal heat flow q_b, the average thermal conductivity of the sediments K, and the internal heat production A (estimated from natural gamma ray logs). The effect of the internal heat generation is greatest at large depths, as can be seen from the third term in eqn. [2.26] (Fig. 10.3). The temperature increase after a time t as a result of the internal heat generation depends on the value of A, but the net temperature change also depends on the rate of conductive heat loss. Over geological time scales of >10 Myr the temperature rise may be considerable

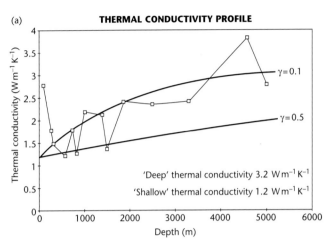

(a) THERMAL CONDUCTIVITY PROFILE

'Deep' thermal conductivity $3.2\,\mathrm{W\,m^{-1}\,K^{-1}}$

'Shallow' thermal conductivity $1.2\,\mathrm{W\,m^{-1}\,K^{-1}}$

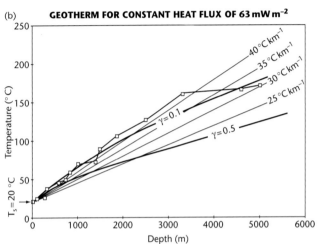

(b) GEOTHERM FOR CONSTANT HEAT FLUX OF $63\,\mathrm{mW\,m^{-2}}$

Fig. 10.2 (a) Thermal conductivity structure of stratigraphy penetrated by borehole 14/20-1 close to Wexford, southeastern Ireland. Two curves represent exponential distribution of thermal conductivity with depth with a depth constant γ of 0.1 and 0.5. The thermal conductivity in the deep section is c.$3.2\,\mathrm{W\,m^{-1}\,K^{-1}}$, and c.$1.2\,\mathrm{W\,m^{-1}\,K^{-1}}$ in the shallow, near-surface section (see eqn. [10.6]). (b) Geotherm using the thermal conductivity structure in (a) for the borehole 14/20-1, exponential distributions of thermal conductivity with γ of 0.1 and 0.5, and a constant basal heat flow of $63\,\mathrm{mW\,m^{-2}}$. Linear geotherms between 25 and $40\,^{\circ}\mathrm{C\,km^{-1}}$ are shown for comparison.

Table 10.2 Typical concentrations of heat-producing elements in various rock types

Rock type	U (ppm)	Th (ppm)	K (%)
Granite	4.7	20	4.2
Shale	3.7	12.0	2.7
Average continental crust	1.42	3.6	1.43
Reference mantle	0.031	0.124	0.031
Chondritic meteorite	0.008	0.029	0.056

(Rybach 1986, p. 317). Internal heat generation in sediments may therefore strongly affect the temperature field in the basin if it is deep (>5 km) or long lived (>10 Myr).

10.3.3 Effects of sedimentation rate and sediment blanketing

The presence of a thick cover of sedimentary rocks with high radiogenic heat production has the effect of 'blanketing' the underlying crust and deeper parts of the sedimentary basin (Karner 1991;

Wangen 1995, 2010 p. 181). This blanketing effect may be important in terms of the temperature-dependent rheology of rocks underlying the basin, but more importantly in the present context in elevating paleotemperatures in the basin-fill.

The effect of sediment blanketing is described in §2.2.4. As a reminder, the deposition of sediments has two effects, first a transient cooling and reduction in the surface heat flow, and, second, a possible long-term warming, dependent on the thermal conductivity and internal heat generation of the new sediment layer, or blanket. The effectiveness of the sediment blanket in causing a temperature transient depends on two main factors: (i) the deposition rate; and (ii) the thermal conductivity of the young sediments comprising the blanket.

A fast deposition rate causes the underlying sediment column to be covered with cool sediments deposited at surface temperatures. During burial, these new sediments do not heat up rapidly enough to offset the cooling effect on the geotherm. This thermal response time is greatest for *instantaneous* deposition and may be >1 Myr for the instantaneous deposition of 1 km of sediment. A deposition rate of >0.1 mm yr^{-1} is required to depart from a stationary state for the geotherm. Rapid deposition of, for example, 1 km of sediments in just 0.1 Myr would reduce the surface heat flow by half. The transient temperature field may extend several kilometres into the basin-fill (see §10.3.5 for discussion of 'skin depth'). This effect of blanketing may be especially relevant in the thick sedimentary successions deposited at the mouths of major river deltas.

The temperature transient is also amplified by the deposition of low thermal conductivity, uncompacted sediments with large volumes of saline pore water ($K_w = 0.6\,\mathrm{W\,m^{-1}\,K^{-1}}$). Consequently, the sediment blanket retards the rapid heating of the newly deposited sediments. Highly porous, uncompacted marine shales, such as may be deposited offshore from suspended load-dominated rivers, may therefore act as strong insulators in this way.

Fig. 10.4 shows the effects of sedimentation rate and thermal conductivity of the sediment blanket on the geotherm and surface heat flow. The solution is given in Appendix 12.

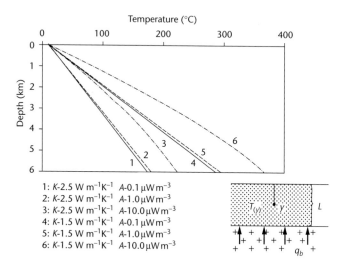

1: K-2.5 W m^{-1}K^{-1} A-0.1 μW m^{-3}
2: K-2.5 W m^{-1}K^{-1} A-1.0 μW m^{-3}
3: K-2.5 W m^{-1}K^{-1} A-10.0 μW m^{-3}
4: K-1.5 W m^{-1}K^{-1} A-0.1 μW m^{-3}
5: K-1.5 W m^{-1}K^{-1} A-1.0 μW m^{-3}
6: K-1.5 W m^{-1}K^{-1} A-10.0 μW m^{-3}

Fig. 10.3 The influence of internal heat generation per unit volume in the sedimentary column A and thermal conductivity K on the distribution of temperature with depth $T(y)$. The different curves were calculated by Rybach (1986) for a thickness of the heat-producing zone of 6 km, a basal heat flux q_b of 70 mW m^{-2} and a surface temperature T_0 of 20 °C. Reproduced with permission from Editions Technip.

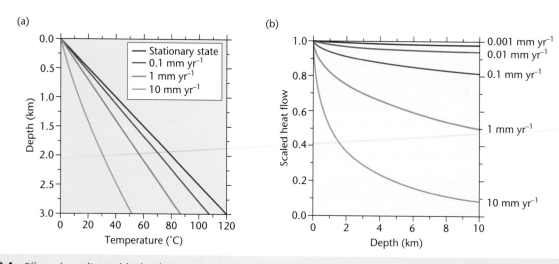

Fig. 10.4 Effect of a sediment blanket for a range of deposition rates. (a) Geotherms in the basin-fill for deposition rates of 0.1, 1 and 10 mm yr^{-1}, at the time when 3 km of sediment has been deposited (30 Myr, 3 Myr and 0.3 Myr respectively), including the steady-state conduction geotherm prior to sedimentation (stationary state). (b) Surface heat flux scaled against the surface heat flow before the onset of sedimentation for a range of deposition rates from 0.001 mm yr^{-1} to 10 mm yr^{-1}, as a function of the thickness of basin sediments. Basin thickness is the product of the sedimentation velocity and time. After Wangen (2010, fig. 6.38). © Magnus Wangen 2010, reproduced with the permission of Cambridge University Press.

10.3.4 Effects of advective heat transport by fluids

The temperatures in sedimentary basins may also be affected by the advective flow of heat through regional aquifers. Such processes may cause anomalously low surface heat flows at regions of recharge, and anomalously high surface heat flows in regions of discharge. The heat flow distributions of the Great Plains, USA (Gosnold & Fischer 1986), and the Alberta Basin (Majorowicz & Jessop 1981; Majorowicz *et al.* 1984) have been explained in this way.

It is important to know the relative contributions of conduction from the interior of the Earth, internal heat production from radiogenic decay (and chemical reactions), and advective transport of fluids through pore space. We perform this heat balance in Appendix 7 and, taking reasonable parameter values for a deeply buried rock in a sedimentary basin, calculate the role of advection in controlling the heat flow (see Appendix 59). Compactionally driven flow is too slow to cause significant thermal effects in a basin, whereas the relatively fast pore water flow velocities associated with groundwater movement in regional aquifers may lead to important heating or cooling.

The likely impact of fluids on the thermal history of the basin-fill is linked to the tectonic evolution of the basin. For example, uplift of rift shoulders during stretching of the continental lithosphere may cause meteoric-derived groundwater flows driven by the topographic elevation of the basin flanks. The meteoric fluxes in this case must displace brines filling pore space within the basin-fill. If the density of pore-filling brine is $1028\,kg\,m^{-3}$ and the density of meteoric water is $1000\,kg\,m^{-2}$, a simple pressure balance indicates that the extent of downward penetration of meteoric water is nearly 40 times the topographic elevation of its influx (Bjorlykke 1983). This implies that for even low topographic basin margin uplifts, meteoric gravitationally driven water is able to displace basinal brines from the entire depth of the sedimentary basin. At the other extreme, fluid movement caused by progressive compaction of basin sediments is slow, with vertical rates of $<10\,mm\,yr^{-1}$, with lower values still as the permeability reduces during compaction (Giles 1987) (Fig. 10.5).

The effects of fluid flow on heat flow in regional-scale systems are well illustrated by the Alberta Basin (Smith & Chapman 1983; Luheshi & Jackson 1986). Using a permeability and thermal conduc-

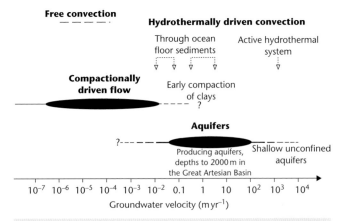

Fig. 10.5 Typical measured pore fluid velocities associated with compactionally driven flow, confined and unconfined aquifers, and hydrothermally driven convection. After Giles (1987, 1997), reproduced with kind permission of Springer Science+Business Media B.V.

tivity structure for the basin, the raised temperatures at discharge points of fluid flow and lowered temperatures at the recharge areas in the fringing hills can be explained (Fig. 10.6). The model results suggest that the temperature distribution is dominated by convection above the Paleozoic succession, while the heat flows within the Precambrian section can be explained simply by conduction. Andrews-Speed *et al.* (1984) similarly found that heat flow measurements strongly suggested a deep-water circulation, possibly controlled by the configuration of faults, in the North Sea failed rift. The implications of detailed studies such as this are that simple one-dimensional conductive heat flow models may be very poor predictors of actual heat flows in some sedimentary basins. The most strongly affected basins are likely to be continental basins with marginal uplifts, such as foreland basins and some intracratonic rifts and sags.

10.3.5 Effects of surface temperature changes

The possible effects of surface temperature changes on the maturation of thermal indicators have been relatively neglected, although the effects of, for example, glacial retreat on near-surface temperatures has been evaluated (Beck 1977). The likelihood of a surface temperature change penetrating an underlying basin-fill is essentially a question of heat diffusion (§2.2), and the problem can therefore be approached in a similar way to the cooling of the oceanic lithosphere (§2.2.7). The amount of time necessary for a temperature change to propagate a distance l in a medium with thermal diffusivity κ is

$$t = \frac{l^2}{\kappa} \qquad [10.9]$$

and the characteristic distance (or thermal diffusion distance) over which a temperature change is felt is

$$L = \sqrt{\kappa t} \qquad [10.10]$$

Taking a thermal diffusivity of $\kappa = 10^{-6}\,m^2\,s^{-1}$, appropriate for the sandstones of a basin-fill, a surface temperature disturbance would be registered at 1 km depth in just over 30 kyr. Expressed differently, the surface temperature change would propagate to a depth of 5.6 km in 1 Myr. We should therefore expect a surface temperature change to rapidly propagate through the upper part of the basin-fill. O'Sullivan and Brown (1998), for example, suggested that the effects of a surface temperature change of c.17°C during the Miocene on the North Slope of Alaska could be recognised in apatite fission tracks in borehole samples. If geotherms derived from apatite fission track analysis were used to calculate denudation in this example, estimates would be significantly in error.

A key point in this argument is that for thermal indicators to be affected by a surface temperature change, the surface temperature change must be sustained for a prolonged period of time. Yet, climatically induced surface temperature changes tend to be cyclic. What would be the effect of a periodic variation in surface temperature on the geotherm? Temperatures must vary cyclically within a surface zone of the Earth whose thickness is determined by the thermal properties of the crust or basin-fill and the period of the temperature fluctuation. If the temperature variation is described by

$$\omega = \frac{2\pi}{f} \qquad [10.11]$$

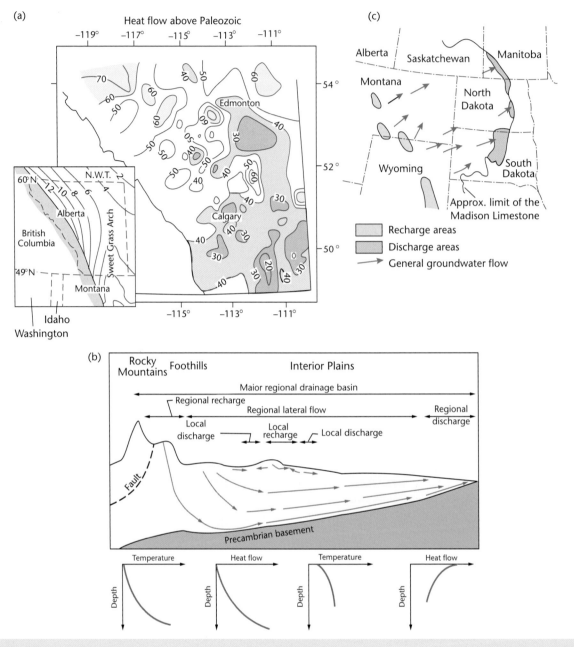

Fig. 10.6   Effects of groundwater flow on surface heat flows in sedimentary basins. (a) Heat flow map of southern and central Alberta, Canada, based on estimated heat flow values (in mW m^{-2}) above the top of the Paleozoic, based on 33,653 bottom hole temperature data from 18,711 wells (Majorowicz *et al.* 1984; reprinted with permission from Elsevier). The heat flows are strongly influenced by groundwater flow from recharge areas in structurally high regions, such as the Sweet Grass Arch (inset), to discharge areas. (b) Pattern of recharge and discharge in a cross-section from the Rocky Mountains to the Great Plains (after Majorowicz *et al.* 1984), and (c) plan view of groundwater flow in the Mississippian (Lower Carboniferous) Madison Limestone aquifer (after Downey 1984).

where *f* is the period of the temperature variation, we can define a *skin depth L* at which the amplitude of the temperature variation is 1/e of that at the surface of the Earth

$$L = \sqrt{\frac{2\kappa}{\omega}} \qquad [10.12]$$

Using $\kappa = 8 \times 10^{-6}\,\text{m}^2\,\text{s}^{-1}$, it is clear that the skin depth for daily temperature variations ($\omega = 7.27 \times 10^{-5}\,\text{s}^{-1}$) is less than 20 cm,

but for Pleistocene climate change variations of frequency 10^5 yr ($\omega = 1.99 \times 10^{-12}\,\text{s}^{-1}$), the skin depth is 1 km. This means that long-period variations in surface temperature may be felt deep within the sedimentary basin-fill.

10.3.6   Heat flow around salt domes

Salt structures are found on many passive margins, such as the Gulf of Mexico, West Africa and the Brazilian margin of the South Atlantic

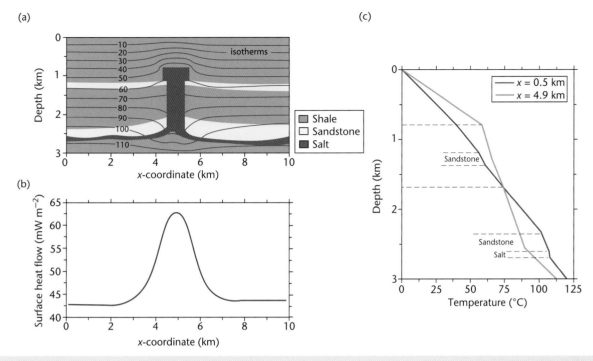

Fig. 10.7 Thermal effects of the presence of a salt diapir in a stratigraphic succession of sandstones and shales. (a) Geometry of salt diapir and isotherms in the subsurface. (b) Surface heat flow (after Wangen 2010, p. 139, fig. 6.15). (c) Geotherms through the centre of the diapir (orange) and to the left side of the model (blue) (after Wangen 2010, fig. 6.16). © Magnus Wangen 2010, reproduced with the permission of Cambridge University Press.

(Hudec & Jackson 2007). Originally deposited as a horizontal sheet, the salt is very mobile because of its rheological weakness, and flows upwards to produce domal diapirs and vertical walls (Gemmer *et al.* 2004). The presence of salt in the subsurface affects heat flow in the sedimentary basin. The reasons for this are straightforward: (i) Salt has a thermal conductivity of ~6 W m^{-1} K^{-1} at surface conditions and essentially zero porosity, whereas a porous shale has a thermal conductivity of ~1 W m^{-1} K^{-1}; (ii) Salt diapirs cause an extreme three-dimensionality to the temperature field, with hotter conditions above the salt bodies and cooler conditions below, compared with adjacent regions in the enclosing shale. The temperature heterogeneity extends a distance laterally approximately the same as the thickness of the salt diapir (Fig. 10.7). For shallowly buried salt, there is an increase in surface heat flow above the diapir. The increase in surface heat flow scales on the average thermal conductivity from the surface down to a characteristic isotherm. Consequently, if the subscript 1 refers to the column to the flank of the diapir, and 2 refers to the vertical column going through the centre of the diapir, the ratio of the surface heat flows scales on the ratio of the thermal conductivities

$$\frac{q_2}{q_1} = \frac{K_1}{K_2}$$ [10.13]

Typical values for the bulk conductivity averaged over depth for the two columns shown in Fig. 10.7 are:

Column 2, comprising salt (4.5 W m^{-1} K^{-1}), shale (1.1 W m^{-1} K^{-1}) and sandstone (2.5 W m^{-1} K^{-1}), has $K_2 = 1.85$ W m^{-1} K^{-1}.

Column 1, comprising shale and sandstone, has $K_1 = 1.17$ W m^{-1} K^{-1}.

K_2/K_1 is therefore 1.58. This explains the variation in surface heat flow from 42 mW m^{-2} on the flank to 63 mW m^{-2} above the salt diapir.

10.3.7 Heat flow around fractures

Heat transport by advection commonly takes place through aquifers, but may also occur through fractures. The strength of advective or convective heat flow relative to that by conduction is measured by a Péclet number (§2.2.5). During transit through a fracture hot fluids lose heat to the rock walls, so it is a two-dimensional problem of (vertical) flow along the fracture plane and lateral heat flow into the surrounding rock (Appendix 60).

Solution of this 2D problem shows that isotherms bend upwards along the fracture and return to horizontal a lateral distance from the fracture. This lateral distance is approximately the vertical extent of the fracture l_0 (Fig. 10.8). The characteristic time for the fracture to heat up its surroundings must depend on the length of the fracture l_0, and the thermal diffusivity of the rock, $t = l_0^2/\kappa$. If the diffusivity is 10^{-6} m^2 s^{-1}, the time scale ranges from weeks to >300 years for fracture lengths l_0 of 1 m to 100 m. In geological terms, therefore, fractures heat their surroundings almost instantaneously, but the thermal effects are relatively close to the fracture. If fractures are long and closely spaced, the entire rock volume would be heated to values close to the temperature of the fluid filling the fracture (Fig. 10.8).

10.3.8 Heat flows around sills, dykes and underplates

Some basins, especially those associated with lithospheric stretching, have been intruded by igneous bodies such as dykes and sills as in the rifts of northeast Africa (Fig. 10.9). Some of these sills and dykes were emplaced at temperatures of c.1000 °C at shallow levels into rocks with temperatures in the range 0–100 °C. It is of some interest, therefore, whether the heating of dykes and sills has a significant

(a)

(b)

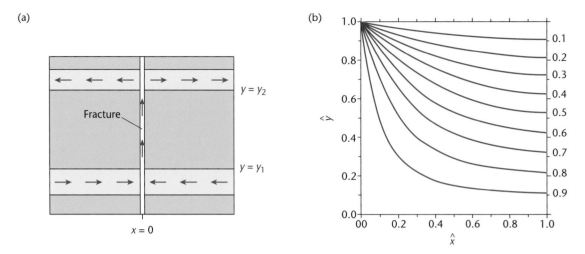

Fig. 10.8 (a) Set-up for a fracture connecting two horizontal aquifers, separated by the vertical distance y_1–y_2, with arrows showing fluid flow; (b) isotherms adjacent to a vertical fracture extending from 0 to 1 along the dimensionless vertical coordinate $\hat{y}$, where the horizontal distance is also scaled by the vertical height of the fracture l_0. Note that the isotherms swing up towards parallelism with the fracture, indicating heating from the fracture fluids, whereas at a distance of $x>l_0$, the isotherms are horizontal, indicating negligible heating from the fracture. For effective heating of the basin-fill, the fracture spacing must therefore be less than the fracture length. After Wangen (2010, figs 6.22 and 6.23). © Magnus Wangen 2010, reproduced with the permission of Cambridge University Press.

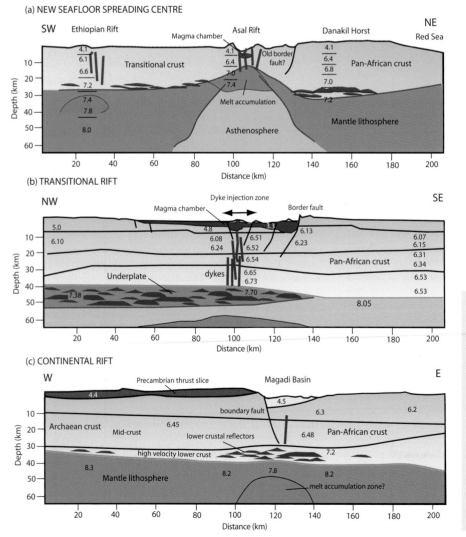

Fig. 10.9 Schematic cross-sections showing the evolutionary development from a continental rift in the south (c), through a transitional rift (b) to a new oceanic spreading centre in the north (a), in the African-Arabian rift system (after Ebinger & Scholz 2012, p. 196, fig. 9.9). Thick red line is Moho, orange is top of thick igneous underplate. Dykes sourced from the mantle and shallow magma chambers are also shown. Dyke injection takes up a considerable amount of horizontal extension and potentially has a strong impact on the thermal state of basin sediments. Numbers are seismic refraction velocities in km s^{-1}. Reproduced with permission of John Wiley & Sons, Ltd.

effect on the temperatures experienced by the sedimentary rocks of the basin-fill.

The heating of igneous bodies can be clearly demonstrated in the increase in vitrinite reflectance values in rocks adjacent to sills (Fig. 10.10). The heating is effective for a distance into the country rock or basin sediments roughly equal to the sill width, but also depends on the emplacement temperature of the sill. Consequently, very thick igneous accretions, such as underplates beneath the Moho, should have significant transient effects on the geotherm through the lithosphere and on surface heat flux (Fig. 10.11).

Let us take the example of a sill with a half-width a, an emplacement temperature of T_0, and let the sill rock and the country rock have the same thermal diffusivity κ. Let the distance from the centre of the sill be y, and let the temperature at any distance from the centre of the sill be T. If we make the temperature dimensionless by introducing $T^* = T/T_0$, the distance dimensionless by introducing $y^* = y/a$, and time dimensionless by introducing a characteristic diffusional time $t^* = t/\tau$, where $\tau = a^2/\kappa$, the temperature field can be solved in a manner similar to the cooling of ocean lithosphere, as discussed in §2.2.7.

The time taken for the sill to cool depends on its thickness, and ranges from days to thousands of years. For the country rock outside of the sill, the temperature first rises, reaches a maximum, and then falls. The dimensionless time t^*_{max} taken to reach the maximum temperature at a dimensionless distance y^* is approximated by

$$t^*_{max} \approx \frac{y^{*2}}{2} \qquad [10.14]$$

and the dimensionless temperature maximum is approximated by

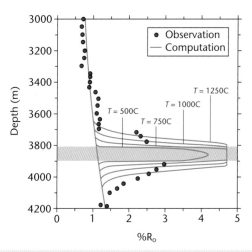

Fig. 10.10 Vitrinite reflectance observations from a borehole penetrating a 92 m-thick sill, showing the effects of heating adjacent to the sill. Vitrinite reflectances calculated based on different sill emplacement temperatures are superimposed. Comparison with measured $R_o\%$ shows a significant discrepancy, with high $R_o\%$ values a long way from the sill margin. This is most likely due to heating by convective fluids. If so, it is likely that the sill was intruded at temperatures above c.1000 °C. The vitrinite above and below the sill is altered over a distance approximately equal to the sill thickness. After Wangen (2010, fig 6.30a). © Magnus Wangen 2010, reproduced with the permission of Cambridge University Press.

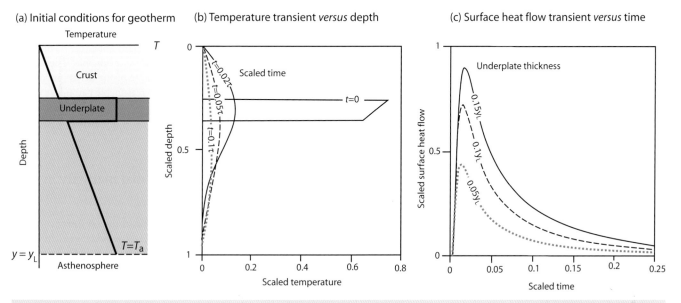

Fig. 10.11 Thermal effects of the intrusion of an igneous underplate below the Moho. (a) Initial conditions and (b) temperature profiles at different times after underplating. Temperature is normalised by the temperature at the base of the lithosphere, depth is normalised by the thickness of the lithosphere, and time by the diffusive lithospheric time constant y_L^2/κ, for an initial lithospheric thickness of 150 km. (c) Transient surface heat flux for different thicknesses of underplate. Flux is scaled by the steady-state solution prior to underplating. After Jaupart & Mareschal (2011, figs 7.18, 7.19 and 7.20). © Claude Jaupart and Jean-Claude Mareschal, reproduced with the permission of Cambridge University Press, 2002.

$$T_{max}^{*} \approx \sqrt{\frac{2}{\pi e}} \frac{1}{y^{*}} \qquad\qquad [10.15]$$

where e is the exponential coefficient (2.73).

With increasing distance from the sill, the time of the achievement of maximum temperatures also increases, but with increasing distance the magnitude of the maximum temperature diminishes. Far from the sill at $y^{*} = y/a = 100$ (10 sill half-widths from the centre line of a 10 m-thick sill), eqn. [10.14] gives $t_{max}^{*} = 5000$, which gives a time of 16 kyr for a diffusivity of 10^{-6} m^2 s^{-1}. From eqn. [10.15] the maximum temperature impact from the sill is c.50 °C. However, close to the sill at $y^{*} = 2$ (20 m from the centreline of a 10 m-thick sill), the max temperature is achieved at a time of just 6 years. The maximum temperature impact from the sill is c.340 °C at this position.

The heating from the sill is clearly extremely important in affecting thermal maturity, but is restricted within a zone approximately one to two sill widths either side of it. As we have seen (Fig. 10.10), thermal indicators such as vitrinite reflectance and apatite fission track ages will be reset by the thermal effects of the sill within this zone.

For igneous underplates, we can assume that the initial temperature of the underplated material is the same as the temperature of the asthenosphere. As the material is molten or partially molten, its solidification will release latent heat (Jaupart & Mareschal 2011, p. 210). The solutions for the temperature field surrounding the underplate and the surface heat flow are shown graphically in Figs 10.11b and 10.11c. The temperature transient (scaled by the asthenospheric temperature) decays rapidly in amplitude over time (scaled by the diffusive time scale y_L^2/κ), and is never large except very close to the underplate.

Inserting reasonable values for the initial lithospheric thickness (y_L=125 km), thermal diffusivity (κ=10^{-6} m^2 s^{-1}), and initial underplate temperature T_0 of 1100 °C, the temperature transient is a maximum of 165 °C 10 Myr after emplacement. But after 50 Myr, the temperature transient has reduced to a maximum of about 55 °C.

The surface heat flux peak occurs some time after underplate emplacement. For a thick underplate of thickness 18.75 km, the surface heat flux transient is nearly equal to the steady-state heat flux through the lithosphere (c.30 mW m^{-2}); it occurs 10 Myr after emplacement and decreases to close to background values after about 100 Myr.

The thermal effects cause an uplift of the overlying Earth's surface due to the effects of thermal expansion over the lithospheric thickness. The amplitude of uplift is likely to be small and depends primarily on the thickness of the underplate. If the volumetric coefficient of thermal expansion $\alpha = 3 \times 10^{-5}$ K^{-1}, and the excess temperature of the underplate compared to the surrounding temperature is 600 °C, the surface uplift is 0.018 times the underplate thickness, that is, it is c.340 m for an underplate thickness of 18.75 km.

10.3.9   Thermal effects of delamination

Delamination, or convective removal, of the mantle lithosphere and its replacement by asthenosphere has been invoked for the surface uplift of orogenic plateaux regions such as Tibet (England & Houseman 1989; Jiménez-Munt & Platt 2006). This is expected to have effects in terms of the surface heat flow as well as uplift of the topographic surface. The effects of the thermal perturbation are shown in Fig. 10.12.

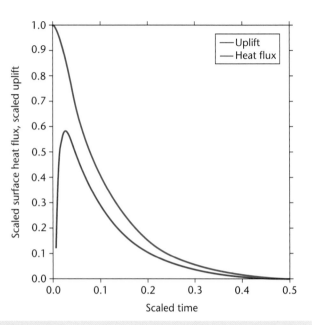

Fig. 10.12   Thermal effects of delamination. Surface flux variation (red line) and surface uplift (blue line) after delamination of the lithospheric mantle. Crustal thickness is ¼ initial lithospheric thickness. Heat flow is scaled by the steady-state solution prior to delamination, and time by the diffusive time scale of the lithosphere y_L^2/κ. After Jaupart & Mareschal (2011, fig. 7.21). © Claude Jaupart and Jean-Claude Mareschal, reproduced with the permission of Cambridge University Press, 2002.

In Fig. 10.12 the surface heat flux scaled by the steady-state heat flow prior to delamination lags behind the surface uplift by c.25 Myr and its amplitude is relatively small, consistent with the generally low surface heat flow values measured in regions such as the Colorado Plateau. For a Moho temperature of 600 °C and an initial lithospheric thickness of 150 km, the surface heat flux transient is just 7.5 mW m^{-2}, so the effect on basin thermal maturity is negligible. The surface uplift, however, instantly peaks at the time of the delamination and its amplitude is of the order of 1 km for reasonable parameter values. Delamination is therefore important in uplifting the Earth's surface but, in the absence of crustal extension, is unimportant in affecting the thermal maturity of basin sediments.

10.4   Measurements of thermal maturity in sedimentary basins

Approaches to understanding the mechanisms of subsidence and uplift in sedimentary basins invariably involve model predictions for burial history, heat flow and paleotemperature that can be compared with observational evidence. But what direct observations are possible in a sedimentary basin that provide information on thermal history and thereby allow a test and calibration of these models? A wide range of techniques are currently available, making use of the changes with temperature of organic particles, clay minerals, geochemical markers, and of the annealing and diffusion histories of certain minerals such as apatite. Most of these techniques generate data on the *maximum* temperature reached by a particle within the basin-fill. This is because thermal reactions are irreversible. Only thermochronological techniques such as apatite fission track and U–Th/He analyses provide any information on thermal evolution. Each technique has its advantages and drawbacks. Taken together, a range of diverse techniques may provide important information on the thermal history of the basin-fill. This information is critically important, not only in understanding the driving mechanisms for basin formation, but also in the evaluation of hydrocarbon prospectivity (Tissot & Welte 1978) (chapter 11).

In the following sections, a number of techniques are presented in outline. For further information and detailed applications, the reader is referred to the references cited. Useful summaries are found in Héroux *et al.* (1979), Gallagher *et al.* (1998), Giles (1997) and Beardsmore and Cull (2001). Thermochronological techniques are covered by Braun *et al.* (2006) and in the compilation edited by Lisker *et al.* (2009).

10.4.1 Estimation of formation temperature from borehole measurements

Formation temperatures from boreholes are used in thermal modelling studies to calculate the geothermal gradient and basal heat flow of the sedimentary section. The temperature in the borehole is recorded on each logging run, using a suite of maximum recording thermometers. Because the circulation of drilling fluid tends to cool the formation, it is necessary to analyse the rate at which temperature restores itself to its original true formation value using temperatures recorded on each successive logging run within a suite of logs. These temperatures may be plotted on a 'Horner'-type plot, as described by Dowdle and Cobb (1975).

The form of the temperature build-up plot is shown in an example from the Gulf Coast in Fig. 10.13. Temperature measured on each logging run is plotted against a dimensionless time factor, $(t_c + \Delta t)/\Delta t$, where t_c is the cooling time (the duration of mud circulation from the time the rock formation opposite the thermometer was drilled to the time circulation of the drilling mud stopped), and Δt is the thermal recovery time (time since mud circulation stopped to the time the logging sonde is in position at the bottom of the borehole). A fully recovered or stabilised formation temperature T_f is obtained by extrapolation to the ordinate, where $(t_c + \Delta t)/\Delta t = 1$. Serra (1984, 1986) discusses the calculation of heat flow from borehole temperature measurements.

10.4.2 Organic indicators

The progressive maturation of organic materials has long been understood in terms of *coal rank*. The coalification process changes peat to anthracite through the intermediate steps of brown coal (lignite and sub-bituminous coal) and bituminous coal. During coalification, the percentage of carbon increases whereas moisture and volatiles are gradually eliminated (Fig. 10.14). Of much greater importance to basin analysis is the quantitative measurement of maturity through the reflectance of the vitrinite maceral, and to a lesser extent the structural changes in distinctive organic molecules known as biomarkers, and the semi-quantitative assessment of spore colour.

Vitrinite reflectance is the most widely used indicator of maturity of organic materials. It is an optical parameter and is denoted by VR or R_o (reflectance in oil). Standard procedures for the measurement of vitrinite reflectance are given in Bostick and Alpern (1977), Bostick (1979), Hunt (1979), Dow and O'Connor (1982), Stach *et al.* (1982),

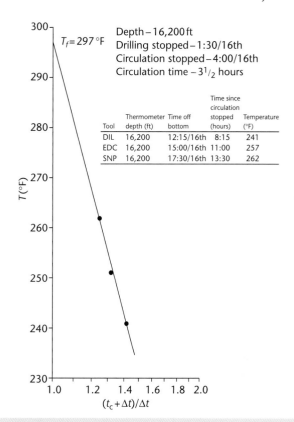

Fig. 10.13 The determination of true formation temperature from a Horner plot (after Dowdle & Cobb 1975). This example is from a high-temperature well in the Gulf Coast, USA. The depth at which measurements were taken was 16,200 ft (c.5 km). Temperature increased from 241 °F at 8 h 15 min after circulation of mud stopped, to 262 °F taken 13 h 30 min after circulation stopped. The estimated formation temperature T_f is 297 °F. © 1975. Reproduced with permission of Society of Petroleum Engineers (SPE).

van Gijzel (1982) and Tissot and Welte (1984). The reflectance of the vitrinite group of macerals appears to vary smoothly and predictably with temperature (Lopatin 1971; Burnham & Sweeney 1989; Sweeney & Burnham 1990).

Drawbacks in the use of vitrinite reflectance measurements are outlined by Héroux *et al.* (1979), Kübler *et al.* (1979) and Durand *et al.* (1986). These arise from a number of problems. Reflectance measurements taken from maceral types other than vitrinite (especially in lacustrine and marine sediments), and even from different macerals within the vitrinite group, may significantly differ (Bensley & Crelling 1994). Other drawbacks are the possibility of reworking of organic material (especially in sandstones), and the lack of higher plants yielding vitrinite in pre-Devonian strata. Vitrinite reflectance tends to be unreliable at low levels of thermal maturity (R_o less than 0.7 or 0.8%). At high temperatures equivalent to depths of >4 km the vitrinite maceral is increasingly anisotropic, making accurate measurement problematical. Nevertheless, with care, reflectance values are a good indicator of maximum paleotemperature within the approximate depth range of 1 to 4 km (Whelan & Thompson-Rizer 1993).

Unfortunately, the distribution of activation energies and value of the frequency factor (see eqn. [10.1]) for the maturation of vitrinite

(a)

Evolution of organic matter and coal

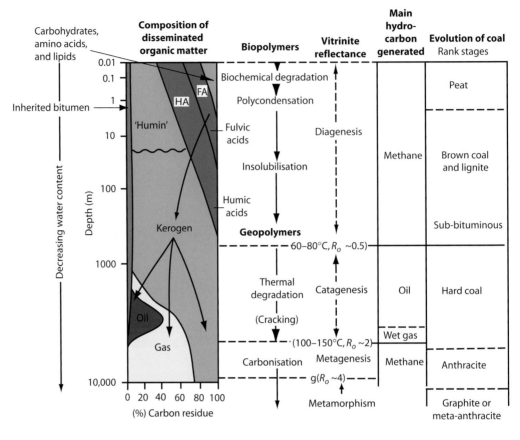

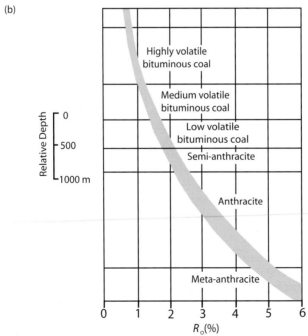

(b)

Fig. 10.14 (a) Evolution of organic matter from organic-rich sediment or peat through the various ranks of coal to meta-anthracite, and the main hydrocarbons generated, correlated with vitrinite reflectance values. After Tissot & Welte (1984), reproduced with kind permission from Springer Science+Business Media B.V. Coal ranks from Stach *et al.* (1982). (b) Correlation of coal rank with vitrinite reflectance R_0 (%).

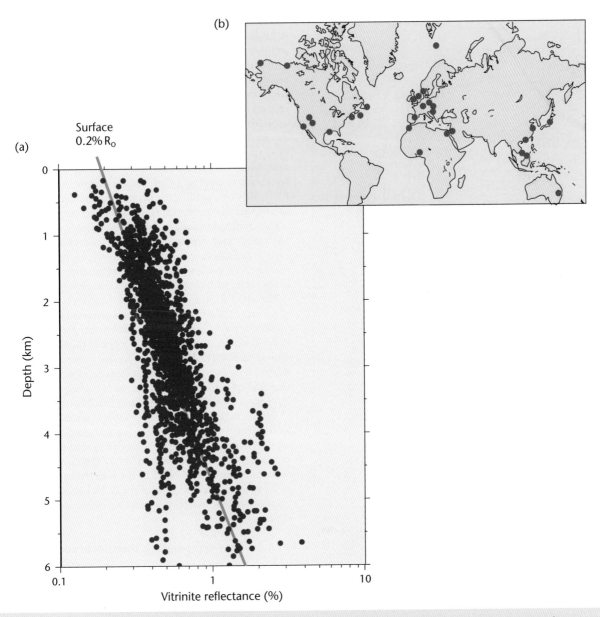

Fig. 10.15 (a) Vitrinite reflectance (logarithmic scale) versus depth for a wide range of selected extensional sedimentary basins (marked on world map in (b)) with predictable subsidence histories. After Rowley & White (pers. comm.).

are not convincingly known (Lerche *et al.* 1984; Burnham & Sweeney 1989; Lakshmanan *et al.* 1991), and a wide range of values has been proposed. The most reliable estimates give a value of E_a in the range 200–300 kJ mol^{-1}, and of A from 2.5×10^{10} to 7.48×10^{18} s^{-1}. It must be emphasised, therefore, that the uncertainties in knowledge of the parameters in the Arrhenius equation make the use of vitrinite reflectance data useful but inexact, and VR measurements should always be compared with paleotemperature estimates from other indices.

Vitrinite reflectance measurements from samples recovered at different depths allow a plot of vitrinite reflectance versus depth to be made (e.g. Corcoran & Clayton 2001). These plots are known as VR or R_o profiles. Examples of their interpretation are given in §10.5.1.

A compilation of VR data from 28 extensional basins shows a relatively well-defined trend (Fig. 10.15), with a surface intercept at 0.2 to 0.4% R_o, and a gradient of $0.15 \pm 0.09\%$ R_o km^{-1} at depths of <4 km.

The Anadarko Basin in western Oklahoma has some of the deepest exploratory wells in the world, penetrating to more than 7900 m (~26,000 ft), and is therefore an excellent case study for thermal maturation. Vitrinite reflectance contours (*isoreflectance lines*) on the Upper Devonian–Lower Mississippian *Woodford Shale* have been constructed from 28 boreholes (Fig. 10.16) (Cardott & Lambert 1985). Isoreflectance maps are useful in combination with structural contours, since cross-cutting relationships give an indication of local thermal anomalies superimposed on the burial-related maturation.

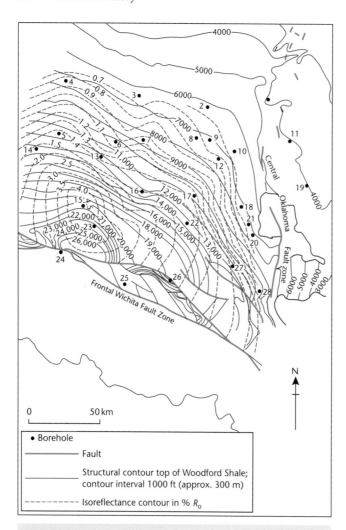

Fig. 10.16 Combined isoreflectance and structure map of the Woodford Shale of the Anadarko Basin of Oklahoma (Cardott & Lambert 1985). Vitrinite reflectance values in general increase with depth of burial, but strong cross-cutting relationships of the isoreflectance and structure contours suggest that there may have been local thermal disturbances superimposed on the burial-related maturation. AAPG © 1985. Reprinted by permission of the AAPG whose permission is required for further use.

Since vitrinite reflectance is a non-reversible thermal indicator, it is important to know whether the distribution of vitrinite reflectance with depth in a sedimentary basin allows a geotherm at a particular time in basin history to be constructed. For example, is it possible that the maximum temperature at different depths in the basin was reached at different times? If so, a R_o profile would not represent a single geotherm that existed at a particular time. This problem can be approached from a forward model of paleotemperature and vitrinite reflectance in a basin with a time-dependent basal heat flow. The heat flow and temperature algorithms are based on the uniform stretching model, but with an additional heat flow originating from the internal heat generation of crustal rocks and sediments due to radioactive decay. The time-temperature history of a number of horizons within a model borehole can then be calculated based on the thermal model and the subsidence history derived from the stratigraphy penetrated by the borehole. Paleotemperatures can be con-

verted to vitrinite reflectance using empirical relationships (Barker & Pawliewicz 1986; Burnham & Sweeney 1989; Sweeney & Burnham 1990) to create a synthetic VR profile (Fig. 10.17).

Two interacting effects control the paleotemperature and vitrinite reflectance of the stratigraphy filling stretched basins. On the one hand, the basal heat flux decreases through time due to thermal relaxation following stretching, causing cooling. On the other hand, subsidence causes a reference horizon to descend to increasing depths over time, causing heating. Fig. 10.17 shows that for low to moderate values of stretching in a basin undergoing continuous subsidence, the present-day temperatures represent the maximum temperatures reached by all but the deepest horizons within the basin-fill. Consequently, the vitrinite reflectance profile can be used to estimate the geotherm. However, model results suggest that in highly stretched basins and for old (especially syn-rift) stratigraphy, VR values may represent maximum paleotemperatures attained early in basin history and not subsequently exceeded. Consequently, measured VR profiles may not accurately reflect the distribution of temperature with depth at any particular instant in time.

The use of R_o profiles in estimating the amount of section removed as a result of basin inversion events is discussed in §10.5.1.

Although vitrinite reflectance has become pre-eminent in its use in basin studies, it is not the only index of thermal maturity. Other optical parameters derived from organic material include sporinite microspectrofluorescence and spore, pollen and conodont colouration scales. Fluorescence and reflectance studies are complementary, fluorescence intensity and reflectance being inversely proportional.

Certain organic molecules (*biomarkers*) undergo transformations with increasing temperature. For example, single *isomers* in biological material are progressively converted to mixtures of isomers with increasing temperature (Abbott *et al.* 1990). *Aromatisation* reactions can also be used, such as the conversion of mono-aromatic to tri-aromatic steroids (Mackenzie & McKenzie 1983). Biomarker transformations take place at rates approximated by the Arrhenius equation.

10.4.3   Low-temperature thermochronometers

Low-temperature thermochronometers are discussed in relation to estimates of erosion in §7.3. The emphasis in this section is on the use of low-temperature thermochronometers in assessing the thermal history of the basin-fill.

Fission track analysis is a very widely used thermochronological technique based on the atomic damage caused to minerals such as apatite, zircon and sphene by their spontaneous decay, almost exclusively by the fission of ^{238}U. The fission of ^{238}U produces a trail of damage to the lattice, known as a *fission track*. Fission tracks become visible when the crystal is chemically etched and viewed at high magnification.

Fission track ages and track length distributions differ according to the time-temperature history of the mineral. For example, *linear heating* (Fig. 10.18a), which may be thought of as characteristic of the trajectory of deposited detrital apatite grains followed by burial in the basin-fill, produces a tightly clustered, unimodal, symmetrical histogram of track length with a short mean track length (MTL). All tracks experience the same maximum temperature. The fission track age bears no relation to any specific or thermal event. Alternatively, basin inversion may cause *heating followed by cooling* (Fig. 10.18b). Old tracks formed during the heating phase have a similar length, with a short MTL. Newer tracks experience different maximum

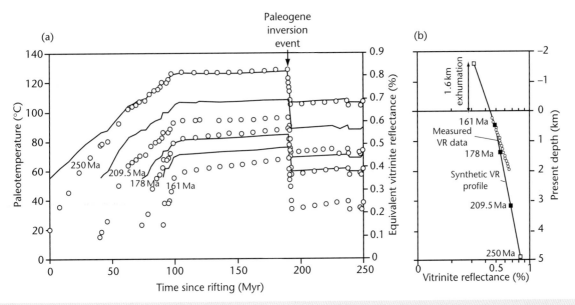

Fig. 10.17 (a) Paleotemperature and equivalent vitrinite reflectance versus time for four different horizons in well 42/21-1 in the extensional Irish Sea Basin ($\beta = 1.5$ at 250 Ma) with internal heat generation ($H_s = 9.6 \times 10^{-10}\,W\,kg^{-1}$; $h_r = 10\,km$) and the thermal conductivity structure shown in Fig. 10.2. Open circles are data points derived by forward modelling temperature from the borehole subsidence history. Solid lines are equivalent vitrinite reflectance. (b) The forward modelled R_0 profile is compared with measured values. Extrapolation of the profile upwards gives an estimate of 1.6 km of denudation since the VR values were set at the maximum temperature at c.60 Ma (c.190 Myr since rifting).

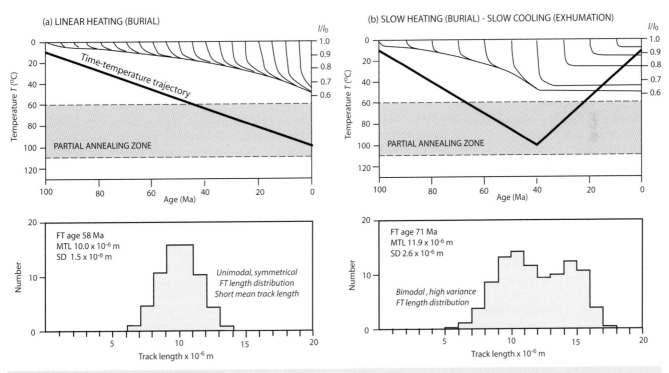

Fig. 10.18 Simple thermal histories and the predicted fission track length parameters using Durango apatite and the annealing model of Laslett et al. (1987), after Gallagher et al. (1998). Each model simulation has 20 tracks formed at equal-time increments over the total duration of the thermal history. Track length reduction (annealing) is shown by the ratio l/l_0, where l is the track length and l_0 is the initial track length. Heavy solid line is time-temperature trajectory. MTL, mean track length. (a) Linear heating as a result of burial, giving a unimodal and symmetrical track length histogram with a short mean track length of $10 \times 10^{-6}\,m$ and a standard deviation of $1.5 \times 10^{-6}\,m$. The fission track age (58 Ma) does not correspond to any distinct tectonic or thermal event; (b) Heating-cooling, due to slow burial and slow exhumation, causes a bimodal track length distribution with a mean track length of $11.9 \times 10^{-6}\,m$ and a large standard deviation of $2.6 \times 10^{-6}\,m$. The fission track age (71 Ma) does not correspond to any distinct tectonic or thermal event.

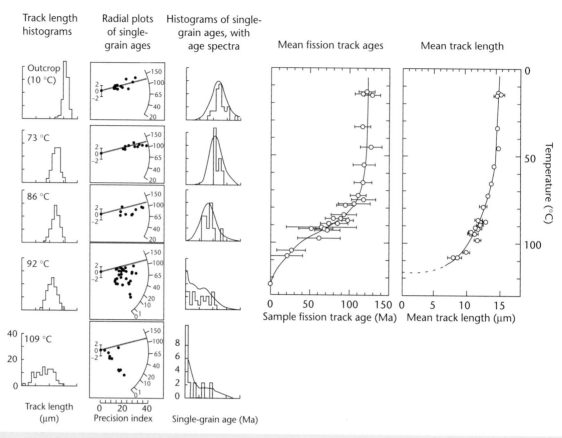

Fig. 10.19 Apatite fission track data from the Otway Basin, southeast Australia (Gleadow & Duddy 1981). The stratigraphic age of the samples is c.120 Ma (shown on the radial plots as a grey band). Note the progressive annealing of fission tracks and reduction of fission track ages of single grains with depth of burial. Reprinted with permission from Elsevier.

temperatures during the cooling phase, causing the histogram to be bimodal, with a high standard deviation. The fission track age does not accurately reflect the timing of the onset of basin inversion.

Apatite fission track analysis has been extensively applied to the thermal history of sedimentary basins (Naeser 1979). This is because the temperature range over which apatite fission track analysis is sensitive (c.50 °C to 120 °C) is also the temperature range over which hydrocarbons are generated (Gleadow *et al.* 1983). The fission track data from the Otway Basin of Australia (Gleadow & Duddy 1981) illustrate the important principles (Fig. 10.19). At shallow depths above the partial annealing zone (PAZ), the fission track lengths are tightly clustered and long (c.14 μm), indicating minimal annealing. The fission track age is more or less equivalent to the stratigraphic age of the samples (120 Ma). With greater depth, fission track length histograms show a wider distribution and a shorter MTL. The fission track ages decrease due to the greater amount of annealing at elevated temperatures, reaching 0 Ma at a temperature of 120 °C (Fig. 10.19).

Detrital grains commonly contain a combination of information on their exhumational history in the source area and their burial history in the basin (Rohrman *et al.* 1996; Carter & Gallagher 2004). Apatite fission track analysis of a number of samples at different depths in a sedimentary basin can be used to calculate geothermal gradients if the effects of the exhumational history can be identified (§10.5.2).

The U–Th/He method is used less in basin studies because the thermal history is reset at relatively low temperatures (50–70 °C) that are common in sedimentary basins. Helium age distributions are known to decrease rapidly down boreholes because of the effect of temperature on diffusion rates (Wolf *et al.* 1998).

There are a number of factors that may affect the distribution of temperature with depth, including variations in thermal conductivity, effects of fluids and internal heat generation. Consequently, calculation of geothermal gradients from apatite fission track data alone are likely to be significantly in error. A combination of paleothermal data from apatite fission track, vitrinite reflectance and any other available techniques is therefore always advised.

10.4.4   Mineralogical and geochemical indices

Mineralogical parameters are controlled by the temperature and chemical properties of the diagenetic environment of the sediment (Fig. 10.20). A number of diagenetic models exist (e.g. Frey *et al.* 1980; Burley *et al.* 1985; Worden & Morad 2003) that allow an interpretation of the sequence of authigenic minerals in terms of their relationship to their depositional environment or surface chemistry (*eogenesis*), the burial or subsurface conditions (*mesogenesis*), and the weathering or re-exposure to surface conditions (*telogenesis*).

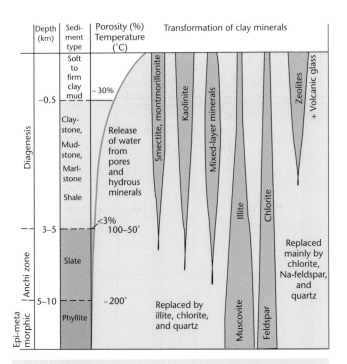

Fig. 10.20 Mineralogical indices. Diagenesis of clay-rich sediments as a function of depth and temperature, showing the most important mineral transformations. After Frey *et al.* (1980); Heling (1988).

The reader is referred to §11.5 for further details on the impact of diagenesis on carbonate and sandstone reservoirs.

Since eogenetic changes are strongly related to depositional environment, climate and associated pore water chemistry, they are of limited use in thermal modelling. However, mesogenesis marks the removal of the sediment from the predominant influence of surface agents in the interstitial pore water. A number of temperature-dependent reactions commonly take place in siliciclastic rocks. Kaolinite transforms to dickite, smectites transform to illite via a process of interlayering, and chloritisation takes place. Physical processes accompany these chemical changes during burial diagenesis. The most important result is *compaction* due to the weight of the overlying sediments (§9.2). In sandstones, compaction brings about a number of porosity-reducing adjustments, including initial mechanical compaction, which simply compresses grains together, rotation, grain slippage, brittle grain deformation and fracturing and plastic deformation of ductile grains.

The best-documented mineral transformations of use in evaluating thermal maturity are from shaly mudstones, where the clay mineral assemblages, the position of the (001) reflection of smectite, the percentage of illite layers in the mixed-layer illite 2:1 expandable, and the illite crystallinity index are used. Derivation of an approximate relationship between mineralogical changes of this type and organic maturity indices has been attempted (Fig. 10.21). One of the fundamental problems is the need for quantitative indices with which to correlate temperature-driven changes. Kübler *et al.* (1979), for example, proposed a quantitative measure of the crystallinity of illite using the width of the illite peak on an X-ray diffractogram measured at half of the peak height. The beginning of oil generation ($R_o = 0.5\%$)

has been correlated with the disappearance of smectite (e.g. Powell *et al.* 1978; Kübler *et al.* 1979).

Quartz cementation and consequent porosity loss is also a temperature-related process acting in sandstones buried to temperatures above 60°C (Walderhaug 1996). It is dealt with in more detail in §10.5.3. At about the same temperature of 60°C, smectite clay starts to react with K-feldspar to form illite (Hower *et al.* 1976; Nadeau & Reynolds 1981). At higher temperatures of 100–120°C, illite precipitates at the expense of kaolinite and K-feldspar (Bjorlykke *et al.* 1986). This can be seen in the percentage of illite in the clay fraction of sandstones penetrated by wells on the Norwegian continental shelf (Fig. 10.22). Quartz cementation and illite precipitation are both temperature-dependent, with activation energies of c.15 kcal mol^{-1} and 20 kcal mol^{-1} respectively. The combined temperature-driven effects of quartz cementation and illite precipitation makes sandstones that have experienced temperatures above 120°C unattractive as potential reservoirs and aquifers (§11.5).

The K content of diagenetic illites allows the age of illite growth to be calculated using the $^{40}K-^{40}Ar$ radiometric technique. K–Ar ages from samples obtained from different depths can then be used to plot age-depth relationships (Hamilton *et al.* 1989). The K–Ar age should, for a continuously subsiding basin, increase with depth, since the deeper horizons should reach the critical temperature for illite cementation first. A study of sandstone reservoir rocks from the Brent province of the North Sea showed that illite growth began at temperatures of 100–110°C (corresponding to a vitrinite reflectance in overlying shales of 0.62%), but that there was considerable scatter in the K–Ar age/depth relationship (Hamilton *et al.* 1992).

The investigation of inclusions of fluids in the rocks and minerals of sedimentary basins forms part of a broader field of paleofluid analysis (Parnell 2010). The inclusions are microscopic bubbles of fluid (or gas) trapped within crystals or fractures. The inclusions preserve their composition and pressure-temperature conditions at the time of entrapment (Fig. 10.23). The fluids analysed commonly come from quartz cements (Walderhaug 1994) and microfractures in quartz grains (Parnell *et al.* 2005). In petroliferous basins, the paleofluids may be hydrocarbons.

The passage of hot fluids through a sedimentary basin is marked by fluid inclusion temperatures far in excess of those expected from the geothermal gradient. In many cases these elevated temperatures must be due to transport of hot fluids up faults and through aquifers (Ziagos & Blackwell 1986; Hulen *et al.* 1994). These high temperatures may represent inversions of the normal temperature profile in the sedimentary basin. Fluid inclusions also allow transient thermal events to be recorded, which are not recorded in other thermochronometers that are based on a time integral for maturity, such as vitrinite reflectance.

10.5 Application of thermal maturity measurements

10.5.1 Vitrinite reflectance (R_o) profiles

Vitrinite reflectance measurements can be plotted as a function of depth to give R_o profiles. The slope of the R_o curves gives an indication of the geothermal gradients in the history of the basin. Although

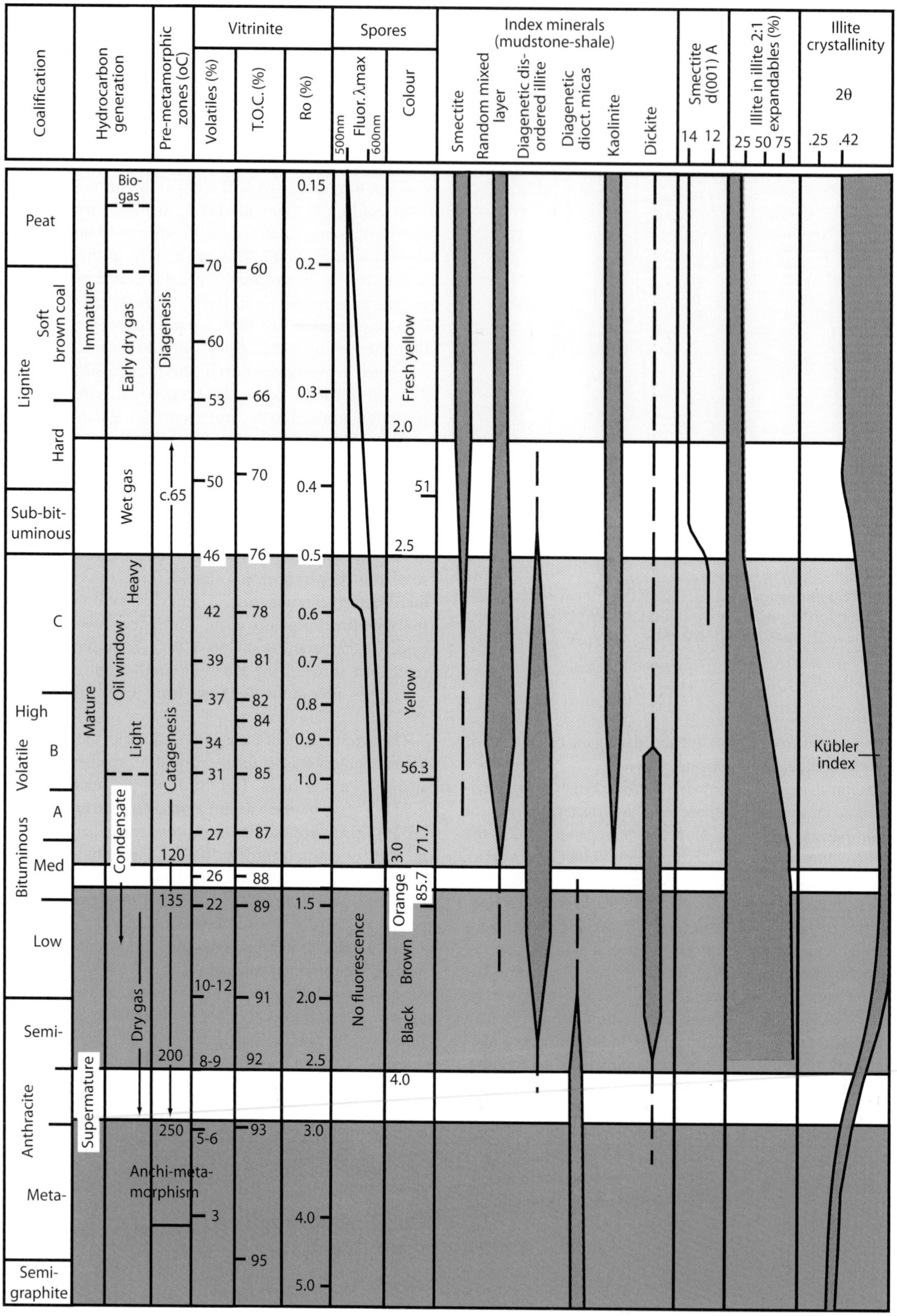

Fig. 10.21 Comparison of a range of thermal indices, modified from Héroux *et al.* (1979). Thermal indices shown are coal rank, vitrinite total organic carbon and % volatiles, vitrinite reflectance, spore fluorescence and colouration, typical clay mineral distributions, position of the (001) reflection of smectite, percent illite in the mixed layer illite 2:1 expandable, and illite crystallinity (Kübler index). These indices are correlated with temperature and hydrocarbon products generated. AAPG © 1979. Reprinted by permission of the AAPG whose permission is required for further use.

(a)

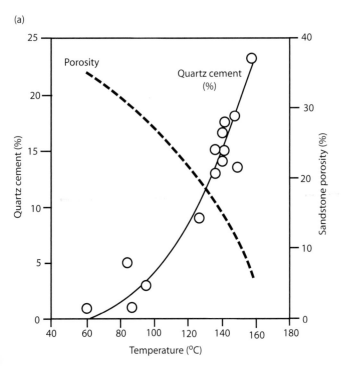

(b)

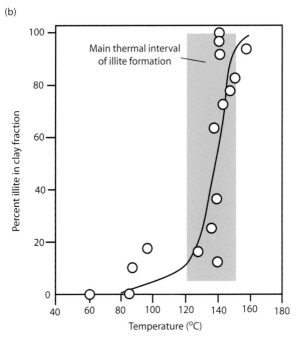

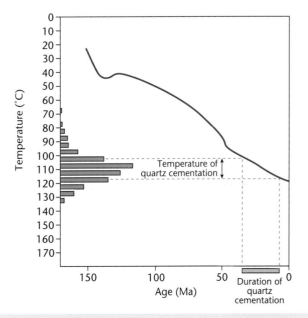

Fig. 10.23 Timing of quartz cementation derived from the combination of temperatures of cementation from fluid inclusion analysis (green bars) with paleotemperature from burial history (red line), Upper Jurassic Brae Formation of Miller Field, northern North Sea. Line shows temperature history of the uppermost part of the Brae Formation in well 16/7b-20 based on a burial history analysis. The temperature range of quartz cementation is based on fluid inclusion analysis of authigenic quartz cements. Quartz cementation took place 35–10 Myr ago. From Gluyas *et al.* (2000), reproduced with permission of John Wiley & Sons, Ltd.

Fig. 10.22 (a) Plot of temperature versus quartz cement content and porosity for wells from the Norwegian continental shelf (Bjorkum & Nadeau 1998), showing an exponential increase in amount of quartz cement with depth. (b) Plot of temperature versus diagenetic illite content measured by X-ray diffraction. Increasing illite content severely reduces permeability. The main thermal interval for illite formation is 120–150°C.

many profile shapes are possible (Fig. 10.24), they generally indicate an exponential evolution of the organic matter with time (Dow 1977), as expected from the kinetics described in §10.2. In basins largely unaffected by major unconformities, young dip-slip faulting and localised igneous activity, there should therefore be a linear relationship between depth and log R_o. Plots of large numbers of reflectance measurements from many sedimentary basins worldwide (Fig. 10.15) (Rowley & White 1998), or from many locations within a basin system (Corcoran & Clayton 2001), show a strong clustering along a roughly linear trend in depth-log R_o space, with an intercept at a depth of zero of 0.2 to 0.4% (Fig. 10.15).

Individual R_o profiles follow a number of different trends, each of which is diagnostic of a particular thermal history. An example of a simple sublinear profile is the Terrebonne Parish well in Louisiana (Heling & Teichmüller 1974) (Fig. 10.24a). R_o is 0.5% at 3 km and 1% at 5 km. It indicates a normal and constant geothermal gradient through time. The Woodford Shale of the Anadarko Basin is another example of a sublinear R_o profile with a surface intercept at $R_o = 0.2\%$, indicating the amount of maturation that the vitrinite had undergone prior to deposition (Fig. 10.24b).

Other R_o profiles are more complex. A dogleg pattern of two linear segments of different slope indicates that two periods of different geothermal gradient have occurred. This may result from a thermal 'event' occurring at the time presented by the break in slope. Such an interpretation is plausible for the R_o profiles from boreholes in the Rhine Graben (Teichmüller 1970, 1982; Robert 1988) (Fig. 10.25).

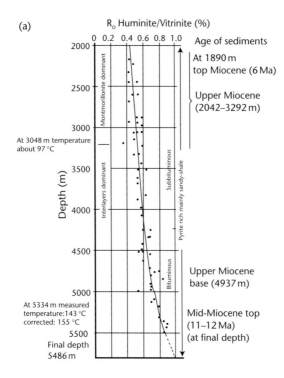

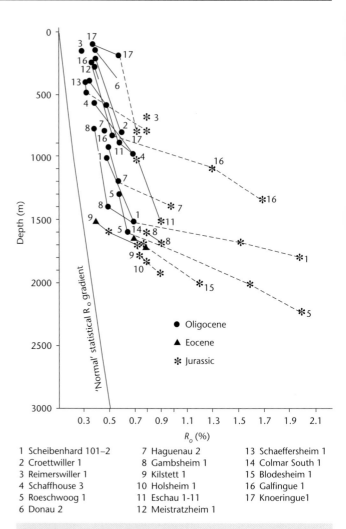

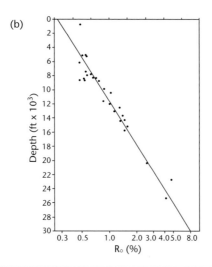

Fig. 10.25 legend table:

1 Scheibenhard 101–2	7 Haguenau 2	13 Schaeffersheim 1
2 Croettwiller 1	8 Gambsheim 1	14 Colmar South 1
3 Reimerswiller 1	9 Kilstett 1	15 Blodesheim 1
4 Schaffhouse 3	10 Holsheim 1	16 Galfingue 1
5 Roeschwoog 1	11 Eschau 1-11	17 Knoeringue1
6 Donau 2	12 Meistratzheim 1	

Fig. 10.25 Reflectance profiles from a number of wells in the Alsace region of the Rhine Graben (after Teichmüller 1970). In general, there are pronounced dog-legs in the R_o profiles at about the age of the Oligocene–Eocene boundary. The post-Eocene history shows a 'normal' gradient, whereas the pre-Oligocene sedimentary rocks have high reflectance values in relation to their depth of burial. This suggests that rifting in the late Eocene caused higher than normal maturity in the older rocks of the basin-fill. Reproduced with kind permission from Springer Science+Business Media B.V.

Fig. 10.24 (a) Vitrinite reflectance profile for Terrebonne Parish, Pont au Fer well in Louisiana. The profile is sublinear and continuous, suggesting a near-constant geothermal gradient through time (after Heling & Teichmüller 1974; reproduced with permission from Springer Science+Business Media B.V.). (b) Woodford Shale of the Anadarko Basin also shows a good sublinear trend (Cardott & Lambert 1985; AAPG © 1985, reprinted by permission of the AAPG whose permission is required for further use).

R_o profiles may consist of two sublinear segments offset by a sharp break or jump in R_o values. The jump may correspond to an unconformity with a large stratigraphic gap. This is well illustrated in the Mazères 2 borehole in the Lacq area of the Aquitaine Basin, France (Fig. 10.26), where R_o values jump from c.0.8% to c.2.4% at the level of an unconformity separating Aptian–Albian rocks from underlying Kimmeridgian.

If there is a known (logarithmic) relationship of R_o with depth, and the subsidence history of a sedimentary basin is known, the R_o values give an indication of the variation of the geothermal gradient through time. This then allows different tectonic histories to be tested (Middleton 1982).

10.5.1.1 Estimation of structural inversion from R_o profiles

R_o profiles can be used to estimate the amount of denudation resulting from a period of basin inversion, since the vitrinite locks in information about the maximum paleotemperature experienced. The technique has been used extensively, particularly in the region of the northwest European continental shelf surrounding the British Isles, where an important crustal uplift event took place in the Early Tertiary (Rowley & White 1998). This uplift event, which was

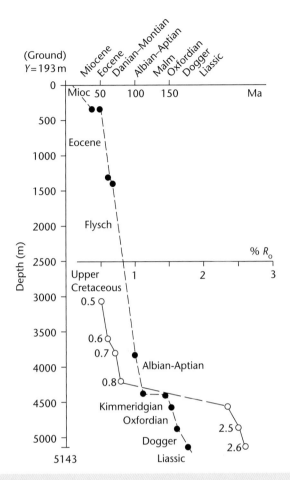

Fig. 10.26 Vitrinite reflectance profile for the Mazères borehole in the Lacq region of southern France (open circles, with reflectance value), and depths of stratigraphic units (black circles). The sharp increase in R_o marks an unconformity between the Lower Cretaceous and the Upper Jurassic. After Robert (1988), reproduced with kind permission of Springer Science+Business B.V.

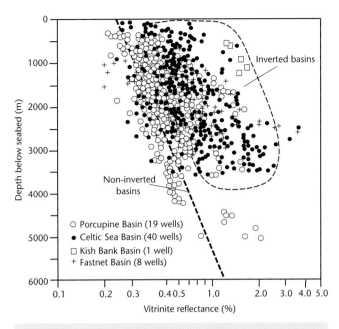

Fig. 10.27 Vitrinite reflectance versus depth for Mesozoic and Cenozoic successions penetrated in boreholes in a number of basins offshore Ireland (Corcoran & Clayton 2001). Basins such as the Porcupine Basin show a 'normal' R_o profile indicative of continual subsidence, whereas those basins that have experienced tectonic uplift and exhumation (such as the Celtic Sea, Kish Bank and Fastnet basins) have displaced R_o profiles indicating higher thermal maturity than expected from the present depth of burial.

most likely related to the thermal, isostatic and dynamic effects of the Icelandic plume (White 1988; Brodie & White 1994) caused an unconformity that cuts down variably into Mesozoic rocks in the form of a paleo-river drainage system (Hartley *et al.* 2011). The paleotemperature profiles from areas such as the Irish Sea basin system are linear and sub-parallel to the present-day geotherm (Duncan *et al.* 1998), suggesting that heating was caused primarily by burial rather than by basal heat flow variations or local magmatic and fluid flow effects. However, most of the R_o values in boreholes in the British-Irish area have elevated values compared to the global data set of non-inverted basins (Fig. 10.27) (Corcoran & Clayton 2001). Calculations of the amount of denudation from the R_o profile can be performed using a number of different methods:

- The curve of log R_o versus depth is extrapolated linearly upwards to a value of approximately 0.2% R_o (Dow 1977; Corcoran & Clayton 2001). The displacement of the depth axis is the amount of denudation. Although this method has been very widely applied, it makes the erroneous assumption that the geothermal gradient is linear. This is a particularly poor assumption in the shallow parts of basins where thermal conductivities vary strongly.

It is more likely that the geotherm is convex near the surface. Secondly, large errors may result from the incorrect choice of surface temperature (and therefore R_o at $y = 0$) at the time of maximum paleotemperatures.

- Maximum depths of burial can be estimated from VR values from empirical relations such as (Barker & Pawlewicz 1986).

$$\ln(R_o) = 0.00096T - 1.4 \qquad [10.16]$$

where T is the temperature, which must be converted to depth using a geothermal gradient. The empirically derived depth can be compared with the present-day depth of the sample. However, there are large uncertainties in the accuracy of the R_o-T relationship.

- Vitrinite reflectance can be calculated from a forward thermal model using the kinetic approaches described in §10.2 (e.g. Middleton 1982). A comparison of observed VR values with the results of a number of forward models allows the most likely parameter values and amount of denudation to be estimated. Alternatively, the amount of denudation can be estimated by inverse methods using the same chemical kinetics.

- Amounts of inversion can be estimated from the trends in sonic velocity versus depth (Giles 1997; Giles & Indrelid 1998; Ware & Turner 2002). For a given lithology, such as shales, the curve between sonic interval transit time and depth reflects the maximum depth (or minimum porosity) attained by a given interval. Inverted successions have smaller porosities and faster sonic velocities than found in continuously buried successions.

Increasingly, vitrinite reflectance data are interpreted alongside other thermal indicators, such as apatite fission track thermochronometry. Both techniques require a conversion between depth and temperature. It is estimated that total fission track annealing in apatites (closure temperature) with typical chlorine content corresponds to a vitrinite reflectance value of 0.7%. This in turn corresponds to a temperature of 110–120 °C.

10.5.2   Fission track age-depth relationships

We saw above that low-temperature thermochronometers such as apatite fission track and U–Th/He are liable to be fully or partially reset during burial in sedimentary basins (Rohrman *et al.* 1996; Carter & Gallagher 2004). Potentially, the source-area fission track signal can be separated from the burial fission track signal by the inversion of track length distributions.

Samples taken as continuously as possible along a stratigraphic depth transect (such as a borehole) are plotted in terms of apatite central age and the peak age of the youngest subpopulation. An example is given from the Miocene–Pliocene Siwalik Group exposed in the Karnali River of western Nepal (Fig. 10.28). Superimposed on the age-depth chart is the depositional age of the samples (Gautam

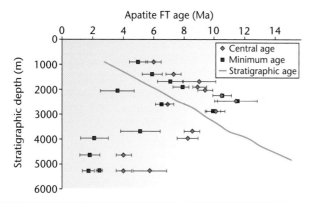

Fig. 10.28 Variations of detrital apatite fission track central age (green diamonds) and the age of the peak of the youngest component in the fission track age histogram ('minimum peak age') (red squares) with stratigraphic depth along the Karnali River section through Miocene–Pliocene (16–5 Ma) foreland basin sediment of the Siwalik Group, western Nepal (after Gautam & Fujiwara 2000). Solid line is stratigraphic age. In the upper part of the section (<2000 m), corresponding to the last 6 Myr, the minimum peak age increases down-section, and there is a nearly constant time lag of ~2 Myr between the stratigraphic age and the minimum apatite fission track (AFT) age, indicating rapid exhumation, transport and burial without annealing since deposition. Using a realistic geothermal gradient, this gives a maximum exhumation rate in the source region of about 1.5 mm yr⁻¹. Below a depth of 2000 m, peak minimum ages decrease down-section and are younger than the stratigraphic age, indicating that these samples have been annealed during burial. The transition from unannealed to partially annealed fission tracks indicates the position of the top of the paleo-partial annealing zone (PAZ), with a temperature of c.60 °C. The base of the paleo-PAZ is at the transition from partially annealed to fully annealed fission tracks. Fully annealed samples should have a constant minimum peak age with depth, with the age marking the age of final exhumation. In the lowest three samples, peak minimum ages are constant at about 2 Ma, suggesting that 2 Myr ago the Siwaliks succession was uplifted along the Main Frontal Thrust of the Himalayas. After Braun *et al.* (2006, p. 149, fig. 9.10). © Jean Braun, Peter van der Beek, Geoffrey Batt 2006, published by University of Cambridge Press.

& Fujiwara 2000). Samples at shallow levels of <2500 m depth have increasing AFT ages and youngest subpopulation ages with depth, and are older than the stratigraphic age. These samples are not annealed and record a thermal history of source-area denudational cooling. They have a lag time of ~2 Myr, indicating rapid exhumation of apatites from the Himalayan mountain belt. At depths greater than 2500 m, however, AFT central ages and youngest subpopulation ages decrease down-section, and are younger than the stratigraphic age. These samples have been partially annealed by the elevated temperatures experienced during burial in the foreland basin. The onset of annealing marks the top of the fossil apatite partial annealing zone (PAZ), representing temperatures of ~60 °C. Since this occurs at a stratigraphic depth of about 2500 m, the geothermal gradient in the basin is c.20 °C km⁻¹ if surface temperature is assumed to be 10 °C. The thermochronological data do not give the geotherm deeper in the sedimentary basin since the base of the PAZ, below which detrital samples should be fully annealed and have a constant fission track age with depth, is not found in the Nepalese samples. Where recognised, the depth of the paleo-PAZ, or partial retention zone for He, would provide an additional constraint of geothermal gradient deeper in the basin.

10.5.3   Quartz cementation

Knowledge of the thermal history of a sedimentary basin enables important changes to the basin-fill to be understood and predicted, such as the occurrence of quartz cementation of sandstones (Fig. 10.23). Strongly cemented sandstones in the subsurface have little pore space left, making them unsuitable as reservoirs and aquifers (§11.5.3).

One suggestion for quartz cementation is that supersaturated pore fluids flow long distances through sandstones and locally precipitate cements, but the high number of pore volumes required makes this unlikely. Instead, dissolution-precipitation models involve: (i) the local derivation of silica by solution, and its local precipitation in adjacent pores. Dissolution takes place in a fluid film at the high stress contact points between two or more grains; and (ii) dissolution in pressure solution seams (*stylolites*) where silica solubility is enhanced by the presence of clay minerals and micas, and precipitation in interstylolite regions. Petrographic studies suggest the second dissolution-precipitation model is the more likely.

The porosity change over time in a sandstone is dependent on the reaction rate, which can be treated as an Arrhenius process, and the surface area available for quartz cementation, which is a function of porosity. For conditions where the temperature does not change with time, the porosity reduces over time to a minimum level where the pores are unconnected, known as the *percolation threshold*. The time scale over which porosity is lost depends on the specific surface (the surface area per bulk volume) available for precipitation, and the temperature. Quartz cementation is known to occur at temperatures in the range 90–120 °C. Laboratory experiments suggest that for a gravel with particles of diameter 1 cm, the specific surface is ~10^2 m² m⁻³. The characteristic time for cementation is 100 Myr at a temperature of 100 °C at this value of specific surface, 75 Myr for a specific surface of 10^3 m² m⁻³ (equivalent to a grain diameter of 1 mm) and 50 Myr for a specific surface of 10^4 m² m⁻³ (finer sands) at the same temperature. These rates of cementation are faster than those calculated with Arrhenius-type kinetics (Walderhaug 1996), but the characteristic time is still very long: at 100 °C, and specific surface of 10^4 m² m⁻³, typical of reservoir sandstones, pore space remains for 100 Myr.

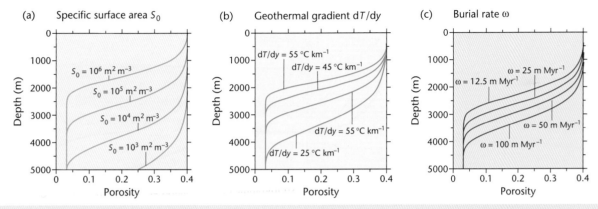

Fig. 10.29 Porosity-depth curve assuming quartz cementation during a constant burial rate (Wangen 2010, figs 11.6 and 11.7). Porosity reduction takes place by cementation only. The porosity curve depends on the initial specific surface S_0 (a), the geothermal gradient dT/dy (b) and the deposition (burial) rate ω (c). For a sandstone with specific surface area (surface area per bulk volume) $S_0 = 10^4\,m^2\,m^{-3}$, cementation starts at 2500 m and is complete by 4500 m. At a higher geothermal gradient, most of the porosity is lost by a depth of 2500 m. The impact of high temperatures is greatest at low burial rates. © Magnus Wangen 2010, reproduced with the permission of Cambridge University Press.

A constant temperature for 100 Myr is unlikely. If there were a constant rate of burial, during which time temperature increased, the quartz cementation could be married with the burial history. Neglecting the effects of compaction and assuming a constant geothermal gradient as well as a constant burial rate, the porosity-depth trajectory for a range of sandstone specific areas is shown in Fig. 10.29. The onset and termination of quartz cementation is clearly seen – from 2500 m to 4500 m for a sandstone with a specific surface of $10^4\,m^2\,m^{-3}$. In reality, compaction of the sandstone would take place before quartz cementation starts, after which cementation would retard further mechanical compaction, and further loss of porosity would be due to progressive quartz cementation. This model with constant burial rate and geothermal gradient can be modified to include a number of periods of different burial rate and geothermal gradient. For example, a period of inversion, causing cooling, would slow the cement precipitation rate, though porosity would continue to be lost (Wangen 2010, p. 375).

10.6 Geothermal and paleogeothermal signatures of basin types

We have previously seen that vitrinite reflectance measurements and apatite fission track and (U–Th)/He analysis can be used to constrain paleotemperatures and paleogeothermal gradients. Particular thermal histories are diagnostic of the formative mechanism of the basin. Robert (1988) suggested three main types of paleogeothermal history:

1 Basins with normal or near-normal paleogeothermal history

Old passive margins have present-day geothermal gradients of c.25–30°C km^{-1} (Congo 27°C km^{-1}; Gabon 25°C km^{-1}; Gulf Coast, USA, 25°C km^{-1}). The Terrebonne Parish well (Fig. 10.24a) shows a vitrinite reflectance of about 0.5% at a depth of 3 km, and the shape of the curve is sublinear. Mature passive margins therefore have near-normal geothermal gradients.

2 Cooler than normal (*hypothermal*) basins

Hypothermal basins include oceanic trenches, outer forearc and foreland basins. Ocean trenches are cold, with surface heat flows often less than 1 HFU (42 mW m^{-2}). In the Japanese archipelago Eocene–Miocene coals occur in two regions, one in Hokkaido in the north along a branch of the present-day Japan trench, and the other in Kyushu in the south is situated in a volcanic arc position relative to the Ryukyu trench. The Hokkaido region is cold, with poorly evolved coals (sub-bituminous coals with $R_o = 0.5\%$ still occurring at a depth of 5 km), whereas the volcanic arc in Kyushu is hot, containing anthracites (>2% R_o). The Mariana Trench, which is a southward continuation of the Japan trench, and its forearc region are also cold, with surface heat flows of less than 1 HFU (42 mW m^{-2}).

Foreland basins are also characterised by low present-day geothermal gradients, 22°C km^{-1} to 24°C km^{-1} being typical of the North Alpine Foreland Basin in southern Germany (Teichmüller & Teichmüller 1975, Jacob & Kuckelkorn 1977). The Anzing 3 well near Munich penetrates the autochthonous Molasse, undisturbed by Alpine tectonic events. At the base of the Tertiary at 2630 m depth the R_o is still only 0.51%. The Miesbach 1 well cuts through about 2 km of thrust sheets of the frontal thrust zone of the Alps (the subalpine zone), before penetrating the autochthonous sediments to a depth of 5738 m (Fig. 10.30). Even at this great depth, the R_o is still only 0.6%, indicating an abnormally low geothermal gradient during the Tertiary. The greater subsidence rate at Miesbach 1 (nearly 0.3 mm yr^{-1}) compared to Anzing 1 (0.1 mm yr^{-1}) may have been responsible for the very low geothermal gradient in the former. In summary, the low present-day geothermal gradients (Anzing 3, 22.8°C km^{-1}; Miesbach 1, 23.5°C km^{-1}) may have been even lower in the past during the phase of rapid subsidence related to continental collision and flexure.

3 Hotter than normal (*hyperthermal*) basins

Hyperthermal basins are found in regions of lithospheric extension such as backarc basins, oceanic and continental rift systems, some strike-slip basins and the internal arcs of zones of B-type subduction. This follows from the mechanics of basin formation in stretched

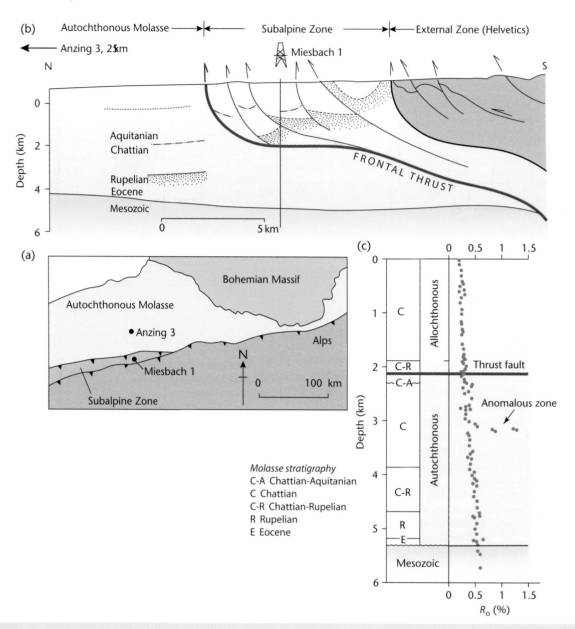

Fig. 10.30 (a) Location of the Bavarian part of the North Alpine Foreland Basin in southern Germany. Anzing 3, near Munich, and Miesbach 1, are boreholes discussed in the text. (b) Cross-section of the southernmost part of the Bavarian section of the North Alpine Foreland Basin, showing the location of Miesbach 1 in the tectonically imbricated subalpine zone (after Teichmüller & Teichmüller 1975, published with kind permission from Springer Science+Business Media B.V.). (c) The R_o profile at Miesbach 1 (Jacob & Kuckelhorn 1977) shows that the autochthonous 'Molasse' under the basal subalpine thrust is poorly evolved, not exceeding 0.6%, even at 5738 m depth. This is indicative of a very low geothermal gradient during the period of rapid sedimentation in the Oligocene.

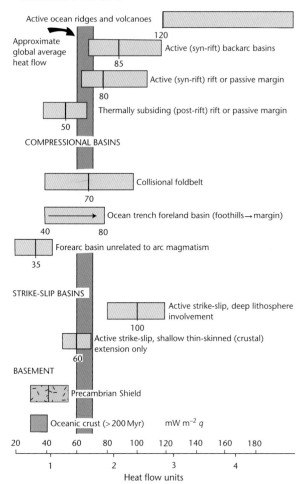

EXTENSIONAL BASINS

Fig. 10.31 Summary of the typical heat flows associated with sedimentary basins of various types.

regions, involving the raising towards the surface of isotherms. *Oceanic rifts* are zones of very high heat flows, 3 to 4 HFU (120–170 mW m^{-2}) being typical, and values occasionally reach 5 to 6 HFU (200–250 mW m^{-2}). Some Californian *strike-slip basins* have very high geothermal gradients (c.200 °C km^{-1} in Imperial Valley), so that very young sediments can be highly mature. Continental rifts have high present-day geothermal gradients (>50 °C km^{-1} in the Red Sea, up to 100 °C km^{-1} in the Upper Rhine Valley) and ancient continental rifts have intense organic maturations in their contained sediments.

Oceanic measurements and deep boreholes in the *Red Sea* (Girdler 1970) suggest that high surface heat flows (generally >3 HFU, >125 mW m^{-2}) occur in a broad band at least 300 km wide, centred on the axis of the rift. The organic maturation shown by R_o profiles and the occurrence of oil, gas and condensate fields suggests that the highest maturity is found in the south of the Red Sea, intermediate values are found in the north of the Red Sea, and the lowest occur in the Gulf of Suez. This can be correlated with different amounts of extension, the largest amount being in the south of the Suez-Red Sea system. The former elevated heat flows in the Oligo–Miocene of the Gulf of Suez have now diminished to near-normal values, while the southern Red Sea, which is still actively rifting, still has very high heat flows.

There are many other examples of high organic maturation in ancient continental rift basins: 2–3% R_o in the Lower Cretaceous of the Congo; 3.3% R_o in the Upper Cretaceous of Cameroon; 3.5% R_o in the Coniacian of the Benue Trough, Nigeria; and 5% R_o in the Permian of the Cooper Basin, Australia.

Internal arc heat flows are elevated because of magmatic activity. The Tertiary anthracites of Honshu, Japan (see above) (2–3% R_o) are an example. Similar patterns are found in ocean–continental collision zones such as the Andean Cordillera, and hyperthermal events may also affect parts of continent–continent collision zones such as the Alps: the 'Black Earths' of southeastern France have R_o values of over 4%, but the precise origin of the thermal event is unknown (Robert 1988, p. 261).

The surface heat flows of the main genetic classes of sedimentary basin are summarised in Fig. 10.31.

PART 4

Application to petroleum play assessment

CHAPTER ELEVEN
Building blocks of the petroleum play

Summary

A play is a perception or model of how a producible reservoir, petroleum charge system, regional topseal and traps may combine to produce petroleum accumulations at a specific stratigraphic level. Prediction of source rocks, reservoirs, topseals and traps requires an understanding of the structural and stratigraphic evolution of the basin-fill. This understanding may be achieved through basin analysis. Correct identification and interpretation of the fundamental tectonic and thermal processes controlling basin formation, and of the geometry and sedimentary facies contained in the basin's depositional sequences, is the first and most important step towards building the geological models that underpin play assessment.

The basic unit of petroleum resource assessment is the play, but the 'petroleum system' concept is also a useful way for the practising petroleum geologist to organise his/her investigations. A petroleum system comprises a pod of mature source rock and all of the migration paths, reservoir rocks, caprocks and traps that can be charged by that source rock to produce oil and gas accumulations. Petroleum systems may be classified to help describe and predict the abundance, geographic location and habitat of petroleum occurrences in a basin. This may be particularly useful in the ranking of relatively immature exploration provinces.

The first requirement for a play is that there is a petroleum charge. The *petroleum charge* system comprises source rocks, which must be capable of *generating* and *expelling* petroleum, and a *migration* pathway into the reservoir unit. Source rocks are sediments rich in organic matter derived from photosynthesising marine or lacustrine algae and land plants that contain chemical compounds known as lipids. Lipids are preserved when sediments are deposited under anoxic conditions. Lakes, deltas and marine basins are the main depositional settings of source beds.

Organic matter buried in sediments is in an insoluble form known as kerogen. Petroleum is *generated* when kerogen is chemically broken down as a result of rising temperature. For typical rates of heating, a stage of oil generation at approximately 100 to 150 °C is followed by a stage of oil cracking to gas (150 to 180 °C) and finally by dry gas generation (150 to 220 °C). Petroleum *expulsion* probably occurs as a result of the build-up of overpressure in the source rock as a consequence of hydrocarbon generation. For lean source rocks, petroleum expulsion is probably very inefficient. *Secondary migration* carries expelled petroleum towards sites of accumulation, and is driven by the buoyancy of petroleum fluids relative to formation pore waters. Migration stops when the capillary pressure of small pore systems exceeds the upward-directed buoyancy force.

Petroleum may be physically and chemically altered while it is in the trap by the processes of biodegradation, water washing, deasphalting, thermal alteration, gravity segregation and dysmigration.

Tertiary migration – the leakage of petroleum from traps towards the Earth's surface – completes the natural life cycle of petroleum that starts with its biological precursors and finishes with its dissipation at the Earth's surface.

A further requirement for a play is a porous and permeable *reservoir rock*. The porosity and mineralogy of potential reservoir rock intervals can be evaluated from wellbore wireline logs and from outcropping correlative units. Reservoir rocks result from deposition in almost any of a very wide range of depositional environments. Reservoir prediction requires careful interpretation of sedimentary facies within each stratigraphic unit. A number of scales of heterogeneity from kilometre-scale to microscopic affect the distribution of porosity and permeability in the gross reservoir unit. Particular basin tectonic settings are associated with particular types of reservoir geometry and composition.

A *regional topseal* or caprock is needed to seal petroleum in the gross reservoir unit. The mechanics of sealing are the same as those that control secondary migration. The ideal caprock comprises a fine-grained lithology, and is ductile and laterally persistent. Thickness and depth of burial do not appear to be critical. Two of the most successful reservoir-caprock associations are where marine shales transgress over gently sloping siliciclastic shelves, and where sabkha evaporites regress over shallow-marine carbonate shelves.

The final requirement for the operation of a petroleum play is the presence of *traps*. Traps are local subsurface concentrations of petroleum, and may be classified into structural, stratigraphic and hydrodynamic traps. Intrusive traps have been proposed as an additional category, based on recent recognition of the potential importance of sand injectites. Structural traps represent the habitat of most of the world's already discovered petroleum, and are formed by tectonic, diapiric and gravitational processes. Gravity-driven foldbelts are an extremely important play type in deep-water provinces. Stratigraphic traps are those inherited from the original depositional morphology of, or discontinuities in, the basin-fill, or from subsequent diagenetic effects.

Basin Analysis: Principles and Application to Petroleum Play Assessment, Third Edition. Philip A. Allen and John R. Allen.
© 2013 John Wiley & Sons, Ltd. Published 2013 by John Wiley & Sons, Ltd.

A systematic petroleum-system-based review of the global distribution of known reserves and undiscovered potential provides a useful statistical top-down perspective on petroleum formation and occurrence, which complements the bottom-up analysis of petroleum plays and their building blocks followed by this book. Empirical observations show the wide variety of geological situations in which oil and gas resources are found worldwide, but also shed light on some general rules and guidelines governing petroleum distribution. One such insight is that the vast majority of the world's conventional oil and gas resources occur within a relatively narrow geothermal range of 60 to 120 °C, an interval that has been termed the 'Golden Zone'. Despite the apparent maturity of the global exploration industry, petroleum-system-based estimates of undiscovered potential indicate there is still a very significant prize to play for.

11.1   From basin analysis to play concept

Basin analysis is critical to the assessment of undiscovered petroleum potential. Assessments of this kind guide the exploration programmes of the petroleum industry. An understanding of the distribution and evolution of depositional sequences and facies allows rational and realistic predictions to be made of petroleum source rocks, reservoir rocks and caprocks – the building blocks of a petroleum play. The associated structural development of the basin is primarily responsible for the formation of petroleum traps.

The bulk of subsurface information in a sedimentary basin is obtained from seismic reflection surveying and well log interpretation. These are fields that are documented elsewhere. The reader is referred to Badley (1985) for a description of seismic interpretation, and to Bacon *et al.* (2003) and Brown (1999) for 3D methods. Rider (1996) provides a summary of well log interpretation.

Play concepts are founded on an understanding of the stratigraphic and structural evolution of the basin. The geological models upon which predictions of source, reservoir and caprocks and their evolution through time are based are outcomes of this understanding. The validity of these models, and therefore of the plays that are generated from them, is dependent on a correct interpretation of the boundaries and overall *geometry* of the genetic stratigraphic packages involved in the play, and on a correct interpretation of the *sedimentary facies* within these stratigraphic units. Basin analysis provides the means of making these interpretations.

The previous chapters of this book have shown that the location and overall form of megasequences and depositional sequences may be understood in terms of the mechanical processes of basin formation. Thus, basins due to lithospheric stretching (Chapter 3), flexure (Chapter 4), mantle dynamics (Chapter 5), and strike-slip deformation (Chapter 6), each exhibit characteristic locations, geometries and evolutions which may be understood in terms of the controlling broad plate tectonic and mantle processes. Knowledge of the underlying basin-forming process also implies a particular tectonic and thermal development for the basin (Chapters 9 and 10), which is an important input to the thermal modelling of potential source rock maturity and reservoir diagenesis.

Correct identification and interpretation of the megasequences present in a province (Chapter 8) is the first step towards building the geological models for play assessment. Each megasequence may be broken down into a series of smaller-scale genetic units and packages representing discrete phases of the basin infill (§8.2). This information forms the basis for the prediction of source, reservoir and caprocks.

The type, amount and quality of data available limits the confidence held in any stratigraphic interpretation. The goal is to achieve a reliable *chronostratigraphic* interpretation of the basin-fill, so that the distribution and nature of sedimentary facies may be understood in terms of geological processes operating at a specific time. The chronostratigraphic interpretation must, however, be built up from interpretations of lithostratigraphy, biostratigraphy and seismic-stratigraphy. Each of these, on its own, is potentially unreliable.

11.2   The petroleum system and play concept

11.2.1   Play definition

A play may initially be defined as a perception or model of how a number of geological factors might combine to produce petroleum accumulations at a specific stratigraphic level in a basin. These geological factors must be capable of providing the essential ingredients of the petroleum play, namely:

- a *reservoir unit*, capable of storing the petroleum fluids and yielding them to the wellbore at commercial rates;
- a *petroleum charge system*, comprising thermally mature petroleum source rocks capable of expelling petroleum fluids into porous and permeable carrier beds, which transport them towards sites of accumulation (traps) in the gross reservoir unit;
- a *regional topseal* or caprock to the reservoir unit, which contains the petroleum fluids at the stratigraphic level of the reservoir;
- petroleum *traps*, which concentrate the petroleum in specific locations, allowing commercial exploitation;
- *the timely relationship* of the above four ingredients so that, for example, traps are available at the time of petroleum charge.

A play may further be defined as a family of undrilled prospects and discovered pools of petroleum that are believed to share a common gross reservoir, regional topseal, and petroleum charge system.

Play ingredients are shown schematically on a chronostratigraphic diagram in Fig. 11.1 (§8.2, Fig. 8.7).

The geographical area over which the play is believed to extend is the *play fairway*. The extent of the fairway is determined initially by the depositional or erosional limits of the gross reservoir unit, but may also be limited by the known absence of any of the other factors. The mapping-out of the play fairway is discussed in §11.2.3.

A play may be considered *proven* if petroleum accumulations (pools or fields) are known to have resulted from the operation of the geological factors that define the play. These geological requirements are thus known to be present in the area under investigation, and the play may be said to be 'working'. In *unproven* plays, there is some doubt as to whether the geological factors actually do combine to produce a petroleum accumulation; the probability of the play working is known as *play chance*.

11.2.2 The petroleum system

The basic unit of petroleum resource assessment is the play, but the 'petroleum system' concept is also a useful way for the practising petroleum geologist to organise his/her investigations. The original work that gave birth to the petroleum system concept was carried out in the Williston Basin of Canada-USA (Dow 1974). Related concepts were subsequently described by Gerard Demaison (the *generative basin concept*, 1984), Meissner *et al.* (the *hydrocarbon machine*, 1984), and Gregory Ulmishek (the *independent petroliferous system*, 1986). The concept was subsequently defined more rigorously by Leslie Magoon of the US Geological Survey and described in AAPG Memoir 60 (Magoon & Dow 1994).

A 'petroleum system' comprises a pod of mature source rock and all of the migration paths, reservoir rocks, caprocks and traps that can be charged by that source rock to produce oil and gas accumulations. The concept, which has seen widespread practical application in the petroleum exploration industry since the mid-1990s, places the source rock as the first and foremost element of the geological system required to produce a petroleum play. When adopting this concept, the practising petroleum geologist goes through an assessment process that mimics the geological process of hydrocarbon generation, migration and entrapment. By starting with the source rock, the petroleum system concept encourages the petroleum geologist to consider all the journeys and destinations of hydrocarbons generated and expelled from a source rock, and is more likely to stimulate ideas on new plays than an exploration approach strongly focused on one or two reservoirs that are already proven in the basin.

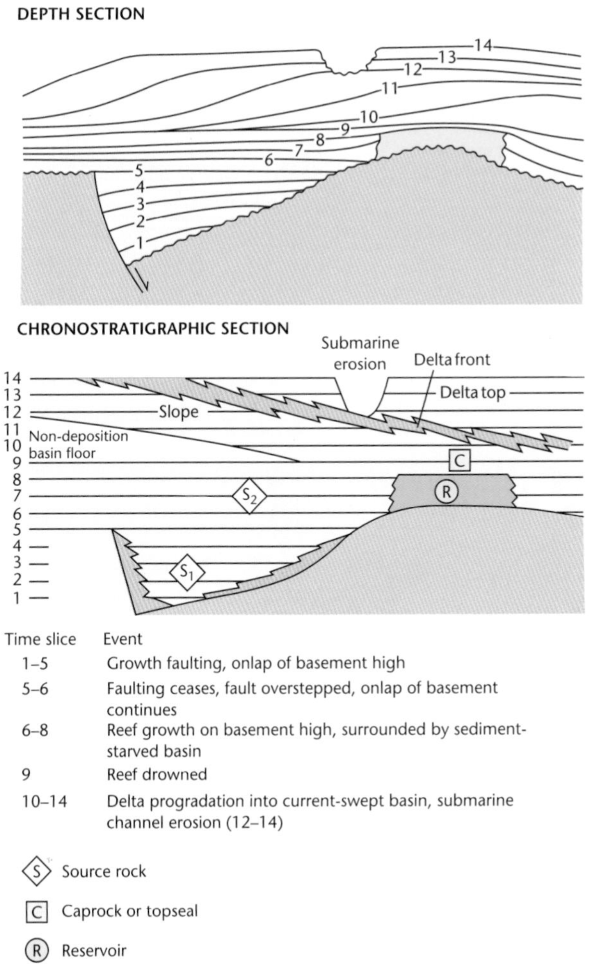

The petroleum system is defined in terms of:

- its *stratigraphic* extent – in the system proposed by Magoon and Dow (1994), the petroleum system takes its name from the source rock, qualified by the main reservoir rock that it charges. This linkage, between source and reservoir, may be: (i) *known* if the petroleum can be conclusively related to the source rock through oil-source or gas-source geochemical correlation; (ii) *hypothetical* if geochemical evidence indicates a source rock but a firm correlation has not been made; and (iii) *speculative* if based only on geological or geophysical evidence. The stratigraphic extent of the petroleum system should also make reference to the other essential elements, namely the carrier bed and caprock, and be illustrated by a cross-section drawn at the *critical moment*.
- its *geographic* extent, shown by a petroleum system map. This map shows the pod of active source rock, together with the associated discoveries, seeps and shows. A table of the discovered accumulations and their field sizes should be included.
- its *temporal* extent, illustrated by a burial history chart and events chart that highlights the critical moment. The *critical moment* is the time at which most of the hydrocarbons were generated, migrated and accumulated in the primary trap type. The events chart shows the time of deposition of the stratigraphic components of the system, and timing of key processes – trap formation, generation-migration and entrapment – and the period over which trapped hydrocarbons are preserved, modified or destroyed.

A genetic classification of petroleum systems (Demaison & Huizinga 1994) helps to describe and predict the abundance, geographic location and habitat of petroleum occurrences in a basin. This may be particularly useful in the ranking of relatively immature exploration provinces. The classification is based on three key factors that may be deduced from basin analysis and geochemical data (Fig. 11.2):

DEPTH SECTION

CHRONOSTRATIGRAPHIC SECTION

Time slice	Event
1–5	Growth faulting, onlap of basement high
5–6	Faulting ceases, fault overstepped, onlap of basement continues
6–8	Reef growth on basement high, surrounded by sediment-starved basin
9	Reef drowned
10–14	Delta progradation into current-swept basin, submarine channel erosion (12–14)

S Source rock
C Caprock or topseal
R Reservoir

Fig. 11.1 Schematic depth and chronostratigraphic diagrams, showing the relationships between sedimentary facies, basin development, and source rocks, reservoirs and topseals.

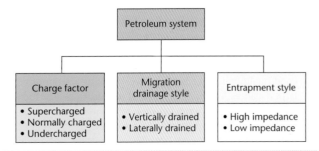

Fig. 11.2 Demaison and Huizinga's (1994) genetic classification of petroleum systems.

1. *Charge factor* – reflects the initial richness and volume of mature source rock, and is therefore a guide to the regional charge potential of the petroleum system. The source potential index (SPI) combines source rock richness and source rock thickness into a single parameter that is known to be positively correlated (in well-explored petroleum provinces) with discovered petroleum reserves. Systems are categorised into supercharged, normally charged, and undercharged. Charge factor is the single most important control on the petroleum richness of a system.

2. *Migration drainage style* – a reflection of the structure and stratigraphy of the basin-fill. Migration drainage style may be either dominantly vertical or dominantly lateral. A *vertical migration style* is assisted by faults and fractures that penetrate regional seals, or by reservoir and carrier beds that interconnect vertically over large distances, and is most common in rifted basins, sandy deltaic sequences, and in highly fractured fold-thrust belts. Examples are the North Sea and Gulf of Suez failed rifts, passive margin basins such as the Lower Congo Basin of Angola, the Campos Basin of Brazil, and the Barrow-Dampier Basin of NW Australia, and the Tertiary deltas of Nigeria and the US Gulf of Mexico. Traps ideally need to be located vertically above the mature pod of source rock in order to be charged. In supercharged vertically drained petroleum systems, abundant surface seepage may occur, as in the San Joaquin Basin of California, the US Gulf Coast salt dome province, the Zagros foldbelt of Iran, and the Magdalena Valley of Colombia. A *lateral migration style* is dominant in basins with extensive reservoir-seal couplets in tectonically stable settings, for example in the foredeeps of foreland basin systems and cratonic basins. Migration may be focused into arches or noses that plunge into the basin. Traps located along these focused migration routes may collect the charge from much of the source kitchen, while other areas are starved. Petroleum occurrences may be located long distances (several hundred kilometres) away from the mature source pod. Good examples are the Oriente/Marañón Basin of Peru and the North Slope of Alaska. In supercharged laterally drained petroleum systems, very large petroleum volumes may migrate as far as the shallow edges of the basin, as in the heavy oil provinces of Western Canada (Athabasca) and eastern Venezuela (Orinoco).

3. *Entrapment style* – reflects the degree to which hydrocarbons are dispersed in moving from source rock to trap. Thermodynamic principles suggest that over geological time, natural geological processes work to ultimately disperse and destroy petroleum in sedimentary rocks. As petroleum migrates from source rock to its ultimate fate of destruction at the Earth's surface, local geological factors (trapping mechanisms) may resist this wholesale process, at least temporarily, and give rise to petroleum accumulations. Physical resistance to petroleum dispersion, or *impedance*, is largely a function of structural and stratigraphic complexity. A *low-impedance entrapment style* indicates a tendency for hydrocarbons to flow efficiently along major migration routes with little resistance or dispersion, and occurs where good-quality carrier beds and regional seals are continuous over large distances. Very large volumes of petroleum may be focused into relatively few ultimate sites of accumulation. A *high-impedance entrapment style* is one in which hydrocarbon migration is dispersed by stratigraphic complexity or structural deformation into many different routes and sites of entrapment, at both the macro and micro scale. Large volumes of petroleum may be lost from commercial exploitation.

11.2.3   Definition and mapping of the play fairway

Play fairway maps show the geographical distribution of the key geological controls on the play fairway. Demaison (1984) introduced the 'generative basin concept', and described how petroleum generative kitchens could be mapped out. White (1988) extended this concept by including the reservoir, topseal and trapping controls on the play.

Plays are essentially reservoir-defined. A single source rock horizon may charge a number of separate reservoir-defined plays, and a single reservoir-defined play may be charged from a variety of separate source rock horizons. The objective of play assessment is to anticipate all of the possible combinations of potential reservoirs, sources and caprocks that may produce petroleum plays in the basin. For each reservoir-defined play, a single map can be produced that shows the distribution of the potential reservoir facies, the source 'kitchen(s)' needed to charge the reservoir, and the potential caprock facies.

Fairways at different stratigraphic levels in a basin may be stacked vertically, as in, for example, the central North Sea, where a number of play fairways are (partially) overlain (Fig. 11.3). Within a single play, all prospects and discovered fields share a common geological mechanism for petroleum occurrence. Petroleum accumulations, discovered or undiscovered, within a single play fairway, can be considered to constitute a naturally occurring population or family of

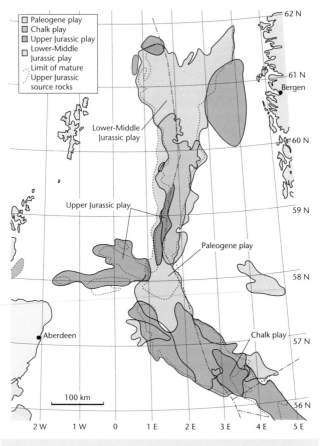

Fig. 11.3 The four main plays of the central and northern North Sea are stacked vertically so that their fairways partially overlie each other. From Gluyas and Swarbrick (2004), reproduced with permission of John Wiley & Sons, Ltd.

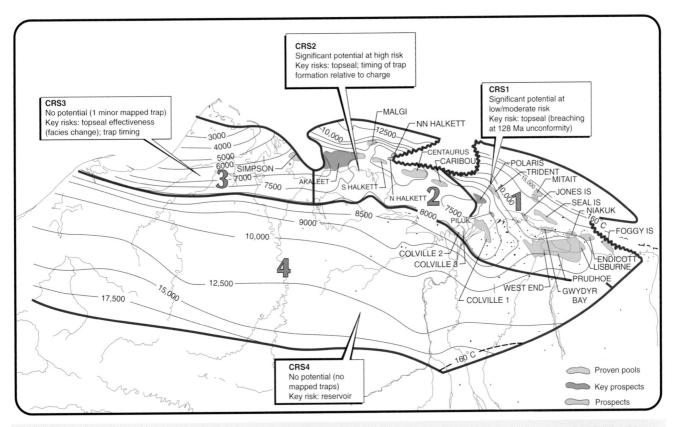

CRS3
No potential (1 minor mapped trap)
Key risks: topseal effectiveness
(facies change); trap timing

CRS2
Significant potential at high risk
Key risks: topseal; timing of trap
formation relative to charge

CRS1
Significant potential at
low/moderate risk
Key risk: topseal (breaching
at 128 Ma unconformity)

CRS4
No potential (no
mapped traps)
Key risk: reservoir

Proven pools

Key prospects

Prospects

Fig. 11.4 An example of play fairway segmentation: Ellesmerian 1 play fairway, Alaska North Slope (from Gluyas and Swarbrick 2004). This 10 billion barrel fairway contains the giant Prudhoe Bay field. Each common-risk segment has a different risk and remaining potential. CRS 1, covering the Prudhoe/Kuparauk High, is proved, and the key risk relates to topseal at the '128 Ma unconformity'; well-defined unbreached dip closures were preferred. To the west, CRS2 was unproven, and a major additional risk was that traps formed after the main period of hydrocarbon charge. CRS3, has an additional top seal (depositional facies) risk and only one mapped structure, while CRS4 has major reservoir risk and no mapped traps at all. Following this regional review, by 1987, BP relinquished its exploration acreage in the central and western Beaufort Sea. Structural contours are believed to be on top Ivishak Formation, in feet. Reproduced with permission of John Wiley & Sons, Ltd.

geological phenomena. Thus, each play can be characterised by a specific field size distribution and drilling success ratio.

In assessing an unproven play, the probability of the play working is known as *play chance*. Owing to the interplay of the critical geological factors over the extent of the fairway, it is normal for play chance to vary spatially. This variation in play chance may be due to hard evidence of adverse geology in different parts of the fairway (determined, for example, from well or seismic data), or to variations in the quantity or quality of the database, which allows greater or lesser confidence in the interpretations made. As a result, an unproven fairway may be subdivided into a mosaic of *common-risk segments*; within each segment play chance is constant. An example of play fairway segmentation (Alaskan North Slope) is shown in Fig. 11.4.

The fairway can also be subdivided into segments if there are strong reasons for believing field sizes are likely to be significantly different (for example, as a result of differing structural development in different parts of the fairway), or drilling success ratios are likely to vary significantly. Thus, the three factors that define fairway segments are variations in: (i) play chance; (ii) expected field sizes; or (iii) expected drilling success ratios. In this way, reservoir-defined plays are sometimes subdivided by trap type.

The following specific items of geological information are typically shown on a play map (Fig. 11.5):

- the depositional or erosional limits of the gross reservoir unit, and the distribution of reservoir facies within it;
- areas where a source rock is present, and where it is mature (the kitchen);
- a migration zone around the kitchen; together with the kitchen itself, it represents the area receiving a petroleum charge;
- areas where there is an effective regional seal;
- areas where traps are present (structural or stratigraphic);
- oil and gas fields, dry holes, and untested prospects, leads and notional prospects;
- drilling success ratios for specific parts of the fairway.

The drilling success ratio is the ratio of the number of *technical successes* to the number of *valid tests* of the fairway. There are normally far more dry holes than technical successes in a play. Within a proven play, dry holes are caused by local geological variations, such as the absence of a lateral seal in a faulted prospect, the existence of a migration shadow, or local diagenetic destruction of reservoir porosity. These factors contribute to *prospect-specific risk*.

Play map

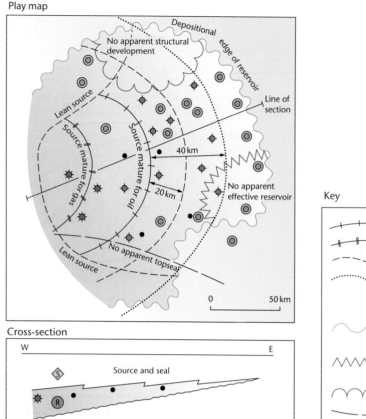

Key

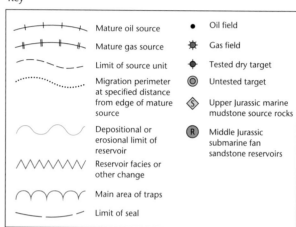

Cross-section

Fig. 11.5 Schematic example of a play map (after White 1988), showing the important geological factors likely to control hydrocarbon accumulation in the fairway. AAPG © 1988. Reprinted by permission of the AAPG whose permission is required for further use.

In the hypothetical case illustrated in Fig. 11.5, an area of mature source in the west of the area has experienced the highest drilling success ratio (75%). As we move outwards, firstly a migration distance of 20 km, and then a migration distance of 40 km, the success ratios drop to 28% and then 20%, as a result of more and more prospects failing to lie on migration routes as the distance from the kitchen increases. Despite these variations in success ratio, however, the play is effectively proven in these areas. Beyond the 40 km migration perimeter, there are no existing discoveries and the play is unproven.

Common-risk segment A is considered already proven (play chance of 1.0). The unproven part of the fairway is divided into the following common-risk segments (Fig. 11.6):

- B: an area lacking structural development in the north (no traps). The seismic grid, however, is coarse and there may be stratigraphic traps, so the possibility of traps cannot be ruled out (play chance assessed at 0.20).
- C: an area beyond the 40 km migration perimeter in the east. Although unlikely, a very focused migration path could charge a prospect in this segment (play chance is assessed at 0.50).
- D: an area beyond the 40 km migration perimeter where reservoir effectiveness may be destroyed by diagenesis (play chance assessed at 0.30).
- E: an area in the northwest, up-dip from a lean source. However, the geochemical database is in fact quite poor and source rock interpretations unreliable (play chance is assessed at 0.5).

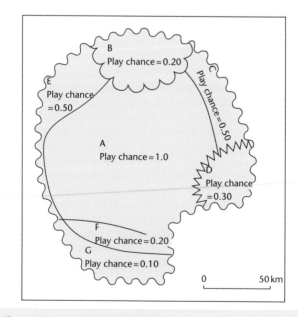

Fig. 11.6 Subdivision of the play fairway into common-risk segments. These are controlled by the distribution of reservoir, charge, topseal and likely trap development shown in Fig. 11.5. A play chance is assessed for each common-risk segment.

- F: an area in the south with no apparent topseal (play chance assessed at 0.20).
- G: an area in the south up-dip from a lean source and also with no apparent regional topseal (play chance assessed at 0.10).

Beyond the depositional or structurally controlled limits of the gross reservoir unit, there is no play at all. This is the outer limit of the fairway.

Play fairway definition and mapping forms the basis for quantitative resource assessment. Quantitative estimates of the undiscovered potential of plays are required by petroleum exploration companies in order to evaluate exploration investment opportunities and guide long-term strategic plans. Estimates of the likely size and timing of future discoveries, and hence future petroleum supply, also form an important element in the planning studies of government organisations. The US Geological Survey's world petroleum assessment employs the petroleum system approach and is available in the public domain (Ahlbrandt *et al.* 2005).

11.3 The source rock

There is now a wealth of geochemical evidence that petroleum is sourced from biologically derived organic matter buried in sedimentary rocks. Organic-rich rocks capable of expelling petroleum compounds are known as *source rocks*.

In order to understand and predict the distribution and type of petroleum source rocks in space and time, it is necessary to consider the biological origin of petroleum. Source beds form when a very small proportion of the organic carbon circulating in the Earth's carbon cycle is buried in sedimentary environments where oxidation is inhibited.

In the world oceans, simple photosynthesising algae (phytoplankton) are the main primary organic carbon producers. Their productivity is controlled primarily by sunlight and nutrient supply. The zones of highest productivity are in the surface waters (euphotic zone) of continental shelves (rather than open ocean) in equatorial and mid-latitudes, and in areas of oceanic upwelling or large river input.

The productivity of land plants is controlled primarily by climate, particularly rainfall. Coals have formed in the geological past predominantly in the equatorial zone and in the cool wet temperate zone centred at about 55° (N and S).

All living organic matter is made up of varying proportions of four main groups of chemical compounds. These are *carbohydrates*, *proteins*, *lipids* and *lignin*. Only lipids and lignin are normally resistant enough to be successfully incorporated into sediment and buried. Lipids are present in both marine organisms and certain parts of land plants, and are chemically and volumetrically capable of sourcing the bulk of the world's oil. Lignin is found only in land plants and cannot source significant amounts of oil, but is an important source of gas. Geochemical studies of coal macerals have shown a very significant oil potential amongst the exinite group, comprising material derived from algae, pollen and spores, resins and epidermal tissue.

The organic compounds provided to sea-bottom sediments by primitive aquatic organisms have probably not changed dramatically over geological time. In contrast, important evolutionary changes have taken place in land plant floras. As a result, a distinction can be made between the generally gas-prone Paleozoic coals, and the coals of the Jurassic, Cretaceous and Tertiary, which may have an important oil-prone component.

Anoxic conditions (depleted in or devoid of oxygen) are required for the preservation of organic matter in depositional environments, because they limit the activities of aerobic bacteria and scavenging and bioturbating organisms that otherwise results in the destruction of organic matter. Anoxic conditions develop where oxygen demand exceeds oxygen supply. Oxygen is consumed primarily by the degradation of dead organic matter; hence, oxygen demand is high in areas of high organic productivity. In aquatic environments, oxygen supply is controlled mainly by the circulation of oxygenated water, and is diminished where stagnant bottom-waters exist. The transit time of organic matter in the water column from euphotic zone to seafloor, sediment grain size, and sedimentation rate also affect source bed deposition.

The three main *depositional settings* of source beds are lakes, deltas and marine basins.

Lakes are the most important setting for source bed deposition in continental sequences (see also §7.5.3). Favourable conditions may exist in deep lakes, where bottom-waters are not disturbed by surface wind stress, and at low latitudes, where there is little seasonal overturn of the water column and a temperature-density stratification may develop. In arid climates, a salinity stratification may develop as a result of high surface evaporation losses. Source bed thickness and quality is improved in geologically long-lasting lakes with minimal clastic input.

Organic matter on lake floors may be autochthonous, derived from freshwater algae and bacteria, which tends to be oil-prone and waxy, or allochthonous, derived from land plants swept in from the lake drainage area, which may be either gas-prone or oil-prone and waxy.

The Eocene Green River Formation of the western USA, and the Paleogene Pematang rift sequences of central Sumatra, Indonesia, are examples of rich, lacustrine source rock sequences.

Deltas may be important settings for source bed deposition (see also §7.5.3). Organic matter is derived from freshwater algae and bacteria in swamps and lakes on the delta top, marine phytoplankton and bacteria in the delta-front and marine pro-delta shales, and, probably most importantly, from terrigenous land plants growing on the delta plain. On post-Jurassic deltas in tropical latitudes, the land plant material may include a high proportion of oil-prone, waxy epidermal tissue. Mangrove material may be an important constituent. Examples of deltaic source rocks include the Upper Cretaceous to Eocene Latrobe Group coals of the Gippsland Basin, Australia.

Much of the world's oil has been sourced from *marine* source rocks. Source beds may develop in *enclosed basins* with restricted water circulation (reducing oxygen supply), or on open shelves and slopes as a result of *upwelling* or impingement of the *oceanic midwater oxygen minimum layer*.

Examples of modern enclosed marine basins include the Black Sea and Lake Maracaibo (Venezuela). Source bed deposition is favoured by a positive water balance, where the main water movement is a strong outflow of relatively fresh surface water, leaving denser bottom-waters undisturbed. The Upper Jurassic Kimmeridge Clay Formation of the North Sea, and the Jurassic Kingak and Aptian–Albian HRZ formations of the North Slope, Alaska, are examples of source rocks deposited in restricted basins on marine shelves.

The upwelling of nutrient-rich oceanic waters may give rise to exceptionally high organic productivity. Oxygen depletion may occur in the underlying bottom-waters as oxygen supply is overwhelmed by the demand created by degradation of dead organic matter. Upwelling coastlines tend to be arid, and the organic matter in upwelling deposits is almost entirely of marine origin and strongly oil-prone. Upwelling may have played a part in the formation of source rocks such as the Permian Phosphoria Formation of the western USA, the Triassic Shublick Formation of the North Slope, the Cretaceous La Luna Formation of Venezuela and Colombia, and the Miocene Monterey Formation of California.

In open oceans whose floors are swept by cold, dense currents originating in the polar regions, an oxygen-deficient layer develops at depths of 100 to 1000 m. At times in the geological past, during periods of warmer climate and higher sea level, this layer may have intensified and impinged on large areas of the continental shelves and slope. The 'global anoxic events' of the Mid-Cretaceous may have resulted from this process. The Toarcian source rocks of western Europe may also be an example.

The organic matter buried in sediments is in a form known as kerogen. *Geochemical measurements* can be used to determine the presence, richness and stage of thermal maturity of a petroleum source rock, as well as the range of compounds likely to be generated and expelled. The richness or petroleum-generating potential of a source rock can be determined by measurements of total organic carbon (TOC) and pyrolysis yield. More sophisticated geochemical techniques, such as gas chromatography and isotope studies can be used to determine likely petroleum products, and for a range of other applications, including the correlation of source rocks with oils. Visual (optical) descriptions of kerogen may also give a useful guide to petroleum potential and petroleum type. From microscopic examination in reflected light, kerogen is classified into the *exinite*, *vitrinite*, and *inertinite* groups. The exinite group comprises macerals with significant oil potential, while the vitrinite group is gas-prone. Inertinites have no petroleum-generating potential.

Measurements of the reflectance of vitrinite are used as an index of thermal maturity (§10.4, 10.5).

11.3.1   The biological origin of petroleum

Since the discovery by Treils in 1934 of a porphyrin 'biological marker' compound in rock material, a wealth of geochemical evidence has accumulated to show that petroleum is sourced from biologically derived organic material buried in sediments. In order to understand the distribution in space and time of source rocks for oil and gas, it is necessary to first consider the characteristics of the biomass from which the organic material is originally derived. This section briefly discusses a number of topics: source bed deposition in the context of the overall carbon cycle; the main components of the biomass; geographical variations in organic productivity in the world's environments at the present day, and the main factors controlling these variations; changes in the composition of the biomass through geological time; and the chemical composition of living organic matter and its likely hydrocarbon products.

11.3.1.1   The carbon cycle

The carbon cycle is initiated by photosynthesising land plants and marine algae, which convert carbon dioxide present in the atmosphere and seawater into carbon and oxygen using energy from sunlight (Fig. 11.7). Carbon dioxide is recycled back in many ways, the most important of which are back to the atmosphere through animal and plant respiration, by bacterial decay and natural oxidation of

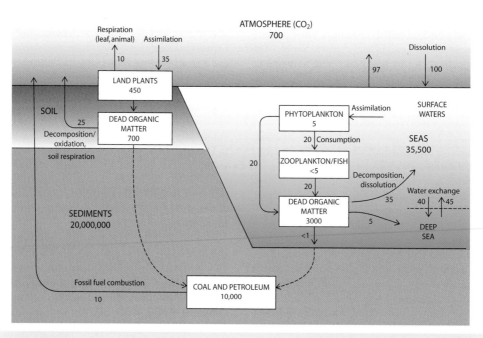

Fig. 11.7   The main elements of carbon circulation in the biosphere. Numbers represent quantities of carbon in billions of metric tons. Numbers in boxes represent estimates of stored or inventoried quantities; numbers beside arrows are yearly transfers or fluxes. Estimates of tonnages are from Bolin (1970), but with the value for fossil fuel consumption increased from 5 to a nominal 10 million tonnes per annum. There are two distinct cycles, one in the seas and the other on land, both interacting with the atmosphere. Living phytoplankton represent a relatively small store of carbon compared to land plants, but owing to their short lives (days/weeks) and high metabolic rates, they are responsible for a very large circulation of carbon in the marine environment. A relatively small amount of organic carbon is thought to escape the carbon cycle through burial in sediments, and an even smaller amount is concentrated into organic-rich sediments which subsequently generate oil and gas. Adapted from Bolin (1970).

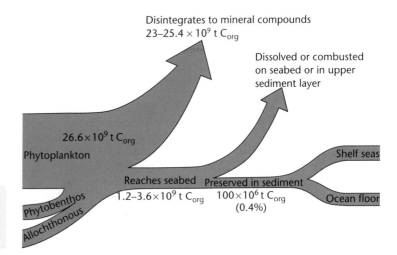

Fig. 11.8 Fate of the 26 billion tons of annual organic carbon (C$_{org}$) production in the world's oceans. Only 0.4% is preserved in the bottom sediments of shelf seas and ocean floors.

dead organic matter, and by combustion of fossil fuels – both naturally and by man.

However, the importance of the carbon cycle to the petroleum geologist is that a small proportion of carbon escapes from the cycle as a result of deposition in environments where oxidation to carbon dioxide cannot occur. These environments are generally depleted in oxygen (for example, some restricted marine basins and deep lakes), or toxic for bacteria (swamps).

The proportion of organic material buried in sediments in this way, relative to that originally produced, is very small (less than 1%), but over geological time is significant. Because it is preferentially concentrated in specific environments, it results in commercially significant petroleum source bed development. Petroleum is sourced, therefore, from organic carbon that has dropped out of the carbon cycle, at least temporarily. It rejoins the cycle when extracted by man and combusted.

11.3.1.2 Organic production

The nature of organic production is quite different in continental and marine ecosystems. Continental ecosystems are dominated by land plants in low-lying coastal plain environments and by freshwater algae in lakes. Marine ecosystems are overwhelmingly dominated by phytoplankton.

Marine ecosystems

Simple photosynthesising algae are the primary organic carbon producers in the world's oceans, and are the start of a complex food chain. Phytoplankton are responsible for over 90% of the supply of organic matter in the world's oceans. The phytoplankton group includes the diatoms, dinoflagellates, blue green algae and nannoplankton.

Fig. 11.8 illustrates the fate of the 26.6×10^9 t supply of organic carbon per year. Only a small percentage (0.4% according to Romankevich 1984) of the net carbon production in the world's seas and oceans is transferred to and preserved in sea-bottom sediment.

Apart from phytoplankton, other organisms such as zooplankton, benthos, bacteria and fish may also be important elements of the biomass. In the Black Sea (Romankevich 1984), annual bacteria production far exceeds even the phytoplankton. The main function of bacteria is to break down dead organic matter, but the bacteria may themselves also contribute to the organic content of the sediment.

Geographical variations in phytoplankton production in the world's marine environments are shown in Table 11.1. Although the open ocean accounts for a large percentage of the organic carbon produced, the concentration of organic carbon per square metre in open ocean water is relatively low (the red pelagic oozes of the deep ocean basins are typically very lean in organic content). In contrast, the continental shelves are very rich, particularly in some specific environments of enhanced organic activity, such as the algally dominated intertidal zone and in reefs and estuaries. Upwelling zones, such as those off the Peruvian coast at the present day, are also areas of relatively high organic productivity.

At a global scale, several trends in organic productivity may be recognised. Primary productivity decreases from coastal/marine shelf

Table 11.1 Global net primary phytoplankton production in the world's marine environments (Woodwell *et al*. 1978, Nienhuis 1981)

Total net primary production				
Ecosystem type	Area 10^6 km^2	10^9 t C$_{org}$yr^{-1}	g C$_{org}$m^{-2}yr^{-1}	Total plant mass of carbon, 10^9 t C$_{org}$
Marine ecosystems including:	361.0	24.7	68.7	1.74
Algal bed and reef	0.6	0.7	1166.7	0.54
Estuaries	1.4	1.0	714.3	0.63
Upwelling zones	0.4	0.1	250.0	0.004
Continental shelf	26.6	4.3	161.6	0.12
Open ocean	332.0	18.7	56.3	0.45

Note: The productivity per square metre of the open oceans is notably low compared to shelf environments.v

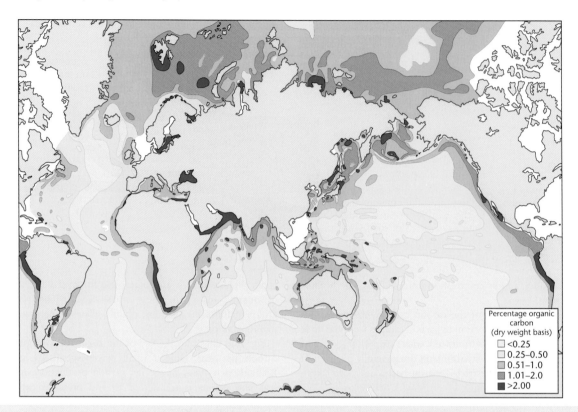

Fig. 11.9   Distribution pattern of organic carbon in the upper sedimentary layer (0–5 cm) beneath the world's oceans, in percentages on a dry weight basis. The highest organic concentrations in bottom sediments are in shelf areas. Organic productivity is an important controlling factor. Reproduced from Romankevich (1984) with kind permission from Springer Science+Business Media B.V.

into open ocean. Mid-latitude humid and equatorial latitudes are more productive than tropical latitudes. Lowest productivity is in polar and tropical areas. The factors controlling organic productivity include:

- *Sunlight.* The zone of highest productivity is the top 200 m of the world seas, especially the upper 60–80 m. This is the photic zone.
- *Nutrient supply.* Nutrients, particularly nitrates and phosphates, are required to sustain high organic productivity. These are supplied by water circulation. Stagnant seas are not very productive. Ocean-bottom currents set up by the sinking of very cold water in the polar regions may cause upwelling along the western coasts of continents in tropical latitudes. The best-known examples are offshore Peru and West Africa. A rich nutrient supply is provided, and organic productivity is very high. Nutrient supply is also locally increased in areas of large river input and coastal abrasion.
- *Turbidity.* Productivity is limited in areas with turbid coastal waters.
- *Salinity.* Extremes of salinity (high or low) reduce the diversity of species present, though productivity of certain groups may still be very high.
- *Temperature.* Temperature also influences the composition of the phytoplankton population, rather than net productivity. Dinoflagellates, for example, require high water temperatures of >25 °C. Diatoms and radiolarians prefer 5–15 °C.

Fig. 11.9 illustrates the concentration of organic carbon in seabottom sediment, and shows that a large proportion is in continental shelf environments. Organic productivity in surface waters is an important, but not sole factor controlling the concentration of organic matter in bottom sediments. Indeed, Demaison and Moore (1980) were unable to find a convincing systematic correlation between these two factors at the present day.

The critical factors for source bed development are the *deposition* and *preservation* of organic matter in significant quantities in sediments, rather than organic productivity *per se*. Factors affecting the deposition and preservation of organic matter in sediment are discussed in §11.3.2.

Continental ecosystems

Organic productivity in continental ecosystems is dominated by land plants and freshwater algae.

The productivity of land plant material is controlled primarily by climate. Land plant material may be swept by rivers into lakes and adjoining marine areas, constituting an allochthonous organic supply. The most important autochthonous land plant deposit is peat, which forms below the water table in swamps or stagnant lakes where a wet climate allows luxuriant plant growth and topography causes poor drainage. A balance is necessary between the rate of accumulation of dead plant matter and the rate of subsidence. Accumulated peat may be preserved where bacterial decay of the dead organic matter is inhibited by anoxic or toxic conditions and where net subsidence takes place.

Peat swamps form in the lower delta plain environment, typically in the lagoonal areas behind coastal spits and barriers, and in bays between vertically accreting distributary channels.

Ancient coal occurrences can be predicted using paleoclimatic maps (Parrish *et al.* 1982), and web-based compilations such as the PALEOMAP project (http://www.scotese.com/Default.htm) and the NOAA Paleoclimatology project (http://www.ncdc.noaa.gov/paleo/paleo.html). Paleolatitude studies show that coals ranging in age from Early Triassic to mid-Miocene are concentrated in the equatorial zone and in the cool wet temperate zone centred on about 55° N and S.

Freshwater algae make an important contribution to the organic matter supply in lakes (see also §7.5.3). An example is the present-day alga, *Botryococcus*. Ancestors of *Botryococcus* have been identified in ancient lake sediments, and derived geochemical compounds have been recognised in many lake-sourced oils.

11.3.1.3   Chemical composition of living organic matter

The chemical compounds that make up all living organic matter fall into four groups:

1. *Carbohydrates* are compounds that function as sources of energy and as supporting tissue in plants and some animals. Examples are sugar, such as glucose and fructose, starch, cellulose and chitin. Cellulose is an important supporting tissue in land plants, while chitin is the material manufactured by crustaceans to form a hard protective exoskeleton.
2. *Proteins* are organic compounds made up of amino acids, and perform a variety of biochemical functions vital to life processes. Examples are enzymes, haemoglobins and antibodies. Proteins also make up most of the organic matter in shells and substances such as hair and nails.
3. *Lipids* are a range of organic substances that are insoluble in water, and include animal fats, vegetable oils and waxes. They are similar in chemical composition to petroleum. Lipids are abundant in marine plankton, and are present in the seeds, fruit, spores, leaf coatings and barks of land plants. A range of lipid-like substances, for example sterols, are important biological markers in crude oils.
4. *Lignin* and *tannins* are compounds common in higher plants. Lignin is the substance that gives strength to plant tissue, for example in trees, providing a much firmer support than cellulose. Tannins are found in some tree barks, seed coats, nut shells, algae and fungi.

Other important organic compounds are *resins* and *essential oils*. Resins are found in the wood and leaf coatings of trees, and are particularly resistant to chemical and biological attack.

The relative amounts of these groups of organic compounds in living organisms varies enormously. Factors such as food supply and overcrowding are also known to affect lipid content. Of the four groups of compounds, proteins and carbohydrates are very susceptible to degradation, and tend to be dissolved, oxidised or bacterially degraded, without being incorporated in sediment beyond its surface layers. In contrast, lipids and lignin are much more resistant to mechanical, chemical and biochemical breakdown, and, under the

conditions to be discussed later, will be buried successfully in sediment.

Lipids are closest in chemical composition to petroleum. A relatively small number of chemical changes are involved in transforming lipids into petroleum, and more petroleum can be produced from lipids than from any of the other substances. The lipid content of organic matter buried in sediments is probably sufficient to source all of the world's known oil.

If the biological precursors of petroleum can be microscopically identified, we can go a long way towards predicting the kind of hydrocarbons, if any, that are likely to be generated. Marine organic matter in sediments tends to be amorphous, but much of our understanding of the generating potential of land plant source rocks is derived from coal petrography. A large number of coal 'macerals' have been identified. The three main geochemical groups of maceral and their hydrocarbon generating potential (Kelley *et al.* 1985) are:

1. *Vitrinite* is derived from the lignin and cellulose component of plant tissues. It is normally the largest constituent of the so-called humic coals, and generates predominantly gas.
2. *Inertinite* is also derived from the lignin and cellulose of plants, but has been oxidised, charred, or biologically attacked. The consensus is that it has negligible hydrocarbon generating potential. What potential it has is for gas. Dispersed inertinite has, however, been proposed as one of the sources of the liquid hydrocarbons in the Permian Gidgealpa Group of the Australian Cooper Basin.
3. *Exinite* is a diverse group of macerals including: (i) *alginite* is derived from algae, and when abundant forms a boghead or cannel coal. It is the main constituent of the torbanites of Scotland. This type of coal is quite rare, being most common in the Permo-Carboniferous. The algae responsible for these alginites are similar to the modern freshwater alga, *Botryococcus*. Alginite is strongly oil-prone. (ii) *Sporinite* is derived from the spores and pollen of plants, and may be abundant in coals from the Devonian through to the present day. The sporinite in Paleozoic coals is derived mainly from spores, whereas in Mesozoic and Tertiary sporinite, pollen predominates. Spores and pollen are extremely lipid-rich (50%) and may give rise to excellent oil-prone source rocks. Since spores and pollen are often transported large distances by wind or water, oxidation before burial in sediment often occurs. The best source rocks are therefore those in which the spores and pollen are autochthonous to the depositional environment, as for example in a coal swamp. (iii) *Resinite* is derived from the resins and essential oils of land plants. It is a prolific source of naphthenic and aromatic hydrocarbons. (iv) *Cutinite* is derived from the protective surface coating or cuticle of higher plants. The cuticle occurs on the outside of the epidermal tissue. It is rich in hydrocarbon waxes, and is thus an important oil source. Poor preservation may result in a gas-prone cutinite, with very little remaining oil potential.

Thus, all of the exinite macerals have oil generating potential. Preservation of the maceral is, however, critical. This is a function of the transport route and distance, and the depositional environment. The oil generating potential of sporinite and cutinite is particularly strongly affected by poor conditions for preservation.

Of the four groups of compounds found in living organic matter, therefore, only lipids and lignins are likely to be substantially incorporated into sediment beyond the surface layer. Lipids are present in both marine organisms and parts of land plants, are chemically suited to sourcing petroleum, and could account for all of the world's known oil. Lignin is found only in land plants. It is unlikely to source oil, but is an important source of gas.

11.3.1.4   Changes in the composition of the biomass through geological time

The ancestors of the primitive aquatic organisms that comprise the phytoplankton, zooplankton and bacteria may be traced back into the Precambrian with little apparent evolutionary change. The kind of aquatic organic matter buried in marine sediment has therefore probably changed very little over geological time.

For land plants, however, important floral changes have taken place since their appearance in the Late Silurian–Devonian that have had a major impact on the hydrocarbon generating characteristics of the source rocks in which they are buried (Fig. 11.10).

The Carboniferous coals of the northern hemisphere and the Permian coals of the southern hemisphere contain floras dominated by early plant groups (mostly lycopods) without extensive foliage. The resulting coal macerals are gas-prone vitrinite and its oxidation product inertinite, with minor amounts or local concentrations of sporinite, resinite, cutinite and alginite. Paleozoic coals, therefore, tend on the whole to be gas-prone.

Important evolutionary changes took place in floras in the Jurassic and Cretaceous. Conifers became dominant in the Jurassic, and angiosperms (flowering plants) appeared in the Cretaceous. Both these plant groups are rich in waxy epidermal tissue and resin, and have significant oil generating potential. Large volumes of gas are also generated from coals of this type, since vitrinite is normally still abundant. Most of Australia's waxy oils have been sourced from coals of this type, most notably the Gippsland Basin oils from the Early Tertiary Latrobe Group, and the Eromanga and Surat Basin oils from the Jurassic (Thomas 1982).

Thus, evolutionary changes in land plant floras through geological time are responsible for the oil-prone component of Mesozoic and Tertiary coals, while Paleozoic sediments are more typically sources solely for gas.

11.3.2   Source rock prediction

11.3.2.1   Introduction: anoxia

Anoxic conditions are critical to the preservation of organic matter in sediments. Source rock prediction is therefore concerned primarily with predicting where and when in the geological past anoxic conditions are likely to have existed. Questions addressed in this section are 'what causes anoxic conditions?', and 'in what geological environments are anoxic conditions likely to develop?'

'Anoxic' means 'devoid of oxygen', but the term is frequently used in the sense of 'depleted in oxygen' – *dysaerobic*. 'Anaerobic' means that insufficient oxygen is available for aerobic biological processes. This critical oxygen concentration is different for different organisms. Below 1.0 millilitres of oxygen per litre of water there is a serious reduction in biomass, but deposit-feeding organisms (those responsible for bioturbation of the sediment) can persist down to concentrations of 0.3 ml/l. As a general guide, 0.5 ml/l can be taken as the oxic/anoxic threshold. Anoxic conditions are critical for source bed deposition because they prevent the scavenging of dead organic matter and bioturbation of the surface sediment by benthic fauna

and degradation of organic matter by bacteria, which would otherwise destroy the organic matter prior to burial.

Anoxic conditions develop where oxygen demand exceeds oxygen supply.

Oxygen demand is caused primarily by the degradation of dead organic matter. Large amounts of organic matter are supplied to the seafloor in areas of high surface organic productivity (photosynthesising algae in the euphotic zones of seas and lakes) and/or where there is a large terrigenous supply of organic matter. Oxygen is consumed as the dead organic matter is degraded.

Oxygen supply is controlled by the circulation of oxygenated water. This may be a downward movement of oxygen-saturated surface waters as a result of mixing by waves, or a movement of cold, oxygen-bearing ocean-bottom currents. Cold water can dissolve more oxygen than warm water. A feature of the world's oceans at the present day is that cold dense water carrying large amounts of oxygen descends in the polar regions and moves over the ocean floor towards the equator, bringing oxygenated conditions to almost all parts of the oceans. It is reasonable to expect, however, that there were times in the past when such a circulation was less well developed. This has important implications for source bed deposition.

A sea or lake floor, therefore, is prone to anoxic conditions primarily under the following two sets of circumstances: (i) when organic productivity in the overlying water column is very high and the system becomes overloaded with organic matter; and (ii) when stagnant bottom water conditions exist, causing a restriction in the supply of oxygen. These factors largely determine the geological settings in which source beds are deposited.

Source beds may, under exceptional circumstances, be deposited under oxic conditions. This sometimes occurs when sedimentation rate is very high. A special case is when mass gravity flows deposit an anoxic sediment almost instantaneously into oxic waters. As a rule, however, a source rock is not expected to be developed under oxic conditions.

The factors that affect the development of anoxia are discussed in more detail in §11.3.2.2, and the depositional environments in which source beds are formed are outlined in §11.3.2.3.

11.3.2.2   Factors affecting source bed deposition

An understanding of anoxic environments and their importance in petroleum exploration has rapidly developed since the 1980s (Demaison & Moore 1980). We have seen in §11.3.2.1 that anoxic conditions are a prerequisite for source bed deposition primarily because they prevent the bacterial degradation of dead organic matter and the scavenging and bioturbation of the surface sediment by benthic fauna. These and other factors are now discussed in greater detail.

Bacterial degradation

Degradation of organic matter by bacteria takes place in both the water column and in sediment pore waters, under both aerobic and anaerobic conditions (Fig. 11.11). Organic matter is oxidised by aerobic bacteria using the available oxygen in the environment until there is no more organic matter to oxidise or there is no more oxygen. If the latter, the environment becomes anoxic. Anaerobic bacteria derive oxygen first of all from nitrates, and then from sulphates. It is thought that they can degrade organic matter just as fast as aerobic bacteria. An important difference, however, is that anaerobic

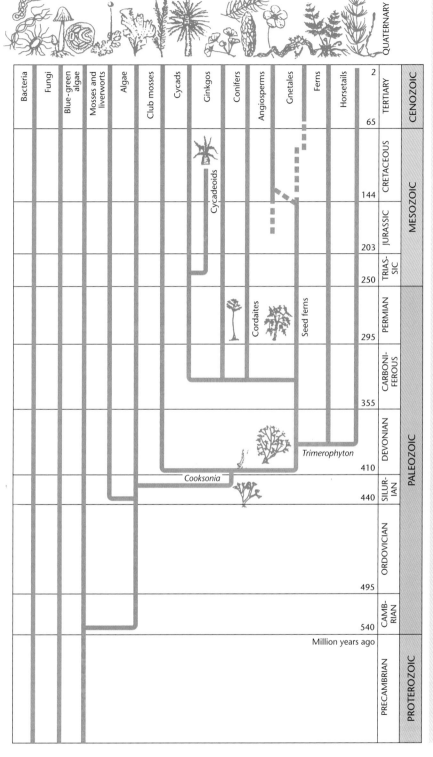

Fig. 11.10 Plant evolution over the last 600 Myr. The first land plants (primitive vascular varieties such as *Cooksonia*) appeared in the Silurian, and a large number of new groups evolved in the Devonian. Important changes in land plant floras have taken place since the Silurian which have affected their hydrocarbon-generating characteristics. Only Mesozoic and Tertiary land plants have significant oil-generating potential. Today, two main groups dominate the land: the *conifers*, mainly in the cooler, drier areas, and the *flowering plants* (angiosperms), which occur everywhere. The conifers may date back as far as the Carboniferous, whereas the oldest flowering plants in the geological record are Early Cretaceous or Late Jurassic. In the sea, organisms adapted to life in water, such as the algae, have undergone their own evolution.

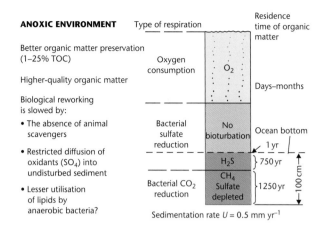

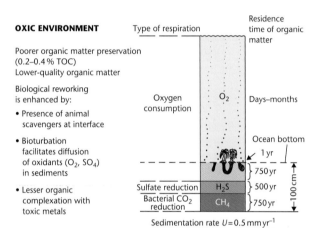

Fig. 11.11   Degradation of organic matter under oxic and anoxic conditions (after Demaison and Moore 1980). Anoxic environments are the primary sites of organic matter preservation because scavenging and bioturbation by benthic fauna and aerobic bacterial degradation are inhibited. AAPG © 1980. Reprinted by permission of the AAPG whose permission is required for further use.

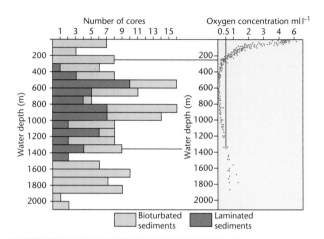

Fig. 11.12   The correlation of laminated sediments with low oxygen concentrations in the bottom waters of the Gulf of California demonstrates that bioturbation is limited by oxygen concentrations of less than 0.5 ml l^{-1}.

degradation appears to result in a greater preservation of lipid-rich, oil-prone material. Furthermore, under anoxic conditions, the bacterial population itself may contribute significantly to the preserved organic matter.

Scavenging and reworking by benthic fauna

The role of benthic metazoans such as worms, bivalves and holothurians is critical to the preservation of organic matter. Their activity is important in two respects. First, they consume particulate organic matter in the water just above the sea or lake floor and in the surface sediment itself. Second, burrowing metazoans churn up the sediment to a depth of 5 to 30 cm, allowing the penetration of oxygen and sulphates into the sediment column, thus promoting bacterial degradation.

Bioturbation seems to take place at all water depths under oxic water columns. Below oxygen concentrations of 0.3 ml/l this activity is virtually eliminated. Sediments remain laminated and organic rich. Even the activity of anaerobic sulphate-reducing bacteria is limited because oxidants cannot easily penetrate the surface. The occurrence of unbioturbated, laminated muds in the Gulf of California is closely correlated with low oxygen concentrations in bottom-waters (Fig. 11.12). For these reasons, organic matter stands a much better chance of being preserved in the absence of benthic fauna, that is, in anoxic environments.

Transit time of organic matter in the water column

Almost all marine organic matter is formed by photosynthesis in the euphotic zone. Before it can accumulate on the seabed it has to fall through a water column of up to 6 km (in deep ocean areas). The smallest particles take the longest to fall, faecal pellets falling the fastest. Organic matter is scavenged by fauna during its transit through the water column. Preservation of organic matter is therefore favoured by shallow water depths and large organic particle size. Scavenging in the water column is probably one of the factors that contributes to the general lack of source bed deposition in deep ocean areas. Another factor is low organic productivity due to remoteness from nutrient supply.

Sediment grain size

The low permeability of fine-grained sediments inhibits the diffusion of oxidants from the water column into the sediment, and, as a result, bacterial activity is lower than in coarse-grained sediment. Coarse-grained sediment is usually associated with high-energy environments that are in any case likely to be well oxygenated.

Sedimentation rate

Under oxic conditions, high sedimentation rates favour source bed deposition because they reduce the period during which organic matter is subject to metazoan grazing, bioturbation and aerobic bacterial attack. Sufficient organic matter may be preserved even under oxic conditions to form a source bed. The organic matter becomes diluted, however, by the large amount of mineral matter in the sediment, and the resulting source rock usually has a low organic matter concentration. It may not be capable of generating sufficient liquid hydrocarbons to saturate the source rock and expel oil. Hydrocarbons may instead be expelled at high maturity as gas. Rapidly deposited sequences, therefore, rarely contain good oil sources. Examples are thick, muddy pro-delta sequences. Under anoxic conditions, high sedimentation rates are likely to have only an adverse diluting effect on organic content. Deposition of rich source beds is therefore favoured by low sedimentation rates.

The ideal conditions for oil source bed deposition are therefore: (i) anoxic conditions with high organic productivity and restricted oxygen supply (poor water circulation); (ii) shallow water depths; and (iii) fine-grained sediment. Under oxic conditions, moderate sedimentation rates favour source bed deposition.

11.3.2.3 Depositional settings of source beds

The main depositional settings of source beds are lakes, deltas, and marine basins. There are a number of other less important settings, including freshwater swamps, non-deltaic shorelines, and continental slopes and rises. These settings appear to have sourced a relatively small proportion of the world's oil and, in terms of source prediction, provide only a relatively low probability of source bed presence. An individual exception to this is paralic source rocks developed along non-deltaic but sheltered tropical shorelines, which may have the same characteristics as the lower delta plain.

Lakes

Lakes are the most important depositional setting for source beds in continental sequences. In order to form volumetrically significant source beds, lakes must be geologically long-lasting. Anoxic conditions develop in 'permanent' lakes when the water column becomes stratified. This is most likely to occur in the following circumstances (Allen & Collinson 1986, Talbot & Allen 1996 for summaries):

- *In deep lakes.* Wind stress causes the mixing of the whole water column in shallow lakes, causing oxygenation of bottom-waters and sediments. Deep lakes are usually tectonically controlled. They frequently develop in rapidly subsiding extensional continental rift systems but may also occur in areas of compressional tectonics. The 1500 m-deep Lake Tanganyika in the East African Rift system is anoxic below 150 m. TOCs of 7 to 11% have been recorded from bottom sediment in the anoxic part of the lake. The shallower Lake Mobutu (Albert) and Lake Victoria are oxic. Beadle (1981) is an excellent source of information on African lakes.
- *At low latitudes.* Wide seasonal variations in weather cause overturn of the water column. Cold, dense river waters carrying large amounts of dissolved oxygen sink to the bottom of temperate lakes, causing oxygenation. All temperate lakes at the present day, even the 1620 m-deep Lake Baikal, are oxygenated for at least part of the year. In warm, tropical, equable climates river water is less dense, does not have a tendency to form high-density flows, and carries less oxygen. These conditions favour the development of anoxic conditions. In addition, the temperature-density behaviour of water (Ragotzkie 1978) means that more work is required to mix two layered water masses at elevated temperatures (e.g. 29 °C and 30 °C) than at low temperatures (e.g. 4 °C and 5 °C), so tropical lakes tend to stratify easily. On the other hand, the slightest cooling in a tropical lake may initiate convection currents that may eventually affect the entire water body, causing mixing and oxygenation of bottom-waters.

- *Abundant water supply*; a wet climate ensures that the lake is kept filled with water. In arid climates, lakes may intermittently dry up, resulting in oxidation of its surface sediment. Provided this does not happen, however, high evaporation losses may encourage anoxia, by producing a *salinity stratification*. Salinity-stratified lakes may form important source environments in low latitudes. Ancient examples are the Devonian Orcadian basin of the UK (review in Allen & Collinson 1986) and the Eocene Green River Formation of western USA (Eugster & Hardie 1975).

Hydrothermal solutions in areas of volcanic activity, and run-off over peralkaline volcanic products, may produce alkaline lakes. Strongly reducing conditions may develop at the lake floor. Distinctive mineral assemblages are characteristic of alkaline lakes. The Lakes Magadi and Natron of East Africa are today precipitating trona, and the Wilkins Peak Member of the Eocene Green River Formation of Colorado and Utah has a similar evaporite mineralogy.

Source beds will be richer and thicker and more likely to develop if a deep lake can be maintained for a long time with the minimum of particulate input. Clastic input to lakes is a complex function of topographic, climatic and bedrock variables in the catchment area of the lake (Chapter 7). Hilly or mountainous relief in tectonically active regions usually causes rapid erosion and the rapid infilling of a lake basin by coarse detritus. Small lakes tend to be completely swamped by clastic input, whereas centres of large lakes may see very little terrigenous influence. If the hinterland is an area of carbonate outcrop, however, much of the weathering is chemical rather than mechanical, and the suspended particulate clastic input to the lake is small. Chemical weathering is dominant in humid climates, rocks quickly breaking down under the combination of high temperatures and abundant water. The transport of such weathering products depends on the rainfall intensity but also on the vegetation type. Lush vegetation, particularly grasses, in the drainage area will tend to reduce the amount of surface erosion, and hence the particulate input to the lake.

Organic matter input to lakes is of two types:

1. *Produced within the lake itself (autochthonous)*, comprising algae and bacteria. It produces strongly oil-prone kerogen. Ancestors of the present-day alga *Botryococcus braunii* have been identified in ancient lake sediments, and its biomarker compound Botryococcane has been recognised in many lake-sourced oils. An example is the oil of the Minas Field in central Sumatra (Williams *et al.* 1985).
2. *Swept into the lake from the drainage area (allochthonous).* This comprises organic matter from higher plants. Much of it will be lignin, which is gas-prone. There may also be a contribution of oil-prone waxy epidermal tissues, spores or pollen. As discussed in §11.3.1.3 and §11.3.1.4, these oil-prone components are only likely to be important in tropical areas since the Jurassic. Terrigenous organic matter is likely to dominate a small lake. In large lakes, it may be concentrated around the margins (particularly around river mouths), leaving the centre the site of algal and bacterial organic matter deposition.

Lake sediments may source oil, gas-condensate or gas, depending on the factors discussed above. Lacustrine oils tend to be very variable in density, low in sulphur, and have a very variable wax content, ranging up to 40%. Wax is derived from land plant cuticles and from wax-secreting freshwater algae.

The lacustrine oil shales of the Eocene Green River Formation of Utah, Wyoming and Colorado (Bradley & Eugster 1969; Surdam & Wolfbauer 1975; Cole & Picard 1981; Smoot 1983) accumulated in paleo-lakes Uinta and Gosiute, now preserved in the Uinta and Piceance Creek sub-basins (Lake Uinta) and the Green River and Washakie sub-basins (Lake Gosiute). A range of lacustrine environments is represented by the members of the Green River Formation. Some of the most important oil source rock units (e.g. Laney Shale, Mahogany Oil Shale) were deposited in deep, anoxic, density-stratified, generally saline lakes, while hypersaline shallow lakes (Wilkins Peak and Parachute Creek members) and their fringes were also sites of significant organic matter accumulation. Hypersalinity and lack of sulphate inhibited bacterial oxidation of the organic matter in these lakes. Arid conditions precluded significant land plant input to the lake systems; organic matter is, as a result, autochthonous and oil-prone.

In contrast to the saline and hypersaline units of the Green River Formation, the Pematang lake sediments of the *Paleogene rifted basins of central Sumatra, Indonesia*, are deposits of low-salinity lakes (Williams *et al.* 1985). The Pematang reaches up to 1800 m in thickness and was deposited in structurally controlled half-grabens, under humid tropical conditions. The most important oil source rock unit is the Brown Shale Formation, which represents the deposits of deep lakes formed by rapid syn-rift subsidence. These are well-laminated, reddish brown to black non-calcareous mudstones. Geochemical correlations demonstrate a convincing match between Brown Shale algal source rocks and crude oils, including those of the giant Minas and Duri fields. In contrast to the deep-lake oil-prone Brown Shale Formation, gas-condensate-prone land plant source rocks predominate in the shallow-lake sequences of the Coal Zone Formation.

Albian–Turonian lacustrine source rocks have been described from the *Songliao Basin of eastern China*. These represent the deposits of deep, thermally stratified freshwater lakes. Small-scale examples of rich oil-prone lacustrine source rocks have been described from the Tertiary Mae Sot and Mae Tip Basins of northwest Thailand (Gibling 1985a, b). They were deposited in shallow, fresh to brackish lakes.

Deltas

Deltas may be an important setting for source rock deposition. In SE Asia and Australasia, deltas sourced a large proportion of the discovered oil. Constructive deltas (fluvial or tide-dominated) are characterised by persistent low-energy environments on the delta top that favour source bed deposition. Destructive or static deltas (wave-dominated) generally provide less favourable environments for source bed deposition. Migration of shoreline bars tends to rework the sediments in the lower delta plain, and organic matter is usually degraded to an inert state.

Organic matter in delta sequences may be of three types:

1. *Freshwater algae (phytoplankton) and bacteria* present in lakes, swamps, and abandoned channels on the delta top. This material is oil-prone. It will only be preserved if anoxic conditions exist at the sediment surface. Fluvial channel migration results in the infilling of lakes in deltaic environments. Lakes are most likely to persist, therefore, in the upper part of the delta plain. High subsidence rates on the lower delta plain may allow lakes to persist despite rapid sedimentation rates, as occurs on the Niger and Nile deltas at the present day.

2. *Marine phytoplankton and bacteria* in the delta-front and pro-delta areas. Abundant nutrients provided by river input frequently stimulate high organic productivity in the marine basin into which the delta debouches, but conditions for preservation are generally poor. Preservation requires anoxicity in the marine basin. The delta front is normally a high-energy, oxic environment. High accumulation rates allow the preservation of some organic matter, but it is strongly diluted with mineral matter, and source rocks are usually lean with potential to expel only gas.

3. *Terrigenous land plant material.* Vegetation growing on the delta plain contributes a large amount of organic matter to depositional environments. On tropical deltas since the Jurassic, the land plant material is likely to comprise both oil-prone (waxy epidermal tissues, resins, spores) and gas-prone (lignin) material. Pre-Jurassic and temperate land plants are predominantly gas-prone. The most prolific sites of accumulation are the inter-distributary peat swamps, where *in situ* coals may form. Terrigenous organic matter may, however, be dispersed across the entire delta top, and into the pro-delta environment where preservation will largely depend on high sedimentation rate. In peat swamps, the high accumulation rates, highly acidic conditions, and presence of bactericidal phenol compounds released from lignin, enhance the preservation of organic material. In the upper part of the delta plain, freshwater swamps dominate. On the lower delta plain, where waters are brackish to saline as a result of some marine influence, mangrove swamps dominate. Mangroves trap plant material drifted in from the freshwater upper delta plain. Terrigenous organic matter may also be reworked onto low-energy tidal flats.

Mangrove-dominated shorelines, such as in Missionary Bay, north Queensland, Australia (Risk & Rhodes 1985), may be important depositional environments for source rocks. Mangrove litter originating on intertidal mudflats is swept into the adjoining anoxic bay-bottom sediments, providing organic matter with a very high lipid content. The mangrove swamps are sites of prolific organic productivity. A thin intertidal strip of mangrove swamp may produce a vast quantity of lipid-rich organic matter and spread it over a large offshore area. The high oxygen demand caused by the abundant influx of mangrove detritus may cause anoxia in the surrounding depositional environments. Mangrove material is also relatively resistant to degradation, both physical and chemical. Under fungal and bacterial attack, the waxy lipid-rich cuticle that coats mangrove epidermal tissue appears to be preferentially preserved. Mangrove material is likely to source oils that have a high wax content.

Reworking of upper delta plain plant material in brackish conditions appears to result in a selective bacterial (or fungal?) degradation of cellulose and lignin (to humic acids), leaving a relative enrichment in oil-prone waxy and resin components (Thomas 1982).

The coals within the Upper Cretaceous–Eocene *Latrobe Group* of the Gippsland Basin of SE Australia (Shanmugam 1985) have sourced about 3000 million barrels of oil. The coals are dominantly vitrinitic but contain up to 15% exinite macerals, comprising cutinite, sporinite and resin. These components are derived from the cones, bark, seeds, leaves and resin bodies originating in the adjacent coniferous forests. The climate was temperate and wet. Oils have been geochemically matched with these coals, and have typical coal-source characteristics, including high wax content (<27%).

Marine basins

Marine source rocks may form in enclosed, silled basins such as the Black Sea, Baltic Sea and Lake Maracaibo, or on open marine shelves and continental slopes. The mechanisms for source bed development in each of these settings are quite different. In enclosed basins, *water stratification* reduces oxygen supply. On open shelves and slopes, source beds may develop as a result of *oceanic upwelling*, causing high organic productivity and hence high oxygen demand, and as a result of impingement of the *oceanic midwater oxygen minimum layer*.

Shallow, large epicontinental seas are favourable settings for the deposition of source rocks, since tidal and wave energies are damped or attenuated, leading to water body stratification (Scourse & Austin 2002). Ancient epicontinental seas would have acted more like vast salty lakes than coastal oceans, with much reduced water circulation (Allison & Wells 2006), potentially leading to anoxia (Wells *et al.* 2005). However, constrictions between islands and shallow platforms are likely to have promoted high bed shear stresses driving sediment transport, as in the Early Jurassic Laurasian seaway of northwest Europe (Mitchell *et al.* 2011).

Large discharges of river water in the coastal region may enhance biological productivity on shallow shelves and semi-enclosed seas by the supply of biolimiting nutrients, promoting anoxia where water circulation is restricted. Climatic forcing of precipitation patterns, and therefore river discharges, may therefore cause alternations of black shale and lean sediment deposition on marine shelves and epicontinental seas, as in the Coniacian–Campanian of tropical Africa (Beckmann *et al.* 2005).

(a) Enclosed basins

Enclosed marine basins are physically restricted, to some extent, by land or by chains of islands, but retain some connection with the open sea. Water exchange is limited, however, and the basin is prone to water stratification and hence anoxic conditions. The nature of the water exchange with the open sea is important, since not all enclosed basins become anoxic.

A *positive water balance* is where the outflow of freshwater (as a surface layer) exceeds the relatively small inflow of deeper saline water. Most of the water movement takes place in the surface layers, allowing stratification of deeper waters. This process characterises the Black Sea, Baltic Sea and Lake Maracaibo. The Black Sea is one of the best-documented anoxic silled marine basins (Fig. 11.13). TOC is up to 15% in sediment 7000 to 3000 years old. At the present day, it is anoxic below 150 to 250 m water depths. In contrast, a *negative water balance* is where the inflow of oceanic water dominates over a relatively meagre freshwater input. This often develops in arid climates where high evaporation losses at the surface cause the sinking of oxygenated waters (Fig. 11.14). Examples of oxic enclosed basins include the Red Sea, Mediterranean Sea and Persian Gulf. The Mediterranean is the world's largest silled marine basin, but it is well oxygenated, and organic contents in bottom sediment are very low.

Size and depth of enclosed basins do not appear to be critical. Lake Maracaibo, for example, is only 30 m deep. The danger of water mixing by wind-generated waves, however, renders shallow basins less favourable source bed environments. Enclosed basins may range in size up to that of the South Atlantic during the Aptian. Small basins tend to be short-lived, particularly when there is high clastic input.

Organic matter type in enclosed marine basins depends on the amount of terrigenous land plant material brought into the basin by rivers. Wet climates imply a positive water balance and hence a ten-

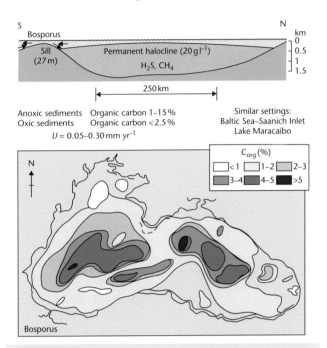

Fig. 11.13 Total organic carbon concentrations in modern sediments of the Black Sea. This is an example of a silled marine basin with a positive water balance. Organic carbon concentrations are locally up to 15% in the deeper parts of the basin at water depths of >1 km below a stratified water body. After Demaison and Moore (1980). AAPG © 1980. Reprinted by permission of the AAPG whose permission is required for further use.

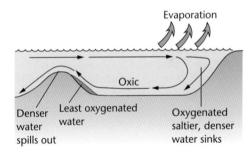

Fig. 11.14 Model for a silled marine basin with a negative water balance (after Demaison and Moore 1980). Oceanic inflow dominates over freshwater fluvial input, a situation that commonly develops in arid climates. Dense, salty, oxygenated waters resulting from surface evaporation may sink and sweep the basin floor, preventing anoxia. AAPG © 1980. Reprinted by permission of the AAPG whose permission is required for further use.

dency to water stratification. However, high terrigenous organic input may produce gas-prone source rocks. In arid climates, the organic matter is made up largely of marine phytoplankton and, when source beds are developed (often in association with carbonates and evaporites), they are predominantly oil-prone. The Devonian of the Western Canada Basin is an example. Source bed deposition is sensitive to changes in the water balance of the basin, and tends to be periodic and of varying lateral extent. Deep, big, enclosed marine basins in areas of wet paleoclimate offer the highest probability of source bed occurrence.

Examples of source rocks deposited in restricted marine basins include the Upper Jurassic Kimmeridge Clay Formation of the North Sea, and the Jurassic Kingak and Aptian–Albian HRZ Formation of the North Slope, Alaska.

(b)   Open marine shelves

Upwelling occurs along coastlines where wind-driven currents flowing parallel to the coast are deflected offshore by the Earth's rotational (Coriolis) force. It is most common on the east side of oceans. Upwelling occurs today along the coasts of Peru–Chile, California, Namibia and Morocco.

The deep ocean water drawn into the upwelling cell to replace the offshore-moving surface water is rich in nutrients such as phosphates and nitrates, and can give rise to exceptionally high organic productivity in the near-surface photic zone. Degradation of dead organic matter creates a high demand for oxygen, and anoxicity may develop in the underlying waters. Underneath the Benguela current offshore Namibia, for example, there is a 340 km by 50 km oxygen depleted zone. Under this zone, the sediment contains up to 26% TOC (Fig. 11.15) (Demaison & Moore 1980). The organic matter is almost entirely made up of marine plankton. There is very little terrestrial input from the arid hinterland; this is a feature of many upwelling coastlines at the present day.

Not all upwelling zones cause anoxic conditions. Oxic examples include off SE Brazil, the northern Pacific and bordering Antarctica. The most likely reason for oxicity is that the upwelling is only seasonal.

A diagnostic feature of sediments deposited under upwelling currents is a distinctive mineral assemblage, including phosphorites and uranium minerals, as in the Neogene deposits of the Californian basins.

Prediction of ancient upwelling zones depends on the accurate reconstruction of atmospheric circulation and paleogeography. Occurrence of upwelling means significantly improved chances of source bed presence. Examples of source beds thought to have been deposited as a result of upwelling include the Permian Phosphoria Formation of western USA, the Triassic Shublik Formation of the North Slope, Alaska, the Cretaceous La Luna Formation of Venezuela/Colombia, the Upper Cretaceous Brown Limestone of Egypt and the Miocene Monterey Formation of California.

The oceanic midwater oxygen minimum layer occurs in the world's oceans at depths of about 100 to 1000 m. It is caused by the degradation of organic matter that has fallen from the overlying highly productive photic zone. Below this midwater zone, oxygen contents rise again because of the influence of cold, dense currents that originate in the polar regions and sweep along the ocean floors towards tropical latitudes. These currents supply oxygen that prevents the midwater oxygen minimum layer becoming anoxic, except in rare examples. The present-day Atlantic Ocean is well oxygenated because there is virtually no obstruction to the passage of cold polar waters from both its northern and southern ends. In contrast, in the eastern Pacific and northern Indian oceans (Fig. 11.16), oxygen levels drop to less than 0.5 ml/l. Ocean floor currents reaching these areas have lost much of their oxygen, and the midwater oxygen minimum layer is free to intensify. TOC values in these areas range up to 11%.

The presence of strong ocean currents derived from the poles is a relatively recent feature of the world's oceans. At the present day, the Earth is in an interglacial phase. At times in the past, for example during most of the Mesozoic (particularly during the Late Jurassic

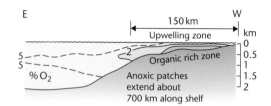
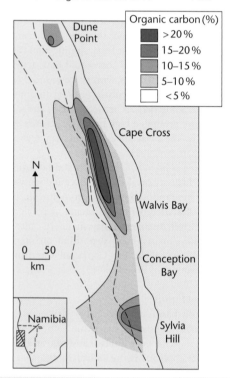

Fig. 11.15 Upwelling zone, offshore Namibia, showing oxygen-depleted zone, and total organic concentrations up to 26% (after Demaison and Moore 1980). Upwelling causes high phytoplankton productivity in surface waters. Sea bottom sediments are anoxic under the highly productive waters. AAPG © 1980. Reprinted by permission of the AAPG whose permission is required for further use.

and Mid-Cretaceous), the global climate was warmer, and the shape of the world's oceans was quite different from that of today. At this time, oceanic circulation may have been much more sluggish, and an intense midwater oxygen minimum zone may have developed. The Mid-Cretaceous 'global anoxic events' probably occurred under such circumstances.

During times of high sea level, the midwater oxygen minimum zone may impinge on the continental shelf over wide areas. As a result, source beds are deposited on the continental shelf in association with reservoir and carrier beds, and a hydrocarbon play may be produced. Oxygen deficiency could be reinforced in bathymetric depressions in the broad epicontinental seas produced at this time. Sediments deposited in midwater anoxic zones may be finely laminated, unbioturbated, organic-rich, diatomaceous mudstones (Gulf of California) or olive grey muds (northern Indian Ocean). Examples of source rocks thought to have been deposited as a result of impinge-

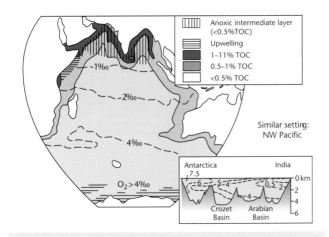

Fig. 11.16 Midwater oxygen concentrations and total organic carbon (TOC) concentrations in the bottom sediment of the Indian Ocean. TOCs are highest around the Indian coastline where oxygen levels in the midwater layer fall to less than 0.5 ml l⁻¹ and impinge on the continental slope and shelf. Inset shows N–S transect of Indian Ocean with water mass contoured in terms of oxygen concentrations in ml l⁻¹.

ment of the midwater oxygen minimum zone on the continental shelf include the Toarcian source rocks of western Europe.

The optimum conditions for marine source bed deposition occur when one of the mechanisms is reinforced by another. For example, a silled geometry may combine with the midwater oxygen minimum zone, or the midwater oxygen minimum may be intensified by upwelling.

The gas- or oil-proneness of a marine source rock depends primarily on the presence or absence of gas-prone terrigenous plant material. Enclosed marine basins close to a major clastic source may be gas-prone. Oil-prone organic matter of truly marine origin occurs in upwelling zones offshore from arid land areas.

11.3.3 Detection and measurement of source rocks

A range of techniques has been developed to identify and measure source rocks and petroleum fluids. These include identification of mature source rock horizons on petrophysical wireline logs, and a range of geochemical and visual microscopic analyses of rock and fluid samples obtained from wells, outcrop and seeps. It is beyond the scope of this book to attempt to describe the detail or the full range of these techniques. However, we outline here a few routine measurements, and some explanations of key terms. These techniques may be used initially to establish simply whether a source rock exists in the sediments being sampled, but a series of other important questions about the petroleum charge system operating in an area may also be addressed. The richness of a source rock, the petroleum composition likely to be expelled, and the thermal maturity of the source rock can be determined. Source rocks and fluids may be correlated geochemically, so that migration routes can be interpreted.

Kerogen is that part of the organic matter in a rock that is insoluble in common organic solvents. It owes its insolubility to its large molecular size. Different types of kerogen can be identified, each with different concentrations of the five primary elements, carbon, hydrogen, oxygen, nitrogen and sulphur, and each with a different potential for generating petroleum.

The organic content of a rock that is extractable with organic solvents is known as *bitumen*. It normally forms a small proportion of the TOC in a rock. Bitumen forms largely as a result of the breaking of chemical bonds in kerogen as temperature rises.

Petroleum is the organic substance recovered from wells and found in natural seepages. Bitumen becomes petroleum at some point during migration. Important chemical differences often exist between source rock extracts (bitumen) and crude oils (petroleum).

Crude oil is naturally occurring petroleum in a liquid form. The term black oil is sometimes used to indicate petroleum that is liquid at both reservoir and surface temperatures and pressures.

Natural gas is petroleum occurring in the gaseous phase. *Wet gas* is differentiated from *dry gas* in that it yields significant volumes of liquid (*condensate*) on changing from reservoir to surface conditions. When the condensate yield is potentially high, the fluid is called a *gas-condensate*.

Under conditions of very low temperature and high pressure, *gas hydrates* may form (§12.2.5).

Natural gas resulting from the thermal breakdown of kerogen is known as *thermogenic*. *Biogenic gas*, however, is a natural gas formed solely as a result of bacterial activity in the early stages of diagenesis (<60–70 °C). It normally occurs at shallow depths, and is always very dry.

Some key measurements routinely made on source rocks and oils are TOC, amount of soluble extract, rock pyrolysis, gas chromatography, visual kerogen descriptions, and vitrinite reflectance.

Total organic carbon (TOC) is a measure of the carbon present in a rock in the form of both kerogen and bitumen. TOC values in source rocks may be quite low, and are frequently less than 2%. Coals, however, may have TOCs of over 50%. 0.5% TOC is frequently taken as the minimum organic content for a shale source rock; a slightly lower value applies for carbonates. Below 0.5% TOC, not enough petroleum can possibly be generated to saturate the source rock, without which expulsion cannot occur. Rocks with greater than 0.5% TOC are not, however, guaranteed as source rocks. If the organic carbon is inert, no amount of it will form a source rock.

The *soluble extract (bitumen)* of source rocks reflects oil content, and shows a strong correlation with thermal maturity. Bitumen content rises with the onset of oil generation, and diminishes as the oil floor is approached.

Rock-Eval pyrolysis is the standard procedure developed by Espitalié for the pyrolysis of rock samples. The amounts of hydrocarbon products evolved through the heating of a rock sample to 550 °C is recorded by a flame ionisation detector (FID) as a function of time. Three peaks are typically recorded, known as the S_1, S_2 and S_3 peaks (Fig. 11.17). The S_1 peak represents hydrocarbons evolved at low temperatures – these represent free or adsorbed hydrocarbons (bitumen) that were already present in the rock before pyrolysis. The S_2 peak is produced at higher temperatures by the thermal breakdown of kerogen. Oxygen-bearing volatile compounds (carbon dioxide and water) are passed to a separate (thermal conductivity) detector, which produces an S_3 response. S_1, S_2 and S_3 are expressed as milligrams per gram of original rock, (mg g⁻¹) or kilograms per tonne (kg t⁻¹). The temperature at which the S_2 generation peak occurs (T_{max}) is also recorded, and is an indicator of source maturity.

Rocks with ($S_1 + S_2$) values of less than 2 kg t⁻¹ are considered as insignificant source rocks. Between 2 kg t⁻¹ and 5 kg t⁻¹ a significant amount of petroleum may be generated in the source rock, but it may be too small to result in expulsion. If the source rock is raised to

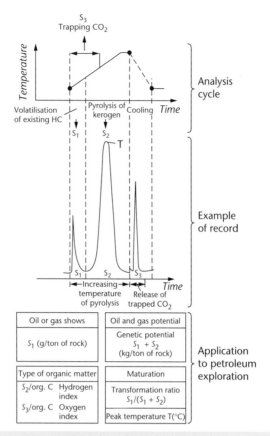

Fig. 11.17   Output obtained from the pyrolysis method of Espitalié *et al.* (1977) (after Tissot and Welte 1978). Three peaks are normally observed. The S_1 peak represents already existing bitumen. S_2 represents hydrocarbons generated by the thermal breakdown of kerogen. The S_3 response is produced by oxygen-bearing compounds released at high temperature. S_1 and S_2 can be used to assess oil-generating potential, while S_2 and S_3 can be used to calculate hydrogen index and oxygen index respectively, which indicate kerogen type. The temperature corresponding to the S_2 peak, and the $S_1/(S_1+S_2)$ ratio, indicate the level of thermal maturation. Reproduced with kind permission from Springer Science+Business Media B.V.

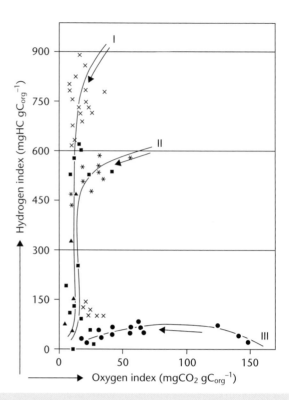

Fig. 11.18   Classification of kerogen types using hydrogen and oxygen indices. The chart is comparable to the van Krevelen diagram plotted from elemental analyses of kerogen (Espitalié *et al.* 1977). Each kerogen type has different hydrocarbon-generating characteristics. Type I is the most oil-prone, whereas Type III produces mostly gas. After Tissot and Welte (1978), reproduced with kind permission from Springer Science+Business Media B.V.

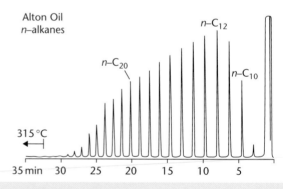

Fig. 11.19   Gas chromatogram of *n*-alkanes of an Australian crude oil. Each *n*-alkane compound in the oil is identified by a peak in the gas chromatogram.

higher maturity, generated oil may be cracked to gas and expelled in the gas phase. Source rocks with potentials of 5–10 kg t^{-1} have the potential to expel a proportion of their generated oil. Source rocks with greater than 10 kg t^{-1} are considered rich; oil generated will almost certainly be in sufficient quantities to ensure expulsion. Exceptionally, yields of several hundred kg t^{-1} are measured; these are usually from coals or oil shale. The type of kerogen in the source rock is indicated by the *hydrogen index*, which combines the S_2 pyrolysis peak with TOC:

$$\text{Hydrogen index} = (S_2/\text{TOC}) \times 100 \text{ mg g}^{-1} \text{ }^{\circ}\text{C}^{-1} \qquad [11.1]$$

It expresses the 'usable' or pyrolysable fraction of the organic content. What is left is inert carbon, which is incapable of sourcing petroleum. Hydrogen indices of <50 imply that the kerogen is made up predominantly of inert kerogen. Values of >200 suggest the presence of significant amounts of hydrogen-rich (oil-prone) kerogen. The hydrogen index may be as high as 900 in strongly oil-prone oil

shales. The ratio of S_3 to TOC is known as the *oxygen index*. Tissot and Welte (1978) cross-plot hydrogen index and oxygen index in order to classify source rocks into three types: I, II and III (Fig. 11.18). Each has different petroleum-generating characteristics.

Gas chromatography is a technique that separates the individual petroleum compounds in a petroleum mixture, according to increasing carbon number (Fig. 11.19). Gas chromatography can be performed on the products from pyrolysis (pyrolysates), on soluble extracts or on crude oil samples. From pyrolysis gas-chromatography (PGC), the broadest application is in estimating the oil versus gas proneness of the kerogen. This can be determined by dividing the pyrolysate into gas (C_1–C_5) and liquid (C_{6+}) fractions. At a greater level of refinement, the gas chromatogram determines the detailed composition of the fluid for use in rock-rock, rock-oil, and oil-oil correlations.

The *visual size, shape, structure and colour of kerogen fragments*, once isolated from the rock, can be microscopically examined in transmitted light, in order to identify *humic* kerogen derived from recognisable higher land plant material, and translucent amorphous *sapropelic* kerogen. The colour of spores, pollen and other micro-fossils is broadly related to thermal maturity: spore colour changes from yellow, through orange to brown and eventually black, as thermal maturity increases (§10.4.2, Fig. 10.21). Humic kerogen was formerly equated with gas-prone source rocks, and sapropelic kerogen with oil-prone source rocks, but this correlation is now known to be exceedingly unreliable.

Examination of kerogen particles in *reflected light* allows the definition of three major maceral groups, each with important implications for petroleum generation potential (§11.4.2), based on their reflectance in oil. The *fluorescence* of organic particles under ultraviolet light allows the identification of the liptinite group, which fluoresce strongly. The vitrinite and inertinite groups do not usually fluoresce.

Vitrinite reflectance is the most widely used indicator of thermal maturity (§10.4.2 and §10.5.1), and therefore of source rock maturity. The vitrinite reflectance scale has been calibrated by other maturity parameters and by field studies in oil and gas provinces, so that R_o (reflectance in oil) may be correlated with the main zones and thresholds of petroleum generation as follows:

$R_o < 0.55$	Immature
$0.55 < R_o < 0.80$	Oil and gas generation
$0.80 < R_o < 1.0$	Cracking of oil to gas (gas-condensate zone)
$1.0 < R_o < 2.5$	Dry gas generation

Vitrinite reflectance is a very good maturity indicator above about 0.7–0.8% R_o.

An important use of vitrinite reflectance measurements in basin analysis is in calibrating thermal and burial history models with present-day maturity data. §10.4 describes a range of thermal maturity indicators, including organic indicators (predominantly vitrinite reflectance), low-temperature thermochronometers (mainly apatite fission tracks), and temperature-dependent mineralogical and geochemical indices (clay mineral assemblages, quartz cementation, and fluid inclusions). §10.5 discusses the application of vitrinite reflectance profiles, apatite fission track age-depth relationships, and analysis of quartz cementation, to the understanding of basin thermal history.

11.4   The petroleum charge

A petroleum charge occurs when petroleum is generated in a source rock, is expelled, and migrates through a carrier bed to a trap.

Petroleum is chemically a mixture of saturated and aromatic hydrocarbons and NSO (nitrogen-sulphur-oxygen) compounds. A wide range of geochemical analyses can be carried out on petroleum and source rock extracts that aim to relate the petroleum back to its original source rock and depositional environment. Important physical properties of petroleum are its density, formation volume factors and boiling points; these influence secondary migration processes, subsurface volume changes and phase behaviour.

Petroleum generation takes place as a result of the chemical breakdown of kerogen with rising temperature. As hydrocarbons are released, the remaining kerogen evolves towards a carbon residue. Temperature and time are the most important factors affecting the breakdown of kerogen. The rate of breakdown can be calculated from the Arrhenius equation given in §10.2. The reactive fraction of kerogen can be subdivided into a labile portion, which yields chiefly oil, and a refractory portion, which yields mainly gas. Labile kerogen breaks down over approximately the 100–150 °C range, followed by refractory kerogen from 150 to 220 °C. Over the 150–180 °C range, oil is rapidly cracked to gas. Thus, a stage of oil generation is succeeded by a stage of wet gas/gas-condensate generation, and finally by a stage of dry gas generation. *Petroleum expulsion* is probably caused by microfracturing of the source rock after overpressure has built up as a result of hydrocarbon generation. Lean source rocks may not generate sufficient oil to cause expulsion. If raised to higher maturity, generated oil may be cracked to gas that will be efficiently expelled. For rich source rocks (>5 kg t^{-1}) efficiency of oil expulsion may be quite high (60–90%). Source rocks can be classified into three types on the basis of their richness and expulsion products.

Secondary migration carries petroleum from the site of expulsion through porous and permeable carrier beds to sites of accumulation (traps) or seepage. The main driving force behind secondary migration is buoyancy, caused by the density difference between oil (or gas) and formation pore waters. The main restricting force is capillary pressure, which increases as pore sizes become smaller. During secondary migration, petroleum flows as slugs through the interconnected network of largest pores in the carrier bed, rather than sweeping its whole volume. Movement is stopped when a smaller pore system is encountered whose capillary pressure exceeds the upward-directed buoyancy of the petroleum column. This pore system constitutes a *seal*. The maximum petroleum column height that can be supported by a seal can be calculated.

Petroleum will tend to move in the true dip direction of the top of the carrier bed. Thus, structural contour maps can be used to model migration pathways. During long-distance migration, for example in some foreland basins, petroleum flow may be strongly focused along regional highs. Losses of petroleum during secondary migration are difficult to quantify.

Petroleum may be physically and chemically altered while it is in the trap by the processes of biodegradation, water washing, deasphalting, thermal alteration, gravity segregation and dysmigration.

The leakage of petroleum from traps towards the Earth's surface is known as tertiary migration: it completes the natural life cycle of petroleum that starts with its biological precursors and finishes with its dissipation at the Earth's surface.

11.4.1   Some chemical and physical properties of petroleum

In order to understand petroleum generation, expulsion and migration, and the chemical changes that may take place in the trap, we need to know a little more about the chemical and physical properties of petroleum. The following discussion is very basic and very brief. Kinghorn (1983), Hunt (1979) and Tissot and Welte (1978) have provided excellent texts on the geochemistry of petroleum.

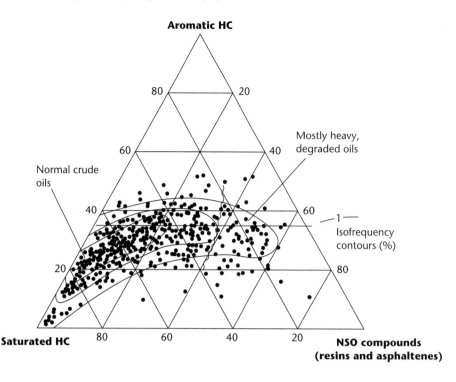

Fig. 11.20 Gross composition of 636 crude oils (from Tissot and Welte 1978) in terms of the three main groups of compounds found in petroleum. Normal (non-degraded) crudes typically contain 60–80% saturates, and less than 20% NSO compounds. Reproduced with kind permission of Springer Science+Business Media B.V.

Hydrocarbons are compounds made up solely of hydrogen and carbon. Petroleum is usually a mixture of hydrocarbon compounds and other compounds containing additional substantial amounts of nitrogen, sulphur and oxygen, and other minor elements.

There are three main groups of compounds found in petroleum (Fig. 11.20):

1. *Saturated hydrocarbons* are compounds in which each carbon atom is completely saturated with respect to hydrogen. Structures include simple straight chains of carbon atoms (the normal paraffins or normal alkanes), branched chains (the isoalkanes), and rings (cyclic hydrocarbons). Methane and ethane are examples of simple normal alkanes, and are commonly referred to as C_1 and C_2.
2. The *aromatic hydrocarbons* are a group of unsaturated hydrocarbons with cyclic structures, and include several important biomarker compounds that allow oils and source rocks to be correlated.
3. *NSO compounds* contain atoms other than carbon and hydrogen, predominantly nitrogen (N), sulphur (S) and oxygen (O). They are known as heterocompounds and are subdivided into the *resins* and the *asphaltenes*.

The composition of source rock bitumens tends to be different from crude oils – they contain fewer aromatic and saturated hydrocarbons and more resins and asphaltenes. These differences are probably due to important chemical changes that take place during migration.

Biomarkers are compounds found in crude oils and source rock extracts that can be unmistakably traced back to living organisms. The nature of the biological input to a sediment, and the chemistry of the depositional environment, give source rock extracts and expelled oils a characteristic 'fingerprint', which is superimposed by the effects of diagenesis and maturation. These fingerprints may be

geochemically recognised. Biomarker compounds may be used to assess thermal maturity (at low levels), and to correlate oils with source rock extracts.

Carbon isotope (δC_{13}) values of crude oils and source rock extracts may be used to distinguish marine from freshwater/terrestrial sources, and biogenic from thermogenic gases.

Crude oils may be classified to enable specific oil types to be directly related back to their source rocks. Such classification schemes use parameters such as oil density, sulphur content, metals content, wax content, carbon isotope value and pristane/phytane ratio. High sulphur contents (>1%) indicate marine sources, high wax content indicates land plant or freshwater algae sources, and high pristane/phytane (C_{17}/C_{18}) ratios (>3) indicate land plant material as the source.

Oil density is normally quoted as an API (American Petroleum Institute) gravity:

$$°API = (141.5/SG_{60}) - 131.5 \qquad [11.2]$$

where SG_{60} is the specific gravity at 60 deg F (15.6 deg C).

From eqn. [11.2], a fluid with a specific gravity of 1.0 g cm⁻³ has an API gravity of 10°. *Heavy oils* are those with API gravities of less than 20° (SG > 0.93). These oils have frequently suffered chemical alteration as a result of microbial attack (biodegradation) and other effects. Not only are heavy oils less valuable commercially, but they are considerably more difficult to extract. API gravities of 20 to 40° (SG 0.83 to 0.93) indicate *normal oils*. Oils of API gravity greater than 40° (SG < 0.83) are *light*.

At surface conditions, normal oils are clearly less dense than water. However, under subsurface conditions, this density difference is much greater. Oil has a great capacity to contain dissolved gas at elevated temperatures. As a result, subsurface oil densities are typically in the 0.5 to 0.9 g cm⁻³ range. Subsurface pore water densities, in contrast, are typically 1.0 to 1.2 g cm⁻³, and largely dependent on salinity. This density difference is the main driving force behind the secondary migration of petroleum.

Gas densities vary markedly between surface and subsurface conditions. At atmospheric pressure, the density of methane (C_1) is only $0.0003\,\text{g cm}^{-3}$, but at subsurface pressures of 5000 psi (equivalent to depths of 3 to 4 km), typical natural gas mixtures have densities of approximately 0.2 to $0.4\,\text{g cm}^{-3}$. In the subsurface, therefore, oils and gases take on more similar physical properties. Methane is the lightest of the hydrocarbon gases, and is normally the most abundant. Dry gases typically have methane concentrations of over 95%.

Gas expands enormously on release of subsurface pressure, and may occupy several hundred times its subsurface volume at surface. Oil shrinks on movement to the surface, owing to the release of dissolved gas. Oil shrinkage factors vary from nearly 1.0 for shallow oils with almost no dissolved gas, to 2.0 or more for extremely gassy oil in deep reservoirs.

The boiling point of a petroleum compound is the temperature above which it is in a vapour state. At temperatures below its boiling point, the compound is in a liquid state. C_1 (methane) to C_4 (butane) are the only hydrocarbon compounds that are gases (vapours) at surface temperature and pressure. The other compounds are liquids.

Phase changes may take place in the subsurface during the processes of hydrocarbon generation, migration and entrapment. Liquids may condense out of a petroleum vapour as it migrates through a carrier bed into areas of lower pressure. These valuable liquids may be lost as a residual oil saturation in the pores of the carrier bed. If an oil accumulation is uplifted, large quantities of gas may be exsolved from the oil. This may cause displacement of oil from the trap.

11.4.2   Petroleum generation

11.4.2.1   Chemical changes to kerogen during source rock maturation

At shallow depths of burial of only a few hundred metres, kerogen remains relatively stable. At greater depths of burial, however, under conditions of higher temperature and pressure, it becomes unstable and rearrangements take place in its structure in order to maintain thermodynamic equilibrium. Structures that prevent the parallel arrangement of cyclic nuclei are progressively eliminated. By this process, a wide range of compounds is generated (heteroatomic compounds, hydrocarbons, carbon dioxide, water, hydrogen sulphide, etc.) as the kerogen evolves towards a highly ordered graphite structure. Petroleum generation is, therefore, a natural consequence of the adjustment of kerogen to conditions of increased temperature and pressure.

Kerogen is a complex macromolecule composed of nuclei linked by heteroatomic bonds or carbon chains that are successively broken as temperature increases. As breakdown occurs, the first products released are heavy heteroatomic compounds, carbon dioxide and water. These are followed by progressively smaller molecules, including hydrocarbons. The kerogen left behind becomes progressively more aromatic and evolves towards a carbon residue.

Mackenzie and Quigley (1988) classify kerogen into *reactive kerogen* and *inert kerogen* (Fig. 11.21). Reactive kerogen is transformed into petroleum at elevated temperatures, whereas inert kerogen rearranges towards graphite-like structures without the generation of petroleum. Reactive kerogen is subdivided into a labile portion, which is transformed into petroleum that is chiefly oil at surface, and a refractory portion that generates chiefly gas.

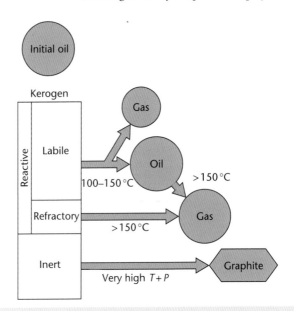

Fig. 11.21 Classification and fate of organic matter in source rocks. *Inert* kerogen rearranges towards graphite-like structures at very high temperatures (*T*) and pressures (*P*) without generating petroleum. *Reactive* kerogen is subdivided into a *refractory* part that yields mainly gas, and a *labile* portion that is transformed into petroleum, chiefly oil at surface conditions. Initial Oil in the diagram represents bitumen in immature source rocks. After Mackenzie and Quigley (1988). AAPG © 1988. Reprinted by permission of the AAPG whose permission is required for further use.

11.4.2.2   Kinetic models of kerogen breakdown

We have seen in §10.2 that temperature and time are the most important factors in controlling the maturation of organic matter. The complex series of consecutive reactions that causes kerogen breakdown proceed at varying rates, governed primarily by temperature and the activation energy of the particular reaction, and expressed in the *Arrhenius equation* (eqn. [10.1], §10.2).

If the constants in the Arrhenius equation are known for a particular petroleum-forming reaction, the rate at which it will proceed can be determined as a function of temperature, and the temperature range over which the bulk of the reaction will take place (before the raw material is used up) can be calculated. Masses of petroleum generated as a result of kerogen breakdown can also be calculated as a function of temperature, and, hence, if the subsidence and thermal history of an area are known, as a function of geological time.

The activation energies of each individual reaction are not known, but for each kerogen type a distribution of activation energies can be established from laboratory and field studies. These distributions, together with the other parameters in the Arrhenius equation, can be used to model kerogen breakdown and the associated generation of petroleum products for each kerogen type.

The kinetic model of Mackenzie and Quigley (1988) shows that kerogen concentrations diminish with increasing temperature for a range of heating rates (Fig. 11.22). Separate sets of curves are shown for labile and refractory kerogen. The heating rate parameter incorporates the time factor in source rock maturation, which is known to be important from the occurrence of generally shallower oil generation thresholds in older basins (Dow 1977). Depending on heating

(a)

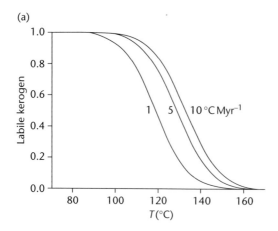

(b)

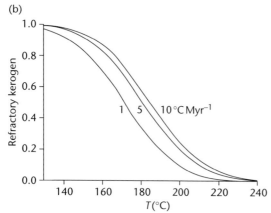

Fig. 11.22 Calculated concentrations of reactive labile (a) and refractory (b) kerogens, relative to initial amount of reactive kerogen, as a function of maximum temperature for a range of heating rates. Mean heating rates of about 0.5 °C Myr⁻¹ occur in old stretched basins, while 10 °C to 50 °C Myr⁻¹ is more typical of young (<25 Myr) stretched basins. Labile kerogen breaks down generally over the 100–150 °C range. Refractory kerogen, however, breaks down at much higher temperatures, from about 150–200 °C. After Mackenzie and Quigley (1988). AAPG © 1988. Reprinted by permission of the AAPG whose permission is required for further use.

rate, kerogen breakdown into petroleum takes place largely over the 100–150 °C range for labile kerogen. For refractory kerogen, the range is approximately 150–220 °C.

Any oil left in a source rock from the breakdown of labile kerogen will be cracked to gas if temperatures continue to rise above 150 °C, and most cracking reactions take place over the 150–180 °C range (Mackenzie & Quigley 1988). The time required to crack half the mass of oil to gas (the 'half-life' of oil), if held at a constant temperature of 180 °C, is less than a million years. Consequently, at temperatures above 160 °C, oil would not be expected to exist for geological periods of time. The cracking process applies, of course, to oil accumulations as well as to oil remaining in source rocks. As a result, oilfields will not exist at depths greater than those corresponding approximately to the 160 °C isotherm.

The kinetic model predicts that the following stages of petroleum formation succeed each other without significant overlap (Figs 11.23, 11.24):

- The *immature* stage precedes petroleum generation (the *diagenesis stage* of Tissot & Welte (1978).
- It is followed by the stage of oil and gas generation from labile kerogen containing lipid material (exinitic macerals), and a stage of wet gas/gas-condensate generation as a result of the cracking of previously generated oil, together making up the *catagenesis* stage of Tissot and Welte (1978).
- Finally, there is a stage of dry gas generation from refractory (vitrinitic) kerogen, called the *metagenesis stage*. Methane is the main petroleum product.

Some heavy heteroatomic (NSO) compounds, together with carbon dioxide and water, are generated in the immature (diagenesis) stage of kerogen evolution, but there is effectively no hydrocarbon generation. Hydrocarbons present in rocks of this maturity are inherited from their precursor organisms (biomarkers, or geochemical fossils) and have not been generated from kerogen.

During the main zone of oil formation, hydrocarbon compounds (normal and isoalkanes, cycloalkanes and aromatics) are generated; the proportions of each depends on kerogen type. As maturity increases, low molecular weight hydrocarbons become most abundant, until only methane is present.

11.4.3   Primary migration: expulsion from the source rock

11.4.3.1   Mechanics

Of the many mechanisms of primary migration debated in the geological literature, the most likely appears to be as a discrete phase, probably through microfractures caused by the release of overpressure. In the case of the Kimmeridge Clay mudstone source rocks sampled at the UK Brae Field (Mackenzie *et al.* 1987), pore radii were big enough for petroleum to be expelled as a pressure-driven petroleum-rich phase through the pore structure of the source rock. However, as source rocks compact during burial, pore sizes may become smaller than the size of some petroleum molecules (Fig. 11.25). In these cases, there is clearly a difficulty in explaining how petroleum migrates out of the source rock without invoking the existence of microfractures. The cause of the overpressure in the source rock may be a combination of oil or gas generation, fluid expansion on temperature increase, compaction of sealed source rock units, or release of water on clay mineral dehydration.

The conversion of kerogen to petroleum results in a significant volume increase. This causes a pore pressure build-up in the source rock. The pressure build-up is sometimes large enough to result in microfracturing (§9.2.2). This releases pressure, and allows the migration of petroleum out of the source rock and into adjoining carrier beds, from which point secondary migration processes take over. Cycles of petroleum generation, pressure build-up, microfracturing, petroleum migration and pressure release continue until the source rock is exhausted. The implication of this theory is that mature source rocks will always expel petroleum as long as they are rich enough. In this sense, primary migration is not a major concern for the practising petroleum geologist. Primary migration clearly takes place both upwards and/or downwards out of the source beds, as governed by local pressure gradients.

A large volume expansion takes place when petroleum liquids are cracked to gas within the source rock. A lean oil-prone source rock

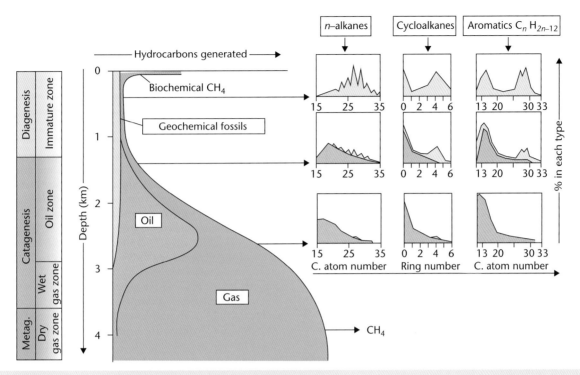

Fig. 11.23 General scheme of hydrocarbon formation as a function of burial, according to Tissot and Welte (1978). As temperatures rise on progressive burial, an immature stage is succeeded by stages of oil generation, oil cracking (wet gas stage), and finally dry gas generation. Typical distributions of *n*-alkanes, cycloalkanes and aromatics at three points in this general evolution are shown. Reproduced with kind permission from Springer Science+Business Media B.V.

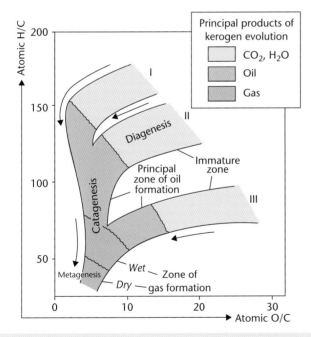

Fig. 11.24 General scheme of kerogen evolution presented on a van Krevelen diagram. The diagenesis, catagenesis and metagenesis stages are indicated and the principal products generated during that time are shown. After Tissot (1973).

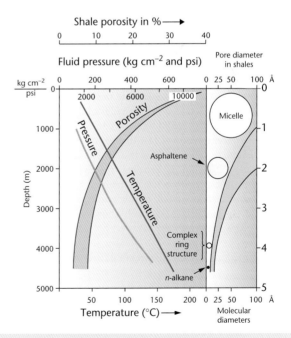

Fig. 11.25 Relationship of shale pore diameters to the molecular diameters of petroleum, with increasing depth of burial. At moderate depths of burial, shale pore diameters typically become very small in relation to the larger petroleum molecules such as the asphaltenes.

may not generate sufficient hydrocarbons to cause microfracturing. As a result, no expulsion will occur. If raised to higher maturity, however, the oil that has remained in the source rock will be cracked to gas. The resulting volume increase and overpressure may allow expulsion to occur. Thus, lean oil-prone source rocks tend to expel gas-condensate once they are raised to sufficiently high maturity.

11.4.3.2   Efficiency of expulsion

How much of the generated (plus initial) petroleum is likely to be expelled from the source rock? Between 120 and 150 °C, petroleum expulsion efficiency is strongly dependent on the original richness of the source rock (Cooles *et al.* 1986). For some rich source rocks (potential greater than 5 kg t^{-1}, TOC>1.5%) oil expulsion may be very efficient with 60–90% of the total petroleum generated being expelled. There is a lag, however, between petroleum generation and petroleum expulsion. It appears that a certain minimum petroleum saturation (probably about 40%) in the source rock is required before efficient expulsion takes place. In leaner source rocks (<5 kg t^{-1}, <1.5% TOC) expulsion efficiency is much lower, and most of the oil generated remains in the source rock. As we have seen, if raised to higher maturity, it may be cracked to gas and expelled. Expulsion appears to be very efficient for gas or gas-condensate, irrespective of original source richness.

Mackenzie and Quigley (1988) classified source rocks into three end-member classes on the basis of initial kerogen concentration and kerogen type. These parameters determine the timing and composition of the petroleum expelled.

*   A *Class 1* source rock has predominantly labile kerogen at concentrations of greater than 10 kg t^{-1}. Generation starts at about 100 °C as the labile kerogen generates an oil-rich fluid. This rapidly saturates the source rock, and between 120 and 150 °C, 60–90% of the petroleum is expelled as oil with dissolved gas. The remaining fluid is cracked to gas at higher temperatures and expelled as a gas phase initially rich in dissolved condensate. Examples of Class I source rocks are the North Sea Kimmeridge Clay, and the Bakken Shale of the Williston Basin.
*   *Class 2* source rocks are a leaner version of Class 1, with initial kerogen concentrations of less than 5 kg t^{-1}. Expulsion is very inefficient up to 150 °C because insufficient oil-rich petroleum is generated. Petroleum is expelled mainly as gas-condensate formed by cracking above 150 °C, followed by some dry gas.
*   *Class 3* source rocks contain mostly refractory kerogen. Generation and expulsion takes place only above 150 °C, and the petroleum fluid is a relatively dry gas.

Some formations contain mixtures of different source rock classes.

11.4.4   Secondary migration: through carrier bed to trap

11.4.4.1   Introduction

Secondary migration concentrates subsurface petroleum into specific sites (traps) where it may be commercially extracted. The main difference between primary migration (out of the source rock) and secondary migration (through the carrier bed) is the porosity, permeability and pore size distribution of the rock through which

migration takes place. These parameters are all much higher for carrier beds. As a result, the mechanics of migration are quite different. The end points of secondary migration are the trap or seepage at surface. If a trap is disrupted at some time in its history, its accumulated petroleum may remigrate either into other traps, or leak to the surface. The same processes of secondary migration apply to the remigration as to the original migration into the trap.

A knowledge of the mechanics of secondary migration is important in the general understanding of active charge systems, but specifically in tracing and predicting migration pathways (hence in defining areas receiving a petroleum charge), in interpreting the significance of subsurface petroleum shows and surface seepages, and in estimating seal capacity in both structural and stratigraphic traps.

The mechanics of secondary migration are now well studied and well understood (Hubbert 1953; Gussow 1954; Berg 1975; Schowalter 1976). In §11.4.4.2 and §11.4.4.3 we describe secondary migration in terms of the main driving forces and main restricting forces. The main driving forces are *buoyancy*, caused by the density difference between oil (or gas) and the pore waters of carrier beds, and *pore pressure gradients,* which attempt to move all pore fluids (both water and petroleum) to areas of lower pressure. The latter is known as a *hydrodynamic* condition. The main restricting force is *capillary pressure,* which increases as pore sizes become smaller. When capillary pressure exceeds the driving forces, entrapment occurs.

11.4.4.2   Driving forces in secondary migration

Buoyancy is a vertically directed force caused by the difference in pressure between some point in a continuous petroleum column and the adjacent pore water. It is a function of the density difference between the petroleum and the pore water, and the height of the petroleum column (Fig. 11.26):

$$\text{Buoyant force } \Delta P = Y_p g (\rho_w - \rho_p) \qquad [11.3]$$

where Y_p is the height of petroleum column, g is the acceleration due to gravity, and ρ_w and ρ_p are the subsurface densities of water and petroleum respectively. Under *hydrostatic conditions,* buoyancy

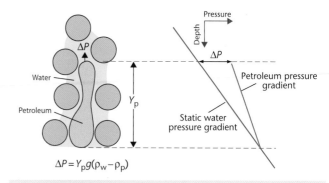

Fig. 11.26 Buoyancy as a driving force in secondary migration. It is a function of the petroleum-water density difference and the height of the petroleum column. A large buoyancy pressure may develop at the tops of large, low-density (gas) petroleum columns. Pressure measurements at points throughout the column define a petroleum pressure gradient, which intersects the hydrostatic gradient at the petroleum–water contact.

is the only driving force in secondary migration. Under *hydrodynamic conditions* (that is, when water flows through a carrier bed), however, the driving force is modified. Hydrodynamics may either assist or inhibit secondary migration, depending on whether it acts with or against the buoyancy force. Hydrodynamics may be important in a number of respects: (i) by affecting the directions and rates of secondary migration; (ii) by increasing or decreasing the driving pressures against vertical or lateral seals, thus reducing or increasing the heights of the petroleum columns that the seals can withstand; and (iii) by tilting petroleum water contacts and displacing petroleum accumulations (for example, off the crests of structural closures).

The effect of hydrodynamics on rates and directions of secondary migration may be safely ignored except in basins where there is good evidence of hydrodynamics operating at the present day. Without this evidence, it is difficult to support a strong argument for hydrodynamics having operated during secondary migration. Geologically long-lasting hydrodynamic conditions are most likely to have existed in foreland basins (§10.3.4).

11.4.4.3   Restricting forces in secondary migration

When a petroleum globule or slug moves through the pores of a rock, work has to be done to distort the globule and squeeze it through the pore throats (Fig 11.27). The force required is called *capillary pressure* (or displacement, or injection pressure), and it is a function of the size (radius) of the pore throat, the interfacial surface tension between the water and the petroleum, and the wettability of the petroleum–water–rock system:

Displacement pressure $P_d = (2\gamma\cos\theta/R)$ [11.4]

where γ is the interfacial tension between petroleum and water (dyne cm^{-1}), θ is the wettability, expressed as the contact angle of the petroleum–water interface against the rock surface (degrees), and R is the radius of the pore (cm). Higher pressures are needed to force petroleum globules through smaller pores.

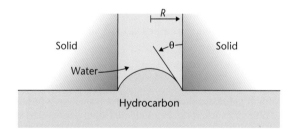

$P_d = \dfrac{2\gamma\cos\theta}{R}$

where P_d = Displacement pressure
γ = Oil–water interfacial tension
θ = Contact angle of oil and water against the solid
R = Radius of the pore throat

As γ increases P_d increases
As θ decreases P_d increases
As R decreases P_d increases

Fig. 11.27 Restricting forces in secondary migration. Higher forces are required to force petroleum globules through smaller pores. After Purcell (1949) in Schowalter (1976).

Interfacial tension is a function primarily of the composition of the petroleum (it is smaller for light, low-viscosity oils), and temperature (interfacial tension generally decreases with increasing temperature). For a given petroleum composition, therefore, interfacial tension may be considered effectively constant over large parts of the migration pathway, unless considerable vertical migration takes place.

Wettability is a function of the petroleum, water and rock. Most rock surfaces are 'water-wet' and θ may be taken to be zero ($\cos\theta = 1$). In carrier beds along secondary migration routes, and in lateral and vertical seals to petroleum accumulations, displacement pressure becomes simply a function of pore size.

Some of the grains of oil-filled reservoir rocks may be oil-wet, and organic-rich source rocks may also be partly oil-wet. Displacement pressures in these cases could, as a result, be considerably smaller than in water-wet rocks. This would assist oil migration.

Pore sizes are the most important control on secondary migration and entrapment. Pore sizes can be estimated visually (in thin section, or by scanning electron microscope, for example). Ideally, displacement pressure can be measured directly by mercury-injection techniques for both reservoir and potential sealing lithologies. Once the displacement pressure has been overcome, and a connected petroleum slug is established in the largest pores of the rock, secondary migration may take place. The petroleum saturation required to produce this connected petroleum slug is surprisingly small (Schowalter 1976): as a result, active secondary migration pathways may be characterised by petroleum saturations of only 10%. Such low saturations provide weak shows that frequently go undetected or are considered of no significance.

So far we have considered only the entry of petroleum into a pore network from an infinite body (Fig. 11.27). Within the pore network of a rock, the pore throat radii at both upper (r_t) and lower (r_b) ends of the oil globule need to be considered, and the capillary pressure equation is modified to:

$$P_c = 2\gamma\left(\frac{1}{r_t} - \frac{1}{r_b}\right)$$ [11.5]

where P_c is the capillary pressure.

11.4.4.4   Petroleum column heights and seal potential

Once a petroleum slug has entered a pore system of a constant size it will continue to move. Its rate of movement is governed by the driving forces and the permeability of the rock. When a smaller pore system is encountered, the driving forces may not be sufficient to overcome the increased capillary pressure. In this case, movement into the smaller pore network will not take place. The slug will either migrate away laterally (in a dipping carrier bed), using the larger pore system, or remain trapped. If joined by a large number of other petroleum slugs, a sufficient vertical column of petroleum may build up to provide a large buoyant force which is enough to cause invasion of the finer pore network. Thus, a seal may be effective only up to a *critical petroleum column height*, at which point it leaks.

The critical petroleum column height (Y_{pc}) is given by:

$$Y_{pc} = \frac{2\gamma\left(\dfrac{1}{r_t} - \dfrac{1}{r_b}\right)}{g\left(\rho_w - \rho_p\right)}$$ [11.6]

where ρ_w and ρ_p are the water and petroleum densities respectively.

When the pore size of a sandstone reservoir r_b is very large in relation to the very small pore size r_t of a shale caprock, the $1/r_b$ term tends towards zero. The equation may then be simplified to:

$$Y_{pc} = \frac{2\gamma}{r_t g (\rho_w - \rho_p)} \qquad [11.7]$$

Since the subsurface density of gas is less than that of oil, it is clear that seals can support much larger oil columns than gas columns. This has important implications for migration and entrapment. For example, it may prevent the formation of gas caps overlying oil columns. It also suggests that gas should migrate vertically more successfully than oil.

In order to calculate seal potential, we need to know the pore radius that is relevant to leakage. This should be the smallest pore throat in a network of large pores that, if penetrated, will allow an interconnected petroleum slug to be established. This may be approximated by a mean hydraulic radius, r_h, where

$$r_h = 2.8 (K/\phi) \qquad [11.8]$$

where K is permeability and ϕ is porosity.

11.4.4.5   Faults and fractures

Fault zones can act as both conduits and barriers to secondary migration (Jones *et al.* 1998). The material crushed by the frictional movement of the fault, the fault gouge, is frequently impermeable and does not allow the passage of petroleum. Similarly, cataclasis of sandstones may cause anastomosing granulation seams that impair permeability (Underhill & Woodcock 1987). Clay smeared along fault planes, as in the growth faults of the Niger Delta (Weber *et al.* 1978) and in the Louisiana Gulf Coast region (Smith 1980; Lopez 1990), also blocks petroleum migration. Faults may also seal as a result of authigenic mineral cementation. Fractures formed in either the footwall or hangingwall, if they remain open, may form effective vertical migration pathways. This is unlikely except at shallow depths, but may occur in the uplifted hangingwalls of contractional (thrust) faults on release of compressive stresses. Tensional fractures in the crestal zones of anticlinal structures may also allow migration of petroleum. Lateral migration will tend to be inhibited by the presence of faults, since they interrupt the lateral continuity of the carrier bed.

The sealing qualities of fault zones are discussed further in §11.6 on petroleum traps. The impact of fractures on fluid flow and the temperature of adjacent rocks is discussed in Appendix 60.

11.4.4.6   Migration pathways

Since the driving force behind secondary migration (in the absence of hydrodynamics) is buoyancy, it is clear that petroleum will tend to move in a homogeneous carrier bed in the direction that has the steepest slope. This is perpendicular to its structural contours, that is, in the true dip direction. When lateral migration is long-distance, as for example in foreland basins, where prospects may be remote from areas of mature source rock, the focusing and de-focusing effects of structural features may strongly influence the pattern of hydrocarbon charge. It is important in play assessment to recognise those parts of the fairway that are located on petroleum migration routes. A petroleum flow may be split when encountering a low, and concentrated along regional highs. The geometry of the kitchen also affects petroleum charge volumes; prospects located close to the ends of strongly elongate source kitchens will receive relatively little charge.

It is important that migration modelling be carried out for the actual time of secondary migration. Present-day structure maps may be used to model present-day migration; otherwise maps of paleostructure must be used. Other factors should also be considered, such as sealing faults, which may deflect petroleum flow laterally, and non-sealing faults, which allow petroleum to flow across the fault plane into juxtaposed permeable units at a different stratigraphic level. From this point a different structure map needs to be used for migration modelling. Communication between carrier beds caused by lateral stratigraphic changes (e.g. by the sanding-out of a shale seal) also needs to be considered. These factors affect the likelihood of petroleum charge into specific segments of the play fairway, and into specific prospects within them.

11.4.4.7   Secondary migration losses

Volume losses occur along secondary migration pathways. These losses are in two distinct habitats:

- In miniature traps – dead-ends along the migration route – produced by faulted and dip-closed geometries, and by stratigraphic changes. The traps may be observable but of no commercial interest, or they may not be observable at all, for example, if they are below the resolution of the seismic tool.
- As a residual petroleum saturation in the pores of the carrier rock, trapped by capillary forces in dead-end pores and absorbed onto rock surfaces. This may represent up to 30% of the pore volume through which the petroleum migrates (and a greater percentage of the hydrocarbon saturation achieved during active migration), so major losses may occur in this way. Losses are minimised when petroleum flows through a relatively small volume of carrier rock, perhaps <10% of its gross volume. This is achieved in high-permeability strata and fracture systems where migration is rapid and takes place without large petroleum columns building up.

The petroleum volumes expelled, lost and trapped can be related by:

$$V_{expelled} = V_{lost} + V_{trapped} \qquad [11.9]$$

The aim of play assessment and prospect evaluation is to estimate $V_{trapped}$. Volumes expelled can be calculated after geochemical source rock evaluation, taking into account source rock richness, thickness, source rock kitchen size and maturity, and expulsion efficiency. However, at the basin or play scale, volumes lost are almost impossible to quantify. Moreover, they are likely to be very large in relation to volumes trapped. As a result, it is very difficult to estimate the volumes trapped in a play fairway through geochemical source rock volumetrics.

It is conceivable that a prospect may lie beyond the 'migration front' of oil generated from a source kitchen. It will not receive a charge, since all of the oil expelled into the carrier bed is lost as a residual saturation (and in small traps) in the carrier bed. In order to determine the position of the migration front, accurate calculations of the volumes of petroleum expelled from the kitchen and the rate of loss in the carrier bed need to be made. Losses between source kitchen and trap may be as little as 5% (England and Fleet 1991). However, the errors involved are considerable and it would be most unwise for a prospect lying just beyond the calculated migration front to be severely downgraded for this reason. The calculations may, however, give useful order-of-magnitude estimates that assist risk estimation.

The focusing of oil migration into specific flow routes, which sweep through only a very small volume of carrier rock, is probably a very important contributing factor in enabling huge oil volumes to migrate very large distances (several hundred kilometres in some foreland basins). These have been previously described in §11.2.2 as petroleum systems with low-impedance entrapment style and a lateral migration style. The heavy oil/asphalt belts on the gentle flanks of these basins (e.g. western Canada) have formed at enormous distances from their source kitchens (§12.2.6).

Flow rates during secondary migration may be estimated using Darcy's law as a function of fluid viscosity, rock permeability, and fluid potential gradient. On this basis, a migration rate of 100 km per million years might be typical for a sandstone carrier bed of 100 mD permeability (England *et al.* 1991).

11.4.5 Alteration of petroleum

Trapped petroleum is not in equilibrium with its environment. Changes may take place in the physical and chemical properties of petroleum while it is in the trap. The longer that petroleum sits in the trap, the more likely it is that physical, chemical and biological processes will significantly alter its original composition. These changes may have an important impact on the recoverable fraction and commercial value of an oil accumulation.

Source rock characteristics are the greatest influence on oil composition prior to trapping. Pressure-volume-temperature (PVT) conditions are the main influence on the original composition of oil in the reservoir. There are a number of secondary alteration processes influencing oil composition after trapping:

1 Biodegradation

Biodegradation is the bacterial alteration of crude oils. Bacteria use any dissolved oxygen present in formation pore waters, or derive oxygen from sulphate ions, in order to selectively oxidise hydrocarbons. Firstly, the light normal alkanes are removed, followed by branched (iso)alkanes, cycloalkanes and finally the aromatics. The physical effect of biodegradation is to increase the density and viscosity of the oil.

Biodegradation appears to take place only at temperatures of less than 60–70 °C. It also appears to require a supply of meteoric water containing dissolved oxygen and nutrients (primarily nitrates and phosphates). These conditions are frequently met in foreland basins, where meteoric water enters the carrier bed/reservoir system in the bordering uplifted thrust belts. Biodegradation may take place in both the thrust belt and in the gentle foreland flank. An example of the former is the Napo Basin, Ecuador; the Athabasca Tar Sands (§12.2.6) are an example of the latter. Biodegradation may also occur at the oil–water contacts of petroleum accumulations, resulting in the formation of tar mats, as in the Burgan Field of Kuwait.

2 Water washing

Water washing commonly accompanies biodegradation. Hydrocarbon-undersaturated meteoric waters may dissolve some hydrocarbons from a reservoired petroleum mixture. Light alkanes and low boiling point aromatics (e.g. benzene, toluene and xylene) are the most soluble and preferentially removed. The net result is a change in composition similar to that caused by biodegradation. Water washing may take place at temperatures greater than the biodegradation threshold of 70 °C. The only requirement is a continued flow of meteoric water.

An example of the profound effects of water washing on petroleum volumes and composition is the northern Bonaparte Basin of Australia (Newell 1999). Long columns of residual oil occurring beneath many of the 11 discovered oilfields had been previously attributed to failure of fault seals. However, aspects of the geochemical composition of the very light oils found in this basin, most importantly the almost complete absence of the highly water-soluble light aromatics benzene and toluene, indicate alteration by water washing. Remarkably, in the order of 70% of the original oil volume in the reservoir has been lost by this process. Subsurface pressure data collected during exploration drilling operations indicate a water flow from the northwest, which can probably be explained by the dewatering of sediments overthrust by the island of Timor during the last 7 Myr.

In assessing the chances of encountering biodegraded or water-washed petroleum in a prospect or play, it is necessary to consider the history of fluid movement in the basin since the time of migration through to the present day. Although *present-day* reservoir temperatures may be greater than 70 °C, this is not a guarantee that biodegradation has not taken place. The effects of strong convective meteoric water flow on geothermal gradients must also be taken into account when reconstructing the thermal history of the petroleum.

3 Deasphalting

Deasphalting is a process whereby reduced solubility and precipitation of heavy asphaltene compounds in a crude oil takes place as a result of the injection of light C_1–C_6 hydrocarbons. This may occur when an oil accumulation experiences a later gas charge as its source kitchen becomes highly mature. It may also occur as a result of oil cracking in the reservoir rock. Deasphalting leads to the formation of light oil and a solid residue rich in asphaltenes. The tar mat in the Norwegian Oseberg Field may have been produced by deasphalting processes (Dahl & Speers 1986). Further examples are tar mats in several oilfields in the Middle East, including Ghawar, and on the North Slope of Alaska, including Prudhoe Bay (Gluyas *et al.* 2004).

4 Thermal alteration

The variations in petroleum composition that take place with increasing thermal maturity of the source rock are described in §11.4.2. Similar compositional changes take place in a reservoired petroleum with rising temperature. This normally occurs when trapped oil is heated through continued burial. Heavy compounds are replaced by progressively lighter ones, until only methane and a solid residue (pyrobitumen) is present. Such bitumens are reported in the Jurassic Norphlet Formation of the Gulf of Mexico (Ajdukiewicz 1995). At high temperatures (greater than 160 °C), oil cracking reactions proceed so rapidly that an oil accumulation may be destroyed within a geologically short period of time.

In addition to the four processes described above, oil composition may also be altered as a result of *gravity segregation* and *dysmigration*. Gravity segregation may take place over long periods of time as heavier molecules sink under gravity toward the base of a hydrocarbon column. The East Painter Field of Wyoming, USA, is an example (Creek *et al.* 1985). Leakage through a caprock (dysmigration) may result from faulting or microfracturing that alters the PVT conditions in the reservoir. A gas phase may form as a result of the pressure drop, and leak through the fractured caprock, leaving a heavier oil accumulation in place.

It is important when assessing exploration plays to block out parts of the fairway that are considered to be susceptible to the petroleum alteration processes described above.

11.4.6   Tertiary migration: leakage to surface

The leakage of petroleum from traps to the Earth's surface has been termed tertiary migration (Gluyas *et al.* 2004). Tertiary migration completes the natural life cycle of petroleum that starts with its biological precursors and finishes with its dissipation at the Earth's surface.

Trap failure may provide petroleum to a carrier bed system at an exceptionally high rate compared to primary migration, but once the trap is breached, the mechanisms of tertiary migration are essentially the same as for secondary migration from source rock to trap. Evidence for seepage to surface includes surface and near-surface phenomena such as soil and water-column geochemical anomalies, water-surface slicks, seabed pock marks, mud volcanoes, gas hydrate mounds, near-surface gas in plumes/chimneys and shallow reservoirs, as well as *in situ* indications of petroleum observed on seismic data (direct hydrocarbon indicators or DHIs) such as flat, bright or dim spots originating from oil, gas and water contacts. A seal may fail if buoyancy pressure is sufficient to force petroleum through the pore structure of the caprock, or if overpressure is sufficient to fracture the seal.

Assessment of such evidence is important not only in regional and local evaluations of the topseal integrity of petroleum prospects and plays, but also in the evaluation of the containment potential of carbon dioxide geological storage plays (§12.3).

11.5   The reservoir

The reservoir has two essential functions in the operation of the play. First, a reservoir rock must be porous enough to constitute a 'tank' of petroleum within the trap, and, second, its pores must be sufficiently interconnected to allow the contained petroleum fluids to flow through the rock towards the wellbore. Thus, the primary considerations in the assessment of reservoir potential are the likely reservoir porosity and permeability. The porosity of the productive part of the reservoir, together with the hydrocarbon saturation, affects the reserves of a prospect or play. Reservoir permeability affects the rate at which petroleum fluids may be drawn off from the reservoir during production.

Reservoir rocks may potentially be of any age, from Precambrian to Neogene, of various lithologies, and result from deposition in a very wide range of environments, from aeolian to pelagic. However, the world's oil and gas resource endowment is strongly dominated by Jurassic and Cretaceous siliciclastics and carbonates, formed in shallow-marine to paralic/nearshore depositional settings. As the industry matures, a number of unconventional reservoir types, in particular tight gas and shale gas reservoirs, are becoming of increasing commercial significance.

Porosity and permeability are the two key measures of reservoir quality. These parameters are influenced by both the depositional pore geometries of the reservoir sediment and the post-depositional

diagenetic changes that take place. Fracturing may significantly enhance reservoir quality, particularly its permeability, as in the low-porosity Chalk reservoir of the Ekofisk Field.

Carbonate reservoirs display a wide range of pore types and geometries that depend on the type of original grain and the subsequent diagenetic alteration of the sedimentary fabric. This complexity is due to the widely differing biological origins of the carbonate grains and their strong chemical reactivity. As a result, porosity-permeability characteristics of carbonate reservoirs may be extremely heterogeneous. A number of carbonate porosity classifications have been suggested. Porosity and permeability of sandstone reservoirs are controlled by depositional factors such as grain size, sorting, and the presence of ductile clasts, plus the subsequent effects of compaction and cementation. Porosity and permeability are simply related in the ideal case of perfectly spherical grains of one size, but the relationship becomes increasingly complex with wide variations in grain size, grain shape and packing arrangement.

A knowledge of hinterland geology and sediment dispersal systems may be helpful in understanding the distribution and quality of reservoir units, particularly in lightly explored basins. At a very broad level, reservoir presence and quality is related to tectonic setting and sediment provenance.

The original depositional fabric of reservoir sediment undergoes early diagenetic changes, then deep burial diagenesis as pressures and temperatures increase on burial. Beyond the reach of surface processes, chemical changes take place in sediments as a result of mixing with basin-derived fluids and rock–water interaction. The hydrology of deep basins depends on their large-scale tectonics. Diagenetic changes to reservoir rocks result from the processes of dissolution, dolomitisation, fracturing, recrystallisation and cementation.

The post-depositional development of secondary porosity is often the key factor in the formation of carbonate reservoirs. Dissolution is the most important process in the formation of secondary porosity. The process of dolomitisation may produce secondary porosity if the source of the CO_3 is local. Secondary porosity associated with breccias can result from evaporite solution collapse, limestone solution collapse, faulting and soil formation, while karstic dissolution associated with exposure and meteoric water influx at major unconformities, and intense fracturing, may also significantly enhance porosity in carbonate rocks.

In sandstones, major influences on reservoir quality during burial diagenesis are the growth of clay minerals and quartz cements, both of which are temperature-dependent. The full diagenetic evolution of a sandstone reservoir is controlled by its original mineralogical composition on deposition, the initial pore water composition, which is itself a function of the depositional environment, the composition of adjacent lithologies which chemically interact with the reservoir sand, its burial and thermal history, and the timing of any cementation relative to petroleum emplacement in the pore space. A number of different diagenetic styles may be observed. The most common is quartz-dominated. Other styles are dominated by clay mineral growth (kaolinite, illite, chlorite and smectite), early diagenetic clay mineral grain coating (such as chlorite), early diagenetic carbonate (commonly siderite) and late diagenetic ferroan dolomite and calcite cementation, and zeolite cementation in association with abundant clay minerals and late diagenetic non-ferroan carbonates. As with carbonates, uplift and exposure to acidic meteoric water may leave a distinctive fingerprint on sandstone reservoirs: feldspars rapidly alter to clay minerals, calcite, dolomite and sulphate cements are dissolved, and iron-bearing cements are oxidised.

Reservoirs are heterogeneous on a number of scales from the >km basin-scale, first-order heterogeneities represented by major, sealing, border fault zones and basin-filling megasequences and tectonostratigraphic units, to the microscopic pore-scale heterogeneities caused by packing arrangements and mineral cements. Reservoir heterogeneity is the main concern of the development geologist. At the critical intermediate scale, permeability heterogeneity is strongly influenced by the aspect ratio and frequency of permeable carbonate and sand bodies and impermeable shale breaks. Such units have characteristic width-thickness relationships that depend, at least in part, on depositional environment. Empirical plots of aspect ratio of reservoir units and shale breaks are therefore indicative of the heterogeneity expected in a poorly known subsurface reservoir if the depositional environment is known from core and wireline log analysis.

Two specific topics are discussed towards the end of this section. Firstly, the depositional and diagenetic histories of carbonate reservoirs are related to their sequence stratigraphic context, namely their setting in relation to periods of relative sea-level lowstand, relative sea-level rise, and relative sea-level highstand. Secondly, the distribution of clay minerals formed in sandstones during early diagenesis (<70 °C, <2 km burial) is shown to be closely related to depositional facies, climate, and relationship to stratigraphic surfaces.

11.5.1 Introduction

The reservoir has two essential functions in the operation of the play. First, a reservoir rock must be porous enough to constitute a 'tank' of petroleum within the trap, and, second, its pores must be sufficiently interconnected to allow the contained petroleum fluids to flow through the rock towards the wellbore. Consequently, the primary considerations in the assessment of reservoir potential are the likely reservoir porosity and permeability. The porosity of the productive part of the reservoir, together with the hydrocarbon saturation, affects the reserves of a prospect or play. Reservoir permeability affects the rate at which petroleum fluids may be drawn off from the reservoir during production. Together, they describe the 'plumbing' of the reservoir.

Reservoir rocks may be any age, from Precambrian to Neogene, and of almost any gross depositional setting, from continental to deep-marine. They may also occur in a wide range of lithologies. However, global statistics collected on the distribution of known oil and gas reserves together with estimated undiscovered petroleum volumes (Ahlbrandt et al. 2005) clearly show: (i) the dominance of Jurassic- and Cretaceous-aged reservoir rocks; (ii) the dominance of shallow-marine and paralic/nearshore depositional environments, despite great success in deep-water reservoirs over the past two decades; and (iii) siliciclastic, carbonate and mixed siliciclastic/carbonate reservoirs essentially make up all of the world's conventional petroleum endowment, with siliciclastics dominating, particularly for the undiscovered component.

A global data set of petroleum fields with carbonate reservoirs (Markello et al. 2008) shows that the majority of producing carbonate reservoirs are situated in intracratonic basins, foreland basins and passive margins. Intracratonic basins, formerly occupied by shallow epeiric seas that promoted anoxia, host the largest fields, such as those of the Middle East. These intracratonic basins were particularly well developed in low latitudes in Paleozoic and Mesozoic times on the northern margins of the Gondwanan and Pangean supercontinents, now located in the Arabian area and North Africa, and also in Paleozoic times on Laurentia, now located in North America.

Basin mechanics has a first-order control on reservoir occurrence and reservoir characteristics (Bosence 2005). Basin-forming mechanisms are important in controlling the supply of particulate sediment from eroding source areas and the delivery to sites of deposition, the spatial extent and duration of deposition of potential reservoir sediments, including chemical sediment such as carbonates, their diagenetic history during burial, and their subsequent tectonic deformation.

This section describes conventional siliciclastic and carbonate reservoir rocks, but it should be acknowledged that other unusual, unconventional or exotic reservoir types are important in specific cases. In particular, unconventional plays dependent on tight or fractured reservoirs are of increasing commercial significance (§12.2). These include the tight gas reservoirs of the Mesaverde Group of the Piceance Basin, USA, which normally require artificial hydraulic fracturing for commercial production, and the Devonian black shales of the Appalachians, USA, in which gas is contained in micropores, naturally occurring fracture networks and absorbed on the surfaces of organic matter. Other unusual reservoir lithologies described in the literature are the cherty mudstones of the Point Arguello Field, California (Mero 1991), and Salym Field, Siberia (Zubkov et al. 1987), and the Precambrian metamorphic rocks forming part of the Clair Field, North Sea (Coney et al. 1993). In addition, unconventional gas may occur as coal seam gas, and in the form of gas hydrates.

Hydrocarbons are known to occur in and around *igneous rocks* in numerous locations worldwide (Schutter 2003), but this is not a major reservoir type. Volcanic reservoirs may have primary intergranular porosity (pyroclastic reservoirs) or vesicular porosity (up to 50%, in basalt and andesite crystalline reservoirs (Chen et al. 1999)). The pores of high-porosity vesicular rocks may be mainly unconnected, and vesicles are also subject to infilling by zeolites. As a result, permeability may be negligible. Secondary porosity, resulting from hydrothermal alteration, fracturing or metamorphism, is often critical to reservoir development. Examples of significant igneous/volcanic reservoirs include the fractured granites of the Cuu Long Basin, offshore Vietnam (Da Hung et al. 2004), the fractured granite of the Suban gas field of South Sumatra, Indonesia (Koning 2003), and the volcanic tuffs of the Jatibarang Field, West Java, Indonesia (Kartanegara et al. 1996). There are also examples of volcanic reservoirs of potential commercial significance that are currently undergoing evaluation (Farooqui et al. 2009). These include the deep gas in the crystalline rhyolites and rhyolitic pyroclastics underneath the Chinese Daqing Field, and the fractured vesicular basalts of the Deccan Traps, underlying the productive conventional reservoirs of the Cambay Basin, India.

Many siliciclastic reservoirs have a strong diagenetic overprinting that modifies the depositional porosities and permeabilities. Diagenesis invariably has a detrimental effect on reservoir porosity and permeability. Many of the advances in studying diagenesis in siliciclastic reservoirs have come through the use of the scanning electron microscope (SEM), X-ray diffraction (XRD), electron microprobe analysis, cathodoluminescence, fluid inclusion analysis and stable isotope analysis, in addition to traditional petrography.

Although depositional systems and facies models (§7.5.3) have clear implications for the occurrence of reservoir rocks within stratigraphic successions, we refer the reader to the many texts dealing with this aspect of reservoir geology (Bathurst 1971; Blatt et al. 1972; Walker 1984; Reading 1996; Emery & Meyers 1996; Einsele 2000; Miall 2010). Similarly, we do not cover the complex subject of reservoir geochemistry and mineral diagenesis in any detail; much information can be found in Cubitt and England (1995), Giles (1997), Kupecz *et al.* (1997) and Gluyas and Swarbrick (2004).

11.5.2   Reservoir properties: porosity and permeability

The porosity and permeability of sediments and sedimentary rocks was discussed in §9.3 in relation to subsidence history and the requirement for porosity-depth relationships in decompaction calculations. §9.3 also describes the various sources of subsurface porosity measurements.

In brief, the pore volume of a sediment can be expressed either as an absolute porosity ϕ_a given by

$$\phi_a = \left(\frac{V_b - V_s}{V_b}\right)100 \qquad [11.10]$$

where V_b and V_s are the bulk and solid volumes respectively, or, as an *effective porosity*

$$\phi_e = \left(\frac{V_i}{V_b}\right)100 \qquad [11.11]$$

where V_i is the interconnected pore volume. Effective porosity is normally measured in studies of reservoirs. Porosity may also be expressed as the void space relative to the solid volume, known as the void ratio e (§9.3):

$$e = \frac{\phi}{(1-\phi)} \qquad [11.12]$$

Different rock types possess different pore geometries, carbonate rocks being very different to siliciclastic rocks in this respect. As described in §9.3.1, porosity can be estimated from a number of downhole wireline logs.

Permeability K or hydraulic conductivity measures the ability of a medium to transmit fluids and is defined according to the Darcy equation, which states

$$Q = KA(dP/dl) \qquad [11.13]$$

where Q is the volume of transmitted flow per unit time (flow rate), A is the cross-sectional area, and dP/dl is the pressure gradient over distance l, or hydraulic gradient. The value of permeability depends not only on rock properties, but also on the medium being transmitted. The *specific permeability*, k, is defined as

$$Q = \frac{kA\gamma}{\mu}\left(\frac{dP}{dl}\right) \qquad [11.14]$$

where γ is the specific weight of the fluid and μ is its absolute viscosity.

Calculations of reservoir porosity and qualitative indications of permeability can be obtained from interpretation of the wireline logs recorded in petroleum boreholes (§9.3.1). Porosity can also be measured from rock (core) material obtained during petroleum drilling operations. The reader is referred to publications by oilfield service companies such as Schlumberger for detailed instruction on the interpretation of lithology, porosity, clay content, fluid content and

other parameters of a reservoir. Good-quality reservoirs typically have porosities of 20 to 30%, but 10 to 20% porosity is not uncommon (Gluyas *et al.* 2004). Some very low-porosity rocks of a few percent may represent productive reservoirs if fractured.

Permeability may be measured by: (i) well testing, which provides an average measurement of permeability across a reservoir interval; (ii) downhole wireline logs (reviewed by Ahmed *et al.* 1991); and (iii) core analysis (sidewall cores, core plugs, and whole core) that allows direct measurement of permeability (and porosity) under controlled laboratory conditions. Rock permeability may range from 0.1 mD to >10 D, although the average permeability of producing fields in the UK and Norwegian North Sea typically ranges from 1 to 30 mD for the chalk and fractured Permian carbonate reservoirs, through hundreds of mD to several darcys for the Jurassic, Cretaceous and Tertiary sandstone reservoirs (Gluyas *et al.* 2004). Gas reservoirs may be productive at 1 mD permeability, while relative permeability and flow rates for light oil will normally require much higher absolute permeability of say >10 mD.

Porosity and permeability are not simply or directly related except in idealised situations of equal-sized spherical grains (§9.3, Appendix 55). Complex pore geometries may, for example, produce highly tortuous paths for transmitted fluids with many dead-ends. This will lower permeability while porosity may be largely unaffected. Similarly, particular pore-filling mineral habits may have different effects on porosity and permeability.

11.5.3   Primary or depositional factors affecting reservoir quality

11.5.3.1   Carbonate pore geometry

Whereas the primary porosity of siliciclastic rocks is mainly intergranular, carbonates display a wide range of pore types and geometries that depend on the type of original grain and the subsequent diagenetic alteration of the sedimentary fabric. This complexity is due to the widely differing biological origins of the carbonate grains and their strong chemical reactivity. As a result, porosity-permeability characteristics of carbonate reservoirs are extremely heterogeneous.

The relationship between pore geometry and permeability in carbonate rocks has been extensively studied (Jardine & Wilshart 1987) and a number of carbonate porosity classifications have been suggested (see Moore 2001 for review). Choquette and Pray (1970) presented a genetic classification that emphasised the importance of depositional and diagenetic fabrics in controlling pore systems (Fig. 11.28a). If porosity is determined by fabric elements, it is said to be *fabric-selective*. Five types of primary fabric-selective porosity were identified: interparticle, intraparticle, fenestral, shelter, and growth framework. Two types of secondary fabric-selective porosity, intercrystal and moldic, were also recognised. Some porosity types are not determined by the fabric of the sediment, such as fracture porosity, channel, vug and cavern types. Other porosity types may be either fabric-selective or not, such as breccia, boring, burrow and shrinkage types. Each of these porosity types may be associated with a particular depositional environment, diagenetic history, and position within a cycle of relative sea-level change.

Lucia (1983, 1995, 1999) placed emphasis on the petrophysical aspects of carbonate pore systems, suggesting that there are two major types of carbonate porosity. The pore space between grains is

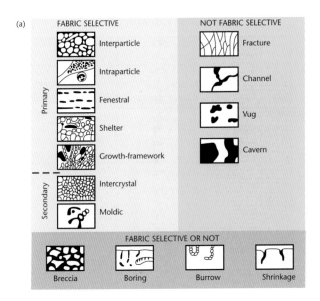

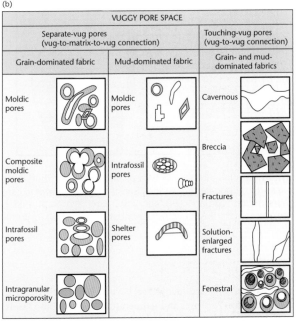

Fig. 11.28 Classification of carbonate porosity: (a) based on the fabric-selective and the non-fabric-selective criteria of Choquette and Pray (1970), and (b) based on the petrophysical classification of vuggy pore space and vug interconnection of Lucia (1995). AAPG © 1970, 1995. Reprinted by permission of the AAPG whose permission is required for further use.

termed interparticle pore space, and all other pore space is termed vuggy, subdivided into separate vugs connected only through the interparticle pore network, and touching vugs that form part of an interconnected pore system (Fig. 11.28b). These three types have different porosity-permeability characteristics and are therefore distinct petrophysically.

11.5.3.2 Primary factors affecting sandstone reservoir quality

Porosity and permeability of sandstone reservoirs are controlled by depositional factors such as grain size, sorting, and the presence of ductile clasts, plus the subsequent effects of compaction and cementation (Cade *et al.* 1994). The dominant pore type is intergranular, although intragranular secondary porosity may be produced by partial dissolution of mineral grains during diagenesis.

The *primary* porosity and permeability of sandstones are mainly dependent on the grain size, sorting and packing of the particulate sediment (see summary in Pettijohn 1975, p. 72–79) and are therefore easier to predict than in carbonate reservoirs. The porosity of artificially packed natural sand is independent of grain size for sand of the same sorting, but porosity varies strongly with sorting, ranging from 28% for very poorly sorted sand to over 42% for extremely well-sorted sand. Permeability is related to pore throat size and the number of interconnected pores. In unconsolidated sands, these parameters are controlled mainly by grain size and sorting:

- *Grain size*: sands with coarser grains have large pore throats and therefore a higher permeability. As a result, permeability may vary over several orders of magnitude between very fine and coarse sand. For example, Hogg *et al.* (1996) measured permeability and porosity on samples obtained while drilling from the Triassic Sherwood Sandstone of the Wytch Farm Field, UK. The intercept of the regression on the permeability axis, representing the notional permeability at zero porosity, ranges from c.0.001 mD for very fine-grained sandstones to >1 mD for coarse-grained sandstones.
- *Sorting*: sands with poorer sorting have smaller mean pore throat diameters and therefore lower permeability than better sorted sediments with the same mean grain size. For well-packed samples, permeability varies over more than an order of magnitude between extremely well-sorted sands and very poorly sorted sands.

Sphericity (grain shape) and *angularity* (grain roundness) may also have an influence on permeability; the effect of low sphericity and high angularity is to increase porosity and permeability of unconsolidated sand.

Apart from the textural properties of the sand, the presence of *ductile clay intraclasts* (Gluyas & Cade 1997), which may compact to form pseudomatrix, and *infiltrated clays* both serve to reduce porosity and permeability.

Compaction and diagenesis also play a major role in controlling the porosity-permeability characteristics of a siliciclastic reservoir rock. Compaction during burial reduces pore throat size and eventually blocks them completely. Although the evolution of porosity with depth in sedimentary rocks of different lithology is relatively well understood (§9.3.2), the impact on permeability is less well known (Ethier & King 1991). *Clay mineral cements* can have varying effects on permeability because they may occupy different positions in the pore space (e.g. Howard 1992). Tangential grain coatings have a much smaller effect on permeability than clays growing perpendicular to grain surfaces or lying within pore throats (Pallatt *et al.* 1984; Kantorowicz 1990). Discrete aggregates, though reducing porosity, may have little effect on permeability, unless they occur as a pseudomatrix, which on compaction severely reduces porosity and permeability.

Finally, porosity and permeability can also be affected positively by *early clay mineral growth*. For example, chlorite rims on grains can inhibit later quartz cementation and pressure dissolution. *Early introduction of oil* into pore space stops or slows clay diagenesis in sandstones, thereby improving porosity and permeability relative to sandstones receiving a late hydrocarbon charge.

11.5.3.3   Effect of sediment provenance on reservoir composition

A knowledge of hinterland geology and sediment routing systems may be helpful in understanding the distribution and quality of reservoir units, particularly in lightly explored basins. By expanding the reservoir study to include questions of provenance and the sediment routing system, the petroleum geologist may be able to develop more sophisticated models for the mineralogical composition of potential reservoir units, possibly assisting with the prediction of reservoir presence and quality across the basin.

Provenance studies classically developed as an analysis of the mineralogy of the light fraction of siliciclastic sediments in the form of ternary diagrams of quartz (Q), feldspars (F), and lithics or rock fragments (L) (e.g. Dickinson & Suczek 1979; Ingersoll & Suczek 1979; Dickinson 1980; Dickinson & Valloni 1980; Lash 1987; Garzanti *et al.* 2007). These techniques are now supplemented by other methods, including heavy-mineral analysis (Mange & Maurer 1991), fission track thermochronology of detrital apatites and zircons, and isotopic studies such as U–Pb dating of detrital zircons and Sm–Nd analysis of basin sedimentary rocks. All of these techniques are directed towards understanding the mineralogy, geochronology, and thermal evolution of the source regions providing basin sediment.

Transport of sediment has a major impact on the resulting composition of a sandstone reservoir (Garzanti *et al.* 2009). Transport in terrestrial systems invariably modifies the composition of the sand, but the effects differ markedly according to climatic zone and type of river system. Franzinelli and Potter (1983) provide an elegant study of the Amazon drainage system in this respect.

At a very broad level, reservoir presence and quality has been related to tectonic setting and sediment provenance (Kingston *et al.* 1983a, b). Continental sags typically contain extensive shallow-marine, fluviatile, aeolian and lacustrine reservoirs. Rifts may contain early spatially restricted and volcanic-rich reservoir units of poor quality, and younger, more extensive, fluviatile, deltaic and marine good-quality reservoirs. Passive margins have very extensive shallow-marine and deltaic sands or thick carbonate reservoirs, and deep-water turbiditic reservoirs. Strike-slip basins have a composition of sedimentary infill determined by the nature of the adjacent plates. Ocean–ocean boundaries generally provide poor reservoirs owing to contamination by pelagic and volcanogenic material; continent–ocean and continent–continent zones have more chance of providing sources of sand. Forearc and trench sediments contain large amounts of volcanogenic material, and porosity and permeability are severely reduced during diagenesis. It must be emphasised, however, that these are very broad generalisations.

11.5.4   Diagenetic changes to reservoir rocks

11.5.4.1   Overview

Depositional fabrics undergo early diagenetic changes and subsequently deep burial diagenesis (*mesogenesis*) (Fig. 11.29). In the

CARBONATE GRAINSTONES

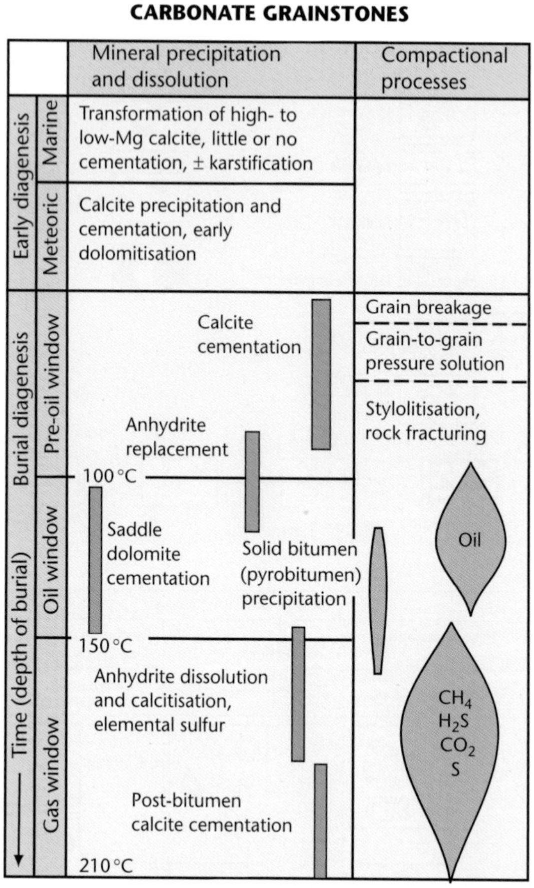

Fig. 11.29 Relationship between early and deep burial diagenesis of carbonate grainstones and the maturation of organic matter. After Heydari (1997a). AAPG © 1997. Reprinted by permission of the AAPG whose permission is required for further use.

burial environment, sediments are beyond the reach of surface-related processes. Pressures and temperatures increase, and fluids are cut off from free exchange with the atmosphere. Chemical changes become dominated by rock-water interaction and mixing with basin-derived fluids.

Diagenetic changes to porosity and permeability in carbonate rocks may generally result from five processes:

1. Dissolution (leaching), which generally improves porosity and permeability.
2. Dolomitisation, which may improve porosity by creating larger pores, or may reduce it by the growth of interlocking mosaics of dolomite crystals. Dolomitisation often increases permeability dramatically, due to the formation of solution vugs and to post-burial fracturing.
3. Fracturing by brecciation, faulting and jointing, which greatly aids permeability.
4. Recrystallisation by aggrading neomorphism of micrite into larger crystal sizes, which enhances porosity.
5. Cementation, which decreases porosity and permeability, the latter especially when cements form in pore throats.

The hydrology of deep basins depends on their large-scale tectonics (Fig. 11.30). For example, passive margin wedges (e.g. northern Gulf of Mexico) (Fig. 11.30a) have a high-velocity, near-surface, gravity-driven flow towards the distal basin, and a deep, moderate-velocity, compaction-driven flow moving from deep to shallow levels (Harrison & Summa 1991). Mechanical and chemical compaction dominate porosity changes, and cementation is relatively unimportant. The major source of $CaCO_3$ for cementation is pressure solution, but the volumes involved are limited. Other sources are dissolution caused by the action of aggressive pore fluids produced during organic diagenesis. Active fold-thrust belt and foreland basin settings (for example the Lower Paleozoic Ouachita-Appalachians) (Fig. 11.30b) are characterised by focused hot fluid flows caused by tectonic loading. The hot fluids react strongly with the rocks forming conduits for migration, causing recrystallisation of early-formed calcite and dolomite, replacement dolomitisation, dissolution of evaporites and precipitation up-dip. The Upper Knox Group carbon-

ates (Lower Ordovician) of the southern Appalachians, USA, have been strongly affected by a regional burial replacement dolomitisation (Montanez 1994). In less active settings (Rocky Mountains of western USA–Canada) (Fig. 11.30c), meteoric waters recharge aquifers exposed in the mountain belt, setting up a gravity-driven flow into the basin. Such meteoric recharge has negligible effects on carbonate porosity unless evaporites are dissolved. If so, the dissolution of gypsum promotes the dissolution of dolomite and precipitation of calcite (*dedolomitisation*), thereby enhancing porosity. The Mississippian Madison aquifer of the Midcontinent, USA, is thought to show this down-flowpath trend of dolomite dissolution and calcite precipitation (Plummer *et al.* 1990). The Madison is the reservoir for a number of fields, including the giant (>1 tcf gas) super-deep (23,000 ft, 7000 m) gas field at Madden Field, Wind River Basin, Wyoming, USA.

11.5.4.2 Secondary porosity development in carbonate reservoirs

Although depositional environment and grain types are important in determining the primary porosity of carbonate sediment, the post-depositional development of secondary porosity is often the key factor in the formation of carbonate reservoirs. Dissolution during burial of carbonate sediment dominates the formation of secondary porosity. Initially, secondary porosity is fabric-selective in the so-called *eogenetic* zone. Later during burial, in the *mesogenetic* zone (beyond the influence of surficial processes) and *telogenetic* zone (exhumation back to the region of surficial processes), porosity generation is generally not fabric-selective. One of the most important processes during diagenesis of carbonate rocks is *dolomitisation*. Dolomitisation may result in cementation due to a net import of CO_3 (Lucia & Major 1994), or if CO_3 is locally sourced, may result in net production of porosity (Moore 1989; Purser *et al.* 1994). Dissolution by ingress of fresh meteoric water during periods of exposure is perhaps the most important process enhancing porosity during dolomitisation.

Secondary porosity associated with breccias can result from evaporite solution collapse, limestone solution collapse, faulting and soil formation (Blount & Moore 1969). For example, solution collapse breccias produced during periods of influx of fresh meteoric waters at times of subaerial emergence contribute significantly to the porosity of the Ordovician Ellenburger carbonate reservoir in the Puckett Field, west Texas (Loucks & Anderson 1985). Major karstic dissolution associated with major unconformities is responsible for a number of important carbonate reservoirs, such as the Mississippian Northwest Lisbon Field, Utah (Miller 1985), and the Permian Yates Field, west Texas (Craig 1988). Intense fracturing commonly enhances porosity in carbonate rocks. A good example is the Oligocene Asmari Limestone of Iran. Despite a matrix porosity of just 9%, the reservoir produces up to 80,000 barrels of oil per day from a fractured reservoir (McQuillan 1985).

11.5.4.3 Diagenesis in sandstone reservoirs

In the burial diagenesis (mesogenesis) zone, temperature is the most important control on the formation of clay minerals. Eogenetic kaolinite, berthierine and smectite are replaced by mesogenetic dickite, illite and chlorite. Vermicular (worm-like) and book-like stacks of kaolinite crystals are progressively replaced by dickite above 70–90 °C (2–3 km burial depth), with a loss of the typical eogenetic kaolinite

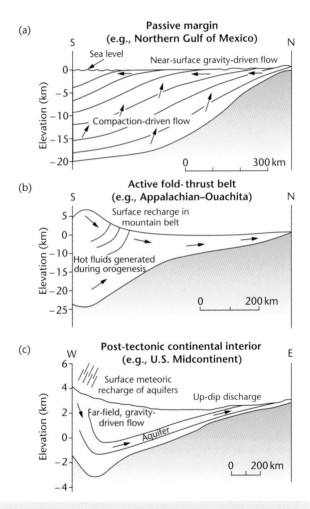

Fig. 11.30 Models of basin-scale fluid movement (after Heydari 1997b; Moore 2001). (a) Passive margin, with near-surface gravity-driven flow and slow upward movement driven by compaction of the deep sedimentary column. (b) Active fold-thrust belt causes the upward migration of hot fluids. (c) Post-tectonic environment, characterised by long-distance fluid transport in regional aquifers, driven by gravity acting on recharged water in the upland. Reprinted with permission from Elsevier.

stacking pattern with increasing temperature and transformation to dickite (Beaufort *et al.* 1998). Transformation is aided by high permeabilities, so cannot be used as a simple paleothermometer. With burial and heating, smectites pass progressively through interlayered forms to illite. Kaolinite also transforms to illite at temperatures above 70 °C, but especially above 130 °C. K-feldspar and kaolinite react to produce illite and quartz, the quartz by-product commonly forming discrete crystals or overgrowths. Dickite also transforms to illite at high temperatures. Finally, mesogenesis is typified by chloritisation. Chlorite grain replacements may form as a result of breakdown of volcaniclastic grains, by transformation of smectite, or by alteration of previous 'green mineral' grain coatings. Chloritisation of kaolinite can also occur at burial depths of 3500–4500 m (165–200 °C) (Boles & Franks 1979).

Burial diagenetic changes are well illustrated by the intensively studied Jurassic sandstone reservoirs of the Brent Group, North Sea. Reservoirs that have experienced only shallow burial have partially dissolved feldspars and authigenic kaolinite. Deeper (>4 km) Brent reservoirs, however, have extensive illite, quartz, and ferroan carbonate cements, severely impairing reservoir quality (Bjorlykke *et al.* 1992; Giles *et al.* 1992). Feldspars are essentially absent in these deep reservoirs (Glasmann 1992).

Based on a wide survey of diagenetic histories of sandstones (Primmer *et al.* 1997), there appear to be five distinct diagenetic styles that are related to original depositional environment, detrital composition and burial history:

- *Quartz-dominated diagenesis*, commonly with smaller quantities of neo-formed clays (kaolinite and/or illite) and late diagenetic ferroan carbonate. This is the most common diagenetic style and is most likely to occur in mineralogically mature sandstones, deposited in high-energy aeolian, deltaic and shallow-marine environments. Quartz cements are temperature-dependent and occur in large volumes at temperatures >75 °C.
- *Clay minerals*, (illite or kaolinite) with smaller quantities of quartz or zeolite and late diagenetic carbonate. Illite rarely forms below 100 °C (Robinson *et al.* 1993). Kaolinite is very common and occurs in more mineralogically mature sandstones. Chlorite is very common in immature sandstones, and smectite only occurs in immature sandstones, such as deep-marine deposits.
- *Early diagenetic (low temperature) grain coating clay mineral* cements such as chlorite, which may inhibit quartz cementation during later burial.
- *Early diagenetic carbonate or evaporite cement*, often localised, which severely reduces porosity at very shallow burial depths. Siderite is a common early carbonate cement in mature sandstones, typically fluviatile and shallow-marine, whereas late diagenetic ferroan dolomite and calcite cements are most common in mineralogically immature sandstones.
- *Zeolites*, which occur over a wide range of burial temperature, in association with abundant clay minerals (usually smectite or chlorite) and late diagenetic non-ferroan carbonates. This style is also likely to occur in mineralogically immature sandstones. Zeolites are common in deep-marine sandstones and especially sandstones derived from volcanogenic terrains.

Telogenetic changes related to uplift and erosion leave a distinctive fingerprint on sandstone reservoirs. Meteoric water fluxes, especially along basin margins and uplifted fault blocks, are commonly dilute, oxidising, saturated with CO_2 and therefore acidic. Feldspars rapidly alter to clay minerals, reduced iron-bearing cements are oxidised, and calcite, dolomite, and sulphate cements are dissolved. The occurrence of kaolinite in the Jurassic Brent Group reservoirs of the North Sea is thought to be due to telogenesis.

The diagenetic evolution of a sandstone reservoir rock is therefore likely to be controlled by the following factors (Gluyas and Swarbrick 2004):

- *the original mineralogical composition of the sand*; a clean quartzose sand is less chemically reactive than a volcaniclastic sand;
- *the initial pore water composition*, controlled by the depositional environment, for example, marine pore waters are richer in dissolved calcium and bicarbonate than aeolian pore waters;
- *the composition of neighbouring lithologies* (for example, evaporites and carbonates), owing to their chemical interaction with the reservoir rock;
- *the burial history*, including maximum burial depth and any subsequent uplift, particularly if it involves exposure and freshwater influx;
- *the thermal history*, since temperature appears to be a control on cementation of certain diagenetic minerals such as illite and quartz;
- *the timing of cementation relative to petroleum accumulation*, since the early presence of a petroleum phase in the pore space may reduce cementation by limiting access to the reservoir by migrating mineral-rich fluids (Gluyas *et al.* 1993).

11.5.5   Reservoir architecture and heterogeneity

So far we have emphasised the factors influencing porosity and permeability on the grain to microscopic scale. All sedimentary deposits, however, have an inhomogeneity caused by the distribution in time and space of sedimentary facies ('architecture'), and by compaction, deformation, cementation and the nature of pore-filling fluids. A classification system of reservoir heterogeneities can be based on their size, origin and influence on fluid flow (early examples are Pettijohn *et al.* 1973; Weber 1986, p. 489). One of the major factors in any classification is the size of the heterogeneities, from the size of a pore to the size of a basin. A scheme of overlapping scales of heterogeneity (Fig. 11.31) is:

- Large-scale heterogeneities at the basin scale (1–100 km scale), including the presence of sealing border fault zones and major unconformities, which define the extent of tectonostratigraphic units or megasequences.
- Large-scale heterogeneities at the field scale (0.1–10 km scale), including the boundaries of stratigraphic genetic units at the depositional sequence scale, individual fault segments, and intrusive bodies such as sills and dykes.
- Medium-scale heterogeneities at the scale of stratigraphic cycles and their characteristic architectural units (1–1000 m), such as fluviatile channel-belts, clinoform wedges, lateral shaling-out of shoreface units, and depositional lobes.
- Medium-scale heterogeneities at the scale of sedimentary bedding (0.01–10 m) such as sets and co-sets of cross-stratification, the interbedding of shales at channel margins, or the permeability contrasts caused by the coarse-fine stratification of a point bar.
- Small-scale heterogeneities at the scale of laminae and laminae-sets, visible at the core-scale (millimetres to tens of centimetres scale). The inclined foresets of cross-sets cause a fine-scale alternation of moderate and high permeabilities, while the toesets

of bedforms introduce essentially horizontal low-permeability 'breaks' or 'baffles' into the reservoir. Fractures and stylolites also fall within this physical scale of heterogeneity.

- Small-scale heterogeneities at the pore scale (<0.1–10 mm) are represented by variations in grain size and sorting, and microscopic heterogeneities caused by the way in which pores are interconnected or blocked.

Large-scale heterogeneities of well-spacing size can often be analysed on the basis of detailed well log correlations and by the use of sedimentological models derived from core descriptions. For smaller-

scale heterogeneities, cores are indispensable, since they provide information on bed thickness, style of cross-stratification, grain size and microscopic features. The correct identification of the environment of deposition of the sediment greatly helps in an assessment of heterogeneity.

A thorough study of the various scales of heterogeneity is essential to the efficient recovery of hydrocarbons from a reservoir but is of less direct concern to the basin analyst. A major objective of the production geologist is to supply engineers with information on the porosity and permeability structure of a reservoir rock. This is commonly done by assigning porosity and permeability values to individual cells in a numerical model of the reservoir. The petroleum engineer then performs a simulation of the flow of fluids, including hydrocarbons, through the reservoir. Where geological information on heterogeneity is poor, it is useful to use scaling relationships between bed thickness, bed width, and bed volume, derived from field studies, as a guide (Figs. 8.43, 8.44). Turbidite successions have in particular been analysed very promisingly in this way (Rothman *et al.* 1994; Malinverno 1997; Talling 2001; Sinclair & Cowie 2003; Felletti & Bersezio 2010) (§8.4.3).

The geometries and scales of reservoir sand bodies can also be assessed with the help of empirical plots of, for example, sand-body width versus mean sand-body thickness. This aspect ratio (width : thickness, or $w:h$) is expected to vary according to the depositional environment (Gluyas & Swarbrick 2004, p. 226), ranging from very small values of 3–70 for small fluvial channel fills, to 10–200 for shallow-marine shoreface units, to 10–1000 for braided and anastomosing fluvial sheets. An alternative visualisation of reservoir sand-body geometry can be obtained by plotting the scaling relationships of shale breaks, as pioneered by Weber (1986, 1987). Different environments are characterised by different lengths and widths of shale break, the aspect ratio varying from 1 to 10000 in field examples (Fig. 11.32) (Gluyas & Swarbrick 2004, p. 226–8). Although empirical compilations of sand-body or shale break aspect ratio may be valuable, the method lacks a firm physical underpinning and neglects the important role of rate of accommodation

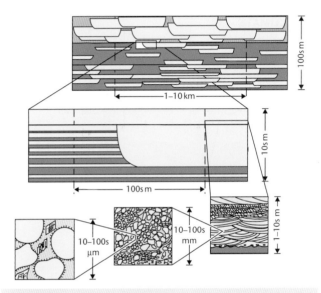

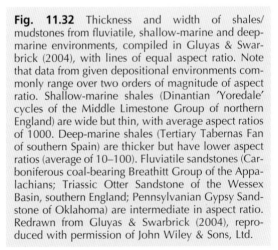

Fig. 11.31 Scheme of overlapping scales of heterogeneity at medium, small and microscopic scales. Modified from Gluyas & Swarbrick (2004, Fig. 5.25), reproduced with permission of John Wiley & Sons, Ltd.

Fig. 11.32 Thickness and width of shales/mudstones from fluviatile, shallow-marine and deep-marine environments, compiled in Gluyas & Swarbrick (2004), with lines of equal aspect ratio. Note that data from given depositional environments commonly range over two orders of magnitude of aspect ratio. Shallow-marine shales (Dinantian 'Yoredale' cycles of the Middle Limestone Group of northern England) are wide but thin, with average aspect ratios of 1000. Deep-marine shales (Tertiary Tabernas Fan of southern Spain) are thicker but have lower aspect ratios (average of 10–100). Fluviatile sandstones (Carboniferous coal-bearing Breathitt Group of the Appalachians; Triassic Otter Sandstone of the Wessex Basin, southern England; Pennsylvanian Gypsy Sandstone of Oklahoma) are intermediate in aspect ratio. Redrawn from Gluyas & Swarbrick (2004), reproduced with permission of John Wiley & Sons, Ltd.

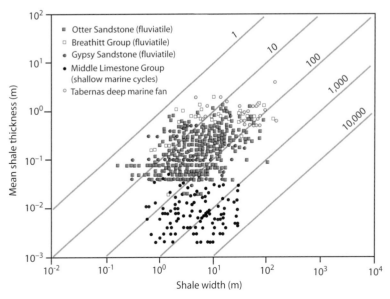

generation on stratigraphic preservation (§8.1), which varies according to basin mechanics.

11.5.6 Carbonate reservoir quality in relation to sea-level change

The methodology of process or genetic stratigraphy facilitates an explanatory approach to the occurrence of carbonate reservoirs. A number of common stratigraphic contexts (Fig. 11.33) can be linked to the occurrence of productive carbonate reservoirs (Saller *et al.* 1994; Harris *et al.* 1999; Moore 2001):

- *Sea-level lowstand (lowstand systems tract, LST).* On carbonate ramps and rimmed carbonate shelves, lowstands of sea level are important because in humid climates chemically unstable marine sediments are potentially subjected to flushing by large volumes of undersaturated meteoric waters. This modifies the porosity and permeability of the host sediment. Marine carbonate systems commonly are exposed and develop karst systems that enhance vertical conductivity of fluids. Gravity-driven meteoric water may move towards the coastline, mixing with marine water, and causing dissolution, secondary porosity and some dolomitisation. On isolated carbonate platforms (such as atolls), there is no active gravity-driven flow beneath the platform, leaving a platform-wide

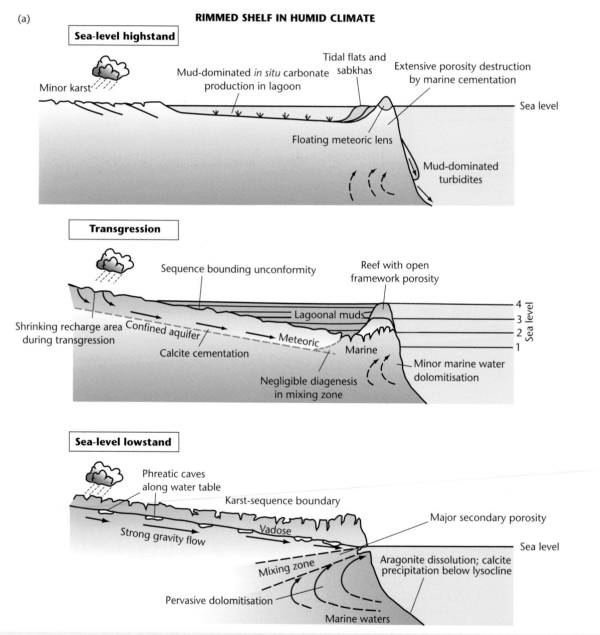

Fig. 11.33 Effects of relative sea level, depositional environment, and climate on the eogenetic changes in carbonate sediments, using a rimmed shelf system as an example (after Moore 2001). (a) Eogenetic processes during relative sea-level lowstand, transgression and highstand for a rimmed shelf in a humid climate. (b) The equivalent processes in an arid climate. Reproduced with permission of John Wiley & Sons, Ltd.

(b)

RIMMED SHELF IN ARID CLIMATE

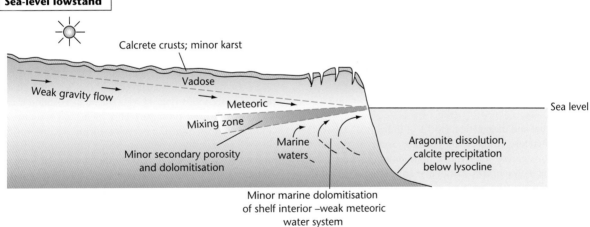

Sea-level highstand

Low-permeability calcrete crusts

Shelf margin reefs

Calcretes on sabkhas

Evaporites

Evaporative brine reflux dolomitisation associated with coastal salinas and sabkhas

Evaporative drawdown-precipitation of gypsum and halite across lagoon, prevents reflux dolomitisation of sediments beneath lagoon

Porosity destruction by marine cementation

Mud-dominated turbidites

Marine

Transgression

Impermeable calcrete crusts reduce meteoric flow

Sequence bounding unconformity

Hypersaline lagoon – low biological productivity

Reef with open framework porosity

Gypsum precipitation

Meteoric

Marine

Confined aquifer

Mixing zone moves up-dip with transgression

Reflux dolomitisation of preceding HST

Marine

4
3
2
1
Sea level

Minor marine water dolomitisation: aragonite dissolution, calcite precipitation below lysocline

Sea-level lowstand

Calcrete crusts; minor karst

Vadose

Weak gravity flow

Meteoric

Mixing zone

Sea level

Minor secondary porosity and dolomitisation

Marine waters

Aragonite dissolution, calcite precipitation below lysocline

Minor marine dolomitisation of shelf interior –weak meteoric water system

Fig. 11.33 *Continued*

meteoric lens floating on marine water at a sea-level lowstand. Cementation may take place along the water table, negatively impacting porosity, but cavernous, karstic porosity may develop along the periphery of the platform. The core of the platform is pervasively dolomitised in a mixing zone between the meteoric lens and marine water drawn into the platform.

- *Sea-level rise (transgressive systems tract, TST).* Diagenesis during sea-level rise is dominated by marine water. During sea-level rise on a ramp, deposition of transgressive deposits seals a confined gravity-driven flow in the highstand deposits of the underlying depositional sequence, often associated with porosity gain in the up-dip recharge area, and porosity loss by cementation in the down-flow area. On rimmed carbonate shelves, reef growth typifies the shelf margin during relative sea-level rise, giving good framework porosity, but marine cementation may reduce porosity along the shelf margin. The interiors of isolated carbonate platforms are generally open to marine waters during a sea-level rise, and there is little meteoric water influence. Reefs may track sea-level rise, particularly on the windward side. Minor dolomitisation of platform margins by marine waters driven by thermal convection, or reflux from the interior of the platform during arid periods, results in porosity loss by cementation.

- *Sea-level highstand (highstand systems tract, HST).* Ramps strongly prograde, and shelf and platform margins slowly accrete during the late highstand as accommodation generation reduces. At a time close to the maximum flooding, the slope and basin offshore ramps receive minimal sediment input, allowing deep-water microbial mud mounds to form. The progradational carbonate shoreline of the ramp contains freshwater lenses during humid periods, but during arid intervals, short-lived meteoric lenses are isolated by saline ponds. Heavy brines reflux downwards through underlying coastal zone sediments causing significant dolomitisation. If evaporites are precipitated in the salinas above, the brines are deficient in the Ca and CO_3 necessary for dolomitisation, so the underlying carbonates are dissolved during dolomitisation, which enhances secondary porosity. On rimmed shelves and isolated platforms, reefs easily keep up with sea level and shelter mud-dominated aggradational and progradational coastal tidal flats, while turbidites are shed into deep water. Reef bodies, if leached by meteoric water, or dolomitised in a marine–meteoric mixing zone, may have good reservoir properties, but otherwise, may suffer extensive porosity-reducing marine cementation.

The depositional and diagenetic histories of individual carbonate reservoirs are therefore complex and unique. Some of the most important carbonate reservoirs result from dolomitisation from brines originating in evaporative marginal marine sabkhas and shelf lagoons. The Permian Capitan reef complex of the Guadaloupe Mountains of west Texas and New Mexico evolved from a progradational build-up on a ramp to a steep-fronted shelf-margin reef and eventually into patch reefs (Kirkland *et al.* 1998). The reef carbonates have a good primary framework porosity. The shelf-margin carbonates of the Capitan pass laterally into extensive Upper Permian shelfal carbonates and evaporites of the Artesia Group, representing deposition in a large evaporative shelf lagoon (Sarg 1981). Dense brines moved from this evaporative lagoon and dolomitised associated carbonate facies, to produce the most prolific oil-producing reservoirs in west Texas (Silver & Todd 1969). The Upper Jurassic Smackover Formation of east Texas, which accumulated as ooidal grainstones on high-energy platforms in a late highstand systems tract, has been extensively dolomitised by reflux from overlying evaporites deposited

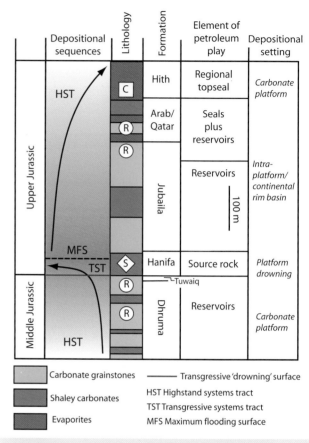

Fig. 11.34 Stratigraphic column of the Middle and Upper Jurassic of Saudi Arabia and the Persian Gulf, modified after Droste (1990). The Hith Formation constitutes an excellent regional topseal. Arab carbonate reservoirs are sealed by Arab evaporitic beds. The Hanifa Formation is an important source rock. Reproduced with permission of Elsevier.

during a lowstand in the overlying depositional sequence (Moore & Heydari 1993). One of the most prolific hydrocarbon systems in the world is found in the Upper Jurassic of the Persian Gulf and Saudi Arabia (Droste 1990) (Fig. 11.34). The Arab Formation carbonates and Hith Anhydrite form part of a highstand systems tract. The dolomitisation of Arab-D reservoirs, including the world's largest field at Ghawar, is due to refluxing brines from the overlying Arab-D evaporative lagoon, as in the case of the Smackover Formation.

Flushing with freshwater has also been identified as an important factor in the development of carbonate reservoirs (Fig. 11.33). The Pennsylvanian carbonate reservoirs of the Aneth Field of the Paradox Basin, USA (Grammer *et al.* 1996), are carbonate mud mounds that have experienced secondary meteoric dissolution during sea-level lowstands. The Ordovician Red River reservoirs of the Williston Basin, USA–Canada (Ruzyla & Friedman 1985) are dolomitised sabkha carbonates associated with evaporites, but secondary porosity is strongly influenced by freshwater flushing. The Ordovician Ellenburger Formation of west Texas consists of shallow-marine shelf to marginal marine supratidal-intertidal deposits that have experienced freshwater flushing during lowstands. The Puckett Field has produced >2.6 tcf gas from dolomites produced by reflux of evaporative brines from adjacent sabkhas (Loucks & Anderson 1985), but secondary porosity has been strongly enhanced by solution collapse breccias related to subaerial exposure. Freshwater flushing of a prograding highstand shoreline succession has been invoked to explain diagenesis in the Jurassic Oaks Field, Louisiana (Moore & Heydari 1993).

11.5.7 Models for clay mineral early diagenesis in sandstone reservoirs

There is now an impressive literature on the formation of clay minerals in sandstones during burial (Haszeldine *et al.* 2000; Morad *et al.* 2000; Ketzer *et al.* 2003; Worden & Morad 2003). The distribution of clay minerals formed in the *eogenetic* regime (<70°C, <2 km burial) is closely related to depositional facies, climate, and stratigraphic surfaces (Fig. 11.35). Consequently, knowledge of the stratigraphic architecture of a basin is valuable for prediction of diagenetic trends and therefore sandstone reservoir quality.

The main eogenetic clay minerals can be grouped under the following headings: (i) kaolinite; (ii) green clay minerals (glauconite, berthierines, verdine); and (iii) smectite, mixed-layer illite/smectite, mixed-layer chlorite/smectite and Mg-clay minerals (e.g. palygorskite). The formation of diagenetic clay minerals in near-surface sandstones and during shallow burial is controlled by depositional environment, detrital composition, and climatic conditions:

- Eogenetic *kaolinite* has a vermicular and book-like habit and forms under humid climatic conditions in continental sediments (Emery *et al.* 1990). It commonly forms during relative sea-level lowstand and relative sea-level fall (forced regression), when large areas of marine sediment are subaerially exposed and subjected to flushing by meteoric water. High rates of flushing are promoted by humid climatic conditions and high permeabilities, as in channel or shoreface sands.
- *Green clay minerals* berthierine and verdine occur as small (<5 μm) lath-shaped grain coatings and crystals, as coats, pellets, ooids, and void-fillings, or as replacements of detrital grains, commonly below the sediment-water interface in deltaic-estuarine deposits, primarily in tropical to subtropical seas. Berthierine is favoured under reducing conditions in volcanogenic sediments in estuarine–coastal plain environments (Jeans *et al.* 2000), whereas verdine forms in mildly reducing conditions in continental shelf sediment off river deltas (<200 m water depths) under low sedimentation rates (Kronen & Glenn 2000). Glauconite forms exclusively in open marine sediments (Odin & Matter 1981) decimetres to metres below the seabed.
- *Smectite* and mixed-layer illite/smectite forms as grain-hugging flakes in semi-arid climates, whereas Mg-smectites and palygorskite form fibres or fibre bundles during near-surface diagenesis of lacustrine, fluvial, and aeolian sediments under evaporative conditions. These minerals therefore commonly form eodiagenetically in arid continental environments.

Ketzer *et al.* (2003) and Worden and Morad (2003) provide summaries of the association of eogenetic clay minerals and systems tracts (Fig. 11.36):

- Lowstand systems tract (LST): sediments on the subaerially exposed continental shelf are subjected to kaolinite eogenesis under humid climatic conditions and formation of Mg-rich clay minerals such as palygorskite under evaporitic arid conditions. Paleosols may form on interfluves between the main incised channels.

- Transgressive systems tract (TST): berthierine may form in the upper part of incised valley-fills as they are flooded during transgression. The transgressive surface may be lined with high amounts of glauconite and verdine intraclasts, giving 'greensands' overlying coastal plain sediments. As water depths deepen and the shelf becomes starved of coarse sediment, authigenic glauconite and verdine forms, with maximum concentrations along the maximum flooding surface, while berthierine forms in estuarine and shallow-marine environments in front of the delta.
- Highstand systems tract (HST): as highstand progradation takes place, the amounts of glauconite and verdine decrease and authigenic berthierine increases. The landward end of the HST may be composed of fluviatile sandstones subjected to kaolinitisation under humid conditions, pedogenesis, or the formation of Mg-rich clay minerals in arid conditions.

11.5.8 Fractures

Fractures are important in enhancing the permeability of otherwise very tight reservoirs, and also in allowing the rapid advection of heat (§10.3.7). Fractures must be open in order to do so, but may subsequently close in response to changing tectonic stress fields. Fractures that have conducted hydrocarbon fluids at some stage are commonly stained with bitumen.

Fractures may cross-cut essentially non-porous lithologies, thereby enabling them to transmit hydrocarbon fluids. Fracture sets are particularly important in reservoir units that have been tectonically deformed, as in fold-thrust belts such as the Zagros province of Iran. Many of the large fields in the Zagros are in fractured carbonate reservoirs (Tertiary Asmari Formation, Cretaceous Ilam and Sarvak Formations). The Tertiary and Cretaceous reservoirs are in pressure communication, despite being separated by several thousand metres of intervening marl, indicating that the marl is highly fractured. One set of fracture orientations is consistent with formation during the NE–SW compression responsible for the main folds of the Zagros, whereas another may be related to localised extension of a pull-apart basin. A key question concerns the precise geological evolution a region must undergo in order for fractures to be open and able to transmit fluids. One explanation is that although the regional state of stress in a fold-thrust belt is compressive, favouring closed fracture systems, localised extension may take place around the outer arc of folds, and in regions of accommodation between major structures in regions of oblique slip (Gluyas & Swarbrick 2004, p. 269).

A different structural setting is presented by the giant fractured Chalk (Maastrichtian–Danian) reservoir of the Ekofisk field in the Norwegian North Sea (Lorenz *et al.* 1997). Although the chalk matrix has very low permeability (1–6 mD), it is crossed by natural interconnecting fracture systems. Effective permeability of the reservoir, which ranges up to 150 mD, correlates with fracture intensity (Lorenz *et al.* 1997; Agarwal *et al.* 2000). Fractures are associated with stylolites and slumps, but the most common type is of tectonic origin. Tectonic fractures are through-going, planar, steep, commonly conjugate features, spaced at 5–100 cm. Slickensides indicate extensional dip-slip movement. The tectonic fractures have orientations consistent with the final stages of extension in the Central Graben of the North Sea Basin, while others reflect radial extension caused by updoming of the anticlinal structure associated with the Ekofisk field over a salt pillow. Stylolite-associated fractures originate from stylolite columns, extend only a short distance from the stylolite seam and do not form an interconnected network throughout the formation.

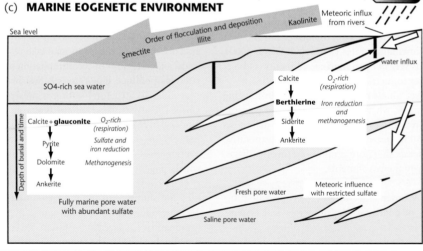

(a) **WARM WET EOGENETIC ENVIRONMENT**

Abandoned channel   Stagnant interdistributary bays   Active crevasse splay   Levee   Fresh water in active river channel

More stagnant, less O_2, more HCO_3^-

Anaerobic bacterial activity: highly reducing, higher Fe^{2+}, salinity and HCO_3^-

Aerobic bacterial activity: low Fe^{2+}, low salinity, low pH

Smectite and 'green clay' growth ± siderite nodules

Freshwater influx

Thick, peaty soil rich in humic acids: extreme silicate weathering: bleached zone + siderite nodules and kaolinite

Kaolinite growth in dilute groundwater

Silty mudstones of floodplains and lakes

(b) **ARID EOGENETIC ENVIRONMENT**

Playa lake

Alluvial fans   Meteoric influx from ephemeral rivers

Calcrete

Increasingly saline groundwater, increasing Mg/Ca ratio

Dolocrete

Evaporites and anhydritic mudstones   Gypcrete

Water table

Pedogenic smectites, infiltrated clays and kaolinite growth from feldspar dissolution

Evaporites

Formation of Mg-rich clays, trioctahedral smectite and palygorskite in Mg-rich groundwaters where evaporation > rainfall

Faulted basin margin

Offshore   Shoreface   Beach

(c) **MARINE EOGENETIC ENVIRONMENT**

Sea level   Meteoric influx from rivers

Order of flocculation and deposition   Kaolinite

Illite

Smectite

water influx

SO4-rich sea water

Calcite O_2-rich (respiration)

Berthierine   Iron reduction and methanogenesis

Depth of burial and time

Calcite + **glauconite** O_2-rich (respiration)

Siderite

Pyrite   Sulfate and iron reduction   Ankerite

Dolomite   Methanogenesis

Ankerite

Fresh pore water

Meteoric influence with restricted sulfate

Fully marine pore water with abundant sulfate

Saline pore water

Fig. 11.35 Relation of eogenetic clay mineral formation to depositional environment. (a) Warm, wet eogenetic environment of river channels, crevasse splays and interdistributary bays; (b) arid eogenetic environment of basin margin fans and evaporitic playas; and (c) marine eogenetic environment offshore from river entry points. For more details see Worden & Morad (2003). Reproduced with permission of John Wiley & Sons, Ltd.

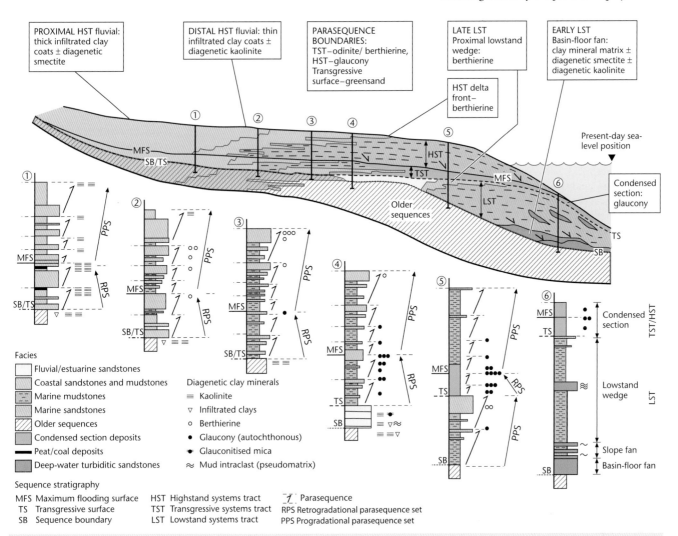

Fig. 11.36 Distribution of diagenetic clay minerals within a depositional sequence, modified after Ketzer *et al.* (2003). Sequence architecture from Van Wagoner *et al.* (1990). Diagenetic clay mineral types are shown in their typical positions within successions 1 to 6. RPS, retrogradational parasequence set during transgressive phase; PPS, progradational parasequence set during highstand phase. Reproduced with permission of John Wiley & Sons, Ltd.

Slump fractures originated as irregular glide planes, later reactivated by tectonic shear.

11.6 The regional topseal

The existence of a petroleum play depends on the presence of an effective regional caprock or topseal.

The basic physical principles governing the effectiveness of petroleum caprocks are the same as those that control secondary migration of petroleum. A caprock is effective if its capillary or displacement pressure exceeds the upward buoyancy pressure exerted by an underlying hydrocarbon column. The capillary pressure of the caprock is largely a function of its pore size. This may be laterally very variable.

The buoyancy pressure is determined by the density of the hydrocarbons and the hydrocarbon column height. A caprock of extremely small pore size is required to prevent the buoyant rise of a tall underlying gas column. Hydrodynamics also affect caprock effectiveness. Loss of gas through caprocks may take place through the process of diffusion.

The most effective caprock lithologies are fine-grained siliciclastics and evaporites. Ductility is an important requirement, particularly in tectonically disturbed areas. Salt and anhydrite are the most ductile, followed by organic-rich shales. Caprocks do not need to be thick to be effective, as long as they are laterally persistent. Similarly, depth of burial does not appear to be critical – seals may be effective at all depths. A good example of a regional caprock is the Upper Jurassic Hith Anhydrite in the Arabian Gulf area, which seals in excess of 100 billion barrels of oil in the underlying Arab reservoirs.

The conditions required for the development of regionally extensive effective caprocks in association with reservoir rocks are frequently met in two particular depositional settings. One of these is where marine shales transgress over gently sloping clastic shelves. An example is the Miocene Telisa Formation shales of the Central

Sumatra Basin. The other is where evaporites in regressive sabkhas regress over shallow-marine carbonate reservoirs, as in the case of the Hith Anhydrite of the Arabian Gulf.

11.6.1   The mechanics of sealing

The basic physical principles governing the effectiveness of petroleum caprocks are the same as those controlling secondary migration (§11.4.4). Compared to the enormous literature devoted to reservoir rocks, and the massive effort devoted to understanding the geochemistry of source rocks, relatively little has been written on caprocks. Downey (1984) and Grunau (1987) provide important reviews. The physical principles of caprock effectiveness are relatively well understood (Schowalter 1976).

In §11.4.4, we divided the forces that control secondary migration into those that drive secondary migration, and those that restrict it. The main driving force is *buoyancy*, caused by the density difference between petroleum fluids and formation pore waters. The main restricting force to the movement of a globule or slug of petroleum through a porous rock is its *capillary* or *displacement pressure*. This depends primarily on the size (radius) of the pore throats. A rock will seal an underlying petroleum accumulation if the displacement pressure of its *largest* pore throats equals or exceeds the upwardly directed buoyancy pressure of the petroleum column. The seal potential or capacity of a caprock can be expressed as the maximum petroleum column height that it will support without leakage. Owing to subsurface density differences of oil and gas, caprocks can support much larger oil columns than gas columns, other things being equal.

The displacement pressure of a piece of caprock can be measured directly in the laboratory by mercury-injection techniques, or can be estimated from the rock porosity and permeability. These data are useful in evaluating seals, but a very severe limitation is caused by the doubtful representativeness of the analysed sample with respect to the entire sealing surface of the trap. Pore sizes are likely to vary considerably over the lateral extent of the caprock, so a core sample tells us little about the sealing capacity of the caprock as a whole. The existence of large pore networks is critical; these represent the weakest points of the seal. As a result of these difficulties, seal capacity calculations are subject to wide ranges of error.

11.6.1.1   Seal failure as a result of fracturing

A caprock that is a very effective capillary or membrane seal due to its small pore throat size, may nevertheless succumb to hydraulic fracturing if pore pressures are sufficient to overcome the tensile strength of the rock and minimum stress acting upon it. In extensional basins where the minimum stress is horizontally directed, vertical fractures may develop in the caprock that destroy its integrity as a seal. Fractures and faults were discussed in §11.4.4 as both potential conduits and barriers to secondary migration. The same mechanisms apply with respect to the seal integrity of a trap. Cross-fault seal depends on the capillary properties of the juxtaposed lithologies and of the fault zone itself. Seal effectiveness along fault planes is enhanced as a result of: (i) the presence of clay smear, if clay-rich rocks have moved past the reservoir zone; (ii) the presence of fine-grained fault gouge material formed by the grinding action of the fault; and (iii) the precipitation of authigenic minerals by migrating fluids. Up-fault seal may be compromised by high pore pressures that episodically exceed the minimum stress and tensile strength of the caprock.

11.6.1.2   The effect of hydrodynamics and overpressured caprock

Under hydrodynamic conditions, the driving forces to migration or leakage are modified. Hydrodynamic flow may either increase or decrease the driving pressure against seals, thus modifying the petroleum column heights the seal can support. When the hydrodynamic force has an upward vector, it acts in support of buoyancy; when it is downward it diminishes the effect of buoyancy on the seal. Hydrodynamic effects on seal capacity may, however, for all practical purposes, be ignored except in those basins with clear evidence of hydrodynamic conditions operating at the present day. In the Powder River Basin of Wyoming, Berg (1975) has shown that hydrodynamic down-dip flow has assisted the lateral sealing of the Recluse Muddy and Kitty Muddy fields.

The existence of overpressure in a shale caprock may create a local pore pressure gradient that greatly assists its capacity to seal adjacent normally pressured reservoirs. Studies in the Niger Delta (Weber *et al.* 1978) and Gulf Coast (Stuart 1970) provide field examples.

11.6.1.3   Loss of petroleum through caprocks by diffusion

Gas may diffuse through water-filled caprocks over geological time scales (Leythaeuser *et al.* 1982). In a study of the 68 billion cubic feet Harlingen gas field in Holland, it is estimated that half of the accumulation would be lost by diffusion through the 400 m-thick shale caprock in 4.5 million years. Thus, gas fields overlain by water-saturated shale caprocks are likely to be ephemeral phenomena unless continuously topped up by active generation in the area. If losses by diffusion are as severe as Leythaeuser *et al.* (1982) suggest, there are some difficulties in explaining the existence of gas fields charged and reservoired in very old stratigraphy, such as the Lower Paleozoic.

11.6.2   Factors affecting caprock effectiveness

11.6.2.1   Lithology

Nederlof and Mohler (1981), in a statistical analysis of 160 reservoirs/seals, found that caprock lithology was of considerable importance in influencing seal capacity. Caprocks need small pore sizes, so the vast majority of caprocks are fine-grained siliciclastics (clays, shales), evaporites (anhydrite, gypsum, halite) and organic-rich rocks. Other lithologies such as argillaceous limestones, tight sandstones and conglomerates, cherts and volcanics may also seal, but they are globally far less important, and are frequently of poor quality and geographically of limited extent.

Grunau (1987) compiled information on the caprock lithologies and ultimately recoverable reserves of the world's 25 largest oil and 25 largest gas fields. The majority of giant oilfields with evaporite caprocks are located in the Middle East and North Africa, while shale caprocks to giant *oil fields* are more ubiquitous, and include Alaska, Western Canada, California, the Gulf Coast, Mexico, Venezuela, the North Sea, the Soviet Union, Indonesia and Brunei. Evaporite caprocks to giant *gas fields* are geographically widely distributed, and apart from the Middle East and North Africa, include the Soviet Union, Netherlands/southern North Sea and Brazil.

Shale caprocks dominate over other lithologies in volume terms, sealing over 900 billion barrels of oil and 500 billion barrels of oil equivalent (boe) of natural gas in known accumulations compiled by

Table 11.2 Ductility of caprocks (Downey, 1984)

Caprock lithology	Ductility
Salt	Most ductile
Anhydrite	
Organic-rich shales	
Shales	
Silty shales	
Calcareous mudstones	
Cherts	Least ductile

the US Geological Survey World Petroleum Assessment (Ahlbrandt *et al.* 2005). Shale caprocks are also expected to dominate undiscovered potential. Evaporites are the next most important caprock lithology in volume terms, and are a critical component of Paleozoic assessment units. A large part of the discovered reserves and undiscovered potential of plays charged in Paleozoic and Mesozoic times are controlled by evaporite seals, and they appear to be particularly effective in extensional passive margin settings.

11.6.2.2 Caprock ductility

Ductile caprock lithologies are less prone to faulting and fracturing than brittle lithologies. Caprocks are placed under substantial stress during periods of structural deformation, including the deformation responsible for trap formation. During the formation of a simple anticline, for example, fractures may occur in brittle caprocks in the crestal parts of the fold that are undergoing tension. Ductility is, therefore, a particularly important requirement of caprocks in strongly deformed areas such as fold-thrust belts.

The most ductile lithologies are evaporites, and the least ductile are cherts (Table 11.2). This may explain the extraordinary success of evaporites as caprocks. A high kerogen content appears to enhance the ductility of shale caprocks. Many source rocks, therefore, also serve as seals. Ductility is also a function of temperature and pressure. Evaporites may be brittle at shallow depths, but very ductile at depths of over 1 km.

11.6.2.3 Caprock thickness

A small thickness of fine-grained caprock may have sufficient displacement pressure to support a large hydrocarbon column. Thin caprocks, however, tend to be laterally impersistent; thus, a thick caprock substantially improves the chances of maintaining a seal over the entire prospect, or even over the entire play fairway or basin. Typical caprock thicknesses range from tens of metres to hundreds of metres (Grunau 1987). Very large volumes of petroleum may be sealed by relatively modest thicknesses of caprock, for example the 30 m-thick Ahmadi shales that seal the 74 billion barrel Burgan Field in Kuwait; the 20 m-thick Arab C-D anhydrite that seals the Arab-D Jubaila main reservoir of the Ghawar Field in Saudi Arabia, the world's largest oilfield (approximately 80 billion barrels); the 33 m-thick Cap-Rock Anhydrite that seals the Asmari oilfields of the Iranian Zagros foldbelt. For gas reservoirs, a thick caprock reduces the risk of substantial losses by diffusion.

11.6.2.4 Lateral seal continuity

In order to provide good regional seals, caprocks need to maintain stable lithological character (and hence capillary pressure and ductil-

ity characteristics) and thickness over broad areas. Most prolific petroleum provinces contain at least one of these regional seals. The search for petroleum in these basins may be focused on the base of the regional seal, rather than on any particular reservoir horizon. The lateral variability of the regional seal may be studied using wireline log information and seismostratigraphic analysis.

Some depositional environments and basin settings are more conducive to the establishment of thick and effective regional caprocks than others. Two of these are discussed in §11.6.3. Of particular importance are the distinctive environments in which evaporites are deposited (§7.5.3).

11.6.2.5 Burial depth of caprocks

The present burial depth of caprocks does not appear to be an important factor in influencing seal effectiveness (Grunau 1987). Seals may be effective at all depths. The requirement always is that a unit of high displacement pressure and ductility is present over wide areas. This has little to do with *present-day* depth of occurrence. However, we know that shale pore diameters do decrease with burial, particularly over the first 2 km (§9.2, §9.3, Fig. 11.25). The *maximum attained depth of burial* of shale caprocks is, therefore, likely to have an influence on sealing capability. Many shallow oil accumulations occur in structures that have undergone significant uplift, bringing well-compacted caprocks close to the surface. Provided these caprocks retain their ductility and avoid brittle deformation through the uplift period, there is no reason why they should not be effective caprocks. In the Duri Field (3.9 billion barrels recoverable) of the Central Sumatra Basin, Indonesia, the oil-bearing lower Miocene Bekasap Formation sands occur at depths of only 100 m below ground surface over the crest of the structure, sealed above by lower to mid-Miocene Telisa Formation shales. Furthermore, the overlying Duri Formation reservoirs in the field occur at depths of less than 30 m. The Minas Field (4.3 billion barrels recoverable), in the same basin, is deeper with the lower Miocene Sihapas A-1 sand at approximately 700 m. This reservoir is sealed from younger water-bearing sands in the Upper Telisa by 70 m of Telisa Formation shales.

A classic example of the importance of regional seals on the location of oil and gas is the Hith Anhydrite regional seal of the central Arabian Gulf (Murris 1980). The Hith Anhydrite is the regional topseal to the prolific Upper Jurassic Arab reservoirs, sealing well in excess of 100×10^9 barrels of recoverable oil reserves. Not only do prolific accumulations in the underlying Arab reservoirs demonstrate the effectiveness of the Hith Anhydrite as a seal, but so also does the general *lack* of accumulations in the overlying Lower Cretaceous where the Hith is present. The Hith is occasionally breached by faulting, allowing upward migration of oil from the upper Oxfordian–lower Kimmeridgian Hanifa source rock. Oils in Arab reservoirs have been confidently traced back to Hanifa source rocks. In the east of the central Gulf area, where the Hith is depositionally absent, hydrocarbons generated in the Hanifa have penetrated upwards into Lower Cretaceous reservoirs.

11.6.3 The depositional settings of caprocks

We have seen in the previous pages that the requirements for good regional caprocks are the maintenance of stable lithology and ductility over broad areas. A stratigraphic unit is not a caprock unless it seals an underlying reservoir; thus, the ideal regional caprock maintains its sealing characteristics over wide areas but also occurs in stratigraphic association with reservoirs. These conditions are liable to be met particularly in two types of depositional setting:

1. *As transgressive marine shales on gently sloping siliciclastic shelves*: these regionally extensive shales may form an excellent seal to basal transgressive sandstone reservoirs. This petroleum play, occurring in the *wedge-base position* of the depositional sequence, as described by White (1980), is a very successful one throughout the world (Fig. 11.37a). The marine transgression may flood wide areas of low-lying coastal flats, and isolate the marine shelf from supplies of coarse clastics. Thus, the *transgressive systems tract* (§8.2.1), extending from the time at which the paleo-shelf begins to be onlapped, to the time of sea-level highstand, is frequently an ideal location for the development of regional caprocks. However, sandy transgressive systems tracts which line the lower depositional sequence boundary may act as 'thief-zones', allowing the up-dip migration of petroleum. The correct prediction of shales in the transgressive systems tract is therefore essential. An example of an excellent and extensive transgressive marine shale caprock is the lower to mid-Miocene Telisa Formation of the highly productive Central Sumatra Basin, Indonesia. Over 12 billion barrels of recoverable oil reserves have been discovered in this basin, much of it as a result of the development of the Telisa shale regional topseal. The Telisa is a marine shelf sequence deposited as a result of the flooding of early Miocene sand-rich deltas (for example, the Bekasap-Duri and Sihapas deltas). In this way, a prolific reservoir-topseal doublet was formed.

2. *As evaporitic deposits in regressive supratidal sabkhas and in evap-oritic interior basins*: in clastic systems, the regressive wedge-top deposits (*sensu* White 1980), or the late highstand systems tract or shelf-margin wedge systems tract (§8.2.1) are of generally poor seal quality, comprising shallow-marine and coastal sands and non-marine deposits. These may form excellent reservoirs, but they do not make good seals. In carbonate systems (Fig. 11.37b), however, extensive evaporitic sabkhas may gently pro-grade across flat marine carbonate platforms, providing exten-sive and excellent-quality seals. The Tithonian Hith Anhydrite of the Arabian Gulf is a good example (Murris 1980). As the climate became arid in the Tithonian, the shallow carbonate platform in which the prolific Arab reservoirs were deposited was replaced by an extensive sabkha (§11.5.6).

Evaporites form through the interaction of paleoclimate and pale-ogeography. They may develop on supratidal sabkhas, in evaporitic continental interior basins, or in wide rifted basins during the early

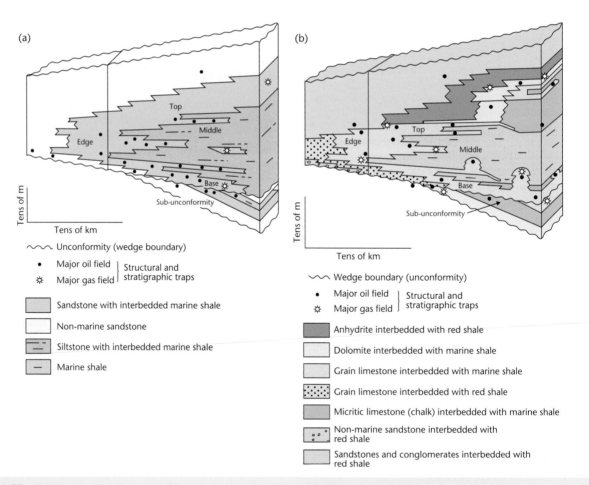

Fig. 11.37   Distribution of facies in (a) the sand-shale wedge of White (1980), corresponding to a continental to marine siliciclastic depositional sequence, and (b) in the carbonate-shale wedge of White (1980), corresponding to an arid continental–peritidal to shallow-marine carbonate-evaporite depositional sequence. In (b), the thick evaporite unit in the wedge-top position may form an excellent regional topseal. AAPG © 1980. Reprinted by permission of the AAPG whose permission is required for further use.

stages of seafloor spreading (e.g. the Aptian salt of the South Atlantic). Clear paleoclimatic and paleogeographic controls can normally be determined. Shale caprocks may be deposited in a wider range of depositional environments, ranging from lacustrine through marine shelf to bathyal.

11.7 The trap

The final requirement for the operation of an effective petroleum play is the presence of traps within the play fairway. A trap represents the location of a subsurface obstacle to the migration of petroleum towards the Earth's surface, which causes a local concentration of petroleum. The petroleum exploration industry is primarily concerned with the recognition of these sites of petroleum accumulation.

Traps are traditionally classified into *structural*, *stratigraphic* and *hydrodynamic* traps. To these may be added *intrusive* traps, a category that has only recently gained recognition.

Structural traps are those caused by tectonic, diapiric, gravitational and compactional processes, and represent the habitat of the bulk of the world's already discovered petroleum resources. The development of most structural traps can be understood in terms of basin-forming mechanics and the ensuing burial history of the basin-fill. Examples of structural traps are the contractional folds of the Zagros foldbelt of Iran and the Wyoming–Idaho fold/thrust belt, the inversion anticlines of Sumatra, the extensional tilted fault blocks of the North Sea, the extensional rollovers and fault traps of the Niger Delta and US Gulf Coast, the compactional drape anticlines of the North Sea, and various traps associated with the salt domes and diapirs of the Gulf Coast and the complex allochthonous salt bodies of the deep-water Gulf of Mexico. A category of contractional structural trap formed by gravitational processes has gained increasing recognition in recent times as the exploration industry has pushed into many deep-water and ultra-deep-water provinces. These gravity-driven foldbelts, associated with décollements in salt or overpressured shale, are found on many deep-water passive margins, and represent a highly successful exploration play in areas such as the Gulf of Mexico and the Atlantic margins of West Africa and Brazil.

Stratigraphic traps are a diverse group in which the trap geometry is essentially inherited from the original depositional morphology of, or discontinuities in, the basin-fill, or from subsequent diagenetic effects. Large volumes of *undiscovered* petroleum may reside in stratigraphic traps, and their discovery will require a very high level of geological expertise. Examples of stratigraphic traps are the fluvial channels and barrier bars in the Cretaceous basins lying along the east flank of the Rockies, the Tertiary submarine fans of the North Sea, the carbonate reefs of the western Canadian Devonian, southern Mexico and Arabian Gulf, the sub-unconformity truncation traps exemplified by the Prudhoe Bay (Alaska) and East Texas fields, and the sub-unconformity paleotopographic traps of the Gulf of Valencia, Spain. Diagenetic traps include those formed by mineral diagenesis, petroleum tar mat formation, and permafrost and gas hydrate formation.

Intrusive traps or sand injectites are dykes, sills and more irregular bodies that have been intruded under high fluid pressure into overlying low-permeability sealing mudstones or shales, and comprise both reservoir and trap. Although observed on an increasingly frequent basis, especially in deep-water clastic systems, their recognition is difficult and still at an early stage. There are numerous well-documented subsurface examples in NW Europe (North Sea), but they are almost certainly much more widespread in occurrence.

Hydrodynamic traps are those formed by the movement of interstitial fluids through basins and, in a worldwide context, tend to be relatively uncommon. Hydrodynamic effects, however, are important in some foreland basins.

Not only must a sealed trap geometry be present for the existence of a petroleum trap, but the timing of its development must also be considered. The geometry must be present prior to the petroleum charge in order to trap petroleum. Thus, an understanding of the history of individual trap growth together with the burial and thermal history of the basin, is essential to the evaluation of petroleum prospects.

11.7.1 Introduction: trap classification

An effective petroleum play requires the presence of traps within the play fairway. A trap exists where subsurface conditions cause the concentration and accumulation of petroleum. After petroleum is generated and expelled from source rocks, it will move from sites of high potential energy to sites of low potential energy. This process ultimately leads to the loss of the petroleum at the Earth's surface. Subsurface traps *en route* may be considered local (and temporary) potential energy minima. In these places, the migration route of petroleum is obstructed.

As a general rule, the commercial exploitation of petroleum resources depends on the concentration and accumulation of petroleum in traps. The exception to this rule is unconventional gas (§12.2). The petroleum industry has been dominated by exploration for specific subsurface geometries that are diagnostic of the presence of a trap. The recognition of these geometries, frequently on seismic sections, has been the goal of explorers for decades. Having identified a trap geometry, however, the petroleum geoscientist must recognise that there are a number of factors that may result in the trap being filled with petroleum to less that its full capacity. This may result from insufficient charge volumes reaching the trap (a charge or trap-timing problem), or leakage through the seal before full capacity is reached (a seal capacity or seal integrity problem).

The same basic physical principles apply to trapping as to secondary migration and seals. A trap is formed where the capillary displacement pressure of a seal exceeds the upward-directed buoyancy pressure of petroleum in the adjoining porous and permeable reservoir rock (§11.4.4 and §11.6).

Both oil and gas may occur in a trap; the gas lies above the oil because it is less dense. If a trap is charged first with oil, and then with gas (for example, as a result of increasing source rock maturity), the expanding gas-cap may displace oil downwards past the spill-point(s) of the trap. The oil may then migrate up-dip to the next available trap. Thus, traps may contain greater proportions of oil relative to gas as the distance from the source kitchen increases. This is the so-called *Gussow principle*.

Traps may be classified into various types. The main purpose of trap classification is to allow comparison between one prospect, or one play, and another. A particular trap type in a basin may be characterised by a distinctive field size distribution and drilling success ratio. Trap classification (Table 11.3) more readily allows the drawing of geological analogies that may be useful in estimation of prospect and play volumes and risk. The main subdivision is between *structural traps*, in which the majority of the world's petroleum resources have been found, and *stratigraphic traps*. The classification is based essentially on the *process* causing the formation of the trap, rather than its geometry. If the geological processes operating in a basin are known, therefore, a particular suite of traps may be predicted. Particular structural traps, for example, can be related to tectonic setting and basin-forming mechanics (Chapters 3 to 6). The detection of stratigraphic traps, on the other hand, is dependent on a good understanding of basin evolution and the stratigraphy of the basin-fill (Chapters 7 and 8).

Structural traps are those caused by tectonic, diapiric, gravitational and compactional processes. The essential point is that movement has occurred in the basin-fill some time after its deposition. *Stratigraphic traps* are those in which the trap geometry is inherited from the original depositional morphology of the basin-fill, or from diagenetic changes that took place subsequently. The best-known stratigraphic traps are caused by facies change or related to unconformities, but we have also included here traps sealed by the up-dip clogging of pore space by biodegraded oil, gas hydrates or permafrost. *Intrusive traps* are caused by the intrusion of fluidised sand into overlying low-permeability sediments. *Hydrodynamic traps* are caused by the flow of water through a reservoir/carrier bed; they may be important in some basins, but are generally rare. More than one process may contribute towards the formation of a trap. Examples are hydrodynamic closures developed on structural noses, onlap, and pinch-out traps combined with structural deformation, and channel sands developed on unconformity surfaces. Furthermore, different trap types may be genetically related. A reef, for example, may be overlain by a compactional (drape) anticline.

Table 11.3   Classification of trap types

Structural	Tectonic	Extensional
		Contractional
	Compactional	Drape structures
	Diapiric	Salt movement
		Mud movement
	Gravitational	Growth faults
		Gravity-driven foldbelts (salt)
		Gravity-driven foldbelts (overpressured shale)
Stratigraphic	Depositional	Reefs
		Pinch-outs
		Channels
		Bars
	Unconformity	Truncation
		Onlap
	Diagenetic	Mineral
		Tar mats
		Gas hydrates
		Permafrost
Intrusive		
Hydrodynamic		

11.7.2   Structural traps

11.7.2.1   Structural traps formed by tectonic processes

Contractional folds and thrust-fault structures

Contractional folds occur in areas undergoing tectonic shortening, and are generally associated with convergent plate boundaries, particularly where continent–continent collision has taken place (Chapter 4). They may also develop where transpression occurs along strike-slip boundaries (Chapter 6).

From a petroleum viewpoint, the most prolific zone of contractional folding is the external zone of the *Zagros Mountains in Iran* (Falcon 1958; Hull & Warman 1970). The main producing reservoir is the lower Miocene Asmari Limestone, which owes its prolific productivity to tectonically induced fracturing. The brittle limestone reservoir is overlain by the ductile evaporite caprocks of the Miocene Lower Fars Group. The folds are up to 60 km long and relatively simple where dramatically exposed in outcrop (Fig. 11.38). At depth,

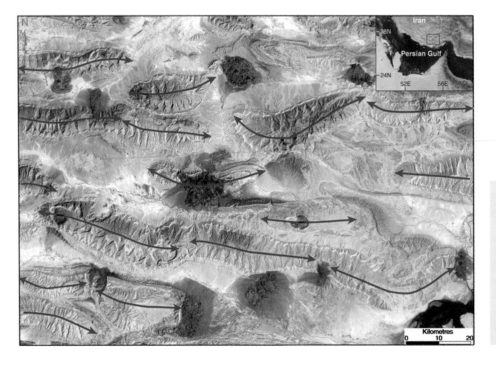

Fig. 11.38   Landsat image of part of the Fars Province of the Iranian Zagros Mountains, showing a series of contractional anticlinal trends, plunging in directions of arrows. Dark areas are diapirs of Late Neoproterozoic to Early Cambrian Hormuz Salt. Image courtesy of K. McClay. Structural interpretation by Rowan and Vendeville (2006). Reprinted with permission from Elsevier.

however, they are considered to be tighter and associated with thrust faults that sole out onto a basal detachment (Fig. 11.39), possibly in the Hormuz Salt.

Further examples of productive contractional fold structures are the *Wyoming thrust belt*, USA, and the *El Furrial* trend of eastern Venezuela. In Wyoming, the Painter reservoir field (Lamb 1980), discovered in 1977, is a large overturned, seismically defined anticline developed in the hangingwall of the Absaroka Thrust (Fig. 11.40). The Painter structure has no surface expression and owes its discovery to an improvement in seismic processing techniques in the mid-1970s.

Surface outcrop lithology and terrain have a large influence on the quality of seismic data in fold-thrust belts. Subsurface accumulations

may be very difficult to find in those areas where surface structure bears little or no relation to subsurface structure, and where surface or subsurface conditions (karstified limestone, volcanics, rugged terrain) prevent the acquisition of good-quality seismic data.

Anticlines may also develop in areas of local contraction along strike-slip systems (Chapter 6). The Wilmington Field in the Los Angeles Basin, California, is such an example (Mayuga 1970). It is developed along the San Andreas Fault system. Transpressional anticlines tend to be arranged *en echelon* and are very strongly faulted: they depend on the presence of thick caprocks to seal reservoirs across fault planes. Other examples of highly complicated faulted anticlines that have developed along predominantly strike-slip fault systems are probably the large Seria and Champion fields

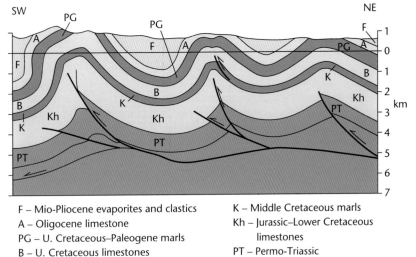

Fig. 11.39 Interpretation of the relationship of the Zagros folds to structure at depth, showing the soling out of the thrust faults onto a basal detachment (after Bailey and Stoneley 1981). Similar listric fault styles typify the thrust belts of the western United States (e.g. Idaho–Wyoming) and the Canadian Rockies of Alberta and British Columbia. Reproduced with permission of John Wiley & Sons, Ltd.

F – Mio-Pliocene evaporites and clastics
A – Oligocene limestone
PG – U. Cretaceous–Paleogene marls
B – U. Cretaceous limestones
K – Middle Cretaceous marls
Kh – Jurassic–Lower Cretaceous limestones
PT – Permo-Triassic

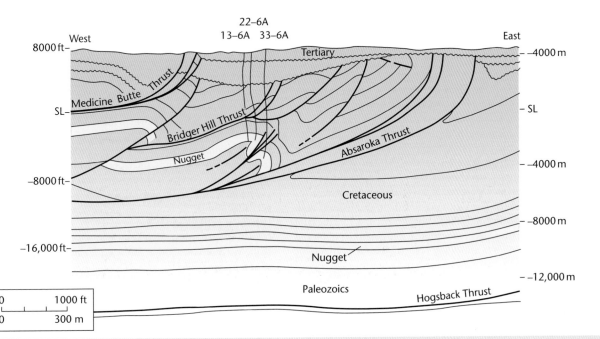
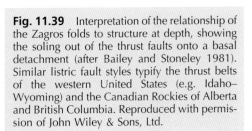

Fig. 11.40 Interpretation of the structure of the Painter Reservoir Field, Idaho–Wyoming thrust belt. The structure has no surface expression, but was identifiable on seismic data. The producing reservoir is the Triassic–Jurassic Nugget Sandstone. AAPG © 1980. Reprinted by permission of the AAPG whose permission is required for further use.

of Brunei. These fields are located on the relatively proximal part of the Miocene to Recent Baram Delta, and are not only subdivided into a large number of fault compartments but also contain numerous stacked deltaic reservoirs. Some component of shale diapirism is present in the development of Champion.

Some contractional anticlines have developed as a result of the reversal of movement along old extensional faults (Cooper & Williams 1989). These are known as *inversion anticlines*. The evidence for the earlier extensional history is usually the thickening of sediments towards the fault plane during its period of growth. Good examples are the so-called 'Sunda' folds of some Indonesian basins, for example, central Sumatra. Inversions tend to occur in areas where relatively subtle changes in the regional stress field cause reversal of movement along faults, and are therefore frequently associated with fault systems that have a strong strike-slip component. Petroleum charge into inversion anticlines may be a problem if a closure did not exist at the time of extension. Migration will tend to be away from the site of the inversion anticline during the extensional phase, and inversion of the basin tends to stop further petroleum generation. Charge may result from the post-inversion remigration of petroleum.

Traps may develop along contractional faults without any element of folding. These may be in the hangingwall or footwall of the fault, and depend for closure on the juxtaposition of sealing lithologies, or on the sealing of the fault zone itself, for example as a result of a finely powdered fault gouge or cemented zone. To complete the trap, closure is also needed in the third dimension. This is frequently produced by slight curvature or obliqueness of the fault plane relative to structural dip, or by the intersection of additional faults.

The subthrust (footwall) has sometimes provided an exploration target in thrust belts. These traps are very difficult to define, mainly due to velocity variations caused by the presence of the overthrust sheet, which make seismic interpretation very difficult. Many dry holes have been drilled on velocity 'pull-ups' in subthrust positions – mere illusions of the presence of a trap.

Extensional structures

Extensional structures form a very important group of traps, being responsible for many of the fields discovered in basins that have experienced a phase of rifting in their geological history (Chapter 3). We will deal in this section only with structural traps resulting from extension of the basement, that is, in stretched rift basins. The extensional structures occurring, for example, in delta sequences that developed in the post-rift stage on passive margins is covered in §11.7.2.2 on gravitational structures.

Rollover anticlines may develop in association with basement-controlled growth faults. The Vicksburg flexure in south Texas is an example. Very large sediment thickness changes occur across the fault zone, particularly in the Oligocene section. Large quantities of oil and gas are trapped in rollover anticlines, fault traps, and stratigraphic pinch-outs.

The most prolific play in the East Shetland Basin of the North Sea province occurs in extensional *tilted fault blocks*. The giant Statfjord Field (approximately 3 billion barrels recoverable, areal closure of 81 km²), the largest in the North Sea, contains Lower Jurassic Statfjord Formation sandstone reservoirs in a large, westward-tilted fault block that has been eroded at its crest to produce a series of Late Jurassic unconformities (Kirk 1980) (Fig. 11.41). Block faulting is responsible for the uplift and erosion of the east flank of Statfjord. The fault block is onlapped by Upper Jurassic Kimmeridgian source rocks, which, together with Lower Cretaceous shales, form the caprocks to the field. The source rocks were deposited in a restricted basin to the west that was bounded to the east by the uplifted Statfjord block, and to the west by the Hutton/Murchison block. The east flank of the field has been complicated by subsidiary downfaulted blocks. The main down-to-the-east bounding fault on the east flank of Statfjord has a total displacement of over 1800 m at Statfjord Formation level, and also controls the location of the giant Brent Field, only 20 km and on-trend to the south. The trap is a result of truncation of the Brent reservoir at the Late Jurassic unconformity surface, and could be considered as stratigraphic. The Statfjord reservoir, however, relies on fault closure. The field is therefore a huge combination structural-stratigraphic trap.

The Ninian Field to the southwest is another of the many examples in the East Shetland Basin of this style of trap – the eroded rotated fault block (Albright et al. 1980). As in the Brent Formation reservoir in the Statfjord Field, the trap at Ninian is produced primarily by the truncation of Mid-Jurassic reservoirs at Upper Jurassic and Cretaceous unconformities (Fig. 11.42).

Extensional fault traps in rifted basins are also responsible for important petroleum provinces in the Egyptian Gulf of Suez, and the Australian Carnarvon and Gippsland Basins.

11.7.2.2   Gravitational structures

An important gravitational structure that forms petroleum traps is the rollover anticline into listric growth faults, occurring particularly in delta sequences. These structures are not caused by extension in the basement, but are due to instability in the sedimentary cover and its movement under gravity. They are most prone to form where a level of undercompacted (overpressured) clays or ductile salt occurs at depth, into which the growth faults sole out, and which is overlain by a thick sequence of more competent rocks. These conditions are commonly created in thick, progradational delta sequences, as in the Gulf of Mexico and Niger Delta.

The dip-closed rollover anticline (Fig. 11.43) is the least risky trap for petroleum, but growth faulting may also give rise to fault-closed traps. The integrity of the fault trap depends on the juxtaposition of a shale seal across the fault plane, so these traps tend to be effective on those parts of the delta where the sand/shale ratio is relatively low (say, less than 50%). Even here, many of the sands may be water-bearing owing to cross-fault leakage; production may be obtained from relatively few sands, interspersed with the water-bearing zones, and distributed over a large gross vertical interval.

The fault zone itself may or may not seal. Slivers of sand tend to get caught up in the fault zones on the Niger Delta, allowing vertical leakage of petroleum (Weber et al. 1978). Although destroying fault traps at deeper levels, this leakage may allow a petroleum charge into shallower reservoirs. Other faults off the Niger Delta are sealing due to the presence of clay smears along the fault plane (Weber et al. 1978). Smith (1980) has investigated fault seal in the Louisiana Gulf Coast, and found that some faults seal even when sandstones are juxtaposed across the fault zone, as long as the sands are of different ages. This is due to the presence of fault-zone material that has formed as a result of mechanical or chemical processes directly or indirectly related to the faulting. Where parts of the *same* sandstone are juxtaposed, the fault tends *not* to seal. Where sand is juxtaposed against shale, a seal is produced.

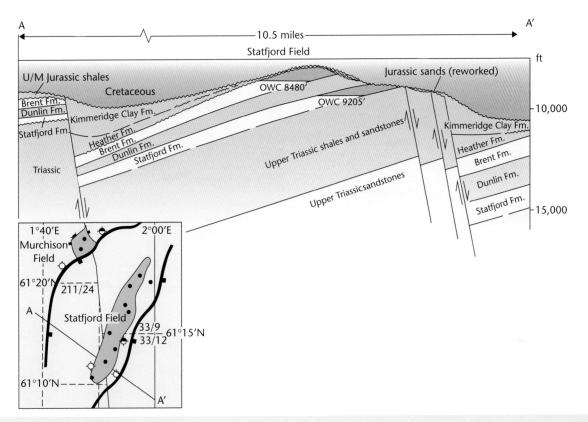

Fig. 11.41 Schematic structural cross section across the Statfjord Field, North Sea (after Kirk 1980). Statfjord is a large westward-tilted fault block that has been strongly eroded at its crest. The trap at Brent Formation level is formed by seal at the erosional unconformity, but the deeper Statfjord Formation reservoir depends on fault closure. AAPG © 1980. Reprinted by permission of the AAPG whose permission is required for further use.

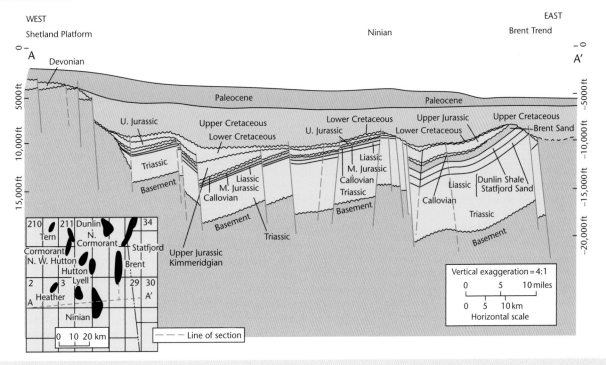

Fig. 11.42 Structural cross section across the eroded, tilted fault blocks of the Ninian area, North Sea (after Albright *et al.* 1980). This is a common and successful trap type in the North Sea. AAPG © 1980. Reprinted by permission of the AAPG whose permission is required for further use.

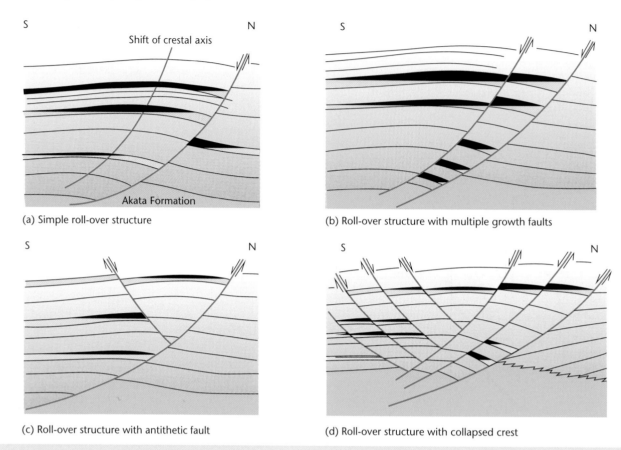

(a) Simple roll-over structure

(b) Roll-over structure with multiple growth faults

(c) Roll-over structure with antithetic fault

(d) Roll-over structure with collapsed crest

Fig. 11.43 Varieties of rollover structure forming petroleum traps in the Niger Delta (after Weber *et al.* 1978). For clarity, only a few reservoir sands are shown in the schematic sections, and the sand thickness has been enlarged.

At shallow depths (a few hundred metres), extensional faults may form open conduits for petroleum; at greater depths, they are likely to be forced closed by overburden pressure. It is reasonable to expect that the likelihood of a fault zone providing a high-permeability conduit for petroleum leakage is enhanced: (i) at shallow depths; (ii) in tensional settings; (iii) during periods of fault movement; and (iv) where reservoirs are overpressured. At depth and in active compressional settings, fault zones are very unlikely to provide a pathway for petroleum migration. In the absence of well-understood local circumstances, it is best for the practising petroleum geologist to evaluate a fault trap on the basis of the juxtaposed lithology alone, that is, assuming the fault zone allows lateral but not vertical migration.

11.7.2.3 Gravity-driven foldbelts

The importance of gravity-driven foldbelts has gained increasing recognition since the 1990s as exploration and development have moved into many deep-water areas around the world. These contractional foldbelts, which only develop where a stratigraphic layer of salt or overpressured mud acts as a good décollement, have emerged as an extremely important play type in deep-water provinces in the Gulf of Mexico and Atlantic margin. This is due to the combination of large structural traps with excellent-quality turbidite sandstone reservoirs. The understanding of driving forces and structural styles in gravity-driven foldbelts has advanced along with the increasing exploration focus (Rowan *et al.* 2004).

Gravity-driven foldbelts are fundamentally different from collisional or accretionary foldbelts: there is no active involvement of basement and there is no net tectonic shortening or extension. Deformation of the sedimentary cover takes place as a result of a combination of *gravity gliding* above a basinward-dipping detachment, and *gravity spreading* of the sedimentary wedge under its own weight. In the proximal areas, deformation is extensional, while shortening occurs in the distal zones. Distal shortening may be accommodated by the development of deep-water foldbelts (Fig. 11.44), by the lateral squeezing of existing salt diapirs, or by the extrusion of salt nappes. Extruded salt may coalesce to form 'allochthonous' salt canopies, as in the Gulf of Mexico. In salt basins, the shortening occurs between the base of slope and the depositional outer edge of the salt. If the detachment is in shale, deformation stops at the point where overpressure is no longer able to overcome the frictional resistance to lateral sliding. If the zone of overpressure advances basinward as the shelf progrades, a forward-propagating sequence of contractional structures may be produced.

For gravity gliding to occur, the primary requirement is for sufficient dip on the basal detachment. For gravity spreading, one of the critical requirements is sufficient dip on the bathymetric surface. Thus, gravitational failure of the margin is driven by increasing basinward tilt (owing to ongoing thermal subsidence), and by shelf-margin progradation, which increases the bathymetric slope. Many margins experience early gravity gliding during the thermal subsidence phase of seafloor spreading. In the case of West Africa, subse-

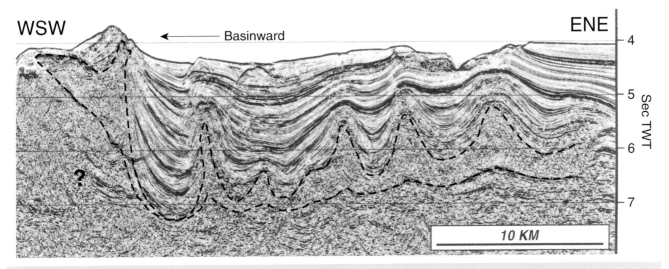

Fig. 11.44 Seismic example of a deep-water gravity-driven foldbelt, detached on salt, Lower Congo/Kwanza Basin, offshore Angola. The profile shows a salt diapir/pillow domain, comprising a regular series of symmetrical folds cored by salt (shown by dashed line), bounded basinward to the WSW by a zone of shallow salt canopies. From Rowan *et al.* (2004), adapted from Marton *et al.* (2000). Reproduced with permission of the American Geophysical Union.

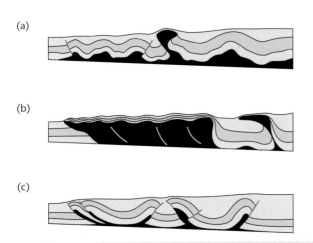

Fig. 11.45 Structural styles in gravity-driven foldbelts developed above salt décollements. (a) Thick salt with symmetrical detachment folds and shortened diapirs. (b) Inflated and thickened salt massifs and nappes. (c) Thin salt with landward and basinward vergent thrusts. From Rowan *et al.* (2004).

Table 11.4 Classification and examples of gravity-driven deepwater foldbelts (Rowan *et al.* 2004)

Foldbelts detached on salt	Foldbelts detached on overpressured shale
Mississippi Fan (Atwater Foldbelt), northern Gulf of Mexico	Mexican Ridges foldbelt, Port Isabel foldbelt, western Gulf of Mexico
Perdido Foldbelt, northern Gulf of Mexico	Sergipe-Alagoas and Para-Maranhao Basins, Brazil
Campos Basin, Santos Basin, Espirito Santo Basin, Brazil	Niger Delta, deepwater Nigeria
Benguela, Kwanza, Congo, Gabon and Rio Muni Basins, West Africa	Mackenzie Delta, arctic Canada
Red Sea, Yemen	
Nile deep sea fan, Egypt	
Rhône deep sea fan, France	
Scotian margin, eastern Canada	

quent uplift of the cratonic interior amplified the basinward tilt and stimulated shelf-margin progradation.

Margin failure is inhibited by processes that produce a landward tilt of the basal detachment or flattening of the bathymetric profile. This may occur when thick shelf-margin deltaic sequences or carbonate build-ups cause flexure in the adjacent deep-water area, or as a result of thick basinal sedimentation that bypasses the slope. Consequently, the development of gravity-driven foldbelts on passive margins is closely linked to the depositional evolution of the margin.

The structural style in the distal zone of shortening depends on a number of factors, including the rheology of the detachment, the thickness of the detachment layer, and the thickness and rheology of the sediment overburden (Rowan *et al.* 2004) (Fig. 11.45). Some

of these structural styles also occur in collisional/accretionary foldbelts.

Structural restorations indicate that the total shortening in gravity-driven foldbelts is usually of the order of tens of kilometres, significantly less than typical of collisional/accretionary foldbelts. Although the gravitational forces involved, and the net deformation, are relatively small compared to collisional foldbelts, the structures and play fairways are of significant scale and of enormous significance to the search for petroleum.

A classification of gravity-driven deep-water foldbelts is shown in Table 11.4 (Rowan *et al.* 2004). The Perdido and Mississippi Fan foldbelts of the deep-water US Gulf of Mexico are examples of gravity-driven salt-cored contractional foldbelts, occurring in front of and beneath the Sigsbee salt nappe (Trudgill *et al.* 1999) (Fig. 11.46). The Perdido foldbelt in the western Gulf of Mexico comprises large northeast–southwest trending concentric box folds, usually bounded by high-angle reverse faults on both flanks (Fig. 11.47),

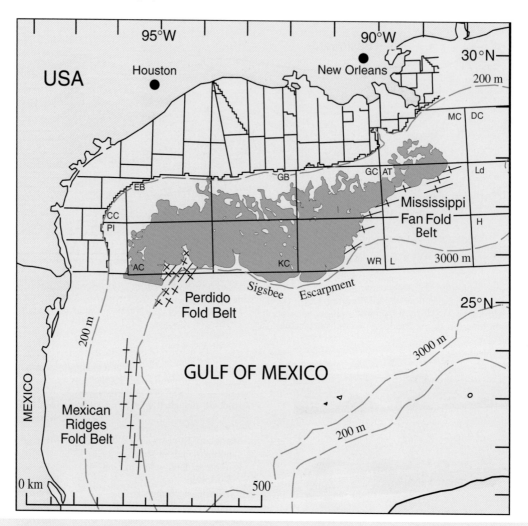

Fig. 11.46 Location map of the northern Gulf of Mexico showing the Mississippi Fan, Perdido and Mexican Ridges foldbelts, and the extent of the allochthonous salt (dark blue). The Perdido and Mississippi Fan foldbelts are gravity-driven, salt-cored Cenozoic foldbelts occurring in front of, and beneath, the Sigsbee salt nappe. After Fiduk *et al.* (1999).

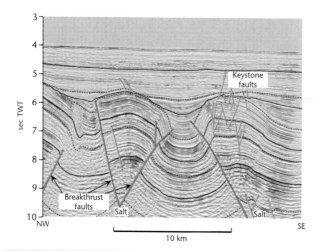

Fig. 11.47 Seismic profile showing contractional box folds in the Perdido foldbelt, bounded by high-angle reverse (Breakthrust) faults that are rooted in autochthonous salt. From Rowan *et al.* (1999).

formed by shortening above the Mid-Jurassic Louann Salt in early Oligocene to earliest Miocene times, and involving strata of Late Jurassic to Eocene age. The reverse faults die or sole out into the autochthonous salt. Post-kinematic sediments gradually onlap and bury the folds (Fig. 11.48). A late stage of regional uplift (5.5 Ma to present) may have been caused by loading of the Louann Salt by the advancing Sigsbee allochthonous salt nappe. At the present day, the Perdido folds disappear northeastwards under the Sigsbee nappe. Pinch-out of the autochthonous Louann Salt occurs just basinward of the frontal folds, demonstrating the linkage between the prior existence of the salt detachment and the geographic extent of deformation. Structural restorations indicate total shortening of 5–10 km (5–10%), and an initial autochthonous salt thickness of 2500 to 3000 m, tapering to zero towards the southeast. As far as the petroleum system is concerned, Paleocene to lower Miocene sand-rich turbidites are the main reservoir targets, derived from major depocentres to the north and northwest that were supplied by sediment routed through the Rio Grande and Houston embayments (Fiduk *et al.* 1999). The main trapping potential is in the large anticlinal folds, and in stratigraphic traps where Oligo-Miocene turbidites onlap the fold flanks. In the mid-Miocene the main sediment routing

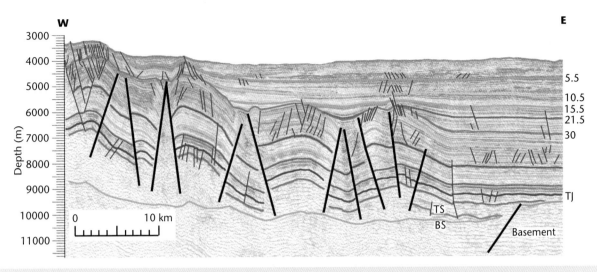

Fig. 11.48 East–west interpreted seismic profile across the Perdido foldbelt, illustrating fold geometries, the progressive westward (landward) onlap of post-kinematic sediments onto the bathymetric relief created by the foldbelt, and the eastward pinch-out of the autochthonous Louann Salt. BS, base salt; TS, top salt; TJ, top Jurassic; 30, 30 Ma horizon, etc. From Fiduk *et al.* (1999).

system shifted east to the Mississippi Embayment on the Louisiana part of the shelf, and turbidite deposition in the Perdido foldbelt since the mid-Miocene is shale-prone.

The structural style of the Mississippi Fan (or Atwater) foldbelt to the east is distinct from the Perdido, comprising basinward-verging anticlines and associated thrust faults, is cored by salt, and has a mid–late Miocene age of deformation (Rowan *et al.* 2000). Also within the Gulf of Mexico, the Mexican Ridges (Fig. 11.46) foldbelt detaches within Cenozoic shales (Buffler *et al.* 1979) and is genetically unrelated to the Perdido and Mississippi Fan foldbelts.

11.7.2.4   Compactional structures

The most important trap type formed by compactional processes is the *drape anticline*, caused by differential compaction. Basement (effectively non-compactible) topography causes significant thickness variations in the overlying highly compactible sediments, which compact and subside most where they are thickest off-structure. If the area above the horst remains elevated relative to surrounding areas, shallower-water sedimentary facies may develop which are less compactible than the surrounding muds. This will exaggerate the differential compaction. The sedimentary facies and diagenetic history of the reservoir unit may be quite different over the crest of the drape anticline than off its flanks.

Drape anticlines form a very successful trap type. They are frequently simple features formed without tectonic disturbance (unless basement faults are reactivated) and frequently persist over a long period of geological time, from shortly after the time of reservoir deposition through to the present day. They are therefore available to trap a petroleum charge over a long time span, and are very forgiving of inaccuracy in estimation of charge timing.

Owing to dependence on the existence of basement topography, drape anticlines commonly form in the passive sedimentary cover to rifted megasequences, particularly over the relatively elevated parts of tilted fault blocks in the pre-rift. In these settings, a petroleum charge is frequently needed from syn-rift or very early post-rift source rocks, since the later post-rift is commonly devoid of a source.

Communication of the reservoir with the source is often, therefore, the critical factor for this play. A second critical factor may also be the presence of a reservoir in the post-rift. Since a considerable amount of crustal thinning is the cause of the basement topography responsible for the formation of the drape anticline, these areas commonly subside rapidly into deep water, potentially limiting reservoir development, which becomes dependent on the presence of deep-sea fans. The Lower Tertiary submarine fans of the North Sea are an example.

Examples of drape anticlines are the Forties and Montrose fields of the North Sea, which occur within very large (90 km² and 181 km² respectively) low-relief domal closures at Paleocene level formed by drape over deeper fault blocks. Drape anticlines may indeed be very large (the world's largest field, Ghawar in Saudi Arabia, comprises Jurassic carbonates draped over a basement high), while the drape structures over the small Devonian pinnacle reefs of the Western Canada sedimentary basin are examples at the other extreme (Gluyas *et al.* 2004). In the Frigg gas field, North Sea, the present-day closure on the Eocene submarine fan reservoir is due not only to compaction processes but also to the rejuvenation of Jurassic faults controlling the deeper structure, and to the original depositional topography on top of the fan (Blair 1975). At the East Brae Field, South Viking Graben, North Sea, compactional drape of the Kimmeridge Clay Formation around the East Brae Fan has accentuated the lobe-like geometry of the Upper Jurassic sandstone reservoir unit (Gluyas *et al.* 2004).

11.7.2.5   Diapiric traps

Diapiric traps result from the movement of salt or overpressured clay. At depths in excess of 600–1000 m, salt is less dense (2200 kg m^{-3}) than its overburden (typically 2500–2700 kg m^{-3}), is rheologically exceptionally weak, and consequently liable to lateral and upward movement. Salt can flow at surprisingly low temperatures and over long periods of geological time. Once a density inversion is present, heterogeneities in either the mother layer of salt or clay, or in the overburden, are sufficient to trigger upward movement. Examples of heterogeneities are lateral changes in thickness, density, viscosity or

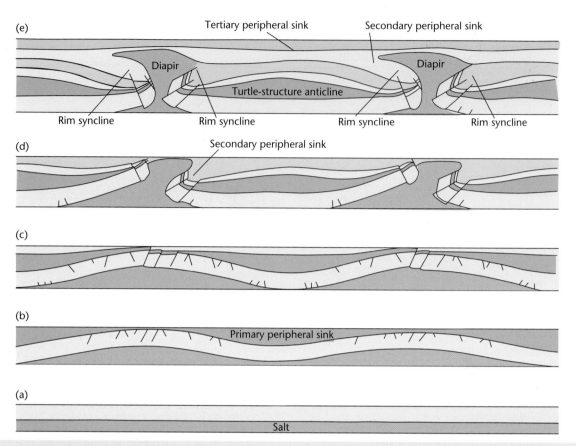

Fig. 11.49 Evolution of salt structures through pillow stage (b) and (c), diapir stage (d), and post-diapir stage (e). Note the lateral migration of peripheral sinks with each stage (from Seni and Jackson 1984, Figure 14). Turtle structures represent the preserved fill of the peripheral sink.

temperature. These changes may be essentially depositional or may be imposed as a result of faulting or folding. In extensional faulted zones, diapirs tend to form through buoyancy where overburden load is most reduced in the footwall. The salt rollers of the US Gulf Coast are examples of this triggering mechanism. Faulting may be basement-involved, or thin-skinned, usually soling out in the ductile layer. This layer may also provide a zone of detachment in contractional areas. Differential loading of a salt layer by thick overlying sediments is a powerful triggering mechanism of diapirism in young shallow delta sequences.

In simple terms, salt structures pass through three stages of growth (Fig. 11.49):

1. *The pillow stage*, characterised by the thinning of sediments over the crest of the pillow, and thickening into the adjacent *primary peripheral sink*. No piercement or intrusion of the overlying sediments has taken place. Depositional facies are affected by pillow growth, with higher-energy facies, perhaps reefs, developing over the crest. Traps formed at this stage are typically broad domes, while sediments that have been channelled into the topographically low peripheral sink may pinch out pillow-wards, forming stratigraphic traps.

2. *The diapir stage*, when the salt body pierces the overburden. These structures are known as salt piercement structures. As the pillow withdraws to form the diapir, *secondary peripheral sinks* may develop close to the diapir, and inside the earlier primary

peripheral sinks of the pillow stage. *Turtle structures* (Fig. 11.50), representing thick lenses of sediment that accumulated in the primary sinks subsequently tilted during diapir growth, may form petroleum traps. Thick clastics may again pinch out onto the flanks of the diapir, forming stratigraphic traps.

3. *The post-diapir stage*. As the diapir grows, a point is reached where the underlying reservoir of salt is depleted, and it can only continue to rise by thinning of its lower trunk or by complete detachment from the mother salt to form an *allochthonous* salt body, separate from the *autochthonous* mother salt. Salt body geometries may become very complex, and overhangs commonly develop. The geometry of the salt diapir and surrounding strata beneath an overhang is generally poorly known owing to the occurrence of a seismic shadow in this zone, and the relatively few penetrations by drilling. The typical thickness of a diapir stem is largely unknown. Piercement of salt may take place through to the surface, forming salt domes that are particularly noticeable on Landsat images and aerial photographs, as in north-central Oman and the Zagros foldbelt of Iran (Fig. 11.38).

Our understanding of the petroleum entrapment possibilities of salt pillows and diapirs has been developed largely through exploration of the Gulf Coast area of the USA, and exploration of the outer shelf and deep-water Gulf of Mexico over the past two to three decades has been responsible for a major advance in the understanding of allochthonous salt bodies and the associated

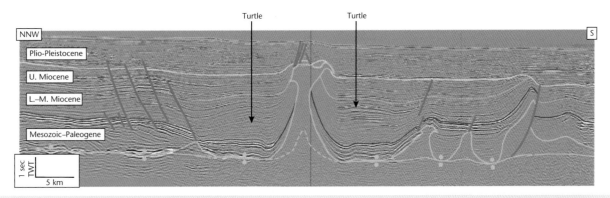

Fig. 11.50 Regional seismic profile in the eastern Mississippi Canyon area, Gulf of Mexico. A salt diapir (delimited by blue lines) appears in the centre of the section, bounded by turtle structures on either side. From Rowan & Vendeville (2006). Reprinted with permission from Elsevier.

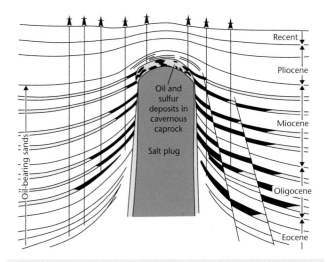

Fig. 11.51 Schematic cross-section through the Spindletop Dome, Texas, showing the distribution of oil reservoirs in both the caprock and on the flanks of the salt plug. After Halbouty (1979). Reproduced with permission from Elsevier.

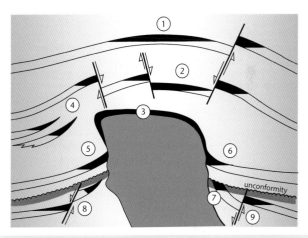

Fig. 11.52 Potential petroleum traps associated with salt diapirs (after Halbouty 1979). A wide variety of structural and stratigraphic traps may develop above and around the diapir, and in its diagenetic caprock. 1, simple domal trap above the diapir with relatively simple or no associated faulting; 2, domal trap faulted into graben structures; 3, diapir caprock reservoir; 4, stratigraphic up-dip pinch-out caused by facies change into the peripheral sink; 5 and 6, reservoirs sealed against the diapir wall (5 is beneath an overhang); 7, unconformity trap formed by erosion towards the diapir crest; 8 and 9, fault traps in the flanking area. Reproduced with permission from Elsevier.

petroleum plays. After the Spindletop discovery (Fig. 11.51) in 1901, traps associated with salt diapirs became one of the most important and prolific plays in this outstandingly successful hydrocarbon province. The piercement salt diapirs of the Houston Salt Basin were formed by the loading of Jurassic salt with a thick sequence of Mesozoic and Tertiary sediments. Most of the diapirs have intruded to shallow depths, and have created complexly faulted structures. Radial fault patterns are common over the diapir flanks. Reservoirs are commonly broken into a very large number of separate fault compartments. Syndepositional diapir growth caused substantial thickness and facies changes, as well as local erosion. Stratigraphic pinch-out and unconformity traps were formed, and local reef limestone reservoirs developed.

Most production comes from Eocene to Pliocene age sediments overlying and surrounding the salt, and from the diagenetic salt dome caprock, which directly overlies and is in contact with the salt. The caprock is a carbonate deposit with zones of gypsum and anhy-

drite, formed by the solution of anhydrite-bearing salt by ground waters. Secondary porosity may be very high (>40%), but may be irregularly distributed.

The petroleum trapping possibilities in sediments above and around a mature diapir deposited during its post-diapir stage are shown schematically in Fig. 11.52. A variety of further traps, mainly stratigraphic, may develop in the diapir- and pillow-stage sediments. These include structural and combined structural-stratigraphic traps on the crests of turtle structures, and a variety of other fault, pinch-out and unconformity traps.

The Greater Burgan Field of Kuwait is an example of a very large, relatively simple elongate domal closure developed over a low-relief swell in the Cambrian Hormuz Salt that began to move in Jurassic

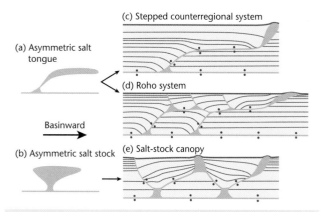

Fig. 11.53 Allochthonous salt geometries. (a) Sub-horizontal asymmetric salt tongues extruded basinward from inclined feeders. (b) Bulb-shaped salt stocks formed by radial spreading from vertical feeders. Individual salt tongues may amalgamate to form salt-tongue canopies, which may evolve as (c) stepped counterregional systems as salt is extruded into leaning diapirs, or (d) roho systems, characterised by basinward-dipping listric growth faults. Individual bulb-shaped salt stocks may coalesce to form (e) salt-stock canopies, in which the salt is gradually replaced by elliptical minibasins. Salt nappes, which are large allochthonous sheets without underlying feeders, are not shown. Dots represent salt welds. From Rowan *et al.* (2001). AAPG © 2001. Reprinted by permission of the AAPG whose permission is required for further use.

times (Gluyas *et al.* 2004). The Cretaceous Chalk fields (such as Ekofisk) in the Norwegian North Sea Central Graben are examples of domal traps formed over salt pillows, in this case the Upper Permian Zechstein Salt (Hancock 1984). The Machar Field (Foster *et al.* 1993) of the Central North Sea is a smaller, highly faulted, structurally complex example of a Zechstein diapiric trap, comprising reservoirs in the overlying Cretaceous Chalk and in onlapping Paleocene submarine fan sandstones.

The northern Gulf of Mexico provides many examples of allochthonous salt geometries (Rowan *et al.* 2001) and their associated fault families (Rowan *et al.* 1999) (Fig. 11.53). Simple asymmetric *salt tongues*, extruded basinward from leaning feeders, may amalgamate into *salt-tongue canopies* and evolve into more complex stepped counterregional systems. Where shelf sediments prograde onto the salt body, these tongues may evolve into salt-based detachment systems characterised by a combination of gravity glides and salt withdrawal structures (roho systems). Individual bulb-shaped *symmetrical salt stocks*, formed by radial spreading from a vertical feeder, may coalesce to form *salt-stock canopies*, where salt is gradually replaced by elliptical *minibasins* above a *salt weld* (the surface remaining after the evacuation of salt). *Salt nappes* are large allochthonous sheets without local underlying feeders.

The evolution of allochthonous salt bodies may be multi-stage and extremely complex. An example is the Mahogany salt body (Rowan *et al.* 2001), located on the Louisiana outer shelf, and associated with the first sub-salt discovery (1993) in the northern Gulf of Mexico. Mahogany is surrounded by many other near-surface salt bodies that overlie a deeper system of salt welds and remnant salt. Although the allochthonous Mahogany salt body does not trap the deeper Mahogany Field, it is a good example of the complex interactions between salt deformation and sedimentation that characterises this trap type.

In summary, diapiric structures may give rise to a very wide range of petroleum traps of both structural and stratigraphic origin. Although the detection of diapirs on seismic sections is not difficult, trap development above and around the diapir is frequently very complex, and individual reservoir units may be quite small, owing to rapid lateral facies changes and complex faulting.

Salt diapirism is likely to occur in quite distinct geological settings. Thick salt deposits may develop in enclosed basins subject to cycles of flooding and desiccation. These may occur in the rift and early post-rift stages of continental margin development. Examples are the Jurassic salt of the Gulf Coast and Aptian salt of the South Atlantic. The occurrence of salt substantially enhances the petroleum resource potential of these provinces.

Mud diapirism

Mud diapirism is most likely to develop in pro-delta clay sequences underneath the thick, rapidly deposited regressive sandy sequences of modern and Tertiary deltas. Examples are the Baram (Borneo), Niger (West Africa), Mississippi (USA) and Mackenzie (Arctic Canada) deltas. Excess pore pressure builds up in these clay sequences because low permeability prevents the expulsion of sufficient pore fluids as the sediment undergoes compaction (§9.2). The overpressure lowers the strength of the sediment and promotes ductile flow.

Although traps associated with salt diapirs have been better studied, the indications are that mud diapir structures offer broadly similar trapping possibilities. Structures may be very complexly faulted, with reservoirs broken into a multitude of separate units. Mud diapirs can be distinguished from salt diapirs on seismic sections on the basis of seismic velocity. Salt has a much higher velocity than overpressured shale. Furthermore, mud diapirs tend to lack the well-developed rim synclines that typically surround salt diapirs. Factors may be the shorter-lived process of mud diapirism, and the smaller area from which mud withdrawal appears to take place (Harding & Lowell 1979).

11.7.3 Stratigraphic traps

A stratigraphic trap is primarily caused by some variation in the stratigraphy of the basin-fill. This variation may be essentially inherited from the original depositional characteristics of the fill, or may result from subsequent diagenetic changes to it. The detailed stratigraphic trap classification of Rittenhouse (1972) (modified slightly in Allen & Allen 1990, p. 390) emphasises the tremendous diversity present amongst stratigraphic traps. Stratigraphic traps may be classified as: (i) *depositional* (in which the trap geometry is related to sedimentary facies changes); (ii) related to *unconformity* surfaces (either above or below); and (iii) *diagenetic*. Among the diagenetic traps, there are not only those caused by mineral diagenesis such as dolomitisation, but also biodegradation of petroleum (tar mats) and phase changes to petroleum gas (gas hydrates) and interstitial water (permafrost).

A large number and wide range of stratigraphic traps have been discovered by over a century of exploration, since the drilling of the first exploration well by Colonel Drake at Titusville, Pennsylvania, in 1857. Many of these stratigraphic discoveries, however, have been made by complete accident or at least by incorrect geological reasoning (Halbouty 1982). The giant East Texas Field, for example, a >5 billion barrel unconformity trap, was drilled as an anticlinal prospect. The detection of stratigraphic traps requires a detailed level of seismic definition, and a high level of geological expertise. Great emphasis must be placed on an understanding of the stratigraphic evolution of the basin, through a detailed sequence-by-sequence analysis. Of particular importance is the understanding of paleogeography and sedimentary facies for each depositional sequence and parasequence. For these reasons, the exploration for stratigraphic traps is perhaps most realistically pursued in near-field and in-field situations where data quality and quantity enable this level of detailed analysis.

11.7.3.1 Pinch-outs

Whatever the geometry and origin of the entire reservoir unit, any porous facies may pinch out laterally, and, when combined with regional structural dip, may give rise to a stratigraphic pinch-out trap. Such traps were previously (§11.7.2.5) described on the flanks of salt diapirs. Pinch-out traps may be extensive and of very low dip. When the depositional pinch-out is gradational, an up-dip transitional 'waste-zone' of very poor-quality reservoir rock may be present. Much or all of the petroleum charge may leak into the poorly producible waste-zone, forming a non-commercial accumulation. Ideally, the up-dip facies change should be rapid and complete in order for an effective lateral seal to be produced (Downey 1984).

Up-dip pinch-out of reservoir units may explain the occurrence of trapped hydrocarbons in apparent synclinal settings, as described at the Scapa Field, Witch Ground Graben, UK (McGann *et al.* 1991), and on the Mutiara/Sanga Sanga Trend, Mahakam Delta, Indonesia (Safarudin *et al.* 1989).

11.7.3.2 Depositional traps

Traps may develop in a wide range of depositional environments, ranging from aeolian dune to submarine fan (§7.5.3). Three of the more common depositional traps discussed here exemplify the variability and complexity of depositional traps.

Fluvial channel traps have been described from the Cretaceous basins along the eastern flanks of the Rockies (Selley 1985). An example is the South Glenrock Field of the Powder River Basin, Wyoming (Curry & Curry 1972), where the productive sand has a very clear meandering channel geometry. The South Glenrock example illustrates two important features of many channel traps: that reservoir sand thicknesses are typically small, limiting the reserves of these accumulations, and that the channel-fill may not be reservoir rock but clay. The geometry of the trap clearly depends on the geometry of the channel. Braided, meandering, anastomosing, delta distributary and tidal channels would have different geometries. Channels may sometimes be detected as a result of differential compaction relative to the surrounding shales. Some degree of structural closure may be needed to produce the trap (for example, a regional tilt or structural nose). Owing to the isolated nature of channel sands, they may not receive a petroleum charge. It is important, therefore, that source rocks are developed within the same depositional sequence. The effectiveness of a channel trap clearly depends on the lithology into which the channel is incised. Thus, although the channel itself may have a very distinct, sharp, lateral boundary, leakage may occur into the adjoining fluvial sediments.

Clay-filled channels may provide a lateral seal to reservoir sands in the adjacent incised sequence. Examples are present in the Sacramento Valley of California (Garcia 1981), and the Pennsylvanian Minnelusa sandstones of the Powder River Basin, Wyoming (Van West 1972).

Submarine fans may be developed on a very much larger scale, and give rise to large petroleum accumulations. An understanding of the sediment routing system is required for accurate prediction of submarine fans: an ability to constrain the timing of delivery of sand to the deep sea from upstream staging areas and river mouths is important.

The Tertiary of the North Sea provides numerous examples of submarine fan reservoirs. An example is the Balder oilfield of the Norwegian sector (Sarg & Skjold 1982). This field is a series of Paleocene-age sand-rich fan lobes, with the trap geometry provided by depositional topography and subsequent submarine erosion. A hemipelagic shale caprock seals the fan complex. The discovery well was drilled by Exxon in 1967 as a test of a structural prospect. The area was reinterpreted in the 1970s using seismostratigraphic principles, and the field found to be a stratigraphic lowstand systems tract trap.

Reefs may provide high-relief stratigraphic traps. Isolated pinnacle reefs may be completely encased in younger marine shale. Barrier reefs on carbonate shelves generally need to be sealed up-dip by tight back-reef facies. As noted for pinch-out traps, there is a risk that the back-reef facies will constitute a waste-zone of non-productive reservoir rock. Although the relief on many reef traps may be considerable, the size of the accumulation is dependent on the presence or absence of permeable strata that terminate against the reef body. Careful seismostratigraphic analysis may be required to detect these zones of leakage. The distribution of reservoir units within reef complexes may be variable and unpredictable, not only owing to depositional facies changes but also to the effects of diagenesis. Reefs have formed very successful petroleum traps around the world, including in the Devonian of the Western Canada Basin, in the Sirte Basin of Libya, in the Tertiary of the Salawati and North Sumatra basins of Indonesia, in the Miocene of Sarawak, in the Permian Basin of west Texas, and in southern Mexico and the Arabian Gulf. Sophisticated exploration techniques, primarily in geophysical data acquisition, processing and interpretation, may be needed to explore successfully for reefs in mature provinces. The undiscovered, smaller reefs in these provinces are extremely subtle features.

11.7.3.3 Unconformity traps

A variety of traps may develop at unconformities, both immediately above and immediately below the unconformity surface. Many of the depositional stratigraphic traps previously described may also develop on unconformities. An example is a pinch-out trap. Overstepping marine shales frequently provide a topseal to shallow shelf or shorezone sands. These traps develop on margins undergoing marine onlap, and transgression, for example, in the transgressive systems tract.

Supra-unconformity sands may be localised by topography in the unconformity surface. Thus, incisive channels in the unconformity surface formed at a relative lowstand may become sand-filled during rising sea level. Valleys developed along the strike of the outcropping pre-unconformity strata may be the location of the first postunconformity fluvial sediment. In both cases, a stratigraphic trap at the unconformity surface may be formed, particularly if regional structural dip is in a suitable direction.

Traps developed beneath an unconformity by *truncation* of reservoir beds may give rise to giant fields. Examples have been previously given of eroded extensional fault blocks in the East Shetland Basin of the North Sea (e.g. the Brent reservoir in the Statfjord Field, Fig.

11.41). Further examples are the Fortescue–Halibut Field of the Gippsland Basin, Australia, the Prudhoe Bay Field of the Alaskan North Slope, and the East Texas Field, USA.

There may be several elements to the closure developed in sub-unconformity traps, including the topography present at the erosional unconformity, and the structural geometry of the pre-unconformity beds. Cross-faulting, for example, may provide closure in the strike direction. Topography on the unconformity may have been provided in the first instance by eroded fault scarps, or eroded carbonate features. Once sealed beneath post-unconformity sediment, such paleotopographic unconformity traps are known as 'buried hills'.

The lithology of the post-unconformity sediments is critical to the effectiveness of sub-unconformity truncation traps. If, for example, a thin marine basal transgressive sand is present above the unconformity, leakage may occur. This sand may be so thin as to be below seismic resolution. Careful seismostratigraphic interpretation and basin analysis is required in order to understand the causes of the unconformity and the evolution of the sequences above and below it, before accurate predictions can be made of sedimentary facies impinging on the unconformity surface. Ideal conditions are created where the unconformity subsides rapidly into deep-water depths, perhaps as a result of rapid initial fault-controlled or thermal subsidence, and the topography is passively infilled by deep-marine muds. Traps developed at unconformities that have been onlapped by shelf, shorezone or non-marine sequences may not be effectively sealed (cf. §11.6.3).

When the sub-unconformity strata are carbonates, exposure at surface may have very important implications for reservoir development in the sub-unconformity trap. The Casablanca Field in the Gulf of Valencia, offshore southern Spain, was formed by Miocene faulting, erosion and leaching of a Mesozoic sequence of tight, dense limestones (Watson 1981). The faulting formed the paleotopography and fractured the brittle carbonates, allowing the penetration of meteoric waters and the development of secondary porosity to depths of up to 50 m. The eroded, elongate limestone ridge is overlain by mid-Miocene Alcanar Formation organic-rich marls, which charge and cap the accumulation. A similar example is the Angila Field in the Sirte Basin of Libya; in this case it is weathered granite that forms the reservoir in the unconformity trap (Williams 1968).

11.7.3.4   Diagenetic traps

There are a number of traps in which diagenesis has played a significant part. Examples are where cementation (for example of dolomite, calcite, clay minerals or quartz) has provided an up-dip seal to an accumulation, or where leaching has produced a local reservoir in an otherwise impermeable sequence. These trapping mechanisms are usually established after the main phase of oil generation, and are relatively difficult to predict.

An interesting form of diagenetic trap is where cementation has taken place below the oil–water contact of an existing accumulation (diagenesis is frequently inhibited by the presence of petroleum in the pore space, §11.5.4.3). This may seal in the accumulation despite the subsequent removal of the original trapping mechanism, for example by tectonic activity.

At temperatures of less than about 70 °C, and in the presence of meteoric water, bacterial degradation of oil may take place (§11.4.5). This can form an impermeable up-dip tar mat that seals subsequently migrated oil. Examples occur in the Orinoco Tar Belt of eastern

Venezuela (§12.2.6), at the Bolivar Coastal Field of Lake Maracaibo, Venezuela, in the Californian San Joaquin Valley (Wilhelm 1945) and in the Russian Volga–Ural region (Vinogradov *et al.* 1983). Similarly, a change of phase from gas or liquid to solid may also provide an unusual kind of trap. At high latitudes, permafrost may provide an up-dip seal to petroleum accumulations, as in the West Siberian Basin. At particular pressure-temperature conditions (low temperature, high pressure) petroleum gases may form *solid hydrates* (§12.2.5). Not only is gas trapped in the hydrates themselves, but may accumulate in reservoirs underlying the zone of hydrate formation (Downey 1984).

11.7.4   Intrusive traps: injectites

Sand injectites have been increasingly observed over the past decade, especially in deep-water clastic systems. When seismically detectable, they form important reservoir targets in their own right. At the sub-seismic scale, they have important implications for recoverable reserves, reservoir productivity and behaviour, as well as for seal integrity and hydrocarbon migration through the basin. The recognition of sand injectites may therefore cause a radical re-think during play, prospect and reservoir assessment.

The trap geometry is formed by the intrusion of highly porous and permeable sand into overlying low-porosity/permeability units. The scale of these intrusions varies from centimetres to hundreds of metres, and geometries range from dykes and sills to more irregular bodies (Hurst and Cartwright 2007) (Fig. 11.54). They may form complex networks of connected sand bodies above the main reservoir unit. Top surfaces may be flat or scalloped, depending on the role of pre- and post-injection faulting. They may also extrude onto the seafloor, forming mound-type structures.

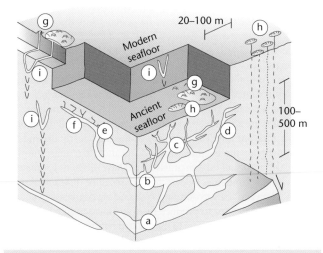

Fig. 11.54   Some common sand injectite geometries and associations identified on seismic data and validated using outcrop data. (a) and (b) Depositional sand bodies, (c) complex of dykes and sills, (d) inclined sills developed along polygonal faults, (e) large irregular intrusive bodies, (f) bedding-concordant sills linked to (d), (g) sand extrusions, (h) gas seeps forming pockmarks, and (i) conical sand intrusions. From Hurst and Cartwright (2007). AAPG © 2007. Reprinted by permission of the AAPG whose permission is required for further use.

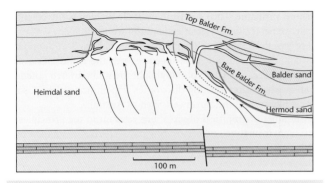

Fig. 11.55 Model for post-depositional sand remobilisation at the Balder Field, North Sea. Contemporaneous sand remobilisation, compaction faulting, sand withdrawal, diapirism and sand injection. From Briedis *et al.* (2007). AAPG © 2007. Reprinted by permission of the AAPG whose permission is required for further use.

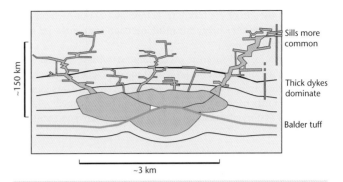

Fig. 11.56 Sketch diagram of the possible geometry of the sand intrusion networks at the Gryphon Field, North Sea. From Lonergan *et al.* (2007). AAPG © 2007. Reprinted by permission of the AAPG whose permission is required for further use.

A number of processes may trigger or induce the formation of sand injectites. Although not yet fully understood, the prerequisites for the generation of sand injectites are likely to be the occurrence of unconsolidated sand encased in low-permeability mudstones, elevated pore pressure (§9.2.2), and a triggering mechanism (Braccini *et al.* 2008). The unconsolidated sands must have sufficiently high volumes of pore fluid to transport the sand upwards. Elevated pore pressure must be enough to induce liquefaction, where the pore pressure rather than grain contacts supports the overburden. Mechanisms for overpressure may include sediment loading, lateral pressure transfer as a result of large-scale slumping, deep pressure transfer from the deep overpressured parts of the basin, the buoyancy effect of migrated hydrocarbons, hydrocarbon generation and expulsion, and rising salt diapirs. A triggering event may be an earthquake, meteorite impact, volcanic eruption or landslide.

Numerous examples of injectites have been described in the UK, Norwegian and Danish North Sea (Hurst and Cartwright 2007; Braccini *et al.* 2008), and in outcrop (Hurst and Cartwright 2007). At the Balder Field (Briedis *et al.* 2007; Fig. 11.55), discovered in 1967, more than 25% of the initial oil in place is now thought to have been located in injectites. At the Alba Field, high-porosity turbidite channel sands were discovered in 1984. Although sand injection features were observed in cores early on, it was a 3D seismic survey in 1998 that revealed dipping reflectors at the margin of the main channel that are now interpreted as sand injectite wings. These have proved to be important additional, high-productivity reservoirs. At Gryphon (Lonergan *et al.* 2007; Fig. 11.56), discovered in 1987, production has been obtained since 2004 from seismic-scale sand injection wings to the main Eocene basin-floor turbidite sandstone reservoirs. The recently discovered Volund Field is an example of a giant sand injection complex (de Boer *et al.* 2007).

11.7.5 Hydrodynamic traps

There are relatively few basins worldwide where hydrodynamics is known to have a significant impact on the entrapment of petroleum. These are typically foreland basins where porous and permeable carrier beds have been uplifted and exposed in adjoining fold-thrust belts, allowing the influx of meteoric water (§10.3.4). If an outlet for the water is available elsewhere in the 'plumbing system' of the basin,

hydrodynamic flow of the basin fluids may take place (Hubbert 1953). Once the necessary conditions are known to be established at the basin scale, individual exploration prospects can be evaluated with a view to hydrodynamic effects.

Under conditions of strong hydrodynamic flow, petroleum–water contacts may be inclined rather than horizontal, petroleum may be completely flushed from structural or stratigraphic closures, or hydrodynamic closures may be produced (for example on structural noses) where there is no other form of closure present. Each prospect needs to be evaluated individually, since prospectivity will depend on a host of regional and local factors that may be difficult to assess. Readers are referred to Hubbert (1953) for the theory and examples of hydrodynamic trapping.

Clearly, an understanding of the structural and stratigraphic evolution of the basin is required for an assessment of the impact, if any, of hydrodynamic conditions on petroleum entrapment. Generally, however, hydrodynamic traps and hydrodynamic effects appear to be relatively rare.

11.7.6 Timing of trap formation

An understanding of the mechanism of trap formation, and therefore the timing of trap formation, is essential to prospect evaluation. A trap that developed too late to receive a petroleum charge will be dry. Each of the structural, stratigraphic, and hydrodynamic trap types discussed in this section has implications for trap timing that are obvious. Depositional and unconformity traps are very early, dating from the time the sealing units became effective. Thus, these traps are ready to receive a charge from a very early stage. Some structural traps, however, are very late in relation to petroleum charge. Each trap needs to be individually evaluated.

The timing problem is perhaps best illustrated by considering a currently active fold-thrust belt. The fold-thrust belt forms as a result of shortening, at least in the sedimentary cover, which results in uplift. Uplift results in cooling of the overthrust sheets, which switches off petroleum generation. Thus, a timing problem may exist, unless generation is maintained in subthrust positions by loading of the allochthonous sheet, or unless earlier trapped petroleum remigrates into the new fold structures. A similar problem exists for inversion structures. Prior to inversion, migration is usually directed away from the site of the future trap. Inversion may switch off new generation, thus the charge into the inversion trap must be from remigrated oil.

Structures formed in areas of continuous subsidence also need careful evaluation, so that the growth of the structure may be closely related to the timing and volume of petroleum charge. The sedimentary section may need to be decompacted and backstripped (§9.4) to the time of first trap formation, and the charge system geochemically modelled.

11.8   Global distribution of petroleum resources

A systematic review of the global distribution of known reserves and undiscovered potential may provide an additional perspective on petroleum formation and occurrence. All estimates of undiscovered petroleum resources are highly uncertain, but the best guide to undiscovered potential is likely to be gained through rigorous petroleum-system, play and prospect based evaluation, applying the principles of basin analysis.

Using a petroleum system approach, the US Geological Survey has assessed undiscovered potential on the basis of geology, and exploration and discovery history, in 128 geological provinces, 149 total petroleum systems and 246 assessment units worldwide, excluding the USA (Ahlbrandt *et al.* 2005). The breakdown of the world's petroleum endowment, excluding unconventional gas and oil sands (Table 11.5) shows that the oil exploration industry is now quite mature; even by 1995, only a quarter (24%) of the world oil endowment of 3 trillion barrels remained to be discovered over a 30-year time frame. Despite this maturity, an undiscovered volume of over 700 billion boe worldwide represents a significant prize. For gas, exploration is less mature; over a third (34%) remained undiscovered by 1995.

For undiscovered oil potential, petroleum systems in the former Soviet Union (FSU) and Middle East/North Africa regions dominate, particularly the West Siberia Basin and Caspian areas, and the Zagros foldbelt of Iran. However, almost half occurs in the other regions, reflecting largely the deep-water potential of the South Atlantic basins in Brazil and West Africa. The USGS attribute significant exploration potential to some frontier areas with no pre-existing discoveries, such as the rift basins of northeast Greenland and

offshore Suriname. The Greenland example highlights the potential of the Arctic as the next great exploration frontier.

For gas, the FSU (West Siberia, Barents Sea, Kara Sea) dominates world undiscovered potential, together with the Middle East and Norwegian Sea. The USGS also highlight the undiscovered potential of areas where large existing discoveries remain undeveloped, such as East Siberia and the Northwest Shelf of Australia.

A number of empirical observations relevant to the principles of petroleum formation and occurrence may be derived from the USGS geological characterisation of known discoveries and future potential (Ahlbrandt *et al.* 2005):

- Source rocks of Jurassic and Cretaceous age, and of marine depositional environment, dominate world petroleum occurrence.
- Young petroleum systems with peak source rock maturation during the Cenozoic dominate, especially for oil, reflecting the risks of trap and seal disruption by tectonic events for earlier-generated oils. Early-generated hydrocarbons, when they occur, are often associated with evaporite seals.
- Mesozoic reservoirs dominate, as do non-marine to shallow-marine depositional environments, despite the important recent successes in deep-water turbidites. Source rocks are not normally present in the non-marine/shallow-marine reservoir sequences, highlighting the importance of migration.
- Siliciclastic reservoir lithologies dominate over carbonates. Carbonate reservoirs are largely restricted to basins that occupied low latitudes in Paleozoic times, and the northernmost margin of Gondwana during the Mesozoic (North Africa and Arabian Peninsula).
- Shale topseals dominate, although evaporites are critical in Paleozoic systems, particularly in extensional and passive tectonic settings.
- No single trap type dominates, although structural traps contain the greatest volumes, especially contractional structural traps. Stratigraphic traps are at a less-advanced stage of exploration maturity.
- Most conventional petroleum systems are dominated by vertical migration, or lateral migration of limited extent (<20 km).
- The vast majority of giant oilfields occur at reservoir depths of less than 15,000 ft (4572 m), at less than 130 °C reservoir temperature, and at less than 12,000 psi reservoir pressure. Fields below 15,000 ft (4572 m) tend to be significantly smaller.

It has also been noted in more recent research (Nadeau *et al.* 2005; Nadeau 2011) that the vast majority of the world's conventional oil and gas resources, and a very high proportion of the world's giant fields, occur within a relatively narrow geothermal range of 60–120 °C, and it is asserted that exploration and production risks are closely controlled by reservoir temperature. This thermal interval spanning 60°C of subsurface temperature has been termed the *Golden Zone*. The rationale that supports this interpretation is derived from mineralogical research, mainly in the North Sea. Firstly, the 60°C temperature threshold is considered critical for the dissolution of clay minerals smectite and kaolinite and the precipitation of diagenetic illite. Illite precipitation in the pore spaces of mudstones and shales is thought to reduce permeability by several orders of magnitude, hence 60°C may be a critical threshold for the effectiveness of vertical and lateral seals. It may also enable the development of significant reservoir overpressure. Secondly, the precipitation rate for diagenetic cements (primarily quartz) is considered to increase

Table 11.5   World petroleum endowment (Ahlbrandt *et al.* 2005)

World (including USA)	Oil (billion barrels)	Natural gas (trillion cubic feet)	Natural gas liquids (billion barrels)
Undiscovered conventional (1995–2025)	732	5196	207
Reserves growth conventional	688	3660	42
Remaining reserves	891	4793	68
Cumulative production	710	1752	7
Endowment	3021	15401	324

Note: 1. US NGLs included in oil statistics, 2. natural gas endowment equivalent to 2567 billion barrels of oil equivalent (Bnboe).

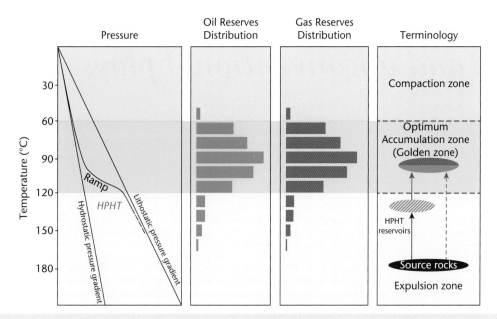

Fig. 11.57 Thermal zonation model of sedimentary basins. The optimum petroleum accumulation zone, or 'Golden Zone', lies between approximately 60 and 120 °C. The underlying high-pressure/high-temperature (HPHT) zone at greater than 120 °C is a zone of petroleum expulsion. After Buller *et al.* (2005).

exponentially with temperature, commencing at 70 °C, resulting in severe porosity loss, particularly in rocks with high quartz surface area such as fine-grained sandstones. Permeability reduction through temperature-related quartz dissolution, transport and cementation enhances the development of isolated overpressured petroleum-bearing cells that ultimately break their seals by hydrofracturing, resulting in the vertical migration of fluids, including petroleum. In fault-segmented basins where lateral drainage is restricted, overpressure may develop dramatically with increasing temperature (and depth). In the North Sea, this critical temperature is 120 °C, roughly equivalent to 4 km depth, below which reservoirs are known as HTHP (high-temperature high-pressure reservoirs). In the HTHP zone, overpressure results in the remigration of petroleum to shallower levels.

In this model, sedimentary basins may be layered into a series of thermal zones (Fig. 11.57) (Buller *et al.* 2005). The optimum petroleum accumulation zone (Golden Zone) between approximately 60 and 120 °C is bounded above by a low-temperature zone characterised by ineffective mudstone seals (the compaction zone), and below by a high-temperature zone characterised by tight reservoirs and

Table 11.6 Distribution of global conventional petroleum resources by subsurface temperature zone (Nadeau 2011)

Reservoir temperature	Oil reserves %	Gas reserves %
<60°C Compaction zone	12	40
60–120°C Golden zone	85	50
>120°C Expulsion zone	3	10

overpressure, from which petroleum fluids have been expelled (fluid expulsion zone).

Remigrated petroleum from HTHP reservoirs supplements indigenously charged and trapped petroleum, ensuring that the Golden Zone contains the vast bulk of the world's conventional resources (Table 11.6).

CHAPTER TWELVE

Classic and unconventional plays

Summary

A small selection of classic petroleum plays are described that collectively illustrate how the building blocks of the play come together to produce petroleum systems that are of enormous economic importance. The examples chosen are the Niger Delta, the Campos Basin of Brazil, the pre-salt of the Brazilian Santos Basin, and the Northwest Shelf of Australia.

In the case of the *Niger Delta*, all the elements of the petroleum system – the organic matter and thermal maturity of the source rocks for the waxy oils, the fluvio-deltaic and deep-water turbidite reservoir rocks, the gravitational tectonics responsible for trap formation – are related to the development of a very thick prograding delta sequence of Eocene to Recent age. The Tertiary delta petroleum system is atypical of other South Atlantic basins in which Cretaceous continental break-up is more critical to petroleum system development. In the Brazilian *Campos Basin*, rift-phase lacustrine source rocks have charged a prolific petroleum system, which in the deep-water sector includes turbidite reservoirs of Late Cretaceous and Oligo-Miocene age. On the same margin, the emergence of the sub-salt petroleum system in the Brazilian *Santos Basin* represents one of the most important new exploration discovery events of the 21st century to date. Key features are the rich lacustrine source rocks that developed in the syn-rift sequence, the 2 km-thick regionally extensive late Aptian evaporitic deposits that covered the syn-rift and sag sequences, providing an excellent regional topseal, and the existence of a regional high at base salt level that served as a focus for hydrocarbon migration. Finally, the *Northwest Shelf of Australia* is one of the world's great gas provinces. Although the petroleum system lacks a world-class source rock, Late Jurassic to Early Cretaceous break-up of the continental margin has produced a number of faulted regional highs that serve as extensive traps for hydrocarbons generated and reservoired in the Triassic pre-rift deltaic sequences, and sealed by the marine Cretaceous post-rift regional topseal.

In addition, a number of unconventional petroleum plays are described, not only because, from a geological perspective, we must understand the genesis of such huge deposits, but also because they are of increasing commercial significance. The oil sands of Western Canada and the heavy oils of the Venezuelan Orinoco Basin represent conventional oils that have been so severely altered that they require unconventional methods of extraction. Unconventional gas plays such as basin-centred tight gas, shale gas and gas hydrates indicate processes at work that vary significantly from conventional models of petroleum formation and occurrence.

Tight gas is produced in a number of North American basins, characterised by extensive or pervasive gas shows, little produced water, low-permeability rocks, abnormally high pressure, and poorly defined traps and seals. Gas accumulations appear to be very large-scale, possibly basin-wide, and locally sourced, rather than a result of secondary migration into discrete defined traps. *Shale gas* production is currently experiencing rapid growth in the USA, and its potential has become of intense interest worldwide. The gas source is commonly indigenous thermogenic, as in the case of the Devonian shales of the Appalachians, where gas distribution is continuous throughout the mature source rock sequence, but it may also be migrated thermogenic or biogenic. Reservoirs are tight and/or fractured. *Coal seam gas* or coal bed methane is gas adsorbed onto the surfaces of organic matter within coal seams. It is an existing form of gas production in many USA basins, and of growing significance around the world, even as a feedstock for liquefied natural gas (LNG) export projects in Queensland, Australia. Finally, *gas hydrates* are accumulations in which gas molecules are encased in a crystal lattice of frozen water, which occurs under certain pressure-temperature conditions found in Arctic permafrost and deep-marine environments. Potential gas hydrate resources may be extremely large, but little is currently known about production technologies and economic viability.

Unconventional oil volumes are also extremely large, and dominated by two provinces, the Alberta oil sands of Western Canada, and the Orinoco Basin heavy oil deposits. In Canada, the oils reservoired in the Cretaceous Mannville Group are the deeply altered remnants of conventional crudes that migrated large distances from source kitchens to the west. Meteoric water influx at the basin edge, where the Mannville sands outcrop, has been responsible for biodegradation of conventional mobile crudes into bitumen and heavy oil. In the Orinoco Basin, the heavy oils found in the Cretaceous Querecual Formation have accumulated on the flexural bulge to the thrust belt and foreland basin formed as a result of oblique plate collision to the north, commencing in Eocene times. Oils generated under the thrust sheets and foreland basin migrated several hundred kilometres to the south into Miocene siliciclastic rocks, where shallow depth of burial and meteoric water influx have contributed to biodegradation.

Finally, the principles of basin analysis, and many of the geological understandings developed in the search for petroleum, can be applied in the new emerging area of carbon dioxide geosequestration. Although the sector is in its infancy, geological assessment has an important role to play in identifying and evaluating potential subsurface sites of CO_2 storage.

Basin Analysis: Principles and Application to Petroleum Play Assessment, Third Edition. Philip A. Allen and John R. Allen.
© 2013 John Wiley & Sons, Ltd. Published 2013 by John Wiley & Sons, Ltd.

12.1 Classic petroleum plays

12.1.1 Introduction

In the following examples of classic petroleum plays, we give a brief description of resource significance, plate tectonic setting and basin-forming mechanisms, the basin-fill and subsidence and thermal history, and the resulting petroleum systems and plays. The examples chosen are the Tertiary Niger Delta, the Brazilian Campos Basin, the emerging sub-salt play of the Brazilian Santos Basin, and the Northwest Shelf of Australia. Common themes are the close linkage between basin analysis and the understanding of hydrocarbon occurrence, the grand scale on which the elements of the petroleum system successfully come together, and their great commercial significance. However, there are some clear contrasts across all the play building blocks described in the previous chapter.

Niger Delta is chosen because all the elements of this petroleum system are contained within a very thick prograding Tertiary delta sequence, atypical of other South Atlantic basins in which Cretaceous continental break-up is more critical to petroleum system development. The Brazilian *Campos Basin*, is a classic example of rich rift-phase lacustrine source rocks that have charged a prolific petroleum system, which includes younger passive margin deep-water turbidite sandstone reservoirs. On the same margin, the emergence of the sub-salt petroleum system in the Brazilian *Santos Basin* represents one of the most important new exploration discovery events of the 21st century to date. Key features are the 2 km-thick regionally extensive late Aptian evaporite sequence that provides an excellent regional topseal, and the existence of a regional high at base salt level that served as a focus for hydrocarbon migration. Finally, the *Northwest Shelf of Australia* is one of the world's great gas provinces. The province appears to lack a world-class source rock. Nevertheless, a world-class gas province has resulted from the Late Jurassic to Early Cretaceous break-up of the continental margin, which produced a number of faulted regional highs that serve as extensive traps for hydrocarbons both generated and reservoired in the Triassic pre-rift deltaic sequences, and sealed by the marine Cretaceous post-rift regional topseal that resulted from drowning of the underlying rift topography.

Space constraints prevent us from providing a wider variety of plays. The importance of many other world-class plays could have justified half a dozen further examples, including orogenic foldbelts and foreland basins, failed rifts, carbonate platforms and reefs, gravity-driven foldbelts in deep water, and salt provinces.

12.1.2 Niger Delta

The Tertiary Niger Delta is a normally charged, vertically drained, high-impedance petroleum system. It has given rise to one of the world's premier petroleum provinces, ranking as twelfth largest in terms of known oil and gas reserves. According to the USGS world petroleum assessment of 2000 (Ahlbrandt *et al.* 2005), the known and potential undiscovered crude oil and natural gas resources of the Niger Delta total 75 billion barrels of oil and over 200 trillion cubic feet (TCF) of gas.

The Niger Delta is situated on the west-facing African passive margin that formed following Late Jurassic to Cretaceous South Atlantic break-up. It is located on the African plate between the West African and southern African cratonic cores. Cretaceous fracture

zones related to mid-Atlantic seafloor spreading subdivide the West African margin and form the boundary faults to the Cretaceous Benue-Abakaliki trough, which penetrates far into the West African basement massif (Tuttle *et al.* 1999). This NE–SW-trending trough represents the failed arm of a rift triple junction. Late Jurassic to Mid-Cretaceous rifting produced a series of strong NE–SW horsts and grabens, parallel to the fracture zone patterns in the deep Atlantic (Haack *et al.* 2000). This provides the basement structure prior to development of the Tertiary delta, which began in the Eocene, well into the drift stage, and continues at the present day.

Passive margin thermal subsidence has been greatly amplified by isostatic compensation to sedimentary loading as the Tertiary delta developed. The great thickness (10 km) of the sedimentary load is due to the prior existence of the Benue-Abakaliki failed rift, which has constrained and focused fluvial drainage and the footprint of the sediment routing system.

The Tertiary delta growth section is separated by décollements from the underlying Cretaceous to Paleocene strata (Haack *et al.* 2000) (Fig. 12.1), and each displays a different structural style. Deposition during delta growth since the mid to late Eocene was controlled by gravity-driven growth faults on a NW–SE trend, 90° to the underlying structural trends. A regionally extensive family of rollover anticlines into these listric growth faults control much of the hydrocarbon accumulation in the delta. In the onshore proximal part of the delta, there is a series of sub-parallel extensional faults that break up hangingwall blocks into many fault compartments. More distally, a listric style of faulting dominates, with hangingwalls more intact. These listric faults are closely associated with mud diapirs in the offshore deep-water area. A contractional thrust belt formed by gravitational tectonics occurs in the toe of the delta (Fig. 12.2).

The Tertiary petroleum system is proven, in the sense that oils have been geochemically linked back to known source rocks in the time-transgressive outer shelf to slope facies beneath the regressive paralic sequence above, and in transgressive shelf facies within the paralic sequence. The primary source of organic matter is terrigenous land plants. A land plant biomarker present in Niger Delta oils indicates lobes of high terrigenous organic matter deposition, while away from these lobes, organic matter deposition was more marine-influenced. In the more oxic depositional environments, where organic matter suffered greater degradation, kerogens are predominantly gas-prone. In parts of the outer shelf and slope, more sub-oxic environments may have been produced by impingement of an oceanic low-oxygen layer. Better organic preservation has allowed the development of a rich Type II kerogen, generating waxy oils.

Other petroleum systems have been proposed based on: (i) a Neocomian lacustrine source (Bucomazi Formation), analogous to the Lower Congo Basin to the south; and (ii) an Upper Cretaceous to lower Paleocene marine source (Haack *et al.* 2000), but these are speculative.

Oils show highest organic maturity in the proximal part of the delta where geothermal gradients are highest. Oil maturity also increases into major bounding faults, suggesting the existence of conduits for vertical migration from deep mature source pods. Different traps and migration pathways would have been available to migrating hydrocarbons at different times. Older traps are more likely to receive an early oil charge prior to oil-prone source rocks passing through the base of the oil generation window. The timing of generation is diachronous as a result of delta growth across the area, post-dating syn-sedimentary faulting. It is as early as

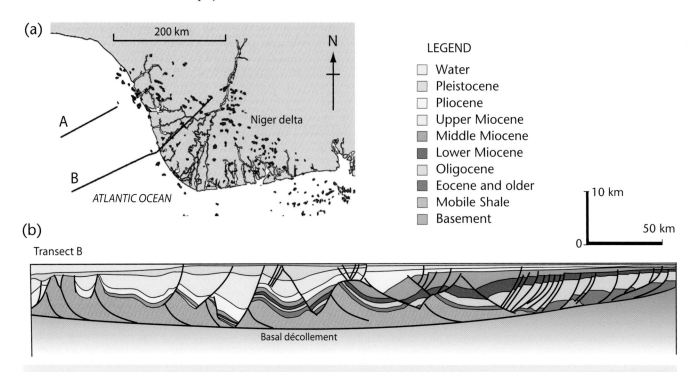

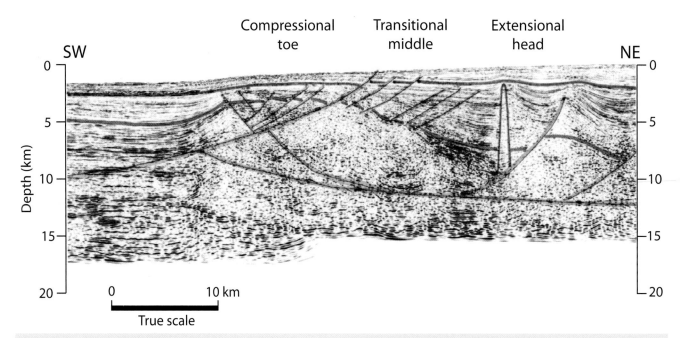

Fig. 12.1 Location map (a) and regional cross-section (b) across the Niger Delta, extending into deep water. The cross-section, shown as transect B on the map, is based on 2D and 3D seismic data, well control, and biostratigraphy, and is depth-converted and structurally-balanced. The Tertiary delta growth section, from which all hydrocarbon production is derived, is separated from the underlying Cretaceous to Paleocene basement by décollements passing through the mobile shale. The map also shows the location of the seismic profile shown in fig 12.2 (transect A). Modified from Haack *et al*. (2000). AAPG © 2000. Reprinted by permission of the AAPG whose permission is required for further use.

Fig. 12.2 Seismic data showing the compressional foldbelt in the distal Niger Delta. The compressional toe forms a positive topographic feature, with considerable stratigraphic thinning over its crest. The location is shown on Fig. 12.1 (transect A). Modified from Haack *et al*. (2000). AAPG © 2000. Reprinted by permission of the AAPG whose permission is required for further use.

late Eocene to early Oligocene in the most proximal parts of the delta, and continues through to the present day. Switching of depositional lobes caused generation to be episodic in any one location. Active growth faulting has redistributed already trapped hydrocarbons; as a result, multiple pay zones are typical. Extensive extensional faulting has focused hydrocarbon migration vertically – it is unlikely long-distance lateral migration would have taken place across major bounding faults.

In the Niger Delta, all production is from the growth section above the regional décollement. The chief reservoir comprises the interbedded delta-front, delta-top and fluvio-deltaic clastic sediments in the Oligocene–Miocene portions of the Agbada Formation. The topseal comprises intraformational shales within the Agbada Formation. These shales also provide lateral seal across faults by juxtaposition and clay smearing. In the lower Agbada, sand/shale ratios are about 50%, but the upper part of the Agbada is a very sand-rich sequence, posing a risk for seal. Traps in the delta sequence are mainly structural, formed by syn-sedimentary deformation, for example rollovers into growth faults.

Since the mid-1990s, exploration in the deep-water distal Niger Delta has pursued plays based on deep-water turbidite reservoirs within the Akata Formation, extending across the shale diapir zone, toe-thrust zone, and frontal deformation zone of the distal delta (Fadahunsi *et al.* 2005). Turbidite systems, comprising channels and fans, are partially controlled by growing shale diapirs and toe thrusts.

12.1.3   Campos Basin, Brazil

The Campos Basin (Lagoa Feia–Carapebus) petroleum system is supercharged, vertically focused, and high impedance, and is the most prolific in Brazil to date. The first discoveries were made in 1974, on the shelf, but the mid-1980s saw globally significant deepwater discoveries in Tertiary turbidite sandstone reservoirs (Marimba, Marlim, Albacora) (Fig. 12.3). By 2000, over 70 discoveries had been made in water depths ranging from 80 m to 2600 m, including seven giant oilfields in deep water. The resource endowment of the system has been estimated at 26 billion barrels of oil and 26 TCF of gas (USGS 2000 World Petroleum Assessment, Ahlbrandt *et al.* 2005).

The petroleum systems of the Campos Basin are described by Guardado *et al.* (2000), while Katz and Mello (2000) provide an overview of the petroleum systems of the entire South Atlantic margin. The Campos Basin formed on the western edge of the South Atlantic passive margin as a result of the break-up of Gondwana in the Early Cretaceous, and contains 9 km of Lower Cretaceous to Holocene basin-fill. The original rift basin structure is seen in a series of NE- and NW-trending horsts and grabens mapped at the level of the Neocomian basalt. Other structural trends are pre-Aptian structures related to the Campos Fault, and a salt dome province in ultra-deep-water.

The basin megasequences represent a rift–drift succession. The rift megasequence comprises lacustrine to hypersaline Barremian

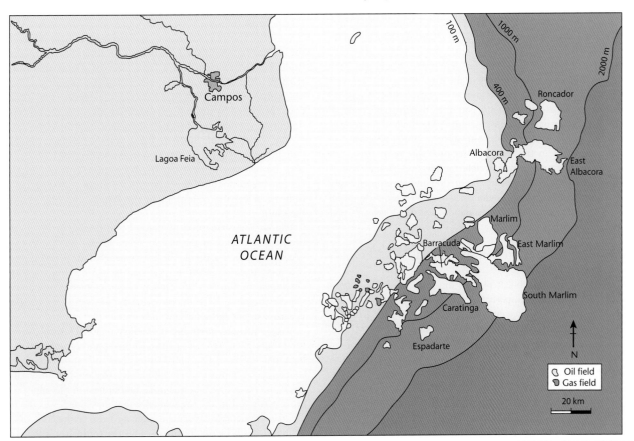

Fig. 12.3   Campos Basin, Brazil: distribution of oil fields discovered to 1999. The nine largest fields are named. These deep-water discoveries in Tertiary turbidite reservoirs have been globally significant. From Guardado *et al.* (2000). AAPG © 2000. Reprinted by permission of the AAPG whose permission is required for further use.

alluvial fans, fan deltas, carbonate banks and lacustrine sediments overlying Neocomian basalts, and includes the Lagoa Feia calcareous shale source rocks. The transitional megasequence is Aptian in age and contains conglomerates and carbonates overlain by evaporites formed during a tectonically quiet period prior to the period of drift. The marine post-rift drift megasequence initially comprises Albian shallow-water carbonates, muds and marls, and grades upwards into an Upper Cretaceous–Paleocene bathyal sequence containing turbidites, in which deposition is strongly influenced by salt tectonics. Sediment provenance is the Precambrian Brazilian platform, with relatively short transport distances.

The main source rock is the excellent-quality Barremian Lagoa Feia organic-rich shale. It is a 100–300 m-thick organic-rich unit (TOC averaging 2–6%) containing oil-prone Type 1 kerogen. Organic material is predominantly lipid-rich, mainly of algal and bacterial origin. Biomarkers suggest lacustrine brackish to saline depositional environments. Oils have 17–37° API gravities, comprising a mix of biodegraded and non-biodegraded oils. More than one pulse of generation and migration took place, combined with biodegradation during successive stages of reservoir filling. High organic content is thought to be a result of blooms of cyanobacteria coinciding with periods of increased water salinity. The depositional environment comprised small alkaline lakes bordered by a large restricted gulf or lagoon with episodic marine incursions.

Reservoir rocks include: (i) Barremian carbonates (coquinas/bivalve grainstones) and fractured and vesicular Neocomian basalts of the rift megasequence; (ii) Albian shallow-water shelf carbonates; (iii) Upper Cretaceous turbidites deposited in gentle bathyal slope troughs formed by salt movement; and (iv) Oligo-Miocene turbidites in extensive basin-floor fans.

Generation began in the Coniacian–Santonian, reaching a peak during late Miocene, and continues to the present day.

Traps occur within the rift sequence on intrabasinal highs; oil migration was short-distance into coquinas flanking the regional highs. The giant Tertiary oilfields, including Marlim and Barracuda, are predominantly stratigraphic, controlled by lateral pinch-out of turbidite reservoirs to the west and regional dip to the east. Salt tectonics also play a major role in determining trap geometry, as at the Maastrichtian Roncador Field. Generation and entrapment were synchronous, resulting in a high efficiency of entrapment. Migration was facilitated by good carrier beds, windows through the salt, and pathways along listric faults.

12.1.4 Santos Basin pre-salt play, Brazil

The emergence of the sub-salt play in the Brazilian Santos Basin represents one of the most important new exploration discovery events of the 21st century to date, ranking in significance alongside the Lower Tertiary discoveries of the ultra-deep-water Gulf of Mexico, unconventional gas developments in the USA, and the giant oil (Kashagan) and gas (Turkmenistan) discoveries of the Caspian area. A number of discoveries in post-rift reservoirs have been made in the Santos Basin since the late 1970s, but exploration success in the pre-salt has been a more recent phenomenon (the Tupi discovery was announced in 2007). Although follow-up discoveries have since been made, at the time of writing the play is at a relatively early stage of maturity, and its petroleum potential is not yet fully understood. Nevertheless, reserves of over 6 billion barrels of oil equivalent of light sweet crude appear to have been discovered in the Tupi Field alone, while the recoverable resource potential of the regional high

on which Tupi is located has been estimated at 60 to 80 billion barrels of oil equivalent (Carminatti *et al.* 2008).

The sub-salt play extends beyond the Santos Basin into the Campos and Espirito Santo basins, but exploration to date has focused on the central part of the Sao Paulo Plateau, a large regional high located in water depths of 2000 to 3000 m in the outer Santos Basin. Initially, the significance of the sub-salt was that it was thought to contain rich lacustrine source rocks, analogous to the black shales and carbonates of the Lagoa Feia Formation of the Campos Basin. The issue addressed by geologists was how to migrate sub-salt generated oil through the barrier of the Aptian evaporites into post-salt reservoirs (Meisling *et al.* 2001). Identification of reservoir and trap potential within the sub-salt itself, that is, recognition of a self-contained sub-salt petroleum system, has been a recent development. Sub-salt exploration has been inhibited by seismic imaging difficulties, owing to the presence of a very thick evaporite sequence overlying the target section, and by extreme water depths and drilling depths. These factors have resulted in very long well durations and extremely high exploration costs (the initial Tupi discovery well has been reported to have cost $240 million and taken a year to drill) (Durham 2009).

The Sao Paulo Plateau is an extensive area of continental crust that remained attached to the Brazilian margin as it separated from Africa during asymmetric South Atlantic break-up. The central part of the plateau is a NE–SW trending horst known as the Santos External High (Carminatti *et al.* 2008) or Outer High of the Santos Basin (Gomes *et al.* 2002) (Fig. 12.4). Bounding the Santos External High to the west is a north–south structural low comprising a wedge of oceanic crust trending northwards into a failed rift system (Meisling *et al.* 2001). The axis of stretching is interpreted to have subsequently shifted during continental break-up to a location east of the Sao Paulo Plateau, leaving it as a large piece of stretched continental crust attached to the South American plate (Carminatti *et al.* 2008). The Santos External High may have initially developed as a thermal dome produced by hotspot-related rifting and regional lithospheric thinning, possibly modified by subsequent local magmatic underplating (Gomes *et al.* 2002). To the south, the Sao Paulo Ridge and Florianopolis High are also important structural features, since they are believed to have separated a restricted marine evaporitic basin in the Santos and other basins to the north during Aptian times from open marine conditions in the Pelotas and other basins to the south.

The regional structural framework and history of the area have been fundamental to the development of the sub-salt petroleum system in the Santos Basin. The key elements of the petroleum system (Carminatti *et al.* 2008) are: (i) rich lacustrine source rocks that developed in the syn-rift sequence, complemented by restricted marine source rocks in the pre-salt sag; (ii) over 2 km of regionally extensive upper Aptian evaporitic deposits that covered the syn-rift and sag sequences, providing an excellent regional topseal; (iii) structural traps associated with rifting, formed on a regional intrabasinal high that acted as a focus for hydrocarbon migration out of pre-salt source rocks (Gomes *et al.* 2002); (iv) shallow-marine carbonate reservoirs (coquinas and microbial limestones) in the upper part of the syn-rift and in the sag sequence that developed over the top of the paleotopographic basement high, far from terrigenous siliciclastic input; and (v) a suitable thermal history for hydrocarbon generation, migration and entrapment.

The Outer High of the Santos Basin forms a 12,000 km² four-way dip closure at Aptian (base salt) level, and contains two individual culminations. The northern Tupi structure extends over an area of

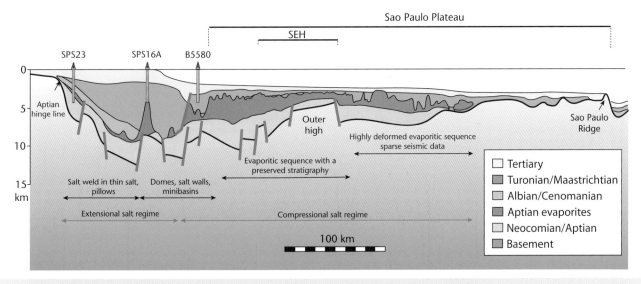

Fig. 12.4 Santos Basin, Brazil: regional geological cross-section extending from the shelf across the Sao Paulo Plateau, showing the prominent Santos External High (SEH), and the thick Aptian salt topseal. From Carminatti *et al.* (2008).

1100 km² at base salt level, and is the location of the Tupi discovery well. The southerly of the two culminations, Sugar Loaf, is larger and represents the crest of the megastructure (Gomes *et al.* 2009).

A critical exploration risk remains reservoir presence and quality. Facies variations take place within the reservoir sequences developed on the evolving carbonate platform during the early Aptian. These result from tectonic and eustatic relative sea-level changes, and from lateral changes between evaporites in the west and shelf-break carbonate build-ups in the east. Subaerial exposure of highs may have led to karstification and reservoir enhancement. On the other hand, erosion of syn-rift volcanics is likely to have given rise to a compositionally immature, poor-quality reservoir. Reservoir complexity represents a major challenge for commercial development of the play (Gomes *et al.* 2009).

12.1.5 Northwest Shelf, Australia (Dampier sub-basin)

The Northwest Shelf (NWS) is a normally charged, vertically drained, high-impedance petroleum system, and is a world-class gas province. Since exploration began in 1953, reserves of 2.6 billion barrels of oil, 2.6 billion barrels of condensate, and over 150 TCF of gas have been discovered in the greater NWS area (by end 2001).

The greater NWS extends from the northern Carnarvon Basin in the south through to the Bonaparte Basin in the north. It developed as a passive margin in response to three phases of continental break-up and seafloor spreading that were initiated: (i) in the Oxfordian in the central area, forming the Argo oceanic basin; (ii) in the Tithonian in the northern area, now subducted under Timor; and (iii) in the Valanginian in the southern area, forming the Cuvier oceanic basin (Longley *et al.* 2003). This pattern of break-up resulted in the drifting away of at least three continental blocks from the Australian craton. This evolution has produced a pre-rift megasequence comprising Permo-Triassic intracratonic sediments, an Oxfordian to Valanginian rift megasequence, and a Cretaceous and younger post-rift megasequence (Fig. 12.5).

The petroleum system description below (Longley *et al.* 2003) refers to the Dampier sub-basin of the northern Carnarvon Basin, home of the existing NWS Gas Project.

The NWS does not have an unambiguous world-class source rock. Oil-source correlation is difficult. There appears to be a range of geochemically similar and generally poor-quality source units that are difficult to identify, or perhaps even remain undrilled, while the oils are thought to have co-mingled during migration. Most sources are gas-prone or lean, consistent with the overwhelming dominance of gas in this area. However, oil-prone source rocks have been identified at various stratigraphic levels in the Dampier sub-basin, associated with the occurrence of small- to medium-sized oilfields. The main oil source is the J40 (Oxfordian–Kimmeridgian) syn-rift anoxic marine shales. The main gas source is the pre-rift deltaic sediments of the Norian TR20 Mungaroo, Rhaetian–Sinemurian J10, and Pliensbachian–Callovian J20 sequences.

A key geological event critical to the maturation of source rocks has been the development of a very thick (up to 4 km) Tertiary prograding carbonate wedge. This late burial has ensured source kitchens are currently at their maximum burial depths and temperatures. In the basinal area of the Dampier sub-basin, the main source units have passed through the oil window in the Late Cretaceous, with Tertiary burial pushing them in to the gas window. Migration out of this relatively unstructured, mud-rich basin has been mainly lateral, at the base of the Cretaceous regional seal. On the adjacent shelf and platform areas though, where the regional topseal becomes much thinner and faulted, migration is vertically focused and the system is of lower impedance. Gas leakage has taken place to shallow stratigraphic levels, where it is visible as bright seismic amplitude anomalies.

The presence and phase of reservoired hydrocarbons is complicated by a number of subsequent processes: (i) gas flushing from a later more mature charge; (ii) biodegradation of light oil into heavy oil and gas (Enfield/Vincent); (iii) water washing (Laminaria, Timor Sea); and (iv) selective gas leakage.

West of the main Oxfordian depocentre, fields occur in large rift-related horst block traps (Goodwyn, North Rankin). Subcrop of

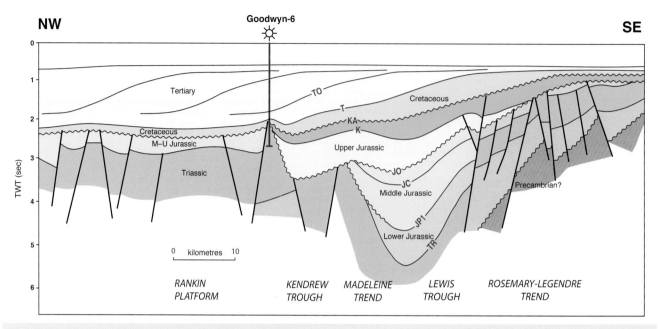

Fig. 12.5 Dampier Basin regional cross-section, NW Australia. A Permo-Triassic pre-rift megasequence is overlain by an Oxfordian to Valanginian rift megasequence, and a Cretaceous to Recent post-rift megasequence. From Longley *et al.* (2003).

tilted TR20 and J10 reservoir units against the Cretaceous regional topseal produces complex fields with many stacked pools with different hydrocarbon–water contacts, and various hydrocarbon phases. Within the central basinal area, oil is reservoired in low-relief drape anticlines over deeper rift structures that sometimes contain gas (Wanaea, Cossack, Lambert, Egret). Fields on the eastern flank of the central basin are less prolific small rollovers, faulted horsts and drape anticlines.

12.2   Unconventional petroleum plays

12.2.1   Introduction

There is no doubt that the production contribution and economic importance of unconventional petroleum is growing as the rate of discovery of conventional petroleum resources has moved into long-term decline. On a global scale, in-place volumes are extremely large, although there are significant challenges to economic production. In the context of basin analysis, we are chiefly interested in unconventional petroleum plays insofar as they shed light on processes of hydrocarbon generation, expulsion, migration, entrapment and alteration that vary from conventional thinking.

In that regard, unconventional oil resources – heavy oils and naturally occurring bitumen – are in essence conventional oils that have achieved such an advanced state of alteration that they require unconventional methods of extraction. Heavy oil differs from light oil on the basis of its high viscosity (>100 centipoises), high density (API gravities of less than 22°), and significant content of asphaltenes (rich in nitrogen, sulphur, oxygen and heavy metal contaminants). Those heavy oils with API gravities of <10° are considered to be 'extra-heavy'. Natural bitumen is chemically similar but is even more viscous (>10,000 cP). The natural bitumen of oil sands is solid at room temperature. Viscosity is extremely sensitive to temperature, so

flow is possible when bitumen is heated; hence the development of specialised extraction technologies such as steam assisted gravity drainage (SAGD) and cyclic steam simulation (CSS). Extraction is either by mining, or by *in situ* methods. Bitumen must be blended or upgraded with diluents (lighter hydrocarbons) before it can be transported by pipeline and used in refineries. We briefly describe the oil sands of Alberta, Canada, and the heavy oils of the Orinoco Basin, eastern Venezuela.

On the other hand, many unconventional gas resources indicate the operation of quite unconventional petroleum systems. They may be classified into the following categories:

- 'Tight gas' is gas contained in low-permeability rocks, often in continuous accumulations in basin-centre settings. The example chosen here is the Mesaverde Group of the Piceance Basin, Colorado.
- 'Shale gas' is gas contained within shales in micropores and fracture networks, and adsorbed onto organic matter surfaces. We describe the example of the Devonian black shales of the Appalachians.
- 'Coal seam gas' or 'coal bed methane' is gas adsorbed onto the surfaces of organic matter within coal seams. We describe this category only very briefly.
- 'Gas hydrates' are accumulations in which gas molecules are encased in a crystal lattice of frozen water, which occur under certain pressure-temperature conditions found in Arctic permafrost and deep-marine environments. Currently, there are no examples of production from gas hydrates.

In addition, the term 'deep gas' is sometimes used to refer to gas contained in rocks beyond conventional drilling depths.

Very large undeveloped resources of unconventional gas occur in the United States, possibly over 300 TCF in the case of tight gas (Energy Information Administration, EIA), over 700 TCF in the case of the Devonian shales of the Appalachian Basin (FERC), and over

160 TCF in coal seam gas (Potential Gas Committee). Domestic natural gas prices in the USA have strengthened significantly over the past decade as conventional supply has waned relative to ongoing demand growth, and this has provided the economic motivation for exploitation of relatively high-cost unconventional sources of gas. As a result, 50% of total USA gas production in 2005 was unconventional (Nehring 2008), most of it tight gas from 'basin-centred' gas plays.

12.2.2 Tight gas

Tight gas is currently produced in a number of North American basins (Fig. 12.6). The 'basin-centred' gas model has evolved to explain a number of observations while drilling in tight gas producing basins (Cumella *et al.* 2008):

- extensive gas shows while drilling, and no gas–water contacts observed on well logs, suggesting gas accumulations that are large-scale and possibly basin-wide;
- little produced water, indicating that reservoirs are close to irreducible water saturation, with much of the produced water deriving from condensation out of the produced gas stream;
- low permeability (less than 0.1 mD), which means that production is more a function of production technology combined with access and gas prices than it is of geology;

- abnormal pressure, indicating formation pore pressures elevated as a result of active hydrocarbon generation within a sealed system. Once generation is switched off, leakage of hydrocarbons by migration or diffusion from the up-dip margin of the accumulation may result in the development of an underpressured rim;
- poorly defined traps and seals, suggesting that conventional gas trapping mechanisms involving the role of gas buoyancy are subtle or not at work.

The requirements for the operation of the play are (Meckel *et al.* 2008): (i) thermally mature gas-prone source rocks, including coals (Rocky Mountain basins), or the thermal cracking of previous oil accumulations (Gulf Coast); (ii) reservoir units in proximity to mature source rocks, precluding the requirement for long-distance migration; and (iii) a volume of gas generated that is large in relation to the reservoir pore space.

12.2.2.1 Cretaceous Mesaverde Group basin-centred gas play, Piceance Basin, Colorado

The gas potential and key geological controls on hydrocarbon distribution in the Mesaverde basin-centred gas play of the Piceance Basin, Colorado, are described by Hood *et al.* (2008) and Yurewicz *et al.* (2008). Today, the Piceance Basin is a strongly asymmetric erosional

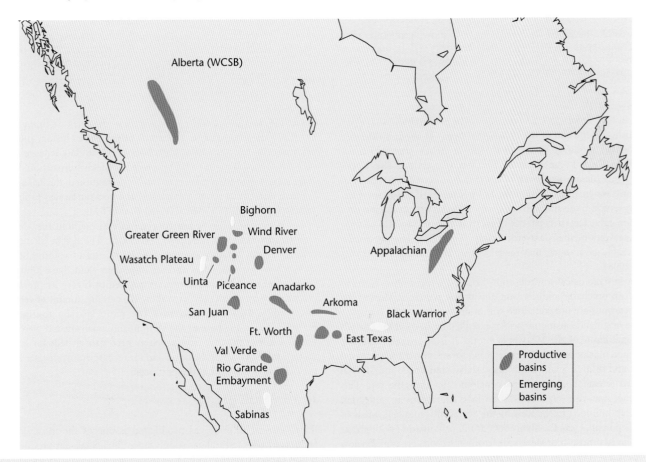

Fig. 12.6 North American basins with production from pervasive tight gas sandstones and emerging tight gas basins at an early stage of evaluation. Modified from Meckel and Thomasson (2008). AAPG © 2008. Reprinted by permission of the AAPG whose permission is required for further use.

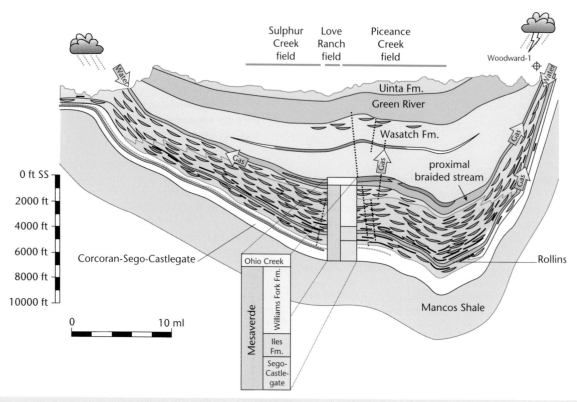

Fig. 12.7 Piceance Basin cross-section, showing regional distribution of gas (orange) and water (blue) within the Mesaverde Group (Yurewicz *et al.* 2008). The Mesaverde is exposed around the perimeter of the basin, and subject to meteoric water influx. Gas is pervasive throughout the Mesaverde, except in the uppermost 300–500 m. AAPG © 2008. Reprinted by permission of the AAPG whose permission is required for further use.

inlier in NW Colorado, USA (Fig. 12.7). The present basin shape was established at the end of the Laramide orogeny at about 40 Ma, and was further modified by regional uplift at 10 Ma, after which the Colorado River began to incise the Tertiary section. The Upper Cretaceous Mesaverde Group is exposed in outcrop and subject to meteoric water influx around the full perimeter of the basin. Currently, several thousand wells in nearly 50 fields are producing over 1 billion cubic feet (bcf) per day of gas from the unconventional tight gas reservoirs of the Mesaverde Group (Fig. 12.8). The potentially recoverable gas resource may range between 100 and 250 TCF (Hood *et al.* 2008).

The key geological controls on hydrocarbon distribution relate to hydrocarbon charge and reservoir quality and geometry. Most of the hydrocarbons are not contained within discrete structural traps.

The original depositional system extended far beyond the current erosional limits of the Mesaverde Group. It is an overall regressive sequence, marine at the base, undergoing a transition through coastal plain and alluvial plain sandstones, shales and coals, into fluvial braided stream sandstones and conglomerates near the top. This sequence was deposited during the Late Cretaceous in a flexural foredeep bounded to the west by the Sevier orogenic belt. Most of the gas potential is in the highly discontinuous Williams Fork Formation fluvial sandstones. Mesaverde Group reservoirs are typically low porosity (2–10%) and very low permeability (<0.1 mD). Hydraulic fracturing, which enhances the naturally occurring fracture set, is critical to successful gas production.

Gas-prone source rocks occur in the marine shales at the base of the section, in the coastal plain coals and in non-marine shales in the

Iles and Williams Fork formations. Of these, the coals are thought to have been the most prolific generators of gas. Burial history modelling indicates gas generation started about 55 Ma (late Paleocene to early Eocene), following rapid burial of the Mesaverde section. Peak generation was probably achieved during deposition of the Paleogene Wasatch and Green River formations, and generation was probably switched off when subsidence ceased about 25 Ma.

Gas is present more or less continuously throughout the Mesaverde Group, except in the upper 300–500 m. The height of the gas column in the Mesaverde appears to be related to volume of gas charge. Low permeabilities and discontinuous sands have inhibited lateral migration. Microfractures propagated by high pore pressures during gas generation may have allowed a certain amount of vertical migration within the Mesaverde. The more laterally continuous marine sands at the base of the section, and amalgamated proximal fluvial sands at the top, have probably been carrier beds for lateral migration of gas out of the basin and for recharge of freshwater from the outcrop areas around the perimeter.

12.2.3   Shale gas

Rapid growth in shale gas production is one of the most salient features of USA natural gas supply since 2005. Shale gas production is now expected by the US Energy Information Administration (EIA) in their Reference Case to grow from 7.8 TCF in 2011 to 16.7 TCF in 2040, representing about 50% of total US domestic dry natural gas production in 2040, and accounting for almost all of the growth in overall US domestic supply to that date (Energy Information Admin-

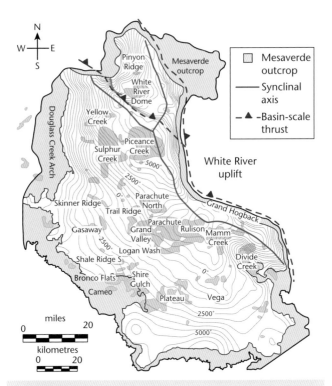

Fig. 12.8 Structure map of the Piceance Basin (top Rollins Member, Iles Formation), showing the wide distribution of fields producing from the Mesaverde Group (Hood and Yurewicz 2008). AAPG © 2008. Reprinted by permission of the AAPG whose permission is required for further use.

istration, 2013a). However, there are major uncertainties over the scale of the resource base, the price level required to sustain its development, and the technical and environmental challenges that need to be overcome.

There are a number of geographically extensive shale gas play types and subtypes in the USA (Fig. 12.9), depending on the source of gas (biogenic, indigenous thermogenic or migrated thermogenic) and type of reservoir (tight or fractured). Examples are the Marcellus shales of the Appalachians, and the Barnett and Eagle Ford (Texas), Haynesville (Texas, Louisiana), Fayetteville (Arkansas), and Woodford (Oklahoma) shale gas plays.

Gas production at economic rates in shale gas plays is generally dependent on horizontal drilling and hydraulic fracturing. Economics may also be improved by associated liquids production.

12.2.3.1 Devonian black shales of the Appalachians

One of the best examples of a North American shale gas petroleum system is the Devonian of the Appalachians (Milici *et al.* 2006). Black shales (for example the Marcellus Shale) were developed during Mid-Devonian to Early Mississippian times in the distal facies of the Catskill Delta, which prograded westwards into a foreland basin during the early stages of the Acadian orogeny. These distal muds are thought to have been deposited in anoxic conditions in the restricted foredeep basin, and in places are extremely organic-rich. Gas is present in micropores in laminated black shales, siltstones and sandstones, and in a network of naturally occurring fractures, as well as adsorbed onto the surfaces of organic matter. Depending on position within the basin, the black shale organic matter passed through the

oil window and became mature for gas generation several hundreds of millions of years ago. This is therefore a highly unconventional petroleum system, in that hydrocarbon distribution is more or less continuous throughout the mature source sequence; reservoir, seal and source are all part of the same self-contained petroleum system; and the hydrocarbon habitat has remained undisrupted for geologically very long periods of time.

12.2.4 Coal seam gas

The thermal maturation of coal source rocks produces a large amount of gas, some of which is stored within the coal. In a similar manner to shale gas, coal bed methane is a self-contained petroleum system where the coal sequence performs the roles of source, reservoir and seal. Most of the gas is trapped on the surfaces of organic matter. Gas content generally increases with the maturity of the coal. Fractures or cleats within the coal are normally filled with water.

Coal seam gas (CSG) or coal bed methane (CBM) is of increasing economic importance. CSG resources are widespread, commonly distributed across the entire coal basin, and in-place resources are very large (estimated at over 700 TCF in USA, of which less than 100 TCF may be economically recoverable). However, production rates are highly variable, and economic success depends on identifying production sweet spots. CSG production also has significant environmental challenges, particularly the disposal of large quantities of saline produced water.

CSG is an established form of gas production in many USA basins, such as the San Juan and Powder River. In Queensland, Australia, in addition to already supplying 80% of the state's domestic gas market, it is emerging as an important gas feedstock for LNG processing and export, mainly around Gladstone.

12.2.5 Gas hydrates

Gas hydrates are formations comprising methane and other small gas molecules encased in a crystal lattice or 'cage' of frozen water. They were first discovered in Arctic permafrost areas, where they occur at depths of 130–2000 m, but they also occur in sediments at the high-pressure/low-temperature conditions (Fig. 12.10) found in deep-marine areas at 100–1100 m below the seafloor, where they are inferred from anomalous seismic reflectors. Current estimates of the volumes of gas contained in the world's hydrate accumulations are highly uncertain but possibly of the order of 20,000 TCF, greatly exceeding conventional gas reserves. While of undoubted geological interest, little is known about the production technologies and economic resource potential of these accumulations. Significant technology and capital investment will be required before the very large in-place resource can be considered exploitable.

By 2001, hydrates had been inferred to occur at about 50 locations worldwide (Fig. 12.11). Some of the best studied are described by Collett (2001). These include Blake Ridge (offshore SE USA), along the Cascadia continental margin off the Pacific coast of the USA, on the North Slope of Alaska, and in the Mackenzie River Delta of northern Canada. Gas hydrates are also considered widespread in West Africa on the continental slope offshore the Niger and Congo River deltas, on the basis of extensive bottom-simulating seismic reflectors (BSRs) (Cunningham *et al.* 2000) (Fig. 12.12). BSRs reported at numerous locations suggest the South Atlantic, in general, may prove to be a rich gas hydrate province. In addition to seismic BSRs, geological evidence for gas hydrates includes seafloor mounds (Fig. 12.13), pockmarks, mud volcanoes, and submarine slumps and

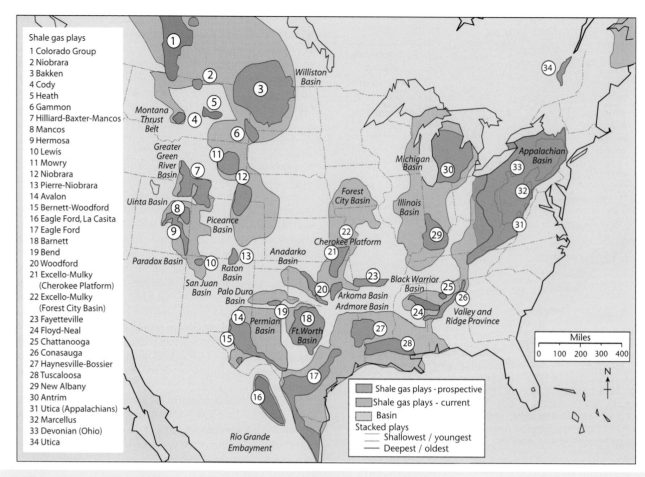

Fig. 12.9 North American shale gas basins and plays (US Energy Information Administration, 2013b). Most US shale gas appears to be located in the Devonian shales (such as the Marcellus) of the Appalachian Basin. The Fayetteville shale of Arkansas and the Texan Barnett and Eagle Ford shales are among other important shale gas plays. Energy Information Administration (EIA) 2013b.

slides, while hydrates have also been directly sampled in piston cores taken at several locations offshore the Niger Delta.

Most gas hydrates are uniformly distributed through sediments as a pore-filling constituent (Booth *et al.* 1996). The formation and occurrence of gas hydrates are controlled by (Collett 1995): (i) formation temperature and formation pore pressure, defining the gas hydrate stability zone; (ii) the availability of gas and water via migration pathways; (iii) the presence of reservoir rocks and seals; and (iv) gas chemistry and pore water salinity. The assessment of gas hydrate potential therefore calls upon many of the concepts and skills used in exploration for conventional petroleum. However, it is also recognised that gas hydrates may form through processes that are not typical for conventional petroleum, particularly in shallow unconsolidated marine sediments where hydrates may mechanically displace sediment to form nodules or massive gas hydrate units.

Elevated organic carbon concentrations (>0.5% TOC) in surficial sediments may be present in areas of high biological productivity such as oceanic upwelling zones or where nutrient supply is elevated offshore major delta systems, such as the Niger and Congo deltas (§11.3). In these areas, gas hydrates may form as a result of the microbial breakdown of organic matter to generate biogenic gas. However, in the case of the Niger and Congo examples, BSRs also appear to be related to areas of recent sediment loading and structural deformation, including the contractional mobile shale belt

off the Niger Delta and the zone of active salt tectonics along the Congo Delta front. This association suggests structuring may have provided a migration focus for biogenic gas generated in and close to the gas hydrate pressure-temperature stability zone, as well as for thermogenic gas generated at deeper levels (Cunningham *et al.* 2000). Carbon isotope analyses indicate that the methane in many oceanic hydrates is biogenic in origin, while molecular and isotopic analyses indicate a thermogenic origin for the methane in several offshore Gulf of Mexico and onshore Alaska examples (Collett 2001).

Enthusiasm over the enormous scale of the gas hydrate resource must be tempered by uncertainties over how it can be technically and economically recovered. Production methods proposed usually involve dissociating the gas by heating or depressurising the hydrate in order to move it outside of its PT stability zone, or by injection of an inhibitor such as methanol or glycol. All current methods appear to be prohibitively expensive, but, nevertheless, the governments of countries such as Japan, India and the USA regard hydrates as a potentially very significant long-term energy resource.

12.2.6 Oil sands and heavy oil

Oil sands (natural bitumen) and heavy oils are conventional oils that have been subsequently degraded. They differ from light oil on the basis of their high viscosity, high density, and significant content of

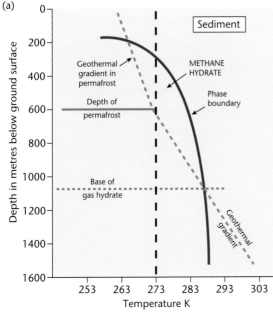

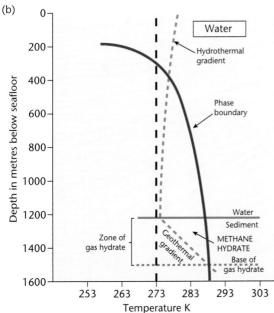

Fig. 12.10 Depth-temperature relationships for gas hydrate stability in (a) a permafrost region, and (b) an outer continental margin marine setting (Collett 2001).

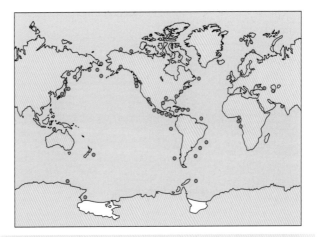

Fig. 12.11 Deep-water distribution of known and inferred gas hydrate deposits (Cunningham & Lindholm 2000). AAPG © 2000. Reprinted by permission of the AAPG whose permission is required for further use.

asphaltenes (rich in nitrogen, sulphur, oxygen and heavy metal contaminants). Heavy oils have API gravities of <22°, and viscosities of greater than 100 cP, while extra-heavy oil is that portion with API gravities of <10. Natural bitumen is chemically similar but is even more viscous (>10,000 cP).

These oils result from the same processes of petroleum generation, expulsion, migration, entrapment and alteration described elsewhere in this chapter. In this sense, they represent conventional petroleum systems. However, heavy oil and oil sands have experienced such an advanced state of alteration that they require unconventional methods of extraction. Volumes are extremely large, in numerous locations worldwide (Attanasi *et al.* 2007). Estimated original reserves

of heavy oil and bitumen in known accumulations total 76 billion barrels and 250 billion barrels respectively at end 2005, as estimated by the US Geological Survey (Meyer *et al.* 2007). The statistics are dominated by two regions in particular. The largest natural bitumen deposit is the Canadian oil sands accumulation of Alberta, making up over 70% of the world's initial recoverable bitumen reserves. Over 90% of the world's extra-heavy oil reserves reside in the Orinoco Basin of eastern Venezuela.

12.2.6.1 Alberta oil sands

The Alberta oil sands represent a supercharged, laterally drained, low-impedance petroleum system. They have been a source of energy in Canada since the 1990s, and crude bitumen production now exceeds conventional crude production in Alberta. By 2015, the government of Alberta expects oil sands production to account for 65% of total Canadian crude oil output, and 17% of total North American supply (Alberta Geological Survey 2010). Volumetric estimates of the oil in place in the Alberta oil sands are vast, almost 1.7 trillion barrels. 170 billion barrels are considered to be recoverable with existing technology, placing Canada second only to Saudi Arabia in terms of proven reserves.

The oil sands deposits of Alberta occur in the Peace River, Athabasca, and Cold Lake regions (Fig. 12.14), mainly in the fluvio-estuarine deposits of the Lower Cretaceous Mannville Group. These sediments lie on a major sub-Cretaceous unconformity that erodes down into strata ranging from the Triassic to the Devonian. Oil sands also occur to a lesser extent in subcropping Devonian carbonates. In addition, there is heavy oil production from the Mannville in the Lloydminster area.

The Mannville Formation sands form part of a thick sequence of deltaic, fluvial and swampland sediments that was deposited in a flexural foredeep basin produced by the loading of thrust sheets in Late Jurassic–Early Cretaceous times during the Columbian orogeny. Sediment supply was from the eroding thrust sheets in the west, and from the Precambrian shield to the east. Further Late Cretaceous–Tertiary shortening in the Cordilleran region during the Laramide orogeny supplied large quantities of clastics to the marginal foredeep, driving further flexural subsidence and sediment loading. This phase

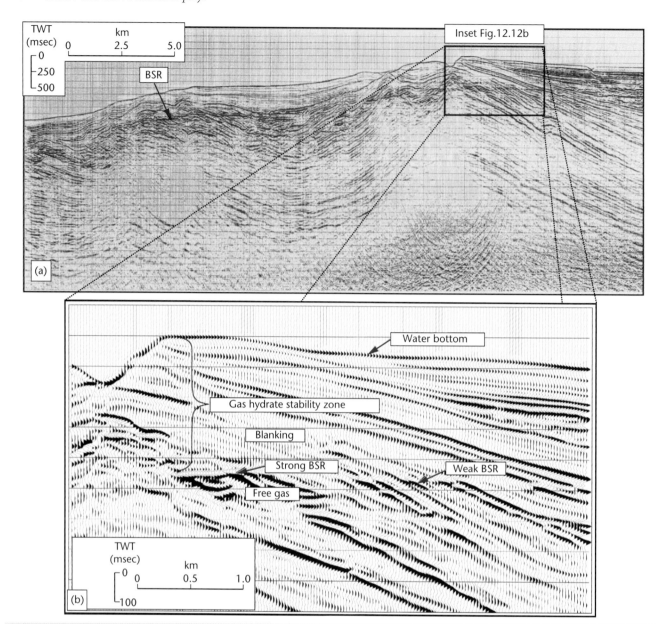

Fig. 12.12 (a) Bottom-simulating seismic reflectors, Nigeria continental slope, and (b) detail showing seismic character of the gas hydrate stability zone (Cunningham & Lindholm 2000). AAPG © 2000. Reprinted by permission of the AAPG whose permission is required for further use.

of basin evolution was the critical moment for the petroleum system, being responsible for the maturation of pre-Cretaceous and Lower Cretaceous source rocks and the long-distance eastward lateral migration of very large quantities of petroleum.

The precise origin of the oils has been the subject of considerable debate over many years (Stanton 2004), but biomarker analysis of oils and source rocks has established the source and migration history of the Western Canada sedimentary basin (Creany *et al.* 1990). It is clear that the Mannville oil sands are the result of alteration of conventional crudes that migrated large distances from source kitchens to the west. According to Creany *et al.* (1990), several pre-Cretaceous source rocks have contributed to the Mannville Group oil sands, but the most important is the Lower Jurassic Nordegg Formation. The Nordegg Formation, although often <20 m thick, comprises an extremely organic-rich source rock containing highly oil-prone

Type II kerogen, with TOCs of up to 27% and hydrogen indices of up to 600. The Nordegg is mature for oil generation over most of its area of occurrence, and overmature in the far west. A liquid hydrocarbon mass balance (Creany *et al.* 1990) suggests that the huge oil sands in-place volumes can be realistically explained by the source rock quality, history and geological setting of the basin. The highly porous and permeable Mannville sands acted as a migration conduit for oil moving through the subcropping pre-Cretaceous rocks, while the Joli Fou shale provides a regional topseal (Fig. 12.15). 2-D modelling studies (Higley *et al.* 2006) indicate northeastward migration of both oil and gas towards the present-day location of the oil sands deposits (Fig. 12.16).

Meteoric water influx into the sandy outcropping Manville Formation is responsible for biodegradation of the conventional crudes into bitumen and heavy oil.

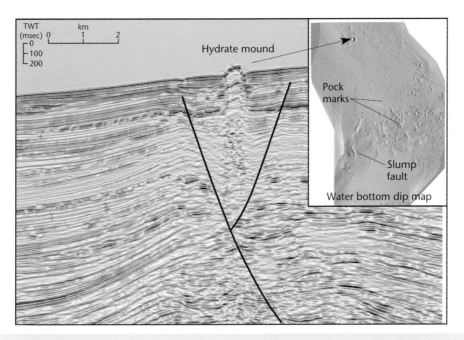

Fig. 12.13　Hydrate mound and pock marks on the continental slope, offshore Nigeria (Cunningham & Lindholm 2000). AAPG © 2000. Reprinted by permission of the AAPG whose permission is required for further use.

12.2.6.2　Orinoco Basin, eastern Venezuela

The Orinoco Basin petroleum system is supercharged, laterally drained, and of low impedance. The US Geological Survey estimates in their 2000 World Petroleum Assessment that the Orinoco oil belt assessment unit of the East Venezuela province contains undiscovered resources of over 500 billion barrels (USGS 2000, 2009). It represents the largest single oil accumulation in the world. The heavy oils have a range of API gravities between 4–16°, and viscosities of 2000–8000 cP.

The source rock for the heavy oils is the Upper Cretaceous Querecual Formation, the stratigraphic equivalent of the La Luna Formation. The reservoirs are the fluvio-deltaic sandstones of the lower Miocene Oficina Group.

As for the Alberta oil sands, the critical moment for the petroleum system was a regional plate tectonic event, which created a thrust belt and adjacent foreland basin and flexural forebulge (Jacome *et al.* 2003) (Fig. 12.17). Initially, the Albian–Turonian Querecual Formation marine pelagic algal-rich calcareous mudstone source rocks were developed on a north-facing passive margin prior to Eocene collision (see §6.3.4). The progressive eastward oblique collision of the Caribbean plate with the passive margin of northern South America, commencing in the Eocene in the west and continuing to the present day, formed a transpressional thrust belt. Lithospheric flexure resulting from the tectonic loading formed an adjoining foredeep together with a flexural forebulge on the Guyana shield to the south. This area of uplifted shield became a source of sediment for rivers draining into the foredeep from the south, and was responsible for the deposition of the chief reservoirs for the heavy oil play. From Oligocene times on, Cretaceous (and possibly older) source rocks buried under the thrust sheets and foreland basin generated large quantities of crude oil that migrated several hundred kilometres to the south through Upper Cretaceous and Miocene carrier beds, until trapped in Miocene clastic rocks in fault traps and stratigraphic pinch-outs against the Guyana shield in the region of the forebulge (Bartok 2003).

Shallow depth of burial and meteoric water influx have contributed to biodegradation of the reservoired oils.

12.3　Geosequestration: an emerging application

As concern in recent times has grown over the increasing concentrations of atmospheric greenhouse gases and their potential impact on climate change, the potential for the geological sequestration of carbon dioxide has grown in interest. CO_2 geosequestration, part of the broader process of carbon capture and storage (CCS), represents one of the options available to reduce the volumes of greenhouse gases that would otherwise be emitted to the atmosphere by power generation plants and industrial processes. A power plant with CCS could typically reduce net CO_2 emissions by 80–90% compared to the same plant without CCS. In effect, CCS provides a means of reducing CO_2 emissions while allowing the continued operation of fossil-fuel-based power generation, thereby buying some time for an orderly and economically non-disruptive transition to renewable and nuclear power generation. As a sign of the increasing focus and investment being made, many field demonstration projects are now underway worldwide to investigate the feasibility and safety of CO_2 geosequestration.

There are a number of potential geological sinks for CO_2 (Fig. 12.18), but the most mature technology for sequestration is associated with enhanced oil recovery (EOR) in existing fields, for which technical and economic feasibility is already proven. CO_2 sequestration in coal beds associated with enhanced coal seam methane recovery is another area currently being demonstrated. Overall, the biggest challenge to CCS is the cost of capture, but there are also a number of important geological challenges impacting subsurface storage (Grobe *et al.* 2009). Firstly, geological identification and evaluation of potential storage sites that have both sufficient storage capacity and proximity to CO_2 sources is critical. Secondly, geological

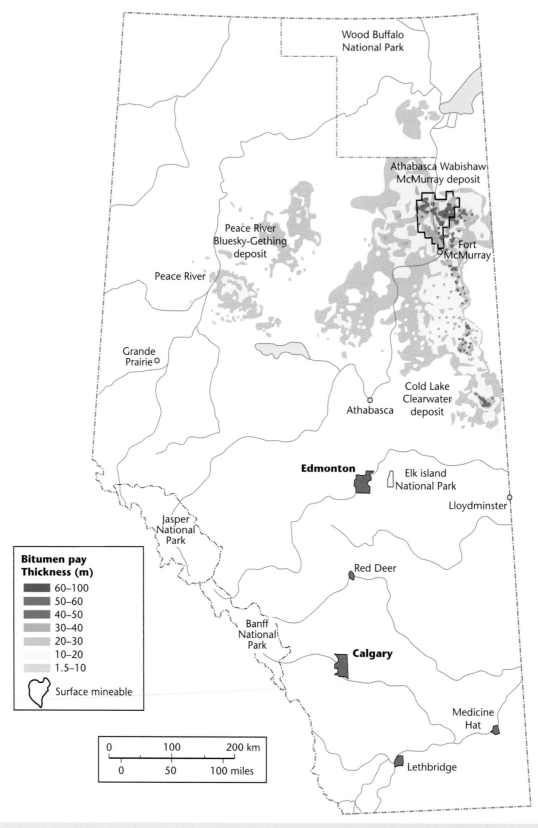

Fig. 12.14 Location of the Athabasca tar sands, Alberta, Canada (courtesy of Energy Resources Conservation Board, Alberta Geological Survey 2010). The contours are bitumen pay thickness, which is greatest in the Fort McMurray and Cold Lake areas.

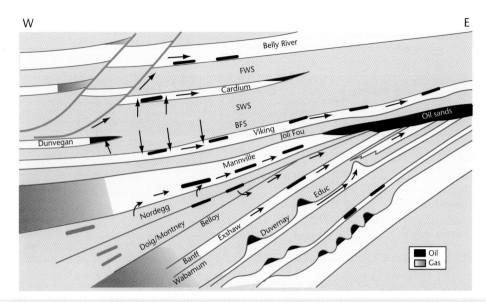

Fig. 12.15 Schematic cross-section of the Western Canada Sedimentary Basin showing migration routes in the pre-Joli Fou section. Modified from Creany & Allan (1990).

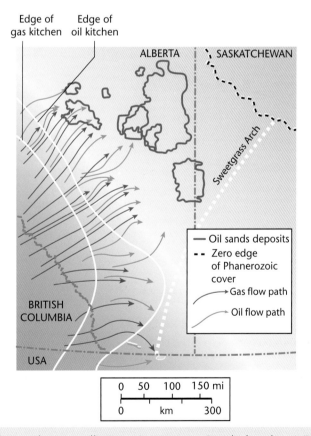

Fig. 12.16 Hydrocarbon flowpaths into the Mannville Formation, Western Canada, based on 2-D modelling studies (modified from Higley *et al.* 2006). The Mannville is buried to over 3000 m below sea level in the western part of the map in British Columbia. Flowpaths are shown emerging from Mesozoic oil and gas kitchens, primarily in a northeasterly direction, towards the known oil sand deposits in northern Alberta. Reproduced courtesy of U.S. Geological Survey.

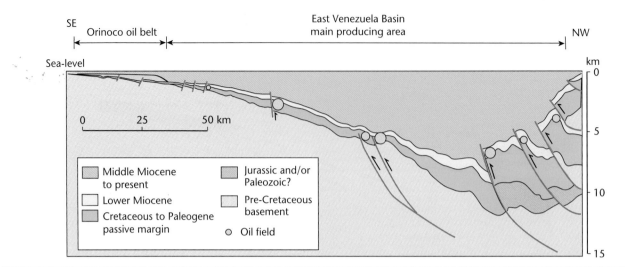

Fig. 12.17 Schematic cross-section of the East Venezuela Basin, showing the up-dip position of the Orinoco Oil Belt relative to the East Venezuela Basin. Oil generated from thermally mature Cretaceous and possibly older source rocks in the deeper part of the basin migrated up-dip to form the accumulations in the Orinoco Oil Belt (Jacome *et al.* 2003). AAPG © 2003. Reprinted by permission of the AAPG whose permission is required for further use.

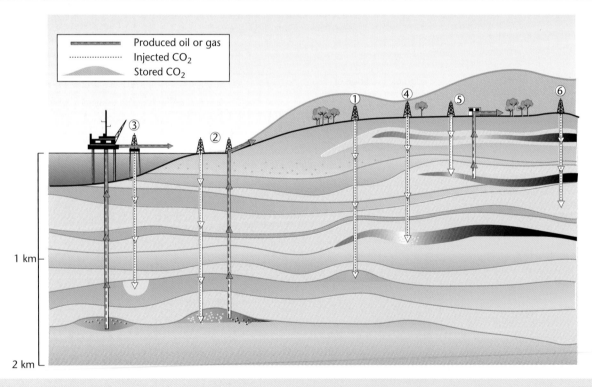

Fig. 12.18 Schematic representation of CO_2 geological storage options (Kaldi *et al.* 2009, courtesy of Australian CO2CRC). 1, depleted oil and gas reservoirs; 2, enhanced oil recovery; 3, deep saline water-saturated aquifers; 4, deep unmineable coal seams; 5, enhanced coalbed methane recovery; 6, other options such as basalts, oil shales, cavities. AAPG © 2009. Reprinted by permission of the AAPG whose permission is required for further use.

assessment has an important role to play in identifying, mitigating and remediating the risks associated with possible CO_2 leakage.

There are a number of key questions that must be addressed by the geological assessment of potential storage sites (Kaldi 2010). Firstly, can we put the CO_2 into the rock? (a question of *injectivity*). Secondly, can we keep the CO_2 in the rock? (a question of *containment*). And, thirdly, what volume of CO_2 can the rock hold? (a question of *capacity*). Fortunately, the geoscience evaluation practices and

skills developed by the petroleum exploration and production industry over many years are highly relevant to addressing these questions. Furthermore, the industry already has considerable relevant experience in natural gas storage (50 years), CO_2-EOR (30 years) and acid gas disposal (20 years).

Potential geological storage options are numerous and large scale. There is likely to be storage capacity of 675–900 gigatonnes of CO_2 in existing oil and gas fields alone (IPCC 2005). The capacity of enhanced coal seam recovery schemes in unmineable coal seams is difficult to quantify but is possibly as much as 200 Gt, while deep saline formations offer an almost unlimited storage capacity (1000–10000 Gt).

Storing CO_2 in existing known depleted oil and gas fields (or still-producing fields in the case of EOR) has many advantages over other storage options. Firstly, the containment potential is already proven effective for oil or gas, over a geological time scale. If the capillary pressure of a seal was effective in supporting a hydrocarbon column over geological time, then it theoretically should support a column of CO_2 at similar temperatures and pressures (Daniel *et al.* 2009). If upward migration of CO_2 were to occur, it would be at extremely low rates. CO_2 injected at the Norwegian Sleipner Field will take over half a million years to reach the seafloor via diffusion (Arts *et al.* 2009). CO_2 injection at Sleipner commenced in 1996, and it is estimated that by 2009 over 8 million tonnes of CO_2 were contained in the reservoir (Fig. 12.19). Secondly, large amounts of geological and engineering data are typically available in existing oil and gas fields for detailed site characterisation. Thirdly, in the case of EOR, there is also an obvious economic benefit from additional oil recovery.

In existing oil and gas fields, and in deep saline formations, the injected CO_2 initially occupies the pore space of the carrier or reservoir rock, and, since an immiscible CO_2 phase is lighter than formation water, it is subject to the same physical principles, and behaves in a similar manner, to oil and gas. In the case of storage in coal seams, the physical process is one of adsorption, and is not discussed further here.

There are a number of trapping mechanisms for CO_2 in the subsurface (Kaldi 2010) (Fig. 12.20): (i) structural or stratigraphic trapping of the free-phase immiscible CO_2; (ii) hydrodynamic traps; (iii) as a low-saturation residual gas, trapped in small pores by capillary forces; (iv) in solution in formation waters; and (v) as a result of permanent precipitation of carbonate minerals through interaction of the CO_2 with formation water and the mineral matrix.

An optimum geosequestration site may combine a number of these trapping mechanisms, as exemplified by the Kingfish area of the Gippsland Basin, Australia (Gibson-Poole *et al.* 2009) (Fig. 12.21). The Gippsland Basin is the site of Australia's premier oil-producing province. The basin has been producing since the 1960s. At the present day, depletion and decommissioning of some of the fields coincides with the desirability for storage of CO_2 generated by anticipated new coal-fired power developments in the adjacent onshore Latrobe Valley. A detailed understanding of CO_2 migration pathways away from potential injection targets is critical to the development of an injection strategy that maximises the storage efficiency of the CO_2. The Kingfish example is a detailed study of stratigraphy, reservoir heterogeneity and seal capacity. Intraformational seals within the upper Latrobe Group are likely to encourage lateral migration of CO_2 into numerous localised traps, reducing dependence on the Lakes Entrance Formation regional topseal. The more tortuous migration path, and increased pore volume accessed, increase the potential for residual gas trapping and dissolution along the migration pathway and provides more time for geochemical precipitation of carbonate minerals. The volume of CO_2 that may be safely stored could significantly exceed the storage capacity suggested by structural/stratigraphic trapping alone.

In identifying and characterising potential storage sites, the principal objective is to understand the distribution of reservoir/seal couplets through a structural and stratigraphic assessment of the area. A good storage site will have reservoir units with attractive

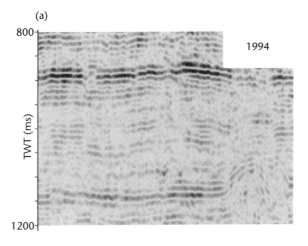

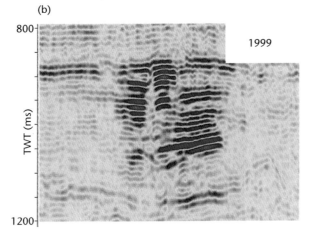

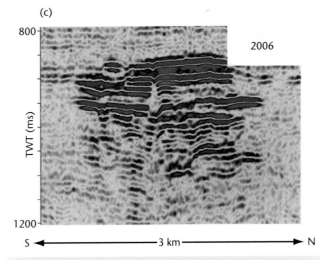

Fig. 12.19 Time-lapse seismic data from the Norwegian Sleipner Field, showing progressive development of the highly reflective CO_2 plume. Panel (a) shows seismic data in 1994, prior to the commencement of CO_2 injection operations in 1996. The other panels show the plume in 1999 (b), a few years after first injection, and in 2006 (c). By 2009, over 8 million tonnes of CO_2 were contained in the reservoir. Modified from Arts *et al.* (2009). AAPG © 2009. Reprinted by permission of the AAPG whose permission is required for further use.

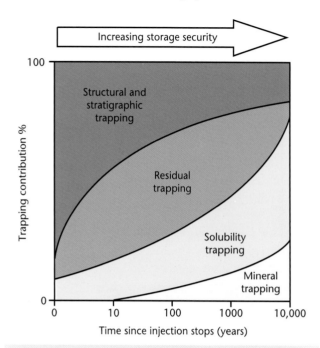

Fig. 12.20 CO$_2$ storage trapping mechanisms (Kaldi *et al.* 2009). In the long term (thousands or tens of thousands of years), trapping through permanent precipitation of carbonate minerals and by solution in formation waters offer the highest storage security. AAPG © 2009. Reprinted by permission of the AAPG whose permission is required for further use.

storage capacity, coupled with extensive and effective seals. Evaluation starts with regional characterisation that assesses the suitability of the basin, and continues through to the identification of prospective sites and detailed site evaluation (Kaldi *et al.* 2009). Detailed site evaluation requires an accurate assessment of the structural and stratigraphic configuration of traps, faults, and reservoir and seal lithologies. This requires the application of sedimentological and genetic stratigraphic principles to determine the distributions of lithologic facies at the basin scale and to characterise the reservoir/seal at specific sites.

Leakage of CO$_2$ from the intended site of storage to other locations where it may contaminate groundwater supplies or interfere with existing petroleum operations is a critical risk in the assessment of geological storage sites. As a result, regional evaluation of containment potential is one of the most important activities that can be undertaken. In a study from Australia's Gippsland Basin, Divko *et al.* (2010) assessed a number of seal attributes (thickness and depth, seal capacity from mercury-injection capillary pressure measurements and seal mineralogy), and assembled empirical evidence of seal failure (soil gas geochemical anomalies, oil slicks, gas chimneys) at the basin scale. The authors found a strong correlation between seal thickness, depth to seal base, and the smectite content of the seal, with seal capacity greatest (>150 m supportable column height) below depths of 700 m and where smectite content exceeds 70%. Hydrocarbon leakage/seepage, onshore and offshore, was clearly related to fault distribution and areas of poor seal capacity.

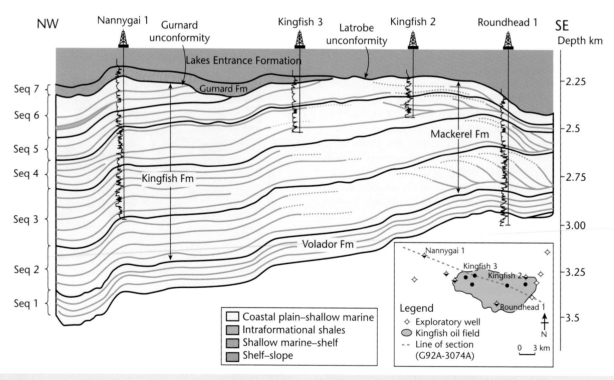

Fig. 12.21 Kingfish area case study, Gippsland Basin, Australia. Cross-section showing intraformational seals, based on sequence stratigraphy, wireline log motifs, and seismic appearance (Gibson-Poole *et al.* 2009). Detailed sequence interpretation is required to understand CO$_2$ trapping possibilities. AAPG © 2000. Reprinted by permission of the AAPG whose permission is required for further use.

Appendices: derivations and practical exercises

1

Rock density as a function of depth

Learning points: rocks increase in density with depth in the Earth due to their compressibility, but they expand due to heating. Knowing the result of this trade-off is important for assessing the internal dynamics of the Earth.

The densities of rocks depend on pressure, so rock densities are commonly expressed as a zero-pressure value (Table A1.1). To work out the increase in density with increasing depth within the Earth, we assume an isotropic state of stress:

$$p = K\Delta = \frac{1}{\beta}\Delta \qquad [A1.1]$$

where p is the pressure, Δ is the dilatation (where each of the principal strains is equal to one third of the dilatation), and K is the bulk modulus – the reciprocal of the compressibility β. Expressed in terms of the material properties used in problems of elasticity,

$$K = \frac{1}{\beta} = \frac{E}{3(1-2v)} \qquad [A1.2]$$

where E is Young's modulus and v is Poisson's ratio.

The density change $\delta\rho$ of a solid element with volume V is related to the fractional change of volume δV, which is equal to the dilatation Δ. Since there is no change of mass,

$$\delta(\rho V) = 0 \qquad [A1.3]$$

which gives

$$\rho\delta V + V\delta\rho = 0 \qquad [A1.4]$$

or

$$\frac{-\delta V}{V} = \Delta = \frac{\delta\rho}{\rho} \qquad [A1.5]$$

Consequently, combining eqns [A1.1] and [A1.5] we obtain an expression for the isothermal change in density in terms of the pressure, the initial density and the compressibility:

$$\delta\rho = \rho\beta p \qquad [A1.6]$$

The compressibility of common rocks is found in Table A1.1.

Table A1.1 Physical properties of common rock types

Rock type	Zero-pressure density $kg\,m^{-3}$	Young's modulus $E\,10^{11}\,Pa$	Poisson's ratio v	Compressibility $\beta\,10^{-11}\,Pa$	Bulk modulus $K\,GPa$
Granite	2650	0.55	0.225	3.0	33.3
Diorite	2800	0.7	0.265	2.0	49.6
Basalt	2950	0.7	0.225	2.4	42.4
Peridotite	3250	1.5	0.25	1.0	100

The density of rock increases with depth below the surface of the Earth. Ignoring initially the effects on density of thermal expansion, at 1 GPa, equivalent to a depth of 38.5 km in granite, the density of granite has increased from 2650 (zero-pressure value) to $2730\,kg\,m^{-3}$. Assuming the mantle lithosphere to be peridotitic in composition, peridotite increases in density from $3282\,kg\,m^{-3}$ at 1 GPa (under 38.5 km of granite) to $3477\,kg\,m^{-3}$ at 7 GPa, equivalent to a depth of 205 km, close to the base of the thermal lithosphere in continental interiors.

Now we incorporate the effects of thermal expansion on rock density. As a reminder, the isothermal compressibility β is a material's fractional change in volume with pressure *at a constant temperature*. The volumetric coefficient of thermal expansion is the fractional change in volume with temperature *at a constant pressure*. Typical values of the volumetric coefficient of thermal expansion are given in Table A1.2.

Table A1.2 Volumetric coefficient of thermal expansion for different lithologies

Peridotite	$2.4 \times 10^{-5}\,K^{-1}$
Granite	$2.4 \times 10^{-5}\,K^{-1}$
Gabbro	$1.6 \times 10^{-5}\,K^{-1}$
Limestone	$2.4 \times 10^{-5}\,K^{-1}$
Sandstone	$3.0 \times 10^{-5}\,K^{-1}$

Basin Analysis: Principles and Application to Petroleum Play Assessment, Third Edition. Philip A. Allen and John R. Allen.
© 2013 John Wiley & Sons, Ltd. Published 2013 by John Wiley & Sons, Ltd.

Since rocks are compressible (they can change in volume), they change in density due to changes in both temperature and pressure. When there is no change in pressure (isobaric), the density of a rock depends on its density at a reference temperature such as 0 °C, the volumetric coefficient of thermal expansion α_v and the temperature relative to the reference temperature

$$\delta\rho = -\rho^* \alpha_v \Delta T \qquad [A1.7]$$

which can be rewritten

$$\rho = \rho^* (1 - \alpha_v \Delta T) \qquad [A1.8]$$

Taking the volumetric coefficient of thermal expansion as $2.4 \times 10^{-5} \, \mathrm{K}^{-1}$ for granite, a linear geothermal gradient of 10 °C km^{-1}, and a density at a reference temperature of 0 °C of 2650 kg m^{-3}, the density at a new temperature of 385 °C, which is approximately the Moho temperature, is 2625 kg m^{-3}. For peridotite, using the same volumetric thermal expansion coefficient of $2.4 \times 10^{-5} \, \mathrm{K}^{-1}$, a density at 0 °C of 3250 kg m^{-3}, the density at a depth of 38.5 km would be 3220 kg m^{-3}, and the density at a depth of about 200 km would be 3094 kg m^{-3}.

The effect of pressure and temperature on density is shown graphically in Fig. 2.15.

These results were obtained with a simple, linear geothermal gradient and will differ with more realistic geothermal gradients incorporating radiogenic heating.

Airy isostatic balance

Learning points: isostatic balances are made in a wide range of problems in the Earth sciences. The simplest is an Airy or 'local' model where it is assumed that the pressure at the base of an imaginary column of rock can be compared with another distant 'reference' column. We mostly perform Airy isostatic balances in evaluating the effect on topography of thickness variations of rock units of known density below the surface.

Continental mountain belts and plateaux, continental platforms, ocean abyssal plains and mid-ocean ridges are all elevated at certain heights or bathymetries dependent on their underlying density struc-ture. Most problems of this type can be approached using a vertical balance of pressure at a level of compensation, known as Airy isostasy.

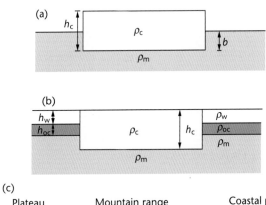

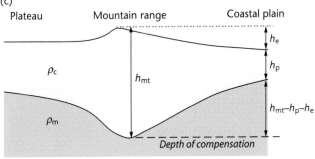

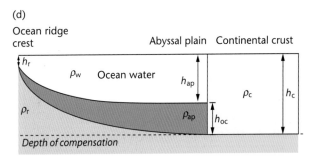

Fig. A2.1 Schematic diagrams illustrating the concepts of Airy isostasy: (a) continental block 'floating' in a fluid mantle, (b) con-tinental block flanked by oceanic crust and overlying water column, (c) continental mountain belt with a root surrounded by a plateau and low-lying plain, and (d) mid-ocean ridge and abyssal plain adjacent to a continental block with its surface at sea level.

Basin Analysis: Principles and Application to Petroleum Play Assessment, Third Edition. Philip A. Allen and John R. Allen.
© 2013 John Wiley & Sons, Ltd. Published 2013 by John Wiley & Sons, Ltd.

The surface force acting on a unit area at the base of a vertical column of rock is given by

$$\sigma_{yy} = \rho g y \qquad [A2.1]$$

where ρ is the density of the rock column, g is the acceleration of gravity and y is the height of the column. For equilibrium to be maintained, we equate the surface forces due to the differing rock columns at a depth of compensation. This depth of compensation is normally chosen as the deepest level at which there are strong lateral density differences in the overlying rock.

For example, the pressure can be calculated under a thick piece of continental crust of thickness h which is 'floating' in the surrounding mantle (Fig. A2.1a). The isostatic balance at a depth equivalent to the base of the continental block is expressed as

$$h\rho_c g = b\rho_m g \qquad [A2.2]$$

where b is the thickness of mantle above the depth of compensation, and ρ_m and ρ_c are the mantle and crustal densities. The elevation of the continental block above the mantle is given by

$$h - b = h\left(1 - \frac{\rho_c}{\rho_m}\right) \qquad [A2.3]$$

Putting $\rho_c = 2700\,\mathrm{kg\,m^{-3}}$, $\rho_m = 3300\,\mathrm{kg\,m^{-3}}$, and $h = 35\,\mathrm{km}$, the elevation of the continental block above the mantle is 6.4 km.

For the case of a continental block with its surface at sea level flanked by oceanic crust and an overlying water column (Fig. A2.1b), the isostatic balance at a depth equivalent to the base of the continental block is

$$h_c\rho_c g = h_w\rho_w g + h_{oc}\rho_{oc} g + (h_c - h_w - h_{oc})\rho_m g$$
$$h_{oc}(\rho_m - \rho_{oc}) + h_w(\rho_m - \rho_w) = h_c(\rho_m - \rho_c) \qquad [A2.4]$$

Putting $\rho_c = 2700\,\mathrm{kg\,m^{-3}}$, $\rho_{oc} = 2950\,\mathrm{kg\,m^{-3}}$, $\rho_m = 3300\,\mathrm{kg\,m^{-3}}$, $h_c = 35\,\mathrm{km}$, and $h_{oc} = 20\,\mathrm{km}$, the water depth is 6 km. This is roughly the depth of the ocean we should expect above the oceanic abyssal plains.

For a mountain belt with a root with an elevation h_e above a coastal plain (Fig. A2.1c), the isostatic balance at a depth equivalent to the base of the continental root is

$$h_{mt}\rho_c g = h_p\rho_c g + (h_{mt} - h_p - h_e)\rho_m g$$
$$h_{mt} = h_p + h_e\left(\frac{\rho_m}{\rho_m - \rho_c}\right) \qquad [A2.5]$$

where h_p is the thickness of the continent below the coastal plain, h_{mt} is the thickness of the continent below the mountain. Putting $h_p = 35\,\mathrm{km}$, $h_e = 5\,\mathrm{km}$, $\rho_m = 3300\,\mathrm{kg\,m^{-3}}$, $\rho_c = 2700\,\mathrm{kg\,m^{-3}}$, the thickness of the crust in the mountain belt h_{mt} is 62.5 km.

If the average density of the hot, young asthenosphere at a mid-ocean ridge is ρ_r, and the average density of the cold, old oceanic lithosphere beneath an oceanic abyssal plain is ρ_{ap}, the depth of the ridge crest below sea level h_r and the depth of the ocean above the abyssal plain h_{ap} can be calculated assuming that these regions are in isostatic balance with a piece of continental crust of thickness h_c with its surface at sea level and with density ρ_c (Fig. A2.1d).

We choose a depth of compensation at the depth of the continental crust h_c. We firstly equate the pressure under the abyssal plain with the pressure under continent:

$$h_{ap}\rho_w g + h_{oc}\rho_{ap} g = \rho_c h_c g \qquad [A2.6]$$

Since $h_{ap} + h_{oc} = h_c$, it can be found that

$$h_{oc} = h_c\frac{(\rho_{ap} - \rho_c)}{(\rho_c - \rho_w)} \qquad [A2.7]$$

Now balancing the columns under the mid-ocean ridge and the continental block, we have

$$(h_c - h_r)\rho_r g + h_r\rho_w g = h_c\rho_c g \qquad [A2.8]$$

from which the depth of water at the ridge crest becomes

$$h_r = h_c\frac{(\rho_r - \rho_c)}{(\rho_r - \rho_w)} \qquad [A2.9]$$

If the continental block is 35 km thick with its surface at sea level and with density of 2700 kg m^{-3}, the average density of the hot asthenosphere at the mid-ocean ridge is 2830 kg m^{-3}, and the average density of the cold oceanic lithosphere beneath the abyssal plain is 3000 kg m^{-3}, the water depth above the abyssal plain is 5.25 km, and the water depth above the mid-ocean ridge is 2.49 km. Consequently, the mid-ocean ridge lies 2.76 km above the abyssal plain. These values are very similar to the typical bathymetric depths of the ocean floor.

3
Deviatoric stress at the edge of a continental block

Learning points: thickness variations and elevation differences generally result in the transmission of lateral forces as well as involving a vertical isostatic balance. Lateral forces result from the vertical integration of the lithostatic pressure over depth. Consequently, a block of elevated continental crust 'pushes' laterally towards the ocean crust, causing the continent to go into tension. The deviatoric stress per unit width of continental block is 70–80 MPa.

Continental blocks may not be in a lithostatic state of stress, since normal forces may be exerted per unit area on the vertical plane defining the block, in which case the horizontal component σ_{xx} (or σ_{zz}) $\neq \sigma_{yy}$. This is also the situation where tectonic forces are transmitted in the plane of the plate. The presence of these horizontal deviatoric stresses strongly affects the state of stress of the plate and therefore its deformation and basin development.

Consider a continental block of density ρ_c surrounded by a piece of ocean crust or mantle lithosphere with density ρ_m. Initially ignore the effect of the ocean water. A force acts on the vertical edge of the block due to the lithostatic stress in the ocean lithosphere, denoted F_m. It has a value given by the integration of the lithostatic pressure over the height of ocean lithospheric column b (Fig. A3.1a).

$$F_m = \int_0^b p \, dy = \rho_m g \int_0^b y \, dy = \frac{1}{2}\rho_m g b^2 \qquad [A3.1]$$

A force also acts on the edge of the continental block due to the integration of the lithostatic pressure over the thickness of the continent h, denoted F_c. Analogously to eqn. [A3.1],

$$F_c = \frac{1}{2}\rho_c g h^2 \qquad [A3.2]$$

The difference between the F_m and F_c is the tectonic deviatoric stress $\Delta\sigma_{xx}$, which can be written:

$$\frac{1}{2}\rho_c g h^2 + \Delta\sigma_{xx} h = \frac{1}{2}\rho_m g b^2 \qquad [A3.3]$$

Since an Airy isostatic balance requires that $\rho_c h = \rho_m b$, we obtain the solution for the deviatoric stress

$$\Delta\sigma_{xx} = -\frac{1}{2}\rho_c g h \left(1 - \frac{\rho_c}{\rho_m}\right) \qquad [A3.4]$$

If $\rho_c = 2750\,\mathrm{kg\,m^{-3}}$ and $\rho_m = 3250\,\mathrm{kg\,m^{-3}}$, and the thickness of the continental block is 30 km, the deviatoric stress is −62 MPa, and the minus sign shows that the stress is extensional in the continental block. This shows that a continental block is naturally in tension when elevated against the ocean lithosphere. For a continental block 70 km thick, as occurs under high plateaux such as Tibet, the deviatoric stress would increase linearly to ≫100 MPa.

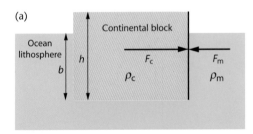

(a)

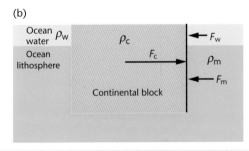

(b)

Fig. A3.1 Set-up for the calculation of the lateral force resulting from the elevation of a light continental block above denser oceanic lithosphere. (a) Force balance between continental block and oceanic lithosphere. (b) Force balance includes effect of ocean water.

Basin Analysis: Principles and Application to Petroleum Play Assessment, Third Edition. Philip A. Allen and John R. Allen.
© 2013 John Wiley & Sons, Ltd. Published 2013 by John Wiley & Sons, Ltd.

Incorporating the effect of the ocean water is straightforward (Fig. A3.1b). The total force against the edge of the continental block from the ocean water and its underlying lithosphere is the sum of $F_w = \rho_w g y_w^2 / 2$ and $F_m = \rho_m g b^2 / 2$.

The isostatic balance incorporating the effect of the ocean water is now

$$\frac{h}{b} = \frac{(\rho_m - \rho_w)}{(\rho_c - \rho_w)} \qquad\qquad [A3.5]$$

Consequently, if $h = 30\,\text{km}$, b is $23.2\,\text{km}$. The force difference is now slightly less, and since the deviatoric stress is $\Delta\sigma_{xx} h$, the deviatoric stress from continent to ocean becomes $-59\,\text{MPa}$. The effect of the incorporation of the ocean water is therefore relatively small.

Lateral buoyancy forces in the lithosphere

Learning points: when the pressure exerted by the weight of overlying rocks (and water) is equal in all three directions it is known as a lithostatic state of stress. But even when lithospheric columns are balanced isostatically, there is commonly a force acting laterally (horizontally). The stress at any depth is made of two components: a lithostatic component (ρgy) and a tectonic component known as a deviatoric stress $\Delta\sigma_{xx}$. The deviatoric stress is assumed to be constant over a given depth range. An elevated continental plateau pushes against a low-elevation coastal plain, and a mid-ocean ridge pushes against an oceanic abyssal plain, causing the plateau and mid-ocean ridge to be in tension.

As a development of the previous exercise in Appendix 3, we wish to know whether there is a horizontal (lateral) force between adjacent columns of lithosphere that are vertically balanced isostatically. This has important implications for shortening or extension between the adjacent lithospheric columns, and plays an important role in determining the lithosphere's state of stress.

We take four lithospheric columns in isostatic equilibrium:

Elevated plateau
Elevation 5 km
Crustal thickness 70 km
Crustal density 2850 kg m^{-3}
Lithospheric thickness 200 km
Mantle lithosphere density 3230 kg m^{-3}

Continental coastal plain
Elevation 0 km
Crustal thickness 30 km
Crustal density 2800 kg m^{-3}
Lithospheric thickness 120 km
Mantle lithosphere density 3200 kg m^{-3}
Asthenospheric mantle density 3330 kg m^{-3}

Mid-ocean ridge
Elevation −2.5 km
Depth-averaged asthenospheric mantle density 3205 kg m^{-3}
Ocean water density 1000 kg m^{-3}

Oceanic abyssal plain
Elevation −5 km
Crustal thickness 25 km

Crustal density 3000 m^{-3}
Lithospheric thickness 60 km
Mantle lithosphere density 3100 kg m^{-3}
Asthenospheric mantle density 3330 kg m^{-3}

Practical exercise

1. Compare the horizontal force acting on the vertical faces of these four columns. This can either be done graphically or numerically. The lateral force F_x (acting in the x-direction) is plotted by integrating over depth y the lithostatic pressure $p = \rho gy$. To calculate this force use

$$F_x = \frac{1}{2}\rho gy^2 \qquad [A4.1]$$

but remember that the density varies between ocean water, crust, mantle lithosphere and asthenosphere.

2. What is the force difference between the elevated continental plateau and the continental coastal plain?

3. What is the force difference between the mid-ocean ridge and the oceanic abyssal plain?

The deviatoric stress (per unit width of the columns) $\Delta\sigma_{xx}$ is the force difference per unit width divided by the height of the overlying rock and water h,

$$F_{1\rightarrow2} = \Delta\sigma_{xx}h \qquad [A4.2]$$

where $F_{1\rightarrow2}$ is the force difference between columns 1 and 2.

Basin Analysis: Principles and Application to Petroleum Play Assessment, Third Edition. Philip A. Allen and John R. Allen.
© 2013 John Wiley & Sons, Ltd. Published 2013 by John Wiley & Sons, Ltd.

Solution

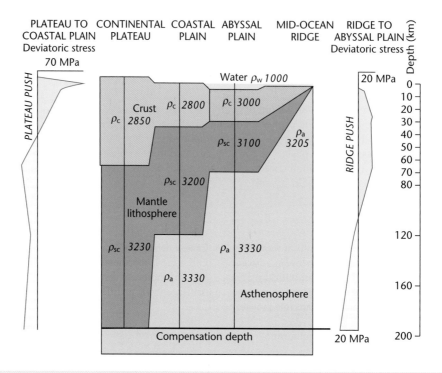

Fig. A4.1 Four lithospheric columns are in an Airy isostatic balance, but there are lateral 'deviatoric' forces between them. The deviatoric stress is calculated as the difference in the horizontal force between two columns averaged over the overlying thickness. The continental plateau pushes against the coastal plain, causing the plateau to go into tension, reaching a maximum of −70 MPa. This explains the normal faults in high plateaux such as Tibet. Note that the maximum tensile deviatoric stress is at shallow depths in the upper crust. The mid-ocean ridge pushes against the oceanic abyssal plain to depths of up to c.100 km, causing the ridge to be in tension and the abyssal plain in compression. This is the familiar ridge-push force. With the parameter values given above, the deviatoric ridge-push is a maximum of 20 MPa.

5

Derivation of flexural rigidity and the general flexure equation

Learning points: flexure takes place in a wide range of complex geological settings, but the general flexure equation is derived for the ideal case of a purely elastic beam of uniform thickness. The coefficient linking the curvature of the beam to the bending moment is called the *flexural rigidity*. It has units of Nm, but is often expressed as a length scale – the *elastic thickness*. The general flexure equation can be solved for particular geodynamical situations by the application of the appropriate boundary conditions.

In the simplest terms, the flexure of a plate depends on its thickness, elastic properties and the nature of the applied load. Imagine a plate of thickness h and width L which is fixed at both ends and which is subjected to a line load at its midpoint (Fig. 2.10). In order to attain a force balance, the vertical line load V_a must be counteracted by vertical forces $V_a/2$ at both ends. Assuming that both plate thickness h and deflection w are small compared to the width of the plate L (as required by linear elastic theory), we can study the forces and torques on a small element of the plate. A downward force per unit area $q(x)$ is exerted on the plate by the applied load, and on the end sections there is a net shear force per unit length V and horizontal force P per unit length, the latter being independent of x (Fig. 2.8). A bending moment M also acts on the end section related to the effects of normal stresses on the cross section, σ_{xx}. These normal stresses are known as *fibre stresses* (Fig. 2.10).

The bending moment M is related to the curvature of the plate, since forces on the end section exert a torque about the midpoint of the plate. If the force on an element of thickness dy on the end of the plate is $\sigma_{xx}dy$, then this force will exert a torque about the midpoint ($y = 0$) of $\sigma_{xx}y\,dy$. Integrated over the entire end section we obtain the bending moment

$$M = \int_{-h/2}^{h/2} \sigma_{xx}y\,\mathrm{d}y \qquad [\text{A5.1}]$$

The fibre stresses σ_{xx} result in longitudinal strains in the plate, ε_{xx}, contractional in the upper half and extensional in the lower half (Fig. 2.10). In the two-dimensional case we assume that the plate is infinite

in the plane normal to the figure, so there is no strain in the direction perpendicular to the xy plane, that is, $\varepsilon_{zz} = 0$. This is the situation of *plane strain* (§2.1.2). If the plate is thin, it is possible to make the further assumption that stresses normal to the plate's surface are zero ($\sigma_{yy} = 0$). The relations between stress and strain can then be restated as (cf. eqn. [2.9])

$$\varepsilon_{xx} = \frac{1}{E}\left(\sigma_{xx} - v\sigma_{zz}\right) \qquad [\text{A5.2}]$$

and from eqn. [2.5]

$$\varepsilon_{zz} = \frac{1}{E}\left(\sigma_{zz} - v\sigma_{xx}\right) \qquad [\text{A5.3}]$$

If the bending is two-dimensional (there is no strain in the direction perpendicular to the page in Fig. 2.10), $\varepsilon_{zz} = 0$, then eqns [A5.2] and [A5.3] give

$$\sigma_{xx} = \frac{E}{\left(1 - v^2\right)}\varepsilon_{xx} \qquad [\text{A5.4}]$$

which is the relation between fibre stresses and longitudinal strains. The bending moment can now be related to longitudinal strains by substitution of eqn. [A5.4] in eqn. [A5.1]

$$M = \frac{E}{\left(1 - v^2\right)}\int_{-h/2}^{h/2} \varepsilon_{xx}y\,\mathrm{d}y \qquad [\text{A5.5}]$$

Basin Analysis: Principles and Application to Petroleum Play Assessment, Third Edition. Philip A. Allen and John R. Allen.
© 2013 John Wiley & Sons, Ltd. Published 2013 by John Wiley & Sons, Ltd.

(a)

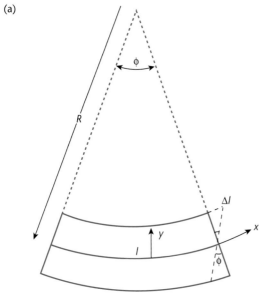

(b)

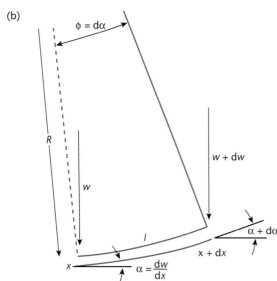

Fig. A5.1 Geometrical aspects of plate bending. (a) Longitudinal strain (extension below the midplane and contraction above the midplane of the plate) is a function of the distance from the midplane of the plate y, and the angle ϕ. (b) Notation to show that the second derivative of the deflection (d^2w/dx^2) gives the rate of change of slope of the plate; this is inversely related to the local radius of curvature R of the plate. From Turcotte & Schubert (2002). © Cambridge University Press, 2002.

The bending moment M can also be related to the deflection w. The local radius of curvature R is inversely proportional to the rate of change in slope of the deflection, or $-d^2w/dx^2$ (if strains are small and $dw/dx \ll 1$, and where the negative sign simply states that w is positive downwards) (Fig. A5.1a). In addition, the longitudinal strain is also related to the radius of curvature. This relation can be derived by geometrical similarity (Fig. A5.1b), since the length of the plate is dependent on the local radius of curvature and ϕ ($l = R\phi$), and the change in length of the plate (Δl) is determined by

the distance from the midline of the plate (y) and ϕ ($\Delta l = \phi y$). Using this result, the longitudinal strains ($\Delta l/l$) can be expressed in terms of the deflection,

$$\varepsilon_{xx} = \frac{\Delta l}{l} = \frac{y}{R} = -y\frac{d^2w}{dx^2} \qquad [A5.6]$$

The bending moment can therefore be rewritten

$$M = \frac{-E}{(1-v^2)}\frac{d^2w}{dx^2}\int_{-h/2}^{h/2}y^2 dy \qquad [A5.7]$$

which when evaluated between the limits $-h/2$ and $h/2$ gives

$$M = \frac{-Eh^3}{12(1-v^2)}\frac{d^2w}{dx^2} \qquad [A5.8]$$

The coefficient of $-d^2w/dx^2$ in eqn. [A5.8] is defined as the *flexural rigidity*, D. Note that the flexural rigidity D is proportional to the cube of the elastic thickness of the plate h. Flexural rigidity is commonly expressed as this length term. When applied to the lithosphere, the term *equivalent elastic thickness* Te is preferred, since the value of Te does not reflect the depth of a real physical discontinuity in the lithosphere; it is simply the equivalent thickness of a purely elastic beam.

The bending moment can therefore be written in terms of the curvature,

$$M = -D\frac{d^2w}{dx^2} = \frac{D}{R} \qquad [A5.9]$$

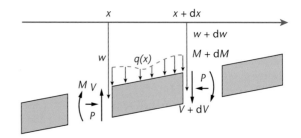

Fig. A5.2 Forces on a small element of a flexed plate. The forces can be balanced vertically and in terms of their tendency to rotate the element (moments or torques). From Turcotte & Schubert (2002). © Cambridge University Press, 2002.

We can now proceed to establish the general flexural equation by carrying out a force balance on a segment of the plate. The force balance (Fig. A5.2) on a flexed plate with a downward force caused by loading $q(x)$ and horizontal forces (P) on the end sections is as follows.

Equating the forces in a vertical direction we have

$$q(x)dx = -dV$$
$$q(x) = -\frac{dV}{dx} \qquad [A5.10]$$

Balancing torques we have in the counterclockwise direction a force P acting over a moment arm of $-dw$ and the torque dM. In the clockwise sense we have $(V + dV)$ acting over a moment arm of dx. Since the term dV is small we can write

$$-P\,dw + dM = V\,dx \qquad\qquad [A5.11]$$

Differentiating eqn. [A5.11] twice, so that eqn. [A5.10] can be substituted into eqn. [A5.11] gives

$$-P\frac{d^2w}{dx^2} + \frac{d^2M}{dx^2} = -q(x) \qquad\qquad [A5.12]$$

Since we have already expressed the bending moment in terms of the flexural rigidity and the local curvature, eqn. [A5.12] can be rewritten as

$$D\frac{d^4w}{dx^4} = q(x) - P\frac{d^2w}{dx^2} \qquad\qquad [A5.13]$$

which is the *general flexure equation* for the deflection of a plate.

In the context of an elastic plate overlying a fluid-like mantle, this equation must be modified to account for a restoring force (buoyancy) acting upwards on the deflected plate (see §2.1.4 and Figs 2.11 and 2.12).

6
Flexural isostasy

Learning points: in contrast to local or Airy isostasy, flexural isostasy involves isostatic compensation by the regional transmission of stresses by flexure. Flexural isostasy must clearly depend on the rigidity of the material being loaded or unloaded, but also on the wavelength of the load. Large wavelengths of load, such as the water in the Pacific Ocean, are fully compensated with an Airy isostatic solution. Small wavelength loads, however, such as individual mountain peaks and valleys, are supported rigidly. Sedimentary basins are at an intermediate wavelength and are expected to be compensated flexurally, with a degree of compensation determined by the flexural rigidity and the density and wave number of the load.

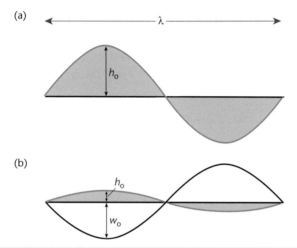

Fig. A6.1 Deflection of the lithosphere under a periodic (sinusoidal) load of wavelength λ. In (a) the wavelength of the load is short and there is no deflection of the lithosphere. In (b) the wavelength of the load is long, leading to isostatic compensation of the load by a deflection of the lithosphere. w_0 is the maximum deflection, h_0 is the maximum elevation of the load (after Turcotte & Schubert 2002, p. 123, fig. 3.25). © Cambridge University Press, 2002.

We start by assuming that topography is represented by a sinusoidal periodic load (Fig. A6.1) given by:

$$h = h_0 \sin(2\pi x/\lambda) \qquad [A6.1]$$

where h_0 is the maximum height of the topography and λ is its wavelength. The load actually exerted on the lithosphere has the form of pressure or stress ρgh, so the distribution of pressure under this periodic topography is

$$q_a(x) = \rho_s gh_0 \sin(2\pi x/\lambda) \qquad [A6.2]$$

where ρ_s is the density of sediments constituting the load. Assuming no horizontal applied forces (i.e. $P = 0$), and returning to the general flexural eqn. [A5.13], we have

$$D\frac{d^4w}{dx^4} + (\rho_m - \rho_s)gw = \rho_s gh_0 \sin(2\pi x/\lambda) \qquad [A6.3]$$

Since the load is sinusoidal and periodic, the deflection of the lithosphere will also be sinusoidal and periodic. The solution for the deflection w can therefore be assumed to be of the type

$$w = w_0 \sin(2\pi x/\lambda) \qquad [A6.4]$$

where w_0 is the maximum deflection. Substituting eqn. [A6.4] into eqn. [A6.3] and rearranging, the amplitude of the deflection of the lithosphere is

$$w_0 = \frac{h_0}{\left\{ D\dfrac{d^4w}{dx^4} \dfrac{1}{\rho_s gw} + \dfrac{\rho_m}{\rho_s} - 1 \right\}} \qquad [A6.5]$$

The fourth differential of w from eqn. [A6.4] is $(2\pi/\lambda)^4 w$, so eqn. [A6.5] simplifies to

$$w_0 = \frac{h_0}{\dfrac{\rho_m}{\rho_s} - 1 + \dfrac{D}{\rho_s g}\left(\dfrac{2\pi}{\lambda}\right)^4} \qquad [A6.6]$$

Basin Analysis: Principles and Application to Petroleum Play Assessment, Third Edition. Philip A. Allen and John R. Allen.
© 2013 John Wiley & Sons, Ltd. Published 2013 by John Wiley & Sons, Ltd.

If the wavelength of the load is short, the deflection is very small compared to the maximum height of the load ($w_0 \ll h_0$). The lithosphere therefore appears to behave very rigidly to loads of this scale. If, however, the wavelength of the load is long, the deflection can be written

$$w = w_{0\infty} = \frac{\rho_s h_0}{(\rho_m - \rho_s)} \tag{A6.7}$$

which is the result obtained for a purely vertical isostatic balance (Airy isostasy) (Appendix 2, §2.1). This means that for sufficiently long wavelengths, the lithosphere appears to have no rigidity and periodic loads should be in hydrostatic equilibrium. It is clearly desirable to be able to predict where we are in this range of complete to no compensation. The degree of compensation C of the load (Fig. 2.13) is the ratio of the actual deflection compared to the maximum (Airy) or hydrostatic deflection

$$C = \frac{w_0}{w_{0\infty}} \tag{A6.8}$$

Substituting eqns [A6.6] and [A6.7] into eqn. [A6.8], the degree of compensation becomes

$$C = \frac{(\rho_m - \rho_s)}{\rho_m - \rho_s + \dfrac{D}{g}\left(\dfrac{2\pi}{\lambda}\right)^4} \tag{A6.9}$$

$(2\pi/\lambda)$ is termed the *wave number* (Watts 1988).

7

The 1D heat conduction equation

Learning points: the transport of heat underpins a wide range of processes in the Earth. Consideration of the transport of heat by conduction provides us with the diffusion equation, which states that the temperature change scales on the second derivative of the temperature field.

Consider a volume element of vertical dimension δy with a cross-sectional area a (Fig. A7.1). A heat flow $Q(y)$ enters the element across its top surface. The element has an internal heat generation A and is composed of material of density ρ and thermal capacity (specific heat) denoted by c. The specific heat is the amount of heat needed to raise the temperature of 1 kg of material by 1 °C (with units therefore of W kg^{-1}°C^{-1}).

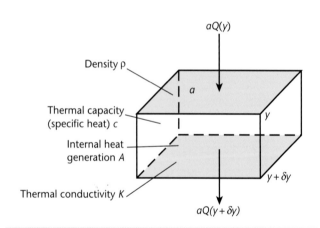

Fig. A7.1 Set-up and notation for conductive heat flow through a volume element of thickness δy, cross-sectional area a, density ρ, thermal capacity (specific heat) c, thermal conductivity K, and internal heat generation A. Heat is conducted across the shaded faces only (it is perfectly one-dimensional).

In a unit of time, the heat entering the volume element is $aQ(y)$ and the heat leaving the element is $aQ(y+\delta y)$.

$Q(y+\delta y)$ can be expanded in a Taylor series to give

$$Q(y+\delta y)=Q(y)+\delta y\frac{\partial Q}{\partial y}+\cdots \qquad [A7.1]$$

Taking the first two terms only of the Taylor series, the net gain or loss of heat is therefore the difference between the heat fluxes through the top of the element and out of the bottom,

$$aQ(y)-a\left\{Q(y)+\delta y\frac{\partial Q}{\partial y}\right\}=-a\delta y\frac{\partial Q}{\partial y} \qquad [A7.2]$$

Let the heat generated internally per unit volume and unit time be A. The heat generated in the element of volume $a\delta y$ is therefore $Aa\delta y$. The total heat gain or loss per unit time is then

$$Aa\delta y-a\delta y\frac{\partial Q}{\partial y} \qquad [A7.3]$$

If the material making up the volume element has specific heat c and density ρ and undergoes a small temperature change δT in a small time δt, the rate at which heat is gained or lost is

$$ca\delta y\rho\frac{\partial T}{\partial t} \qquad [A7.4]$$

Equating eqns [A7.3] and [A7.4],

$$ca\delta y\rho\frac{\partial T}{\partial t}=Aa\delta y-a\delta y\frac{\partial Q}{\partial y} \qquad [A7.5]$$

which when simplified gives

$$c\rho\frac{\partial T}{\partial t}=A-\frac{\partial Q}{\partial y} \qquad [A7.6]$$

Basin Analysis: Principles and Application to Petroleum Play Assessment, Third Edition. Philip A. Allen and John R. Allen.
© 2013 John Wiley & Sons, Ltd. Published 2013 by John Wiley & Sons, Ltd.

Substituting Fourier's law (eqn. [2.14]) into eqn. [A7.6],

$$\frac{\partial T}{\partial t} = \frac{A}{\rho c} + \frac{K}{\rho c}\frac{\partial^2 T}{\partial y^2} \qquad \text{[A7.7]}$$

which is the *1D heat conduction equation*. We can define the thermal diffusivity as $\kappa = K/\rho c$. The thermal diffusivity expresses the ability of a material to gain or lose heat by conduction.

Certain boundary conditions can be applied to the 1D heat conduction equation in order to simplify it for certain physical situations in the lithosphere:

(a) steady-state solution where there is no temperature change with time, that is, $\partial T/\partial y = 0$,

$$\frac{\partial^2 T}{\partial t^2} = -\frac{A}{K} \qquad \text{[A7.8]}$$

(b) no internal heat generation, that is, $A=0$,

$$\frac{\partial T}{\partial t} = \frac{K}{\rho c}\frac{\partial^2 T}{\partial y^2} = \kappa\frac{\partial^2 T}{\partial y^2} \qquad \text{[A7.9]}$$

which is the *diffusion equation*. An equation of this form is used in a wide range of physical problems.

The volume element may be in motion relative to its surroundings and may therefore move through a region where the temperature varies with depth. If the volume element moves at a velocity u_y in the y direction (vertically in the lithosphere), its depth after time t will be $y+u_y t$, and the associated temperature change will be $u_y\partial T/\partial y$. The 1D heat conduction equation therefore needs to be modified. The temperature change after a small time period δt is given by

$$\frac{\partial T}{\partial t} = \frac{K}{\rho c}\frac{\partial^2 T}{\partial y^2} + \frac{A}{\rho c} - u_y\frac{\partial T}{\partial y} \qquad \text{[A7.10]}$$

The temperature change of a piece of lithosphere is therefore seen to be made of three components: a *basal heat flow* term, an *internal heat generation* term and an *advective* term. The advective term may be the movement towards the surface of a volume element of rock associated with the downcutting action of erosion. Or, with a different sign, it could be the velocity of deposition.

8

Derivation of the continental geotherm

Learning points: the heat conduction equation can be used to calculate the variation of temperature with depth, the *geotherm*. Here, we provide the solution of conductive heat flow with a contribution from radiogenic heating. Inclusion of radiogenic heat production causes the geotherm to be curved, so near-surface temperature gradients are not a close reflection of the deeper geotherm in the continental lithosphere.

In the case where there is no advective term and a constant heat flux, we have the equilibrium situation where $\partial T/\partial y = 0$. In this case

$$\frac{\partial^2 T}{\partial y^2} = -\frac{A}{K} \qquad [A8.1]$$

where A is the internal heat generation and $K/\rho c$ is the diffusivity κ. We can calculate the geotherm as long as we can provide some boundary conditions. Two boundary conditions are that the surface temperature is T_0 ($T = T_0$ at $y = 0$) and that the surface heat flow is $-Q_0$ at $y = 0$. Integrating eqn. [A8.1] gives

$$\frac{\partial T}{\partial y} = -\frac{A}{K}y + c_1 \qquad [A8.2]$$

where c_1 is a constant of integration. Since $\partial T/\partial y = Q_0/K$ at $y = 0$, the constant c_1 must be equal to Q_0/K. Eqn. [A8.2] after further integration therefore becomes

$$T = -\frac{A}{2K}y^2 + \frac{Q_0}{K}y + c_2 \qquad [A8.3]$$

Since $T = T_0$ at $y = 0$, c_2 must be equal to T_0. The temperature as a function of depth is therefore

$$T = T_0 + \frac{Q_0}{K}y - \frac{A}{2K}y^2 \qquad [A8.4]$$

There are instances where it would be more useful to calculate the geotherm with knowledge of the heat flow from the mantle rather than the surface heat flow (the reason for this is that the basal heat flow may be provided by a forward model of lithospheric stretching – Chapter 3). In this case, we have a pair of new boundary conditions. Firstly, $T = T_0$ at $y = 0$ and, secondly, $Q = -Q_m$ at $y = y_c$, where y_c is the depth to the base of the crust, and Q_m is the basal heat flow from the mantle. Since $\partial T/\partial y = Q_m/K$ at $y = y_c$, the constant of integration c_1 in eqn. [A8.2] is given by

$$c_1 = \frac{Q_m}{K} + \frac{A}{K}y_c \qquad [A8.5]$$

Substituting into eqn. [A8.2] and integrating again,

$$T = -\frac{A}{2K}y^2 + \frac{(Q_m + Ay_c)}{K}y + c_2 \qquad [A8.6]$$

Since $T = T_0$ at $y = 0$, $c_2 = T_0$. Consequently, the geotherm in the region $0 < y < y_c$ is given by

$$T = T_0 + \frac{(Q_m + Ay_c)}{K}y - \frac{A}{2K}y^2 \qquad [A8.7]$$

Eqns [A8.4] and [A8.7] give the geotherms where we have information of either the surface heat flow or the basal heat flow from the mantle.

Basin Analysis: Principles and Application to Petroleum Play Assessment, Third Edition. Philip A. Allen and John R. Allen.
© 2013 John Wiley & Sons, Ltd. Published 2013 by John Wiley & Sons, Ltd.

9

Radiogenic heat production

Learning points: there are various ways of dealing with radiogenic heat production. Radiogenic heat production can be viewed as exponentially reducing with depth, or to be distributed in a number of layers or 'slabs' with different heat production values. Radiogenic heat production on average contributes to about half of the surface heat flow and causes geotherms to be curved.

Two models for radiogenic heat production are considered: (i) radiogenic heat production is uniform within slabs of variable thickness and depth; or (ii) radiogenic heat production reduces from surface values exponentially with depth.

Slab models for radiogenic heat generation

The simplest slab model is to have a single layer of radiogenic crust above a non-radiogenic mantle. The solution is given by eqn. [A8.7] where y_c is the thickness of the crust. In a multilayer model, the same equations would be used but individual layers treated separately. For a two-layer model with a highly radiogenic upper crust, a less radiogenic lower crust and a non-radiogenic mantle (Fig. 2.17), we let

$A = A_1$ over the depth range $0 \le y < y_1$
$A = A_2$ over the depth range $y_1 \le y < y_2$
$Q = -Q_2$ at $y = y_2$
$T = T_0$ at $y = 0$

The solutions for temperature with depth are:
for the depth range between $y = 0$ and $y = y_1$,

$$T = T_0 + \left\{ \frac{Q_2}{K} + \frac{A_2(y_2 - y_1)}{K} + \frac{A_1 y_1}{K} \right\} y - \frac{A_1}{2K} y^2 \qquad \text{[A9.1]}$$

and for the depth range between y_1 and y_2,

$$T = T_0 + \left\{ \frac{Q_2}{K} + \frac{A_2 y_2}{K} \right\} y + \left\{ \frac{A_1 - A_2}{2K} \right\} y_1^2 - \frac{A_2}{2K} y^2 \qquad \text{[A9.2]}$$

Using this two-layer model, with $A_1 = 2.5 \times 10^{-6}$ W m^{-3} in the upper crust ($y_1 = 20$ km) and $A_2 = 0.8 \times 10^{-6}$ W m^{-3} in the lower crust ($y_2 = 35$ km), and the basal heat flow from the mantle $Q_2 = 30 \times 10^{-3}$ W m^{-2}, the geotherm is as sketched in Fig. 2.17. The two-layer solution is compared to a one-layer crustal model with $A = 1.25 \times 10^{-6}$ W m^{-2}.

Exponential distribution of radiogenic heat generation

In an alternative model, the distribution of radiogenic heat production can be assumed to be exponential with depth (Fig. 2.17). If A_0 is the surface heat generation per unit volume, we can introduce a length scale for the exponential decrease in radiogenic heat production a_r such that the radiogenic heat generation per unit volume A at any depth is given by

$$A = A_0 \exp(-y/a_r) \qquad \text{[A9.3]}$$

where a_r is between 4.5 km and 16 km for a range of continental heat flow provinces (Sclater *et al.* 1980a, table 2.2) (see also Appendix 10 showing how a_r can be estimated).

One of the consequences of the exponential distribution of radiogenic heat is a linear relationship between the surface heat flow and the radiogenic heat content of surface rocks. This relationship has been demonstrated from many measurements in continental regions (Appendix 10). The reason for this predicted linear relationship is as follows. The steady-state solution for one-dimensional heat conduction (eqn. [A8.1]) can be modified to account for an exponential radiogenic heat production to give

$$0 = K \frac{\partial^2 T}{\partial y^2} + A_0 \exp(-y/a_r) \qquad \text{[A9.4]}$$

Integrating this once gives

$$c_1 = K \frac{\partial T}{\partial y} - A_0 a_r \exp(-y/a_r) \qquad \text{[A9.5]}$$

Using Fourier's law (eqn. [2.14]), eqn. [A9.5] can be modified slightly to give

$$c_1 = -Q - A_0 a_r \exp(-y/a_r) \qquad \text{[A9.6]}$$

Basin Analysis: Principles and Application to Petroleum Play Assessment, Third Edition. Philip A. Allen and John R. Allen.
© 2013 John Wiley & Sons, Ltd. Published 2013 by John Wiley & Sons, Ltd.

The constant of integration can be found be applying an appropriate boundary condition. Let us assume that as the depth becomes large ($y \rightarrow$ infinity), the heat flow is the basal heat flow from the mantle Q_m. Consequently, eqn. [A9.5] shows that $c_1 = Q_m$. The heat flux at any depth is therefore

$$Q = -Q_m - A_0 a_r \exp(-y/a_r)$$ [A9.7]

or in terms of the surface heat flow, where $Q_0 = -Q$ at $y = 0$, is

$$Q_0 = Q_m + A_0 a_r$$ [A9.8]

showing that there is a linear relationship between the surface heat flow and the radiogenic heat production of surface rocks, with an intercept equal to the mantle heat flow and a slope of a_r. The example shown in Fig. 2.19 from granite plutons in eastern North America (Roy *et al.* 1968) gives a mantle heat flow, or *reduced heat flow*, Q_m of 30 mW m^{-2}, and a length scale a_r of 7.5 km.

A further integration of eqn. [A9.6] gives the temperature profiles at any depth. Using the boundary condition $T = T_0$ at $y = 0$,

$$T = T_0 + \frac{Q_m}{K}y + \frac{A_0 a_r^2}{K}\{1 - \exp(-y/a_r)\}$$ [A9.9]

or in terms of the surface heat flow, by integrating eqn. [A9.7]

$$T = T_0 + \frac{Q_m}{K}y + \frac{(Q_0 - Q_m)a_r}{K}\{1 - \exp(-y/a_r)\}$$ [A9.10]

A typical geotherm, with $A_0 = 2.5 \times 10^{-6}$ W m^{-3}, $Q_0 = 70$ mW m^{-2}, $K = 3$ W m$^{-1\circ}$C^{-1}, $T_0 = 10\,^{\circ}$C, and $a_r = 10$ km, is shown in Fig. 2.17.

Practical exercise

Calculate and plot the geotherm where there is a radiogenic heat production that can be approximated by an exponential function with a depth constant and basal heat flow for the Sierra Nevada taken from Appendix 10. Use the following parameter values:

Surface temperature $T_0 = 0\,^{\circ}$C
Thermal conductivity $K = 3$ W m^{-1} K^{-1}
Surface heat production $A_0 = 1$ to 4×10^{-6} W m^{-3}

Heat production depth constant $a_r = 8.4$ km
Invariant (reduced) heat flow $Q_m = 22$ mW m^{-2}

Solution

The geotherm is given by:

$$T(y) = T_0 + \frac{Q_m}{K}y + \frac{A_0 a_r^2}{K}\{1 - \exp(-y/a_r)\}$$ [A9.11]

The geotherm for different values of A_0 is plotted in Fig. A9.1.

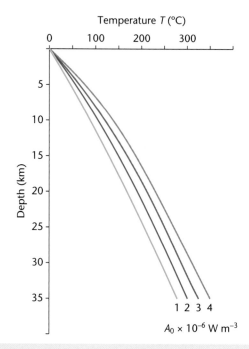

Fig. A9.1 Geotherms for different values of the surface radiogenic heat production A_0 using an exponential model of distribution of heat-producing elements (HPEs).

Surface heat flow and the radiogenic contribution

Learning points: plots of surface heat flow versus radiogenic heat production allow the non-radiogenic, 'basal', 'reduced' or 'invariant' heat flow and the depth distribution of radiogenic heat production to be estimated. Each geological province has its own linear trend between Q_0 and A. The basal heat flow is generally <40 mW m^{-2} in tectonically inactive provinces, and higher in regions of lithospheric stretching and magmatic activity. Heat-producing elements are concentrated in the upper 10 km of the crust.

Practical exercise

This practical exercise demonstrates the wide range of surface heat flow values caused by variable concentrations of radiogenic heat-producing elements (HPEs) in the continental crust. *Conversion of units*: For surface heat flow, note that 1 cal cm^{-2} s^{-1} × 10^{-6} is 1 heat flow unit or 41.84 mW m^{-2}. For radiogenic heat production, 1 cal cm^{-3} s^{-1} × 10^{-13} is 0.4184 μW m^{-3}.

Using the data in Table A10.1, plot the heat production versus surface heat flow and draw best-fit linear regressions to obtain: (i) the 'invariant', 'basal' or 'reduced' heat flow, which is the intercept on the Q_0 axis; and (ii) the radiogenic heat production depth constant a_r, which is the slope.

The solutions are given in Figure A10.1.

Table A10.1 Radiogenic heat production and surface heat flow values from a range of geological provinces. Data from Roy *et al.* (1968).

Locality	A Radiogenic heat production 10^{-13} cal cm^{-3} s^{-1} (μW m^{-3})	Q$_0$ Surface heat flow 10^{-6} cal cm^{-2} s^{-1} (mW m^{-2})	Locality	A Radiogenic heat production 10^{-13} cal cm^{-3} s^{-1} (μW m^{-3})	Q$_0$ Surface heat flow 10^{-6} cal cm^{-2} s^{-1} (mW m^{-2})
New England			19	6.4 (2.7)	1.06 (44.4)
1	20.7 (8.7)	2.27 (95.0)	20	4.7 (2.0)	0.83 (34.7)
2	21.2 (8.8)	2.15 (90.0)	21	3.2 (1.3)	0.73 (30.5)
3	17.6 (7.4)	1.89 (79.1)	22	1.8 (0.75)	0.62 (25.9)
4	12.9 (5.4)	1.80 (75.3)	23	2.2 (0.92)	0.60 (25.1)
5	11.6 (4.9)	1.63 (68.2)			
6	9.6 (4.0)	1.63 (68.2)	Basin and Range province		
7	7.8 (3.3)	1.34 (56.1)	24	10.7 (4.5)	2.40 (100.4)
8	3.8 (1.6)	1.08 (45.2)	25	6.0 (2.5)	3.40 (142.3)
			26	7.9 (3.3)	2.30 (96.2)
Central Stable region			27	10.3 (4.3)	2.22 (92.9)
9	7.6 (3.2)	1.46 (61.1)	28	7.1 (3.0)	2.20 (92.0)
10	5.8 (2.4)	1.22 (51.0)	29	6.7 (2.8)	2.20 (92.0)
11	5.5 (2.3)	1.17 (49.0)	30	2.0 (0.84)	2.14 (89.6)
12	1.4 (0.59)	0.82 (34.3)	31	7.7 (3.2)	2.00 (83.7)
13	<0.4 (0.17)	0.81 (33.9)	32	5.7 (2.4)	1.90 (79.5)
14	<0.4 (0.17)	0.79 (33.1)	33	5.3 (2.2)	1.88 (78.7)
15	<0.4 (0.17)	0.81 (33.9)	34	7.7 (3.2)	1.82 (76.1)
			35	3.8 (1.6)	1.78 (74.5)
Sierra Nevada			36	3.1 (1.3)	1.65 (69.0)
16	8.8 (3.7)	1.30 (54.4)	37	6.6 (2.8)	1.64 (68.6)
17	4.0 (1.7)	1.25 (52.3)	38	3.1 (1.3)	1.60 (66.9)
18	9.6 (4.0)	1.25 (52.3)			

Basin Analysis: Principles and Application to Petroleum Play Assessment, Third Edition. Philip A. Allen and John R. Allen.
© 2013 John Wiley & Sons, Ltd. Published 2013 by John Wiley & Sons, Ltd.

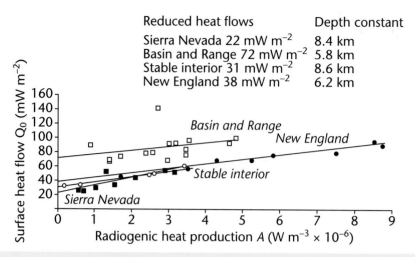

Fig. A10.1 Four heat flow provinces, data from Roy *et al.* (1968). Values of intercept and slope are slightly different to those in Roy *et al.* (1968, table 3), which were calculated after some data selection. Note that depth constants a_r are <10 km, and that reduced (basal) heat flows are in the range 20–40 mW m^{-2} with the exception of the tectonically active Basin and Range province where the reduced heat flow is ~70 mW m^{-2}.

11

Radiogenic heat production of various rock types

Learning points: radiogenic heat production can be estimated from the measurement of the isotopic abundances of the main heat producers, uranium, thorium and potassium, combined with knowledge of the energy release from these different isotopic systems. The technique can also be used with measurements from a natural gamma ray spectrometer deployed as a downhole tool.

Practical exercise

This practical exercise illustrates the ways in which the radiogenic heat production can be estimated from data on the isotopic abundances of HPEs in a rock, and how A_0 may be estimated from spectral gamma ray logs measured continuously from boreholes. 98% of the total radiogenic heat production comes from the decay of ^{238}U, ^{232}Th and ^{40}K, but energy (and therefore heat) released during their decay varies. The heat generation is the product of isotopic abundance in a rock (ppm or %), the energy release per unit mass of HPE (μW kg^{-1}), and the density of the rock ρ (kg m^{-3}).

Calculate the total heat production (W m^{-3}) of a (granitic) rock of density 2700 kg m^{-3} with the properties given in Table A11.1.

Table A11.1 Radiogenic heat production from the three main heat-producing elements (HPEs)

Isotope	Concentration/ abundance	Heat generation per unit mass
Uranium (^{238}U)	4.3 ppm	96.7 μW kg^{-1}
Thorium (^{232}Th)	12.2 ppm	26.3 μW kg^{-1}
Potassium (^{40}K)	5.1%	3.5×10^{-3} μW kg^{-1}

Solution

The total heat generation A from the sample of granite is the sum of abundance $\times$ density $\times$ heat generation from each element

Uranium: $4.3 \times 10^{-6} \times 2700 \times 96.7 \times 10^{-6} = 1.123 \, \mu$W m^{-3}
Thorium: $12.2 \times 10^{-6} \times 2700 \times 26.3 \times 10^{-6} = 0.866 \, \mu$W m^{-3}
Potassium: $5.1 \times 10^{-2} \times 2700 \times 3.5 \times 10^{-9} = 0.482 \, \mu$W m^{-3}
Total heat production $= 2.471 \, \mu$W m^{-3}

Heat generation derived from well logs

The empirical relationship between heat generation A (in μW m^{-3}) and natural gamma ray log values GR (in API units) from Rybach (1986) and Bücker and Rybach (1996) is:

$$A = 0.0158(GR - 0.8) \quad [A11.1]$$

Heat generation can also be obtained from the ppm uranium, ppm thorium and %potassium from the natural gamma spectrometer, and rock bulk density can be obtained from the compensated density log.

Practical exercise

Calculate A using the following values from black shales in borehole East Yeeda 1 in the Canning Basin, Australia (Beardsmore & Cull 2001, p. 30–33) using the data in Table A11.2.

Table A11.2 Abundances of HPEs and bulk density from black shales in the Canning Basin, Australia

From natural gamma spectrometer (NGS)			From CDL
Uranium ppm	Thorium ppm	Potassium %	Bulk density (kg m^{-3})
8	14	2.3	2600

Solution

There are two steps:

1. Calculate the total GR count from the elemental abundances
2. Calculate the heat generation from the total GR and density

The total gamma ray count GR is proportional to the abundances of potassium, thorium and uranium, but the proportionality coefficient varies down the borehole depending on the depth into the

Basin Analysis: Principles and Application to Petroleum Play Assessment, Third Edition. Philip A. Allen and John R. Allen.
© 2013 John Wiley & Sons, Ltd. Published 2013 by John Wiley & Sons, Ltd.

formation that gamma rays are detected, which is affected by formation density, mud weight, hole diameter, gamma ray energy and tool type (Serra 1984). If the radius of influence is the only parameter controlling the proportionality coefficient, then the total gamma ray count becomes

$$GR = c_{GR} \times (K + 0.13Th + 0.36U) \qquad [A11.2]$$

where c_{GR} can be obtained as the gradient of the best-fit line from a plot of GR against (K+0.13Th+0.36U) from the depth range of interest in the borehole. c_{GR} is typically 0.005.

The total heat generation can be obtained from

$$A = 3.5.10^{-2} \times \rho_b \times GR \times \frac{c_A}{c_{GR}} \qquad [A11.3]$$

where

$$c_A = \frac{(K) + (0.751Th) + (2.76U)}{(K) + (0.13Th) + (0.36U)} \qquad [A11.4]$$

and bulk density is in g cm^{-3} and GR is in API units.

Calculations of A should be done for many points in the depth range of interest, but taking just one set of data (Table A11.2), c_A is 4.98, c_A/c_{GR} is 0.143, GR is 175 API units, and A is 2.3 μW m^{-3}. This is compared with 2.8 μW m^{-3} calculated using the Rybach regression of eqn. [A11.1].

Effects of erosion and deposition on the geotherm

Learning points: rapid rates of sedimentation or erosion have a strong effect on the geotherm in the underlying lithosphere. Rapid erosion quickly cools the lithosphere, but there is a period of transient adjustment to the new surface temperature during which rocks are hotter than their steady-state temperature. Sedimentation causes rocks to be buried to greater depths and to heat up, but with a transient stage during which they are cooler than their steady-state temperature. The length of the transient stage is a diffusive time scale. Effects on temperature and heat flow of variations in sedimentation velocity may be stronger than the effects of temporal changes in basal heat flow.

Instantaneous erosion

We start by considering the effect on subsurface temperatures of an instantaneous removal of rock by erosion. The instantaneous removal of rock causes warm crust to be subjected to a cold surface temperature T_0. Over time, the crust cools by conduction to the surface of the Earth. The time scale of this cooling depends on the diffusivity of crustal rocks, and the depth to which the temperature perturbation is felt is the diffusion distance of the form $\sqrt{\kappa t}$.

The temperature following the erosion event is given by

$$T(y,t) = T_0 + Gy + Gl\,erf\left(\frac{y}{2\sqrt{\kappa t}}\right)$$ [A12.1]

where G is the steady-state geothermal gradient, l is the thickness of crust removed by erosion and *erf* is the error function (available as ERF in Excel and tabulated in texts such as Turcotte & Schubert 2002, p. 155). Eqn. [A12.1] simply states that the temperature at any depth y is made up of a steady-state linear solution (second term on right) and a transient solution (third term on right).

The thermal effects of instantaneous erosion and deposition are provided in the main text, § 2.2.5.

Practical exercise: transient effects of rapid sedimentation

Consider a passive margin where there is an abrupt increase in sediment accumulation rate caused by switching in the sediment discharge from a large river. What are the thermal effects of the deposition of cold sediment in the rapidly accumulating sediment column and underlying basin-fill?

The problem of the transient temperature field resulting from a period of high deposition rate can be solved by considering the basin as a half space that has a moving boundary. We keep the temperature at $T_0 = 0\,°C$ at the surface, and $y = 0$. The steady-state geotherm before the onset of rapid deposition is therefore $T(y, t = 0) = Gy$.

The solution for the evolution of the temperature field over time, at $t > 0$, is provided by Jaupart and Mareschal (2011, p. 84):

$$T(y,t) = G(y - vt) + \frac{G}{2}(y + vt)\exp\left(\frac{vy}{\kappa}\right)erfc\left(\frac{y+vt}{2\sqrt{\kappa t}}\right)$$
$$+ \frac{G}{2}(y - vt)\,erfc\left(\frac{y-vt}{2\sqrt{\kappa t}}\right)$$ [A12.2]

where v is the steady deposition rate, κ is the thermal diffusivity ($10^{-6}\,m^2\,s^{-1}$), $G = 40\,°C\,km^{-1}$ is the steady-state linear geothermal gradient, and *erfc* is the complementary error function (available as ERFC in Excel, and tabulated in Turcotte & Schubert 2002, p. 155). Make a plot of temperature versus depth for a basin with a sediment thickness of 3 km deposited at the rate v. Plot transient geotherms for $v = 0.01\,mm\,yr^{-1}$, $0.1\,mm\,yr^{-1}$, $1\,mm\,yr^{-1}$ and $10\,mm\,yr^{-1}$. The times t for these sedimentation velocities are therefore 300 Myr, 30 Myr, 3 Myr and 0.3 Myr.

Solution

The chart showing the four geotherms is given in Fig. A12.1. Note that the temperature deficit compared to the linear geotherm at $t = \infty$ increases with depth for each transient geotherm. The temperature deficit is, however, very low for the slow sedimentation rate. At the base of the sediment column deposited at the rate v, the temperature is 87 °C at $v = 1\,mm\,yr^{-1}$, and 52 °C at $v = 10\,mm\,yr^{-1}$, compared to the steady-state temperature of 120 °C. Sedimentation rate therefore has a strong effect on the temperature of the entire sediment column.

Basin Analysis: Principles and Application to Petroleum Play Assessment, Third Edition. Philip A. Allen and John R. Allen.
© 2013 John Wiley & Sons, Ltd. Published 2013 by John Wiley & Sons, Ltd.

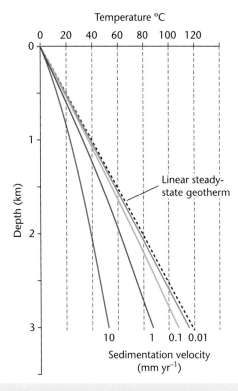

Fig. A12.1   Geotherms for different rates of sediment deposition v after the deposition of 3 km of sediment, which implies that the time scale varies from 0.3 Myr to 300 Myr for sedimentation velocities between 10 mm yr^{-1} and 0.01 mm yr^{-1}.

Table A12.1   The geothermal gradient scaled against the steady-state geotherm for different sedimentation velocities

Sedimentation velocity v (m yr^{-1})	Scaled geothermal gradient	Geothermal gradient where $G = 0.04\,°\text{C m}^{-1}$
0.01	0.724	0.029
0.001	0.867	0.034
0.0001	0.950	0.038
0.00001	0.983	0.039

For rapid sedimentation rates, the surface geothermal gradient is strongly reduced compared to the steady-state value.

We take the surface heat flow as $Q = K\delta T/\delta y$ and a thermal conductivity K of $3\,\text{W m}^{-1}\,\text{K}^{-1}$. The range of values in Table A12.1 correspond to surface heat flow values of between $87\,\text{mW m}^{-2}$ and $117\,\text{mW m}^{-2}$. But if the rapidly deposited sediments are dominated by porous shales, where the thermal conductivity is closer to $1.5\,\text{W m}^{-1}\,\text{K}^{-1}$, the surface heat flow would reduce to $43\,\text{mW m}^{-2}$ at the high sedimentation velocity of $10\,\text{mm yr}^{-1}$. The key learning point is that variations in sedimentation velocity are potentially more important than basal heat flow variations in basin evolution.

Practical exercise: characteristic time scale and length scale for heating

The characteristic time scale is κ/v^2.
The characteristic length scale is κ/v.
Calculate these values for the sedimentation velocities above.

Solution

Table A12.2   The characteristic length (depth) and time of the temperature transient caused by sedimentation

Sedimentation velocity v (m yr^{-1})	Characteristic length (km)	Characteristic time (Myr)
0.01	3.153	0.315
0.001	31.53	31.54
0.0001	315.4	3154.0
0.00001	3154	315×10^3

Practical exercise: effect of rapid sedimentation on the geothermal gradient and heat flow at the surface

The scaled (versus G) geothermal gradient at the surface where $y = 0$ is

$$\frac{1}{G}\frac{\partial T}{\partial y} = \left(1 + \frac{v^2 t}{2\kappa}\right) erfc\left(\frac{v}{2}\sqrt{\frac{t}{\kappa}}\right) - v\sqrt{\frac{t}{\pi\kappa}}\exp\left(-\frac{v^2 t}{4\kappa}\right) \qquad [\text{A12.3}]$$

Calculate the scaled geothermal gradient for the different sedimentation velocities used above.

Solution

The scaled geothermal gradient at the surface is provided in Table A12.1.

The time scale of thermal response to sedimentation is very long, so the speed of adjustment to the deposition of cold sediment is slow. At a sedimentation rate of $0.1\,\text{mm yr}^{-1}$ ($0.0001\,\text{m yr}^{-1}$), the characteristic time is about $3 \times 10^9\,\text{yr}$ for the geothermal gradient to recover from the effects of sedimentation. This has major implications for the impact of variable sedimentation rates on the thermal maturity of the basin-fill.

Effects of variable radiogenic heating and thermal conductivity on the geotherm in the basin-fill

Learning points: sedimentary basins contain radiogenic detrital minerals derived by erosion of adjacent radiogenic upper crust. Consequently, strong variations in the geotherm may exist between a full-thickness upper crustal layer overlain by a sedimentary basin, an adjacent column with an eroded radiogenic upper crustal layer, and an adjacent column with a thickened crustal radiogenic layer. This may result in a lateral heat flows between the three columns. The choice of geothermal model has profound implications for the paleotemperatures experienced by stratigraphic horizons during burial.

Geotherm for crust blanketed by a radiogenic sedimentary basin

Firstly, we consider the geotherm resulting from heat flow through a two-layer model composed of a sedimentary basin overlying a radiogenic crustal layer. This solution is valuable for calculating the geotherms through two adjacent columns of continent, one eroded and the other blanketed with sediment. For a uniform radiogenic heat production A_1 in an upper layer representing the sedimentary basin in the depth range $0 < y < y_1$, and a uniform radiogenic heat production A_2 in the depth range $y_1 < y < y_2$, we have the two solutions:

$$T = T_0 + \left\{ \frac{Q_m}{K} + \frac{A_2(y_2 - y_1)}{K} + \frac{A_1 y_1}{K} \right\} y - \frac{A_1}{2K} y^2 \quad 0 \le y < y_1 \quad \text{[A13.1]}$$

and

$$T = T_0 + \left\{ \frac{Q_m}{K} + \frac{A_2 y_2}{K} \right\} y + \left\{ \frac{A_1 - A_2}{2K} \right\} y_1^2 - \frac{A_2}{2K} y^2 \quad y_1 \le y < y_2 \quad \text{[A13.2]}$$

The effect of the difference in heat production of the basin-fill overlying a radiogenic upper crustal layer compared to the adjacent eroded upper continental crust can now be evaluated using the following parameter values:

Radiogenic basin heat production, $A_1 = 5 \times 10^{-6} \, \text{W m}^{-3}$
Radiogenic upper crustal layer heat production $A_2 = 2.5 \times 10^{-6} \, \text{W m}^{-3}$
Depth to base of basin $y_1 = 5 \, \text{km}$

Depth to base of upper crustal layer $y_2 = 15 \, \text{km}$
Surface temperature $T_0 = 0 \, ^\circ\text{C}$
Basal heat flow $Q_m = 30 \, \text{mW m}^{-2}$
Basin-fill and upper crustal thermal conductivity $K = 3 \, \text{W m}^{-1} \text{K}^{-1}$

Geotherm for an eroded upper crustal radiogenic layer

We make the following modifications:

Depth to base of radiogenic upper crustal layer $y_2 = 5 \, \text{km}$ (5 km eroded)
Radiogenic upper crustal layer heat production $A_2 = 2.5 \times 10^{-6} \, \text{W m}^{-3}$
Upper crustal thermal conductivity $K = 3 \, \text{W m}^{-1} \text{K}^{-1}$
Surface temperature $T_0 = 0 \, ^\circ\text{C}$
Basal heat flow $Q_m = 30 \, \text{mW m}^{-2}$

Geotherm for a thickened upper crustal radiogenic layer (no sedimentary basin)

We make the further modifications to account for a doubling of the thickness of the upper crustal radiogenic layer (as may result from convergent tectonic deformation), with no erosion and no basin development:

Depth to base of radiogenic upper crustal layer $y_2 = 20 \, \text{km}$
Radiogenic upper crustal layer heat production $A_2 = 2.5 \times 10^{-6} \, \text{W m}^{-3}$
Upper crustal thermal conductivity $K = 3 \, \text{W m}^{-1} \text{K}^{-1}$

Basin Analysis: Principles and Application to Petroleum Play Assessment, Third Edition. Philip A. Allen and John R. Allen.

Solution

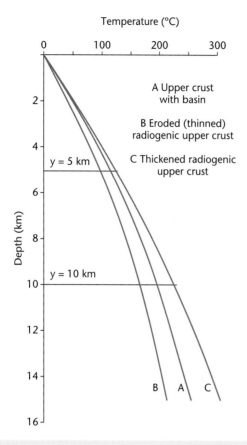

Fig. A13.1 Comparison of geotherms using a two-layer model, for (A) a sedimentary basin overlying a radiogenic upper crustal layer, (B) an eroded (thinned) upper crust with no sedimentary basin, and (C) a thickened upper crustal radiogenic layer with no sedimentary basin. At 5 km depth, the temperature range is from 81 to 123 °C, whereas at 10 km depth, the temperature range is from 142 to 225 °C. These temperature differences are large enough to cause lateral heat flows and different tectonic deformation through the impact on temperature-dependent rheologies.

Temperatures are strongly elevated in and under the sedimentary basin with its radiogenic infill and normal thickness upper crustal radiogenic slab compared to the adjacent eroded crust. Clearly, the radiogenic structure of the crust and basin-fill are critical to the prediction of the continental geotherm. The deposition of a radiogenic, self-heating sedimentary infill has an important impact on basin-fill and crustal temperatures, and thereby influences the thermal maturation of lithologies within the basin, and potentially the deformation of upper crustal rocks. The highest temperatures are, however, where the radiogenic upper crustal layer has been thickened tectonically.

Geotherm in a basin with variable thermal conductivity and self-heating

For the case of radiogenic self-heating and a variable thermal conductivity structure in the basin-fill, we start by considering, for sim-

plicity, a uniform radiogenic heat production A extending to the base of the crust at depth y_c, and a surface temperature T_0. As a reminder, for a constant thermal conductivity K, the geotherm is given by

$$T_y = T_0 + \left(\frac{Q_m + Ay_c}{K}\right)y - \frac{A}{2K}y^2 \qquad [\text{A13.3}]$$

where T is the temperature at any depth y and Q_m is the basal heat flow. With a variable thermal conductivity structure in the basin-fill, we have the contribution to the temperature field of the different layers:

$$T_y = T_0 + (-Q)\left\{\frac{l_1}{K_1} + \frac{l_2}{K_2} + \cdots \frac{l_n}{K_n}\right\} \qquad [\text{A13.4}]$$

which can be modified to:

$$T_y = T_0 + \left\{\left(\frac{Q_m + Ay_c}{K_1}\right)l_1 + \left(\frac{Q_m + Ay_c}{K_2}\right)l_2 + \cdots n\right\} \\ - \left\{\frac{A}{2K_1}l_1^2 + \frac{A}{2K_2}l_2^2 + \cdots n\right\} \qquad [\text{A13.5}]$$

where the basin-fill is made of n layers of thickness $l_1, l_2 \ldots l_n$ with thermal conductivities K_1, K_2 to K_n. If the geotherm is calculated for the basin-fill only, y_c becomes the sum of the n layers $l_1 + l_2 + \ldots . l_n$.

Finally, if n different layers in the basin-fill have their own radiogenic heat production A_1 to A_n, eqn. [A13.5] becomes

$$T_y = T_0 + \left\{\left(\frac{Q_m + A_1y_c}{K_1}\right)l_1 + \left(\frac{Q_m + A_2y_c}{K_2}\right)l_2 + \left(\frac{Q_m + A_3y_c}{K_3}\right)l_3 + \ldots n\right\} \\ - \left\{\left(\frac{A_1}{2K_1}\right)l_1^2 + \left(\frac{A_2}{2K_2}\right)l_2^2 + \left(\frac{A_3}{2K_3}\right)l_3^2 + \ldots n\right\} \qquad [\text{A13.6}]$$

Practical exercise

Compare the geotherms for the following two cases:

1. Two 'reference' geotherms caused by: (i) a basal heat flow of 60 mW m^{-2} but no radiogenic heating; and (ii) radiogenic heat production $A = 2.5 \times 10^{-6}$ W m^{-3} in the crust to a depth $y_c = 30$ km, and a basal heat flow of 25 mW m^{-2}. In both cases use a constant thermal conductivity of 3.0 W m^{-1} C^{-1}. Assume the surface temperature T_0 is 15 °C.

2. Variable radiogenic heat production and thermal conductivity in the basin-fill as follows:

Table A13.1

Depth range (m) [Layer thickness l_i (m)]	Layer radiogenic heat production A_i (W m$^{-3} \times 10^{-6}$)	Lithology	Layer bulk conductivity K_i (W m^{-1} C^{-1})
0–1000 [1000]	2.0	Anhydrite	6.0
1000–1500 [500]	1.0	Dolomite	3.35
1500–3000 [1500]	5.0	Shales	1.6
3000–4500 [1500]	0.3	Quartzite	7.0
4500–5000 [500]	5.0	Granite	3.1

Note the different temperatures at the base of the sedimentary basin (5 km) between the three models (Fig. A13.1). The gotherm for variable radiogenic heat production and thermal conductivity is shown in Fig. A13.2.

Solution

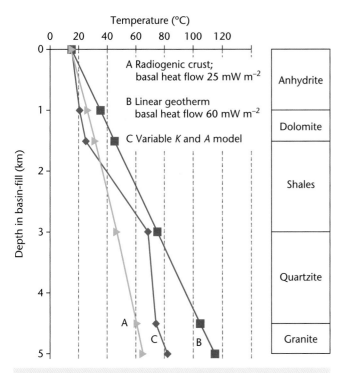

Fig. A13.2 Geotherm within the basin-fill with a variable thermal conductivity and radiogenic heat production structure (see Table A13.1) (labelled C), compared with a geotherm for a basal heat flow only (Q_m 60 mW m^{-2}) (B), and geotherm with a radiogenic crust ($A = 2 \times 10^{-6}$ W m^{-3}) and basal heat flow ($Q_m = 25$ mW m^{-2}) (A). A and B have a constant thermal conductivity of $K = 3$ W m^{-1} K^{-1}. Note how the geothermal gradient varies strongly with model C, showing strong departures from the two other models.

Practical exercise: the time-temperature trajectory

Appendix 58 gives a step-by-step practical example of the calculation of decompacted depths and paleotemperatures. Using the dataset above, we use decompacted depths and layer thicknesses as given in Table A13.2.

Table A13.2 Decompacted depths and thicknesses of stratigraphic layers in the basin-fill

Depth of top of layer (m)					
Age (Ma)	150	120	100	50	0
Layer					
Anhydrite	–	–	–	–	0
Dolomite	–	–	–	0	1000
Marine shales	–	–	0	600	1500
Quartzite	–	0	2550	2850	3000
Basement	0	1900	4200	4380	4500
Thickness of layer (m)					
Anhydrite	–	–	–	–	1000
Dolomite	–	–	–	600	500
Marine shales	–	–	2550	2250	1500
Quartzite	–	1900	1650	1530	1500

Solution

The decompacted depths of chosen horizons in the basin-fill are given in Fig. A13.3, and the time-temperature trajectory for three different thermal models is shown in Fig. A13.4.

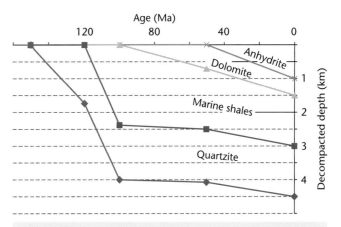

Fig. A13.3 Decompacted depths for different stratigraphic layers in the basin-fill. The different layers reduce in thickness over time due to compaction.

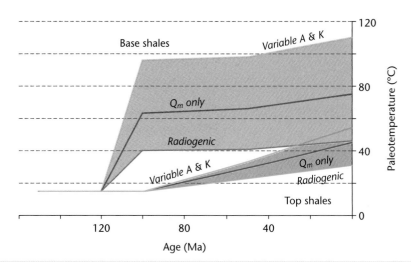

Fig. A13.4 Time-temperature trajectories for the top and base of the marine shales (deposited 120–100 Ma) for three geothermal models, as explained in text. Model A (labelled 'Q_m only') has a basal heat flow of $Q_m = 60\,\mathrm{mW\,m^{-2}}$, a uniform thermal conductivity of $3\,\mathrm{W\,m^{-1}\,K^{-1}}$, and no radiogenic heating. Surface temperature is 15 °C. Model B (labelled 'radiogenic') has a basal heat flow of $25\,\mathrm{mW\,m^{-2}}$, a uniform radiogenic heat production of $1 \times 10^{-6}\,\mathrm{W\,m^{-3}}$, and a uniform thermal conductivity of $3\,\mathrm{W\,m^{-1}\,K^{-1}}$. Model C (labelled 'Variable A and K') has a basal heat flow Q_m of $25\,\mathrm{mW\,m^{-2}}$ and a variable thermal conductivity structure and radiogenic heat production as given above. Note that the exercise is simplified by assuming that K_i and A_i values do not vary over time. For more accurate results, K_i values should be updated as the porosity of layers in the basin-fill evolves over time (Appendix 58). The key learning point is that the choice of geothermal model has a large influence on the time-temperature trajectory of units in the basin-fill.

14

The mantle adiabat and peridotite solidus

Learning points: the temperature gradient in the mantle is adiabatic, due to the progressive increase in pressure with depth. If the adiabatic geotherm intersects the solidus for peridotite, melting starts to occur. This may be promoted by lithospheric stretching or by anomalously high temperatures in the asthenosphere.

The mantle adiabat

Since the mantle is convecting, it can be assumed that a small volume of mantle rock does not exchange heat by conduction to its surroundings as it takes part in this convective motion. Instead, the temperature of the small volume or rock changes in response to the pressure it experiences since mantle rocks are compressible. A conservation of energy gives the temperature change with a change in pressure or depth (see Wangen 2010, p. 192)

$$\frac{dT}{dp} = \frac{T\alpha_v}{\rho c_p} \quad or \quad \frac{\partial T}{\partial y} = \frac{T\alpha_v g}{c_p} \qquad \text{[A14.1]}$$

where ρ is the rock density, c_p is the heat capacity at constant pressure, T is the absolute temperature in Kelvin, α_v is the volumetric coefficient of thermal expansion, and the pressure is assumed to be equal to $p = \rho g y$.

The temperature gradient can be expressed as

$$\frac{dT}{T} = \frac{\alpha_v g}{c_p} dy \qquad \text{[A14.2]}$$

Letting the characteristic length scale $l_0 = c_p/(\alpha_v g)$, eqn. [A14.2] can be integrated to obtain

$$T = T_0 \exp\left(\frac{y - y_0}{l_0}\right) \qquad \text{[A14.3]}$$

where T_0 is the temperature at the reference depth y_0. If the magnitude of $y-y_0$ is small compared to l_0, the adiabat becomes a linear approximation given by

$$T = T_0\left(1 + \frac{y - y_0}{l_0}\right) \qquad \text{[A14.4]}$$

If $T_0 = 1300\,°C$ (or 1573 K), $\alpha_v = 3 \times 10^{-5}\,K^{-1}$, and $c_p = 1\,kJ\,kg^{-1}\,K^{-1}$, l_0 must be of order $3 \times 10^5\,km$, and if the reference temperature is at the base of the thermal lithosphere at $y_0 = 125\,km$, the linear adiabatic geotherm in the mantle can be simply obtained from eqn. [A14.4]. The mantle temperature gradient is approximately $0.5\,K\,km^{-1}$ using the parameter values above, giving a temperature of $1427\,°C$ or 1700 K at a depth of $400\,km$.

The peridotite solidus

The solidus for mantle rock (peridotite) is approximated by a linear function

$$T_s(y) = T_{s,0} + Ay \qquad \text{[A14.5]}$$

where $T_{s,0}$ is the solidus temperature at surface conditions ($1100\,°C$) and A is a constant with a value of approximately $3\,°C\,km^{-1}$.

The peridotite solidus and the mantle adiabat are compared in Fig. A14.1. It can be seen that melting is unlikely to take place with the standard parameter values chosen.

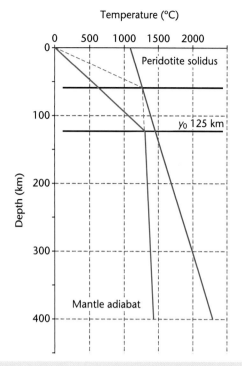

Fig. A14.1 Comparison of lithospheric conduction profile and mantle adiabat (blue) and solidus for peridotite (red). The closest approach of the two curves is at the base of the lithosphere. Stretching of the lithosphere by a factor of ~2 would raise the geotherm in the lithosphere (dashed line) and cause the base of the lithosphere to melt.

The onset of melting near the base of the lithosphere should take place at a minimum stretch factor (see Chapter 3) given by

$$\beta = \frac{y_L A}{T_a - T_{s,0}}$$ [A14.6]

where y_L is the initial lithospheric thickness and T_a is the temperature at the base of the lithosphere. If y_L is 125 km and T_a is 1300 °C, the minimum stretch factor for the onset of melting is 1.875.

15

Lithospheric strength envelopes

Learning points: the layering of the Earth in terms of its strength is a fundamental property that controls its deformation or flow. Two behaviours are recognised – a brittle domain and a temperature-dependent ductile domain. Conventionally, there is thought to be two brittle and two ductile layers in the continental crust and mantle lithosphere, but the presence and thickness of these different rheological layers depends of the geotherm, amount of volatiles and composition.

The relationship between stress and strain rate in ductile (power-law) creep is temperature-dependent and given by

$$\dot{\varepsilon} = (\sigma_1 - \sigma_3)^n A \exp\left(-\frac{E}{RT}\right) \qquad [A15.1]$$

where σ_1 and σ_3 are the largest and least principal stresses respectively, $A\exp(-E/RT)$ is the Arrhenius factor, and n is the power exponent on the stress difference. The strain rate during stretching of the lithosphere by a factor β over a time period t is

$$\dot{\varepsilon} = \frac{\ln \beta}{t} \qquad [A15.2]$$

so the strain rate derived from the stretch factor can be used to solve for the state of stress over a lithospheric profile of initial thickness y_L for a known temperature. The *lithospheric strength* S_l is then the integral of the stress difference over the lithospheric thickness:

$$S_l = \int_0^{yL} (\sigma_1 - \sigma_3) \mathrm{d}y \qquad [A15.3]$$

The least principal stress is assumed to be vertical and equal to gravity acting on the bulk density of the lithosphere ρ_b, and the greatest principal stress is assumed to be in the direction of stretching, assumed to be horizontal. The force F needed to stretch the lithosphere is then the equivalent to the lithospheric strength plus the gravity term

$$F = \int_0^{yL} \sigma_1 \mathrm{d}y = S_l + \frac{1}{2}\rho_b g y_L^2 \qquad [A15.4]$$

The strength of rock in the brittle field is given by Byerlee's law. The ratio of principal stresses is

$$\frac{\sigma_1}{\sigma_3} = \left(\sqrt{1+\mu^2} + \mu\right)^2 \approx 5 \qquad [A15.5]$$

where the coefficient of friction μ is taken as 0.85, and $\sigma_1 = \rho g y$ is the largest principal stress and σ_3 is the least principal stress in the direction of extension.

To plot the strength profile for the lithosphere, it is necessary to treat the brittle and ductile domains of the crust and mantle lithosphere separately.

Practical exercise

Use the parameter values in Table A15.1 (mostly from Stüwe 2002).

Basin Analysis: Principles and Application to Petroleum Play Assessment, Third Edition. Philip A. Allen and John R. Allen.
© 2013 John Wiley & Sons, Ltd. Published 2013 by John Wiley & Sons, Ltd.

Table A15.1   Rheological, thermal and mechanical parameters of the lithosphere for calculation of strength profiles

Parameter		Value	Units
Temperature T	Surface	273	K
	Base lithosphere	1593	K
Crustal thickness y_c		15	km
Lithosphere thickness y_L		120	km
Stretch factor β		2	[-]
Duration of stretching t		20	Myr
Coefficient of friction μ		0.85	[-]
Arrhenius prefactor A	Lower crust	100	$MPa^{-n}\,s^{-1}$
	Mantle	7×10^4	$MPa^{-n}\,s^{-1}$
Activation energy E	Lower crust	300	$kJ\,mole^{-1}$
	Mantle	540	$kJ\,mole^{-1}$
Arrhenius exponent n	Lower crust	3	[-]
	Mantle	3	[-]
Universal gas constant R		8.31451	$J\,mole^{-1}\,K^{-1}$
Strain rate $\dot{\varepsilon}$		1.1×10^{-15}	s^{-1}
Bulk density ρ	Crust	2650	$kg\,m^{-3}$
	Mantle	3300	$kg\,m^{-3}$
Basal heat flow Q_m		0.33	$W\,m^{-2}$
Thermal conductivity K	Lithosphere	3	$W\,m^{-1}\,K^{-1}$
Radiogenic heat production A	Crust	0, 2×10^{-6} and 4×10^{-6}	$W\,m^{-3}$

1. Plot the stress difference versus depth for the crust and mantle lithosphere with the properties given above.

   The strength envelope of the continental lithosphere is affected by its thermal state and its strain rate.

2. Plot the strength envelopes for variable radiogenic heat production of $0\,\mu W\,m^{-3}$, $2\,\mu W\,m^{-3}$ and $4\,\mu W\,m^{-3}$. For radiogenic heat production confined to the crust, use $y_c=15\,km$, thermal conductivity $K = 3\,W\,m^{-1}\,K^{-1}$ and basal heat flow $Q_m = 33\,mW\,m^{-2}$, and the following equation for the geotherm

$$T(y) = T_0 + \frac{Q_m + Ay_c}{K}\,y - \frac{A}{2K}\,y^2 \qquad [A15.6]$$

   for $0 < y < y_c$. For $y > y_c$, put $A = 0$ in eqn. [A15.6] and use $T(y = y_c)$ as T_0.

3. If the strain rate varies with values of $10^{-16}\,s^{-1}$, $10^{-14}\,s^{-1}$ and $10^{-12}\,s^{-1}$, what is the effect on the depth of the upper and lower brittle–ductile transitions (BDT1 and BDT2)?

Solutions

1. The lithosphere has two brittle–ductile transitions, BDT1 and BDT2, one in the lower crust between 10 and 15 km, and the other in the mantle lithosphere at a depth of c.25 km (Fig. A15.1). The brittle failure criterion is for tension.

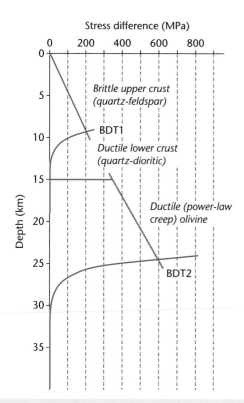

Fig. A15.1   Strength envelope for a strain rate of $\sim\!10^{-15}\,s^{-1}$, a geothermal gradient of 22 °C km^{-1}, zero radiogenic heat production, using the rheological information in Table A15.1 for the brittle upper crust and mantle lithosphere, and the ductile lower crust and mantle lithosphere. Crust is 15 km thick in a 120 km-thick lithosphere. Basal heat flow is 33 mW m^{-2} and thermal conductivity 3 W m^{-1} K^{-1} in order to have a temperature at the base of the lithosphere of 1593 K (1320 °C).

2. The geotherms for different values of radiogenic heat production show curvature in the crust ($y_c = 15\,\text{km}$), and since the basal heat flow Q_m is held constant at $33\,\text{mW}\,\text{m}^{-2}$, the temperature at the base of the lithosphere varies between geotherms (Fig. A15.2).

3. Table A15.2 summarises the impact of variable strain rate on the depth of BDT1 and BDT2. Depths are to nearest half kilometre.

Table A15.2 Results of calculations of the depth to the upper and lower brittle–ductile transitions

Strain rate	Depth of BDT1 (km)	Depth of BDT2 (km)
$10^{-16}\,\text{s}^{-1}$	8.5	23.5
$10^{-14}\,\text{s}^{-1}$	10	25
$10^{-12}\,\text{s}^{-1}$	11.5	27.5

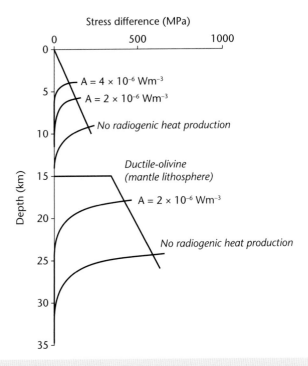

Fig. A15.2 Strength envelopes for variable radiogenic heat production, from zero to $2 \times 10^{-6}\,\text{W}\,\text{m}^{-3}$ to $4 \times 10^{-6}\,\text{W}\,\text{m}^{-3}$. Parameter values in Table A15.1. Note that the zero radiogenic heat production curve uses a basal heat flow of $33\,\text{mW}\,\text{m}^{-2}$ to obtain a temperature at the base of the lithosphere of $1593\,\text{K}$ ($1320\,^\circ\text{C}$) with a geothermal gradient of $22\,^\circ\text{C}\,\text{km}^{-1}$. The same basal heat flow is used for the curves using radiogenic heat production.

Rift zones: strain rate, extension velocity and bulk strain

Learning points: there are simple relationships between parameters concerning the stretching (or shortening) of the lithosphere. Some rigour is required in using these terms. Rift basins show wide variations in strain rate, degree of localisation of brittle faulting, total stretch and width of zone undergoing stretching.

Simple relationships between the initial width of a piece of lithosphere l_0, the amount of extension Δl, the extension velocity v, the stretch factor β and the extensional strain rate $\dot{\varepsilon}$ are as follows.

The stretch factor is the extended length compared to the initial length:

$$\beta = \frac{l_0 + \Delta l}{l_0} \qquad [\text{A16.1}]$$

The velocity of extension averaged over a time interval t is:

$$v = \frac{\Delta l}{t} \qquad [\text{A16.2}]$$

The stretch factor β evolves over time since it is the integral over time of the strain rate. The stretch factor increases exponentially for a constant strain rate over time. Consequently, the total stretch factor resulting from a constant strain rate over a time interval t is given by the relation

$$\dot{\varepsilon} = \frac{\ln \beta}{t} \qquad [\text{A16.3}]$$

which can be rearranged as

$$\dot{\varepsilon} = \frac{v \ln \beta}{\Delta l} \qquad [\text{A16.4}]$$

These relationships can be applied to the 'reference' uniform stretching model. Although the 'reference' uniform stretching model involves instantaneous stretching, the model is applicable to finite durations of stretching, so there must be a strain rate given by

$$\dot{\varepsilon}(t) = \frac{1}{l_0} \frac{dl}{dt} \qquad [\text{A16.5}]$$

where l_0 is the original length and dl/dt is the change in length over a time period t. The expression can be modified to describe the thinning in the vertical direction y:

$$\dot{\varepsilon}_y(t) = -\frac{1}{y} \frac{dy}{dt} \qquad [\text{A16.6}]$$

The strain rate is associated with horizontal and vertical velocities v_x and v_y at any depth y. When the vertical velocity is set to zero at the surface $y = 0$, the vertical velocity at any depth is simply

$$v_y = -\dot{\varepsilon}_y y \qquad [\text{A16.7}]$$

and the horizontal lateral velocity is

$$v_x = \dot{\varepsilon}(x - x_{ref}) \qquad [\text{A16.8}]$$

where x_{ref} is an arbitrary reference position where the lateral velocity is zero.

The relationship between extensional strain rate, stretch factor and extension velocity is shown graphically in Fig. A16.1.

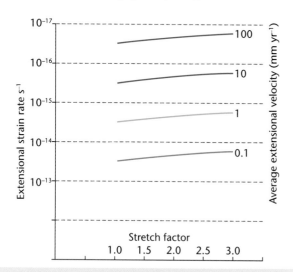

Fig. A16.1 Relationship between stretch factor and extensional strain rate for different values of the extensional velocity. Strain rate is assumed to be constant over the duration of stretching.

Basin Analysis: Principles and Application to Petroleum Play Assessment, Third Edition. Philip A. Allen and John R. Allen.
© 2013 John Wiley & Sons, Ltd. Published 2013 by John Wiley & Sons, Ltd.

Practical exercise

A piece of continental lithosphere originally 100 km across is extended to 120 km in a period of 5 Myr. What is the strain rate, average horizontal velocity and stretch factor? Assume that the strain rate is constant.

Solution

1. The stretch factor after 5 Myr of stretching, from eqn. [16.1], is 1.2.
2. The horizontal velocity averaged over the duration of stretching, given by eqn. [16.2], is 0.004 m yr^{-1}.
3. The extensional strain rate, assumed constant, given by eqn. [A16.3] or [A16.4] is 3.65×10^{-8} yr^{-1}, or 1.16×10^{-15} s^{-1}.

The 'reference' uniform extension model

Learning points: although the boundary conditions and assumptions can be debated, the uniform stretching model serves as a reference model with clear predictions for heat flow and subsidence. The model involves an Airy isostatic balance of a piece of continental lithosphere that is instantaneously stretched by pure shear, experiences heat flow by 1D conduction, and subsequently cools. The basal temperature is held constant, and there is no radiogenic heat production. There is therefore a steady-state and transient component to the temperature field after stretching.

The lithostatic column before rifting is made up of two components (see §2.1)

$$y_c \rho_c g + (y_L - y_c)\rho_{sc} g \qquad [\text{A17.1}]$$

where y_c and y_L are the thicknesses of the crust and lithosphere respectively, ρ_c and ρ_{sc} are the average densities of the crust and subcrustal lithosphere, and g is the gravitational acceleration, which, because it is common to all lithospheric columns, can be ignored in the following analysis. These densities are assumed to have a linear relationship to temperature, and the geotherm is also assumed to be linear from T_m at the base of the lithosphere to T_0 at the surface (Fig. 3.19). As a result, the average densities are given by

$$\rho_c = \rho_c^* (1 - \alpha_v T_c) \qquad [\text{A17.2}]$$

and

$$\rho_{sc} = \rho_m^* (1 - \alpha_v T_{sc}) \qquad [\text{A17.3}]$$

where ρ_c^* and ρ_m^* are crustal and mantle densities at 0 °C, α_v is the volumetric coefficient of thermal expansion, and T_c and T_{sc} are the average crustal and subcrustal temperatures. Since the geotherms are linear, T_c and T_{sc} can be easily obtained as

$$T_c = T_0 + \left(\frac{T_m - T_0}{2}\right)\left(\frac{y_c}{y_L}\right) \qquad [\text{A17.4}]$$

and

$$T_{sc} = \frac{1}{2}\left\{T_m + T_0 - (T_m - T_0)\frac{y_c}{y_L}\right\} \qquad [\text{A17.5}]$$

By assuming that the surface temperature is 0 °C, these expressions simplify to

$$T_c = \frac{T_m}{2}\frac{y_c}{y_L} \qquad [\text{A17.6}]$$

and

$$T_{sc} = \frac{T_m}{2}\left(1 - \frac{y_c}{y_L}\right) \qquad [\text{A17.7}]$$

After rifting there are four components to the lithostatic stress at the depth of the original lithospheric thickness:

$$y_s \rho_s + (y_L - y_c)\rho_{sc} + \left\{\frac{y_L}{\beta_L} - \frac{y_c}{\beta_c}\right\}\rho_{sc} + \left\{y_L - \frac{y_L}{\beta_L} - y_s\right\}\rho_m \qquad [\text{A17.8}]$$

where the new terms introduced are y_s, the thickness of sediment, water or air filling the rift axis; β_c and β_L, the stretch factors for the crust and lithosphere (for uniform stretching $\beta_c = \beta_L$); ρ_s, the average sediment, water, or air bulk density; and ρ_{sc}, the subcrustal mantle density (at a temperature of T_m), equal to $\rho_m^*(1 - \alpha T_m)$ where ρ_m^* is the mantle density at 0 °C.

Balancing the columns before and after uniform stretching ($\beta_c = \beta_L = \beta$)

$$y_c \rho_c + (y_L - y_c)\rho_{sc} = y_s \rho_s + \left(\frac{y_c}{\beta}\right)\rho_c$$
$$+ (y_L - y_c)\frac{1}{\beta}\rho_{sc} + \left(y_L - y_s - \frac{y_L}{\beta}\right)\rho_m \qquad [\text{A17.9}]$$

Regrouping the terms

$$y_s = \frac{(1 - 1/\beta)}{(\rho_m - \rho_s)}\{\rho_m y_L - y_c \rho_c - (y_L - y_c)\rho_{sc}\} \qquad [\text{A17.10}]$$

Substituting the correct terms for ρ_c, ρ_{sc}, ρ_s and ρ_m into eqn. [A17.10] we have after rearrangement

$$y_s = \frac{y_L\left\{(\rho_m^* - \rho_c^*)\frac{y_c}{y_L}\left(1 - \alpha_v \frac{T_m}{2}\frac{y_c}{y_L}\right) - \frac{\alpha_v T_m \rho_m^*}{2}\right\}(1 - 1/\beta)}{\rho_m^*(1 - \alpha_v T_m) - \rho_s} \qquad [\text{A17.11}]$$

Basin Analysis: Principles and Application to Petroleum Play Assessment, Third Edition. Philip A. Allen and John R. Allen.
© 2013 John Wiley & Sons, Ltd. Published 2013 by John Wiley & Sons, Ltd.

where β is the stretch factor, y_L is the initial thickness of the lithosphere, y_c is the initial thickness of the crust, ρ_m^* is the density of the mantle (at 0°C), ρ_c^* is the density of the crust (at 0°C), ρ_s is the average bulk density of sediment or water filling the rift, α_v is the thermal expansion coefficient of both crust and mantle, T_m is the temperature of the asthenosphere, and y_s is positive for subsidence, negative for uplift. Typical values for the above parameters, mostly derived from oceanic lithosphere studies (Parsons & Slater 1977), are $y_L = 125\,km$, $\rho_m^* = 3330\,kg\,m^{-3}$, $\rho_c^* = 2800\,kg\,m^{-3}$, $\alpha_v = 3.28 \times 10^{-5}\,°C^{-1}$ and $T_m = 1333\,°C$.

For all values of the stretch factor, the initial subsidence S_i is positive for values of $y_c/y_L > 0.12$, corresponding to a crustal thickness $y_c > 15\,km$ within a lithosphere of 125 km (parameter values as above, and sediment density of 2000 $kg\,m^{-3}$). The reference model therefore predicts an initial subsidence with no uplift where continental crusts are 'normal' in thickness (Fig. 3.20).

The initial subsidence caused by isostatic adjustment to mechanical stretching is followed by a long-term gradual subsidence caused by cooling and thermal contraction of the lithosphere following extension. This *thermal subsidence* is dependent on β alone.

The stretching of the lithosphere causes two responses: (i) the thinning of the crust and the fault-controlled subsidence is *permanent*, i.e. the brittle crust cannot regain its original thickness; and (ii) the thinning of the mantle lithosphere and any elevation changes caused by the presence of hot asthenosphere are *transient*.

In order to predict the cooling of the lithosphere following rifting that causes thermal subsidence, we must know the heat flow through time. Following the *instantaneous* increase in heat flow accompanying rifting, the heat flow decreases exponentially with time. We know two boundary conditions and we make two assumptions:

Boundary conditions:

(1) $T = 0$ at $y = y_L$
(2) $T = T_m$ at $y = 0$

Assumptions:

(1) lateral temperature gradients are much smaller than vertical gradients.

$$\left(\frac{\partial T}{\partial x} \approx 0 \approx \frac{\partial T}{\partial z}\right) \qquad [A17.12]$$

(2) internal heat production from radioisotopes is ignored.

The one-dimensional unsteady (that is, time-dependent) heat flow equation is

$$\frac{\partial T}{\partial t} = K\frac{\partial^2 T}{\partial y^2} \qquad [A17.13]$$

where the second derivative gives the curvature of the geotherm as it relaxes to its pre-stretching gradient. If the surface temperature $T_0 = 0°C$, the temperature at any depth and time $T(y, t)$ is made up of a steady-state solution $s(y) = T_m(1-y/y_L)$ which applies to a linear geotherm in the lithosphere, and an unsteady-state component $u(y, t)$. The general solution for the unsteady term is

$$u(y,t) = \sum_{n=0}^{\infty} An\sin\left(\frac{n\pi y}{y_L}\right)\exp\left(\frac{-n^2\pi^2\kappa t}{y_L^2}\right) \qquad [A17.14]$$

where A is a constant and n is an integer that expresses the order of the harmonic of the Fourier transform, and κ is the thermal diffusivity. Therefore, as n increases, the negative exponential becomes very small and ceases to contribute to the unsteady temperature field. As a gross approximation it is frequently satisfactory to consider only $n = 1$ (but see Appendix 23 for a more accurate solution using higher values of n). The constant A depends on the asthenospheric temperature (T_m) and the amount of stretching.

At $t = 0$, $An = \left\{\frac{2}{n}(-1)^{n+1}\frac{\beta}{n\pi}\sin\left(\frac{n\pi}{\beta}\right)T_m\right\}$ [A17.15]

so if $n = 1$, An simplifies to

$$\frac{2}{\pi}\frac{\beta}{\pi}\sin\left(\frac{\pi}{\beta}\right)T_m \qquad [A17.16]$$

The full solution for $T(y,t)$ is therefore the sum of the steady and unsteady components,

$$T(y,t) = T_m\left(1-\frac{y}{y_L}\right) + \left\{\frac{2}{\pi}\frac{\beta}{\pi}\sin\left(\frac{\pi}{\beta}\right)T_m\right\}\exp\left(-\frac{\pi^2\kappa t}{y_L^2}\right)\sin\left(\frac{\pi y}{y_L}\right) \qquad [A17.17]$$

or regrouping and simplifying

$$\frac{T(y,t)}{T_m} = \left(1-\frac{y}{y_L}\right) + \frac{2}{\pi}\frac{\beta}{\pi}\sin\left(\frac{\pi}{\beta}\right)\exp\left(-\frac{t}{\tau}\right)\sin\left(\frac{\pi y}{y_L}\right) \qquad [A17.18]$$

where $\tau = y_L^2/\pi^2\kappa$ and is known as the *thermal time constant* of the lithosphere.

The surface heat flux is given by Fourier's law (§2.2.1), which states that the flux is the temperature gradient times the thermal conductivity ($Q_{y=L} = K\,dT/dy$). For $n = 1$, this is

$$Q = \frac{KT_m}{y_L}\left\{1 + \frac{2\beta}{\pi}\sin\left(\frac{\pi}{\beta}\right)e^{-t/\tau}\right\} \qquad [A17.19]$$

The subsidence caused by the thermal contraction is then given by

$$S(t) \approx E_0\frac{\beta}{\pi}\sin\left(\frac{\pi}{\beta}\right)\left(1-e^{-t/\tau}\right) \qquad [A17.20]$$

where

$$E_0 = \frac{4y_L\rho_m^*\alpha_v T_m}{\pi^2(\rho_m^*-\rho_s)} \qquad [A17.21]$$

and ρ_s is the average density of the water or sediment filling the basin, ρ_m^* is the mantle density at 0°C, α_v is the volumetric coefficient of thermal expansion, and the remaining terms are defined above.

The result of eqn. [A17.19] is that heat flux is strongly dependent on the amount of stretching, the dependency becoming less after about one thermal time constant (τ), or 50 Myr, and insignificant after two thermal time constants. At times younger than ~30 Myr, higher harmonics ($n = 2, 3\ldots$) should be used in eqn. [A17.14]. Heat flux and subsidence are shown in Fig. 3.18. If the subsidence history of a basin is known, it is therefore possible to estimate β from thermal subsidence curves (Appendices 19, 20).

18

Boundary conditions for lithospheric stretching

Learning points: the boundary conditions for any quantitative model are critical. With reference to lithospheric stretching, a constant velocity BC requires the extensional strain rate to decrease with time, whereas a constant strain rate requires an accelerating velocity of extension. A constant tectonic force causes both the strain rate and the extension velocity to increase over time. There are two thermal BCs. If the basal heat flux is constant, the temperature at the base of the plate must decrease over time, whereas, if the basal temperature is held constant, the basal heat flux must increase as a result of the stretching, as in the reference uniform stretching model.

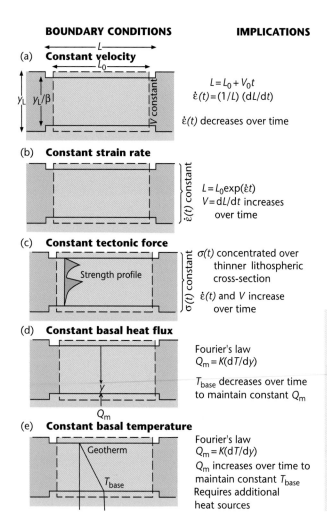

BOUNDARY CONDITIONS | **IMPLICATIONS**

(a) **Constant velocity**

$L = L_0 + V_0 t$
$\dot{\varepsilon}(t) = (1/L)\,(dL/dt)$

$\dot{\varepsilon}(t)$ decreases over time

(b) **Constant strain rate**

$\dot{\varepsilon}(t)$ constant

$L = L_0 \exp(\dot{\varepsilon} t)$
$V = dL/dt$ increases over time

(c) **Constant tectonic force**

Strength profile

$\sigma(t)$ constant

$\sigma(t)$ concentrated over thinner lithospheric cross-section
$\dot{\varepsilon}(t)$ and V increase over time

(d) **Constant basal heat flux**

Q_m

Fourier's law
$Q_m = K(dT/dy)$

T_{base} decreases over time to maintain constant Q_m

(e) **Constant basal temperature**

Geotherm

T_{base}

Fourier's law
$Q_m = K(dT/dy)$

Q_m increases over time to maintain constant T_{base}
Requires additional heat sources

Fig. A18.1 Boundary conditions for dynamical models of lithospheric extension. (a) Constant velocity of extension over time, which implies that the strain rate must decrease over time. (b) Strain rate constant over time, which implies that the extension velocity must increase over time. (c) Constant tectonic force (deviatoric stress) over time, which causes the same force to be concentrated over a thinner lithospheric cross-section over time. This force becomes concentrated in the strong layers in the lithosphere, which undergo large-scale necking. The strain rate therefore increases over time. (d) Constant basal heat flux over time, which implies, via Fourier's law, that the temperature at the base of the lithosphere must decrease over time. (e) Constant basal temperature over time, which implies that the basal heat flux must increase with time, as in the uniform stretching model of McKenzie (1978a). This requires additional heat sources to increase the basal heat flux.

Basin Analysis: Principles and Application to Petroleum Play Assessment, Third Edition. Philip A. Allen and John R. Allen.
© 2013 John Wiley & Sons, Ltd. Published 2013 by John Wiley & Sons, Ltd.

Dynamic models require *boundary conditions* on the margins of the extending lithosphere (Fig. A18.1). The choice of boundary conditions has important implications for the evolution through time of parameters in the model. As examples, the implications of five different boundary conditions are given below.

1. *Constant velocity boundary condition*: if L_0 is the initial width and v_0 is the initial (constant) extension velocity, the width of the extending zone after time t is

$$L = L_0 + v_0 t \qquad [A18.1]$$

and the extensional strain rate is

$$\dot{\varepsilon}(t) = (1/L)(dL/dt) \qquad [A18.2]$$

The extensional strain rate clearly decreases with time for a constant velocity boundary condition. A constant velocity boundary condition might be applicable to a situation where the extension is driven by relative plate motion, which over the duration of the extension has a constant vector.

2. *Constant strain rate boundary condition*: the width of the extending zone must vary in time as

$$L = L_0 \exp(\dot{\varepsilon}t) \qquad [A18.3]$$

and the extension velocity $v = (dL/dt)$ must increase as a function of time. For a constant strain rate boundary condition, the extensional velocity would accelerate unreasonably.

3. *Constant stress boundary condition*: the result of a constant tectonic force boundary condition is to concentrate the stress on a progressively thinner lithosphere, especially on those parts of the lithosphere that are most resistant to deformation, such as the strong upper-mid crust or lithospheric mantle (Kusznir 1982). Tensile deviatoric stresses would therefore increase with time at the site of lithospheric necking, leading to accelerating strain rate, unless a 'hardening' process prevents it. Such a 'hardening' process might be cooling, causing an increase in viscosity.

4. *Constant basal heat flux boundary condition*: if it assumed that there are no additional heat sources in the asthenosphere, such as plume heads and secondary convective cells, the heat flux at the base of the lithosphere can be assumed to be constant as extension proceeds. If so, since $Q = K(dT/dy)$, and the thickness of the lithosphere decreases by extension, the temperature at the base of the lithosphere should decrease by the stretch factor β with time.

5. *Constant basal temperature boundary condition*: in similar fashion, a constant basal temperature implies that the basal heat flow increases (by a factor β) through time. Since the base of the lithosphere is usually taken as the solidus temperature of peridotite (Pollack & Chapman 1977), which is a constant temperature, this is a commonly applied boundary condition, and is implied in the uniform stretching model of McKenzie (1978a). However, it requires additional heat fluxes from the mantle, such as those caused by secondary convection, as stretching proceeds.

Subsidence as a function of the stretch factor

Learning points: subsidence is made of two components, a fault-related syn-rift subsidence and a post-rift thermal subsidence. The thermal subsidence over time has the form of a negative exponential, with the amplitude increasing with the stretch factor. The time scale of the reduction of the subsidence rate in the post-rift stage is a diffusive time scale known as the lithospheric time constant, which is about 50 Myr.

The equations for calculating the syn-rift and post-rift subsidence using the uniform stretching model are found in Appendix 17.

Practical exercise

Calculate the syn-rift and post-rift subsidence for values of the stretch factor of 1.2, 1.5, 2.0 and 4.0. Assume that the sedimentary basin is filled with water, in which case $\rho_s = 1000\,\text{kg m}^{-3}$. Use the parameter values in Table A19.1.

Table A19.1 Parameter values used to calculate thermal subsidence

Parameter	Notation	Value	Units
Initial crustal thickness	y_c	35	km
Initial lithosphere thickness	y_L	125	km
Density of mantle at 0 °C	$\rho_m^{\bullet}$	3330	kg m^{-3}
Density of crust at 0 °C	$\rho_c^{\bullet}$	2800	kg m^{-3}
Volumetric coefficient of thermal expansion	α_v	3.28×10^{-5}	°C^{-1}
Temperature of the asthenosphere	T_m	1333	°C
Thermal diffusivity	κ	10^{-6}	m^2 s^{-1}

Solution

The *water-loaded* syn-rift and post-rift subsidence using the uniform stretching model for stretch factors of 1.2, 1.5, 2.0 and 4.0 are as illustrated in Fig. A19.1:

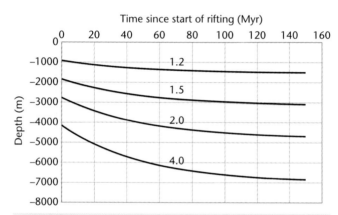

Fig. A19.1 Subsidence versus time since rifting for values of the stretch factor of 1.2. 1.5, 2.0 and 4.0. Subsidence values are derived from the reference uniform stretching model using water as the infilling material. Syn-rift subsidence takes place instantaneously at $t = 0$.

Basin Analysis: Principles and Application to Petroleum Play Assessment, Third Edition. Philip A. Allen and John R. Allen.
© 2013 John Wiley & Sons, Ltd. Published 2013 by John Wiley & Sons, Ltd.

Inversion of the stretch factor from thermal subsidence data

Learning points: accurate estimates of the stretch factor are invaluable for basin modelling, particularly of the heat flow history of a basin, which cannot be directly measured. One way of estimating the stretch factor is from the shape of thermal subsidence curves. This is best done by firstly removing the effects of the sediment load to obtain the water-loaded thermal subsidence, and then finding a solution for the stretch factor.

It is possible to calculate the stretch factor from the (post-rift) thermal subsidence. The underlying theory and the method for inverting for β are found in Appendices 17 and 19.

To calculate the stretch factor β we must first identify the beginning of the post-rift phase (from, for example, seismic reflection data showing the end of extensional faulting, the end of fault-controlled sedimentation, the beginning of regional sag-type subsidence).

The beginning of thermal subsidence is taken at 125 Ma in the practical example below. From the decompaction and backstripping exercise (see Appendices 56 to 58 for further details), we know that the water-filled tectonic subsidence Y is as given in Table A20.1:

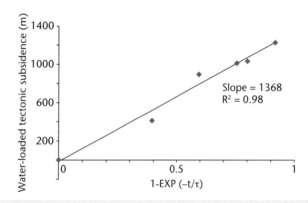

Fig. A20.1 Linear regression used to calculate stretch factor.

Table A20.1 Water-loaded tectonic subsidence versus time before present used in the practical exercise

Age (Ma)	125	100	80	55	45	0
Y (m)	0	0.410	0.892	1.009	1.037	1.231

Practical exercise

Plot a graph of the water-loaded thermal subsidence versus $1 - \exp(-t/\tau)$, where t is the time since the end of rifting, and τ is the lithospheric time constant (50 Myr).

Solution

The slope of the linear regression is equal to $E_0(\beta/\pi)\sin(\pi/\beta)$ and E_0 is given by $E_0 = (4y_L\rho_m\alpha_v T_m)/(\pi^2(\rho_m-\rho_w))$ (Fig. A20.1). The solution for the stretch factor is $\beta = \mathbf{1.52}$ (E_0 is 3212 m).

Maximum thermal subsidence

The maximum thermal subsidence caused by stretching can be obtained analytically using

$$S_{t,max} = \frac{1}{2}\left(1-\frac{1}{\beta}\right)\frac{\rho_m^*}{\rho_m^*-\rho_s}\alpha_v T_a y_L \qquad [A20.1]$$

where the asterisk superscript denotes the density as a reference temperature of $0\,^{\circ}C$, α_v is the volumetric coefficient of thermal expansion, T_a is the temperature of the top of the asthenosphere, and y_L is the initial thickness of the lithosphere. Eqn. [A20.1] can be modified for the water-loaded case by the substitution of ρ_w for ρ_s.

Basin Analysis: Principles and Application to Petroleum Play Assessment, Third Edition. Philip A. Allen and John R. Allen.
© 2013 John Wiley & Sons, Ltd. Published 2013 by John Wiley & Sons, Ltd.

Practical exercise

Take the following parameter values:

Water density	$1030\,kg\,m^{-3}$
Bulk sediment density	$2300\,kg\,m^{-3}$
Reference mantle density	$3300\,kg\,m^{-3}$
Asthenospheric temperature	$1300\,°C$
Initial lithospheric thickness	$125\,km$
Thermal expansion coefficient	$3 \times 10^{-5}\,K^{-1}$

What is the stretch factor if a maximum of 3.5 km of sediment accumulates in a sediment-filled basin during the post-rift stage?

What is the stretch factor if a maximum of 2 km of water-loaded subsidence takes place during the post-rift stage?

Solution

The maximum thermal subsidence for a sediment-filled and a water-filled basin as a function of the stretch factor is given in Fig. A20.2.

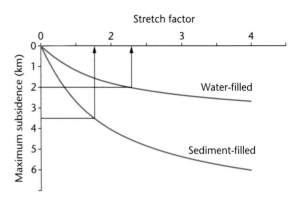

Fig. A20.2 Plot of maximum thermal subsidence versus stretch factor for sediment- and water-filled basins.

The stretch factor for a maximum thermal subsidence of 3.5 km in a sediment-filled basin is 1.75.

The stretch factor for a maximum thermal subsidence of 2 km in a water-filled (backstripped) basin is 2.3.

21

Calculation of the instantaneous syn-rift subsidence

Learning points: the syn-rift subsidence results from an isostatic balance before and after lithospheric stretching and is a trade-off between subsidence due to crustal thinning and uplift due to the thermal effects of lithospheric thinning. Syn-rift subsidence depends on the stretch factor and on the ratio between crustal and lithospheric thicknesses. Crustal stretching is permanent, but is spatially highly variable, depending on the localisation of brittle deformation in prominent fault zones.

The syn-rift subsidence can be calculated from the isostatic balance shown in Fig. A21.1, where the thickness of sediment is small enough to not alter the linear conduction geotherm. It is important to note that crustal thinning reduces the thickness of relatively light crust, causing subsidence, whereas the thinning of the subcrustal lithosphere is accompanied by heating, thermal expansion and consequently uplift. We therefore need to know what the effects of temperature changes will be on the densities of the crust and, especially, the subcrustal lithosphere. The initial geothermal gradient is linear and given by

$$\frac{dT}{dy} = (T_a - T_s)/y_L \qquad [A21.1]$$

where T_a is the temperature at the base of the lithosphere and y_L is the initial lithospheric thickness.

Uniform stretching of crust and lithosphere

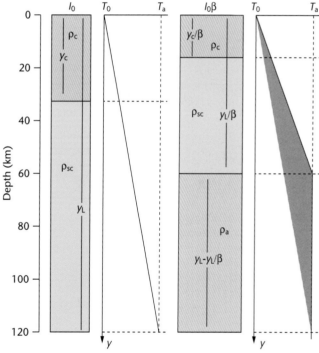

Fig. A21.1 Isostatic balance before and after instantaneous stretching. Shaded area shows temperature transient immediately after instantaneous stretching.

Basin Analysis: Principles and Application to Petroleum Play Assessment, Third Edition. Philip A. Allen and John R. Allen.
© 2013 John Wiley & Sons, Ltd. Published 2013 by John Wiley & Sons, Ltd.

Immediately after instantaneous stretching, the geothermal gradient is increased by a factor β, giving

$$\frac{dT}{dy} = \beta \frac{(T_a - T_s)}{y_L}$$
[A21.2]

We will simplify by assuming that the surface temperature T_s is zero °C.

After stretching, the lithosphere cools back to its initial position. The upwelled asthenosphere, with temperature T_a, therefore cools conductively to the surface and becomes lithosphere again, whereas the crust is permanently thinned by mechanical stretching.

We know (Chapter 2, Appendix 1) that the density of a rock changes with its temperature

$$\rho(T) = \rho^* \left(1 - \alpha_v \left(T - T^*\right)\right)$$

where ρ^* is the density at a reference temperature of T^*, usually taken as 0 °C, and α_v is the volumetric coefficient of thermal expansion, which is ~3×10^{-5} K^{-1} for mantle rock.

An isostatic balance, where a thin layer of sediment y_s of density ρ_s is included at the top, gives the syn-rift subsidence

$$y_s = y_L \left(1 - \frac{1}{\beta}\right) \frac{\left(\rho_m^* - \rho_c^*\right)\left(y_c / y_L\right)\left(1 - \frac{1}{2}\alpha T_a \frac{y_c}{y_L}\right) - \left(\frac{1}{2}\rho_m^* \alpha T_a\right)}{\rho_m^* \left(1 - \frac{1}{2}\alpha T_a\right) - \rho_s}$$
[A21.3]

Practical exercise

Assume the following:

Temperature at base of lithosphere T_a	1300 °C
Thermal expansion coefficient crust α_v	3×10^{-5} K^{-1}
Thermal expansion coefficient mantle α_v	3×10^{-5} K^{-1}
Density of crust at 0 °C ρ_c^*	2800 kg m^{-3}
Density of mantle at 0 °C, ρ_m^*	3300 kg m^{-3}
Initial crustal thickness y_c	32 km
Initial lithosphere thickness y_L	120 km

1. Calculate the syn-rift subsidence for a range of stretch factors between 1 and 4.
2. Experiment with a range of y_c/y_L between 0.1 and 0.4.
3. At what value of y_c/y_L does the syn-rift subsidence become negative, indicating uplift?

Solution

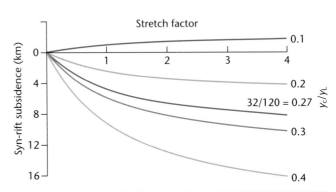

Fig. A21.2 Syn-rift subsidence for a sediment-filled basin as a function of the stretch factor and the crustal/lithosphere thickness ratio. The curve for $y_c/y_L = 0.27$ represents standard crust and lithosphere thicknesses.

The syn-rift subsidence as a function of the stretch factor is shown for a range of values of y_c/y_L in Figure A21.2.

The elevation change at the surface of the Earth becomes zero and uplift occurs when

$$\frac{y_c}{y_L} > \frac{1}{2} \frac{\rho_m^*}{\left(\rho_m^* - \rho_c^*\right)} \alpha_v T_a$$
[A21.4]

which is when $y_c/y_L \sim 0.13$.

The transient temperature solution

Learning points: cooling following stretching is responsible for post-rift thermal subsidence. The transient temperature field, which is controlled mainly by the stretch factor, has a characteristic half-life (or decay time). The temperature transient causes an exponentially decreasing subsidence rate over time.

The temperature during conductive cooling following stretching is made of two components: the initial steady-state temperature plus an unsteady (transient) term that varies in both depth and time (Fig. A22.1).

We firstly make depth dimensionless by $\hat{y} = y/y_L$, and time dimensionless by $\hat{t} = t/t_0$, where t_0 is a characteristic diffusive time scale equal to y_L^2/κ.

The solution for transient (unsteady) temperature T_u is given by

$$T_u(\hat{y}, \hat{t}) = 2\beta \sum_{n=1}^{\infty} \frac{\sin(n\pi/\beta)}{(n\pi)^2} \sin(n\pi\hat{y}) \exp\left(-(n\pi)^2 \hat{t}\right) \quad [A22.1]$$

The half-life of the temperature transient is

$$\hat{t}_{1/2} = \frac{\ln 2}{\pi^2} = 0.07 \quad [A22.2]$$

which means that the temperature transient has decayed to half of its initial value after 7% of the characteristic time. The geothermal temperature is then the sum of the steady ($T_a y/y_L$) and the transient temperature, where y is measured positive downwards from the surface. The scaled version of the steady state temperature is simply $\hat{y}$, so the sum of the steady state and transient solutions is

$$\hat{T}(\hat{y}, \hat{t}) = \hat{y} + T_u(\hat{y}, \hat{t}) \quad [A22.3]$$

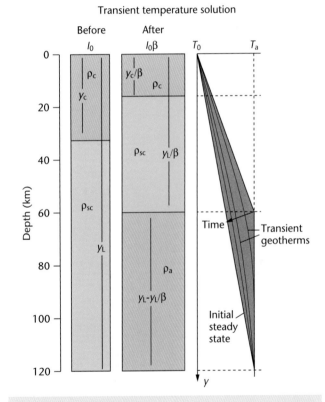

Fig. A22.1   Transient temperature solution for cooling following instantaneous stretching, shown for a stretch factor of 2.

Basin Analysis: Principles and Application to Petroleum Play Assessment, Third Edition. Philip A. Allen and John R. Allen.
© 2013 John Wiley & Sons, Ltd. Published 2013 by John Wiley & Sons, Ltd.

Practical exercise

1. Calculate the temperature $T(y,t)$ for a stretch factor of 2, using an initial lithospheric thickness y_L of 120 km and a thermal diffusivity κ of $10^{-6}\,\mathrm{m^2\,s^{-1}}$
2. What is the half-life in millions of years?

Solution

The transient temperature is given in Fig. A22.2.
    The characteristic time t_0 is 457 Myr, so the half-life is $0.07 \times 457 = 32$ Myr.

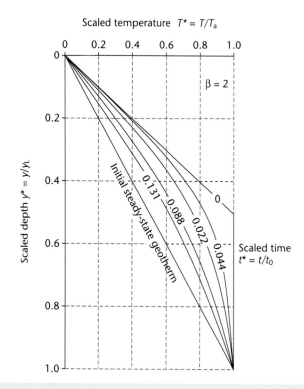

Fig. A22.2   Temperature as a function of depth and time, for a stretch factor of 2, using the Fourier series solution given in eqn [A22.1] ($n = 10$). Curves are labelled with values for dimensionless time.

Heat flow during uniform stretching using a Fourier series

Learning points: the transient heat flow following stretching is solved using a Fourier series expansion, where the accuracy of the solution improves with the number of terms in the expansion n. But how large should n be to get an acceptable solution? At short times following rifting (<20 Myr), the Fourier series needs to be expanded to $n > 10$, whereas the first term only ($n = 1$) is adequate for long times since rifting.

The surface heat flow immediately after lithospheric stretching relative to the steady-state heat flow is denoted by $\hat{q}$.

The steady-state heat flow (that is, prior to stretching) is given by

$$q_0 = \frac{KT_a}{y_L} \qquad [A23.1]$$

where K is the thermal conductivity, T_a is the temperature at the top of the asthenosphere and y_L is the initial lithospheric thickness.

The surface heat flow q_s relative to the steady-state heat flow q_0 can be expressed as a Fourier series

$$\hat{q} = \frac{q_s}{q_0} = 1 + 2\beta \sum_{n=1}^{\infty} \frac{\sin(n\pi/\beta)}{n\pi} \exp\left(-(n\pi)^2 \hat{t}\right) \qquad [A23.2]$$

where $\hat{t}$ is a scaled time equal to t/t_0, and t_0 is the characteristic diffusive time y_L^2/κ for the lithosphere, and β is the stretch factor.

The scaled surface heat flow immediately after instantaneous stretching is the solution in eqn. [A23.2] at time zero, which is

$$\hat{q} = 1 + 2\beta \sum_{n=1}^{\infty} \frac{\sin(n\pi/\beta)}{n\pi} \qquad [A23.3]$$

Since $\hat{q} = \beta$ at $t = 0$, the sum of the Fourier series must be

$$\sum_{n=1}^{\infty} \frac{\sin(n\pi/\beta)}{n\pi} = \frac{\beta - 1}{2\beta} \qquad [A23.4]$$

The value of the sum of the Fourier series at $t = 0$ is $(\beta - 1)/2\beta$, which for $\beta = 2$, is equal to 0.25, so the scaled surface heat flow becomes $1 + 0.25(2\beta) = 2$. That is, the surface heat flow increases by the factor β immediately after instantaneous stretching.

Practical exercise

Carry out the following:

1. Calculate the accuracy of the solution in eqn. [A23.2] based on the number of terms in the Fourier series expansion.
2. Plot the scaled surface heat flow $\hat{q}$ versus scaled time $\hat{t}$ for a stretch factor of $\beta = 2$.
3. What is the half-life of the scaled surface heat flow for a stretch factor of $\beta = 2$?

Solution

1. The accuracy of the solution as a function of the length of the Fourier series is shown in Fig. A23.1.

Basin Analysis: Principles and Application to Petroleum Play Assessment, Third Edition. Philip A. Allen and John R. Allen.
© 2013 John Wiley & Sons, Ltd. Published 2013 by John Wiley & Sons, Ltd.

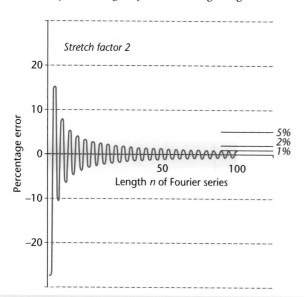

Fig. A23.1   Percentage error on the scaled surface heat flow based on the value of *n* in the Fourier series solution. Shaded area shows ±5% error. Significant errors are found at *n* < 10.

2. The scaled surface heat flow versus scaled time is shown in Fig. A23.2.

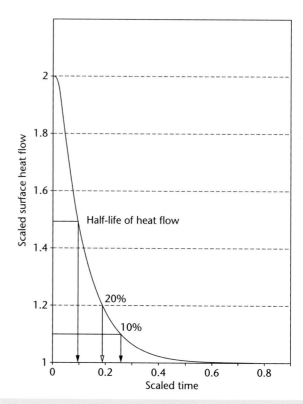

Fig. A23.2   Surface heat flow scaled by the steady-state heat flow versus time scaled by the characteristic diffusive time scale for the lithosphere, for a stretch factor of 2.

3. The scaled surface heat flow reduces to half of its value immediately after stretching at $\hat{t} \sim 0.1$, that is, at a time of approximately 50 Myr after instantaneous stretching.

24

The stretch factor for extension along crustal faults

Learning points: the thermal subsidence of a stretched basin gives an estimate of the *lithospheric* stretch factor. The geometry of normal faults provides an estimate of the *crustal* stretch factor. Some extension may take place on faults that are not imaged on seismic reflection profiles, so β_c derived from this method is a minimum, and should be compared where possible with β_c estimated from mapping of the Moho.

The brittle crust may deform during extension by the formation of tilted 'domino-style' fault blocks (Fig. A24.1). The stretch factor can be obtained from the dip of the faults defining the domino-like pieces of crust:

$$\beta = \frac{1}{\sin\phi} \qquad \text{[A24.1]}$$

where ϕ is the angle that the blocks make with the horizontal.

Practical exercise

What is the stretch factor for domino blocks inclined at 60° to the horizontal?

Solution

Using eqn. [A24.1], $\beta = 1.16$.

(a)

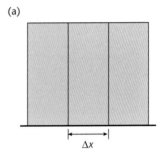

(b)   (c)

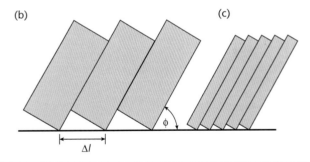

Fig. A24.1 Extension and thinning by a rotation of domino-like fault blocks (following Wangen 2010, p. 219, fig 7.16). The horizontal separation of the faults is determined by the angle of rotation (b) and the initial spacing (and displacement) of the faults (c). © Magnus Wangen, published by Cambridge University Press, 2010.

Basin Analysis: Principles and Application to Petroleum Play Assessment, Third Edition. Philip A. Allen and John R. Allen.
© 2013 John Wiley & Sons, Ltd. Published 2013 by John Wiley & Sons, Ltd.

Extension may also take place along a low-angle detachment that crosses the entire lithosphere (Fig. A24.2). The stretch factor for the crust, where $y_{c,0}$ is the original crustal thickness, y_c is the new crustal thickness, d is the horizontal offset of the plates, and θ is the dip of the detachment fault, is given by

$$\beta_c = \frac{y_{c,0}}{y_c} = \frac{y_{c,0}}{y_{c,0} - d\tan\theta} \qquad [\text{A24.2}]$$

Mantle rocks will be exhumed in the footwall after a vertical offset (throw) equal to the original crustal thickness, corresponding to a horizontal offset of $y_{c,0}/\tan\theta$.

The isostatic subsidence resulting from crustal thinning on a low-angle detachment fault is given by

$$y_s = \frac{(\rho_m - \rho_c)}{(\rho_m - \rho_s)} d\tan\theta \qquad [\text{A24.3}]$$

Practical exercise

Take the following parameter values: $y_{c,0} = 35\,\text{km}$; $\theta = 20°$; $d = 50\,\text{km}$.

1. What is the crustal stretch factor?
2. At what horizontal offset will mantle rocks in the footwall start to be exhumed?
3. What thickness of sediments will potentially accumulate on the hangingwall block at a horizontal offset of 50 km? Use mantle, crustal and sediment densities of $3300\,\text{kg m}^{-3}$, $2750\,\text{kg m}^{-3}$ and $2500\,\text{kg m}^{-3}$.

Solution

1. $\beta_c = 2.1$.
2. The critical horizontal offset for the exhumation of mantle is 96 km.
3. The sediment thickness from an isostatic balance is 12.5 km.

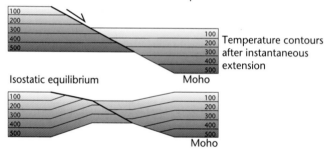

Fig. A24.2   Simple shear stretching of the crust by a low-angle detachment fault (a) and its isostatic compensation (b). Contours of temperature in °C, immediately after instantaneous stretching, redrawn from Wernicke (1985) in Wangen (2010, p. 220).

Protracted rifting times during continental extension

Learning points: the reference uniform stretching model assumes instantaneous stretching, which provides a convenient boundary condition for thermal calculations, but the geological record suggests that periods of rifting may be long (protracted) in duration. The importance of protracted rifting from a thermal point of view can be gauged by a Péclet number – expressing the relative importance of conductive cooling to the surface, and the effect of the upward advection of rock during stretching. Protracted stretching, as is common in 'slow' rifts and cratonic basins, requires modification of the heat flow calculations compared to the reference 'instantaneous' model.

The applicability of the assumption of instantaneous stretching can be gauged by considering the time scale for the decay of thermal transients by heat conduction, which has the usual diffusive form:

$$t_s = \frac{\ln 2 l_0^2}{\pi^2 \kappa} \qquad [A25.1]$$

where t_s is the half-life for the decay, l_0 is the characteristic length scale, which is the thickness of the stretched lithosphere equal to y_L/β, and κ is the thermal diffusivity. The time scale of rifting must be shorter than t_s for heat flow due to upward advection to dominate over conductive heat flow causing cooling – a requirement for the instantaneous stretching model.

Practical exercise

What is the critical duration of rifting for which the instantaneous model applies, assuming $\kappa = 10^{-6}\,\mathrm{m^2\,s^{-1}}$, and $y_L = 125\,\mathrm{km}$, expressed in terms of the stretch factor β?

Solution

Substituting $l_0 = y_L/\beta$ into eqn. [A25.1], and recalling that the lithospheric time constant is $\tau = y_L^2/\pi^2\kappa$, we have

$$t_s = \frac{\ln 2 y_L^2}{\pi^2 \kappa \beta^2} = \frac{\tau}{\beta^2} \qquad [A25.2]$$

which gives a critical duration of rifting in Myr of $50.2/\beta^2\,\mathrm{Myr}$. For $\beta = 1.5$, this gives $t_s = 22\,\mathrm{Myr}$, and for $\beta = 2$, gives $12.5\,\mathrm{Myr}$.

The Péclet number gives an indication of the dominance of advection over conduction. For vertical flow, $\mathbf{Pe} = v_y y_L/\kappa$, where

the vertical velocity at the base of the lithosphere v_y is $\dot{\varepsilon} y_L$, and the characteristic time constant for the lithosphere is $\tau = y_L^2/\kappa$. For convection to dominate, at $\mathbf{Pe} \gg 1$, we have

$$Pe = \frac{v_y y_L}{\kappa} = \frac{\dot{\varepsilon} y_L^2}{\kappa} = \ln \beta \frac{\tau}{t_s} \gg 1 \qquad [A25.3]$$

where the strain rate $\dot{\varepsilon} = \ln\beta/t_s$ is that when the stretching has reached the value β after a time interval t_s.

Practical exercise

1. Use eqn. [A25.3] to find the critical stretch factor at which advection dominates.
2. What is the required vertical strain rate $\dot{\varepsilon}_y$?

Solution

1. Since the decay half-life t_s depends on the stretch factor, eqn. [A25.3] is solved by iteration, and $\beta_{cr} = \exp(1/\beta^2)$. The critical stretch factor at which $\mathbf{Pe} > 1$ is 1.53.
2. The vertical strain rate required is $0.02 \times 10^{-6}\,\mathrm{yr^{-1}}$, or $6 \times 10^{-16}\,\mathrm{s^{-1}}$. The vertical velocity at the base of the lithosphere is $2.5\,\mathrm{mm\,yr^{-1}}$.

This vertical strain rate and critical stretch factor are commonly exceeded in the stretching of continental lithosphere, suggesting that the instantaneous solution may be widely applicable. But the Péclet number approach suggests that slowly stretching lithosphere, as under cratonic basins, may lose heat by conduction during the period of extension.

Basin Analysis: Principles and Application to Petroleum Play Assessment, Third Edition. Philip A. Allen and John R. Allen.
© 2013 John Wiley & Sons, Ltd. Published 2013 by John Wiley & Sons, Ltd.

Lithospheric extension and melting

Learning points: lithospheric stretching may cause the geotherm to intersect the peridotite solidus, causing melting by decompression. For a given initial lithospheric thickness and asthenospheric temperature, the amount of stretching required to initiate decompression melting, and the amount of melt generated as a function of the stretch factor, can be calculated.

During extension, mantle rocks may be brought to sufficiently shallow depths that their temperature intersects the peridotite solidus. They start to melt because of the reduction of pressure during vertical advection associated with lithospheric thinning, that is, the melting is caused by *decompression*. Melting takes place when the temperature at the base of the lithosphere T_a is brought to the depth where it crosses the solidus T_s by stretching at β_{min}.

$$T_{s,0} + A \frac{y_L}{\beta_{min}} = T_a \qquad [A26.1]$$

where A is the slope of the adiabatic solidus curve and y_L is the initial thickness of the lithosphere.

Practical exercise

Calculate the minimum stretch factor required to cause the start of adiabatic decompression of mantle rocks, with parameter values as follows:

Slope of solidus A	$3\,°C\,km^{-1}$
Temperature of base of lithosphere T_a	$1300\,°C$
Initial thickness of lithosphere y_L	$125\,km$
Solidus temperature at surface $T_{s,0}$	$1100\,°C$
Temperature of liquidus at the surface $T_{l,0}$	$1600\,°C$

Solution

The minimum stretch factor is 1.875, but if the asthenosphere is cooler (1200 °C), the stretch factor increases to 3.75, and if the asthenosphere is hotter (1400 °C), the stretch factor becomes 1.25.

The total amount of melt produced M depends on the stretch factor β according to:

$$M(\beta) = \frac{T_a y_L^2 A \beta}{2(T_{l,0} - T_{s,0})(\beta T_a - y_L A)} \left(\frac{1}{\beta_{min}} - \frac{1}{\beta} \right)^2 \qquad [A26.2]$$

Practical exercise

Calculate the melt generated (in terms of thickness) for values of the stretch factor of β_{min} (1.875), $\beta = 2$, $\beta = 3$ and $\beta = 4$.

Now increase the value of T_a to 1400 °C, and repeat the calculations.

Solution

The amount of melt as a function of the stretch factor is as given in Table A26.1.

Table A26.1 Melt generation for different stretch factors with an asthenospheric temperature of 1200 °C

Stretch factor β	β_{min} 1.875	2	3	4	5
Melt generation M	0	0.061 km	2.074 km	4.055 km	5.527 km

To generate large thickness of melt requires high stretch factors if the temperature at the base of the plate is 1300 °C. However, if the asthenospheric temperature is raised to 1400 °C, as would occur over a mantle upwelling, the amount of melt generated increases markedly (Table A26.2).

Table A26.2 Melt generation for different stretch factors with an asthenospheric temperature of 1400 °C

Stretch factor β	β_{min} 1.25	2	3	4	5
Melt generation M	0	4.871 km	11.21 km	15.20 km	17.83 km

This explains the large thicknesses of igneous products in large igneous provinces (LIPs) such as the Siberian, Deccan and Parana traps.

Basin Analysis: Principles and Application to Petroleum Play Assessment, Third Edition. Philip A. Allen and John R. Allen.
© 2013 John Wiley & Sons, Ltd. Published 2013 by John Wiley & Sons, Ltd.

Igneous underplating – an isostatic balance

Learning points: melt formed by decompression may be underplated beneath the crust, with a density intermediate between that of the crust and mantle lithosphere. Underplating therefore causes isostatic uplift of the Earth's surface. Note that this is a purely isostatic effect and does not involve dynamic topography generated by flow in the mantle (§5.2).

The density of igneous rock generated by adiabatic decompression of the mantle depends on the potential temperature at the site of partial melting, and ranges between 2990 and 3070 kg m^{-3} (at a depth of 10 km) for potential temperatures of 1280 °C and 1480 °C respectively. This range of density is mid-way between the density of the continental crust and the density of the mantle, so there is a strong likelihood that melts will be trapped and underplated beneath the crust. Since these igneous additions replace lithospheric mantle (density 3300 kg m^{-3}), the net effect is uplift relative to the depth expected from uniform stretching of the lithosphere without magmatic activity.

Assuming Airy isostasy, the thickness of underplate X required to generate an amount of surface uplift U can be found by balancing two lithospheric columns (Fig. A27.1). In the undisturbed column, crust of thickness y_c and density ρ_c occurs within a lithosphere of thickness y_L and subcrustal density ρ_m. In the underplated lithospheric column, crust of the same thickness is underlain by a layer of basaltic underplate with thickness X and density ρ_x. The total thickness of this lithospheric column is y_L plus the surface uplift U.

Under the normal lithosphere with its surface at sea level, the pressure at the depth of y_L is

$$y_c \rho_c g + (y_L - y_c)\rho_m g \qquad \text{[A27.1]}$$

Under the uplifted lithosphere with an underplate beneath the crust the pressure at the same depth y_L is

$$y_c \rho_c g + X\rho_x g + (y_L - y_c - X + U)\rho_m g \qquad \text{[A27.2]}$$

Balancing the pressures at a depth of compensation at the base of the lithosphere yields

$$U = X\frac{(\rho_m - \rho_x)}{\rho_m} = X\left(1 - \frac{\rho_x}{\rho_m}\right) \qquad \text{[A27.3]}$$

The density of the mantle lithosphere can be taken as 3300 kg m^{-3} and the density of the basaltic underplate as 3000 kg m^{-3}. The rock uplift for an underplate thickness of 5 km is therefore 455 m, or approximately one tenth of the underplate thickness. This is the surface uplift without any erosion. We can incorporate erosion in the simplest way, as follows (Brodie & White 1994). Consider a piece of lithosphere underplated with basalt of thickness X that is eroded so that the surface elevation change is zero. The crust therefore has a thickness of y_c minus the denudation D. We can once again perform an isostatic balance, giving

$$D = X\left(\frac{\rho_m - \rho_x}{\rho_m - \rho_c}\right) \qquad \text{[A27.4]}$$

where D is the total denudation of crust caused by the emplacement of a basaltic underplate of thickness X. The total denudation for an underplate thickness of 5 km if there is no change in the surface elevation is 2.5 km, if crustal rocks of density 2700 kg m^{-3} are eroded. We therefore have the maximum erosion case and the no-erosion case. Reality is probably in between. Clearly, magmatic underplating can have significant effects on surface uplift and on the delivery of erosional products to neighbouring sedimentary basins (White & Lovell 1997). The uplift caused by magmatic underplating is permanent.

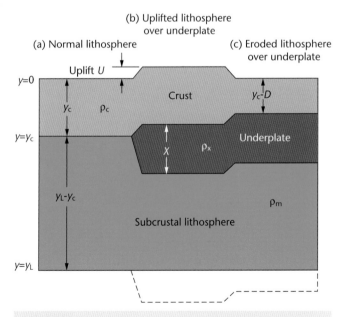

Fig. A27.1 Isostatic balance for lithosphere with a crust underplated with igneous rocks.

Basin Analysis: Principles and Application to Petroleum Play Assessment, Third Edition. Philip A. Allen and John R. Allen.
© 2013 John Wiley & Sons, Ltd. Published 2013 by John Wiley & Sons, Ltd.

28

Uniform stretching at passive margins

Learning points: passive margins occur when stretching proceeds to a point where the continental lithosphere is strongly attenuated and a new ocean basin forms. Cooling of the highly stretched passive margin potentially lasts $>10^7$ yr, resulting in deep oceanic basins where sediment supplies are low. Where sediment supplies are high, thick seaward-prograding sediment wedges $>10\,$km thick may accumulate.

The initial fault-controlled subsidence for the reference uniform stretching model is given by the solution provided in Appendix 21:

$$y_s = \frac{y_L\left\{(\rho_m^* - \rho_c^*)\dfrac{y_c}{y_L}\left(1-\alpha_v\dfrac{T_m}{2}\dfrac{y_c}{y_L}\right) - \dfrac{\alpha_v T_m \rho_m^*}{2}\right\}(1-1/\beta)}{\rho_m^*(1-\alpha_v T_m) - \rho_s^*} \qquad [\text{A28.1}]$$

where y_s is the syn-rift subsidence, y_L is the initial lithospheric thickness, y_c is the initial crustal thickness, ρ_m^*, ρ_c^* and ρ_s^* are the densities of mantle, crust and material infilling the basin, respectively, referenced to 0 °C, α_v is the volumetric coefficient of thermal expansion, T_m is the mantle temperature at the top of the asthenosphere, and β is the stretch factor.

But since we are now dealing with passive margins, it is important to investigate the subsidence $S(\infty)$ at time $t = \infty$. The initial subsidence S_i (eqns. [A21.3, A28.1]) is a linear function of $(1 - 1/\beta)$, and $S(\infty)$ can also be expressed as a linear function of $(1 - 1/\beta)$. By introducing new parameters ρ_L' and ρ_c' as the average densities of the mantle part of the lithosphere and crust at time $t = \infty$, the final subsidence $S(\infty)$ can be expressed

$$S_\infty = \frac{y_L(\rho_L' - \rho_L) + y_c\{\rho_L - (\rho_L'/\beta) + (\rho_c'/\beta) - \rho_c\}}{(\rho_m - \rho_w)} \qquad [\text{A28.2}]$$

The average densities of the mantle part of the lithosphere and the crust at time $t = \infty$ are

$$\rho_L' = \rho_m^*\left(1 - \frac{\alpha_v}{2}T_m - \frac{\alpha_v}{2}T_m\frac{y_c}{\beta y_L}\right) \qquad [\text{A28.3}]$$

and

$$\rho_c' = \rho_c^*\left(1 - \frac{\alpha_v}{2}T_m\frac{y_c}{\beta y_L}\right) \qquad [\text{A28.4}]$$

where the infilling material on this sediment-starved margin is the density of the seawater, ρ_w.

From eqn. [A28.2] to eqn. [A28.4],

$$S_\infty = y_c(1-1/\beta)\left\{\frac{\rho_L - \rho_c + \rho_m^*(\alpha_v/2)T_m + \varepsilon}{\rho_m - \rho_w}\right\} \qquad [\text{A28.5}]$$

where

$$\varepsilon = \left(\frac{\rho_m^* - \rho_c^*}{\beta}\right)\left(\frac{\alpha_v}{2}T_m\frac{y_c}{y_L}\right) \qquad [\text{A28.6}]$$

or by slight rearrangement

$$S_\infty = y_c(1-1/\beta)\left\{\frac{(\rho_m^* - \rho_c^*)[1-(\alpha_v/2)T_m(y_c/y_L)]\varepsilon}{(\rho_m - \rho_w)}\right\} \qquad [\text{A28.7}]$$

In fact, neglect of the additional term ε introduces an error of less than 0.5%.

Basin Analysis: Principles and Application to Petroleum Play Assessment, Third Edition. Philip A. Allen and John R. Allen.
© 2013 John Wiley & Sons, Ltd. Published 2013 by John Wiley & Sons, Ltd.

Flexure of continuous and broken plates

Learning points: the general flexure equation (see Appendix 5) requires boundary conditions in order to be solved for particular geodynamical situations. The most common boundary conditions are for a continuous (infinite) plate subjected to a narrow load (point or line load) far from its edge, and for a broken (semi-infinite) plate subjected to a point or line load at its free end. Using these flexural models, characteristic deflections of the plate can be calculated dependent on the magnitude of the applied load and the material properties of the plate, especially its flexural rigidity.

There are different boundary conditions for a continuous plate loaded far from its edge, and a plate loaded at its free end. Whereas the former is representative of the loading of a seamount chain, the latter is representative of an ocean trench or a pro-foreland basin in the early stages of collision.

Comparing the key boundary conditions, for the continuous plate:

w = maximum at $x = 0$
$w \to 0$ as $x \to \infty$
$dw/dx = 0$ at $x = 0$

and for the broken plate:

w = maximum at $x = 0$
$w \to 0$ as $x \to \infty$
$d^2w/dx^2 = 0$ at $x = 0$

Deflection of a continuous plate under a point or line load

We assume that a vertically acting load V_0 is emplaced at $x = 0$, and that the deflection goes to zero ($w \to 0$) a long way from the load ($x \to \infty$). It is also assumed that there are no horizontal applied loads ($P = 0$), in which case the general flexural equation is

$$D\frac{d^4w}{dx^4} + \Delta\rho gw = 0 \qquad [A29.1]$$

where w is the deflection, x is the horizontal coordinate, D is the flexural rigidity and $\Delta\rho$ in this case is the difference in density between mantle ρ_m and infilling material ρ_i.

The general solution for a fourth-order differential equation such as this requires four boundary conditions.

The solution using these boundary conditions is

$$w(x/\alpha) = w_{max}\exp(-x/\alpha)(\cos x/\alpha + \sin x/\alpha) \qquad [A29.2]$$

where the characteristic flexural length scale (or *flexural parameter*), is given by

$$\alpha = \left\{\frac{4D}{\Delta\rho g}\right\}^{1/4} \qquad [A29.3]$$

and the maximum deflection in terms of the vertical load is

$$w_{max} = \frac{V_0}{2}\frac{1}{\Delta\rho g\alpha} \qquad [A29.4]$$

From this solution, several useful and simple expressions emerge for the geometry of the deflection.

The half-width of the downward deflection (x_0) can be found since it is defined by the horizontal distance from the maximum deflection (at $x = 0$) to the point where the deflection is zero ($w = 0$). Setting $w = 0$, eqn. [A29.2] gives

$$x_0 = \frac{3\pi\alpha}{4} \qquad [A29.5]$$

The distance from the line load ($x = 0$) to the highest part of the forebulge (x_b) can be found since at the forebulge crest, the slope of the deflection is zero. That is, $dw/dx = 0$ at $x = x_b$. Eqn. [A29.2] then gives

$$x_b = \pi\alpha \qquad [A29.6]$$

The height of the forebulge w_b above the datum of zero deflection can also be found using the condition that $x = \pi\alpha$ at the point where $w = w_b$. As a result eqn. [A29.2] reduces to

$$w_b = -w_0\exp(-\pi) = -0.0432w_0 \qquad [A29.7]$$

Basin Analysis: Principles and Application to Petroleum Play Assessment, Third Edition. Philip A. Allen and John R. Allen.
© 2013 John Wiley & Sons, Ltd. Published 2013 by John Wiley & Sons, Ltd.

Practical exercise

Plot the shape of the deflection for flexural rigidities of $D = 10^{22}$, 10^{23} and 10^{24} Nm, using the following parameter values:

$V_0 = 10^{13}$ kg s^{-2}
$\rho_m = 3300$ kg m^{-3}
$\rho_i = 1030$ kg m^{-3} (sea water)

Solution

The solution is shown in Fig. A29.1.

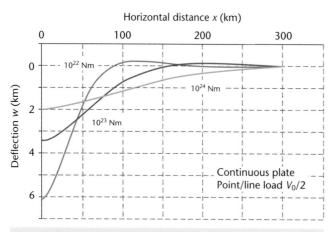

Fig. A29.1 Comparison of the deflections of a continuous plate for different values of the flexural rigidity.

Deflection of a broken plate under a point or line load

For the broken plate, the deflection is given by

$$w(x) = w_{max} \exp(-x/\alpha)\cos(x/\alpha) \qquad \text{[A29.8]}$$

where the maximum deflection is

$$w_{max} = \frac{V_0}{\Delta \rho g \alpha} \qquad \text{[A29.9]}$$

and the flexural parameter is

$$\alpha = \left\{ \frac{4D}{\Delta \rho g} \right\}^{1/4} \qquad \text{[A29.10]}$$

Practical exercise

Compare the continuous plate and broken plate profiles for the following parameter values:

Mantle density ρ_m	3300 kg m^{-3}
Infill density (water) ρ_i	1030 kg m^{-3}
Flexural rigidity D	10^{23} Nm
Vertical load V_0	10^{13} kg s^{-2}

Solution

The solution is given in Fig. A29.2.

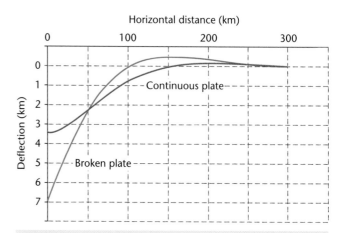

Fig. A29.2 Comparison of the deflection of a broken and a continuous plate with identical flexural rigidities and applied loads.

Note that the broken plate has a deeper maximum deflection, a narrower basin, and a higher forebulge than the continuous plate.

The time scale of flexural isostatic rebound or subsidence

Learning points: the transient response of the lithosphere to the rapid removal or addition of a load is short – geologically instantaneous. The rate of the response, as seen in the isostatic rebound from the last glaciation, tells us a lot about the viscosity of the upper mantle.

When a load is applied or removed from the lithosphere, it does not instantaneously achieve its full elastic deflection. It takes time for the lithosphere to fully adjust to the new load, like a shock absorber in a car. Geological applications are the rebound of the lithosphere upon melting of an ice cap load, or upon rapid erosion.

Following Wangen (2010, p. 301), we assume that the lithosphere has an initial periodic deflection

$$w_e(x) = \frac{q_0 \cos(kx)}{Dk^4 + \Delta\rho g} \qquad [\text{A30.1}]$$

where the initial periodic load is $q(x) = q_0\cos(kx)$ and k is the wave number $k = 2\pi/\lambda$. When the load is rapidly removed, the flexural equation is solved for $q(x) = 0$ and $w_e(x)$ as the initial condition. We should expect the time scale of the flexural response to be determined principally by the flexural rigidity of the lithosphere D, the wavelength of the periodic load expressed by the wave number k, and the viscosity of the mantle given by η, the ratio of viscosity over a mantle length scale μ/l. The deflection as a function of time is

$$w(x,t) = w_e(x)\exp\left(-\frac{t}{t_0(k)}\right) \qquad [\text{A30.2}]$$

where the characteristic time t_0 is a function of the wave number k given by

$$t_0(k) = \frac{\eta}{Dk^4 + \Delta\rho g} \qquad [\text{A30.3}]$$

The half-life of the flexural response for a given wave number is

$$\ln 2\, t_0(k) \qquad [\text{A30.4}]$$

indicating that the response is slowest for a long-wavelength load and fastest for a short wavelength load.

Practical exercise

Assume that the initial condition is of an ice cap with a half-wavelength of 2000 km and a maximum thickness 4 km. Take the density of the ice as 900 kg m^{-3} and the density of the mantle as 3300 kg m^{-3}.

Calculate the initial deflection $w_e(x)$.

1. Plot the deflection as a function of time (eqn. [A30.2]) using a value of $\eta = \mu/l$ of 10^{21} Pa s$/10^6$ m $= 10^{15}$ P s m^{-1}.
2. What is the half-life of the flexural response?

Solution

The solution for the practical exercise is given in Fig. A30.1.

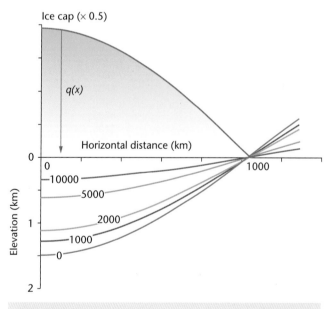

Fig. A30.1 Flexural isostatic rebound following the rapid melting of an ice cap, showing the initial condition at $t = 0$, and successive stages at $t = 1000$ yr, 2000 yr, 5000 yr and 10,000 yr. The ice load is shown in terms of ice thickness (vertical exaggeration 0.5). $q(x)$ is ρgh. The flexural rigidity of the plate is 5×10^{23} Nm, the viscosity of the mantle is 5×10^{21} Pa s, and the mantle length scale is 10^3 km.

Basin Analysis: Principles and Application to Petroleum Play Assessment, Third Edition. Philip A. Allen and John R. Allen.
© 2013 John Wiley & Sons, Ltd. Published 2013 by John Wiley & Sons, Ltd.

The half-life of the flexural response is 4,667 years. In geological terms this is instantaneous.

Postglacial rebound of Scandinavia

The elevation of the mouth of the Angerman River, Sweden, is known as a function of time (from ^{14}C dating of river-mouth deposits) for the last 10,000 years, corrected for eustatic change (Table A30.1). A reasonable wavelength for the Scandinavian ice cap is 3000 km. The highest elevation river mouth is at c.300 m above present-day sea level.

Practical exercise

Calculate the rebound relaxation time $t_0(k)$ for different values of the mantle viscosity, taking the mantle length scale as 300 km, and the flexural rigidity as 5×10^{24} Nm. Make a chart of the elevation of the river mouth over time in order to find the most reasonable value of mantle viscosity.

Table A30.1 Data for uplift and age of the Angerman river mouth, rounded from graph in Turcotte & Schubert (2002, p. 241, fig. 6.16)

Elevation (m)	Time since melting (yr)
310	900
250	1000
220	1200
160	1300
110	3200
95	3700
90	4000
88	4200
50	5300
45	5800
40	6000
30	6400
20	6700
10	8100
10	8700
5	10000
3	10300

Solution

The solution is given in Fig. A30.2.

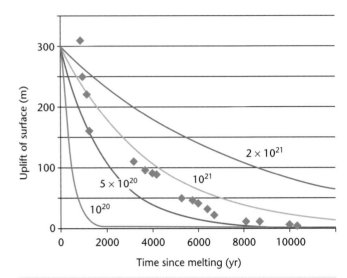

Fig. A30.2 Uplift of the mouth of the River Angerman, Sweden, as a function of time since melting at c.10,000 years ago. Observational data shown as diamonds. Model curves are for different values of mantle viscosity. With the parameter values given, the best fit for mantle viscosity is slightly less than 10^{21} Pa s. The characteristic time for rebound with this value of mantle viscosity is c.4000 years.

Flexural rigidity derived from uplifted lake paleoshorelines

Learning points: the rebound from the last glaciation can be seen in the pattern of uplifted paleoshorelines found around the edges of and around islands in ancient Lake Algonquin. This enables the flexural rigidity of the Canadian lithosphere to be estimated.

Lake Algonquin

The extent of former Lake Algonquin is known from the distribution of old lake shorelines, which are now at variable elevations due to flexural rebound (see §4.1.1 and Fig. 4.2). What is the most plausible range of flexural rigidity for the North American craton to explain the pattern of uplifted shorelines? Assume that the ice cap was 3 km in maximum height and that the rebound following late Pleistocene–Holocene warming is close to complete.

The heights of former lake shorelines of glacial Lake Algonquin, Canada, as a function of distance from the former ice front are given in Table A31.1.

Table A31.1

Location	Distance from former ice front (km)	Present-day elevation of lake shoreline (m)
Bayfield	255	0
Port Elgin	185	15
	150	25
Alliston	110	40
	100	45
	90	55
Beaverton	80	60
	60	70
	45	75
	40	80
	30	110
	20	120
Huntsville	0	130
	−5	140
	−10	150
	−20	175
Fossmill	−40	190

Since the shorelines are 12,000 years old, we can safely assume that postglacial rebound is complete. We can therefore use a steady-state solution for the deflection under a line load, with a maximum deflection (equivalent to rebound) of w_{max} equal to 190 m. We firstly transform the x coordinate to x/α, where α is the flexural parameter, given by

$$\alpha = \left(\frac{4D}{\Delta\rho g}\right)^{1/4}$$ [A31.1]

where D is the flexural rigidity (Nm), and $\Delta\rho$ is the density difference between the mantle and the infilling material, in this case water.

The solution of the general flexure equation for a line load on a continuous plate is

$$w(x/\alpha) = w_{max}e^{-x/\alpha}(\cos x/\alpha + \sin x/\alpha)$$ [A31.2]

Plot the flexural rebound for different values of flexural rigidity. Which is the best value to explain the uplifted shorelines around ancient Lake Algonquin?

Solution

The solution is shown in Fig. A31.1.

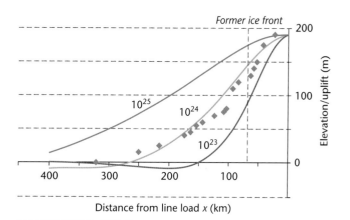

Fig. A31.1 Model curves for different values of flexural rigidity in Nm. Observational data points are diamonds. The best-fitting flexural rigidity is close to 10^{24} Nm, equivalent to an elastic thickness of 50 km (E = 70 GPa, v = 0.25).

Basin Analysis: Principles and Application to Petroleum Play Assessment, Third Edition. Philip A. Allen and John R. Allen.
© 2013 John Wiley & Sons, Ltd. Published 2013 by John Wiley & Sons, Ltd.

Deflection under a distributed load – Jordan (1981) solution

Learning points: geological loads are commonly distributed spatially rather than being point or line loads. An example is an orogenic wedge loading a foreland plate. The practical exercise illustrates how distributed loads cause wide deflections (basins) to form whose dimensions no longer scale closely on the flexural rigidity. When the load is distributed very widely, the lithospheric response approaches that of Airy compensation.

Method of Jordan (1981): the deflection produced by each load i is given by the following set of equations (Hetényi 1979; Jordan 1981).

If x is situated in the basin outboard from the load:

$$w_i = \frac{h}{2}\frac{\Delta\rho_1}{\Delta\rho_2}\left\{\begin{array}{l}\exp(-\lambda(x-s+a))\cos(\lambda(x-s+a))\\ -\exp(-\lambda(x-s-a))\cos(\lambda(x-s-a))\end{array}\right\} \qquad [A32.1]$$

where
$$\Delta\rho_1 = \rho_c - \rho_{air}$$
$$\Delta\rho_2 = \rho_m - \rho_{air}$$

λ is an inverse flexural parameter $(1/\alpha)$ (see eqn. [4.3] in which $\Delta\rho$ is the density difference between mantle and air), and s is the position of the centre of the load, a is its half-width and h its height.

If x is under the load:

$$w_i = -\frac{h}{2}\frac{\Delta\rho_1}{\Delta\rho_2}\left\{\begin{array}{l}2-\exp(-\lambda(x-s+a))\cos(\lambda(x-s+a))\\ -\exp(-\lambda(-x+s+a))\cos(\lambda(-x+s+a))\end{array}\right\} \qquad [A32.2]$$

If x is situated behind (to the left of) the load block:

$$w_i = -\frac{h}{2}\frac{\Delta\rho_1}{\Delta\rho_2}\left\{\begin{array}{l}\exp(-\lambda(-x+s-a))\cos(\lambda(-x+s-a))\\ -\exp(-\lambda(-x+s+a))\cos(\lambda(-x+s+a))\end{array}\right\} \qquad [A32.3]$$

The total deflection resulting from the distributed load is found by summing the individual deflections w_i, as shown in Figure A32.1.

(a) **LOAD BLOCK CONFIGURATION**

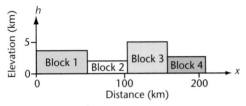

(b) **INDIVIDUAL LOAD BLOCK CONTRIBUTIONS TO DEFLECTION**

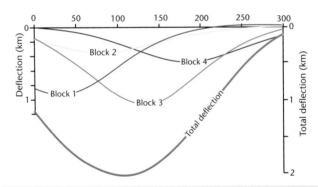

Fig. A32.1 (a) The spatially distributed load is approximated by a series of load blocks of a given height h, half-width a and position of centre s. (b) Individual deflections from load blocks 1 to 4, including the summation of all the individual loads. As the load system becomes more distributed, individual deflections mutually interfere.

Basin Analysis: Principles and Application to Petroleum Play Assessment, Third Edition. Philip A. Allen and John R. Allen.
© 2013 John Wiley & Sons, Ltd. Published 2013 by John Wiley & Sons, Ltd.

33

Deflection under a distributed load – numerical solution of Wangen (2010)

Learning points: the numerical solution for the deflection under a distributed load system is simpler to implement than the procedure given in Appendix 32. The principle is the same, that the total deflection is the summation of all the individual deflections from loads distributed in the x-direction. The load blocks have a height equal to their elevation above a regional datum, such as sea level.

The method outlined in Appendix 32 is difficult to use efficiently in a spreadsheet. A semi-numerical solution is provided by Wangen (2010, p. 293).

The distributed load $q(x)$ can be treated as a series of columns of height h_i, extending from x_i to x_{i+1}, with a load magnitude V_i.

The centre of each load block is at $x_{i+1} - x_i$.

Each individual load (weight) is given by:

$$V_i = \int_{x_i}^{x_{i+1}} q(x)dx = \frac{1}{2}\{q(x_i)+q(x_{i+1})\}(x_{i+1}-x_i) \qquad [A33.1]$$

The deflection $w(x)$ can be calculated as the sum of all of the individual deflections under $i = N$ load blocks. We first make use of the function

$$f(u) = \exp(-u)(\cos u + \sin u) \qquad [A33.2]$$

where u is given by

$$u = |x - x_{c,i}|/\alpha \qquad [A33.3]$$

where $x_{c,i}$ is the position of the centre of each load block, and α is the flexural parameter.

The deflection at any position x is then the sum of all the individual deflections

$$w(x) = \sum_{i=1}^{N} \frac{V_i}{2\Delta\rho g\alpha} f(u) \qquad [A33.4]$$

where the grouping before $f(u)$ in the summation is the maximum deflection for each load V_i.

Practical exercise

Use the numerical solution above to calculate the deflection from the sum of load blocks given in Table A33.1. Assume the loads are made of crustal rocks.

Table A33.1 Position, width and height of 16 load blocks used in the practical exercise

x_c (centre) in m	$x_{i+1} - x_i$ (width) in m	Height of load in m
5000	10000	1000
15000	10000	1500
22500	5000	2000
30000	10000	1500
40000	10000	1000
60000	30000	2000
80000	10000	2500
90000	10000	1500
105000	20000	1000
125000	20000	1000
150000	30000	1500
200000	70000	2000
250000	30000	1500
275000	20000	1000
290000	10000	500
300000	10000	200

Flexural rigidity	5×10^{23} Nm
Gravity	9.81 m s^{-2}
Mantle density	3300 kg m^{-3}
Crustal density	2750 kg m^{-3}
Flexural parameter	138,755 m

Calculate the deflection at x = 0, 100, 200, 300, 400, 500, 600 and 700 km. Make a chart of deflection w versus horizontal distance x.

Solution

The solution is shown in Fig. A33.1.

Basin Analysis: Principles and Application to Petroleum Play Assessment, Third Edition. Philip A. Allen and John R. Allen.
© 2013 John Wiley & Sons, Ltd. Published 2013 by John Wiley & Sons, Ltd.

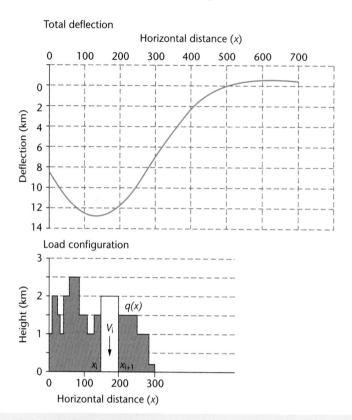

Fig. A33.1 Deflection calculated for a distributed load, with parameters given in Table A33.1.

Deflection under a periodic distributed load

Learning points: periodic topography, such as that of an ice cap, produces a periodic isostatic response, but the wavelength of the periodic load is critical to the isostatic support. For long wavelengths, the lithosphere compensates fully for the topography with a solution identical to that of Airy isostasy, but at smaller wavelengths the lithosphere only partially compensates and behaves flexurally. Information on both flexural rigidity and load wavelength is therefore required in order to predict the deflection under a spatially distributed or periodic load.

The total deflection can also be calculated where the load has the form of an analytical expression. The reader is also referred to Appendix 6 on flexural isostasy. An example is the loading of an ice cap approximated by a periodic (sine) function. We assume an elastic lithosphere, and neglect time-dependent problems caused by mantle viscosity (Appendix 30).

First, we can write for the periodic applied load

$$q_a(x) = \rho_i g h_0 \sin \frac{2\pi x}{\lambda} \qquad [A34.1]$$

where ρ_i is the density of the periodic topography, in this case ice, λ is the wavelength of the periodic topography, x is the horizontal coordinate, and h_0 is the maximum elevation of the ice cap. The sinusoidal applied load produces a deflection in the x-direction that is also periodic. It has the form

$$w = w_0 \sin \frac{2\pi x}{\lambda} \qquad [A34.2]$$

where w_0 is the maximum elevation, which is given by

$$w_0 = \frac{h_0}{\dfrac{\rho_m}{\rho_i} - 1 + \dfrac{D}{\rho_i g}\left(\dfrac{2\pi}{\lambda}\right)^4} \qquad [A34.3]$$

where ρ_m and ρ_i are the mantle and ice cap densities, and D is the flexural rigidity in Nm. The flexure under the ice cap is dependent on the wavelength of the load.

Practical exercise

1. What is the maximum deflection of the lithosphere for an ice cap with maximum height 3 km, average density 900 kg m^{-3}, for flexural rigidities between 10^{22} and 10^{24} Nm, and for wavelengths of 250 km and 500 km?

Solution

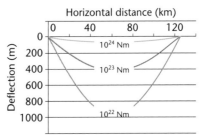

Wavelength 250 km, maximum ice elevation 3 km

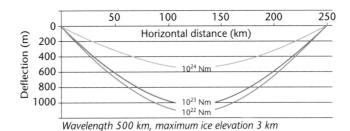

Wavelength 500 km, maximum ice elevation 3 km

Fig. A34.1 Deflection under a periodic ice load as a function of load wavelength and flexural rigidity. Note that shorter wavelengths produce smaller deflections for the same flexural rigidity.

Basin Analysis: Principles and Application to Petroleum Play Assessment, Third Edition. Philip A. Allen and John R. Allen.
© 2013 John Wiley & Sons, Ltd. Published 2013 by John Wiley & Sons, Ltd.

Flexural unloading from a distributed load – the cantilever effect

Learning points: large-scale extension of the crust causes a distributed negative load in the form of a hangingwall basin. The flexural unloading associated with this negative load produces uplift of the basin margin by a so-called flexural cantilever effect.

A number of problems involve the unloading of a distributed load, such as erosion in a mountain belt or at the edge of a high plateau, the desiccation of water in a deep basin, or the melting of an ice cap. Of particular interest in basin analysis is the flexural unloading that takes place due to the replacement of crustal rocks by light sediments in extensional hangingwall basins. Erosional unloading of the footwall block is ignored for simplicity.

The load block configuration is summarised in Table A35.1: note that the replacement of crustal density rocks with water represents a negative load.

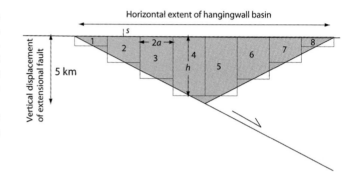

Fig. A35.1 Distributed load of basin sediments in an extensional hangingwall basin.

Table A35.1 Load block configuration for the unloaded hangingwall basin shown in Fig. A35.1

Load block	Centre s (km)	Half-width a (km)	Height h (km)
1	150	10	0.5
2	170	10	1.5
3	190	10	2.5
4	210	10	3.5
5	230	10	3.5
6	250	10	2.5
7	270	10	1.5
8	290	10	0.5

Over what distance from the fault plane does the footwall experience uplift due to the flexural cantilever effect? Where should be the maximum erosion? Use the parameter values given in Table A35.2.

Table A35.2 Parameter values for flexural cantilever exercise

Parameter	Notation (units)	Value
Flexural rigidity	D (Nm)	10^{23}
Density of mantle	ρ_m (kg m^{-3})	3300
Density of crust	ρ_c (kg m^{-3})	2600
Density of infill	ρ_i (kg m^{-3})	1000

Practical exercise

Calculate the flexural effect of unloading the lithosphere by extension along a major basin-bounding fault, as shown in Fig. A35.1.

Solution

The individual deflections from the load blocks 1–8 are shown in Fig. A35.2. Note that the fault plane intersects the surface at $x = 140$ km. To simplify, only the footwall ($0 < x < 140$ km) is shown; all positions of x are to the left of the load blocks, so eqn. [A32.3] can be used throughout. Eqn. [A32.3] calculates the deflection from positive supracrustal loads, so the deflection of the footwall from negative loads in the hangingwall will have the opposite sign.

Basin Analysis: Principles and Application to Petroleum Play Assessment, Third Edition. Philip A. Allen and John R. Allen.
© 2013 John Wiley & Sons, Ltd. Published 2013 by John Wiley & Sons, Ltd.

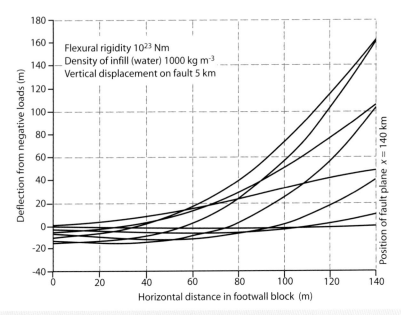

Fig. A35.2   Individual deflections from the eight load blocks shown in Fig. A35.1.

The sum of all of these individual deflections gives the uplift or subsidence of the footwall (Fig. A35.3).

Note that uplift increases towards the fault plane, reaching 600 m at $x = 140$ km. The region of uplift extends over 60 km from the fault plane.

It should be recognised that the distribution of footwall uplift will be affected by:

- the flexural rigidity of the plate, here assumed to be 10^{23} Nm;
- the dip of the fault plane, here assumed to be 45 degrees;
- the density of the infilling material, here assumed to be water;
- the size of the hangingwall basin, here 160 km width, 3.5 km maximum depth.

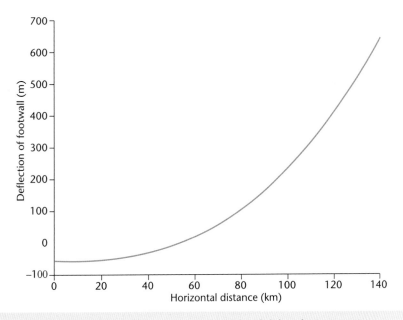

Fig. A35.3   The total deflection caused by the cantilever effect associated with lithospheric extension.

Bending from multiple loads: the Hellenides and Apennines in central Italy–Albania

Learning points: some geological scenarios involve flexure under complex load systems. One example is the flexure of the 'Italian' or Adriatic plate under two oppositely verging loads, in the Hellenides of Albania–Greece, and the Apennines of Italy. In this case, unusually, the two forebulges coincide, producing the Puglia High.

The E–W crustal cross-section across the southern Adriatic Basin shows two depocentres separated by the Puglia High (Fig. A36.1). It appears that the Hellenides and the Apennines are both loading the Adriatic plate.

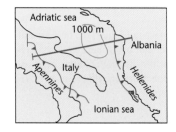

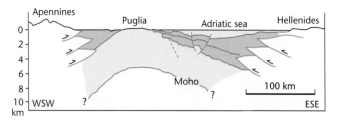

Fig. A36.1 The South Adriatic area is unusual in being loaded on two sides by supracrustal loads – the Apennines and the Hellenides. Puglia is thought to be the uplifted forebulge region, with a filled Apenninic foreland basin in the west and a partially filled Hellenic foreland basin in the east. Modified from De Alteriis (1995). Reprinted with permission from Elsevier.

The deflection of the Adriatic plate can be modelled by making the following assumptions and simplifications:

- at the western edge of the profile, a line load at $x = 0$ represents the Apennines load, with a maximum deflection of basement w_0 of 10 km.

- at the eastern edge of the profile, a line load at $x = 400$ km represents the Hellenides load, with a maximum deflection of basement w_0 of 10 km.

Practical exercise

1. Using the expression for the deflection of a broken plate, make a chart of the deflection of the Adriatic plate from the two tectonic loads, taking note of the location, width and shape of the forebulge region. Does this fit with the position of the Puglia High, centred on $x = 130$ km?
2. Experiment with changing the flexural rigidity of the Adriatic plate. To simulate the correct position of the Puglia High, experiment with different values of flexural rigidity for the Adriatic plate beneath the Hellenides and Apennines. Take note of the effects that this variable flexural rigidity has on the position of the forebulge and its elevation.

Solution

1. A chart giving the total deflection from the two loads for different values of the flexural rigidity is given in Fig. A36.2. Observe how the deflections from the two tectonic loads interfere. This has a major effect on the position of the forebulge, its width and its shape.
2. When the Apennines load deflects a weaker plate than the Hellenides load, the forebulge is driven towards the Apennines. But notice again how the two deflections interfere. Using the line load approximations, and the simplified way of dealing with a variable flexural rigidity, you can see how the forebulge position, height and width are sensitive parameters, especially when the plate is loaded from two sides. A chart is shown in Fig. A36.3 with a 'solution' that shows the forebulge at sea level in approximately the correct position.

A further possibility is that the Apennines load is deflecting the end of a broken plate, and the Hellenides load is deflecting a continuous plate. To investigate this possibility, the appropriate solutions for the deflection (Eqn A29.2 and Eqn A29.8) should be used.

Basin Analysis: Principles and Application to Petroleum Play Assessment, Third Edition. Philip A. Allen and John R. Allen.
© 2013 John Wiley & Sons, Ltd. Published 2013 by John Wiley & Sons, Ltd.

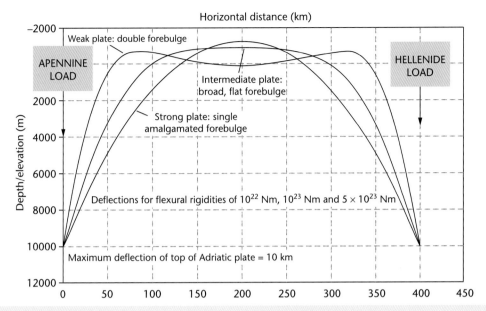

Fig. A36.2 Deflections from the oppositely vergent loads of the Hellenides and Apennines for different flexural rigidities of the Adriatic plate. The shape of the Puglia High is best produced with flexural rigidity of between 10^{23} and 5×10^{23} Nm, but the modelled forebulge is incorrectly positioned if the flexural rigidity is uniform across the Adriatic plate.

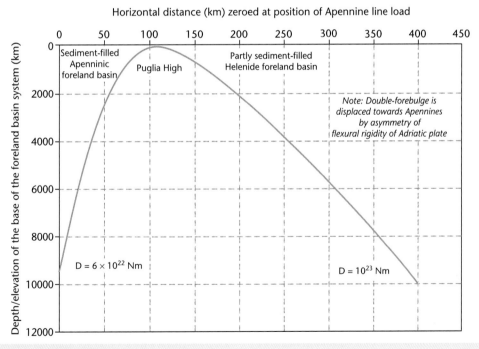

Fig. A36.3 The match with the observed position of the Puglia High is improved by taking a variable flexural rigidity of 6×10^{22} for the Apennines and 10^{23} Nm for the Hellenides.

Flexural profiles, subsidence history and the flexural forebulge unconformity

Learning points: tectonic loads commonly move in position relative to the deflected plate, particularly in pro-foreland basins. The translation of the load system across the plate generates new accommodation that may be filled with sediment. It also forces ahead of it a forebulge region that migrates as a wave of uplift across the foreland plate, producing a flexural forebulge unconformity. Locations on the pro-foreland plate experience an accelerating subsidence rate over time due to this translation of the orogenic load.

In pro-foreland basins, the characteristic subsidence history is caused by the migration of the tectonic load comprising the orogenic wedge. Let us assume that the deflection of the foreland plate can be modelled as due to flexure of a continuous elastic beam subjected to a vertical line load (Appendix 29).

Practical exercise

Use the following input parameters:

Magnitude of line load V_0 ($10^{13}\,\mathrm{kg\,s^{-2}}$)
Flexural rigidity of foreland plate D ($10^{23}\,\mathrm{Nm}$)
Advance rate of orogenic wedge A ($0.01\,\mathrm{m\,yr^{-1}}$)
Density of basin infill ρ_I ($2400\,\mathrm{kg\,m^{-3}}$)
Density of mantle ρ_m ($3300\,\mathrm{kg\,m^{-3}}$)

Use

$$w(x/\alpha) = w_{max}\exp(-x/\alpha)(\cos x/\alpha + \sin x/\alpha) \qquad [A37.1]$$

where the characteristic flexural length scale (or *flexural parameter*), is

$$\alpha = \left\{\frac{4D}{\Delta\rho g}\right\}^{1/4} \qquad [A37.2]$$

and the maximum deflection in terms of the vertical load is

$$w_{max} = \frac{V_0}{2}\frac{1}{\Delta\rho g\alpha} \qquad [A37.3]$$

1. Initially calculate the deflection at $t = 0$. Then assuming that the load advances at a steady rate A, construct subsidence histories for the foreland plate at distances of 150 km, 200 km, 250 km, 300 km, 350 km, 400 km and 450 km from the initial position ($x = 0$) of the line load at time $t = 0$. Hint: the horizontal coordinate is displaced by a horizontal distance At for each increment of movement of the load.
2. What is the duration of stratigraphic gap at the flexural forebulge unconformity for each of the horizontal distances x?

Solution

1. The subsidence histories for locations initially 150 to 450 km from the line load at $x = 0$ are shown in Fig. A37.1.

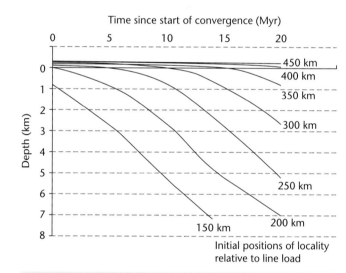

Fig. A37.1 Subsidence history for localities at different initial distances from the line load, under a steady advance rate of $10\,\mathrm{mm\,yr^{-1}}$, flexural rigidity of $10^{23}\,\mathrm{Nm}$, and sediment infill with ρ_i of $2400\,\mathrm{kg\,m^{-3}}$. Note the diachroneity of flexural uplift and subsidence. Points initially close to the line load experience subsidence, whereas points initially far from the line load experience prolonged flexural uplift, causing an unconformity to form at the base of the foreland basin megasequence.

2. The characteristic time of uplift in the forebulge region is $\pi\alpha/A = 25.8\,\mathrm{Myr}$. However, the actual time uplifted varies in the x-direction because of progressive advance of the wedge, from $0\,\mathrm{Myr}$ at $x = 0$ (at $t = 0$) to a maximum of $25.8\,\mathrm{Myr}$ for locations initially distant (>400 km) from the line load.

Basin Analysis: Principles and Application to Petroleum Play Assessment, Third Edition. Philip A. Allen and John R. Allen.
© 2013 John Wiley & Sons, Ltd. Published 2013 by John Wiley & Sons, Ltd.

Bending stresses in an elastic plate

Learning points: curvature of an elastic plate causes fibre or normal stresses that are extensional on the outer arc of the deflected plate. These stresses due to bending may be considerably in excess of the yield strength of rocks where the upper surface of the plate is near the surface, as in the forebulge region. Bending stresses may be responsible for joints and low-displacement faults seen in the uplifted rocks of the forebulge region, and for the extensional fault systems on the outer ramp margin of the flexural basin imaged on seismic reflection profiles.

The distribution of fibre stresses (normal stresses due to bending) for a continuous plate and a broken plate subjected to a vertical load applied at $x = 0$ is shown in Fig. A38.1 (Watts 2001).

For the continuous plate, note that the maximum tensile stress (negative by convention) is situated at $\pi\alpha/2$. Now compare this with the deflection of a continuous plate – the 'basin' extends to the first node at $3\pi\alpha/4$, so the maximum tensile stress is in the outer part of the basin, where the plate dips down towards the maximum deflection. This is where we would expect to see down-to-the-basin normal faults, or normal reactivation of older structures.

For the broken plate, note that the tensile stress is at a maximum at $\pi\alpha/4$. The first node is at $\pi\alpha/2$, so you can see that the maximum tensile stress is located at half way between the maximum deflection and the first node.

Normal (fibre) stresses, continuous plate

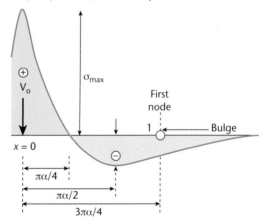

Normal (fibre) stresses, broken plate

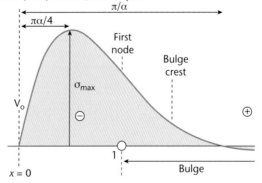

Fig. A38.1 Bending stresses for continuous and broken plates subjected to a line load, modified from Watts (2001). © Cambridge University Press.

Basin Analysis: Principles and Application to Petroleum Play Assessment, Third Edition. Philip A. Allen and John R. Allen.
© 2013 John Wiley & Sons, Ltd. Published 2013 by John Wiley & Sons, Ltd.

The maximum fibre stress of a continuous flexed elastic beam σ_{xx} is found at its upper surface ($y = h/2$) and its lower surface ($y = -h/2$) where y is zero along the midplane of the plate. We are primarily concerned with the upper surface, where fibre stresses may cause the crust to fault during flexure. The solution for any distance from the midplane ($y = 0$) to the upper surface of the plate ($y = h/2$) is

$$\sigma_{xx} = -\frac{EV_0}{(1-v^2)\Delta\rho g\alpha^3}e^{-x/\alpha}\left(\cos\frac{x}{\alpha}-\sin\frac{x}{\alpha}\right)y \qquad [A38.1]$$

which, since $V_0 = 2w_{max}\alpha\Delta\rho g$, can be expressed in terms of the maximum deflection w_{max}

$$\sigma_{xx} = -\frac{E}{(1-v^2)}\frac{2w_{max}}{\alpha^2}e^{-x/\alpha}\left(\cos\frac{x}{\alpha}-\sin\frac{x}{\alpha}\right)y \qquad [A38.2]$$

Using the parameter values below, the distribution of the bending stress in the upper 10% of the plate is shown in Fig. A38.2. The bending stress is at a maximum tension at $x = \pi\alpha/2$.

Maximum deflection	8 km
Mantle density	3300 kg m^{-3}
Sediment density	2500 kg m^{-3}
Poisson's ratio	0.25
Young's modulus	70 GPa
Flexural rigidity	10^{22} Nm
Flexural parameter	47.514 km
Equivalent elastic thickness	11.702 km

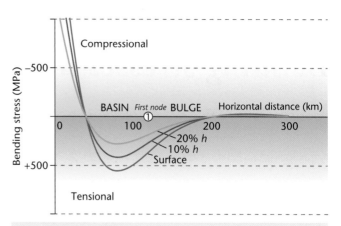

Fig. A38.2   Bending stress due to flexure of an elastic beam with parameter values given above, for three depths in the upper part of the plate: at the surface, depth of 10% plate thickness and depth of 20% plate thickness, showing large region in which tensional bending stresses are found. These bending stresses are much greater than the strength of near-surface rocks.

The strength of rocks in the upper 20% of the flexed plate is <20 MPa. This is exceeded by the bending stress across a broad arc of the flexed plate, suggesting that at shallow depths the rocks of the flexed plate should be faulted and jointed (Billi & Salvini 2003).

In-plane forces and surface topography during orogenesis

Learning points: lateral forces result from the vertical integration of stresses between different lithospheric columns. These lateral buoyancy forces are related to topographic elevation, but in addition reflect density variations with depth. Lateral buoyancy forces add or subtract from the tectonic driving forces caused by tectonic extension or convergence (Sonder & England 1989; Zhou & Sandiford 1992) to give an effective driving force. Use of an elevation-dependent erosion law reveals that the lateral buoyancy forces and erosion velocity vary through an orogenic lifetime, with important implications for orogenic progradation and collapse, and for sediment effluxes to neighbouring basins.

The surface topography of a mountain region relative to an adjacent lowland is a result of the isostatic balance of the two lithospheric columns down to a compensation depth y_k. For each column

$$\sigma_{yy}\big|_{y=y_k} = \int_0^{y_k} \rho(y) g \, dy \qquad [\text{A39.1}]$$

where σ_{yy} is the vertically acting pressure or stress, and ρ is the density. In contrast, the gravitational potential energy is the vertically integrated vertical normal stresses. Integrating eqn. [A39.1] gives

$$E_p = \int_0^y \int_0^{y_k} \rho(y) g \, dy \, dy \qquad [\text{A39.2}]$$

The difference between the gravitational potential energy of two lithospheric columns is the lateral or in-plane force.

The solutions for the topography H and the in-plane force F_h were derived by Sandiford and Powell (1990) and applied by Stüwe and Barr (2000). If the vertical tectonic strains of the subcrustal lithosphere and the crust are ε_L and ε_c respectively, the elevation of the mountain belt H is given by

$$H = \delta y_c (\varepsilon_c - 1) - \xi y_L (\varepsilon_L - 1) \qquad [\text{A39.3}]$$

where δ is the density contrast between the mantle and the crust equal to $(\rho_m - \rho_c)/\rho_m$, and ξ is the average thermal expansion of the lithospheric column where $\xi = \alpha(T_m - T_s)/2$, ε_c and ε_L are the thickening strains of the crust and subcrustal lithosphere respectively, α is the coefficient of thermal expansion, and T_m and T_s are the mantle and surface temperatures (Fig. A39.1).

Similarly, the horizontal in-plane force F_h is given by

$$\frac{F_h}{\rho_m g y_c^2} = \frac{\delta(1-\delta)}{2}\left(\varepsilon_c^2 - 1\right) - \frac{\alpha T_m}{6(y_c/y_L)^2}\left\{\varepsilon_L^2 - 1 - 3\delta(\varepsilon_c \varepsilon_L - 1)\right\} \\ + \frac{\alpha^2 T_m^2}{8(y_c/y_L)^2}\left(1 - \varepsilon_L^2\right) \qquad [\text{A39.4}]$$

Stüwe and Barr (2000, p. 1058) also provide solutions for the lateral in-plane force and surface topography during the evolution of a model orogen by incorporating a vertical erosion velocity proportional to elevation, and a vertical strain rate.

Basin Analysis: Principles and Application to Petroleum Play Assessment, Third Edition. Philip A. Allen and John R. Allen.
© 2013 John Wiley & Sons, Ltd. Published 2013 by John Wiley & Sons, Ltd.

Strain trajectories

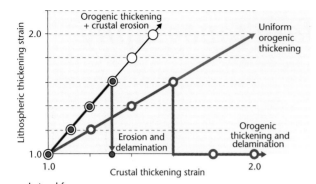

Lateral force

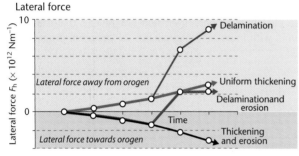

Topography

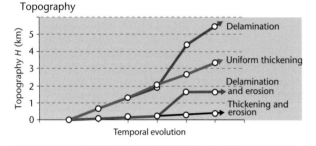

Fig. A39.1 Lateral buoyancy force and topography based on the crustal and lithospheric strains, for four imagined scenarios: (1) the uniform thickening of the crust and the lithosphere; (2) uniform thickening by tectonics is accompanied by erosion of the crust, thereby reducing the crustal thickening strain; (3) uniform thickening followed by delamination of the lithosphere, thereby reducing the lithospheric thickening strain; and (4) thickening accompanied by erosion and delamination.

The lithosphere is homogeneously thickened over time at a constant strain rate, and erosion rate is treated as proportional to elevation, $v_e = H/E$, where E is an erosion constant. The surface elevation as a function of time is then

$$H = \frac{\dot{\varepsilon}A}{\dot{\varepsilon} - B/E} \{\exp(-t(\dot{\varepsilon} - B/E)) - 1\} \qquad [A39.5]$$

where A and B are constants describing the initial density structure, $A = \delta y_c - \xi y_L$, and $B = \delta - \xi$, and $\dot{\varepsilon}$ is the strain rate.

Practical exercise

Plot the surface elevation over time (from $t = 0$ to $t = 50\,\mathrm{Myr}$) for erosion constants E of $B/2\dot{\varepsilon}$ (equivalent to 2.17 Myr) and for $B/3\dot{\varepsilon}$ (equivalent to 1.5 Myr) for a convergent orogen with the following parameter values:

Strain rate	$10^{-15}\,\mathrm{s}^{-1}$
Asthenospheric temperature T_m	1280 °C
Surface temperature T_s	0 °C
Mantle density ρ_L	$3200\,\mathrm{kg\,m}^{-3}$
Crustal density ρ_c	$2700\,\mathrm{kg\,m}^{-3}$
Coefficient thermal expansion α_v	$3 \times 10^{-5}\,\mathrm{K}^{-1}$
Initial crustal thickness y_c	35 km
Initial lithosphere thickness y_L	120 km

Solution

The topographic and erosional history during orogenesis is shown in Fig. 4.33. The surface elevation of the mountain belt becomes asymptotic over time, first for rapid erosion rates (erosion constant E of 1.5 Myr) at 30–40 Myr after the beginning of shortening, and later for lower erosion rates (erosion constant E of 2.17 Myr) after >60 Myr. The erosion velocity increases during the early stage of orogenesis, but reaches an asymptotic value of c. 1 mm yr⁻¹. Such an asymptotic behaviour has implications for the exhumation of orogenic belts estimated from thermochronological methods and for the calculation of sediment effluxes.

The onset of convection

Learning points: the Rayleigh number is a dimensionless grouping that gives the ratio of factors tending to cause thermal instability and convection, versus resisting forces. From a Rayleigh number analysis, it is clear that the mantle must be convecting vigorously.

The onset of convection is given by the Rayleigh number. It can be derived for different conditions of heating (Fig. 5.3). If we take the simple situation of heating from below, with no internal heating (Fig. 5.3A), and we keep the top and bottom temperatures isothermal, the vigour of convection resulting from a temperature difference between the top and bottom of the layer ΔT depends on the fluid density ρ, coefficient of thermal expansion α_v, viscosity μ, thermal diffusivity κ, gravitational acceleration g, and the layer thickness d. Increases in ρ, α_v and g facilitate convection by allowing more buoyancy. However, increases in μ and κ impede convection by resisting motion and diffusing heat away faster. Combining these facilitating and impeding properties into a ratio, and introducing d^3 in the denominator to give units of temperature, we have

$$\frac{\mu\kappa}{\rho g \alpha_v d^3} = \frac{[ML^{-1}T^{-1}][L^2T^{-1}]}{[ML^{-3}][LT^{-2}][T^{-1}][L^3]} = [T] \qquad [A40.1]$$

Eqn. [A40.1] gives the temperature difference between the top and bottom boundaries that must be reached in order to cause convection. The Rayleigh number is the dimensionless ratio of these temperatures

$$Ra = \frac{\rho g \alpha_v \Delta T d^3}{\mu\kappa} \qquad [A40.2]$$

The critical Rayleigh number for the onset of convection is in the region 600 to 2000, and can be taken as $\cong 10^3$. For vigorous convection, **Ra** must be of the order of $>10^5$.

The Rayleigh number given above in eqn. [A40.2] applies to the case of isothermal top and bottom boundaries and heating from below. Since Fourier's law gives the basal heat flux Q_m as proportional to the temperature gradient by the thermal conductivity K (eqn. [2.14]), we can substitute $Q_m d/K$ for ΔT in eqn. [A40.2] giving

$$Ra = \frac{\rho g \alpha_v Q_m d^4}{K\mu\kappa} \qquad [A40.3]$$

We can also allow for some internal heating from radiogenic decay (Fig. 5.3b). If so, we make use of the equation for the geotherm where there is a basal heat flow Q_m and an internal heat generation A. Setting $y = y_c = d$, the temperature difference over the layer thickness d is $(d/K)(Q_m + Ad/2)$, so eqn. [A40.2] becomes

$$Ra = \frac{\rho g \alpha_v \dfrac{d}{K}\left(Q_m + \dfrac{A}{2}d\right)d^3}{\mu\kappa} \qquad [A40.4]$$

which simplifies to

$$Ra = \frac{g\alpha_v d^4\left(Q_m + \dfrac{A}{2}d\right)}{K\kappa\nu} \qquad [A40.5]$$

where α_v is the volumetric coefficient of thermal expansion, g is gravity, d is the thickness of the fluid layer, Q_m is the heat flow through the lower boundary, A is the internal heat generation, κ is the thermal diffusivity, K is the thermal conductivity and ν is the kinematic viscosity (μ/ρ). For the upper mantle, let us initially put $A = 0$, $d = 700$ km, $\alpha = 2 \times 10^{-5}\,°C^{-1}$, $\kappa = 10^{-6}\,m^2\,s^{-1}$, $\nu = 10^{17}\,m^2\,s^{-1}$, $K = 4\,W\,m^{-1}\,°C^{-1}$, $Q/K = 10^{-3}\,°C^{-1}\,m^{-1}$, and g $= 10\,m\,s^{-2}$, which gives a Rayleigh number of c.5×10^5, indicating that even without radiogenic heating the upper mantle should be convecting. Substituting $d = 2000$ km, $\nu = 10^{16}\,m^2\,s^{-1}$, and $Q/K = 10^{-4}\,°C^{-1}\,m^{-1}$ for the lower mantle gives **Ra** $= 3 \times 10^7$, again indicating that the lower mantle should be convecting. However, allowing an internal heat generation in the upper mantle of $A = 3.6 \times 10^{-8}\,W\,m^{-3}$, would raise the Rayleigh number for the upper mantle to c.2×10^6. For the whole mantle, using $d = 2000$ km and $A = 3.6 \times 10^{-8}\,W\,m^{-3}$ and otherwise identical values to those in the upper mantle calculation, the Rayleigh number is 3.2×10^9. Despite the simplifications, this indicates that the mantle should be vigorously convecting.

Basin Analysis: Principles and Application to Petroleum Play Assessment, Third Edition. Philip A. Allen and John R. Allen.
© 2013 John Wiley & Sons, Ltd. Published 2013 by John Wiley & Sons, Ltd.

A global predictor for sediment discharge: the BQART equations

Learning points: a global regression using parameters influencing sediment discharge of rivers to the ocean, corrected for human activities, provides an empirical basis for predictions of sediment discharge. Estimates from this predictor can be compared with values derived from theory, and with values inverted from mass/volumetric budgets.

Interrogation of a dataset of sediment discharge in rivers from a wide variety of climatic and topographic settings (Milliman & Syvitski 1992; Mulder & Syvitski 1996) revealed a relationship between sediment mass discharge Q_s [M T^{-1}], basin area A [L^2] and maximum relief R [L]. Since rivers transport sediment and water at the surface of the Earth under the influence of gravity, we should expect sediment density ρ_s [M L^{-3}] and gravity g [L T^{-2}] to occur in any dimensionless analysis. This gives three dimensions and five parameters, so there should be two dimensionless groups. We can make the sediment mass discharge dimensionless by dividing by $R^{5/2}\rho_s\sqrt{g}$, so one of the dimensionless groups is

$$\left[\frac{Q_s}{R^{5/2}\rho_s\sqrt{g}}\right] \qquad [\text{A41.1}]$$

where Q_s is a mass transport rate. The other group must be a dimensionless drainage area,

$$\left[\frac{A}{R^2}\right] \qquad [\text{A41.2}]$$

A global regression analysis (nearly 500 rivers) gives

$$Q_s = \alpha A^m R^n \qquad [\text{A41.3}]$$

where m and n are area and relief exponents and α is a coefficient dependent on climatic and lithologic effects. When Q_s is in Mt yr^{-1}, area is in km$^2 \times 10^6$, and relief is in km, the regression gives $m = 0.41$ and $n = 1.3$. The median from 488 rivers suggests that $\alpha = 17$, but this value is likely to change with variations in climate and lithology.

Eqn. [A41.3] can be expressed in terms of the dimensionless groups as follows:

$$\left[\frac{Q_s}{R^{5/2}\rho_s\sqrt{g}}\right] = \alpha\left[\frac{A}{R^2}\right]\left(\frac{A^{1-m}R^{n-0.5}}{\rho_s\sqrt{g}}\right) \qquad [\text{A41.4}]$$

This dimensionless form of the global sediment discharge predictor is for pristine rivers unaffected by man. Syvitski *et al.* (2005) therefore introduced an anthropogenic factor B, as follows

$$B = IL(1 - T_E)E_h \qquad [\text{A41.5}]$$

where I is a glacial erosion factor calculated as related to the percentage of the drainage area occupied by ice, L is an average basin-wide lithology factor, T_E is the trapping efficiency of lakes or man-made reservoirs, and E_h is a human-induced soil erosion factor. Eqn. [A41.4] applies to $B = 1$, for pristine conditions. The slope of the regression produced by plotting the two dimensionless groups is therefore equal to

$$\alpha B\left(\frac{A^{1-m}R^{n-0.5}}{\rho_s\sqrt{g}}\right) \qquad [\text{A41.6}]$$

Syvitski and Milliman (2007) provide information on the likely values of the parameters in eqn. [A41.5]. The means of the two databases in Milliman and Syvitski (1992) and Milliman *et al.* (2008) totalling 488 rivers, are given in Table A41.1, allowing a global mean of **B** to be calculated.

Table A41.1 Average parameter values from global databases in Milliman & Syvitski (1992) and Milliman *et al.* (2008)

	M&S92	M et al. 08
No. of rivers	294	194
Area A (km^2)	215400	22400
Relief R (km)	2.0	1.5
Temp (°C)	12.3	14.2
Run-off (m yr^{-1})	0.7	0.6
Discharge Q_w (m^3 s^{-1})	2300	274
Sediment load Q_s (kg yr^{-1})	8.4×10^9	0.4×10^9
Glaciers A_g (%)	1.0	1.0
$I = 1 + 0.09Ag$	1.09	1.09
Lithology L	1.5	1.1
Trapping $(1 - T_E)$	0.8	0.8
Soil erosion E_h	1.3	1.1
Calculated B	**1.70**	**1.06**

Basin Analysis: Principles and Application to Petroleum Play Assessment, Third Edition. Philip A. Allen and John R. Allen.
© 2013 John Wiley & Sons, Ltd. Published 2013 by John Wiley & Sons, Ltd.

A dimensionless plot of the 488 rivers in the Syvitski and Milliman (2007) data set is given in Fig. 7.19. For consistency with Syvitski and Milliman (2007), Q_s is in Mt yr^{-1}, area A is in Mkm^2, and relief R is in km. There is clearly a correlation between dimensionless discharge and dimensionless area scaled by relief, but there is considerable scatter (note that it is a log-log chart).

Practical exercise

Calculate the sediment discharge to the ocean for the two sediment routing systems below, using: (i) the global regression of 488 rivers in eqn. [A41.4]; and (ii) the power law

$$Q_s = cA^d \qquad [A41.7]$$

where Q_s is in Mt yr^{-1}, A is in Mkm^2, and c and d are given below. Coefficients are chosen to represent relief classes of Milliman and Syvitski (1992), with

Sediment routing system	[1] Po	[2] Zambezi
Catchment area A (km^2)	72,183	1,400,000
Catchment relief R (km)	4.8	2.35
Mean elevation H (km)	0.793	1.033
Basin length L_b (km)	480	2040
River length L_r (km)	691	2660
Average basin slope S_b	0.01	0.0012
Average river channel slope S_r	0.007	0.0009
Prefactor in power law c	95	12
Power law Area exponent d	0.6	0.42

Use the following values for coefficients:

Area exponent m	0.41
Relief exponent n	1.3
αB from regression (488 rivers)	11.3

With the above coefficients and exponents, Q_s is in mega tonnes per year (Mt yr^{-1}). The value of α has been adjusted to account for human influences on discharge data ($B = 1.5$).

Assuming that the effective diffusivity κ_e of the sediment routing system is a constant, what is its value in m^2 s^{-1} using the two estimates of Q_s? Recall that $Q_s = \kappa_e S$.

Solution

Table A41.2 Calculated parameters of the Po and Zambezi river systems for the practical exercise

Parameter	Value	Units
Sediment routing system [1] Po		
Q_s (Area-relief equation with m and n constrained from global regression)	29.6	Mt yr^{-1}
q_s unit width sediment discharge	71	m^2 yr^{-1}
Equivalent unit width diffusivity κ_e	7×10^3	m^2 yr^{-1}
Q_s (power law of Area with c and d for elevation class of 1–3 km)	19.6	Mt yr^{-1}
q_s unit width sediment discharge	47.5	m^2 yr^{-1}
Equivalent unit width diffusivity κ_e	4.7×10^3	m^2 yr^{-1}
Sediment routing system [2] Zambezi		
Q_s (Area-relief equation with m and n constrained from global regression)	39.4	Mt yr^{-1}
q_s unit width sediment discharge	27.2	m^2 yr^{-1}
Equivalent unit width diffusivity κ_e	31×10^3	m^2 yr^{-1}
Q_s (power law of Area with c and d for elevation class of 0.5–1.0 km)	13.8	Mt yr^{-1}
q_s unit width sediment discharge	9.5	m^2 yr^{-1}
Equivalent unit width diffusivity κ_e	11×10^3	m^2 yr^{-1}

We can compare these values obtained from regressions with the actual measurements from the two river systems. Taking the values for total sediment discharge in Mt yr^{-1} from, first, the global regression of the Area-relief relationship and, second, the power law of area (Table A41.3):

Table A41.3

River Q_s (Mt yr^{-1})	Area-relief regression Q_s (Mt yr^{-1})	Area power law Q_s (Mt yr^{-1})	Observed
[1] Po	29.6	19.6	17.2
[2] Zambezi	39.4	13.8	19.9

Estimates are highly dependent on the values of coefficients and exponents, indicating that careful calibration of Area-power law and Area-relief relationships is important.

Modelling hillslopes

Learning points: the slow creep on hillslopes can be modelled as a diffusive process, whereby the sediment flux scales on the topographic slope, and the elevation change of the surface over time scales on the curvature. This produces a parabolic hillslope profile. Diffusion can be used in a wide range of geoscientific applications.

The sediment continuity equation can easily be modified for the case of the topography of a hillslope:

$$\frac{\partial q}{\partial x} = -\rho_b \frac{\partial y}{\partial t} \qquad [A42.1]$$

where q is the discharge of mass per unit width of hillslope, ρ_b is the bulk density of the mobile regolith, x is the horizontal distance from the ridge crest, y is the vertical coordinate and t is time (Fig. 7.32).

The mass discharge per unit width is assumed to be proportional to the local topographic slope by a transport coefficient k, encompassing all of the geomorphic processes acting on the hillslope, such as rainsplash, soil creep, bioturbation, overland flow, rilling and gullying:

$$q = -k \frac{\partial y}{\partial x} \qquad [A42.2]$$

The discharge of sediment per unit width is given by eqn. [A42.2)] as long as the regolith production rate by weathering (§7.2.2) is sufficiently large. Combining eqn. [A42.1] and eqn. [A42.2] gives the familiar diffusion equation

$$\frac{\partial y}{\partial t} = \kappa \frac{\partial^2 y}{\partial x^2} \qquad [A42.3]$$

where the *diffusivity* $\kappa = k/\rho_b$. This states the familiar result that the erosion or deposition scales on the topographic curvature. The form of a hillslope profile therefore depends on the diffusivity, but the amplitude of the profile is set by the rate of channel incision. Let the incision rate be $\dot{e}$, and the length of the hillslope from ridge crest to valley bottom be L. The steady-state profile of the hillslope is obtained by integrating the diffusion equation twice:

$$y = \frac{\dot{e}}{2\kappa}(L^2 - x^2) \qquad [A42.4]$$

which shows that the hillslope profile is parabolic. The slope of the hillslope at the ridge crest is zero, and the maximum slope is at the boundary with the river channel (at $x = L$), where it has the value

$$\left|\frac{\partial y}{\partial x}\right|_{\max} = \frac{\dot{e}L}{\kappa} \qquad [A42.5]$$

The time constant for the hillslope system is given by

$$\tau = \frac{L^2}{\kappa} \qquad [A42.6]$$

The diffusion equation (eqn. [A42.3]) can be modified to include an advective term to account for the vertical uplift rate of rock:

$$\frac{\partial y}{\partial t} = \kappa \left(\frac{\partial^2 y}{\partial x^2}\right) + V(x) \qquad [A42.7]$$

where $V(x)$ is the vertical tectonic uplift rate of rock. This diffusion-advection equation can be integrated twice to obtain the height of the hillslopes. We firstly assume that the landscape is in steady state, so $\partial y/\partial t \rightarrow 0$. Consequently, eqn. [A42.7] becomes

$$\kappa \frac{\partial^2 y}{\partial x^2} = -V(x) \qquad [A42.8]$$

which when integrated is

$$\frac{\partial y}{\partial x} = -\frac{Vx}{\kappa} + c_1 \qquad [A42.9]$$

Since $\partial y/\partial x = 0$ at $x = 0$, $c_1 = 0$. Integrating eqn. [A42.9] once again we obtain

$$y = -\frac{Vx^2}{2\kappa} + c_2 \qquad [A42.10]$$

Basin Analysis: Principles and Application to Petroleum Play Assessment, Third Edition. Philip A. Allen and John R. Allen.
© 2013 John Wiley & Sons, Ltd. Published 2013 by John Wiley & Sons, Ltd.

The constant of integration can be found by applying the boundary condition that $y = 0$ at $x = L$. Consequently, $c_2 = VL^2/2\kappa$. Eqn. [A42.10] therefore becomes

$$y = -\frac{V}{2\kappa}(L^2 - x^2) \qquad [\text{A42.11}]$$

The maximum height of the hillslope y_{max} is the height at $x = 0$ minus the height at $x = L$. Consequently,

$$y_{max} = \frac{VL^2}{2\kappa} \qquad [\text{A42.12}]$$

An example of the profile of hillslopes between two channels spaced 100 m apart, which incise at 0.5 mm yr^{-1}, where the hillslope diffusivity is 50×10^{-3} m^2 yr^{-1} is given in Fig. 7.32. In this case the maximum height of the hillslope is 12.5 m. The maximum gradient $\tan^{-1}$ (0.5 radians) = 27 degrees.

If the incision rate of the channels is doubled to 1 mm yr^{-1}, the hillslopes attain a maximum height of 25 m, and the maximum slope becomes $\tan^{-1}$ (1 radian) = 45 degrees, which is close to the limit for landsliding.

The sediment continuity (Exner) equation

Learning points: the sediment continuity equation (commonly referred to as the Exner equation) is the basic statement of the conservation of mass associated with sediment transport. There is an elevation change of the bed when the downstream sediment transport rate changes in the x-direction and/or there is tectonic uplift or subsidence. The change in sediment transport rate may be due to a change in the bedload flux or in the suspended load concentration.

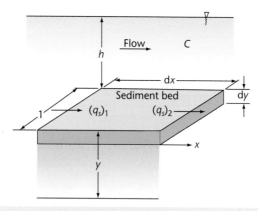

Fig. A43.1 Set-up and notation for the derivation of the sediment continuity equation. Notation explained in text.

Consider an area of stream bed with unit width normal to the flow and length parallel to the flow of dx, with a height of y above some reference datum (Fig. A43.1). The area of this piece of stream bed is therefore dx. A fluid flow with suspended sediment volume concentration C has a sediment transport rate (per unit width of stream bed) at the upstream end of the reference area $(q_s)_1$ and at the downstream end of the reference area $(q_s)_2$. We can denote the difference in the transport rate at the upstream and downstream ends of the reference area as dq_s. Any change in the bed elevation due to either erosion or deposition must be reflected in a downstream change in the sediment transport rate as long as the sediment concentration C does not change. The amount of sediment deposited over the reference area associated with a downstream decrease in the sediment transport rate is simply the product of the difference between the upstream and downstream sediment transport rates (dq_s) and the time interval dt. This sediment produces a layer of thickness dy with a porosity λ over an area dx. Expressed mathematically

$$dy\,dx = -\frac{1}{1-\lambda}dq_s dt \qquad [A43.1]$$

A change in bed elevation may also result from a temporal change in the volume concentration of suspended sediment throughout the flow. For example, if the suspended sediment concentration throughout the flow is decreasing, the bed must be accreting, irrespective of downstream variations in the sediment transport rate. The bed elevation change due to this effect in a flow of depth h is

$$dy\,dx = -\frac{1}{1-\lambda}(dC \cdot h)dx \qquad [A43.2]$$

The net change in bed elevation over a time interval dt is therefore

$$\frac{dy}{dt} = -\frac{1}{1-\lambda}\left(\frac{dq_s}{dx} + h\frac{dC}{dt}\right) \qquad [A43.3]$$

In other words, the change of bed elevation in time (erosion or deposition) is related to the downstream change of the sediment transport rate and to the change of the suspended sediment concentration in time. If the suspended sediment concentration does not vary in time, that is, $dC/dt = 0$, or if the suspended sediment concentration is zero, then the sediment continuity equation simplifies to

$$\frac{dy}{dt} = -\frac{1}{1-\lambda}\frac{dq_s}{dx} \qquad [A43.4]$$

The change in the bed elevation is affected by the subsidence (or uplift) of the alluvial basin. If $\sigma(x)$ is the subsidence rate as a function of the downstream distance x, the sediment continuity equation becomes

$$\sigma(x) + \frac{dy}{dt} = -\frac{1}{1-\lambda}\frac{dq_s}{dx} \qquad [A43.5]$$

Basin Analysis: Principles and Application to Petroleum Play Assessment, Third Edition. Philip A. Allen and John R. Allen.
© 2013 John Wiley & Sons, Ltd. Published 2013 by John Wiley & Sons, Ltd.

Use of the stream power rule

Learning points: the conventional framework for considering the erosion by bedrock channels is the stream power rule – a rule of thumb that relates the elevation change in a bedrock channel to a power law of water discharge and slope. It is commonly used in landscape evolution models. The two exponents in the stream power rule combine to make a quotient *m/n*, known as concavity.

Consider a volume V of water, density ρ_w, in a river that moves downstream, changing its elevation by dy over a lateral distance dx in a time period dt. There is a change in potential energy caused by this downstream motion. If we assume that the mass of water ($V\rho_w$) does not change, this change in potential energy is $V\rho_w g dy$. The rate of change of energy loss, or power, is simply $V\rho_w g dy/dt$. Now the volume of water moving through the river per interval of time dt is Q_w/dt, where Q_w is the water discharge. Consequently, the stream power P per length of stream channel dx is given by

$$P/dx = kQ_w \frac{dy}{dx} \qquad [A44.1]$$

where the coefficient $k = \rho_w g$. The stream power (per length of stream channel) is therefore proportional to a discharge-slope product. The rate of bedrock incision is conventionally assumed to have the form of a power-law version of eqn. [A44.1]

$$\frac{\partial y}{\partial t} = -c_b Q_w^m S^n \qquad [A44.2]$$

where Q_w is the total discharge of water, S is the slope, m and n are empirical coefficients, and c_b is a bedrock incision coefficient. The parameters contributing to the value of the bedrock erosion coefficient c_b remains debatable. In the simplest case, let us assume that c_b is constant and determined by lithology, ranging from 7×10^{-3} for mudstones to 6×10^{-6} for granite and basalt (Stock & Montgomery 1999).

The discharge of water in the channel depends on the drainage area upstream of the stream channel A, the average precipitation over the area P, and a run-off coefficient C_r that accounts for transmission losses through the hillslope-channel system ($Q_w = C_r AP$). If the long-term effective discharge varies linearly with drainage area, then $m = n = 1$, and the incision rate is then linearly proportional to stream power (Seidl & Dietrich 1992).

The drainage area upstream of a particular point in the channel increases with distance down the channel, and on the basis of empirical results (Montgomery & Dietrich 1992) is well described by

$$A = \frac{x^2}{3} \qquad [A44.3]$$

Consequently, the rate of channel incision can be expressed as

$$\frac{\partial y}{\partial t} = -c_1 x^2 S \qquad [A44.4]$$

where c_1 is a coefficient equal to $c_b PC_r/3$. It can be seen, therefore, that in a bedrock valley where the width of the stream does not change significantly in the downstream direction, the rate of bedrock incision varies by the square of the distance from the drainage divide x and the single power of the slope S. These two effects are traded off against each other. The result is that the rate of valley incision increases to a maximum at a certain distance from the drainage divide.

However, theory and empirical data suggest that m/n is ~0.5, and analysis of the topography of the Zagros fold and thrust belt (Tucker 1996) indicates that m and n are 1/3 and 2/3 respectively. In such a case, the rate of bedrock channel incision becomes

$$\frac{\partial y}{\partial t} = -c_2 x^{2/3} S^{2/3} \qquad [A44.5]$$

where c_2 is equal to $c_b(PC_r/3)^{1/3}$.

It is also possible that the bedrock erosion coefficient is a more complex function of hydrological variables, including the flow resistance of the bed. Inspection of eqn. [A44.5] shows that c_2 has the units of $[T]^3$. By dividing c_1 by the units of discharge ($m^3 \ s^{-1}$), we can express the channel incision rate in terms of an *'erosional velocity'* c_v (units of $m \ s^{-1}$), so that

$$\frac{\partial y}{\partial t} = -c_v \left(\frac{Q_w}{Q_*} \right)^m S^n \qquad [A44.6]$$

Basin Analysis: Principles and Application to Petroleum Play Assessment, Third Edition. Philip A. Allen and John R. Allen.
© 2013 John Wiley & Sons, Ltd. Published 2013 by John Wiley & Sons, Ltd.

where Q_* is a characteristic channel discharge (equal to the total area of the catchment times the average precipitation rate, Tucker & Slingerland 1996).

Practical exercise

What is the bedrock incision rate along the course of an upland river flowing over relatively easily eroded sedimentary rocks, from $x = 0$ at the headwaters to $x = 350$ km, using the stream power rule. Take m and n as 1/3 and 2/3, bedrock incision coefficient $c_b = 10^{-4}$, run-off coefficient $C_r = 0.9$, and average rate of precipitation over the catchment $P = 1$ m yr^{-1}. Measurement of the stream profile gives the following information on slope:

x (km)	slope
350	0.00061
300	0.00069
250	0.00079
200	0.00090
100	0.00126
60	0.00163
40	0.00195
30	0.00220
20	0.00250

Solution

The maximum incision rate of 0.14 mm yr^{-1} is at c.250 km from the source (Fig. A44.1).

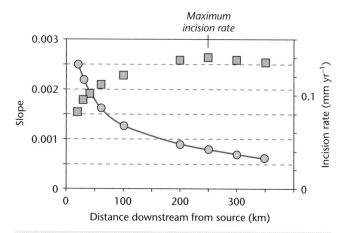

Fig. A44.1   Incision rate along a river profile, reaching a maximum at approximately $x = 250$ km from source.

Effects of tectonic uplift on stream longitudinal profiles

Learning points: the longitudinal profile of a river can be modelled using the stream power rule. A tectonic displacement field causing uplift of rocks has impact on the long profile of the river, including on parameters such as concavity and steepness. Tectonic uplift commonly results in knickzones that propagate upstream as convex reaches.

The analysis of longitudinal stream profiles is useful in assessing landscape response to forcing mechanisms such as variations in tectonic uplift rate. We build on the stream power law in Appendix 44 by including a term for the tectonic uplift rate of rock. As a reminder, according to the stream power rule for detachment-limited incision of channels into bedrock, the erosion rate scales on the area and slope

$$\frac{dy}{dt} = k_1 A^m S^n \qquad [A45.1]$$

where k_1 is a coefficient of erosion that contains information on the density of water, gravity, the erodibility of bedrock, and the precipitation in terms of channel discharge; A is the contributing drainage basin area serving as a proxy for water discharge; and S is channel slope. This erosion rate balances the tectonic uplift rate of rocks $U(x,t)$ in *steady-state conditions*, giving

$$U(x,t) = k_1 A^m S^n \qquad [A45.2]$$

U is considered initially to be spatially uniform, and m and n are constants, so eqns [A45.1] and [A45.2] become after rearrangement

$$S = \left(\frac{U}{k_1}\right)^{1/n} A^{-m/n} \qquad [A45.3]$$

Eqn. [A45.3] has the form of a power law (Sklar & Dietrich 2001). The coefficient $(U/k_1)^{1/n}$ is known as the *steepness*, and $m/n = \theta$ is known as the *concavity* (Tucker & Whipple 2002). If incision rate is linearly proportional to shear stress, $n = 2/3$. If incision rate is linearly proportional to stream power, $n = 1$. Concavity values are likely to vary, though within a relatively narrow band from 0.35 to 0.6. The value of concavity m/n in stream profiles thought to be in steady state in the Siwalik Hills of Nepal is ~0.46 (Kirby & Whipple 2001) compared with the theoretical value of 0.5 for the case of shear stress proportionality. The erosion coefficient k_1 may vary over orders of

magnitude (10^{-3} to 10^{-7} m$^{0.2}$ yr^{-1}), with a mean value of 4.3×10^{-4} m$^{0.2}$ yr^{-1} in the same Nepalese rivers cutting into relatively easily erodible material.

For the uniform rock uplift case, the straight line regression of log S plotted against log A has a slope of $-m/n$, and an intercept of $(U/k_1)^{1/n}$. Consequently, values for concavity and steepness can be derived from plots of field data from modern rivers (Snyder *et al.* 2000).

Where the tectonic rock uplift rate is variable, concavities and steepness values are modified from the uniform case above. An uplift function that yields a power-law relationship between slope and area is $U = U_0 x^\alpha$, where U_0 is the rock uplift rate at the edge of the model, x is the horizontal coordinate, and α is a constant describing the downstream rate of change in tectonic uplift rate. Assuming Hack's law for drainage area versus downstream distance, $A = k_2 x^h$, and steady-state conditions, this power-law relationship has gradient and intercepts giving the steepness k_s and the concavity θ, as follows (Kirby & Whipple 2001)

$$k_s = \left(\frac{U_0}{k_1}\right)^{1/n} k_2^{-(\alpha/hn)};$$
$$\theta = (m/n) - (\alpha/hn) \qquad [A45.4]$$

Eqn. [A45.4] indicates that where the rock uplift rate increases downstream ($\alpha > 0$), streams should have low concavity or even convexity ($\theta < 0$), whereas where the rock uplift rate decreases downstream ($\alpha < 0$), streams should have high concavity ($\theta > 0$). A study of the concavity of streams crossing the Siwalik Hills (Kirby & Whipple 2001) showed strong variations in θ dependent on the rates of change in tectonic uplift rate associated with differential movement across the Main Frontal Thrust (Lavé & Avouac 2001). Channels crossing the tectonically uplifting anticline from north to south experience increasing tectonic uplift rate and have convex-up longitudinal profiles (concavity < 0). Tectonic uplift rates and their lateral variability therefore can potentially be inverted from stream profiles if parameter values can be well calibrated.

Basin Analysis: Principles and Application to Petroleum Play Assessment, Third Edition. Philip A. Allen and John R. Allen.
© 2013 John Wiley & Sons, Ltd. Published 2013 by John Wiley & Sons, Ltd.

Practical exercise

Plot stream profiles over the horizontal distance of 10 km for different distributions of tectonic uplift rate using the parameter α: (i) $\alpha = +0.25$, indicating a downstream increase in tectonic uplift rate as would be experienced by a river occupying a hinterland catchment associated with a downstream dip-slip normal fault; (ii) $\alpha = -0.25$, indicating a downstream decrease in tectonic uplift rate as would be experienced by a river flowing down the flanks of a dynamically supported uplift or a flank of an orogenic belt; (iii) a plateau-like uniform spatial distribution of uplift, in which case $\alpha = 0$. For all three cases, use $k_2 = 2$ and $h = 0.5$ for the coefficient and exponent of Hack's relationship between drainage basin area and downstream distance, $m = 1/3$, $n = 2/3$ and $k_1 = 4 \times 10^{-4}\,\mathrm{m}^{1/3}\,\mathrm{yr}^{-1}$.

Solution

Concavity values range from (i) -0.25 to (ii) $+1.25$, to (iii) 0.5, showing that with the downstream increasing tectonic uplift rate, stream profiles are convex-up. This is commonly observed upstream from active extensional faults, as in the Central Apennines of Italy (Whittaker *et al.* 2010).

Estimation of the uplift rate from an area-slope analysis

Learning points: erosion can be approximated by a combined advective-diffusive rule, which can be calibrated using a plot of the logarithms of the local slope versus the contributing upstream drainage area. Assuming steady-state (erosion is balanced by tectonic uplift), the uplift rate can be inverted from river long profiles.

The change of elevation of the bed of a river is the balance between the erosional downcutting of the river and the tectonic uplift of rock. The erosion rate is controlled by the long-term water discharge of the river catchment, which varies over time and reflects run-off from precipitation and the contributing catchment size. The variation of erosion rate E along the river profile, and over time, is then based on a combination of diffusive and stream power (discharge-slope) relationships, that is,

$$E(x,t) = -vA^m\left(\frac{\partial y}{\partial x}\right)^n + \kappa\frac{\partial^2 y}{\partial x^2} \qquad [A46.1]$$

where v is an advective erosion coefficient (units of m^{1-2m} Myr^{-1}), A is the contributing (upstream) catchment area, κ is an erosional diffusivity, and m and n are area and slope exponents respectively (m is typically 0.35–0.6, and n is 1–2).

If the diffusive term (second term on right-hand side of eqn. [A46.1]) is very small compared to the advective (first term on right-hand side), and if the river is in steady state ($dh/dt = 0$), then erosion is balanced by tectonic uplift, and eqn. [A46.1] simplifies to

$$U = -vA^m\left(\frac{\partial y}{\partial x}\right)^n \qquad [A46.2]$$

which after some rearrangement gives

$$\log\left(\frac{\partial y}{\partial x}\right) = -\left(\frac{m}{n}\right)\log A + \frac{1}{n}\log\left(\frac{U}{v}\right) \qquad [A46.3]$$

In other words, when log of the topographic slope is plotted against log of the area, the slope is $-m/n$ (also known as concavity) and the intercept on the slope axis is $(1/n)\log(U/v)$. Consequently, the unknowns in eqn. [A46.2] can be derived from a slope-area analysis (Whipple 2004; Wobus *et al.* 2006), allowing U/v to be calculated. The advective erosion coefficient v can be estimated from rates of incision in modern rivers (Karlstrom *et al.* 2008). Although this slope-area method is potentially useful, the advective erosion coefficient is not independent of m and n, and the assumption of steady state may not be appropriate during a landscape's transient response to tectonic or climatic change.

Practical exercise

The dating of basalts in Colorado Canyon allows the incision rate at a given location along the river to be constrained at $118\,m\,Myr^{-1}$ (Karlstrom *et al.* 2007), where the local slope is 1/350 and the upstream drainage area is $3.6 \times 10^{11}\,m^2$. If $m = 0.2$ and $n = 1$, find the value of the advective erosion coefficient v, assuming the river to be in steady state ($dy/dt = 0$).

Solution

Using eqn. [A46.2], v is $149\,m^{0.6}\,Myr^{-1}$.

Basin Analysis: Principles and Application to Petroleum Play Assessment, Third Edition. Philip A. Allen and John R. Allen.
© 2013 John Wiley & Sons, Ltd. Published 2013 by John Wiley & Sons, Ltd.

Uplift history from stream profiles characterised by knickpoint migration

Learning points: phases of accelerated rock uplift may be recognised by the presence of knickzones in the longitudinal profile of a river. Knickzones move erosionally upstream in a wave-like manner. Compilation of long profile information from many streams in a region experiencing uplift may reveal otherwise unrecognised long-wavelength doming of the continental surface.

Appendices 45 and 46 showed the potential for inverting tectonic uplift rate from information contained in stream profiles. A similar goal can be reached by considering the elevation profile of the river $y(x)$ to reflect the migration of knickpoints generated by spatially uniform tectonic activity $U(t)$ (Pritchard *et al.* 2009; Roberts *et al.* 2010). Erosion is assumed to be made of two components, one a diffusive lowering, and the other a result of knickpoint migration. Knickpoints migrate upstream with a celerity C, which causes a change in the elevation of the stream channel. The knickpoint migrates upstream at a rate controlled by the water discharge and the erodibility of rocks comprising the stream bed. The change of elevation of the bed resulting from the migration of the knickpoint is

$$\frac{\partial y}{\partial t} = C\frac{\partial y}{\partial x} = (k_e \rho g)q_w \frac{\partial y}{\partial x} \qquad [A47.1]$$

where the celerity is proportional to the water unit discharge q_w, and k_e is the erodibility coefficient, dependent on rock strength. Since water discharge varies with the downstream distance x, eqn. [A47.1] can be expressed

$$\frac{\partial y}{\partial t} = v_0 x^m \left(\frac{\partial y}{\partial x}\right)^n \qquad [A47.2]$$

where v_0 is a knickpoint migration parameter, equal to $k_e \rho g$, which equals the celerity C when $m = 0$ and $n = 1$. Assuming diffusion to be slow compared to advection (knickpoint migration), the tectonic uplift is given by

$$U(t) = \frac{\partial y}{\partial t} + v_0 x^m \left(-\frac{\partial y}{\partial x}\right)^n \qquad [A47.3]$$

where the first term is the change of bed elevation over a time increment, and the second term is a wave-like advection describing the upstream migration of a knickpoint.

The inversion of uplift rate history from the variation of the stream gradient is given in full in Pritchard *et al.* (2009) and not repeated here. Following rescaling of the horizontal coordinate and the stream gradient, and introducing a time variable τ to describe the time since the origination of the wave-like knickpoint at $x = L$,

the uplift rate as a function of time can be inverted from topographic information on river longitudinal profiles (such as provided by a digital elevation model). The tectonic uplift history as a function of time is then given by

$$U(\tau) = v_0 x^m \left(-\frac{\partial y}{\partial x}\right)^n \qquad [A47.4]$$

where the time variable is

$$\tau = -\frac{L^{1-m}}{(n-m)v_0}\left\{\frac{1-(x/L)^{1-m/n}}{(x/L)^{m(n-1)/n}}\right\}\left(-\frac{\partial y}{\partial x}\right)^{1-n} \qquad [A47.5]$$

There is clearly a need for m, n and v_0 to be calibrated. In their study of the uplift of the Bié Dome, Namibia, Pritchard *et al.* (2009) used $m = 0.5$ and $n = 1$, in which case eqn. [A47.5] simplifies to

$$\tau = -\frac{\sqrt{L}-\sqrt{x}}{0.5v_0} \qquad [A47.6]$$

Consequently, if the advective erosion coefficient v_0 can be approximated, the time scales for the propagation of a knickzone up the river profile starting at $x = L$ can be found, and the uplift history can be calculated from stream gradients extracted from a digital elevation model. Pritchard *et al.* used $v_0 = 50 \text{ m}^{1-2m} \text{Myr}^{-1}$ from their study of the Bié Dome, whereas Roberts *et al.* (2012) used $v_0 = 200 \text{ m}^{1-2m} \text{Myr}^{-1}$ for the Colorado River, USA.

Practical exercise

The Colorado River has a total length of $L = 2300 \text{ km}$. Two knickzones are found at $x = 200 \text{ km}$ and 1000 km. What are the characteristic time scales for the migration of these knickzones to their present position?

Solution

Using $m = 0.25$, $n = 1$, and $v_0 = 84 \text{ m}^{0.5} \text{Myr}^{-1}$, derived from an area-slope analysis, and using the knickpoint positions at 200 km and 1000 km, characteristic time scales are 25 and 12 Myr respectively.

Basin Analysis: Principles and Application to Petroleum Play Assessment, Third Edition. Philip A. Allen and John R. Allen.
© 2013 John Wiley & Sons, Ltd. Published 2013 by John Wiley & Sons, Ltd.

48
Sediment deposition using the heat equation

Learning points: the solution for temperature resulting from the instantaneous heating of a semi-infinite half space, widely applied to the temperature field of cooling oceanic lithosphere, can be applied to diffusive depositional profiles after a rapid generation of accommodation by tectonic changes, or to the progradation of clinoforms into a large water body. Diffusive profiles due to deposition are characteristically concave-up.

The progradation of depositional surfaces can be solved by using a steady sediment discharge, and treating the problem as identical to the cooling of a semi-infinite half space (§2.2.7). The elevation of the bed h is then given by

$$h = \frac{2q_{s,0}}{\kappa}\left\{ \left(\frac{\kappa t}{\pi}\right)^{1/2} \exp\left(-\frac{x^2}{4\kappa t}\right) - \frac{x}{2}\,erfc\left(\frac{x}{2\sqrt{\kappa t}}\right) \right\} \qquad [A48.1]$$

where $q_{s,0}$ is the unit width sediment discharge at the sediment source ($x = 0$), t is the time elapsed, κ is the diffusivity, and $erfc$ is the complementary error function. This solution cannot be used unless there is information on the sediment discharge at the source, and on the sediment transport coefficient or diffusivity. These parameter values can be constrained by taking field measurements of the slope at $x = 0$, and the elevation h_0 of the apex above the base-level over which the clinoforms migrate. We then make use of an expression for the maximum elevation

$$h_0 = 2q_{s,0}\left(\frac{t}{\pi\kappa}\right)^{1/2} \qquad [A48.2]$$

and for the maximum slope at $x = 0$

$$\left(\frac{\partial h}{\partial x}\right)_{x=0} = -\frac{q_{s,0}}{\kappa} \qquad [A48.3]$$

First, solve for $q_{s,0}$ using

$$q_{s,0} = -\frac{\pi h_0^2}{4t\,(\partial h/\partial x)} \qquad [A48.4]$$

then use this value for $q_{s,0}$ to find κ.

Practical exercise

Plot a graph of clinoform elevation versus horizontal distance from the shoreline.

Use the following input data, applicable to an alluvial fan in the Basin and Range province, USA:

Elevation of fan or clinoform apex, h_0	400 m
Maximum slope at $x = 0$	−0.0075
Time t	10^6 yr

Now use the same approach for the Mississippi Delta with the following input parameters:

Elevation of clinoform apex, h_0	107 m
Maximum slope at $x = 0$	−0.0075
Time t	400 yr

Solution

For the Basin and Range fan, the sediment discharge $q_{s,0}$ is 1.68 m^2 yr^{-1} and diffusivity 22.3 m^2 yr^{-1}. The profile of the fan or clinoform is given in Fig. A48.1. The downstream length of the clinoform or fan is approximately 15 km. Note that the slope is determined by the ratio of sediment discharge to diffusivity.

For the Mississippi Delta, the sediment discharge $q_{s,0}$ is c.2000 m^2 yr^{-1} and diffusivity 2×10^5 m^2 yr^{-1} – much higher than the values obtained from the Basin and Range fan. The length scale of the clinoforms is also considerably greater.

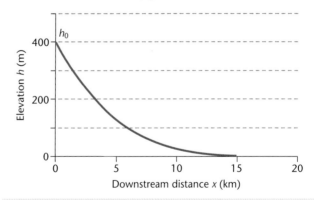

Fig. A48.1 Longitudinal profile of a fan with an apex elevation of 400 m, built by diffusive processes with a sediment discharge $q_{s,0}$ at $x = 0$.

Basin Analysis: Principles and Application to Petroleum Play Assessment, Third Edition. Philip A. Allen and John R. Allen.
© 2013 John Wiley & Sons, Ltd. Published 2013 by John Wiley & Sons, Ltd.

Axial versus transverse drainage

Learning points: the drainage of rivers, and the dispersal of sediment, may be either axial-longitudinal or transverse depending on the sediment fluxes from mountain catchments and the spatial distribution of accommodation. The simplest accommodation pattern is of a half-graben tilting down to a vertical normal or reverse fault. The flux from each margin versus the generation of accommodation determines fan lengths and fluxes to axial systems. High supply compared to low or reducing accommodation favours large transverse megafans (Burbank 1992).

Consider a sedimentary basin that undergoes subsidence by a simple tilting. There are two sediment sources, one from a tectonically uplifting region in the extensional footwall or contractional hangingwall, and another from the basin shoulder (Fig. A49.1). These sediment sources build fans from each margin. Using the terminology for extensional systems (Fig. A49.1a), the geometrical expressions below give the lengths of hangingwall- and footwall-derived fans based on a mass balance. For hangingwall-derived fans

$$w_h = \sqrt{2\frac{W_b}{S_{max}}(q_h - q_{ha})} \qquad \text{[A49.1]}$$

Similarly, for the footwall-derived fans

$$w_f = 2\frac{(q_f - q_{fa})}{S_{max}} \qquad \text{[A49.2]}$$

where w_h and w_f are the lengths of hangingwall and footwall-derived fans, W_b is the total basin width, S_{max} is the maximum subsidence rate along the vertical fault, q_h and q_f are the unit width sediment discharges from the hangingwall and footwall, and q_{ha} and q_{fa} are the sediment discharges from the hangingwall-derived fan to the axial system and from the footwall-derived fan to the axial system (Fig. A49.1).

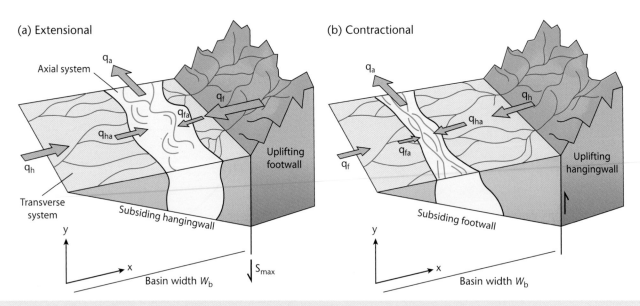

Fig. A49.1 Block diagrams to illustrate the dual supply of sediment from opposing hangingwall and footwall source areas in the geometrically similar extensional half-graben (a) and wedge-top basin tilting down towards thrust hangingwall (b).

Basin Analysis: Principles and Application to Petroleum Play Assessment, Third Edition. Philip A. Allen and John R. Allen.
© 2013 John Wiley & Sons, Ltd. Published 2013 by John Wiley & Sons, Ltd.

Practical exercise

1. Plot the length of footwall-derived and hangingwall-derived alluvial fans in an extensional basin as a function of tectonic subsidence rate (S_{max}). Assume that the sediment discharges are constants. Take the following parameter values

Maximum subsidence rate S_{max}	0.0001, 0.0005 and 0.001 m yr^{-1}
Basin width W_b	20 km
Sediment efflux of footwall q_f	0.75 m^2 yr^{-1}
Sediment efflux of hangingwall q_h	0.25 m^2 yr^{-1}
Sediment discharge from footwall to axial system q_{fa}	0.10 m^2 yr^{-1}
Sediment discharge from hangingwall to axial system q_{ha}	0.15 m^2 yr^{-1}

2. What must the discharges be in order to completely fill the available accommodation at these different maximum tectonic subsidence rates? Assume initially that the sediment discharges to footwall and hangingwall fans are identical.

Solution

1. The fan lengths are given in Table A49.1.

Table A49.1 Fan lengths for different values of the maximum subsidence rate in a 20 km-wide half-graben, with sediment discharges from the footwall and hangingwall of 0.75 and 0.25 m^2 yr^{-1}, and sediment discharges to the axial system from the footwall-derived and hangingwall-derived fans of 0.10 and 0.15 m^2 yr^{-1}

Maximum tectonic subsidence rate S_{max} (mm yr^{-1})	0.1	0.5	1
Length of footwall-derived fan w_f (km)	12.000	2.400	1.200
Length of hangingwall-derived fan w_h (km)	7.746	3.464	2.449
Width of axial system w_a (km)	0.254	14.136	16.351

2. The sediment discharges necessary to perfectly fill accommodation must satisfy

$$\frac{S_{max}}{2} W_b = q_f + q_h \qquad \text{[A49.3]}$$

and $w_f + w_h$ must be less than or equal to W_b. A range of sediment discharges is possible.

If we initially set $q_f = q_h$, then $q_f = q_h = S_{max}W_b/4$. The lengths of the footwall-derived and hangingwall-derived fans also depends on the fluxes into the axial system. Fig. A49.2 shows results for discharges to the axial system of 20% of the discharge to the fan, that is, $q_{fa} = 0.2\, q_f$, and $q_{ha} = 0.2\, q_h$. Note that if the ratio of discharge to axial system over source discharge is constant, the fan lengths do not change despite variations in the maximum tectonic subsidence rate.

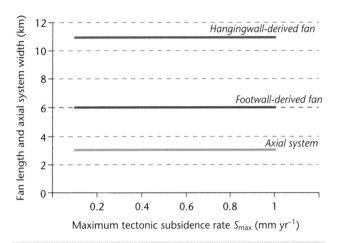

Fig. A49.2 Fan lengths and width of the axial system remain constant over a range of tectonic subsidence rate where the discharge to the axial system is a fixed percentage of the discharge from the source regions to the fans (20%).

Changing the relative contribution of the transverse fans to the axial drainage causes the fan lengths and axial system width to change as a function of maximum tectonic subsidence rate. Fig. A49.3 shows a perfectly filled basin evolving into an underfilled stage as sediment discharge is unable to keep up with a step increase in accommodation generation, to a final overfilled stage.

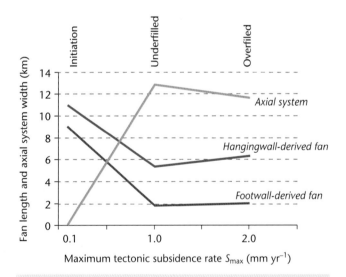

Fig. A49.3 Hypothetical evolution of fan lengths and axial system width from basin initiation to underfilled stage to overfilled stage.

Parameter values used to generate the chart in Fig. A49.3 are in Table A49.2.

Table A49.2 Parameter values used to generate the chart in Fig. A49.3 as the basin evolves through an underfilled to overfilled stage. Oversupply of the half-graben cross-section results in sediment export to down-system locations along the longitudinal system.

Parameter	Basin initiation	Underfilled	Overfilled
Total sediment supply q_T (m^2 yr^{-1})	1.0	1.8	20
Discharge from footwall q_f (m^2 yr^{-1})	0.6	1.0	12
Discharge from hangingwall q_h (m^2 yr^{-1})	0.4	0.8	8
Discharge from footwall-derived fan to axial system q_{fa} (m^2 yr^{-1})	0.15	0.1	10
Discharge from hangingwall-derived fan to axial system q_{ha} (m^2 yr^{-1})	0.1	0.8	6
Total rate of accommodation generation (m^2 yr^{-1})	1	5	10
Filling ratio [-]	1	0.36	2
Axial discharge exported (m^2 yr^{-1})	0	0	10
Footwall-derived fan length w_f (km)	9.000	1.800	2.000
Hangingwall-derived fan length w_h (km)	10.954	5.367	6.325
Axial system width (km)	0.045	12.833	11.675

Learning points are that the increase in tectonic subsidence rate is accompanied by a marked backstepping of footwall-generated fans, a widening of the axial fluvial belt and underfilling. Later overfilling witnesses a small progradation of fans and the export of large sediment discharges down-system. Fan lengths in the overfilled stage depend on the mass transfers to the axial system. Under high streamflood-dominated conditions q_{fa} and q_{ha} are likely to be high, enabling axial sediment export to take place. Under arid conditions dominated by debris flows, this process is likely to be far less efficient.

Downstream fining of gravel

Learning points: the commonly observed downstream fining of grain-size in stratigraphy is the result of the selective extraction of certain grain sizes from the mobile sediment flux. The grain-size trend in stratigraphy is different to that of the mobile surface flux. Analysis of regional grain-size trends in stratigraphy has the potential to aid the understanding of the role of sediment discharge and accommodation generation in basin filling.

The impact of deposition is to reduce the sediment discharge q_s in the downstream direction from its initial value $q_s(0)$ at $x = 0$. Allowing for the porosity of deposited sediment, and defining the depositional length L_d as the point at which the sediment discharge is exhausted,

$$q_s(x) = q_s(0)(1-\varphi)\int_0^{Ld} \sigma(x)dx \qquad [\text{A50.1}]$$

where φ is the porosity of the deposited sediment and $\sigma(x)$ is the spatial distribution of accommodation due to subsidence.

This equation can easily be adapted to consider the gravel fraction only f_g where $q_{\text{gravel}}(0) = f_g q_s(0)$. At each point in the downstream direction $x < L_d$ the solid sediment volume extracted by deposition as a fraction of the surface sediment flux is $(1-\varphi)\sigma(x)/q_s(x)$. When normalised by transforming the x-axis by setting $x^* = x/L_d$, and non-dimensionalised by introducing the length term L_d in the numerator, the dimensionless parameter for the extraction from the surface flux by deposition R^* becomes

$$R^*(x^*) = (1-\varphi)L_d \frac{\sigma(x^*)}{q_s(x^*)} \qquad [\text{A50.2}]$$

The integration of $R^*(x^*)$ over the normalised downstream distance x^* gives the cumulative downstream sediment deposition in relation to the available surface flux, $y^*(x^*)$, where

$$y^*(x^*) = \int_0^{x^*} R^*(x^*)dx^* \qquad [\text{A50.3}]$$

The dimensionless parameter $y^*(x^*)$ can be used to calculate the variation of mean grain size $D(x^*)$ from an initial mean grain size D_0 and variation ψ_0 in the sediment supply (at $x = 0$),

$$D(x^*) = D_0 + \psi_0 \frac{C_2}{C_1}\{\exp(-C_1 y^*)-1\} \qquad [\text{A50.4}]$$

where C_1 and C_2 are constants related to the variation in the grain-size distribution (details in Fedele & Paola 2007; Duller *et al.* 2010;

Whittaker *et al.* 2011). C_1 is 0.5 and C_2 is 0.45 in the geological application of Duller *et al.* (2010).

The grain-size trend therefore reflects the grain-size distribution of the sediment supply, the sediment discharge from the upstream source, and the spatial distribution of accommodation. An example is provided in Fig. A50.1 of the trend in mean grain-size for a starting grain-size of 40 mm, and a unit width sediment discharge of gravel from the source region of 4.6 m² yr⁻¹, which becomes exhausted at the gravel front at 50 km. The accommodation decreases exponentially down-system from a maximum of 0.2 mm yr⁻¹ at $x = 0$.

Exponential Accommodation Model

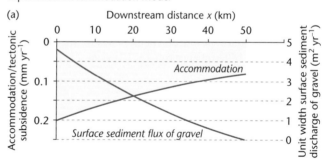

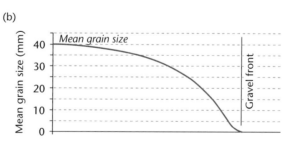

Fig. A50.1 Surface sediment flux $q(x)$, tectonic subsidence (accommodation) $\sigma(x)$ for an exponential accommodation model, and the resulting regional grain-size trend for an initial mean grain size of 40 mm.

Basin Analysis: Principles and Application to Petroleum Play Assessment, Third Edition. Philip A. Allen and John R. Allen.
© 2013 John Wiley & Sons, Ltd. Published 2013 by John Wiley & Sons, Ltd.

Sinusoidal eustatic change superimposed on background tectonic subsidence

Learning points: the packaging of stratigraphy results from a combination of sediment supply, background tectonic subsidence and base-level change. This balance can be initially considered by using analytical expressions for the tectonic subsidence and eustatic change. This simple theory allows predictions to be made of the generation of unconformities during falling relative sea level and of the diachroneity of points of maximum flooding.

We initially consider the variation of relative sea level through a cycle of eustatic change with wavelength λ and amplitude h_0 in a basin with a linear tectonic subsidence rate a, as originally envisaged by Barrell (1917). (Fig. 8.2). The elevation of sea level caused by a sinusoidal eustatic fluctuation is

$$h = h_0 \sin\left(\frac{2\pi t}{\lambda}\right) \qquad [A51.1]$$

Adding the tectonic subsidence rate, assumed linear, to obtain the relative sea level h_{rel}, we have

$$h_{rel} = h_0 \sin\left(\frac{2\pi t}{\lambda}\right) + at \qquad [A51.2]$$

where t is time.

The point at which the inflexion point in relative sea level occurs is at $dh_{rel}/dt = 0$. Differentiating eqn. [A51.2] gives

$$\frac{dh_{rel}}{dt} = h_0 \frac{2\pi}{\lambda}\cos\left(\frac{2\pi t}{\lambda}\right) + a \qquad [A51.3]$$

Setting $dh_{rel}/dt = 0$ gives the solution

$$t = \frac{\lambda}{2\pi}\cos^{-1}\left(-\frac{\lambda}{2\pi}\frac{a}{h_0}\right) \qquad [A51.4]$$

Eqn. [A51.4] shows that for a given eustatic period λ, the time period to the relative sea level inflexion point decreases with increasing values of the dimensionless parameter $\psi = a\lambda/h_0$.

At what critical value of a does relative sea level increase monotonically, that is, dh_{rel}/dt is always positive? This is when

$$a > h_0 \frac{2\pi}{\lambda}\cos\left(\frac{2\pi t}{\lambda}\right) \qquad [A51.5]$$

for all t, which is when $\lambda a/2\pi h_0 > 1$, or $\psi/2\pi > 1$.

We now add the effects of sediment supply. Sediment supply controls how much of the accommodation is filled (Fig. 8.3). In the simplest scenario, the sediment supply rate s could be considered a constant velocity, in which case water depth as a function of time $w(t)$ is given by

$$w(t) = h_0 \sin\left(\frac{2\pi t}{\lambda}\right) + at - st \qquad [A51.6]$$

However, it is reasonable to assume that sediment supply rate is coupled in some way to the rate of change of relative sea level. If sediment supply rate $s(t)$ peaks at the maximum rate of relative sea-level fall, and is zero at the maximum rate of relative sea-level rise, we can write

$$s(t) = s_0 \sin\left(\frac{\pi t}{\lambda}\right) \qquad [A51.7]$$

It would also be simple to formulate sediment supply as coupled to the rate of relative sea-level variation but with a certain time lag to account for the response time of sediment delivery systems to any base-level change (§7.6). However, to obtain the sediment accumulated S at any point within a relative sea-level cycle simply using eqn. [A51.7], we integrate it to give

$$S(t) = -s_0 \frac{\lambda}{\pi}\cos\left(\frac{\pi t}{\lambda}\right) + c \qquad [A51.8]$$

Since the total accumulated sediment is zero at $t = 0$, c must be $s_0\lambda/\pi$. The thickness of accumulated sediment is

$$S(t) = \frac{s_0\lambda}{\pi}\left\{1 - \cos\left(\frac{\pi t}{\lambda}\right)\right\} \qquad [A51.9]$$

and the water depth is therefore

$$w(t) = h_0 \sin\left(\frac{2\pi t}{\lambda}\right) + at + s_0\frac{\lambda}{\pi}\left\{\cos\left(\frac{\pi t}{\lambda}\right) - 1\right\} \qquad [A51.10]$$

where the first term is the eustatic effect, the second the tectonic subsidence effect and the third the sediment supply effect. The evolution of water depth through a relative sea-level cycle with different rates of sediment supply is sketched in Fig. 8.3.

Basin Analysis: Principles and Application to Petroleum Play Assessment, Third Edition. Philip A. Allen and John R. Allen.
© 2013 John Wiley & Sons, Ltd. Published 2013 by John Wiley & Sons, Ltd.

Isostatic effects of absolute sea-level change

Learning points: sea-level change commonly results in a change of water depth, but the addition or subtraction of water load results in an isostatic compensation in the bathymetry of the seabed. Using Airy isostasy, the isostatic subsidence of the seabed is about 40% of the sea-level change. The 'freeboard' is the sea-level change after isostatic compensation, and is not the same as the change in water depth.

Consider an ocean basin of depth h_1 that deepens to a depth h_2. The greater depth of the ocean is made up of two components: the subsidence of the ocean floor S, and the sea-level rise Δ_{SL}.

The balance of pressure at a depth of compensation C_1 below the surface of the ocean crust (Fig. 8.21a) is

$$\rho_w g h_1 + \rho_m g C_1 = \rho_w g h_2 + \rho_m g (C_1 - S) \qquad [A52.1]$$

where ρ_w and ρ_m are the water and ocean crust densities respectively. Rearrangement gives

$$\Delta_{SL} = S\left(\frac{\rho_m - \rho_w}{\rho_w}\right) \qquad [A52.2]$$

If rocks comprising the ocean crust have a density (ρ_m) of $3300\,kg\,m^{-3}$ and ocean water a density (ρ_w) of $1000\,kg\,m^{-3}$, the isostatic subsidence of the ocean floor is approximately 0.4 of the sea-level change. Expressed slightly differently, the sea-level change is 0.7 of the increase in the water depth of the ocean ($h_2 - h_1$). These figures represent approximate upper bounds on the water loading of the oceanic lithosphere since they result from a purely local (vertical) isostasy. A flexural response of the lithosphere would reduce the water-loaded subsidence. However, the approximation is reasonable, since the wavelength of the water load (the ocean) is very large compared to lithospheric thickness.

Basin Analysis: Principles and Application to Petroleum Play Assessment, Third Edition. Philip A. Allen and John R. Allen.
© 2013 John Wiley & Sons, Ltd. Published 2013 by John Wiley & Sons, Ltd.

53

Sea-level change resulting from sedimentation

Learning points: sedimentation in the ocean depresses the ocean floor isostatically but elevates the seabed by deposition. A one-dimensional balance results in a sea-level rise of about a quarter of the sediment layer thickness.

Consider a piece of ocean crust of density ρ_m overlain by water of density ρ_w and depth h_w. A layer of sediment of density ρ_s and thickness h_s is deposited on the seafloor (Fig. 8.21b). An isostatic balance at a depth C_1 below the surface of the ocean crust is:

$$\rho_m g C_1 + \rho_w g h_w = \rho_m g C_2 + \rho_s g h_s + \rho_w g h_w \qquad [A53.1]$$

where C_2 is the thickness of ocean crust above the depth of compensation after sediment loading, and $C_1 - C_2 = h_s - \Delta_{SL}$. The solution is

$$\Delta_{SL} = h_s\left(1 - \frac{\rho_s}{\rho_m}\right) \qquad [A53.2]$$

If ρ_m is $3300\,\mathrm{kg\,m^{-3}}$ and the density of ocean sediment ρ_s is $2500\,\mathrm{kg\,m^{-3}}$, the sea-level change is 0.24 of the sediment thickness h_s. If 25 mm of sediment is deposited every 1000 years, the sea-level change is 6 mm per 1000 years.

Basin Analysis: Principles and Application to Petroleum Play Assessment, Third Edition. Philip A. Allen and John R. Allen.
© 2013 John Wiley & Sons, Ltd. Published 2013 by John Wiley & Sons, Ltd.

The consolidation line

Learning points: the void ratio of a soil or porous sediment changes when the medium is compressed, as when a new layer of sediment is deposited on top. Compression leads to a reduction of the thickness of the layer and therefore a vertical strain. The change in the void ratio as a function of the effective stress is known as the normal consolidation line.

All soils compact by losing void space caused by a rearrangement and crushing of grains, so the compressibility of the mineral grains is negligible compared to the consolidation responsible for the reduction of void space. Compaction in soils is measured by compressing a sample axially, the results being shown in a semi-log plot of effective vertical stress σ' versus void ratio e (Fig. 9.1). The void ratio is related to porosity ϕ since

$$e = \frac{V_p}{V_t - V_p} = \frac{\phi}{1 - \phi} \qquad [A54.1]$$

where V_p is the volume of pore space and V_t is the total volume (both solid framework and pore space). This expression can also be inverted to give

$$\phi = \frac{e}{e + 1} \qquad [A54.2]$$

The *normal consolidation line* at relatively low vertical effective stresses is

$$e = e_0 - C \ln\left(\frac{\sigma'}{\sigma'_0}\right) \qquad [A54.3]$$

where e_0 is the void ratio at a reference effective stress σ'_0 (0.1 MPa in Fig. 9.1, see fig 4.2, p. 89, Wangen 2010) and C is called the *compression index*.

A change in the void ratio results in a change in the layer thickness. Treating the solid volume as a thickness ζ, which does not vary during compaction, and a porosity that does vary

$$h = \frac{\zeta}{1 - \phi} = (1 - e)\zeta \qquad [A54.4]$$

then a change in the void ratio Δe leads to a change in the thickness of the sample Δh. The fractional change in the layer thickness or strain in the vertical direction is given by

$$\frac{\Delta h}{h} = \frac{\Delta e}{1 + e} \qquad [A54.5]$$

The strain $\Delta h / h$ is caused by a change in effective stress. Since stress divided by strain gives Young's modulus E for elastic materials, we can write

$$E' = \frac{\Delta \sigma'}{\Delta h / h} = \frac{1 + e}{C} \sigma' \qquad [A54.6]$$

The porous layer (soil) therefore has a higher Young's modulus at higher levels of compression. It becomes 'stiffer' with loading.

Consider a clay-rich layer of thickness $h = 4\,\mathrm{m}$ subjected to a change in vertical stress $\Delta \sigma' = 20\,\mathrm{kPa}$ caused by a weight placed on the top of the layer, which overlies an incompressible substrate. The reference void ratio for clay is $e_0 = 1.5$, corresponding to a porosity of 0.6, the compression index $C = 0.2$ and the reference effective vertical stress $\sigma'_0 = 1\,\mathrm{kPa}$. The change in void ratio Δe is given by a rearrangement of the consolidation relationship (eqn. [A54.3])

$$\Delta e = C \ln\left(\frac{\Delta \sigma'}{\sigma'_0}\right) \qquad [A54.7]$$

which with the parameter values above gives $\Delta e = 0.60$. This change in void ratio causes a subsidence of the clay-rich layer of $\Delta h = h \Delta e / (1 + e_0) = 0.96\,\mathrm{m}$. The new layer thickness is therefore 3.04 m.

Basin Analysis: Principles and Application to Petroleum Play Assessment, Third Edition. Philip A. Allen and John R. Allen.
© 2013 John Wiley & Sons, Ltd. Published 2013 by John Wiley & Sons, Ltd.

Relation between porosity and permeability – the Kozeny-Carman relationship

Learning points: porosity and permeability are the key properties of a porous or fractured rock in terms of its ability to transmit fluids. The relationship between porosity and permeability depends on the packing arrangement of grains.

For a medium of well-sorted grains, in which pore throats are the same size, and assuming laminar flow, the form of the permeability function $k(\phi)$ is known as the Kozeny-Carman relationship:

$$k(\phi) = \frac{\phi^3}{2S^2}$$

$$k(\phi) = \frac{\phi^3}{2S_s^2(1-\phi)^2} \qquad \text{[A55.1a, b]}$$

where $S_s = S/(1 - \phi)$ is the *specific surface area* of the pore space with respect to the volume of solid instead of the total volume. For spherical grains, S depends on grain radius a, the grain density ρ (number of grains per unit volume) and the porosity ϕ, giving $S = 4\pi a^2 \rho \phi$. S therefore reflects the arrangement or packing of the grains. As a rule of thumb, the Kozeny-Carman permeability scales on the *cube* of the porosity.

An empirical relationship based on measurements of porosity and permeability has the simpler form

$$k(\phi) = k_c \phi^m \qquad \text{[A55.2]}$$

where k_c is a constant with units of permeability and m is an exponent that ranges between 2 and 7. If $m = 3$ and k_c is $1/(2S_s^2)$,

the empirical expression is the same as the Kozeny-Carman relationship.

Measurements of porosity and permeability show that there is commonly a very wide variation of permeability for a given porosity (Bloch *et al.* 2002; Dutton *et al.* 2003) (Fig. 9.7). Permeability is plotted on a $\log_{10}$ scale against porosity, so that the linear least-squares fit is of the form

$$\log_{10} k = a\phi + b \qquad \text{[A55.3]}$$

where a is the gradient and b the intercept (Fig. 9.7), which can be rewritten

$$k(\phi) = k_c \exp(a\phi) \qquad \text{[A55.4]}$$

where k_c is the permeability at zero porosity, $k_c = \exp(b)$. We could also write eqn. [A55.4] in terms of a reference porosity rather than a zero porosity, giving

$$k(\phi) = k_0 \exp(a(\phi - \phi_0)) \qquad \text{[A55.5]}$$

where $k_0 = \exp(b + a\phi_0)$. Log-linear expressions fit many porosity-permeability measurements for a range of different lithologies.

Basin Analysis: Principles and Application to Petroleum Play Assessment, Third Edition. Philip A. Allen and John R. Allen.
© 2013 John Wiley & Sons, Ltd. Published 2013 by John Wiley & Sons, Ltd.

Decompaction

Learning points: decompaction is an important procedure to remove the progressive effects of compaction and reduction of layer thicknesses and average porosities during basin subsidence. Decompaction allows the depths of certain stratigraphic horizons to be tracked as a function of time. The decompaction technique essentially restores present-day thicknesses to those at a given time in the burial history by use of porosity-depth relationships for each lithological type. Variations in layer thickness during burial can also be calculated by changing the coordinate system so that the depth term is the *porosity-free* vertical coordinate. A stratigraphic layer has a constant porosity-free thickness over time, so its true thickness as a function of time is the solid (porosity-free) thickness plus the amount due to porosity.

The decompaction method proposed by Sclater and Christie (1980) is summarised below. Consider a sediment layer at present depths of y_1 and y_2 that is to be moved vertically to new shallower depths y_1' and y_2' (Fig. A56.1).

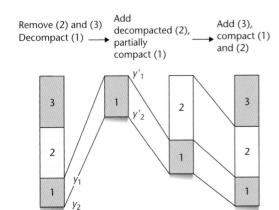

Fig. A56.1 Concept of the successive stages in a decompaction exercise.

From eqn. [9.24] the amount of water-filled pore space V_w between depths y_1 and y_2 is simply the porosity integrated over the depth interval,

$$V_w = \int_{y1}^{y2} \phi_0 e^{-cy} dy \qquad [A56.1]$$

where ϕ_0 is the surface porosity, and c is a depth constant for the downward decrease in porosity (see §9.3.2), which on integration gives

$$V_w = \frac{\phi_0}{c}\{\exp(-cy_1) - \exp(-cy_2)\} \qquad [A56.2]$$

Since the total volume of the sediment layer (V_t) is the volume due to pore-filling water (V_w) and the volume of the sediment grains (V_s),

$$V_s = V_t - V_w \qquad [A56.3]$$

and from eqn. [A56.2], considering a unit cross-sectional area,

$$y_s = y_2 - y_1 - \frac{\phi_0}{c}\{\exp(-cy_1) - \exp(-cy_2)\} \qquad [A56.4]$$

On decompaction, the sediment volume remains the same, only the volume of water expanding. The height of the water in a unit area sedimentary column lying between depths y_1' and y_2' is from eqn. [A56.2]

$$y_w' = \frac{\phi_0}{c}\{\exp(-cy_1') - \exp(-cy_2')\} \qquad [A56.5]$$

The new decompacted thickness of the sediment layer is the sum of the thickness due to the sediment grains (eqn. [A56.4]) and that due to the water (eqn. [A56.5]). That is,

$$y_2' - y_1' = y_s + y_w' \qquad [A56.6]$$

which becomes

$$y_2' - y_1' = y_2 - y_1 - \frac{\phi_0}{c}\{\exp(-cy_1) - \exp(-cy_2)\}$$
$$+ \frac{\phi_0}{c}\{\exp(-cy_1') - \exp(-cy_2')\} \qquad [A56.7]$$

This is the *general decompaction equation*. It represents mathematically the exercise of sliding the sediment layer up the exponential porosity-depth curve. Its solution is by numerical iteration, which makes it ideal for solving by computer.

Basin Analysis: Principles and Application to Petroleum Play Assessment, Third Edition. Philip A. Allen and John R. Allen.
© 2013 John Wiley & Sons, Ltd. Published 2013 by John Wiley & Sons, Ltd.

The average porosity of the layer at any depth is

$$\phi = \frac{\phi_0}{c} \cdot \frac{\exp(-cy_1') - \exp(-cy_2')}{y_2' - y_1'} \qquad [A56.8]$$

Practical exercise: calculation of decompacted depths, layer thicknesses and average porosities

Consider the three-layer dataset in Table A56.1.

Table A56.1

Layer i	Present-day thickness Δy_i (m)	Present-day base y_i (m)	Porosity-depth coefficient c (km^{-1})	Surface porosity ϕ_0
3 Shales	2000	2000	0.51	0.63
2 Sandstones	1000	3000	0.27	0.49
1 Chalk	1000	4000	0.71	0.70

Solution

To decompact Layer 1 (Chalk) so that its upper surface is at sea level, we use eqn. [A56.7], where $y_2 = 4000$ m, $y_1 = 3000$ m, $y_1' = 0$, $\phi_0 = 0.70$, and $c = 0.71$. y_2' appears on both the right-hand and left-hand sides of the equation, which therefore needs to be solved by iteration (Hint: choose a range of values of y_2' and calculate the left-hand side and right-hand side of equation [A56.7]. The correct value for y_2' is where the LHS equals the RHS). The result is $y_2' = 1612$ m, and the average porosity of the decompacted layer 1 from eqn. [A56.8] is 0.417.

To track the depth of the base of Layer 1 over time, it is necessary to decompact Layer 2 by bringing it to the surface, so that the decompacted base of Layer 2 becomes the top of Layer 1. The decompacted base of Layer 2 is at 1280 m and the average porosity of decompacted Layer 2 is 0.414. The depth to the top of Layer 1 is therefore 1280 m. To calculate the base of Layer 1 under the decompacted Layer 2, we set $y_1' = 1280$ m. The base becomes $y_2' = 2446$ m. The average porosity of Layer 1 under the decompacted Layer 2 is 0.192.

Layer thicknesses and average porosities calculated through the decompaction process are shown in Fig. A56.2.

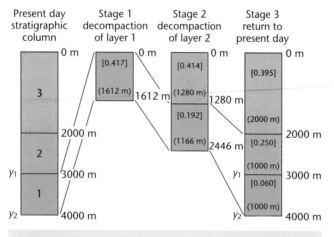

Fig. A56.2 Result of decompaction exercise, showing decompacted depths, layer thicknesses (curved brackets) and average porosities (straight brackets).

Lagrangian method: porosity-free and real depths

An alternative procedure for decompaction is a Lagrangian method (modified from Wangen 2010). A Lagrangian coordinate system is one that moves – in this case the vertical coordinate system moves with the compacting sedimentary basin layers. The set-up is shown diagrammatically in Fig. A56.3. During burial, the net (porosity-free) thicknesses of individual layers do not change over time, since there are no solid additions or subtractions to the stratigraphic layer. The stratigraphic section is divided by i number of horizons, the top surface of the basin being at $i = N$. The porosity-free depth from basement to a chosen horizon is ζ_i and the porosity-free depth measured downwards from the surface in η_i. The true depth of each horizon is y_i, and the layer thicknesses are $\Delta \zeta_i$ in terms of the net porosity-free thickness and $\Delta y_i = y_{i+1} - y_i$ in terms of the real depth. The total ζ-coordinate thickness of the stratigraphic column is the sum of the layer thicknesses for every i from 1 to N. We approximate the average porosity of the layer by the porosity at the centre of the layer – suitable if layer thicknesses are small.

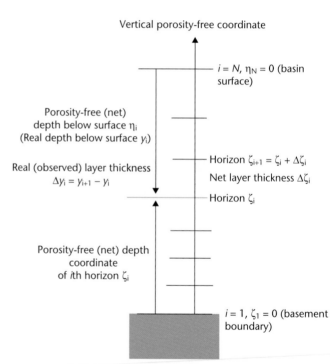

Fig. A56.3 Lagrangian porosity-free coordinate system for a basin made of N layers (modified from Wangen 2010, p. 98, fig. 5.3). © Magnus Wangen 2010, reproduced with the permission of Cambridge University Press.

Porosity-free and real depths

The vertical Lagrangian coordinate system used is measured as a thickness of sediment with porosity removed above a horizon in the basin, such as basement. The relationship between the real thickness and the porosity-free thickness is of course governed by the average porosity of the sedimentary layer.

If the porosity at depth is close to zero, the porosity varies exponentially with net (porosity-free) depth as

$$\phi = \phi_0 \exp\left(-\frac{(\zeta^* - \zeta)}{\zeta_0}\right) \qquad \text{[A56.9]}$$

where ζ^* is the current height or total thickness of the sedimentary basin, ζ is the porosity-free height of a chosen horizon above the basement, and ζ_0 is a porosity-depth coefficient equivalent to the reciprocal of c in the Athy relationship.

The real depth of sediments from the basin surface is a function of the ζ-depth from the surface $(\zeta^* - \zeta)$ given by

$$y = (\zeta^* - \zeta) + \zeta_0 \ln\frac{1-\phi}{1-\phi_0} \qquad \text{[A56.10]}$$

which can be rewritten in terms of the layer thickness

$$\Delta y = \Delta\zeta + \ln\left\{\frac{(1-\phi_i)}{(1-\phi_{i+1})}\right\} \qquad \text{[A56.11]}$$

where $\Delta y = y_{i+1} - y_i$ is the layer thickness in terms of real depth, and $\Delta\zeta = \zeta_{i+1} - \zeta_i$ is the layer thickness in terms of porosity-free depth.

Practical exercise: porosity versus depth in the vertical porosity-free coordinate

Use $\zeta^* = 3000$ m as the porosity-free thickness of the basin, $\zeta_0 = 1408$ m (the reciprocal of $c = 0.71\,\text{km}^{-1}$) and surface porosity ϕ_0 of 0.70.

Use eqn. [A56.10] and eqn. [A56.11] to calculate the porosity in terms of porosity-free and real depths.

Solution

The results are shown in Fig. A56.4. The blue line shows the porosity of sediments with a surface porosity of 0.70 (Chalk) as a function of true depth below the surface. The red line shows the porosity for the same sediments *versus* the net (porosity-free) height above basement.

Practical exercise: real depth versus time (decompacted burial history)

To simulate burial of a given sedimentary layer, the net (solid) thickness remains a constant over time. To track a horizon over time, we firstly set porosity-free height above basement as, for example, $\zeta = 1000$ m. We then assume a constant porosity-free sedimentation rate S of $0.1 \times 10^{-3}\,\text{m yr}^{-1}$, so during basin subsidence the basement height is $\zeta^* = St$, where t is time in years.

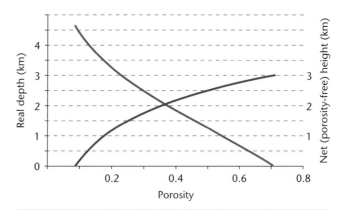

Fig. A56.4 Porosity versus net (porosity-free) height and real depth, with parameter values for Chalk, for a basin of 3 km porosity-free depth. The difference between real and net depths is due to the progressive loss of porosity with subsidence and burial.

Solution

The plot of real depth versus time for two stratigraphic horizons with a porosity-free height of 1 km and 2.5 km is shown in Fig. A56.5, for a steady solid sedimentation rate of $0.1\,\text{mm yr}^{-1}$, in a basin with the porosity-depth behaviour of Chalk.

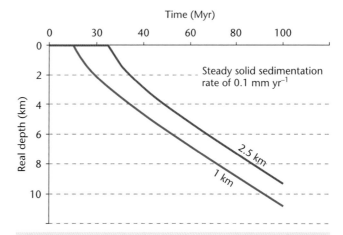

Fig. A56.5 Decompacted burial history for two horizons in a sedimentary basin made of one lithology, undergoing steady solid sediment accumulation, using a Lagrangian coordinate system.

The same procedure can be followed for other horizons in the basin, with their own porosity-depth relationships.

Layer thicknesses

In summary, starting with the present-day stratigraphic thicknesses, the average porosity of each layer at the present-day is found from an exponential (Athy-type) relationship, which allows the net (porosity-free) thickness of each layer to be found. The net (porosity-free) thickness of each layer can be found from

$$\Delta\zeta_i = (1-\phi)\Delta y_i \qquad [A56.12]$$

where the porosity is

$$\phi = \phi_{0,i}\exp\{-c_i(\bar{y}-y_N)\} \qquad [A56.13]$$

and the present-day real depth of the centre-point of the ith layer is

$$\bar{y} = (y_i + y_{i+1})/2 \qquad [A56.14]$$

where y_N is the depth of the present-day surface of the sedimentary basin, and Δy_i is the real thickness of the ith layer.

The net (porosity-free) thickness of each layer at the present-day can then be used to calculate the true layer thicknesses at other times in the burial history.

Practical exercise: layer thicknesses using the Lagrangian method

As an example of the method, consider the same three-layer dataset as we used in the previous exercise. From these data, $y_N = 0$ m.

Taking layer $i = 1$, which is currently between real depths of 3000 (y_2) and 4000 (y_1) m, the centre of the layer $\bar{y}$ is at 3500 m, and taking the surface of the sedimentary basin y_N to be at 0 m elevation, the average porosity of layer 1 at present-day depth is 0.058, so the solid porosity-free thickness becomes 942 m. The mean porosity of the layer if brought to the surface with a centre at half of Δy_i is 0.401. This can be used to calculate by iteration the real layer thickness when at the surface, giving Δy_i of 1571 m. This gives the depth of the horizon below layer 1 at the end of its deposition.

Table A56.2

Layer i	Centre-point of layer $\bar{y}$ (m)	Average present-day porosity ϕ	Porosity-free layer thickness $\Delta\zeta_i$ (m)
3 Shales	1000	0.38	1243
2 Sandstones	2500	0.25	751
1 Chalk	3500	0.06	942

57

Backstripping

Learning points: the depths of a stratigraphic horizon over time is partly a result of the loading by overlying sediment. Backstripping is the removal of this sediment loading effect to reveal the driving force for tectonic subsidence, with the assumption that it is taking place in a water-filled basin. Backstripping allows different boreholes or stratigraphic sections to be compared in terms of their water-loaded tectonic subsidence.

At its simplest, following Sclater and Christie (1980), the influence of the sediment load can be evaluated as follows. The porosity of the sediment layer at its new depth is

$$\phi = \frac{\phi_0}{c} \cdot \frac{\exp(-cy_1') - \exp(-cy_2')}{y_2' - y_1'} \qquad [A57.1]$$

Since the bulk density of the new sediment layer (ρ_b) depends on the porosity and the density of the sediment grains (ρ_{sg})

$$\rho_b = \phi\rho_w + (1-\phi)\rho_{sg} \qquad [A57.2]$$

the bulk density of the entire sedimentary column ($\bar{\rho}_b$) made up of i layers is

$$\bar{\rho}_b = \sum_i \left\{ \frac{\bar{\phi}_i\rho_w + (1-\bar{\phi}_i)\rho_{sgi}}{S} \right\} y_i' \qquad [A57.3]$$

where $\bar{\phi}_i$ is the mean porosity of the ith layer, ρ_{sgi} is the sediment grain density of the same layer, y_i is the thickness of the ith sediment layer, and S is the total thickness of the column corrected for compaction.

The loading effect of the sediment can then be treated as a problem of a local (Airy) isostatic balance. Where sediment is replacing a column of water,

$$Y = S\left(\frac{\rho_m - \bar{\rho}_b}{\rho_m - \rho_w} \right) \qquad [A57.4]$$

where Y is the depth of the basement corrected for sediment load, and ρ_m, $\bar{\rho}_b$ and ρ_w are mantle, mean sediment column and water densities.

Incorporating the various effects of paleobathymetry, eustatic sea-level change and sediment loading gives the Airy compensated tectonic subsidence

$$Y = S\left(\frac{\rho_m - \bar{\rho}_b}{\rho_m - \rho_w} \right) - \Delta_{SL}\left(\frac{\rho_w}{\rho_m - \rho_w} \right) + (W_d - \Delta_{SL}) \qquad [A57.5]$$

where Δ_{SL} is the paleo-sea level relative to the present, and W_d is the paleo-water depth. This is the subsidence relative to a stationary datum (today's sea level) that would have occurred in an entirely water-filled basin.

Backstripping using the void ratio

The same backstripping exercise can be carried out making use of the void ratio e and the porosity-free ζ-coordinate system (Wangen 2010). Two values are required for the isostatic balance to remove the effects of the sediment load, the sediment thickness $y(t)$ and the average sediment bulk density ($\bar{\rho}_b$):

$$y(t) = \sum_i \{1 + e_i(t)\} d\zeta_i \qquad [A57.6]$$

where e_i is the void ratio of layer i, and $d\zeta_i$ is the net (porosity-free) thickness of layer i, and

$$\bar{\rho}_b(t) = \frac{\sum_i \{e_i(t)\rho_w + \rho_{s,i}\} d\zeta_i}{\sum_i \{1 + e_i(t)\} d\zeta_i} \qquad [A57.7]$$

where the sediment framework or grain density $\rho_{s,i}$, water density ρ_w and the net porosity-free thickness $d\zeta_i$ are all constant over time, which means that the bulk sediment density depends on paleo-porosity alone. Porosity for each layer ϕ_i can be found from the relationship

$$\Delta y(t)_i = \Delta\zeta_i / \{1 - \phi_i(t)\} \qquad [A57.8]$$

The required values of sediment bulk density and decompacted sediment thickness can therefore be obtained using the Lagrangian coordinate system and used in the isostatic balance to solve for the tectonic subsidence $Y(t)$, as given in eqn. [9.32]:

$$Y(t) = W_d(t) + y(t)\left(\frac{\rho_m - \bar{\rho}_b(t)}{\rho_m - \rho_w} \right) - \frac{\rho_m}{\rho_m - \rho_w}\Delta_{SL}(t) \qquad [A57.9]$$

Basin Analysis: Principles and Application to Petroleum Play Assessment, Third Edition. Philip A. Allen and John R. Allen.
© 2013 John Wiley & Sons, Ltd. Published 2013 by John Wiley & Sons, Ltd.

58

From decompaction to thermal history

Learning points: a common workflow in basin analysis is aimed at understanding the time-temperature history of chosen stratigraphic intervals. This has relevance to the prediction of rock properties (porosity, permeability, bulk density), temperature-sensitive diagenetic changes, deformation, and thermal maturity of organic constituents. The workflow involves distinct stages starting with decompaction and ending with time-temperature trajectories.

This exercise starts with stratigraphic information derived from a borehole or outcropping stratigraphic succession and goes through the various stages towards a paleotemperature history for chosen horizons.

Step 1. Decompaction: stratigraphic columns derived from field measurements and from well penetrations record the thicknesses of stratigraphic units at the present day, in a compacted state (Table A58.1). Consequently, they give a false idea of the rate of sediment accumulation over time. Present-day stratigraphic thicknesses therefore need to be decompacted. Conventionally, this is done by assuming an exponential relationship between porosity and depth. The parameters required are given in Table A58.1.

The exponential porosity-depth relationship is

$$\phi(y) = \phi_0 \exp(-cy) \qquad \text{[A58.1]}$$

where $\phi(y)$ is the porosity at any depth y, ϕ_0 is the porosity at the surface, and c is the porosity-depth coefficient with units of km^{-1}.

The decompaction procedure sequentially restores the thicknesses of stratigraphic units at time intervals corresponding to the ages of the unit boundaries (Appendix 56). The lowest (1st) unit is restored to a position with its top at zero depth and its decompacted thickness calculated. Then in step 2, Unit 1 is buried by the decompacted Unit 2, which causes it to be partially re-compacted. This is repeated for each stratigraphic unit. The mathematics is given by:

$$y_2' - y_1' = y_2 - y_1 - \frac{\phi_0}{c}\{e^{-cy_1} - e^{-cy_2}\} + \frac{\phi_0}{c}\{e^{-cy_1'} - e^{-cy_2'}\} \qquad \text{[A58.2]}$$

where y_1 and y_2 are the top and bottom of the stratigraphic unit at the present day, and y_1' and y_2' are the top and bottom of the stratigraphic unit at the decompacted or partially re-compacted depth.

Table A58.1 Parameter values used in decompaction procedure

Lithology (unit)	Surface porosity	Exponential decay constant c (km^{-1})	Grain density (kg m^{-3})	Age (Ma)	Present depths (m)
Shales (10)	0.63	0.51	2720	45–0	0–200
Sandstones (9)	0.49	0.27	2650	55–45	200–450
Shales (8)	0.63	0.51	2720	80–55	450–1200
Chalk (7)	0.70	0.71	2710	100–80	1200–2500
Sandstones (6)	0.49	0.27	2650	125–100	2500–3400
Limestones (5)	0.40	0.60	2710	145–125	3400–3600
Dolomite (4)	0.20	0.60	2870	160–145	3600–4000
Sandstone (3)	0.49	0.27	2650	210–160	4000–4250
Anhydrite (2)	0.05	0.20	2960	245–210	4250–5000
Quartzite (1)	0.20	0.30	2650	260–245	5000–5400

Step 2. Corrections for paleobathymetry and eustasy: paleobathymetric and eustatic changes over time at the site of the stratigraphic column are shown in Table A58.2.

Table A58.2 Paleobathymetry and eustatic sea level relative to present-day sea level W_d (m) from 260 Ma to present

260 Ma	245	210	160	145	125	100	80	55	45	0 Ma
−20	0	20	10	20	20	200	300	350	325	300

Sea level relative to today Δ_{SL} (m)

260 Ma	245	210	160	145	125	100	80	55	45	0 Ma
10	0	0	−20	−40	+70	+80	+100	+50	+40	0

Basin Analysis: Principles and Application to Petroleum Play Assessment, Third Edition. Philip A. Allen and John R. Allen.
© 2013 John Wiley & Sons, Ltd. Published 2013 by John Wiley & Sons, Ltd.

We incorporate these paleobathymetric and eustatic corrections to the decompacted depths (S) using

$$S^* = S - \Delta_{SL}\left(\frac{\rho_w}{\rho_m - \rho_w}\right) + (W_d - \Delta_{SL}) \qquad [\text{A58.3}]$$

where S^* is the corrected decompacted depths, and ρ_m and ρ_w are the mantle and water densities.

Step 3. Removal of effect of sediment load: the sediment thicknesses in a basin are amplified by the effect of sediment loading. This load can be accounted for using flexural isostasy, but a simpler method is to assume Airy isostasy. Backstripping requires two procedures: (i) the calculation of the bulk density of the sediment column as it evolves over time; and (ii) removal of the sediment loading effect using Airy isostasy.

The bulk density of an individual layer i with grain density ρ_g is

$$\rho_b = \phi\rho_w + (1-\phi)\rho_{gi} \qquad [\text{A58.4}]$$

The bulk density of a sedimentary column made of i layers of thickness $y'_2 - y'_1$ is given by:

$$\bar{\rho}_b = \sum_i \left\{\frac{\bar{\phi}_i\rho_w + (1-\bar{\phi}_i)\rho_{gi}}{S}\right\}(y'_2 - y'_1) \qquad [\text{A58.5}]$$

and the porosity of a stratigraphic unit at any depth can be found from

$$\bar{\phi} = \frac{\phi_0}{c}\left\{\frac{\exp(-cy'_1) - \exp(-cy'_2)}{y'_2 - y'_1}\right\} \qquad [\text{A58.6}]$$

The sediment infill of a basin is denser than the water that it replaces and therefore loads the underlying lithosphere, causing further subsidence. The tectonic driving subsidence is given by

$$Y = S^*\left(\frac{\rho_m - \bar{\rho}_b}{\rho_m - \rho_w}\right) \qquad [\text{A58.7}]$$

where S^* is the decompacted subsidence corrected for paleobathymetry and eustasy (eqn. A58.3).

Solution: decompaction and backstripping

Table A58.3 Decompacted thicknesses (km) of a stratigraphic column comprising 10 lithological units

Stratigraphic unit	245 Ma	210 Ma	160 Ma	145 Ma	125 Ma	100 Ma	80 Ma	55 Ma	45 Ma	0 Ma
Shales (Unit 10)	0	0	0	0	0	0	0	0	0	0.200
Sandstones (Unit 9)	0	0	0	0	0	0	0	0	0.263	0.250
Shales (Unit 8)	0	0	0	0	0	0	0	0.889	0.800	0.750
Chalk (Unit 7)	0	0	0	0	0	0	1.747	1.373	1.330	1.300
Sandstones (Unit 6)	0	0	0	0	0	1.204	0.959	0.917	0.907	0.900
Limestones (Unit 5)	0	0	0	0	0.300	0.232	0.205	0.201	0.200	0.200
Dolomite (Unit 4)	0	0	0	0.475	0.458	0.423	0.404	0.401	0.400	0.400
Sandstone (Unit 3)	0	0	0.392	0.356	0.339	0.300	0.260	0.253	0.251	0.250
Anhydrite (Unit 2)	0	0.770	0.768	0.765	0.763	0.758	0.752	0.750	0.750	0.750
Quartzite (Unit 1)	0.470	0.450	0.442	0.433	0.429	0.416	0.404	0.401	0.400	0.400
Stratigraphic thickness (km)	0.470	1.220	1.602	2.029	2.289	3.333	4.731	5.185	5.301	5.400

Table A58.4 Decompacted depths in km corrected for paleobathymetry and eustasy

Unit	260	245	210	160	145	125	100	80	55	45	0
10	0	0	0	0	0	0	0	0	0	0.267	0.500
9	0	0	0	0	0	0	0	0	0.277	0.530	0.750
8	0	0	0	0	0	0	0	0.155	1.166	1.330	1.500
7	0	0	0	0	0	0	0.084	1.902	2.539	2.660	2.800
6	0	0	0	0	0	0.082	1.288	2.861	3.456	3.567	3.700
5	0	0	0	0	0.078	0.218	1.520	3.066	3.657	3.767	3.900
4	0	0	0	0.039	0.553	0.676	1.943	3.470	4.058	4.167	4.300
3	0	0	0.020	0.431	0.909	1.015	2.243	3.730	4.311	4.418	4.550
2	0	0	0.790	1.199	1.674	1.778	3.001	4.482	5.061	5.168	5.300
1	0.035	0.470	1.240	1.641	2.107	2.207	3.417	4.886	5.462	5.568	5.700

Table A58.5 Porosities (to nearest percent) as a function of time for 10 lithological units undergoing progressive compaction during burial

Time before present (Ma)	260	245	210	160	145	125	100	80	55	45	0
Quartzite (Unit 1)	0	0.19	0.16	0.13	0.11	0.10	0.09	0.06	0.04	0.04	0.04
Anhydrite (Unit 2)	0	0	0.05	0.04	0.04	0.04	0.03	0.02	0.02	0.02	0.02
Sandstone (Unit 3)	0	0	0	0.46	0.41	0.38	0.32	0.22	0.17	0.15	0.15
Dolomite (Unit 4)	0	0	0	0	0.17	0.14	0.09	0.04	0.02	0.02	0.02
Limestones (Unit 5)	0	0	0	0	0	0.37	0.24	0.10	0.05	0.04	0.04
Sandstones (Unit 6)	0	0	0	0	0	0	0.42	0.28	0.21	0.19	0.18
Chalk (Unit 7)	0	0	0	0	0	0	0	0.37	0.15	0.11	0.10
Shales (Unit 8)	0	0	0	0	0	0	0	0	0.45	0.33	0.31
Sandstones (Unit 9)	0	0	0	0	0	0	0	0	0	0.44	0.41
Shales (Unit 10)	0	0	0	0	0	0	0	0	0	0	0.52

Table A58.6 Bulk density (kg m^{-3}) for ten lithological units undergoing progressive compaction during burial

Time before present (Ma)	260	245	210	160	145	125	100	80	55	45	0
Quartzite (Unit 1)	0	2346	2398	2439	2465	2480	2509	2555	2581	2588	2590
Anhydrite (Unit 2)		0	2870	2880	2887	2891	2900	2914	2922	2925	2926
Sandstone (Unit 3)			0	1902	1987	2037	2136	2293	2381	2405	2413
Dolomite (Unit 4)				0	2560	2615	2698	2794	2831	2838	2840
Limestones (Unit 5)					0	2094	2300	2534	2620	2637	2642
Sandstones (Unit 6)						0	1971	2193	2311	2342	2352
Chalk (Unit 7)							0	2087	2464	2524	2540
Shales (Unit 8)								0	1951	2156	2202
Sandstones (Unit 9)									0	1938	1982
Shales (Unit 10)										0	1843
Bulk density column		2346	2696	2514	2523	2493	2379	2371	2455	2492	2396

Table A58.7 Water-loaded tectonic subsidence (km) as a function of time for two selected horizons in the basin-fill

Time before present (Ma)	260	245	210	160	145	125	100	80	55	45	0
Tectonic subsidence 260 Ma horizon	−0.050	0.197	0.330	0.568	0.721	0.784	1.385	1.999	2.033	1.981	2.269
Y (km) 140 Ma horizon					0.027	0.078	0.616	1.254	1.361	1.340	1.553
Y (km) 125 Ma horizon						−0.029	0.522	1.170	1.287	1.269	1.473

Decompacted depths versus age for the 10 stratigraphic units are shown in Table A58.3. Decompacted depths corrected for the effects of water depth and eustatic variations are given in Tabe A58.4 and shown graphically in Fig. A58.1. The average porosities of the decompacted units are given in Table A58.5, allowing the average bulk densities to be calculated (Table A58.6). Following backstripping, the water-loaded tectonic subsidence as a function of time is given in Table A58.7. The water-loaded tectonic subsidence for the base of Unit 1 and for the 125 Ma horizon are shown in Fig. A58.2.

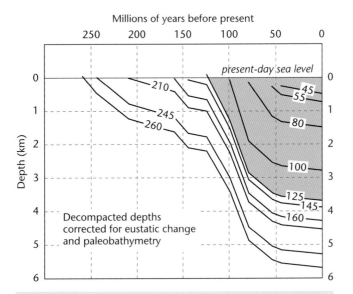

Fig. A58.1 Chart of decompacted depths corrected for changes in paleowater depth and eustatic sea level.

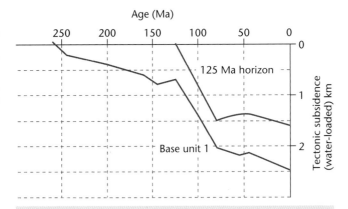

Fig. A58.2 Tectonic subsidence Y after decompaction and backstripping. Values are shown for the top of basement (base of quartzite) and for the 125 Ma horizon, close to the level of source rocks.

Step 4. Thermal conductivity of the basin-fill: our aim is to calculate the paleotemperature of chosen horizons in the basin-fill, such as source-rock horizons. To do this we require:

- an estimate of the basal heat flow (Q_m) variation over time;
- the decompacted subsidence history (S);
- the radiogenic heat production (A);
- the thermal conductivity of crustal rocks and sedimentary infill (K).

Data for the calculation of the bulk thermal conductivity of stratigraphic layers 1 to 10 are provided in Table A58.8. The bulk conductivity for two-phase media is calculated from the thermal conductivities of solid matrix and pore-filling fluid:

$$K_{bulk} = K_s(1-\theta) + K_w\theta \qquad [A58.8]$$

where K_s and K_w are the thermal conductivities of sediment and water respectively, and θ is the porosity.

The conductivity structure of the basin-fill changes over time as the different stratigraphic units evolve in their porosities (Table A58.9).

Step 5. Calculation of the geotherm with variable thermal conductivity: we firstly use a thermal model with a constant heat flow from the mantle, so the geotherm obeys Fourier's law:

$$T(y) = T_0 + (-Q_m y/K) \qquad [A58.9]$$

where $T(y)$ is the temperature at depth y, T_0 is the surface temperature, Q_m is the basal heat flow, and K is the thermal conductivity. In the case of variable thermal conductivity, we use values of thermal conductivity K_i for n layers of thickness l_i

$$T(y) = -Q\{l_1/K_1 + l_2/K_2 + l_3/K_3 + ...\} \qquad [A58.10]$$

and we let $T_0 = 0$ for simplicity, and $l_1 + l_2 + l_3 ....$ must equal y.

It is recommended, though laborious, to construct a geotherm for each time period from 260 to 0 Ma. The technique can be illustrated by taking the present-day situation, 0 Ma – we can therefore see the effect on the geotherm of the thermal conductivity structure of the entire basin-fill.

In constructing the geotherm in Step 5, it is important to understand the assumptions:

- we keep the heat flow constant and allow the geothermal gradient to vary as heat passes through the different stratigraphic units;
- surface temperature is 0 °C and does not vary over time;
- there is no radiogenic heat production.

Step 6. The radiogenic heat contribution: now we introduce a radiogenic heat source to generate a more realistic geotherm. Eqn. [A58.10] is modified to:

$$T_y = T_0 + \left\{ \left(\frac{Q_m + Ay_c}{K_1} \right)l_1 + \left(\frac{Q_m + Ay_c}{K_2} \right)l_2 + \cdots n \right\}$$
$$- \left\{ \frac{A}{2K_1}l_1^2 + \frac{A}{2K_2}l_2^2 + \cdots n \right\} \qquad [A58.11]$$

where y_c is the total thickness of the basin-fill (or crust), and the bulk thermal conductivities are as in Table A58.8. Initially, we keep the radiogenic heat production constant through the basin-fill, and incorporate a variable thermal conductivity structure to account for the heterogeneity of basin sedimentary rocks in Step 9.

Table A58.8 Porosity, sediment framework thermal conductivity (at surface temperatures), bulk thermal conductivity (accounting for pore fluid) and radiogenic heat production data, all at the present-day depths of the stratigraphic layers. Pore fluid has a thermal conductivity of $0.6\,\mathrm{W\,m^{-1}\,K^{-1}}$ at surface temperatures

Lithology	Porosity	K_i framework $(\mathrm{W\,m^{-1}\,K^{-1}})$	K_i bulk $(\mathrm{W\,m^{-1}\,K^{-1}})$	$A_i \times 10^{-6}$ $(\mathrm{W\,m^{-3}})$	l_i (m)
10. Shale	0.52	2.0	1.27	1.7	200
9. Sandstone	0.41	5.5	3.48	1.2	250
8. Shale	0.31	2.0	1.57	1.7	750
7. Chalk	0.10	3.3	3.03	0.4	1300
6. Sandstone	0.18	5.5	4.60	1.2	900
5. Limestone	0.04	3.3	3.17	0.5	200
4. Dolomite	0.02	5.3	5.23	0.4	400
3. Sandstone	0.15	5.5	4.78	1.2	250
2. Anhydrite	0.02	6.3	6.20	0.1	750
1. Quartzite	0.04	7.7	7.44	0.5	400
Basement	0.00	3.0	3.00	2.8	–

Step 7. Calculation of the stretch factor: it is possible to calculate the stretch factor from the (post-rift) thermal subsidence. To calculate the stretch factor β we must first identify the beginning of the post-rift phase (from, for example, seismic reflection data showing the end of extensional faulting, the end of fault-controlled sedimentation, the beginning of regional sag-type subsidence, the break-up unconformity).

The beginning of thermal subsidence is taken at 125 Ma in the practical example. Use the (water-loaded) tectonic subsidence for the 125 Ma horizon derived from the decompaction and backstripping exercises. Plot a graph of the water-loaded thermal subsidence versus $1 - \exp(-t/\tau)$, where t is the time since the end of rifting, and τ is the lithospheric time constant (c. 50 Myr) (Fig. A58.3).

The slope of the linear regression is equal to $E_0(\beta/\pi)\sin(\pi/\beta)$ and E_0 is given by $E_0 = (4y_L\rho_m\alpha_v T_m)/(\pi^2(\rho_m - \rho_w))$.

Obtain the slope of the regression and solve for the value of the stretch factor.

The value of β is approximately 1.71.

Step 8. Calculation of paleotemperatures using the reference uniform stretching model: the time-temperature trajectory of a particular stratigraphic horizon determines the thermal maturity of sedimentary rocks in the basin-fill.

A common procedure is to assume a certain geothermal gradient and to calculate the time-temperature trajectory based on the decompacted depths of a chosen stratigraphic horizon over time. An improvement on this is to invoke a geotherm that varies over time controlled by the stretch factor. In this case, the solution for uniform stretching is given by:

$$T(y,t) = T_m \frac{K_b}{K_s} \frac{S^*}{y_L}\left\{1 + \frac{2\beta}{\pi}\sin\left(\frac{\pi}{\beta}\right)\exp\left(-\frac{t}{\tau}\right)\right\} \qquad [A58.12]$$

where S^* is the corrected decompacted depth, y_L is the initial lithospheric thickness, τ is the lithospheric time constant, β is the stretch factor, and K_s and K_b are the thermal conductivities of sedimentary rock and basement respectively.

We can calculate the time-temperature trajectory of selected horizons (such as the Top Jurassic horizon, 145 Ma), using a stretch factor inverted from thermal subsidence data (Step 7). Parameter values are:

Thermal conductivity of sedimentary rocks $K_s = 1.25\,\mathrm{W\,m^{-1}\,K^{-1}}$
Thermal conductivity of basement $K_b = 3\,\mathrm{W\,m^{-1}\,K^{-1}}$
Asthenosphere temperature $T_m = 1330\,°C$
Initial lithosphere thickness $y_L = 125\,\mathrm{km}$
Lithospheric time constant $\tau = 50.2\,\mathrm{Myr}$
Stretch factor $\beta = 1.71$

The results for paleotemperature are shown in Fig. A58.4. The peak paleotemperature is in excess of $100\,°C$ and was attained at 55 Ma. The present-day temperature of the 145 Ma horizon is $97\,°C$.

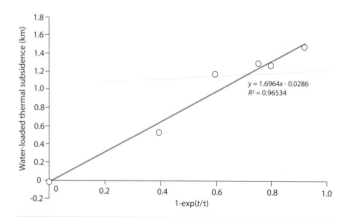

$y = 1.6964x - 0.0286$
$R^2 = 0.96534$

Fig. A58.3 Chart of the thermal subsidence for the 125 Ma horizon versus $1 - \exp(t/\tau)$. The slope gives $E_0(\beta/\pi)\sin(\pi/\beta)$, from which β can be calculated.

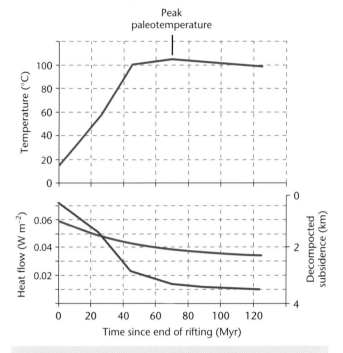

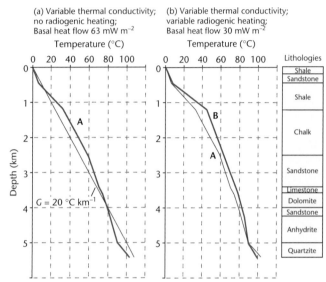

Fig. A58.4 Paleotemperature of the 145 Ma horizon (above), and the heat flow (blue) and decompacted subsidence (red). Note that peak paleotemperature occurs at c.55 Ma.

basal heat flow Q_m over time, the geotherm at the present-day is shown in Fig. A58.5.

The procedure should be repeated for each time step of the burial history, with the basal heat flow Q_m updated at each step based on the basal heat flow calculated from the uniform stretching model (Appendix 17).

Fig. A58.5 Comparison of linear geotherm (20 °C km⁻¹) with geotherms due to variable thermal conductivity with a basal heat flow of 63 mW m⁻² (a) and both radiogenic heating and variable thermal conductivity with a basal heat flow of 30 mW m⁻² (b).

Step 9. Calculation of paleotemperatures using the geotherm calculated using variable thermal conductivity and variable radiogenic heat production: the geotherm for a variable thermal conductivity and radiogenic heat production in the basin-fill is given in Appendix 13, eqn. A13.6. For the $n = 10$ layers of the basin-fill, and with a constant

Table A58.9 Thermal conductivity of stratigraphic layers as a function of time, in W m⁻¹ K⁻¹

Age (Ma)	260	245	210	160	145	125	100	80	55	45	0
Quartzite	6.37	6.60	6.78	6.89	6.96	7.09	7.29	7.40	7.43	7.44
Anhydrite		6.04	6.07	6.09	6.10	6.12	6.16	6.19	6.20	6.20
Sandstone			3.24	3.49	3.65	3.95	4.42	4.69	4.76	4.78
Dolomite				4.51	4.65	4.86	5.11	5.20	5.22	5.23
Limestones					2.31	2.50	2.95	3.12	3.16	3.17
Sandstones						3.45	4.12	4.48	4.57	4.60
Chalk							2.30	2.91	3.00	3.03
Shales								1.36	1.53	1.57
Sandstones									3.35	3.48
Shales										1.27

Advective heat transport by fluids

Learning points: slow conductive heat transport may be supplemented by a more rapid heat flow caused by advection of magmatic or hydrous fluids. Advective heat transport depends on the velocity of fluid flow, and is particularly important in the flow of groundwater through regional aquifers. Flow rates depend on the permeability of the aquifer, which is strongly influenced by lithology and burial history.

Consider a subsiding and compacting volume of porous rock with a cross-sectional area a and thickness δy (Fig. A59.1). The total heat gain or loss across this volume of porous rock of thickness δy is made of three components: a change in conductive heat flow q_c, an internal heat generation A, and a change in advective heat flow q_a

$$(a\partial y)\frac{\partial q_c}{\partial y} + (a\partial y)A + (a\partial y)\frac{\partial q_a}{\partial y} \qquad [\text{A59.1}]$$

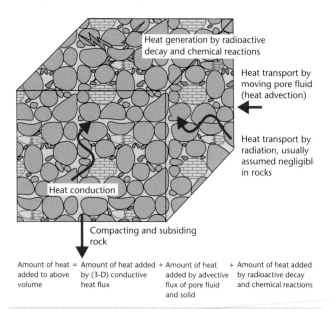

Amount of heat = Amount of heat added + Amount of heat + Amount of heat added
added to above   by (3-D) conductive   added by advective   by radioactive decay
volume   heat flux   flux of pore fluid   and chemical reactions
                                             and solid

Fig. A59.1 Schematic representation of a cube of subsiding and compacting sedimentary rock as a basis for a heat balance (after Giles 1997). Reproduced with kind permission from Springer Science+Business Media B.V.

The mass of the rock volume is $\rho_r a\delta y$, where ρ_r is the density of the porous rock. The heat stored in this rock volume is therefore the product of the specific heat c_r and its mass, $c_r\rho_r a\delta y$. If this rock volume undergoes a temperature change δT over a short time period δt, the rate of heat loss or gain is $c_r\rho_r a\delta y\delta T/\delta t$. Consequently,

$$c_r\rho_r a\partial y\frac{\partial T}{\partial t} = (a\partial y)\frac{\partial q_c}{\partial y} + (a\partial y)A + (a\partial y)\frac{\partial q_a}{\partial y} \qquad [\text{A59.2}]$$

The mass of fluid occupying the pore space is $(a\delta y\rho_f\phi)$. Its heat content is therefore $(a\delta y\rho_f\phi)c_f T$, and the advective heat flow is

$$q_a = (a\partial y)\rho_f\phi c_f v_f T \qquad [\text{A59.3}]$$

Substituting eqn. [A59.3] into eqn. [A59.2], making use of Fourier's law (eqn. [2.14]), simplifying and rearranging, eqn. [A59.3] can be written

$$\frac{\partial T}{\partial t} = \frac{1}{c_r\rho_r}\left\{ K_r\frac{\partial^2 T}{\partial y^2} + A + c_f\rho_f\phi v_f\frac{\partial T}{\partial y} \right\} \qquad [\text{A59.4}]$$

where K_r is the thermal conductivity of the rock volume, and it is assumed that the pore fluid velocity v_f does not vary with depth in the rock volume.

One of the most important parameters in an analysis of the temperature changes caused by advection of pore waters is the flow velocity of the moving groundwater. Using the parameter values for a volume of rock deeply buried in a sedimentary basin (Table 10.1) eqn. [A59.3] gives an advective heat flow of $16\,\text{mW}\,\text{m}^{-2}$ for a pore fluid velocity of $10^{-10}\,\text{m}\,\text{s}^{-1}$, and $2\,\text{mW}\,\text{m}^{-2}$ for a pore fluid velocity of $10^{-11}\,\text{m}\,\text{s}^{-1}$, typical of compaction-driven flow (Giles 1997, p. 278). Advective heat flows are therefore likely to be important in porous sedimentary rocks (such as uncemented aeolian sandstones) with compaction-driven groundwater velocities, and to dominate in aquifers with high groundwater velocities (10^{-6} to $10^{-7}\,\text{m}\,\text{s}^{-1}$).

Flow velocity can be estimated from the *Darcy equation*

$$U = k\frac{\partial P}{\partial x} \qquad [\text{A59.5}]$$

where k is the permeability and $\delta P/\delta x$ is the pressure gradient. Flow velocities are therefore strongly dependent on lithology. With compaction during burial, sedimentary units become less permeable and therefore less able to transmit fluids.

Basin Analysis: Principles and Application to Petroleum Play Assessment, Third Edition. Philip A. Allen and John R. Allen.
© 2013 John Wiley & Sons, Ltd. Published 2013 by John Wiley & Sons, Ltd.

Heat flow in fractured rock

Learning points: open fractures allow the advective and convective transmission of heat, which heats up the intervening rock volume. Long, wide fractures may heat the surrounding rock to the temperature of the fracture fluid on time scales that are geologically relatively short. Proximity to a fracture transmitting hot fluids may cause thermochronometers to be reset.

The heat flows in fractures is a 2D problem with both a convection-dominated component in the fracture and a conduction-dominated component in the surrounding rock. We should therefore expect to be able to find a Péclet number describing the balance of the two components. An energy balance (Wangen 2010, p. 152) reveals that this Péclet number $\mathbf{Pe}_f$ has the form

$$Pe_f = \frac{c_f \rho_f v_0 w}{K} \qquad \text{[A60.1]}$$

where c_f is the heat capacity of the fluid, ρ_f is the density of the fluid, K_f is its thermal conductivity, v_0 is the average fluid velocity in the fracture, and w is the width of the fracture. $\mathbf{Pe}_f$ is much less than 1 in the conduction-dominated regime, and much greater than 1 in the convection-dominated regime.

The solution for the temperature field in the convection-dominated case is as follows. The vertical fracture located at $x = 0$ has a base at y_1 and a top at y_2; $y_1 - y_2$ is therefore a characteristic length scale l_0. Fluid is supplied to the base of the fracture by a lower aquifer, and is discharged from the top of the fracture into an upper aquifer. The boundary condition for temperature along the fracture ($x = 0$) is T_2, and the boundary condition for temperature along the horizontal sides of the model is T_1 ($y = y_1$) and T_2 ($y = y_2$). A dimensionless temperature solution can be expressed in terms of two dimensionless coordinates, as follows:

Dimensionless temperature difference:

$$T^* = (T - T_1)/(T_2 - T_1)$$

Dimensionless horizontal distance:

$$x^* = x/l_0$$

Dimensionless vertical distance:

$$y^* = (y_2 - y)/l_0$$

The temperature field can be expressed as a conductive temperature component $T^* = (1 - y^*)$ and a convective component giving the effect of the hot fluid-filled fracture as a Fourier series (Fig. 10.8):

$$T^*\left(x^*, y^*\right) = 1 - y^* + \sum_{n-1}^{\infty} a_n \exp\left(-n\pi x^*\right)\sin\left(n\pi y^*\right) \qquad \text{[A60.2]}$$

where the Fourier coefficients a_n are

$$a_n = -\frac{2(-1)^n}{n\pi} \qquad \text{[A60.3]}$$

As can be seen from Fig. 10.8b, the isotherms bend upwards along the fracture but become horizontal at large distances from the fracture. A time scale τ_0 for the heating up of the surrounding rock by the fluid in the vertical fracture is

$$\tau_0 = \frac{l_0^2 c_b \rho_b}{K_b} \qquad \text{[A60.4]}$$

where c_b and ρ_b are the heat capacity and bulk density of the surrounding sedimentary rock, and K_b is its thermal conductivity, where $K_b/\rho_b c_b$ is the thermal diffusivity κ.

Taking $\kappa = 10^{-6}\,\mathrm{m^2\,s^{-1}}$, and a fracture (fault) 1 km long, the time scale is ~32 kyr – quick by geological standards. At time scales greater than this value, the stationary temperature field given by eqn. [A60.2] applies. To heat a significant volume of a sedimentary basin, long fractures would need to be closely spaced at $<l_0$.

How far away from the fracture does the hot fluid heat up the surrounding rock? Taking $x^* = 1$, at the base of the fracture, the temperature of the surrounding rock is the same as the temperature of the fluid. At the level of the top of the fracture at $x^* = 1$, the temperature of the surrounding rock is unaffected by the hot fluid.

To calculate the Péclet number, let us assume that the 1 km-long fault is 0.1 m wide and that the thermal diffusivity is $3 \times 10^{-7}\,\mathrm{m^2\,s^{-1}}$. The Péclet number is unity for a flow velocity through the fracture of approximately $3 \times 10^{-6}\,\mathrm{m\,s^{-1}}$, or ~100 m yr^{-1}. At velocities less than this, conductive heat flow should dominate, and, above this, convection should dominate. But if the fracture is shorter (100 m) and narrower (0.01 m), the heat flow would be by conduction only at this flow velocity, and the dimensionless temperature would be $T^* = (1 - y^*)$ and vary in the vertical dimension only.

Basin Analysis: Principles and Application to Petroleum Play Assessment, Third Edition. Philip A. Allen and John R. Allen.
© 2013 John Wiley & Sons, Ltd. Published 2013 by John Wiley & Sons, Ltd.

References

Abbott, G.D., Wang, G.Y., Eglington, T.I. and Petch, G.S. (1990) The kinetics of sterane biological marker release and degradation during the hydrous pyrolysis of vitrinite kerogen. *Geochimica et Cosmochimica Acta*, 54, 2451–2461.

Adam, C. and Bonneville, A. (2005) Extent of the South Pacific Superswell. *Journal of Geophysical Research*, 110, B09408.

Adams, R.D. and Grotzinger, J.P. (1996) Lateral continuity of facies and parasequences in Middle Cambrian platform carbonate, Carrara Formation, southeastern California, USA. *Journal of Sedimentary Research*, 66, 1079–1090.

Agarwal, B., Hermansen, H., Sylte, J.E. and Thomas, L.K. (2000) Reservoir characterization of Ekofisk Field: a giant, fractured Chalk reservoir in the Norwegian North Sea – history match. *Society of Petroleum Engineers*, 3(6), 534–543.

Aharon, P. (1983) 140,000 yr isotope climatic record from raised coral reef in New Guinea. *Nature*, 304, 720–723.

Ahern, J.L. and Dikeou, P.J. (1989) Evolution of the lithosphere beneath the Michigan Basin. *Earth and Planetary Science Letters*, 95, 73–84.

Ahlbrandt, T.S., Charpentier, R.R., Klett, T.R., Schmoker, J.W., Schenk, C.J. and Ulmishek, G.F. (2005) *Global Resource Estimates from Total Petroleum Systems*. American Association of Petroleum Geologists Memoir, 86.

Ahmed, U., Crary, S.F. and Coates, G.R. (1991) Permeability estimation: the various sources and their interrelationships. *Journal of Petroleum Technology*, 42, 578–587.

Ahnert, F. (1970) Functional relationships between denudation, relief and uplift in large mid-latitude drainage basins. *American Journal of Science*, 270, 243–263.

Ahrens, J.L. and Schubert, G. (1975) Gabbro–eclogite reaction rate and its geophysical significance. *Reviews of Geophysics and Space Physics*, 13, 383–400.

Aigner, T., Brandenburg, A., van Vliet, A., Doyle, M., Lawrence, D. and Westrich, J. (1990) Stratigraphic modelling of epicontinental basins: two applications. *Sedimentary Geology*, 69, 167–190.

Aigner, T., Doyle, M., Lawrence, D., Epting, M. and van Vliet, A. (1989) Quantitative modeling of carbonate platforms: some examples. In: *Controls on Carbonate Platform and Basin Development* (ed. by P.D. Crevello *et al.*), Society of Economic Paleontologists and Mineralogists Special Publication, 44, 323–338.

Ajdukiewicz, J.M. (1995) A model for quartz cementation in the Norphlet Formation, Mobile Bay, offshore Alabama. *American Association Petroleum Geologists Annual Convention*, Houston, Texas, May 5–8, 1995.

Alberta Geological Survey (2010) Alberta Oil Sands, http://www.ags.gov.ab.ca/energy/oilsands.

Albright, W.A., Turner, W.L. and Williamson, K.R. (1980) Ninian field, UK sector, North Sea. In: *Giant Oil and Gas Fields of the Decade 1968–1978* (ed. by M.T. Halbouty), *American Association of Petroleum Geologists Memoir*, 30, 173–194.

Al-Hajri, Y., White, N. and Fishwick, S. (2010) Scales of transient convective support beneath Africa. *Geology*, 883–886.

Allen, P.A. (1984) Reconstruction of ancient sea conditions with an example from the Swiss Molasse. *Marine Geology*, 60, 455–473.

Allen, P.A. (1997) *Earth Surface Processes*, Blackwell Publishing Ltd., Oxford, 404pp.

Allen, P.A. (2008) Time scales of tectonic landscapes and their sediment routing systems. In: *Earth's Dynamic Surface: Catastrophe and Continuity in Landscape Evolution* (ed. by K. Gallagher, S.J. Jones and J. Wainwright), *Geological Society, London, Special Publication*, 296, 7–28.

Allen, P.A. and Allen, J.R. (1990) *Basin Analysis: Principles and Applications*, Blackwell Publishing Ltd., Oxford, 451pp.

Allen, P.A. and Allen, J.R. (2005) *Basin Analysis: Principles and Applications, Second Edition*, Blackwell Publishing Ltd., Oxford, 549pp.

Allen, P.A. and Armitage, J.J. (2012) Cratonic basins. In: *Tectonics of Sedimentary Basins: Recent Advances* (ed. by C. Busby and A. Azor), Wiley-Blackwell, 602–620, 647pp.

Allen, P.A., Armitage, J.J., Carter, A., Duller, R.A., Michael, N.A., Sinclair, H.D., Whitchurch, A.L. and Whittaker, A.C. (2013) The Qs problem: Sediment volumetric balance of proximal foreland basin systems. *Sedimentology*, 60, 102–130.

Allen, P.A. and Collinson, J.D. (1986) Lakes. In: *Sedimentary Environments and Facies* (ed. by H.G. Reading), Blackwell Scientific Publications, Oxford, 63–94.

Allen, P.A. and Densmore, A.L. (2000) Sediment flux from an uplifting fault block. *Basin Research*, 12, 367–380.

Allen, P.A. and Heller, P.L. (2012) The timing, distribution and significance of tectonically generated gravels in terrestrial sediment routing systems. In: *Tectonics of Sedimentary Basins: Recent Advances* (ed. by C. Busby and A. Azor Pérez), Wiley-Blackwell, 111–130, 647pp.

Allen, P.A. and Hovius, N. (1998) Sediment supply from landslide-dominated catchments: implications for basin-margin fans. *Basin Research*, 10, 19–35.

Allen, P.A., Armitage, J.J., Carter, A., Duller, R.A., Michael, N.A., Sinclair, H.D., Whitchurch, A.L. and Whittaker, A.C. (2013) The Qs problem: Sediment volumetric balance of proximal foreland basin systems. *Sedimentology*, 60, 102–130.

Allen, P.A., Burgess, P.M., Galewsky, J. and Sinclair, H.D. (2001) Flexural-eustatic numerical model for drowning of the Eocene perialpine carbonate ramp and implications for Alpine geodynamics. *Geological Society of America Bulletin*, 113, 1052–1066.

Allen, P.A., Crampton, S. and Sinclair, H.D. (1991) Inception and early evolution of the North Alpine Foreland Basin, Switzerland. *Basin Research*, 3, 143–163.

Allen, P.A., Verlander, J.E., Burgess, P.M. and Audet, D.M. (2000) Jurassic giant erg deposits, flexure of the United States continental interior, and timing of the onset of Cordilleran shortening. *Geology*, 28, 159–162.

Basin Analysis: Principles and Application to Petroleum Play Assessment, Third Edition. Philip A. Allen and John R. Allen.
© 2013 John Wiley & Sons, Ltd. Published 2013 by John Wiley & Sons, Ltd.

Alley, N.F. (1998) Cainozoic stratigraphy, palaeoenvironments and geological evolution of the Lake Eyre Basin. *Palaeogeography, Palaeoclimatology, Palaeoecology*, 144, 239–263.

Allison, P.A. and Wells, M.R. (2006) Circulation in large ancient epicontinental seas: what was different and why? *Palaios*, 21, 513–515.

Alvarez, W. (1999) Drainage on evolving fold-thrust belts: a study of transverse canyons in the Apennines. *Basin Research*, 11, 267–284.

An Yin (2010) Cenozoic tectonic evolution of Asia: a preliminary synthesis. *Tectonophysics*, 488, 293–325.

Anderson, R.S. (1982) Hotspots, polar wander, Mesozoic convection and the geoid. *Nature*, 297, 391–393.

Anderson, R.S. (1994) Evolution of the Santa Cruz Mountains, California, through tectonic growth and geomorphic decay. *Journal of Geophysical Research*, 99, 20161–20179.

Anderson, R.S. and Anderson, S.P. (2010) *Geomorphology: The Mechanics and Chemistry of Landscapes*, Cambridge University Press, 637pp.

Anderson, R.S. and Humphrey, N.F. (1990) Interaction of weathering and transport processes in the evolution of arid landscapes. In: *Quantitative Dynamic Stratigraphy* (ed. by T.A. Cross), Prentice Hall, Englewood Cliff, NJ, 349–361.

Andrews-Speed, C.P., Oxburgh, E.R. and Cooper, B.A. (1984) Temperatures and depth-dependent heat flow in western North Sea. *American Association Petroleum Geologists Bulletin*, 68, 1764–1781.

Antonov, J.I., Levitus, S. and Boyer, T.P. (2005) Thermosteric sea level rise, 1955–2003. *Geophysical Research Letters*, 32, L12602.

Arenas, C., Millan, H., Pardo, G. and Pocovi, A. (2001) Ebro basin continental sedimentation associated with late compressional Pyrenean tectonics (northeastern Iberia): controls on basin-margin fans and fluvial systems. *Basin Research*, 13, 65–89.

Armitage, J.J. and Allen, P.A. (2010) Cratonic basins and the long-term subsidence history of continental interiors. *Journal of the Geological Society*, 167, 61–70.

Armitage, J.J., Duller, R.A., Whittaker, A.C. and Allen, P.A. (2011) Transformation of tectonic and climatic signals from source to sedimentary archive. *Nature Geoscience*, 4, 231–235.

Artemieva, I.M. (2006) Global 1x1 thermal model TC1 for the continental lithosphere: implications for lithosphere secular evolution. *Tectonophysics*, 416, 245–277.

Artoni, A. (2007) Growth rates and two-mode accretion in the outer orogenic wedge-foreland basin system of Central Apennine (Italy). *Italian Journal of Geosciences*, 126, 531–556.

Arts, R.J., Trani, M., Chadwick, R.A., Eiken, O., Dortland, S. and van der Meer, L.G.H. (2009) Acoustic and elastic modelling of seismic time-lapse data from the Sleipner CO_2 storage operation. In: *Carbon Dioxide Sequestration in Geological Media – State of the Science* (ed. by M. Grobe, J.C. Pashin and R.L. Dodge), *American Association of Petroleum Geologists Studies in Geology*, 59, 391–403.

Ashby, M.F. and Verall, R.A. (1977) Micromechanisms of flow and fracture and their relevance to the rheology of the upper mantle. *Philosophical Transactions of the Royal Society of London*, A288, 59–95.

Athy, L.F. (1930) Density, porosity and compaction of sedimentary rocks. *American Association Petroleum Geologists Bulletin*, 14, 1–24.

Atkinson, B.K. (1987) *Fracture Mechanics of Rock*, Academic Press, New York.

Attanasi, E.D. and Meyer, R.F. (2007) Natural bitumen and extra-heavy oil. In: *2007 Survey of Energy Resources* (ed. by J. Trinnaman and A. Clarke), World Energy Council, 119–143.

Audemard, F.E. and Serrano, I. (2001) Future petroliferous provinces of Venezuela. In: *Petroleum Provinces of the 21st Century* (ed. by M. Downey, J. Threet and W. Morgan), *American Association Petroleum Geologists Memoir*, 74, 353–372.

Audet, D.M. and McConnell, J.D.C. (1992) Forward modelling of porosity and pore pressure evolution in sedimentary basins. *Basin Research*, 4, 147–162.

Audet, D.M. and McConnell, J.D.C. (1994) Establishing resolution limits for tectonic subsidence curves by forward basin modelling. *Marine and Petroleum Geology*, 11, 400–411.

Audet, P. and Bürgmann, R. (2011) Dominant role of tectonic inheritance in supercontinental cycles. *Nature Geoscience*, 4, 184–187.

Avouac, J.P. (1993) Analysis of scarp profiles: evaluation of errors in morphologic dating. *Journal of Geophysical Research*, 98, 6745–6754.

Avouac, J.P. and Burov, E.B. (1996) Erosion as a driving mechanism for intracontinental mountain growth. *Journal of Geophysical Research*, 101, 17747–17769.

Aydin, A. and Nur, A. (1982) Evolution of pull-apart basins and their scale independence. *Tectonics*, 1, 91–105.

Aydin, A. and Nur, A. (1985) The types and roles of stepovers in strike-slip tectonics. In: *Strike-Slip Deformation, Basin Formation and Sedimentation* (ed. by K.T. Biddle and N. Christie-Blick), *Society of Economic Paleontologists and Mineralogists Special Publication*, 37, 35–44.

Bacon, M., Simm, R. and Redshaw, T. (2003) *3D Seismic Interpretation*, Cambridge University Press.

Badley, M.E. (1985) *Practical Seismic Interpretation*, Prentice Hall.

Badley, M.E., Price, J.D., Rambech Dahl, C. and Agdestein, T. (1988) The structural evolution and the northern Viking Graben and its bearing upon extensional modes of basin formation. *Journal of the Geological Society*, 145, 455–472.

Bailey, R.J. and Stoneley, R. (1981) Petroleum: entrapment and conclusions. In: *Economic Geology and Geotectonics* (ed. by D.H. Tarling), Blackwell Scientific Publications, Oxford, 73–97.

Baird, D.J., Knapp, J.H., Steer, D.N., Brown, L.D. and Nelson, K.D. (1995) Upper-mantle reflectivity beneath the Williston basin, phase-change Moho, and the origin of intracratonic basins. *Geology*, 23, 431–434.

Baker, B.H., Mohr, P.A. and Williams, L.A.J. (1972) *Geology of the Eastern Rift System of Africa. Geological Society of America Special Paper*, 136.

Baldwin, B. and Butler, C.O. (1985) Compaction curves. *American Association Petroleum Geologists Bulletin*, 69, 622–626.

Bally, A.W. (1975) A geodynamic scenario for hydrocarbon occurrences. *Proceedings 9th World Petroleum Congress*, Tokyo, 33–44, Vol. 2 (Geology), Applied Science Publishers, Barking.

Bally, A.W. and Snelson, S. (1980) Realms of subsidence. In: *Facts and Principles of World Petroleum Occurrence* (ed. by A.D. Miall), *Canadian Society Petroleum Geologists Memoir*, 6, 9–75.

Banks, R.J. and Swain, C. (1978) The isostatic compensation of East Africa. *Proceedings of the Royal Astronomical Society*, A364, 331–352.

Barazangi, M. and Dorman, J. (1969) World seismicity map compiled from ESSA Coast and Geodetic Survey epicenter data, 1961–1967. *Bulletin of the Seismological Society of America*, 59, 369–380.

Barker, C.E. and Pawlewicz, M.J. (1986) The correlation of vitrinite reflectance with maximum temperature in humic organic matter. In: *Lecture Notes* (ed. by G. Buntebarth and L. Stegena), *Earth Sciences, 5. Palaeogeothermics*, Springer-Verlag, Berlin, 79–93.

Barrell, J. (1917) Rhythms and the measurement of geologic time. *Geological Society of America Bulletin*, 28, 745–904.

Barrier, L., Proust, J.-N., Nalpas, T., Robin, C. and Guillocheau, F. (2010) Control of alluvial sedimentation at foreland-basin active margins: a case-study from the northeastern Ebro Basin (southeastern Pyrenees, Spain). *Journal of Sedimentary Research*, 80, 728–749.

Bartlett, W.L., Friedman, M. and Logan, J.M. (1981) Experimental folding and faulting of rocks under confining pressure. Part IX. Wrench faults in limestone layers. *Tectonophysics*, 79, 255–277.

Bartok, P. (2003) The peripheral bulge of the Interior Range of the Eastern Venezuela Basin and its impact on oil accumulations. In: *The Circum-Gulf of Mexico and Caribbean – Hydrocarbon Habitats, Basin Formation, and Plate Tectonics* (ed. by C. Bartolini, R.T. Buffler and J. Blickwede), *American Association of Petroleum Geologists Memoir*, 79, 925–936.

Barton, P.J. (1986) Comparison of deep reflection and refraction structures in the North Sea. *Reflection Seismology: A Global Perspective* (ed. by M. Barganzi and L. Brown), *American Geophysical Union Series*, 13, 297–300.

Basile, C. and Brun, J.-P. (1999) Transtensional faulting patterns ranging from pull-apart basins to transform continental margins: an experimental investigation. *Journal of Structural Geology*, 21, 23–37.

Bastow, I.D. and Keir, D. (2011) The protracted development of the continent–ocean transition in Afar. *Nature Geoscience*, 4, 248–250.

Bastow, I.D., Nyblade, A.A., Stuart, G.W., Rooney, T.O. and Benoit, M.H. (2008) Upper mantle seismic structure beneath the Ethiopian hotspot: rifting at the edge of the African Low Velocity Anomaly. *Geochemistry, Geophysics, Geosystems*, 9, Q12022.

Bathurst, R.G.C. (1971) *Carbonate Sediments and their Diagenesis. Developments in Sedimentology*, 12, Elsevier, Amsterdam, 620pp.

Bayona, G. and Thomas, W.A. (2003) Distinguishing fault reactivation from flexural deformation in the distal stratigraphy of the peripheral Blountain Foreland Basin, southern Appalachians, USA. *Basin Research*, 15, 503–526.

Beadle, L.C. (1981) *The Inland Waters of Tropical Africa*, Longman, London, 365pp.

Beamud, E., Garces, M., Cabrera, L., Muñoz, J.A. and Almar, Y. (2003) A new middle to late Eocene continental chronostratigraphy from NE Spain. *Earth and Planetary Science Letters*, 216, 501–514.

Beardsmore, G.R. and Cull, J.P. (2001) *Crustal Heat Flow: A Guide to Measurement and Modelling*, Cambridge University Press, 324pp.

Beaufort, D., Cassagnabere, A., Petit, S. *et al.* (1998) Kaolinite-to-dickite reaction in sandstone reservoirs. *Clay Minerals*, 33, 297–316.

Beaumont, C. (1978) The evolution of sedimentary basins on a viscoelastic lithosphere: theory and examples. *Geophysical Journal Royal Astronomical Society*, 55, 471–497.

Beaumont, C. (1981) Foreland basins. *Geophysical Journal Royal Astronomical Society*, 65, 291–329.

Beaumont, C., Ellis, S. and Pfiffner, O.A (1999) Dynamics of sediment subduction-accretion at convergent margins: short-term modes, long-term deformation, and tectonic implications. *Journal of Geophysical Research – Solid Earth*, 104, 17573–17601.

Beaumont, C., Ellis, S., Hamilton, J. and Fullsack, P. (1996a) Mechanical model for subduction-collision: tectonics of Alpine-type compressional orogens. *Geology*, 24, 675–678.

Beaumont, C., Fullsack, P. and Hamilton, J. (1992) Erosional control on active compressional orogens. In: *Thrust Tectonics* (ed. by K.R. McClay), Chapman and Hall, New York, 1–18.

Beaumont, C., Kamp, P.J.J., Hamilton, J. and Fullsack, P. (1996b) The continental collision zone, South Island, New Zealand; comparison of geodynamic models and observations. *Journal of Geophysical Research*, 101, 3333–3359.

Beaumont, C., Keen, C.E. and Boutilier, R. (1982) On the evolution of rifted continental margins; comparison of models and observations for Nova Scotian margin. *Geophysical Journal of the Royal Astronomical Society*, 70, 667–715.

Beaumont, C., Kooi, H. and Willett, S.D. (2000) Coupled tectonic-surface process models with applications to rifted margins and collisional orogens. In: *Geomorphology and Global Tectonics* (ed. by M.A. Summerfield), John Wiley & Sons, Inc., New York, 29–55.

Bechstädt, T., Brack, P., Preto, N., Rieber, H. and Zülke, R. (2003) *Field Trip to Latemar, A Guidebook, Triassic Geochronology and Cyclostratigraphy – A Field Symposium*, St. Christina/Val Gardena, Dolomites, Italy, September 11–13, 2003.

Bechtel, T., Forsyth, D. and Swain, C. (1987) Mechanisms of isostatic compensation in the vicinity of the East African rift, Kenya. *Geophysical Journal of the Royal Astronomical Society*, 90, 445–465.

Beck, A.E. (1977) Climatically perturbed temperature gradients and their effect on regional and continental heat flow means. *Tectonophysics*, 41, 17–39.

Beck, A.E., Anglin, F.M. and Sass, J.H. (1971) Analysis of heat flow data – *in situ* thermal conductivity measurements. *Canadian Journal of Earth Sciences*, 8, 1–19.

Beck, R.A., Vondra, C.G., Filkinds, J.E. and Olander, J.D. (1988) Syntectonic sedimentation and Laramide basement thrusting, Cor-

dilleran foreland: timing of deformation. In: *Interaction of the Rocky Mountain Foreland and the Cordilleran Thrust Belt* (ed. by C.A. Schmidt and W.J. Perry Jr.), *Geological Society of America Memoir*, 171, 465–487.

Beckmann, B., Flögel, S., Hofmann, P., Schulz, M. and Wagner, T. (2005) Orbital forcing of Cretaceous river discharge in tropical Africa and ocean response. *Nature*, 437, 241–244.

Bedle, H. and van der Lee, S. (2009) S-velocity variations beneath North America, *Journal of Geophysical Research*, 114, B07308.

Belknap, D.F. *et al.* (1987) Late Quaternary sea level changes in Maine. In: *Sea Level Fluctuation and Coastal Evolution* (ed. by D. Nummedal, O.H. Pilkey and J.D. Howard), *Society of Economic Paleontologists and Mineralogists Special Publication*, 41, 71–85.

Belt, E.S. (1968) Carboniferous continental sedimentation, Atlantic provinces, Canada. *Geological Society of America Special Paper*, 106, 127–176.

Ben-Avraham, Z. (1985) Structural framework of the Gulf of Elat (Aqaba), northern Red Sea. *Journal of Geophysical Research*, 90, 703–726.

Ben-Avraham, Z. and Zoback, M.D. (1992) Transform-normal extension and asymmetric basins: an alternative to pull-apart models. *Geology*, 20, 423–426.

Ben-Avraham, Z., Almagor, G. and Garfunkel, Z. (1979) Sediments and structure of the Gulf of Elat (Aqaba) – northern Red Sea. *Sedimentary Geology*, 23, 239–267.

Ben-Avraham, Z., Hanel, R. and Villinger, H. (1978) Heat flow through the Dead Sea rift. *Geology*, 28, 253–269.

Bennett, R.A., Wernicke, B.P., Davis, J.L., Elósegui, P., Snow, J.K., Abolins, M.J., House, M.A., Stirewalt, G.L. and Ferrill, D.A. (1997) Global Positioning System constraints on fault slip rates in the Death Valley region, California and Nevada. *Geophysical Research Letters*, 24, 3073–3076.

Bensley, D.F. and Crelling, J.C. (1994) The inherent heterogeneity within the vitrinite maceral group. *Fuel*, 73, 1306–1316.

Bentham, P.A., Burbank, D. W. and Puigdefabregas, C. (1993) Temporal and spatial controls on alluvial architecture in an axial drainage system, Upper Eocene Campodarbe Group, southern Pyrenean foreland basin, Spain. *Basin Research*, 4, 335–352.

Beranek, L., Link, P. and Fanning, C. (2006) Miocene to Holocene landscape evolution of the western Snake River Plain region, Idaho: using the SHRIMP detrital zircon provenance record to track eastern migration of the Yellowstone hotspot. *Geological Society of America Bulletin*, 118, 1027–1050.

Bercovici, D., Ricard, Y. and Richards, M.A. (2000) The relation between mantle dynamics and plate tectonics: a primer. In: *The History and Dynamics of Global Plate Motions* (ed. by M.A. Richards, R.G. Gordon and R.D. van der Hilst), *American Geophysical Union, Geophysical Monograph*, 121, 5–46.

Berg, R.R. (1975) Capillary pressures in stratigraphic traps. *American Association of Petroleum Geologists Bulletin*, 59, 939–956.

Berner, R.A. (2004) A model for calcium, magnesium and sulfate in seawater over Phanerozoic time. *American Journal of Science*, 304, 438–453.

Berner, R.A. and Berner, E.K. (1997) Silicate weathering and climate. In: *Tectonic Uplift and Climate Change* (ed. by W. Ruddiman), Plenum Press, New York, 353–365.

Bernet, M., Brandon, M.Y., Garver, J.I. and Molitor, B. (2004) Fundamentals of detrital zircon fission track analysis for provenance and exhumation studies with examples from the European Alps. In: *Detrital Thermochronology – Provenance Analysis, Exhumation and Landscape Evolution of Mountain Belts* (ed. by M. Bernet and C. Spiegel), Geological Society of America, Boulder Colorado, 25–36.

Bernet, M., van der Beek, P., Pik, R. *et al.* (2006) Miocene to Recent exhumation of the central Himalaya determined from combined detrital zircon fission track and U/Pb analysis of Siwalik sediments, western Nepal. *Basin Research*, 18, 393–412.

Bernoulli, D. and Jenkyns, H.C. (1974) Alpine, Mediterranean and Central Atlantic Mesozoic facies in relation to the early evolution of the Tethys. In: *Modern and Ancient Geosynclinal Sedimentation* (ed. by R.H. Dott and R.H. Shaver), *Society of Economic Paleontologists and Mineralogists Special Publication*, 19, 129–160.

Bertotti, G., Picotti, V., Chilovi, C., Fantoni, R., Merlini, S. and Mosconi, A. (2001) Neogene to Quaternary sedimentary basins in the south Adriatic (Central Mediterranean): foredeeps and lithospheric buckling. *Tectonics*, 20, 771–787.

Bice, D.M. (1991) Computer simulation of carbonate platform and basin systems. In: *Sedimentary Modeling: Computer Simulations and Methods for Improved Parameter Definition* (ed. by E.K. Franseen, W.L. Watney, C.G.S.C. Kendall and W. Ross), *Kansas Geological Survey Bulletin*, 233, 431–447.

Biddle, K.T., Uliana, M.A., Mitchum, R.M. Jr., Fitzgerald, M.G. and Wright, R.C. (1986) The stratigraphic and structural evolution of the central and eastern Magallanes Basin, southern South America. In: *Foreland Basins* (ed. by P.A. Allen and P. Homewood), *Special Publications of the International Association of Sedimentologists*, 8, 41–61.

Bierman, P.R. (1994) Using *in situ* produced cosmogenic isotopes to estimate rates of landscape evolution. A review from the geomorphic perspective. *Journal of Geophysical Research*, 99, 13885–13896.

Bierman, P.R. and Caffee, M. (2002) Cosmogenic exposure and erosion history of Australian bedrock landforms. *Geological Society of America Bulletin*, 114, 787–803.

Bilham, R., Larson, K., Freymüller, J., and Project Idylhim members (1997) GPS measurements of present-day convergence across the Nepal Himalaya. *Nature*, 386, 61–64.

Billi, A. and Salvini, F. (2003) Development of systematic joints in response to flexure-related fibre stress in flexed foreland plates: the Apulian forebulge case history, Italy. *Journal of Geodynamics*, 36, 523–536.

Bjorkum, P.A. and Nadeau, P.H. (1998) Temperature controlled porosity/permeability reduction, fluid migration, and petroleum exploration in sediment basins. *APPEA Journal*, 38, 453–465.

Bjorlykke, K. (1983) Diagenetic reactions in sandstones. In: *Sediment Diagenesis* (ed. by A. Parker and B.W. Sellwood), Reidel, 169–213.

Bjorlykke, K., Aagaard, P.A., Dypvik, H., Hastings, D.S. and Harper, A.S. (1986) Diagenesis and reservoir properties of Jurassic sandstones from the Haltenbanken area, offshore mid-Norway. In: *Habitat of Hydrocarbons on the Norwegian Continental Shelf* (ed. by A.M. Spencer), Graham and Trotman, London, for The Norwegian Petroleum Society, 275–286.

Bjorlykke, K., Nedkvitne, T., Ramm, M. and Saigal, G.C. (1992) Diagenetic processes in the Brent Group (Middle Jurassic) reservoirs of the North Sea: an overview. In: *Geology of the Brent Group* (ed. by A.C. Morton, R.S. Haszeldine, M.R. Giles and S. Brown), *Geological Society, London, Special Publication*, 61, 263–287.

Blackwell, D. and Steele, J. (1989) *Thermal Conductivity of Sedimentary Rocks: Measurement and Significance*, Springer, New York, 13–36.

Blair, D.G. (1975) Structural styles in North Sea oil and gas fields. In: *Petroleum and the Continental Shelf of NW Europe*, Vol. 1 (ed. by A.W. Woodland), Applied Science Publishers, Barking, 327–338.

Blair, T.C. and McPherson, J.G. (1994) Alluvial fan processes and forms. In: *Geomorphology of Desert Environments* (ed. by A.D. Abrahams and A.J. Parsons), Chapman and Hall, London, 354–402.

Blanckenburg, von F., Hewawasam, T. and Kubik, P.W. (2004) Cosmogenic nuclide evidence for low weathering and denudation in the wet, tropical highlands of Sri Lanka. *Journal of Geophysical Research – Earth Surface*, 109, F03008.

Blasband, B., White, S., Brooijmans, S., De Broeder, H. and Visser, W. (2000) Late Proterozoic extensional collapse in the Arabian-Nubian Shield. *Journal of the Geological Society*, 157, 615–628.

Blatt, H., Middleton, G. and Murray, R. (1972) *Origin of Sedimentary Rocks*, Prentice Hall, Englewood Cliffs, New Jersey, 634pp.

Bloch, S., Lander, R. and Bonnell, L. (2002) Anomalously high porosity and permeability in deeply buried sandstone reservoirs: origin and predictability. *American Association Petroleum Geologists Bulletin*, 86, 301–328.

Blount, D. and Moore, C.H. (1969) Depositional and non-depositional carbonate breccias, Chiantla Quadrangle, Guatemala. *Geological Society of America Bulletin*, 80, 429–442.

Blum, M.D. and Hattier-Womack, J. (2009) Climate change, sea-level change, and fluvial sediment supply to deepwater depositional systems. In: *External Controls on Deepwater Depositional Systems, Society Economic Paleontologists and Mineralogists (Society for Sedimentary Geology)*, 92, 15–39.

Boles, J.R. and Franks, S.G. (1979) Clay diagenesis in Wilcox sandstones of southwest Texas: implications of smectite diagenesis on sandstone cementation. *Journal of Sedimentary Petrology*, 49, 55–70.

Bolin, B. (1970) The Carbon cycle. *Scientific American* V223(3), 124–132.

Bond, G. (1976) Evidence for continental subsidence in North America during the Late Cretaceous global submergence. *Geology*, 4, 557–560.

Bond, G. (1978) Speculations on real sea level changes and vertical motions of continents at selected times in the Cretaceous and Tertiary periods. *Geology*, 6, 247–250.

Bond, G. (1979) Evidence for some uplifts of large magnitudes in continental platforms. *Tectonophysics*, 61, 285–305.

Bond, G., Kominz, M.A. and Sheridan, R.E. (1995) Continental terraces and rises. In: *Tectonics of Sedimentary Basins* (ed. by C. Busby and R. Ingersoll), Blackwell Publishing Ltd., Oxford, 149–178.

Bondevik, S., Svendson, J.I., Johnsen, G., Mangerud, J. and Kaland, P.E. (1997) The Storegga tsunami along the Norwegian coast, its age and run-up. *Boreas*, 26, 29–53.

Bonnet, C., Malavieille, J. and Mosar, J. (2008) Surface processes versus kinematics of thrust belts: impact on rates of erosion, sedimentation, and exhumation – insights from analogue models. *Bulletin Geological Society France*, 179, 297–314.

Boote, D.R.D., Clark-Lowes, D.D. and Traut, M.W. (1998) Palaeozoic petroleum systems of North Africa. In: *Petroleum Geology of North Africa* (ed. by D.S. MacGregor, R.T.J. Moody and D.D. Clark-Lowes), *Geological Society, London, Special Publication*, 132, 7–68.

Booth, J.S., Rowe, M.M. and Fischer, K.M. (1996) *Offshore Gas Hydrate Database with an Overview and Preliminary Analysis. US Geological Survey Open-File Report*, 96-272, 17pp.

Bosence, D. (2005) A genetic classification of carbonate platforms based on their basinal and tectonic settings in the Cenozoic. *Sedimentary Geology*, 175, 49–72.

Bosscher, H. and Schlager, W. (1992) Computer simulation of reef growth. *Sedimentology*, 39, 503–512.

Bostick, N.H. (1979) Microscopic measurement of the level of catagenesis of solid organic matter in sedimentary rocks to aid exploration for petroleum and to determine former burial temperatures – a review. In: *Aspects of Diagenesis* (ed. by P.A. Scholle and P.R. Schluger), *Society of Economic Paleontologists and Mineralogists Special Publication*, 26.

Bostick, N.H. and Alpern, B. (1977) Principles of sampling, preparation and constituent selection for microphotometry in measurement of maturation of sedimentary organic matter. *Journal of Microscopy*, 109, 41–47.

Bott, M.H.P. (1992) Modelling the loading stresses associated with active continental rift systems. *Tectonophysics*, 215, 99–115.

Boyd, R., Dalrymple, R.W. and Zaitlin, B.A. (1992) Classification of clastic coastal depositional environments. *Sedimentary Geology*, 80, 139–150.

Boyer, S.E. (1995) Sedimentary basin taper as a factor controlling the geometry and advance of thrust belts. *American Journal of Science*, 295, 1220–1254.

Braccini, E., de Boar, W., Hurst, A., Huuse, M., Vigorito, M. and Templeton, G. (2008) Sand injectites. *Schlumberger Oilfield Review*, summer 2008.

Brace, W.H. and Kohlstedt, D.L. (1980) Limits on lithospheric stress imposed by laboratory experiments. *Journal of Geophysical Research*, 85, 6248–6252.

Brack, P., Mundil, R., Oberli, F. *et al.* (1996) Biostratigraphy and radiometric age data question the Milankovitch characteristics of the Latemar cycles (southern Alps, Italy). *Geology*, 24, 371–375.

Bradley, D.C. (1983) Tectonics of the Acadian orogeny in New England and adjacent Canada. *Journal of Geology*, 91, 381–400.

Bradley, D.C. and Kidd, W.S.F. (1991) Flexural extension of the upper continental crust in collisional foredeeps. *Geological Society of America Bulletin*, 103, 1416–1438.

Bradley, W.H. and Eugster, H.P. (1969) Geochemistry and paleolimnology of the trona deposits and associated authigenic minerals of the Green River Formation of Wyoming. *US Geological Survey Professional Paper*, 469-B, 71pp.

Braile, L.W., Hinze, W.J., Keller, G.R., Lidiak, E.G. and Sexton, J.L. (1986) Tectonic development of the New Madrid rift complex, Mississippi Embayment, North America. *Tectonophysics*, 131, 1–21.

Brandon, M.T. and Vance, J.A. (1992) New statistical methods for analysis of fission track grain-age distributions with applications to detrital grain ages from the Olympic subduction complex, western Washington State. *American Journal of Science*, 292, 565–636.

Braun, J. (1993) Three-dimensional numerical modelling of compressional orogenies: thrust geometry and oblique convergence. *Geology*, 21, 153–156.

Braun, J. (2005) Quantitative constraints on the rate of landform evolution derived from low-temperature thermochronology. *Reviews in Mineral Geochemistry*, 58, 351–374.

Braun, J. and Beaumont, C. (1987) Styles of continental rifting: results from dynamic models of lithospheric extension. In: *Sedimentary Basins and Basin-Forming Mechanisms* (ed. by C. Beaumont and A.J. Tankard), Canadian Society of Petroleum Geologists, Calgary, 12, 241–258.

Braun, J. and Beaumont, C. (1995) Three-dimensional numerical experiments of strain partitioning at oblique plate boundaries: implications for contrasting tectonic styles in the southern Coast Ranges, California, and central South Island, New Zealand. *Journal of Geophysical Research*, 100, 18059–18074.

Braun, J. and Sambridge, M. (1997) Modelling landscape evolution on geological time scales: a new method based on irregular spatial discretization. *Basin Research*, 9, 27–52.

Braun, J., van der Beek, P. and Batt, G. (2006) *Quantitative Thermochronology: Numerical Methods for the Interpretation of Thermochronological Data*, Cambridge University Press.

Brewer, I.D., Burbank, D.W. and Hodges, K.V. (2003) Modelling detrital cooling-age populations: insights from two Himalayan catchments. *Basin Research*, 15, 305–320.

Bridge, J.S. (2003) *Rivers and Floodplains: Forms, Processes and Sedimentary Record*, Blackwell Publishing Ltd., Oxford, 491pp.

Briedis, N.A., Bergslien, D., Hjellbakk, A., Hill, R.E. and Moir, G.J. (2007) Recognition criteria, significance to field performance, and reservoir modelling of sand injections in the Balder filed, North Sea. In: *Sand Injectites: Implications for Hydrocarbon Exploration and Production* (ed. by A. Hurst and J. Cartwright), *American Association of Petroleum Geologists Memoir*, 87, 91–102.

Brigaud, F. and Vasseur, G. (1989) Mineralogy, porosity and fluid control on thermal conductivity of sedimentary rocks. *Geophysical Journal*, 98, 525–542.

Brocklehurst, S.H. and Whipple, K.X. (2004) Hypsometry of glaciated landscapes. *Earth Surface Processes and Landforms*, 29, 907–926.

Brocklehurst, S.H. and Whipple, K.X. (2007) Response of glacial landscapes to spatial variations in rock uplift rate. *Journal of Geophysical Research – Earth Surface*, 112, F02035.

Brodie, J. and White, N. (1994) Sedimentary basin inversion caused by igneous underplating: Northwest European continental shelf. *Geology*, 22, 147–150.

Brown, A.R. (1999) *Interpretation of 3-D Seismic Data, Memoir 42*, Geological Society of London, Bath.

Brown, E.T., Stallard, R.F., Larsen, M.C., Raisbeck, G.M. and Yiou, F. (1995) Denudation rates determined from the accumulation of in situ-produced ^{10}Be in the Luquillo experimental forest, Puerto Rico. *Earth and Planetary Science Letters*, 129, 193–202.

Brown, L.F. and Fisher, W.L. (1977) Seismic-stratigraphic interpretation of depositional systems. In: *Seismic Stratigraphy – Applications to Hydrocarbon Exploration* (ed. by C.E. Payton), *American Association of Petroleum Geologists Memoir*, 26, 213–248.

Bruhn, R.L., Stern, C.R. and De Wit, M.J. (1978) Field and geochemical data bearing on the development of a Mesozoic volcano-tectonic rift zone and back-arc basin in southernmost South America. *Earth and Planetary Science Letters*, 41, 32–46.

Brun, J.P. (1999) Narrow rifts *versus* wide rifts: inferences for the mechanics of rifting from laboratory experiments. *Philosophical Transactions of the Royal Society, London, A*, 357, 695–712.

Buck, W.R. (1984) Small-scale convection and the evolution of the lithosphere. Unpublished PhD thesis, MIT, 256pp.

Buck, W.R. (1986) Small-scale convection induced by passive rifting: the cause for uplift of rift shoulders. *Earth and Planetary Science Letters*, 77, 362–372.

Buck, W.R. Steckler, M.S. and Cochran, J.R. (1988) Thermal consequences of lithospheric extension: pure and simple. *Tectonics*, 7, 213–234.

Bücker, C. and Rybach, L. (1996) A simple method to determine heat production from gamma logs. *Marine and Petroleum Geology*, 13, 373–375.

Buffler, R.T., Shaub, F.J., Watkins, J.S. and Worzel, J.L. (1979) Anatomy of the Mexican Ridges, southwestern Gulf of Mexico. In: *Geological and Geophysical Investigations of Continental Margins* (ed. by J.S. Watkins, L. Montadert and P.W. Dickerson), *American Association of Petroleum Geologists Memoir*, 29, 319–327.

Buiter, S.J.H. Steinberger, B., Medvedev, S. and Tetreault, J.L. (2011) Could the mantle have caused subsidence of the Congo Basin? *Tectonophysics*, 514–517, 62–80.

Bull, W.B. (1962), Relations of alluvial fan size and slope to drainage basin size and lithology in western Fresno County, California. *US Geological Survey Professional Paper*, 450-B, 51–53.

Bull, W.B. (1977) The alluvial fan environment. *US Geological Survey Professional Paper*, 353-E, 89–129.

Buller, A.T., Bjorkum, P.A., Nadeau, P.H., and Walderhaug, O. (2005) *Distribution of hydrocarbons in sedimentary basins. Research and Technology Memoir*, 7, Statoil ASA, Stavanger, 15pp.

Bullimore, S. (2004) Shelf-edge and shoreline trajectories: effect upon facies, environments and stratigraphic architecture in wave-dominated regressive depositional systems. Unpublished PhD dissertation, University of Bergen, Norway.

Bunge, H.P., Richards, M.A. and Baumgartner, J.R. (1996) Effect of depth-dependent viscosity on the planform of mantle convection. *Nature*, 379, 436–438.

Burbank, D.W. (1992) Causes of recent Himalayan uplift deduced from deposited patterns in the Ganges basin. *Nature*, 357, 680–682.

Burbank, D.W. and Anderson, R.S. (2001) *Tectonic Geomorphology*. Blackwell Publishing Ltd., Oxford, 274pp.

Burbank, D.W. and Vergés, J. (1994) Reconstruction of topography and related depositional systems during active thrusting. *Journal of Geophysical Research*, 99, 20281–20297.

Burbank, D.W., Beck, R.A. and Mulder, T. (1996) The Himalayan foreland basin. In: *The Tectonic Evolution of Asia* (ed. by A. Yin and T.M. Harrison), Cambridge University Press, Cambridge, 149–188, 678pp.

Burchfiel, B.C. and Stewart, J.H. (1966) 'Pull-apart' origin for the central segment of Death Valley, California. *Geological Society of America Bulletin*, 77, 439–441.

Burg, J.-P. and Podladchikov, Y. (2000) From buckling to asymmetric folding of the continental lithosphere: numerical modelling and application to the Himalayan syntaxes. In: *Tectonics of the Nanga Parbat Syntaxis and the Western Himalaya* (ed. by M.A. Khan, P.J. Treloar, M.P. Searle and M.Q. Jan), *Geological Society, London, Special Publication*, 170, 219–236.

Burg, J.-P., Davy, P. and Martinod, J. (1994b) Shortening of analogue models of the continental lithosphere: new hypothesis for the formation of the Tibetan plateau. *Tectonics*, 13, 475–483.

Burg, J.-P., Guiraud, M., Chen, G.M. and Li, G.C. (1984) Himalayan metamorphism and deformations in the North Himalayan Belt (southern Tibet, China). *Earth and Planetary Science Letters*, 69, 391–400.

Burg, J.-P., Sokoutis, D. and Bonini, M. (2002) Model-inspired interpretation of seismic structures in the Central Alps: crustal wedging and buckling at mature stage of collision. *Geology*, 30, 643–646.

Burg, J.-P., Van Den Driessche, J. and Brun, J.-P. (1994a) Syn- to post-thickening extension in the Variscan Belt of Western Europe: mode and structural consequences. *Géologie de la France*, 3, 33–51.

Burgess, P.M. and Allen, P.A. (1996) A forward modelling analysis of the controls on sequence stratigraphic geometries. In: *Sequence Stratigraphy in British Geology* (ed. by S.P. Hesselbo and W. Parkinson), *Geological Society, London, Special Publication*, 103, 9–24.

Burgess, P.M. and Gurnis, M. (1995) Mechanisms for the formation of cratonic stratigraphic sequences. *Earth and Planetary Science Letters*, 136, 647–663.

Burgess, P.M. and Hovius, N. (1998) Rates of delta progradation during highstands: consequences for timing of deposition in deep marine systems. *Journal of the Geological Society*, 155, 217–222.

Burgess, P.M. and Moresi, L.N. (1999) Modelling rates and distribution of subsidence due to dynamic topography over subducting slabs: is it possible to identify dynamic topography from ancient strata? *Basin Research*, 11, 305–314.

Burgess, P.M., Gurnis, M. and Moresi, L. (1997) Formation of sequences in the cratonic interior of North America by interaction between mantle, eustatic, and stratigraphic processes. *Geological Society of America Bulletin*, 108, 1515–1535.

Burgess, P.M., Paola, C. and Mountney, N. (eds.) (2002) Numerical and physical experimental modelling of stratigraphy. *Basin Research*, Thematic Set, 14.

Burgess, P.M., Wright, V.P. and Emery, D. (2001) Numerical forward modelling of peritidal carbonate parasequence development: implications for outcrop interpretation. *Basin Research*, 13, 1–16.

Burke, K. (1976) The Chad Basin: an active intracontinental basin. *Tectonophysics*, 36, 197–206.

Burke, K. (1977) Aulacogens and continental breakup. *Annual Review of Earth and Planetary Sciences*, 5, 371–396.

Burke, K. and Dewey, J.F. (1973) Plume-generated triple junctions: key indicators in applying plate tectonics to old rocks. *Journal of Geology*, 81, 406–433.

Burke, K. and Gunnell, Y. (2008) *The African Erosion Surface: A Continental-Scale Synthesis of Geomorphology, Tectonics, and Environmental Change Over the Last 180 Million Years. Geological Society of America Memoir*, 201, 72pp.

Burke, K. and Torsvik, T. (2004) Derivation of large igneous provinces of the past 200 million years from long-term heterogeneities in the deep mantle. *Earth and Planetary Science Letters*, 227, 531–538.

Burke, K., Steinberger, B., Torsvik, T.H. and Smethurst, M.A. (2008) Plume generation zones at the margins of large low shear velocity provinces on the core-mantle boundary. *Earth and Planetary Science Letters*, 265, 49–60.

Burke, K.C. and Wilson, J.T. (1976) Hotspots on the Earth's surface. *Scientific American*, 235, 46–57.

Burkhard, M. and Sommaruga, A. (1998) Evolution of the Swiss Molasse basin: structural relations with the Alps and Jura belt. In: *Cenozoic Foreland Basins of Western Europe* (ed. by A. Mascle, C. Puigdefàbregas,

H.P. Luterbacher and M. Fernàndez), *Geological Society, London, Special Publication*, 134, 279–298.

Burland, J.B. (1990) On the compressibility and shear strength of natural clays. *Géotechnique*, 40, 329–378.

Burley, S.D., Kantorowicz, J.D. and Waugh, B. (1985) Clastic diagenesis. In: *Sedimentology: Recent Developments and Applied Aspects* (ed. by P.J. Brenchley and B.P.J. Williams), *Geological Society, London, Special Publication*, 18, 189–226.

Burnham, A.K. and Sweeney, J.J. (1989) A chemical kinetic model of vitrinite reflectance maturation. *Geochimica et Cosmochimica Acta*, 53, 2649–2657.

Burov, E.B. and Diament, M. (1995) The effective elastic thickness (Te) of continental lithosphere: what does it really mean? *Journal of Geophysical Research*, 100, 3905–3927.

Burov, E.B. and Watts, A.B. (2006) The long-term strength of continental lithosphere: 'jelly sandwich' or 'crème brulée'? *GSA Today*, January 2006, 16, 4–10.

Burov, E.B., Lobkovsky, L.I., Cloetingh, S. and Nikishin, A.M. (1993) Continental lithosphere folding in Central Asia (part 2): constraints from gravity and topography. *Tectonophysics*, 226, 73–87.

Busse, F.H. and Whitehead, J.A. (1971) Instabilities of convection rolls in high Prandtl number fluid. *Journal of Fluid Mechanics*, 47, 305–320.

Byerlee, J. (1977) A review of rock mechanics studies in the United States pertinent to earthquake prediction. In: *Proceedings of Conference II; Experimental Studies of Rock Friction With Application to Earthquake Prediction*, US Geological Survey, Office of Earthquake Studies, Menlo Park, California, 559–590.

Byerlee, J.D. (1978) Friction of rocks. *Pure and Applied Geophysics*, 116, 615–626.

Cade, C.A., Evans, J. and Bryant, S.L. (1994) Analysis of permeability controls: a new approach. *Clay Minerals*, 29, 491–501.

Cadek, O., Kyvalova, H. and Yuen, D.A. (1995) Geodynamical implications from the correlation of surface geology and seismic tomographic structure. *Earth and Planetary Science Letters*, 136, 615–627.

Caldwell, J.G., Haxby, W.F., Karig, D.E. and Turcotte, D.L. (1976) On the applicability of a universal elastic trench profile. *Earth and Planetary Science Letters*, 31, 239–246.

Cardott, B.J. and Lambert, M.W. (1985) Thermal maturation by vitrinite reflectance of Woodford Shale, Anadarko Basin, Oklahoma. *American Association of Petroleum Geologists Bulletin*, 69, 1982–1998.

Cardozo, N. and Jordan, T. (2001) Causes of spatially variable tectonic subsidence in the Miocene Bermejo Foreland Basin, Argentina. *Basin Research*, 13, 335–358.

Carey, S.W. (1976) *The Expanding Earth. Developments in Geotectonics*, 10, Elsevier, Amsterdam, 488pp.

Carlson, J. and Grotzinger, J.P. (2001) Submarine fan environment inferred from turbidite thickness distributions. *Sedimentology*, 48, 1331–1351.

Carminatti, M., Wolff, B. and Gamboa, L. (2008) New exploratory frontiers in Brazil. *19th World Petroleum Congress*, Spain 2008.

Carslaw, H.S. and Jaeger, J.C. (1959) *Conduction of Heat in Solids, Second Edition*, Oxford University Press, Oxford.

Carson, R.J. (1969) *Water, Earth and Man*, Methuen, London.

Carter, A. and Bristow, C.S. (2000) Detrital zircon geochronology: enhancing the quality of sedimentary source information through improved methodology and combined U–Pb and fission track techniques. *Basin Research*, 12, 47–57.

Carter, A. and Gallagher, K. (2004) Characterising the significance of provenance on the inference of thermal history models from apatite fission track data – a synthetic data study. In: *Detrital Thermochronology – Provenance Analysis, Exhumation and Landscape Evolution of Mountain Belts* (ed. by M. Bernet and C. Spiegel), Geological Society of America, Boulder, Colorado, 7–23.

Carter, R.M. (1998) Two models: global sea-level change and sequence stratigraphic architecture. *Sedimentary Geology*, 22, 23–36.

Carter, R.M. (2005) A New Zealand climatic template back to c.3.9 Ma: ODP Site 1119, Canterbury Bight, southwest Pacific Ocean, and its relationship to onland successions. *Journal Royal Society New Zealand*, 35, 9–42.

Carvajal, C. and Steel, R.J. (2009) Shelf-edge architecture and bypass of sand to deep water: influence of sediment supply, sea level, and shelf-edge processes. *Journal of Sedimentary Research*, 79, 652–672.

Carvajal, C. and Steel, R.J. (2012) Source-to-sink sediment volumes within a tectonostratigraphic model for a Laramide shelf-to-deep-water basin: methods and results. In: *Tectonics of Sedimentary Basins: Recent Advances* (ed. by C. Busby and A. Azor Pérez), Blackwell Publishing Ltd., Oxford, 131–151.

Castelltort, S. and Van Den Driessche, J. (2003) How plausible are high-frequency sediment supply-driven cycles in the stratigraphic record? *Sedimentary Geology*, 157, 3–13.

Cattaneo, A., Correggiari, A., Langone, L. and Trincardi, F. (2003) The late Holocene Gargano subaqueous delta, Adriatic shelf, sediment pathways and supply fluctuations. *Marine Geology*, 193, 61–91.

Catuneanu, O. (2006) *Principles of Sequence Stratigraphy*, Elsevier, Amsterdam, 375pp.

Catuneanu, O., Abreu, V., Bhattacharya, J.P. and 25 others (2009) Towards the standardization of sequence stratigraphy. *Earth Science Reviews*, 92, 1–33.

Catuneanu, O., Beaumont, C. and Waschbusch, P. (1997) Interplay of static loads and subduction dynamics in foreland basins: reciprocal stratigraphies and the 'missing' peripheral bulge. *Geology*, 25, 1087–1090.

Cazenave, A., Sourian, A. and Domink, K. (1989) Global coupling of the Earth surface topography with hotspots, geoid and mantle heterogeneity. *Nature*, 340, 54–57.

Cermak, V. (1979) Review of heat flow measurements in Czechoslovakia. In: *Terrestrial Heat Flow in Europe* (ed. by V. Cermak and L. Rybach), Springer-Verlag, New York, 152–160.

Cerveny, P.F., Naeser, N.D., Zeitler, P.K., Naeser, C.W. and Johnson, N.M. (1988) History of uplift and relief of the Himalaya during the past 18 million years: evidence from fission track ages of detrital zircons from sandstones of the Siwalik Group. In: *New Perspectives in Basin Analysis* (ed. by K.L. Kleinspehn and C. Paola), Springer-Verlag, New York, 43–61.

Chapple, W.M. (1978) Mechanics of thin-skinned fold-and-thrust belts. *Geological Society of America Bulletin*, 89, 1189–1198.

Charreau, J., Gumiaux, C., Avouac, J.P. *et al.* (2009) The Neogene Xiyu Formation, a diachronous prograding gravel wedge at the front of the Tien Shan: climatic and tectonic implications. *Earth and Planetary Science Letters*, 287, 298–310.

Chase, C.G. (1979) Subduction, the geoid, and lower mantle convection. *Nature*, 282, 464–468.

Chase, T.E., Menard, H.W. and Mammerickx, J. (1970) *Bathymetry of the North Pacific*, Chart 8 of 10, Scripps Institution of Oceanography and Institute of Marine Resources.

Cheadle, M.J., Czuchra, B.L., Byrne, T., Ando, C.J., Oliver, J.E., Brown, L.D., Kaufman, S., Malin, P.E. and Phinney, R.A. (1985) The deep crustal structure of the Mojave Desert, California, from COCORP seismic reflection data. *Tectonics*, 5, 293–320.

Chen, W.P. and Molnar, P. (1983) Focal depths of intracontinental and interplate earthquakes and their implication for the thermal and mechanical properties of the lithosphere. *Journal of Geophysical Research*, 88, 4183–4214.

Chen, Z., Yan, H., Li, J., Zhang, G. and Liu, B. (1999) Relationship between Tertiary volcanic rocks and hydrocarbons in the Liaohe Basin, People's Republic of China. *American Association of Petroleum Geologists Bulletin*, 6, 1004–1014.

Choquette, P.A. and Pray, L.C. (1970) Geologic nomenclature and classification of porosity in sedimentary carbonates. *American Association of Petroleum Geologists Bulletin*, 54, 207–250.

Christiansson, P., Faleide, J.I. and Berge, A.M. (2000) Crustal structure in the northern North Sea: an integrated geophysical study. In: *Dynamics of the Norwegian Margin* (ed. by A. Nottvedt), *Geological Society, London, Special Publication*, 167, 15–40.

Christie-Blick, N. and Biddle, K.T. (1985) Deformation and basin formation along strike-slip faults. In: *Strike-slip Deformation, Basin Formation and Sedimentation* (ed. by K.T. Biddle and N. Christie-Blick), *Society of Economic Paleontologists and Mineralogists Special Publication*, 37, 1–34.

Clark, I.R. and Cartwright, J.A. (2009) Interaction between submarine channel systems and deformation in deepwater fold belts: examples from the Levant Basin, eastern Mediterranean Sea. *Marine and Petroleum Geology*, 26, 1465–1482.

Clark, J.A., Farrell, W.E. and Peltier, W.R. (1978) Global changes of post-glacial sea level: a numerical calculation. *Quaternary Research*, 9, 265–287.

Clarke, P.J., Davies, R.R., England, P.C. *et al.* (1998) Crustal strain in central Greece from repeated GPS measurements in the interval 1989–1997. *Geophysical Journal International*, 135, 195–214.

Clevis, Q., de Boer, P. and Maarten, W. (2003) Numerical modelling of drainage basin evolution and three-dimensional fan stratigraphy. *Sedimentary Geology*, 163, 85–110.

Clevis, Q., de Boer, P.L. and Nijman, W. (2004) Differentiating the effect of episodic tectonism and sea level fluctuations in foreland basins filled by alluvial fans and axial deltaic systems; insights from a three-dimensional stratigraphic forward model. *Sedimentology*, 51, 809–835.

Clift, P.D. (2006) Controls on the erosion of Cenozoic Asia and the flux of clastic sediment to the ocean. *Earth and Planetary Science Letters*, 241, 571–580.

Clift, P.D., Lee, J.I., Hildebrand, P. *et al.* (2002) Nd and Pb isotope variability in the Indus River system: implications for sediment provenance and crustal heterogeneity in the Western Himalaya. *Earth and Planetary Science Letters*, 200, 91–106.

Clift, P.D., Shimizu, N. and Layne, G.D. *et al.* (2001) Development of the Indus Fan and its significance for the erosional history of the western Himalaya and Karakorum. *Geological Society of America Bulletin*, 113, 1039–1051.

Cloetingh, S. and Burov, E. (1996) Thermomechanical structure of European continental lithosphere: constraints from rheological profiles and EET estimates. *Geophysical Journal International*, 124, 695–723.

Cloetingh, S., Burov, E. and Poliakov, A. (1999) Lithospheric folding: primary response to compression? (From Central Asia to Paris Basin). *Tectonics*, 18, 1064–1083.

Cloetingh, S., McQueen, H. and Lambeck, K. (1985) On a tectonic mechanism for regional sea level variations. *Earth and Planetary Science Letters*, 75, 157–166.

Cloetingh, S., Van Wees, J.D., Van der Beck, P.A. and Spadini, G. (1995) Role of pre-rift rheology in kinematics of basin formation: constraints from thermo-mechanical modelling of Mediterranean basins and intracratonic rifts. *Marine and Petroleum Geology*, 12, 793–808.

Cloos, E. (1955) Experimental analysis of fracture patterns. *Geological Society of America Bulletin*, 66, 241–256.

Coakley, B.C. and Gurnis, M. (1995) Far field tilting of Laurentia during the Ordovician and constraints on the evolution of a slab under an ancient continent. *Journal of Geophysical Research*, 100, 6313–6327.

Cobbold, P.R., Davy, P., Gapais, D., Rossello, E.A., Sadybakasov, E., Thomas, J.C., Tondji Biyo, J.J. and de Urreiztieta, M. (1993) Sedimentary basins and crustal thickening. *Sedimentary Geology*, 86, 77–89.

Coblentz, D.D. and Sandiford, M. (1994) Tectonic stresses in the African plate: constraints on the ambient lithospheric stress state. *Geology*, 22, 831–834.

Coblentz, D.D., Zhou, S., Hillis, R.R., Richardson, R.M. and Sandiford, M. (1998) Topography, boundary forces, and the Indo-Australian intraplate stress field. *Journal of Geophysical Research*, 103, 919–931.

Cochran, J.R. (1983) Effects of finite extension times on the development of sedimentary basins. *Earth and Planetary Science Letters*, 66, 289–302.

Coe, A.L. (2003) *The Sedimentary Record of Sea-level Change.* Cambridge University Press, New York, 287pp.

Cole, R.D. and Pickard, M.D. (1981) Sulfur-isotope variations in marginal lacustrine rocks of the Green River Formation, Colorado and Utah. In: *Recent and Ancient Nonmarine Depositional Environments: Models for Exploration* (ed. by F.G. Ethridge and R.M. Flores), *Society of Economic Paleontologists and Mineralogists Special Publication*, 31, 261–275.

Collella, A. and Prior, D.B. (eds.) (1990) *Coarse Grained Deltas, International Association of Sedimentologists Special Publication*, 10, Blackwell, Oxford.

Collett, T.S. (1995) Gas hydrate resources of the United States. In: *1995 National Assessment of Oil and Gas Resources* (ed. by D.L. Gautier, G.L. Dolton, K.I. Takahashi and K.L. Varnes), on CD-ROM, *US Geological Survey Digital Data Series*, 30.

Collett, T.S. (2001) Natural gas hydrates: resource of the twenty-first century? In: *Petroleum Provinces of the 21st Century* (ed. by M.W. Downey, J.C. Threet and W.A. Morgan), *American Association of Petroleum Geologists Memoir*, 74, 85–108.

Colman, S.M., Clark, J.A., Clayton, L., Hansel, A.K. and Larsen, C.E. (1994) Deglaciation, lake levels, and meltwater discharge in the Lake Michigan Basin. *Quaternary Science Reviews*, 13, 879–890.

Condie, K.C. (1998) Episodic continental growth and supercontinents: a mantle avalanche connection? *Earth and Planetary Science Letters*, 163, 97–108.

Condie, K.C. (2004) Supercontinents and superplume events: distinguishing signals in the geologic record. *Physics of the Earth and Planetary Interiors*, 146, 319–332.

Coney, D., Fyfe, T.B., Retail, P. and Smith, P.J. (1993) Clair appraisal: the benefits of a co-operative approach. In: *Petroleum Geology of Northwest Europe, Proceedings of the 4th Conference of the Geological Society, London* (ed. by J.R. Parker), 1409–1420.

Connan, J. (1974) Time-temperature relation in oil genesis. *American Association of Petroleum Geologists Bulletin*, 58, 2516–2521.

Conrad, C.P. and Gurnis, M. (2003) Seismic tomography, surface uplift, and the breakup of Gondwanaland: integrating mantle convection backwards in time. *Geochemistry, Geophysics, Geosystems*, 4.

Conrad, C.P. and Lithgow-Bertelloni, C. (2007) Faster sea floor spreading and lithosphere production during the mid-Cenozoic. *Geology*, 35, 29–32.

Conrad, C.P., Lithgow-Bertelloni, C. and Louden, K.E. (2004) Iceland, the Farallon slab, and dynamic topography of the North Atlantic. *Geology*, 32, 177–180.

Cooles, G.P., Mackenzie, A.S. and Quigley, T.M. (1986) Calculations of masses of petroleum generated and expelled from source rocks. In: *Advances in Organic Geochemistry* (ed. by D. Leythaeuser and J. Rullkotter), Pergamon Press, Oxford, 235–246.

Cooper, M.A. and Williams, G.D. (1989) *Inversion Tectonics, Geological Society, London, Special Publication*, 44.

Copeland, P. and Harrison, T.M. (1990) Episodic rapid uplift in the Himalaya revealed by $^{40}Ar/^{39}Ar$ analysis of detrital K-feldspar and muscovite, Bengal fan. *Geology*, 18, 354–357.

Corcoran, D.V. and Clayton, G. (2001) Interpretation of vitrinite reflectance profiles in sedimentary basins, onshore and offshore Ireland. In: *The Petroleum Exploration of Ireland's Offshore Basins* (ed. by P.M. Shannon, P.D.W. Haughton and D.V. Corcoran), *Geological Society, London, Special Publication*, 188, 61–90, 473pp.

Correggiani, A., Trincardi, F., Langone, L. *et al.* (2001) Styles of failure in late Holocene highstand prodelta wedges on the Adriatic shelf. *Journal of Sedimentary Research*, 71, 218–236.

Courtillot, V., Davaille, A., Besse, J. *et al.* (2003) Three distinct types of hotspot in the Earth's mantle. *Earth and Planetary Science Letters*, 205, 295–308.

Courtney, R.C. and White, R.S. (1986) Anomalous heat flow and geoid across the Cape Verde Rise: evidence for dynamic support from a thermal plume in the mantle. *Geophysical Journal of the Royal Astronomical Society*, 87, 815–867.

Covault, J.A., Romans, B.W., Graham, S. *et al.* (2011) Terrestrial source to deep sea sink sediment budgets at high and low sea levels: insights from tectonically active Southern California. *Geology*, 39, 619–622.

Covault, J.A., Romans, B.W. and Fildani, A. *et al.* (2010) Rapid climate signal propagation from source to sink in a southern California sediment routing system. *Journal of Geology*, 118, 247–259.

Covey, M. (1986) The evolution of foreland basins to steady state: evidence from the western Taiwan foreland basin. In: *Foreland Basins* (ed. by P.A. Allen and P. Homewood), *International Association of Sedimentologists Special Publication*, 8, Blackwell Scientific Publications, Oxford, 77–90.

Coward, M.P. (1986) Heterogeneous stretching, simple shear and basin development. *Earth and Planetary Science Letters*, 80, 325–336.

Cowie, P.A., Attal, M., Tucker, G.E., Whittaker, A.C., Naylor, M., Ganas, A. and Roberts, G.P. (2006) Investigating the surface processes response to fault interaction and linkage using a numerical modelling approach. *Basin Research*, 18, 231–266.

Cowie, P.A., Gupta, S. and Dawers, N.H. (2000) Implications of fault array evolution for synrift depocentre development: insights from a numerical fault growth model. *Basin Research*, 12, 241–261.

Cox, K.G. (1989) The role of mantle plumes in the development of continental drainage patterns. *Nature*, 342, 873–877.

Craig, D.H. (1988) Caves and other features of Permian karst in San Andres Dolomite, Yates field reservoir, West Texas. In: *Paleokarst* (ed. by N.P. James and P.W. Choquette). Springer-Verlag, New York, 342–363.

Crampton, S.E. and Allen, P.A. (1995) Recognition of forebulge unconformities associated with early stage foreland, basin development: example from the North Alpine Foreland Basin. *American Association of Petroleum Geologists Bulletin*, 79, 1495–1514.

Creany, S. and Allan, J. (1990) Hydrocarbon generation and migration in the Western Canada sedimentary basin. In: *Classic Petroleum Provinces* (ed. by J. Brooks), *Geological Society, London, Special Publication*, 50, 189–202.

Creek, J.L. and Schraeder, M.L. (1985) *East Painter Reservoir: An Example of a Compositional Gradient.* SPE 14411, Society of Petroleum Engineers, Las Vegas.

Crittenden, M.D. Jr. (1963) Effective viscosity of the Earth derived from isostatic loading of Pleistocene Lake Bonneville. *Journal of Geophysical Research*, 68, 5517–5530.

Crosby, A.G., Fishwick, S. and White, N. (2010) Structure and evolution of the intracratonic Congo Basin. *Geochemistry, Geophysics, Geosystems*, Q06010.

Cross, T.A. (1986) Tectonic controls of foreland basin subsidence and Laramide style deformation, western United States. In: *Foreland Basins* (ed. by P.A. Allen and P. Homewood), *International Association of Sedimentologists Special Publication*, 8, 15–39.

Cross, T.A. (ed.) (1990) *Quantitative Dynamic Stratigraphy*, Prentice Hall, New Jersey, 625 pp.

Cross, T.A. *et al.* (1993) Applications of high resolution sequence stratigraphy to reservoir analysis. In: *Subsurface Reservoir Characterization from Outcrop Observations* (ed. by R. Eschard and B. Doligez), Editions Technip, Paris, 11–33.

Crough, S.T. and Jurdy, D.M. (1980) Subducted lithosphere, hotspots and the geoid. *Earth and Planetary Science Letters*, 48, 15–22.

Crowell, J.C. (1974a) Origin of late Cenozoic basins in southern California. In: *Tectonics and Sedimentation* (ed. by W.R. Dickinson), *Society of Economic Paleontologists and Mineralogists Special Publication*, 22, 190–204.

Crowell, J.C. (1974b) Sedimentation along the San Andreas fault, California. In: *Modern and Ancient Geosynclinal Sedimentation* (ed. by R.H. Dott Jr. and R.H. Shaver), *Society of Economic Paleontologists and Mineralogists Special Publication*, 19, 292–303

Crowell, J.C. (1982) The tectonics of Ridge Basin, southern California. In: *San Andreas Fault in Southern California* (ed. by J.C. Crowell), *California Division of Mines and Geology Special Report*, 118, 208–218.

Crowell, J.C. and Link, M.H. (1982) Ridge Basin, Southern California. In: *Geologic History of Ridge Basin, Southern California* (ed. by J.C. Crowell and M.H. Link), Society of Economic Paleontologists and Mineralogists Pacific Section, 1–4, 304 pp.

Cubitt, J.M. and England, W.A. (1995) *The Geochemistry of Reservoirs, Geological Society, London, Special Publication*, 86, 321pp.

Cumella, S.P., Shanley, K.W. and Camp, W. (2008) Introduction. In: *Understanding, Exploring, and Developing Tight-Gas Sands* (ed. by S.P. Cumella, K.W. Shanley and W.K. Camp), *2005 Vail Hedberg Conference, American Association of Petroleum Geologists Hedberg Series*, 3, 1–4.

Cunningham, R. and Lindholm, R.M. (2000) Seismic evidence for widespread gas hydrate formation, offshore west Africa In: *Petroleum Systems of South Atlantic Margins* (ed. by M.R. Mello and B.J. Katz), *American Association of Petroleum Geologists Memoir*, 73, 93–105.

Curry, W.H. and Curry, W.H. III (1972) South Glennock oilfield, Wyoming: a pre-discovery thinking and post-discovery description. In: *Stratigraphic Oil and Gas Fields* (ed. by R.E. King), *American Association of Petroleum Geologists Memoir*, 16, 415–427.

Da Hung, N. and Van Le, H. (2004) Petroleum geology of Cuu Long Basin – offshore Vietnam, *Search and Discovery*, article 10062.

Dade, W.B. and Friend, P.F. (1998) Grain size, sediment-transport regime and channel slope in alluvial rivers. *Journal of Geology*, 106, 661–675.

Dadson, S.J., Hovius, N. and 10 others (2003) Erosion of the Taiwan orogen. *Nature*, 426, 648–651.

Dahl, B. and Speers, G.C. (1986) Geochemical characterization of a tar mat in the Oseberg Field, Norwegian Sector, North Sea. In: *Advances in Organic Geochemistry, 1985; Part I, Petroleum Geochemistry* (ed. by D. Leythaeuser and J. Ruellkotter), *Organic Geochemistry*, 10, 547–558.

Dahlen, F.A. (1981) Isostasy and the ambient state of stress in the oceanic lithosphere. *Journal of Geophysical Research*, 86, 7801–7807.

Dahlen, F.A. (1984) Non-cohesive critical Coulomb wedges: an exact solution. *Journal of Geophysical Research*, 89, 10125–10133.

Dahlstrom, C.D.A. (1970) Structural geology in the eastern margin of the Canadian Rocky Mountains. *Bulletin of Canadian Petroleum Geology*, 18, 332–406.

Dalmayrac, B. and Molnar, P. (1981) Parallel thrust and normal faulting in Peru and constraints of the state of stress. *Earth and Planetary Science Letters*, 55, 473–481.

Dalrymple, R.W., Zaitlin, B.A. and Boyd, R. (1992) Estuarine facies models: conceptual basis and stratigraphic implications. *Journal of Sedimentary Petrology*, 62, 1130–1146.

Daniel, R.F. and Kaldi, J.G. (2009) Evaluating seal capacity of caprocks and intraformational barriers for CO_2 containment. In: *Carbon Dioxide Sequestration in Geological Media – State of the Science* (ed. by M. Grobe, J.C. Pashin and R.L. Dodge), *American Association of Petroleum Geologists Studies in Geology*, 59, 335–345.

Davies, D.R. and Davies, J.H. (2009) Thermally driven mantle plumes reconcile multiple hot-spot observations. *Earth and Planetary Science Letters*, 278, 50–54.

Davies, G.F. (1988) Ocean bathymetry and mantle convection, 1, large-scale flow and hotspots. *Journal of Geophysical Research*, 93, 10467–10480.

Davies, G.F. (1999) *Dynamic Earth: Plates, Plumes and Mantle Convection*, Cambridge University Press, Cambridge, 458pp.

Davies, G.F. and Pribac, F. (1993) Mesozoic seafloor subsidence and the Darwin Rise, past and present. In: *The Mesozoic Pacific* (ed. by M. Pringle, W. Sager, W. Sliter and S. Stein), American Geophysical Union, Washington D.C., 39–52.

Davies, G.F. and Richards, M.A. (1992) Mantle convection. *Journal of Geology*, 100, 151–206.

Davies, J.H. and Davies, D.R. (2010) Earth's surface heat flux. *Solid Earth*, 1, 5–24.

Davis, D., Suppe, J. and Dahlen, F.A. (1983) Mechanics of fold-and-thrust belts and accretionary wedges. *Journal of Geophysical Research*, 88, 1153–1172.

Davy, P. and Cobbold, P.R. (1991) Experiments on shortening of a 4-layer model of the continental lithosphere. *Tectonophysics*, 188, 1–25.

Dawers, N.H. and Anders, M.H. (1995) Displacement-length scaling and fault linkage. *Journal of Structural Geology*, 17, 607–614.

De Alteriis, G. (1995) Different foreland basins in Italy: examples from the central and southern Adriatic Sea. *Tectonophysics*, 252, 349–373.

De Boer, W., Rawlinson, P.B. and Hurst, A. (2007) Successful exploration of a successful sand injection complex: Hamsun prospect, Norway block 24/9. In: *Sand Injectites: Implications for Hydrocarbon Exploration and Production* (ed. by A. Hurst and J. Cartwright), *American Association of Petroleum Geologists Memoir*, 87, 65–68.

DeCelles, P.G. and Giles, K.A. (1996) Foreland basin systems. *Basin Research*, 8, 105–124.

DeCelles, P.G., Garzione, C.N., Copeland, P., Upreti, B.N., Robinson, D.M., Quade, J. and Ohja, T.P. (2001) Stratigraphy, structure and tectonic evolution of the Himalayan fold-thrust belt in western Nepal. *Tectonics*, 20, 487–509.

DeCelles, P.G., Gehrels, G.E., Najman, Y., Martin, A.J., Carter, A. and Garzanti, E. (2004) Detrital geochronology and geochemistry of Cretaceous–Early Miocene strata of Nepal: implications for timing and diachroneity of initial Himalayan orogenesis. *Earth and Planetary Science Letters*, 227, 313–330.

DeCelles, P.G., Gehrels, G.E., Quade, J., Ojha, T.P., Kapp, P.A. and Upreti, B.N. (1998) Neogene foreland basin deposits, erosional unroofing, and the kinematic history of the Himalayan fold-thrust belt, western Nepal. *Geological Society of America Bulletin*, 110, 2–21.

Demaison, G. (1984) The generative basin concept. In: *Petroleum Geochemistry and Basin Evaluation* (ed. by G. Demaison and R.J. Murris). *American Association of Petroleum Geologists Memoir*, 35, 1–14.

Demaison, G. and Huizinga, B.J. (1994) Genetic classification of petroleum systems using three factors: charge, migration and entrapment. In: *The Petroleum System – From Source to Trap* (ed. by L.B. Magoon and W.G. Dow), *American Association of Petroleum Geologists Memoir*, 60.

Demaison, G.J. and Moore, G.T. (1980) Anoxic environments and oil source bed genesis. *American Association of Petroleum Geologists Bulletin*, 64, 1178–1209.

Demicco, R.V. (1998) CYCOPATH 2D – A two-dimensional forward model of cyclic sedimentation on carbonate platforms. *Computers and Geosciences*, 24, 405–423.

Demicco, R.V. and Spencer, R.J. (1989) Maps – a basic program to model accumulation of platform sediments. *Computers and Geosciences*, 15, 95–105.

Densmore, A.L. and Hovius, N. (2000) Topographic fingerprints of bedrock landslides. *Geology*, 28, 371–374.

Densmore, A.L., Allen, P.A. and Simpson, G. (2007) Development and response of a coupled catchment-fan system under changing tectonic and climatic forcing. *Journal of Geophysical Research – Earth Surface*, 112, F01002.

Densmore, A.L., Anderson, R.S., McAdoo, B.G. and Ellis, M.A. (1997) Hillslope evolution by bedrock landslides. *Science*, 275, 369–372.

Densmore, A.L., Ellis, M.A. and Anderson, R.S. (1998) Landsliding and the evolution of normal-fault-bounded mountains. *Journal of Geophysical Research*, 103, 15203–215219.

Dewey, J.F. (1977) Suture zone complexities: a review. *Tectonophysics*, 40, 53–67.

Dewey, J.F. (1980) Episodicity, sequence, and style at convergent plate boundaries. In: *The Continental Crust and its Mineral Deposits* (ed. by D.W. Strangeway), *Geological Association of Canada Special Paper*, 20, 553–573.

Dewey, J.F. (1982) Plate tectonics and the evolution of the British Isles. *Journal of the Geological Society*, 139, 371–414.

Dewey, J.F. (1988) Extensional collapse of orogens. *Tectonics*, 7, 1123–1139.

Dewey, J.F. and Bird, J.M. (1970) Mountain belts and the new global tectonics. *Journal of Geophysical Research*, 75, 2625–2647.

Dewey, J.F. and Pindell, J.L. (1985) Neogene block tectonics of eastern Turkey and northern South America: continental applications of the finite difference method. *Tectonics*, 4, 71–83.

Dezes, P. and Ziegler, P.A. (2002) Moho depth map of western and central Europe. http://www.unibas.ch/eucor-urgent.

Dibblee, T.W. Jr. (1977) Strike-slip tectonics of the San Andreas Fault and its role in Cenozoic basin development. In: *Late Mesozoic and Cenozoic Sedimentation and Tectonics in California, San Joaquin Geological Society Short Course*, 26–38.

DiCaprio, L., Gurnis, M. and Müller, R. (2009) Long-wavelength tilting of the Australian continent since the Late Cretaceous. *Earth and Planetary Science Letters*, 278, 175–185.

Dickinson, W.R. (1974) Plate tectonics and sedimentation. In: *Tectonics and Sedimentation* (ed. by W.R. Dickinson), *Society of Economic Paleontologists and Mineralogists Special Publication*, 22, 1–27.

Dickinson, W.R. (1980) Plate tectonics and key petrologic associations. In: *The Continental Crust and its Mineral Deposits* (ed. by D.W. Strangeway), *Geological Association of Canada Special Paper*, 20, J.T. Wilson Volume, 341–360.

Dickinson, W.R. (1995) Forearc basins. In: *Tectonics of Sedimentary Basins* (ed. by C. Busby and R. Ingersoll), Blackwell Publishing Ltd., Oxford, 221–262.

Dickinson, W.R. and Seely, D.R. (1979) Structure and stratigraphy of forearc regions. *American Association of Petroleum Geologists Bulletin*, 63, 2–31.

Dickinson, W.R. and Suczek, C.A. (1979) Plate tectonics and sandstone compositions. *American Association of Petroleum Geologists Bulletin*, 63, 2164–2182.

Dickinson, W.R. and Valloni, R. (1980) Plate tectonics and provenance of sands in modern ocean basins. *Geology*, 8, 82–86.

Dietz, R.S. (1963) Collapsing continental rises, an actualistic concept of geosynclines and mountain building. *Journal of Geology*, 71, 314–333.

Dingle, R.V. (1980) Large allochthonous sediment masses and their role in the construction of the continental slope and rise off southwestern Africa. *Marine Geology*, 37, 333–354.

Divins, D.L. (2003) Total sediment thickness of the world's oceans and marginal seas, NOAA National Geophysical Data Center, Boulder, CO. http://www.ngdc.noaa.gov/mgg/sedthick/sedthick.html.

Divko, G.D., O'Brien, G.W., Harrison, M.L., and Hamilton, P.J. (2010) Evaluation of the regional top seal in the Gippsland Basin: implications for geological carbon storage and hydrocarbon prospectivity. *APPEA Journal*, 2010, 463–486.

Doglioni, C., Mongelli, F. and Pialli, G. (1998) Boudinage of the Alpine Belt in the Apenninic back-arc. In: *Results of the CROP03 Deep Seismic Reflection Profile* (ed. by G. Pialli, M. Barchi and G. Minelli), *Memorie della Societa Geologica Italiana*, 52, 457–468.

Donovan, D.T. and Jones, E.J.W. (1979) Causes of world-wide changes in sea level. *Journal of the Geological Society*, 136, 187–192.

Dooley, T. and McClay, K. (1997) Analog modelling of pull-apart basins. *American Association of Petroleum Geologists Bulletin*, 81, 1804–1826.

Dooley, T., Monastero, F., Hall, B., McClay, K.R. and Whitehouse, P. (2004) Scaled sandbox modelling of transtensional pull-apart basins: applications to the Coso geothermal system. *Geothermal Research Council Transactions*, 28, 637–641.

Doré, A.G., Cartwright, J.A., Stoker, M.S., Turner, J.P. and White, N. (2002) *Exhumation of the North Atlantic Margin: Timing, Mechanisms and Implications for Petroleum Exploration*, Geological Society, London, Special Publication, 196, 494pp.

Dorman, L.M. and Lewis, B.T.R. (1972) Experimental isostasy. 3. The isostatic Green's function and lateral density changes. *Journal of Geophysical Research*, 77, 3068–3077.

Dorobek, S.L. (1995) Synorogenic carbonate platforms and reefs in foreland basins: controls on stratigraphic evolution and platform/reef morphology. In: *Stratigraphic Evolution of Foreland Basins* (ed. by S.L. Dorobek and G.M. Ross), *Society for Sedimentary Geology Special Publication*, 52, 127–147.

Dow, W.G. (1974) The application of oil correlation and source rock data to exploration in the Williston Basin. *American Association of Petroleum Geologists Bulletin*, 58, 7, 1253–1262.

Dow, W.G. (1977) Kerogen studies and geological interpretations. *Journal of Geochemical Exploration*, 7, 79–99.

Dow, W.G. and O'Connor, D.I. (1982) Kerogen maturity and type by reflected light microscopy applied to petroleum exploration. In: *How to Assess Maturation and Paleotemperatures, Society of Economic Paleontologists and Mineralogists Short Course Notes*, 7, 133–157.

Dowdle, W.L. and Cobb, W.M. (1975) Static formation temperature from well logs – an empirical method. *Journal of Petroleum Technology*, Nov. 1975, 1326.

Downey, J.S. (1984) Geohydrology of the Madison and associated aquifers in parts of Montana, North Dakota, South Dakota, and Wyoming. *US Geological Survey Professional Paper*, 1273-G.

Downey, N.K. and Gurnis, M. (2009) Instantaneous dynamics of the cratonic Congo basin. *Journal of Geophysical Research*, 114, B06401.

Drinkwater, N.J. and Pickering, K.T. (2001) Architectural elements in a high-continuity, sand-prone turbidite system, late Precambrian Kongsfjord Formation, northern Norway: application to hydrocarbon reservoir characterization. *American Association of Petroleum Geologists Bulletin*, 85, 1731–1757.

Droste, H. (1990) Depositional cycles and source rock development in an epeiric intra-platform basin: the Hanifa Formation of the Arabian peninsula. *Sedimentary Geology*, 69, 281–296.

Drummond, C.N. and Wilkinson, B.H. (1993) Aperiodic accumulation of cyclic peritidal carbonate. *Geology*, 21, 1023–1026.

Duerto, L. and McClay, K.R. (2009) The role of syntectonic sedimentation in the evolution of doubly vergent thrust belts and foreland folds. *Marine and Petroleum Geology*, 26, 1051–1069.

Dühnforth, M., Densmore, A.L., Ivy-Ochs, S. and Allen, P.A. (2008) Controls on sediment evacuation from glacially modified and unmodified catchments in the eastern Sierra Nevada, California. *Earth Surface Processes and Landforms*, 33, 1602–1613.

Duller, R.A., Whittaker, A.C., Fedele, J.J. *et al.* (2010) From grain size to tectonics. *Journal of Geophysical Research – Earth Surface*, 115, F03022.

Dunbar, C.O. and Rogers, J. (1957) *Principles of Stratigraphy*, John Wiley & Sons, New York, 356 pp.

Dunbar, J.A. and Sawyer, D.S. (1989) How pre-existing weaknesses control the style of continental breakup. *Journal of Geophysical Research*, 94, 7278–7292.

Duncan, R.A. and Richards, M.A. (1991) Hotspots, mantle plumes, flood basalts, and true polar wander. *Reviews of Geophysics*, 29, 31–50.

Duncan, W.I., Green, P.F. and Duddy, I.R. (1998) Source rock burial history and seal effectiveness: key facets to understanding hydrocarbon exploration potential in the East and Central Irish Sea Basins. *American Association of Petroleum Geologists Bulletin*, 82, 1401–1415.

Durand, B., Alpern, B., Pittion, J.L. and Pradier, B. (1986) Reflectance of vitrinite as a control of thermal history of sediments. In: *Thermal Modeling in Sedimentary Basins* (ed. by J. Burrus), *1st IFP Exploration Research Conference*, Carcans, France, June 3–7, 1985, Editions Technip, Paris, 441–474, 600pp.

Durham, L.S. (2009) Brazil dancing the pre-salt salsa. *American Association of Petroleum Geologists Explorer*, Nov. 2009.

Dutton, S., Flanders, W. and Barton, M. (2003) Reservoir characterization of a Permian deep-water sandstone, Earl Ford field, Delaware basin, Texas. *American Association of Petroleum Geologists Bulletin*, 87, 609–627.

Duval, B., Cramez, C. and Jackson, M.P.A. (1992) Raft tectonics in the Kwanza Basin, Angola. *Marine and Petroleum Geology*, 9, 389–390.

Dzevanshir, R.D., Buryakovskiy, L.A. and Chilingarian, G.V. (1986) Simple quantitative evaluation of porosity of argillaceous sediments at various depths of burial. *Sedimentary Geology*, 46, 169–173.

Dziewonski, A.M. and Woodhouse, J.H. (1987) Global images of the Earth's interior. *Science*, 236, 37–48.

Ebinger, C. and Casey, M. (2001) Continental breakup in magmatic provinces: an Ethiopian example. *Geology*, 29, 527–530.

Ebinger, C. and Scholz, C.A. (2012) Continental rift basins: an East African perspective. In: *Tectonics of Sedimentary Basins: Recent Advances* (ed. by C. Busby and A. Azor), Wiley-Blackwell, 185–208, 647pp.

Ebinger, C. and Sleep, N. (1998) Cenozoic magmatism throughout East Africa resulting from impact from a single plume. *Nature*, 395, 788–791.

Ebinger, C.J., Bechtel, T.D., Forsyth, D.W and Bowin, C.O. (1989) Effective elastic plate thickness beneath the East African and Afar plateaus and dynamic compensation of the uplifts. *Journal of Geophysical Research*, 94, 2883–2901.

Ebinger, C.J., Karner, G.D., and Weissel, J.K. (1991) Mechanical strength of extended continental lithosphere: constraints from the Western Rift System. *Tectonics*, 10, 1239–1256.

Ehlers, J. (1996) *Quaternary and Glacial Geology*, John Wiley & Sons Ltd., Chichester.

Einsele, G. (2000) *Sedimentary Basins: Evolution, Facies and Sediment Budget, Second Edition*, Springer, Berlin.

Einsele, G. and Hinderer, M. (1997) Terrestrial sediment yield and the lifetimes of reservoirs, lakes and larger basins. *Geologische Rundschau*, 86, 288–310.

Einsele, G., Ratschbacher, L. and Wetzel, A. (1996) The Himalaya-Bengal fan denudation-accumulation system during the past 20 Ma. *Journal of Geology*, 104, 163–184.

Elder, J. (1976) *The Bowels of the Earth*, Oxford University Press, 222pp.

Elliott, D. (1976) The motion of thrust sheets. *Journal of Geophysical Research*, 81, 949–963.

Ellis, M.A. and Densmore, A.L. (2006) First order topography over blind thrusts. In: *Tectonics, Climate and Landscape Evolution* (ed. by S.D. Willett, N. Hovius, M.T. Brandon *et al.*), *Geological Society of America Penrose Conference, Taiwan, Geological Society of America Special Papers*, 398, 251–266.

Ellis, M.A., Densmore, A.L. and Anderson, R.S. (1999) Evolution of mountainous topography in the Basin and Range Province. *Basin Research*, 11, 21–42.

Ellis, S. (1996) Forces driving continental collision: reconciling indentation and mantle subduction tectonics. *Geology*, 24, 699–702.

Ellis, S., Fullsack, P. and Beaumont, C. (1995) Oblique convergence of the crust driven by basal forcing: implications for length-scales of deformation and strain partitioning in orogens. *Geophysical Journal International*, 120, 24–44.

Embry, A.F. and Johannessen, E.P. (1992) T-R sequence stratigraphy, facies analysis and reservoir distribution in the uppermost Triassic–Lower Jurassic succession, western Sverdrup Basin, Arctic Canada. In: *Arctic Geology and Petroleum Potential* (ed. by T.O. Vorren, E. Bergseger, O.A. Dahl-Stamnes *et al.*), *Special Publication*, 2, Norwegian Petroleum Society, 121–146.

Emery, D. and Myers, K.J. (1996) *Sequence Stratigraphy*, Blackwell Publishing Ltd., Oxford, 297pp.

Emery, D., Myers, R.J. and Young, R. (1990) Ancient subaerial exposure and freshwater leaching in sandstones. *Geology*, 18, 1178–1181.

Emery, K.O. (1977) Structure and stratigraphy of divergent continental margins. In: *Geology of Continental Margins* (ed. by H. Yarborough *et al.*), *American Association of Petroleum Geologists Continuing Education Course Notes Series*, 5, B1–B20, Washington D.C.

Emter, D. (1971) Ergebnisse seismischer Untersuchungen der Erdkruste und des obersten erdmantels in Südwestdeutschland. Dissertation, Universität Stuttgart, 108pp.

Energy Information Administration (EIA) (2013a) Annual Energy Outlook 2013 Early Release, Report Number DOE/EIA-0383ER(2013) www.eia.gov/forecasts/aeo/er.

Energy Information Administration (EIA) (2013b), Energy in Brief article "What is shale gas and why is it important?", http://www.eia.gov/energy_in_brief/article/about_shale_gas.cfm.

England, P. and Houseman, G. (1986) Finite strain calculations of continental deformations, 2, Comparison with the India–Asia collision. *Journal of Geophysical Research*, 91, 3664–3676.

England, P., Houseman, G. and Sonder, L. (1985) Length scales for continental deformation in convergent, divergent and strike-slip environments: analytical and approximate solutions for a thin viscous sheet model. *Journal of Geophysical Research*, 90, 4797–4810.

England, P.C. (1983) Constraints on extension of continental lithosphere. *Journal of Geophysical Research*, 88, 1145–1152.

England, P.C. (1987) Diffuse continental deformation – length scales, rates and metamorphic evolution. *Philosophical Transactions of the Royal Society, London*, A321, 1557, 3–22.

England, P.C. and Houseman, G.A. (1988) Uplift and extension of the Tibetan plateau. *Journal of Geophysical Research*, 326, 301–320.

England, P.C. and Houseman, G.A. (1989) Extension during continental convergence, with application to the Tibetan Plateau. *Journal of Geophysical Research*, 94, 17561–17579.

England, P.C. and McKenzie, D.P. (1982) A thin viscous sheet model for continental deformation. *Geophysical Journal of the Royal Astronomical Society*, 70, 295–322.

England, P.C. and McKenzie, D.P. (1983) Correction to a thin viscous sheet model for continental deformation. *Geophysical Journal of the Royal Astronomical Society*, 73, 523–532.

England, P.C. and Molnar, P. (1990) Surface uplift, uplift of rocks, and exhumation of rocks. *Geology*, 18, 1173–1177.

England, W.A. and Fleet, A.J. (1991) *Petroleum Migration, Geological Society, London, Special Publication*, 59.

England, W.A., Mann, A.L. and Mann, D.M. (1991) Migration from source to trap. In: *Source and Migration Processes and Evaluation Techniques* (ed. by R.K. Merrill), *Treatise of Petroleum Geology Handbook of Petroleum Geology*, American Association of Petroleum Geologists, Tulsa, Oklahoma, Chapter 3, 23–46.

Ericksen, M.C. and Slingerland, R. (1990) Numerical simulations of tidal and wind-driven circulation in the Cretaceous Interior Seaway of North America. *Geological Society of America Bulletin*, 102, 1499–1516.

Escalona, A. and Mann, P. (2011) Tectonics, basin subsidence mechanisms, and palaeogeography of the Caribbean–South American plate boundary zone. *Marine and Petroleum Geology*, 28, 8–39.

Espitalié, J., Laporte, J.L., Madec, M., Marquis, F., Leplat, P., Paulet, J. and Boutefeu, A. (1977) Méthode rapide de characterisation des roches mères, de leur potentiel pétrolier et de leur degré d'évolution. *Revue de l'Institut Francais du Petrole et Annales des Combustibles Liquides*, 32, 23–42.

Ethier, V.G. and King, H.R. (1991) Reservoir quality evaluation from visual attributes on rock surfaces: methods of estimation and classification from drill cuttings or cores. *Bulletin Canadian Petroleum Geologists*, 39, 233–251.

Eugster, H.P. and Hardie, L.A. (1975) Sedimentation in an ancient playa-lake complex: the Wilkins Peak member of the Green River Formation of Wyoming. *Geological Society of America Bulletin*, 86, 319–334.

Eugster, H.P. and Kelts, K. (1983) Lacustrine chemical sediments. In: *Chemical Sediments and Geomorphology* (ed. by A.S. Goudie and K. Pye), Academic Press, London, 321–368.

Evans, A.E. (1988) Neogene tectonic and stratigraphic events in the Gulf of Suez area, Egypt. *Tectonophysics*, 153, 235–247.

Eyles, C.H., Eyles, N. and Miall, A.D. (1985) Models of glaciomarine sedimentation and their application to the interpretation of ancient glacial sequences. *Palaeogeography, Palaeoclimatology, Palaeoecology*, 51, 15–84.

Fadahunsi, O.O., Lawrence, S.R., Richards, M. and Bray, R. (2005) Plays, traps, and petroleum systems in the Deepwater Niger Delta. *Search and Discovery*, article 90037.

Fairbanks, R.G. and Matthews, R.R. (1978) The marine oxygen isotope record in Pleistocene corals, Barbados, West Indies. *Quaternary Research*, 10, 181–196.

Fairbridge, R.W. (1961) Eustatic changes in sea level. In: *Physics and Chemistry of the Earth* (ed. by L.H. Ahrens), 4, Pergamon Press, London, 99–185.

Falcon, N.L. (1958) Position of oil fields of southwest Iran with respect to relevant sedimentary basins. In: *Habitat of Oil* (ed. by L.G. Weeks), Symposium Volume, American Association of Petroleum Geologists, Tulsa, 1279–1293.

Falvey, D.A. (1974) The development of continental margins in plate tectonic theory. *Australian Petroleum Production and Exploration Association Journal*, 14, 95–106.

Falvey, D.A. and Middleton, M.F. (1981) Passive continental margins: evidence for a pre-breakup deep crustal metamorphic subsidence mechanism. In: *Colloquium on Geology of Continental Margins*, C3, Paris, 7–17 July 1980, *Oceanologica Acta*, 4 (Supplement), 103–114.

Farooqui, M.Y., Hou, H., Li, G., Machin, N., Neville, T., Pal, A., Shrivastva, C., Wang, Y., Yang, F., Yin, C., Zhao, J. and Yang, X. (2009) Evaluating volcanic reservoirs. *Oilfield Review* Spring 2009, 21, no. 1.

Favre, P. and Stampfli, G.M. (1992) From rifting to passive margin – the examples of the Red Sea, Central Atlantic and Alpine Tethys. In: *Geodynamics of Rifting Symposium, Tectonophysics*, 215, 69–97.

Fedele, J.J. and Paola, C. (2007) Similarity solutions for fluvial sediment fining by selective deposition. *Journal of Geophysical Research – Earth Surface*, 112, F02038.

Felletti, F. and Bersezio, R. (2010) Quantification of the degree of confinement of a turbidite-filled basin: a statistical approach based on bed thickness distribution. *Marine and Petroleum Geology*, 27, 515–532.

Fernandez, M. and Ranalli, G. (1997) The role of rheology in extensional basin formation modelling. *Tectonophysics*, 2282, 129–145.

Fernandez-Lozano, J., Soukoutis, D., Willingshofer, E. *et al.* (2011) Cenozoic deformation of Iberia: a model for intraplate mountain building and basin development based on analogue modelling. *Tectonics*, 30, TC1001.

Fiduk, J.C., Weimer, P., Trudgill, B.D., Rowan, M.G., Gale, P.E., Phair, R.L., Korn, B.E., Roberts, G.R., Gafford, W.T., Lowe, R.S. and Queffelec, T.A. (1999) The Perdido Foldbelt, northwestern deep Gulf of Mexico, Part 2: seismic stratigraphy and petroleum systems. *American Association of Petroleum Geologists Bulletin*, 83, no. 4, 578–612.

Fielding, E.J. (2000) Morphotectonic evolution of the Himalayas and Tibetan Plateau. In: *Geomorphology and Global Tectonics* (ed. by M.A. Summerfield), John Wiley & Sons, 201–222, 367pp.

Finlayson, D.P., Montgomery, D.R. and Hallet, B. (2002) Spatial coincidence of rapid inferred erosion with young metamorphic massifs in the Himalayas. *Geology*, 30, 219–222.

Fitzgerald, P.G., Sorkhabi, R.B., Redfield, T.F. and Stump, E. (1995) Uplift and denudation of the central Alaska Range: a case-study in the use of apatite fission track thermochronology to determine absolute uplift parameters. *Journal of Geophysical Research*, 100, 21075–21091.

Fleitout, L. and Froideveaux, C. (1982) Tectonic stresses in the lithosphere. *Tectonics*, 2, 315–324.

Flemings, P.B. and Jordan, T.E. (1989) A synthetic stratigraphic model of foreland basin development. *Journal of Geophysical Research*, 94, B3851–B3866.

Flemings, P.B. and Jordan, T.E. (1990) Stratigraphic modelling of foreland basins: interpreting thrust deformation and lithospheric rheology. *Geology*, 18, 430–434.

Flint, R.F. (1971) *Glacial and Quaternary Geology*, John Wiley & Sons, New York, 892pp.

Ford, M. (2004) Depositional wedge-tops: interaction between low basal friction external orogenic wedges and flexural foreland basins. *Basin Research*, 16, 361–375.

Forsyth, D.W. (1985) Subsurface loading and estimates of the flexural rigidity of the continental lithosphere. *Journal of Geophysical Research*, 90, 38–47.

Forte, A.M., Mitrovica, J.X., Moucha, R., Simmons, N.A. and Grand, S.P. (2007) Descent of the ancient Farallon slab drives localized mantle flow below the New Madrid seismic zone. *Geophysical Research Letters*, 34, L04308.

Fossen, H. (2010) *Structural Geology*, Cambridge University Press, 463pp.

Foster, P.T. and Rattey, P.R. (1993) The evolution of a fractured chalk reservoir, Machar Oilfield, UK North Sea. In: *Petroleum Geology of Northwest Europe, Proceedings of the 4th Conference of the Geological Society, London* (ed. by J.R. Parker), 1445–1452.

Fournier, F. (1960) *Climat et érosion; la relation entre erosion du sol par l'eau et les precipitations atmosphériques*, Presses Universitaires de France, Paris, 201pp.

Fowler, C.M.R. (1990) *The Solid Earth: An Introduction to Global Geophysics*, Cambridge University Press.

France-Lanord, C. and Derry, L.A. (1997) Organic carbon burial forcing of the carbon cycle from Himalayan erosion. *Nature*, 390, 65–67.

Franzinelli, E. and Potter, P.E. (1983) Petrology, chemistry and texture of modern river sands, Amazon River system. *Journal of Geology*, 91, 23–39.

Frazier, D.E. (1974) Depositional episodes: their relationship to the Quaternary stratigraphic framework in the northwestern portion of the Gulf Basin. *University of Texas at Austin, Bureau of Economic Geology, Geological Circular*, 4(1), 28pp.

Frenz, M., Wynn, R.B. and Georgiopoulou, A. *et al.* (2009) Provenance and pathways of late Quaternary turbidites in the deep-water Agadir Basin, northwest African margin. *International Journal of Earth Sciences*, 98, 721–733.

Freund, R. (1970) Rotation of strike-slip faults in Sitan, southeast Iran. *Journal of Geology*, 78, 188–200.

Freund, R. (1971) The Hope Fault, a strike-slip fault in New Zealand. *Bulletin of the New Zealand Geological Survey*, 86, 1–49.

Frey, M., Teichmüller, M., Teuichmüller, R., Mullis, J., Kunzi, B., Breitschmid, A., Gruner, U. and Schwizer, B. (1980) Very low grade metamorphism in the external parts of the Central Alps: illite crystallinity, coal rank and fluid inclusion data. *Eclogae Geologicae Helvetiae*, 73, 173–203.

Friedmann, S.J. and Burbank, D.W. (1995) Rift basins and supradetachment basins: intracontinental extensional end-members. *Basin Research*, 7, 109–127.

Frisch, W., Dunkl, I. and Kuhlemann, J. (2000) Post-collisional orogen-parallel large-scale extension in the Eastern Alps. *Tectonophysics*, 327, 239–265.

FrOG Tech Pty Ltd. (2005) OZ SEEBASE™ Study, 2005, Public Domain Report to Shell Development Australia by FrOG Tech Pty Ltd., viewed 19 January 2006, http://www.frogtech.com.au/.

Fu, B., Lin, A. and Kano, K., Maruyama, T. and Guo, J. (2003) Quaternary folding of the eastern Tian Shan, northwest China. *Tectonophysics*, 369, 79–101.

Fuis, G.S., Mooney, W.D., Healey, J.H., McMechan, G.A. and Lutter, W.J. (1982) Crustal structure of the Imperial Valley region. *US Geological Survey Professional Paper*, 1254, 145–154.

Fuis, G.S., Mooney, W.D., Healy, J.H., McMechan, G.A. and Lutter, W.J. (1984) A seismic refraction survey of the Imperial Valley region, California. *Journal of Geophysical Research*, B, 89, 1165–1189.

Fullsack, P. (1995) An arbitrary Lagrangian–Eulerian formulation for creeping flows and its application to tectonic models. *Geophysical Journal International*, 120(1), 1–23.

Furman, T., Bryce, J., Hanan, B., Yirgu, G. and Ayalew, D. (2006) Heads and tails: 30 years of the Afar plume. In: *The Structure and Evolution of the East African Rift System in the Afar Volcanic Province* (ed. by G. Yirgu, C.J. Ebinger and P.K.H. Maguire), *Geological Society, London, Special Publication*, 259, 95–119.

Gaillardet, J., Dupré, B., Louvat, P. and Allègre, C.J. (1999) Global silicate weathering and CO₂ consumption rates deduced from the chemistry of large rivers. *Chemical Geology*, 159, 3–30.

Galewsky, J. (1998) The dynamics of foreland basin carbonate platforms: tectonic and eustatic controls. *Basin Research*, 10, 409–416.

Gallagher, K. and Brown, R. (1997) The onshore record of passive margin evolution. *Journal of the Geological Society*, 154, 451–457.

Gallagher, K., Brown, R. and Johnson, C. (1998) Fission track analysis and its application to geological problems. *Annual Reviews of Earth and Planetary Sciences*, 26, 519–572.

Gallagher, K., Hawkesworth, C.J. and Mantovani, M.J.M. (1994) The denudation history of the onshore continental margin of SE Brazil inferred from apatite fission track data. *Journal of Geophysical Research*, 99, 18117–18145.

Galloway, W.E. (1975) Process framework for describing the morphologic and stratigraphic evolution of deltaic depositional systems. In: *Deltas, Models for Exploration* (ed. by M.L. Broussard), Houston Geological Society, Houston, 87–98.

Galloway, W.E. (1989) Genetic stratigraphic sequences in basin analysis: architecture and genesis of flooding surface bounded depositional units. *American Association of Petroleum Geologists Bulletin*, 73, 125–142.

Galy, A., France-Lanord, C., Hutrez, J.E. and Lucazeau, F. (1996) Mass transfer during Himalayan erosion during the monsoon: mineralogical and geochemical constraints. *Eos, Transactions, American Geophysical Union*, 77, 236.

Galy, V., France-Lanord, C., Beyssac, P., Faure, P., Kudrass, H. and Palhol, F. (2007) Efficient organic carbon burial in the Bengal fan sustained by the Himalayan erosional system. *Nature*, 450, 407–410.

Garcia, R. (1981) Depositional systems and their relation to gas accumulation in Sacramento Valley, California. *American Association of Petroleum Geologists Bulletin*, 65, 653–673.

Garcia-Castellanos, D. (2002) Interplay between lithospheric flexure and river transport in foreland basins. *Basin Research*, 14, 89–104.

Garcia-Castellanos, D. and Cloetingh, S. (2012) Modelling the interaction between lithospheric and surface processes in foreland basins. In: *Tectonics of Sedimentary Basins, Recent Advances* (ed. by C. Busby and A. Azor), Wiley-Blackwell, Oxford, 152–181.

Garfunkel, Z. (1981) Internal structure of the Dead Sea leaky transform (rift) in relation to plate kinematics. *Tectonophysics*, 80, 81–108.

Garfunkel, Z. and Bartov, Y. (1977) The tectonics of the Suez rift. *Bulletin of the Geological Survey of Israel*, 71, 44pp.

Garfunkel, Z. and Ben-Avraham, Z. (1996) The structure of the Dead Sea basin. *Tectonophysics*, 266, 155–176.

Garfunkel, Z., Zak, I. and Freund, R. (1981) Active faulting in the Dead Sea Rift. *Tectonophysics*, 80, 1–26.

Garzanti, E., Ando, S. and Vezzoli, G. (2009) Grain-size dependence of sediment composition and environmental bias in provenance studies. *Earth and Planetary Science Letters*, 277, 422–432.

Garzanti, E., Doglioni, C. and Vezzoli, G. *et al.* (2007) Orogenic belts and orogenic sediment provenance. *Journal of Geology*, 115, 315–334.

Gautam, P. and Fujiwara, Y. (2000) Magnetic polarity stratigraphy of Siwalik Group sediments of Karnali River section in western Nepal. *Geophysical Journal International*, 142, 812–824.

Gawthorpe, R.L. and Hurst, J.M. (1993) Transfer zones in extensional basins: their structural style and influence on drainage development and stratigraphy. *Journal of the Geological Society*, 150, 1137–1152.

Gawthorpe, R.L. and Leeder, M.R. (2000) Tectono-sedimentary evolution of active extensional basins. *Basin Research*, 12, 195–218.

Gemmer, L., Ings, S.J., Medvedev, S. and Beaumont, C. (2004) Salt tectonics driven by differential sediment loading: stability analysis and finite element experiments. *Basin Research*, 16, 199–218.

Gibbs, R. J. (1970) Mechanisms controlling world water chemistry. *Science*, 170, 1088–1090.

Gibling, M.R., Tantisukrit, C., Utamo, W., Thanasuthipitak, T. and Haraluck, M. (1985a) Oil shale sedimentology and geochemistry in Cenozoic Mae Sot basin, Thailand. *American Association of Petroleum Geologists Bulletin*, 69, 767–780.

Gibling, M.R., Ukakimaphan, Y. and Srisuk, S. (1985b) Oil shale and coal in intermontane basins of Thailand. *American Association of Petroleum Geologists Bulletin*, 69, 760–766.

Gibson-Poole, C.M., Svendsen, L., Watson, M.N., Daniel, R.F., Ennis-King, J. and Rigg, A.J. (2009) Understanding stratigraphic heterogeneity: a methodology to maximise the efficiency of the geological storage of CO₂. In: *Carbon Dioxide Sequestration in Geological Media – State of the Science* (ed. by M. Grobe, J.C. Pashin and R.L. Dodge), *American Association of Petroleum Geologists Studies in Geology*, 59, 347–264.

Gilchrist, A.R. and Summerfield, M.A. (1990) Differential denudation and flexural isostasy in formation of rifted-margin upwarps. *Nature*, 346, 739–742.

Gildner, R.F. and Cisne, J.L. (1990) Quantitative modelling of carbonate stratigraphy and water-depth history using depth-dependent sediment accumulation function. In: *Quantitative Dynamic Stratigraphy* (ed. by T.A. Cross), Prentice Hall, Englewood Cliffs, New Jersey, 417–432.

Giles, M.R. (1987) Mass transfer and the problems of secondary porosity creation in deeply buried hydrocarbon reservoirs. *Marine and Petroleum Geology*, 4, 188–204.

Giles, M.R. (1997) *Diagenesis: A Quantitative Perspective. Implications for Basin Modeling and Rock Property Prediction*, Kluwer Academic Publishers, Dordrecht, 526pp.

Giles, M.R. and Indrelid, S.L. (1998) Divining burial and thermal histories from indicator data: application and limitations. In: *Advances in Fission Track Geochronology* (ed. by P. van den Haute and F. De Corte), Kluwer Academic Publishers, Dordrecht, 115–150.

Giles, M.R., Stevenson, S., Martin, S.V., Cannon, S.J.C., Hamilton, P.J., Marshall, J.D. and Samways, G.M. (1992) The reservoir properties and diagenesis of the Brent Group: a regional perspective. In: *Geology of the Brent Group* (ed. by A.C. Morton, R.S. Haszeldine, M.R. Giles and S. Brown), *Geological Society, London, Special Publication*, 61, 289–327.

Giltner, J.P. (1987) Application of extensional models to the northern Viking graben. *Norsk Geologisk Tidsskrift*, 67, 339–352.

Ginsburg, R.N. (1971) Landward movement of carbonate mud: new model for regressive cycles in carbonates. *American Association Petroleum Geologists Annual Meeting, Abstract Programs*, 55, 340.

Ginsburg, R.N. and James, N.P. (1974) Holocene carbonate sediments of continental margins. In: *The Geology of Continental Margins* (ed. by C.A. Burke and C.L. Drake), Springer-Verlag, New York, 137–155.

Girdler, R.W. (1970) A review of Red Sea heat flow. *Philosophical Transactions Royal Society London*, A267, 191–203.

Glasmann, J.R. (1992) The fate of feldspar in Brent Group reservoirs, North Sea: a regional synthesis of diagenesis in shallow, intermediate, and deep burial environments. In: *Geology of the Brent Group* (ed. by A.C. Morton, R.S. Haszeldine, M.R. Giles and S. Brown), *Geological Society, London, Special Publication*, 61, 329–350.

Gleadow, A.J.W. and Duddy, I.R. (1981) A natural long term annealing experiment for apatite. *Nuclear Tracks*, 5, 169–174.

Gleadow, A.J.W., Duddy, I.R. and Lovering, J.F. (1983) Fission track analysis: a new tool for the evaluation of thermal histories and hydrocarbon potential. *Australian Petroleum Production and Exploration Association Journal*, 23, 93–102.

Glotzbach, C., Bernet, M. and van der Beek, P. (2011) Detrital thermochronology records changing source areas and steady exhumation in the Western European Alps. *Geology*, 39, 239–242.

Gluyas, J. and Cade, C.A. (1997) Prediction of porosity in compacted sands. In: *Reservoir Quality Prediction in Sandstones and Carbonates* (ed. by J.A. Kupecz, J. Gluays and S. Bloch), *American Association of Petroleum Geologists Memoir*, 69, 19–28.

Gluyas, J.G. and Swarbrick, R. (2004) *Petroleum Geoscience*, Blackwell Publishing Ltd., Oxford, 359pp.

Gluyas, J.G., Garland, C.R., Oxtoby, N.H. and Hogg, A.J.C. (2000) Quartz cement; the Miller's tale. In: *Quartz Cementation in Sandstones* (ed. by R.H. Worden and S. Morad), *International Association Sedimentologists Special Publication*, 29, Blackwell Publishing Ltd., Oxford, 199–218.

Gluyas, J.G., Robinson, A.G., Emery, D., Grant, S.M. and Oxtoby, N.H. (1993) The link between petroleum emplacement and sandstone cementation. In: *Petroleum Geology of Northwest Europe: Proceedings of the 4th Conference* of the *Geological Society, London* (ed. by J.R. Parker), 1395–1402.

Goes, S. and Van der Lee, S. (2002) Thermal structure of the North American uppermost mantle inferred from seismic tomography. *Journal of Geophysical Research*, 107(B3).

Goetze, C. and Evans, B. (1979) Stress and temperature in the bending lithosphere as constrained by experimental rock mechanics. *Geophysical Journal of the Royal Astronomical Society*, 59, 463–478.

Goldhammer, R.K., Dunn, P.A. and Hardie, L.A. (1990) Depositional cycles, composite sea level changes, cycles stacking patterns, and the hierarchy of stratigraphic forcing: examples from Alpine Triassic platform carbonates. *Geological Society of America Bulletin*, 102, 535–562.

Goldhammer, R.K., Lehmann, P.J. and Dunn, P.A. (1993) The origin of high-frequency platform carbonate cycles and third-order sequences (Lower Ordovician El Paso Group, West Texas): constraints from outcrop data and stratigraphic modelling. *Journal Sedimentary Petrology*, 63, 318–359.

Gomberg, J. and Ellis, M. (1994) Topography and tectonics of the central New Madrid Seismic Zone: results of numerical experiments using a 3-dimensional boundary element program. *Journal of Geophysical Research*, 99, 20299–20310.

Gomes, P.O., Kilsdonk, W., Minken, J., Grow, T. and Barraga, R. (2009) The Outer High of the Santos Basin, southern Sao Paulo Plateau, Brazil: pre-salt exploration outbreak, paleogeographic setting, and evolution of the syn-rift structures. *Search and Discovery*, article 10193.

Gomes, P.O., Parry, J. and Martins, W. (2002) The Outer High of the Santos Basin, southern Sao Plateau, Brazil: tectonic setting, relation to volcanic events and some comments on hydrocarbon potential. *American Association of Petroleum Geologists Hedberg Conference on Hydrocarbon Habitat of Volcanic Rifted Passive Margins*, Sept. 8–11, 2002, Stavanger, Norway.

Gordon, M.B. and Hempton, M.R. (1986) Collision-induced rifting: the Greenville orogeny and the Keweenawan rift of North America. *Tectonophysics*, 127, 1–25.

Gosnold, W.D. and Fischer, D.W. (1986) Heat flow studies in sedimentary basins. In: *Thermal Modeling in Sedimentary Basins* (ed. by J. Burrus), *1st IFP Exploration Research Conference*, Carcans, France, June 3–7, 1985, Editions Technip, Paris, 199–218, 600pp.

Goto, S. and Matsubayashi, O. (2008) Inversion of needle-probe data for thermal properties of the eastern flank of the Juan de Fuca Ridge. *Journal of Geophysical Research*, 113 (B08105), 1–17.

Gould, S.J. (1989) *Wonderful Life: The Burgess Shale and the Nature of History*, W.W. Norton and Co., New York.

Grammer, G.M., Eberli, G.R., Van Buchem, F.S.P., Stegvenson, G.M. and Homewood, P. (1996) Application of high-resolution sequence stratigraphy to evaluate lateral variability in outcrop and subsurface-Desert Creek and Ismay intervals, Paradox Basin. In: *Paleozoic Systems of the Rocky Mountain Region* (ed. by M.W. Longman and M.D. Sonnenfeld), *Rocky Mountain Section of the Society of Economic Paleontologists and Mineralogists*, 235–266.

Grand, S. (2002) Mantle shear-wave tomography and the fate of subducted slabs. *Royal Society London Philosophical Transactions*, 360, 2475–2491.

Granger, D.E., Kirchner, J.W. and Finkel, R. (1996) Spatially averaged long-term erosion rates measured from *in situ*-produced cosmogenic nuclides in alluvial sediments. *Journal of Geology*, 104, 249–257.

Granjeon, D. and Joseph, P. (1999) Concepts and applications of a 3-D multiple lithology, diffusive model in stratigraphic modeling. In: *Numerical Advances in Stratigraphy: Recent Advances in Stratigraphic and Sedimentologic Simulations* (ed. by J.W. Harbaugh et al.), *Society for Sedimentary Geology Special Publication*, 62, Tulsa, Oklahoma, 197–210.

Griffiths, R.W. and Campbell, I.H. (1990) Interaction of mantle plume heads with the Earth's surface and onset of small-scale convection. *Journal of Geophysical Research*, 96, 18, 295–310.

Griggs, D.T., Turner, F.J. and Heard, H.C. (1960) Deformation of rocks at 500 °C to 800 °C. In: *Rock Deformation* (ed. by D.T. Griggs and J. Handin), *Geological Society of America Memoir*, 79, 21–37, 382pp.

Grigné, C., Labrosse, S. and Tackley, P.J. (2007) Convection under a lid of finite conductivity: heat flux scaling and application to continents. *Journal of Geophysical Research*, 112, B08402.

Grobe, M., Pashin, J.C. and Dodge, R.L. (2009) Carbon dioxide sequestration in geological media – state of the science: introduction. In: *Carbon Dioxide Sequestration in Geological Media – State of the Science* (ed. by M. Grobe, J.C. Pashin and R.L. Dodge), *American Association of Petroleum Geologists Studies in Geology*, 59, 1–2.

Grocott, J. and Watterson, J. (1980) Strain profile of a boundary within a large ductile shear zone. *Journal of Structural Geology*, 2, 111–117.

Grotzinger, J.P. (1986) Cyclicity and paleoenvironmental dynamics, Rocknest platform, northwest Canada. *Geological Society of America Bulletin*, 97, 1208–1231.

Grunau, H.R. (1987) A worldwide look at the caprock problem. *Journal of Petroleum Geology*, 10, 245–266.

Guardado, L.R., Spadini, A.R., Brandao, J.S.L. and Mello, M.R. (2000) Petroleum system of the Campos Basin, Brazil. In: *Petroleum Systems of South Atlantic Margins* (ed. by M.R. Mello and B.J. Katz), *American Association of Petroleum Geologists Memoir*, 73, 317–324.

Guillou, L. and Jaupart, C. (1995) On the effect of continents on mantle convection. *Journal of Geophysical Research*, 100, 24217–24238.

Gunnell, Y. (1998) The interaction between geological structure and global tectonics in multistoreyed landscape development: a denudation chronology of the South Indian shield. *Basin Research*, 10, 281–310.

Gupta, S. (1997) Himalayan drainage patterns and the origin of fluvial megafans in the Ganges foreland basins. *Geology*, 25, 11–14.

Gupta, S. & Allen, P.A. (1999) Fossil shore platforms and drowned gravel beaches: evidence for high-frequency sea-level fluctuations in the distal Alpine foreland basin. *Journal of Sedimentary Research*, 69, 394–413.

Gupta, S. and Allen, P.A. (2000) Implications of foreland paleotopography for stratigraphic development in the Eocene distal Alpine foreland basin. *Geological Society of America Bulletin*, 112, 515–530.

Gupta, S., Underhill, J.R., Sharp, I.R. and Gawthorpe, R.L. (1999) Role of fault interactions in controlling synrift sediment dispersal patterns; Miocene, Abu Alaqa Group, Suez Rift, Sinai, Egypt. *Basin Research*, 11, 167–189.

Gürbüz, A. (2010) Geometric characteristics of pull-apart basins. *Lithosphere*, 2, Geological Society of America, 199–206.

Gurnis, M. (1988) Large-scale mantle convection and the aggregation and dispersal of supercontinents. *Nature*, 332, 695–699.

Gurnis, M. (1990a) Bounds on global dynamic topography from Phanerozoic flooding of continental platforms. *Nature*, 344, 754–756.

Gurnis, M. (1990b) Ridge spreading, subduction and sea level variations. *Science*, 250, 970–972.

Gurnis, M. (1992) Rapid continental subsidence following the initiation and evolution of subduction. *Science*, 255, 1556–1558.

Gurnis, M. (1993) Depressed continental hypsometry behind ocean trenches: a clue to subduction controls on sea-level change. *Geology*, 21, 29–32.

Gurnis, M. and Hager, B.H. (1988) Controls of the structure of subducted slabs. *Nature (London)*, 335, 317–321.

Gurnis, M., Eloy, C. and Zhong, S. (1996) Free-surface formulation of mantle convection; II, Implication for subduction-zone observables. *Geophysical Journal International*, 127, 719–727.

Gussow, W.C. (1954) Differential entrapment of oil and gas: a fundamental principle. *American Association of Petroleum Geologists Bulletin*, 38, 816–853.

Gust, D.A., Biddle, K.T., Phelps, D.W. and Uliana, M.A. (1985) Associated Middle to Late Jurassic volcanism and extension in southern South America. *Tectonophysics*, 116, 223–253.

Haack, R.C., Sundararaman, P., Diedjomahor, J.O., Xiao, H., Gant, N.J., May, E.D. and Kelsch, K. (2000) Niger Delta petroleum systems, Nigeria. In: *Petroleum Systems of South Atlantic Margins* (ed. by M.R. Mello and B.J. Katz), *American Association of Petroleum Geologists Memoir*, 73, 213–231.

Haack, U. (1982) Radioactivity of rocks. In: *Physical Properties of Rocks* (ed. by G. Angenheister), Vol. 1b, Springer-Verlag, Berlin, 433–481.

Hack, J.T. (1973) Drainage adjustment in the Appalachians. In: *Fluvial Geomorphology* (ed. by M. Morisawa), SUNY Publications in Geomorphology, Binghampton, N.Y., 51–69.

Hager, B.H. (1980) Eustatic sea level and spreading rate are not simply related. *Eos, Transactions, American Geophysical Union*, 61, 374.

Hager, B.H. (1984) Subducted slabs and the geoid: constraints on mantle rheology and flow. *Journal of Geophysical Research*, 89, B7, 6003–6015.

Hager, B.H. and Clayton, R.W. (1989) Constraints on the structure of mantle convection using seismic observations, flow models, and the geoid. In: *Mantle Convection* (ed. by R.W. Peltier), Gordon and Breach, New York, 657–763.

Hager, B.H., Clayton, R.W., Richards, M.A., Comer, R.P. and Dziewonski, A.M. (1985) Lower mantle heterogeneity, dynamic topography and the geoid. *Nature*, 313, 541–545.

Halbouty, M.T. (1979) *Salt Domes, Gulf Region, United States and Mexico, Second Edition*, Gulf Publishing, Houston, Texas.

Halbouty, M.T. (1982) The time is now for all explorationists to purposely search for the subtle trap. In: *The Deliberate Search for the Subtle Trap* (ed. by M.T. Halbouty), *American Association of Petroleum Geologists Memoir*, 32, 1–10.

Hallet, B., Hunter, L. and Bogen, J. (1996) Rates of erosion and sediment evacuation by glaciers. A review of field data and their implications. *Global and Planetary Change*, 12, 213–235.

Halley, R.B. and Schmoker, J.W. (1983) High-porosity Cenozoic carbonate rocks of South Florida: progressive loss of porosity with depth. *American Association of Petroleum Geologists Bulletin*, 67, 1191–1200.

Hambrey, M.J. (1994) *Glacial Environments*, UCL Press, London.

Hamilton, P.J., Giles, M.R. and Ainsworth, P. (1992) K–Ar dating of illites in Brent Group reservoirs: a regional perspective. In: *Geology of the Brent Group* (ed. by A.C. Morton, R.S. Haszeldine, M.R. Giles and S. Brown), *Geological Society, London, Special Publication*, 61, 377–400.

Hamilton, P.J., Kelley, S. and Fallick, A.E. (1989) K–Ar dating of illite in hydrocarbon reservoirs. *Clay Minerals*, 24, 215–231.

Hampson, G.J. (2000) Discontinuity surfaces, clinoforms and facies architecture in a wave-dominated, shoreface-shelf parasequence. *Journal of Sedimentary Research*, 70, 325–340.

Hancock, G.S., Anderson, R.S., Whipple, K.X. and Wohl, E.E. (1998) Beyond power: bedrock river incision process and form. Rivers over rock, fluvial processes in bedrock channels. *Geophysical Monograph*, 107, 35–60.

Hancock, J.M. (1984) Cretaceous. In: *Introduction to the Petroleum Geology of the North Sea, Second Edition* (ed. by K.W. Glennie), Blackwell Scientific Publications, Oxford, 133–150.

Hand, M. and Sandiford, M. (1999) Intraplate deformation in central Australia, the link between subsidence and fault reactivation. *Tectonophysics*, 305, 121–140.

Hanks, T.C., Bucknam, R.C., LaJoie, K.R. and Wallace, R.E. (1984) Modification of wave-cut and faulting controlled landforms. *Journal of Geophysical Research*, 89, 5771–5790.

Hanne, D., White, N., Butler, A. and Jones, S. (2004) Phanerozoic vertical motions of Hudson Bay. *Canadian Journal of Earth Science*, 41, 1181–1200.

Hansen, S. (1996) A compaction trend for Cretaceous and Tertiary shales on the Norwegian Shelf based on sonic transit times. *Petroleum Geoscience*, 2, 159–166.

Haq, B.U. and Al-Qahtani, A.M. (2005) Phanerozoic cycles of sea-level change on the Arabian Platform. *GeoArabia*, 10, 127–160.

Haq, B.U., Hardenbol, J. and Vail, P.R. (1987) Chronology of fluctuating sea levels since the Triassic (250 Myr ago to present). *Science*, 235, 1156–1167.

Haq, B.U., Hardenbol, J. and Vail, P.R. (1988) Mesozoic and Cenozoic chronostratigraphy and cycles of sea-level change. In: *Sea-level Changes: An Integrated Approach* (ed. by C.K. Wilgus, B.S. Hastings, C.G.St.C. Kendall, H.W. Posamentier, C.A. Ross and J.C. Van Wagoner), *Society of Economic Paleontologists and Mineralogists Special Publication*, 42, 40–45.

Harbaugh, J.W., Watney, L., Rankey, G., Slingerland, R., Goldstein, R. and Franseen, E. (eds.) (1999) *Numerical Experiments in Stratigraphy. Recent Advances in Stratigraphic and Sedimentologic Computer Simulations. Society for Sedimentary Geology Special Publication*, 62, Tulsa, Oklahoma, 362pp.

Hardie, L.A. and Shinn, E.A. (1986) Carbonate depositional environments, modern and ancient – 3, tidal flats. *Quarterly Journal Colorado School of Mines*, 81.

Hardie, L.A., Bosellini, A. and Goldhammer, R.K. (1986) Repeated subaerial exposure of subtidal carbonate platforms, Triassic, northern Italy: evidence for high frequency sea level oscillations on a 10^4 year scale. *Paleoceanography*, 1, 447–457.

Hardie, L.A., Dunn, P.A. and Goldhammer, R.K. (1991) Field and modelling studies of Cambrian cycles, Virginia Appalachians – discussion. *Journal of Sedimentary Petrology*, 61, 636–646.

Harding, T.P. and Lowell, J.D. (1979) Structural styles, their plate tectonic habitats and hydrocarbon traps in petroleum provinces. *American Association of Petroleum Geologists Bulletin*, 63, 1016–1058.

Harris, L.B. and Cobbold, P.R. (1984) Development of conjugate shear bands during simple shearing. *Journal of Structural Geology*, 7, 37–44.

Harris, P.M., Saller, A.H. and Simo, J.A. (1999) Introduction. In: *Advances in Carbonate Sequence Stratigraphy: Applications to Reservoirs, Outcrops and Models* (ed. by P.M. Harris, A.H. Saller and J.A. Simo), Society of Economic Paleontologists and Mineralogists, Tulsa, Oklahoma, 1–10.

Harrison, C.G.A. (1990) Long-term eustasy and epeirogeny in continents. In: *Sea-Level Change* (ed. by R. Revelle), National Academy Press, Washington, D.C., 141–160.

Harrison, T.K., Copeland, P., Hall, S.A. *et al.* (1993) Isotopic preservation of Himalayan/Tibetan uplift, denudation and climatic histories in two molasse deposits. *Journal of Geology*, 100, 157–173.

Harrison, W.J. and Summa, L.L. (1991) Paleohydrology of the Gulf of Mexico Basin. *American Journal of Science*, 291, 2181–2187.

Hartley, R.A., Roberts, G.G., White, N.J. and Richardson, C. (2011) Transient convective uplift of an ancient buried landscape. *Nature Geoscience*, 4, 562–565.

Hartley, R.W. and Allen, P.A. (1994) Interior cratonic basins of Africa: relation to continental break-up and role of mantle convection. *Basin Research*, 6, 95–113.

Haszeldine, R.S., Macaulay, C.I., Marchands, A. *et al.* (2000) Sandstone cementation and fluids in hydrocarbon basins. *Journal of Geochemical Exploration*, 69–70, 195–200.

Hawkesworth, C., Cawood, P., Kemp, T., Storey, C. and Dhuime, B. (2009) A matter of preservation. *Science*, 323, 49–50.

Haxby, W.F. Turcotte, D.L. and Bird, J.M. (1976) Thermal and mechanical evolution of the Michigan Basin. *Tectonophysics*, 36, 57–75.

Hay, W.W. (1998) Detrital sediment fluxes from continents to oceans. *Chemical Geology*, 145, 287–323.

Hay, W.W., Sloan, J.L. and Wold, C.N. (1988) Mass/age distribution and composition of sediments on the ocean floor and the global rate of sediment subduction. *Journal of Geophysical Research*, 93, 14933–14940.

Hayes, M.O. (1979) Barrier island morphology as a function of tidal and wave regime. In: *Barrier Islands from the Gulf of St. Lawrence to the Gulf of Mexico* (ed. by S.P. Leatherman), Academic Press, New York, 1–27.

Hays, J.D. and Pitman, W.C. (1973) Lithospheric plate motion, sea level changes, and climatic and ecological consequences. *Nature*, 246, 18–22.

Hays, J.D., Imbrie, J. and Shackleton, N.J. (1976) Variations in the Earth's orbit: pacemaker of the ice ages. *Science*, 194, 2212–2232.

Hedberg, H.D. (1936) Gravitational compaction of clays and shales. *American Journal of Science*, 31, 241–287.

Heffern, E.L., Reiners, P.W., Naeser, C.W. and Coates, D.A. (2008) Geochronology of clinker and implications for evolution of the Powder River Basin landscape, Wyoming and Montana. *Geological Society of America Reviews in Engineering Geology*, 18, 155–175.

Heidbach, O., Tingay, M., Barth, A. *et al.* (2010) Global crustal stress pattern based on the World Stress Map database release 2008. *Tectonophysics*, 482, 3–15.

Heine, C., Muller, R., Steinberger, B. and Torsvik, T. (2008) Subsidence in intracontinental basins due to dynamic topography. *Physics Earth and Planetary Interiors*, 171, 252–264.

Heine, C., Müller, R.D., Steinberger, B. *et al.* (2010) Integrating deep Earth dynamics in palaeogeographic reconstructions of Australia. *Tectonophysics*, 483, 135–150.

Heling, D. and Teichmüller, M. (1974) Die Grenze Montmorillonit/ Mixed Layer Minerale und ihre Beziehung zur Inkohlung in der grauen Schichtenfolge des Oligozäns im Oberrheingraben. *Fortschritte in der Geologie von Rheinland und Westfalen, Krefeld*, 24, 113–128.

Helland-Hansen, W. and Gjelberg, J.G. (1994) Conceptual basis and variability in sequence stratigraphy: a different perspective. *Sedimentary Geology*, 92, 31–52.

Helland-Hansen, W. and Hampson, G.J. (2009) Trajectory analysis: concepts and applications. *Basin Research*, 21, 454–483.

Helland-Hansen, W. and Martinsen, O.J. (1996) Shoreline trajectories and sequences: description of variable depositional-dip scenarios. *Journal of Sedimentary Research*, B66, 670–688.

Heller, P.L. and Paola, C. (1992) The large scale dynamics of grain-size variation in alluvial basins; 2, application to syntectonic conglomerate. *Basin Research*, 4, 91–102.

Heller, P.L., Angevine, C.L. and Winslow, N.S. (1988) Two-phase stratigraphic model of foreland basin sequences. *Geology*, 16, 501–504.

Hellinger, S.J. and Sclater, J.G. (1983) Some comments on two-layer extensional models for the evolution of sedimentary basins. *Journal of Geophysical Research*, 88, 8251–8270.

Hempton, M.R. and Dunne, L.A. (1984) Sedimentation in pull-apart basins: active example in eastern Turkey. *Journal of Geology*, 92, 513–530.

Henriksen, S., Hampson, G.J., Helland-Hansen, W., Johannessen, E.P. and Steel, R.J. (2009) Shelf-edge and shoreline trajectories, a dynamic approach to stratigraphic analysis. *Basin Research*, 21, 445–453.

Henriksen, S., Helland-Hansen, W. and Bullimore, S. (2011) Relationships between shelf-edge trajectories and sediment dispersal along depositional dip and strike: a different approach to sequence stratigraphy. *Basin Research*, 23, 3–21.

Herb, R. (1988) Eocaene palaeogeographie and palaeotektonic des Helvetikums. *Eclogae Geologicae Helvetiae*, 81, 611–657.

Héroux, Y., Chagnon, A. and Bertrand, R. (1979) Compilation and correlation of major thermal maturation indicators. *American Association of Petroleum Geologists Bulletin*, 63, 2128–2144.

Hetényi, M. (1979) *Beams on Elastic Foundations*, University of Michigan Press, Ann Arbor, Michigan, 255pp.

Heward, A.P. (1981) A review of wave-dominated clastic shoreline deposits. *Earth-Science Reviews*, 17, 223–276.

Heydari (1997a) The role of burial diagenesis in hydrocarbon destruction and H$_2$S accumulation, upper Jurassic Smackover Formation, Black Creek Field, Mississippi. *Bulletin American Association of Petroleum Geologists* 81, 26–45.

Heydari, E. (1997b) Hydrotectonic models of burial diagenesis in platform carbonates based on formation water geochemistry in North American sedimentary basins. In: *Basin-wide Diagenetic Patterns: Integrated Petrologic, Geochemical and Hydrological Considerations* (ed. by I.P. Montañez, J.M. Gregg and K.L. Shelton), *Society for Sedimentary Geology Special Publication*, 57, 53–79.

Higley, D.K., Lewan, M., Roberts, L.N.R. and Henry, M.E. (2006) *Petroleum System Modeling Capabilities for Use in Oil and Gas Resource Assessments*, US Geological Survey Open-File Report, 1024.

Hill, R.I. (1991) Starting plumes and continental break-up. *Earth and Planetary Science Letters*, 104, 398–416.

Hillgärtner, H., van Buchem, F.S.P., Gaumet, F., Razin, P., Pittet, B., Grötsch, J., and Droste, H. (2003) The Barremian–Aptian evolution of the eastern Arabian carbonate platform margin (northern Oman). *Journal of Sedimentary Petrology*, 73, 756–773.

Hillis, R.R. and Reynolds, S.D. (2000) The Australian stress map. *Journal of the Geological Society*, 157, 915–921.

Hilton, R.G., Galy, A. and Hovius, N. (2008) Riverine particulate organic carbon from an active mountain belt: Importance of landslides. *Global Biogeochemical Cycles*, 22.

Hilton, R.G., Galy, A. and Hovius, N. (2011) Efficient transport of fossil organic carbon to the ocean by steep mountain rivers: an orogenic carbon sequestration mechanism. *Geology*, 39, 71–74.

Hinderer, M. (2001) Late Quaternary denudation of the Alps, valley and lake fillings and modern river loads. *Geodinamica Acta*, 14, 231–263.

Hinnov, L.A. and Goldhammer, R.K. (1991) Spectral analysis of the Triassic Latemar Limestone. *Journal of Sedimentary Petrology*, 61, 1173–1193.

Hinsch, R., Decker, K. and Peresson, H. (2005) 3-D seismic interpretation and structural modelling in the Vienna Basin: implications for Miocene to recent kinematics. *Austrian Journal of Earth Science*, 97, 38–50.

Hirn, A. and Perrier, G. (1974) Deep seismic sounding in the Limagne graben. In: *Approaches to Taphrogenesis* (ed. by J.H. Illies and K. Fuchs), E. Schweizerbartsche Verlagsbuchhandlung, Stuttgart, 329–340.

Hirst, J.P.P. and Nichols, G.J. (1986) Thrust tectonic controls on Miocene alluvial distribution patterns, southern Pyrenees. In: *Foreland Basins* (ed. by P.A. Allen and P. Homewood), *International Association of Sedimentology Special Publication*, 8, Blackwell Scientific Publications, Oxford, 247–258.

Hoffman, A.W. and White, M.W. (1982) Mantle plumes from ancient oceanic crust. *Earth and Planetary Science Letters*, 57, 421–436.

Hogg, A.J.C., Mitchell, A.W. and Young, S. (1996) Predicting well productivity from grain-size analysis and logging while drilling. *Petroleum Geoscience*, 2, 1–15.

Holland, H.D. (1981) River transport to the oceans. In: *The Sea*, 7, *The Oceanic Lithosphere* (ed. by C. Emiliani), John Wiley & Sons, N.J., 763–800.

Holt, P., Allen, M.B. and van Hunen, J. (2010) Lithospheric cooling and thickening as a basin forming mechanism. *Tectonophysics*, 495.

Homewood, P., Allen, P.A. and Williams, G.D. (1986) Dynamics of the Molasse Basin of western Switzerland. In: *Foreland Basins* (ed. by P.A. Allen and P. Homewood), *International Association of Sedimentology Special Publication*, 8, Blackwell Scientific Publications, Oxford, 199–217.

Homewood, P.W., Guillocheau, F., Eschard, R. and Cross, T.A. (1992) Corrélations haute résolution et stratigraphie génétique: une démarche intégrée. *Bulletin des Centre de Recherches Exploration-Production Elf Aquitaine*, 16, 357–381.

Homewood, P.W., Mauriaud, P. and Lafont, F. (2000) *Best Practices in Sequence Stratigraphy*, ELF EP Edition, Mémoire 25, 81.

Hood, A., Gutjahr, C.C.M. and Heacock, R.L. (1975) Organic metamorphism and the generation of petroleum. *American Association of Petroleum Geologists Bulletin*, 59, 986–996.

Hood, K.C. and Yurewicz, D.A. (2008) Assessing the Mesaverde basin-centred gas play, Piceance Basin, Colorado. In: *Understanding, Exploring, and Developing Tight-Gas Sands, 2005 Vail Hedberg Conference* (ed. by S.P. Cumella, K.W. Shanley and W.K. Camp), *American Association of Petroleum Geologists Hedberg Series*, 3, 87–104.

Horton, B.K. and DeCelles, P.G. (1997) The modern foreland basin system adjacent to the central Andes. *Geology*, 25, 895–898.

Horton, B.K. and DeCelles, P.G. (2002) Modern and ancient fluvial megafans in the foreland basin system of the central Andes, southern Bolivia: implications for drainage network evolution in fold-thrust belts. *Basin Research*, 13, 43–63.

Horton, B.K., Hampton, B.A. and Waanders, G.L. (2001) Paleogene synorogenic sedimentation in the Altiplano plateau and implications for initial mountain building in the central Andes. *Geological Society of America Bulletin*, 113, 1387–1400.

Hoth, S., Adam, J., Kukowski, N. *et al.* (2006) Influence of erosion on the kinematics of bivergent orogens: results from scaled sandbox simulations. In: *Tectonics, Climate, and Landscape Evolution, Book Series, Geological Society of America Special Papers*, 398, 201–225.

Hoth, S., Hoffmann-Rothe, A. and Kukowski, N. (2007) Frontal accretion: an internal clock for bivergent wedge deformation and surface uplift. *Journal of Geophysical Research – Solid Earth*, 112, B6, B06408.

Hoth, S., Kukowski, N. and Oncken, O. (2008) Distant effects in bivergent orogenic belts – How retro-wedge erosion triggers resource formation in pro-foreland basins. *Earth and Planetary Science Letters*, 273, 28–37.

Houseman, G. and England, P.C. (1986) A dynamical model of lithosphere extension and sedimentary basin formation. *Journal of Geophysical Research*, 91, 719–729.

Hovius, N. (1995) Macro-scale process systems of mountain belt erosion and sediment delivery to basins. Unpublished DPhil. thesis, University of Oxford.

Hovius, N. (1996) Regular spacing of drainage outlets from linear mountain belts. *Basin Research*, 8, 29–44.

Hovius, N. (1998) Controls on sediment supply by large rivers. In: *Relative Role of Eustasy, Climate and Tectonics in Continental Rocks* (ed. by K.W. Shanley and P.J. McCabe), *Society of Economic Paleontologists and Mineralogists Special Publication*, 59, 3–16.

Hovius, N. (2000) Macroscale process systems of mountain belt erosion. In: *Geomorphology and Global Tectonics* (ed. by M.A. Summerfield), John Wiley & Sons Ltd., Chichester. 77–105.

Hovius, N., Stark, C.P. and Allen, P.A. (1997) Sediment flux from a mountain belt derived by landslide mapping. *Geology*, 25, 231–234.

Hovius, N., Stark, C.P., Chu, H.T. and Lin, J.C. (2000) Supply and removal of sediment in a landslide-dominated mountain belt: Central Range, Taiwan. *Journal of Geology*, 108, 73–89.

Howard, A.D. (1987) Modelling fluvial systems: rock-, gravel- and sandbed channels. In: *River Channels* (ed. by K. Richards), Basil Blackwell, New York, 69–94.

Howard, A.D. (1994) A detachment limited model of drainage basin evolution. *Water Resources Research*, 30, 739–752.

Howard, J.J. (1992) Influence of authigenic clay minerals on permeability. In: *Origin, Diagenesis, and Petrophysics of Clay Minerals in Sandstones* (ed. by D.W. Houseknecht and E.D. Pittman), *Society of Economic Paleontologists and Mineralogists Special Publication*, 47, 257–264.

Howell, D.G. and Vedder, J. (1981) Structural implications of stratigraphic discontinuities across the southern California borderland. *Journal of Sedimentary Petrology*, 49, 517–540.

Howell, D.G., Crouch, J.K., Greene, H.G., McCulloch, D.S. and Vedder, J.G. (1980) Basin development along the late Mesozoic and Cainozoic California margin: a plate tectonic margin of subduction, oblique subduction and transform tectonics. In: *Sedimentation in Oblique Slip Mobile Zones* (ed. by P.F. Ballance and H.G. Reading), *International Association of Sedimentology Special Publication*, 4, Blackwell Scientific Publications, Oxford, 43–62.

Hower, J., Eslinger, E.V., Hower, M. and Perry, E.A. (1976) Mechanism of burial metamorphism in argillaceous sediments. I. Mineralogical and chemical evidence. *Geological Society of America Bulletin*, 87, 725–737.

Hsu, K.J. (ed.) (1983) *Mountain Building Processes*, Academic Press, Orlando, Florida.

Hubbard, R.J., Pape, J. and Roberts, D.G. (1985) Depositional sequence mapping as a technique to establish tectonic and stratigraphic framework and evaluate hydrocarbon potential on a passive continental margin. In: *Seismic Stratigraphy II: An Integrated Approach* (ed. by O.R. Berg and D. Woolverton), *American Association of Petroleum Geologists Memoir*, 39, 79–91.

Hubbert, M.K. (1953) Entrapment of petroleum under hydrodynamic conditions. *American Association of Petroleum Geologists Bulletin*, 37, 1954–2026.

Hubbert, M.K. and Rubey, W.W. (1959) Role of fluid pressure in mechanics of overthrust faulting. *Geological Society of America Bulletin*, 70, 115–166.

Hudec, M.R. and Jackson, M.P.A. (2007) Terra infirma: understanding salt tectonics. *Earth Science Reviews*, 82, 1–28.

Huff, K.F. (1978) Frontiers of world oil exploration. *Oil and Gas Journal*, 76(40), 214–220.

Huismans, R.S. and Beaumont, C. (2002) Asymmetric lithospheric extension: the role of frictional plastic strain softening inferred from numerical experiments. *Geology*, 30, 211–214.

Huismans, R.S. and Beaumont, C. (2008) Complex rifted continental margins explained by dynamical models of depth-dependent lithospheric extension. *Geology*, 36, 163–166.

Hulen, J.B., Goff, F., Ross, J.R., Bortz, L.C. and Bereskin, S.R. (1994) Geology and geothermal origin of Grant Canyon and Bacon Flat oil fields, Railroad Valley, Nevada. *American Association of Petroleum Geologists Bulletin*, 78, 596–623.

Hull, C.E. and Warman, H.R. (1970) Asmari oil fields of Iran. In: *Geology of Giant Petroleum Fields, American Association of Petroleum Geologists Memoir*, 14, 428–437.

Humphrey, N.F. and Heller, P.L. (1995) Natural oscillations in coupled geomorphic systems – an alternative origin for cyclic sedimentation. *Geology*, 23, 499–502.

Hunt, D. and Tucker, M.E. (1992) Stranded parasequences and the forced regressive wedge systems tract: deposition during base level fall. *Sedimentary Geology*, 81, 1–9.

Hunt, J.M. (1979) *Petroleum Geochemistry and Geology*, W. H. Freeman, San Francisco, 617pp.

Hurford, A.J. (1986) Cooling and uplift patterns in the Lepontine Alps, south-central Switzerland, and an age of vertical movement on the Insubric fault line. *Contributions to Mineralogy and Petrology*, 92, 413–427.

Hurford, A.J. and Carter, A. (1991) The role of fission track dating in discrimination of provenance. In: *Developments in Sedimentary Provenance Studies* (ed. by A. Morton, S.P. Todd and P.D.W. Haughton), *Geological Society, London, Special Publication*, 57, 67–78.

Hurford, A.J., Flisch, M. and Jäger, E. (1989) Unravelling the thermotectonic evolution of the Alps: a contribution from fission track analysis and mica dating. In: *Alpine Tectonics* (ed. by M.P. Coward, D. Dietrich and R.G. Park), *Geological Society, London, Special Publication*, 45, 369–398.

Hurst, A. and Cartwright, J. (2007) Relevance of sand injectites to hydrocarbon exploration and production. In: *Sand Injectites: Implications for Hydrocarbon Exploration and Production* (ed. by A. Hurst and J. Cartwright), *American Association of Petroleum Geologists Memoir*, 87, 1–19.

Huuse, M. and Clausen, O.R. (2001) Morphology and origin of major Cenozoic sequence boundaries in the eastern North Sea Basin: top Eocene, near-top Oligocene and the mid-Miocene unconformity. *Basin Research*, 13, 17–41.

Huyghe, P., Galy, A., Mugnier, J.-L. and France-Lanord, C. (2001) Propagation of the thrust system and erosion in the Lesser Himalaya: geochemical and sedimentological evidence. *Geology*, 29, 1007–1010.

Illies, J.H. (1977) Ancient and recent rifting in the Rhine graben. *Geologie en Mijnbouw*, 56(4), 329–350.

Illies, J.H. and Greiner, G. (1978) Rhine graben and the Alpine system. *Geological Society of America Bulletin*, 89, 770–782.

Imbrie, J. (1982) Astronomical theory of the Pleistocene ice ages: a brief historical review. *Icarus*, 50, 408–422.

Ingersoll, R.V. (1988) Tectonics of sedimentary basins. *Geological Society of America Bulletin*, 100, 1704–1719.

Ingersoll, R.V. (2011) Tectonics of sedimentary basins, with revised nomenclature. In: *Tectonics of Sedimentary Basins, Second Edition* (ed. by C.J. Busby and A. Azor), Wiley-Blackwell.

Ingersoll, R.V. and Busby, C.J. (1995) Tectonics of sedimentary basins. In: *Tectonics of Sedimentary Basins* (ed. by C.J. Busby and R.V. Ingersoll), Blackwell Publishing Ltd., Oxford, 1–52, 579pp.

Ingersoll, R.V. and Suczek, C.A. (1979) Petrology and provenance of Neogene sand from Nicobar and Bengal fans, DSDP 211 and 218. *Journal of Sedimentary Petrology*, 49, 1217–1228.

Intergovernmental Panel on Climate Change (IPCC) (2005) *IPCC Special Report on Carbon Dioxide Capture and Storage, Prepared by Working Group III of the IPCC* (ed. by B. Metz, O. Davidson, H.C. de Connick, M. Loos and L.A. Meyer), Cambridge University Press, New York, 442pp.

Isaaks, B.L. (1988) Uplift of the central Andean plateau and bending of the Bolivian orocline. *Journal of Geophysical Research*, 93, 3211–3231.

Isacks, B. and Molnar, P. (1971) Distribution of stresses in the descending lithosphere from a global survey of focal-mechanism solutions of mantle earthquakes. *Reviews of Geophysics and Space Physics*, 9, 103–174.

Isacks, B., Oliver, J. and Sykes, L. (1969) Seismology and the new global tectonics. *Journal of Geophysical Research*, 73, 5855–5899.

Jackson, J. (2002) Strength of the continental lithosphere: time to abandon the jelly sandwich? *GSA Today*, 12, 4–10.

Jackson, J. (2005) Mountain roots and the survival of cratons. *The Harald Jeffreys Lecture 2005, Royal Astronomical Society, A and G*, 46, 2.33–2.36.

Jackson, J., McKenzie, D.P., Preistley, K. and Emmerson, B. (2008) New views on the structure and rheology of the lithosphere. *Journal of the Geological Society*, 165, 453–465.

Jackson, J., Norris, R. and Youngson, J. (1996) The structural evolution of active fault and fold systems in central Otago, New Zealand: evidence revealed by drainage patterns. *Journal of Structural Geology*, 18, 217–234.

Jacob, H. and Kuckelhorn, K. (1977) Das Inkohlungsprofil der Bohrung Miesbach 1 und seine erdolgeologische Interpretation. *Erdöl-Erdgas Z.*, 4, 115–124.

Jacome, M.I., Kusznir, N., Audemard, F. and Flint, S. (2003) Tectonostratigraphic evolution of the Maturin foreland basin – Eastern Venezuela. In: *The Circum-Gulf of Mexico and Caribbean – Hydrocarbon Habitats, Basin Formation, and Plate Tectonics* (ed. by C. Bartolini, R.T. Buffler and J. Blickwede), *American Association of Petroleum Geologists Memoir*, 79, 925–936.

Jamieson, R.A. and Beaumont, C. (1998) Orogeny and metamorphism – A model for deformation and pressure-temperature-time paths with applications to the central and southern Appalachians. *Tectonics*, 7, 417–445.

Jansa, L.F. and Wade, J.A. (1975) Geology of the continental margin off Nova Scotia and Newfoundland. In: *Offshore Geology of Eastern Canada* (ed. by W.J.M. Van der Linden and J.A. Wade), *Geological Survey of Canada*, 2, 51–105.

Jardine, D. and Wilshart, J.W. (1987) Carbonate reservoir description. In: *Reservoir Sedimentology* (ed. by R.W. Tillman and W.J. Weber), *Society of Economic Paleontologists and Mineralogists Special Publication*, 40, 129–152.

Jarvis, G.T. and McKenzie, D.P. (1980) Sedimentary basin formation with finite extension rates. *Earth and Planetary Science Letters*, 48, 42–52.

Jaupart, C. and Mareschal, J.-C. (2011) *Heat Generation and Transport in the Earth*, Cambridge University Press, Cambridge, 464pp.

Jeans, C.V., Wray, D.S., Merriman, R.J. and Fisher, M.J. (2000) Volcanogenic clays in Jurassic and Cretaceous strata of England and the North Sea basin. In: *Mineral Diagenesis and Reservoir Quality, the Way Forward* (ed. by D.C. Bain, P.L. Hall, H.F. Shaw and D.A. Spears), *Clay Minerals*, 35, 25–55.

Jennings, C.W. (compiler) (1975) Preliminary fault and geologic map of southern California. In: *San Andreas Fault in Southern California* (ed. by J.E. Crowell), *California Division of Mines and Geology Special Report*, 118, Scale 1:750,000.

Jerolmack, D.J. and Sadler, P. (2007) Transience and persistence in the depositional record of continental margins. *Journal of Geophysical Research – Earth Surface*, 112, F03S13.

Jervey, M.T. (1988) Quantitative geological modeling of siliciclastic rock sequences and their seismic expressions. In: *Sea Level Changes: An Integrated Approach* (ed. by C.K. Wilgus, B.S. Hastings, C.G.St.C. Kendall, H.W. Posamentier, C.A. Ross and J.C. Van Wagoner), *Society of Economic Paleontologists and Mineralogists Special Publication*, 42, 47–69.

Jessop, A.M. and Lewis, T. (1978) Heat flow and the heat generation in the Superior Province of the Canadian Shield. *Tectonophysics*, 50, 55–77.

Jiménez-Munt, I. and Platt, J.P. (2006) Influence of mantle dynamics on the topographic evolution of the Tibetan Plateau: results from numerical modeling. *Tectonics*, 25, TC6002.

Johnson, D.D. and Beaumont, C. (1995) Preliminary results from a planform kinematic model of orogen evolution, surface processes and the development of clastic foreland basin stratigraphy. In: *Stratigraphic Evolution of Foreland Basins* (ed. by S.L. Dorobek and G.M. Ross), *Society of Economic Paleontologists and Mineralogists Special Publication*, 52, 3–24.

Johnson, H.D. and Baldwin, C.T. (1986) Shallow siliciclastic seas. In: *Sedimentary Facies and Environments* (ed. by H.G. Reading), Blackwell Scientific Publications, Oxford, 229–282.

Johnson, J.G. and Murphy, M.A. (1984) Time-rock model for Siluro-Devonian continental shelf, western United States. *Geological Society of America Bulletin*, 95, 1349–1359.

Jolley, E.T., Turner, P., Williams, G.D., Hartley, A.J. and Flint, S. (1990) Sedimentological response of an alluvial system to Neogene thrust tectonics, Atacama Desert, northern Chile. *Journal of the Geological Society*, 147, 769–784.

Jones, C.H., Unruh, J.R. and Sonder, L.J. (1996) The role of gravitational potential energy in active deformation in the southeastern United States. *Nature*, 381 (6577), 37–41.

Jones, G., Fisher, G.J. and Knipe, R.J. (eds.) (1998) *Faulting, Fault Sealing and Fluid Flow in Hydrocarbon Reservoirs, Geological Society, London, Special Publication*, 147, 319pp.

Jones, M.A., Heller, P.L., Roca, E., Garcés, M. and Cabrera, L. (2004) Time lag of syntectonic sedimentation across an alluvial basin: theory and example from the Ebro Basin, Spain. *Basin Research*, 16, 467–488.

Jones, S.M., White, N. and Lovell, B. (2001) Cenozoic and Cretaceous transient uplift in the Porcupine Basin and its relationship to a mantle plume. In: *Petroleum Exploration of Ireland's Offshore Basins* (ed. by P.M. Shannon, P.D.W. Haughton and D.V. Corcoran), *Geological Society, London, Special Publication*, 188, 345–360.

Jordan, R.E. and Flemings, P.B. (1991) Large-scale stratigraphic architecture, eustatic variation, and unsteady tectonism – a theoretical evaluation. *Journal of Geophysical Research – Solid Earth and Planets*, 96, B4, 6681–6699.

Jordan, T.E. (1981) Thrust loads and foreland basin evolution, Cretaceous, western United States. *American Association of Petroleum Geologists Bulletin*, 65, 2506–2520.

Jordan, T.E. and Allmendinger, R.W. (1986) The Sierras Pampeanas of Argentina; a modern analogue of Rocky Mountain foreland deformation. *American Journal of Science*, 286, 737–764.

Jordan, T.E., Flemings, P.B. and Beer, J.A. (1988) Dating thrust-fault activity by use of foreland basin strata. In: *New Perspectives in Basin Analysis* (ed. by K.L. Kleinspehn and C. Paola), Springer-Verlag, New York, 307–330.

Jordan, T.H. (1979) Mineralogies, densities and seismic velocities of garnet lherzolites and their geophysical implications. In: *The Mantle Sample: Inclusions in Kimberlites and other Volcanics* (ed. by F.R. Boyd and H.O.A. Meyer), *Proceedings Second International Kimberlites Conference*, American Geophysical Union, Washington D.C., 2, 1–14.

Joyner, W.B. (1967) Basalt–eclogite transition as a cause for subsidence and uplift. *Journal of Geophysical Research*, 72, 4977–4998.

Junger, A. (1976) Tectonics of the southern California borderlands. In: *Aspects of the Geological History of the California Continental Borderland Miscellaneous Publication, 24* (ed. by D.G. Howell), American Association of Petroleum Geologists, Pacific Section, Bakersfield, California, 486–498.

Kaldi, J.G. (2010) Evaluating storage capacity in saline aquifers and depleted oil and gas fields: geological and engineering issues. *China-Australia Geological Storage of CO2 Workshop*, Canberra, Jan. 19–21.

Kaldi, J.G., Gibson-Poole, C.M. and Payenberg, T.H.D. (2009) Geological input to selection and evaluation of CO2 geosequestration sites. In: *Carbon Dioxide Sequestration in Geological Media – State of the Science* (ed. by M. Grobe, J.C. Pashin and R.L. Dodge), *American Association of Petroleum Geologists Studies in Geology*, 59, 5–16.

Kaminski, E. and Jaupart, C. (2000) Lithospheric structure beneath the Phanerozoic intracratonal basins of North America. *Earth and Planetary Sciences Letters*, 178, 139–149.

Kantorowicz, J.D. (1990) The influence of variations in illite morphology on the permeability of Middle Jurassic Brent Group sandstones, Cormorant Field, UK North Sea. *Marine and Petroleum Geology*, 7, 66–74.

Karáson, H. and van der Hilst, R.D. (2000) Constraints on mantle convection from seismic tomography. In: *The History and Dynamics of Global Plate Motions* (ed. by M.A. Richards, R.G. Gordon and R.D. van der Hilst), *American Geophysical Union, Geophysical Monograph*, 121, 277–288.

Karlstrom, K.E., Crow, R., Crossey, L.J., Coblentz, D. and Van Wijk, J.W. (2008) Model for tectonically driven incision of the younger than 6 Ma Grand Canyon. *Geology*, 36, 835–838.

Karlstrom, K.E., Crow, R.S., Peters, L., McIntosh, W., Raucci, J., Crossey, L.J., Umhoefer, P. and Dunbar, N. (2007) $^{40}Ar/^{39}Ar$ and field studies of Quaternary basalts in the Grand Canyon: quantifying the interaction of river incision and normal faulting across the western edge of the Colorado Plateau. *Geological Society of America Bulletin*, 119, 1283–1312.

Karner, G.D. (1986) Effects of lithospheric in-plane stress on sedimentary basin stratigraphy. *Tectonics*, 5, 573–588.

Karner, G.D. (1991) Sediment blanketing and the flexural strength of the extended continental lithosphere. *Basin Research*, 3, 177–185.

Karner, G.D. and Watts, A.B. (1983) Gravity anomalies and flexure of the lithosphere at mountain ranges. *Journal of Geophysical Research*, 88, 10449–10477.

Kartanegara, A.L., Baik, R.N. and Ibrahim, M.A. (1996) Volcanics oil-bearing in Indonesia. *American Association of Petroleum Geologists Bulletin*, 80, no. 13, A73.

Katz, B.J. and Mello, M.R. (2000) Petroleum systems of South Atlantic marginal basins: an overview. In: *Petroleum Systems of South Atlantic Margins* (ed. by M.R. Mello and B.J. Katz), *American Association of Petroleum Geologists Memoir*, 73, 1–13.

Katzman, R., ten Brink, U.S. and Lin, J. (1995) Three-dimensional modeling of pull-apart basins: implications for the tectonics of the Dead Sea Basin. *Journal of Geophysical Research*, 100, B4, 6295–6312.

Kaufman, P., Grotzinger, J.P. and McCormick, D.S. (1991) Depth-dependent diffusion algorithm for simulation of sedimentation in shallow marine depositional systems. In: *Sedimentary Modeling: Computer Simulations and Methods for Improved Parameter Definition* (ed. by E.K. Franseen, W.L. Watney, C.G.S.C. Kendall and W. Ross), *Kansas Geological Survey Bulletin*, 233, 489–508.

Kaus, B.J.P., Connolly, J.A.D., Podladchikov, Y.Y. and Schmalholz, S.M. (2005) Effect of mineral phase transitions on sedimentary basin subsidence and uplift. *Earth and Planetary Science Letters*, 233, 213–228.

Keen, C.E. (1985) The dynamics of rifting: deformation of the lithosphere by active and passive driving forces. *Geophysical Journal of the Royal Astronomical Society*, 80, 95–120.

Keen, C.E. (1987) Some important consequences of lithospheric extension. In: *Continental Extensional Tectonics* (ed. by M.P. Coward, J.F. Dewey and P.L. Hancock), *Geological Society, London, Special Publication*, 28, 67–73.

Keen, C.E. and Boutilier, R.R. (1995) Lithosphere–asthenosphere interactions below rifts. In: *Rifted Ocean–Continent Boundaries* (ed. by E. Banda *et al.*), Kluwer, Dordrecht, 17–30.

Keen, C.E. and Dehler, S.A. (1997) Extensional styles and gravity anomalies at rifted continental margins. Some North Atlantic examples. *Tectonics*, 16, 744–754.

Keen, C.E., Stockmal, G.S., Welsink, H., Quinlan, G. and Mudford, B. (1987) Deep crustal structure and evolution of the rifted margin northeast of Newfoundland. Results from Lithoprobe East. *Canadian Journal of Earth Sciences*, 24, 1537–1550.

Keller, E.A., Bonkowski, M.S., Korsch, R.J. and Shlemon, R.J. (1982) Tectonic geomorphology of the San Andreas fault zone in the southern Indio Hills, Coachella Valley, California. *Geological Society of America Bulletin*, 93, 46–56.

Kelley, P.A., Bissada, K.K., Burda, B.H., Elrod, L.W. and Pheifer, R.N. (1985) Petroleum generation potential of coals and organic-rich deposits: significance in Tertiary coal-rich basins. *Proceedings of the Indonesian Petroleum Association, 14th Annual Convention*, 1985.

Kempf, O. and Pfiffner, O.A. (2004) Early Tertiary evolution of the North Alpine Foreland Basin of the Swiss Alps and adjoining areas. *Basin Research*, 16, 549–567.

Kendall, C.G.S.C., Moore, P., Strobel, J., Cannon, R., Perlmutter, M., Bezdek, J. and Biswas, G. (1991a) Simulation of the sedimentary fill of basins. In: *Sedimentary Modeling: Computer Simulations and Methods for Improved Parameter Definition* (ed. by E.K. Franseen, W.L. Watney, C.G.S.C. Kendall and W. Ross), *Kansas Geological Survey Bulletin*, 233, 489–508.

Kendall, C.G.S.C., Strobel, J., Cannon, R., Bezdek, J. and Biswas, G. (1991b). The simulation of the sedimentary fill of basins. *Journal of Geophysical Research*, 96, 6911–6929.

Kennett, J.P. and Shackleton, N.J. (1976) Critical development in evolution of deep-sea waters 38 m.y. ago; oxygen isotopic evidence from deep-sea sediments. *Eos, Transactions, American Geophysical Union*, 57, 256.

Kerr, R.A. (1985) Plate tectonics goes back 2 billion years. *Science*, 230, 1364–1367.

Ketzer, J.M., Morad, S. and Amorosi, A. (2003) Predictive diagenetic clay-mineral distribution in siliciclastic rocks within a sequence stratigraphic framework. In: *Clay Mineral Cements in Sandstones* (ed. by R.H. Wordon and S. Morad), *International Association of Sedimentologists Special Publication*, Blackwell Publishing, 34, 43–61.

Khain, V.Y. (1992) The role of rifting in the evolution of the Earth's crust. *Tectonophysics*, 215, 1–7.

Kim, W., Connell, S.D., Steel, E., Smith, G.A. and Paola, C. (2011) Mass-balance control on the interaction of axial and transverse channel systems. *Geology*, 39, 611–614.

Kim, W., Paola, C., Swenson, J.B. and Voller, V.R. (2006) Shoreline response to autogenic processes of sediment storage and release in the fluvial system. *Journal of Geophysical Research*, 111, F04013.

King, B.C. and Williams, L.A.J. (1976) The East African rift system. In: *Geodynamics; Progress and Prospects* (ed. by C.L. Drake), American Geophysical Union, Washington D.C., 63–74.

King, G.C.P., Stein, R.S. and Rundle, J.B. (1988) The growth of geological structure by repeated earthquakes. 1. Conceptual framework. *Journal of Geophysical Research*, 93, 13307–13318.

King, S.D. (2007) Hotspots and edge-driven convection. *Geology*, 35, 223–226.

King, S.D. and Anderson, D.L. (1995) An alternative mechanism of flood basalt formation. *Earth and Planetary Science Letters*, 136, 269–279.

King, S.D. and Anderson, D.L. (1998) Edge-driven convection. *Earth and Planetary Science Letters*, 160, 289–296.

Kinghorn, R.R.F. (1983) *An Introduction to the Physics and Chemistry of Petroleum*, John Wiley & Sons Ltd., Chichester, 420 pp.

Kingston, D.R., Dishroon, C.P. and Williams, P.A. (1983a) Global basin classification. *American Association of Petroleum Geologists Bulletin*, 67, 2175–2193.

Kingston, D.R., Dishroon, C.P. and Williams, P.A. (1983b) Hydrocarbon plays and global basin classification. *American Association of Petroleum Geologists Bulletin*, 67, 2194–2198.

Kinsman, D.J.J. (1975) Rift valley basins and sedimentary history of trailing continental margins. In: *Petroleum and Global Tectonics* (ed. by A.G. Fischer and S. Judson), Princeton University Press, 83–126.

Kirby, E. and Whipple, K.X. (2001) Quantifying differential rock-uplift rates via stream profile analysis. *Geology*, 29, 415–418.

Kirby, J.F. and Swain, C.J. (2009) A reassessment of spectral Te estimation in continental interiors: the case of North America. *Journal of Geophysical Research – Solid Earth*, 114, B08401.

Kirk, R.H. (1980) Statfjord field: a North Sea giant. In: *Giant Oil and Gas Fields of the Decade 1968–1978* (ed. by M.T. Halbouty), *American Association of Petroleum Geologists Memoir*, 30, 95–116.

Kirkland, B.L., Dickson, J.A.D., Wood, R.A. and Land, L.S. (1998) Microbialite and microstratigraphy: the origin of encrustations in the middle and upper Capitan Formation, Guadalupe Mountains, Texas and New Mexico. *Journal of Sedimentary Research*, 68, 956–969.

Klemme, H.D. (1980) Petroleum basins: classification and characteristics. *Journal of Petroleum Geology*, 3, 187–207.

Klemperer, S.L. (1988) Crustal thinning and nature of extension in the northern North Sea from deep seismic reflection profiling. *Tectonics*, 7, 803–821.

Kohlstedt, D.L., Evans, B. and Mackwell, S.J. (1995) Strength of the lithosphere; constraints imposed by laboratory experiments. *Journal of Geophysical Research*, 100, 17587–17602.

Koide, H. and Bhattacharji, S. (1977) Geometric patterns of active strike-slip faults and their significance as indicators for areas of energy release. In: *Energetics of Geological Processes* (ed. by S.K. Saxena), Springer-Verlag, New York, 46–66.

Kolata, D.R. and Nelson, W.J. (1990) Tectonic history of the Illinois Basin. In: *Interior Cratonic Basins* (ed. by M.W. Leighton, D.R. Kolata, D.F. Oltz and J.J. Eidel), *American Association of Petroleum Geologists Memoir*, 51, 263–285.

Kominz, M.A. (1984) Oceanic ridge volumes and sea level change: an error analysis. In: *Interregional Unconformities and Hydrocarbon Accumulation* (ed. by J. Schlee), *American Association of Petroleum Geologists Memoir*, 36, 109–127.

Kominz, M.A. and Pekar, St.F. (2001) Oligocene eustasy from two-dimensional sequence stratigraphic backstripping. *Geological Society of America Bulletin*, 113, 291–304.

Kominz, M.A., Miller, K.G. and Browning, J.V. (1998) Long-term and short-term global Cenozoic sea-level estimates. *Geology*, 26, 311–314.

Koning, T. (2003) Oil and gas production from basement reservoirs: examples from Indonesia, USA and Venezuela. In: *Hydrocarbons in Crystalline Rocks* (ed. by N. Petford and K.J.W. McCaffrey), *Geological Society, London, Special Publication*, 214, 83–92.

Konstantinovskaya, E. and Malavieille, J. (2005) Erosion and exhumation in accretionary orogens: experimental and geological approaches. *Geochemistry, Geophysics, Geosystems*, 6, Q02006.

Kooi, H. and Beaumont, C. (1994) Escarpment evolution on high-elevation rifted margins: insights derived from a surface processes model that combines diffusion, advection and reaction. *Journal of Geophysical Research*, 99, 12191–12209.

Kooi, H. and Beaumont, C. (1996) Large-scale geomorphology: classical concepts reconciled and integrated with contemporary ideas via a surface processes model. *Journal of Geophysical Research*, 101, 3361–3386.

Kooi, H., Cloetingh, S. and Burrus, J. (1992) Lithospheric necking and regional isostasy at extensional basins; 1. Subsidence and gravity modeling with an application to the Gulf of Lions margin (SE France). *Journal of Geophysical Research*, 97, 17553–17571.

Koons, P.O. (1989) The topographic evolution of collisional mountain belts: a numerical look at the Southern Alps, New Zealand. *American Journal of Science*, 289, 1041–1069.

Koons, P.O. (1990) Two-sided orogen: collision and erosion from the sandbox to the Southern Alps, New Zealand. *Geology*, 18, 679–682.

Koons, P.O. (1995) Modelling the topographic evolution of collisional belts. *Annual Reviews of Earth and Planetary Sciences*, 23, 375–408.

Koons, P.O., Norris, R.J., Craw, D. and Cooper, A.F. (2003) Influence of exhumation on the structural evolution of transpressional plate boundaries: an example from the Southern Alps, New Zealand. *Geology*, 31, 3–6.

Kreemer, C., Holt, W.E. and Haines, A.J. (2003) An integrated global model of present-day plate motions and plate boundary deformations. *Geophysical Journal International*, 154, 8–34.

Kronen, J.D. Jr. and Glenn, C.R. (2000) Pristine to reworked verdine: keys to sequence stratigraphy in mixed carbonate-siliciclastic forereef sediments (Great Barrier Reef). In: *Marine Authigenesis: From Global to Microbial* (ed. by C.R. Glenn, J. Lucas and L. Lucas), *Society of Economic Paleontologists and Mineralogists Special Publication*, 65, 387–403.

Kübler, B., Pittion, J.-L., Héroux, Y., Charolais, J. and Weidmann, M. (1979) Sur le pouvoir réflecteur de la vitrinite dans quelques roches du Jura, de la Molasse et des Nappes préalpines, helvétiques et penniques (Suisse occidentale, Haute Savoie). *Eclogae Geologicae Helvetiae*, 72, 347–373.

Kuehl, S.A., Levy, B.M., Moore, W.S. and Allison, M.A. (1997) Subaqueous delta of the Ganges-Brahmaputra river system. *Marine Geology*, 144, 81–96.

Kuhlemann, J., Frisch, W., Szekeley, B., Dunkl, I. and Kazmer, M. (2002) Post-collisional sediment budget history of the Alps: tectonic versus climatic control. *International Journal of Earth Sciences*, 91, 818–837.

Kupecz, J., Gluyas, J.G. and Bloch, S. (1997) *Reservoir Quality Prediction in Sandstones and Carbonates. American Association of Petroleum Geologists Memoir*, 69.

Kusznir, N.J. (1982) Lithosphere response to externally and internally derived stresses: a viscoelastic stress guide with amplification. *Geophysical Journal of the Royal Astronomical Society*, 70, 399–414.

Kusznir, N.J. and Egan, S.S. (1990) Simple-shear and pure-shear models of extensional sedimentary basin formation: application to the Jeanne d'Arc Basin, Grand Banks of Newfoundland. In: *Extensional Tectonics of the North Atlantic Margins* (ed. by A.J. Tankard and H.R. Balkwill), *American Association of Petroleum Geologists Memoir*, 46, 305–322.

Kusznir, N.J. and Morley, C. (1990) The Lake Tanganyika rift, East Africa. Application of the flexural cantilever model of continental extension. *Geophysical Journal International*, 101, 275.

Kusznir, N.J. and Park, R.G. (1987) The extensional strength of the continental lithosphere: its dependence on geothermal gradient, and crustal composition and thickness. In: *Continental Extensional Tectonics* (ed. by M.P. Coward, J.F. Dewey and P.L. Hancock), *Geological Society, London, Special Publication*, 28, 35–52.

Kusznir, N.J. and Ziegler, P.A. (1992) The mechanics of continental extension and sedimentary basin formation; a simple-shear/pure-shear flexural cantilever model. *Tectonophysics*, 215, 117–131.

Kusznir, N.J., Karner, G.D. and Egan, S. (1987) Geometric, thermal and isostatic consequences of detachments in continental lithosphere extension and basin formation. In: *Sedimentary Basins and Basin-Forming Mechanisms* (ed. by C. Beaumont and A.J. Tankard), *Canadian Society of Petroleum Geologists Memoirs*, 12, 185–203.

Kusznir, N.J., Marsden, G. and Egan, S.S. (1991) A flexural-cantilever simple-shear/pure-shear model of continental lithosphere extension: applications to the Jeanne d'Arc Basin, Grand Banks and Viking Graben, North Sea. In: *The Geometry of Normal Faults, Geological Society, London, Special Publication*, 56, 41–60.

Lachenbruch, A.H. and Sass, J.H. (1978) Models of extending lithosphere and heat flow in the Basin and Range province. In: *Cenozoic Tectonics and Regional Geophysics of the Western Cordillera, Geological Society of America Memoir*, 152, 209–250.

Lachenbruch, A.H. and Sass, J.H. (1980) Heat flow and energetics of the San Andreas fault zone. *Journal of Geophysical Research*, 85, 6185–6222.

Lachenbruch, A.H. and Sass, J.H. (1988) The stress-heat flow paradox and preliminary thermal results from Cajon Pass. *Geophysical Research Letters*, 15, 981–984.

Lakshmanan, C.C., Bennett, M.L. and White, N. (1991) Implications of multiplicity in kinetic parameters to petroleum exploration; distributed activation energy models. *Energy and Fuels*, 5, 110–117.

Lal, D. (1991) Cosmic ray labeling of erosion surfaces. *In situ* nuclide production rates and erosion. *Earth and Planetary Science Letters*, 104, 424–439.

Lamb, C.F. (1980) Painter reservoir field: giant in the Wyoming thrust belt. *American Association of Petroleum Geologists Bulletin*, 64, 638–644.

Lambeck, K. (1983) Structure and evolution of intracratonic basins in central Australia. *Geophysical Journal of the Royal Astronomical Society*, 74, 843–886.

Larsen, P.H. (1988) Relay structures in a Lower Permian basement-involved extension system, East Greenland. *Journal of Structural Geology*, 10, 3–8.

Lash, G.G. (1987) Longitudinal petrographic variations in a Middle Ordovician trench deposit, central Appalachian orogen. *Sedimentology*, 34, 227–235.

Laske, G., and Masters, G. (1997) A global digital map of sediment thickness. *Eos, Transactions, American Geophysical Union*, 78, F483.

Laslett, G.M., Green, P.F., Duddy, I.R. and Gleadow, A.J.W. (1987) Thermal annealing of fission tracks in apatite, 2. A quantitative analysis. *Chemical Geology*, 65, 1–13.

Latin, D. and White, N. (1990) Generating melt during lithospheric extensions: pure shear vs. simple shear. *Geology*, 18, 327–331.

Lavé, J. and Avouac, J.P. (2001) Fluvial incision and tectonic uplift across the Himalayas of central Nepal. *Journal of Geophysical Research*, 106, 25561–26591.

Lavier, L.L. and Steckler, M.S. (1997) The effect of sedimentary cover on the flexural strength of continental lithosphere. *Nature*, 389, 476–479.

Lawrence, D.T., Doyle, M. and Aigner, T. (1990) Stratigraphic simulation of sedimentary basins: concepts and calibration. *American Association of Petroleum Geologists Bulletin*, 74, 273–295.

Le Pichon, X. and Francheteau, J. (1978) A plate tectonic analysis of the Red Sea – Gulf of Aden area. *Tectonophysics*, 46, 369–406.

Le Pichon, X. and Sibuet, J.-C. (1981) Passive margins: a model of formation. *Journal of Geophysical Research*, 86, 3708–3720.

Le Pichon, X., Fournier, M. and Jolivet, L. (1992) Kinematics, topography, shortening, and extrusion in the India–Eurasia collision. *Tectonics*, 11, 1085–1098.

Le Stunff, Y. and Ricard, Y. (1995) Topography and geoid due to lithospheric mass anomalies. *Geophysical Journal International*, 122, 982–990.

Lee, Y.-H. and Harbaugh, J.W. (1992) Stanford's Sedsim project: dynamic three-dimensional simulation of geological processes that affect clastic sediments. *Computer Graphics in Geology*, 41, 113–127.

Leeder, M.R. (1999) *Sedimentology and Sedimentary Basins: From Turbulence to Tectonics*, Blackwell Publishing Ltd., Oxford.

Leeder, M.R. (2011) *Sedimentology and Sedimentary Basins: From Turbulence to Tectonics, Second Edition*, Wiley-Blackwell, 768pp.

Leeder, M.R. and Gawthorpe, R.L. (1987) Sedimentary models for extensional tilt-block/half-graben basins. In: *Continental Extensional Tectonics* (ed. by M.P. Coward, J.F. Dewey and P.L. Hancock), *Geological Society, London, Special Publication*, 28, 139–152.

Leeder, M.R. and Mack, G.H. (2001) Lateral erosion ('toe-cutting') of alluvial fans by axial rivers: implications for basin analysis and architecture. *Journal of the Geological Society*, 158, 885–893.

Leeder, M.R., Harris, T. and Kirkby, M.J. (1998) Sediment supply and climate change: implications for basin stratigraphy. *Basin Research*, 10, 7–18.

Leeds, A.R., Knopoff, L., and Kausel, E.G., (1974) Variations of upper mantle structure under the Pacific Ocean. *Science*, 186, 141–143.

Lees, A. and Buller, A.T. (1972) Modern temperate-water and warm-water shelf carbonate sediments contrasted. *Marine Geology*, 13, M67–M73.

Leighton, M.W. and Kolata, D.R. (1990) Selected interior cratonic basins and their place in the scheme of global tectonics: a synthesis. In: *Interior Cratonic Basins, American Association of Petroleum Geologists Memoir*, 51, 729–797.

Leighton, M.W., Kolata, D.R., Oltz, D.R. and Eidel, J.J. (1990) *Interior Cratonic Basins, American Association of Petroleum Geologists Memoir*, 51, 797pp.

Lemoine, F.G., Kenyon, S.C., Factor, J.K., Trimmer, R.G., Pavlis, N.K. Chinn, D.S., Cox, S.M. Klosko, S.M., Luthcke, S.B., Torrence, M.H. Wang, Y.M. Williamson, R.G., Pavlis, E.C., Rapp, R.H. and Olson, T.R. (1998) The development of the joint NASA GSFC and the National Imagery and Mapping Agency (NIMA) geopotential model EGM96. *NASA Technical Paper*.

Lenardic, A. and Moresi, L.N. (1999) Some thoughts on the stability of cratonic lithosphere: effects of buoyancy and viscosity. *Journal of Geophysical Research*, 104, 12747–12758.

Lerch, F.J., Klosko, S.M. and Patch, G.B. (1983) A refined gravity model from LAGEOS (GEM-L2), *NASA Technical Memorandum*, 84986.

Lerche, I., Dromgoole, E., Kendall, C.G.S.C., Walter, L.M. and Scaturo, D. (1987) Geometry of carbonate bodies: a quantitative investigation of factors influencing their evolution. *Carbonates and Evaporites*, 2, 15–42.

Lerche, I., Yarzab, R.F. and Kendell, C.G. St. (1984) Determination of paleoheat flux from vitrinite reflectance data. *American Association of Petroleum Geologists Bulletin*, 69, 1709–1717.

Leythaeuser, D., Schaefer, R.G. and Yukler, A. (1982) Role of diffusion in primary migration of hydrocarbons. *American Association of Petroleum Geologists Bulletin*, 66, 408–429.

Li, Y. and Yang, J. (1998) Tectonic geomorphology in the Hexi Corridor, north-west China. *Basin Research*, 10, 345–352.

Lin, J.C. (2000) Morphotectonic evolution of Taiwan. In: *Geomorphology and Global Tectonics* (ed. by M.A. Summerfield), John Wiley & Sons Ltd., 135–146.

Lindsay, J.F. (2002) Supersequences, superbasins, supercontinents – evidence from the Neoproterozoic–Early Palaeozoic basins of central Australia. *Basin Research*, 14, 207–223.

Link, M.H. (1982) Provenance, palaeocurrents and palaeogeography of Ridge Basin, southern California. In: *Geologic History of Ridge Basin, Southern California* (ed. by J.C. Crowell and M.H. Link), Society of Economic Paleontologists and Mineralogists, Pacific Section, 265–276.

Link, M.H. and Osborne, R.H. (1978) Lacustrine facies in the Pliocene Ridge Basin Group: Ridge Basin, California. In: *Modern and Ancient Lake Sediments* (ed. by A. Matter and M.E. Tucker), *International Association of Sedimentology Special Publication*, 2, Blackwell Scientific Publications, Oxford, 169–187.

Lisker, F., Ventura, B. and Glasmacher, U.A. (2009) Apatite thermochronology in modern geology. In: *Thermochronological Methods: From Paleotemperature Constraints to Landscape Evolution Models* (ed. by F. Lisker, B. Ventura and U.A. Glasmacher), *Geological Society, London, Special Publication*, 324, 1–23.

Lister, C.R.B., Sclater, J.G., Davis, E.E., Villinger, H. and Nagihara, S. (1990) Heat flow maintained in ocean basins of great age: investigations in the north-equatorial west Pacific. *Geophysical Journal International*, 102, 603–630.

Lister, G.S. and Davis, G.A. (1989) The origin of metamorphic core complexes and detachment faults formed during Tertiary continental extension in the northern Colorado River region, USA. *Journal of Structural Geology*, 11, 65–94.

Lister, G.S., Banga, G. and Feenstra, A. (1984) Metamorphic core complexes of Cordilleran type in the Cyclades, Aegean Sea, Greece. *Geology*, 12, 221–225.

Lister, G.S., Etheridge, M.A. and Symonds, P.A. (1986) Detachment faulting and the evolution of passive continental margins. *Geology*, 14, 890–892.

Lithgow-Bertelloni, C. and Gurnis, M. (1997) Cenozoic subsidence and uplift of continents from time-varying dynamic topography. *Geology*, 25, 735–738.

Lithgow-Bertelloni, C. and Richards, M.A. (1998) The dynamics of Mesozoic and Cenozoic plate motions. *Journal of Geophysical Research*, 36, 27–78.

Lithgow-Bertelloni, C. and Silver, P. (1998) Dynamic topography, plate driving forces and the African superswell. *Nature*, 395, 269–272.

Livingstone, D.A. (1963) Chemical composition of rivers and lakes. Data of geochemistry. *US Geological Survey Professional Paper*, 440-G, 1–64.

Lonergan, L. and Johnson, C. (1988) Reconstructing orogenic exhumation histories using synorogenic detrital zircons and apatites: an example from the Betic Cordillera, SE Spain. *Basin Research*, 10, 353–364.

Lonergan, L., Borlandelli, C., Taylor, A., Quine, M. and Flanagan, K. (2007) The three-dimensional geometry of sandstone injection complexes in the Gryphon field, United Kingdom North Sea. In: *Sand Injectites: Implications for Hydrocarbon Exploration and Production* (ed. by A. Hurst and J. Cartwright), *American Association of Petroleum Geologists Memoir*, 87, 103–112.

Longley, I.M., Buessenschuett, C., Clydsdale, L., Cubitt, C.J., Davis, R.C., Johnson, M.K., Marshall, N.M., Murray, A.P., Somerville, R., Spry, T.B. and Thompson, N.B. (2003) The Northwest Shelf of Australia – A Woodside perspective. In: *The Sedimentary Basins of Western Australia 3* (ed. by M. Keep and S.J. Moss), *Proceedings of the Petroleum Exploration Society of Australia Symposium*, Perth, West Australia.

Lopatin, N.V. (1971) Temperature and geologic time as factors in coalification (in Russian). *Akad. Nauk SSSR Izvestiya, Seriya Geologicheskaya*, 3, 95–106.

Lopez, J.A. (1990) Structural styles of growth faults in the US Gulf Coast Basin. In: *Classic Petroleum Provinces* (ed. by J. Brooks). *Geological Society, London, Special Publication*, 50, 203–219.

Lopez-Blanco, M. (2002) Sediment response to thrusting and fold growing on the SE margin of the Ebro Basin (Palaeogene, NE Spain). *Sedimentary Geology*, 146, 133–154.

Lorenz, J.C., Farrell, H.E., Hanks, C.L., Rizer, W.D. and Sonnenfeld, M.D. (1997) Characteristics of natural fractures in carbonate strata. In: *Carbonate Seismology* (ed. by I. Palaz and K.J. Marfut), *Geophysical Developments Series*, 6, Society of Exploration Geophysicists, 179.

Loucks, R.G. and Anderson, J.H. (1985) Depositional facies, diagenetic terranes, and porosity development in Lower Ordovician Ellenburger Dolomite, Puckett Field, west Texas. In: *Carbonate Petroleum Reservoirs* (ed. by P.O. Roehl and P.W. Choquette), Springer-Verlag, New York, 19–37.

Louden, K.E. and Chian, D. (1999) The deep structure of non-volcanic rifted continental margins. *Philosophical Transactions of the Royal Society, London*, A357, 767–804.

Loutit, T.S., Hardenbol, J., Vail, P.R. and Baum, G.R. (1988) Condensed sections: the key to age dating of continental margin sequences. In: *Sea Level Changes: An Integrated Approach* (ed. by C.K. Wilgus, B.S. Hastings, C.G.St.C. Kendall, H.W. Posamentier, J.C. van Wagoner and C.A. Ross), *Society of Economic Paleontologists and Mineralogists Special Publication*, 42, 183–213.

Lowell, J.D. and Genik, G.J. (1972) Sea floor spreading and structural evolution of southern Red Sea. *American Association of Petroleum Geologists Bulletin*, 56, 247–259.

Lowry, A.R. and Smith, R.B. (1994) Flexural rigidity of the Basin and Range–Colorado Plateau–Rocky Mountain transition from coherence analysis of gravity and topography. *Journal of Geophysical Research*, 99, B10, 20123–20140.

Lucia, F. (1983) Petrophysical parameters estimated from visual descriptions of carbonate rocks: a field classification of carbonate pore space. *Journal of Petroleum Technology*, 35, 629–637.

Lucia, F.J. (1995) Rock-fabric/petrophysical classification of carbonate pore space for reservoir characterization. *American Association of Petroleum Geologists Bulletin*, 79, 1275–1300.

Lucia, F.J. (1999) *Carbonate Reservoir Characterization*, Springer-Verlag, Berlin.

Lucia, F.J. and Major, R.P. (1994) Porosity evolution through hypersaline reflux dolomitization. In: *Dolomites* (ed. by B. Purser, M. Tucker and D. Zenger), *International Association of Sedimentologists Special Publication*, 21, 325–341.

Luheshi, M.N. and Jackson, D. (1986) Conductive and convective heat transfer in sedimentary basins. In: *Thermal Modeling in Sedimentary Basins* (ed. by J. Burrus), *1st IFP Exploration Research Conference*, Carcans, France, June 3–7, 1985, Editions Technip, Paris, 219–234, 600pp.

Lundin, E.R. and Doré, A.G. (1997) A tectonic model for the Norwegian passive margin with implications for the NE Atlantic Early Cretaceous to break-up. *Journal of the Geological Society*, 154, 545–550.

Lvovich, M.I., Karasik, G.Ya., Bratseva, N.L., Medvedeva, G.P. and Maleshko, A.V. (1991) *Contemporary Intensity of the World Land Intracontinental Erosion*, USSR Academy of Sciences, Moscow.

Lynch, H.D. and Morgan, P. (1987) The tensile strength of the lithosphere and the localization of extension. In: *Continental Extensional Tectonics* (ed. by M.P. Coward, J.F. Dewey and P.L. Hancock), *Geological Society, London, Special Publication*, 28, 53–65.

Lynch, H.D. and Morgan, P. (1990) Finite-element models of continental extension. *Tectonophysics*, 174, 115–135.

Lyon-Caen, H. and Molnar, P. (1985) Gravity anomalies, flexure of the Indian plate, and the structure, support and evolution of the Himalaya and Ganga Basin. *Tectonics*, 4, 513–538.

Mack, G.H. and Seager, W.R. (1990) Tectonic control on facies distribution of the Camp Rice and Palomas Formations (Pliocene–Pleistocene) in the southern Rio Grande Rift. *Geological Society of America Bulletin*, 102, 45–53.

Mackenzie, A.S. and McKenzie, D. (1983) Isomerization and aromatization by hydrocarbons in sedimentary basins formed by extension. *Geological Magazine*, 120, 417–528.

Mackenzie, A.S. and Quigley, T.M. (1988) Principles of geochemical prospect appraisal. *American Association of Petroleum Geologists Bulletin*, 72, 399–415.

Mackenzie, A.S., Price, I., Leythaeuser, D., Muller, P., Radke, M. and Schaefer, R.G. (1987) The expulsion of petroleum from Kimmeridge Clay source rocks in the area of the Brae Oilfield, UK continental shelf. In: *Petroleum Geology of North West Europe* (ed. by J. Brooks and K. Glennie), Graham and Trotman, London, 865–877.

Mackey, S.D. and Bridge, J.S. (1995) Three-dimensional model of alluvial stratigraphy: theory and application. *Journal of Sedimentary Research*, 65B, 7–31.

Mackwell, M.J., Zimmerman, M.E. and Kohlstedt, D.L. (1998) High temperature deformation of dry diabase with application to the tectonics of Venus. *Journal of Geophysical Research*, 103, 975–984.

Magara, K. (1976) Thickness of removed sedimentary rocks, paleopore pressure and paleotemperatures, southwestern part of western Canada Basin. *American Association of Petroleum Geologists Bulletin*, 60, 554–565.

Maggi, A., Jackson, J.A., McKenzie, D. and Preistley, K. (2000) Earthquake focal depths, effective elastic thickness, and the strength of the continental lithosphere. *Geology*, 28, 495–498.

Magoon, L.B. and Dow, W.G. (1994) The petroleum system. In: *The Petroleum System – From Source to Trap* (ed. by L.B. Magoon and W.G. Dow), *American Association of Petroleum Geologists Memoir*, 60, 3–24.

Majorowicz, J.A. and Jessop, A.M. (1981) Present heat flow and a preliminary geothermal history of the central Prairies Basin, Canada. *Geothermics*, 10, 81–93.

Majorowicz, J.A., Jones, F.W., Lam, H.L. and Jessop, A.M. (1984) The variability of heat flow both regional and with depth in southern Alberta, Canada: effect of groundwater flow? *Tectonophysics*, 106, 1–29.

Malavieille, J. (1984) Modélisation expérimentale des chevauchements imbriqués: application aux chaînes des montagnes. *Société Géologique de France, Bulletin*, 26, 129–138.

Malinverno, A. (1997) On the power-law size distribution of turbidite beds. *Basin Research*, 9, 263–274.

Malinverno, A. and Ryan, W.B.F. (1986) Extension in the Tyrrhenian Sea and shortening in the Apennines as result of arc migration driven by sinking of the lithosphere. *Tectonics*, 5, 227–245.

Malmon, D.V., Dunne, T. and Reneau, S.L. (2003) Stochastic theory of particle trajectories through alluvial valley floors. *Journal of Geology*, 111, 525–542.

Mancktelow, N.S. (1985) The Simplon Line: a major displacement zone in the western Lepontine Alps. *Eclogae Geologicae Helvetiae*, 78, 73–96.

Mancktelow, N.S. and Grasemann, B. (1997) Time-dependent effects of heat advection and topography on cooling histories during erosion. *Tectonophysics*, 270, 167–195.

Mange, M.A. and Maurer, H.F.W. (1991) *Heavy Minerals in Colour*, Chapman and Hall, London, 147pp.

Mann, C.D. and Vita-Finzi, C. (1988) Holocene serial folding in the Zagros. In: Gondwana and Tethys (ed. by M. Audley-Charles and A. Hallam), *Geological Society, London, Special Publication*, 37, 51–59.

Mann, P., Draper, G. and Burke, K. (1985) Neotectonics of a strike-slip restraining bend system, Jamaica. In: *Strike-Slip Deformation, Basin Formation, and Sedimentation* (ed. by K.T. Biddle and N. Christie-Blick), *Society of Economic Paleontologists and Mineralogists Special Publication*, 37, 211–226.

Mann, P., Hempton, M.R., Bradley, D.C. and Burke, K. (1983) Development of pull-apart basins. *Journal of Geology*, 91, 529–554.

Manspeizer, W. (1985) The Dead Sea rift: impact of climate and tectonism on Pleistocene and Holocene sedimentation. In: *Strike-Slip Deformation, Basin Formation, and Sedimentation* (ed. by K.T. Biddle and N. Christie-Blick), *Society of Economic Paleontologists and Mineralogists Special Publication*, 37, 143–158.

Marcelja, S. (2010) The timescale and extent of thermal expansion of the global ocean due to climate change. *Ocean Science*, 6, 179–184.

Markello, J.R., Koepnick, R.B., Waite, L.E. and Collins, J.L. (2008) The Carbonate Analogs Through Time (CATT) hypothesis and the global atlas of carbonate fields: a systematic and predictive look at Phanerozoic carbonate systems. In: *Controls on Carbonate Platform and Reef Development* (ed. by J. Lukasik and J.A. Simo), *Society for Sedimentary Geology Special Publication*, 89, 15–46.

Marr, J.G., Swenson, J.B., Paola, C. and Voller, V.R. (2000) A two-diffusion model of fluvial stratigraphy in closed depositional basins. *Basin Research*, 12, 381–398.

Marsaglia, K.M. (1995) Interarc and backarc basins. In: *Tectonics of Sedimentary Basins* (ed. by C. Busby and R. Ingersoll), Blackwell Publishing Ltd., Oxford, 299–329.

Marsden, G., Yielding, G., Roberts, A.M. and Kusznir, N.J. (1990) Application of a flexural cantilever simple-shear/pure-shear model of continental lithosphere to the formation of the northern North Sea Basin. In: *Tectonic Evolution of the North Sea Rifts* (ed. by D.J. Blundell and A.D. Gibbs), Oxford University Press, 236–257.

Martin, J., Palola, C., Abreu, V., Neal, J. and Sheets, B. (2009) Sequence stratigraphy of experimental strata under known conditions of differential subsidence and variable base level. *American Association of Petroleum Geologists Bulletin*, 93, 503–533.

Marton, L.G., Tari, G.C. and Lehmann, C.T. (2000) Evolution of the Angolan passive margin, West Africa, with emphasis on post-salt structural styles. In: *Atlantic Rifts and Continental Margins* (ed. by M. Webster and M. Talwani), American Geophysical Union, Geophysical Monograph Series 115, 129–149.

Masson, D.G. (1996) Catastrophic collapse of the volcanic island of Hierro 15 ka ago and the history of landslides in the Canary Islands. *Geology*, 24, 231–234.

Massonnet, D., Rossi, M., Carmona, C., Adragna, F., Peltzer, G., Feigl, K. and Rabaute, T. (1993) The displacement field of the Landers earthquake mapped by radar interferometry. *Nature*, 364, 138–142.

Matter, A., Homewood, P., Caron, C., van Stuijvenberg, J., Weidmann, M. and Winkler, W. (1980) Flysch and molasse of western and central Switzerland. In: *Geology of Switzerland, a Guide Book* (ed. by R. Trümpy), Schweizerische Geologische Kommission, 261–293.

Matthews, R.K. (1974) *Dynamic Stratigraphy*, Prentice Hall, New Jersey, 370pp.

Mayall, M. Lonergan, L., Bowman, A., James, S. *et al.* (2010) The response of turbidite slope channels to growth-induced seabed topography. *American Association of Petroleum Geologists Bulletin*, 94, 1011–1030.

Mayuga, M.N. (1970) Geology and development of California's giant Wilmington oil field. In: *Geology of Giant Petroleum Fields*, American Association of Petroleum Geologists Memoir, 14, 158–184.

McClay, K. and Dooley, T. (1995) Analog models of pull-apart basins. *Geology*, 23, 711–714.

McConnell, R.B. (1977) East African rift system dynamics in view of Mesozoic apparent polar wander. *Journal of the Geological Society*, 134, 33–39.

McConnell, R.B. (1980) A resurgent taphrogenic lineament of Precambrian origin in eastern Africa. *Journal of the Geological Society*, 137, 483–489.

McConnell, R.K. (1968) Viscosity of the mantle from relaxation time spectra of isostatic adjustment. *Journal of Geophysical Research*, 73, 7089–7105.

McGann, G.J., Green, S.C.H., Harker, S.D. and Romani, R.S. (1991) The Scapa Field, Block 14/19, UK North Sea. In: *United Kingdom Oil and Gas Fields* (ed. by I.L. Abbotts), American Association of Petroleum Geologists, Tulsa, Oklahoma, 27–58.

McKenzie, D. and Bickle, M.J. (1988) The volume and composition of melt generated by extension of the lithosphere. *Journal of Petrology*, 29, 625–679.

McKenzie, D., Watts, A., Parsons, B. and Roufosse, M. (1980) Planform of mantle convection beneath the Pacific Ocean. *Nature*, 288, 442–446.

McKenzie, D.P. (1967) Some remarks on heat flow and gravity anomalies. *Journal of Geophysical Research*, 72, 6261–6273.

McKenzie, D.P. (1978a) Some remarks on the development of sedimentary basins. *Earth Planetary Science Letters*, 40, 25–32.

McKenzie, D.P. (1978b) Active tectonics of the Alpine–Himalayan belt: the Aegean and surrounding regions. *Geophysical Journal of the Royal Astronomical Society*, 55, 217–254.

McKenzie, D.P. (1983) The Earth's mantle. *Scientific American*, 249, 66–113.

McKenzie, D.P., Jackson, J. and Preistley, K. (2005) Thermal structure of oceanic and continental lithosphere. *Earth and Planetary Science Letters*, 233, 337–349.

McKenzie, D.P., Roberts, J.M. and Weiss, N.O. (1974) Convection in the Earth's mantle: towards a numerical simulation. *Journal of Fluid Mechanics*, 62, 465–538.

McLennan, J. and Lovell, B. (2002) Control of regional sea level by surface uplift and subsidence caused by magmatic underplating of Earth's crust. *Geology*, 30, 675–678.

McLeod, A.E., Dawers, N.H. and Underhill, J.R. (2000) The propagation and linkage of normal faults; insights from the Strathspey–Brent–Statfjord fault array, northern North Sea. *Basin Research*, 12, 263–284.

McNutt, M.K. (1984) Lithospheric flexure and thermal anomalies. *Journal of Geophysical Research*, 89, 11180–11194.

McNutt, M.K. (1998) Superswells. *Reviews of Geophysics*, 36, 311–344.

McNutt, M.K. and Fischer, K.M. (1987) The South Pacific superswell. In: *Seamounts, Islands and Atolls* (ed. by B.H. Keating, P. Fryer, R. Batiza and G. Boehlert), American Geophysical Union, Washington D.C., 25–34.

McNutt, M.K. and Kogan, M.G. (1987) Isostasy in the USSR, 2, Interpretation of admittance data. In: *The Composition, Structure, Dynamics of the Lithosphere–Aesthenosphere System* (ed. by K. Fuchs and C. Froideveaux), *American Geophysical Union Geodynamics Series*, Washington D.C., 16, 309–327.

McNutt, M.K., Diament, M. and Kogan, M.G. (1988) Variations of elastic plate thickness at continental thrust belts. *Journal of Geophysical Research*, 93, 8825–8838.

McQuillan, H. (1985) Fracture-controlled production from the Oligo-Miocene Asmari Formation in Gachsaran and Bibi Hakimeh Fields, southwest Iran. In: *Carbonate Petroleum Reservoirs* (ed. by P.O. Roehl and P.W. Choquette), Springer-Verlag, New York, 511–523.

Meckel, L D. and Thomasson, M.R. (2008) Pervasive tight-gas sandstone reservoirs: an overview. In: *Understanding, Exploring, and Developing Tight-Gas Sands* (ed. by S.P. Cumella, K.W. Shanley and W.K. Camp), *2005 Vail Hedberg Conference, American Association of Petroleum Geologists Hedberg Series*, 3, 13–27.

Medvedev, S.E. and Podlachikov, Y.Y. (1999a) New extended thin-sheet approximation for geodynamic applications – I. Model formulation. *Geophysical Journal International*, 136, 567–585.

Medvedev, S.E. and Podlachikov, Y.Y. (1999b) New extended thin-sheet approximation for geodynamic applications – II. Two-dimensional examples. *Geophysical Journal International*, 136, 586–608.

Meisling, H.E., Cobbold, P.R. and Mount, V.S. (2001) Segmentation of an obliquely rifted margin, Campos and Santos basins, southeastern Brazil. *American Association of Petroleum Geologists Bulletin*, 85/11, 1903–1924.

Meissner, F.F., Woodward, J. and Clayton, J.L. (1984) Stratigraphic relationships and distribution of source rocks in the greater Rocky Mountain region. In: *Hydrocarbon Source Rocks of the Greater Rocky Mountain Region* (ed. by J. Woodward, F.F. Meissner and J.L. Clayton), Rocky Mountain Association of Geologists, Denver, Colorado, 1–34.

Mellere, D., Plink-Bjorklund, P. and Steel, R. (2002) Anatomy of shelf deltas at the edge of a prograding Eocene shelf margin, Spitzbergen. *Sedimentology*, 49, 1181–1206.

Mero, W.E. (1991) Point Arguello Field. In: *Structural Traps V, Treatise of Petroleum Geology Atlas of Oil and Gas Fields* (ed. by E.A. Beaumont and N.H. Foster), American Association of Petroleum Geologists, Tulsa, 27–58.

Métivier, F. and Gaudemer, Y. (1999) Stability of output fluxes of large rivers in South and East Asia during the last 2 million years: implications for floodplain processes. *Basin Research*, 11, 293–304.

Métivier, F., Gaudemer, Y., Tapponier, P. and Klein, M. (1999) Mass accumulation rates in Asia during the Cenozoic. *Geophysical Journal International*, 137, 280–318.

Meybeck, M. (1979) Concentration des eaux fluviales en éléments majeurs et apports en solution aux océans. *Revue de Géologie Dynamique et de Géographie Physique*, 21, 215–246.

Meybeck, M. (1987) Global chemical weathering of surficial rocks estimated from river dissolved loads. *American Journal of Science*, 287, 401–428.

Meyer, R.F., Attanasi, E.D. and Freeman, P.A. (2007) *Heavy Oil and Natural Bitumen Resources in Geological Basins of the World, US Geological Survey Open-File Report*, 2007-1084.

Meyers, J.H., Suttner, L.J., Furer, L.C., May, M.T. and Soreghan, M.J. (1992) Intrabasinal tectonic control on fluvial sandstone bodies in the Cloverly Formation (Early Cretaceous) east-central Wyoming, USA. *Basin Research*, 4, 315–333.

Miall, A.D. (1991) Stratigraphic sequences and their chronostratigraphic correlation. *Journal of Sedimentary Petrology*, 61, 497–505.

Miall, A.D. (1992) The Exxon global cycle chart: an event for every occasion. *Geology*, 20, 787–790.

Miall, A.D. (1994) Sequence stratigraphy and chronostratigraphy: problems of definition and precision in correlation, and their implications for global eustasy. *Geoscience Canada*, 21, 1–26.

Miall, A.D. (2000) *Principles of Sedimentary Basin Analysis, Third Edition*, Springer-Verlag, New York, 616pp.

Miall, A.D. (2010) *The Geology of Stratigraphic Sequences, Second Edition*, Springer Verlag, 522pp.

Michael, N. (2013) Functioning of an ancient routing system, the Escanilla Formation, South Central Pyrenees. Unpublished PhD thesis, Imperial College London.

Middleton, M.F. (1982) Tectonic history from vitrinite reflectance. *Geophysical Journal of the Royal Astronomical Society*, 68, 121–132.

Milici, R.C. and Swezey, C.S. (2006) *Assessment of Appalachian Basin Oil and Gas Resources: Devonian Shale – Middle and Upper Palaeozoic Total Petroleum System. US Geological Survey Open-File Report*, 2006-1237.

Miller, J.A. (1985) Depositional and reservoir facies of the Mississippian Leadville Formation, Northwest Lisbon Field, Utah. In: *Carbonate Petroleum Reservoirs* (ed. by P.O. Roehl and P.W. Choquette), Springer-Verlag, New York, 161–173.

Miller, K.G., Kominz, M.A., Browning, J.V. *et al.* (2005) The Phanerozoic record of global sea level change. *Science*, 310, 1293–1298.

Miller, K.G., Mountain, G.S., Browning, J.V., Kominz, M., Sugarman, P.J., Christie-Blick, N., Katz, M.E. and Wright, J.D. (1998) Cenozoic global sea level, sequences, and the New Jersey transect: results from coastal plain and continental slope drilling. *Reviews of Geophysics*, 36, 569–601.

Milliman, J.D. and Meade, R.H. (1983) Worldwide delivery of river sediment to the oceans. *Journal of Geology*, 91, 1–21.

Milliman, J.D. and Syvitski, J.P.M. (1992) Geomorphic/tectonic control of sediment discharge to the ocean: the importance of small mountainous streams. *Journal of Geology*, 100, 525–544.

Milliman, J.D., Farnsworth, K.L., Jones, P.D., Xu, K.H. and Smith, L.C. (2008) Climatic and anthropogenic factors affecting river discharge to the global ocean, 1951–2000. *Global and Planetary Change*, 62, 187–194.

Minster, J.B., Jordan, T.H., Molnar, P. and Haines, E. (1974) Numerical modelling of instantaneous plate tectonics. *Geophysical Journal of the Royal Astronomical Society*, 36, 553–562.

Mitchell, A.J., Allison, P.A., Gorman, G.J. *et al.* (2011) Tidal circulation in an ancient epicontinental sea: The early Jurassic Laurasian Seaway. *Geology*, 39, 207–210.

Mitchell, S.G. and Reiners, P.W. (2003) Influence of wildfires on apatite and zircon (U–Th)/He ages. *Geology*, 31, 1025–1028.

Mitchum, R.M. Jr., Vail, P.E. and Thompson, S. III (1977) The depositional sequence as a basic unit for stratigraphic analysis. In: *Seismic Stratigraphy: Applications to Hydrocarbon Exploration* (ed. by C.E. Payton), *American Association of Petroleum Geologists Memoir*, 26, 53–62.

Mitrovica, J.X. (1996) Haskell [1935] revisited. *Journal of Geophysical Research*, 101, 555–569.

Mitrovica, J.X. and Jarvis, G.T. (1985) Surface deflections due to transient subduction in a convecting mantle. *Tectonophysics*, 120, 211–237.

Mitrovica, J.X., Beaumont, C. and Jarvis, G.T. (1989) Tilting of continental interiors by the dynamical effects of subduction. *Tectonics*, 8, 1079–1094.

Molnar, P. (1988) Continental tectonics in the aftermath of plate tectonics. *Nature*, 335, 131–137.

Molnar, P. (1992) Brace-Goetze strength profiles, the partitioning of strike-slip and thrust faulting at zones of oblique convergence and the stress-heat flow paradox of the San Andreas Fault. In: *Fault Mechanics and Transport Properties of Rocks* (ed. by T.F. Wong and B. Evans), San Diego, California, 435–459.

Molnar, P. (2004) Late Cenozoic increase in accumulation rates of terrestrial sediment: how might climate change have affected erosion rates? *Annual Review of Earth and Planetary Sciences*, 32, 67–89.

Molnar, P. and Atwater, T. (1978) Interarc spreading and Cordilleran tectonics as alternates related to the age of subducted oceanic lithosphere. *Earth Planetary Science Letters*, 41, 330–340.

Molnar, P. and Chen, W.P. (1982) Seismicity and mountain building. In: *Mountain Building Processes* (ed. by K.G. Hsü), Academic Press, London, 41–57.

Molnar, P. and England, P. (1990) Late Cenozoic uplift of mountain ranges and global climate change: chicken or egg? *Nature*, 346, 29–34.

Molnar, P. and Lyon-Caen, H. (1988) Some simple physical aspects of the support, structure, and evolution of mountain belts. In: *Processes in Continental Lithospheric Deformation* (ed. by S.P. Clark Jr. *et al.*), *Geological Society of America Special Paper*, 218, 179–207.

Molnar, P. and Tapponnier, P. (1975) Cenozoic tectonics of Asia: effects of a continental collision. *Science*, 189, 419–426.

Molnar, P.H., Anderson, R.S and Anderson, S.P. (2007) Tectonics, fracturing of rock, and erosion. *Journal of Geophysical Research*, 112, F03014.

Montadert, L., Roberts, D.G., de Charpal, O. and Guennoc, P. (1979) Rifting and subsidence of the northern continental margin of the Bay of Biscay. *Deep Sea Drilling Project Initial Report*, 48, 1025–1060.

Montanez, I.P. (1994) Late diagenetic dolomitization of Lower Ordovician, Upper Knox carbonates: a record of the hydrodynamic evolution of the southern Appalachian Basin. *American Association of Petroleum Geologists Bulletin*, 78, 1210–1239.

Montelli, R., Nolet, G., Dahlen, F.A., Masters, G., Engdahl, E.R. and Hung, S.H. (2003) Finite-frequency tomography reveals a variety of plumes in the mantle. *Science*, 303, 338–343.

Montgomery, D.R. (1994) Valley incision and uplift of mountain peaks. *Journal of Geophysical Research*, 99, 13913–13921.

Montgomery, D.R. and Brandon, M.T. (2002) Topographic controls on erosion rates in tectonically active mountain ranges. *Earth and Planetary Science Letters*, 201, 481–489.

Montgomery, D.R. and Dietrich, W.E. (1988) Where do channels begin? *Nature*, 336, 232–234.

Montgomery, D.R. and Dietrich, W.E. (1992) Channel initiation and the problem of landscape scale. *Science*, 255, 826–830.

Moore, C.H. (1989) *Carbonate Diagenesis and Porosity*, Elsevier, New York.

Moore, C.H. (2001) *Carbonate Reservoirs*, Elsevier Science, Amsterdam.

Moore, C.H. and Heydari, E. (1993) Burial diagenesis and hydrocarbon migration in platform limestones: a conceptual model based on the Upper Jurassic of Gulf Coast of USA. In: *Diagenesis and Basin Development* (ed. by A.D. Horbury and A.G. Robinson), *American Association of Petroleum Geologists Studies in Geology*, 36, 213–229.

Moore, D.G. (1969) *Reflection Profiling Studies of the California Continental Borderlands*. *Geological Society of America Special Paper*, 107, 138pp.

Morad, S., Ketzer, J.M. and De Ros, L.F. (2000) Spatial and temporal distribution of diagenetic alterations in siliciclastic rocks. Implications for mass transfer in sedimentary basins. *Sedimentology*, 47 (Millennium Reviews), 95–120.

Morelli, C., Gantar, G. and Pisani, M. (1975) Bathymetry, gravity and magmatism in the Strait of Sicily and in the Ionian Sea. *Bolletino di Geoftsica, Teorica ed Applicala*, 11, 3–190.

Moretti, I. and Turcotte, D.L. (1985) A model for erosion, sedimentation, and flexure with application to New Caledonia. *Journal of Geodynamics*, 3, 155–168.

Morgan, J.P., Morgan, W.J. and Proce, E. (1995) Hotspot melting generates both hotspot volcanism and a hotspot swell. *Journal of Geophysical Research – Solid Earth*, 100, 8045–8062.

Morgan, P. and Baker, B.H. (1983) Introduction: processes of continental rifting. *Tectonophysics*, 94, 1–10.

Morgan, W.J. (1981) Hotspot tracks and the opening of the Atlantic and Indian Oceans. In: *The Sea*, 7 (ed. by C. Emiliani), John Wiley & Sons Inc., New York, 443–487.

Morton, A.C., Haszeldine, R.S., Giles, M.R. and Brown, S. (eds.) (1992) *Geology of the Brent Group, Geological Society*, London, *Special Publication*, 61, 506pp.

Moucha, R., Forte, A.M., Mitrovica, J.X., Rowley, D.B., Quéré, S., Simmons, N.A. and Grand, S.P. (2008) Dynamic topography and long-term sea-level variations: there is no such thing as a stable continental platform. *Earth and Planetary Science Letters*, 271, 101–108.

Mulder, T. and Syvitski, J.P.M. (1996) Climatic and morphologic relationships of rivers: implications of sea-level fluctuations on river loads. *Journal of Geology*, 104, 509–523.

Müller, R.D., Sdrolias, M., Gaina, C. and Roest, W.R. (2008a) Age, spreading rates and spreading symmetry of the world's ocean crust. *Geochemistry, Geophysics, Geosystems*, 9, Q04006.

Müller, R.D., Sdrolias, M., Gaina, C., Steinberger, B. and Heine, C. (2008b) Long-term sea-level fluctuations driven by ocean basin dynamics. *Science*, 319, 1357–1362.

Müller, S., Peterschmitt, E., Fuchs, K., Emter, D. and Ansorge, J. (1973) Crustal structure of the Rhine graben area. *Tectonophysics*, 20, 381–391.

Mundil, R., Brack, P., Meier, M., Rieber, H. and Oberli, F. (1996) High resolution U–Pb dating of Middle Triassic volcaniclastics: time-scale calibration and verification of tuning parameters for carbonate sedimentation. *Earth and Planetary Science Letters*, 141, 137–151.

Murphy, J.B and Nance, R.D. (2008) The Pangea conundrum. *Geology*, 36, 703–706.

Murris, R.J. (1980) Middle East: stratigraphic evolution and oil habitat. *American Association of Petroleum Geologists Bulletin*, 64, 597–618.

Mussett, A.E., Dagley, P. and Skelhorn, R.R. (1988) Time and duration of activity in the British Tertiary Igneous Province. *Early Tertiary Volcanism and the Opening of the North Atlantic. Geological Society, London, Special Publication*, 39, 201–214.

Muto, T. and Steel, R.J. (2002) In defense of shelf-edge delta development during falling and lowstand of relative sea level. *Journal of Geology*, 110, 421–436.

Nadeau, P.H. (2011) Earth's energy 'Golden Zone': a synthesis from mineralogical research. The 2010 George Brown Lecture. *Clay Minerals*, 46, 1–24.

Nadeau, P.H. and Reynolds, R.C. (1981) Burial and contact metamorphism in the Mancos shale. *Clays and Clay Minerals*, 29, 249–259.

Nadeau, P.H., Bjorkum, P.A. and Walderhaug, O. (2005) Petroleum system analysis: impact of shale diagenesis on reservoir fluid pressure, hydrocarbon migration, and biodegradation risks. In: *Petroleum Geology: Northwest Europe and Global Perspectives* (ed. by A.G. Doré and B.A. Vining), *Proceedings of the 6th Petroleum Geology Conference*, Geological Society, London, 1267–1274.

Naeser, C.W. (1967) The use of apatite and sphene for fission track determinations. *Geological Society of America Bulletin*, 78, 705–710.

Naeser, C.W. (1979) Thermal history of sedimentary basins: fission track dating of subsurface rocks. In: *Aspects of Diagenesis* (ed. by P.A. Scholle and P.R. Schluger), *Society Economic Paleontologists and Mineralogists Special Publication*, 26, 109–112.

Najman, Y. and Garzanti, E. (2000) Reconstructing early Himalayan tectonic evolution and palaeogeography from Tertiary foreland basin sedimentary rocks, northern India. In: *Special Focus on the Himalaya* (ed. by J.W. Geissmann and A.F. Glazner), *Geological Society America Bulletin*, 112, 435–449.

Najman, Y., Pringle, M.S., Johnson, M.R.W., Robertson, A.H.F. and Wijbrans, J.R. (1997) Laser $^{40}Ar/^{39}Ar$ dating of single detrital muscovite grains from early foreland basin sedimentary deposits in India: implications for early Himalayan evolution. *Geology*, 25, 535–538.

Nakiboglu, S.M. and Lambeck, K. (1983) A re-evaluation of the isostatic rebound of Lake Bonneville. *Journal of Geophysical Research*, 88, 10439–10448.

Nance, R.D., Worsley, T.R. and Moody, J.B. (1988) The supercontinent cycle. *Scientific American*, 256, 72–79.

Nance, R.D. and Murphy, J.B. (2003) Do supercontinents introvert or extrovert? Sm–Nd isotopic evidence. *Geology*, 31, 873–876.

Naylor, M. and Sinclair, H.D. (2008). Pro- versus retro-foreland basins. *Basin Research*, 20, 285–303.

Nederlof, M.H. and Mohler, H.P. (1981) Quantitative investigation of trapping effect of unfaulted caprock (abstr.). *American Association of Petroleum Geologists Bulletin*, 65, 964.

Nehring, R. (2008) Growing and indispensable: the contribution of production from tight-gas sands to US gas production. In: *Understanding, Exploring, and Developing Tight-Gas Sands* (ed. by S.P. Cumella, K.W. Shanley and W.K. Camp), *2005 Vail Hedberg Conference, American Association of Petroleum Geologists Hedberg Series*, 3, 5–12.

Nemec, W. and Steel, R.J. (eds.) (1988) *Fan Deltas: Sedimentology and Tectonic Settings*, Blackie, London.

Neugebauer, H.J. (1983) Mechanical aspects of continental rifting. *Tectonophysics*, 94, 91–108.

Newell, N.A. (1999) Water washing in the northern Bonaparte Basin. *Australian Petroleum Production and Exploration Association Journal*, 227–247.

Newell, N.D. (1962) Paleontological gaps and geochronology. *Journal of Paleontology*, 36, 592–610.

Newman, R. and White, N. (1999) The dynamics of extensional sedimentary basins: constraints from subsidence inversion. *Philosophical Transactions of the Royal Society of London*, A 357, 805–834.

Nicholson, C., Seeber, L., Williams, P. and Sykes, L.R. (1986a) Seismic evidence for conjugate slip and block rotation within the San Andreas fault system, southern California. *Tectonics*, 5, 629–648.

Nicholson, C., Seeber, L., Williams, P. and Sykes, L.R. (1986b) Seismicity and fault kinematics through the eastern Transverse Ranges, California: block rotation, strike-slip faulting and shallow-angle thrusts. *Journal of Geophysical Research*, 91, 4891–4908.

Niedoroda, A.M., Reed, C.W., Swift, D.J.P., Arata, H. and Hoyanagi, K. (1995) Modeling shore-normal large-scale coastal evolution. *Marine Geology*, 126, 181–199.

Nienhuis, P.H. (1981) Distribution of organic matter in living marine organisms. In: *Marine Organic Chemistry* (ed. by E.K. Duursma and R. Dawson), 31–69, Elsevier, Amsterdam.

Nikishin, A.M. & Kopaevich, L.F. (2009) Tectonostratigraphy as a basis for paleotectonic reconstructions. *Moscow University Geology Bulletin*, 64, 65–74.

Nilsen, T.H. and McLaughlin, R.J. (1985) Comparison of tectonic framework and depositional patterns of the Hornelen strike-slip basin of Norway and the Ridge and Little Sulphur Creek strike-slip basins of California. In: *Strike-Slip Deformation, Basin Formation and Sedimentation* (ed. by K.T. Biddle and N. Christie Block), *Society of Economic Paleontologists and Mineralogists Special Publication*, 37, 79–103.

Nilsen, T.H. and Sylvester, A.G. (1995) Strike-slip basins. In: *Tectonics of Sedimentary Basins* (ed. by C.J. Busby and R.V. Ingersoll), Blackwell Publishing Ltd., Oxford, 425–457, 579pp.

Nittrouer, C.A., Kuehl, S.A., Figueiredo, A.G., Allison, M.A., Sommerfield, C.K., Rine, J.M., Faria, E.C. and Silveira, O.M. (1996) The geological record preserved by Amazon shelf sedimentation. *Continental Shelf Research*, 16, 817–841.

Nolet, G. (2008) *A Breviary of Seismic Tomography*, Cambridge University Press, Cambridge.

Nolet, G. (ed.) (2008) *Seismic Tomography: With Applications in Global Seismology and Exploration*, D. Reidel Publishing Company.

Noller, J.S., Sowers, J.M. and Lettis, W.R. (2000) *Quaternary Geochronology: Methods and Applications*, American Geophysical Union, Washington, D.C., 582pp.

Normark, W.R. and Gutmacher, C.E. (1988) Sur submarine slide, Monterey Fan, central California. *Sedimentology*, 35, 629–648.

Norris, R.J., Koons, P.O. and Cooper, A.F. (1990) The obliquely-convergent plate boundary in the South Island of New Zealand: implications for ancient collision zones. *Journal of Structural Geology*, 12, 715–725.

Nunn, J.A. and Sleep, N.H. (1984) Thermal contraction and flexure of intracratonic basins: a three dimensional study of the Michigan Basin. *Geophysical Journal of the Royal Astronomical Society*, 79, 587–635.

Nyblade, A.A. and Robinson, S.W. (1994) The African Superswell. *Geophysical Research Letters*, 21, 765–768.

O'Connor, J.M., Stoffers, P. and Wijbrans, J.R. (2002) Pulsing of a focused mantle plume: evidence from the distribution of Foundation Chain hotspot volcanism. *Geophysical Research Letters*, 29, 10.1029/2002GL014681.

O'Hara, M.J. (1975) Is there an Icelandic mantle plume? *Nature*, 253, 708–710.

O'Neill, C., Lenardic, A., Jellinek, A.M. and Moresi, L. (2009) Influence of supercontinents on deep mantle flow. *Gondwana Research*, 15, 276–287.

O'Sullivan, P.B. and Brown, R.W. (1998) Effects of surface cooling on apatite fission track data: evidence for Miocene climate change, North Slope, Alaska. In: *Advances in Fission Track Geochronology* (ed. by P. van den Haute and F. De Corte), Kluwer Academic Publishers, The Netherlands, 255–267.

Oberlander, T.M. (1985) Origin of drainage transverse to structures in orogens. In: *Tectonic Geomorphology* (ed. by M. Morisawa and J.T. Hack), *Binghampton Symposia in Geomorphology, International Series*, 15, Allen and Unwin, London, 155–182.

Odin, G.S. and Matter, A. (1981) De glauconiarum origine. *Sedimentology*, 28, 611–641.

Odinsen, T., Reemst, P., van der Beek, P., Faleide, J.I. and Gabrielsen, R.H. (2000) Permo-Triassic and Jurassic extension in the northern North Sea: results from tectonostratigraphic forward modelling. In: *Dynamics of the Norwegian Margin* (ed. by A. Nøttvedt), *Geological Society, London, Special Publication*, 167, 83–103.

Odonne, F. and Vialon, P. (1983) Analogue models of folds above a wrench fault. *Tectonophysics*, 99, 31–46.

Ogniben, L., Parotto, M. and Praturion, A. (eds.) (1975) *Structural Model of Italy*. Consiglio Nazionale delle Ricerche, 90.

Okay, A., Demirbag, E., Kurt, H., Okay, N. and Kuscu, I. (1999) An active, deep marine strike-slip basin along the North Anatolian fault in Turkey. *Tectonics*, 18, 129–147.

Olaiz, A.J., Muñoz-Martin, A., Vicente, G. De, Vegas, R. and Cloetingh, S. (2009) European continuous active tectonic strain-stress map. *Tectonophysics*, 474, 33–40.

Oliver, H.W., Chapman, R.H., Biehler, S., Robbins, S. L., Hanna, W. F., Griscom, A., Beyer, L. and Silver, E.A. (1981) *Gravity map of California and its continental margin*, Geol. Data Map 3, 2 sheets, scale 1:750,000, Calif. Div. of Mines and Geol., Sacramento.

Olsen, H.W. (1960) Hydraulic flow through saturated clays. *Clays and Clay Minerals*, 11, 131–161.

Olson, H.C. and Leckie, R.M. (2003) *Micropaleontologic Proxies for Sea-Level Change and Stratigraphic Discontinuities*. Society for Sedimentary Geology Special Publication, 75, DOI: 10.2110/pec.03.75.

Ori, G.G. and Friend, P.F. (1984) Sedimentary basins, formed and carried piggyback on active thrust sheets. *Geology*, 12, 475–478.

Ori, G.G., Roveri, M. and Valloni, F. (1986) Plio-Pleistocene sedimentation in the Apenninic-Adriatic foredeep (central Adriatic Sea, Italy). In: *Foreland Basins* (ed. by P.A. Allen and P. Homewood), *International Association of Sedimentology Special Publication*, 8, Blackwell Scientific Publications, Oxford, 183–198.

Osleger, D.A. (1990) Subtidal carbonate cycles: implications for allocyclic versus autocyclic controls. *Geology*, 19, 917–920.

Overeem, J., Weltje, G.J., Bishop-Kay, C. and Krooenenberg, S.B. (2001) The late Cenozoic Eridanos delta system in the southern North Sea Basin: a climate signal in sediment supply. *Basin Research*, 13, 293–312.

Palciauskas, V.V. (1986) Models for thermal conductivity and permeability in normally compacting basins. In: *Thermal Modeling in Sedimentary Basins* (ed. by J. Burrus), *1st IFP Exploration Research Conference*, Carcans, France, June 3–7, 1985, Editions Technip, Paris, 323–336, 600pp.

Pallatt, N., Wilson, M.J. and McHardy, W.J. (1984) The relationship between permeability and the morphology of diagenetic illite in reservoir rocks. *Journal of Petroleum Technology*, 36, 2225–2227.

Paola, C. (2000) Quantitative models of sedimentary basin filling. *Sedimentology*, 47 (Suppl. 1), 121–178.

Paola, C. and Martin, J. (2012) Mass balance effects in depositional systems. *Journal of Sedimentary Research*, 82, 435–450.

Paola, C. and Seal, R. (1995) Grain-size patchiness as a cause of selective deposition and downstream fining. *Water Resources Research*, 31, 1395–1407.

Paola, C., Heller, P.L. and Angevine, C.L. (1992) The large-scale dynamics of grain size variation in alluvial basins. 1. Theory. *Basin Research*, 4, 73–90.

Paola, C., Straub, K., Mohrig, D. and Reinhardt, L. (2009) The 'unreasonable effectiveness' of stratigraphic and geomorphic experiments. *Earth Science Reviews*, 97, 1–43.

Parker, G. (1978a) Self-formed straight rivers with equilibrium banks and mobile bed. Part 1. The sand-silt river. *Journal of Fluid Mechanics*, 89, 109–125.

Parker, G. (1978b) Self-formed straight rivers with equilibrium banks and mobile bed. Part 2. The gravel river. *Journal of Fluid Mechanics*, 89, 127–146.

Parker, G., Paola, C., Whipple, K.X. and Mohrig, D.C. (1998) Alluvial fans formed by channelized fluvial and sheet flow. I: Theory. *Journal of Hydraulic Engineering*, 124, 985–995.

Parkinson, N. and Summerhayes, C. (1985) Synchronous global sequence boundaries. *American Association of Petroleum Geologists Bulletin*, 69, 685–687.

Parnell, J. (2010) Potential of palaeofluid analysis for understanding oil charge history. *Geofluids*, 10, 73–82.

Parnell, J., Green, P.F., Watt, G. and Middleton, D. (2005) Thermal history and oil charge on the UK Atlantic margin. *Petroleum Geoscience*, 11, 99–112.

Parrish, J.T., Ziegler, A.M. and Scotese, C.R. (1982) Rainfall patterns and the distributions of coals and evaporites in the Mesozoic and Cenozoic. *Palaeogeography, Palaeoclimatology, Palaeoecology*, 40, 67–101.

Parsons, A.J., Michael, N., Whittaker, A.C., Duller, R.A. and Allen, P.A. (2012) Grain-size trends reveal the late orogenic tectonic and erosional history of the south-central Pyrenees, Spain. *Journal of the Geological Society*, 169, 111–114.

Parsons, B. (1982) Causes and consequences of the relation between area and age of the ocean floor. *Journal of Geophysical Research*, 87, 289–302.

Parsons, B. and Daly, S. (1983) The relationship between surface topography, gravity anomalies, and the temperature structure of convection. *Journal of Geophysical Research*, 88, 1129–1144.

Parsons, B. and Sclater, J.G. (1977) An analysis of the variation of ocean floor bathymetry and heat flow with age. *Journal of Geophysical Research*, 82, 803–827.

Partington, M.A., Mitchener, B.C., Milton, N.J. and Fraser, A.J. (1993) Genetic sequence stratigraphy of the North Sea Late Jurassic and Early Cretaceous; distribution and prediction of Kimmeridgian–Late Ryazanian reservoirs in the North Sea and adjacent areas. In: *Petroleum Geology of Northwest Europe* (ed. by J.R. Parker), *Proceedings of the 4th Conference of the Geological Society, London*, Bath, 347–370.

Paterson, M.S. (1958) Experimental deformation and faulting of Wombeyan marble. *Geological Society of America Bulletin*, 69, 465–476.

Pawlak, A., Eaton, D.W., Bastow, I.D., Kendall, J-M., Helffrich, G., Wookey, J. and Snyder, D. (2011) Crustal structure beneath Hudson Bay from ambient-noise tomography: implications for basin formation. *Geophysical Journal International*, 184, 65–82.

Payton, C.E. (ed.) (1977) *Seismic Stratigraphy: Applications to Hydrocarbon Exploration. American Association of Petroleum Geologists Memoir*, 26, 516pp.

Pazzaglia, F.J. and Brandon, M.T. (1996) Macrogeomorphic evolution of the post-Triassic Appalachian mountains determined by deconvolution of the offshore basin sedimentary record. *Basin Research*, 8, 255–278.

Peltier, W.R. (1980) Models of glacial isostasy and relative sea level. In: *Dynamics of Plate Interiors* (ed. by A.W. Ball, P.L. Bender, T.R. McGetchin and R.I. Walcott), *American Geophysical Union Geodynamics Series*, 1, 111–127.

Peltier, W.R. (2007) Mantle dynamics and the D-double prime layer implications of the post-perovskite phase. In: *Post-Perovskite: The Last Mantle Phase Transition* (ed. by K. Hirose, J. Brodholt, T. Lay and D. Yuen), *American Geophysical Union Geophysical Monographs*, 174, 217–227.

Peltzer, G. and Tapponnier, P. (1988) Formation and evolution of strike-slip faults, rifts and basins during India–Asia collision – an experimental approach. *Journal of Geophysical Research*, 93, 15085–15117.

Pépin, E., Carrétier, S. and Herail, G. (2010) Erosion dynamics modelling in a coupled catchment-fan system with constant external forcing. *Geomorphology*, 122, 78–90.

Perez-Arlucea, M., Mack, G. and Leeder, M. (2000) Reconstructing the ancestral (Plio-Pleistocene) Rio Grande in its active tectonic setting, southern Rio Grande rift, new Mexico, USA. *Sedimentology*, 47, 701–720.

Perez-Gussinye, M., Lowry, A.R., Watts, A.B. *et al.* (2004) On the recovery of effective elastic thickness using spectral methods: examples from synthetic data and from the Fennoscandian Shield. *Journal of Geophysical Research – Solid Earth*, 109, B10409.

Perez-Gussinye, M., Swain, C.J., Kirby, J.F. *et al.* (2009) Spatial variations of the effective elastic thickness Te using multitaper spectral estimation and wavelet methods: examples from synthetic data and application to South America. *Geochemistry, Geophysics, Geosystems*, 10, Q04005.

Petersen, K.D., Nielsen, S.B., Clausen, O.R., Stephenson, R. and Gerya, T. (2010) Small-scale mantle convection produces stratigraphic sequences in sedimentary basins. *Science*, 329, 827–831.

Petrunin, A.G. and Sobolev, S.V. (2006) What controls thickness of sediments and lithospheric deformation at a pull-apart basin? *Geology*, 34, 389–392.

Petrunin, A.G. and Sobolev, S.V. (2008) Three-dimensional numerical model of the evolution of pull-apart basins. *Physics of the Earth and Planetary Interiors*, 171, 387–399.

Pettijohn, F.J. (1975) *Sedimentary Rocks, Third Edition*, Harper and Row, New York.

Pettijohn, F.J., Potter, P.E. and Siever, R. (1973) *Sand and Sandstone*, Springer-Verlag, New York.

Pfiffner, O.A. (2000) Collision tectonics of the Swiss Alps: insight from geodynamic modelling. *Tectonics*, 19, 1065–1094.

Pfiffner, O.A., Schlunegger, F. and Buiter, S.J.H. (2002) The Swiss Alps and their peripheral foreland basin: stratigraphic response to deep crustal processes. *Tectonics*, 21, 1009.

Philippi, G.T. (1965) On the depth, time and mechanism of petroleum generation. *Geochimica et Cosmochimica Acta*, 29, 1021–1049.

Phillips, B.R. and Bunge, H.-P. (2007) Supercontinent cycles disrupted by strong mantle plumes. *Geology*, 35, 847–850.

Phillips, B.R. and Coltice, N. (2010) Temperature beneath continents as a function of continental cover and convective wavelength. *Journal of Geophysical Research*, 115, B04408.

Pickup, S.L.B., Whitmarsh, R.B., Fowler, C.M.R. and Reston, T.J. (1996) Insight into the nature of the ocean-continent transition off West Iberia from a deep multichannel seismic reflection profile. *Geology*, 24, 1079–1082.

Pilger, R.H. (1984) Cenozoic plate kinematics, subduction and magmatism, South American Andes. *Journal of the Geological Society*, London, 141, 793–802.

Pinet, C., Jaupart, C., Mareschal, J.-C., Gariépy, C., Bienfait, G. and Lapointe, R. (1991) Heat flow and lithospheric structure of the eastern Canadian shield. *Journal of Geophysical Research*, 96, 19923–19941.

Pinet, P. and Souriau, M. (1988) Continental erosion and large-scale relief. *Tectonics*, 7, 563–582.

Pirmez, C., Pratson, L.F. and Steckler, M.S. (1998) Clinoform development by advection-diffusion of suspended sediment: Modeling and comparison to natural systems. *Journal of Geophysical Research*, 103, 24141–24157.

Pitman, W.C. (1978) The relationship between eustasy and stratigraphic sequences of passive margins. *Geological Society of America Bulletin*, 89, 1389–1403.

Pitman, W.C. (1979) The effect of eustatic sea level changes on stratigraphic sequences at Atlantic margins. In: *Geological and Geophysical Investigations of Continental Margins, American Association of Petroleum Geologists Memoir*, 29, 453–460.

Platt, J.P. (1986) Dynamics of orogenic wedges and the uplift of high-pressure metamorphic rocks. *Geological Society of America Bulletin*, 97, 1037–1053.

Plummer, N., Busby, J., Lee, R. and Hanshaw, B. (1990) Geochemical modeling of the Madison Aquifer in parts of Montana, Wyoming, and South Dakota. *Water Resources Research*, 26, 1981–2014.

Poag, C.W. (1980) Foraminiferal stratigraphy, palaeoenvironments and depositional cycles in the Outer Baltimore Canyon Trough. In: *Geological Studies of the COST B-3 Well, US Mid-Atlantic Continental Slope Area* (ed. by P.A. Scholle), *US Geological Survey Circular*, 833, 44–65.

Podladchikov, Y.Y., Poliakov, A.N.B. and Yeun, D.A. (1994) The effect of lithospheric phase transitions on subsidence of extending continental lithosphere. *Earth and Planetary Science Letters*, 124, 95–103.

Pollack, H.N. and Chapman, D.S. (1977) On the regional variation of heat flow, geotherms and lithospheric thickness. *Tectonophysics*, 38, 279–296.

Pollack, H.N., Hurter, S.J., and Johnson, J.R. (1993) Heat flow from the Earth's interior: analysis of the global data set. *Reviews of Geophysics*, 31(3), 267–280.

Pomar, L. (2001) Types of carbonate platforms: a genetic approach. *Basin Research*, 13, 313–334.

Poore, H., White, N. and Maclennan, J. (2011) Ocean circulation and mantle melting controlled by radial flow of hot pulses in the Iceland plume. *Nature Geoscience* 4, 558–561.

Posamentier, H.W. and Chamberlain, C.J. (1993) Sequence stratigraphic analysis of Viking Formation lowstand beach deposits in the Joarcam field, Alberta. In: *Sequence Stratigraphy and Facies Associations* (ed. by H.W. Posamentier, C.P. Summerhayes, B.U. Haq and G.P. Allen), *International Association of Sedimentologists Special Publication*, 18, Blackwell Scientific Publications, Oxford, 469–486.

Posamentier, H.W. and James, D.P. (1993) An overview of sequence stratigraphic concepts: Uses and abuses. In: *Sequence Stratigraphy and Facies Associations* (ed. By H.W. Posamentier, C.P. Summerhayes, B.U. Haq and G.P. Allen), *International Association of Sedimentologists Special Publication*, 18, Blackwell Scientific Publications, Oxford, 3–18.

Posamentier, H.W. and Vail, P.R. (1988) Eustatic controls on clastic deposition, II, Sequence and systems tract models. In: *Sea Level Changes: An Integrated Approach* (ed. by C.K. Wilgus, B.S. Hastings, C.G.St.C. Kendall, H.W. Posamentier, C.A. Ross and J.C. Van Wagoner), *Society of Economic Paleontologists and Mineralogists Special Publication*, 42, 125–154.

Posamentier, H.W., Allen, G.P., James, D.P. and Tesson, M. (1992) Forced regressions in a sequence stratigraphic framework: concepts, examples and sequence stratigraphic significance. *American Association of Petroleum Geologists Bulletin*, 76, 1687–1709.

Posamentier, H.W., Jervey, M.T. and Vail, P.R. (1988) Eustatic controls on eustatic deposition, I, Conceptual framework. In: *Sea Level Changes: An Integrated Approach* (ed. by C.K. Wilgus, B.S. Hastings, C.G.St.C. Kendall, H.W. Posamentier, C.A. Ross and J.C. Van Wagoner), *Society of Economic Paleontologists and Mineralogists Special Publication*, 42, 109–124.

Postma, G. (1990) Depositional architecture and facies of river and fan deltas: a synthesis. In: *Coarse-grained Deltas* (ed. by A. Colella and D.B. Prior), *International Association of Sedimentologists Special Publication*, 10, 13–27.

Powell, T.G., Foscolos, A.E., Gunther, P.R. and Snowdon, L.R. (1978) Diagenesis of organic matter and fine clay minerals: a comparative study. *Geochimica et Cosmochimica Acta*, 42, 1181–1197.

Pratt, B.R. and James, N.P. (1986) The St. George Group (Lower Ordovician) of western Newfoundland: tidalflat island model for carbonate sedimentation in shallow epeiric seas. *Sedimentology*, 33, 313–343.

Preistley, K. and McKenzie, D.P. (2006) The thermal structure of the lithosphere from shear wave velocities. *Earth and Planetary Science Letters*, 244, 285–301.

Preto, N., Hinnov, L.A., Hardie, L.A. and De Zanche, V. (2001) Middle Triassic orbital signature recorded in the shallow marine Latemar carbonate build-up (Dolomites, Italy). *Geology*, 29, 1123–1126.

Price, R.A. (1973) Large-scale gravitational flow of supracrustal rocks, southern Canadian Rockies. In: *Gravity and Tectonics* (ed. by K. de Jong and R. Scholten), John Wiley & Sons Inc., New York, 491–502.

Price, R.A. and Hatcher, R.D. (1983) Tectonic significance of similarities in the evolution of the Alabama–Pennsylvania Appalachians and the Alberta–British Columbia Canadian Cordillera. In: *Contributions to the Tectonics and Geophysics of Mountain Chains* (ed. by R.D. Hatcher, H. Williams and I. Zietz), *Geological Society of America Memoir*, 158, 149–160.

Primmer, T.J., Cade, C.A., Evans, J., Gluyas, J.G., Hopkins, M.S., Oxtoby, N.H., Smalley, P.C., Warren, E.A. and Worden, R.H. (1997) Global patterns in sandstone diagenesis: their application to reservoir quality prediction for petroleum exploration. In: *Reservoir Quality Prediction in Sandstones and Carbonates* (ed. by J.A. Kupecz, J. Gluyas and S. Bloch), *American Association of Petroleum Geologists Memoir*, 69, 61–77.

Pritchard, D., Roberts, G.G., White, N.J., and Richardson, C.N. (2009) Uplift histories from river profiles. *Geophysical Research Letters*, 36, L24301, DOI: 10.1029/2009GL040928.

Prosser, S. (1993) Rift-related linked depositional systems and their seismic expression. In: *Tectonics and Seismic Sequence Stratigraphy* (ed. by G.D. Williams and A. Dobbs), *Geological Society, London, Special Publication*, 71, 35–66.

Puigdefàbregas, C., Muñoz, J.A. and Marzo, M. (1986) Thrust belt development in the eastern Pyrenees and related depositional sequences in the southern foreland basin. In: *Foreland Basins* (ed. by P.A. Allen and P. Homewood), *International Association of Sedimentologists Special Publication*, 8, Blackwell Scientific Publications, Oxford, 229–246, 453pp.

Puigdefàbregas, C., Muñoz, J.A. and Vergès, J. (1992) Thrusting and foreland basin evolution in the southern Pyrenees. In: *Thrust Tectonics* (ed. by K.R. McClay), 247–254.

Purcell, W.R. (1949) Capillary pressures – their measurement using mercury and the calculation of permeability therefrom. *AIME Petroleum Transactions*, 186, 39–48.

Purser, B.H., Brown, A. and Aissaoui, D.M. (1994) Nature, origins and evolution of porosity in dolomites. In: *Dolomites* (ed. by B. Purser, M. Tucker and D. Zenger), *International Association of Sedimentologists Special Publication*, 21, 283–308.

Quinlan, G.M. and Beaumont, C. (1984) Appalachian thrusting, lithospheric flexure, and the Paleozoic stratigraphy of the eastern interior of North America. *Canadian Journal Earth Science*, 21, 973–996.

Ragotzkie, R.A. (1978) Heat budgets of lakes. In: *Lakes: Chemistry, Geology, Physics* (ed. by A. Lerman), Springer-Verlag, Berlin, 1–20.

Rahl, J.M., Reiners, P.W., Campbell, I.H., Nicolescu, S. and Allen, C.M. (2003) Combined single grain (U–Th)/He and U/Pb dating of detrital zircons from the Navajo Sandstone, Utah. *Geology*, 31, 761–764.

Raiverman, V., Kunte, S.V. and Mukherjea, A. (1983) Basin geometry, Cenozoic sedimentation and hydrocarbons in north western Himalaya and Indo-Gangetic plains. *Petroleum Asia Journal*, 6, 67–92.

Ramberg, I.B. (1978) Comparison of the Oslo Graben with the Rio Grande Rift. *Conference Proceedings: Los Alamos Scientific Laboratory*, 7487, 68–71.

Ramos, E., Busquets, P. and Vergés, J. (2002) Interplay between longitudinal fluvial and transverse alluvial fan systems and growing thrusts in a piggyback basin (SE Pyrenees). In: *Sedimentary Geology of Growth Strata* (ed. by M. Marzo, J.A. Muñoz and J. Vergés), *Sedimentary Geology*, 146, 105–131.

Ranalli, G. (1995) *Rheology of the Earth, Second Edition*, Chapman and Hall, London, 413pp.

Ratcliff, J.T., Tackler, P.J., Schubert, G. and Zebib, A. (1997) Transitions in thermal convection with strongly variable viscosity. *Physics of the Earth and Planetary Interiors*, 102, 201–212.

Raymo, M.E and Ruddiman, W.F. (1992) Tectonic forcing of late Cenozoic climate, *Nature*, 359, 117–122.

Raymo, M.E., Ruddiman, W.F. and Froehlich, P.N. (1988) Influence of late Cenozoic mountain building on ocean geochemical cycles. *Geology*, 16, 649–653.

Read, J.F., Osleger, D. and Elrich, M. (1991) Two-dimensional modelling of carbonate ramp sequences and component cycles. In: *Sedimentary Modeling: Computer Simulations and Methods for Improved Parameter Definition* (ed. by E.K. Franseen, W.L. Watney, G.G.S.C. Kendall and W. Ross), *Kansas State Geological Survey Bulletin*, 233, 473–488.

Reading, H.G. (1980) Characteristics and recognition of strike-slip fault systems. In: *Sedimentation in Oblique-Slip Mobile Zones* (ed. by P.F. Ballance and H.G. Reading), *International Association of Sedimentologists Special Publication*, 4, 7–26.

Reading, H.G. (ed.) (1996) *Sedimentary Environments: Processes, Facies and Stratigraphy, Third Edition*, Blackwell Publishing Ltd., Oxford, 688pp.

Reading, H.G. and Collinson, J.D. (1996) Clastic coasts. In: *Sedimentary Environments: Processes, Facies and Stratigraphy, Third Edition* (ed. by H.G. Reading), Blackwell Publishing Ltd., 154–231.

Reasenberg, P. and Ellsworth, W.L. (1982) Aftershocks of the Coyote Lake, California earthquake of August 6, 1979: a detailed study. *Journal of Geophysical Research*, 87, 10637–10655.

Reineck, H.-E. (1960) Uber Zeitlücken in rezenten Flachsee-Sedimenten. *Geologisches Rundschau*, 48, 149–161.

Reiners, P.W. (2002) (U–Th)/ He chronometry experiences a renaissance. *Eos, Transactions, American Geophysical Union*, 83, 26–27.

Reiners, P.W., Ehlers, T.A. and Zeitler, P.K. (2005) Past, present and future of thermochronology. In: *Thermochronology* (ed. by P.W. Reiners and T.A. Ehlers) *Reviews in Mineralogy and Geochemistry*, 58, 1–18.

Repke, J.L., Anderson, R.S. and Finkel, R.C. (1997) Cosmogenic dating of fluvial terraces, Fremont River, Utah. *Earth and Planetary Science Letters*, 152, 59–73.

Ribberink, J.S. and van der Sande, J.T.M. (1985) Aggradation in rivers due to overloading. *Journal of Hydraulic Research*, 23, 273–284.

Ricci-Lucchi, F. (1986) The Oligocene to Recent foreland basins of the northern Apennines. In: *Foreland Basins* (ed. by P.A. Allen and P. Homewood), *International Association of Sedimentologists Special Publication*, 8, Blackwell Scientific Publications, Oxford, 105–140.

Richards, M.A. and Hager, B.H. (1984) Geoid anomalies in a dynamic Earth. *Journal of Geophysical Research*, 89, 5487–6002.

Richards, M.A., Duncan, R.A. and Courtillot, V.E. (1989) Flood basalts and hot-spot tracks: plume heads and tails. *Science*, 246, 103–107.

Richardson, R.M. (1992) Ridge forces, absolute plate motions, and the intraplate stress field. *Journal of Geophysical Research*, 97, 11739–11749.

Richter, F.M. (1973) Convection and the large-scale circulation of the mantle. *Journal of Geophysical Research*, 78, 8735–8745.

Rider, M.H. (1996) *The Geological Interpretation of Well Logs*, Rider-French Publications, 288pp.

Riedel, W. (1929) Zur Mechanik geologischer Brucherscheinungen. *Zentralblatt fur Mineralogie, Geologie und Palaeontologie*, 1929B, 354–368.

Risk, M.J. and Rhodes, E.G. (1985) From mangroves to petroleum precursors: an example from tropical N.E. Australia. *American Association of Petroleum Geologists Bulletin*, 69, 1230–1240.

Ritsema, J. and van Heijst, H.J. (2000) New seismic model of the upper mantle beneath Africa. *Geology*, 28, 63–66.

Rittenhouse, G. (1972) Stratigraphic trap classification. In: *Stratigraphic Oil and Gas Fields: Classification, Exploration Methods and Case Histories* (ed. by R.E. King), *American Association of Petroleum Geologists Memoir*, 16, 14–28.

Rivenaes, J.C. (1992) Application of a dual-lithology, depth-dependent diffusion equation in stratigraphic simulation. *Basin Research*, 4, 133–146.

Rivenaes, J.C. (1997) Impact of sediment transport efficiency on large-scale sequence architecture: results from stratigraphic computer stimulation. *Basin Research*, 9, 91–105.

Robert, P. (1988) *Organic Metamorphism and Geothermal History*, Elf-Aquitaine and Reidel Publishing, Dordrecht, 311pp.

Roberts, G.G. and White, N. (2010) Estimating uplift rate histories from river profiles using African examples. *Journal of Geophysical Research*, 115, B02406.

Roberts, G.G., White, N.J., Martin-Brandis, G.L. *et al.* (2012) An uplift history of the Colorado Plateau and its surroundings from inverse modelling of longitudinal river profiles. *Tectonics*, 31, TC4022.

Robinson, A.G., Coleman, M.L. and Gluyas, J.G. (1993) The age of illite cement growth, Village Fields area, southern North Sea: evidence from K–Ar ages and $^{18}O/^{16}O$ ratios. *American Association of Petroleum Geologists Bulletin*, 77, 68–80.

Rodgers, D.A. (1980) Analysis of pull-apart basin development produced by en echelon strike-slip faults. *International Association of Sedimentologists Special Publication*, 4, Blackwell Scientific Publications, Oxford, 27–41.

Rogers, N.R., Macdonald, R., Fitton, J., George, R., Smith, R. and Barreiro, B. (2000) Two mantle plumes beneath the East African Rift system: Sr, Nd and Pb isotope evidence from Kenya rift basalts. *Earth and Planetary Science Letters*, 176, 387–400.

Rohrman, M., Andriessen, P.A.M. and van der Beek, P.A. (1996) The relationship between basin and margin thermal evolution assessed by fission track thermochronology: an application to offshore southern Norway. *Basin Research*, 8, 45–63.

Romankevich, E.A. (1984) *Geochemistry of Organic Matter in the Ocean*, Springer-Verlag, Berlin, 334pp.

Romans, B.W., Normark, W.R., McGann, M.M., Covault, J.A. and Graham, S.A. (2009) Coarse-grained sediment delivery and distribution in the Holocene Santa Monica Basin, California: implications for evaluating source-to-sink flux at millennial time scales. *Geological Society of America Bulletin*, 121, 1394–1408.

Rona, P.A. (1982) Evaporites at passive margins. In: *Dynamics of Passive Margins* (ed. by R.A. Scrutton), *American Geophysical Union and Geological Society of America Geodynamics Series*, 6, 116–132.

Rosenbloom, N.A. and Anderson, R.S. (1994) Hillslope and channel evolution in a marine terraced landscape, Santa Cruz, California. *Journal of Geophysical Research*, 99, 14013–14029.

Rosendahl, B.R., Reynolds, D.J., Lorber, P.M., Burgess, C.F., McGill, J., Scott, D., Lambiase, J.J. and Derksen, S.J. (1986) Structural expressions of rifting: lessons from lake Tanganyika, Africa. In: *Sedimentation in the African Rifts* (ed. by L. E. Frostick *et al.*), *Geological Society, London, Special Publication*, 25, 29–43.

Ross, W.C., Watts, D.E. and May, J.A. (1995) Insights from stratigraphic modelling: mud-limited versus sand-limited depositional systems. *American Association Petroleum Geologists Bulletin*, 79, 231–258.

Rothman, D.H., Grotzinger, J.P. and Flemings, P. (1994) Scaling in turbidite deposition. *Journal of Sedimentary Research*, 64, 59–67.

Rotstein, Y. and Schaming, M. (2011) The Upper Rhine Graben (URG) revisited: Miocene transtension and transpression account for the observed first-order structures. *Tectonics*, 30, TC3007.

Rowan, M.G. and Vendeville, B.C. (2006) Foldbelts with early salt withdrawal and diapirism: physical model and examples from the northern Gulf of Mexico and the Flinders Ranges, Australia. *Marine and Petroleum Geology*, 23, 871–891.

Rowan, M.G., Jackson, M.P.A. and Trudgill, B.D. (1999) Salt-related fault families and fault welds in the northern Gulf of Mexico. *American Association of Petroleum Geologists Bulletin*, 83/9, 1454–1484.

Rowan, M.G., Peel, F. and Vendeville, B.C. (2004) Gravity-driven foldbelts on passive margins. In: *Thrust Tectonics and Hydrocarbon Systems* (ed. by K.R. McClay), *American Association of Petroleum Geologists Memoir*, 82, 157–182.

Rowan, M.G., Ratliff, R.A., Trudgill, B.D. and Duarte, J.B. (2001) Emplacement and evolution of the Mahogany salt body, central Louisiana outer shelf, northern Gulf of Mexico. *American Association of Petroleum Geologists Bulletin*, 85/6, 947–969.

Rowan, M.G., Trudgill, B.D. and Fiduk, J.C. (2000) Deepwater salt-cored foldbelts: lessons from the Mississippi Fan and Perdido foldbelts, northern Gulf of Mexico. In: *Atlantic Rifts and Continental Margins* (ed. by W. Mohriak and M. Talwani), *Geophysical Monograph Series*, 115, American Geophysical Union, Washington, 173–191.

Rowley, D.B. (2002) Rate of plate creation and destruction: 180 Ma to present. *Geological Society of America Bulletin*, 114, 927–933.

Rowley, D.B. and Sahagian, D. (1986) Depth-dependent stretching: a different approach. *Geology*, 14, 32–35.

Rowley, E. and White, N. (1998) Inverse modelling of extension and denudation in the East Irish Sea and surrounding areas. *Earth and Planetary Science Letters*, 161, 57–71.

Roy, R.F., Blackwell, D.D. and Birch, F. (1968) Heat generation of plutonic rocks and continental heat flow provinces. *Earth and Planetary Science Letters*, 5, 1–12.

Royden, L. (1993) The tectonic expression of slab pull at convergent plate boundaries. *Tectonics*, 12, 303–325.

Royden, L. and Karner, G.D. (1984) Flexure of the continental lithosphere beneath Apennine and Carpathian foredeep basins: evidence for an insufficient topographic load. *American Association of Petroleum Geologists Bulletin*, 68, 704–712.

Royden, L. and Keen, C.E. (1980) Rifting processes and thermal evolution of the continental margin of eastern Canada determined from subsidence curves. *Earth and Planetary Science Letters*, 51, 343–361.

Royden, L., Patacca, E. and Scandone, P. (1987) Segmentation and configuration of subducted lithosphere in Italy; an important control on thrust-belt and foredeep-basin evolution. *Geology*, 15, 714–717.

Royden, L., Sclater, J.G. and Von Herzen, R.P. (1980) Continental margin subsidence and heat flow: important parameters in formation of petroleum hydrocarbons. *American Association of Petroleum Geologists Bulletin*, 64, 173–187.

Royden, L.H. (1985) The Vienna Basin: a thin-skinned pull-apart basin. In: *Strike-Slip Deformation, Basin Formation and Sedimentation* (ed. by K.T. Biddle and N. Christie-Blick), *Society of Economic Paleontologists and Mineralogists Special Publication*, 37, 319–338.

Rudge, J.F., Champion, M.E.S., White, N. *et al.* (2008) A plume model of transient diachronous uplift at the Earth's surface. *Earth and Planetary Science Letters*, 267, 146–160.

Ruhl, K.W. and Hodges, K.V. (2005) The use of detrital mineral cooling ages to evaluate steady state assumptions in active orogens: an example from the central Nepalese Himalaya. *Tectonics*, 24, TC4015.

Ruiz, G.M.H., Seward, D. and Winkler, W. (2004) Detrital thermochronology – a new perspective on hinterland tectonics, an example from the Andean Amazon Basin, Ecuador. *Basin Research*, 16/3, 413–430.

Runcorn, S.K. (1967) Flow in the mantle inferred from the low degree harmonics of the geopotential. *Journal of the Royal Astronomical Society*, 14, 375–384.

Rutter, E.H. (1976) The kinetics of rock deformation by pressure solution. *Philosophical Transactions of the Royal Society, London*, A283, 203–219.

Rutter, E.H. (1983) Pressure solution in nature, theory and experiment. *Journal of the Geological Society*, 140, 725–740.

Rutter, E.H. and Brodie, K.H. (1988) The role of tectonic grain size reduction in the rheological stratification of the lithosphere. *Geologische Rundschau*, 77, 295–308.

Ruzyla, K. and Friedman, G.M. (1985) Factors controlling porosity in dolomite reservoirs of the Ordovician Red River Formation, Cabin Creek Field, Montana. In: *Carbonate Petroleum Reservoirs* (ed. by P.O. Roehl and P.W. Choquette), Springer-Verlag, New York, 39–58.

Rybach, L. (1986) Amount and significance of radioactive heat sources in sediments. In: *Thermal Modelling in Sedimentary Basins* (ed. by J. Burrus), *1st IFP Exploration Research Conference*, Careans, France, June 3–7, 1985, Editions Technip, Paris, 311–322, 600pp.

Rybach, L. and Cermak, V. (1982) Radioactive heat generation in rocks. In: *Physical Properties of Rocks* (ed. by G. Angenheister), Vol. 1b, Springer-Verlag, Berlin, 353–371.

Sadler, P.M. (1981) Sediment accumulation rates and the completeness of stratigraphic sections. *Journal of Geology*, 89, 569–584.

Sadler, P.M. and Strauss, D.J. (1990) Estimation of completeness of stratigraphical sections using empirical data and theoretical models. *Journal of the Geological Society*, 147, 471–485.

Safarudin, X. and Manulang, M.H. (1989) Trapping mechanism in Mutiara Field, Kutei Basin, East Kalimantan. In: *Proceedings of the Indonesian Petroleum Association 8th Annual Convention*, 1, 31–54.

Safran, E.B. (2003) Geomorphic interpretation of low-temperature thermochronologic data: insights from two-dimensional thermal modeling. *Journal of Geophysical Research*, 108, 2189.

Sahagian, D. (1980) Sublithospheric upwelling distribution. *Nature (London)*, 287, 217–218.

Sahagian D.L. (1993) Structural evolution of African basins; stratigraphic synthesis. *Basin Research*, 5, 41–54.

Saller, A.H., Buded, D.A. and Harris, P.M. (1994) Unconformities and porosity development in carbonate strata: ideas from a Hedberg conference. *American Association of Petroleum Geologists Bulletin*, 78, 857–872.

Sambrook-Smith, G.H. and Ferguson, R.I. (1995) The gravel–sand transition along river channels. *Journal of Sedimentary Research*, 65, 423–430.

Sandiford, Lithospheric Dynamics On-line. http://jaeger.earthsci.unimelb.edu.au/msandifo/Publications/Geodynamics/geodynamics.html.

Sandiford, M. and Hand, M. (1998) Controls on the locus of Phanerozoic intraplate deformation in central Australia. *Earth and Planetary Science Letters*, 162, 97–110.

Sandiford, M. and McLaran, S. (2002) Tectonic feedback and the ordering of heat producing elements within the continental lithosphere. *Earth and Planetary Science Letters*, 204, 133–150.

Sandiford, M. and Powell, R. (1990) Some isostatic and thermal consequences of the vertical strain geometry in convergent orogens. *Earth and Planetary Science Letters*, 98, 154–165.

Sanford, B.V. (1987) Paleozoic geology of the Hudson Platform. In: *Sedimentary Basins and Basin-Forming Mechanisms* (ed. by C. Beaumont and A.J. Tankard), *Canadian Society Petroleum Geologists Memoir*, 12, 483–505.

Sarg, J.F. (1981) Petrology of the carbonate-evaporite facies transition of the Seven Rivers Formation (Guadalupian, Permian), southeast New Mexico. *Journal of Sedimentary Petrology*, 51, 73–95.

Sarg, J.F. (1988) Carbonate sequence stratigraphy. In: *Sea Level Changes: An Integrated Approach* (ed. by C.K. Wilgus, B.S. Hastings, C.G.St.C. Kendall, H.W. Posamentier, C.A. Ross and J.C. Van Wagoner), *Society of Economic Paleontologists and Mineralogists Special Publication*, 42, 155–181.

Sarg, J.F. and Skjold, L.J. (1982) Stratigraphic traps in Palaeocene sands in the Balder area, North Sea. In: *The Deliberate Search for the Subtle Trap* (ed. by M.T. Halbouty), *American Association of Petroleum Geologists Memoir*, 32, 197–206.

Sass, J.H., Lachenbruch, A.H. and Munroe, R.J. (1971) Thermal conductivities of rocks from measurements on fragments and its application to heat flow determination. *Journal of Geophysical Research*, 76, 3391–3401.

Saunders, A.D., England, R.W., Reichow, M.K. and White, R.V. (2005) A mantle plume origin for the Siberian traps: uplift and extension in the East Siberian Basin, Russia. *Lithos*, 79, 407–424.

Schaller, M., von Blanckenburg, G., Hovius, N. *et al.* (2004) Paleoerosion rates from cosmogenic Be-10 in a 1.3 Ma terrace sequence: response of the River Meuse to changes in climate and rock uplift. *Journal of Geology*, 112, 127–144.

Schaller, M., von-Blanckenburg, F., Hovius, N. and Kubik, P.W. (2001) Large-scale erosion rates from *in situ*-produced cosmogenic nuclides in European river sediments. *Earth and Planetary Science Letters*, 188, 3–4.

Schedl, A. and Wiltschko, D.V. (1984) Sedimentological effects of a moving terrain. *Journal of Geology*, 92, 273–287.

Scherer, M. (1987) Parameters influencing porosity in sandstones: a model for sandstone porosity prediction. *American Association of Petroleum Geologists Bulletin*, 71(5), 485–491.

Schlager, W. (1981) The paradox of drowned reefs and carbonate platforms. *Geological Society of America Bulletin*, 92, 197–211.

Schlager, W. (1992) *Sedimentology and Sequence Stratigraphy of Reefs and Carbonate Platforms. American Association of Petroleum Geologists Continuing Education Course Notes Series*, 34, Tulsa, Oklahoma.

Schlager, W. (2005) Carbonate sedimentology and sequence stratigraphy. *Society for Sedimentary Geology, Concepts in Sedimentology and Paleontology*, 8, 200pp.

Schlische, R.W. (1991) Half-graben basin filling models: new constraints on continental extensional basin development. *Basin Research*, 3, 123–141.

Schluger, P.R. (ed.) (1979) *Diagenesis as it Affects Clastic Reservoirs. Society of Economic Paleontologists and Mineralogists Special Publication*, 26, 443pp.

Schlunegger, F. and Willett, S. (1999) Spatial and temporal variations in exhumation of the central Swiss Alps and implications for exhumation mechanisms. *Geological Society, London, Special Publication*, 154, 157–179.

Schlunegger, F., Jordan, T.E. and Klaper, E.M. (1997a) Controls of erosional denudation in the orogen on foreland basin evolution: the Oligocene central Swiss Molasse Basin as an example. *Tectonics*, 16, 823–840.

Schlunegger, F., Leu, W. and Matter, A. (1997b) Sedimentary sequences, seismofacies, subsidence analysis, and evolution of the Burdigalian Upper Marine Molasse Group (OMM) of central Switzerland. *American Association of Petroleum Geologists Bulletin*, 81, 1185–1207.

Schneider, F., Burrus, J. and Wolf, S. (1993) Modelling overpressures by effective stress/porosity relationships in low permeable rocks: empirical artifice or physical reality? In: *Basin Modelling; Advances and Applications, Norwegian Petroleum Society, Special Publication*, 3, Elsevier, New York, 333–341.

Scholz, C.H. (1988) The brittle–plastic transition and the depth of seismic faulting. *Geologische Rundschau*, 77, 319–328.

Scholz, C.H. (1990) *The Mechanics of Earthquakes and Faulting*, Cambridge University Press.

Schowalter, T.T. (1976) The mechanics of secondary hydrocarbon migration and entrapment. *Wyoming Geological Association Earth Science Bulletin*, 9, 1–43.

Schröder, K.W. and Theune, C. (1984) Festoffabtrag und Stauraumsverlandung in Mitteleuropa. *Wasserwirstschaft*, 74, 374–379.

Schubert, C. (1982) Origin of Cariaco Basin, southern Caribbean Sea. *Marine Geology*, 47, 345–360.

Schumer, R. and Jerolmack, D.J. (2009) Real and apparent changes in sediment deposition rates through time. *Journal of Geophysical Research*, 114, F00A06.

Schumer, R., Jerolmack, D. and McElroy, B. (2011) The stratigraphic filter and bias in measurement of geological rates. *Geophysical Research Letters*, 38, L11405.

Schutter, S.R. (2003) Hydrocarbon occurrence and exploration in and around igneous rocks. In: *Hydrocarbons in Crystalline Rocks* (ed. by N. Petford and K.J.W. McCaffrey), In: *Geological Society, London, Special Publication*, 214, 7–33.

Schwartz, R.K. and DeCelles, P.G. (1988) Cordilleran foreland basin evolution in response to interactive Cretaceous thrusting and foreland partitioning, southwestern Montana. In: *Geological Society of America Memoir*, 171, 489–513.

Schwarzacher, W. (1987) Astronomically controlled cycles in the Lower Tertiary of Gubbio (Italy). *Earth and Planetary Science Letters*, 84, 22–26.

Sclater, J.G. and Christie, P.A.F. (1980) Continental stretching: an explanation of the post Mid-Cretaceous subsidence of the central North Sea basin. *Journal of Geophysical Research*, 85, 3711–3739.

Sclater, J.G., Jaupart, C. and Galson, D. (1980a) The heat flow through oceanic and continental crust and the heat loss of the Earth. *Reviews of Geophysics and Space Physics*, 18, 269–311.

Sclater, J.G., Royden, L., Horvath, F., Burchfiel, B.C., Semken, S. and Stegena, L. (1980b) The formation of the intra-Carpathian basins as determined from subsidence data. *Earth and Planetary Science Letters*, 51, 139–162.

Scourse, J.D. and Austin, W.E.N. (2002) Quaternary shelf sea palaeoceanography: recent developments in Europe. *Marine Geology*, 191, 87–94.

Segall, P. and Pollard, D.D. (1980) Mechanics of discontinuous faults. *Journal of Geophysical Research*, 85, 4337–4350.

Seidl, M.A. and Dietrich, W.E. (1992) The problem of channel erosion into bedrock. *Catena Supplement*, 23, 101–124.

Seidl, M.A., Weissel, J.K. and Pratson, L.F. (1996) The kinematics and pattern of escarpment retreat across the rifted continental margin of SE Australia. *Basin Research*, 12, 301–316.

Selley, R.C. (1972) Diagnosis of marine and non-marine environments from the Cambro-Ordovician sandstones of Jordan. *Journal of the Geological Society*, 128, 135–150.

Selley, R.C. (1985) *Elements of Petroleum Geology*, W.H. Freeman, New York.

Selley, R.C. (1997) The basins of northwest Africa: structural evolution. In: *African Basins: Sedimentary Basins of the World* (ed. by R.C. Selley), Elsevier, Amsterdam, 17–26.

Sengör, A.M.C. and Burke, K. (1978) Relative timing of rifting and volcanism on Earth and its tectonic implications. *Geophysical Research Letters*, 5, 419–421.

Sengör, A.M.C., Burke, K. and Dewey, J.F. (1978) Rifts at high angles to orogenic belts: tests for their origin and the Upper Rhine Graben as an example. *American Journal of Science*, 278, 24–40.

Sengör, A.M.C., Gorur, N. and Saroglu, F. (1985) Strike-slip faulting and related basin formation in zones of tectonic escape: Turkey as a case study. In: *Strike-Slip Deformation, Basin Formation and Sedimentation* (ed. by K.T. Biddle and N. Christie-Blick), *Society of Economic Paleontologists and Mineralogists Special Publication*, 37, 227–264.

Seni, S.J. and Jackson, M.P.A. (1984) Sedimentary record of Cretaceous and Tertiary salt movement, East Texas Basins. *Report of Investigations, Texas University, Bureau of Economic Geology*, 139, 89pp.

Serra, O. (1984) *Fundamentals of Well-Log Interpretation*. 1. *Acquisition of Logging Data*, Elsevier, 423pp.

Serra, O. (1986) *Fundamentals of Well-Log Interpretation*. 2. *The Interpretation of Logging Data*, Elsevier, 684pp.

Shackleton, N.J. (1977) Oxygen isotope and palaeomagnetic evidence for early northern hemisphere glaciation. *Nature*, 270, 216–219.

Shackleton, N.J. and Opdyke, N.D. (1973) Oxygen isotope and palaeomagnetic stratigraphy of equatorial Pacific core V28–V2381: oxygen isotope temperatures and ice volumes on a 10^5 year and 10^6 year scale. *Quaternary Research*, 3, 339–355.

Shanmugam, G. (1985) Significance of coniferous rain forests and related organic matter in generating commercial quantities of oil, Gippsland Basin, Australia. *American Association of Petroleum Geologists Bulletin*, 69, 1241–1254.

Sharp, I.R., Gawthorpe, R.L., Armstrong, B. and Underhill, J.R. (2000) Propagation history and passive rotation of mesoscale normal faults: implications for syn-rift stratigraphic development. *Basin Research*, 12, 285–306.

Sharp, R.V. (1976) Surface faulting in Imperial Valley during the earthquake swarm of January–February, 1975. *Bulletin of the Seismological Society of America*, 66, 1145–1154.

Shaw-Champion, M.E., White, N.J., Jones, S.M. and Lovell, J.P.B. (2008) Quantifying transient mantle convective uplift: an example from the Faroe–Shetland Basin. *Tectonics*, 27, TC1002.2008.

Shephard, G. *et al.* (2009) Contribution of mantle convection to shifting South American coastlines during the Tertiary. *Eos, Transactions, American Geophysical Union*, 90, 52.

Shephard, G.E., Muller, R.D., Liu, L. and Gurnis, M. (2010) Miocene drainage reversal of the Amazon River driven by plate–mantle interaction. *Nature Geoscience*, 3, 870–875.

Shibaoka, M. and Bennett, A.J.R. (1977) Patterns of diagenesis in some Australian sedimentary basins. *Journal Australian Petroleum Production and Exploration Association*, 17, 58–63.

Shillington, D., White, N., Minshull, T.A., Edwards, G.R.H., Jones, S.M., Edwards, R.A. and Scott, C.L. (2008) Cenozoic evolution of the eastern Black Sea: a test of depth-dependent stretching models. *Earth and Planetary Science Letters*, 265, 360–378.

Shinn, E.A. (1986) Modern carbonate tidal flats; their diagnostic features. *Quarterly Journal Colorado School of Mines*, 81, 7–35.

Shor, G.G. and Pollard, D.D. (1964) Mohole site selection studies north of Maui. *Journal of Geophysical Research*, 69, 1627–1637.

Shuster, M.W. and Aigner, T. (1994) Two-dimensional synthetic seismic and log cross-sections from stratigraphic forward models. *American Association of Petroleum Geologists Bulletin*, 78, 409–431.

Sibson, R.H. (1983) Continental fault structure and the shallow earthquake source. *Journal of the Geological Society*, 140, 741–768.

Silver, B.A. and Todd, R.G. (1969) Permian cyclic strata, northern Midland and Delaware Basins, west Texas and southeastern New Mexico. *American Association of Petroleum Geologists Bulletin*, 53, 2223–2251.

Silver, P.G., Carlson, R.W. and Olson, P. (1988) Deep slabs, geochemical heterogeneity, and the large-scale structure of mantle convection; investigation of an enduring paradox. *Annual Review of Earth and Planetary Sciences*, 16, 477–541.

Simons, F.J., Zuber, M.T. and Korenaga, J. (2000) Isostatic response of the Australian lithosphere: estimation of effective elastic thickness and anisotropy from multitaper spectral analysis. *Journal of Geophysical Research – Solid Earth*, 105, B8, 19163–19184.

Simpson, G. and Schlunegger, F. (2003) Topographic evolution and morphology of surfaces evolving in response to coupled fluvial and hillslope sediment transport. *Journal of Geophysical Research – Solid Earth*, 108(B6), 16pp.

Simpson, G.D.H. (2004a) Dynamic interactions between erosion, deposition, and three-dimensional deformation in compressional fold belt settings. *Journal of Geophysical Research*, 109, F03007.

Simpson, G.D.H. (2004b) Role of river incision in enhancing deformation. *Geology*, 32, 341–344.

Simpson, G.D.H. (2006a) A dynamic model to investigate coupling between erosion, deposition, and three-dimensional (thin-plate) deformation. *Journal of Geophysical Research*, 109, F02006.

Simpson, G.D.H. (2006b) How and to what extent does the emergence of orogens above sea level influence their tectonic development? *Terra Nova*, 18, 447–451.

Simpson, G.D.H. (2006c) Modelling interactions between fold-thrust belt deformation, foreland flexure and surface mass transport. *Basin Research*, 18, 1–19.

Sinclair, H.D. (1996) Plan-view curvature of foreland basins and its implications for the palaeostrength of the lithosphere underlying the western Alps. *Basin Research*, 8, 173–182.

Sinclair, H.D. (1997) Tectono-stratigraphic model for underfilled peripheral foreland basins: an Alpine perspective. *Geological Society of America Bulletin*, 109, 324–346.

Sinclair, H.D. (2012) Thrust wedge/foreland basin systems. In: *Tectonics of Sedimentary Basins: Recent Advances* (ed. by C. Busby and A. Azor), Wiley-Blackwell, 522–537.

Sinclair, H.D. and Allen, P.A. (1992) Vertical versus horizontal motions in the Alpine orogenic wedge: stratigraphic response in the foreland basin. *Basin Research*, 4, 215–232.

Sinclair, H.D. and Cowie, P.A. (2003) Basin-floor topography and the scaling of turbidites. *Journal of Geology*, 111, 277–299.

Sinclair, H.D., Coakley, B.J., Allen, P.A. and Watts, A.B. (1991) Simulation of foreland basin stratigraphy using a diffusion model of mountain belt uplift and erosion: an example from the central Alps, Switzerland. *Tectonics*, 10, 599–620.

Sinclair, H.D., Gibson, M., Naylor, M. and Morris, R.G. (2005) Asymmetric growth of the Pyrenees revealed through measurement and modelling of orogenic fluxes. *American Journal of Science*, 305, 369–406.

Sixsmith, P.J., Hampson, G.J., Gupta, S., Johnson, H.D. and Fofana, J.J. (2008) Facies architecture of a transgressive sandstone reservoir analogue: the Cretaceous Hosta Sandstone, New Mexico, USA. *American Association Petroleum Geologists Bulletin*, 92, 513–547.

Sklar, L.S. and Dietrich, W.E. (2001) Sediment and rock strength controls on river incision into bedrock. *Geology*, 29, 1087–1090.

Skogseid, J., Planke, S., Faleide, J.I., Pedersen, T., Eldholm, O. and Neverdal, F. (2000) NE Atlantic continental rifting and volcanic margin formation. In: *Dynamics of the Norwegian Margin* (ed. by A. Nottvedt), *Geological Society, London, Special Publication*, 167, 295–326.

Sleep, N.H. (1971) Thermal effects of the formation of Atlantic continental margins by continental break-up. *Geophysical Journal of the Royal Astronomical Society*, 24, 325–350.

Sleep, N.H. (1990) Hotspots and mantle plumes: some phenomenology. *Journal of Geophysical Research*, 95, 6715–6736.

Sleep, N.H. (2009) Stagnant lid convection and the thermal subsidence of sedimentary basins with reference to Michigan. *Geochemistry, Geophysics, Geosystems*, 10, Q12015.

Sleep, N.H. and Snell, N.S. (1976) Thermal contraction and flexure of Mid-continent and Atlantic marginal basins. *Geophysical Journal of the Royal Astronomical Society*, 45, 125–154.

Sloss, L.L. (1950) Paleozoic stratigraphy in the Montana area. *American Association of Petroleum Geologists Bulletin*, 34, 423–451.

Sloss, L.L. (1962) Stratigraphic models in exploration. *Journal of Sedimentary Petrology*, 32, 415–422.

Sloss, L.L. (1963) Sequences in the cratonic interior of North America. *Geological Society of America Bulletin*, 74, 93–114.

Sloss, L.L. (1988) Forty years of sequence stratigraphy. *Geological Society of America Bulletin*, 100, 1661–1665.

Sloss, L.L. (1990) Epilog. In: *Interior Cratonic Basins* (ed. by M.W. Leighton, D.R. Kolata, D. Oltz and J.J. Eidel), *American Association of Petroleum Geologists Memoir*, 51, 799–805.

Sloss, L.L. and Speed, R.C. (1974) Relationships of cratonic and continental-margin tectonic episodes. In: *Tectonics and Sedimentation* (ed. by W.R. Dickinson), *Society of Economic Paleontologists and Mineralogists Special Publication*, 22, 98–119.

Small, E.E. and Anderson, R.A. (1998) Pleistocene relief production in Laramide mountain ranges, western United States. *Geology*, 26, 123–136.

Small, E.E. and Anderson, R.S. (1995) Geomorphically driven Late Cenozoic rock uplift in the Sierra Nevada, California. *Science*, 270, 277–280.

Small, E.E., Anderson, R.S., Finkel, R.C. and Repka, J. (1997) Erosion rates of summit flats using cosmogenic radionuclides. *Earth and Planetary Science Letters*, 150, 413–425.

Small, E.E., Anderson, R.S., Hancock, G.S. and Finkel, R.C. (1999) Estimates of regolith production from ^{10}Be and ^{26}Al: evidence for steady state alpine hillslopes. *Geomorphology*, 27, 131–150.

Smith, D.A. (1980) Sealing and non-sealing faults in Louisiana Gulf Coast Salt Basin. *American Association of Petroleum Geologists Bulletin*, 64, 145–172.

Smith, G.A. and Landis, C.A. (1995) Intra-arc basins. In: *Tectonics of Sedimentary Basins* (ed. by C. Busby and R. Ingersoll), Blackwell Publishing Ltd., Oxford, 263–298.

Smith, L. and Chapman, D.S. (1983) On the thermal effects of groundwater flow, 1. Regional scale systems. *Journal of Geophysical Research*, 88, 593–608.

Smith, T.R. and Bretherton, F.P. (1972) Stability and the conservation of mass in drainage basin evolution. *Water Resources Research*, 8, 1506–1529.

Smoot, J.P. (1983) Depositional subenvironments in an arid closed basin; the Wilkins Peak Member of the Green River Formation (Eocene), Wyoming, USA. *Sedimentology*, 30, 801–828.

Snyder, N.P., Whipple, K.X., Tucker, G.E. and Merrits, D.J. (2000) Landscape to tectonic forcing: digital elevation model analysis of stream profiles in the Mendocino triple junction region, northern California. *Geological Society of America Bulletin*, 112, 1250–1263.

Somme, T.O., Helland-Hansen, W., Martinsen, O.J. and Thurmond, J.B. (2009) Relationships between morphological and sedimentological parameters in source-to-sink systems: a basis for predicting semiquantitative characteristics in subsurface systems. *Basin Research*, 21/4, 361–388.

Somme, T.O., Piper, D.J.W., Deptuck, M.E. and Helland-Hansen, W. (2011) Linking onshore–offshore sediment dispersal in the Golo source-to-sink system (Corsica, France) during the Late Quaternary. *Journal of Sedimentary Research*, 81, 118–137.

Sonder, L. and England, P.C. (1989) Effects of a temperature-dependent rheology on large-scale continental extension. *Journal of Geophysical Research*, 94, 7603–7619.

Spasojevic, S., Liu, L. and Gurnis, M. (2009) Adjoint models of mantle convection with seismic, plate motion, and stratigraphic constraints. *Geochemistry, Geophysics, Geosystems*, 10, Q05W02.

Spasojevic, S., Liu, L., Gurnis, M. and Müller, R.D. (2008) The case for dynamic subsidence of the U.S. east coast since the Eocene. *Geophysical Research Letters*, 35, L08305.

Spiegel, C., Siebel, W., Kuhlemann, J. and Frisch, W. (2004) Towards a comprehensive provenance analysis: a multi-method approach and its implications for the evolution of the central Alps. In: *Detrital Thermochronology – Provenance Analysis, Exhumation and Landscape Evolution of Mountain Belts* (ed. by M. Bernet and C. Spiegel), Geological Society of America, Boulder, Colorado, 37–50.

Spohn, L. and Schubert, G. (1983) Convective thinning of the lithosphere: a mechanism for rifting and midplate volcanism on Earth, Venus and Mars. *Tectonophysics*, 94, 67–90.

Stacey, F.D. (1992) *Physics of the Earth*, Brookfield Press, Brisbane, 513pp.

Stach, E., MacKowsky, M.-Th., Teichmüller, M., Teichmüller, R., Taylor, G.H. and Chandra, D. (1982) *Stach's Textbook of Coal Petrology*, Third Edition, Gebrüder Borntraeger, Berlin-Stuttgart, 535pp.

Stallard, R.F. (1995) Tectonic, environmental and human aspects of weathering and erosion: a global review using a steady-state perspective. *Annual Review of Earth and Planetary Sciences*, 23, 11–39.

Stallard, R.F. and Edmond, J.M. (1983) Geochemistry of the Amazon; 2, The influence of geology and weathering environment on the dissolved load. *Journal of Geophysical Research, Oceans and Atmospheres*, 88, 9671–9688.

Stanton, M.S. (2004) Origin of the Lower Cretaceous heavy oils (Tar Sands) of Alberta. *Search and Discovery*, article 10067.

Steckler, M.S. (1981) Thermal and mechanical evolution of Atlantic-type margins. Unpublished PhD thesis, Columbia University, New York.

Steckler, M.S. (1985) Uplift and extension at the Gulf of Suez: indications of induced mantle convection. *Nature*, 317, 135–139.

Steckler, M.S. (1999) High resolution sequence stratigraphic modeling: 1. The interplay of sedimentation, erosion and subsidence. In: *Numerical Experiments in Stratigraphy* (ed. by J. Harbaugh, L. Watney, G. Rankey, R. Slingerland, R. Goldstein and E. Franseen), *Society of Economic Paleontologists and Mineralogists Special Publication*, 62, 139–150.

Steckler, M.S., Watts, A.B. and Thorne, J.A. (1988) Subsidence and basin modelling at the U.S. Atlantic continental margin. In: *The Atlantic Continental Margin, U.S., Vol. 1–2, The Geology of North America* (ed. by R.E. Sheridan and J.A. Grow), Geological Society of America, 199–416.

Steel, R.J., Carvajal, C., Petter, A. and Urosa, C. (2008) Shelf and shelf-margin growth in scenarios of rising and falling sea level. In: *Recent Advances in Models of Siliciclastic Shallow-marine Stratigraphy* (ed. by G.J. Hampson, P.M. Burgess, R.J. Steel *et al.*), *Society of Economic Paleontologists and Mineralogists Special Publication*, 90, 47–71.

Steel, T.J. and Olsen, T. (2002) Clinoforms, clinoform trajectories and deepwater sands. In: *Sequence Stratigraphic Models for Exploration and Production: Evolving Methodology, Emerging Models and Application Histories* (ed. by J.M. Armentrout and N.C. Rosen), *Proceedings 22nd Annual Bob F. Perkins Research Conference*, Gulf Coast Section, Society Economic Paleontologists Mineralogists, 367–381.

Stefanik, M. and Jurdy, D.M. (1994) The distribution of hotspots. *Journal of Geophysical Research*, 89, 9919–9125.

Stein C.A. and Stein S. (1992) A model for the global variation in oceanic depth and heat flow with lithospheric age. *Nature*, 359, 123–129.

Stein, R.W., King, G.C.P. and Rundle, J.B. (1988) The growth of geological structure by repeated earthquakes 2. Field examples of continental dip-slip faults. *Journal of Geophysical Research*, 93, 13319–13331.

Steinberger, B. (2007) Effects of latent heat release at phase boundaries on flow in the Earth's mantle, phase boundary topography and dynamic topography at the Earth's surface. *Physics of the Earth and Planetary Interiors*, 164, 2–20.

Steinberger, B., Schmeling, H. and Marquart, G. (2001) Large-scale lithospheric stress field and topography induced by global mantle circulation. *Earth and Planetary Science Letters*, 186, 75–91.

Stel, H., Cloetingh, S., Heeremans, M. and van der Beek, P. (1993) Anorogenic granites, magmatic underplating and the origin of inter-cratonic basins in a non-extensional setting. *Tectonophysics*, 226, 285–299.

Stern, R.J. (2002) Subduction zones. *Reviews of Geophysics*, 40, DOI 10.1029/2001RG000108.

Stern, T.A., Quinlan, G.M. and Holt, W.E. (1992) Basin formation behind an active subduction zone: 3-dimensional flexural modelling of Wanganui Basin, New Zealand. *Basin Research*, 4, 197–214.

Stevenson, C.J and Turner, J.S. (1977) Angle of subduction. *Nature (London)*, 270, 334–336.

Stewart, J. and Watts, A.B. (1997) Gravity anomalies and spatial variations of flexural rigidity at mountain ranges. *Journal of Geophysical Research*, 102, B3, 5327–5352.

Stewart, R.J. and Brandon, M.T. (2004) Detrital zircon fission track ages for the 'Hoh Formation': implications for late Cenozoic evolution of the Cascadia subduction wedge. *Geological Society of America Bulletin*, 116, 60–75.

Stock, G.M., Ehlers, T.A and Farley, K.A. (2006) Where does sediment come from? Quantifying catchment erosion with detrital apatite (U–Th)/He thermochronometry. *Geology*, 34, 725–728.

Stock, J.D. and Montgomery, D.R. (1999) Geologic constraints on bedrock river incision using the stream power law. *Journal of Geophysical Research*, 104, 4983–4993.

Stöckli, D.F., Farley, K.A. and Dumitru, T.A. (2000) Calibration of the apatite (U–Th)/He thermochronometer on an exhumed fault block, White Mountains, California. *Geology*, 28, 983–986.

Stockmal, G.S., Beaumont, C. and Boutilier, R. (1986) Geodynamic models of convergent tectonics: the transition from rifted margin to overthrust belt and consequences for foreland basin development. *American Association of Petroleum Geologists Bulletin*, 70, 181–190.

Storti, F. and McClay, K. (1995) Influence of syntectonic sedimentation on thrust wedges in analog models. *Geology*, 23, 999–1002.

Strakhov, N.M. (1967) *Principles of Lithogenesis, Vol. 1*, Oliver and Boyd, Edinburgh.

Straub, C. and Kahle, H.G. (1995) Active crustal deformation in the Marmara Sea region, NW Anatolia, inferred from GPS measurements. *Geophysical Research Letters*, 22, 2533–2536.

Strong, N., Sheets, B., Hickson, T. and Paola, C. (2005) Mass balance framework for quantifying downstream changes in fluvial architecture. *International Association Sedimentologists Special Publication*, 35, 243–253.

Stuart, C.A. (1970) *Geopressures, Proceedings Second Symposium Abnormal Subsurface Pressure*, Louisiana State University, Baton Rouge, Louisiana, Supplement.

Stumm, W. and Morgan, J.J. (1996) *Aquatic Chemistry: Chemical Equilibria and Rates in Natural waters, 3rd edition*, New York, Wiley-Interscience, 1013 pp.

Stüwe, K. (2002) *Geodynamics of the Lithosphere*, Springer, Berlin.

Stüwe, K. and Barr, T.D. (1998) On uplift and exhumation during convergence. *Tectonics*, 17, 80–88.

Stüwe, K. and Barr, T.D. (2000) On the relationship between surface uplift and gravitational extension. *Tectonics*, 19, 1056–1064.

Stüwe, K., White, L. and Brown, R. (1994) The influence of eroding topography on steady state isotherms; applications to fission track analysis. *Earth and Planetary Science Letters*, 124, 63–74.

Styron, R.H., Taylor, M.H. and Murphy, M.A. (2011) Oblique convergence, arc-parallel extension, and the role of strike-slip faulting in the High Himalaya. *Geosphere*, 7, 582–596.

Su, W.J. and Dziewonski, A.M. (1992) On the scale of mantle heterogeneity. *Physics of the Earth and Planetary Interiors*, 74, 29–54.

Su, W.J., Woodward, R.L. and Dziewonski, A.M. (1994) Degree 12 model of shear velocity heterogeneity in the mantle. *Journal of Geophysical Research*, 99, 6945–6980.

Summerfield, M.A. (1991) *Global Geomorphology*, Longman, London, 537pp.

Summerfield, M.A. and Hulton, N.J. (1994) Natural controls on fluvial denudation rates in major world drainage basins. *Journal of Geophysical Research*, 99, 13871–13883.

Suppe, J. (1983) Geometry and kinematics of fault-bend folding. *American Journal of Science*, 283, 684–721.

Surdam, R.C. and Wolfbauer, C.A. (1975) Green River Formation, Wyoming: a playa-lake complex. *Geological Society of America Bulletin*, 86, 335–345.

Sweeney, J.J. and Burnham, A.K. (1990) Evaluation of a simple model of vitrinite reflectance based on chemical kinetics. *American Association Petroleum Geologists Bulletin*, 74, 1559–1570.

Swenson, J.B., Paola, C., Pratson, L., Voller, V.R. and Murray, A.B. (2005) Fluvial and marine controls on combined subaerial and subaqueous delta progradation: morphodynamic modeling of compound clinoform development. *Journal of Geophysical Research*, 110, F02013.

Swenson, J.B., Voller, V.R., Paola, C., Parker, G. and Marr, J.G. (2000) Fluvio-deltaic sedimentation: a generalized Stefan problem. *European Journal of Applied Mathematics*, 11, 433–452.

Swift, D.J.P. (1974) Continental shelf sedimentation. In: *The Geology of Continental Margins* (ed. by C.A. Burk and C.L. Drake), Springer-Verlag, Berlin, 117–135.

Swift, D.J.P., Han, G. and Vincent, C.E. (1986) Fluid processes and sea-floor responses on a modern storm-dominated shelf: Middle Atlantic shelf of North America. Part I. The Storm Current Regime. In: *Shelf Sands and Sandstones* (ed. by R.J. Knight and J.R. McLean), *Memoir Canadian Society Petroleum Geologists* Calgary, 11, 99–119.

Sylvester, A.G. (1988) Strike-slip faults. *Geological Society of America Bulletin*, 100, 1666–1703.

Sylvester, Z. (2007) Turbidite bed thickness distributions: methods and pitfalls of analysis and modelling. *Sedimentology*, 54, 847–870.

Syvitski, J.P.M. and Milliman, J.D. (2007) Geology, geography, and humans battle for dominance over the delivery of sediment to the coastal ocean. *Journal of Geology*, 115, 1–19.

Syvitski, J.P.M. and Morehead, M.D. (1999) Estimating river-sediment discharge to the ocean: application to the Eel margin, northern California. *Marine Geology*, 154, 13–28.

Syvitski, J.P.M., Vorosmarty, C.J., Kettner, A.J. *et al.* (2005) Impact of humans on the flux of terrestrial sediment to the global coastal ocean. *Science*, 308, 376–380.

Szulc, A.G., Najman, Y., Sinclair, H.D. *et al.* (2006) Tectonic evolution of the Himalaya constrained by detrital ^{40}Ar–^{39}Ar, Sm–Nd and petrographic data from the Siwalik foreland basin succession, SW Nepal. *Basin Research*, 18, 375–392.

Talbot, M.R. and Allen, P.A. (1996) Lakes. In: *Sedimentary Environments: Processes, Facies and Stratigraphy, Third Edition* (ed. by H.G. Reading), Blackwell Publishing Ltd., 83–124.

Talling, P.J. (2001) On the frequency distribution of turbidite thickness. *Sedimentology*, 48, 1297–1331.

Talling, P.J., Lawton, T.F., Burbank, D.W. and Hobbs, R.S. (1995) Evolution of latest Cretaceous–Eocene nonmarine deposystems in the Axhandle piggyback basin of central Utah. *Geological Society of America Bulletin*, 107, 297–315.

Talling, P.J., Wynn, R.B., Masson, D.G. *et al.* (2007) Onset of submarine debris flow deposition far from original giant landslide. *Nature*, 450, 541–544.

Tamaki, K. and Honza, E. (1991) Global tectonics and formation of marginal basins: role of the Western Pacific. *Episodes*, 14, 224–230.

Tankard, A.J. (1986) On the depositional response to thrusting and lithospheric flexure: examples from the Appalachian and Rocky Mountain basins. In: *Foreland Basins* (ed. by P.A. Allen and P.

Homewood), *International Association of Sedimentologists Special Publication*, 8, 369–394.

Tankard, A.J., Jackson, M.P.A., Eriksson, K.A., Hobday, D.K., Hunter, D.R. and Minter, W.E.L. (1982) *Crustal Evolution of Southern Africa*, Springer-Verlag, New York, 523pp.

Tapponnier, P. and Molnar, P. (1976) Slip-line field theory and large-scale continental tectonics. *Nature*, 264, 319–324.

Taylor, M., Yin, A., Ryerson, F., Kapp, P. and Ding, L. (2003) Conjugate strike-slip faulting along the Bangong-Nujiang suture zone accommodates coeval east–west extension and north-south shortening in the interior of the Tibetan Plateau. *Tectonics*, 22, 1044.

Taylor, S.R. and McLennan, S.M. (1995) The geochemical evolution of the continental crust. *Reviews of Geophysics*, 33, 241–265.

Taymaz, T., Jackson, J. and McKenzie, D. (1991) Active tectonics of the north and central Aegean Sea. *Geophysical Journal International*, 106, 433–490.

Tchalenko, J.S. (1970) Similarities between shear zones of different magnitudes. *Geological Society of America Bulletin*, 81, 1625–1640.

Tchalenko, J.S. and Ambraseys, N.N. (1970) Structural analyses of the Dasht-e Baÿaz (Iran) earthquake fractures. *Geological Society of America Bulletin*, 81, 1625–1640.

Teichmüller, M. (1970) Bestimmung des Inkohlungsgrades von kohligen Einschlussen in Sedimenten des Oberrheingrabens: ein Hilfsmittel bei der Klärung geothermischer Fragen. In: *Graben Problems* (ed. by J.H. Illies and S. Müller), *Upper Mantle Project Scientific Report*, 27, 124–142.

Teichmüller, M. (1982) *Fluoreszenz von Liptiniten und Vitriniten in Beziehung zu Inkohlungsgrad und Verkokungsverhalten. Geologisches Landesamt Nordrhein-Westfalen Special Paper*, 119pp.

Teichmüller, M. and Teichmüller, R. (1975) Inkohlungsuntersuchungen in der Molasse des Alpenvorlands. *Geologica Bavarica*, 73, 123–142.

ten Brink, U.S. and Ben-Avraham, Z. (1989) The anatomy of a pull-apart basin: Seismic reflection observations of the Dead Sea basin. *Tectonics*, 8, 333–350.

ten Brink, U.S., Ben-Avraham, Z., Bell, R.E., Hassouneh, M., Coleman, D.F., Andreasen, G., Tibor, G. and Coakley, B.J. (1993) Structure of the Dead Sea pull-apart basin from gravity analyses. *Journal of Geophysical Research*, B 98, 21877–21894.

ten Brink, U.S., Katzman, R. and Lin, J. (1996) Three-dimensional models of deformation near strike-slip faults. *Journal of Geophysical Research*, 101, 16205–16220.

ter Voorde, M. and Cloetingh, S. (1996) Numerical modelling of extension in faulted crust: effects of localized and regional deformation on basin stratigraphy. *Geological Society Special Publication*, 99, 283–296.

Terres, R.R. and Sylvester, A.G. (1981) Kinematic analysis of rotated fractures and blocks in simple shear. *Bulletin of the Seismological Society of America*, 71, 1593–1605.

Terzaghi, K. (1936) The shearing resistance of saturated soils. In: *Proceedings of the 1st International Conference on Soil Mechanics*, Harvard, 1, 54–56.

Terzaghi, K. and Peck, R.B. (1948) *Soil Mechanics in Engineering Practice*, John Wiley & Sons, 566pp.

Thiessen, R., Burke, K. and Kidd, W.S.F. (1979) African hotspots and their relation to the underlying mantle. *Geology*, 7, 263–266.

Thomas, B.M. (1982) Land plant source rocks for oil and their significance in Australian basins. *Australian Petroleum Production and Exploration Association Journal*, 22, 164–178.

Thybo, H. and Nielsen, C. (2009) Magma-compensated crustal thinning in continental rift zones. *Nature*, 457, 873–876.

Tissot, B. (1969) Premières données sur les mécanismes et le cinétique de la formation du pétrole dans les sédiments; simulation d'un schéma reactionnel sur ordinateur. *Revue de l'Institut Français du Pétrole*, 24, 470–501.

Tissot, B. (1973) Vers l'evaluation quantitative du petrole forme dans les basins sedimentaires. *Petrole et Techniques*, 222, 27–31.

Tissot, B. and Espitalié, J. (1975) L'evolution thermique de la matière organique des sediments: applications d'une simulation mathématique. *Revue de l'Institut Français du Pétrole*, 30, 743–777.

Tissot, B.P. and Welte, D.H. (1978) *Petroleum Formation and Occurrence: A New Approach to Oil and Gas Exploration*, Springer-Verlag, Berlin, 538pp.

Tissot, B.P. and Welte, D.H. (1984) *Petroleum Formation and Occurrence, Second Edition*, Springer-Verlag, New York, 699pp.

Toksöz, M.N. and Bird, P. (1977) Formation and evolution of marginal basins and continental plateaus. In: *Island Arcs, Deep Sea Trenches and Back-arc Basins* (ed. by M. Talwani and W.C. Pitman), *American Geophysical Union, Maurice Ewing Series*, 1, 379–393.

Torsvik, T.H., Smethurst, M.A., Burke, K. and Steinberger, B. (2008) Long-term stability in deep mantle structure: evidence from the ca. 300 Ma Skagerrak-centered Large Igneous Province (the SCLIP). *Earth and Planetary Science Letters*, 267, 444–452.

Trudgill, B.D., Rowan, M.G., Fiduk, J.C., Weimer, P., Gale, P.E., Korn, B.E., Phair, R.L., Gafford, W.T., Roberts, G.R. and Dobbs, S.W. (1999) The Perdido Foldbelt, northwestern Deep Gulf of Mexico, Part 1: Structural geometry, evolution and regional implications. *American Association of Petroleum Geologists Bulletin*, 83/1, 88–113.

Tucker, G.E. (1996) Modelling the large-scale interaction of climate, tectonics, and topography. PhD dissertation, Pennsylvania State University, University Park, PA.

Tucker, G.E. and Slingerland, R.L. (1994) Erosional dynamics, flexural isostasy, and long-lived escarpments: a numerical modelling study. *Journal of Geophysical Research*, 99, 12229–12243.

Tucker, G.E. and Slingerland, R.L. (1996) Predicting sediment flux from fold and thrust belts. *Basin Research*, 8, 329–349.

Tucker, G.E. and Whipple, K.X. (2002) Topographic outcomes predicted by stream erosion models: sensitivity analysis and intermodel comparison. *Journal of Geophysical Research*, 107(B9), 2179.

Turcotte, D.L. (1983) Driving mechanisms of mountain building. In: *Mountain Building Processes* (ed. by K.J. Hsu), Academic Press, Orlando, Florida, 141–146.

Turcotte, D.L. and Oxburgh, E.R. (1967) Finite amplitude convective cells and continental drift. *Journal of Fluid Mechanics*, 28, 29–42.

Turcotte, D.L. and Schubert, G. (1982) *Geodynamics: Applications of Continuum Mechanics to Geological Problems*, Wiley, New York.

Turcotte, D.L. and Schubert, G. (2002) *Geodynamics, Second Edition*, Cambridge University Press, Cambridge, 456pp.

Tuttle, M.L.W., Charpentier, R.R. and Brownfield, M.E. (1999) *The Niger Delta Petroleum System: Niger Delta Province, Nigeria, Cameroon, and Equatorial Guinea, Africa. US Geological Survey Open-File Report*, 99-50-H.

Ulmishek, G. (1986) Stratigraphic aspects of petroleum resource assessment. In: *Oil and Gas Assessment: Methods and Applications* (ed. by D.D. Rice), *American Association of Petroleum Geologists Studies in Geology*, 21, 59–68.

Underhill, J.R. and Woodcock, N.H. (1987) Emplacement-related fault patterns around the Northern Granite, Arran, Scotland. *Geological Society of America Bulletin*, 98, 515–527.

Underwood, M.B. and Moore, G.F. (1995) Trenches and trench slope basins. In: *Tectonics of Sedimentary Basins* (ed. by C. Busby and R. Ingersoll), Blackwell Publishing Ltd., Oxford, 179–220.

Upcott, N.M., Mukasa, R.K., Ebinger, C.J. and Karner, G.D. (1996) Along-axis segmentation and isostasy in the Western Rift, East Africa. *Journal of Geophysical Research*, 101, 3247–3268.

US Geological Survey (2000) World Petroleum Assessment 2000; Orinoco heavy oil and tar belt; assessment unit 60980104. http://energy.usgs.gov/OilGas/AssessmentsData/WorldPetroleumAssessment.

US Geological Survey (2009) *An estimate of recoverable heavy oil resources of the Orinoco oil belt, Venezuela, Fact Sheet*, 2009-3028.

Ussami, N., Shiraiwa, S. and Landin-Dominguez, J.M. (1999) Basement reactivation in a sub-Andean foreland flexural bulge: the Pantanal wetland, SW Brazil. *Tectonics*, 18, 25–39.

Uyeda, S. and Kanamori, H. (1979) Back-arc opening and the mode of subduction. *Journal of Geophysical Research*, 84, 1049–1061.

Vail, P.E. Mitchum, R.M. Jr. and Thompson, S. III (1977a) Relative changes of sea level from coastal onlap. In: *Seismic Stratigraphy: Applications to Hydrocarbon Exploration* (ed. by C.E. Payton), *American Association of Petroleum Geologists Memoir*, 26, 63–82.

Vail, P.R., Mitchum, R.M. Jr. and Thompson, S. (1977b) Seismic stratigraphy and global changes of sea level, Part 4: Global cycles of relative changes of sea level. In: *Seismic Stratigraphy: Applications to Hydrocarbon Exploration* (ed. by C.E. Payton), *American Association of Petroleum Geologists Memoir*, 26, 83–97.

Van Balen, R.T., Van der Beek, P.A. and Cloetingh, S. (1995) The effect of rift shoulder erosion on stratal patterns at passive margins: implications for sequence stratigraphy. *Earth and Planetary Science Letters*, 134, 532–544.

van Gijzel, P. (1982) Characterization and identification of kerogen and bitumen and determination of thermal maturation by means of qualitative and quantitative microscopical techniques. In: *How to Assess Maturation and Paleotemperatures, Society of Economic Paleontologists and Mineralogists Short Course Notes*, 7, 159–216.

van Hinte, J.E. (1978) Geohistory analysis: application of micropalaeontology in exploration geology. *American Association of Petroleum Geologists Bulletin*, 62, 201–222.

Van Wagoner, J.C., Mitchum, R.M. Jr., Campion, K.M. and Rahmanian, V.D. (1990) *Siliciclastic Sequence Stratigraphy in Well Logs, Cores and Outcrop: Concepts for High Resolution Correlation of Time and Facies. American Association of Petroleum Geologists Methods in Exploration Series*, 7, Tulsa, 55pp.

Van Wagoner, J.C., Posamentier, H.W., Mitchum, R.M., Vail, P.R., Sarg, J.F., Loutit, T.S. and Handenbol, J. (1988) An overview of the fundamentals of sequence stratigraphy and key definitions. In: *Sea Level Changes: An Integrated Approach, Society of Economic Paleontologists and Mineralogists Special Publication*, 42, 39–45.

Van Wees, J.D. and Cloetingh, S. (1996) 3-D modelling of stress-induced subsidence in the North Sea basin. *Tectonophysics*, 266, 343–359.

Van West, F.P. (1972) Trapping mechanisms of Minnelusa oil accumulations, northeastern Powder River Basin, Wyoming. *Mountain Geologist*, 9, 3–20.

Vauchez, A., Tommasi, A. and Barruol, G. (1998) Rheological heterogeneity, mechanical anisotropy and deformation of the continental lithosphere. *Tectonophysics*, 296, 61–86.

Vaughan, S., Bailey, R.J. and Smith, D.G. (2011) Detecting cycles in stratigraphic data: spectral analysis in the presence of red noise. *Paleoceanography*, 26, PA4211.

Veevers, J.J. (1981) Morphotectonics of rifted continental margins in embryo (East Africa), youth (Africa–Arabia) and maturity (Australia). *Journal of Geology*, 89, 57–82.

Veevers, J.J. (2000) *Billion Year Earth History of Australia and Neighbours in Gondwanaland*, GEMOC Press, Sydney, 388pp.

Veevers, J.J., Jones, J.G and Powell, C.McA. (1982) Tectonic framework of Australia's sedimentary basins. *Australian Petroleum Production and Exploration Association Journal*, 22, 283–300.

Vening-Meinesz, F.A. (1941) *Gravity over the Hawaiian Archipelago and Over the Madcira Area, Proceedings Netherlands Academy Wetensia*, 44pp.

Vening-Meinesz, F.A. (1948) *Gravity Expeditions at Sea*, 1923–1938, Netherlands Geodetic Commission, Waltman, Delft.

Vergés, J. (2007) Drainage responses to oblique and lateral ramps: a review. In: *Sedimentary Processes, Environments and Basins: A Tribute to Peter Friend* (ed. by G. Nichols, E. Williams and C. Paola), John Wiley & Sons, 29–47.

Vergés, J., Marzo, M., Santaeulària, T. *et al.* (1998) Quantified vertical motions and tectonic evolution of the SE Pyrenean foreland basin: In: *Cenozoic Foreland Basins of Western Europe* (ed. by A. Mascle, C. Puigdefàbregas, H.P. Luterbacher and M. Fernàndez), *Geological Society, London, Special Publication*, 134, 107–134.

Vincent, S.J. (2001) The Sis palaeovalley: a record of proximal fluvial sedimentation and drainage basin development in response to Pyrenean mountain building. *Sedimentology*, 48, 1235–1276.

Vinogradov, L.S., Averyanova, I.S. and Nigmati, I.S. (1983) Catagenetically sealed oil pools of the Volga–Ural region. *Petroleum Geology*, 19, 266–268.

Vogt, P.R. (1991) Bermuda and Appalachian–Labrador rises. *Geology*, 19, 41–44.

Vyssotski, A.V., Vyssotski, V.N. and Nezhdanov, A.A. (2006) Evolution of the West Siberian Basin. *Marine and Petroleum Geology*, 23, 93–126.

Walcott, R.I. (1970) Flexural rigidity, thickness and viscosity of the lithosphere. *Journal of Geophysical Research*, 75, 3941–3954.

Walderhaug, O. (1994) Temperatures of quartz cementation in Jurassic sandstones from the Norwegian continental shelf – evidence from fluid inclusions. *Journal of Sedimentary Research*, 64, 311–323.

Walderhaug, O. (1996) Kinetic modeling of quartz cementation and porosity loss in deeply buried sandstone reservoirs. *American Association Petroleum Geologists Bulletin*, 80, 731–745.

Walford, H., White, N. and Sydow, J.C. (2005) Solid sediment load history of the Zambezi delta. *Earth and Planetary Science Letters*, 238, 49–63.

Walker, R.G. (ed.) (1984) *Facies Models, Second Edition*, Geoscience Canada Reprint Series, 1.

Walling, D.E. and Webb, B.W. (1983) Patterns of sediment yield. In: *Background to Palaeohydrology* (ed. by K.J. Gregory), John Wiley & Sons Ltd., Chichester, 69–100.

Walling, D.E. and Webb, B.W. (1996) Erosion and sediment yield: a global overview. In: *Erosion and Sediment Yield: Global and Regional Perspectives* (ed. by D.E. Walling and B.W. Webb), *International Association of Hydrological Sciences Publication*, 236, 3–19.

Walter, M.R., Veevers, J.J., Calver, C.R. and Grey, K. (1995) Neoproterozoic stratigraphy of the Centralian Superbasin, Australia. *Precambrian Research*, 73, 173–195.

Wangen, M. (1995) The blanketing effect in sedimentary basins. *Basin Research*, 7, 283–298.

Wangen, M. (2010) *Physical Principles of Sedimentary Basin Analysis*, Cambridge University Press, 527pp.

Wangen, M., Fjeldskaar, W. and Faleide, J. *et al.* (2007) Forward modeling of stretching episodes and palaeo heat flow of the Voring margin, NE Atlantic. *Journal of Geodynamics*, 45, 83–98.

Waples, D.W. (1980) Time and temperature in petroleum formation: application of Lopatin's method to petroleum exploration. *American Association of Petroleum Geologists Bulletin*, 64, 916–926.

Waples, D.W. (1981) *Organic Geochemistry for Exploration Geologists*, Burgess Publishing, Minneapolis, Minnesota.

Ware, P.D. and Turner, J.P. (2002) Sonic velocity analysis of the Tertiary denudation of the Irish Sea basin. In: *Exhumation of the North Atlantic Margin: Timing, Mechanisms and Implications for Petroleum Exploration* (ed. by A.G. Doré, J.A. Cartwright, M.S. Stoker, J.P. Turner and N. White), *Geological Society, London, Special Publication*, 196, 355–370.

Warrick, J.A. and Rubin, D.M. (2007) Suspended-sediment rating curve response to urbanization and wildfire, Santa Ana River, California. *Journal of Geophysical Research – Earth Surface*, 112, F02018, DOI: 10.1029/2006JF000662.

Warrick, R.A. and Oerlemans, J. (1990) Sea Level Rise, Chapter 9, 260–281, In: *Climate Change – The IPCC Scientific Assessment (1990)* (ed. by J.T. Houghton, H.J. Jenkins and J.J. Ephraums), Cambridge University Press, 410pp.

Waschbusch, P.J. and Royden, L.H. (1992) Spatial and temporal evolution of foredeep basins: lateral strength variations and inelastic yielding in continental lithosphere. *Basin Research*, 4, 179–196.

Watson, H.J. (1981) Casablanca field, offshore Spain, a palaeogeomorphic trap. *American Association of Petroleum Geologists Bulletin Abstract*, 65, 1005–1006.

Watts, A.B. (1978) An analysis of isostasy in the world's oceans: 1. Hawaiian–Emperor seamount chain. *Journal of Geophysical Research*, 83, 5989–6004.

Watts, A.B. (1982) Tectonic subsidence, flexure and global changes in sea level. *Nature*, 297, 469–474.

Watts, A.B. (1988) Gravity anomalies, crustal structure and flexure of the lithosphere at the Baltimore Canyon Trough. *Earth and Planetary Science Letters*, 89, 221–238.

Watts, A.B. (1992) The effective elastic thickness of the lithosphere and the evolution of foreland basins. *Basin Research*, 4, 169–178.

Watts, A.B. (2001) *Isostasy and Flexure of the Lithosphere*, Cambridge University Press, Cambridge, 458pp.

Watts, A.B. and Cochran, J.R. (1974) Gravity anomalies and flexure of the lithosphere along the Hawaiian–Emperor seamount chain. *Geophysical Journal of the Royal Astronomical Society*, 38, 119–141.

Watts, A.B. and Ryan, W.B.F. (1976) Flexure of the lithosphere and continental margin basins. *Tectonophysics*, 36, 25–44.

Watts, A.B. and Steckler, M.S. (1979) Subsidence and eustasy at the continental margin of eastern North America. *American Geophysical Union, Maurice Ewing Series*, 3, 218–239.

Watts, A.B. and Talwani, M. (1974) Gravity anomalies seaward of trenches and their tectonic implications. *Geophysical Journal of the Royal Astronomical Society*, 36, 57–90.

Watts, A.B. and ten Brink, U.S. (1989) Crustal structure, flexure and subsidence history of the Hawaiian Islands. *Journal of Geophysical Research*, 94, 10,473–20,500.

Watts, A.B. and Thorne, J. (1984) Tectonics, global changes in sea level and their relationship to stratigraphic sequences at the US Atlantic continental margin. *Marine and Petroleum Geology*, 1, 319–339.

Watts, A.B. and Torné, M. (1992) Subsidence history, crustal structure and thermal evolution of the Valencia Trough: a young extensional basin in the western Mediterranean. *Journal of Geophysical Research*, 97, 20021–20041.

Watts, A.B., Karner, G.D. and Steckler, M.S. (1982) Lithospheric flexure and the evolution of sedimentary basins. *Philosophical Transactions of the Royal Society, London*, A305, 249–281.

Watts, A.B., Peirce, C., Collier, J., Dalwood, R., Canales, J.P. and Henstock, T.J. (1997) A seismic study of lithospheric flexure in the vicinity of Tenerife, Canary Islands. *Earth and Planetary Science Letters*, 146, 431–447.

Watts, A.B., Rodger, M., Pierce, C., Greenroyd, C.J. and Hobbs, R.W. (2009) Seismic structure, gravity anomalies, and flexure of the Amazon continental margin, NE Brazil. *Journal of Geophysical Research – Solid Earth*, 114, B07103.

Weaver, P.P.E., Rothwell, R.G., Ebbing, J., Gunn, D. and Hunter, P.M. (1992) Correlation, frequency of emplacement and source directions of megaturbidites on the Madeira Abyssal Plain. *Marine Geology*, 109, 1–20.

Weber, K.J. (1986) How heterogeneity affects oil recovery. In: *Reservoir Characterization* (ed. by L.W. Lake and H.B. Carroll, Jr.), Academic Press, Orlando, 487–544.

Weber, K.J. (1987) Computation of initial well productivities in aeolian sandstone on the basis of a geological model, Leman Gas Field, U.K. In: *Reservoir Sedimentology* (ed. by R.W. Tillmann and K.J. Weber), *Society of Economic Paleontologists and Mineralogists Special Publication*, 40, 333–354.

Weber, K.J., Mandl, G., Pilaar, W.F., Lehner, F. and Precious, R.G. (1978) The role of faults in hydrocarbon migration and trapping in Nigerian growth fault structures. *Offshore Technology Conference*, Houston, paper OTC 3356, 2643–2651.

Weber, V.V. and Maximov, S.P. (1976) Early diagenetic generation of hydrocarbon gases and their variations dependent on initial organic composition. *American Association of Petroleum Geologists Bulletin*, 60, 287–293.

Wegmann, K.W., Zurek, B.D., Regalla, C.A., Bilardello, D. *et al.* (2007) Position of the Snake River watershed divide as an indicator of geodynamic processes in the greater Yellowstone region, western North America. *Geosphere*, 3, 272–281.

Weissel, J.K. and Karner, G.D. (1989) Flexural uplift of rift flanks due to mechanical unloading of the lithosphere during extensions. *Journal of Geophysical Research*, 94, 13919–13950.

Wells, M.R., Allison, P.A., Piggott, M.D., Pain, C.C., Hampson, G.J. and De Oliviera, C.R.E. (2005) Large sea, small tides: the Late Carboniferous seaway of NW Europe. *Journal of the Geological Society*, 162, 417–420.

Wen, L. and Anderson, D.L. (1995) The fate of slabs inferred from seismic tomography and 130 million years of subduction. *Earth and Planetary Science Letters*, 133, 185–198.

Werner, B.T. (1999) Complexity in natural landform patterns. *Science*, Viewpoint, 284, 102–104.

Wernicke, B. (1981) Low-angle normal faults in the Basin and Range province: nappe tectonics in an extending orogen. *Nature*, 291, 645–648.

Wernicke, B. (1985) Uniform-sense normal simple shear of the continental lithosphere. *Canadian Journal of Earth Sciences*, 22, 108–125.

Wernicke, B. and Axen, G.J. (1988) On the role of isostasy in the evolution of normal fault systems. *Geology*, 16, 848–851.

Wesson, R.L., Helley, E.J., Lajoie, K.R. and Wentworth, C.M. (1975) Faults and future earthquakes. In: *Studies for Seismic Zonation of the San Francisco Bay Region* (ed. by R.D. Borchardt), *US Geological Survey Professional Paper*, 941A, 5–30.

Wheeler, H.E. (1958) Time stratigraphy. *American Association of Petroleum Geologists Bulletin*, 42, 1047–1063.

Wheeler, H.O. (1964) Base level, lithosphere surface, and time-stratigraphy. *Geological Society of America Bulletin*, 75, 599–610.

Wheeler, P. and White, N. (2000) Quest for dynamic topography: observations from southeast Asia. *Geology*, 28, 963–966.

Whelan, J.K. and Thomson-Rizer, C.L. (1993) Chemical methods for assessing kerogen and protokerogen types and maturity. In: *Organic Geochemistry* (ed. by M.H. Engel and S.A. Macko), Plenum Press, New York, 289–346.

Whipple, K.X. (2004) Bedrock rivers and the geomorphology of active orogens. *Annual Review of Earth and Planetary Sciences*, 32, 151–185.

Whipple, K.X. (2009) The influence of climate on the tectonic evolution of mountain belts. *Nature Geoscience*, 2, 97–104.

Whipple, K.X. and Trayler, C.R. (1996) Tectonic control on fan size: the importance of spatially-variable subsidence rates. *Basin Research*, 8, 351–366.

Whitchurch, A.L., Carter, A., Sinclair, H.D., Duller, R.A., Whittaker, A.C. and Allen, P.A. (2011) Sediment routing system evolution within a diachronously uplifting orogen: insights from detrital zircon thermochronological analyses from the south-central Pyrenees. *American Journal of Science*, 311, 442–482.

White, A.F. and Blum, A.E. (1995) Effects of climate on chemical weathering in watersheds. *Geochimica et Cosmochimica Acta*, 59, 1729–1747.

White, D.A. (1980) Assessing oil and gas plays in facies cycle wedges. *American Association of Petroleum Geologists Bulletin*, 64, 1158–1178.

White, D.A. (1988) Oil and gas play maps in exploration and assessment. *American Association of Petroleum Geologists Bulletin*, 72, 944–949.

White, N. (1993) Recovery of strain rate variation from inversion of subsidence data. *Nature*, 366, 449–452.

White, N. (1994) An inverse method for determining lithospheric strain rate variation on geological timescales. *Earth and Planetary Science Letters*, 122, 351–371.

White, N. and Lovell, B. (1997) Measuring the pulse of a plume with the sedimentary record. *Nature*, 387, 888–891.

White, N. and McKenzie, D.P. (1988) Formation of the steer's head geometry of sedimentary basins by differential stretching of the crust and mantle. *Geology*, 16, 250–253.

White, N.M., Pringle, M., Garzanti, E., Bickle, M., Najman, Y., Chapman, H. and Friend, P.F. (2002) Constraints on the exhumation and erosion of the High Himalayan Slab, NW India, from foreland basin deposits. *Earth and Planetary Science Letters*, 195, 29–44.

White, R.S. and McKenzie, D. (1989) Magmatism at rift zones: the generation of volcanic continental margins and flood basalts. *Journal of Geophysical Research*, 94, 7685–7729.

Whittaker, A.C., Attal, M. and Allen, P.A. (2010) Characterising the origin, nature and fate of sediment exported from catchments perturbed by active tectonics. *Basin Research*, 22, 809–828.

Whittaker, A.C., Duller, R.A., Springett, J. *et al.* (2011) Decoding downstream trends in stratigraphic grain size as a function of tectonic subsidence and sediment supply. *Geological Society of America Bulletin*, 123, 1363–1382.

Wilcox, R.E., Harding, T.P. and Seely, D.R. (1973) Basin wrench tectonics. *American Association of Petroleum Geologists Bulletin*, 57, 74–96.

Wilgus, C.K., Hastings, B.S., Kendall, C.G. St. C., Posamentier, H.W., Ross, C.A. and Van Wagoner, J.C. (eds.) (1988) *Sea-level Changes: An Integrated Approach, Society of Economic Paleontologists and Mineralogists Special Publication*, Tulsa, Oklahoma, 42, 407pp.

Wilhelm, O. (1945) Classification of petroleum reservoirs. *American Association of Petroleum Geologists Bulletin*, 29, 1537–1580.

Wilkinson, B.H., Diedrich, N.W. and Drummond, C.N. (1996) Facies successions in peritidal carbonate sequences. *Journal of Sedimentary Research*, 66, 1065–1078.

Wilkinson, B.H., Diedrich, N.W., Drummond, C.N. and Rothman, E.D. (1998) Michigan hockey, meteoric precipitation, and rhythmicity of accumulation on peritidal carbonate platforms. *Geological Society of America Bulletin*, 110, 1075–1093.

Wilkinson, B.H., Drummond, C.N., Rothman, E.D. and Diedrich, N.W. (1997) Stratal order in peritidal carbonate sequences. *Journal of Sedimentary Research*, 67, 1068–1082.

Willenbring, J.K. and von Blanckenburg, F. (2010) Long-term stability of global erosion rates and weathering during late Cenozoic cooling. *Nature*, 465, 211–214.

Willett, S.D. (1992) Dynamic and kinematic growth and change of a Coulomb wedge. In: *Thrust Tectonics*, Chapman and Hall, London, 19–31.

Willett, S.D. (1999) Orogeny and orography: the effects of erosion on the structure of mountain belts. *Journal of Geophysical Research*, 104, 28957–28981.

Willett, S.D. and Brandon, M.T. (2002) On steady states in mountain belts. *Geology*, 30, 175–178.

Willett, S.D. and Schlunegger, F. (2010) The last phase of deposition in the Swiss Molasse Basin: from foredeep to negative-alpha basin. *Basin Research*, 22, 623–639.

Willett, S.D., Beaumont, C. and Fullsack, P. (1993) Mechanical model for the tectonics of doubly vergent compressional orogens. *Geology*, 21, 371–374.

Willett, S.D., Chapman, D.S. and Neugebauer, H.J. (1985) A thermomechanical model of continental lithosphere. *Nature*, 314, 520–523.

Willett, S.D., Slingerland, R.J. and Hovius, N. (2001) Uplift, shortening and steady-state topography in active mountain belts. *American Journal of Science*, 301, 455–485.

Williams, C.A., Connors, C., Dahlen, F.A., Price, E.J. and Suppe, J. (1994) Effect of the brittle–ductile transition on the topography of compressive mountain belts on Earth and Venus. *Journal of Geophysical Research*, 99, B10, 19947–19974.

Williams, H.H., Kelley, P.A., Janks, J.S. and Christensen, R.M. (1985) The Palaeogene rift basin source rocks of Central Sumatra. *Proceedings of the Indonesia Petroleum Association 14th Annual Convention*, October, 1985.

Williams, J.J. (1968) The stratigraphy and igneous reservoirs of the Angila field, Libya. In: *Geology and Archaeology of Northern Cyrenaica, Libya* (ed. by F.T. Barr), 197–206.

Wilson, D., Aster, R., West, M., Ni, J., Grand, S., Gao, W., Baldridge, W.S., Semken, S. *et al.* (2005) Lithospheric structure under the Rio Grande Rift. *Nature*, 433, 851–855.

Wilson, J.T. (1966) Did the Atlantic close and then re-open? *Nature*, 211, 676–681.

Wobus, C.W., Crosby, B.T. and Whipple, K.X. (2006) Hanging valleys in fluvial systems: controls on the occurrence and implications for landscape evolution. *Journal of Geophysical Research*, 111, F02017.

Wolf, R., Farley, K. and Silver, L. (1996) Assessment of (U–Th)/He thermochronometry: the low temperature history of the San Jacinto Mountains, California. *Geology*, 25, 65–68.

Wolf, R., Farley, K.A. and Kass, D.M. (1998) Modelling the temperature sensitivity of the apatite U–Th/He thermochronometer. *Chemical Geology*, 148, 105–114.

Wood, R.J. and Barton, P.J. (1983) Crustal thinning and subsidence in the North Sea. *Nature*, 302, 134–136.

Woodcock, N.H. (1986) The role of strike-slip fault systems at plate boundaries. *Philosophical Transactions of the Royal Society, London*, A317, 13–29.

Woodcock, N.H., 2004, Life span and fate of basins. *Geology*, 32, 685–688.

Woodside, W. and Messmer, J.H. (1961) Thermal conductivity of porous media (parts I and II). *Applied Physics*, 32(9), 1688–1707.

Woodwell, G.M., Whittaker, R.H., Reiners, W.A., Likens, G.E., Delwiche, C.C. and Botkin, D.B. (1978) The biota and the world carbon budget. *Science*, 199, 141–146.

Wooler, D.A., Smith, A.G. and White, N.J. (1992) Measuring lithospheric stretching on Tethyan passive margins. *Journal of the Geological Society*, 149, 517–532.

Worden, R.H. and Morad, S. (2003) Clay minerals in sandstones: controls on formation, distribution and evolution. In: *Clay Mineral Cements in Sandstones* (ed. by R.H. Worden and S. Morad), *International Association of Sedimentologists Special Publication*, 34, Blackwell Publishing Ltd., Oxford, 3–41.

Wright, L.D. (1977) Sediment transport and deposition at river mouths: a synthesis. *Geological Society of America Bulletin*, 88, 857–868.

Wright, L.D. and Coleman, J.M. (1974) Mississippi River mouth processes; effluent dynamics and morphologic development. *Journal of Geology*, 82, 751–778.

Wu, J.E., McClay, K., Whitehouse, P. and Dooley, T. (2009) 4D analogue modelling of transtensional pull-apart basins. *Marine and Petroleum Geology*, 26, 1608–1623.

Xie, X. and Heller, P.L. (2009) Plate tectonics and basin subsidence history. *Geological Society of America Bulletin*, 121, 55–64.

Xu, X., Lithgow-Bertelloni, C. and Conrad, C.P. (2006) Global reconstructions of Cenozoic seafloor ages: implications for bathymetry and sea level. *Earth and Planetary Science Letters*, 243, 552–564.

Yong, L., Allen, P.A., Densmore, A.L. and Qiang, X. (2003) Evolution of the Longmen Shan Foreland Basin (Western Sichuan, China) during the Late Triassic Indosinian orogeny. *Basin Research*, 15, 117–138.

Yurewicz, D.A, Bohacs, K.M, Kendall, J., Klimentidis, R.E., Kronmueller, K., Meurer, M.E., Ryan, T.C. and Yeakel, J.D. (2008) Controls on gas and water distribution, Mesaverde basin-centred gas play, Piceance Basin, Colorado. In: *Understanding, Exploring, and Developing Tight-Gas Sands* (ed. by S.P. Cumella, K.W. Shanley and W.K. Camp), *2005 Vail Hedberg Conference, American Association of Petroleum Geologists Hedberg Series*, 3, 105–136.

Zak, I. and Freund, R. (1981) Asymmetry and basin migration in the Dead Sea rift. *Tectonophysics*, 80, 27–38.

Zeitler, P.K. *et al.* (2001a) Crustal reworking at Nanga Parbat, Pakistan: metamorphic consequences of thermal-mechanical coupling facilitated by erosion. *Tectonics*, 20, 712–728.

Zeitler, P.K., Herczig, A.L., McDougall, I. and Honda, M. (1987) U–Th–He dating of apatite: a potential thermochronometer. *Geochimica et Cosmochimica Acta*, 51, 2865–2868.

Zeitler, P.K., Meltzer, A.S., Koons, P.O., Craw, D., Hallet, B., Chamberlain, C.P., Kiss, W.S.F, Park, S.K. and Seeber, L. (2001b) Erosion, Himalayan geodynamics, and the geomorphology of metamorphism. *GSA Today*, 11/1, 4–9.

Zhang, P., Molnar, P. and Downs, W.R. (2001) Increased sedimentation rates and grain sizes 2–4 Myr ago due to the influence of climate change on erosion rates. *Nature*, 410, 891–897.

Zhong, S., Gurnis, M. and Moresi, L. (1996) Free-surface formulation of mantle convection. I. Basic theory and application to plumes. *Geophysical Journal International*, 127, 708–718.

Zhou, S. and Sandiford, M. (1992) On the stability of isostatically compensated mountain belts. *Journal of Geophysical Research*, 97, 14207–14221.

Ziagos, J.P. and Blackwell, D.D. (1986) A model for the transient temperature effects of horizontal fluid flow in geothermal systems. *Journal of Volcanology and Geothermal Research*, 27, 371–397.

Ziegler, P.A., Cloetingh, S. and van Wees, J.D. (1995) Dynamics of intraplate compressional deformation: the Alpine foreland and other examples. *Tectonophysics*, 252, 7–60.

Zimmerman, R., Somerton, W. and King, M. (1986) Compressibility of porous rocks. *Journal of Geophysical Research*, 91, 12765–12777.

Zoback, M.L. (1992) First- and second-order patterns of stress in the lithosphere: The World Stress Map Project. *Journal of Geophysical Research*, 97, 11703–11728.

Zoback, M.L., Zoback, M.D., Mount, V.S. *et al.* (1987) New evidence on the state of stress of the San Andreas fault system. *Science*, 238, 1105–1111.

Zubkov, M.Y. and Mormyshev, V.V. (1987) Correlation and formation conditions in the Bazhenov Suite in the Salym deposit. *Lithology and Mineral Resources*, 22, 1167–1174.

Zuffa, G.G. (1985) *Provenance of Arenites*, D. Reidel Publishing Company, Dordrecht, 408pp.

Zühlke, R., Bechstaedt, T. and Mundil, R. (2003) Sub-Milankovitch and Milankovitch forcing on a model Mesozoic carbonate platform – the Latemar (Middle Triassic, Italy). *Terra Nova*, 15, 69–80.

Zweigel, J. and Zweigel, P. (1998) Plan-view curvature of foreland basins and its implications for the palaeostrength of the lithosphere underlying the Western Alps. Discussion of Sinclair (1996), with reply. *Basin Research*, 10, 271–278.

Index

Basin Analysis: Principles and Application to Petroleum Play Assessment, Third Edition. Philip A. Allen and John R. Allen.
© 2013 John Wiley & Sons, Ltd. Published 2013 by John Wiley & Sons, Ltd.